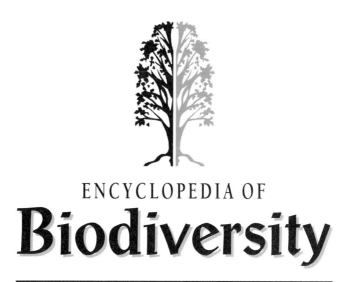

ENCYCLOPEDIA OF
Biodiversity

VOLUME 2

D – Fl

EDITORIAL BOARD

ENCYCLOPEDIA OF
Biodiversity

VOLUME 2
D–Fl

Editor-in-Chief

Simon Asher Levin

Moffett Professor of Biology
Princeton University
Princeton, New Jersey, USA

ACADEMIC PRESS

A Harcourt Science and Technology Company

SAN DIEGO   SAN FRANCISCO   BOSTON   NEW YORK   LONDON   SYDNEY   TOKYO

OCLC: 45279099

Copyright © 2001 by ACADEMIC PRESS

All Rights Reserved.
No part of this publication may be reproduced or transmitted in any form or by any means, electronic or mechanical, including photocopy, recording, or any information storage and retrieval system, without permission in writing from the publisher.

Requests for permission to make copies of any part of the work should be mailed to: Permissions Department, Harcourt Inc., 6277 Sea Harbor Drive, Orlando, Florida 32887-6777

Academic Press
A Harcourt Science and Technology Company
525 B Street, Suite 1900, San Diego, California 92101-4495, USA
http://www.academicpress.com

Academic Press
Harcourt Place, 32 Jamestown Road, London NW1 7BY, UK
http://www.academicpress.com

Library of Congress Catalog Card Number: 00-105903

International Standard Book Number: 0-12-226865-2 set
International Standard Book Number: 0-12-226866-0 Volume 1
International Standard Book Number: 0-12-226867-9 Volume 2
International Standard Book Number: 0-12-226868-7 Volume 3
International Standard Book Number: 0-12-226869-5 Volume 4
International Standard Book Number: 0-12-226864-4 Volume 5

PRINTED IN THE UNITED STATES OF AMERICA
00 01 02 03 04 05 MM 9 8 7 6 5 4 3 2 1

Contents

Contents of Other Volumes

CONTENTS OF VOLUME 4

CONTENTS OF VOLUME 5

Contents by Subject Area

ENVIRONMENTAL CONDITIONS AND EFFECTS

ACID RAIN AND DEPOSITION

AIR POLLUTION

ATMOSPHERIC GASES

CARBON CYCLE

CLIMATE CHANGE AND ECOLOGY, SYNERGISM OF

CLIMATE, EFFECTS OF

DESERTIFICATION

DISTURBANCE, MECHANISMS OF

ECONOMIC GROWTH AND THE ENVIRONMENT

ECOTOXICOLOGY

ENERGY FLOW AND ECOSYSTEMS

ENERGY USE, HUMAN

ENVIRONMENTAL ETHICS

EUTROPHICATION AND OLIGOTROPHICATION

FIRES, ECOLOGICAL EFFECTS OF

GRAZING, EFFECTS OF

GREENHOUSE EFFECT

INSECTICIDE RESISTANCE

MARINE AND AQUATIC COMMUNITIES, STRESS FROM EUTROPHICATION

NITROGEN, NITROGEN CYCLE

PESTICIDES, USE AND EFFECTS OF

POLLUTION, OVERVIEW

SOIL BIOTA, SOIL SYSTEMS, AND PROCESSES

STRESS, ENVIRONMENTAL

ULTRAVIOLET RADIATION

EVOLUTION

ADAPTATION

ADAPTIVE RADIATION

BIODIVERSITY, EVOLUTION AND

BIODIVERSITY GENERATION, OVERVIEW

BIODIVERSITY, ORIGIN OF

COEVOLUTION

COMPLEMENTARITY

DARWIN, CHARLES

DEFENSES, ECOLOGY OF

DIAPAUSE AND DORMANCY

EUKARYOTES, ORIGIN OF

EVOLUTION, THEORY OF

FOSSIL RECORD

GEOLOGIC TIME, HISTORY OF BIODIVERSITY IN

LIFE HISTORY, EVOLUTION OF

MUTUALISM, EVOLUTION OF

PHYLOGENY

PLANT COMMUNITIES, EVOLUTION OF

EXTINCTIONS

DINOSAURS, EXTINCTION THEORIES FOR

ENDANGERED BIRDS

ENDANGERED ECOSYSTEMS

ENDANGERED FRESHWATER INVERTEBRATES

ENDANGERED MAMMALS

ENDANGERED MARINE INVERTEBRATES

ENDANGERED PLANTS

ENDANGERED REPTILES AND AMPHIBIANS

ENDANGERED TERRESTRIAL INVERTEBRATES

EXTINCTION, CAUSES OF

EXTINCTION, RATES OF

EXTINCTIONS, MODERN EXAMPLES OF

HUMAN EFFECTS ON ECOSYSTEMS, OVERVIEW

LATENT EXTINCTIONS: THE LIVING DEAD

LOSS OF BIODIVERSITY, OVERVIEW

MAMMALS (LATE QUATERNARY), EXTINCTIONS OF

MAMMALS (PRE-QUATERNARY), EXTINCTIONS OF

MARINE MAMMALS, EXTINCTIONS OF

MASS EXTINCTIONS, CONCEPT OF

MASS EXTINCTIONS, NOTABLE EXAMPLES OF

NATURAL EXTINCTIONS (NOT HUMAN-INFLUENCED)

GENETICS

BACTERIAL GENETICS

ECOLOGICAL GENETICS

GENE BANKS

GENES, DESCRIPTION OF

POPULATION ISSUES

PUBLIC POLICIES AND ATTITUDES

SPECIES INTERACTIONS AND INTERRELATIONSHIPS

SYSTEMATICS AND SPECIES CONCEPT

Foreword

Biodiversity is the totality of the inherited variation of all forms of life across all levels of variation, from ecosystem to species to gene. Soon after the term was introduced at the first National Forum on Biodiversity in 1986, and after it began its rapid spread around the world, there occurred a reconfiguration in the way much of the science of biology is conceived. Where previously comparative biology had been almost entirely focused on the fundamentals of classification, evolution, behavior, and ecology, now it was augmented by a wide range of analyses from the social sciences. Where taxonomy and biogeography had been marginalized through the middle half of the twentieth century, now they moved back toward center stage. And where extinction had been little more than a phenomenon recognized and lamented, now it became a major concern of science. Much of ecology shifted toward the study of biodiversity's role in the assembly and maintenance of ecosystems. A growing number of economists, political scientists, and bioethicists took up the issue as part of their scholarly agenda. From this mix the discipline of conservation biology was born, and the Society of Conservation Biology became one of the fastest growing organizations in modern science. In 1992 the Rio Summit catapulted biodiversity to global prominence, from which most of the nations of the world endorsed the Convention on Biodiversity and have since used it as a guideline for conservation programs.

The new biodiversity initiative gave organismic and evolutionary biologists a global mission worthy of their science. It confirmed for those who labored in the vineyards that, as medicine is to molecular and cellular biology, the environment is to organismic and evolutionary biology. The first is responsible for personal health, and the second for planetary health. The additional evidence adduced moreover made clear that we are in the midst of an episode of massive extinction, unprecedented since that closing of the Mesozoic Era—and that scientists must lead the attempt to save the Creation.

The articles in the *Encyclopedia of Biodiversity* are unusually eclectic, yet organized by a set of easily articulated goals. They are the following: to carry the systematics and biogeography of the world fauna and flora toward completion; map the hot spots where conservation will save the most biodiversity; orient studies of natural history to understand and save threatened species; advance ecosystems studies and biogeography to create the needed principles of community assembly and maintenance; acquire the knowledge of resource use, economics, and polity to advance conservation programs based on sustainability; and enrich the ethic of global conservation in terms persuasive to all.

The road ahead, down which we must urgently travel, will be smoothed by the exponential growth of information and a growing public awareness and support. The *Encyclopedia of Biodiversity* will serve as an important knowledge base to guide this supremely important effort.

Edward O. Wilson
Museum of Comparative Zoology
Harvard University

Preface

The science of biodiversity has become the science of our future. Our awareness of the disappearance of biodiversity has brought with it a long-overdue appreciation of the magnitude of our loss, and a determination to develop the tools to protect our future. This encyclopedia brings together, for the first time in its completeness, study of the dimensions of diversity with examination of the services that biodiversity provides, and measures to protect it.

The entries in the encyclopedia have been arranged alphabetically, but the coverage is designed functionally. At the core is a comprehensive survey of biodiversity, across taxonomic groups and ecological regions. The emergence of biodiversity is then placed in an evolutionary perspective, as background for an understanding of current trends. Particular attention is given to the loss of services—for example, in fisheries, forestry and climate mediation—that are derived from natural systems. These are placed in an economic framework through a comprehensive set of papers that address problems of valuation, costs, and benefits, and develop a framework for prioritizing actions. Finally, a review is given of institutions and other mechanisms that exist and are needed for the preservation of biodiversity and, with it, the services that humans derive from nature.

The background for understanding biodiversity is to be found in the fossil record, and in the evolutionary patterns and trends that it reveals. The encyclopedia hence discusses these patterns, the origins of biodiversity, the effects of geological events, the mechanisms of evolution, and the uniqueness of the evolutionary process, with implications for conservation and restoration. The essential processes in macroevolution are those of speciation and extinction, which together govern the dynamics of diversity at higher levels of organization. These are given extensive coverage, both from a mechanistic and from a historical perspective, and provide an essential context for understanding the rest of the contributions.

The classification of organisms into species and higher taxa, and the elucidation of the mechanisms of natural selection, were the essential intellectual advances that allowed the development of the science of biodiversity. Carl Linnaeus introduced a systematic framework for understanding phylogenies, which continues to provide the foundation for evolutionary studies today; and Charles Darwin's great legacy—the theory of evolution by natural selection—is the essential organizing principle for understanding the processes that gave rise to the patterns Linnaeus recognized. The encyclopedia provides unmatched taxonomic coverage of the organization of diversity into taxonomic groups and complements that with an extensive examination of ecosystems by biogeographic region and by functional type. These chapters elucidate latitudinal trends, life zones, species–area relationships and the distribution of diversity within and among ecological communities.

Throughout the core chapters, there is a healthy balance between empirical facts and conceptual theories. Such theories help to illuminate principles that cross systems and levels of organization, and transform the study of biodiversity into a science. Basic ecological constructs, such as the habitat and the niche, are given extensive treatment, as are key ecological mechanisms such as competition, predation, herbivory, parasitism and mutualism. These treatments are complemented by exploration of fundamental evolutionary mechanisms related to local differentiation, aspect diversity, sex, and recombination, and especially theories of extinction.

With these foundational chapters in hand, one can turn to the contemporary problems in biodiversity and compare today's rapid rates of change to the historical patterns. Key chapters examine agriculture, fisheries, and forests, their importance to human needs, and their status and trends in response to changing land-use patterns, population growth, overexploitation, and climate change. Threatened and endangered species are discussed in detail, with relation to the consequences of the spread of invading species.

The utilization of nature's bounty for food, fiber and fuel provides some of the most obvious benefits of biodiversity to humanity. Equally important, however, are the things that are less well appreciated: the potential for the discovery of new pharmaceuticals that can improve human health, the role of biodiversity in pollinating crops and wild species, and the importance of natural systems in regulating climate, mediating nutrient fluxes, and sequestering carbon as well as toxic materials. Each of these services provides humans with direct and indirect benefits, and somehow we must find ways to weigh these benefits, along with the ethical and aesthetic values we place on natural systems and biodiversity, to provide priorities for action. Only recently have economists recognized the importance of such issues as intellectual challenges essential to our survival on the planet. Much of biodiversity is exploited by humanity as part of a global commons, in which one does not pay in fair measure for extracting parts, or affecting the commons otherwise through land use or pollution. Economists have come to realize, along with ecologists and others, the magnitude of the externalities involved. When such externalities are involved, the market does not function as it must to maintain the resource, and new measures are needed if the sustainability and resilience of these resources are to be preserved. There is as yet, however, no ecological equivalent to the power of financial institutions, such as the Federal Reserve Board in the United States, to modify individual incentives sufficiently to maintain regional or global stability in the system of interest. In this encyclopedia, some of the most enlightened and thoughtful economists turn their attention to the economic challenges, and discuss the mechanisms and institutions that might be needed.

Together, the state-of-the-art entries in this encyclopedia tell an exciting story of how biodiversity arose, continues to arise, and is maintained. It is a story of a complex, self-organizing system—the biosphere—whose pieces can be examined individually, but cannot be understood outside the context of the whole. It is also a story of the coevolution of the biosphere and *Homo sapiens*, the first species whose own activities can feed back to influence the evolution of the biosphere on time scales that could lead to its own demise. The articles in the encyclopedia can be used as material for a wide spectrum of courses, tracing the history of the emergence of biodiversity from its origins to the challenges we face today.

This has been a massive effort, but one of the most rewarding I have ever undertaken. So many people have played a role that it is difficult to know where to begin. The project began through the initiative of Scott Bentley at Academic Press, and then was managed flawlessly by Chris Morris at AP. I cannot recall ever having dealt with an editor who operated more professionally than Chris, who combined a true vision and enthusiasm for the project with a sense of economic realities and the energy and insight to make the whole project work. In this he was ably assisted by outstanding Academic Press staff, especially Naomi Henning, Nick Panissidi, and Ann Marie Martin. At every step, it has been a pleasure to work with Academic Press, and I especially single out Chris for his fantastic and scholarly efforts.

At the next stage, the Editorial Boards were terrific in generating and commenting on ideas, suggesting authors, and critiquing contributions. More than 400 authors then adopted our view of the importance of the project, accepted the task of writing, and produced timely and comprehensive articles that make this Encyclopedia like no other source available today. To all of these, I extend my thanks and congratulations.

And finally, special acknowledgment and gratitude are due my wife, Carole, and my assistant, Amy Bordvik. Carole put up with the late nights and obsessiveness that were essential to the process, and Amy worked tirelessly and without complaint through the whole long process. To them, any expression of thanks is insufficient.

Simon Levin
Princeton University

References

Simon A. Levin (1999). *Fragile Dominion*, Perseus Books. Reading, Massachusetts
Edward O. Wilson (1992). *The Diversity of Life*. Norton. New York, New York.

Guide to the Encyclopedia

The *Encyclopedia of Biodiversity* is a comprehensive study of the topic of diversity in the natural world, contained within the covers of a single unified work. It consists of five volumes and includes 313 separate full-length articles by leading international authors.

Each article in the encyclopedia provides a comprehensive overview of the selected topic to inform a broad spectrum of readers, from research professionals to students to the interested general public. In order that you, the reader, will derive the greatest possible benefit from the *Encyclopedia of Biodiversity,* we have provided this Guide. It explains how the encyclopedia was developed, how it is organized, and how the information within it can be located.

ENTRY SELECTION

This encyclopedia was conceived with the goal of providing a complete description of all the issues contained within, or impacting upon, the field of biodiversity. To that end, a thorough and systematic method of entry selection was devised for the work.

To begin the selection process, the project's chief editor, Simon Levin, prepared a bibliography of leading source materials in the field, including books, journal articles, conference proceedings, Websites, and so on. Then the reference staff of Academic Press combed through these materials to develop a list of potential article topics for the encyclopedia. This preliminary list was refined and approved by Prof. Levin; at this point the number of possible entries was approximately twice as large as the eventual total in the published encyclopedia.

The entry list was then provided to all the associate editors and the international editorial board for their evaluation. Their mandate was to read through the list and rate each topic on a numerical scale according to how important they deemed it to be for inclusion in the encyclopedia. The editors were also encouraged to recommend new topics not on the existing list, and to make other comments on the list as appropriate. A number of additions to the entry list emerged from this process.

The editors' ratings and comments were returned to Academic Press for scoring, and an overall tabulation was created that indicated the consensus of the group as to the priority of each topic. Then the list was sent to Prof. Levin for a final evaluation in which he made "tie-breaker" decisions for certain topics on which the editors' vote was split, and also other adjustments based on his expert judgment. The result was a working entry list of about 325 topics which, after some attrition and the combining of related topics, resulted in the final table of contents of 313 articles.

ORGANIZATION

The *Encyclopedia of Biodiversity* is organized in a single alphabetical sequence by title. Articles whose titles begin with the letters A to C are in Volume 1, articles with titles from D through Fl are in Volume 2, then Fo through Man in Volume 3, Mar through Q in Volume 4, and R to Z in Volume 5.

Volume 5 also includes a complete subject index for the entire work, an alphabetical list of the authors who contributed to the encyclopedia, and a glossary of key terms used in the articles.

TABLE OF CONTENTS

A complete table of contents for the *Encyclopedia of Biodiversity* appears at the front of each volume. This alphabetical list of article titles (see p. vii) is followed by a second contents list (p. xix) in which the titles are listed according to their subject area within the overall field of biodiversity.

Articles are classified in 20 different subject areas, including not only core disciplines of biodiversity such as evolution, speciation, populations, extinction, and ecosystems, but also areas that link biodiversity to other disciplines, such as environmental science, agriculture, public policy, and economics.

ARTICLE TITLES

Article titles generally begin with the key term describing the topic, and have inverted word order if necessary to begin the title with this term. For example, "Archaea, Origin of" is the article title rather than "Origin of Archaea," "Grazing, Effects of" is the title rather than "Effects of Grazing," and so on with other titles such as "Species, Concepts of," "Mammals, Biodiversity of," "Pollinators, Role of," and so on. This is done so that the reader can more easily locate a desired topic. For example, eight different articles on endangered groups (e.g., endangered birds) appear in succession in the "En-" section of the encyclopedia.

INDEX

The index appears as the last element of Volume 5. Subjects are listed alphabetically and indicate the volume and page number where information on this topic can be found. In addition, the table of contents by subject area also functions as an index, since it lists all the topics covered in a given area; e.g., the encyclopedia has 26 different articles dealing with invertebrates.

ARTICLE FORMAT

Articles in the *Encyclopedia of Biodiversity* are arranged in a standard format, as follows:

- Title and Author
- Outline
- Glossary
- Defining Statement
- Main Body of the Article
- Cross-References
- Bibliography

OUTLINE

Entries in the encyclopedia begin with a topical outline that indicates the general content of the article. This outline serves two functions. First, it provides a preview of the article, so that the reader can get a sense of what is contained there without having to leaf through the pages. Second, it serves to highlight important subtopics that are discussed within the article. For example, the article "Greenhouse Effect" includes subtopics such as "Climatic Consequences: Global Warming" and "Climate Change and Biodiversity."

The outline is intended as an overview and thus it lists only the major headings of the article. In addition, extensive second-level and third-level headings will be found within the article.

GLOSSARY

The Glossary section contains terms that are important to an understanding of the article and that may be unfamiliar to the reader. Each term is defined in the context of the article in which it is used. The same term may appear as a glossary entry in different articles, with the details of the definition varying slightly from one article to another. The encyclopedia includes approximately 2,500 glossary entries. For example, the article "Mangrove Ecosystems" has the following entry:

aerenchyma A spongy plant tissue composed largely of air spaces enabling gas exchange to take place by diffusion in underground mangrove roots.

In addition, Volume 5 has a comprehensive glossary that presents the core vocabulary of biodiversity in one A-Z list. This section can be consulted for definitions of unfamiliar terms not found in the individual glossary for a given article.

DEFINING STATEMENT

The text of each article in the encyclopedia begins with a single introductory paragraph that defines the topic under discussion and summarizes the content of the article. For example, the article "Agriculture, Sustainable" begins with the following statement:

> Sustainable agriculture describes a food and fiber production system that is economically viable, environmentally safe, and socially acceptable over long periods.

CROSS REFERENCES

The entry list for *Encyclopedia of Biodiversity* has been constructed so that each entry is supported by one or more other entries that provide additional information. Therefore all articles in the encyclopedia have references to other articles. These cross references appear at the conclusion of the article text. They indicate articles that can be consulted for further information on the same issue, or for pertinent information on a related issue. The encyclopedia includes a total of about 1,750 cross references to other articles. For example, the article "Biodiversity-Rich Countries" contains the following list of references:

> Biodiversity as a Commodity • Deforestation • Economic Growth and the Environment • Indigenous Peoples, Biodiversity and • Social and Cultural Factors • Tropical Ecosystems

BIBLIOGRAPHY

The Bibliography section appears as the last element in an article. Entries in this section include not only print sources but relevant Websites as well.

The bibliography entries in this encyclopedia are for the benefit of the reader and do not represent a complete list of all the materials consulted by the author in preparing the article. Rather, the sources listed are the author's recommendations of the most appropriate materials for further research on the given topic. For example, the article "Fires, Ecological Effects of" lists as references (among others) the works *Fire and Plants, Fire in the Environment, Fire in the Tropical Biota,* and *The Role of Fire in Mediterranean Ecosystems.*

COMPANION WORKS

Encyclopedia of Biodiversity is part of a continuing program of multivolume reference works published by Academic Press. This program encompasses many different areas of science, ranging from organismal biology (e.g., *Encyclopedia of Dinosaurs, Encyclopedia of Microbiology*) to biomedical topics (*Encyclopedia of Reproduction, Encyclopedia of Stress*), to physical science (*Encyclopedia of the Solar System, Encyclopedia of Volcanoes*) to social and political issues (*Encyclopedia of Applied Ethics, Encyclopedia of Creativity, Encyclopedia of Nationalism, Encyclopedia of Violence, Peace, and Conflict*).

For information on these and other Academic Press reference titles, please see the Website at:

www.academicpress.com/reference/

DARWIN, CHARLES (DARWINISM)

Michael T. Ghiselin
California Academy of Sciences

GLOSSARY

biota Fauna and flora; the organisms in a given place.

coevolution Evolution where two species have a mutual influence on each other.

genetic drift Change in gene frequencies due to random fluctuations in a finite population.

heterostyly Having two or more kinds of flowers with stamens and pistils of different lengths.

mimicry Defensive mechanism in which organisms of two species resemble each other, ordinarily because at least one of them is distasteful.

morphology The study of form.

phylogeny The history of a species or lineage.

pleiotropy Phenomenon of a gene affecting more than one trait.

sexual selection Kind of selection that Darwin distinguished from natural and artificial selection in which reproductive success depends on monopolizing mates.

taxon Group of organisms in a formal system of classification.

CHARLES DARWIN (1809–1882) DISCOVERED the principle of natural selection and laid the foundations for modern evolutionary biology. The term "Darwinian" is applied to his theory, and others like it, in which natural selection is considered the main, although not the only, mechanism. Terms such as "Lamarckian" suggest alternatives in which other mechanisms, such as use and disuse, are considered the main cause of change. "Darwinism" has come to mean both Darwin's version of evolutionary theory and those elaborated by his followers. Modified versions of Darwinism that emphasize selection are often called "neo-Darwinism." Among these are the neo-Darwinism of August Weismann and the Synthetic Theory that took shape in the middle of the twentieth century and still remains dominant. Darwinians have always acknowledged the existence of mechanisms other than natural selection. Darwin himself invoked Lamarckian and quasi-Lamarckian mechanisms that are no longer accepted by his followers. He also accepted auxiliary mechanisms that are still considered valid. These include sexual selection, pleiotropy, and developmental constraints. Evolution as a result of sampling error (genetic drift) is "non-Darwinian" in the sense that Darwin did not know about it. The "neutral theory" that invokes sampling error as a case of essentially non-adaptive change at the molecular level is best considered part of mainstream neo-Darwinism.

I. INTRODUCTION

Although natural selection theory is generally considered Darwin's most important and original contribution to knowledge, he also deserves credit for providing compelling evidence that evolution is a fact and for making its fundamental implications clear. Indeed, Darwin began an intellectual revolution of the first magnitude, one that has affected the whole scope of modern life and culture. The emphasis here will be on those aspects of Darwin's contribution and of Darwinism in general that are particularly relevant to the study of biodiversity. Within biology it has been the sciences of systematics and ecology that have been most revolutionized by Darwinian thinking, because it enables us to understand both how and why diversity originates and is maintained. Some of the larger philosophical and social issues will also be considered, especially as they relate to the central theme. These include the rejection of teleology, the replacement of typological thinking with population thinking, the recognition of the importance of historical contingency, and a competitive view of the natural economy.

II. DARWIN'S LIFE AND WORK

A. Childhood and Education

Charles Darwin was born on February 12, 1809, in Shrewsbury, England. His father was a wealthy physician and his mother was a member of the Wedgwood family, famous for the manufacture of pottery. She died when he was eight years old. Young Darwin was not very successful in school and at the age of sixteen he was sent to study medicine at Edinburgh University. There he was able to cultivate his interest in natural history, largely on an extra-curricular basis. A decision to abandon medicine in favor of preparation for a career as a clergyman led him to enroll in Cambridge University on October 15, 1827. During the years at Cambridge he continued his largely extracurricular studies of natural history and received much encouragement from both faculty and students. In those days there was nothing unusual about a clergyman also being a scientist, a university professor, or both. He passed the examina-

tions for his B.A. degree on January 22, 1831, but remained there to complete his residency requirements. He took up the study of geology, and the following summer obtained some field experience in the company of Professor Adam Sedgwick (1785–1873).

B. The *Beagle* Voyage

On September 1, 1831, Darwin accepted an invitation to join a surveying expedition on H.M.S. *Beagle*, as an unofficial naturalist and gentlemanly companion for the captain of the ship, Robert FitzRoy (1805–1865). The *Beagle* left for South America on December 27, 1831, and after sailing around the world finally got back to England on October 2, 1836. During the voyage, Darwin was able to observe and collect on a vast scale, especially in South America, on the coast of which most of the survey work was done. He also visited Australia, the Cape of Good Hope, and many islands along the way. At first his work emphasized the marine invertebrates that had interested him since his days at Edinburgh. But he took in increasing interest in geology. He read the first volume of the *Principles of Geology* by Charles Lyell (1797–1875) not long after the ship left port, and began to read the second volume the following November. Darwin became an enthusiastic supporter of Lyell's uniformitarian methodology, and was also influenced by the discussions of biogeography and Lamarckism in the *Principles*. Darwin developed a theory of coral reef morphology while still working in South America. Later in the voyage he began to test it by means of observations on reefs. It was the coral reef theory, coupled with other geological work, that first established his high reputation as a scientist.

Observations during the voyage on the geographical distribution of animals and plants ultimately led Darwin to believe in evolution. The differences among geographically separated biotas and the peculiar character of insular ones were crucial in his thinking. It must be stressed, however, that these evolutionary insights did not happen instantaneously and that his commitment to evolution was a later development. Darwin began to take the possibility of evolution seriously only toward the end of the voyage, and the compelling evidence was not really in place until shortly after it was over, and some of his specimens had been examined by specialists. During his visit to the Galapagos Islands (from September 15 to October 20, 1835) he still assumed that species were immutable. He did, however, expect them to be somewhat adaptable, and to exist as local varieties. Some of the birds that he had supposed were just varieties turned out to be distinct species, implying that the inhabitants had diversified more than he had assumed was possible, and furthermore that they had done so locally. They were related by community of descent. This was just the beginning of many of his biogeographical insights. Facts such as the lack of frogs and other amphibians on islands far distant from land could easily be explained in terms of evolution, but otherwise remained a mystery. Contrary to the impression that one sometimes gets from text books, the most compelling evidence for evolution comes from biogeography, not the study of fossils.

C. Natural Selection

The fossils that Darwin collected in South America turned out to be recently extinct members of groups of animals that were still living in the same area. Again this suggested the possibility of local diversification. Around March, 1837, he began to search for an evolutionary mechanism. After much reading and theorizing, he discovered natural selection. This happened toward the end of September, 1838, when he read the *Essay on the Principle of Population* by Thomas Robert Malthus (1766–1834), who was among other things the world's first professor of economics. Darwin now understood how reproductive competition between organisms of the same species could produce evolutionary change.

The influence of Malthus on Darwin requires some elaboration here because it is so important for biodiversity studies. Malthus's ideas (or what have been represented as his ideas) have been controversial and have often entered into discussions on population planning and conservation policy. His basic point was that the growth of population tends to outstrip the growth of the food supply. His notion that the growth of population is exponential whereas that of the food supply is linear is well known, although it never was more than just a crude approximation. Malthus based his arguments on empirical evidence. After plagues and other catastrophes have reduced the numbers of human population there are fewer mouths to feed and the amount of food available per capita increases. Consequently people can and do support larger families and population grows rapidly. As time passes, the amount of food per capita decreases and population growth levels off, as people marry later and have smaller families. The standard of living tends to drop as well. A comparable situation could be observed where new land had been brought into cultivation, as in the United States, following which a high living standard was accompanied by rapid population growth.

Malthus saw a problem in this scenario with respect to the future economic prosperity of society: population growth would tend to depress the standard of living. There was a serious question in his mind, as well as those of his critics, whether his model of population growth need have those particular consequences. "Prudential constraints" (as he called sexual abstinence) would have some mitigating effect. The growth of technology has of course at least delayed the sort of famine that might be envisaged. On the other hand, the Malthusian model applies quite well to non-human populations. It contributed to Darwin's understanding of population dynamics, which he incorporated into his evolutionary and ecological theory. Darwin recognized that there was reproductive competition between individuals of the same species, and because these individuals varied, those that were able to utilize resources more effectively would have more offspring. Natural selection follows when it is realized that the properties of those individuals that have more offspring are differentially transmitted by inheritance to the next generation. Extrapolating further, it made sense that natural selection would operate differently depending on whether or not population levels had been lowered by the effects of weather or other density-independent factors.

Given this kind of population dynamics, it became clear to Darwin that variation was far more important than had previously been understood. Natural selection would not work without it. This shift in emphasis is often discussed under the rubric of "population thinking." In contrast, the older "typological" approach, which conceived of groups of organisms in terms of stereotypes, viewed the differences between individuals as a kind of departure from a norm, and these were screened out as unimportant. To the extent that variability had been recognized, it was assumed that something held it within definite limits. It was believed that species might vary, but not so far as to give rise to new ones. At a deeper philosophical level, treating species are reproductive populations implied that they were very different from the "natural kinds" of chemistry and other physical sciences. The new Darwinian thinking accepted that species were concrete, that they had a history, and that they could evolve. Treating species as something more than abstractions was a major conceptual innovation.

D. The Long Delay

Following this shift in his thinking, Darwin immediately began elaborating a comprehensive theory of evolution together with its many implications. However, he was busy with other things, and publication of his theory was long delayed. One factor in the delay was the decline of his health, which, if not brought on by stress, was at least exacerbated by it. Ill health was a major consideration in his decision to leave London and live in the country. Darwin had married his cousin Emma Wedgwood on January 29, 1839, and the first of their many children was born on December 27th of that year. He moved to the village of Down, not far south of London, and resided there from September, 1842, until his death on April 19, 1882.

He had already published a semi-popular account of his travels in 1839. Usually called *The Voyage of the Beagle*, it was a great literary success. He also edited the scientific results of the voyage and published three books on the geology of the voyage, including one on coral reefs. He began to publish a few papers on zoology. One of these grew into a vast monograph on barnacles (Cirripedia). It was well over a thousand pages long, and consumed eight years on his working time.

E. The *Origin of Species*

At last, in 1854, the work on barnacles was completed and Darwin began to write a long book on evolution. It was to incorporate much original research as well as a detailed review of the literature. As it happened, Alfred Russel Wallace (1832–1913) was interested in the possibility of evolution too, and spent many years in the field collecting animals and plants and studying their geographical distribution, first in South America (1848–1852) and then in the region that is now mostly part of Indonesia (1845–1862). They entered into correspondence, and in a letter dated May 1, 1857, Darwin informed Wallace that he had an evolutionary theory but did not let on what it was. Wallace may have gained valuable hints from his readings of Darwin's publications, especially the book on the voyage, which contains among other things a brief discussion on Malthus. Nonetheless, he is generally credited with an independent discovery of the principle of natural selection.

A manuscript by Wallace on that topic reached Darwin in June, 1858, and colleagues arranged for a joint publication, which was read at the Linnean Society on July 1. It became available in print the following month, but it was not well understood at the time. Almost immediately Darwin began an "abstract" of the longer book that he had been writing. This "abstract" became *On the Origin of Species by Means of Natural Selection, or the Preservation of Favoured Races in the Struggle for Life*. Published on November 24, 1859, it became the focus of a controversy such as the world has never seen before or since.

As a supporter, Darwin already had his friend and confidant the botanist Joseph Dalton Hooker (1817–1911), and others soon joined them. There were of course public debates, including the one at Oxford on June 9, 1860. Although the exchange there between Bishop Samuel Wilberforce (1805–1873) and Thomas Henry Huxley (1825–1895) is better known, Hooker was really more effective in suport of Darwin's ideas. Darwin himself furthered his interests through behind the scenes negotiation and, more importantly, through research and publication. There were a total of six editions of the *Origin of Species*, the last of which was published in 1872.

F. Later Publications

Plans to complete and publish the big were abandoned, but part of it, somewhat revised, appeared in 1868 as *The Variation of Animals and Plants under Domestication*. Concurrently Darwin had an active empirical research program under way, much of which was conducted on plants. His book of 1862, entitled *The Various Contrivances by which British and Foreign Orchids Are Fertilised by Insects, and on the Good Effects of Intercrossing*, dealt with the problems of adaptation. It was a seminal document of pollination ecology and the study of coevolution. Later works on floral biology, *The Effects of Cross and Self-Fertilisation in the Vegetable Kingdom* (1876) and *The Different Forms of Flowers on Plants of the Same Species* (1877) addressed mainly the effects of inbreeding and outbreeding.

Darwin was a first-rate plant physiologist, and his book *The Movement and Habits of Climbing Plants* had first appeared as a journal article in 1865; it was followed by *The Power of Movement in Plants* (1880), which treated the underlying mechanisms of plant behavior. The book on climbing plants and *Insectivorous Plants* (1875) are early treatises in what later came to be called ethology. They deal with behavior from a comparative and evolutionary point of view. Treating humans, other animals, and even plants as part of an evolutionary continuum was a major contribution to comparative psychology, as well as a serious challenge to traditional philosophical ideas such as mind–body dualism. There was also a lot of behavior in *The Descent of Man, and Selection in Relation to Sex* (1871) and its sequel. *The Expression of the Emotions in Man and Animals* (1872), and in the last of his books, *The Formation of Vegetable Mould through the Action of Worms, with Observations on Their Habits* (1881).

Darwin's, work on sexual selection and the evolution of social behavior as treated in *The Descent of Man* is strikingly modern in its approach. In retrospect, his efforts to solve the problems of heredity seem much less successful, though they stimulated others to work on them. It is unfortunate that his theory of pangenesis was appended to the 1868 book *Variation*, with the result that it is rarely read because that theory was superseded. Although pangenesis explained the inheritance of acquired characters, it was more a theory of development than of heredity. In fact the book presents a remarkably sophisticated treatment of the relationships between embryology and evolution.

III. DARWIN'S CONTRIBUTIONS TO THE STUDY OF BIODIVERSITY

A. The Competitive Natural Economy

Darwin succeeded, where others had failed, in convincing the scientific community that evolution had in fact occurred, and that it was responsible for the patterns of diversity that may be observed in both fossil and living organisms. Biogeography, paleontology, and classification took on new significance, becoming profoundly historical in nature. He was less successful at gaining immediate support for his more fundamental (or at least original) contribution, which was the basic mechanism for evolution and the new way of thinking that went along with it. Evolution by natural selection made it possible to reject all sorts of older ideas about the world, especially the notion that organisms and the "natural economy" had somehow been designed; or produced by a supernatural agency. Darwin had demonstrated that there was much less "purpose" in the world than had been supposed, that there was nothing inherently "good" about the way things are, and that there was no necessary progression from lower to higher levels of organization.

Darwin's theory was immediately recognized as having profound philosophical significance. From time immemorial it had seemed natural to interpret the living world and the universe in general as if they were the product of something like human intelligence. Such notions, and the anthropomorphic way of thinking that underlies them, are called "teleology." The idea that somehow the world had been designed or was otherwise the product of supernatural action was of course fundamental of much religious thinking, even among those who had come to accept early versions of evolution. The remarkable adaptations of living organisms were taken as evidence for the existence of a supernatural order. Although many of the features of inanimate nature could be explained in terms of laws of nature and matter in motion, a more subtle purposefulness was widely attributed to the universe as a whole. Indeed, the great German philosopher Immanuel Kant (1724–1804) had proclaimed that there never could be a "Newton of the grass-blades" able to explain adaptation in strictly naturalistic terms. But that is precisely what Darwin accomplished.

Darwin also provided a way of thinking about adaptation that allows us to avoid the naive and anthropomorphic thought processes that lead to the illusion of design. Rather than ask what things are for, we ask how they might have evolved, and explain the facts in terms of what has happened and why. When this is done properly, it becomes clear that much ascription of adaptive significance has been mere guesswork and that there is far less "optimality" in the world than is sometimes supposed even today. On the other hand, research has repeatedly revealed that supposedly "non-adaptive" features turn out to be very important in the lives of organisms. Defensive metabolites are an excellent example.

Opponents of Darwinism of course include many persons who have been reluctant to abandon teleological interpretations of nature. It has not always been a matter of denying evolution or natural selection outright. Often evolution has been envisaged as a means of achieving some supernaturally ordained plan. And often the teleology is merely assumed or presupposed rather than being advocated explicitly. "Holistic" notions that treat the world as if it were a super-organism have a long history in myth and pseudoscience that goes back to antiquity. They retain their popularity among religious persons and advocates of occult metaphysics. Because such older ways of thinking are still widespread, it is important to avoid giving the appearance of endorsing them. Functional explanations should be called "teleonomic" (i.e., where the apparent purposefulness in organisms is derived from evolutionary adaptation) rather than "teleological," and metaphors such as "design" should be avoided.

The idea of progress has rarely been disassociated from teleological thinking. For that reason, among others, Darwin's ideas about progress have often been misinterpreted and but rarely understood. They are often confused with quite different ideas, especially those of philosophers, social scientists, and non-Darwinian evolutionists. In pre-evolutionary biology it was widely accepted that organisms could be placed in a single series from lower to higher, connecting to non-living matter at the bottom and to the various ranks of supernatural beings at the top. This great chain of being, or *scala naturae*, was later modified to make the order an historical or quasi-evolutionary one. Some authors treated the history of life as something like the development of an embryo, and the succession of organisms would become increasingly more complicated because such change was in a sense built in to the ancestral germ. Others attributed the same basic pattern to laws of nature, laws that were supposedly pre-ordained so as to have such an effect. In either case, it was Divine Providence on a geological scale.

Darwin forthrightly rejected that kind of progressionist thinking, especially as it had been presented by his predecessor Jean-Baptiste Lamarck (1744–1829) and his contemporaries Richard Owen (1804–1892) and Herbert Spencer (1820–1903). There was nothing like a cosmic escalator carrying species upward toward higher levels of organization. Nonetheless, Darwin believed that evolutionary progress is real and that his theory could explain it. He was well aware of the progress that had been made during his own life-time in science and technology. Artificial selection was one technique by means of which domesticated breeds of plants and animals had been much improved. His study of the fossil record clearly indicated to him that analogous directional changes had taken place, even though Lyell and others had long denied that. Natural selection, he reasoned, would lead to the continued accumulation of adaptations that made their possessors increasingly able to compete with those around them. But he saw nothing truly inevitable about that kind of progressive change, for it depends on environmental and other circumstances that are not always realized.

Darwin replaced the older notion of a largely harmonious and cooperative natural economy with that of a fundamentally competitive one. The organisms within populations reproduce differentially as a consequence of how they utilize the resources in their environments, and change occurs in the direction of more effective use of those resources. That accounts for adaptation and for such progressive and regressive changes as may in fact occur. But it greatly limits the kind of adaptation that can evolve, because it is organisms, and to some extent families, that compete as reproductive units. Darwin was profoundly aware of the point that his mechanism would not produce adaptations at the species level or at the level of the community or ecosystem. This point, however, was often neglected by his contemporaries and received due emphasis only in the latter half of the twentieth century.

Darwin's idea of what a species is lacked a clear conception of reproductively isolated populations with distinct gene pools. Furthermore, he did not fully ap-

preciate the difference between genotypic and phenotypic variability. Because his mechanism depended on differences between individual organisms, he needed to show that the necessary variability does in fact exist. He was able to show that species are indeed far more variable than his predecessors had believed. He also discovered that sexual reproduction and outcrossing are much more prevalent than even he had at first suspected. Therefore, his contributions to the study of population structure in nature were the foundation for a great deal of work along such lines.

Variation within and between the component populations of a species is one important form of biodiversity within species. Another is the sort of differences that may be observed between the sexes, and here again Darwin's contribution was fundamental. His theory of sexual selection helped to account for sexual dimorphism on the basis of males competing for the females, either through fighting with one another or through efforts to attract the females. He also discovered heterostyly in plants. This is a kind of polymorphism in which hermaphrodite flowers of the same species have different morphologies, furthering outcrossing and reducing inbreeding.

Darwin also was able to explain the global pattern of diversity between species. Of course the splitting up of species and their ability to undergo an indefinite amount of change allows diversification through geological time. However, that by itself does not tell us why or how the diversification takes place. To account for that Darwin invoked what he called a "principle of divergence of character." According to this principle, diversification allows the organisms to exploit a wider range of resources, so that groups of them can expand their numbers. As he put it, they can occupy more places in the natural economy, or as we would say, they can occupy more Eltonian niches. It is important not to view this diversity in a strictly negative way, as if it were merely a matter of "avoiding competition." Rather, new resources come to be exploited, and the original ones may be exploited more effectively.

Darwin was also one of the first scientists to study competition experimentally. In *The Origin of Species* he described field experiments that document changes in diversity as a result of competition. He believed that competition, is most severe between closely related organisms, and this tends both to drive intermediate forms to extinction and to bring about divergent evolution among the more modified ones. At lower taxonomic levels this means that higher biotic diversity should develop where the competition is most severe, notably in the tropics. At higher taxonomic levels, it means that there is adaptive radiation within such groups as families, orders, and classes. The lack of intermediate forms among extant taxa is the consequence of those forms that were close to the ancestral ones having been driven to extinction. On a global scale, whole biotas would tend gradually to diversify through time. Darwin specified the conditions under which such diversification would lead to the evolution of competitively effective organisms. These included a large area and a diverse topography, coupled with a long period of competitive interaction among the inhabitants. He realized that the biotas of separate biotic provinces have not undergone the same amount or kind of modification, and that similar considerations apply to insular biotas. His insights were thus fundamental to our understanding of invasions, faunal interchanges, and many aspects of extinction.

Darwin contributed a great deal toward our understanding of the interdependencies among species by emphasizing how complicated such ecological relationships may be. He was particularly interested in pollination symbiosis, and his work in that area initiated the study of coevolution between animals and plants. Although his research was focused on showing the importance of outcrossing, it also revealed the ecological significance of pollination.

B. Classification

Systematics is often defined as the study of biodiversity, and Darwin made fundamental contributions to both the theory and the content of that science. Classification is part of everybody's language and thinking, and is necessary for the organization of knowledge, whether it be the knowledge of the lay-person or the professional scientist. The objects of our experience can be referred to by more or less general terms, such as "food," "pasta," and "spaghetti." Because what is true of the more general groups is true of all the groups that fall under them, we have a powerful instrument for thinking about more than just single objects. If pasta is made of grain, and all spaghetti is pasta, then spaghetti is made of grain.

Scientific classification may be much more sophisticated than that of the lay person. Scientists attempt to discover groups of things that share properties that are of theoretical importance. A good example of a scientific classification system is the periodic table of the elements, in which chlorine and bromine are both halogens and undergo similar chemical reactions. The groups in question may be worked out on an empirical basis, and only later understood in theoretical terms. This was approximately the case with Darwin's work

on biological systematics. Human beings could be assigned to more general groups, such as Mammalia, Vertebrata, and Animalia, but it was not obvious why. Sometimes it was attributed to God having some plan in mind when He created organisms, and sometimes unknown laws of nature were thought to be responsible.

Darwin's solution to this puzzle was straight-forward and evolutionary. According to his biodiversity model, species have repeatedly undergone a kind of splitting, giving rise to separate lineages that evolve more and more differences as time since common ancestry becomes greater. The more general the group, the earlier the common ancestry. Once the theoretical reason for traditional classification was understood, a new approach could be developed in which the groups were explicitly recognized as lineages sharing a common ancestor. As Darwin put it: "Our classifications will come to be, so far as they can be so made, genealogies."

Putting that prognostication into practice was not easy and, generally speaking, classifications have been only rough approximations to genealogies. They also generally attempt to express the amount of difference that separates the various groups, and sometimes the criteria conflict. Be this as it may, Darwin was writing from experience. On October 1, 1846, he began to study the anatomy of barnacles in order to write a short paper on an aberrant species that he had discovered. The project expanded to the point that he spent eight years revising and monographing the entire subclass, both living and fossil, and treating all aspects of their biology in great detail. Although he did not explain what he was doing at the time, it is clear that his classification is based on his understanding of the phylogeny of the group. Subsequent workers have modified and enlarged his sysetm, but the basic arrangement remains the same. Why Darwin delayed publishing his evolutionary theory in order to undertake a major research program in systematics has been something of a puzzle to historians. One possible reason, however, is quite clear. He was interested in topics such as the evolution of organisms with separate sexes from ancestral hermaphrodites. The problems were solvable, but only if the sequence of changes could be documented, and that meant reconstructing the pattern of branching in the phylogenetic tree. Without good systematics, much of evolutionary biology would consist of untested hypotheses and conjectural history. This is as true today as it was for Darwin.

Because, as was already mentioned, Darwin did not have an adequate species concept, modern systematists have made some important changes in Darwin's barnacle classification. In particular, many of the groups that he treated as mere varieties are now classified as species. In many other respects, however, his approach was very advanced for his day. He worked with extensive series of specimens and went out of his way to document all kinds of variation.

IV. DARWINISM AFTER DARWIN

A. Alternatives to Darwinism

Darwin maintained that natural selection is the "main, but not exclusive" evolutionary mechanism. This seems about as good a criterion as any for distinguishing "Darwinians" from a wider range of scientists who accepted evolution but either rejected natural selection or treated it as an unimportant relative to other causes. Among these scientists, the followers of Jean-Baptiste Lamarck (1744–1829), who emphasized use and disuse, were particularly numerous. However, there were other alternatives. Étienne Geoffroy Saint-Hilaire (1772–1844), whose views are often confused with those of Lamarck, believed that the environment directly induces changes in structure. Indeed, Darwin was strongly influenced by Geoffroy, and accepted the direct action of the environment as a minor cause of change. There were also various versions of "orthogenesis" or evolution that is directed in a particular direction. Some versions of orthogenesis were based on the notion that evolution is like embryological development, and change has been built in from the start. In other versions, laws of nature, similar to those that determine the structure of minerals, were invoked. Sometimes the appeal was to unknown causes, perhaps of a supernatural character. Many scientists were, or at least claimed to be, agnostic with respect to what causes evolution.

B. Early Darwinians

Given such a range of alternatives and the small amount of research that had been done, it makes sense that from the outset there were few Darwinians other than Darwin himself. Because he was largely responsible for getting evolutionary thinking in general accepted by the intellectual community, a lot of evolutionists who did not accept natural selection nonetheless considered themselves his followers. The degree to which Darwinism, in the sense that we use that term here, was a minority position has sometimes been exaggerated. We can identify quite a number of important contemporaries of Darwin who established successful research programs based on the study of natural selection. Foremost

among these of course was Alfred Russel Wallace (1832–1913), its co-discoverer. Also very distinguished was Wallace's traveling companion, Henry Walter Bates (1825–1892), the discoverer of Batesian mimicry. Second only to Darwin in his mastery of the theory was Fritz Müller (1822–1897). He is best remembered for his discovery of Müllerian mimicry, but he also was the first to propound the idea that developmental stages may recapitulate evolutionary ones. Both Fritz Müller and his brother Hermann (1829–1883) conducted magisterial research on pollination symbiosis under Darwin's influence. It is worth emphasizing that these scientists were outstanding for their performance as naturalists in the field. The kind of research that they did has been fundamental to our understanding of biodiversity because it documents how natural selection takes place in real environments.

Darwin also had important followers whose work was more focused in the museum and the laboratory. He had a close circle of followers in the botanist Joseph Dalton Hooker (1817–1911) and the zoologists George John Romanes (1848–1894) and John Lubbock (1834–1913). There was also August Weismann (1834–1914), whose ideas about the continuity of the germ plasm made natural selection seem a much more plausible explanation for evolution than Lamarckism. Neo-Darwinism is rightly associated with the name of Weismann, whose basic position was that natural selection is not just the main, but the exclusive evolutionary mechanism. Actually he admitted two minor ones that had been invoked by Darwin: sexual selection and pleiotropy.

C. Genetics and the Modern Synthesis

Beginning in 1900, Mendelism, developed from the revolutionary research of Gregor Johann Mendel (1822–1884), began to supply the genetics support that Darwinism needed. However, it did not have the immediate effect of gaining adherents to Darwin's theory. Instead it gave rise to alternatives such as mutationism. Advocates of mutationism were over impressed by conspicuous mutants and erroneously concluded that their sudden appearance in a single generation could be extrapolated to the origin of new species and even higher taxa. However, the study of genetics soon moved away from mutationism and gradually undermined some of the other alternatives, especially Lamarckism and orthogenesis.

The reconciliation of Darwinism with genetics depended in part on the elimination of such alternatives, in part on the growth of theoretical population genetics, and in part on the study of natural populations in the field. Studies of natural populations by systematists like Ernst Mayr (b. 1904) and geneticists like Theodosius Dobzhansky (1900–1975) were crucial to the "evolutionary synthesis" that began in the 1930s and matured by around 1950. A new and more sophisticated biological species concept emerged, and the richness of genetic diversity within populations became much better understood. The result was a theoy very much like Darwin's, but without the minor elements of Lamarckism and Geoffroyism, and integrated with modern genetics.

This modernized Darwinism continued to be expanded and refined. Regarding biodiversity, there were two important developments, both of which involved a return to Darwin's original theory. First, there was Darwin's ecology, which was based on individualistic competition and which rejected the notion that organisms do things "for the good of the species" or "in order to benefit the ecosystem." Ecologists had commonly treated species and ecosystems as if they were super-organisms and had attributed a kind of adaptation to them that could only have resulted from selection of populations. The rejection of "group selection" in its unsophisticated and uncritical form was largely carried out in the 1960s and 1970s.

Second, there was a return to Darwin's idea that classification should be genealogical. Although both the extent and the manner to which this program should be carried out have been controversial, it is generally agreed that comparative biology is best conducted within the framework of phylogenetic relationships. Ill-defined "similarity" has proved too subjective and not rigorous enough for modern biodiversity research. This development has partly been the result of new methodologies (especially cladistic analysis). It has also been strengthened by new kinds of evidence, such as isozyme data and genomic sequencing, that supplement the more traditional data.

D. Contemporary Developments

From time to time Darwinism and neo-Darwinism are challenged by what are purported to be genuine alternatives. Some of these challenges, such as efforts to resurrect Lamarckism or orthogenesis, have simply not measured up. Others, such as developmental constraints, are components of Darwinism that may not have re-

ceived as much attention as they deserve. Still others, such as the notion of punctuated equilibria, with its occasional periods of rapid change interspersed with long interludes of stasis, are by no means incompatible with the Darwinian tradition.

Darwinism has never been so monolithic or devoid of major unanswered questions as to preclude a broad range of possibilities for new developments. A present it is by no means clear to what degree processes at various levels need to be invoked to provide a satisfactory account of evolution. The "reductionist" view that would explain everything in terms of genes and essentially ignore organisms and species is obviously too simplistic. The models used in theoretical population genetics make all sorts of unrealistic assumptions, and more realistic ones might give surprising results. There is still no really satisfactory explanation for the prevalence and ecological patterns of sexuality. In spite of much recent progress, many of the tarditional problems of phylogenetics remain highly debatable. Much work remains to be done in turning branching phylogenetic diagrams into explanatory historical narratives. The vast majority of extant species remain undescribed, and the fossil record consists largely of "roadkills" strung together with gaps. There are plenty of opportunities for important new discoveries within the Darwinian research tradition.

See Also the Following Articles

ADAPTATION • ADAPTIVE RADIATION • BIODIVERSITY, ORIGIN OF • COEVOLUTION • PHYLOGENY • SPECIATION, PROCESS OF • SPECIES, CONCEPT(S) OF • SYSTEMATICS, OVERVIEW

Bibliography

Browne, J. (1995). *Charles Darwin Voyaging*. Knopf, New York.

Gayon, J. (1998). *Darwinism's Struggle for Survival: Heredity and the Hypothesis of Natural Selection*. Cambridge University Press, Cambridge, United Kingdom.

Goldie, P., and Ghiselin, M. T. (1997). The Darwin CD ROM. Lightbinders, San Francisco.

Mayr, E. (1991). *One Long Argument: Charles Darwin and he Genesis of Modern Evolutionary Thought*. Harvard University Press, Cambridge, Massachusetts.

Mayr, E., and Provine, W. B. (1980). *The Evolutionary Synthesis: Perspectives on the Unification of Biology*. Harvard University Press, Cambridge, Massachusetts.

DEFENSES, ECOLOGY OF

Phyllis D. Coley* and John A. Barone[†]
*University of Utah and [†]Mississippi State University

GLOSSARY

chemical defenses Compounds used by plants to deter or poison herbivores and pathogens.
constitutive defenses Defenses that are manufactured and maintained by a plant, regardless of whether it has been attacked by an herbivore or pathogen.
endophyte Fungus or other organisms residing or growing within plant tissues.
herbivory Damage to plant tissues by herbivores or pathogens.
induced defenses Plant defenses, including both chemical and physical defenses, that are produced, at least in their final form, only after the plant has been damaged by herbivores or pathogens.
mutualism Interactions between organisms of different species that increase the fitness of both participants.
secondary compounds A synonym for chemical defenses in plants; contrasts with chemical compounds used in primary metabolism, such as photosynthesis and cellular respiration.

PLANT DEFENSES ARE ADAPTATIONS that reduce the damage and mortality caused by herbivores and pathogens. Here we describe the diverse array of features that have a defensive role in plants. Most prominent are chemical defenses that plants use to deter or poison their natural enemies. Also important are physical defenses, such as spines and trichomes, that prevent herbivores from feeding on plant tissues, and mutualisms, in which plants feed or house ants, other arthropods, or fungi in exchange for their defensive help. All of these defenses are presumed to cost plants in terms of energy and nutrients, and they only evolve when their benefits outweight these costs. Selection, however, has favored different investments in defenses across habitats, reflecting both underlying costs of defenses and the potential for plant damage. We conclude with a discussion of how humans have used plant defenses, primarily chemical compounds, in their everyday lives, underscoring their importance in both traditional and Western societies.

I. DIVERSITY AND FUNCTION OF CHEMICAL DEFENSES

Plants contain a tremendous diversity of chemical compounds that have no function in any aspect of primary

TABLE I

Types of Chemical Defenses in Plants

Class of compound	Activity	Examples
Alkaloids	Variety of mechanisms; often disrupt nervous system function	Caffeine, nicotine, morphine
Nonprotein amino acids	Disrupt proteins when incorporated in herbivores	Canavanine
Phenolics	May bind proteins, reducing the nitrogen available to herbivores	Hydrolyzable tannis and condensed tannins
Saponins	May reduce uptake of sterol from herbivore digestive system	Digitonin
Terpenes	Various; often toxic	Pyrethroids, hormone mimics
Toxic proteins	Variety of mechanisms; highly toxic compounds	Ricin

metabolism, such as in cellular respiration or photosynthesis. Therefore, these compounds are frequently called "secondary compounds" and were originally thought to be either waste products or storage molecules. In the 1950s, however, it was realized that these compounds were in fact defenses against herbivores and pathogens. There is an impressive variety of forms of these compounds, and we begin with a brief description of the more common types and how they act. For additional details, see Rosenthal and Berenbaum (1991) and Harborne (1988) (Table I).

A. Phenolic Compounds

Perhaps the most common chemical defenses in plants are composed of phenols. Simple phenols consist of a benzene ring with a hydroxyl group attached, along with varying functional groups. More than 200 different simple phenols have been described, and they have been found to deter feeding of both invertebrate and vertebrate herbivores. Several studies have shown that plants with higher concentrations of phenolics were avoided by species of herbivorous birds (Harborne, 1988). For example, the Canada goose selects plant species that are low in phenolic content, and in the subarctic phenolic resins deter the feeding of ptarmigans and grouse on several plant species. In the latter case, the resins may be effective defenses because they have antimicrobial effects and may disrupt disgestion by microbes in the birds' caeca.

Compounds composed of multiple phenolic units are called tannins, most of which belong to either of two groups. Hydrolyzable tannins have relatively simple structures that are made up of phenolic acids, whereas condensed tannins (also called proanthocyanidins) have more complicated structures and are made from the condensation of hydroxyflavanol units. Condensed tannins are also more common across plant taxonomic

groups, occurring in 54% of angiosperm and 74% of gymnosperm genera that have been examined. In contrast, hydrolyzable tannins are found in only 13% of angiosperm genera (all dicots) and are absent from gymnosperms. Although the actual mechanism of tannin action is controversial, tannins may function by binding proteins. As a herbivore chews a leaf, the tannins stored in the leaf vacuole are released and mix with the proteins, binding them and rendering them indigestible. Since proteins are a primary source of nitrogen for herbivores, this chemical reaction makes leaves less nutritious, although it is unclear whether this reaction is effective in the digestive tract of herbivores. Nonetheless, evidence suggests that tannins do deter feeding and are even toxic to some herbivores.

B. Alkaloids

Though a structurally and biochemically diverse group of secondary compounds, all alkaloids have a heterocyclic ring that either contains a nitrogen atom or has one attached to it (Roberts and Wink, 1998). Some of the best known include atropine, the main poison in deadly nightshade (*Atropa belladonna*), quinine, caffeine, cocaine, morphine, and nicotine. It has been estimated that between 15 and 20% of all vascular plants contain alkaloids, and that they are found in one-third of angiosperm families, primarily the dicots. The distribution of alkaloids in relation to climate or geographic distribution remains in dispute, but in general it appears that alkaloids are a more common form of defense in tropical habitats than in the temperate zone. About 16% of temperate species in one survey contained alkaloids versus 35% of tropical species. The mechanisms by which alkaloids affect herbivores (including humans) are highly varied and specific and often effective at small doses, making them effective poisons (see Section VII). Many that have been described affect the functioning of the autonomic nervous system by competing with neurotransmitters. An alkaloid found in the genus *Erythrina* (Fabaceae) blocks acetylcholine receptors in animals, whereas caffeine, an alkaloid found in *Coffea arabica* (Rubiaceae), causes the release of calcium from neural receptors. Both types of disruption can lead to the death of susceptible herbivores.

C. Terpenes

These are an extremely heterogeneous group of chemicals derived from a common metabolic pathway that produces polymers of isoprene units, a five-carbon compound. The isoprene units may then be arranged in a variety of ways, generating a diverse range of nonpolar compounds. Monoterpenes, which have 10 carbon atoms, are often volatile and include pyrethroids from the genus *Chrysanthemum* that are highly toxic to insects and used as commercial insecticides. These neurotoxins disrupt sodium channels in neurons, causing uncoordinated movement and paralysis. Sesquiterpenes, diterpenes, and triterpenes are larger polymers containing 15, 20, and 30 isoprene units, respectively. Terpenes are especially common in pine and fir trees, which secrete the chemicals in resins that reduce insect attack, although monoterpenes can act as feeding attractants for some species of beetles.

Terpenes can be combined with other molecules to form additional types of secondary compounds, for example, sesquiterpene lactones. In this case, a lactone ring (composed of three carbon atoms, with an oxygen and a carbonyl group) is attached to a sesquiterpene. These compounds are found predominantly in the family Asteraceae. Studies have shown that they are toxic to a variety of herbivorous insects and mammals, particularly livestock. Currently, the cause of the toxicity of sesquiterpene lactones is not known, although they cause tissue lesions in mammals and may disrupt hormone function in insects.

Another group of compounds that are chemically related to triterpenes have the remarkable feature that they can act as analogs to hormones found in animals. These analogs do not appear to have any function in plant metabolism, but their similarity to animal hormones is so striking that it is easy to believe that they evolved primarily as a defense against herbivores. Among these plant hormones are mimics of human female sex hormones, and these plant "phytoestrogens" are being used as alternative medicines.

Some plant species also produce analogs that mimic insect hormones. Two critical hormones in insects are juvenile hormone and molting hormone, both of which are necessary for the development of larvae or juveniles. Molting hormone mimics (or phytoecdysones) are common in ferns and gymnosperms (but rare in flowering plants) and, when consumed by insect larvae, disrupt hormonal control of development, in some cases causing death by preventing the insect from shedding its cuticle. Juvabione and other analogs of juvenile hormone are less common than mimics of molting hormone but appear to have similar, fatal effects in insects.

Three other important groups of terpenes are cucurbitacins (found in the family Cucurbitaceae), limonoids (found in Rutaceae, among others), and saponins, which are discussed in the following section.

D. Saponins

A type of triterpenoid glycoside, saponins are complicated molecules composed of a polycyclic structure (with either 27 or 30 carbon atoms) attached to a carbohydrate group. Consequently, these molecules have both hydrophobic and hydrophilic properties, making them water soluble. Taxonomic surveys suggest that saponins are widespread; one study found that they are present in 500 species from 80 plant families. Saponins have been shown to be toxic to many arthropods, including mites, beetles, and lepidopterans. In insects, saponins may reduce the uptake of sterol from the digestive system. Since sterol is necessary for insects to molt, saponins apparently interfere with insect development. There is little evidence, however, that saponins are toxic to most mammals.

E. Nonprotein Amino Acids

Many plant species produce amino acids that are not used in making proteins but instead are involved in defenses against natural enemies. More than 600 of these nonprotein amino acids have been identified from a wide range of plant species. For example, canavanine, a structural analog of the amino acid arginine, has been found in 1500 species in the legume family. Fed to sensitive insect herbivores, canavanine causes dramatic and often fatal defects in development. This effect occurs because canavanine becomes incorporated into newly synthesized proteins in place of arginine, thus altering their structure and function. A species that produces canavanine, the woody vine *Dioclea megacarpa*, has seeds that are 8% canavanine by dry weight. Not surprisingly, these seeds are largely free from attack except by one specialized beetle (Rosenthal and Berenbaum, 1991).

F. Toxic Proteins

Plants also make toxic defensive proteins that produce their effects through a variety of different mechanisms (Kiowa *et al.*, 1997). One such protein, ricin, is produced by the castor bean (*Ricinus communis*) and functions by inhibiting ribosomes in herbivores. By weight, it is among the most toxic compounds known. In general, there is considerable interest in identifying genes that code for toxic proteins in wild plants. If such genes could then be used to modify agricultural crops, they would be protected from herbivores without the use of chemical insecticides.

II. PHYSICAL AND STRUCTURAL DEFENSES

A. Spines

In addition to chemical defenses, plants also employ a wide range of physical or structural features that either reduce the accessibility of plant tissue to herbivores or deter herbivore feeding. The most apparent of these are spines or thorns on plants. There is little doubt that the primary function of these structures is as a defense against herbivorous animals, particularly vertebrates, although in some cases spines may help plants to thermoregulate or, in the case of some vines, to climb.

B. Trichomes

In some ways resembling small spines, trichomes are structures found on the leaves and stems of many plant species that inhibit attack by herbivores and pathogens, either physically or chemically. The shapes and sizes of trichomes vary considerably across species from slender hairs to stout spikes, and many are armed with recurved barbs. Furthermore, glandular trichomes of many species produce or store chemicals that act as deterrents to herbivores. There are countless examples from plants in both natural and agricultural settings of how trichomes can kill or deter herbivores (Levin, 1973). Acting as physical barriers, trichomes can reduce egg laying by herbivores on some varieties of wheat and increase plant resistance to leaf hoppers and mites in some varieties of cotton. The chemicals secreted by many trichomes include sticky gums that immobilize small insects, as found in some varieties of wild potato and tomato. On other species trichomes secrete chemicals such as alkaloids and terpenes as well as waxes, fatty acids, and alcohols. These chemicals are frequently toxic to herbivores, for example, nicotine (an alkaloid) secreted by tobacco plants (*Nicotiana* in the Solanaceae) that kills aphids on contact. Thus, trichomes not only act as physical defenses but also provide a way for plants to use chemical defenses against their natural enemies without first having to suffer tissue damage.

C. Toughness

Leaves with thick cell walls, consisting of lignin and fiber, are tough. In one assessment of both physical and chemical defenses of mature leaves in tropical forests, leaf toughness and fiber content were the two most important factors in reducing herbivory. Toughness is an effective defense against insect herbivores, particularly small or immature ones, because they have a difficult time cutting or chewing the leaves. Furthermore, cell walls are largely indigestible; therefore, tough leaves are nutritionally poor.

III. BIOTIC DEFENSES AND THE THIRD TROPHIC LEVEL

In addition to using chemical and physical defenses against herbivores and pathogens, many plant species have evolved complex, mutualistic interactions with other groups of organisms that act as a type of defense against herbivores. In such interactions, the plants typically provide food, shelter, or both, whereas the other organisms defend the plant from its natural enemies.

A. Ants

The best studied defensive mutualisms in plants are those with ants (Huxley and Cutler, 1991). Given the incredible abundance of ants in most habitats and their propensity to forage on leaves and stems, it is not surprising that these mutualisms are so common. Ant–plant interactions vary from loose, facultative associations in which the plant offers rewards to any ants in its vicinity to more tightly coevolved relationships in which both partners display highly specialized traits. In facultative associations, a plant offers nectar, food bodies, and other rewards to lure ants that nest elsewhere to patrol its leaves and remove any herbivores they encounter. These associations are especially common in tropical forests.

There are also many obligate ant–plant mutualisms, in which one or both participants require the other to survive. The best known is the interaction between the tree *Acacia cornigera* (Fabaceae) and the ant *Pseudomyrmex ferruginea* (Pseudomyrmecinae) in Central America. The plants in this pair have numerous adaptations for playing host to the ants, including swollen thorns on which the ants have their colonies, enlarged nectaries on the leaves that provide the ants with sugars, and modified leaflet tips on which the ants can feed. The young queen ant establishes a colony by landing on a young tree and setting up a nest in one of the swollen thorns. After the colony is established, the worker ants patrol the tree constantly, removing any other insects on the tree and cutting away any other plants that touch it. In one study, when the ants were experimentally removed the plants suffered high rates of defoliation, leading to a reduction in growth and a doubling of mortality during the 11 months of the study (55 vs 28% for control plants with ants). Thus, there is little doubt that ants can play an important role in protecting plants in these tightly obligate mutualisms.

The clear defensive effect of ants on their host plants has been demonstrated for other species as well. Less easily measured is the cost of the ants as a defense: Presumably the benefits outweigh the costs, but what are they? Recent studies show that *Cecropia* trees produce different types of food rewards for their ants in relation to the nutrients available—in effect producing rewards that use the less limiting nutrients. Circumstantial evidence also suggests that the ants are a costly defense since on islands on which native ants are rare or absent, few plant species have either extrafloral nectaries or food bodies.

B. Domatia, Mites, and Other Predators

The mutualistic relationship between ants and plants is apparently paralleled by similar but looser associations between plants and small arthropods. Many plant species have structures on their leaves called domatia that can serve as homes for mites and small, predaceous insects. Although there are a variety of forms, most domatia are pits, pockets, or tufts of hairs on the undersides of leaves. Domatia are widespread taxonomically among dicots, occurring in 28% of 290 plant families, and they are quite common, at least in some forests. Recent work has supported the hypothesis that domatia provide shelter to arthropods, which in turn attack herbivores on the plant. For example, when artificial do-

matia were placed on the leaves of cotton plants, the density of predaceous bugs increased, the number of herbivores decreased, and the number of fruit produced increased by 30%. Although preliminary, this work suggests that domatia are a part of loose mutualisms used by plants to defend themselves.

C. Endophytic Fungi

In addition to using animals as a means of defense against herbivores and pathogens, there is considerable evidence that plants can employ fungi in a similar manner. Most plants are infected asymptomatically by endophytic fungi—that is, fungi that live within the plant, usually between the plant cells. These endophytes are a potential defense since they produce chemical compounds not made by plants.

Although endophytes are common in an impressive diversity of plant species, most research has focused on their relationship with grasses. In fact, their defensive role in plants was only appreciated after it was discovered that cattle suffered toxic symptoms after eating tall fescue grass infected with a fungal endophyte. Since then, grasses infected with fungal endophytes have been shown to be toxic to other domestic and wild mammalian herbivores as well as herbivorous insects. At the community level, research has shown that pastures dominated by endophyte-infected grasses have lower populations of small mammals. Thus, it seems likely that plants can gain some protection against herbivores from fungal endophytes.

D. Predators and Parasitoids as Plant Defenses

Plant defenses have evolved within ecological communities in which predators and parasitoids that attack their herbivores are often common. Evidence suggests that some plant species may directly or indirectly elicit the help of these predators and parasitoids as a form of defense. Recent research has found that some plants release volatile compounds when damaged, and that these are used as location cues by parasitoids and predators (Turlings and Benrey, 1998). Frequently, these volatile chemicals are emitted only when damage is accompanied by oral secretions of herbivores, meaning that the signals are specifically released when a herbivore is present and not when the plants incur other types of damage. By facilitating the discovery of herbivores by their predators and parasites, plants may enjoy reduced herbivory.

Other defenses, such as tannins and toughness, do not kill herbivores outright but reduce their growth and prolong the time until pupation. As a result, herbivores may actually consume more leaf tissue than if the plants lacked these defenses. This seems paradoxical, but by slowing growth these defenses increase the chance that herbivores will be preyed on while still in the earlier instars (Benrey and Denno, 1997). Since the majority of leaf damage occurs in the final instar, this could greatly reduce herbivory. Thus, the effectiveness of some defenses may depend critically on help from parasitoids and predators.

IV. PHENOLOGICAL STRATEGIES

New, expanding leaves are generally more vulnerable to attack by herbivores and pathogens than mature leaves because they are tender and nutritious. In addition to the defenses already described, many plant species reduce new leaf damage by altering the timing of new leaf production. These phenological strategies are of two general types. In the first, plants produce new leaves during times of the year when herbivores and pathogens are rare. In temperate forests, the spring flush of leaves occurs during the short window of opportunity between the end of winter and the recovery of herbivore populations in the late spring. Similarly, studies conducted in tropical forests in India, Ghana, and Panama have shown that when plants produce new leaves during the dry season, when herbivores are rare, herbivory is reduced. Obviously, this seasonal escape strategy would not work in nonseasonal climates, and because of coincident changes in light and water availability with season it is difficult to demonstrate conclusively that the timing of leaf production is mainly an adaptation to reduce herbivore and pathogen damage.

The second phenological strategy to decrease damage to new leaves is for the plants of a given species to produce new leaves synchronously. By flushing simultaneously, plants may be able to overwhelm their specialist herbivores with an abundance of new leaves so that the chance that any particular new leaf will be discovered and eaten is decreased. One test of this strategy involved following new leaf production and herbivory in a moist tropical forest for a year. The results showed that highly synchronous species experienced lowered herbivory than those that were continuous in production. This phenological strategy may be more important in less seasonal forests in which a seasonal escape strategy may be less effective.

V. PLANT INVESTMENTS IN DEFENSES

A. Assumptions

In their evaluations of plant defenses, ecologists typically make the following assumptions: (i) Herbivory and other tissue damage by natural enemies is bad for plants, reducing their fitness; (ii) defenses reduce damage by natural enemies; and (iii) defenses cost the plants in terms of energy and nutrients.

Surprisingly, the first assumption about the negative effect of herbivory on plants has been controversial. Several studies have shown that some plant species, in certain circumstances, are able to "overcompensate" for tissue damage, achieving higher fitness than those that were never damaged. However, in general the evidence is quite clear that herbivory, for most plant species, reduces fitness. The billions of dollars spent annually throughout the world on pesticides to protect crops are a testament to the negative effects of herbivores and pathogens. Experimental studies have also demonstrated that damage reduces plant fitness. In a study by Marquis, for example, shrubs of the species *Piper arieianum* (Piperaceae) were artificially defoliated and their growth and seed production followed for 2 years. The results showed that individuals that had lost at least 30% of their leaf area suffered a 50% decrease in growth and produced approximately half the seeds of control plants. Clearly, damage affects plant fitness.

B. Costs and Benefits of Defenses

Plant defenses may be costly to plants for several reasons (Fritz and Sims, 1992). First, in order to invest in defenses, plants must divert resources from other aspects of growth and reproduction. For example, nitrogen that a plant invests in alkaloids as a defense cannot be used in making proteins necessary for greater photosynthesis. A second reason why defenses may be costly to plants is that some chemical defenses, although effective deterrents against natural enemies, may also be toxic to the plant, so the plant has to expend additional energy protecting itself from its own defenses. Tannins must be sequestered in the vacuole to avoid cell damage, and many terpenes and resins are restricted to specialized canals in plants.

Third, a defense that reduces plant damage from one herbivore or pathogen species may inadvertently make the plant more attractive to another. For example, Eisner and coworkers showed that the trichomes on the leaves and stems of *Mentzelia pumila* (Loasaceae) frequently ensnare a variety of insects, including many damaging herbivores. However, the trichomes do not affect a damaging aphid species, whereas they kill one of its predators. Thus, the trichome defense protects the plant from some herbivores but makes it more vulnerable to others.

Despite a variety of direct and indirect costs of defense, plants generally benefit from being defended from herbivores and pathogens. One instance was already discussed: When the ants were removed from ant-defended *Acacias*, herbivory and plant mortality increased. Other studies of chemical defenses have shown the benefits of chemical defenses as well as their costs. One such study was conducted by Berenbaum and associates, who examined whether wild parsnips benefited from being chemically defended in the presence of the herbivore parsnip webworms. They showed that plants with greater chemical defenses had higher fitness in the presence of the webworms, meaning that the defenses had a clear benefit. The chemicals were costly, however, because in the absence of herbivores in the greenhouse plants that were better defended chemically had lower seed production. Under natural conditions, however, the benefits of reduced herbivore damage outweighed the costs to these plants.

C. Induced Defenses

The chemical defenses described so far are usually constitutive—that is, they are produced and maintained regardless of whether herbivores or pathogens have damaged the plant. Many secondary compounds, however, are induced, with production (or at least the final stages of production) only occurring after the plant has been attacked (Karban and Baldwin, 1997). Usually, an aspect of damage, such as partially eaten cell walls, leads to a transduction process that causes the cell, tissue, or whole plant to begin synthesis of defensive compounds. This entire process can take place in less than 1 hr. In field and laboratory conditions, plants that have been induced suffer less herbivory and have higher fitness than controls. However, most species do not appear to have inducible defenses.

Plants should have induced defenses instead of constitutive ones if defenses are costly to plants, and energy and nutrients that are allocated for secondary compounds or other compounds cannot be used for growth or reproduction. Therefore, by producing defenses only when they are needed—during an attack by herbivores or pathogens—the plant is able to divert these resources to growth and reproduction. However, most species

do not appear to have inducible defenses, suggesting that plants may frequently not be in the position to predict when it would be advantageous to induce defenses.

D. Theories of Plant Defense

Both across and within habitats, the amount of damage that plants suffer from herbivores and pathogens varies enormously. In a tropical forest, for instance, rates of herbivory to mature leaves can vary more than two orders of magnitude, from species that receive virtually no damage to those that lose more than 0.6% of their leaf area per day. This huge range implies that plants differ in both the types and amount they invest in defenses.

Much research has focused on trying to correlate defensive investment with other plant life history traits and habitat preferences. It is generally accepted that short-lived, fast-growing species, which typically are found in resource-rich habitats, are not as well defended as slower growing, longer lived woody species. However, the underlying reasons for these patterns are not understood, and numerous theories have been proposed. Two important elements appear in many of these theories: value and risk. First, selection should favor high investments in defense if the risk of herbivory is high. Second, effective defenses should be favored in tissues with a high value or replacement cost.

Apparency theory was the first influential attempt to provide a theoretical framework for plant defenses (Feeny, 1976). It emphasized differences in the risk of herbivory, arguing that defense investment should depend on how apparent plants are to herbivores. Species that are ephemeral, such as annuals, may not be readily apparent to herbivores, whereas long-lived trees and shrubs should be readily found. Thus, unapparent plants should be able to largely escape specialist herbivores and, as a result, should invest less in defenses and use defenses that are toxic at low concentrations to generalist herbivores. These "qualitative" defenses include low-molecular-weight compounds such as alkaloids, cyanogenic glycosides, and monoterpenes. In contrast, apparent plants are certain to be discovered by host-specialist herbivores and should produce leaves that are generally less palatable. The theory predicts that they should invest in "quantitative" defenses, such as tannins and toughness, that are not poisonous per se but have dosage-dependent effects, making leaves harder to digest. One test of this prediction found that herbaceous plants had a greater abundance of toxic alkaloids than more apparent, woody plants. Several authors have also suggested that the risk of discovery by herbivores may depend on the frequency of other more palatable species in the community.

Other theories, although acknowledging that herbivore pressure is important, also suggest that the value of the plant tissue should influence the evolution of defense. For example, species adapted to resource-poor environments, such as those with low light or poor soils, have inherently slow growth rates and long leaf lifetimes. The resource availability hypothesis (Coley et al., 1985) argues that slower growing species should invest more in defenses than fast-growing ones because the cost of replacing damaged leaves is higher when resources are limited. Moreover, in slow growers, the decrement of lost growth for a given investment in defenses is small because the species are already growing slowly. However, in a fast growing species, the investment in defenses leads to a higher percentage decrease in plant growth. This means that the relative or opportunity cost of plant defenses is greater for fast-growing species than for slow growers. Thus, for a given abundance of natural enemies, plant species adapted for growth in high resource environments should maximize their growth by investing less in defense and suffering greater herbivory. The optimal level of defense for species adapted to low resource environments should be much higher.

The resource availability theory also predicts the general types of defenses that plants should use. Species with long-lived leaves should invest in defensive compounds that, despite high initial costs due to their high concentrations, have low turnover rates. These defenses include fibers, which increase leaf toughness, and tannins, both of which would be quantitative defenses in the plant apparency theory. In fast-growing species with short-lived leaves, these defenses would not be profitable because the leaves would be dropped before the plants would have an opportunity to recoup their investment. Instead, the hypothesis predicts that species with short-lived leaves should invest in compounds that are effective at low concentrations even though they may have high turnover rates. The combination of low concentration (or low incremental cost) and high turnover rates means that these defenses are relatively less costly to plants that have short-lived leaves, especially because the components of these defenses can be withdrawn from leaves before they senesce. These defenses would be very costly for species with long-lived leaves since they would have to be continually regenerating these defenses over a long period of time.

VI. PLASTICITY IN DEFENSES

The theories of plant defenses presented in the last section attempt to explain why selection would favor different amounts and types of defenses in different species. An underlying assumption is that defense strategies will be optimized for the predominant conditions experienced by each species. However, there is variation around this optimum, depending on environmental perturbations. In general, if a particular resource is available in excess of normal growth requirements, the extra is shunted into making defenses (Bryant et al., 1983). For example, if a plant is fertilized but retained in low light, it will shunt the extra nitrogen into nitrogen-containing defenses such as alkaloids. Of course, this occurs only in species that have the ability to make alkaloids or other nitrogen-based defenses. In contrast, if a plant is placed in a high light environment, the high rates of photosynthesis will lead to an excess of carbon in the form of starch. This extra carbon will be used to produce higher levels of carbon-based defenses such as tannins and terpenes. Note that these patterns are the opposite of the interspecific trends in which high resources select for lower defense (see Section V,D). Thus, shifts in the ratio of carbon and nitrogen available to an individual plant will cause phenotypic changes in defense allocation patterns that do not necessarily reflect the optimal trends seen across species.

VII. HUMAN USES OF PLANT DEFENSES

In their daily lives, humans have long made use of plant defenses, especially plant secondary compounds. This use has been one focus of ethnobotany, which is the general study of the relationship between humans and plants. Here, we briefly review a few examples of how these plant chemicals have been employed by people. More detailed information can be found in Johns (1996).

A. Hunting and Fishing

Undoubtedly, one of the first uses of plant secondary compounds by humans was as poisons on the tips of arrows. As described by Neuwinger (as cited in Roberts and Wink, 1998), such poisons were formerly used with arrows throughout the world (except perhaps in Australia and New Zealand) and are still used in some areas, both in hunting and in warfare. Poisons for hunting must be highly toxic in small quantities and either easy to process during hunting trips or persistent while on arrow tips so they do not have to be replaced often. For these reasons, cardiac glycosides and alkaloids are the most common arrow poisons.

Cardiac glycosides, which are secondary compounds common in the families Asclepiadaceae and Apocynaceae, were mostly used as arrow poisons in Africa. They were often the primary active ingredient in concoctions of several different plant extracts, with the other plants added to enhance the effectiveness of the poison, to help it adhere better to the arrows, or to fulfill a magical purpose. In northern South America curare poisons were used in blowgun darts. These were typically derived from plants in the genus *Strychnos* (Loganiaceae; also used in Africa) or from species in the Menispermaceae. Chemically, curare poisons all contain quaternary alkaloids, albeit with different structures. Once in the bloodstream, these block the transmission of neural impulses to the skeletal muscles by competing for receptors with the neurotransmitter acetylcholine. Progressive paralysis of the skeletal muscles follows, eventually leading to respiratory failure in the animal (Roberts and Wink, 1998).

In many traditional societies, poisons from plant secondary compounds have also been used in fishing. These poisons are normally put into small ponds or slow-moving streams, in which the concentration of poison will be high enough to kill or at least stun the fish. A variety of compounds have been used as fish poisons, including isoflavonoids, saponins, cardiac glycosides, alkaloids, tannins, and cyanogenic compounds. Some of the better known are flavonoids derived from the tropical legumes in the genera *Derris*, *Lonchocarpus*, *Mundulea*, and *Tephrosia*

B. Medicines

As has been noted many times, the difference between a poison and a medicine is often only a matter of dosage, and this is certainly true of products from plant secondary compounds. One example is a curare poison derived from the vine *Chondrodendron tomentosum* that contains the alkaloid tubocurarine. Although it works like other curare poisons in paralyzing muscles, it was introduced to Western medicine in 1939 and is now used as a muscle relaxant in anesthesia.

It would be difficult to exaggerate the importance of plant secondary compounds in the history of medicine, both traditional and Western. It has been estimated that 25% of all prescriptions written in North America are from plant-derived drugs, and the majority of medicines

in traditional cultures are derived from plants. However, only about one-half of 1% of the approximately 265,000 flowering plant species have been comprehensively evaluated for biological active compounds, the first step in determining whether a chemical can be effective as a drug. Of course, most plants do not contain medically useful compounds; therefore, an important task is to determine how to screen species in a more effective manner than simply collecting and testing plants at random. One approach, pioneered by ethnobotanists, has been to document the plants used by healers in traditional cultures. In a preliminary study, plants used by a healer in Belize produced four times as many positive results in laboratory assays than did species collected at random. A second approach is to collect plants based on ecological information. For example, as noted previously, an important defense of mature leaves in tropical forests in toughness, resulting from increased lignin and cellulose in the leaves. However, new, expanding leaves cannot be defended by toughness and so are more likely to use chemical defenses. For this reason, one way to increase the effectiveness of plant screenings for drugs may be to focus on young leaves instead of mature ones.

Regardless of the screening process, plant secondary compounds have been an important source of drugs. In particular, alkaloids have provided many noteworthy medicines in both traditional and Western societies (Schmeller and Wink as cited in Roberts and Wink, 1998). These include atropine, codeine, colchicine, ephedrine, morphine, reserpine, taxol, and theobromine. Codeine, for example, is derived from the fruit of *Paper somniferum* (Paperveraceae) and is a component of opium. Currently, more than 200 pharmaceuticals contain codeine, and it is used as a cough suppressant, a pain killer, and a sedative. In the West, ephedrine is used as a nasal decongestant in cold medicines and is a treatment for asthma in both Western and traditional societies. Its source is *Ephedra sinica* and *E. shunnungiania* (Ephedraceae), and it works in part by raising blood pressure and respiration while causing an opening of air passages. Taxol is a recently discovered diterpene alkaloid in the tree *Taxus brevifolia* (Taxaceae) that has therapeutic use as a treatment for breast and ovarian cancer. This compound disrupts microtubule assembly in cells, making it an effective weapon against rapidly dividing tumor cells.

C. Self-Medication in Animals

Mounting evidence suggests that humans are not the only animals that use plant chemical defenses as medicines. Naturalists have long observed that some animals occasionally eat or rub leaves or other plant parts on themselves in an apparent attempt to cure either diseases or kill off parasites (Clayton and Wolfe, 1993). Demonstrating that this behavior is truly self-medication requires showing that the behavior is deliberate, that the plant substance used kills parasites or disease-causing organisms, and that the self-medication leads to increased fitness in the animal. Since in many cases the behavior is rare, gathering evidence to meet these criteria is difficult, but some instances are suggestive. For example, kodiak bears chew the roots of the *Ligusticum* spp. and then rub their saliva through their fur. Since this species is also used by humans as a medicine, it is possible that the bears are also using the plant to cure or ward off infections. Other vertebrates, including chimpanzees, white-faced monkeys, and birds, also use leaves of particular plant species in unusual ways or only when they are sick, suggesting that they may have medicinal value. However, it remains difficult to demonstrate conclusively that animals are truly using plant defenses as medicines.

D. Spices in Foods

In addition to their uses in treatment of diseases, plant secondary compounds may also be used to prevent illness. Billing and Sherman (1998) argued that people use spices in food preparation to inhibit or kill pathogenic microorganisms. They marshaled several lines of evidence to support this "antimicrobial hypothesis." First, they noted that numerous studies have demonstrated that many spices, such as onions, garlic, pepper, and chili peppers, can inhibit the growth or even kill many species of bacteria. Second, by comparing recipes of traditional cuisines from 36 countries, they found that both the proportion of recipes calling for spices and the number of spices per recipe were positively correlated with the mean annual temperature of the country. In warmer climates food should spoil more quickly, they suggested, so spices may be an important way to preserve food. Third, in warmer countries the spices that are commonly used in cooking more strongly inhibit bacterial growth, again reflecting the greater chance of food going bad. For example, despite its strongly antimicrobial properties, garlic is not part of any traditional dishes in Norway, but it is in 80% of the recipes from Indonesia. Billing and Sherman also noted that spices are used in small quantities and thus the alternative explanation that they are used for nutritional reasons seems unlikely. Although other factors may come into play, including the greater diversity of plants in equatorial countries and the relatively slow incorporation of newly available spices into traditional cuisines, the correlations observed by Billing and Sherman strongly support their hypothesis.

VIII. SUMMARY AND CONCLUSIONS

Ultimately, the theme of this article, like that of this book, is biodiversity. Plants have evolved a stunning diversity of defenses and strategies to reduce damage by herbivores and pathogens. The most notable of these are chemical defenses, which plants use either to deter feeding (such as tannins) or to poison their natural enemies (such as nonprotein amino acids.) Plant defenses also include physical defenses such as thorns, and many species have evolved elaborate mutualistic relationships with ants, mites, and fungi as an additional means of defense. All defenses have some cost, and a defense will only be favored by natural selection if its benefits in reduced damage and higher fitness outweigh the costs of producing and maintaining the defense. Why particular species invest what they do in defenses, and why they invest in some defenses and not others, remains a subject of controversy, but there are patterns across habitats and successional seres, making it clear that the evolution of plant defenses correlates with at least some abiotic factors. Finally, humans (and perhaps other animals) make use of plant defenses in their everyday lives. In food and with medicine, the lives of people have been greatly enhanced by the diverse wealth of plant defenses. Understanding how to make a fuller, more effective use of the diversity of plant defenses found in the natural world is an important, practical goal emerging from this type of research.

See Also the Following Articles

ADAPTATION • PARASITISM

Bibliography

Benrey, B., and Denno, R. F. (1997). Slow-growth—high-mortality hypothesis: A test using the cabbage butterfly. *Ecology* 78, 87–99.
Billing, J., and Sherman, P. W. (1998). Antimicrobial functions of spices: Why some like it hot. *Q. Rev. Biol.* 73, 3–49.
Bryant, J. P., Chapin, F. S., III, and Klein, D. R. (1983). Carbon/ nutrient balance of boreal plants in relation to vertebrate herbivory. *Oikos* 40, 357–368.
Clayton, D. H., and Wolfe, N. D. (1993). Adaptive significance of self-medication. *Trends Ecol. Evol.* 8, 60–63.
Feeny, P. (1976). Plant apparency and chemical defense. *Recent Adv. Phytochem.* 10, 1–40.

Fritz, R. S., and Sims, E. L. (1992). *Plant Resistance to Herbivores and Pathogens: Ecology, Evolution and Genetics.* Univ. of Chicago Press, Chicago.

Harborne, J. B. (1988). *Introduction to Ecological Chemistry.* Academic Press, San Diego.

Huxley, C. R., and Cutler, D. F. (1991). *Ant–Plant Interactions.* Oxford Univ. Press, Oxford.

Johns, T. (1996). *Origins of Human Diet and Medicine.* Univ. of Arizona Press, Tucson.

Karban, R., and Baldwin, I. T. (1997). *Induced Responses to Herbivory.* Univ. of Chicago Press, Chicago.

Kiowa, H., Bressan, R. A., and Hasegawa, P. M. (1997). Regulation of protease inhibitors and plant defense. *Trends Plant Sci.* **2,** 379–384.

Levin, D. A. (1973). Role of trichomes in plant defense. *Q. Rev. Biol.* **48,**3–15.

Roberts, M. F., and Wink, M. (1998). *Alkaloids: Biochemistry, Ecology and Medicinal Applications.* Plenum, New York.

Rosenthal, G. A., and Berenbaum, M. R. (1991). *Herbivores: Their Interactions with Secondary Plant Metabolites.* Academic Press, San Diego.

Turlings, T. C. J., and Benrey, B. (1998). Effects of plant metabolites on the behavior and development of parasitic wasps. *Ecoscience* **5,** 321–333.

DEFORESTATION AND LAND CLEARING

Jaboury Ghazoul and Julian Evans
Imperial College of Science Technology and Medicine, University of London

I. Introduction
II. Quantifying Global Deforestation
III. Causes of Deforestation
IV. Consequences of Deforestation
V. Alternatives for Sustainable Development
VI. Conclusion

GLOSSARY

afforestation The planting of previously unforested land with trees.

agroforestry The practice of combining agricultural crops or animal husbandry with the maintenance and cultivation of trees on the same patch of land.

criteria and indicators A monitoring system for the assessment of the economic, social, or ecological data about land management practices as an aid to improving sustainability.

deforestation The complete or almost complete removal of tree cover and conversion of forested land to other uses as a result of human activities.

desertification The development of desert conditions as a result of human activity, frequently by overuse of trees and overgrazing, or climate change.

forest degradation Damage to forest ecosystems through human activities that does not result in the total elimination of forest cover.

natural forest An area of land which supports a mini-

mum of 20% tree cover that has arisen as a result of natural processes of establishment and succession.

plantation An area of land that supports planted forest, usually for commercial exploitation.

remote sensing The measurement or acquisition of information of landscape pattern by a recording device that is not in physical or intimate contact with the landscape under study.

secondary forest An area of previously forested land that was subsequently degraded or deforested through human or natural action but which now supports regenerating or mature natural forest.

shifting cultivation Any temporally or spatially cyclical agricultural system that involves the clearing of land followed by cultivation and fallow periods.

AT THE END OF THE TWENTIETH CENTURY, there were an estimated 1700 million ha of tropical forest and 1600 million ha of temperate forests worldwide. These figures represent about 60% of the original forest cover that is estimated to have existed approximately 8000 years ago. Much of this loss can be directly attributed to human impacts during the past three millennia with increased clearance in recent centuries, and even more recently in tropical regions. Forests provide important resources and a multitude of natural services, and their recent rapid destruction is causing increasing concern due to environmental, social, and economic problems throughout the world. However, developing solutions is proving to be a highly complex task due to the variety of causes of deforestation and conflicting stakeholder interests.

I. INTRODUCTION

Deforestation is the complete or almost complete removal of tree cover and conversion of the land to other uses. Technically, deforestation may be defined as the semipermanent depletion of tree crown cover to less than 10%. In this respect, a distinction needs to be made between deforestation and forest degradation, which is the significant damage to forest ecosystems but without the total elimination of forest cover.

This article begins by describing historic changes in forest cover and proceeds to current deforestation trends and their measurement. Section II discusses the causes of deforestation in recent years and section III considers the consequences of deforestation for a variety of environmental parameters. Solutions to the deforestation problem are presented in section IV, and potential future trends are described in section V with a brief discussion of the impact of projected climate change.

A. Historical Deforestation and Land Clearance

At the advent of agriculture, approximately 8000 years ago, forests are thought to have covered approximately 40% of the world's land area, or approximately 6000 million ha. Up to 1500 AD, the spread of agriculture across the globe resulted in the clearance of many forests, particularly those on the most accessible and fertile land. The areas most affected were the Middle East, the Mediterranean, South Asia, Europe, and the Far East.

Deforestation in the New World was underway long before the arrival of Europeans, but much of this clearance was patchy or localized. Settlers from Europe started to clear accessible forests of coastal Brazil and the Caribbean for sugar plantations. In North America, settlers cleared large areas of temperate forests for cattle ranches and food crops. Wood was used as fuel and for construction.

In the late eighteenth century the industrial revolution put tremendous pressure on the remaining forests to supply fuel for industrial development, and by the end of the nineteenth century many of Europe's ancient woodlands had been cleared. However, it is also clear

that demand for wood and charcoal also led to conservation of resources to ensure optimality of supply.

In the past 200 years, deforestation rates have increased greatly. Between 1850 and 1980, 15% of the world's forests and woodlands were cleared (Rowe *et al.*, 1992). The world forest area has shrunk to 3500 million ha as a consequence of human exploitation, most of which occurred in the latter half of the twentieth century [Food and Agriculture Organization (FAO), 1999].

B. Contemporary Deforestation

Deforestation and land clearance in the twentieth century have increased greatly, with the highest rates of clearance occurring since 1960. Most current deforestation occurs in the tropical regions, whereas in temperate countries there has been a net increase of forest cover by 0.1% due to reforestation and regeneration policies. In Canada, the area of land under tree cover increased by 1.4 million ha to 417.6 million ha in the late 1980s.

FAO has estimated annual rates of forest clearance in developing countries at 15.5 million ha for the period 1980–1990 and 13.7 million ha for 1990–1995. Thus, the total area of forest cleared during this 15-year period is approximately 200 million ha.

The causes of deforestation vary among regions. In Africa, FAO reports that the major direct cause of deforestation is clearance by farmers driven by increasing population pressures. In Latin America, settlement and infrastructure projects in forested areas result in clearance of land for cattle ranching and permanent agriculture, often combined with financial incentives in the form of subsidies and favorable tax policies. In Asia, intensive timber harvesting and shifting cultivation as well as the expansion of large-scale agricultural projects and plantation crops, such as oil palm and rubber, and to a lesser extent transmigration projects, all contribute to deforestation. Forest land is often not suitable for sustainable agricultural development, and as soils become exhausted new areas of forest have to be cleared. For example, 80% of the Amazon basin is ill suited to sedentary farming. Desertification through unsustainable agricultural development has contributed to much deforestation in drier regions of Africa and Asia.

Most deforestation is concentrated in relatively few tropical countries. Fifty percent of global deforestation occurs in 10 developing countries (Table I). Brazil is typical of tropical countries in that deforestation rates were low until the 1970s. Since then, extensive spread of agriculture and ranching, encouraged by government subsidies, and clearance by landless farmers have re-

TABLE I

"Top 10" Deforesting Countries in Terms of Total
Forest Loss (1995)[a]

Country	Ranking	Annual loss (ha)
Brazil	1	2,550,000
Indonesia	2	1,080,000
Congo	3	740,000
Bolivia	4	580,000
Mexico	5	510,000
Venezuela	6	500,000
Malaysia	7	400,000
Myanmar	8	390,000
Sudan	9	350,000
Thailand	10	330,000

[a] Source: FAO (1999).

sulted in very rapid deforestation (Table II). Large-scale development and industrial projects such as mining and hydroelectric plants have contributed to the high deforestation rates in the tropics. Thus, deforestation is largely a tropical issue. Because a large proportion of the world's biodiversity is found within tropical forests, tropical deforestation also has very great relevance to global biodiversity.

In temperate countries there is no overall deforestation in terms of net area but a small increase in forest cover owing to policies of regeneration and afforestation. However, this statement disguises a steady and continuing transition from natural forest formations to managed and plantation forests. In this sense, there is some clearance of natural forest in temperate regions with the con-

TABLE II

Important Deforesting Countries and Regions in
Terms of Annual Rate of Loss (1995)[a]

Country	Ranking	% Annual loss
Philippines	1	−3.5
Sierra Leone	2	−3.0
Pakistan	3	−2.9
Thailand	4	−2.6
Paraguay	5	−2.6
Central America	6	−2.1
Caribbean Islands	7	−1.7
Cambodia	8	−1.6
Ecuador	9	−1.6
Myanmar	10	−1.6

[a] Source: FAO (1999).

sequent loss of biodiversity. This trend is likely to decrease as conservation priorities assert themselves, except in eastern Europe and Russia, where the importance of forest resources as an accessible and tradeable commodity takes priority. Nevertheless, in Europe as a whole there is very little natural forest undisturbed by human intervention; the forest of Bialowiezca in Poland is one of the few examples of reasonably extant forest not dissimilar to the ancient "wild wood."

II. QUANTIFYING GLOBAL DEFORESTATION

A. Improving Assessment Accuracy

Accurate measures of deforestation are increasingly needed for the development of appropriate forest management policies and for estimating global carbon pools and gas emissions. Conflicting estimates lead to confusion on the part of policymakers, the media, and the public, but obtaining accurate estimates is difficult due to the lack of reliable time-sequence maps, varying standards of forest and nonforest classification, and inadequate ground-truthing of satellite imagery. FAO, the United Nations Food and Agriculture Organisations, the body responsible for the collecting and interpretation of forest cover statistics, is largely dependent on the information supplied to it by national forest departments. These data are often based on out-of-date inventory records and frequently reflect political and institutional biases.

The objectivity of forest cover measurement has increased since the 1970s, when estimates varied considerably due to differences in the methods used. In 1976, FAO produced the first estimates of global deforestation rates extrapolated from information derived from 13 countries. Although these estimates had a large degree of uncertainty associated with them, they did serve to draw attention to the increasing problem of tropical deforestation. The Tropical Forest Resources Assessment Project carried out by FAO and UNEP, the United Nations Environment Programme, in the early 1980s was the first major step to measuring the extent of tropical forests and their deforestation. Seventy-six countries were surveyed, with estimates for 31 countries based on remote sensing data. Models were constructed to estimate deforestation rates through the late 1970s and to estimate projected deforestation trends through 1985. In 1989, Norman Myers attempted to determine global deforestation rates for the late 1980s using data from 37 countries accounting for 97% of tropical forests and relying primarily on nongovernment sources and remote sensing.

By 1993, FAO had produced the Tropical Forest Assessment for 1990, a computerized database of forest areas that was used to develop deforestation models to estimate forest cover in 1990 and rates of change in the previous decade. The database covers 90 tropical countries divided into 644 subnational units. Satellite data were available for 71 of these countries, whereas information for the remaining 19 was of generally poor quality and based on coarse resolution and often out-of-date vegetation maps and land-use surveys. Data were particularly poor for many countries in Africa. Multi-year observations are required to generate reliable estimates, but only 24 of the 90 countries satisfied this criterion. Estimates of deforestation for the remaining 66 countries are based almost entirely on models that relate forest cover change to population density, population growth, ecological zone, or initial forest cover. The 1993 assessment estimates the earth's total area of tropical rain forest and moist deciduous forest to have been 13.06 million km^2 in 1990 with an annual deforestation rate of 106,000 km^2 during 1981–1990 (i.e., 7.5% net loss of these forest types during this decade).

B. Remote Sensing Developments

The development of remote sensing technology in recent decades has vastly improved the capabilities for the measurement of forest cover and deforestation. Aerial photography, developed in the 1950s, provides high resolution and detail, and despite its relative costs it is still used for ground truthing of small areas. Airborne radar was used extensively in the 1960s and 1970s and has the advantagof being able to provide information in almost any weather conditions. Satellite imaging began in the 1970s with the launch of the first LANDSAT satellite. Of the six LANDSAT satellites successfully launched, three continue to operate, with each providing digital data "images" covering a ground area of about 185 km^2 with a resolution of either 80 m or 30 m. Radar satellite sensing capable of penetrating cloud cover is being developed and will improve assessment in cloud-covered tropical areas.

C. Conflicting Estimates

Despite the development and use of remote sensing for measuring deforestation, there remain large discrepancies for regional deforestation rates due to differences in the definitions used, geographical areas, types of satellite photos, and different methods of projecting trends.

Definitions of deforestation are sometimes confused. "Deforestation" is confused with "degradation," which refers to significant damage to forests without actually removing forest cover entirely. Degraded forests may regenerate to secondary forest formations. Furthermore, given the continuous gradations between vegetation types, the classification of vegetation types into forest and non-forest categories may not be consistent across different studies. Open savanna forest, for example, may be included in some definitions of forest but excluded from others. Worse, the total land area often includes water bodies or open savannas, and the deforested area, which is expressed as a percentage of the total land area, is therefore underestimated.

Satellite data may be interpreted by computer or manually. Manual interpretation is less expensive and may give better results, but observers often differ in their interpretation of the data. Computer-aided interpretation of low-resolution images eliminates inconsistencies of judgment but may overestimate the area deforested if deforestation has occurred in a piecemeal manner. Low-resolution images have also tended to overestimate the amount of burning forest because even very small fires are capable of saturating a 1.1-km^2 pixel.

Variations in the models used for projection of future deforestation rates explain much of the variation among different projections. Exponential models based on rapidly accelerating deforestation in the 1970s have tended to give excessively high projected rates of deforestation for the 1980s. With improved models and more accurate interpretation of statistics on the ground, increasingly accurate estimates of deforestation are expected in the near future.

III. CAUSES OF DEFORESTATION

There is no single cause of deforestation but rather it is the result of the interaction of social, economic, political, and cultural forces with the environment. Several underlying socioeconomic causes create conditions that favor forest clearance by readily identifiable direct causes.

A. Underlying Causes of Deforestation

The underlying causes of deforestation are the factors that give rise to conditions in which forest clearance becomes a rational or necessary behavior. They may be local or national socioeconomic or political forces, or they may be external global forces such as the state of the global market economy. They are generally beyond the control of an individual but strongly influence the decisions individuals make regarding the management and use of forests and forest resources.

1. Population Growth and Poverty

Population growth is one of the most publicized underlying factors for tropical deforestation. The global population currently grows by about 1000 million every decade, and most of this increase occurs in developing countries in which deforestation is greatest. Increasing populations place pressure on forests and the resources they supply for food, energy, water, wood, paper, and a variety of other products. The importance of population pressure as an underlying cause for deforestation can be misleading in that it is dependent on the ability of the land to support the population, the importance of forest for supplying goods and services to the people, and the level of control by national or local institutions.

An estimated 2.8 billion people live in rural areas and are dependent on agriculture to meet their basic needs. The rural poor have very few economic options and are often forced to seek short-term solutions to their economic problems. These solutions include clearing forested land to grow subsistence crops. Opportunities for improving livelihoods by other means are limited due to the lack of rural capital, low capacity of subsistence farming to generate income, and the lack of infrastructure and education.

2. Development Policies and Tax Incentives

Debt repayments constitute a large proportion of the national budget of many tropical countries, and structural adjustment programs introduced as a result often favor the maximization of foreign exchange through direct and unsustainable exploitation of forest capital and by conversion of forests to agriculture for export crops. Large-scale extensive agricultural development, frequently at the expense of small farmers as well as forest cover, is further encouraged through the provision of state subsidies for agriculture and livestock expansion, reduction in income and corporate taxes, and tax breaks on imports of equipment for new industries. Expansion of agricultural crops for export or to satisfy national demands destroys forest directly but also causes the displacement of subsistence farmers who are forced to relocate and clear new and often marginal lands elsewhere.

The privatization of public resources, advocated by the World Bank and other bilateral donor agencies, favors management strategies that maximize the short-term economic gain for the new owners, whereas non-

monetary forest services, such as soil conservation and watershed protection, are not valued highly in a market-driven environment. Government incentives and subsidies have allowed some otherwise uneconomical industries to prosper at the expense of forest cover, whereas development projects often fail to account fully for the value of forest capital lost.

A lack of understanding of the real value of forests' goods and services results in poor policies. The institutional weakness of the national forest department or corruption within the government can lead to policy decisions that favor private interests at the expense of the benefits to society as a whole. In recent years there has been an improvement in the reformulation of forest policies of several tropical countries. Subsidies that promote cattle ranching have been withdrawn in Brazil, whereas Costa Rica is now beginning to account for the destruction of forest capital in its national economic accounts.

3. Resettlement Programs

Government-sponsored resettlement programs, such as the transmigration programs of Indonesia and Brazil, encourage landless farmers or the urban poor to clear areas of forest for subsistence agriculture or for the cultivation of cash crops for export. These schemes relieve the pressure of urban crowding and allow governments to avoid the difficult issue of land reform by providing new agricultural land from forest. However, many forested lands are unsuitable for permanent agriculture and resettlement programs have invariably failed as a result. Transmigrants are thus forced to resort to shifting cultivation, leading to further forest degradation and deforestation.

4. Tenurial Policies

Much of the agricultural land of tropical countries is owned by large land-owners or corporations and is therefore not accessible to the majority of the farming population. Improved agricultural production is gained through the use of chemical fertilizers and pesticides which, together with mechanization of labor, is most efficient on large-scale agricultural systems. This favors large farmers who have the capital to invest in such innovations and the land area to benefit from economies of scale. As the large landowners become more economically successful and powerful, small farmers who may not have legal title to their land are frequently displaced or forced to sell their land because of increasing debts. These farmers often move to the forest frontier to clear a new plot of land, and it is usually politically easier for governments to ignore deforestation than to deal with the difficult issues of land redistribution or job creation.

Legal ownership of land has a great effect on the attitude people have of the land. Without some legal land title there is little incentive to invest in the land to make it more productive. It becomes more economically logical to pursue short-term gain and to move on to clear new forest land once productivity declines. Because most tropical forest lands are owned by the state, clearance is often illegal and governments are unwilling to grant legal title to small farmers for land acquired in this way. Lack of ownership of the land excludes farmers from obtaining credit to purchase seed or fertilizers and pesticides and discourages any long-term investment.

In countries in which there is rapid growth of the rural population, individual farm plots that pass through inheritance to subsequent generations become smaller as they are divided among the inheritor, and eventually they become too small to be economically viable. Consequently, the growing rural population is forced to clear new land from the remaining reserves of unused forest.

In many Pacific rim countries, customary land ownership prevails. Precise boundaries are frequently unsurveyed, and local communities and groups know only from tradition what is their land. Although attachment to such land is deeply held, wholesale allocation of logging rights can be conveyed either by the local people with relatively little outside control or, conversely, imposed from outside by governments that fail to take account of local peoples' interests.

In either case, the land tenure system is a weak instrument in preventing unplanned deforestation and land clearance.

5. Market Demands

As populations grow and become more affluent, the demand for forest products increases, particularly for industrial timber and pulpwood for making paper. However, although it can be readily demonstrated that some countries have significant exports of forest products (Table III), the extent to which international markets contribute to deforestation varies greatly from country to country. In the top 10 deforesting countries, it is the national demand for forest products that accounts for most industrial deforestation. Furthermore, there is no strong link between increasing international demand for forest products generally and deforestation. This is because almost all pulp and paper comes from temperate forests or specially established tropical plantations since paper manufactured from mixed tropical forest is generally expensive and has poor quality owing to the lack of wood uniformity. Second, much of the world's industrial-grade timber is softwood (i.e., coniferous), and tropical forests are overwhelming broad-leaved. Supplies of industrial lumber in tropical countries increasingly come from tropical plantations, notably pines, cypress, eucalyptus, and teak.

International markets for forest-derived products are important in terms of volume only for Malaysia, which exports approximately 26% of its industrial forest products by volume. Indonesia and Brazil export large quantities of paper and paperboard, but much of this comes

TABLE III

The Importance of Exports to the Forest-Based Economies of the Top 10 Deforesting Countries in 1996[a]

Country	Timber products[b]			Paper and paperboard[c]			Charcoal and fuelwood			Total (%)
	Production	Export	%	Production	Export	%	Production	Export	%	
Brazil	107,360	5,019	4.7	12,110	3,396	28.0	135,652	63	0.05	3.3
Indonesia	64,711	9,414	14.5	7,021	2,342	33.4	153,540	1,039	0.7	5.7
Congo, D.R.	3,554	227	6.4	3	0	0	45,142	0	0	0.5
Bolivia	1,070	150	14.0	2	0	0	1,419	0	0	6.0
Mexico	9,063	557	6.1	3,558	328	9.2	16,731	136	0.8	3.5
Venezuela	1,790	52	2.9	900	50	5.6	918	0	0	2.8
Malaysia	50,923	16,143	31.7	777	41	5.3	10,035	186	1.9	26.5
Myanmar	3,399	689	20.3	24	0	0	20,612	53	0.3	3.1
Sudan	2,321	0	0	3	0	0	14,600	0	0	0
Thailand	3,636	614	16.9	2,744	335	12.2	36,894	53	0.1	2.3

[a] Note. All values in thousand m³. Source: FAO (1999).
[b] Includes industrial roundwood, sawnwood, and wood-based panels.
[c] Includes pulp for paper.

from plantations of fast-growing species that were established on nonforest land, previously cleared land, or logged forest. In some cases, market demand for wood-based products has led to clearance of natural forests and subsequent establishment of plantations, but most of this production has been to satisfy domestic rather than international markets. In countries in which deforestation is most severe, it is the national demand for forest products that is the more important cause of deforestation. The importance to tropical deforestation of the international market varies from region to region, but it is likely to diminish as populations in developing countries grow and become more affluent while those in developed countries demand products from demonstrably sustainable sources.

Clearance of forest for agricultural production also appears to be driven by increasing national demand for agricultural crops. All the important deforesting countries listed in Table I remain net importers of rice or maize except for Thailand, which exports approximately 40% of its rice production. Also, self-sufficiency in agricultural production has been a primary developmental goal of many of these countries and has led to policies that encourage conversion of forests to fields. Growth in the domestic market can far outstrip that in the export market. During the past decade, for example, the demand for palm oil in Indonesia has led to the widespread establishment of plantations even though palm oil exports have remained at between 6 and 8%. Similarly, rising production of beef in Central and South America to feed a growing domestic market has resulted in extensive deforestation by ranchers, farmers, and land speculators. Livestock exports from Brazil decreased from about 15 to 5% of production during the past two decades, and in a similar period Central American exports declined from 45% to approximately 20% of production.

Deforestation due to agricultural and livestock production is therefore largely a result of growth in the domestic markets and only partially attributable to markets in the developed world. Although the importance of export markets should not be underestimated, national market forces appear to be more important than international trade in determining the rates and extent of tropical deforestation. Consequently, it is likely that international trade offers only limited scope for reducing deforestation rates in most tropical countries.

6. Undervaluation of Forests and Forest Products

Where logging has preceded and been the first stage in wholesale land clearance leading to deforestation, it is often because the value placed on the timber is no more than the cost of extraction and marketing. In this sense, logging is not only unsustainable forest management but also becomes a mining activity in which timber value reflects the ease or otherwise of obtaining the raw material. The value does not reflect cost of replacement or cost of growing beyond what is often a nominal payment of royalty to the owner. If timber was valued to reflect its true cost of replacement, then growing trees to produce timber would become economically worthwhile and hence potentially a sustainable option. While it is not worthwhile, deforestation, especially in the tropics, is likely to continue because clearance and conversion is perceived as more profitable.

However, undervaluation has a further dimension. The non-timber benefits and services trees and forests provide are often far more important than their timber products. Environmental benefits, including soil protection, shelter, micro-climate amelioration, contribution to regional and global hydrological and carbon cycles, are rarely quantified and hardly ever incorporated in economic assessments. Only when massive downstream flooding is traced back to wanton deforestation in the catchment are such connections made.

A major underlying cause of deforestation is the widespread failure to value sufficiently both forest products and the many environmental benefits forests bring. Regrettably, this is largely because simple and widely accepted approaches to such valuations in economic terms do not exist. For example, stumpage, which is the charges that governments demand from loggers for state-owned timber, often undervalues the resource which encourages waste and makes other land uses more economically attractive.

7. Weak Government Institutions

Although almost all countries have explicit forest laws and policies designed to conserve forest, two features of forest conditions in tropical countries exacerbate the risk of deforestation. First, extensive forests by their very geography are likely to be remote from towns and cities and hence far from the rule of law. It is easy for illegal logging and clearance to continue unseen and unchecked. Second, forest services are frequently the "Cinderella" organization of government, being viewed as inferior to agriculture and even wildlife and tourism. Few resources are attracted and poorly paid staff often have difficulty both in ensuring that sustainable management practices are implemented and in imposing their authority on perhaps large private sector interests. Quite apart from the risk of corruption that these circumstances afford, many staff once trained simply dislike the remoteness of forest management and supervision and prefer the white-collar work of the city office.

B. Direct Causes of Deforestation

1. Shifting Cultivation

The contribution of small-scale shifting agriculture to tropical deforestation remains unresolved due to the widely variable agricultural practices that are encompassed by this term. Some types of small-scale agriculture undoubtedly cause deforestation, but Brown and Schreckenberg (1998) indicate that the inherent stability and long-term viability of many shifting cultivation systems are unlikely to result in long-term forest clearance. The least destructive form of shifting cultivation is when land cultivated for 2 or 3 years is then left for a long fallow period. This long fallow shifting cultivation occurs only under conditions of very low population density. Where land is scarce, traditional shifting cultivators using short fallow cultivation make decisions as to which crops to plant, what techniques to use, and how long to fallow the land. These management decisions have allowed this land use to continue in forest areas for centuries without undue forest loss.

Expanding populations, land scarcity, and government policies have also created shifted cultivators who are the typical slash-and-burn farmers of recent decades. Unlike the traditional farmers who have practiced shifting cultivation for decades, these shifted agriculturists have been forced by circumstances or government policy to cultivate habitats that are unfamiliar to them. Government resettlement and transmigration schemes attract migrants for whom forest cultivation is an unfamiliar means of generating a livelihood and income. Similarly, in the Amazon, migrant cultivators are attracted to the forest frontier, where they clear and cultivate land for a few years. This land quickly becomes exhausted due to unsuitable soils or farming techniques and the land is sold to cattle ranchers and the "farmers" move on to clear more land.

2. Commercial Agriculture

Large-scale commercial agriculture is most frequently practiced by large corporations or state enterprises. These large operations can dispossess local landowners and farmers of the best and most fertile agricultural land, indirectly leading to deforestation in areas in which the farmers relocate. The establishment of oil palm plantations in valleys of Honduras in the 1970s displaced thousands of farmers who were forced to clear forests from steep slopes to establish new farms.

Commercial agriculture often leads to direct conversion of large tracts of forest to plantation estates and rice fields. This has been particularly prevalent in Indonesia and other regions of Southeast Asia where oil palm, coconut, or rubber plantations have been established on cleared forest land. In Indonesia, oil palm plantations increased from approximately 4 million ha in 1980 to 5.8 million ha in 1995. Land clearance for agricultural development is often subsidized by governments, and because the owners of the agribusiness companies are politically well appointed there is little interest in forest protection. Areas for conversion are frequently burnt because this is the least expensive method of clearance, and natural events such as the 1997–1998 El Niño are used as an opportunity to do so.

3. Cattle Ranching and Livestock Grazing

Intensive clearing of forest land in South and Central America occurred as a result of the expansion of cattle ranching, which was economically attractive due to relatively low risk, little labor, well-established markets, and the availability of various government subsidies. Cattle ranching expanded greatly initially in response to the opening of large markets in North America but has been sustained by the development of the domestic markets for beef. Ranchers cleared forest land by purchasing it directly and employing laborers to clear the forest cover or by purchasing land or dispossessing it from slash-and-burn farmers that was then converted to grasslands. The shifting cultivators would move deeper into the forest to repeat the cycle. The area of pasture in Central America is estimated to have increased from 3.9 million ha in 1955 to 13.4 million ha in 1995 and this increase has largely occurred at the expense of the region's tropical forests. Thus, deforestation in Latin America due to ranching is also associated with slash-and-burn agriculture and land speculation.

Livestock grazing can be a serious cause of deforestation in Africa where livestock herds exceed the carrying capacity of the area. Such pressure is acute in the drier tropics such as the Sahel region of Africa and in the Middle East, where large flocks of sheep and goats are maintained. The history of deforestation around the Mediterranean is linked to grazing regimes, especially by goats, but past simplistic assumptions have given way to the recognition that climatic, sociological, and agrarian factors have also contributed to forest clearance.

4. Infrastructural Development and Industrial Projects

During the 1970s and early 1980s, development of the Amazon, largely illustrative of similar strategies throughout the tropics, was actively encouraged by the Brazilian government through the building of roads, tax incentives and subsidies, massive resettlement programs, and large-scale development programs. The Trans-Amazonian highway opened up millions of square kilometers of previously inaccessible forest to colonization and allowed further expansion of the cattle industry. Such roads improve access to poorly developed areas and therefore tend to increase the adjacent land value for non-forest uses and encourage land speculation and deforestation. Recent slowing of deforestation is due to Brazil's economic recession and has been aided by changes in government policies on tax incentive and subsidies and increased enforcement of environmental regulations. Logging roads in Asia also facilitate deforestation by allowing access to farmers, and illegal loggers who follow may deforest an area that is otherwise merely degraded through selective logging.

Mining and oil exploration have a localized deforestation impact that has been widely publicized. Large mines consume large amounts of wood for fuel, whereas oil companies clear forests to create lines for exploration which are later used by colonizers following the exploration teams. Other large-scale development projects include hydro-electric power plants, whose reservoirs flood forests and transmission line paths are cut through the forest causing permanent loss of forest cover.

5. Plantations

Much forest has been cleared for commercial plantation crops such as rubber, oil palm, and the beverage crops of cocoa, coffee, and tea. Huge areas of dipterocarp forest in peninsular Malaysia have been converted to oil palm or rubber plantations, and although such perennial woody crops offer some soil protection, the loss in biodiversity is enormous. Indeed, it can be argued that the great bulk of the world's tropical plantation crops, approximately 26.5 million ha (FAO, 1999), are on former forest land.

Conversion of natural forest to forest plantations has also occurred, although currently such practice is deprecated and is less common (Evans, 1992). Indeed, a key principle of good forest stewardship is that forest plantations are only located on already cut-over, abandoned, or wasteland and in this way can actually help deflect pressure away from natural forest. However, the subject is not quite so simple since many forests have enjoyed enrichment by planting or have arisen through tree planting operations, such as many "natural" forests in France, Germany, and elsewhere.

Illustrative examples of significant conversion of natural forest formations to forest plantations include the following: conversion to exotic pines, *Gmelina*, and eucalypts at Jari in the drier east Amazon for pulp and paper production, conversion on the hills around Bulolo in Papua New Guinea—albeit with the species of *Araucaria* native to the area—for high-quality softwood veneer logs for making plywood, and clearance of miombo and savanna woodland for pine plantations in Zambia for industrial constructional material.

6. Fuelwood Collection and Charcoal Production

Fuelwood accounts for approximately 80% of all wood use in developing countries. Dependence on fuelwood is expected to decrease gradually with the introduction of electricity, kerosene, and propane, but heavy dependence on fuelwood by the rural poor is expected to continue well into the twenty-first century. Fuelwood collection as an agent of deforestation is particularly marked near urban centers and in villages in which continuous collection results in the gradual degradation and eventual deforestation of accessible areas. This is critical in the dry tropics, along with domestic use of wood for other uses such as construction and fencing material.

7. Logging

The most optimistic independent estimates of the amount of sustainably managed productive forest in the tropics are no more than 2% of the productive forest area. Most tropical logging consists of short-term exploitation of timber products with little or no concern for the future potential of the forest to provide additional resources and revenues. This is largely due to the insecurity of tenure and short concession periods. Although the intensity of logging in the tropics is usually low, removal of only 10% of the timber trees can result in damage to 55% or more of the remaining trees. Nevertheless, logging operations in the tropics usually result in degradation of the forest rather than its complete elimination. Deforestation does occur along logging roads, where forest is cleared for several meters on either side of the road, allowing the sun to dry the road surface. Poorly designed roads can result in severe erosion and landslides as well as facilitate movement into the area by illegal loggers and slash-and-burn cultivators. Illegal loggers may cause much greater damage to forest cover than the initial logging operations. Logging continues to be one of the most important causes of forest degradation but not deforestation, although intensive logging in Southeast Asia has resulted in the conversion of thousands of hectares of forest to alang-

alang (*Imperata cylindrica*) grassland that excludes almost all other vegetation.

Following industrial extraction of timber, farmers, agribusinesses, ranchers, fuelwood collectors, and illegal loggers move in along logging roads to clear the land for other uses. Management plans and government policies oblige industrial foresters to prevent encroachment of this sort, but these rules are inadequately enforced due to lack of will or manpower. The failing of forestry departments and companies is the inability to maintain forest cover in a single area.

The length of concessions is very often short term, sometimes less than 10 years, and very rarely more than the length of time for a rotation of the crop. In the absence of a long-term commitment, the logging company has little incentive to invest in long-term forest management. Concessions are also granted for timber with little regard for the other resources provided by the forest and the impact of the logging on local people.

8. Fire

Serious losses in forest cover in Southeast Asia and South America have been reported as a result of forest fires in 1997 and 1998. The causes of these fires are new large-scale commercial agricultural projects (including plantations) and shifting cultivation. The fires were exacerbated by the dry, coarse, woody debris left after logging operations and the very dry climatic conditions caused by the El Niño phenomenon. The area of forest consumed by fires in 1997 and 1998 has not been accurately documented, but estimates vary from 170,000 to more than 2 million ha (Byron and Shepherd, 1998). Extensive fires in tropical moist forests have been previously associated with the El Niño phenomenon, as in 1982, but the underlying causes are clearance of forest to establish plantations of oil palm, pulp wood, and rice, and, in South America, cattle pastures and shifting cultivation.

IV. CONSEQUENCES OF DEFORESTATION

A. Biodiversity

Perhaps the most severe consequence of the loss of tropical forest cover is the associated loss of biodiversity. It is not known how many species become extinct through the loss of tropical forest habitats, but estimates range up to a yearly loss of 50,000 species. Small forest fragments that often remain after clearance usually support only a small fraction of the original forest biodiversity.

B. Ecosystem Services

The loss or disruption of ecosystem services following deforestation is well documented, particularly for regulation of water flow through watersheds and maintenance of soil structure and fertility. Watersheds become susceptible to extremes of water flow following removal of tree cover, with which erosion of topsoil is associated. Thus, forest clearance often results in erratic water flow with periodic droughts punctuated by flood events which put communities living downstream at risk. Loss of water purity and topsoil through erosion deprive rural communities of safe potable water and fertile soils. Tropical forest soils rapidly become marginal for farming following loss of vegetation cover due to compaction, loss of organic material, and leaching of soil nutrients. Sedimentation of coastal fisheries can deprive communities of their sources of food and income far from the sites of deforestation.

Undisturbed forested land exhibits low to negligible rates of soil erosion. The layers of forest canopy, composed of trees and shrubs, intercept rainfall and thereby reduce raindrop impact while the network of tree roots adds mechanical strength. The accumulated decaying debris branches, twigs, leaves, and other organic material on the forest floor—the litter layer—is effective at protecting the soil from direct water erosion and facilitates infiltration of water into the soil. Removal of this litter layer may contribute more to severe erosion than any other single factor.

Clearance of forests disrupts these soil-conserving roles and subsequent rapid soil erosion is frequently observed, as the muddy red, yellow, and brown streams and rivers of so much of the tropics testify. Typical erosion rates under different land uses are presented in Table IV.

Owing to higher evapotranspiration of forests than most other vegetation types, forests tend to consume more water and therefore lead to reduced runoff and diminished stream flow. In dry countries, this can have a negative impact where water is at a premium, but in general undisturbed forest or sensitively managed plantations reduce sediment loads and flooding and may, in some instances, prolong dry season flows owing to improved infiltration. Overall, less water flows from forested catchments, but it is usually of better quality with less sediment, and streams may show more persistent flow.

Land degradation through loss or overuse of forest

TABLE IV

Soil Erosion Rates[a]

Land use	$t\ ha^{-1}y^{-1}$
Undisturbed natural forest	Negligible
Cut-over forest with litter and organic matter intact	1–5
Cut-over forest with litter and organic matter removed	30–100
Forest plantations with litter and undergrowth	0.5–3
Forest plantations with no undergrowth and litter removed	Up to 100
Undisturbed or lightly grazed grassland	2–10
Cultivated arable land (depending on slope, terracing, soil type, etc.)	20–400

[a] Source: Ambar (1986) as cited in Evans (1992).

cover in dry forest zones can result in desertification. Moderate to severe desertification affects between 3000 and 3500 million ha worldwide, about one-fourth of the world's land area, and threatens the livelihoods of 900 million people in approximately 100 developing countries. Firewood consumption is a significant contributing factor, but this is further compounded by overgrazing and overcultivation. The combination of these factors is resulting in widespread desertification of the Sahel in Africa, causing massive losses of topsoil and soil fertility and declines in productivity.

C. Global Climate Change

It is generally agreed that global climate is warming, with an increase in average global temperatures of approximately 0.3°C each decade. This is due to the emissions of carbon dioxide and other greenhouse gases from burning of fossil fuels mainly in industrialized countries. However, it is estimated that approximately 25% of total CO_2 emissions are a consequence of deforestation and forest fires as the carbon stocks in forest are liberated. The consequences of global warming are difficult to predict, but increased drought and desertification leading to crop failures and reduced productivity seem likely in many tropical regions. Higher temperature and reduced rainfall coupled with more frequent and severe El Niño events are predicted to result in increased frequency and intensity of fires in even moist lowland tropical forests. Geographic and altitudinal shifts in the distribution of vegetation types can also be expected.

D. Economic Losses

Annual loss of forest capital is estimated at $45 billion (World Commission on Forests and Sustainable Development, 1999). Destruction of forest represents opportunity costs through the loss of future revenues and employment that might be derived from the sustainable management of the forest for timber and non-timber products as well as potential income from ecotourism and other such enterprises. Economic losses can be much greater if the costs of lost ecosystem services are factored into the equation. If the non-timber benefits and services tropical forests provide are even partially costed, then the value per hectare of extant forest increases enormously. This is now being quantified. For example, the carbon sequestration role of such forests is valued in ways simply ignored in the past.

E. Social Consequences

Deforestation leads to the loss of traditional lifestyles associated with forested regions. Individual and collective rights to forest land have frequently been ignored by developers, ranchers, transmigrants, and governments. The loss of traditional lifestyles, customs, and religious beliefs of indigenous communities in the Amazon and Borneo has paralleled the loss and degradation of forests in these regions. Approximately 200 million people in the developing world live in or at the forest margins and wholly depend on forests for their livelihoods. Rarely are their concerns heeded or their rights even acknowledged.

V. ALTERNATIVES FOR SUSTAINABLE DEVELOPMENT

A. Protection and Management of the Remaining Forests

1. Protected Area Systems

Protected area systems are needed to establish areas that conserve habitat and biodiversity from encroachment and poaching. The majority of countries have not placed 12% of their land area within a protected area system as endorsed by the United Nations Conference on the Environment and Development.

To be effective, protected areas need to be sufficiently large to conserve all the biodiversity they contain and to include a broad range of naturally occurring forest types and all stages of natural succession. Protected areas need to be established in areas that harbor the best examples of the representative vegetation and in which biodiversity values are greatest or most at risk.

To be successful, protected area systems need to be supported by adequate funding and legislation and managed by strong institutional departments. However, forest protection must be developed with the cooperation of the local communities that use forest resources because alienation of the local communities has led to failure or inadequate protection. Environmentally sensitive development such as ecotourism, sustainable agriculture, or agroforestry should be permitted adjacent to, or even within, forest reserves, allowing local people to benefit directly from them. Protected area staff need to be adequately trained and equipped, and they should be sensitive to the needs of the local population. Environmental education programs should be promoted in and around protected forest reserves.

2. Joint Forest Management

New approaches to managing forest resources involve partnerships between local communities and local or national governments. These partnerships provide for sensitive management that acknowledges the needs of all stakeholders by providing a wide variety of benefits in a sustainable manner. Forest management decisions are made at a local level and are informed by state-supported science and developmental technologies. Providing local stakeholders with access to and benefits from forest resources and empowering them with the management of these resources encourages sustainable development and investment in forests.

3. Sustainable Timber Harvest Procedures

Timber harvest practices that minimize damage to forests are well-known but rarely implemented due to the perceived high cost of doing so. Indeed, over a short time scale reduced impact logging procedures are more costly, but over the longer term they are cheaper due to increased efficiency of extraction (by eliminating extraction of low-value trees and damage to high-value timber) and reduced damage to the remaining forest stand. National standards for logging operations, where they exist, are often flouted and enforcement is weak.

To encourage sustainable timber production, most timber-producing countries have been involved in developing criteria and indicator systems for sustainable forest management. These systems provide tools for assessing the state of forests which can be used to promote and inform sustainable forest management. The development of such systems has been hampered by the complexities of the definition of sustainable forest

management and how to interpret the information generated by criteria and indicator systems.

Coupled with criteria and indicator systems is the idea of certification that aims to encourage producers to adopt sustainable practices which, after verification, may be certified as such. However, there are several difficulties associated with certification as a means to reduce forest degradation and deforestation, the principal one being the need to ensure that certified wood is kept separate from non-certified wood from the source through the chain of custody and to the retailer. The success of certification also depends on the market demand for certified forest products, which is currently greatest in Europe. The domestic markets of timber-producing tropical countries account for the majority of tropical timber consumption in these countries, in which the market demand for certified wood remains low.

B. Socioeconomic and Agricultural Development

1. Improving the Productivity of Subsistence Agriculture and Ranching

Agricultural expansion is the most significant cause of tropical deforestation. Large numbers of people are involved in poorly productive subsistence agriculture or extensive cattle ranching into which there is very little investment to improve production and efficiency. State-supported investment in deforested marginal lands, fair credit schemes, and educational development are all needed to provide extension services to improve the efficiency of land use. Greater productivity from and improved use of existing agricultural land will lessen the pressure for clearance of new forested lands and will promote private investment for currently occupied land which in turn encourages sustainable use.

Agroforestry, the practice of combining agricultural crops or animal husbandry with the maintenance and cultivation of trees on the same patch of land, has been demonstrated to increase crop production by as much as 25–200% through the use of multipurpose trees. These trees can reduce soil erosion, increase soil fertility, provide favorable microclimates for crops and livestock, as well as provide firewood or other tree products such as rubber or fruit.

An estimated 200 million ha of degraded lands exist in upland watersheds of tropical countries as a result of deforestation. Restoration of these lands into productive use through tree planting or clearing for agricultural production will further alleviate the pressure on forest lands from agricultural expansion. Multiple-use species provide benefits directly from the products they provide, but they also serve to renew a watershed's ability to regulate groundwater and reduce soil erosion.

2. Tree Plantations

Tree and forest plantations are not a substitute for natural forest, but appropriately sited they can alleviate deforestation pressures. Plantations on already degraded land can be an excellent source of industrial-grade timber, pulpwood for paper making, and fuel—solid firewood and charcoal. Thus, they provide an alternative source of such products. In the tropics and subtropics, there are estimated to be approximately 70 million ha of industrial plantations, of which probably 55 million ha are reasonably well stocked. This area is slowly increasing, is often more productive than temperate plantations, and represents a major new wood resource. Considering the world as a whole, nearly half of all industrial timber is likely to be sourced from plantations by about 2020 (Fisher and de Fegely, 1999).

Small-scale plantings on farms, around villages, and in compounds helps satisfy domestic demand for building and fencing materials, fuel, and local constructional uses. The estimated 20–30 million ha of such "woodland" further diverts pressure away from natural forest.

Profitable plantations risk undermining the perceived value of natural forest and may accelerate conversion of these to plantations. If the primary cause behind deforestation is the conversion of land for agricultural development, it is unlikely that plantations will reduce deforestation. However, tree plantations do offer an opportunity to reduce the pressure on natural forests if the latter are exploited primarily for wood products.

C. Policy and Institutional Reform

1. Forest Valuation and the Reformation of Government Policies and Institutions

National policies need to promote the necessary framework that encourages sustainable forest use before community-level projects can have an effect. However, under-valuation of forests by governments and people undermines forestry institutions nationally and internationally, making it difficult to enact the necessary changes. The lack of forest-user taxes and low stumpage fees discourage sustainable management of forests by undervaluing the resource. Subsidies to competing land uses should be removed and effective natural resource accounting should be established to reflect true forest values. Opportunity costs and externalities associated with conversion need to be factored into assessments of forest values. Thus, the value of the ecosystem services as well as alternative income-generating businesses, such as tourism, should be considered. Furthermore, forest land value can be greatly increased by more efficient use of a greater variety of forest resources and by improved marketing of these products.

Forestry institutions are weak not only because of the undervaluation of forests but also because they lack information about forest impacts and resources. The absence of clear policy guidelines and trends in forests and forest resources leads to a confusion of strategies among national and international organizations. Strengthening of these institutions and developing mutually agreed frameworks that provide for more effective policy development and monitoring are needed if deforestation is to be arrested.

Finally, more collaboration, coordination, and communication between government forestry institutions, non-governmental organizations, and local stakeholders are needed to produce national and local development plans that are relevant to the needs of the society. Ecologically based land-use planning needs to include environmental and social impact assessments, and additional forest research is needed to document the impact of public policies on deforestation.

2. Provide Education

Educating the public and political and economic decision makers about environmental and socioeconomic issues related to forests is crucial if policies promoting forest preservation through sustainable use are to be heeded. Education must be based on thorough and demonstrable arguments about the economic, social, and biological benefits of using forests wisely and the costs associated with deforestation.

VI. CONCLUSION

Deforestation is largely a tropical issue. The quality of data on deforestation rates has improved but remains poor and the source of contention and debate. The causes of deforestation are complex and multi-faceted, although socioeconomic factors and trade are foremost. Deforestation impacts the environment through loss of biodiversity and disruption of ecosystem processes and the economy by affecting ecosystem services and because of inefficient squandering of resources. Perhaps the most tragic consequence is the loss of traditional beliefs and customs and the displacement of forest-dependent communities. Solving the problem of defor-

estation requires a suite of strategies that include establishing an effective and global protected area network
to preserve forest biodiversity and the genuine implementation of good practice guidelines in the use of
forest resources. Ultimately, a reformulation of policy
and a change in the attitudes of decision makers are
needed to ensure that forests are correctly valued economically, socially, and biologically.

See Also the Following Articles

AGRICULTURE, INDUSTRIALIZED • AGRICULTURE,
SUSTAINABLE • BIODIVERSITY-RICH COUNTRIES • CATTLE,
SHEEP, AND GOATS • DESERTIFICATION • FOREST
ECOLOGY • LAND-USE PATTERNS, HISTORIC •
REFORESTATION • TIMBER INDUSTRY

Bibliography

Brown, D., and Schreckenberg, K. (1998, April). Shifting cultivators as
agents of deforestation: Assessing the evidence. Natural Resources
Perspective No. 29. Overseas Development Institute, London.

Brown, K., and Pierce, D. (1994). *The Causes of Tropical Deforestation.*
Univ. of British Columbia Press, Vancouver, Canada.

Bryant, D., Nielsen, D., and Tangley, L. (1997). *The Last Frontier
Forests: Ecosystems and Economies on the Edge.* World Resources
Institute, Washington, DC.

Byron, N., and Shepherd, G. (1998, April). Indonesia and the 1997–98
El Niño: Fire problems and long-term solutions. Natural Resource
Perspective No. 28. Overseas Development Institute, London.

Downton, M. W. (1995). Measuring tropical deforestation: Development of the methods. *Environ. Conservation* **22**, 229–240.

Evans, J. (1992). *Plantation Forestry in the Tropics.* Oxford Univ.
Press, Oxford.

Fisher, B. S., and de Fegely, R. (1999). A study on the global outlook
for plantations. Technical paper for the Intergovernmental Forum
on Forests. Commonwealth of Australia.

Food and Agriculture Organization (FAO) (1999). *State of the World's
Forests 1999.* FAO, Rome.

Myers, N. (1989). *Deforestation Rates in Tropical Forests and Their
Climatic Implications.* Friends of the Earth, London.

Rowe, R., Sharma, N. P., and Browder, J. (1992). Deforestation: Problems, causes and concerns. In *Managing the World's Forests: Looking for Balance between Conservation and Development* (N. P.
Sharma, Ed.), pp. 33–45. Kendall/Hunt, Dubuque, IA.

World Commission on Forests and Sustainable Development (1999).
Our Forests, Our Future. Cambridge Univ. Press, Cambridge, UK.

DESERT ECOSYSTEMS

James A. MacMahon
Utah State University

GLOSSARY

alluvial fan Fine sediments deposited by water that form a conical fan when laid down by streams or rivers

annuals A plant, sometimes called an ephemeral, that survives year to year as a seed rather than as a "plant."

convergence or convergent evolution Species develop similar behavioral, physiological, or morphological characteristics in response to living in similar environments in different places.

desertification Process of changing a nondesert area into what appears to be a desert, regardless of the climate. This process often has a negative connotation relating it to overgrazing by domestic animals, or general human misuse; however, factors, not related to humans, may cause desertification.

Hadley cell A pattern of air circulation that causes deserts at about 30° N and 30° S of the equator. Sun heats the equator, air rises, pressure decreases, moist air moves north and south, cooling as it goes and falls onto the spinning earth, drying as it falls. The falling dry air increases pressure preventing the incursion of moist air.

life-form A system of classification of the plants in communities based on the position of their perennating structures relative to the ground.

mesic Moist, but not wet, environment.

perennating structures The tissue that carries a plant into a new growing season. For example, buds of trees, shrubs, etc., bulbs of some plants (e.g., tulips) or seeds of annuals where the "plant" dies each year.

physiognomy The general appearance of an area. For example, if you are in any desert in the world it has a certain recognizable "look," regardless of species composition.

saltation Sand particles jumping across the surface due to wind movement.

torper A condition, usually in birds and mammals, when the body temperature falls a few degrees below ambient for a short period of time. Differs from hibernation in (a) the amount the body temperatures fall and (b) the length of the resting state.

xeric A dry environment.

DEFINING "DESERT" is difficult because the literature contains so many definitions, using quite different criteria. A widely used approach developed for UNESCO includes consideration of rainfall and temperature (Fig. 1). Others calculate aridity indices using a variety of data. One system recognizes extremely arid (<100 mm [<4 in] annual precipitation), arid (100–250 mm annual precipitation [<10 in]), semiarid (250–500 mm precipitation [up to 20 in]), and several more mesic subdivisions. While overly simple, this system is easy to understand. Other workers define arid lands by the ratio of precipitation to potential evapotranspiration. In this system arid = annual precipitation is 33% of the potential evapotranspiration and semiarid = 33–67%. Using this standard, Africa has the greatest absolute area of arid lands and Europe has the least. The largest single area of desert, the Sahara of North Africa, covers nearly 9×10^6 km², an area about the size of the United States.

I. WHAT IS A DESERT?

This essay discusses extremely arid, arid, and some drier semiarid areas (transitions), calling them, collectively, deserts. This area has been calculated as 19 to 23% (29–32 million km²) of the earth's surface. The discussion will not include semiarid grasslands. Deserts range from areas devoid of any conspicuous vegetation to areas moderately well vegetated with shrubs and subtrees, a scattering of grasses, and a variety of annuals and succulents. They have scant and unpredictable rainfall and poorly developed soils. Deserts represent a highly variable group of ecosystems found in areas ranging from temperate to tropical zones around the world. I exclude those cold, high-latitude areas termed polar deserts and those barren areas of high elevations termed mountain deserts.

II. WHAT CAUSES DESERTS?

Generally, deserts are caused by one of four phenomena. First is the rain shadow effect whereby moisture is lost from air as it moves inland over mountains. Moisture-laden air condenses to form rain as it ascends. As the air crosses the mountains and descends it becomes drier and creates arid conditions (rain shadow). The Mojave and Great Basin deserts of North America are predominantly rain shadow deserts, as are some deserts of Central Asia. A second source of drying air involves cooling and condensation of water in air over cool coastal cur-

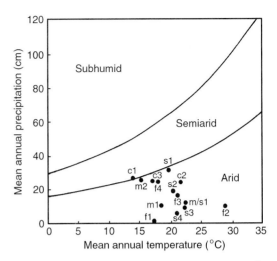

FIGURE 1 Depiction of arid and semiarid moisture-province boundaries as presented by Bailey (1979, in A. E. Hall *et al.*, eds., Agriculture in Semi-Arid Environments, Springer-Verlag), with data from some desert localities superimposed. Mojave Desert, m1 (Las Vegas, Nevada); m2 (St. George, Utah); m/s1 (Needles, California, a Mojave-Sonoran transition site); Sonoran Desert, s1 (Tucson, Arizona); s2 (Phoenix, Arizona), s3 (Yuma, Arizona); s4 (Brawley, California), c2 (Ojinaga, Chihuahua), c3 (El Paso, Texas); sites on other continents, f1 (Atacama Desert, Antofagasta, Chile), f2 (southern Sahara Desert, Tessalit, Mali), f3 (northern Sahara Desert, Biskra, Algeria), f4 (Karoo Desert, Beaufort West, South Africa). Note that despite the common assertions that North American deserts are not really deserts, their aridity is of a magnitude similar to that for many "true" deserts throughout the world. [From Figure 8.4, Chapter 8, Warm Deserts, by James A. MacMahon, 2000, p. 290 in *North American Terrestrial Vegetation*, Michael G. Barbour and William Dwight Billings (eds.). Reprinted with the permission of Cambridge University Press, New York.]

rents. Examples of this occur in Africa's Namib Desert, in Peru's Atacama Desert, and in the Baja California portion of the Sonoran Desert. The available moisture in these coastal deserts is often in the form of fog rather than rain. The third possibility occurs in the interior of a continent where moisture-laden air is rare because of the distance from sources of moisture. Some deserts of central Australia and of China are caused by their continental climates. Finally, there is the effect of high-pressure zones that occur at about 30° north and 30° south latitude that are caused by Hadley Cells. These convection entities are driven by solar energy and the spinning of the earth on its axis. Air lifted at the equator produces rain. As the air moves both north and south of the equator it descends, dries, and creates high-pressure areas that preclude inward movement of moist air. These subtropical deserts include the Sahara and Kalahari, in northern and southern Africa. Disruption of any of these climate patterns can disturb deserts, altering both their inhabitants and their geomorphology. For detailed introductions to the specific deserts of the world, their causes, and biological and geological characteristics, refer to the publications in the bibliography.

III. DESERT CLIMATES

Typical desert climates include low and unpredictable precipitation, relatively cloudless days, and high summer and moderate to cold winter temperatures that vary greatly from day to night.

The clarity of desert air occurs because it has so little moisture. Thus, during the day the ground surface heats to very high temperatures because there is little water to intercept the radiation. The ground re-radiates heat upward causing an increase in the air temperature. At night, when there is no incoming radiation, the ground continues to radiate toward outer space, thus lowering the temperatures significantly. Representative climograms for a variety of deserts are presented in Figure 2.

A. Rain Shadow Deserts

The Sonoran, Mojave, and Great Basin deserts of North America, some areas of Australia, and much of the Patagonian Desert are rain shadow deserts. They tend to have highly variable temperatures and, depending on latitude, precipitation can occur in the form of snow or rain. There are one or two periods of annual precipitation (Fig. 3). Because these deserts cover a wide latitudinal range it is difficult to generalize about their climates. See specific deserts in Section VI.

B. Cool Coastal Deserts

Cool coastal deserts include the Namib, the Atacama, the Baja California section of the Sonoran, and a small area in northwestern Australia. These deserts are dominated by fog that causes them to have relatively cool daytime temperatures ranging from 15 to 18°C. Condensation occurs at night, so that vegetation is wet many days of the year. Seasons are similar with winter being only 3 to 6°C below summer temperatures. Temperature extremes increase inland where summer maxima may be near 30°C. These tend to be extremely dry deserts, although in El Niño years the Atacama and Peruvian deserts may be subject to rare, devastating rainstorms.

C. Continental Deserts

Two large areas of continental deserts are the Great Basin of North America and the deserts of central Asia. Because they are far from the moderating influences of large bodies of water (usually oceans), temperature extremes are pronounced. Thus, continental deserts are extremely cold in the winter and extremely hot in the summer, where temperatures often exceed 38°C and may reach 44°C. Similarly, winter minima can be extremely low, regularly reaching −30°C and occasionally −50°C in the Caspian Lowlands. These deserts appear to have more water available because of the extended periods of snow cover, however, much of the water in the snow is lost by ablation (water goes from the solid state of snow directly to gas vapor without going through an intermediate liquid stage).

D. Subtropical Deserts

Subtropical deserts include the Sahara and Kalahari of Africa, several deserts of Australia, the Monte of Argentina, and the deserts of Arabia, the Middle East, India, west Pakistan, and the Sonoran Desert of North America. Freezing temperatures seldom occur with lows averaging above 5°C. Daytime temperatures reach 38°C on about one-third of the days each year and maximums may exceed 46°. The Kalahari, because of its elevation, is an exception to the low temperature rule in that subfreezing temperatures regularly occur over large areas in the winter. Subtropical deserts also have very low humidity, with little dew, frost, or fog, except following some rainy periods. Precipitation may come from either summer storms with high intensity, low areal extent, and short duration or from winter storms of low intensity, high areal extent, and long duration. Some areas (e.g., the Sonoran Desert) may receive approximately equal proportions of both storm types.

Any one desert may be typified by one of the four climate patterns, however, one might also have influences of several patterns. For example, the Mojave Desert is a subtropical desert, in a rain shadow, with some continental influences.

IV. DESERT GEOMORPHOLOGY

The geological substrates of deserts are fully exposed because there are so few plants. Generally, desert topography falls into two categories: (a) shield and platform

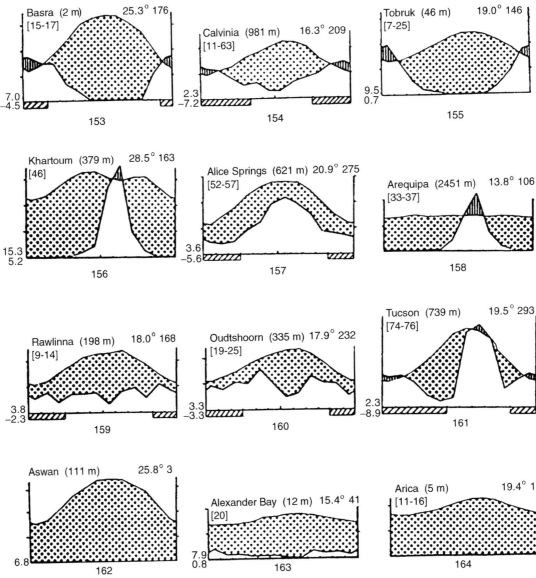

FIGURE 2 Climatic diagrams of desert regions. (Upper row) deserts with winter rains; (second row) those with summer rains; (third row) those with rains throughout the year or with two rainy seasons; (bottom row) those almost rain-free. Data are, left to right across the top: station name (below station name in brackets is number of observation years [occasionally first figure for temperature, second for precipitation]); height above sea level (m); mean annual temperature (°C); mean annual precipitation (mm). Lower left side: mean daily minimum temperature of the coldest month; below that, the absolute minimum temperature (the lowest ever recorded). The two curves show monthly mean temperatures (one scale interval, 10°C) and mean monthly precipitation (scale 10°C, 20 mm); drought periods (dotted); humid period (vertical hatching). [Used by permission from Figs. 153–164, p. 269 in *Ecology of Tropical and Subtropical Vegetation,* by Heinrich Walter, 1971, J. H. Burnett (ed.), Oliver & Boyd, Edinburgh.]

deserts that are associated with the stable plains of areas such as the Sahara, Central Asia, and Australia and (b) basin and range deserts characterized by large basin areas dotted by mountain ranges that cover various proportions of the total area and that are frequently the product of tectonic activity. The southwestern United States, Iran, Afghanistan, Pakistan, and parts of central Asia have large areas of basin and range topography. In both desert topographies, one finds alluvial fans, pediments (impressive plains cut into bedrock), playas, dunes, and a variety of incisions caused by moving water.

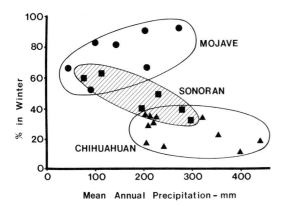

FIGURE 3 Percentage of winter rainfall plotted against mean annual precipitation for a variety of warm desert sites. Site assignment to a desert type was by knowledge of vegetation at each site. [Used by permission from Figure 5.6, Chapter 5, The Mojave, Sonoran and Chihuahuan deserts of North America, by James A. MacMahon and Frederic H. Wagner, 1985, p. 112 in *Hot Deserts and Arid Shrublands, A,* Michael Evenari, Imanuel Noy-Meir, and David W. Goodall (eds.), Elsevier, Amsterdam.]

A. Alluvial Fans

At their upper ends, alluvial fans have coarse soils, steep slopes and deep deposits. Fans gradually change down slope to areas of fine soils, gentle (generally 1% or less) slopes, and shallow deposits. Coalesced alluvial fans, forming bajadas, may occupy nearly 100% of valley floors in some areas. The soil particle size changes that occur down a bajada have significant influences on vegetation composition.

B. Playas

In places where watercourses drain to internal basins without outlets, ephemeral lakes, termed playas, are formed. These are characteristic of the world's deserts with nearly 50,000 playas worldwide (Fig. 4), most seldom exceeding 100 km^2 (Neal, 1969).

Playas often have commercially significant deposits of a variety of salts. They are common worldwide and are especially conspicuous in deserts of Australia, Africa (north and south), Argentina, Chile, and the United States.

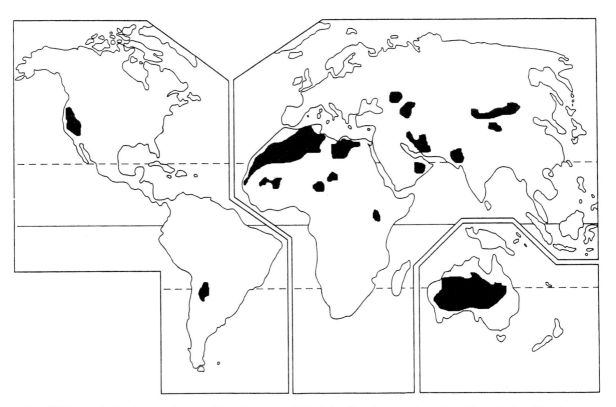

FIGURE 4 World distribution of zones of playa formation. [Modified and used by permission from Figure 5.12, p. 35 in *Desert Geomorphology,* by Ron Cooke, Andrew Warren, and Andrew Goudie, 1993, UCL Press, London.]

Playas usually lack vegetation except on their perimeter. Many animals use playas and some (e.g., frogs of several genera) seem reasonably adjusted to the unpredictable filling of playas with water.

C. Dunes

Dunes are usually composed of relatively fine sands that are mobile or may have become stabilized either through geological processes or by vegetation. Some dunes are formed of gypsum and are so different from those of silica sand that they contain endemic plants.

Fine particles that form dunes are generally moved across the surface by either saltation or reptation (short jumps or creeping). This movement is caused by the shearing action of wind that is often restricted by sand grain size and shape, the slope of the ground surface, moisture content, and the surface roughness of the ground, especially that caused by vegetation. While there are many types of dunes, the various morphologies repeat themselves across dune fields of the world (Fig. 5). Dune form is controlled by a combination of the availability of sediments, the local wind regime, and vegetation.

D. Arroyos

Arroyos, stream channels formed by periodic water flow, characterize most world deserts, especially in the United States, Africa, and the Middle East (where they are termed wadis), Australia, and parts of South America. Arroyo vegetation is different from that of adjacent desert surfaces in species composition and stature. Many of these species are phreatophytes, drought avoiders that draw water from deep in the soil. Many animals use arroyos as "highways" and for hunting. Vegetation characteristics change as the arroyo goes from deeply incised with straight sides to wider, flat fans.

V. DESERT SOILS

A. General Characteristics

Soils of arid lands are generally low in organic matter and range from slightly acidic to strongly alkaline on the surface. In many places the accumulation of calcium carbonate in the upper parts of the profile may create

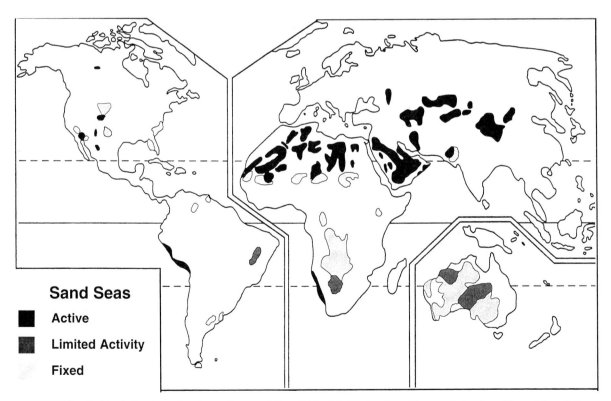

Sand Seas

■ Active

▦ Limited Activity

▫ Fixed

FIGURE 5   Worldwide locations of major sand seas and dune fields. [Modified and used by permission from Figure 1.1, p. 2, in *Geomorphology of Desert Dunes*, by Nicholas Lancaster, 1995, Routledge, London.]

hardpans of cemented calcium carbonate, but also of silica or iron compounds. The calcium carbonate layers can be up to 90 m thick and in places are virtually impervious to water.

In most desert areas, with the exceptions of valley floors or gentle slopes, desert soils are thin and bedrock is exposed. Because soil development is relatively poor in desert areas, the parent materials tend to dominate the chemistry of the soil, except where significant plant activity has added organic matter and all of its constituent elements.

B. Pavements

Soil surfaces in deserts often have one of two unusual characteristics—pavements and crusts. A superficial layer of stones, termed desert pavement, gibber, hammada, or reg occurs in many deserts. Pavements can be caused by the removal of fine materials by wind or surface water or by the upward migration of stones caused by several physical processes. Most likely involved are cycles of freezing and thawing or of wetting and drying. Often the stones of pavements develop a thin patina referred to as desert varnish.

C. Desert Varnish

In some deserts (e.g., southwestern United States), 75% of the rock may be covered with desert varnish. The color of varnish ranges from red to black depending on manganese content. There is considerable debate about the process of varnish formation. Some workers believe only physicochemical processes are involved, while others promote a biological model involving lichens and bacteria. Both processes probably contribute to the variety of colors of varnishes. Surprisingly, desert varnish may form in as little as 13 yr.

D. Crusts

The second unusual soil surface in deserts is a thin crustlike layer. When viewed from the side, these crusts are full of vesicles caused by rainfall or perhaps the expansion of soil gases. Crusts that are formed in association with algae and lichens have increased infiltration rates, whereas others may be somewhat water repellant. Most biological crusts involve cyanobacteria, green algae, lichens, and mosses. In many areas of the world, disturbance of crusts decreases nitrogen and organic matter available to plants. Off-road vehicles, cattle, and human foot traffic can all disturb such crusts. In the Great Basin, the winter grazing of cattle had little impact on the soil crusts containing mosses.

VI. DISTRIBUTION AND CHARACTERISTICS OF WORLD DESERTS

The world's deserts occur on five continents (Fig. 6). Any one continent may contain deserts caused by several factors. Similarly, a continent may contain both warm (precipitation comes as rain) and cool deserts (precipitation comes mainly as snow). The floristic affinities of world deserts are depicted in Figure 7.

A. North America

In North America, four desert types are recognized including one cool desert, the Great Basin Desert, and three warm deserts, the Mojave, Sonoran, and Chihuahuan deserts. These four deserts differ in climate and in the plant species and life forms that dominate them. The Great Basin Desert receives over 65% of its precipitation as snow and is dominated by low growing shrubs of a few genera (e.g., *Atriplex, Artemisia, Tetradymia*), some bunch grasses, few native annuals, but many alien species, and a few low-stature cacti.

The Mojave Desert, often considered a transitional desert between the Great Basin and the Sonoran, is characterized by winter precipitation in the form of rain, shrubby vegetation of low-form physiognomy, with the exception of the joshua tree (*Yucca brevifolia*) and a few other yuccas and mesquite (*Prosopis*). The Mojave has an abundance of annual species, and in wet years they may number $1000/m^2$.

The Sonoran Desert receives both summer and winter rain in approximately equal amounts and has the greatest variety of life forms, including subtrees, many species of shrubs, and a wide variety of cacti ranging in size from tall (over 12 m) columnar cacti such as saguaro (*Carnegiea gigantea*) (Fig. 8) down to a variety of small (a few cm high) species. Sonoran sites typically contain few grasses and have both summer and winter annuals, but only locally in the profusion of some Mojave Desert areas.

The Chihuahuan Desert has predominately summer rainfall and is dominated by low shrubs and species in the genera *Agave* and *Yucca*. The Chihuahuan has more dominant grasses than other North American deserts. Summer annuals and a number of cacti, but mostly of relatively low physiognomic form, occur.

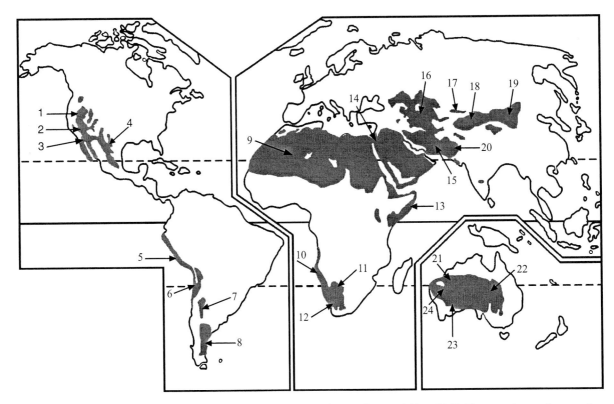

FIGURE 6 The arid areas of the world (dark) as defined by Peveril Meigs for UNESCO in 1953. The general area of many of the deserts mentioned in this article are indicated. 1, Great Basin; 2, Mojave; 3, Sonoran; 4, Chihuahuan; 5, Peruvian; 6, Atacama; 7, Monte; 8, Patagonian; 9, Sahara; 10, Namib; 11, Kalahari; 12, Karoo; 13, Somali-Chabli; 14, Arabian; 15, Iranian; 16, Turkestan; 17, Dzungaria; 18, Takla-Makan; 19, Gobi; 20, Thar; 21, Great Sandy; 22, Simpson; 23, Great Victoria; 24, Gibson.

All three of the warm deserts have a common plant, creosotebush (*Larrea tridentata*), that dominates their landscapes, especially on flats and on the lower slopes of bajadas with calcareous soils. Other species of *Larrea* occur in South American deserts.

B. South America

South America contains two general areas of deserts, a strip along the west coast of Chile and Peru from about 5° to 32°S that includes the Atacama and Peruvian deserts and a section east of the Andes on the Patagonian Plateau that includes both the Patagonian Desert 40° to 55°S, a cool semidesert, and the Monte of Argentina 28° to 35°S, a warm desert/semidesert.

The northern Peruvian Desert is a mosaic of vegetation and substrates ranging from sand dunes to rocky soils and from areas devoid of vegetation to those containing trees (*Prosopis*) and large cacti (*Neoraimondia*). Some extensive areas are covered with relatives of Spanish moss. These ground-dwelling species of *Tillandsia* absorb water from the humid air.

The Atacama is among the driest deserts in the world

and in places like Iquique, the average annual rainfall is 3 mm. It is characterized by winter fogs that form over the cold Humboldt currents. The sparse vegetation of the Atacama consists of a number of succulents, low cushion-like shrubs and dwarf shrubs with some grasses.

Many basins have salt pans or pavements or, at the least, gravelly surfaces. Vegetation is least developed near the coast where the fog effect is limited. Many areas are dominated by blue-green algae and lichens. As the fog effect increases inland, grasses (*Pennisetum*) may dominate some sites. Some areas have a variety of cacti ranging from low barrel-like species to taller ones with some branching. This is especially so at somewhat higher elevations.

The Patagonian Plateau has vegetation bracketing the full range from somewhat developed grasslands to areas that are clearly arid and support desert vegetation. In those areas, vegetation ranges from some cushion-shaped shrubs in the genera *Brachyclados* and *Azorella*, among others, to normal shrubs including the genera *Mulinum*, *Nassauvia*, and *Verbena*. Many of the plants belong to the Asteraceae, Verbenaceae, and Rubiceace.

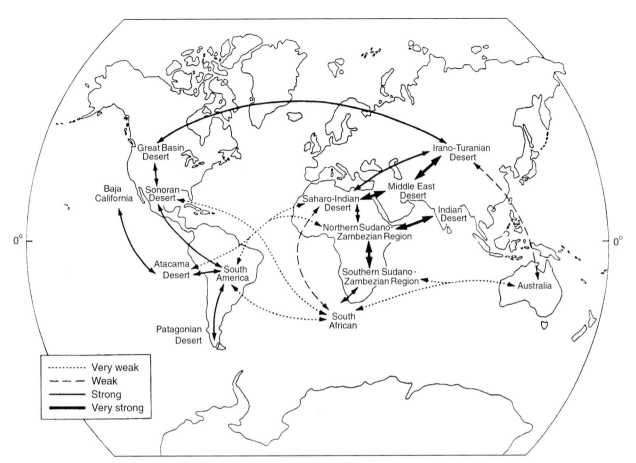

FIGURE 7 Floristic affinities between the major desert regions of the world. The width of the arrow indicates the degree of overlap in types (genera) of plants in each region. [From Fig. 4.8, p. 101, in *Ecology of World Vegetation*, by O. W. Archibold, Chapman & Hall, London, with kind permission from Kluwer Academic Publishers.]

Significant grass grazing by sheep is common, an ongoing practice since at least 1885.

Moving from Chile to Argentina, we encounter the Monte Desert. The Monte is well known because of comparative studies between it and the Sonoran Desert of Arizona. This comparison was motivated by the similar life-forms and many shared genera of the two deserts. Dominant plant genera include *Acacia, Larrea, Opuntia,* and *Prosopis* that occur in the Sonoran Desert. Additionally, there are a number of genera that resemble those of North America, including subtrees (*Bulnesia*) and columnar cacti (*Trichocereus*). The Monte has soils that range from clay to quite rocky on upper slopes but does not include extensive sand dunes.

C. Africa

Africa has desert areas at both the north and south ends of the continent. The Sahara Desert covers nearly 7 million km², mostly covered by gravelly or pebbly sur-

faces or sand dunes. Rainfall ranges from near zero to 200 mm. Plants that occur in the Sahara often reoccur across the deserts of the Middle East in similar microhabitats. On some sand dunes, shrub genera such as *Retama* and *Ephedra* occur along with a variety of grasses. Stony deserts may be occupied by *Ziziphus* and some perennial grasses, as well as a few other shrubs. Saline areas contain several salt-tolerant succulents. Where more moisture is available, several acacias, and a variety of grasses predominate.

The deserts of southern Africa are unrelated floristically to those of northern Africa (see Fig. 7). These three deserts—the Kalahari, the Karoo, and the Namib—differ from each other. The Kalahari is an inland area that ranges in precipitation from 250 to 500 mm; it represents an extensive semidesert, although some areas of the Kalahari include well-developed sand dunes that are stabilized by the presence of plants. Where there are extensive surface depressions, hardpans of calcareous materials form. Plants dominating

FIGURE 8   Typical scene in the Sonoran Desert of Arizona. Saguaro cacti, subtrees, and shrubs dominate the scene.

the arid portions are shrubs and some scattered trees, along with a variety of grasses.

The Karoo is often divided into two sections. The Nama Karoo is an area covered by dwarf shrubs and a variety of low-growing grasses. This low-form vegetation occupies 600,000 km². The succulent Karoo is about one-fourth the size of the Nama Karoo and has a richness of plants in the Mesembryanthemaceae (Fig. 9). This family and several others contain a variety of succulent plants that offer a rich and unusual flora. Many of the species are sold worldwide to collectors of succulent plants because of their unusual forms, including the widely known "stone plants," *Lithops* (Fig. 10). This desert has few tall shrubs or trees.

The Namib Desert is the most extreme of the desert areas, with precipitation ranging from 5 to about 80 mm. Much of the area is dominated by moving sand dunes up to 300 m in height. It is watered mainly by coastal fog that in some places occurs on half of the days each year. Annuals that respond to the sparse rainfall may blanket the area when rain occurs. Among the other plants in this landscape, some of the most famous are not considered true desert plants (e.g., *Wel-*

witschia). This unusual plant with large strap-shaped leaves, which is often pictured in desert books, resides in areas along water courses that have high amounts of available soil water.

Two other areas of Africa are sometimes identified as arid. The first is the Somali-Chalbi area that extends along the African coast from the Red Sea down the Indian Ocean to about the equator. In essence this is an extension of the Sahara. It includes zones with <100 mm precipitation supporting low shrubs and dwarf trees. During rainy periods there is good cover by annuals. There are very few areas with no vegetation. The second region is the southern and southwestern portion of Madagascar, which receives ~350 mm of rainfall/yr, but none in an obvious rainy season (i.e., precipitation occurs throughout the year). This area is dominated by a variety of woody plants that have such incredible forms that they are often pictured. These include the genera *Pachypodium, Didierea, Alluadia,* and baobabs, *Adansonia.*

Vegetation of north Africa is similar to that stretching all the way to the Gobi of China and Mongolia. This whole area is essentially a desert broken by some semi-

FIGURE 9 General view of vegetation in the Nama Karoo Desert of South Africa.

arid patches or mountains. People award different names to the dissected desert areas, even though they are not dramatically different in many cases. The commonality of the Sahara and parts of Asia is exemplified by the recognition of the Saharo-Sindian floristic region that covers this zone (see also Fig. 7). A floristic region is one that is defined by an internally consistent group of plant species or genera.

FIGURE 10 Stone plant (*Lithops*) in the Nama Karoo of South Africa.

D. Middle East

The Arabian deserts cover the Arabian Peninsula and Iran occupying some 2,600,000 km². This area has very close affinities to the Sahara and contains some areas of extreme aridity. The northern portion of the Arabian desert has winter precipitation, in contrast to the southern portion which has scant rainfall year round. The northern area is characterized by the presence of *Zygophyllum-Raeumuria* and *Artemisia-Gymnocarpus* communities and winter annuals. The southern region is characterized by a sparse vegetation with many annuals. The Rub'al Khali (empty quarter) is an extremely dry portion of the Arabian desert that contains the largest continuous body of sand in the world, covering 650,000 km².

E. Asia

The Iranian Desert, covering Iran, Afghanistan, and part of Pakistan, has a continental climate with winter precipitation. Vegetation types vary dramatically with specific substrates. Soils are generally coarse with some developing pavements. There are vast areas of saline plains that are virtually devoid of plants. Where plants occur, genera such as *Tamarix* with numerous species, *Artemisia*, *Atriplex*, *Zygophyllum*, *Calligonum*, *Ephedra*, and *Salsola* are common. These cool deserts reach their northern limit at about 30°N. The northern sections, dominated by *Artemisia*, appear similar to the Great Basin Desert of North America.

The Thar Desert of western India and Pakistan varies in climate from its northern and western sections where winter precipitation occurs, to the eastern portions where summer precipitation dominates. There are sandy areas broken by zones of rocky soils and plateaus. The soils differ between the north and the east with larger soil particles in the north and small gravel and sand in the east. Dominant genera include *Calligonum*, *Capparis*, *Ziziphus*, and *Prosopis*. Salt-laden playas occur throughout.

The Eurasian continent has a band of cool deserts that span 70° of longitude between approximately 40 and 50°N latitude. These are often divided into the Middle Asian deserts, mainly in the former Soviet Union, and the Central Asian deserts, those occurring in Mongolia and China. The Dzungaria Desert forms a transition between the two. The middle Asian deserts grade into the Iranian deserts to the south.

The Turkestan Desert in the western portion of middle Asia has a variety of subdivisions including the Kyzl-

Kum, the Kara-Kum, and Ust-Urt. This vast area has rainfall ranging from 75 to 200 mm/yr falling in the spring and autumn in the north but tending toward winter rains in the south. Summers reach a maximum of 50°C while winter reaches −26°C. The Kyzl-Kum (red sands) and Kara-Kum (black sands) describe the nature of the substrate, vast sandy basins with a variety of plant genera that are widely distributed in Asia, for example, *Haloxylon* (Fig. 11), a genus often used as firewood, shade for domesticated animals, or even as fodder. Some areas across the Turkestan Desert have highly saline soils in the basins with virtually no vegetation. Less saline, nonsandy areas contain a variety of genera including *Salsola*, *Capparis*, *Artemisia*, *Ephedra*, *Ceratoides*, and *Calligonum*. The transitional area of Dzungaria has precipitation ranging from 96 to 247 mm/yr. The flora is more like middle Asia than central Asia, including genera such as *Haloxylon*, *Anabasis*, and *Nanophyton*.

The deserts of central Asia range east to west across China straddling the southern portion of Mongolia. The Takla Makan covers part of central, western China in the Tarim Basin, the most arid part of central Asia. This is a vast sand desert with many areas nearly devoid of vegetation. Where vegetation occurs, the genera *Ephedra*, *Nitraria*, *Tamarix*, *Zygophyllum*, and *Haloxylon* are locally abundant. The Tsaidam is southeast of the Takla Makan in China. Here July temperatures seldom reach 18°C and the mean annual temperature is near 3°C. These cool areas are dominated by *Artemisia-Salsola* communities with representatives of the genera *Ephedra* and *Tamarix* on a variety of soils. Some areas are covered by saline flats that are essentially vegetationless. The Ala Shan is just south of the western half of the Gobi Desert in China. Sand covers nearly 65% of the area with much of it shifting. Rainfall ranges from 39 to 220 mm/yr over a gradient from west to east toward the base of the Ala Shan Mountains. Common plant genera include *Nitraria*, *Lycium*, *Kalidium*, and *Haloxylon* (tree forms), especially in sandy washes. The Ordos lies south of the eastern half of the Gobi Desert in China and is a basin dominated by widespread genera such as *Artemisia* on sandy sites, and *Zygophyllum*, *Raeumuria*, and *Caragana* on a variety of soil types. The basins of the Ordos have saline soils and have either salt-adapted plants or they are vegetationless.

The famous Gobi Desert runs east-west straddling the border of northern China and southern Mongolia. Summer rains range from 100 to 150 mm, although there are years with no rain. In contrast to many of the other central Asian deserts, the soils are mostly rocky,

FIGURE 11   A sandy area of the Kara-Kum Desert of Turkestan dominated by the sand-adapted tree *Haloxylon*.

in many places forming pavements that are covered with varnish. *Haloxylon* occurs in sandy areas, but perhaps the most widespread and characteristic plant is baglur (*Anabasis brevifolia*). Other genera that are widespread in these central Asian deserts include *Caragana, Raeumuria, Kalidium,* and *Nanophyon*. Two other areas that are often referred to as deserts occur on high elevation plateaus. The Tibetan Plateau receives less than 100 mm of precipitation per year and its flora is limited to about 50 species of low shrubs, grasses, and sedges. The Pamir, occurring in part of the former Soviet Union and Afghanistan, has a somewhat richer vegetation but has similar features. Neither of these is a typical desert.

F. Australia

The arid center of Australia covers more than 20% of the continent. If additional scattered arid areas are included the coverage is nearly 4.2 million km². The deserts of Australia are generally broken into five or six zones within the arid center. The Great Sandy Desert occupies an area in the northern part of Western Australia. South of there, the Gibson Desert occupies east

central Western Australia and southwestern Northern Territory. Even further south, the Great Victoria Desert occupies southeastern Western Australia and west central South Australia. In the center of the Northern Territory the Tanami Desert is sometimes recognized. South of it, in the southeastern corner of the Northern Territory and northeastern South Australia, lies the Simpson Desert. Some authors include the Stuart Desert of east central south Australia.

Most areas of Australian desert have reasonably well-developed vegetation composed of scattered subtrees, shrubs, and a variety of perennial grasses. Especially common are areas that contain hummock grasses and low shrublands dominated by members of the Chenopodiaceae. Australia's deserts fall in the same 30°S latitude high pressure belt as the Kalahari of Africa. The lowest precipitation is near 20 mm/yr and the boundary of the desert is usually set at about 254 mm/yr. Some areas receive more than two times that on occasion. Summer (January) has temperatures ranging to 49°C, but more commonly they are 25 to 34°C. In winter (July) the lowest temperature is 4.4°C with the mean ranging from 10 to 19°. Sandy areas that are generally stabilized cover nearly 2 million km² and these support several types of vegetation. The main vegetation types occur in more than one of the deserts. Arid mallee covers 416,000 km². This vegetation contains eucalyptus trees and members of the genera *Cassia, Atriplex,* and *Maireana* as dominants. Hummock grass vegetation is the most widespread in the arid zone occupying ~1.6 million km². These areas are dominated by spinifex (*Triodea spp.*) (Fig. 12). Hummocks of these species are often widely scattered and after rainfalls are interspersed with a variety of annuals.

Areas dominated by *Acacia* shrublands occur throughout. *Acacia* shrublands contain several species of acacias, especially *A. aneura.* These communities are extremely complex, varying significantly from site to site, based on soil characteristics. Grasses form an important portion of the vegetation. *Cassia* occurs in some places as do other perennial grasses. This vegetation is especially common in the Gibson and Simpson Deserts.

A tussock grass vegetation that covers ~500,000 km² dominated by *Astrebla* represents a small but significant portion of the arid zone. *Astrebla* grasslands further east in Australia are of significant economic importance.

The general distribution of arid tussock grasses is in the eastern third of the arid zone, while arid hummock grasses are in the central zone.

VII. DESERT PLANTS

Plants in deserts are exposed to extreme and unpredictable values of environmental variables, especially temperature, precipitation, and wind. Under these selection pressures, numerous morphological and physiological adaptations have developed. Current knowledge about these is changing our view of their functions compared to that postulated as recently as 20 years ago. Interestingly, the adaptations of desert plants may occur in other extreme environments such as those with saline soils, extreme cold, or at high elevation sites with high solar radiation and low temperatures. One common adaptation of plants in all ecosystems is to converge in their general structure (life-forms) in areas of similar climate. Table I compares the percentage of species in various life-forms in several deserts, a temperate forest

FIGURE 12   Ayers Rock in the central desert of Australia. The sandy soils are dominated by spinifex grass (*Triodea* spp.).

TABLE I

Lifeforms (% of total flora) for Several Desert Sites

Locality	No of spp.	Life forms							
		S	MM	N	Ch	H	G	HH	Th
Hot desert climate:									
Death Valley, California	294	3	2	21	7	18	2	5	42
El Golea, Central Sahara	169	0	0	9	13	15	5	2	56
Oudjda Desert, North Africa	49	0	0	0	4	17	6	0	73
Ooldea, Australia	188	4	19	23	14	4	1	0	35
Cool desert climate:									
Transcaspian lowlands	768	0	0	11	7	27	9	5	41
Hot Steppe climate:									
Tucson, Arizona	266	0	0	18	11	—	—	0	47
Whitehill, South Africa	428	1	1	8	42	2	18	0	23
Temperate forest:									
Connecticut	—	0		—	2	49	22	0	12
Tropical wet forest:									
Queensland	—	0		—	2	0	2	0	0

Lifeforms are S = Stem succulents, plants with aboveground water storage; MM = Mega- and Mesophanerophytes, trees over 8 m; N = Nanophanerophytes, shrubs under 2 m; Ch = Chamaephytes, plants with buds just above or at surface of ground; H = Hemicryptophytes, plants with buds in the ground surface; G = Geophytes, plants with bulbs, rhizomes, tubers; HH = Helo- and Hydrophytes, marsh and water plants; Th = Therophytes, annuals. From *Deserts of the World: An Appraisal of Research Into Their Physical and Biological Environments,* edited by William G. McGinnies, Bram J. Goldman, and Patricia Paylore. Copyright © 1969, 1988 The Arizona Board of Regents. Reprinted by permission of the University of Arizona Press.

and a rain forest. It is obvious that species of annuals dominate deserts.

A common classification of plant responses to desert conditions includes (a) drought-escaping plants—those that bloom only during periods of high available moisture (e.g., annuals), (b) drought-evading plants—those nonsucculent plants that conduct their physiological activities only when water is available, (c) drought-enduring plants—those adapted to high water stresses (e.g., many dominant shrubs), and (d) drought-resisting plants—species that store water in some manner to use for photosynthesis during periods of otherwise low water availability (e.g., cacti). These four strategies highlight the complex of adaptations that can occur.

A. Morphology and Behavior

Desert plants have a variety of structural adaptations, for example, succulent stems or leaves, reduced leaf size, and a more linear leaf shape. It was long thought that some of these structural characteristics prevented water loss. For example, having a low leaf area presents less total surface for transpiration. Recently, it has been argued that small leaf size reduces water loss by maximizing photosynthesis because such a leaf is cooler, not simply because the leaf is smaller and transpires less.

Belowground portions of plants are often equal to or exceed the biomass of aboveground portions. In a Great Basin Desert plant, *Atriplex confertifolia*, there are 133 g C/m^2 aboveground but 682 g C/m^2 belowground. This investment in roots provides sites for storage of energy as well as a high surface area for the uptake of water when it is available. Additionally, because many plants are exposed to wind, they are well anchored by extensive root systems. Some plants produce extremely long tap roots, sometimes exceeding 30 m, allowing plants to access the water table or water stored as a deep soil moisture.

B. Physiological Processes

Desert plants need to obtain and retain water and produce photosynthate under extreme conditions. There are three photosynthetic pathways: C_3, C_4, and CAM. C_4 plants form a four-carbon molecule during photosynthesis, while C_3 plants form a three-carbon molecule in the early stages of photosynthesis. C_3 and C_4 plants differ morphologically in that the leaves of the C_4 species are often subdivided into leaflets or have leaves with serrated margins, while C_3 species more often have simple leaves with no incisions. There are also internal anatomical differences in the arrangements of cells of each type.

Most summer annuals are C_4 plants, while most winter annuals are C_3 plants. Compared to C_3 plants under similar conditions, C_4 plants are more effective at photosynthesizing at high temperatures and in high light intensities, and they use water more efficiently to produce photosynthate. C_3 plants germinate during cool periods, sometimes forming rosettes that allow the plant to develop at the ground surface under temperature conditions not conducive for C_4 plants. Often there is a gradient in the percentages of the C_3 and C_4 plants in a community coinciding with measures of aridity, although this is an imperfect relationship.

The third kind of metabolism, CAM (short for crassulacean acid metabolism), derives its name from the Crassulaceae, a plant family widely distributed in some desert areas of the world but also occurring under cool, nondesert conditions and along the edges of oceans. CAM plants take up carbon dioxide at night but they cannot complete the light-mediated steps of photosynthesis. Instead they produce malic acid that is later used as a substrate to complete the C_3 pathway when light is available. CAM plants, under favorable conditions, can switch to using only the C_3 pathway.

VIII. DESERT ANIMALS

Animals in deserts are exposed to limited quantities of food, low and unpredictable water supplies, high and widely varying temperatures, loose substrates, high incident radiation, and lack of concealment because of scant vegetation.

A. Scant Food

There are several adaptations to the relative scarcity of food. One is to collect and store food when there are abundant supplies in order to have food available for long periods when food is scarce. Increasing body reserves is another form of storage. One obvious case of food caching is the storage of seeds by rodents and ants in deserts across the world. Both ants and rodents vary in the degree to which they rely on seeds. Some are specially adapted to handling seeds and may be nearly obligate granivores, (e.g., some ants in the genus *Pogonomyrmex* in North American deserts). Other species use seeds according to availability as part of their diet. Beetles in the Namib store body fat very readily, as do a variety of lizards in several deserts across the world.

Another way to deal with food shortages, lack of water, or low temperatures is to become dormant. An extreme case is anhydrobiosis, where organisms lose all of their body water and metabolism halts (e.g., some nematodes). A variety of species of invertebrates, including collembolans in Australia and the eggs and immature stages of many insects in several deserts, exhibit this adaptation. A less dramatic approach is to simply lower metabolic rate, so the expenditure of energy is less. This can be done by changing behavior patterns or going into a state of rest, such as torpor, that occurs in heteromyid rodents weighing ~40 g, for example, *Perognathus*. Another avoidance strategy is simply to migrate away from the site. Many birds that cannot withstand the summer conditions of the deserts use them heavily in winter to obtain food and then migrate when food is less available.

B. Water

Water scarcity and unpredictability in deserts causes a host of adaptations. Amphibians become active only when water is available. It takes ~30 mm of rainfall to cause frogs to come out of the ground where many of them are protected by a cocoon of their shed skins that remains around their bodies. Similarly, birds may be able to breed throughout the year in some areas as long as there is sufficient water. Again, it takes ~30 mm of precipitation to assure that seeds or insects are available to feed their young. Some species are able to store water when it is available, for example, the genus *Cyclorana*, an Australian frog that can increase its normal body weight by over 50% by acquiring additional water. Other species conserve body water. For example, many frogs store urea as a nitrogen waste product rather than producing urine which requires a lot of water. Many species withstand higher degrees of dehydration than relatives from more mesic areas. Birds have high mobility, and thus they can travel to water sources when many other animals cannot. In the Kalahari, sand grouse (*Pterocles*) may fly up to 80 km, one way, to obtain water. When birds drink water they can endure salt concentrations up to about 50% of that in seawater. Salt is also a problem for herbivorous lizards and mammals. Lizards often excrete salt through special nasal glands, while some mammals use their highly adapted kidneys to concentrate their urine up to 17 to 25 times the concentration of their blood (e.g., sand rats [*Psammomys*]) in North Africa and hopping mice (*Notomys*) in Australia. Several small mammals in three families have a different solution to salt problems. These species, *Dipodomys microps* in North America, *Psammomys obesus* in North Africa, and *Tympanoctomys barrerae* in the Monte of Argentina, scrape the outer salty layers from leaves with specialized incisors and discard them, leaving less salty material to eat. *T. barrerae* has a unique bundle of bristles on the side of its mouth that assists the process. As is the case for many of the extremes in deserts, both behavioral and physiological adaptation solutions may be necessary.

C. Temperature

During periods of low temperatures, most animals simply retreat to burrows or other refugia and may go into torpor. This also occurs at high temperatures, however, some species remain active because they have higher critical thermal maxima than related species in cooler environments. Birds tolerate an increase in their body temperature of up to about 4°C above normal, as can a number of diurnal rodents (e.g., *Ammospermophilus*) and some large mammals like camels. Interestingly, birds have little other capacity to deal with temperature other than the normal range of cooling behaviors that birds have everywhere.

D. Loose Substrates

Loose substrates create problems for locomotion and for burrowing. Many adaptations for moving across sand surfaces occur. Some lizards in Asia and Australia and North America have enlarged scales on their toes that allow them to move across sand surfaces by dissipating the weight of their body over a larger area. Many species become bipedal on loose substrates so that the force is directed downward rather than laterally. Specialized forms of locomotion include the side-winding behavior of rattlesnakes in North America and sand-dwelling vipers in Africa. Many species swim through loose sand substrates, rather than burrow, and have the capacity to extract air from between the sand grains (Fig. 13). In loose substrates, soil particles could get into the eyes, ears, or other orifices of small animals, and frequently there are hairs, scales, or other protective means of preventing this.

E. High Radiation

Most species seem to lack techniques for dealing with the high radiation in deserts other than avoiding it by seeking shade or burrowing. The fur and feathers of birds and mammals insulates them to a degree. In contrast, most desert lizards have dark pigment covering their peritoneum that absorbs genetically harmful ultraviolet radiation. Interestingly, the genus *Cnemidophorus*

FIGURE 13   A fringe-toed sand lizard (*Uma notata*) from the northern Sonoran Desert of Mexico. Note the enlarged scales on the hind foot. This helps these swift lizards run across loose sand.

in North America lacks such a peritoneum, but it does have dark pigment in its skin.

Another problem faced in deserts is the lack of cover, and thus, camouflage or cryptic coloration, is common. There has been some argument as to whether or not this is truly the reason that desert animals match their background so well, but most workers tend to accept the cryptic coloration hypothesis rather than some of its alternatives. Examples range from birds that match the color of the sand that they live on, with the same species having different colors in other areas, to grasshoppers and lizards responding in a similar manner.

Desert animals provide some dramatic examples of convergent evolution—that is, the evolution of similar body forms or physiological adaptations in unrelated organisms under the same environmental conditions worldwide. Bipedality in a variety of rodents—for example, springhares (*Pedetes*) in Africa, jerboas in Asia (*Jaculus*), and kangaroo rats (*Dipodomys*) in North America, is a good example of morphological adaptation in mammals. Similar adaptations occur in herbivorous lizards in Australia (*Amphibolurus*), North America (*Dipsosaurus*), Asia (*Liolepis*), and Israel (*Agama*). One of the most striking cases of convergence is the two ant-eating lizards, the Australian thorny devil (*Moloch*) and the North American horned toads (*Phrynosoma*) that are unusually similar in body form and habits.

IX. DESERT ECOSYSTEM PROCESSES

In such extreme environments, it is no surprise that the functioning of desert systems is often at one end of a spectrum when compared to more mesic systems such as forests. Nonetheless, the processes that occur in deserts are categorized as they are elsewhere for other ecosystems; it is generally only the values of these processes that differ. It is not possible to cover all ecosystem processes, but it is informative to look at primary productivity, biogeochemical cycling of nutrients, the spatial pattern of resources in deserts, succession, the process of repair after disturbance, and, finally, the role of animals in controlling desert ecosystem processes.

A. Primary Productivity

For the entire world, the annual net terrestrial primary productivity is 56.4 Pg/yr (one petagram = 10^{15}g). Of that total productivity, 0.5 Pg/yr (about 1% of the total) occurs in deserts. This is unimpressive given that arid areas cover 22% of the terrestrial world. There is a strong correlation between annual net primary produc-

tivity and precipitation in deserts. Extreme deserts are 0 to 10 g/m^2/yr; in well-developed shrubby deserts, this value is 10 to 250 g/m^2/yr. These values are much lower than those of temperate grasslands (200–1500 g/m^2/yr) and forests (600–1500 g/m^2/yr). In North America, the Mojave Desert is the lowest producing area at 25 g/m^2/yr. Chihuahuan Desert values are close to 120/g^2/yr.

B. Nutrients

Nitrogen (N), phosphorus (P), and potassium (K) are important nutrients for plant growth. In arid areas, these nutrients are usually positively associated with shrubs. In a creosotebush community there were 0.253 mol/kg N beneath the canopy and 0.148 and 0.10 at the canopy edge and interspace, respectively. There may be several reasons for this association, the most general one relating to the concept of "islands of fertility." Once established, a shrub sheds leaves (19–53 g/m^2/yr) and portions of its belowground roots die adding organic matter to the soil. Additionally, the roots may extend into the barren area between shrubs and extract nutrients from those areas, concentrating them under shrubs. Rodents often make burrows at the bases of shrubs and the addition of their nesting material and digging activities can increase soil nutrients.

In desert areas, the amount of carbon (C) available is generally lower than in adjacent grasslands, e.g., in a creosotebush stand in Chihuahuan Desert, there were 102 g C/m^2 in aboveground biomass, and a total of 159 g C/m^2 both above and belowground in biomass. The total carbon for that area is 2,088 g C/m^2 including the biomass of plants (above and belowground) plus the stored carbon in the soil. In contrast, the same values for a grassland are 131, 190, and 2,302 g C/m^2, respectively.

Nitrogen is generally scarce in deserts. It enters the system through fixation by organisms in soil crusts, through fixation by the nodules on the roots of legumes, and through airfall deposition. In the past it was thought that low quantities of nitrogen limited primary productivity. More recently, it has been suggested that phosphorus may be more limiting because it is relatively unavailable in the alkaline soils of deserts. When nutrients exist in desert soils, they may not be available because decomposition occurs more slowly in the absence of water, and because the microfauna and microflora that usually affect decomposition are at much lower densities than in forest or grassland systems. None of these decomposers can continue activity when soils are very dry. In the Chihuahuan Desert, termites are extremely important in decomposition, but seem less so in drier areas.

C. Succession

Ecosystem repair is referred to as succession. It is controversial as to whether succession occurs in deserts because there is usually not an obvious group of plant species occupying recently disturbed sites (pioneers) and that is ultimately replaced by a climax vegetation. Rather, in many cases following a disturbance, both climax and pioneering species may co-occur. In Israel, 16 of 20 species that were typical of mature communities germinated during the first year following a disturbance. In mesic areas, plants begin to occupy disturbed sites immediately following the disturbance. In arid areas, it may take much longer before establishment occurs. This is probably due to the episodic nature of the conditions necessary for establishment. One good year of rainfall is usually insufficient for establishment, and some species require the presence of other species for establishment. In one common example, some desert cacti must germinate and begin to develop in the shade of a shrub or subtree. This "nurse-plant" phenomenon (Fig. 14) is known from North American deserts and the Karoo where mound-building members of the succulent Mesembryanthemaceae establish early and are later invaded by woody species. In other cases, it is merely a matter of a statistical probability as to what sequence of plants will occur following disturbance.

D. Role of Animals in Ecosystem Processes

Animals can alter the species composition as well as the environmental conditions of ecosystems. Observations in deserts around the world indicate that animals will enhance particular species and hinder others. When the local extinction of the plain Vizcacha (*Lagostomus*) occurred in Argentina, the number of forb species declined on colony sites as the cover of grasses increased. Five years later richness of grasses had changed again.

In almost all deserts of the world, one group of three types of organisms are influential. These include granivorous species of harvester ants, small mammals, and birds. These organisms disperse or consume seeds, turn over soil at very high rates, change the physical conditions of the soil through digging activities, incorporate organic matter, and, in the case of ants, remove vegetation from the sites of their nests. We can predict the area disturbed by digging rodents by knowing

FIGURE 14 Two columnar cacti, the saguaro (*Carnegiea gigantea*) and the organ pipe cactus (*Stenocereus thurberi*), have germinated and grown in the shade of a nurse plant, in this case, a foothill palo verde (*Cercidium microphyllum*) in the Sonoran Desert.

their longevity and applying the formula: longevity in years $= 9.33$ Area$^{0.735}$.

Interestingly, the relative importance of these three groups differs around the world. In North America and Israel, small mammals are most important and remove the greatest number of seeds, followed by ants, and then by birds. In the Karoo, ants take more seeds than small mammals as they do in the Monte of Argentina. Birds, while generally taking fewer seeds than the other two groups, may be quite important seasonally. For some species of plants, nearly 100% of the seeds are removed while other species lose a very small portion. This differential feeding can clearly alter the composition of the plant community. Additionally, for species like the harvester ants that carry seeds back to their nests and belowground, there are dramatic changes in soil properties where these nests exist. The plants that establish on mounds are often not the same as would establish in the disturbed area adjacent to mounds. Clearly, these animals, along with many others, have an important role in structuring desert ecosystems.

Another interesting facet of desert systems is that they seem to operate on what is called the "pulse-reserve paradigm." For long periods, when water is not available or temperatures are too high, plants and animals in deserts, to a great extent, shut down either physiologically or behaviorally. If temperature moderates and water is available, there is a sudden pulse of breeding activity by animals, primary production by plants, and nutrient cycling and decomposition mediated by a variety of soil organisms. The gains made during these favorable periods allow accumulation of material that will carry the organisms through the next period of extreme environmental factors. The adaptations discussed here help with this pulse-reserve approach to survivorship.

X. HUMANS AND DESERTS

A. Human Uses of Deserts

Deserts are home to about 15% of the world's human population. People use deserts and their transitions with grasslands for a wide variety of purposes. Sometimes, when vegetation is destroyed or significantly altered, the process of desertification begins (i.e., transitional grasslands are replaced by desert vegetation). While estimates vary, it is thought that nearly 10×10^6 ha are subjected to desertification each year. One can imagine that anything that humans do might be exacerbated by natural disturbances (e.g., drought). Desertification does not have a single cause, rather it occurs because of people, climate, and environments of particular areas.

In addition to using desert areas for livestock grazing and browsing, firewood collection, habitation, and mining, humans also use deserts as sites for recreation. In some places human impact relates to unusual activities. For example, in North America, people with off-road vehicles frequently traverse desert areas at great speeds, tearing up vegetation, altering soil crusts, and changing the soil characteristics in other ways.

The collection of succulent plants around the world by hobbyists has become a serious threat for many species in at least five plant families. Many human activities can have indirect effects. For example, irrigated agriculture can change the vapor pressure deficit over relatively large areas, thus impacting the thermal and light environment of desert plants, often to their detriment.

An unusual use of deserts is to harvest water. In the Atacama large screens are erected to intercept and condense the fog so that it can be used by humans.

Interestingly, deserts seem to have the capacity to rebound from short duration, high-impact disturbances, even the testing of atomic bombs, and yet may be more negatively influenced by long duration, lower level assaults such as grazing and browsing.

B. Uses of Desert Organisms and Their Products

The list of desert animals and plants that are of direct value to human beings is as long as the uses to which they are put. A few examples should suffice: in North America guayale (*Parthenium argentatum*) was grown as a substitute for rubber for a period of time; jojoba (*Simmondsia*) oil appears in a variety of beauty products; in the Sudan *Blepharis* has wound healing properties; *Agave* is used in Mexico to produce an alcoholic beverage that is known worldwide; *Schinus*, from the Atacama Desert, has proven to have insecticidal activity, as does *Calotropis* in India.

Many species have turned out to be good forage plants, especially some in the genus *Atriplex*. *Larrea*, occurring in both the North and South American warm deserts, has great potential for commercialization because of its antifungal properties and nutritious forage. Similarly, *Acacia* is used for food and fuel but may be more important as a source of commercial gums and tannins. Other species worldwide have been used as sources of anti-diabetics, the basis for cosmetics, fibers for weaving and other uses, plant growth regulators, surfactants, waxes and many other uses.

Desert animals have been used in a variety of ways. Animals ranging from beetle larvae to large antelope have been harvested for food, clothing, weapons, and medicines. Recently, as a greater knowledge of the chemistry of some compounds produced by animals has occurred, they have found even greater use. The venoms of cobras, rattlesnakes, and gila monsters all have medicinal properties under certain circumstances. Some desert animals (e.g., species of camelids) have been domesticated in Africa, Asia, and South America.

XI. DISTURBANCE IN DESERTS

A. Natural

Natural disturbances include the effects of all environmental factors exceeding those that plants and animals are generally exposed to. Deserts are systems that occur in extreme environments, so periodic change in a natural factor that exceeds tolerance of desert organisms is highly likely. Because the time to recovery from such perturbations can be long, deserts are especially susceptible to disturbance and may, in fact, seldom be in equilibrium with their environment.

While this is not the place to discuss every environmental disturbance, a few will be informative. Temperature can vary dramatically in deserts (see Table I). High temperatures have the capacity to kill both plants and animals if they are exposed for long periods. However, there are few data suggesting that established plants die because of high temperature. In contrast, young plants or seedlings may die upon short exposure to high temperatures. Low temperatures, however, are known to kill a variety of desert plants. Catastrophic freezes, where temperatures may reach −8.4 to −5.6°C and last for about a day, can wreak havoc as they did in the Mexican state of Sonora in 1937 when a variety of genera were killed following a rather short cold spell.

Temperature may control the life-forms of plants in deserts (see Table I). It has been suggested that a major limiting factor for the succulent life history strategy is the inability of these life-forms to withstand low temperatures during the growing season. This might explain the relative lack of succulents in the deserts of central Asia, the Great Basin of North America, and the Patagonian Desert of Argentina.

Too little or too much water can always be a disturbance in deserts. In a very real sense, water is the environmental factor that drives desert ecosystems. Different attributes of water (e.g., availability, predictability, and flooding) have different influences on desert species. Deserts are already characterized by low annual values and high year-to-year variation in rainfall. In years of sequential drought, some species may be edited out of the community, although others have surprising resilience. The rebound of the Sahel following a long period of drought, and perceived devastation, is a good example of such resiliency. High rainfall amounts can cause flooding which, in turn, can mechanically remove plants or animals or cover them for long enough periods that respiration is not possible.

Wind is ever present in deserts. Many familiar land-forms (e.g., plants perched on pedestals) are caused, at least in part, by the downcutting of soil by wind in interspaces and deposition beneath plants. If wind carries a substantial load of geological materials, the abrasion can damage or kill plants and animals; the desiccating power of the air, partly due to wind, can bring organisms to critically low levels of tissue water. In the Patagonian Deserts, much of the dryness is due to the effect of wind rather than directly to the amount of rainfall.

Contrary to expectations, fire can be an important factor in deserts, especially in years of above average precipitation when a significant fuel load develops. In an interesting case in the Sonoran Desert, fuel load accumulated over a few years, and a single fire destroyed 68% of all the individuals of saguaro cactus (*Carnegiea gigantea*). In similar habitats, species like creosotebush (*Larrea tridentata*) were burnt to the ground but sprouted back a short time later.

In some systems, the activities of animals may form a disturbance for plants. This might take the form of plagues of grasshoppers in some African deserts or the soil turning activities of fossorial mammals, ants, or termites.

B. Anthropogenic Disturbance

Humans alter deserts dramatically. Semideserts hold an even larger portion of the world's population and nearly every desert and semidesert in the world has been influenced by the grazing and browsing activities of domesticated ungulates.

There are good historical records of the uses of deserts for thousands of years, for example, the early Egyptian cultures in Africa and by the Seri Indians for at least 2000 years in the Sonoran Desert of Mexico. The result of human use is often to change areas that are transitional with grasslands into deserts, at least in part, because of the effect of grazing by domestic livestock or removal of vegetation for fuel. Similar changes occur because of climate variations. Vegetation removal may actually alter some climatic patterns in deserts because of changes in the albedo (reflectivity) of the earth's surface.

Humans have introduced non-native plants, often to the detriment of deserts. In North America, the introduction of tamarisk (*Tamarix*) has created problems in riparian areas of deserts where these aliens alter water regimes and animal and plant community composition. Annuals and grasses have been moved around the world and often out-compete native plants. In North America,

TABLE II

Significance of Different Land-Use Management Objectives in the Different Regions

	Israel	Southwestern United States	Mexico	Australia	LDC[a]
Grazing	Low	Medium	High	High	High
Conservation	High	High	High	High	Low
Recreation	Medium	High	Medium	Medium	Low
Agroforestry	Low	High	High	Low	Medium
Forestry	Low	Medium	Low	Low	Low
Military	High	Medium	Low	Low	Medium
Agriculture	High	Medium	Medium	Low	Low
Waste Dump	Low	Medium	Medium	Low	Low
Mining	Low	Medium	Low	Medium	Medium
Land enhancement					
Urban	Medium	Medium	Medium	Low	Low
Rural	High	Low	Medium	Medium	Low

Note: Precise values are debatable, but note that spatially intensive uses tend to become more important in countries with higher population densities and that land uses other than primary production tend to become more important in countries with higher levels of affluence.

[a] LDC = less-developed countries. This column was included to compare likely patterns with countries around the edge of the Sahara, for example; it is even more generalized than the region-specific columns since there is a huge diversity in LDCs.

These results were assessed by participants in an Arid Lands Management Workshop. From Fisher *et al.*, Land use and management: Research implications from three arid and semiarid regions of the world. From *Arid Lands Management* (T. W. Hoekstra and M. Shachak, Tech. Eds.), p. 159. Copyright 1999 by the Board of Trustees of the University of Illinois. Used with permission of the University of Illinois Press.

the Great Basin Desert is dominated by non-native annuals that actually cause an increase in the fire frequency in those systems compared to natural systems that do not contain a large proportion of annuals. These fires kill native shrubs not adapted to fire (e.g., *Artemisia*) thus changing the entire community structure.

Obviously, building highways, large irrigation projects, and human settlement in what are often considered to be the favorable climates of deserts has dramatically diminished the area that can, by any sense of the imagination, be considered natural. The potential effect of different land uses is likely to differ dramatically in different desert areas (Table II).

XII. CONSERVATION ISSUES AND THE FUTURE OF DESERTS AND THEIR INHABITANTS

Misconceptions about deserts inhibit efforts to conserve them. First, because they are so severe, many believe that they are not much utilized by human beings. Second, because many animals and plants are inconspicuous it is thought that, compared to rain forests, there are very few species that require conservation efforts. Both of these are incorrect.

As mentioned earlier, the deserts are used, and at an increasing rate, for a variety of purposes by humans, often to their detriment. Second, the rich diversity of desert species are of interest in their own right and are potential sources for materials (food, medicine, clothing) of value to humans.

Even the United Nations, in its 1992 Conference on Environment and Development, called for a "Convention on Desertification" that would be similar to those that have been developed for rain forests.

Setting aside areas that currently represent sustainable desert communities is one form of conservation. Unfortunately, many areas have been degraded to the point where they require restoration, a process more challenging in deserts because of the extreme conditions than it is in more mesic areas. Additionally, certain species have to be restored or even reintroduced. For example, a program in Abu Dhabi was developed to restore the Houbara bustard (*Chlamydotis undulata*) that seems to have suffered from the effects of some range practices. In the Kyzylkum desert of Uzbekistan, studies are underway to reintroduce Przewalski's horse. Animals that had been zoo reared seem to adapt well when turned out into semiwild, desert conditions.

We must remember that the deserts of the world have been populated by indigenous people who are highly adapted to a desert way of life. While such people have probably altered deserts by their presence, they have come into a sort of equilibrium with the altered deserts and their way of life must be protected, just as any rare plant or animal should be.

One problem with attempting to conserve desert areas is that vast tracts are often required to maintain viable breeding populations of rare species. This need is antithetical to the needs of ever-expanding populations that require spaces to live, as well as places to obtain resources to sustain themselves.

In cases where nature reserves have been established in deserts, many of them are suffering challenges to their boundaries. These often relate to the encroachment of human populations or to the movement of materials or alien species from populated areas into reserves. It has been suggested that large buffer areas are needed and care must be taken to prevent leakage of organisms and effects into or out of the reserve.

Clearly, deserts deserve as much attention as rain forests but they have not captured the imagination of the world's peoples in the same way. They are not desolate lands that require no attention. Rather, they are amazing places housing plants and animals of indescribable beauty and complexity and humans with folkways that enrich the human spirit.

I have often said that the very starkness of the desert allows one to have an almost religious experience while in awe of the capacity of living organisms to adapt to such extreme and unpredictable environments. For that reason alone, I believe they should be preserved. Everyone deserves that experience.

Acknowledgments

John Mull, Weber State University, and an anonymous reviewer toiled over the manuscript to its benefit. Linda Finchum, as always, worked feverishly to keep me from making too many errors.

See Also the Following Articles

AFRICA, ECOSYSTEMS OF • AUSTRALIA, ECOSYSTEMS OF • C$_4$ PLANTS • DESERTIFICATION • NEAR EAST ECOSYSTEMS • SUCCESSION, PHENOMENON OF

Bibliography

Abrahams, A. D., and Parsons, A. J. (Eds.) (1994). *Geomorphology of Desert Environments*. Chapman & Hall, Cambridge.

Allan, T., and Warren, A. (Eds.) (1993). *Deserts: The Encroaching Wilderness. A World Conservation Atlas*. Oxford University Press, New York.

Cloudsley-Thompson, J. L. (Ed.) (1984). *Sahara Desert*. Pergamon Press, Oxford.

Cooke, R., Warren, A., and Goudie, A. (1993). *Desert Geomorphology*. UCL Press, London.

Cowling, R. M., Richardson, D. M., and Pierce, S. M. (Eds.) (1997). *Vegetation of Southern Africa*. Cambridge University Press, Cambridge.

Dean, W. R. J., and Milton, S. J. (1999). *The Karoo: Ecological Patterns and Processes*. Cambridge University Press, Cambridge.

Evenari, M., Noy-Meir, I., and Goodall, D. W. (1986). Hot deserts and arid shrublands, A. *Ecosystems of the World 12A*. Elsevier, Amsterdam.

Evenari, M., Noy-Meir, I., and Goodall, D. W. (1986). Hot deserts and arid shrublands, B. *Ecosystems of the World 12B*. Elsevier, Amsterdam.

Lovegrove, B. (1993). *The Living Deserts of Southern Africa*. Fernwood Press, Vlaeberg.

Man, J. 1997. *Gobi: Tracking the Desert*. Yale University Press, New Haven.

Mares, M. A. (Ed.) (1999). *Encyclopedia of Deserts*. University of Oklahoma Press, Norman.

McGinnies, W. G., Goldman, B. J., and Paylore, P. (1968). *Deserts of the World*. University of Arizona Press, Tucson.

Neal, J. T. (Ed.) (1975). Playas and dried lakes: Occurrence and development. *Benchmark Papers in Geology*, **20**. Dowden, Hutchinson and Ross, Stroudsburg, PA.

Nobel, P. S. (1988). *Environmental Biology of Agaves and Cacti*. Cambridge University Press, Cambridge.

Petrov, M. P. (1976). *Deserts of the World*. John Wiley & Sons, New York.

Sen, D. N. (1982). *Environment and Plant Life in Indian Desert*. Bharat Printers, Jodphur.

Smith, S. D., Monson, R. K., and Anderson, J. E. (1997). *Physiological Ecology of North American Desert Plants*. Springer-Verlag, Berlin.

Von Willert, D. J., Eller, B. M., Werger, M. J. A., Brinckmann, E., and Ihlenfeldt, H.-D. 1992. *Life Strategies of Succulents in Deserts: With Special Reference to the Namib Desert*. Cambridge University Press, Cambridge.

Walter, H., and Breckle, S.-W. (1986). *Ecological Systems of the Geobiosphere. 2. Tropical and Subtropical Zonobiomes*. Springer-Verlag, Berlin.

West, N. E. (Ed.) (1983). Temperate deserts and semi-deserts. *Ecosystems of the World 5*. Elsevier, Amsterdam.

Zahran, M. A., and Willis, A. J. (1992). *The Vegetation of Egypt*. Chapman & Hall, London.

DESERTIFICATION

James F. Reynolds
Duke University

GLOSSARY

combating desertification Activities specifically aimed at prevention and/or reduction of land degradation, rehabilitation of partly degraded land, and reclamation of desertified land.

desertification Land degradation in arid, semiarid, and dry subhumid areas resulting from various factors, including climatic variations and human activities.

dimensions of desertification The interactions and feedbacks of meteorological, ecological, and human components of land degradation.

drylands Arid and semiarid croplands, pastures, rangelands, and subhumid woodlands in which the index of aridity is less than 0.65; drylands cover about two-fifths of the land surface of the earth and are home to more than 20% of the human population.

index of aridity Ratio of mean annual precipitation to mean annual potential evapotranspiration.

land The terrestrial ecosystem that encompasses soils, vegetation, other biota, and the ecological, biogeochemical, and hydrological processes that operate therein.

land degradation Reduction or loss of the biological and economic productivity and complexity of terrestrial ecosystems, including soils, vegetation, other biota, and the ecological, biogeochemical, and hydrological processes that operate therein.

myths of desertification Controversy stemming from failure to consider all dimensions (meteorological, ecological, and human) of the problem, and alarmist tone connoted by the word desertification, which incorrectly suggests the action of deserts "moving" across the landscape, engulfing fertile lands and leaving starving people in their wake.

THE PHENOMENON OF DESERTIFICATION involves the loss of biological or economic productivity and biodiversity in arid and semiarid croplands, pastures, rangelands, and subhumid woodlands due mainly to nonsustainable human activities, such as overcultivation, fuel gathering, overgrazing by domestic animals, deforestation, and poor irrigation practices, and is often triggered or exacerbated by climate variability, mainly drought.

I. INTRODUCTION

A. Land Degradation in Global Drylands

Drylands cover about two-fifths of the land surface of the earth and are home to more than 1 billion people, or 20% of the human population. The fate of rural people in these regions is ultimately dependent on the effective use of natural resources, e.g., water, soils, plants, and wildlife. However, rapid growth of populations in dryland regions, which have some of the highest birth rates in the world, has contributed to rapid land degradation. This degradation—which encompasses soils, vegetation, other biota, and the ecological, biogeochemical, and hydrological processes that operate therein—has occurred over large areas of drylands in Asia, the Mediterranean, Africa, Oceania, and the Americas. Natural vegetation has been eliminated or severely reduced, soils are eroding at accelerated rates, and the

capacity of the land to support human populations, livestock, and wild herbivores has been substantially reduced (Figs. 1a–1f).

Can land degradation be reversed? What is the impact of deteriorated drylands on global climate systems? Do natural factors, such as climate variability, play a role in land degradation? Do ecologically sound land-use practices lie at the level of recognizing the rights and environmental knowledge of local communities? Such questions exemplify the suite of concerns underlying the general concept of land degradation in drylands, or desertification. Desertification involves the loss of biological or economic productivity and biodiversity in

FIGURE 1   (a–f) Images of land degradation in arid and semiarid lands. (a) Cattle disturbance around a watering point in northern Australia (photo by J. Ludwig). (b) Soil erosion following logging in dry subhumid region in south China (photo by J. Wu). (c) Fence line contrast between inside and outside grazed rangeland on Santa Rita Experimental Range, southern Arizona (1932). (d) Woody plant encroachment in native grasslands of central Argentina (photo by R. Distel). (e) Denuded grassland in South Africa (photo by S. Milton, courtesy of *Bioscience*). (f) Extensive soil erosion due to cultivation of marginal lands in central Argentina (photo by R. Distel). (g–i) Rehabilitation efforts to curtail or reverse land degradation. In southern New Mexico, (g) mechanical shrub removal and root plowing, which is extremely expensive and leads to high soil erosion (photo by R. Gibbens) and (h) manure spreading to restore soil fertility (photo by J. Herrick). (i) Pile of mulga (*Acacia aneura*) branches used to trap soil and runoff on a bare slope in Australian rangeland, creating a new vegetation patch (photo by J. Ludwig).

arid and semiarid croplands, pastures, rangelands, and subhumid woodlands. It is due mainly to nonsustainable human activities, such as overcultivation, fuel gathering, overgrazing by domestic animals, deforestation, and poor irrigation practices, and often is triggered or exacerbated by climate variability, mainly drought.

B. Reality vs Myths

The topic of desertification has evoked much controversy throughout the years in scientific, social, and political arenas (Table I). This controversy stems in part from the lack of scientific rigor in the preparation of many of the databases complied in the 1970s, misinformation disseminated by various organizations, the lack of a consensus definition, and the alarmist tone connoted by the word itself, which suggests the action of deserts "moving" across the landscape, engulfing fertile lands and leaving starving people in their wake. These images have their roots in a series of papers written in the 1930s by the British forester Stebbing, who used titles such as "The Encroaching Sahara" and "The Man-Made Desert in Africa." Furthermore, a series of major droughts in sub-Saharan Africa in the 1960s and 1970s, combined with weak economies and civil strife, resulted in widespread famine and—along with reports that the

Sahara desert was moving southward into the Sahel at a rapid rate (and which was referred to as "desertification")—only served to reinforce this image. Subsequent evidence using satellite images of plant cover on the margins of the Sahara desert has clearly demonstrated that its boundary is actually dynamic (Fig. 2)—there are periods of both "expansion" and "retreat," coinciding with the severity and duration of droughts. As findings such as these were reported in the scientific literature, news articles appeared with provocative titles such as "Threat of Encroaching Deserts May Be More Myth Than Fact" (*New York Times*, January 18, 1994) and "The Sahara Is Not Marching Southward" (Science, July 31, 1998), illustrating that desertification remains a controversial topic.

There are several themes underlying misunderstandings of the concept, which have given rise to numerous myths throughout the years. Current estimates of the rates and the spatial extent of global desertification are based on limited data and thus are generally rough at best or (at worse) inaccurate (see Section IV,C). In some instances, gross exaggerations about the rates and extent of desertification have been made, e.g., areas that had been arid for centuries have been claimed to have become "desertified." In other cases, observa-

TABLE I

Some of the Major Issues of the Desertification Debate[a]

- What is land degradation?
- What is the precise meaning of desertification?
- Does desertification refer to the state of a system, or is it a process?
- Can land degradation be quantitatively measured?
- What is the amount of land affected or at risk?
- How can a desert become "desertified?"
- What are the causes of land degradation? For example, what is the relative importance of natural (climate-driven) vs man-made (e.g., due to overgrazing by domestic animals) processes?
- Does climate variability cause desertification, or does desertification cause climate variability?
- Are short-term ecosystem changes indicative of desertification? When are these changes considered "permanent"?
- Does land degradation really matter?
- Is land degradation reversible?
- What are the relative roles of scientific vs grassroot efforts in abatement? In remediation?

[a] Historically, desertification has been a contentious issue involving ecological, meteorological, and human dimensions but usually in isolation from each other.

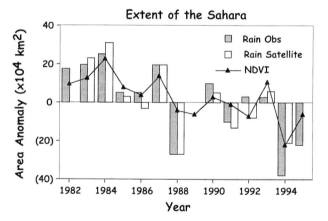

FIGURE 2 Extent of the Sahara Desert, shown in terms of annual deviations from average size defined by a 200-mm rainfall isohyet extending to the north of 25°N. Calculations are based on rainfall data collected at 141 locations throughout west Africa (Rain Obs), from METEOSAT satellite data (Rain Satellite), and from the normalized difference vegetation index (NDVI; a simple ratio between the red and near-infrared reflectance bands obtained from the NOAA polar-orbiting satellite). The NDVI correlates well with percentage vegetation cover and biomass, and thus the interannual fluctuations of the desert boundary, as assessed from NDVI, generally mimic those of rainfall. This work by Sharon Nicholson and coworkers at Florida State University shows considerable agreement between estimates based on rainfall and those based on NDVI and demonstrates that there is no "march" of the desert over West Africa (plotted from data in Nicholson *et al.*, 1998).

tions made on short-term ecosystem dynamics have been cited as evidence of desertification. With regard to short-term ecosystem dynamics, a key misunderstanding stems from the fact that drylands are fundamentally fragile ecosystems and, hence, highly susceptible to disturbance. Thus, the loss of vegetation cover (a short-term response to drought, e.g., that captured by satellite images in the Sahara) and soil degradation (a longer term response resulting from chronic and severe disturbances) are distinct, and not necessarily related, phenomena.

C. Dimensions of the Problem

The popular image of "creeping deserts moving like a series of waves of sand" continues to distract scientific and public attention from the real issue of desertification, which is indeed occurring globally, often far from the so-called true "deserts" such as the Sahara and Atacama. Desertification consists of three dimensions: (i)

meteorological, (ii) ecological, and (iii) human (Fig. 3). Each of these components is complex, difficult to predict, and highly interdependent. A failure to recognize the simultaneous role of—and feedbacks between—these different components has led to many of the controversies and misconceptions alluded to previously. In this article, I provide a brief background to drylands, present regional examples of desertification, and discuss international efforts to "combat" desertification. Finally, I explore the concept of desertification from each of its critical dimensions.

II. WHAT ARE DRYLANDS?

A. Index of Aridity

As the term suggests, annual precipitation in drylands is quite low. Rainfall often consists of high-intensity events that occur over short periods of time, it is highly

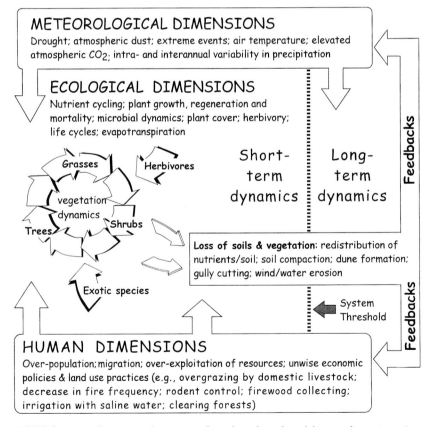

FIGURE 3 Desertification involves meteorological, ecological, and human dimensions. Past failures to recognize the interdependencies of these components has led to controversies and misconceptions about the concept.

variable throughout the year, and there is high overland runoff. Most, if not all, of the annual precipitation is lost from the system via evapotranspiration (evaporated from the soil or transpired by plants) rather than via drainage to ground water or surface runoff. This is due to the high evaporative demand of the atmosphere resulting from high air temperatures, low humidity, and abundant solar radiation, i.e., the potential for evapotranspiration is high. Therefore, drylands may be defined as regions where the ratio of mean annual precipitation (P) to mean annual potential evapotranspiration (PET), known as the index of aridity, is less than 0.65.

B. Global Extent and Classification

Drylands account for 47% of the global land mass (Table II). If we restrict the aridity index to the range of 0.05–0.65, drylands consist of arid, semiarid, and dry subhumid regions, which cover approximately 5.2 billion ha (Fig. 4). This definition of drylands omits 0.98 billion

ha (~7.5% of the global land area) occupied by hyperarid regions (P/PET < 0.05), such as the Atacama, Arabian, and Sahara deserts. These hyperarid regions have very low rainfall and high rates of potential evaporation that restrict (or preclude) plant growth and settlement by human populations and are not included in this discussion of desertification.

The greatest concentrations of drylands are in Asia and Africa, which account for ~30% of the total land area of the world and 64% of all drylands. Hence, total dryland area elsewhere—when expressed on a global basis—appears relatively small (5% in Europe and 9% in South America). However, these numbers are deceptive. For example, although Australia contains only ~11% of the world's drylands, they cover 75% of the Australian continent; similarly, both Europe and South America are about 30% drylands on a continental basis. In terms of a general land-use classification, arid (1569 million ha) and semiarid (2305 million ha) lands consist of rangelands and rain-fed farmlands, whereas the dry subhumid lands (1295 million ha) consist of woodlands, farmlands, and pasturelands.

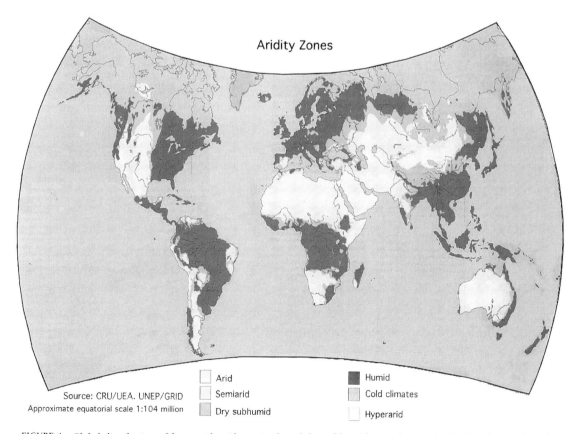

FIGURE 4 Global distribution of hyperarid, arid, semiarid, and dry subhumid areas (source: UNEP, 1997). See also color insert, Volume 1.

III. REGIONAL OVERVIEWS

The following brief overviews from select parts of the world illustrate the great diversity of issues and challenges presented by desertification. The causes and consequences vary from region to region, mainly in terms of localized intensity, as do programs to solve problems of land degradation, which vary regionally as a function of ecosystem characteristics, culture, economics, and political will. However, there exists some similarities. For example, land degradation resulting from human settlement in the rangelands of the United States, Argentina, and Australia share many of the same ecological consequences and social issues.

A. Africa: Extensive Problems

Desertification is perhaps most often associated with Africa, due largely to the great droughts and famines of the 1970s and 1980s. The drought in the Sudano–Sahel region in the early 1970s claimed more than 250,000 lives, left millions destitute, and resulted in mass migrations of rural people to urban areas in search of food, housing, and work. Similarly, in 1982–1986 the combined effects of severe drought and civil war killed more than 1 million Ethiopians. These high-profile events received much attention from the world media and served to cement the association of desertification with Africa. However, the true problems of desertification in Africa are much more complex than these tragic, but relatively short-lived, localized events.

More than one-third of the world's drylands are found in Africa (Table II) and one-third of Africa is considered to suffer from some degree of land degradation, including more than 70% of all agricultural lands. The African regions most affected by desertification include the northern portion of the continent, the Sudano–Sahelian (equatorial) region, and the Kalahari–Namib in the south. In the arid and semiarid zones of Algeria, Libya, Morocco, and Tunisia, overgrazing of rangelands has led to severe soil erosion and loss of vegetation. This loss of valuable topsoil, combined with a cycle of droughts, has resulted in declining production and intensified food insecurity. Rain-fed croplands in this region have been degraded by the use of inappropriate heavy machinery, salinization, and cultivation of dry, sandy soils. Woodlands have been cleared for fuel or by fires for agriculture and grazing. In the Sudano–Sahelian region the story is basically the same: fragile soils degraded through improper cultivation practices, deforestation, and overgrazing, leaving the land more vulnerable to drought and soil erosion. In southern Africa, overgrazing has led to severe land degradation and a decrease in productivity.

Michael Darkoh of the University of Botswana recently reviewed the large number of programs and organizations established in Africa during the past 30 years to help arrest and mitigate the effects of drought and

TABLE II

Aridity Zones as Defined by P/PET Ratios and World Drylands in Millions of Hectares[a]

		Aridity zones				All aridity zones	% of world drylands	% of continental land area	% of total land area of world
	Land mass	Hyperarid <0.05[b]	Arid 0.05–<0.20	Semiarid 0.20–<0.50	Dry subhumid 0.50–<0.65				
Africa	2,965.6	672.0	503.5	513.8	268.7	1959	31.9	66.0	15
Asia	4,255.9	277.3	625.7	693.4	352.7	1949	31.7	45.8	15
Australasia	882.2	0	303.0	309.0	51.3	663	10.8	75.2	5
Europe	950.5	0	11.0	105.2	183.5	300	4.9	31.5	2
North America	2,190.9	3.1	81.5	419.4	231.5	736	12.0	33.6	6
South America	1,767.5	25.7	44.5	264.5	207.0	543	8.8	30.6	4
Total	13,012.6	978.1	1569.2	2305.3	1294.7	6150	100	47.2	47.2
% of world drylands		16	26	38	21	100			
% of total land area of world		7.5	12.1	17.7	9.9	47.2			

[a] Based on UNEP (1997).

[b] P/PET ratio, where P is the mean annual precipitation and PET is the mean annual potential evapotranspiration.

desertification. The list is long, and these programs involve a plethora of action plans and activities and usually result in the formation of various types of organizations at the local, regional, and national levels (see Section IV). Unfortunately, many of these efforts have had limited success. The reasons for these failures vary from region to region but include such factors as lack of political will, shortage of funds, lack of participation of the people directly affected, inappropriate land-tenure policies, civil strife, misdirected research priorities, failure to include desertification control activities in national development plans, and a failure of governments to make desertification a priority. Poor countries have found it difficult to finance environmental programs aimed to protect natural resources, and often foreign financial and technical assistance has been used to support urban populations at the expense of rural ones. Of course, none of these problems and concerns are necessarily unique to Africa.

B. Mediterranean Region: Integrative Assessment

Land degradation is not a new issue in the Mediterranean. Humans have cut, grazed, and burned these lands for thousands of years, leaving virtually no natural vegetation and resulting in extensive soil erosion. However, rapidly changing social and economic conditions—along with the potential for climate change—pose many serious challenges to the Mediterranean region. In particular, the availability, distribution, and use of water underlies many of the contemporary themes of land degradation. Extensive forest fires throughout the Mediterranean in the early 1990s, which resulted in extensive soil erosion, runoff, and flooding, had a major impact on the public perception of land degradation. As a result, there has been a renewed emphasis on national programs of abatement in numerous Mediterranean countries.

The European Union (EU) recently established several major projects, such as Mediterranean Desertification and Land Use (MEDALUS). The aim of this project was to understand, predict, and mitigate the effects of desertification in Mediterranean countries of the EU, including the southern and eastern parts of the Iberian Peninsula of Spain, Mediterranean France, Italian Mezzogiorno, and almost all of Greece. This was motivated by many concerns, particularly increasing human populations and widespread land degradation (particularly soil erosion), air and water pollution, and water shortages. For example, as new EU economic policies are adopted, many pastoral and drylands farmers are being

marginalized and moving to urban areas; social–economic conditions are rapidly changing, e.g., the rise of tourism, intensification of high-tech agriculture, and the shifting of populations to urban environments. Such changes have resulted in the abandonment of land for traditional agriculture and the increase in demand for water for urban expansion, tourism, and irrigation, resulting in increased land-use conflicts.

The Mediterranean, although small on a global scale, is a complex mosaic of landscapes, soils, different cultures, localized climates, and land-use histories. Although many of the problems grouped under the general heading of desertification are similar throughout the region (e.g., rainfall variability, growing urbanization, loss of farmland and agricultural potential, increased tourism and human population density, forest fires, soil erosion, and flooding), it is doubtful that any one approach for solving these problems will apply. However, programs such as MEDALUS have made significant advances in integrating desertification research across all levels of concerns (Fig. 3), which may eventually provide general tools for linking science and policy in other regions of the world.

C. Australia: Landcare and Satellites

More than 70% of Australia is arid or semiarid. It is estimated that about 50% of these drylands are severely degraded or desertified to some extent. Most of this land has traditionally been used for pastoralism and the most obvious form of desertification is the encroachment of shrubs into grasslands and savannas. This encroachment is preceded by a loss of native perennial grasses and increased rates and extent of soil and wind erosion.

In contrast to the Mediterranean, which has experienced centuries of chronic land degradation, Europeans only began settling in Australia in the late 1880s. However, the landscape underwent rapid change over a very short period of time. Large herds of domestic livestock were put in the rangelands in order to maximize short-term economic profits but with little or no view toward long-term sustainability. These new pastoralists overstocked herds, suppressed fire (unaware of its importance in maintaining grasslands), and built fences that concentrated the impact of grazing animals. Exotic animals, such as the rabbit, were introduced with disastrous impact on native plant and animal communities. Furthermore, recurrent droughts tended to accelerate this chronic and systematic degradation of the land.

As in the Mediterranean, several instances of extreme drought, massive flooding, and soil erosion brought this problem to the attention of the general public. Recent national programs of desertification management are relying on a grassroots approach to help resolve these problems, although this is complicated by the extreme isolation of many pastoralists. For example, federal funds are being used in Landcare Australia to encourage pastoralists and farmers to form local "landcare" groups and to employ a variety of proven techniques to restore degraded lands. The National Drought Alert Program employs satellite imagery to detect trends in greenness (similar to the NDVI index in Fig. 2) to forecast potential drought conditions in pastoral and farming regions; media alerts can then be issued to encourage appropriate strategies of land management, e.g., reduction of stocking rates.

D. China: Green Great Wall

China currently suffers from extensive land degradation, which has resulted in "colossal environmental detriment and economic loss" and threatens the lives of more than 400 million people (Zha and Gao, 1997). It is estimated that more than one-third of the total land area of China is desertified (at least to some extent), especially in the arid and semiarid northern areas of the country. Desertification has caused reduced soil fertility, degraded soil structure and massive erosion, and loss of native vegetation. Not surprisingly, China has experienced centuries of land degradation.

As in other areas of the world, the causes and consequences of desertification in China are varied but all can be attributed to various combinations of human and meteorological factors acting out in concert with localized ecological conditions (Fig. 3). There are some unique ecological settings in China that are conducive to desertification. Many northern regions of the country contain sandy, loose surficial sediment deposits that are readily transported. These regions are also characterized by the coincidence of droughts with the windy season. All of this combines to produce much wind erosion and transport of sand, particularly when the native vegetation is removed or decreased.

China has made an enormous effort to assess the causes and consequences of desertification, including programs to rehabilitate desertified land into productive uses. The Three North project (named after the three northern regions that suffer desertification) involves constructing a forested shelter belt that spans across 13 northern provinces. The hope is that this Green Great Wall, which is expected to be completed by 2025, will help stabilize sand dunes in the region, thus protecting cropland and pastures and abating soil erosion.

IV. "COMBATING" DESERTIFICATION

A. United Nations Efforts

Because of the global magnitude of desertification and its current and potential effects on human populations, the United Nations (UN) has taken a leading role in providing an international framework for studying and resolving this problem. Hence, much current desertification research (across a broad spectrum of disciplines) is conducted within the auspices of former and existing UN programs and related efforts. These programs are involved in educational and outreach efforts, basic research, and the coordination of disparate international and local projects. Although some of these programs have been subjected to criticism over the years, they provide a valuable context for addressing this problem. A brief overview is provided in the following sections.

1. Earth Summit

In June 1992, the UN convened the Earth Summit in Rio de Janeiro, Brazil, to address—among a variety of critical global issues—the problem of desertification. The Earth Summit was not the first UN attempt to tackle the problem of desertification. In the late 1970s, the term desertification was first brought to the public attention when the UN announced that this process was affecting more than 3500 million ha of land globally and that more than one-third of the terrestrial land surface of the earth was under similar risk. At the same time, the aforementioned great drought and famine of the Sudano–Sahel was in the public spotlight. The UN held a high-profile conference on desertification in Nairobi in 1977 at which a Plan of Action to Combat Desertification (PACD) was adopted "to curb this onslaught." However, despite this and numerous other international efforts, by 1991 it was concluded that the PACD was not working and the problem of land degradation was actually intensifying, rather than abating, globally. Africa was identified as a particularly serious region of concern. As a result, desertification remained a major issue for the UN when it convened in Rio de Janeiro 15 years later.

The Earth Summit produced a significant, revamped program of desertification, including the establishment of a separate UN Convention on Desertification, an action framework designed to establish programs to counter the threat of dryland degradation throughout the world, particularly in arid, semiarid, and dry subhumid regions. The UN General Assembly was called on to establish an Intergovernmental Negotiating Committee to prepare a Convention to Combat Desertification (CCD) in "those countries experiencing serious drought and/or desertification, particularly in Africa." In December 1992, the general assembly adopted Resolution 47/188. The convention was adopted in Paris in June 1994 and on October 14, 1994, it was opened for signature; as of October 1999, the convention had been ratified or acceded by 159 countries (see *http://www.unccd.de/* for updates). Meanwhile, many of the convention's provisions, particularly those relating to action programs in Africa, are being implemented.

2. Combating Desertification

What are the essential elements involved in combating desertification? The CCD adopted the following definition: Combating desertification includes activities that are part of the integrated development of land in arid, semiarid, and dry subhumid areas for sustainable development. These activities are specifically aimed at (i) prevention and/or reduction of land degradation, (ii) rehabilitation of partly degraded land, and (iii) reclamation of desertified land.

Although there are language and contextual differences, many of the goals and objectives of the CCD are not unlike those contained in the 1977 PACD. Hence,

it is crucial to understand why the PACD failed. Three major reasons are most often cited: (i) Many processes and issues leading to land degradation (e.g., overcultivation of the land, deforestation, overgrazing by livestock, and poverty) have their roots in socioeconomic and political realms, and thus they are not amenable to technical or science "solutions" as proposed in the PACD; (ii) local populations were not included in drawing up solutions; and (iii) there was a failure to integrate the PACD into other existing development programs. Note that these problems reflect the failure to recognize the simultaneous importance of the three fundamental dimensions of desertification (Fig. 3).

Recognizing these shortcomings, the Rio conference supported a new, integrated approach to the desertification problem, emphasizing programs and strategies to promote sustainable development at the community level in an effort to overcome some of the failings of the previous PACD. Consequently, the CCD places considerable emphasis on the social dimensions of the problem and the role of local peoples and nongovernmental organizations in solving local problems. There is a conspicuous emphasis on the human dimensions (a "grassroot" effort), as evident in the prologue to the convention (Table III). This shift of emphasis from scientific solutions to social dimensions to solve prob-

TABLE III

Excerpts from the Prologue of the 1994 United Nations Convention on Desertification, Classified as to Their Emphasis on Ecological, Meteorological, and Human Dimensions of the Problem

Parties to this convention	Ecological aspects	Meteorological aspects	Human dimensions
Affirm that human beings … are at the center of concerns to combat desertification and mitigate the effects of drought.			•
Are aware that arid, semiarid and dry subhumid areas together account for a significant proportion of the earth's land area and are the habitat and source of livelihood for a large segment of its population.	•		•
Note the high concentration of developing countries, notably the least developed countries … particularly … in Africa.			•
Note that desertification is caused by complex interactions among physical, biological, political, social, cultural, and economic factors.	•	•	•
Are conscious that sustainable economic growth, social development, and poverty eradication are priorities of affected developing countries, particularly in Africa, and are essential to meeting sustainability objectives.			•
Stress the important role played by women in regions affected by desertification and/or drought, particularly in rural areas of developing countries, and the importance of ensuring the full participation of both men and women at all levels in programs to combat desertification and mitigate the effects of drought.			•
Believe that strategies to combat desertification and mitigate the effects of drought will be most effective if they are based on sound systematic observation and rigorous scientific knowledge and if they are continuously reevaluated.	•	•	

lems of desertification is another reason it remains somewhat of a contentious issue (see Section I,B). There have also been criticisms that some developing countries are using the convention to assist in their economic development rather than combating desertification per se. Despite these concerns, the CCD is making significant progress and there is a clear recognition that regardless of emphasis, dryland degradation—given its ecological, climatic, economic, and social implications and complexities—will require contributions from a multitude of disciplines to understand and identify useful resolutions.

B. Defining Desertification

More than 100 formal definitions of desertification have been proposed in the literature. Although most restrict its usage for drylands per se (see Section II), these definitions cover a gamut of issues, involve a multitude of spatial and temporal scales of concerns, and often connote disparate meanings. As a result, some have called for eliminating the use of the term altogether. The main criticism is that it has been used in so many different ways throughout the years, and has been applied to such an extensive range of environmental problems, that it often lacks precise meaning. Does it refer to the state of a system, or is it a process? For example, is soil salinization a cause or a symptom? At what stage does land degradation constitute desertification? Issues such as these have contributed to the controversy surrounding the concept (see Table I). In general, the various definitions of desertification differ in their emphasis on the three distinct dimensions of the problem (Table IV).

TABLE IV

Select Examples of Definitions of Desertification from the Literature That Differ in Their Emphasis on Ecological, Meteorological, and/or Human Dimensions of the Problem

Definition	Ecological aspects	Meteorological aspects	Human dimensions
The spread of desert-like conditions in arid or semiarid areas due to man's influence or to climate change.		•	•
Diminution or destruction of the biological potential of the land that can lead ultimately to desert-like conditions. It is an aspect of the widespread deterioration of ecosystems, and has diminished or destroyed the biological potential, i.e., plant and animal production, for multiple-use purposes at a time when increased productivity is needed to support growing populations in quest of development.	•		•
An aspect of the widespread deterioration of ecosystems under the combined pressure of adverse and fluctuating climate and excessive exploitation.	•	•	•
The process of environmental degradation in nonsandy areas where the fragile ecology is disturbed by excessive human activities.	•		•
The impoverishment of terrestrial ecosystems under the impact of man ... the process of deterioration ... that can be measured by reduced productivity of desirable plants, undesirable alterations in the biomass and the diveristy of the micro and macrofauna and flora, accelerated soil deterioration, and increased hazards for human occupancy.	•		•
The irreversible, sustained decline of the biological productivity of arid and semiarid land resulting from pressures caused both by people (e.g., increased population) and by abiotic factors (e.g., variable rainfall and long-term climate changes).	•	•	•
The spread of desert-like conditions of low biological productivity due to human impact under climatic variations.	•	•	•
The expansion of desert-like conditions and landscapes to areas where they should not occur climatically.		•	
Land degradation ... resulting from adverse human impact.			•
Desertification, revealed by drought, is caused by human activities in which the carrying capacity of land is exceeded; it proceeds by exacerbated natural or man-induced mechanisms, and is made manifest by intricate steps of vegetation and soil deterioration which results, in human terms, in an irreversible decrease or destruction of the biological potential of the land and its ability to support population.	•	•	•
Land degradation in arid, semiarid, and dry subhumid areas resulting from various factors, including climatic variations and human activities.	•	•	•

The French forester Aubréville is credited with being the first to coin the term desertification. Aubréville's usage was in the context of observations he and colleagues made in 1949 of indiscriminate felling and burning of forest and woodlands in subhumid Africa and the accelerated soil erosion that resulted, i.e., the "creation of deserts." In 1977, the UN PACD defined desertification as the "diminution or destruction of the biological potential of the land [that] can lead ultimately to desert-like conditions." This definition, which identified human activities as the sole causal mechanism, was heavily criticized as being overly simplistic, alarmist, and ignoring the role of climatic factors. A more precise definition was needed—one that distinguished between true desertification and the shorter term phenomenon of cyclic oscillations of vegetation at desert fringes (i.e., desert "expansion and contraction" as observed by satellite data and which is related to natural climate fluctuations; Fig. 2).

The 1994 CCD addressed these concerns and adopted the following definition: land degradation in arid, semiarid, and dry subhumid areas resulting from various factors, including climatic variations and human activities. This definition distinguishes natural factors (i.e., climate) from anthropogenic ones, which is especially important since climate (both short and longer term) may play a substantial role in exacerbating the degradation of semiarid lands, particularly in marginal areas at the transitions between semiarid and arid ecosystems.

C. Global Extent of Desertification

1. UNEP Estimates

How widespread is land degradation in drylands? The most comprehensive study to date to estimate the global extent and distribution of desertification is that of the UN Environmental Programme (UNEP). Drylands (as in Table II) were subdivided into three dominant types of human land-use categories: irrigated agricultural cropland (146 million ha), rain-fed agricultural cropland (458 million ha), and rangelands (4556 million ha) (Fig. 5). Hence, from the point of view of agricultural land use, the overwhelming majority of drylands are rangelands (88%) and only 3% are irrigated croplands and 9% rain-fed cropland. It was estimated that 70% of all susceptible drylands suffered from some form of land degradation. By land-use category, this was 73% of all rangelands, 47% of rain-fed cropland, and 30% of irrigated cropland.

2. Limitations of Estimates

How accurate are these numbers? The UNEP numbers cited previously were compiled from a diversity of sources, including heavy reliance on the subjective

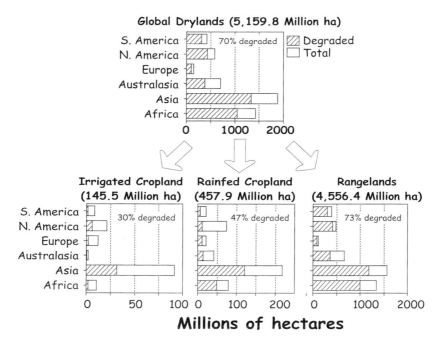

FIGURE 5   Estimates of dryland degradation (source: UNEP, 1992).

judgments of scientists and laypersons, and from surveys completed by local governments. Although they represent the most thorough estimates to date, they must be considered initial approximations. Although it is generally acknowledged that extensive areas of the earth have experienced varying degrees of chronic land degradation during the past century, there is a paucity of rigorous, consistent data. This is not surprising given that land degradation is a complex phenomenon (see Section V) and thus not amenable to practical and standardized methodologies for assessing, mapping, and quantifying on large scales. There are many perplexing challenges. For example, it is an exceptionally difficult task to quantify any single indicator from the suite of components that may reflect the rates and magnitudes of desertification, such as the degree of soil erosion, plant primary production, soil organic matter content, or socioeconomic conditions (Table V). This is true even at small spatial scales when each variable is analyzed in isolation, let alone consideration of key interactions with other variables. Many challenges remain. Most existing monitoring programs have been implemented for relatively short time frames, which makes it difficult to decipher short- vs long-term trends; meth-

TABLE V

Partial List of Potential Indicators of Land Degradation in Drylands[a]

Indicator/process	Ecological aspects	Meteorological aspects	Human dimensions
Soils			
Chemical degradation: salinization, contamination (heavy metals), acidification (aluminum, pH)			•
Physical degradation: compaction, soil depth, infiltration, percolation, temperature, shear strength	•		
Water erosion: runoff rates, soil loss, rill and gully density, sediment loading	•		
Wind erosion: soil loss, dune formation	•		
Soil organic matter: nutrient concentration, bulk density, soil texture	•		
Depth to water table	•	•	
Albedo (surface reflectance)		•	
Ratio of soil carbon to nitrogen	•		
Biological soil crusts: cyanobacteria, green algae, lichens, mosses	•		
Climate			
Precipitation: spatial distribution, amounts, intensity, frequency, variability, duration		•	
Wind speed		•	
Evaporation; evapotranspiration	•	•	
Vegetation			
Structure: cover, density, spatial distribution, plant morphology, rooting depth	•		
Agriculture: genetic diversity, species, harvest	•	•	•
Composition: species richness, diversity, endemism, exotic species, life forms	•		
Function: net primary productivity, standing biomass, leaf area index, growth rates, mortality, disease susceptibility	•		
Litter accumulation	•		
Socioeconomics			
Human population density: changes in numbers, migrations			•
Income per capita			•
Land-use changes: Ownership, clearing for firewood/cultivation			•
Water: availability, consumption, distribution, storage	•	•	•
Grazing: pressure, wildlife restricted to game reserves, domestic cattle/sheep	•		•
Tourism: game ranching, off-road vehicle use, road development			•
Agroforestry practices: irrigation water quality, pesticides, fertilizer-use rates			•

[a] Indicators classified based on dominant driver (Fig. 3) of the processes, although all are highly interdependent.

ods and models that may work for one region may not be appropriate for another; and selecting appropriate indicators of desertification (e.g., to separate natural vs social factors) is difficult.

Current estimates of rates and extent of desertification are constantly being updated and improved as evident in the revised UN *World Atlas of Desertification* (UNEP, 1997). However, the process is difficult and slow. Although modern techniques of satellite-based remote sensing (combined with geographic information systems) provide some new opportunities for obtaining improved data—at least for some indicators of land degradation such as primary production and land-use changes—currently there are no integrative approaches or tools that can incorporate critical aspects of climate, ecology, and human dimensions into a single integrative framework.

V. CONCEPTUAL MODEL OF DESERTIFICATION

A. Ecological Aspects

1. Threshold of Response

Nearly all drylands are characterized by extreme year-to-year precipitation fluctuations; hence, it is often difficult to distinguish between short-term variability and long-term changes in ecosystem appearance as well as between temporary and permanent changes. Short-term variability in precipitation tends to affect the range and frequency of resource pulses, whereas long-term change alters the resource base: that is, the entire system moves beyond some threshold (dashed line in Fig. 3). Once this threshold has been crossed, it is unlikely that the system can be returned to its predisturbed state without expensive, and often extensive, remediation efforts (see Section V,D). All of this depends on the interactions of numerous climatic, edaphic, and biological factors in combination with the economic feasibility of rehabilitation.

I have stressed the inherent difficulties in teasing out underlying causal factors that give rise to short- vs long-term ecosystem dynamics in drylands. A recent case study from the Serengeti in Tanzania provides an excellent illustration (Morell, 1997). In the 1970s, there was much concern that the Serengeti was degrading and turning into a desert. The large bush and acacia forests that characterized the Serengeti were declining at an alarming rate. Elephants, which fed on these trees, were largely considered responsible and many believed

that a culling program was the only way to save these "pristine" woodlands. In fact, these woodlands were not pristine but instead a relatively recent feature of the Serengeti reflecting a sequence of complex interactions between ecological factors (e.g., the life span of the acacias and the number of wildebeest), human dimensions (cattle grazing and fire setting), and meteorological factors (precipitation patterns) (Fig. 6).

Woodlands in the Serengeti started appearing after an outbreak of rinderpest disease in the 1890s that devastated wildebeest herds and local cattle populations and which led to mass human starvation. This decline in human populations meant a decrease in fires, which were set to create new pasture for cattle, and fewer fires meant higher establishment of acacia seedlings and the eventual establishment of a forest in the midst of grassland. By the 1920s and 1930s, the practice of setting fires was becoming reestablished as the human population recovered; however, the new fires were more frequent and hotter since conditions existed that favored grass growth—low herbivory (wildebeest herds were still low) and a relatively wet climate in the 1960s. These fires resulted in high mortality of acacias seedlings and, by the 1960s, the oldest trees (which have a life span of about 60–70 years) started to die off. However, concerns about the Serengeti turning into a desert were abated by the late 1990s as the Serengeti was experiencing a "return of the forest." Heavy poaching of elephants, the recovery of the wildebeest populations, and fewer fires set by humans all favored high recruitment of tree seedlings.

This example illustrates several key elements of the conceptual model of desertification: (i) Short-term dynamics may not necessarily be indicative of long-term phenomena (i.e., system thresholds are extremely difficult to establish without long-term studies), (ii) natural ecosystems are complex and difficult to predict, and (iii) humans have major impacts on natural systems in often unknown and unpredictable ways. This serves to reinforce the view that land degradation is a complex phenomena that frequently evades simple characterizations.

2. Ecological Dimensions of Land Degradation

Drylands typically have many distinguishing ecological characteristics that contribute to their susceptibility to disturbance and, ultimately, to desertification. The natural vegetation is composed of various mixtures of grasslands, shrublands, and savannas, with trees either scattered or concentrated along watercourses. Since vegetation cover is usually relatively sparse, much of

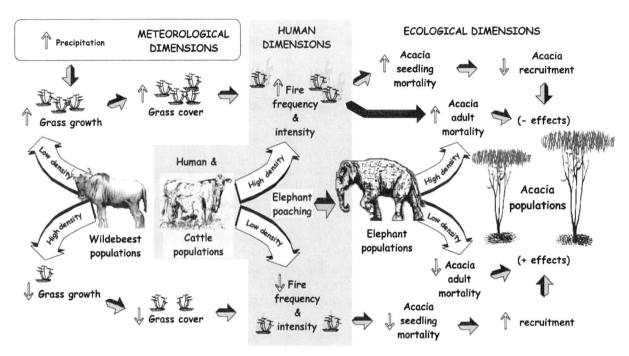

FIGURE 6   Dynamics of acacia forests in the Serengeti grasslands are a function of complex interactions between meteorological, ecological, and human dimensions. See also color insert, Volume 1.

the soil is exposed directly to rain, overland flow, sunlight, and wind. Many dryland soils are sensitive to disturbances because they contain small amounts of organic matter and have low aggregate strength. Both tillage and grazing by domesticated animals can have profound effects—in a very short period of time—on these soils, including lowering their permeability to water (thus decreasing infiltration), disturbing their surface integrity (thus increasing susceptibility to erosion and sedimentation), and decreasing their quality (decreased nutrient status) for plant growth.

Although land degradation is often equated with soil degradation, it is a more general phenomenon that involves whole ecosystems. The UN CCD defines land degradation as the reduction or loss of the biological or economic productivity and complexity. This reduction in the complexity of the land is the major physical restructuring of a system that is symptomatic of land degradation. In drylands, this includes erosion and sedimentation by both water and wind, often resulting in a redistribution of topsoil, compacting of the soil, loss of soil silt fraction, dune formation, and arroyo cutting. There may be shifts in natural fire cycles with a disruption of biogeochemical cycling, including the redistribution of essential nutrients, decreased efficiency of nutrient cycling, and increased nutrient losses from

the system. Both native perennial plants (cover and biomass) and many associated microbial and animal populations are reduced, whereas exotic species increase in dominance. All of this results in a decrease in the biological and hence economic potential of the land.

B. Meteorological Aspects

Williams and Balling (1996) concluded that substantial improvements have been made in our understanding of the causes of interannual variability in dryland climates, including the natural causes of droughts. For example, it is accepted that variations in annual precipitation levels are related to natural variations within global-scale climate systems. However, great uncertainties remain and the relationship between desertification and climate resembles the proverbial "chicken and egg" problem. To what extent does climate induce desertification? How do changes in land surface properties due to land degradation in turn affect climate? Since the two processes often work together, it is virtually impossible to separate the impact of drought from that of desertification.

Drylands are particularly vulnerable to climate variability, of which precipitation is the most important component. For example, a slight shift in seasonal precipitation and/or frequency of extreme rain events could potentially lead to the overexploitation of the meager resources of drylands and contribute to the further degradation of the very resource base on which human populations are so dependent. Preliminary studies with general circulation models (GCMs) in the 1980s projected that a doubling of atmospheric carbon dioxide (due to the rapidly expanding human population and associated activities) would result in lower precipitation, as well as shifts in the timing and frequency of rains, in the interior of large continents. Recent GCM studies also predict increases in rainfall intensity and longer dry periods in many dryland regions of the world. In the long run, global climate change may further exacerbate the already high natural variability of drylands, leading to permanent degradation of their productive potentials, particularly since there is a lack of "buffering" by large reserves of organic matter in the soils or in woody vegetation.

The array of impacts of climate on land and the implications of degraded land surface for the climate system are varied and complex (Fig. 3). Human activities impact surface characteristics (and hence albedo) and the atmospheric composition of dryland regions, including the breakdown of soil structure, reductions in soil moisture, increased surface runoff, changes in species composition and cover, increased dust in the atmosphere, and increases in aerosol and trace gas emissions from burning. In response to such human impacts, dryland climate is greatly influenced mainly via changes in energy balance. Changes in albedo affect the amount of solar radiation absorbed by the surface and changes in soil moisture levels affect the portion of energy used in evaporation and transpiration processes. Changes in surface roughness influence wind speeds and turbulence, which have a bearing on evapotranspiration. Finally, changes in atmospheric composition may affect atmospheric temperature profiles and thus influence the capacity to generate precipitation over land.

C. Human Dimensions

Rapidly changing social and economic conditions— along with the potential for climate change—pose serious challenges to many dryland regions of the world. Globally, there are differences in socioeconomic factors (e.g., human population growth rates) and biogeography (e.g., natural vegetation) that play a large role in the type of major human activities in any given area. Moreover, there are differences in how human interference is affecting biodiversity and ecosystem functioning in poor and richer countries. Taking these differences

into account, there are approximately six major classes of human interference in drylands: (i) the loss of habitat; (ii) the fragmentation of crucial habitat; (iii) overexploitation (mainly overgrazing by domestic animals); (iv) the spread of exotic organisms (aliens); (v) air, soil, and water pollution; and (vi) climate change. Although it is difficult to generalize about the vulnerability of terrestrial ecosystems, some problems virtually exist everywhere (e.g., habitat loss, habitat fragmentation, and the direct and indirect effects of exotic species), whereas pollution and climate change tend to be more significant threats in richer, temperate-zone nations. Key ecosystem "goods and services" (e.g., food, construction materials, water purification, flood control, climate regulations, soil maintenance, carbon sequestration, nutrient recycling, wildlife habitat, erosion control, and tourism/recreation) are being seriously affected.

Given that all three of the developing regions of the world (Africa, Asia, and Latin America) have similar percentages of land degradation, and the highest levels of population growth, desertification is a growing problem with major ecological, economic, and social implications. However, to understand the human dimensions of desertification, it is important to compare and contrast different regions of the world to seek out generalities, particularly in richer vs poorer countries. As new economic policies are adopted, many pastoral and dryland farmers are marginalized and must move to urban areas; socioeconomic conditions are rapidly changing, e.g., the rise of tourism, intensification of high-tech agriculture, and the shifting of populations to urban environments. Often, such changes result in the abandonment of land for traditional agriculture and the increase in demand for water for urban expansion, tourism, and irrigation, resulting in increased land- and resource-use conflicts.

The human dimensions of desertification (Fig. 3) represent some of the most complicated and perplexing issues in identifying the causes and solutions to land degradation. For example, in large areas of Africa, civil strife and government policies are key factors influencing resources (land, water, and wildlife), land degradation, and food security. Civil strife displaces people, often to other marginal lands and in great numbers; their abandoned land is left unattended, indigenous management systems are lost, and they are forced to use methods of ranching and farming that are often not appropriate in these new areas. In some countries, government policies have encouraged planned human settlements in the wetter margins of arid and semiarid lands or nearer to water supplies. This has led to land-use conflicts between agriculture, livestock, wildlife, and human settlements as a result of the intrusion of agriculture into lands traditionally used for domestic stock.

D. Stepwise Degradation Model

The conceptual model of desertification presented in Fig. 3 suggests that the process of chronic land degradation is directional—that is, it increases in severity until a threshold is eventually surpassed whereby the ecological state of the land (e.g., loss of soils and vegetation) is irreversible. An example is provided by a study of land degradation of rangelands in South Africa by Sue Milton and colleagues at the FitzPatrick Institute. Mil-

ton and colleagues showed that grazing-induced desertification is a stepwise phenomena and that the potential for recovery at any given step is related to the function of the affected component. Their model is summarized in Table VI, adapted to the scheme proposed in Fig. 3. In this particular case, the human dimensions component emphasizes management options and that part of the ecosystem specifically targeted for management.

At step zero, the biomass and composition of vegetation varies as a function of natural climatic cycles and stochastic events (e.g., fire, drought, and disease). An understanding of the relationships between these processes can be used effectively as a management tool. For example, livestock densities (secondary producers) can be manipulated depending on rangeland condi-

TABLE VI

Ecological, Meteorological, and Human Dimensions of Grazing Induced Rangeland Desertification as Illustrated in Fig. 3[a]

Step No.	Description	Ecological dimensions	Meteorological dimensions	Human dimensions	
				Management option	Management level
0	Biomass and composition of vegetation vary with climatic cycles and stochastic events	Biomass and composition of vegetation vary with weather	Natural climatic cycles and stochastic events	Adaptive management	Secondary producers
1	Herbivory reduces recruitment of palatable plants, allowing populations of unpalatable species to expand	Plant demography, increase in exotic or undesirable species		Strict grazing controls (rotation schemes, intensity, type of animals, brower: grazer ratios, etc.)	Secondary producers
2	Plant species that fail to recruit are lost, as are their specialized predators and symbionts	Loss of biodiversity (plants, animals, specialized predators, facilitators, symbionts), reduced primary and secondary productivity		Manage vegetation (e.g., seeding, plant removal), removal of livestock, culling abundant herbivores	Primary and secondary producers
3	Biomass and productivity of vegetation fluctuate as ephemerals benefit from loss of perennial cover	Perennial biomass reduced (short-lived plants and instability increase), resident birds decrease, monads increase	Changes in surface albedo and soil moisture; do land surface feedbacks affect regional climate? Increase in variability of precipitation?	Manage soil cover (e.g., mulching, erosion barriers, roughen soil surface)	Physical environment
4	Denudation of all vegetation cover; changes in ecosystem structure and function	Bare ground, accelerated wind and water erosion, aridification, soil salinization		?	?

[a] Symptoms characterize the state of plant and animal assemblages, management options refer to actions to improve the condition of the range, and management level refers to the system (level of the food chain) in which management could be focused. The "system threshold" (Fig. 3) may be somewhere between steps 2 and 4 (based on conceptual model for stepwise degradation of arid and semiarid rangelands by Sue Milton and coworkers in South Africa).

tions. The critical importance of this is illustrated by an example from Mexico. It was recently authorized that communally owned Mexican rangelands could be privatized in hopes of improving resource conditions and productivity. However, a follow-up analysis showed no differences between private and communal tenure systems, and instead the amount of annual precipitation was the most important factor explaining rangeland conditions (Coronado-Q., 1998).

The first step of degradation concerns changes in the age structure of the plant populations (Table VI). Those species most frequently defoliated by grazers fail to set seed, whereas toxic or distasteful plants, which are generally unaffected by grazers, are very successful. The type of management that might be used at this step is a function of the type of objective (e.g., game viewing and meat production) and vegetation type (annual or perennial grassland, shrubland, or savanna). Options to restore certain desirable forage species include shifting or varying the grazing season, stocking intensity, or animal type. The second step of degradation involves a decrease in biodiversity and both plant and animal productivity. There are fewer ecosystem "goods and services" (see Section V,C). Reversal of degradation at this stage is usually not cost-effective, particularly for arid regions in which it would involve removal of domestic livestock, culling of other abundant herbivores, and manipulating the vegetation (reseeding, herbicide treatments, bush cutting, etc.; see Figs. 1g and 1h). The third step involves processes associated with the reduction of perennial plant vegetation cover and an increase in ephemeral and weedy species. This includes accelerated wind and water erosion and various land surface impacts, such as increased albedo and surface temperatures, reduced soil moisture storage, and cloudiness. Rangelands in this condition are not profitable to ranchers and restoration must focus on the physical environment, e.g., reducing erosion, increasing water infiltration, protecting the soil surface from sun and frost, and creating microsites suitable for the establishment of perennial seedlings (Fig. 1i). The final step in the degradation process is characterized by a complete loss of vegetation cover, accelerated erosion, and soil salinization—a true "human-made" desert. Such rangelands are usually abandoned due to the high costs of—and low probability of successful—restoration and rehabilitation.

VI. CONCLUSIONS

Desertification or land degradation is the loss of biological and economic productivity and biodiversity in arid and semiarid croplands, pastures, rangelands, and sub-humid woodlands of the world. It is due mainly to nonsustainable human activities, such as overcultivation, overgrazing, deforestation, and poor irrigation practices, and is often triggered or exacerbated by climate variability, mainly drought. A new integrated assessment paradigm for rational decision making is needed to solve this problem. Such a paradigm must recognize the simultaneous roles of, and complex feedbacks between, the meteorological, ecological, and human dimensions of this problem. The failure to recognize and include the interdependencies of these dimensions has slowed progress in developing the necessary synthetic framework for tackling the enormous problem of dryland degradation. International researchers—representing a wide range of disciplines, such as ecology, atmospheric science, social sciences, policy, and integrated assessment—must work together to define new interdisciplinary approaches for addressing this pressing global problem.

Acknowledgments

The author thanks Roberto Distel, Roberto Fernández, Bob Gibbens, Jeff Herrick, John Ludwig, Kiona Ogle, Sue Milton, Sharon Nicholson, Colin Saunders, John Tenhunen, and Jianguo Wu for helpful input and acknowledges support from the Alexander von Humboldt-Stiftung during the preparation of the manuscript. This article is a contribution to the Carnegie Mellon Center for the Integrated Study of the Human Dimensions of Global Change (NSF SBR 95-21914).

See Also the Following Articles

AFRICA, ECOSYSTEMS OF • AUSTRALIA, ECOSYSTEMS OF • DEFORESTATION • DESERT ECOSYSTEMS • MEDITERRANEAN-CLIMATE ECOSYSTEMS

Bibliography

Coronado-Q., J. A. (1998). Relationship between range condition and the land tenure system in Sonora. PhD dissertation, University of Arizona, Tucson.
Darkoh, M. B. K. (1998). The nature, causes and consequences of desertification in the drylands of Africa. Land Degradation Dev. 9, 1–20.
Grainger, A. (1992). Characterization and assessment of desertification processes. In Desertified Grasslands: Their Biology and Management (G. P. Chapman, Ed.), Linnean Society Symposium Series No. 13, pp. 17–33. Academic Press, London.
Ludwig, J. A., and Tongway, D. J. (1995). Desertification in Australia: An eye to grass roots and landscapes. Environ. Monitoring Assessment 37, 231–237.
Milton, S. J., Dean, W. R. J., Duplessis, M. A., and Siegfried, W. R.

(1994). A conceptual model of arid rangeland degradation—the escalating cost of declining productivity. *Bio Science* **44**, 70–76.

Morell, V. (1997). Counting creatures of the Serengeti, great and small. *Science* **278**, 2058–2060.

Nicholson, S. E., Tucker, C. J., and Ba, M. B. (1998). Desertification, drought, and surface vegetation: An example from the West African Sahel. *Bull. Am. Meteorol. Soc.* **79**, 815–829.

Thomas, D. S. G., and Middleton, N. (1994). *Desertification: Exploding the Myth.* Wiley, Chichester, UK.

United Nations Environmental Programme (UNEP) (1992). *World Atlas of Desertification* (editorial commentary by N. Middleton and D. S. G. Thomas). Arnold, London.

United Nations Environmental Programme (UNEP) (1997). *World Atlas of Desertification* (N. J. Middleton and D. S. G. Thomas, Coordinating Eds.), 2nd ed. UNEP/Arnold, New York.

Williams, M. A. J., and Balling, R. C. Jr. (1996). *Interactions of Desertification and Climate.* World Meteorological Organization, United Nations Environmental Programme/ Arnold, London.

Zha, Y., and Gao, J. (1997). Characteristics of desertification and its rehabilitation in China. *J. Arid Environ.* **37**, 419–432.

DIAPAUSE AND DORMANCY

Nelson G. Hairston, Jr.
Cornell University

I. What Are Dormancy and Diapause?
II. Dormancy and Diapause as Adaptive Traits
III. Variation in Dormancy and Diapause
IV. Time Travelers: Dormancy and Diapause as "Migration from the Past"
V. The Storage Effect, Prolonged Dormancy and Diapause, and the Maintenance of Biodiversity
VI. Summary

GLOSSARY

bet-hedging strategy A trait of an organism, living in a variable environment, that leads to low variation in fitness. In general, such a trait provides an organism greater net fitness over a range of environmental conditions than would a trait specialized for any single environment. A bet-hedging trait is expected to evolve when the environment in which a species lives fluctuates over a fixed range of conditions that is sufficiently broad that fitness varies significantly, and when precisely which state the environment will take in the immediate future is unpredictable.

diapause A state of dormancy in some animals that is induced by a "token" environmental cue, such as day length. The token cue serves as a reliable indicator of a coming onset of harsh environmental conditions, but is not by itself harsh.

dormancy Any state of reduced metabolic activity of an organism. Typically, dormant organisms have associated characteristics such as cessation of development, the absence of reproduction, and enhanced resistance to harsh environmental conditions. Some disciplines have distinct meanings for dormancy that add further constraints to its meaning.

egg bank An accumulation of long-lived diapausing eggs (i.e., those eggs that persist in diapause for longer than a single growing season) of aquatic invertebrates in the sediments of marine or freshwater habitats.

seed bank An accumulation of dormant plant seeds that persist in dormancy for longer than a single growing season.

storage effect A general mechanism for the maintenance of biodiversity within a single habitat based on differences between competing species in their responses to environmental conditions. A resistant life-history stage can allow coexistence by the storage effect if each species reproduces successfully under the conditions favorable for that species and can survive through unfavorable periods (e.g., when a competing species dominates) in the resistant stage. Often, the resistant stage has prolonged dormancy. This mechanism can also serve to promote the maintenance of genetic diversity within a single population.

temporal dispersal The emergence of individuals from dormancy over a range of years (or other time interval), when those individuals entered dormancy in a single year. Often, the years (or other time intervals) have different environmental qualities for growth and reproduction.

temporal migration The avoidance of harsh environmental conditions in an environment by an individual organism that enters dormancy before conditions become harsh and emerges from dormancy when favorable conditions return.

————————————————————————————————

DORMANCY IS CRITICAL for surviving stressful environmental periods for a great many species. Some form of this trait is expressed by organisms ranging from bacteria to trees, from protozoa to vertebrates, and for a great diversity of organisms in between. In many cases, the period of dormancy lasts only as long, or slightly longer than, the duration of harsh conditions and simply determines the likelihood that a species can persist until the arrival of the next favorable period. For a variety of other species, however, dormancy can last for periods extending well beyond the typical duration of harsh environmental conditions. When this "prolonged dormancy" lasts long enough for multiple generations to have occurred in the active life history stage, powerful conditions exist for fostering the coexistence of multiple competing species within a community or genotypes within a species. Thus, both short-duration and prolonged dormancy have important implications for understanding biodiversity.

I. WHAT ARE DORMANCY AND DIAPAUSE?

Dormancy is a very general term that encompasses a wide variety of different physiological states. The names applied to each of these states depend on the scientific traditions accompanying particular taxon-based disciplines. Only a small amount of this terminological diversity can be covered here, but it is important to recognize that the biological similarities (homologies) among the various physiological conditions called "dormancy" in different kinds of organisms are often questionable. In its most general sense, a dormant organism is simply one with a reduced metabolic rate. For bacteria, in which life cycle and cell cycle are synonymous, dormancy is defined as a temporary loss of the ability to reproduce (Henis, 1987), although many single-celled organisms also produce hardened cases at the time of

dormancy (Whitton *et al.* and Bradbury in Henis, 1987). At the other end of the phylogenetic tree, vertebrate dormancy is most clearly expressed in "adaptive hypothermia" which includes both the daily torpor exhibited by a variety of small mammals and birds and the longer seasonal hibernation or aestivation seen in many taxa (Bartholomew, 1972). These latter states are often induced in part by "token" day-length cues that initiate physiological responses (e.g., fat storage), in advance of the onset of harsh conditions. In higher plants, dormancy is divided into "seasonal" (induced by external day-length cues that foreshadow a change in season) and "opportunistic" (imposed by direct exposure to harsh conditions) categories (Harper, 1977). Similarly, in arthropods, Tauber *et al.* (1986) distinguish between "aseasonal quiescence" ("a reversible state of suppressed metabolism," similar to the "opportunistic dormancy" of plants) and "dormancy" ("a seasonally recurring period ... during which growth, development, and reproduction are suppressed"). By their definition, dormancy is then further divided as "diapause-mediated dormancy," which is anticipatory, being induced by token cues (similar to seasonal dormancy of plants), and "non-diapause dormancy," which is essentially seasonal quiescence. The common theme here is that organisms from bacteria to vertebrates exhibit dormancy as an adaptation for survival in temporally varying environments. Over a remarkable range of organismal complexity and phylogenetic origin, species have evolved broadly similar mechanisms for avoiding harsh conditions either by direct physiological responses to the imposition of harsh conditions or through distinct anticipatory responses to seasonally predictable changes. These mechanisms are generally lumped under the broad umbrella of dormancy.

Depending on the organism and the environment in which it resides, the duration of dormancy or diapause can be brief or can last for extraordinary periods of time. For microbial cysts, dormant seeds, and diapausing eggs, there are reports of remarkable abilities to survive for centuries in dormancy: There are examples of bacteria becoming active after 200 years, of plant seeds germinating after thousands of years, and of crustacean eggs hatching after 300 years (Hairston *et al.*, 1996). The diapausing eggs of *Artemia* (fairy shrimp from salt pans along San Francisco Bay) can survive more than 4 years of complete anoxia with no measurable amounts of stored carbohydrates used during this period and no detectable metabolic rate, even down to 1/50,000th of the aerobic respiration rate (Clegg and Jackson in Brendonck *et al.*, 1998). Thus, at least some organisms can apparently survive as "living dead"—that is, in a state without detectable metabolism but still capable of becoming active when exposed to favorable environmental conditions.

Dormancy can occur at many points in an organism's life cycle, from embryo to juvenile and adult stages, depending on the species. Indeed, for some organisms, dormancy may occur at several different life history stages in a single population. Perennial plants may express both bud and seed dormancy (Harper, 1977). Some invertebrates, such as cyclopoid copepods, can enter diapause at several different stages ranging from immatures to adults (Elgmork in Alekseev and Fryer, 1996) and even as fertilized females (Alekseev and Fryer, 1996). Some fishes can aestivate as adults, whereas others such as the annual killifishes make diapausing eggs capable of surviving several years in desiccation. Some mammals can exhibit both adult hibernation and delayed implantation, a kind of embryonic diapause. In general, short-term dormancy (from a few weeks to a few years) is characteristic of species that are dormant as immature individuals or adults, whereas long-term dormancy (years to centuries) is only found in species that possess embryonic (i.e., seed or egg) dormancy. However, many seeds and eggs actually only remain dormant for brief time periods.

II. DORMANCY AND DIAPAUSE AS ADAPTIVE TRAITS

It is axiomatic that if individuals that possess dormancy within a population survive a harsh period, whereas others that lack dormancy do not survive, then genotypes expressing dormancy will be favored by natural selection. It is less obvious, however, precisely where the trade-off lies if the harsh environment is not fatal to all non-dormant individuals. Dormancy represents a reproductive delay (i.e., a lengthening of generation time) and therefore a reduction in potential rate of population increase. Cohen (1970), modeling insect diapause, showed that all individuals in a population should enter dormancy when the arithmetic mean of potential reproduction in the active stage drops below the harmonic mean of survival in the dormant stage. The "means" in his analysis represent long-term probabilities of successful reproduction or survival. Likewise, Cohen (1966), this time taking plant seed dormancy as his inspiration, showed that the optimal fraction of seeds germinating in any given year depends on the probability that an individual will contribute to long-term fitness through reproduction as a growing plant

compared with the contribution it would make by simply surviving another season as a dormant seed. His important contribution in these two papers was to show that dormancy is not simply a response to the certain extermination of active individuals in a highly seasonal environment but also that it is expected to evolve as a response to environmental uncertainty. It is because not all harsh periods are fatal, and not all growing seasons are bountiful, that variation in the expression of dormancy exists both within and among species.

III. VARIATION IN DORMANCY AND DIAPAUSE

There is often variation within a group of species living a in single environment either in whether any given species makes dormant eggs or seeds or in the time of year that these dormant stages are produced [see Harper (1977) for examples in plants and Tauber et al. (1986) for examples in insect diapause]. This may be expected if different species react to the environment in different ways. One species may perceive a change as highly unfavorable, whereas another may be relatively immune to the change. There are many examples of populations in which only a fraction of individuals enter dormancy (Venable in Leck et al., 1989; Tauber et al., 1986). In these cases, it seems likely that a kind of "bet-hedging" strategy has evolved in response to variation through time, typically among years. In the models previously reviewed, dormancy is favored or not depending on the relative expectations of growth and reproduction in the active stage and survival in the dormant stage. If these expectations vary substantially over time, then the most successful phenotype may be one in which some individuals remain active (in case conditions remain favorable) and others enter dormancy (in case conditions become too harsh). For the same reason, there may be variation in the time of year that individuals within a population enter dormancy or diapause: In some years, the seasonal onset of harsh conditions occurs early, and in other years the onset occurs later. An example of this variation is a population of freshwater crustaceans (the calanoid copepod, *Diaptomus sanguineus*, living in a small lake in Rhode Island) in which one subpopulation switches from making eggs that hatch immediately to making diapausing eggs significantly earlier in the season compared with the second subpopulation (Ellner et al. in Brendonck et al., 1998).

IV. TIME TRAVELERS: DORMANCY AND DIAPAUSE AS "MIGRATION FROM THE PAST"

Many organisms migrate seasonally away from environments that become uninhabitable at one time of year (usually winter). These include both vertebrates (birds, mammals, and fish) and invertebrates (e.g., some butterfly and lobster species). There is a large group of organisms, however, that are unable to undertake long-distance movement and therefore must survive the seasonally harsh environments in which they live. Microbes, fungi, plants, and a great many invertebrates have this life history constraint, as do some vertebrates that hibernate rather than migrate. In a very real sense, the organisms that cannot migrate spatially have evolved dormancy as a kind of temporal migration from one favorable season to the next. Migration is a directed movement between spatially separated habitats. Dormancy is especially analogous to migration when an organism not only begins dormancy before conditions become harsh but also emerges only when conditions become favorable again. For plant seeds, this is known as "predictive germination" (Venable in Leck *et al.*, 1989), but many other types of dormancy termination have a predictive character, such as spring arousal from hibernation in mammals or the hatching of crustacean diapausing eggs only when their temporary pond habitat refills with water.

Equally analogous are spatial and temporal dispersal in unpredictably varying environments. Spatial dispersal contrasts with migration in that it is a non-directional movement of individuals away from a source population. It is often interpreted as an adaptation for survival when habitat patches become uninhabitable asynchronously. An individual that produces many offspring, each of which disperses to a new habitat, plays a kind of probabilistic game with many young landing in unfavorable habitats in which they die but with at least some landing at sites in which they can successfully grow and reproduce. For organisms lacking effective spatial dispersal in an environment that is unpredictable through time, prolonged dormancy is an effective alternative mechanism for continuing success. An individual that produces many dormant seeds, cysts, or eggs, each of which is capable of surviving long periods of time (i.e., over what would be multiple generations or seasons in the active life history stage) before germinating or hatching, exhibits a kind of temporal dispersal. This is particularly true if the dormant propagules also distribute their emergence from dormancy over multiple seasons. Then, as in spatial dispersal, the dormant offspring awaken in a range of different conditions. Many will emerge at times when the habitat is harsh, but some will emerge and have successful growth and reproduction, producing new long-lived dormant propagules. Seeds with prolonged dormancy are said to belong to a "seed bank" (Leck *et al.*, 1989), and eggs in prolonged diapause are, by analogy, said to belong to an "egg bank" (Hairston *et al.*, 1996).

Spatial dispersal and prolonged dormancy are predicted in theory to be alternative life history adaptations (Hairston and also McPeek and Kalisz in Brendonck *et al.*, 1998). Organisms that have excellent spatial dispersal capabilities should not experience strong selection for long-term survival in dormancy, and vice versa, organisms with effective prolonged dormancy should not also evolve mechanisms for spatial dispersal. Consistent with this theory, there are significant negative relationships between spatial dispersal ability and seed dormancy in plants (Venable in Leck *et al.*, 1989; Rees, 1993). Similar considerations may explain why winged insects, although often exhibiting single-season diapause (Tauber *et al.*, 1986), very rarely possess diapause that extends over longer periods.

For a variety of species that make dormant propagules (i.e., either dormant seeds or diapausing eggs), there is evidence that these stages can facilitate both spatial and temporal dispersal. The resistant stages that permit dormant organisms to survive stressful periods *in situ* also make it possible for them to withstand unfavorable conditions during transport, whether by physical forces such as wind or water currents or as hitchhikers attached to more mobile animals. There is evidence, however, that trade-offs still exist in which function is most important for a given organism: Plant species with elaborations attached to their seeds that promote wind dispersal have shorter term dormancy than those that lack such structures (Rees, 1993).

Limitations inherent in spatial dispersal for invertebrates living in lakes and other inland pools very likely explain why diapause and dormancy are much more prevalent among taxa living in this habitat than in related animals found in marine environments in which ocean currents provide for passive dispersal. In a survey of 167 species of crustaceans, Hairston and Cáceres (in Alekseev and Fryer, 1996) found that more than 55% of those living in inland water bodies possessed a long-lived diapausing stage, whereas less than 10% of marine species possessed this trait. More broadly, prolonged dormancy occurs much more frequently in invertebrate phyla with species that occur in nonmarine habitats than in invertebrate phyla that are exclusively marine.

Indeed, no exclusively marine phylum, and only 1 of 28 exclusively marine classes, is known to exhibit prolonged dormancy (Cáceres, 1997a). Several studies suggested that possession of prolonged dormancy or diapause may facilitate invasion of inland waters by aquatic invertebrates (Hairston and Cáceres in Alekseev and Fryer, 1996; Cáceres, 1997a; Hairston and Bohonak, 1998). Furthermore, those taxa that have persisted the longest in freshwater habitats, over geological time, are the ones with prolonged diapause (Alekseev and Starobogatov in Alekseev and Fryer, 1996).

V. THE STORAGE EFFECT, PROLONGED DORMANCY AND DIAPAUSE, AND THE MAINTENANCE OF BIODIVERSITY

Long-lived dormant stages that spread their germination or hatching over an extended period of time can play a major role in maintaining the coexistence of both species within a community and genotypes within a population. This is especially true when the environment varies through time with some species (or genotypes) favored under some environmental conditions and others favored under other conditions. For example, consider two competing plant species for which the environment varies through time so that each species has years in which it does well and others in which it does poorly. Also suppose that in those years when one species does well, the other does poorly. In the absence of seeds with prolonged dormancy, one species would be expected to be competitively superior on average, and thus to eliminate the other over time. However, if both species have long-lived seeds, then the one that is on average the poorer performer can persist by producing seeds when conditions are favorable and survive in dormancy through years of poor growth and reproduction. This scenario can be extended to communities of many species: If for each species there is an environmental condition in which it does better than its competitors, and if each has some seeds that germinate in each year, then all species can take advantage of years favorable to their own growth and reproduction while not suffering serious fitness loss in years that are unfavorable. Theoretical studies have shown that many species can coexist in this scenario, which is one example of a more general mechanism called the "storage effect" (Chesson, 1994). Its importance in natural communities has been demonstrated for two species of competing

freshwater crustaceans in the genus Daphnia that make long-lived diapausing eggs (Cáceres, 1997b) and for desert annual plants with long-lived seeds (Philippi in Brendonck et al., 1998).

A similar process can maintain biodiversity of genotypes within a single species if individuals with different genotypes perform best in years of differing environmental quality (Hairston et al., 1996). The major difference between the theory for species competing within a community and that for genotypes competing within a population is evolution. In a community of competing species, we need only ask if those species present can coexist. With evolution, however, it is theoretically possible for natural selection to produce a single genotype that has a greater long-term fitness than any other genotype. The question now becomes one of whether other genotypes with lower long-term fitnesses can coexist in the presence of this most-fit type. The answer is "no" if dormancy is absent or only short term but "yes" if dormant stages are sufficiently long lived, and if natural selection fluctuates sufficiently to favor different genotypes at different times (Hairston et al., 1996). This has been shown to be the case for a population of freshwater crustacean copepods, Diaptomus sanguineus, with long-lived diapausing eggs that experience fluctuating natural selection due to year-to-year changes in fish predation pressure (Ellner et al. 1999).

The characteristics of organisms and their environments that make it likely that dormancy plays an important role in maintaining biodiversity are (i) an environment that fluctuates through conditions that favor a diversity of types of organisms, and (ii) dormancy or diapause of sufficient duration to span the time it takes for favorable conditions to recur for each organism type. A literature review on these topics by Hairston et al. (1996) shows that both of these conditions are quite common in nature for a wide variety of taxa and habitats.

VI. SUMMARY

Dormancy is a term that covers a variety of physiological states in a wide range of kinds of organisms. In each of its forms, dormancy plays a role in the ability of species to live where they do and is thus important in explaining what organisms are found in which environments. The presence of prolonged dormancy as a character in a taxonomic group of organisms can be critical to the ability of that group to invade and colonize new habitats. This is particularly true for habitats that vary greatly through time and for groups with restricted abilities to disperse spatially. Finally, prolonged dormancy, when combined with a temporally varying environment, can be a significant factor in maintaining both genetic and community biodiversity within a habitat.

Acknowledgments

S. P. Ellner and C. E. Cáceres provided helpful comments. My research on diapause and dormancy is supported by grants from the U.S. National Science Foundation and the U.S. Environmental Protection Agency.

See Also the Following Articles

ADAPTATION • COMPETITION, INTERSPECIFIC

Bibliography

Alekseev, V. R., and Fryer, G. (Eds.) (1996). *Diapause in the Crustacea. Developments in Hydrobiology*, Volume 114. Kluwer, Boston.

Bartholomew, G. A. (1972). Energy metabolism. In *Animal Physiology: Principles and Adaptations* (M. S. Gordon, Ed.), 2nd ed. Macmillan, New York.

Brendonck, L., De Meester, L., and Hairston, N. G., Jr. (Eds.) (1998). Evolutionary and ecological aspects of crustacean diapause. In *Advances in Limnology*, Vol. 52. Schweizerbart'sche, Stuttgart.

Cáceres, C. E. (1997a). Dormancy in invertebrates. *Invert. Biol.* **116**, 371–383.

Cáceres, C. E. (1997b). Temporal variation, dormancy, and coexistence: A field test of the storage effect. *Proc. Natl. Acad. Sci. USA* **94**, 9171–9175.

Chesson, P. L. (1994). Multispecies competition in variable environments. *Theor. Popul. Biol.* **45**, 227–276.

Cohen, D. (1966). Optimizing reproduction in a randomly varying environment. *J. Theor. Biol.* **12**, 119–129.

Cohen, D. (1970). A theoretical model for the optimal timing of diapause. *Am. Nat.* **104**, 389–400.

Ellner, S. P., Hairston, N. G., Jr., Kearns, C. M., and Babaï, D. (1999). The roles of fluctuating selection and long-term diapause in microevolution of diapause timing in a freshwater copepod. *Evolution* **53**, 111–122.

Hairston, N. G., Jr., and Bohonak, A. J. (1998). Copepod reproductive strategies: Life-history theory, phylogenetic pattern and invasion of inland waters. *J. Marine Syst.* **15**, 23–34.

Hairston, N. G., Jr., Ellner, S., and Kearns, C. M. (1996). Overlapping generations: The storage effect and the maintenance of biotic diversity. In *Population Dynamics in Ecological Space and Time* (O. E. Rhodes, R. K. Chesser, and M. H. Smith, Eds.), pp. 109–145. Univ. of Chicago Press, Chicago.

Harper, J. L. (1977). *Population Biology of Plants*. Academic Press, New York.

Henis, Y. (Ed.) (1987). *Survival and Dormancy of Microorganisms*. Wiley Interscience, New York.

Leck, M. A., Parker, V. T., and Simpson, R. L. (Eds.) (1989). *Ecology of Soil Seed Banks*. Academic Press, New York.

Rees, M. (1993). Tradeoffs among dispersal strategies in British plants. *Nature (London)* **366**, 150–152.

Tauber, M. J., Tauber, C. A., and Masaki, S. (1986). *Seasonal Adaptations of Insects*. Oxford Univ. Press, Oxford.

DIFFERENTIATION

Nicholas H. Barton
Institute of Cell, Animal and Population Biology, University of Edinburgh, Scotland

GLOSSARY

adaptive landscape A graph of average fitness against the state of the population, represented by allele frequencies or trait means.

allele A particular variant of a gene.

balancing selection Selection that acts to maintain two or more alleles in a stable equilibrium.

character displacement The displacement of characters caused by competition between species that live in the same place.

coalescent process The merging of lineages traced backward in time from a sample of genes.

fitness The total number of offspring produced over one generation.

frequency-dependent selection Selection in which fitnesses vary with the frequency of different genotypes.

gene flow The movement of genes from place to place.

genotype The particular combination of genes carried by an individual organism.

hybrid zone A narrow region in which genetically distinct populations meet, mate, and produce hybrids.

neutral Having no effect on fitness.

QTL Quantitative trait locus.

random genetic drift The random fluctuation in allele frequency caused by random variation in fitness between individuals.

stabilizing selection Selection on a continuously varying trait toward an intermediate optimum.

sympatric speciation The separation of a single population living in one place into two species.

transposable element A DNA sequence that can move from one place in the genome to another.

IN EVOLUTIONARY BIOLOGY, THE TERM DIFFERENTIATION usually refers to the accumulation of genetic differences between populations or species. However, the term can be applied more broadly to the diversification of genes, organisms, and populations. The extraordinary conservation of the basic genetics and biochemistry across all living organisms implies that life evolved from one most recent common ancestor; the obstacles to the chemical evolution of the first replicator suggest that this ancestor itself evolved from a single origin—most likely a short self-replicating RNA sequence. Thus, we must explain how past and present biodiversity has differentiated from this simple beginning. We can ask how the first replicating molecules diversified to produce the many kinds of gene in present-day organisms, how the first cells evolved into these diverse organisms, and how populations have come to contain so much genetic diversity. This chapter sets out to explain how elementary evolutionary and ecological processes lead to differentiation in both the narrow and broad senses.

I. INTRODUCTION

If we regard species as simply groups of organisms with similar morphology, then the differentiation of populations becomes equivalent to the process of speciation. However, under the biological species concept, speciation consists of the evolution of genetic differences that prevent successful interbreeding; under the phylogenetic species concept, speciation consists of the separation of lineages. Thus, at least in the sexually reproducing populations to which these ideas of species apply, population differentiation is distinct from speciation.

This chapter introduces successively more complex processes involved in differentiation. In the simplest kind of evolution, organisms and populations differentiate under mutation and random reproduction. If there is inherited variation in the rate of reproduction (i.e., in fitness), then natural selection will shape variation; populations in different environments will become differentiated in a straightforward way. This divergence may be opposed by the flow of genes from place to place, but will only be prevented over the smallest spatial scales. Just as diverse ecological niches are required if species are to coexist, so selection can maintain differentiation within a population only if genotypes exploit different limiting resources. Finally, the chapter examines how far these genetical and ecological processes can account for differentiation in the broadest sense and over the largest scales.

II. NEUTRAL EVOLUTION

Even if all copies of a gene are equivalent, some will by chance leave more descendants than others, leading to random drift of gene frequencies. As we trace their descendant lineages forward in time, some will become extinct, while others will grow in numbers. Eventually, it is inevitable that one gene will become the ancestor of the entire population (Fig. 1). Therefore, if we trace the ancestry of all present-day genes backward in time, the lineages will at some point coalesce in a single common ancestor. In an ideal population of N diploid

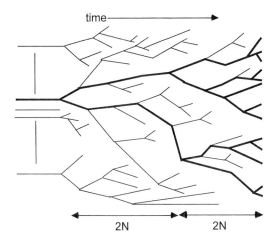

FIGURE 1 A sketch of the random reproduction of genes. Moving forward in time (left to right), genes leave random numbers of descendants. Eventually, only one family of descendants survives (in bold). Moving backward in time (right to left), the ancestry of the population takes on average 2N generations to coalesce to two lineages and then the same time, on average, to reach the single common ancestor.

individuals, the chance that any two lineages coalesce is just

$$\frac{1}{2N}$$

per generation. A large sample of genes will take about $2N$ generations to coalesce into two ancestral lineages, and these two will then take, on average, another $2N$ generations to coalesce in one ancestor. Much attention has focused on identifying "mitochondrial Eve," the single female who (assuming strictly maternal inheritance) was the most recent common ancestor of all human mitochondria. The key issue here is when this ancestor lived: a recent Eve would imply a small effective size for the human population.

This coalescent process determines the simplest form of differentiation of genes and populations. Suppose that the gene in question consists of a long stretch of DNA, which is always passed on as a unit. Mutations will occur at a rate μ, at random times and at random sites in the sequence; to a good approximation, mutations never occur at the same site twice (the infinite sites model). If we examine sequences that shared an ancestor a time T in the past, they will on average differ by $2\mu T$ mutations. If these gene sequences come from two well-separated species, then we can date the divergence of those species by the steady ticking of a molecular clock. Indeed, it was the striking observation that any given protein accumulates mutations at a steady rate that stimulated Kimura to develop the neutral theory of molecular evolution. Across a range of eukaryote lineages, sequences that have no effect on fitness mutate and diverge at a rate of $\sim 10^{-8}$ per base pair per year.

Variation within a population can be understood in a similar way. If two gene sequences are chosen at random, they will share an ancestor on average $T = 2N$ generations back; therefore, they will differ by, on average, $4N\mu$ mutations. This simple prediction is in qualitative agreement with observations on molecular genetic diversity within populations. However, variation in extremely large populations is not vastly greater than in much less abundant species: for example, about 5% of enzyme loci in humans are heterozygous, while about 25% are heterozygous in the bacterium *Escherichia coli*. There are two possible explanations for the weak relation between within-population diversity and census numbers. First, the effective population size may be much smaller than the actual census number, because of occasional bottlenecks and local extinctions or because of selection on linked genes. For example, in *Drosophila melanogaster*, $\sim 1\%$ of noncoding base pairs are heterozygous, which, with $\mu = 10^{-8}$ per base per year, implies that a random pair of bases share a common ancestor

$$T \sim \frac{0.01}{2\mu} = 5 * 10^{5}$$

years back. Assuming 4 generations per year, this implies an effective population number of $\sim 10^{6}$ flies—many fewer than the census number. Second, variation in coding and regulatory regions of the DNA may be maintained by selection, regardless of population size (discussed later).

Neutral evolution does not lead to a homogeneous pattern of differentiation. Imagine that a process of random reproduction produces lineages as in Figure 1. (We might think of asexual organisms or of species originating and going extinct.) If morphology changes by random increments, then the variance among a group reflects the age of that group. Now, random genealogies typically have a hierarchical structure, with most family groups being closely related compared with the age of the whole. Morphology is therefore expected to fall into well-separated clusters (Fig. 2). Of course, the clustering that we see in the living world, which allows us to classify organisms into species, genera, and higher taxa, is not merely due to random reproduction. However, it is important to realize that what Darwin termed "descent with modification" tends naturally to produce clusters of distinct morphologies.

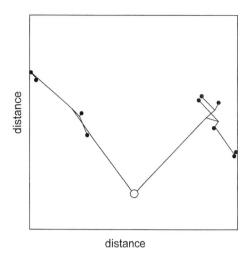

FIGURE 2 An example of the clustered pattern produced by random reproduction. Each dot represents an asexual organism, or a separate species. The position of each point represents its morphology, reduced to two dimensions; this evolves in a random walk, following a normal distribution. Each lineage becomes extinct with constant probability and splits at a rate that decreases with the total number of lineages. The open circle represents the ancestor.

The recent availability of DNA-based markers, which for the most part can be presumed to have negligible effects on fitness, has transformed evolutionary biology. These provide a molecular clock against which to measure divergence times, an objective and quantitative means of reconstructing phylogeny, and a baseline level of differentiation within and between populations against which to measure other processes. Having established this straightforward benchmark, we can consider how selection, gene flow, and ecological interactions influence differentiation.

III. NATURAL SELECTION

A difference in relative fitness of s (the selection coefficient) causes change over a time

$$\sim \frac{1}{s}$$

It dominates over random genetic drift if

$$s > \frac{1}{N}.$$

and over mutation if $s > \mu$. Thus, even slight fitness differences of $s > 10^{-5}$ will be the predominant force in a population of effective size larger than 10^5 individuals and for genes consisting of, say, 10^3 base pairs, each with mutation rate 10^{-8} per generation. Such minute fitness differences cannot be measured directly. However, a multitude of studies show that selection on morphological traits and on individual genes can be strong ($s > 10\%$, say). Nevertheless, we will see that this does not imply that differentiation among populations is a simple consequence of selection alone.

In this section, I consider the straightforward effects of selection in the simplest case, where each genotype (that is, each set of genes) has a fixed effect on fitness. Variation within populations is then maintained by a balance between mutation generating deleterious variants and selection eliminating them. (For the moment, we leave aside cases in which heterozygotes are fitter than either homozygote.) While mutation rates on individual bases are extremely low, the net mutation rate over the whole genome may be high, and so may sustain substantial genetic variation. For example, Eyre-Walker and Keightley (1999) used rates of DNA sequence divergence among primates to show that the human lineage has been subject to several deleterious mutations per genome per generation. At the level of the whole organism, the variance of continuous traits such as growth rate or skeletal proportions increases by roughly 0.1% every generation as a result of new mutations; thus, much of the response to long-term artificial selection is due to mutation, rather than to standing variation. The genetic basis of such mutations is beginning to be understood. For example, Lai *et al.* (1994) found that ~10% of the variance in bristle number in a natural population of Drosophila melanogaster was due to the insertion of transposable elements around the *scabrous* gene; these insertions are clearly deleterious, since they were all found to be of relatively recent origin, and hence rapidly eliminated by selection.

There is a long-standing debate over the processes that maintain genetic variation. While this is as yet unresolved, it is at least plausible that most variation, in both discrete genes and continuous traits, is due to deleterious mutations. On this view, differentiation within populations is merely a fluctuation around some optimal state, while differentiation between populations is a direct consequence of changes in that optimum. The ability of artificial selection to rapidly change almost any aspect of an organism was Darwin's strongest argument for the efficacy of natural selection, and remains so today. Almost all traits show heritable variation, and therefore they respond to selection; what is remarkable is that this response continues apparently without limit

(at least, over the span of human experiments). For example, Weber (1992) selected on the angle between two small veins on Drosophila wings. Lines selected in opposite directions diverged steadily and differed by 20 standard deviations after 15 generations. We now understand that at the genetic level, this divergence is due to the establishment of individual alleles, either present in the original base population or generated by new mutations. Weber's lines, for example, differed by at least 11 genes on one chromosome, each of small effect.

There has been much dispute over whether adaptive differences between populations and species are due to the accumulation of minute differences or to a few genes of large effect. Darwin emphasized the former view, on the grounds that complex and finely adjusted adaptations could only be built up by slight variations. In contrast, the geneticists who rediscovered Mendelian heredity emphasized the importance of major mutations. After a century of research, the answer is only beginning to become clear. On the one hand, sustained, steady responses to selection must be based on multiple genes of small effect, and genetic analysis of species differences usually shows that many genes are involved in maintaining their separation. On the other hand, differentiation in particular traits is often based on one or a few genes. Insecticide resistance usually evolves through mutations in particular target genes (for example, in the acetyl cholinesterase enzyme that is the target of organophosphates). To take a natural example, the striking differences between warning color patterns of Heliconius butterflies are due to only a handful of genes. Much effort is now being devoted to identification of the quantitative trait loci (QTL) responsible for human variation related to disease, for differences between varieties of domesticated plants and animals, and for adaptive differences between species. Such methods do often identify genes of major effect: indeed, the whole enterprise of genetic engineering rests on the possibility of obtaining useful results by manipulating a few key genes. However, there is a statistical bias toward finding genes of large effect, and it is too early to know the answer to this long-standing debate over the genetic basis of adaptation.

Given that selection can rapidly adapt populations to changing conditions, and given that selection is often the strongest evolutionary force, one might suppose that differentiation would simply reflect the optimal genotype that is favored by the local environment. We could then say rather little about what patterns of differentiation to expect, since they would be the outcome of the particular interactions between organism and environment that determine reproductive success. However, matters are not so simple. There may be alternative combinations of genes that are locally optimal, and so populations may adapt to essentially the same conditions in different ways. Sewall Wright devised an influential way of thinking about this issue, the "adaptive landscape" (Fig. 3). This is a surface that represents the average fitness of a population, plotted against a set of axes that describe the state of a population (for example, gene frequencies or trait means). Under the simple assumptions of this section, natural selection pushes populations uphill on this landscape, toward a local "adaptive peak." In general, there will be many such peaks, and so populations will differentiate under selection even if they all experience the same environment.

Divergence may be observable at the phenotypic level or may be cryptic. For example, suppose that a trait is determined as the sum of effects of many genes, each carrying either "+" or "−" alleles. If selection favors a single intermediate optimum, any combination of genes that achieves that optimum can be established $(+ + - -, - - + +, \ldots)$. Such cryptic divergence is revealed when crosses between populations produce progeny with reduced fitness and increased variance; over the largest scales, it is reflected by sibling species, which are morphologically indistinguishable and yet reproductively incompatible.

Divergence toward alternative "adaptive peaks" can occur in several ways. Random genetic drift in small populations may knock populations into the domain of attraction of different peaks. This process is especially likely if there are flat ridges on the adaptive landscape along which divergence is not opposed by natural selection. For example, if only the double recessive homozygote $P_1P_1P_2P_2$ is unfit, then populations initially containing Q_1 and Q_2 can fix either P_1 or P_2, with no effect on fitness. However, when crossed they have reduced fitness. Genetic analysis of hybrid inviability and sterility in Drosophila show that incompatibilities are largely recessive, which supports this model of quasi-neutral divergence. A different line of evidence comes from studies of functional RNA molecules (for example, the transfer RNAs that translate the genetic code). Many changes in individual bases cause no appreciable change in three-dimensional structure, so that extensive divergence can occur without disruption of function.

Random processes can drive divergence even in large populations. If adaptation depends on new mutations, then different mutations may arise first in different places. For example, humans have evolved resistance to malaria by a variety of different mutations in the β-hemoglobin molecule: the S allele in Africa, the C allele in Asia, and various thalassemias in the Mediterranean. These alternative alleles tend to exclude each other, so that chance differences in first occurrence of mutations leads to permanent divergence. Finally, divergence may be driven by fluctuations in selection. Suppose that the "adaptive landscape" fluctuates as conditions change but maintains overall distinct peaks. Then, populations may end up on different peaks, even if they experience the same mix of environments: the outcome depends on the particular sequence of environments that they encounter.

Although the "adaptive landscape" is a helpful metaphor, it gives a misleading picture of the geometry of evolution. Evolution can proceed in very many dimensions, rather than just the few that can be visualized by the human mind. This dimensionality is nicely illustrated by the model experiments of Eigen and coworkers, in which RNA molecules ~100 bases long are replicated by a polymerase derived from the Qβ virus. Under given conditions, some definite sequence (and its mutational variants) is picked out by selection from among $4^{100} \sim 10^{60}$ possibilities. This is possible because any sequence is connected to any other by no more than 100 mutational steps. At any stage, 300 variants are available from a single mutation, since each base can mutate to three alternatives. Provided that there is some sequence of steps that

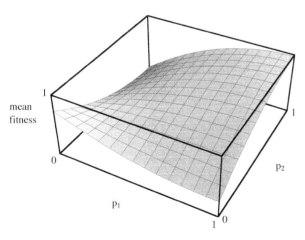

mean fitness

FIGURE 3 A simple "adaptive landscape" for two loci, each carrying two alleles. One allele at each locus is dominant, with frequency denoted by p_1 and p_2. The double dominant (P_1P_2) and double recessive (Q_1Q_2) genotypes are assumed to be most fit, with $W_{PP} = 1$, $W_{QQ} = 0.8$, $W_{PQ} = 0.4$ and $W_{QP} = 0.2$.

leads uphill toward the optimal sequence, selection will effectively search from among the astronomical numbers of possibilities. This multidimensional view emphasizes that the outcome depends on which mutations are available: while evolution could take any of a large number of directions at any stage, this number is very much smaller than the total space of possibilities, and it sets a strong constraint.

IV. GENE FLOW

The most obvious impediment to differentiation is gene flow—that is, the mixing of genes from different populations caused by movement of the individuals that carry them. At least since the mid-19th century, there has been controversy over whether such mixing prevents differentiation. Darwin held that populations could adapt to gradually varying conditions across their range and that this could lead to the origin of new species. In contrast, Wagner emphasized the importance of barriers to migration as prerequisites for speciation. This view has been strongly propounded in more recent years by Ernst Mayr. However, both population genetic theory and observations of geographic divergence strongly support the Darwinian view, at least if one considers adaptations that can be built up by individual alleles or by continuous changes in quantitative traits.

Gene flow has straightforward and well-understood effects. The rate of gene flow into a single population is measured by m, the fraction of genes derived by immigration in each generation. Mixing then occurs over a timescale

$$\sim \frac{1}{m}$$

In a broadly continuous habitat, gene flow can be well approximated as a diffusion: in the same way that gas molecules diffuse as a result of random collisions, so genes diffuse through their habitat as a result of the random movement of the organisms that carry them. The rate of diffusion is measured by σ^2, the variance of distance between parent and offspring along any particular axis. Over a time T, this diffusion causes genes to spread over a distance $\sim \sqrt{2\pi\sigma^2 T}$. For example, the annual grasshopper *Chorthippus parallelus* is divided into two subspecies *C. p. parallelus* and *C. p. erythropus,* which met in the Pyrenees after the last Ice Age, $\sim 10^4$ years ago. These grasshoppers move by about 30 meters in a year, so that $\sigma^2 \sim 900m^2$ per year. Enzymes that are fixed for different alleles in the two subspecies mix over a region ~ 15 Km wide, which is of the same order as the distance predicted by simple diffusion, $\sqrt{2\pi\sigma^2 T} \sim 7.5$ Km (Hewitt, in Harrison, 1993).

Whether populations can adapt to local conditions, despite mixing, depends essentially on the relative rates of gene flow and selection. An allele with selective advantage s in the heterozygote can be established in a single population despite gene flow, provided that the rate of immigration of individuals carrying the alternative allele is lower than the rate of selection ($m < s$). If genes diffuse across a continuous habitat, then an allele can become established provided that it is favored in a region larger than some critical distance proportional to the characteristic scale $\sigma/\sqrt{s}$. New mutations have a chance of fixation approaching $2s$ if they arise within the appropriate region, but the probability falls to zero outside this region. Similar considerations apply to adaptations based on continuously varying traits. Suppose that fitness falls away in a Gaussian function with variance V_s (i.e., stabilizing selection); the optimal trait value varies from place to place. The characteristic scale over which populations can respond to local conditions is now $\sigma\sqrt{V_s/V_g}$, where V_s is a measure of the strength of stabilizing selection, and V_g is the additive genetic variance.

Because the dispersal range σ is typically much smaller than the species' range, these theoretical considerations imply that weak selection can allow a species to adapt to diverse local conditions. This can be seen directly. For example, the grass *Agrostis tenuis* has evolved tolerance to heavy metals on mines only a few meters across. Similarly, narrow clines ~ 10 Km wide separate races of *Heliconius* butterflies, which carry alternative warning color patterns (Mallet, in Harrison, 1993). Such examples leave little doubt that gene flow need not prevent divergence over very short scales.

If variation in selection can cause differentiation over local scales, why do species not fall apart into a myriad of separate forms? Mayr was led to argue that gene flow prevents differentiation of populations by his observation that species remain morphologically homogeneous over much of their contiguous range: it is this homogeneity that makes taxonomy practicable. How can we explain this apparent conflict with the potential power of selection to cause differentiation?

Much variation in protein or DNA sequence may have such small effects on fitness that selection is negligible relative to gene flow. Such neutral variation will differentiate as a result of random genetic drift. However, because drift is weak in even moderately abundant populations, weak gene flow homogenizes neutral variation. In a single population, the number of migrants, Nm, is crucial. The variance of allele frequency within the local population is

$$\frac{\bar{p}(1 - \bar{p})}{1 + 4\,Nm}.$$

where is the allele frequency in the source population; hence, if more than one migrant per generation enters, there will be little differentiation. This result carries over to populations spread over two dimensions, either in an array of subpopulations or in a continuous habitat. (In the latter case, the crucial parameter is $\rho\sigma^2$, where ρ is the population density.) The effectiveness of gene flow in overcoming random drift has an important consequence: even if selection within local populations is ineffective relative to drift ($Ns \ll 1$), the cumulative effect of selection across the whole population of N_{tot} individuals dominates if $N_{tot}s > 1$. For many purposes, a population with $Nm > 1$ can be treated as a single gene pool.

Many surveys have been made of geographic variation in gene frequency, using both protein and DNA variation. The usual finding is that the variance in allele frequency between locations is small

$$\frac{\text{var}(p)}{\bar{p}(1 - \bar{p})} < 5\%.$$

This ratio is termed F_{ST}. This implies that if this variance is at an equilibrium between random drift and gene flow, the number of migrants is large ($Nm > 5$). There are, of course, exceptions: for example, the salamander *Ensatina eschscholtzii* has exceptionally low dispersal, and correspondingly, $F_{ST} \sim 0.7$ (Wake and Yanev, 1986). However, the general conclusion is that gene flow is sufficient to homogenize differentiation generated by random drift.

Surveys of geographic variation often aim at estimating the number of migrants (Nm). However, even if large-scale differentiation is generated by random drift, its geographic pattern may be dominated by past history rather than by present-day rates of gene flow (see Whitlock and McCauley, 1998). If a species is extinguished from a large region (as happened in temperate regions during the Ice Ages) and then reinvades, the pattern of variation reflects the pattern of recolonization. In European human populations, Cavalli-Sforza and co-workers have reconstructed the history of population movements from the detailed genetic data that is available: many loci show traces of the introduction of distinct alleles as agriculture spread from the Near East.

When the genealogical relationships among DNA sequences can be inferred, historical events can be seen more directly. For example, Avise (1994) has identified in many species a cryptic subdivision that runs across the southwestern United States; this is a relict of a separation into different Pleistocene refugia. Such reconstructions should be based on several unlinked genetic loci, however: within a sexual species, different loci may have different histories.

Much molecular differentiation may have slight effects on fitness, allowing it to be used to reconstruct the history of gene movements. However, we cannot explain the apparent homogeneity of species simply by an absence of selection. There are many examples of geographic variation at enzyme loci that are maintained by selection for adaptation to different environments. Evidence of selection may come from direct measurement, from close correlation with environment, or from discordance in pattern between different loci. (For example, the geographic pattern at several enzymes in European humans differs significantly from the majority of genes, whose pattern reflects population history.) Morphological traits are usually strongly selected, and their geographic patterns therefore do not match those predicted from putatively neutral markers. (For a thorough comparison of morphological and molecular patterns of differentiation in *Daphnia obtusa*, see Lynch *et al.*, 1999.) Because selection can cause rapid divergence, even in the face of gene flow, extinctions and recolonizations can only generate a transient homogeneity in selected traits. Thus, we expect to see morphological patterns that follow recent selection imposed by environmental features and cut across patterns at neutral markers that reflect population history. For example, Schneider et al. (1999) found that in the skinks of northeastern Australia, mitochondrial DNA and microsatellite markers tend to change along the Black Mountain Corridor, a biogeographic feature marking past range changes. In contrast, skeletal morphology changes in a quite different pattern, which parallels the transition from open to closed forest.

The evident power of selection to cause divergence remains hard to reconcile with the apparent similarity of species across large areas that Mayr and others have emphasized. The problem is close to that posed by examples of stasis in the fossil record, in which morphology remains similar over long times rather than large distances. One possibility is that in sexual species, different traits or genes can evolve independent spatial patterns, so that overall morphology is more stable than any one component and varies more smoothly. In any case, the supposed homogeneity may be exaggerated. Sharp transitions ('hybrid zones') do often occur, prompting classification of the taxa on either side as species or subspecies. These may arise either when many traits respond to a sharp environmental discontinuity (as in the skink example used earlier) or when populations that have diverged to become partially incompatible are brought together by a range expansion. In the latter case, divergence remains permanently because the hybridizing populations are at two different "adaptive peaks."

V. ECOLOGICAL DIFFERENTIATION

A simple form of natural selection, in which each gene combination has a fixed fitness, can explain much of evolution: it accounts for differentiation between populations and species as a result of changing conditions, mutation, and random drift, and despite gene flow. However, at the very least absolute fitness (that is, the number of offspring) must decrease with density if the population is to remain bounded. We expect the relative fitnesses of different genotypes to change with their abundance in the population: selection is in general frequency dependent. If the reproductive success of an allele decreases as it becomes more common, then genetic diversity can be maintained. This is known as balancing selection.

Maintenance of genetic diversity by frequency-dependent selection arises from competition for different limiting resources, in just the same way as maintenance of species diversity. If the concept of a "limiting resource" is interpreted broadly, then all examples of balancing selection can be seen in this way. Evidence for a heterogeneity of resources, cited by Darwin and exploited by plant breeders, comes from the increased yield of a mixture of varieties compared with a pure stand. Here, the limiting resources may consist of different nutrients, though other mechanisms are possible. A classic example of frequency-dependent selection is Batesian mimicry, in which a palatable species avoids predation by mimicking a distasteful model species. As the mimic becomes common, predators fail to associate their pattern with unpalatability, and protection is lost. Batesian mimics such as swallowtail butterflies therefore evolve diverse mimetic patterns, each of which remains rare. The limiting resources here are the various different model species. In diploid species, variation can be maintained by heterozygote advantage. The classic example is the polymorphism for the β-hemoglobin S allele, where the heterozygote gains resistance to malaria, while the homozygote suffers from sickle-cell anemia. The relative fitnesses of diploid individuals are fixed, but from the gene's point of view, fitness decreases with frequency: a common allele is more likely to find itself in a homozygote. The limiting resources here are the two kinds of allele with which each gene can find itself paired in a diploid.

The existence of heterogeneous limiting resources is, in a broad sense, necessary for maintaining diverse species and may be responsible for much genetic differentiation within species. However, we do not know how much genetic variation is maintained by balancing selection, as opposed to mutation and gene flow, and we do not know how far organisms respond to environmental heterogeneity by evolving diverse genotypes or diverse species. A tradeoff between variation within and between species is revealed by character displacement, where coexisting species tend to diverge in morphology and behavior. The *Anolis* lizards of the Caribbean provide a classic example. Where one species lives on an island, it has a broad distribution of body size; where two species share an island, they take up nonoverlapping distributions and so partition the available prey. A similar phenomenon can be seen where asexual clones compete with their sexually reproducing relatives. The freshwater fish *Poeciliopsis monacha* and *P. lucida* occasionally mate and produce a parthenogenetically reproducing clone. Vrijenhoek (1994) found that in streams with many different asexual clones, there tend to be fewer of the sexual progenitor species. This suggests that the space of available ecological niches can be filled either by a set of distinct asexual clones or by members of the genetically diverse sexual population.

If multiple forms, each exploiting a different resource, coexist in a sexual population, then matings between them may produce maladapted progeny. To make this point in another way, imagine that resource use depends on a set of continuously varying traits. If these traits are determined as the sum of effects of several genes, then sexual reproduction produces an approximately normal trait distribution. There is no reason why this should match the distribution of resources that are available. The problem can be avoided if the genetic system can somehow produce the appropriate distribution despite the random shuffling of genes that occurs with sexual reproduction. For example, swallowtail butterflies produce many distinct wing patterns, each mimicking a different distasteful model. These are determined by several genes, which are so tightly linked that they behave as a single genetic locus. If such genetic tricks do not evolve, then selection favors mating behaviors that cause like to mate with like. Ultimately, such selection could lead to complete reproductive isolation between the two forms. This splitting of a single population into two separate species in the absence of spatial separation is termed sympatric speciation. It is almost impossible to demonstrate directly, since any present pattern of divergence might have originated with some spatial isolation. However, there are several plausible cases. For example, sticklebacks from the sea have colonized postglacial lakes within the past $\sim 10^4$ years and have repeatedly evolved distinct benthic and limnetic feeding morphs. These differ in morphology and also prefer to mate with their own kind; hybrids between them have reduced fitness in the wild (Hatfield and Schluter, 1999). A key question is how often ecological differentiation among genotypes evolves into differentiation between species in this way.

Darwin emphasized that natural selection is driven by competition for limited resources (the "struggle for existence"). Moreover, he argued that the lack of intermediate forms—the evident clustering of organisms that makes taxonomy possible—can be explained largely by the diversifying effects of competition. There is a "division of labor" among organisms, which (as with market economics) results from competition between individuals rather than any optimization of species or community productivity. Similar processes account for the differentiation of the $\sim 10^5$ different genes in complex organisms. Often, one gene may acquire multiple functions. For example, arginosuccinate lyase in birds acts as a lens crystallin as well as having a catalytic function. Similarly, most of the genes responsible for the establishment of segmentation in early *Drosophila* development are also involved much later in specifying various organs. If by chance a multifunctional gene duplicates, then each copy may specialize to become more efficient at one of the two original functions. The sequencing of complete genomes has made clear the extent of gene duplication and differentiation: most yeast genes are members of closely related families. There are close analogies between the division of labor among different species within an ecological community, different genotypes within a population, and different genes within an organism.

VI. EVOLUTION OVER LARGE SCALES

The processes of selection, mutation, and random drift that lead to differentiation are well understood, at least over the short time scales that are accessible to experiment and observation. How far do they account for the differentiation of genes and organisms over the whole span of evolution? First, consider the overall pattern. The first fossil evidence of multicellular organisms is found in the Edicaran fauna, about 560 million years ago. An extraordinary diversity of animals with skeletons evolved soon after, at the

beginning of the Cambrian, and included representatives of almost all present phyla. Molecular evidence suggests that the ancestral lineages of these phyla may trace back much further; nevertheless, there was a rapid diversification of body plans during this "Cambrian explosion." The subsequent fossil record shows an overall increase in diversity, as measured by numbers of species, but interspersed with occasional mass extinctions. Organismal complexity is hard to quantify, and trends in complexity are controversial: while the most complex organisms may become more complex as diversity increases, it is not clear that there is (or should be) any trend toward more elaborate organisms. Rates of morphological evolution can be quantified more readily, at least for hard parts. Morphology changes much more slowly over the long time span of the geological record than in current populations (Gingerich, 1983). Only weak selection pressures need be invoked to explain observed rates, and indeed it is surprising that morphology has often changed little over millions of years.

If we suppose that organisms differentiate primarily as a result of selection to fill different ecological niches (more precisely, to exploit different limiting resources), then we might imagine that these macroevolutionary patterns could be explained by diversification to fill a set number of preexisting niches. At equilibrium, there might be a balance between extinction and species formation, leaving many niches unfilled; nevertheless, diversity of species and their morphology would be expected to rise toward some definite value. This view is supported by statistical analysis of the fossil record, which suggests saturation of species numbers. Over shorter time scales, remote islands such as the Hawaiian archipelago are occupied by the descendants of a few colonizing species, which radiated to fill the roles normally occupied by other taxa. However, new organisms generate new niches. One host species is often plagued by several specialist parasites, which are restricted to that host by specific adaptations to evade its defenses. The evolution of novel features (multicellularity, bony skeletons, flight, etc.) are followed by a rapid diversification into novel ways of living. While the maintenance of diverse genotypes and species can be understood formally in terms of competition for distinct limiting resources, their number is not fixed and seems inherently unpredictable.

VII. CONCLUSION

The pattern of diversity that we see may not be entirely a matter of historical accident. Even though there is no difficulty in explaining the rapid evolution of any one gene or trait, there are fundamental limits on the numbers of genes, the rates at which new genes are substituted for old, and the number of different genotypes that can be maintained within a population. The total mutation rate cannot greatly exceed one per generation, since most mutations are deleterious, and each deleterious mutation requires roughly one selective death to be eliminated. The human mutation rate approaches this upper bound, which may limit the number of functional genes that we carry. Each substitution of one allele by another requires that individuals carrying that allele leave more offspring than those that do not: Haldane (1957) showed that this implies a "cost of natural selection," which limits the rate at which new alleles are substituted. Again, this argument places a strong constraint on the number of adaptive changes that can have occurred during human evolution. Finally, the diversity that can be maintained within a finite population is limited: the maximum number of genes that vary independently can be no greater than the effective population size. These constraints give some hope that we can understand the overall features of organismal differentiation, even if the detailed pattern is a conglomeration of unique historical events.

See Also the Following Articles

ADAPTATION • ADAPTIVE RADIATION • DARWIN, CHARLES • GENETIC DIVERSITY • NUCLEIC ACID BIODIVERSITY • SPECIATION, PROCESS OF

Bibliography

Avise, J. C. (1994). *Molecular markers, natural history and evolution.* Chapman & Hall, New York.
Cavalli-Sforza, L. L., Menozzi, P., and Piazza, A. (1994). *The History and Geography of Human Genes.* Princeton University Press, Princeton, NJ.
Eigen, M. (1992). *Steps Towards Life.* Oxford University Press, Oxford.
Eyre-Walker, A., and Keightley, P. D. (1999). High genomic deleterious mutation rates in hominids. *Nature* **397**, 344–347.
Gingerich, P. D. (1983). Rates of evolution: Effect of time and temporal scaling. *Science* **222**, 159–161.
Haldane, J. B. S. (1957). The cost of natural selection. *J.Genet.* **55**, 511–524.
Harrison, R. G. (1993). *Hybrid Zones and the Evolutionary Process.* Oxford University Press, Oxford.
Hatfield, T., and Schluter, D. (1999). Ecological speciation in sticklebacks: Environment-dependent hybrid fitness. *Evolution* **53**, 866–873.
Lai, C., and Mackay, T. F. C. (1993). Mapping and characterization of P-element induced mutations at quantitative trait loci in *Drosophila melanogaster. Genet. Res.* **61**, 177–194.
Lynch, M., Pfrender, M., Spitze, K., Lehman, N., Hicks, J., Allen, D., Latta, L., Ottene, M., Bogue, F., and Colbourne, J. (1999). The quantitative and molecular genetic architecture of a subdivided species. *Evolution* **53**, 100–110.
Schneider, C. J., *et al.* (1999). A test of alternative models of diversification in tropical rainforests: Ecological gradients vs rainforest refugia. *Proc. Natl. Acad. Sci.* (USA) **96**, 13869–13873.
Vrijenhoek, R. C. (1994). Unisexual fish: Model systems for studying ecology and evolution. *Ann. Rev. Ecol. & Syst.* **25**, 71–96.
Wake, D. B., and Yanev, K. P. (1986). Geographic variation in allozymes in a ring species, the Plethodontid salamander Ensatina eschscholtzii of western North America. *Evolution* **40**, 702–715.
Weber, K. E. (1992). How small are the smallest selectable domains of form? *Genetics* **130**, 345–353.
Whitlock, M. C., and McCauley, D. E. (1998). Indirect measures of gene flow and migration: F_{ST} not equal to $1/(4Nm+1)$. *Heredity* **82**, 117–125.

DINOSAURS, EXTINCTION THEORIES FOR

J. David Archibald
San Diego State University

GLOSSARY

actinopterygia Ray-finned bony fishes.
amniota Taxon of vertebrates including mammals and reptiles (including birds).
champsosauridae Extinct taxon (Choristodera) of reptiles superficially resembling crocodilians
dinosauria Taxon of reptiles including Ornithischia (bird-hipped dinosaurs) and Saurischia (reptile-hipped dinosaurs, including birds).
elasmobranchii Sharks and their relatives, such as skates and rays.
K/T Abbreviation for Cretaceous/Tertiary, usually in reference to the K/T boundary.
multituberculata Extinct taxon of mammals superficially resembling rodents.
nonavian dinosaurs Dinosauria excluding birds.
squamata Lizards and snakes.

THE LOSS OF NONAVIAN DINOSAURS is the most widely recognized extinction either past or present. Although the dinosaurs that we commonly call birds survived and thrived, no nonavian dinosaurs are known beyond 65 mya (million years ago). Why all nonavian dinosaurs became extinct after some 145 my (million years) of evolutionary success as the world's largest land vertebrates is a much debated topic. Here the most recent extinction theories are presented and tested against the known record of nonavian dinosaurs and their contemporaries.

I. EXTINCTION AS COUNTERPOINT TO EVOLUTION

Extinction is commonly perceived as a rare, negative process. This is the case because much of the extinction that we recognize as occurring in historical times has been at the hands of humans, resulting in a loss to overall biodiversity. In this context, extinction certainly is a negative process, and unfortunately it is becoming more and more common.

Taking the much longer geological perspective, however, extinction is an integral part of the processes that have shaped the global biota ever since the appearance of life at least 3.5 billion years ago. Of all the species that have existed throughout Earth history, it is commonly estimated that over 90%, possibly over 99% of them have gone extinct. At first surprising, this figure is comprehensible when we realize that species have been constantly added through evolution and winnowed by extinction ever since life appeared. For the Phanerozoic era as a whole, there has been a general increase in species diversity. This trend can be further divided into several major increases in species diversity, along with at least one long interval when species diversity remained at a steady state. It is the balance between originations of new species and losses of existing species that maintains such a steady state of species numbers. Thus, extinction, considered a rare and negative process in human time frames and sensibilities, is an integral counterpoint to evolution. The vast majority of extant species, including *Homo sapiens*, would probably have never evolved had not many other species become extinct.

The vast majority of these extinctions, possibly as much as 95%, occurred as so-called normal or background extinctions. This is because rates and especially numbers of extinctions during most intervals of time do not stand out from extinctions in other such intervals. Although it is difficult to provide a specific figure because of vagaries of the fossil record, differences between environments, and differences between organisms, it can be safely said that normal extinction probably affects much less than 50% of existing species in a given time period. Similarly, there is no general agreement on the time frame over which these normal extinctions occur, but it is almost certainly for intervals of less than one million years. There are, however, at least five almost universally accepted intervals during the Phanerozoic when the numbers of extinctions far surpassed those of background extinctions. These are the so-called mass extinctions. During a mass extinction, well over 50%, in one case possibly reaching as high as 95%, of all extant species became extinct. This horrendously high level of extinction occurred at the end of the Permian period some 250 million years ago. All five known mass extinctions, from oldest to youngest, were in (or at the end of) the Late Ordovician, Late Devonian, Late Permian, Late Triassic, and Late Cretaceous. There is an unresolved debate as to whether these five mass extinctions represent a separate class of extinction from normal extinctions or instead form a continuum with normal extinctions in both rate and cause.

The most recent of the five mass extinctions near the end of the Late Cretaceous includes that of all the nonavian dinosaurs. The level of extinction for all species during the Cretaceous/Tertiary (or K/T) transition has been frequently estimated at about 75%, although no studies directly document this level. For backboned animals or vertebrates, the level of extinction hovers around 50% for species, but this is based on only one area, the Western Interior of North America.

II. MISCONCEPTIONS ABOUT DINOSAUR EXTINCTION

A. Quality of the Dinosaur Record

The belief that the nonavian dinosaurs disappeared from Earth instantaneously is not new, but the proposition in 1980 that an asteroid impact caused this very rapid decline and extinction gave new life to this idea. Some proponents of this theory have explicitly stated that these extinctions were essentially instantaneous around the world. Such explicit assertions about global records of nonavian dinosaurs are patently false.

All one need do is examine the fossil record of nonavian dinosaurs to see that this belief is unsupported. Figure 1 shows that for the Late Cretaceous as a whole, an interval of some 35 my, the record is reasonably good. Every continent, and even smaller areas such as New Zealand, produced Late Cretaceous nonavian dinosaurs. By comparison, Fig. 2 shows that the K/T record of nonavian dinosaurs is very incomplete. In only one region do we currently have a reasonably large sample of nonavian dinosaurs and contemporary vertebrates extending to or near the K/T boundary (and have a fossil record of similar quality above this boundary). This is in the Western Interior of North America,

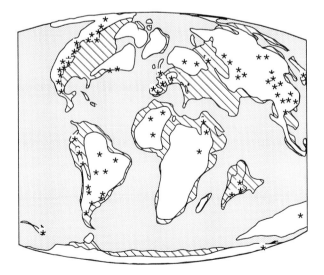

FIGURE 1   Record of nonavian dinosaurs for the Late Cretaceous (modified from Archibald, 1996).

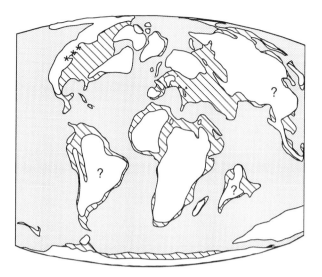

FIGURE 2   Record of nonavian dinosaurs at the K/T boundary (asterisks) and new areas (question marks) possibly preserving a similar record (modified from Archibald, 1996).

notably in the eastern part of Montana and into southern Canada. This region formed the eastern coast of a continental-sized landmass (Laramidia) that bordered the Pierre Seaway, a great inland sea that split North America in half in the Late Cretaceous. Although extensive, this region does not represent the entire globe regarding questions of extinction. A global record of latest Cretaceous nonavian dinosaurs may emerge in the next few years. Especially promising are new finds in several sedimentary basins in China, localities in central South America, and possibly sites on the Indian subcontinent. Until we have a more global record, arguments about the pace of dinosaur extinction on a global scale remain unsubstantiated speculation. We simply have no record of nonavian dinosaurs that permits us to show clearly whether dinosaur extinction was catastrophically fast or glacially slow. Rather, the data we do have are more regional in scope and only permit us to examine questions of the magnitude and selectivity of these extinctions, but nothing of its pace.

B. Ecological and Evolutionary Diversity among Dinosaurs

Although our perceptions of nonavian dinosaurs have changed during the past century, only now is a clearer picture emerging of them as living and breathing animals. Nevertheless, we must be cautious in assessing how these animals might have reacted to various stresses that could lead to extinction. Almost certainly

the various groups of nonavian dinosaurs would not have reacted in monotonously similar ways to various ecological stresses. A short summary can provide a sense of this ecological and evolutionary diversity.

The two great clades of Dinosauria, Ornithischia and Saurischia, both appear in the Late Triassic, over 200 million years ago. Late Cretaceous Ornithischia ("bird-hipped" dinosaurs) includes the considerable diversity of ceratopsians ("horned" dinosaurs), hadrosaurians ("duck-billed" dinosaurs) and their relatives, the low-built ankylosaurs, and some lesser known clades. Saurischia most importantly includes sauropods (the largest dinosaurs) and Theropoda (all the medium- to large-sized carnivorous, bipedal dinosaurs, plus Aves, or birds).

All of these dinosaurs were present in the Western Interior of North America leading up to the end of the Cretaceous. They ranged from small (about 1–2 kg) birds up to ponderous, many-tonned sauropods known from the southern areas of the continent. They also ranged from probable insectivores through carnivores, as well as a variety of herbivores. In short, although not great in numbers of species (e.g., compared to extant mammals), they were ecologically diverse. This means that any theory of extinction must explain how at least all the nonavian dinosaurs succumbed while many other vertebrates did not.

The cladogram in Fig. 3 shows the relationship of birds to other dinosaurs, as well as the relationships of other vertebrates known from the Late Cretaceous. The inclusion of birds within Dinosauria becomes stronger all the time with new discoveries, notably of feathered

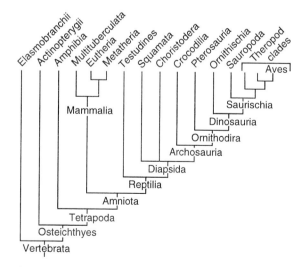

FIGURE 3   Cladogram of vertebrates from the Late Cretaceous (modified from Archibald, 1996).

nonavian dinosaurs, and ever more sophisticated phylogenetic analyses.

The most important concern regarding K/T boundary extinctions comes from workers that object to including birds within Dinosauria, even if they accept the very robust hypothesis of dinosaur–bird relationships as true. This objection comes from those who do not wish to accept the fact that dinosaurs survived the K/T boundary.

III. CURRENT THEORIES

It is estimated that more than 80 dinosaur extinction theories have been proposed, more than for any other group of animals. This list of theories ranges from the absurd (e.g., overhunting by aliens) to the highly reasonable. This section briefly outlines the three primary causes of extinction that are well enough articulated to be testable with the fossil record: marine regression (and habitat fragmentation), massive eruptions of flood basalts, and asteroid impact. In Section V these are tested using the vertebrate fossil record.

Marine regression refers to the draining of epicontinental seas. One of the greatest such regressions is recorded in rocks near the end of the Cretaceous period some 65 mya. Estimates suggest that 29 million km² of land were exposed during this interval (Fig. 4), more than twice the next largest such addition of land during the past 250 million years. This landmass is approximately the size of Africa. There were marked proximate effects of this regression—major loss of low coastal plain habitats, fragmentation of the remaining coastal plains, establishment of land bridges, extension of fresh-

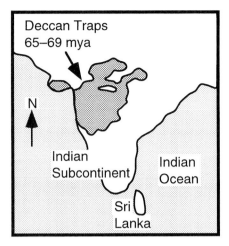

FIGURE 5 Geographic extent of latest Cretaceous and earliest Paleocene Deccan Traps (modified from Archibald, 1996).

water systems, and climatic change with a general trend toward cooling on the newly emerged landmasses.

Massive eruptions of flood basalts called the Deccan Traps on the Indian subcontinent occurred over a longer interval than marine regression, perhaps as much as 4 million years surrounding the K/T boundary (Fig. 5). The volume of material estimated to have been erupted over this interval would cover both Alaska and Texas to a depth of 610 m. Proximate effects resulting from such massive volcanism have not been as well studied as for either marine regression or asteroid impact, but are argued to be similar to those from impact. Climatic changes caused by massive eruptions would have spanned a longer time than changes caused by impact.

The original asteroid impact theory argued that a 10-km-diameter asteroid struck Earth 65 mya, producing ejecta and a plume reaching far enough into the atmosphere to spread around the globe, blocking the Sun. The cessation of photosynthesis resulted in the death and extinction of many plants, the herbivores that fed upon them, and the carnivores that in turn fed upon the herbivores. The probable crater, named Chicxulub, has been located in the Gulf of Mexico near the tip of the Yucatan Peninsula (Fig. 6). At 170 km across, it is thought to be one of the largest such structures on Earth. Interestingly, the 100 km Popigai Crater in Siberia formed almost 36 mya at the same time as the 80-km-wide Chesapeake Bay impact crater. Thus we have 100- and 80-km-wide impact craters that formed at the same time (within present margins of error) that caused no identified increase in extinction levels. Further, there is as yet no clear evidence of an impact for the terminal Triassic and Permian mass extinctions. Nonetheless,

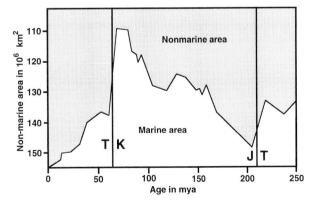

FIGURE 4 Comparison of marine and nonmarine areas for the 250 million years. Note marked additions of nonmarine areas surrounding the Triassic/Jurassic and Cretaceous/Tertiary boundaries (modified from Smith *et al.*, 1994).

FIGURE 6   Possible extent of Chicxulub crater in Yucatan Peninsula (modified from Archibald, 1996).

two other important pieces of physical evidence supporting an impact are an increase in the element iridium at the K/T boundary and minerals, especially quartz grains, showing shocked lamellae in two directions. A high level of iridium at Earth's surface and double lamellae are both more indicative of an impact than volcanism. Some of the more proximate effects of an asteroid impact that have been suggested are acid rain with a pH as low as battery acid, global wildfire consuming some 25% of aboveground biomass, sudden subfreezing temperatures lasting from 45 days to six months, tsunamis, and superhurricanes.

IV. PATTERNS OF VERTEBRATE EXTINCTION AND SURVIVAL AT THE K/T BOUNDARY

Although the vertebrate record of the K/T boundary is almost exclusively limited to the Western Interior, the uppermost Cretaceous Hell Creek Formation in eastern Montana has yielded a taxonomically rich sample of well over 100 vertebrate species, 107 of which are well enough known to be used in this analysis. There are 12 major vertebrate clades—5 species of elasmobranchs, 15 of actinopterygians, 8 of amphibians, 10 of multituberculates, 6 of placental mammals, 11 of marsupials, 17 of turtles, 10 of squamates, 1 of champsosaurs, 5 of crocodilians, 10 of ornithischians, and finally 9 species of saurischians (not including birds)

(Table I). Of these 107 vertebrate species, 49% (52 of 107) survived across the K/T boundary in the Western Interior. This is the *minimum* percentage survival of vertebrate species across the K/T boundary, because some of the very rare species may have survived undetected. Twenty of the 107 are quite rare species, represented by fewer than 50 identifiable specimens out of 150,000 specimens estimated to have been recovered from the Hell Creek Formation. Although any estimate is speculative, certainly some of these very rare species did survive. The extreme and very improbable scenario would be if all 20 survived, which would yield 67% (72 of 107 species) survival. This represents the extreme maximum percentage survival in the region. If I were to hazard an educated guess, I would say that no more than 60% of vertebrate species survived the K/T boundary.

When examined in greater detail, a very interesting pattern emerges within the record of survival and extinction for these Hell Creek Formation vertebrates. This is a pattern of differential species survival and extinction among the 12 major clades. Only five of the groups—elasmobranchs, marsupials, squamates, ornithischians, and saurischians—contributed to 75% of the extinctions. What do elasmobranchs, squamates, marsupials, and the two lineages of nonavian dinosaurs have in common in these faunas, other than that each suffered at least 70% or more species extinction at the K/T boundary in western North America? If we are to understand causes of extinction at the K/T boundary we must explain this differential pattern of extinctions.

TABLE I

Percent Survival of Vertebrate Species Across the K/T Boundary

Major clades from the Upper Cretaceous Hell Creek Formation, eastern Montana	Number and % species survival
ELASMOBRANCHII	0/5 (0%)
ACTINOPTERYGII	9/15 (60%)
AMPHIBIA	8/8 (100%)
MAMMALIA	
MULTITUBERCULATA	5/10 (50%)
EUTHERIA (placentals)	6/6 (100%)
METATHERIA	1/11 (9%)
REPTILIA	
TESTUDINES	15/17 (88%)
SQUAMATA	3/10 (30%)
CHORISTODERA (Champsosauridae)	1/1 (100%)
CROCODILIA	4/5 (80%)
ORNITHISCHIA	0/10 (0%)
SAURISCHIA	0/9 (0%)
TOTAL NUMBER & % SURVIVAL	52/107 (49%)

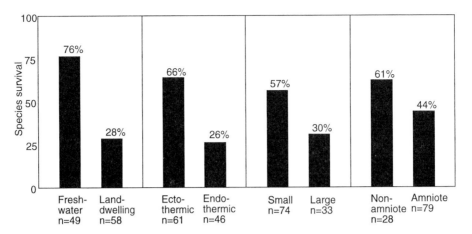

FIGURE 7   Four factors showing pattern of differential vertebrate species survival and extinction at the K/T boundary. Chi-square tests show that all of the comparisons, except possibly the last, are significant (freshwater vs land-dwelling comparison and ectothermic vs endothermic comparison, $P \ll 0.005$, small vs large comparison, $0.010 < P < 0.025$, nonamniote vs amniote comparison $0.150 < P < 0.100$) (modified from Archibald, 1996).

Any theories about the cause(s) of extinction of dinosaurs and their contemporary vertebrates must be able to explain why elasmobranchs, squamates, marsupials, ornithischians, and saurischians suffered very high levels of or even total extinction, whereas actinopterygians, amphibians, multituberculates, placentals, turtles, champsosaurs, and crocodilians suffered 50% or often much less extinction.

There are at least four biological factors that may account for much of this pattern of differential species survival and extinction: freshwater vertebrates survived better than land-dwelling vertebrates, ectotherms did better than endotherms, small vertebrates did better than large species, and nonamniotes did better than amniotes. Except for possibly the last comparison, chi-square tests confirm the differences among these factors (Fig. 7). The biological significance of these differences is more difficult to demonstrate, but these comparisons emphasize that the K/T vertebrate pattern of extinction and survival was indeed differential.

V. TESTING K/T EXTINCTION THEORIES USING THE VERTEBRATE FOSSIL RECORD

Of the three testable hypotheses of K/T extinction—the impact, volcanism, and marine regression theories—only the impact and marine regression theories have been relatively thoroughly tested against the vertebrate fossil record. Proponents of the volcanism theory, how-

ever, suggest that many of the biotic responses to an impact would also be found with massive volcanism. One method to examine the efficacy of these theories is to examine them in the context of the known K/T vertebrate record, starting with the impact theory.

A. The Impact Theory

The original impact theory paper in 1980 still offers the basic mechanism of how such an impact might cause extinction among both animals and plants, including vertebrates. The impact would create a dust cloud enveloping the globe for a few months to a year, and darkness would shroud the world as long as the dust remained in the atmosphere. Photosynthesis in the sea and on land would cease. As the plants died or became dormant, herbivores would soon starve, followed by the carnivores. Anoxia would rise in the oceans. Some of the best physical evidence of such an impact is the discovery of anomalously high levels of the rare earth element iridium at the K/T boundary in strata around the world and the probable remains of an impact crater. Neither elevated levels of iridium nor an impact crater, however, is direct evidence for specific causes of extinction at the K/T boundary.

An incorrect assumption often made in testing the impact theory and its possible corollaries is that all major taxa show very high levels of extinction across the ecological spectrum on a global scale. As discussed earlier, for vertebrates, there is no K/T global record at the species level and there is a differential pattern of extinction. The impact-generated scenario of extremely high levels of catastrophic extinction across most environments is so broad spectrum and tries to explain so much that it is difficult to test. The burden of proof for sweeping, catastrophic extinction scenarios rests with the proposers of the theory. The various corollaries of the impact theory, such as a sudden cold snap, highly acidic rain, or global wildfires, are more easily tested using the known K/T vertebrate record.

Although a short, sharp decrease in temperature was not emphasized in the originally proposed hypothesis, it soon became an important corollary of the impact theory. It is argued that if tremendous amounts of dust were injected into the atmosphere after a large impact, the darkness would not only suppress photosynthesis but would also produce extremely cold temperatures. This hypothesized condition has become known as "impact winter." Accordingly, following a large impact, ocean temperatures would decrease only a few degrees because of the huge heat capacity of the oceans, but on the continents temperatures would be subfreezing from 45 days up to six months. The temperature would remain subfreezing for about twice the time of darkness caused by the dust.

If a suddenly induced, prolonged interval of subfreezing temperatures occurred in subtropical and tropical regions today, among vertebrates the ectothermic tetrapods would suffer most. Ectotherms, as the name suggests, heat or cool themselves using the environment. Endotherms such as mammals and birds generate their heat through metabolic activity. In endotherms, about 80% of the energy derived from food consumption goes toward thermoregulation. Fishes, which are by and large ectothermic, would be generally more tolerant of a severe temperature drop.

The northern limits of extant turtles and crocodilians today are controlled by temperature. These animals cannot tolerate freezing, and become sluggish or immobile at 10–15°C. Various amphibians and reptiles inhabit areas with low winter temperatures or severe drought, but they have evolved methods of torpor (estivation and hibernation) to survive. These are the exceptions, however, because species diversity for ectothermic tetrapods is far higher in warmer climates. More importantly, we should not assume that Late Cretaceous ectothermic tetrapods living in subtropical to tropical climates such as in eastern Montana were capable of extended torpor. Torpor is most often preceded by decreases in ambient temperature, changes in light regimes, and decreases in food supply. The ectotherms in eastern Montana could not have anticipated a short, sharp decrease in temperature. This is true even if the impact had occurred during a Northern Hemisphere winter, when temperatures would have been slightly lower. We must remember that this was a subtropical to tropical setting, and thus the extended, subfreezing temperatures advocated by proponents of this corollary would have been devastating to ectotherms even during a terminal Cretaceous winter in Montana.

Except for a 70% decline in lizards, the other ectothermic tetrapods (frogs, salamanders, turtles, champsosaurs, and crocodilians) did very well across the K/T boundary. The corollary of a sudden temperature decrease simply does not fit with the vertebrate data at the K/T boundary. A latest (but not terminal) Cretaceous vertebrate fauna from northern Alaska strengthens the evidence that a hypothesized sudden temperature drop was not a likely cause of K/T boundary extinctions. A comparison of Late Cretaceous vertebrate faunas from Alaska and eastern Montana reveals a striking difference. Although the Alaskan fauna is decidedly smaller and with fewer species than that from eastern Montana, both have elasmobranchs, actinopterygians, dinosaurs, and mammals. The Alaskan fauna, however, completely lacks amphibians, turtles, lizards, champsosaurs, and crocodilians. These taxa comprise 41 of the 107 (38%) eastern Montana species. If even the fairly balmy temperature range of 2–8°C for Late Cretaceous Alaska was enough to exclude ectothermic tetrapods, a severe temperature drop to below subfreezing temperatures at the K/T boundary should have devastated the rich ectothermic tetrapod faunas at midlatitudes. Instead, these species flourished.

A second prominent corollary of the impact theory is the creation of highly acidic rain. The most commonly cited acids as products of an impact are nitric and sulfuric acid. It is argued that nitric acid would be produced by the combination of atmospheric nitrogen and oxygen as a result of the tremendous energy released by an impact. Sulfuric acid would be produced as large amounts of sulfur dioxide are vaporized from rock at the impact site. These acids would be precipitated in the form of rain.

Estimates of the pH of these acid rains vary, but go as low as 0.0–1.5. It is suggested that global effects could have caused the pH of near-surface marine water and freshwater to drop below 3. In today's environment, rain below a pH of 5.0 is considered unnaturally acidic. Rain as low as 2.4 has been recorded, but annual averages in areas affected by acid rain range from 3.8 to 4.4. Acid fogs and clouds from 2.1 to 2.2 have been recorded in southern California and have been known to bathe spruce–fir forests in North Carolina. The biological consequences of such low pH values vary from one vertebrate group to another but are always detrimental. Aquatic species (fishes, amphibians, and some reptiles) are the first and most drastically affected, with those reproducing in water being the first to suffer. If pH drops lower than about 3.0, adults often die. The effects on aquatic vertebrates across the K/T boundary would have been very bad if a pH of 3.0 were reached and truly horrendous if it hit 0.0 as suggested by some proponents. Some advocates of K/T acid rain argue that the surrounding soils or bedrock would have buffered the aquatic systems, even suggesting that limestone caves could have been important refugia for birds, mammals, amphibians, and small reptiles. The major problem with this scenario is that these kinds of buffering soils or bedrock and limestone caves did not exist in eastern Montana in the latest Cretaceous. From what we know of the present-day biota's reaction to acid rain, aquatic animals should have been devastated by acid rain at the K/T boundary. Of all the aquatic species, only elasmobranchs show a drastic drop in eastern Montana. Thus, the likelihood of low pH rain is highly implausible.

A third corollary of the impact theory that receives various levels of support is global wildfire resulting from the aftermath of the impact. Soot and charcoal have been reported from several sites at the K/T boundary coincident with the enrichment of iridium noted earlier. It was argued that this pattern is unique and must come from the extremely rapid burning of vegetation equivalent to half of all present-day forests. Other scenarios posit that some 25% of the aboveground biomass burned at the end of the Cretaceous.

Such a global conflagration is really beyond our comprehension. To grasp the magnitude of this scenario, imagine one-quarter to half of all structures on the globe engulfed in flames within a matter of days or weeks. This still would be only a fraction of what is argued to have been burned at the K/T boundary. In such an apocalyptic global wildfire, much of the aboveground biomass all over the world would have been reduced to ashes. In freshwater, those plants and animals not boiled outright would have faced a rain of organic and inorganic matter unparalleled in human experience. These organisms would have literally choked on the debris or suffocated as oxygen was suddenly depleted with the tremendous influx of organic matter. The global wildfire scenario is so broad in its killing effects that it could not have been selective, yet as discussed earlier, the vertebrate pattern of extinction and survival is highly selective. Thus it is no surprise that this scenario of equal opportunity losers does not show any significant agreement with the pattern of vertebrate extinction and survival at the K/T boundary.

Not only is there almost no fossil evidence supporting global wildfire, but the physical basis for such an event is suspect. Proponents argue that a global charcoal and soot layer coincides with the K/T boundary, whose emplacement is measured in months. This also assumes that the sedimentary layer encasing the charcoal and soot was also deposited in only months. This is demonstrably not the case for at least one K/T section that continues to be cited in these studies—the Fish Clay of the Stevns Klint section on the coast of Denmark. The Fish Clay is a laterally discontinuous, complexly layered and burrowed clay reflecting the conditions at the time of its deposition. It is not the result of less than a year of deposition caused by an impact-induced global wildfire. Thus, carbon near the K/T boundary at Stevns Klint, as well as in other sections, is likely the result of much longer accumulation during normal sedimentation.

When all of the corollaries of an impact of an asteroid or comet are compared to the pattern of extinction and survival for vertebrates at the K/T boundary in eastern Montana, there is relatively poor agreement. Without special pleading, these corollaries as currently proposed are unlikely causes of vertebrate extinction. This does not mean that all corollaries of impact should be rejected. For example, the argument for a period of extended darkness and concomitant suppression of photosynthesis seems to be supported by the fossil record of lower and midlatitude plants. It is imperative, however, that researchers proposing the various corollaries of impact separate those that are supported by the vertebrate fossil record from those that are not.

B. The Volcanism Theory

Although some proponents of the impact theory do not agree, many advocates of the impact and volcanism theories feel that a number of the same physical events would have occurred at the K/T boundary if either extensive volcanism or an impact took place. These advocates also say that the biological results would be similar. In light of the foregoing discussion of how most of the corollaries of the impact theory do not test well against the vertebrate fossil record, the volcanism and impact theories are equally weak in their biological predictions. The major difference in these two theories lies in their timing. The impact theory measures most of the cataclysmic effects in months or years, with physical effects possibly lingering for a few hundred or a few thousand years, whereas the volcanism theory measures effects into the millions of years. The effects of many volcanic eruptions, such as that of Mt. St. Helens in the United States, linger for a few months or a few years. Many other episodes of volcanism are very prolonged and these are flood basalt eruptions.

The best-known flood basalt eruption in the United States is the 16-million-year-old Columbia River flows in the Pacific Northwest. In the past 250 million years, arguably one of the biggest flood basalt eruptions occurred on the Indian subcontinent. This was occurring during (and is probably related to) the collision of the subcontinent with the remainder of Asia. The most obvious manifestation today is the tallest mountain range in the world, the Himalayas. As discussed earlier, these flood basalts, known as the Deccan Traps, cover an immense part of modern-day India and Pakistan. On the basis of radiometric dating, paleomagnetics, and vertebrate fossils, the bulk of the eruptions was centered around the K/T boundary during a reversal in Earth's magnetic poles known as 29R or 29 reversed. The number 29 represents the 29th reversal of the magnetic field counting backward from the present, which today has normal polarity by definition. The K/T boundary happens to fall in 29R.

What would the effect have been on the global biota if something of the magnitude of the Deccan Traps erupted for tens of thousands of years or longer? One of the greatest effects would have been to increase and maintain a much higher level of particulate matter in the atmosphere. Whether it would have caused warming through a greenhouse effect, cooling because of less light, or simply prettier sunsets is not certain. The amount of CO_2 pumped into the atmosphere by the eruptions may have been a boon for green plants that require CO_2 for photosynthesis, but a reduction in the light that reached Earth's surface because of particulate matter may have canceled the effects of increased CO_2. If I must choose, I would say that the effects of added particulate matter would prevail for no other reason than because they would linger longer after eruptions ceased, whereas the release of CO_2 would diminish much more rapidly after each eruption stopped.

If the latter scenario is correct, the longer-term effects over a million years or more would be to precipitate a global cooling. Most estimates suggest that the climate cooled regionally if not globally through the K/T transition. Because the time frame is moderately long, many species on land or in the sea, especially smaller ones, could have adapted to changes, whereas larger species such as dinosaurs may not have been fortunate. Although I doubt that the cooling across the K/T boundary can be said to be a cause of extinction for most species, it would have been an added stress.

A final possible long-term effect from eruption of

the Deccan Traps is reduced hatching success for eggs of herbivorous dinosaurs. Volcanic activity can release elements such as selenium that are highly toxic to developing embryos. Increased levels of selenium in the eggshells of dinosaurs are reported from near the K/T boundary in southern France. The poisoning of eggs has also been suggested for dinosaur eggs near the K/T boundary in Nanxiong Basin in southeastern China.

C. The Marine Regression Theory

The final hypothesis that has been tested with the vertebrate fossil record is the marine regression/habitat fragmentation theory, or marine regression theory for short. Many areas of the modern terrestrial realm were repeatedly inundated by shallow epicontinental seas throughout geologic history. The term "epicontinental" refers to the occurrence of these very shallow seas on the continental shelves and platforms rather than in deep ocean basins. Epicontinental seas reached depths of only 1500 to 2000 ft, very shallow relative to most large present-day marine bodies. Epicontinental seas are almost nonexistent today, except for bodies of water such as Hudson Bay. It is known that during the Late Cretaceous, large areas of modern continents were submerged under warm, shallow, epicontinental seas.

It became clear only recently just how dramatic the regression of these seas was leading up to the K/T boundary. There is absolutely no mistaking that the K/T loss of shallow seas (or increase in nonmarine area) is greater than at any time in the past 250 million years. An equivalent of the land area of all of Africa, the second largest continent today, was added during this time. The second largest increase in continental area in the past 250 million years occurred across the Triassic/Jurassic boundary. Like the K/T transition, this is also during one of the five universally recognized mass extinctions during the Phanerozoic or last 550 million years.

Some of the most dramatic additions of nonmarine areas at or near the K/T boundary occurred in North America. Near the end of the Cretaceous, maximum marine transgression divided North America into two continents (Fig. 8). As regression continued until at or near the K/T boundary, coastal plains decreased in size and became fragmented; stream systems multiplied and lengthened; and as sea level fell, land connections were established or reestablished (Fig. 9).

The driving force for these repeated inundations or transgressions of the lower-lying portions of continents is still not fully understood. The general consensus is

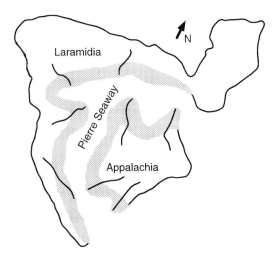

FIGURE 8 North America, divided into two continents by seaway throughout much of the Late Cretaceous. Exact positions of most streams and coastal plains is not known (modified from Archibald, 1996).

that it was related to plate tectonics, that is, rises in sea level and inundations began as the motion of the plates increased. As this occurs, the margins along which the colliding plates converge are subducted or pushed downward into the depths of the Earth. This causes inundation by the seas over shallow continental shelves and platforms. Whatever the geophysical factors driving these processes, the physical manifestations of marine regression, like those of impacts and volcanism, are important ultimate causes of extinction.

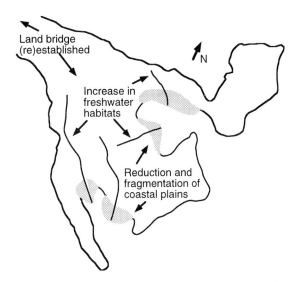

FIGURE 9 North America, reunited as single continent in latest Cretaceous or early Tertiary, showing some results of this event. Exact positions of most streams and coastal plains is not known (modified from Archibald, 1996).

Although these processes of marine transgression and regression were global in extent, a closer examination of North America is instructive because this is where we have the best vertebrate data at the K/T boundary. North America was split into two continents—a western continent, Laramidia, and an eastern continent, Appalachia—by the Pierre Seaway for almost 40 million years during the Late Cretaceous. Most of our latest Cretaceous vertebrate fossils come from the east coast of Laramidia. The west coast of Appalachia, as well as the eastern seaboard of Appalachia, have also produced some specimens.

In the last few million years of the Cretaceous the Pierre Seaway began to regress from both Laramidia and Appalachia. At or just shortly before the K/T boundary, the seaway reached its nadir. The positions of the receding coastlines both north and south have not been well established, but we know the southern coastline reached well into Texas. There is no question that there was a dramatic reduction in coastal plains. This is exactly the kind of environment from which we are sampling the last of the Late Cretaceous vertebrate community. A common refrain is that because the total amount of land increased with the regression, dinosaurs should have had more, not less, area and environments in which to live. We know with considerable certainty that dinosaurs did live in other environments such as the higher, drier Gobi Desert in Mongolia during part of the Late Cretaceous. At present, however, the only well-known vertebrate communities that preserve dinosaurs at the K/T boundary are coastal. Thus arguments about what dinosaurs and other vertebrates may or may not have done in other environments are moot. It is simply incorrect to say the dinosaurs and other vertebrates may have survived elsewhere, when we have little or no information about other environments. We must test our theories using the data we have, not what we wish we had.

The drastic reduction of coastal plains put tremendous pressure on some vertebrate species, especially large ones. Reduction of habitat, for example, in the Rift Valley System of modern-day East Africa, first affects larger vertebrates, especially mammals. In the shrinking coastal plains of latest Cretaceous North America, the equivalent large vertebrates first affected were the dinosaurs. An additional problem, whether in East Africa today or the coastal plains of latest Cretaceous North America, is the fragmenting of the remaining habitat. This present-day consequence as a result of human activity has become known as habitat fragmentation by ecologists. In larger, undisturbed habitats, animals (and plants) can move more freely from one area to another. If the habitat is fragmented, although the amount of habitat may not have been greatly altered, it will reduce the flow of species from one fragment to another.

For some species, even seemingly small barriers such as two-lane roads can be insurmountable, and the consequences can be disastrous if viable populations cannot be maintained in the various fragments. This fragmentation can lead to extinctions. Barriers also arise in nature even among animals that would seem easily capable of dispersing. Although I have no doubt that such barriers often result in extinction, we usually see only what survives in the form of differences between closely related species. Small arboreal primates in the rain forests of both South America and Africa form small fragmented groups that are sometimes isolated from each other by rivers only tens of yards wide. Another example is the Kaibab squirrel on the North Rim of the Grand Canyon. Unlike its nearest relative, Abert's squirrel, which is found on the south side of the Grand Canyon and in the western United States and Mexico, the Kaibab squirrel is restricted to an area of only 30 by 60 miles. Fragmentation, in this case the development of the Grand Canyon, helped produce the differences, but the margin for survival for the Kaibab squirrel is slim.

The idea of habitat fragmentation extends not only to natural processes operating today, but to processes operating in the geological past. Although habitat fragmentation is not well understood among earth scientists, it is all too real for biologists studying the effects of human activity in present-day rain forests and in urban settings. Declines of bird and mammal populations have been well documented in the city of San Diego as urban development divides and isolates the remaining habitats in canyon areas. One would not expect that habitat fragmentation would be easily preserved in geological strata. Yet the forcing factor for fragmentation in the latest Cretaceous—marine regression—is a thoroughly documented fact during the waning years of the Late Cretaceous in North America. Globally, marine regression occurred within this same general time frame, although how close in time it occurred in various regions is a matter of debate.

Theory predicts that large species would be the most severely affected by habitat fragmentation, for the reasons discussed earlier. During the K/T transition in eastern Montana, only 8 of 30 large species survived and these were all or partially aquatic (two fishes, one turtle, one champsosaur, and four crocodilians), whereas all 22 large terrestrial species (and one aquatic species) became extinct (one turtle, one lizard, one crocodilian, and 19 dinosaurs). Thus, predictions from

habitat fragmentation studies fit the observed data very well.

As noted in the preceding discussion, two other major physical events in addition to habitat fragmentation occurred with marine regression: stream systems multiplied and lengthened, and as sea level fell, land connections were established or reestablished. Only a few of the K/T boundary stream systems have been studied in detail in the Western Interior, and thus we do not know the exact drainage patterns for most latest Cretaceous and early Tertiary stream systems in the eastern part of Laramidia. Nevertheless, we are certain that as new land was added following marine regression in the early Tertiary, stream systems increased and lengthened. This process is another major corollary of marine regression. When freshwater habitats were increased following marine regression, most aquatic vertebrates did well except those with close marine ties—sharks and some bony fishes. Such fishes may need to spend at least a portion of their life in a marine environment, in some instances to reproduce. The major group most likely to suffer would have been the sharks and their relatives. In fact, all five species of sharks disappeared. It is not clear, however, whether these disappearances from the Western Interior were actually extinctions at the K/T boundary or whether they survived elsewhere in marine environments into the earliest Paleocene. The problem is that the definitively oldest marine sediments that postdate the K/T boundary in the Western Interior are at most late early Paleocene in age. This means that there is a gap in marine sedimentation in the Western Interior of possibly one million years or more. This pattern of disappearance and reappearance strongly suggests that as the Pierre Seaway regressed farther and farther away from eastern Montana, all sharks and relatives departed because connections to the sea became attenuated. New species of elasmobranchs did not occur in the area until a smaller marine transgression reached the Western Interior at or just before middle Paleocene times. This is known as the Cannonball Sea, which was a smaller seaway than the Pierre. The total disappearance of sharks and their relatives is the only prediction that can be made with any certainty as a result of the loss of marine connections and the lengthening of stream systems. The increase of stream systems helped to mitigate other stresses that may have been put on the freshwater system.

New land areas were exposed as the sea level lowered. In some cases this included the establishment or reestablishment of intercontinental connections. One such connection was the Bering Land Bridge joining western North America and eastern Asia. At various times during the Late Cretaceous this bridge appeared and then disappeared. This is suggested by similarities in parts of the Late Cretaceous vertebrate faunas in Asia and North America, especially the better-studied turtles, dinosaurs, and mammals. Competition and extinction often result from biotic mixing, but predicting the fates of various taxonomic groups is usually not possible. An exception may have been the fate of marsupials in North America near the K/T boundary.

The oldest marsupials are known from approximately 100-million-year-old sites in western North America. By some 85 mya, we know of about 10 species of marsupial. This number rose to about 15 species and remained stable from around 75 mya until the K/T boundary some 65 mya, when it plummeted to a single species. These were all quite small mammals, from the size of a mouse up to a very well fed opossum or raccoon. Their teeth resembled those of modern opossums, with slicing crests and well-developed but relatively low cusps (compared to contemporary placental mammals) for poking holes in insect carapaces, seeds, or whatever they found. Most did not appear to be specialists on any particular food. With the reestablishment of the Bering Land bridge (or at least closer islands) near the K/T boundary, a new wave of placental mammals appeared in western North America. These mammals (traditionally known as condylarths) were the very early relatives of modern ungulates and whales. Their appearance in North America coincides with the very rapid decline of marsupials near the K/T boundary. Within a million years of the K/T boundary, 30 species of these archaic ungulates are known in North America, and their numbers kept on rising. Our best guess now is that the lineage that gave rise to these mammals first appeared in middle Asia between about 80–85 mya and reached North America near the K/T boundary. What is of interest is that the archaic ungulate invaders had dentitions very similar to those of the then contemporary marsupials and presumably ate similar things. It seems more than coincidence that marsupials did well in North America for about 20 million years, only to nearly disappear with the appearance of the ungulate clade. It is ironic that both marsupials and ungulates were joint invaders of South America very soon after the K/T boundary. Their dentitions were already beginning to show differentiation, with the marsupials headed toward carnivory and ungulates headed toward herbivory. It shows what a little cooperation can do.

These various physical events accompanying marine regression fit very well the pattern of extinction and survival described earlier for the 107 vertebrate species from near the K/T boundary in eastern Montana. In fact, the patterns of extinction and survival for 11 of 12 of the major vertebrate groups agree quite well with predictions from the marine regression theory.

VI. A SCENARIO FOR DINOSAUR EXTINCTION

Some scenarios argue that dinosaur extinction was basically caused by one factor, such as the impact of an asteroid with its various related effects. The vertebrate fossil record, however, clearly points to multiple causes that may include volcanism, marine regression, and asteroid impact. Each cause might have been necessary for these extinctions to have occurred, but none was sufficient to cause the decidedly differential pattern of extinctions seen in the fossil record of vertebrates from western North America.

One must start at least 10 million years before the end of the Cretaceous in order to understand this biotic upheaval. Because the record of vertebrate change during this time is largely limited to the Western Interior, the scenario must be similarly limited in scope. How much more widely it can be applied remains a paleontological puzzle.

The scene some 75 million years ago is of an open plain with scattered trees on the eastern shore of Laramidia (Western Interior of North America). Rivers of moderate size meander across the landscape. In the distance, the plains are lapped gently by a shallow sea stretching to the eastern horizon. The scene is dominated by vast herds of several species of duck-billed and horned dinosaurs, much like the hoofed mammals that dominate Africa's Serengeti Plain of today. Other ornithischian dinosaurs and the occasional meat-eating theropods cross the landscape. The streams are populated by numerous species of turtles, amphibians, crocodilians, and fishes, including the occasional skate or shark swimming up from the nearby sea. It is daytime, mouse- and rat-sized mammals are asleep in their dens.

Fast forward to about 66 million years ago. The shallow seaway has begun to slip rapidly away to the south and east. As the exiting seaway reaches lower-lying, flatter terrain, the rate of exodus quickens, with the final stages of withdrawal occurring in at most tens of thousands of years. The great herds of duck-billed and horned dinosaurs have followed the retreating seas.

But as their remaining refugia of low coastal plains rapidly decrease, first one and then another of the species dwindles until the great herds are reduced to at most two or three remaining species, much like the herds of bison that once roamed North America. Dinosaurs, like large vertebrates everywhere and at every time, are the first to experience biotic stresses leading to their decline and disappearance. Thus, as with the bison, the slide to extinction of the few remaining species can occur quickly.

We do not know what was happening to vertebrates in more inland areas just before the K/T boundary, because few such areas are well preserved and none has been studied. The coastal plains dinosaurs certainly were capable of migrating from one shrinking coastal habitat to another, but finally even this could not stop further declines in population size—just as the relentless encroachments of increasing human populations are causing many biotas to shrink today. Other large vertebrates also suffered. The Komodo Dragon-sized lizards and the single exclusively terrestrial turtle, *Boremys,* experienced declines. Populations of smaller terrestrial vertebrates were also declining, but perhaps because they had shorter life spans and quicker turnover rates, they adapted more quickly to the environmental stresses caused by the loss and fragmentation of the coastal plains.

Marsupials had flourished for some 25 million years in North America. Newly emerging land bridges appeared as the seas retreated. Invaders appeared. In North America these were the diminutive archaic ungulates, probably from Asia. In the Western Interior, at least, they out–competed the marsupials for dwindling resources. In South America events were different. Both groups of mammals appeared in South America soon after the K/T boundary, but here they divided the guilds, with marsupials becoming the carnivores and the ungulates the herbivores. This coevolutionary arrangement lasted for almost 50 million years in South America, with only an infusion of rodents and primates from the outside world.

Unlike the terrestrial vertebrates, freshwater species faced far less stress, especially because the size of their habitat was at least holding its own as the lengthening streams followed the retreating seas. Yet not all aquatic vertebrates fared so well. With the loss of close ties to the seas in areas like eastern Montana, sharks and skates ventured into the rivers in the area less and less frequently, as the distance to the sea expanded from tens to thousands of miles, eventually reaching Texas to the south. Plants and nearshore species also showed added

stresses as their respective habitats shrunk. Certainly, some species must have done fine as new habitats were formed as the seas regressed. As with vertebrates, however, we do not have any clear record of these environments away from the coastal areas.

Even before the seas began to retreat, the waxing and waning of the eruptions of the Deccan Traps added further stresses. One such stress was increase in particulate matter in the atmosphere that very slowly began to cool and dry some areas of the globe.

Suddenly, a literally Earth-shattering event magnified the differences between the have and have-not species. A 10-km-wide asteroid struck what today we call Yucatan. Material injected into the upper atmosphere formed a cover of darkness, blanketing the Sun to the point that photosynthesis ceased or diminished for many weeks, depending on where you were. The effects were especially acute at lower latitudes and closer to the impact, such as in North America. Plants unaccustomed to lower light regimes caused by seasonal changes in the Sun's position were especially hard hit. Higher-latitude plants accustomed to seasonally lower light regimes survived much better, as did the animals that fed upon them. The effects on higher-latitude plants and animals were tempered by which season they were experiencing when impact occurred. Extinction rates for coastal plants in North America soared because of the cumulative effects of continued habitat loss, drought, and loss of sunlight.

Except for the elasmobranchs, who had already departed or become extinct as the seas regressed, all ectothermic, aquatic vertebrate species (bony fishes, amphibians, turtles, champsosaurs, and crocodilians) weathered the impact quite well in their still flourishing freshwater habitats (80% survival, or 37 out of 46 species). Although some species can be resilient, they are also susceptible to certain kinds of shocks to their environment—just like the proverbial canary in the coal mine. If drastic temperature spikes, global wildfires, or battery acid rains had accompanied the Chicxulub impact as is often argued, these species would have been decimated because they lacked the torpor mechanisms or places to hide from such hypothesized devastation.

With the added loss of more plant species, and the reduction of biomass that the impact brought in the already highly stressed terrestrial ecosystem, other vertebrate species rapidly succumbed. Most notable were the last of the large herbivorous nonavian dinosaurs. The remaining predaceous, nonavian dinosaurs followed very soon, with the larger species going first. In some places on the globe the great creatures may have lingered a while longer, but finally, for the first time in more than 150 million years, no large land vertebrates graced Earth. The landscape was now open and waiting for evolution's next gambit—mammals.

See Also the Following Articles

EXTINCTION, CAUSES OF • MASS EXTINCTIONS, CONCEPT OF • MASS EXTINCTIONS, NOTABLE EXAMPLES OF

Bibliography

Alvarez, W. L. (1997). *T. rex and the Crater of Doom.* Princeton University Press, Princeton, New Jersey.

Archibald, J. D. (1996). *Dinosaur Extinction and the End of an Era: What the Fossils Say.* Columbia University Press, New York.

Dingus, L., and Rowe, T. (1998). *The Mistaken Extinction: Dinosaur Evolution and the Origin of Birds.* W. H. Freeman and Co., New York.

Fastovsky, D. E., and Weishampel, D. B. (1996). *The Evolution and Extinction of the Dinosaurs.* Cambridge University Press, Cambridge, United Kingdom.

MacLeod, N., and Keller, G. (eds.). (1996). *The Cretaceous–Tertiary Mass Extinction: Biotic and Environmental Changes.* W. W. Norton and Co., New York.

DISEASES, CONSERVATION AND

Sonia Altizer, Johannes Foufopoulos, and Andrea Gager
Princeton University

GLOSSARY

disease The debilitating effects of parasites on their hosts.

endemic Refers to a pathogen that is consistently present in a region and does not exhibit large fluctuations in prevalence.

epidemic Characterizes a sudden increase in parasite prevalence or intensity beyond what is normally present.

helminth One of several classes of parasitic worms: nematodes, cestodes, trematodes (monogeneans and digeneans), and acanthocephalans.

intensity Either the mean number of parasites within the subset of infected hosts or the average parasite load of the entire population.

morbidity Host weakness or lethargy caused by disease.

pathogen A microbial agent that causes disease.

prevalence The proportion of hosts in a population that are infected or diseased.

R_0 The basic reproductive ratio of disease; a parameter that describes the number of new infections generated by a single infected host entering an entirely susceptible population.

vector An animal that transmits parasites among definitive hosts; for example, mosquitos are vectors of malaria.

vertical transmission The process of infection from parents to offspring (as opposed to horizontal transmission, in which parasites are transferred by direct contact or vectors).

virulence Disease-induced mortality rate, or the severity of infection on individual hosts (note that in plant pathology this term specifically refers to the range of host genotypes that can be attacked rather than pathogen aggressiveness).

PARASITES AND INFECTIOUS DISEASES can greatly influence the population dynamics and evolution of their hosts. Collectively, they comprise a major component of biodiversity (perhaps up to one-half of all living species) and are a ubiquitous component of natural ecosystems. Because of their effects on host abundance and diversity, diseases also pose an unusual problem for conservation biologists. From one perspective, they may have positive effects on biodiversity, facilitating species coexistence and maintaining genetic variation in natural populations. On the other hand, disease outbreaks can threaten biodiversity by catalyzing or accelerating species or population declines and extinctions, often in conjunction with human modifications of the environment or the accidental introduction of virulent pathogen strains. Therefore, tracking epidemics and measuring the effects of diseases in natural ecosystems are central to developing conservation strategies and managing biodiversity.

I. TYPES OF PARASITES AND CAUSES OF DISEASE

Parasites and pathogens obtain nutrients from one or a few individuals, usually to the detriment of their hosts. Disease refers to the signs and symptoms of infection, whereas parasites and pathogens are the disease-causing organisms. Infectious diseases can be caused either by microparasites (such as viruses, bacteria, protozoa, and fungi) or by larger macroparasites, including worms and arthropods. The distinction between macro- and microparasites is particularly useful to ecologists and epidemiologists because these groups differ in the degree of within-host replication, their ability to generate a lasting host immune response, and how they are quantified in natural populations (Table I).

Disease outbreaks require the presence of a susceptible host population, an infectious pathogen, and favorable environmental conditions; these three factors are often referred to as the disease triangle. Endemic parasites can often persist in host populations with few

or no negative impacts, and infected hosts may never develop signs or symptoms of disease. Epidemics, on the other hand, are increases in the prevalence or intensity of infection and may result from the introduction of a new pathogen or from changing environmental conditions that favor pathogen transmission or increase host susceptibility. In some cases, these environmental modifications are human-induced changes that lead to host crowding, habitat degradation and host stress, or shifts in the geographic distribution of host, pathogen, or vector species. Diseases and parasites can also pose grave problems in captive populations in which plants and animals are maintained at high density, may be stressed, and are likely to be exposed to cross-species transmission.

II. BASIC EPIDEMIOLOGICAL PRINCIPLES

Since the pioneering work of Anderson and May in the late 1970s, ecologists have shown increasing interest in the effects of diseases at the population and community level. A general understanding of parasite ecology and epidemiology can illuminate problems in conservation biology ranging from detecting disease threats for endangered species to using parasites as control agents for invasive exotic species. Epidemiologists quantify and model disease spread over time and space to identify parameters that influence the prevalence and popula-

TABLE I

Characteristics, Examples, and Ecological Properties of Micro- and Macroparasites

	Microparasites	Macroparasites
Representative taxa	Viruses, bacteria, protozoa, fungi, microsporidians	Helminths (e.g., nematodes, cestodes, and acanthocephalans), arthropods (e.g., mites, ticks, and lice)
Size and reproduction	Small, unicellular, short generation times within individual hosts	Large, multicellular, longer generation times, usually no direct replication within hosts
Transmission of infective stages	Transmission via direct contact (e.g., venereal and vertical), vectors, or contaminated air/soil/water	Complex life cycles and intermediate hosts, vector transmission, or direct transmission (fecal/oral or host contact)
Effects on host immunity	Lasting host immunity develops quickly	Antigenic diversity of parasites usually too high for host to mount effective immune response
Effects on host fitness	Disease can be acute or chronic, may have strong effects on host survival or fecundity	Depends on the number of parasites within host (can affect mortality or fecundity, usually chronic infection with sublethal effects)
Quantification in host populations	Prevalence, seroprevalence, incidence	Prevalence, degree of aggregation in individual hosts
Frequency of epidemics	Common	Rare

tion-level effects of disease. Prevalence refers to the proportion of hosts harboring infection or expressing disease and allows for useful comparisons among geographic locations or over time. Parasite intensity refers to the average number of parasites within infected hosts and is important in quantifying the distribution and effects of macroparasites in host populations.

Models are used in epidemiology to detail how processes operating at the level of individuals (such as transmission, recovery, and death) translate into population-level phenomena. Mathematical models of host–parasite interactions differ from models of other antagonistic interactions (such as predators and prey) because diseases do not necessarily kill their hosts, and recovered hosts can develop a long-term immunity to reinfection. It is useful to address models designed for micro- vs macroparasites separately to simplify the analysis of

disease dynamics and quantify their effects on host populations.

A. Microparasites

Mathematical models for microparasites divide the host population into susceptible (S), infected (I), and recovered/immune (R) classes and track changes in the number of hosts within each category (Fig. 1). This type of compartment model (often called a SIR model) is commonly used in modeling directly transmitted microparasites, and has been developed and analyzed extensively through the work of Anderson and May (1991) and others from classical approaches of Kermack and McKendrick (1927). This model makes many assumptions, including that hosts are uninfected at birth, the disease does not affect host fecundity, and host

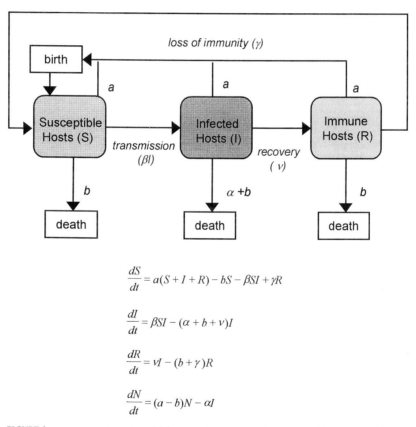

$$\frac{dS}{dt} = a(S + I + R) - bS - \beta SI + \gamma R$$

$$\frac{dI}{dt} = \beta SI - (\alpha + b + \nu)I$$

$$\frac{dR}{dt} = \nu I - (b + \gamma)R$$

$$\frac{dN}{dt} = (a - b)N - \alpha I$$

FIGURE 1 SIR compartment model for directly transmitted microparasitic disease. This diagram depicts a population containing susceptible hosts (S), infected hosts that can transmit the parasite to others (I), and recovered or immune hosts that are no longer infected (R). Total host population size (N) = S + I + R. Susceptible hosts arise from birth or immunity loss, at per capita rates a and γ, respectively. Individuals leave the susceptible class through natural mortality (rate b) or by acquiring the parasite at rate β after encountering an infected host. Hosts leave the infected category through natural death or disease-induced mortality (rates b and α, respectively) or through recovery (rate ν) to an immune state.

populations are large enough that stochastic processes can be ignored. For diseases for which hosts do not acquire immunity to re-infection (e.g., many plant and insect pathogens and vertebrate diseases such as tuberculosis and brucellosis), the resistant class is eliminated and the equations simplify to a SI model. The simple SIR model in Fig. 1 is useful for parasites with density-dependent transmission, a mass-action process in which transmission increases directly with host population density.

Many complications can be added to the simple compartment model in Fig. 1. For example, a disease may reduce the fecundity of infected hosts or be associated with a long latent period. Age or social structure may complicate among-host contact rates and parasite transfer. In addition, the density-dependent mixing assumed by the equations in Fig. 1 is often inappropriate to describe the transmission dynamics of many pathogens. Other transmission modes, such as venereal, vector, or vertical transmission, can have profound effects on the invasion, persistence, and temporal dynamics of disease, and their consequences have been explored in many theoretical and comparative studies (Getz and Pickering, 1983; Thrall *et al.*, 1993). Additional factors that increase the realism and complexity of host–parasite interactions lead to spatially explicit models to track disease epidemics and stochastic models that simulate disease dynamics in small populations.

Nevertheless, the basic SIR model gives rise to several key principles that characterize interactions between hosts and microparasites and have important consequences for diseases in wild populations. R_o is the basic reproductive ratio of disease and specifies the conditions for which pathogens can increase in prevalence when the disease is initially rare. This is estimated by multiplying the expected number of new infections from a single infected host by the average duration of infectiousness. For the SIR model in Fig. 1,

$$R_o = \frac{\beta S}{\alpha + b + \nu} \qquad (1)$$

This quantity must exceed 1.0 for the disease to invade. The form of Eq. (1) suggests that diseases with high transmission rates (β), low virulence (α), and low host recovery (ν) are most likely to establish in host populations.

The establishment and persistence of many directly transmitted parasites can be related to a population density of susceptible hosts in which the parasite can just maintain itself, often called N_T, or the disease's threshold host population density. Assuming that the population is homogeneously mixed, this value is

$$N_T = \frac{\alpha + b + \nu}{\beta} \qquad (2)$$

Pathogens that are highly virulent (high α) or have lower transmission rates (low β) are likely to require much higher host densities for establishment than those that are highly transmissible and relatively benign. It is important to note that this host density threshold disappears if parasites are transmitted by a frequency-dependent rather than a density-dependent process (Getz and Pickering, 1983). Instead of ever-increasing with host density, frequency-dependent transmission remains constant or saturates at high host densities. This type of transmission characterizes vector- and sexually transmitted diseases, and in theory parasites with frequency-dependent transmission should be able to invade and persist at arbitrarily low host densities.

Microparasitic diseases will regulate or depress total host population size (N) through their effects on host survival or fecundity. If the intrinsic growth rate of uninfected hosts is $r = a - b$, and the prevalence of disease is $y = I/N$ then the change in host population size can be written as

$$\frac{dN}{dt} = (r - \alpha y)N \qquad (3)$$

Equation (3) suggests that the mechanism by which parasites regulate their hosts is through disease-induced mortality (α) compensating for the host's intrinsic growth rate. For example, with a SIR model modified to include density-dependent host regulation in the absence of disease, parasites can reduce host density below the disease-free carrying capacity if β is sufficiently high. If pathogens affect host mortality alone, those with intermediate virulence will depress host density to the greatest degree (Fig. 2) because extremely virulent parasites are likely to kill their hosts before new transmissions occur, and parasites with negligible effects on host survival usually have small population-level effects. In addition, disease prevalence at equilibrium ($y = r/\alpha$) is inversely related to pathogenicity— that is, the more virulent a parasite, the lower its expected prevalence.

Analysis of simple microparasite models can generate important insights for managing and identifying disease risks in wild or captive populations. For example, they suggest that parasites that cause high host mortality in captive or laboratory populations are less

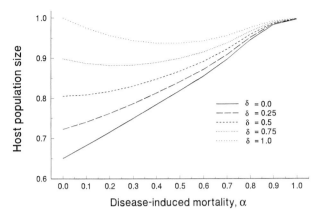

FIGURE 2 Host population size as a function of disease-induced mortality and sterility. The fecundity of infected hosts, δ, ranges from 0 (100% sterility) to 1.0 (no effect of disease). If pathogens affect mortality alone, diseases of intermediate virulence will have the largest impact on host abundance. If pathogens affect host fecundity alone, then diseases with high virulence can continually depress host density. Host population size is measured as proportions relative to the carrying capacity in the absence of disease. Results are derived from a modification of the SIR model in Fig. 4, with density-dependent host mortality in the absence of disease (host mortality = $b_0 + b_1 N$). Parameters used are $a = 1.5$, $b_0 = 0.2$, $b_1 = 0.01$, $\beta = 0.02$, $\nu = 0.2$, and $\gamma = 0.5$.

likely to cause problems in low-density natural populations because infected hosts may die before transmitting the disease. The goal of vaccination and disease eradication programs is usually to drive the number of susceptible hosts in a population below N_T, which in theory leads to extinction of the pathogen. In addition, considerations of threshold densities are important for zoo collections or captive populations in which animal densities are artificially high and the introduction of one infected animal (or the transfer of diseases among species) can devastate captive breeding programs.

It is important to note that these conclusions assume that parasites have a narrow host range and cannot rely on a reservoir host for persistence. Diseases with a wide host range that are relatively benign in reservoir hosts can have severe consequences for endangered or rare species (McCallum and Dobson, 1995). Predictions concerning intermediate virulence also assume that parasites do not affect host fecundity: if parasites decrease host reproduction but have no effect on host survival, then extremely virulent (essentially sterilizing) pathogens can persist at high prevalence and have devastating effects on host abundance (Fig. 2). Finally, these results are contingent on host populations being homogeneously mixed. If hosts are spatially segregated then

invasion thresholds should be considered on a local scale, and if contact rates vary among age, sex, or social classes, then models of disease transmission can be altered to reflect population heterogeneity.

B. Macroparasites

In contrast to microparasites, macroparasites typically cause chronic and persistent infections. Disease severity and the reproductive output of adult macroparasites usually depend on the number of parasites harbored by individual hosts, and a small fraction of the host population may accommodate most of the parasites. Therefore, models of macroparasite dynamics must account for this variation and aggregation in parasite abundance. Fundamental macroparasite models developed by Anderson and May (1991) have been modified by Dobson and Hudson (1992) and others to consider the presence of free-living infective stages, arrested parasite development, and parasites with complex life cycles. These models typically track the density of the entire host population, the abundance of adult parasites within hosts, and the number of free-living parasite stages in the external environment.

As in the microparasite model, host birth and death rates are denoted by a and b (Fig. 3). Adult parasites can induce host sterility and mortality at rates δ and α, respectively: these are per capita rates induced by each individual parasite and assume that overall host death rate increases linearly with parasite burden. Adult parasites give birth to free-living infective stages at rate λ and die as a result of three different processes—parasite background mortality (μ), host background mortality (b), and parasite-induced host mortality (α). Thus, the model assumes that when hosts die, so do their parasites. Free-living egg and larval stages die in the external environment at rate γ and are eaten by hosts at rate β, thus giving rise to new adult infections. The model in Fig. 3 also assumes that parasites are aggregated within hosts according to the negative binomial distribution, where the degree of aggregation varies inversely with k. As indicated by the equations (Fig. 3), the mortality of adult parasites is affected by within-host clustering, with parasite mortality increasing when k is small (and parasites are highly aggregated).

The basic reproductive ratio of macroparasites is the product of the mean number of new infections produced by a single adult parasite and the average life expectancy of adult and larval stages:

$$R_o = \frac{\beta \lambda H}{(\mu + b + \alpha)(\gamma + \beta H)} \qquad (4)$$

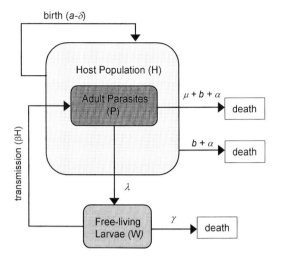

$$\frac{dH}{dt} = (a - b)H - (\alpha + \delta)P$$

$$\frac{dP}{dt} = \beta WH - (\mu + b + \alpha)P - \alpha \frac{P^2}{H}\frac{(k+1)}{k}$$

$$\frac{dW}{dt} = \lambda P - \gamma W - \beta WH$$

FIGURE 3 Schematic representation of host and parasite life cycle for macroparasitic infections. This diagram depicts a host population of size H, collectively harboring an adult parasite population of size P. As in the microparasite model, per capita host birth and death rates are denoted by a and b, respectively. δ and α are the per capita host sterility and mortality rates induced by each individual parasite. Adult parasites give birth at rate λ and die at rates μ (background mortality), b (host mortality), and α (disease-induced mortality). Thus, the model assumes that when hosts die, so do their parasites. Free-living egg and larval stages die at rate γ, and transmission occurs when hosts eat these stages at rate β, thus giving rise to new adult infections.

As with microparasites, Eq. (4) must exceed 1.0 for the parasite to establish when rare. Therefore, parasite invasion and persistence depend strongly on the rate of production of eggs or larval stages (λ), the rate at which parasites are consumed by hosts (β), and the survival of free-living infective stages ($1 - \mu$). The threshold host population necessary to sustain infection is

$$H_T = \frac{\gamma(\mu + b + \alpha)}{\beta(\lambda - (\mu + b + \alpha)} \tag{5}$$

Because the mortality rate of adult and larval parasites is likely to be low (larval parasites often have long-

lived resistant stages and adult worms can live for years within their hosts), and the transmission rate of macroparasites is relatively high (especially when infective stages actively seek out their hosts), macroparasites will likely persist at lower host population densities than will directly transmitted microparasites.

For macroparasite models, the effects of parasite virulence on host thresholds and the ability of parasites to regulate hosts depend on the degree of parasite aggregation and the strength of within-host density dependence. In general, highly aggregated parasite distributions tend to stabilize host–macroparasite interactions, and random or regular parasite distributions tend to destabilize them. The dynamic properties of this model indicate that parasites regulate host density when they have strong negative effects on host survival, provided that within-host density-dependent effects are strong. Parasite-induced sterility, on the other hand, destabilizes the interaction and increases the probability of parasite-induced host population cycles.

Macroparasites can cause conservation problems for several reasons. Parasites adapted to low-density host populations often have high fecundity and long-lived infectious stages. When hosts are restricted in range or achieve high densities (as in zoos and small game parks), parasite burdens can rapidly increase. Macroparasites with negative effects on host fecundity and long-lived infectious stages are also likely to generate dramatic cycles in host abundance over time, causing problems for managing wild populations. Finally, these simple macro- and microparasite models can be used to quantify the impact of parasites and pathogens on host populations, provided that data are available on the effects of disease at the individual level and on patterns of infection in wild populations.

III. DISEASES AND THE MAINTENANCE OF BIODIVERSITY

A. Parasites and the Diversity of Ecological Communities

Although the risks they pose to endangered species are apparent, diseases and other natural enemies can play an important role in maintaining biodiversity and driving successional dynamics. The stability of ecological communities is thought to depend on the number of species they contain and the strengths of interactions between them (Elton, 1958; but see May, 1974). One way that diversity can be maintained at any given trophic level is through ecological tradeoffs between species with respect to factors such as competitive ability and resistance to pathogens and other natural enemies. Therefore, the effects of diseases on individuals or species may facilitate species coexistence and increase the diversity of ecological communities. For example, one explanation for why species diversity in the tropics is so high is that because there are no seasonal climate changes to reduce pest populations, tropical species face greater pressure from parasites and diseases (Connell and Orias, 1964; Gilbert and Hubbell, 1996). Perpetual parasite populations may prevent any single species or group of species from dominating communities, allowing many species to coexist at relatively low densities. If more diverse and complex communities are more stable, then the goal of conservation should be not only to increase the abundance of rare or threatened species but also to maintain a complex and diverse array of interspecies interactions.

Clarifying the role of pathogens in community structure requires extending models of host–parasite dynamics to multi-species systems (Fig. 4). A range of mathematical models have been developed that explore the roles of pathogens in modifying interactions among species (Holt and Pickering, 1985). These systems fall into several classes: (i) a single pathogen species infecting multiple, potentially competing hosts; (ii) a host species with several natural enemies, at least one of which is a pathogen or parasite; or (iii) tri-trophic interactions in which the consumers (herbivores or predators) are susceptible to parasitism or disease. Theoretical studies of host–parasite interactions involving more

than two species can yield complex dynamical outcomes and often support the proverb that "my enemy's enemy is also my friend" (Dobson and Crawley, 1994). In the case of host–host–pathogen systems, parasites can reverse the outcome of competition between species sharing the same resource if the dominant competitor is more susceptible to infection. Apparent competition is a related phenomenon whereby two or more hosts not directly competing for resources are affected by the same parasite to different degrees (Hudson and Greenman, 1998). In this case, generalist pathogens that are relatively benign to one host species may depress the density of other hosts for which they are more pathogenic. Diseases that are transmitted via vectors or long-lived infective stages are more likely to play a role in mediating host competition due to increased transmission opportunities among species. Although generalist pathogens can decrease diversity by threatening rare or endangered species, specialist diseases are more likely to facilitate coexistence by attacking common species that would otherwise dominate communities.

Several empirical observations illustrate the role of pathogens in determining plant and animal community structure and modifying ecosystems. Parasite-mediated competition has been shown to allow coexistence between *Anolis* lizard species on the Caribbean island of St. Maarten and in laboratory studies reversed the outcome of competition between two species of *Tribolium* beetles. Introduced pathogens can dramatically alter the species composition of ecological communities, particularly when they cause epidemics in dominant species. For example, fungal blights that removed chestnut trees from eastern deciduous forests of the United States and elm trees from much of western Europe favored less competitive species or those from earlier successional stages. Soil-borne diseases have been implicated in the rate and direction of succession in plant communities and in determining patterns of seedling recruitment in tropical rain forests. Fungal endophytes of grasses that limit seed production may also be responsible for enhancing the survival and dominance of tall fescue in open-field communities.

Pathogens that attack herbivore species can in turn produce major changes in plant recruitment and abundance (Dobson and Crawley, 1994) and in the density of predators and other natural enemies (Dobson and Hudson, 1986). One example is furnished by the myxoma virus epidemic in rabbits in south England. Although a high abundance of rabbits in the mid-1900s prevented the regeneration of woody plants in grassland habitats, myxoma virus (introduced in the 1950s) led

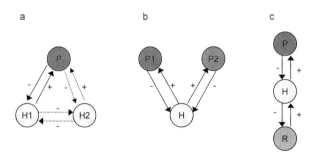

FIGURE 4 Diagrammatic illustrations of host–parasite assemblages involving more than two species. (a) Host–host pathogen interactions (pathogen infecting multiple host species within the same trophic level). Hosts may or may not be directly competing for resources, and one or both species may be affected by the pathogen. (b) Host–pathogen–predator interactions (host is susceptible to a pathogen or parasite in addition to other natural enemies). (c) Resource–consumer–parasite interactions (pathogen affects predator or other consumer feeding on one or several types of prey).

to a scarcity of rabbits for the next 15 years. In areas where rabbit grazing had been so intense as to prevent tree establishment, a cohort of oak seedlings grew into forests following the initial epidemic (Dobson and Crawley, 1994). Similar cases can be found in east Africa, where viral and bacterial pathogens of herbivores have caused changes in herbivore abundance and radically altered the structure of plant communities. Although these examples are cases in which diseases have generated striking changes in community structure, a vast majority of host–parasite interactions are likely to yield more subtle effects on host growth, survival, or fecundity that underlie the persistence of species and the assembly of ecological communities. When preserving intact and functioning ecosystems is the goal of conservation strategies, attention must focus on the structure of parasite communities and the role they play in maintaining biodiversity.

B. Pathogens as Biological Control Agents

Another more optimistic side to parasites and pathogens is that they can be used as biological control agents to regulate the densities of introduced or pest species that pose major threats to endangered species or ecosystems. Alternate strategies for the eradication of exotic organisms usually involve trapping and culling, building exclosures, or using chemical agents to attract, repel, or disrupt the reproductive biology of invasive species. However, these strategies can be time-consuming, expensive, and labor-intensive. Biological control strategies involving parasites and other pathogens have been proposed and successfully implemented for several vertebrate pest species and may offer long-term regulation with minimal input of labor or resources. For example, viral pathogens (myxomatosis and calicivirus) have been successfully employed against rabbit populations in Australia, and feral cats have been dramatically reduced on the sub-Antarctic Marion Island by a combination of feline parvovirus and culling. Microbial pathogens have also been proposed as biocontrol agents to regulate brown tree snakes on the Pacific island of Guam and fire ants in North America. Characteristics of promising biological control agents include a high specificity for the target host species, high R_0 (determined by high transmission rates or low host recovery), and the ability to retain infectivity for long periods of time within hosts or in the external environment. Epidemiological models also suggest that pathogens with large effects on host fecundity (δ) or moderate effects on host mortality (α) will provide the greatest degree of sustainable host regulation (Fig. 2). Clearly,

the introduction of novel pathogens to island and other ecosystems requires careful feasibility studies to ensure the pathogens will not infect native species or escape to other areas and cause devastating epidemics. Therefore, this method of pest control should probably be restricted to oceanic islands to avoid the escape of novel pathogens into mainland ecosystems (Dobson, 1988).

C. Disease and Host Genetic Diversity

In many ways, the maintenance of genetic polymorphism within species parallels the coexistence of multiple species in ecological communities and will presumably enable hosts to evolve in response to changing environments and resist ecological invasions and perturbations. Coevolution between hosts and parasites is likely to generate host genetic diversity and may provide selective advantages for host recombination and sexual reproduction (Lively and Apanius, 1995). Although examples from wild animal populations are rare (Read *et al.*, 1995), parasites infecting Soay sheep, snails, and insects have been implicated in the maintenance of allelic diversity or sexual recombination in their hosts. In contrast, one of the most distinctive features of natural plant populations is the staggering abundance of genetic polymorphism for resistance to fungal diseases.

Models of host–parasite coevolution demonstrate that advantages held by rare alleles can lead to the maintenance of genetic variation in host populations. These models are based on reproductive costs for hosts that acquire infection (or parasites that "attempt" to infect incompatible host genotypes), leading to cycles in both host and parasite allelic frequencies. In this way, frequency-dependent selection favors rare host genotypes that are resistant to diseases attacking the majority of individuals. Modern agriculture and animal husbandry provide examples of how parasites exploit monocultures or genetically homogeneous host populations. The phenomenon of parasites tracking common host genotypes is also critical for arguments concerning the role of parasites in generating advantages to genetic recombination through sexual reproduction (Lively and Apanius, 1995). Key insights from models of agonistic coevolution are that genetic diversity can be maintained even in the absence of stable equilibria, and that genetic polymorphism is more likely when fitness differences among host genotypes are small rather than large.

Parasites are likely to be powerful selective agents in natural populations, and host species that are continually exposed to a diverse array of parasites should harbor a variety of resistance alleles or repertoire of inducible defenses. The major histocompatibility locus

(MHC) in vertebrates provides a notable example of extreme polymorphism and diversity of rare alleles (Nei and Hughes, 1991). Polymorphism at this locus determines the variety of foreign antigens the host immune system can recognize and attack, and it may be the ultimate response to selection in the face of unpredictable and frequent disease outbreaks. Gene-for-gene co-evolution between plants and pathogens has also been demonstrated to promote a high diversity of resistance and virulence alleles. For example, long-term field studies of the interaction between wild flax and flax rust in natural populations in Australia indicate that many alleles can persist among metapopulations, and the distribution of genotypes can change rapidly during individual epidemics (Burdon and Jarosz, 1991). From a broader perspective, coevolution between hosts and parasites may be a major force determining the earth's biodiversity. Conservation strategies that fail to recognize this potential and restrict disease spread may ultimately rob host populations of genetic diversity needed to respond to future ecological changes.

IV. DISEASES AS THREATS TO BIOLOGICAL DIVERSITY

A. Introduced Pathogens and Species Declines

Exotic diseases and parasites are increasingly recognized as important factors driving population declines, geographic range contractions, or even terminal extinctions in many plant and animal species. Although native parasites can play important ecological roles in natural ecosystems, a great number of introduced or exotic pathogens are directly implicated in catastrophic declines of rare or endangered species. In such "virgin soil epidemics," which occur after new pathogens are introduced into previously unexposed host populations, the disease progresses rapidly through immunologically naive hosts and can cause mass mortalities. Because of the high rate of spread and the potential to devastate a host population, parasitic organisms are now considered the greatest threats to the survival of some endangered species (Primack, 1998).

Parasitic organisms have been shown to impact small populations in a variety of ways. In the most direct manner, they can reduce population sizes below a threshold necessary for maintenance and growth. This can happen through direct disease mortality or indirect mortality due to reduced competitive ability or increased susceptibility to predation. For example, red

grouse in Scotland that carry heavy infestations of the nematode *Trichostrongylus tenuis* are more susceptible to predation by red foxes (Hudson *et al.*, 1992). In very small populations, differential mortality between male and female hosts can sufficiently distort sex ratios and hamper future reproduction. In the long term, host populations infected with a disease tend to be far less stable than otherwise similar (but uninfected) populations (Dobson and Hudson, 1992). This is of great conservation importance because fluctuating host populations are more likely to decrease to low numbers, at which point they are susceptible to extinction from other factors.

In recent years, a plethora of case studies have documented the catastrophic effects of introduced parasites in small host populations (Table II). In general, because simple ecosystems are more susceptible to invasion by new diseases, many examples come from island ecosystems. One of the best known examples is the introduction of both the mosquito vector and the exotic malarial parasite *Plasmodium relictum* into the Hawaiian archipelago. Although most non-native birds were apparently resistant to this infection, this parasite was extremely pathogenic to native bird species. The epidemics that raged through the Hawaiian archipelago caused massive mortalities in the native birds, essentially clearing the lower island elevations of the native avifauna and contributing to the extinctions of several endemic bird species (Van Riper *et al.*, 1986). On other island ecosystems, the extinction of native small mammals (such as three species of rats and shrew on Christmas Island) has been linked to an unidentified disease apparently derived from non-native rats. Furthermore, there is evidence that the thylacine, the largest marsupial carnivore in modern Oceania, went extinct in the early twentieth century from a combination of overhunting and a disease epidemic of unknown origin.

In addition to single species extinctions, introduced epidemics can generate secondary effects that ripple through natural communities and, in severe cases, trigger secondary species extinctions termed "extinction cascades." Several examples of such community-wide effects have been reported in the literature from both marine and terrestrial ecosystems. For example, the pathogen-caused disappearance of long-spined sea urchins (*Diadema antillarum*) throughout the Caribbean Sea triggered numerous secondary community changes and in some cases caused a shift from coral- to algae-dominated reef communities (Harvell *et al.*, 1999). In a similar manner, the spread of rinderpest (caused by an introduced morbillivirus) through sub-Saharan Africa not only devastated ungulate populations but also generated complicated shifts in the communities of both plants and mammalian predators occurring in the same area. Furthermore, after the chestnut blight (caused by the introduced fungal pathogen *Endothia parasitica*) led to the virtual disappearance of the American chestnut, eight species of Lepidopterans feeding predominantly on this tree species also succumbed to extinction (Meffe and Carroll, 1997).

New and potentially dangerous pathogens can be introduced into wildlife populations in several ways. Shifts in the geographic distributions of species can result in two previously separated host species contacting each other and exchanging new diseases. Epidemics of introduced pathogens can also result from contact between domestic species and wildlife (Table II). Finally, new diseases may be introduced into the wild by captive-bred hosts that acquired an infection while in human care (Table III).

It is important to keep in mind that for most species,

endangered populations are not likely to sustain any pathogens in the long run simply because they are too small. This is particularly true for either specialized or highly virulent parasites. In fact, it is often generalist (i.e., nonspecific) parasite species that are of greatest conservation concern because they can survive for long periods in alternate host species. Past experience suggests that it is mostly directly transmitted microparasites that cause most severe problems for endangered species. Nonetheless, more research is needed to understand the potential impacts of parasites with different life cycles or natural history requirements on small host populations.

B. Endemic Diseases

In addition to introduced pathogens, conservation biologists regularly face outbreaks of parasites that are well

TABLE II

Selected Disease Outbreaks in Natural Populations

Host species	Parasite/pathogen	Location	Comments
Plants			
American chestnut (*Castanea dentata*)	Chestnut blight (*Endothia parasitica*)	Eastern North America	Pathogen introduced by ornamental trees
Flowering dogwood (*Cornus florida*)	*Discula destructiva* (fungus)	Eastern North America	Introduced pathogen
Several native plant species	*Phytophthora cinnamoni* (fungus)	Western Australia	Responsible for large-scale diebacks and permanent plant community shifts
American elm (*Ulmus americana*)	Dutch elm disease *Ceratocystis ulmi* (fungus)	North America	Introduced pathogen
Invertebrates			
Long-spined sea urchin (*Diadema antillarum*)	Unidentified	Caribbean Sea	Epidemic spread across the entire Caribbean Sea
Fish			
Rainbow trout, salmon (*Salmo*)	*Myxobolus cerebralis* (Protozoa; whirling disease)	Montana	Introduced with stocked fish
Aral Sea sturgeon (*Acipenser nudiventis*)	*Nitzschia sturionis* (Monogenean)	Aral Sea, former USSR	Introduced with stocked Caspian sturgeon
Amphibians			
Various species of frogs and toads	*Batrachochytrium dendrobatidis* (Fungus)	Australia, Central America, Arizona, Sierra Nevada	Highly lethal pathogen; additional contributing stressors involved
Reptiles			
Desert tortoise (*Gopherus agassizii*)	Upper respiratory tract syndrome	Mojave Desert	Introduction through released pets
Birds			
Hawaiian honeycreepers	*Plasmodium relictum capistranoae*	Hawaii	Implicated in the extinction of several Hawaiian bird species
Various species of ducks and geese	Duck plague (herpesvirus)	North America	Introduced with domestic waterfowl

TABLE III

Examples of Diseases-Related Problems in Captive Breeding Programs

Species	Pathogen	Comments
Elephants (both African and Asian) (*Loxodonta* and *Elephas*)	Herpesvirus	Cross-species infection between the two taxa
Black-footed ferrets (*Mustela nigripes*)	Canine distemper virus (morbillivirus)	High sensitivity to disease
Cranes (several species)	Herpesvirus	Outbreak at the International Crane Foundation facilities in Wisconsin
Mauritius pink pigeon (*Columba mayeri*)	Herpesvirus	Infected from domestic pigeons (foster parents)
Bali mynah	*Isospora* sp. (Coccidia)	Reduces reproduction in captivity
Various species of primates and large cats	Bovine tuberculosis (*Mycobacterium bovis*)	Pathogen infects wide range of mammals
Nene (*Anser sandwitchensis*)	*Mycobacterium*	Captive-bred birds cannot be released into the wild because of infection

established in a regional wildlife population but may erupt because of unusual environmental conditions. As discussed earlier, the occurrence of disease depends not only on the characteristics of both the parasite and the host but also on the prevailing environmental circumstances. As a result, specific changes in environmental conditions can precipitate disease outbreaks in wildlife populations. This may occur either because the new conditions increase host population density above the threshold needed for rapid spread of the pathogen or because they directly favor parasite transmission. Hence, such conditions could either enhance the survival of infective stages between hosts or stress hosts to the point that they become particularly susceptible to infection. A good example in which environmental conditions caused increased host population size and eventually a disease epidemic is the outbreak of the Sin Nombre virus in the deer mouse populations in the American Southwest. In the early 1990s, El Niño–Southern Oscillation event (ENSO) associated rains led to an explosive increase in the rodent populations in Arizona and New Mexico. These enormously high population densities triggered an epidemic of the Sin Nombre virus in deer mice that eventually spilled over into human populations. In contrast, apparently stable frog and toad populations in the American West suffered from previously established pathogens such as *Aeromonas hydrophila* (redleg disease) and the *Saprolegnia* fungus after some new environmental stressors (perhaps in the form of increased UV radiation) appeared in the area (Meffe and Caroll, 1997).

C. Concerns for Captive Breeding Programs

Conservation efforts rely increasingly on captive breeding programs to augment and restore free-living populations. Because captive animals are particularly susceptible to infections, disease has emerged as a serious threat to these programs, and pathogens have become one of the principal concerns of captive breeding managers. Disease outbreaks are now being taken seriously not only because of their unpredictable nature and the potentially catastrophic loss of breeding stock but also because of increased costs of screening and treatment, concerns for housing and animal care, and the erosion of public support in the event of outbreaks. In addition, diseases propagated in captive animals, when introduced into the wild with released hosts, can jeopardize the health and demography of the very same populations the programs are trying to save.

Captive animals may acquire novel infections from a wide range of infected hosts, including unrelated species kept in the same pen, wildlife living around outdoor cages, foster parents, or individuals from the same species but different subspecies. For example, captive African elephants kept in mixed collections have been infected with a lethal herpesvirus that occurs without disease symptoms in their Asian elephant pen mates. Furthermore, many captive-bred hatchlings of the endangered Mauritius pink pigeon contracted and succumbed to a herpesvirus infection that their foster parents (domestic rock doves) were carrying without ill effects (Primack, 1998). Finally, humans pose a serious threat to many captive animals. Several human diseases, such as measles, tuberculosis, or influenza, are highly virulent for wildlife; because of the large numbers of visitors that pass through zoos every year, captive animals may become exposed to a very large pool of potentially infected humans. For example, captive-bred orangutans that were infected with tuberculosis from humans could not be released into the wild due to fear of introducing the pathogen into the wild population.

Many animals in captive breeding programs are often held close together, a practice that poses two disease-related risks. First, animals are likely to be stressed and hence more susceptible to infection (particularly those that are territorial or normally persist at low densities). Second, crowding in pens or cages can elevate host densities above the threshold necessary for virulent pathogens to invade and will also increase transmission rates (e.g., hosts may reinfect themselves by ingesting the eggs of their own parasitic nematodes released into their pen). Interestingly, parasites with complex life cycles may be less of a concern to captive breeding programs because of the likely absence of intermediate hosts (or vectors) that are necessary for transmission.

Finally, additional complications exacerbating disease problems in zoos stem from inbreeding depression, or the genetic impoverishment of a captive colony due to loss of diversity and the expression of deleterious recessive alleles. This loss of genetic variability leads to homogeneous captive populations that can be very susceptible to a variety of pathogens. Such loss of genetic heterozygosity has been implicated in the high mortality that captive cheetah populations experienced due to a feline infectious peritonitis virus (Soulé, 1986). Hence, genetic and ecological problems can operate in synergy to reduce population size and diminish heterozygosity, leading populations toward increased disease susceptibility and possible extinction. In conclusion, as captive breeding programs expand, disease-related problems are likely to become even more prevalent than they are now. Because of their unpredictability and their potential to cause great harm, serious precautions are needed to prevent future disasters.

V. INTERACTIONS BETWEEN DISEASE, WILDLIFE, AND HUMAN ACTIVITIES

Wildlife disease outbreaks are usually directly or indirectly facilitated by human activities. In particular, economic development in conjunction with social pro-

cesses can influence disease emergence through changes in land use or demographic patterns. These activities can lead to quantitative habitat changes (i.e., causing direct habitat destruction or fragmentation) or qualitative changes (such as pollution or other forms of habitat degradation).

A. Quantitative Changes in Natural Habitats

Humans destroy natural habitats in a plethora of ways, many of which have been summarized under the term habitat fragmentation. Habitat fragmentation is the process of reduction and subsequent isolation of a continuous natural habitat into smaller patches; today, it is an extremely widespread phenomenon affecting every major terrestrial ecosystem. Habitat fragmentation is of great conservation importance because it affects native communities both directly (by removing individuals and their resources) and indirectly (through changes in community composition and interactions). Fragmentation of a continuous habitat can affect the spread of parasitic organisms in a variety of ways, all of which depend on specific community characteristics or idiosyncrasies of individual host–parasite systems.

Habitat fragmentation results in both reduction and subdivision of a host population. Depending on the degree of isolation between the different habitat patches, parasites or pathogens may infect some but not all host subpopulations. As a result, many of these subpopulations escape a disease epidemic that can devastate other infected patches. Furthermore, if host subpopulations are relatively small they may fall below the critical host threshold required for disease persistence (N_T). Consequently, one possible result of habitat fragmentation may be the decline or extinction of a parasitic species.

Habitat fragmentation will also increase the contact between endangered native hosts and other domesticated and wild species living in adjacent habitats, facilitating the introduction of new diseases into previously unexposed populations. As human activities degrade continuous natural habitat into progressively smaller habitat fragments, the percentage of habitat edge—defined as the habitat close to the border with the disturbed matrix—steadily increases. As a result, increasingly more individuals of an endangered species living in this habitat are likely to encounter other host species living along the edge of this matrix. The increased likelihood of such chance encounters also raises the probability of cross-species infection. Indeed, there is evidence that in many cases host switching between domestic and wild animals is facilitated by habitat fragmentation and penetration by exotic disease-carrying hosts. For example, such degradation appears to have preceded the spillover of rabies and canine distemper virus from domestic dogs to African wild dogs and Ethiopian wolf populations.

Fragmentation-caused disease problems are further exacerbated by wildlife crowding. As increasingly more habitat is destroyed, surviving native hosts will increasingly move into the last few remaining patches. Such crowding is particularly severe in waterfowl populations, which in many areas of the world are forced to overwinter or migrate through wetland areas greatly reduced by human activities. Such large numbers of birds not only degrade vegetation cover and water quality but also compete for reduced food resources and may hence become nutritionally stressed. High bird densities also increase the probabilities of disease transmission through either direct contact or fecal contamination and can further facilitate a disease epidemic. In line with this argument, wildlife managers have seen massive outbreaks of disease in crowded waterfowl populations during the past several years.

B. Qualitative Changes and Pollution

Degradation of natural habitats takes many forms and has frequently been associated with outbreaks of new diseases in both wildlife and humans. Human assaults on natural ecosystems as varied as logging, changes in vegetation cover, overexploitation of game and fisheries, erosion, and pollution can trigger pathogen epidemics. Direct injury due to logging and other anthropogenic factors can cause increased levels of disease in native plant communities (Gilbert and Hubbell, 1996). Many scientists suggest that the widespread 1988 phocine distemper epidemic that spread through the seal and small cetacean populations of the North Sea can also be attributed to human activities. Depletion of the fish stocks in the North Atlantic ocean because of commercial overfishing forced harp seal populations carrying the virus to move southward; there, they infected immunologically naive harbor seals which were subsequently ravaged by this epidemic (Grenfell and Dobson, 1995).

Pollution from anthropogenic sources is frequently associated with outbreaks of disease in natural ecosystems. Because pollutants are important stressors for many species of wild animals and plants, they can shift the dynamic balance between the immune defenses of the host and an invading parasite and determine whether wild animals and plants become infected and

ultimately die during the course of an epidemic. For example, in aquatic ecosystems, eutrophication (pollution with various nutrients) and the resulting stress on the natural communities can trigger disease outbreaks in native species. Furthermore, high nutrient input may lead to algal and dinoflagellate blooms that promote the survival and proliferation of various marine microparasites. Across the eastern seaboard of the United States and in the Gulf of Mexico, increased eutrophication due to excess runoff from agricultural fertilizers has created conditions leading to outbreaks of *Pfiesteria piscicida,* an aggressive protozoan that attacks and kills large numbers of fish. In addition to the obvious economic and conservation ramifications of such massive fish die-offs, *Pfiestaria* also poses direct human risks, such as skin lesions, memory loss, and respiratory problems.

In freshwater ecosystems, the spread and duration of avian cholera outbreaks in waterfowl populations is greatly facilitated by pollution from nutrient-rich runoff into wetlands. This disease, caused by the introduced bacterium *Pasteurella multocida,* has become the second most important cause of waterfowl die-offs in North America. Increased turbidity and high organic content in a wetland enhance pathogen survival and extend the duration of epidemics. In a similar manner, freshwater habitat degradation due to polluted runoff from intensive agriculture or animal farming practices appear to be associated with disease outbreaks in amphibian populations in the continental United States. Hence, pollution and habitat degradation should be viewed as contributing both directly and indirectly (i.e., through increased incidence of disease) to the worldwide decline of amphibian populations.

C. Global Climate Change and Disease

There is increasing evidence that the changes wrought by human activities on global climate are affecting the distribution, prevalence, and severity of pathogenic organisms in natural ecosystems. A growing body of research demonstrates that the infusion of anthropogenic greenhouse gases such as methane, CFCs, and CO_2 is changing both temperature and precipitation patterns throughout the planet. Because the distribution and prevalence of parasites often depends crucially on environmental conditions, it is not surprising that changing weather patterns affect the prevalence of wildlife diseases.

The best documented examples of such climate–disease interactions come from marine ecosystems. Increased water temperatures have been shown to trigger coral bleaching (expulsion of symbiotic algae from the coral) and a suite of associated opportunistic infections that have led to massive coral die-offs (Harvell *et al.*, 1999). Although such bleaching events can occur naturally in association with ENSO events, both their frequency and severity have increased during the past 20 years. Because such die-offs have profound community consequences in coral reef ecosystems (because affected reefs take many years to recover), this has become an issue of great conservation importance.

The link between global climate change and animal pathogens is less well documented for terrestrial systems. Although evidence exists that the expanding distribution of some human diseases (such as malaria and dengue) is tied to recent warming trends, much less is known about animal parasites. In the human transmission cycle, the distributions of both the *Plasmodium* parasites and their *Anopheles* mosquito vectors are limited by low environmental temperatures. As a result, malaria does not exist in high latitudes or at high elevations in the tropics. In recent years, however, malaria has been occurring at progressively higher elevations, potentially because of warmer ambient temperatures (Patz, 1997). Similar advances of animal diseases (such as the previously discussed *P. relictum*) could presumably spread into high-elevation refugia such as the mountains of Hawaii, leading to the extinction of many endangered bird species surviving there.

In addition to changes in temperature and precipitation regimes throughout the world, greenhouse gases also contribute to the depletion of the ozone layer and a concomitant increase in harmful UV radiation. Such radiation can affect negatively many species of vertebrates and may facilitate the spread of disease. There is indeed evidence suggesting that such increases impact amphibian populations by raising the susceptibility of egg masses to fungal infections (Kiesecker and Blaustein, 1995). In summary, there is accumulating evidence that global climate change will emerge as an important factor facilitating the spread of animal pathogens. Therefore, exploring the potential consequences of future changes in global climate should be an important future research goal for scientists and wildlife managers involved in conservation.

D. Accidental Introductions of Disease via Worldwide Commerce and Travel

Worldwide commerce and animal trafficking are probably the most important causes of new disease outbreaks in wildlife populations. This traffic, which has dramatically increased in the past few decades, takes a variety of forms. It includes international trade of live animal and plant stock for commercial breeding purposes, for zoos and animal parks, for the pet trade and hunting, and for laboratory research. The situation is further exacerbated by the staggering amounts of international transport of fruit, vegetables, and various animal parts, all of which facilitate the spread of pathogens. For example, most of the major pathogens that cause disease epidemics in wild birds in North America were probably introduced by humans. Both avian cholera (a highly virulent disease caused by the bacterium *Pasteurella multocida*) and duck plague (caused by a herpesvirus) appeared first in North American domestic birds and spread to wild waterfowl. Today these diseases are responsible for the deaths of tens of thousands of wild waterfowl each year in the United States alone. Similarly, avian malaria was introduced to Hawaii with imported birds, and brucellosis (caused by the bacterium *Brucella abortus*) was introduced to North America by domestic livestock (and has since spread to wild deer, elk, and buffalo populations). Although brucellosis is controlled in domestic livestock today, it is of great conservation concern for large ungulates in the greater Yellowstone area.

The spread of disease into wildlife populations is further facilitated by the continual breakdown of barriers between livestock and wild animals. For example, today there is a wide spectrum of waterfowl populations, all with varying degrees of dependency on humans (such as traditionally farmed animals, captive collections of exotic waterfowl, birds bred in game farms and released into the wild, feral birds, nonmigratory birds, and fully migratory wild birds). Many of these populations transfer pathogens as a result of their frequent contact with each other.

Finally, international travel and tourism has emerged as an important factor that facilitates the spread of pathogens and their vectors into new populations. The constant shuttling of infected humans between different continents can potentially allow parasites to spread throughout the world in a matter of days. Although this point is not lost to epidemiologists dealing with human infections, it probably needs to be taken more seriously by both ecologists and conservation biologists.

VI. MANAGING DISEASES AND BIODIVERSITY IN THE FUTURE

A. Preventing Disease Outbreaks

Pathogens that pose significant threats to conservation programs were discussed in detail earlier. In summary, the sources of these diseases can be divided into two categories. The overwhelming majority of epidemics begin with the introduction of a disease into a naïve population, either via captive-bred individuals carrying a pathogen or via the invasion of a host species into new areas. Also of concern are environmental changes that affect host physiology and shift relationships with endemic parasites. Anthropogenic effects on global and local environments (such as habitat loss and fragmentation) and predicted changes in global temperature and weather patterns will have a cascade of consequences for wild populations, including the threats from diseases.

To prevent future disease outbreaks it is imperative to monitor the sources of disease introduction. Disease threats to wild and agricultural populations are managed through screening imported animals, plants, and products made from their tissues. Frequently, wild and domestic animals must undergo quarantine before release into native populations. Furthermore, disease screening is increasingly included in captive breeding programs to prevent the inadvertent introduction of infected individuals into a susceptible population. Ultimately, the majority of disease-related threats to conservation arise from human-related disturbances, and it is difficult to find an example of a disease-induced wildlife crisis in which human activity is not the origin of the problem.

B. Monitoring Populations for Infection

Most diseases are discovered after epidemics have already spread through wild populations. A powerful tool in the management of diseases is thus to monitor threatened populations for signs of infection before overt mortality occurs. Several such tools exist to both monitor unaffected populations and track the spread of an epidemic. The efficacy of a monitoring program will increase with the number of host animals included in the screening.

Common sources of information that can be used to monitor parasite prevalence and intensity in animals are blood, tissue, and fecal samples. The presence of blood parasites (e.g., trypanosomes, malaria, and filarial nematodes), anemia, elevated leukocyte levels, and pathogen-specific antibodies can all be detected from relatively small volumes of blood. Antibody assays (e.g., enzyme-linked immunosorbent assay) can show evidence of both current and past infection. Many microbial pathogens can be cultured directly from tissue showing signs of infection or from swabs taken from the mouth, ears, eyes, nose, genitals, or anus. Scans of feces will also provide information on the diversity of species of intestinal parasites in an individual as well as the number of worms per host or intensity of infection.

Hosts that are hunted or culled for other purposes are frequently examined for internal and external parasites. Parasitic arthropods (e.g., ticks, mites, fleas, and biting dipterans) are also monitored because they can decrease the condition of their hosts by drawing on resources. High parasite burdens can indicate a decrease in immune function due to other causes, and arthropods should be examined for internal microparasites because they often serve as vectors for other diseases. For both internal and external macroparasites, the distribution of parasites among hosts is particularly critical in assessing the impact of infection on individual hosts and at the population level.

Molecular techniques have recently been employed as methods for screening free-living populations with surprising results. For example, polymerase chain reaction tests can reveal the presence of infection in host blood or tissue by identifying pieces of pathogen genetic material. Such a study on native Hawaiian birds indicated that the prevalence of avian malaria was higher than original estimates based on microscopic examination of blood (Feldman et al., 1995).

C. Assessing Disease Threats

Screening programs will verify the presence or absence of a particular pathogen in a wild population. However, this information alone does not provide a definitive indication of severe population consequences resulting from infection. Epidemiological models discussed earlier indicate that this determination can only be made when information on both the prevalence of a disease (y) and its effects on individual hosts (α and δ) have been collected. In estimating prevalence, care should be taken to sample a large number of hosts throughout their entire range. To determine the effects of disease on individual hosts, postmortem examinations are often performed. However, the presence of a pathogen in dead or dying animals does not necessarily indicate the population-level impact of the disease, nor does it offer a measure of disease effects on host life span or fecundity. Ideally, captive or wild hosts should be monitored throughout the course of infection to compare survival and fecundity between infected and uninfected hosts. Experimental manipulation of parasite loads in natural populations remains the most direct way of assessing the effects of pathogens on host populations (McCallum and Dobson, 1995).

Epidemiological models suggest that if parasites affect host mortality alone, then those of intermediate virulence are the most likely to pose problems for wild populations (Fig. 2). These models also predict that a highly virulent pathogen may be maintained at a high prevalence in a host population if it infects multiple host species and is more benign in some hosts than in others. For example, in the late nineteenth century, Rinderpest devastated much of the native ungulate fauna of sub-Saharan Africa because it persisted in domestic cattle as a less virulent infection. Essentially, this notion explains most disease introductions into managed populations because the carrier host remains asymptomatic and undetected. Therefore, complete screening efforts and disease management plans should include the determination of the infection status and disease history of closely related taxa that overlap in range.

It is worth noting that correlations have been demonstrated between the degree of inbreeding in individuals (indicated by high levels of homozygosity or from known pedigrees) and the likelihood of contracting a serious infection. Given this association, measuring inbreeding coefficients as a part of population screening and avoiding inbreeding in captive populations may restrict future epidemics by increasing the genetic variation in managed populations.

D. Intervention Methods

Historically, diseases in wild populations have drawn the attention of wildlife managers only after an epidemic severely threatens the host population or when the disease is perceived at a threat to valuable agricultural crops or livestock. In general, the types of management regimes used to limit disease spread vary depending on the type of pathogen, the threat to the host population or to other neighboring species, available financial resources, and the existence of agricultural or veterinary information and tools. The use of veterinary and agricultural designed tools to treat domestic species is particularly important because very little information exists on disease treatment in native wildlife and plants. However, direct application of domestic animal medicine on wildlife can have disastrous effects, as illustrated by the mortality caused by a canine distemper vaccine given to endangered black-footed ferrets. Some of the more common intervention methods available to conservation biologists are discussed in the following sections.

1. Vaccination

Viral infections and, less frequently, bacterial infections are most effectively controlled by vaccinations. However, the cost of vaccinations may severely limit the success of population-wide disease control plans. Programs intended to eradicate or control the spread of a virus must consider basic epidemiological principles and models in determining where to concentrate efforts and what proportion of the population to target to limit or eradicate the disease. Both oral vaccines and intramuscular injections have been used in wild populations. Vaccinations have been administered to prevent the spread of rabies in European foxes, and vaccination of domesticated reservoir hosts (cattle and domestic dogs) has reduced the threats of morbillivirus outbreaks in wild ungulates and carnivores in east Africa.

2. Chemotherapy

Drugs are most frequently administered for bacterial, fungal, helminth, and ectoparasitic infections. This method of disease control least effectively addresses the ultimate cause of a disease and may be extremely costly for population-wide control measures (e.g., systemic fungicides to counter tree blights). Interestingly, treatment of helminth infections in red grouse in Scotland and in Soay sheep in St. Kilda reduces dramatic host population cycling. A more stable population may be easier to manage; therefore, the treatment of endemic macroparasite infections can have great conservation value.

3. Culling

The removal of infected individuals by lethal means is analogous to intensifying parasite-induced mortality (α; Fig. 1). This strategy can effectively decrease disease prevalence and lower R_0, and it has been implemented to counter the spread of certain tree diseases (e.g., Dutch elm disease). Although less frequently applied to vertebrate populations, Cape buffalo in some South African national parks are screened for tuberculosis and killed if infected. The culling of individuals of a host species, regardless of infection status, has also been suggested as a means for reducing the numbers of susceptible hosts in population below the threshold required for parasite persistence.

4. Quarantine

The costly removal and care of infected individuals from a susceptible population has been carried out in extreme circumstances. For example, when host populations are dangerously low in numbers, then every member may be of great value and worth rehabilitating. This was the case during a distemper epidemic in Wyoming black-footed ferrets, and a similar situation may arise if rabies invades endangered Ethiopian wolf or African wild dog populations.

5. Biological Control

The use of one pathogen to attack another is certainly a controversial measure and should be included in management regimes with much caution. There are instances in which natural parasites of pathogenic organisms have been proposed as potential tools in slowing or controlling the spread of infection. For example, hypovirulence in the fungal pathogen responsible for chestnut blight, *Cryphonectria (Endothia) parasitica,* is caused by a virus that is transmitted via direct hyphal contact. This virus debilitates the pathogen and allows chestnut trees to counter and recover from fungal infections.

6. Hybridization

In cases in which variability in pathogen resistance in a host population is negligible, management plans may resort to crossing highly susceptible species with more resistant organisms. For example, crossing black-footed ferrets with domestic European ferrets has been proposed to decrease susceptibility to viral pathogens. Hybrids of American elm trees and Eurasian elm species (which are resistant to Dutch elm disease) are already available on the commercial market.

7. Habitat Alteration

Management regimes that address the ultimate causes of disease outbreaks have the greatest potential for removing disease threats but are also the most difficult to implement. Indeed, the ultimate causes of marine invertebrate diseases, namely, pollution and ocean temperature changes, are so global and diffuse in origin as to be impossible to confront in any single species management plan. However, alterations in terrestrial habitats have been suggested to alleviate disease problems in conservation. Two such examples are the proposed creation of a bovid-free land zone around Yellowstone National Park to prevent contact between bison and cattle and the removal of feral pigs from the Hawaiian Islands (because their activities increase mosquito-breeding areas and elevate the transmission of avian malaria).

An understanding of disease ecology is also pertinent to the design of habitat reserves. For example, how large should reserves be to prevent host crowding? Do corridors between reserves increase the threat of pathogen transfer among locations, or does host dispersal among habitats facilitate the spread of resistance genes or aid in parasite avoidance? Maintaining species richness and genetic diversity within reserves is also critical to limiting threats from disease. For example, high species diversity may buffer natural communities from dev-astating epidemics, and habitats that are restored with genetically homogeneous stock may be much more susceptible to pathogen invasion.

Ultimately, no one method of intervention will ever be sufficient to conserve endangered species, eliminate disease threats, and satisfy local farmers, ranchers, and residents. Cost–benefit analyses of intervention methods along with the use of epidemiological models to predict effectiveness of any proposed measures are vital components to an effective disease management plan. The most expedient measures may not always address the true causes of an outbreak, such as habitat disturbance, invasive species encroachment, or crowding. Plans based in community ecology with a multispecies approach are more likely to address the presence and effects of reservoir species. Finally, it is important to note that parasites are natural elements of all communities and their presence alone may not always warrant intervention. In fact, parasites can act as selective forces in host populations, and maintaining variability in host resistance may be important for future population responses to disease.

See Also the Following Articles

CAPTIVE BREEDING AND REINTRODUCTION • CLIMATE CHANGE, SYNERGISM OF • COEVOLUTION • DIVERSITY, COMMUNITY/REGIONAL LEVEL • GENETIC DIVERSITY • MICROORGANISMS, ROLE OF • PARASITISM

Bibliography

Anderson, R. M., and May, R. M. (1991). *Infectious Diseases of Humans: Dynamics and Control.* Oxford Univ. Press, Oxford.

Burdon, J. J., and Jarosz, A. M. (1991). Host–pathogen interactions in natural populations of *Linum maringale* and *Melampsora lini:* I. Patterns of resistance and racial variation in a large host population. *Evolution* 45, 205–217.

Connell, J. H., and Orias, E. (1964). The ecological regulation of species diversity. *Am. Nat.* 13, 791–833.

Dobson, A. P. (1988). Restoring island ecosystems: The potential of parasites to control introduced mammals. *Conserv. Biol.* 2, 31–39.

Dobson, A. P., and Crawley, M. (1994). Pathogens and the structure of plant communities. *TREE* 9(10), 393–398.

Dobson, A. P., and Hudson, P. J. (1986). Parasites, disease, and the structure of ecological communities. *TREE* 1(1), 11–15.

Dobson, A. P., and Hudson, P. J. (1992). Regulation and stability of a free-living host-parasite system, *Trichostrongylus tenuis* in red grouse. II. Population models. *J. Anim. Ecol.* 61, 487–500.

Elton, C. S. (1958). *The Ecology of Invasions.* Wiley, New York.

Feldman, R. A., Freed, L. A., and Cann, R. L. (1995). A PCR test for avian malaria in Hawaiian birds. *Mol. Ecol.* 4, 663–673.

Getz, W. M., and Pickering, J. (1983). Epidemic models: Thresholds and population regulation. *Am. Nat.* 121, 892–898.

Gilbert, G. S., and Hubbell, S. P. (1996). Plant diseases and the conservation of tropical forests. *Bioscience* **46**, 98–106.

Grenfell, B. T., and Dobson, A. P. (1995). *Ecology of Infectious Diseases in Natural Populations.* Cambridge Univ. Press, Cambridge, UK.

Harvell, C. D., Kim, K., Burkholder, J. M., Colwell, R. R., Epstein, P. R., Grimes, D. J., Hofmann, E. E., Lipp, E. K., Osterhaus, A. D. M. E., Overstreet, R. M., Porter, J. W., Smith, G. W., and Vasta, G. R. (1999). Emerging marine diseases—Climate links and anthropogenic factors. *Science* **285**, 1505–1510.

Holt, R. D., and Pickering, J. (1985). Infectious disease and species coexistence: A model in Lotka–Volterra form. *Am. Nat.* **126**, 196–211.

Hudson, P. H., and Greenman, J. (1998). Competition mediated by parasites: Biological and theoretical progress. *TREE* **13**(10), 387–390.

Hudson, P. J., Dobson, A. P., and Newborn, D. (1992). Do parasites make prey more vulnerable to predation? Red grouse and parasites. *J. Anim. Ecol.* **61**, 681–692.

Kermack, W. O., and McKendrick, A. G. (1927). A contribution to the mathematical theory of epidemics. *Proc. R. Soc. London A* **115**, 700–721.

Kiesecker, J. M., and Blaustein, A. R. (1995). Synergism between UV-B radiation and a pathogen magnifies amphibian embryo mortality in nature. *Proc. Natl. Acad. Sci. USA* **92**(24), 11049–11052.

Lively, C. M., and Apanius, V. (1995). Genetic diversity in host-parasite interactions. In *Ecology of Infectious Diseases in Natural Populations* (A. P. Dobson and B. T. Grenfell, Eds.), pp. 421–449. Cambridge Univ. Press, Cambridge, UK.

Lyles, A. M., and Dobson, A. P. (1993). Infectious disease and intensive management: Population dynamics, threatened hosts and their parasites. *J. Zoo Wildlife Med.* **24**(3), 315–326.

May, R. M. (1974). *Stability and Complexity in Model Ecosystems*, 2nd ed. Princeton Univ. Press, Princeton, NJ.

McCallum, H., and Dobson, A. P. (1995). Detecting disease and parasite threats to endangered species and ecosystems. *TREE* **10**(5), 190–193.

Meffe, G., and Carroll, R. (1997). *Principles of Conservation Biology*, 2nd ed. Sinauer, Sunderland, MA.

Nei, M., and Hughes, A. L. (1991). Polymorphism and evolution of the major histocompatibility complex loci in mammals. In *Evolution at the Molecular Level* (R. K. Selander, A. G. Clark, and T. G. Whittam, Eds.), pp. 249–271. Sinauer, Sunderland, MA.

Patz, J. A. (1997). Public health repercussions of climate and ecological change. *Bull. Ecol. Soc. Am.* **78**(4 Suppl.), 28.

Primack, R. B. (1998). *Essentials of Conservation Biology*, 2nd ed. Sinauer, Sunderland, MA.

Read, A. F., Albon, S. D., Antonovics, J., Apanius, V., Dwyer, G., Holt, R. D., Judson, O., Lively, C. M., Martin-Lof, A., Mclean, A. R., Metz, J. J., Schmid-Hempel, P., Thrall, P. H., Via, S., and Wilson, K. (1995). Group report: Genetics and evolution of infectious diseases in natural populations. In *Ecology of Infectious Diseases in Natural Populations* (A. P. Dobson and B. T. Grenfell, Eds.), pp. 450–477. Cambridge Univ. Press, Cambridge, MA.

Soulé, M. (1986). *Conservation Biology. The Science of Scarcity and Diversity.* Sinauer, Sunderland, MA.

Thrall, P., Antonovics, J., and Hall, D. (1993). Host and pathogen coexistence in sexually-transmitted and vector-borne diseases characterized by frequency dependent disease transmission. *Am. Nat.* **142**, 543–552.

Van Riper, C. III, van Riper, S. G., Goff, M. L., and Laird, M. (1986). The epizootiology and ecological significance of malaria in Hawaiian land birds. *Ecol. Monogr.* **56**, 327–344.

DISPERSAL BIOGEOGRAPHY

Ran Nathan
Princeton University

GLOSSARY

dispersal The movement of disseminules away from their parent source. Because of the variety of disciplines involved with the study of movements of organisms, the term is often misused and confused with a plethora of terms. The term *dispersion* refers to the fine-scale spatial distribution pattern of organisms, and *migration* is generally equivalent to dispersal, though it typically refers to a particular (directional and synchronized) type of dispersal. The term *immigration* is dispersal from the perspective of the recipient locality regardless of the source—the arrival of new individuals to an area previously not occupied by that species. It should be emphasized that dispersal does not deal with events that occur after a movement has been completed. The term *colonization* should be used for incidents of immigra-

tion that are followed by establishment of a viable population.

dispersal barrier The species-specific physical and biological restrictions a propagule must overcome to accomplish successful colonization.

dispersal biogeography Usually refers specifically to "a branch of historical biogeography that attempts to account for present-day distributions based on the assumption that they resulted from differences in the dispersal abilities of individual lineages" (Brown and Lomolino, 1998; p. 628). More generally, it is the study of the relationships between dispersal and the geographical distribution of organisms. I adopt the latter definition for this article.

dispersal route The particular set of physical and biological conditions that allow organisms to cross dispersal barriers.

disseminule The unit of dispersal; any part or stage in the life cycle of an organism that is used for dispersal.

propagule The unit of colonization; a disseminule, or group of disseminules, which has the full potential to establish a new population.

range expansion The spatiotemporal process in which successive colonization events increase the size of a species' geographical range. The terms *invasion*, *migration* (as used in the paleobotanical literature), and *spatial spread* are roughly synonyms for range expansion. *Invasion*, in particular, is often used to describe colonization outside the species range.

DISPERSAL IS A PROCESS COMMON to all organisms, and it is highly variable in details among organisms and across space and time. At large spatial or temporal scales, this process has broad implications for biogeography through both long- and short-distance dispersal events. The study of the relationships between dispersal and the geographic distribution of organisms, advanced by various methodological approaches, is critical to our understanding of forces affecting the world's biodiversity.

I. INTRODUCTION

In Chapters XII and XIII of *The Origin of Species*, Charles Darwin (1859) discussed how patterns of geographical distribution of species are associated with his ideas of natural selection. He argued that "all the grand leading facts of geographical distribution are explicable on the theory of migration, together with subsequent modification and the multiplication of new forms." Darwin illustrated this argument through many examples, by explaining different biogeographic patterns, emphasizing the role of *dispersal barriers* (see glossary) and the various means of *long-distance dispersal* through which organisms are able to cross them. Although Carl Linnaeus (1707–1778) and Augustin-Pyramus de Candolle (1778–1841) had already expressed similar ideas about the importance of dispersal in determining species distributions, Darwin (1809–1882) and Alfred Wallace (1823–1913) are those most identified with the *dispersalist* school. This school asserts that disjunction and range expansion occur through rare long-distance dispersal events across preexisting barriers. Joseph Hooker (1817–1911), one of Darwin's closest confidants, viewed long-distance dispersal across barriers as extremely unlikely and argued that range expansion was possible only across ancient land bridges that once connected currently disjoint continents. He thus argued that biogeographic disjunction can be explained by vicariant (splitting) events in which new barriers have subdivided a previously continuous range (see *Vicariance Biogeography*). Evidence for Hooker's land bridges never emerged; however, the two basic opponent ideas of dispersal versus vicariant history were already brought into the front stage of biogeography. In *The Secular Ark*, Janet Browne (1983) describes the 19th century discussion on this issue involving Darwin, Wallace, Hooker, and others.

The 20th century has seen a further polarization of the vicariance/dispersal debate. Plant biogeographers such as Evgenii Wulff (1885–1941) and Stanley Cain (1902–1995) reiterated serious doubts about the ability of long-distance dispersal to account for present plant disjunctions, in light of the extreme rarity and lack of direct empirical evidence of these events. Others, like Philip Darlington (1904–1983), advocated Darwin's dispersalist view, but without Darwin's reliance on specific *centers of origin*. In particular, the *equilibrium theory of island biogeography* (MacArthur and Wilson, 1967) ascribes a critically important role to *immigration* (see glossary) in determining the number of species on islands. This landmark theory, which had a major impact on both ecology and biogeography, modeled the probability of immigration of a new species as roughly a negative exponential function of the number of species on an island, with the rate also decreasing with distance from the mainland [see *Island Biogeography*].

At approximately the same time that Robert MacArthur and Edward Wilson published their theory, the important theory of *continental drift*, or more generally of *plate tectonics*, first proposed early in the 20th century, was finally accepted. This theory provided a vicariant mechanism to explain biogeographic disjunctions without long-distance dispersal events. In the 1970s and early 1980s, the concept of vicariance biogeography received much attention among biogeographers, as reflected in three books by Nelson and Rosen (1981), Nelson and Platnick (1981), and Wiley (1981). The main argument raised against dispersalists' explanations, following the similar criticism by Leon Croizat (1894–1982), was that long-distance dispersal is a unique, highly stochastic and unpredictable phenomenon, hence its relationships with species distributions are not testable. However, truth and testability remain distinct: obstacles to proving the truth of an idea do not imply its falsehood. The explanations of biogeographic patterns by dispersalists were condemned as "storytelling" scenarios which "in explaining everything really explain nothing." However, solutions of complex biogeographic patterns by vicariant events alone, if taken to their limits, are also speculative. There are, indeed, examples of extreme cases in which disjunctions are exclusively explained by dispersal (e.g., colonization of remote newly created islands) or by vicariance (e.g., postglacial relicts). However, none of these extreme arguments provides a general explanation; in fact, as we shall see, not only do both phenomena exist, but they also frequently co-occur.

The dispersal-vicariance debate reflects a more general controversy over the biogeographic relevance of dispersal: *to what extent does dispersal influence species distribution?* This debate concerns whether disjunct distributions have resulted from dispersal across a barrier or from a barrier splitting a continuous range. But dispersal and vicariant events are not the sole explanations for disjunctions, which may also result from differential extinction within a species range. The major problem in limiting *dispersal biogeography* to its currently narrow definition (see glossary) is that such a definition reduces the entire phenomenon of dispersal to long-distance dispersal across barriers, thereby ignoring the various biogeographic patterns to focus solely on disjunct distributions. As this discussion will show, there are many other distributional patterns that cannot be explained by either vicariant or extinction events but only by long- and short-distance dispersal events. Rich evidence gathered from diverse independent sources indicates not only that long-distance dispersal exists, but also that it is the only possible explanation for the geographical distributions of many species, such as the Hawaiian biota (Carlquist, 1974). Further, long-distance dispersal is not necessarily as random or unpredictable as it is often considered. The relative importance of dispersal varies between species, biogeographic regions, and times. Thus, to assess the relative biogeographic importance of dispersal, we need to explore the nature of dispersal and its relationships with the geographical distribution of species. The broader view of *dispersal biogeography* (see glossary) propounded here aims to understand how dispersal influences, and is influenced by, the geographical distribution of organisms.

II. PERSPECTIVES ON DISPERSAL

A. The Ecological Perspective

As a general rule, most *disseminules* (see glossary) travel relatively short distances from their parent source. The basic description of dispersal, called the *dispersal curve*, is a graph plotting the distribution of disseminules against the distance from their parent source (Fig. 1). Dispersal curves, derived for a great number of organisms, show tremendous variability, but also a right-skewed leptokurtic distribution with overwhelming regularity. That is, they typically describe a rapid decline in disseminule density with distance from their parent source, with more disseminules both near the source and along the tail than a corresponding normal distribution (Section IV.C.1). The restricted spatial extent of dispersal has important implications for the dynamics of populations, metapopulations, and communities and for other topics addressed in ecological studies of dispersal (Nathan and Muller-Landau, 2000).

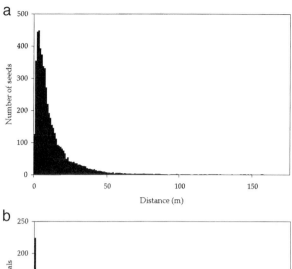

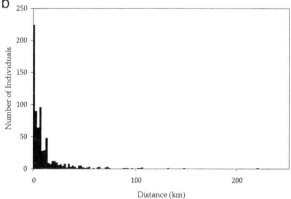

FIGURE 1 Dispersal diagrams representing the distribution of dispersal distances, for the winged seeds of Aleppo pine (*Pinus halepensis*) (a) and for first-year woodpigeons (*Columba palumbus*) (b). Dispersal of pine seeds was estimated by combining mechanistic simulation of wind dispersal with seed-trap data collected during six dispersal seasons in an isolated pine stand on Mt. Carmel, Israel; bird natal dispersal was estimated from ringing and recovery data gathered in Britain from 1909 to 1994. Note the 3-order of magnitude difference between abscissas. Pine data from Nathan *et al.*, (2000); the bird diagram was redrawn from Paradis *et al.*, (1998) *J. Anim. Ecol.* 67, 518–536 with permission from Blackwell Science Ltd.

The ecological perspective of dispersal is concerned not only with small spatial scales, but also with relatively short temporal scales. Dispersal is often restricted to early stages during an organism's life cycle (seeds, eggs, larvae, and juvenile stages in animals); the term *natal dispersal* denotes dispersal of young animals while *breeding dispersal* indicates dispersal of reproductive adults. Disseminules of many plants, nematodes, rotifers, desert butterflies, copepods, and other invertebrates can persist for years—sometimes hundreds of years—as dormant seeds or eggs, providing the means for a temporal exploration for favorable *establishment* conditions. From another temporal perspective, dispersal is typically a seasonal phenomenon, and synchronization of dispersal in different individuals, populations, and species is fairly common in both animal and plant communities. These temporal patterns have important effects at the ecological scale, but they are unlikely to exert a significant effect at the biogeographic level.

B. The Evolutionary and Population Genetic Perspectives

The great variety of dispersal-determining traits presents clear evidence that dispersal is of important adaptive value for many life forms on earth. From the evolutionary perspective, the study of dispersal is concerned with the source and significance of this variation: the factors and processes that affect the advantage of dispersal. Typically dispersal is advantageous for escaping the high mortality near the source, reducing competition with close relatives, minimizing the fitness cost of inbreeding depression by reducing the chances of mating with close relatives (in sexual species), and for locating new resources for establishment, growth, and reproduction. Theoretical models (see Johnson and Gaines, 1990, for a review) show that even in environments in which the physical conditions are stable in time and space, surprisingly high proportions of dispersing progeny may be adaptive (Hamilton and May, 1977). Since mutations beneficial to successful dispersal are typically rare, evolutionary effects are observable only at relatively long time scales (but see Section III.C). However, there is no characteristic spatial scale for the evolutionary perspective of dispersal: adaptive advantages of dispersal are expressed at both small and large spatial scales.

Dispersal provides the means for *gene flow* within and among populations; thus studies of population genetics explore how dispersal, in conjunction with the breeding system and selection regime, affect the genetic structure of populations. In general, high rates of dispersal lead to mixing of genotypes and thus lower regional genetic diversity, while low rates of dispersal (*isolation*) typically result in low local genetic diversity, but high regional diversity. Although these effects may also occur at small spatial and temporal scales, they are notably more pronounced in large ones.

C. The Biogeographic Perspective

Dispersal becomes important at the biogeographic level only at large temporal or spatial scales, when and where propagules disperse outside the species range and successfully establish viable populations (Section III.B). The identification of locations as "within" or "outside" the species range (and obviously the definition of a species range itself) is problematic since sampling of any species is typically incomplete. Even if we assume complete information on the exact location of each individual at any particular time, the issue of species boundaries is difficult because each species exhibits a profound variation in its local density and, in fact, is entirely absent from most locations. In biogeography, this distinction is less problematic when dealing with larger spatial and temporal scales.

Spatial spread can occur at different rates and can generate different spatiotemporal patterns, from slow gradual expansion to abrupt long-distance jumps. Both rate and pattern depend on the two basic components of spatial progression—the movement itself (dispersal) and local population growth (reproduction)—and their interplay (Section IV.C). The biogeographic literature provides numerous examples showing a continuum between these extremes, and it is also clear that gradual spread and long jumps can follow each other. However, there are basic differences between these two modes: gradual range expansion is the outcome of multiple ordinary (i.e., short-distance) dispersal events over a relatively long time period; jumps are rare individual events that abruptly change the species' range.

1. Gradual Range Expansion

Gradual range expansion is the continuous spread of populations through multiple ordinary (short-distance) dispersal events, usually over quite long time scales. Very slow expansion has been termed *secular dispersal* (or *secular migration*); more rapid (though still gradual) expansion is referred to as *diffusion* (Pielou, 1979). Examples of gradual range expansion are the spread of mollusks, worms, crabs, and fishes from the Red Sea to the Mediterranean (Section IV.A.3) and of muskrats in Europe (Section IV.A.5).

2. Jump Range Expansion (Jump Dispersal)

Although most dispersal is limited to relatively short distances and therefore covered appropriately by the ecological scale (Section II.A), rare long-distance events have crucial implications for biogeography. There is enormous variation in the ability to disperse disseminules over large distances within and, more significantly, between species. Specific morphological characteristics are associated with long-distance dispersal ability (Box 1). When such an event ends in establishment, the spatial spread is clearly discontinuous. The frequently used term for this type of spatial spread, *jump dispersal*, does not fully describe the process: spatial spread is not restricted to the movement phase (dispersal) but also necessitates establishment. Hence, the term *jump dispersal* should be replaced with *jump range expansion*. Note also that while the original term is appropriate for describing long-distance dispersal over barriers, the term *jump range expansion* is also applicable to situations in which suitable habitats are as common in the gap as in the recipient area but are unoccupied (see Sections IV.A.4.b and IV.C.1).

Box 1

Mechanisms and Adaptations for Dispersal

Previous workers have suggested many methods of classifying the various means of dispersal. The most fundamental distinguishes between organisms that are actively dispersed (by their own power) and those that are passively dispersed (by a physical agent or by other organisms). Organisms unusually capable of *active dispersal* are said to be *vagile*, while those unusually capable of *passive dispersal* are called *pagile*. Another common classification, mostly used for plants, is based on the primary dispersal agent, using the suffix *chory* (to disperse) at different levels: for example, *zoochory* is dispersal by animals, *ornithochory* by birds, and *endozoochory* is within an animal. Neither classification should be taken as an absolute: many vagile organisms also disperse passively, and many pagile organisms can move actively over short distances. Seeds of many plants are actually dispersed by multiple agents, rather than by a single one.

(a) Shy albatross (*Thalassarche cauta*)—a vagrant seen off the shore of California in 1999. Photo by Luke Cole.

(b) Squirting cucumber (*Ecballium elaterium*). Photo by Pietro Pavone; drawing from Hegi (1915).

Relatively few animals have the capacity to travel long distances under their own power. Most of these are volant (flying) species such as birds (a), bats, and butterflies. Although many migrate seasonally [see *Migration*], this large-scale movement usually does not affect the geographical distribution of the breeding range. However, extra-limital sightings are well documented for many taxa, and occasionally may result in the establishment of a new colony outside the species range (Section IV.A.4.a). Nonvolant animals cannot vault dispersal barriers as do flying animals, rendering their active dispersal generally less effective. Despite this, large-scale active dispersal of large mammals, reptiles, and fishes is frequently observed. In plants, active dispersal is very rare and is restricted to the phenomenon of *ballistic dispersal*, in which seeds are explosively discharged from their base (b) and are dispersed at most tens of meters.

Passive dispersal is the main, or sole, means of movement for the vast majority of organisms. In plants, morphological appendages such as wings or hairs aid the passive wind dispersal of diaspores (c); barbs or spines adhere to mobile animals; fleshy fruits attract frugivores that may digest the fruit and defecate the seeds (d). Seed dispersal by volant vertebrates (especially birds and bats), either external (epizoochory) or internal (endozoochory), is considered an efficient adaptation for long-distance dispersal.

Passive dispersal is also very frequent among organisms other than plants. Many marine organ-

(c) Seeds of sticky fleabane (*Dittrichia viscosa*). Photo by Avinoam Danin.

(d) Fruits of thorny smilax (*Smilax aspera*). Photo by Avinoam Danin.

(e) Veliger larva of neapolitan triton *Cymatium parthenopeum*, a telepanic gastropod (Renellidae). Photo by Rudolf S. Scheltema, used with permission.

(f) Floating island of 14 m long, composed mainly of papyrus, in Lake Malawi, Africa. Photo by Kenneth R. McKaye.

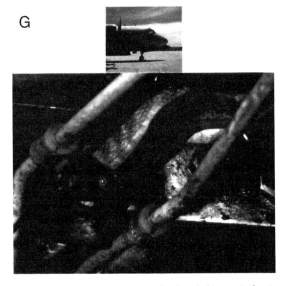

(g) Brown tree snake (*Boiga irregularis*) coiled in an airplane's wheelwell in Guam. Photo by Gad Perry.

isms have free-living juvenile stages in the form of cell, eggs, or larvae that are carried passively by ocean currents over large distances. Being concentrated near the water surface, these tiny planktonic propagules (e) are dispersed by the winds that drive the oceanic circulation. Microbes, spores, light seeds, and tiny insects, spiders, and mites constitute the so-called *aerial plankton* known to be an extremely effective source of colonization of remote islands. Some spiders, for example, climb to an eminence, release silk parachutes into the air, and are subsequently carried off by wind. Dispersal of larger disseminules by winds is usually very local, but strong updrafts and extreme storms can carry them over large distances. Another passive mechanism for dispersal of relatively large organisms is by *floating islands* or *rafts* (f), which may disperse entire trees, rodents and reptiles. However, the most influential dispersal agent at the biogeographic scale—at least during the past 10,000 years—is man. *Anthropogenic dispersal* has moved many species across the globe (Section IV.A.5), either intentionally or not. The arrival of the brown tree snake to Guam is considered to be unintentional (Section V.B.); years later, a snake was found in an airplane's wheelwell just before leaving Guam for Japan (g).

The probability of long-distance dispersal is inherently very low, and jump range expansion also requires the improbable discovery of a suitable site and establishment of a viable population in a remote location. Jump range expansion is thus highly improbable (Section III.B). This improbability might lead one to doubt the biogeographic importance of long-distance dispersal (Section I). However, even extremely low (but nonzero) probabilities will eventually be realized given enough attempts. In fact, there are large numbers of examples of jump range expansion, including, but not restricted to, the colonization of newly formed volcanic islands (Section IV.A.2).

The low probability of jump range expansion implies that this phenomenon is highly stochastic and unpredictable; although this generalization is true to a large extent, there are several predictable mechanisms of jump range expansion. Such mechanisms include the dispersal of seeds and fruits by migrating birds, of aerial plankton (Box 1) by wind, of marine plankton by sea currents, and others.

III. DETERMINANTS OF DISPERSAL AND COLONIZATION SUCCESS

A. Mechanisms and Adaptations for Dispersal

Most (if not all) organisms are able to move or be transported at some stage during their life cycle, exhibiting a diverse array of dispersal mechanisms (Box 1). Henry Ridley (1930), Sherwin Carlquist (1974), Leendert van der Pijl (1982), and Hugh Dingle (1996) provide many examples of dispersal-determining traits in animals and plants. Overall, organisms vary greatly in their *dispersal ability*, the capacity for dispersal, which is determined by a complex set of interacting factors. High dispersal ability can be achieved by either active or passive dispersal (Box 1), and it is often associated with low competitive ability and low longevity (Tokeshi, 1999).

B. Components of Colonization Success

Successful colonization necessitates the physical presence of a propagule followed by establishment. Both dispersal and establishment require physiological, morphological, and behavioral preadaptations or adaptations enabling them to disperse far from the source, survive during dispersal, and become established in a remote location. In general, the larger the distance between any two locations, the more dissimilar the environmental conditions. Therefore, long-distance dispersal typically takes propagules to environments that are very different from their usual habitats. Thus, successful colonization in remote locations is generally extremely rare, because it is the product of the very low probabilities of these sequential processes. However, given enough time and multiple attempts, long-distance dispersal may result in long-distance colonization.

1. Preadaptations

Species that often combine high dispersal ability with broad ecological tolerance are those adapted to exploit disturbed or newly created environments and those that inhabit ephemeral or uncertain environments. Such species are thus preadapted for successful colonization outside their species range. MacArthur and Wilson (1967) called these species *r*-strategists, characterized by small size, early maturity, rapid growth, high fecundity, high dispersal ability, and broad niche. In contrast, *K*-strategists are better adapted for local competition through the efficient use of limited resources (Pianka, 1970). The dichotomous r/K scheme is obviously an oversimplification and the concept is rather controversial (see Williamson, 1996); yet empirical studies generally, but not always, do support the theoretical expectation that r-strategy preadaptations characterize effective colonizers (Brown and Lomolino, 1998; see Section IV.A.3).

2. Physical Access: Dispersal Barriers and Dispersal Routes

Physical access to a remote location depends not only on specific traits that enhance long-distance dispersal (Box 1), but also on the environmental conditions in the area between the source and the deposition location. Crucial to the problem of physical access is that of *dispersal barriers*, which propagules must overcome during jump range expansion (Section II.C.2). The manner of overcoming barriers defines the *dispersal route* (see glossary), the other key term in the context of physical access; some routes may be transversed with ease, while others rely upon chance. Examples of dispersal barriers and dispersal routes are given in Section IV.A.3 and in other sections that follow.

It is important to emphasize that *dispersal barriers* are species-specific; that is, conditions that may prevent dispersal of one species may not affect the dispersal of another. They may even be specific for a particular stage in the species' life cycle. Dispersal barriers can be physiological, ecological, or behavioral, with organisms subjected to physiological stresses, ecological hazards, or behavioral difficulties in the environments they traverse.

Dispersal barriers become traversable via *dispersal routes*. George Simpson (1902–1984) made a distinction between *corridors, filters,* and *sweepstakes routes,* which roughly cover the entire range of probabilities of crossing a barrier from highly probable to highly improbable, respectively (Simpson, 1965). *Corridor routes* allow dispersal of many species from one region to another, conferring a considerable degree of similarity between the biotas of the two sides. Climatic changes and plate tectonics have repeatedly connected previously separated land or water masses, generating dispersal corridors such as those between Africa and Eurasia (Section IV.A.4.c, Fig. 2). The passage through *filter routes* is more selective than through corridors: some species pass rather easily, while others do not; as a result, each biota represents a biased subset of the other. The Suez Canal between the Red Sea and the Mediterranean Sea (Section IV.A.3) is an example of a filter route. *Sweepstakes routes* in which crossing

occurs as a chance event at a very low frequency, such as by dispersal on floating islands (Box 1), represent the most improbable way to cross a barrier (Section IV.A.2).

3. Ecological Opportunity: Postdispersal Establishment

Given that physical access has been attained, nevertheless the probability of successful colonization in remote areas is still rather low. As indicated above, both the physical access and the establishment of a viable population require physiological and morphological preadaptations; often each characteristic is restricted to either dispersal or establishment, and plays little role in the other. The importance of post-dispersal establishment for successful colonization is illustrated by the many examples in Section IV.

Establishment success is determined in part by the ecological conditions at the new site. For passively dispersed species, only the conditions in the particular arrival site matter, while actively dispersed propagules may exhibit *habitat choice* (also called *habitat preference*) due to their ability to reach preferred habitats among those locally available. Passively dispersed dormant propagules like seeds and invertebrate eggs may exhibit temporal habitat choice by becoming dormant until conditions favorable for establishment become available. Meeting the physical conditions for establishment may not be sufficient: colonizers may still be excluded by biotic interactions such as competition and predation. Thus establishment also requires traits specific to the recipient community, particularly the resident competitors and predators.

Successful colonization in remote areas requires that colonists be able to reproduce and establish a viable population. Given that long-distance dispersal is likely to bring only one or very few reproductive individuals close together in a remote area, the initial population size of a naturally colonizing species is typically very low. Very small populations are prone to extinction due to *demographic, genetic,* and *environmental stochasticity*—the influence of random events on population persistence—which are intensified as population size decreases. These risks may be reduced with repeated immigration to the site, by so-called *rescue effects* (Section V.B); hence dispersal ability itself is a factor that determines not only arrival but also, in part, establishment probability. Small, newly established isolated populations exhibit reproductive difficulties due to *Allee effects* such as difficulty in finding a mate. Such effects are obviously graver for obligate sexual reproducers (higher animals and plants) than for asexual or self-

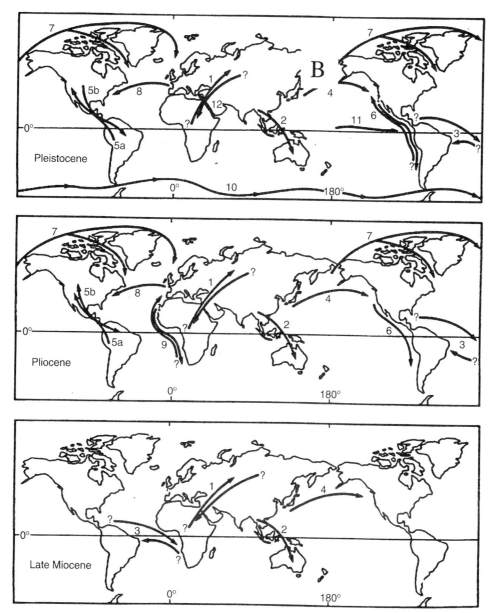

FIGURE 2   Major episodes of biotic interchange during the late Miocene, Pliocene, and Pleistocene-Recent. Roughly symmetrical interchanges are marked by two-headed arrows, asymmetrical ones by single-headed arrows pointing at the predominant direction, and uncertain direction by question marks. Reprinted with permission from Vermeij *Science* 253, 1099–1104 copyright 1991, American Association for the Advancement of Science.

compatible organisms that can reproduce without gametes from another individual. However, the severity of Allee effects is reduced in obligate sexual species by certain reproductive mechanisms, such as *sperm storage* and *delayed implantation*, which assure sufficient gametes of both types and by social behavior such as flocking.

C. The Geography of Dispersal

The main issue addressed so far is the effect of dispersal on the geographic distribution of organisms. The complementary issue of geography's effects on dispersal entered the discussion of dispersal barriers and dispersal routes (Section III.B.2). This topic is concerned with geographic variation in dispersal explained by geographic variation in the external factors affecting dispersal; it typically neglects geographic effects on the dispersal traits themselves. Theoretical models predict that dispersal is favored during colonization but selected against when a population is strongly isolated. The degree of isolation is a relevant factor: (1) when isolation is low, colonization is easily achieved by dispersal. Thus, elevated levels of isolation initially select for greater dispersal abilities; (2) at higher levels of isolation colonization is unlikely, and dispersal is selected against. While the second prediction has been long recognized and is supported by overwhelming evidence, there have been only a few recent tests of the first.

The prediction that very high degrees of isolation select against dispersal is evident in the loss of dispersal ability on oceanic islands, as in many flightless birds, land snails, insects, and plants (see Carlquist, 1974). Reduced dispersal is also evident in organisms in unique isolated habitats as diverse as subterranean caves, mountaintops, deep-sea trenches, thermal vents, and hot springs. The observation of this phenomenon raises an intriguing paradox, which Darwin (1859) explained through counteracting selective forces acting before and after colonization of a remote island. Martin Cody and Jacob Overton (1996) demonstrated the strength of the selection for reduced dispersal potential in weedy, short-lived wind-dispersed plants of inshore islands in Canada. Over the course of just a few generations, island populations of two of three species with sufficiently large sample sizes showed marked reduction in dispersal potential relative to mainland populations or young island populations. Aphids, leafhoppers, crickets, and many other insects have the potential to develop into a normal or flightless adult, depending on the environmental conditions encountered during development (Harrison, 1980).

Habitat fragmentation has substantial effects on dispersal at mesoscales by creating dispersal barriers and disrupting dispersal routes. Recent empirical studies on British butterflies have taken the first step toward experimental investigation of the effect of landscape structure on dispersal-related traits (Van Dyck and Matthysen, 1999). These studies show that the relative investment in the thorax, which mainly contains flight muscles, is higher when patches are >40 km apart; similar selection appears to have occurred in a region where population expansion was most rapid. No differences were found in the relative investment in flight between sites <5 km apart. Since flight is associated with other functional needs of butterflies such as foraging and mate-location, further study is needed to test the relationship between morphological traits and dispersal ability.

IV. METHODOLOGICAL APPROACHES AND CASE STUDIES

A. Observing Dispersal and Colonization in Action

1. Observations of Dispersal

Direct observation is potentially a powerful method of quantifying dispersal. Peter Turchin (1998) provided an extensive review and explanation of the various methods used to track the movements of individuals and to estimate population redistribution in both animals and plants, including records of movement paths and mark-recapture methods. These methods are typically laborious and, from a biogeographic perspective, where large spatial or temporal scale are most important (Section II.C), nearly impossible. The indirect approach, in which dispersal is estimated from its outcome (recruitment), is more feasible. Therefore, all methodological approaches described here are based on indirect evidence. The greater feasibility of indirect methods is a result of the larger size and more conspicuous nature of organisms in relation to their propagules; moreover, it is usually easier to observe a pattern (recruitment) than a process (dispersal). As emphasized in Section III.B, dispersal is meaningless without subsequent recruitment; therefore, indirect evidence of dispersal is indeed important. Nevertheless, it does not tell us much about the dispersal process itself, since the distances dispersed, the number of dispersal events, and the mechanisms of unsuccessful recruitment are unknown. These are important parameters that should be incorporated into biogeographic models in order to enhance their generality and predictive ability.

Data on dispersal events are best obtained by following movements of individuals for which both source and end points are known. Unfortunately, for practical reasons, these kind of data are relatively rare. In many cases only the end location is known and there are multiple potential sources. Although our knowledge of long-distance dispersal continues to increase, the *paucity of direct observations of these dispersal events is one of the largest obstacles to the development of dispersal biogeography as a mature scientific discipline.* One possible solution is the development of methods to track individual movements at both small and large spatial scales. Recent advances in developing molecular and remote sensing techniques promise significant progress in the near future. Note that in genetic studies the term *direct evidence* is used for evidence of gene flow obtained by comparing the genotypes of recruits and their putative parents (and then applying methods such as *paternity analysis* to identify the most probable parent): it would be considered as *indirect* evidence under the preceding definition. However, for the study of long-distance dispersal this approach is not very valuable; evaluating putative parents becomes impractical as the distance from the offspring increases. In contrast, the so-called indirect genetic approach, which compares the genetic structure between populations, has provided compelling evidence for jump range expansions (Section IV.A.4.b).

There are two types of methods for estimating the dispersal distances when there are multiple possible sources (reviewed in Nathan and Muller-Landau, 2000). The first makes a priori assumptions about the source location (the simplest assumes the nearest possible parent is the source), while the second uses inverse methods to fit a particular *dispersal kernel* (a functional form of the probability distribution of dispersal distances—see Section IV.C.1). Both methods have been applied to study dispersal at the ecological scale. Since short-distance dispersal governs gradual range expansion, it is also important from the biogeographic perspective. However, using these methods to estimate long-distance dispersal is problematic since the exact position along the dispersal kernel tail is typically very sensitive to small differences in probability—sampling errors are likely to obscure the actual probabilities. Furthermore, the number of endpoints in typically very small (in most cases there is only one), which prohibits inverse modeling techniques. With a better understanding of the process of long-distance dispersal itself, we might expect that mechanistic dispersal models (Section IV.C.1) would help surmount this difficulty.

2. Colonization of Virgin Habitats

Volcanic eruptions that eliminate an entire local biota provide unique opportunities to study patterns of dispersal and colonization. Although there are some well-studied cases of recolonization in terrestrial systems after volcanic eruptions, such as the 1980 eruption of Mt. St. Helens (western North America), the best case studies are on oceanic islands. A well-known example is the island of Surtsey near Iceland that emerged from the sea in 1963. The best example, however, has its roots 80 years earlier: on August 27, 1883, the island of Krakatau, located between Sumatra and Java (Fig.

3a), exploded with tremendous force, obliterating all traces of life. Less than 2 months after the eruption, a Dutch expedition found no evidence for any surviving life form; the first animal, a single tiny spider that had dispersed as aerial plankton (Box 1), and the first plant were detected 9 and 13 months after the eruption, respectively. Later biological surveys documented the rapid recolonization of the island by various life forms, in a complex *successional* process, which began with fungi and lichens breaking down the volcanic rock to form soil, enabling seeds to germinate. The possible sources of immigrants were the mainland areas of Java and Sumatra, both approximately 40 km away, and two small islands, 12 and 20 km away (Fig. 3a). More precise identification of the sources for colonization in Krakatau by any current available method is highly speculative. Furthermore, most species were discovered after they had already established populations, with no data on their actual arrival; obviously, there are very few data on arrivals not followed by recruitment. Yet, even with these restrictions, the main insights from this unique long-term classical case study are extremely important; a recent book by Ian Thornton (1996) tells the story of the reassembly of this island ecosystem.

The immigration data collected on Krakatau, albeit incomplete, provides important information on the dispersal process itself. One of the most important modes of dispersal was by sea, whether passively (e.g., fruits and seeds) or actively (e.g., the water monitor *Varanus salvator*). There was a striking example of *rafting* (Box 1) in a 20-m²-area floating island with 3 to 4 m high palms observed in the archipelago in 1986. Dispersal by air, either passively (spores, light seeds, and other aerial plankton) or actively (butterflies, birds, and bats), was another important mode. Birds and bats served as dispersal agents for plant seeds over short (within the archipelago) and long (from the mainland to the archipelago) distances.

One of the most remarkable patterns observed in the Krakatau system is that of unequal colonization. This pattern was largely clarified for plants when dispersal agents were classified into three primary types (wind, sea, and animals) and examined in three time phases since the eruption (Fig. 3b). Sea-dispersed species were the most important early colonizer but became the least important by the third phase; animal-dispersed species generally showed the opposing trend; the relative importance of wind-dispersed species remained fairly constant in all phases. The mode of dispersal also explains why several kinds of taxa were underrepresented: species with large and heavy seeds dispersed either by wind (wings) or by bats and terrestrial mammals (Fig. 3c).

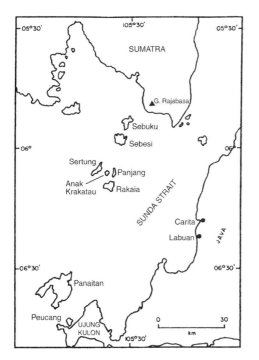

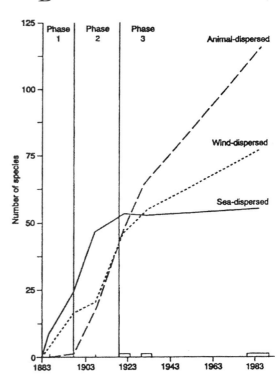

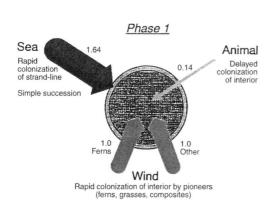

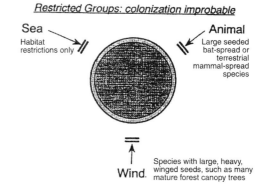

Moreover, some species reached, but did not colonize, the archipelago. Over the period of eight years (1984–1992), 54 species of animals were recorded as new arrivals to Anak Krakatau (a small island in the center of the archipelago, which emerged from the sea in 1928; Fig. 3a): 14 were new to the archipelago, and more than 20 did not become established. The absence of several sea-dispersed plant species is attributed to post-dispersal conditions. Similarly, several species such as the estuarine crocodile (*Crocodylus porosus*) were seen in Krakatau but have not yet colonized, presumably due to the lack of suitable habitats. The role of establishment conditions is also reflected in the decrease of the colonization rate with species richness.

3. Removal of Barriers

Dissolution of barriers is typically accompanied by species invasion in both directions. Although the term *biotic interchange* might be interpreted as a balanced exchange, the pattern is often highly asymmetric (Vermeij, 1991). Climatic changes and tectonic plate movements in geological times have repeatedly created and destroyed dispersal barriers, and both kinds of events have had important biogeographic and evolutionary consequences (Fig. 2). Such events include the closure of the Tethyan seaway and the formation of a wide land bridge between Africa and Asia during the Miocene 18 million years ago (mya). Likewise, the reoccurring Bering land bridge and strait have intermittently served as dispersal route between the terrestrial biota of northeastern Asia and northwestern North America, and between the marine biota of the north Pacific and Arctic-Atlantic Basins. Another well-studied example is the current land bridge between South and North America which was formed 3.5 Mya; it provided a dispersal route for a massive passage—the *Great American Interchange*—of terrestrial organisms, and a dispersal barrier for the marine biota of the Atlantic and Pacific oceans. These examples demonstrate how dispersal and vicariant events can take place simultaneously.

Since the fossil record of mammals is richer than that of other vertebrates, the passage of mammals during the Great American Interchange has been the most thoroughly documented. It is clearly asymmetric, with many more North American taxa passing through the filter and diversifying in South America than doing the converse. In other terrestrial vertebrates, however, the interchange was either roughly balanced (e.g., birds) or biased in the opposite direction—that is, predominantly from south to north (reptiles, amphibians, and even freshwater fishes). Indeed, the differential response to the creation of the land bridge shows large variation also at lower taxonomic levels; in general, most taxa in both continents actually did *not* expand their ranges to the other continent.

Of the several episodes of biotic interchange that have occurred recently, perhaps the one most intensively studied is that through the Suez Canal, which connects the Mediterranean Sea to the Red Sea. Before the canal was opened for navigation in 1869, only a very few euryhaline species were able to cross the hypersaline lakes that have occasionally connected the two seas (Por, 1978). The creation of the Suez Canal led to an almost unidirectional colonization named *Lessepsian migration* after Ferdinand de Lesseps, the designer and chief constructor of the canal. It involves the colonization of the Mediterranean by at least 14 species of worms (Polychaeta), 91 mollusks, 15 crabs, and 44 fish species from the Red Sea. Two main explanations for the unidirectional colonization have been proposed. The first, that species from the Gulf of Suez, the saline arm of the Red Sea at the southern end of the canal, were better preadapted physiologically to cross the hypersaline canal and lakes than were species from the northern end of the Mediterranean. The second postulates that species from the more diverse biota of the Red Sea are superior in competition, defense, or reproduction to species of the impoverished biota of the eastern Mediterranean. A similar explanation has also been raised to explain the asymmetry in the passage of mammals during the Great American Interchange. Comparisons between species that colonize and closely related species that do not colonize show, in general, a tendency for *r*-selection attributes (Section III.B.1) among colonizers, especially higher dispersal ability and fast growth (Safriel and Ritte, 1983).

FIGURE 3 Colonization of plant species to the Krakatau archipelago between Sumatra and Java (a). The number of species colonizing the island of Rakata has increased over the period 1883–1989. Species with different primary dispersal mechanisms (sea, animal, and wind) exhibit dissimilar colonization dynamics over three phases of this period (b). A model of island colonization derived from these empirical data (c) incorporates the dispersal mechanisms and temporal phases with a distinction between strand-line (outer circle) and interior habitats, and emphasizes colonization constraints (the lower right figure). Arrow widths are proportional to the cumulative number of species in species year⁻¹ (values are given by each arrow). From Bush and Whittaker (1991) with permission from Blackwell Science Ltd. and Whittaker and Jones (1994) with permission.

4. Natural Range Expansion

a. Recent Range Expansion

Well-documented major range expansions provide a unique opportunity to quantify and model the process of spatial spread at large scales. Charles Elton (1900–1991) and J. Skellam laid the foundations for the dynamic progress recently experienced in this field. Most range expansions at present are related to human activity (Section IV.A.5), but some that are presumably natural range expansion are of particular interest to explore how species expand their range without human-aided transportation (see Hengeveld, 1989; Shigesada and Kawasaki, 1997; Williamson, 1996; also see *Plant Invasions*).

Range expansion can be very abrupt and may be followed by a rapid contraction, a phenomenon called *boom-and-bust* (Williamson, 1996). The most remarkable examples are three African locusts (*Schistocerca gregaria*, *Locusta migratoria*, and *Nomadacris septemfasciata*), which usually inhabit limited areas, but which reproduce rapidly, change their morphology and behavior, and disperse in huge swarms to forage over much larger areas under particularly favorable weather and food conditions. Populations of several rodent species such as lemmings (*Lemmus* spp.) also exhibit marked (cyclic) fluctuations, but due to their lower dispersal capacities, the resulting geographical impact is much more limited. Typically, these outbreaks do not result in expanding the *breeding* distribution of these species, as occasionally occurs after outbreaks of some bird species such as crossbills (*Loxia curvirostra*) and Pallas's sandgrouse (*Syrrhaptes paradoxus*).

The accidental occurrence of individuals far away from the breeding sites has been documented in many taxa, especially in volant animals. For example, in 1979 I observed an accidental arrival to Elat (South Israel) of a female long-tailed duck (*Clangula hyemalis*) whose southernmost breeding and wintering regions in Europe are in Sweden and Northern France, respectively. Fifteen months later, at the very same salt pond, I collected an exhausted shy albatross (Box 1), a rare seabird that breeds on islands near Tasmania and New Zealand. Some accidental vagrants may establish viable populations (jump range expansion), as did the purple gallinule (*Gallinula martinica*), a colorful waterfowl, in the Tristan da Cunha Islands at the Mid-Atlantic Ridge after years as a vagrant visitor. This tropical species has been reported as a rare vagrant as far north as Nova Scotia in Canada.

Events of boom-and-bust and of vagrant occurrence illustrate three important points: first, that species differ in their dispersal ability (Section III.A, Box 1); second, that the current breeding distributions of these species is not limited by dispersal (Section III.B.2), and third, that postdispersal factors are important determinants of colonization success (Section III.B.3). However, these events are episodic and thus difficult to study. In contrast, some persistent cases of recent broad-scale range expansions that have been documented in detail are important contributions to the study of dispersal and population spread. As in the previous case, the best examples come from highly mobile animals. One well-studied case is the collared dove (*Streptopelia decaocto*), which since 1900 has expanded its range from Turkey through the Balkans to the rest of Europe. Another is the cattle egret (*Bubulcus ibis*), which expanded its range in all continents since the late 1800s. It naturally colonized South America from its breeding grounds in Africa by the mid-1930s (jump range expansion) and rapidly expanded throughout the New World. See Sections IV.A.5 and IV.C for a more detailed discussion of range expansion.

b. Holocene Postglacial Spread

The considerable fluctuations in the global climate during the Pleistocene (the geological epoch between 1.8 million years ago [Mya] and 11 thousand years ago [Kya]) have led to several glacial-interglacial cycles. The last glacial period reached its maximum about 18 Kya, when ice sheets covered extensive parts of Eurasia and North America. Species responded in various independent ways, depending on their physiology, life history strategies, and dispersal abilities. Many species were able to track the shift of their most favorable climate and habitats, experiencing range contraction or range shift southward. From about 15 Kya, the climate warmed, the ice retreated, and species expanded their range northward from refugia. Valuable data on this range expansion during the Holocene (the last 11,000 years) come from insect exoskeletons, plant macrofossils and pollen cores collected from many localities.

The postglacial expansion of trees in North America and Europe has been studied extensively based on fossil pollen cores from lake sediments across a wide geographical range, dated by radiocarbon (^{14}C) analysis (Davis, 1976; Huntley and Birks, 1983; Delcourt and Delcourt, 1987; Birks, 1989). The results are often expressed as isochrone maps in which contours join sites where similar pollen-stratigraphical events occurred at the same time and as rates of range expansion during a given period (Table I). These interpretations, however, should be made with care since pollen of many tree species are effectively dispersed over large distances;

pollen data also contains a nonlocal component. Estimation of this component is possible through data on other plant remains (seeds, fruits, leaves, twigs, wood), so-called *plant macrofossils*, that are less likely to be transported over large distances. The relatively few studies in which macrofossils have been well documented showed that the main conclusions based on pollen data alone were fairly robust. Genetic studies also strongly support these conclusions.

The paleoecological pollen record shows an individualistic response of tree species, in the timing, direction, and speed of their spread. It also shows that species were able to cross dispersal barriers as large as the North American Great Lakes, the English Channel, and the Baltic Sea. In particular, it provides compelling evidence that range expansion of trees can be very rapid with estimated *average* rates of spread of a few hundred meters per year (Table 1) and more. These estimated rates greatly exceed the rates expected from the observed dispersal rates of these species, collected at ecological spatial and temporal scales (Section II.A). The discrepancy has been a long-standing puzzle for plant ecologists, and termed *Reid's paradox* after Clement Reid who described it for British oaks. Even more puzzling than the rate of expansion of oaks—whose acorns may occasionally be dispersed over large distances by birds—is the rapid rate of expansion of woodland herbs that mostly propagate clonally and possess no obvious mechanism for long-distance dispersal (Cain *et al.*, 1998). The discussion of this paradox has recently been revived because of the need to predict biological responses to human-induced climate warming. In fact, evidence on population expansions and retractions from butterflies (Parmesan, 1996; Parmesan *et al.*, 1999) and breeding birds (Thomas and Lennon, 1999) indicate that global climatic warming is already reflected in considerable poleward shift of species' distributions. Consequently, scientists have developed new mathematical models for seed dispersal that incorporate the long-distance component as well (Section IV.C).

Another recent advance in the study of postglacial range expansion follows the progress in molecular genetics that enables assessment of present genetic structure; that is, in turn, used to infer past colonization events (see Section IV.B). Godfrey Hewitt (1999) summarized the results of studies dealing with the postglacial colonization of the European biota (both animals and plants) from three main refugia at the eastern (Balkans), central (Italian), and western (Iberian) peninsulas of southern Europe. He proposed three main expansion patterns, which were termed, after their exemplars, "hedgehog," "grasshopper," and "bear" (Fig. 4c): the hedgehog (*Erinaceus* spp.) expanded with three genomes advancing north from the three refugia; the grasshopper (*Chorthippus parallelus*) colonized mostly from a eastern refugium, with other genomes blocked at the Pyrenees and Alps; the bear (*Ursus arctos*) colonized from the east and the west, with its central refugium blocked at the Alps.

c. Spread of Early Hominids

Modern man (*Homo sapiens sapiens*) is currently the species with the most widespread geographical distribution, and other globally distributed species are mostly human commensalists. The dispersal biogeography of humanity is a very intriguing subject; it is also an excellent case study that demonstrates many essential aspects of dispersal biogeography. The following discussion briefly summarizes the patterns of dispersal and range expansion of early hominids, mostly based on Roy Larick and Russell Ciochon (1996) and Ian Tattersall (1997); note that there still is some argument over this topic.

The ancestral source location of early hominids is considered to be the plains of the eastern Rift Valley in Africa (now Ethiopia, Kenya, and Tanzania) where fossils of the first, woodland-dwelling, hominids were found (*Australopithecus anamensis*, 4.2–3.9 Mya, and *Australopithecus afarensis*, 3.0–3.9 Mya). The cooling period during the Middle Pliocene (3.0–2.4 Mya) marks the emergence and dispersal of the genus *Homo*. The fossil record shows that during this period many mammalian species shifted from forests to open landscapes, including the earliest *Homo* species, omnivorous scavengers who lived in open habitats between 2.5 and 2.0 Mya. One of these species, *Homo ergaster* (closely related to *Homo erectus* found in Java) had long torso and limbs, narrow hips, a large brain, reduced dentition, and elementary stone technology. The findings of many bone and stone refuse near lakes and streams suggest that early hominids congregated within catchment basins and utilized local resources until they were exhausted, and then moved to another catchment, and so on. All of these attributes served as morphological and behavioral preadaptations (Section III.B.1) for successful dispersal to and colonization of similar habitats outside Africa during the Pliocene (Fig. 5). At the same period some large bovids dispersed from Africa to Eurasia; perhaps the two groups emerged and dispersed together, with the bovids being hunted or scavenged by *Homo*.

The earliest fossil of *Homo* (species indeterminate) outside Africa, dated 1.9 Mya, was found in Longgupo (China). Early *Homo erectus* skulls dated 1.8 and 1.7

TABLE I

Observed Rates of Range Expansion, Arranged in Decreasing Order within Each Taxonomic Group

Species	Period	Region	Average expansion rate ($km\ y^{-1}$)	Reference
Mammals				
European rabbit	1870–1900	Australia	54	Myres, 1970
Oryctolagus cuniculus				
Muskrat		Czechoslovakia		Andow et al., 1990
Ondatra zibethicus	1905–1947	to east-southeast	25	
	1905–1948	to north	12	
	1932–1954	France	0.9–4.6	Andow et al., 1990
		3 localities		
	1923–1938	Finland	3.5–6.7	Andow et al., 1990
		4 localities		
Grey squirrel	1965–1981	Britain	7.7	Okubo et al., 1989
Sciurus carolinensis				
Red deer	1861–1910	New Zealand	1–1.6	Shigesada and Kawasaki, 1997
Cervus elaphus	1910–1920		4.3	
California sea otter	1938–1972	U.S., California		Lubina and Levin, 1988
Enhydra lutris		to south	3.1	
		to north	1.4	
Himalayan tahr	1936–1966	New Zealand	0.7	Shigesada and Kawasaki, 1997
Hemitragus jemlahicus				
Birds				
Cattle egret	1950–1970	S. America to N. America	106	Van den Bosch et al., 1992
Bubulcus ibis				
House sparrow	1864–1888	U.S.	66	Okubo, 1988
Passer domesticus				
European Starling	1900–1915	U.S.	11	Okubo, 1988
Sturnus vulgaris	1915–1950		51	
Collared dove	1928–1963	Europe	44	Van den Bosch et al., 1992
Streptopelia decaocto				
House finch	1947–1962	U.S.	3.5	Shigesada and Kawasaki, 1997
Carpodacus mexicanus	1962–1979		21	
Insects				
Africanized bee	1957–1992	S. America to N. America	300–500	Winston, 1992
Apis melifera scutellata				
Rice water weevil	1976–1986	Japan		Andow et al., 1993
Lissorhoptrus oryzophilus		to northeast	28–470	
		to west	47–250	
Cabbage white butterfly	1860–1883	Canada, Quebec	15–170	Andow et al., 1990
Pieris rapae				
Cereal leaf beetle	1962–1979	U.S., Michigan	27–90	Andow et al., 1990
Oulema melanopus				
Gypsy moth	1900–1916	N. America	9.5	Liebhold et al., 1992
Lymantria dispar	1916–1965		2.8	
	1966–1990		21	
Vascular plants[a]				
Alder	8.5–5.5 Kya	British Isles	0.33	Birks, 1989
Alnus glutinosa	8.5–7 Kya		0.55	
	7–6 Kya		0.15	

continues

Continued

Species	Period	Region	Average expansion rate (km y^{-1})	Reference
Aspen	20 Kya–present	N. America	0.26	Delcourt and Delcourt, 1987
Populus spp.	14–12 Kya		0.54	
	10–8 Kya		0.15	
Oak	9.5–6 Kya	British Isles	0.21	Birks, 1989
Quercus spp.	9.5–8.5 Kya		0.48	
	7–6 Kya		0.04	
	20 Kya–present	N. America	0.13	Delcourt and Delcourt, 1987
	16–14 Kya		0.18	
	10–8 Kya		0.04	
Bacteria and viruses				
Bubonic plague (black death)	1347–1350	Europe	320–650	Noble, 1974
Yersinia pestis				
Rabies	1939–1983	Europe	30–60	Shigesada and Kawasaki, 1997
Lyssavirus				

[a] Postglacial expansion rates of trees during the late Quaternary, based on fossil pollen records. The first row in each data set shows the overall average rate during the entire period examined, the second row shows the period with the fastest rate, the third shows data from the same period (in North America and British Isles separately), for comparison.

Mya were found in Mojokerto and Sangiran (Java, Indonesia). Thus, the departure from Africa occurred roughly 2 Mya. During this period there were wide land bridges between Africa and Eurasia at the Sinai Peninsula, and at the straits of Bab-el-Mandab between the Red Sea and the Indian Ocean. Probably both served as dispersal routes for bovids and *Homo*; fossils and stone-tools dated 1.96 to 1.4 Mya were found along the possible paths to China and Indonesia from 'Ubeidiya and Erq-el-Ahmar (Israel), and through Dmanisi (Georgia), Riwat, and Pabbi hills (Pakistan). Before the findings from eastern Asia, the consensus about the earliest hominid "out-of-Africa" dispersal was that it occurred roughly 1 Mya later. Since this early departure from Africa 2 Mya, early hominids successively dispersed from Africa to Asia, and later to Europe, where the earliest *Homo* fossil found (in Atapuerca hills, Spain) is dated 0.78 Mya. Compelling evidence shows that *Homo sapiens*, our own species, also originated in Africa from where it eventually colonized a much broader range than any of its extinct predecessors.

5. The Spread of Alien Species

Humans can facilitate the physical access of organisms from one region to another by breaking dispersal barriers or by providing dispersal routes. Species that have colonized nonnative regions due to human activities are called *alien species*, and those that have been deliberately transferred are called *introduced species*. Many closely related terms, including *adventive, exotic*, and *invasive species*, have been used confusingly in the literature. Though examples are numerous, the spread of alien species represents merely a small, though growing, fraction of the total number of incidents in which humans have intentionally or accidentally facilitated physical access. Further, as a general rule—termed *propagule pressure* by Mark Williamson (1996), which is principally equivalent to the *rescue effects* (Section V.B)—single or very few propagules are unlikely to invade. Increasing the number of propagules (e.g., by repeated introduction attempts) increases the chance of an alien species to establish and spread. The literature provides a wide array of case studies of various taxa across the entire globe, from zebra mussels and starlings in North America to rabbits and myxoma virus in Australia (Table I). I have chosen the classical case study of the muskrat (*Ondatra zibethicus*) to illustrate the implications from such events to dispersal biogeography; the chapter *Plant Invasions* provides more examples of the spread of alien plants.

The muskrat is a large semiaquatic rodent native to North America (Fig. 6). In 1905, five muskrats escaped from a fur-breeding farm near Prague in Czechoslovakia. Within 50 years, their descendants and others inhabited large parts of the European continent (Fig. 6), despite eradication programs initiated due to muskrat

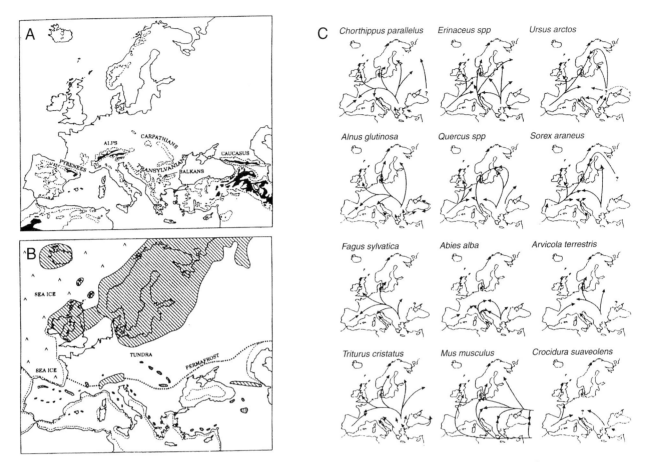

FIGURE 4 Postglacial recolonization of European biota. The physical geography of Europe (a) is characterized by mountain ranges running from the Pyrenees eastward through the Alps to Caucasus, marked as black regions (>2000 m) and dashed lines (>1000 m altitude). By the end of the Pleistocene ice age (b) extensive ice sheets (hatched) covered northern Europe, and a permafrost plain extended southward to the mountain ranges. The rapid postglacial expansion out of the three southern peninsular refugia (c), as deduced from molecular data and fossil evidence, followed three main patterns as illustrated by four examples in each column (see text for details). From Hewitt (1999).

damage to roads, dikes, and agriculture. Expansion of the muskrat appeared to be rapid and of constant rate (see Section IV.C.2 for quantification of this case study). At finer scales, however, expansion was fairly variable in space and, to a lesser extent, in time (Table I). Overall, the expansion was closely related to the geographical structure of the species' favorable wet habitats. River valleys, one kind of favorable habitat, provided dispersal routes, while mountain ranges acted as dispersal barriers, forming a starlike expansion pattern. Interestingly, favorable habitats do not necessarily increase but may also decrease the rate of spread, as happened in regions with abundant wet habitats where muskrats settled and built up huge populations. Understanding the interplay between the processes of dispersal and of population growth is critical to the understanding of range expan-

sion (Section IV.C). Temporal changes should also be taken into account, since expansion rates were lower in dry years when the species' favorable habitats were more restricted.

B. Inference from Current Distributions

Early biogeographers have frequently compared species composition of different regions to evaluate their biogeographic affinities. This analysis, later scrutinized to include various quantitative methods, has revealed a common nonrandom pattern called *nestedness*, in which small biotas contain a nonrandom subset of the species in richer ones (Patterson and Atmar, 1986). Found in various taxonomic groups, nestedness is generally attributed to differential extinction and colonization,

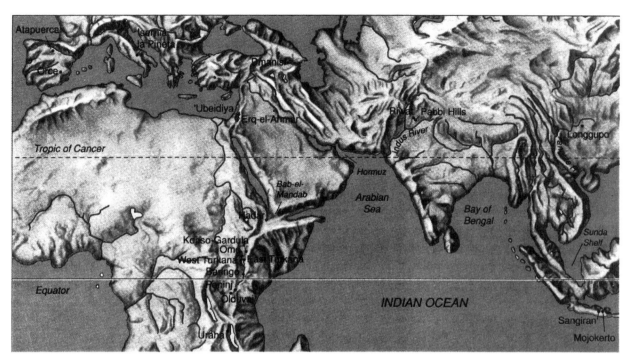

FIGURE 5 Dispersal routes that allowed passage from Africa across the Middle-East into South and East Asia during the late Pliocene and the main sites where human remains were discovered. From Larick and Ciochon (1996) with permission.

or a combination of both, though other factors may also be responsible. Differential extinction is arguably the prevalent mechanism of nestedness, especially in fragmented biotas. However, several biotas assembled by colonization also exhibited nested patterns. Some studies suggested that nestedness reflects differential dispersal ability coupled with geographical isolation. This suggestion can be tested by ranking the subsets (e.g., islands) according to their distance from the presumed source (e.g., mainland), and by examining the correlation between individual species contributions to the observed nestedness and species dispersal ability. Ronen Kadmon (1995) used this approach for woody plant species in seven man-made islands in a reservoir in eastern North America and confirmed that strong nestedness was attributable to geographic isolation and was correlated with dispersal ability. Wind-dispersed species showed no evidence of nestedness and species without particular adaptations for long-distance dispersal showed the strongest nestedness. An equivalent pattern is seen in much broader spatial scales, where high dispersal abilities are associated with low levels of endemism, for example in marine benthic invertebrates with teleplanic larvae such as the gastropod shown in Box 1 (Scheltema, 1971). Similarly, the biotas of isolated islands such as Krakatau (Section IV.A.2), Galápa-

gos, and Hawaii, are overrepresented by taxa having long-distance dispersal mechanisms (but see Section III.C), a characteristic termed *disharmony* to denote the lack of biotic elements that are conspicuously present on continents (Carlquist, 1974).

During the 1970s and early 1980s, the approach used in the study of historical biogeography shifted dramatically from explanations of the unique responses of individual groups to a quantitative and more rigorous analysis of general patterns. This progress was achieved mainly by the scholars of vicariance biogeography (Section I; Myres and Giller, 1988) who, based on Hennig's phylogenetic systematics and Croizat's panbiogeography, developed the *area cladogram* as a general technique for the analysis of biogeographic patterns (see *Vicariance Biogeography*). Area cladograms are branching diagrams representing history, similar to traditional phylogenetic trees, but with the geographic areas replacing taxa at the terminal tips. Correspondence between phylogeny and independently derived geological history is taken to imply vicariance; dispersal is generally invoked to explain lack of correspondence. Congruence among area cladograms of different taxonomic groups from the same region more strongly implies a vicariant event and may be used to infer geological history in the absence of independent data.

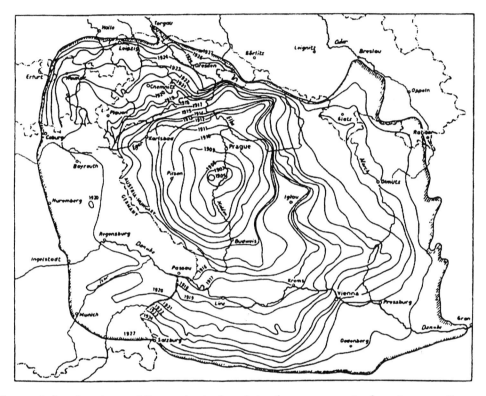

FIGURE 6   The spread of muskrats in central Europe, showing boundaries of apparent expansion for various years. From Kluwer Academic Publishers, *Biological Invasions*, Williamson, Figures 4.3 and 4.4 © (1996) with kind permission from Kluwer Academic Publishers.

Juan Morrone and Jorge Crisci (1995) classified the historical biogeography methods into five groups: dispersalist, phylogenetic biogeography, panbiogeography, cladistic biogeography, and parsimony analysis of endemicity. The methods of cladistic biogeography (see *Vicariance Biogeography*) are the most diverse and most widely used; important ones include *component analysis*, *Brooks parsimony analysis*, and *reconciled trees* (Morrone and Crisci, 1995). These approaches construct an area cladogram assuming vicariance as a null hypothesis. An alternative interpretation is presented in the book edited by Warren Wagner and V. A. Funk (1995) where simple area cladograms were produced for 25 taxonomic groups, including insects, spiders,

birds, and flowering plants, of the Hawaiian Islands. The cladistic analysis was undertaken using various attributes, including morphological and anatomical characters, banding patterns on chromosomes, and DNA sequences from chloroplast, mitochondria, and nuclear ribosomes. Area cladograms were compared with several hypothetical patterns that represent historical scenarios that putatively generated the observed patterns, based on the geological history of the islands. The most frequent pattern was of dispersal from older to younger islands; several patterns were interpreted either as recent colonization from the mainland or as vicariant events within the archipelago. A book on the historical biogeography of Southeast Asia edited by

Robert Hall and Jeremy Holloway (1998) provides another exemplary integration of ecological and evolutionary processes that account for the impact of dispersal and vicariance along with a profound analysis of the complex geological history of this region.

Marrone and Crisci (1995) classified Kåre Bremer's (1992) *ancestral areas* method as a dispersalist approach. This method is aimed at identifying the ancestral area of a group from the topological information of its cladogram by estimating the likelihood that different areas were part of the ancestral area for the group. More recently, Fredrik Ronquist (1997) developed the *dispersal-vicariance analysis (DIVA)*, which reconstructs the ancestral distributions of the species in a given phylogeny without any prior assumption about the form of area relationships and allowing dispersal and extinction. The optimal ancestral distributions are those that minimize the number of dispersal and extinction events required to explain the pattern. The number of dispersal events is underestimated because similar distributions between species are interpreted as sharing vicariant history, though this is not necessarily the case. Nevertheless, this method disentangles some important methodological and conceptual restrictions of alterna-

tive methods and is so far the biogeographic method that most realistically incorporates the effects of dispersal. Gary Voelker (1999) has recently applied the two methods to reconstruct the ancestral distributions and the direction of dispersal events in the pipits (*Anthus* spp.), small songbirds taxonomically related to the wagtails that occur in all continents. The results show that 16 dispersal events are required to explain the biogeographic pattern of the 40 species analyzed. The reconstructed map of colonizations from the postulated ancestral area in central Asia (Fig. 7) shows broad intercontinental movements, suggesting that dispersal played an important role in the distribution of this group.

C. Modeling

Mathematical models are useful tools for exploring the role of dispersal in the process of range expansion and for predicting expansion dynamics. A comprehensive study from various fields such as genetics, epidemiology, and biological control has led to the development of highly diverse mathematical approaches of varying complexity to modeling movement and spatial popula-

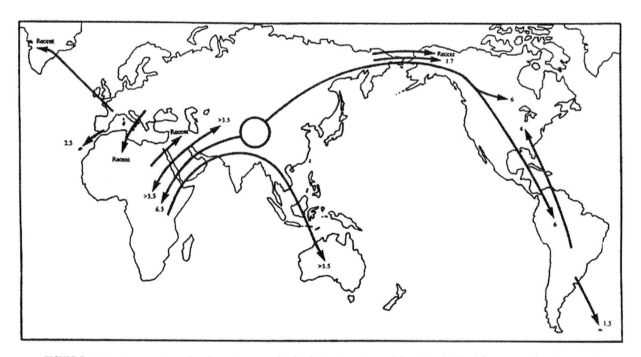

FIGURE 7   Major intercontinental and continent-to-island colonizations (arrows) by pipits (*Anthus*) form a postulated ancestral area in central Asia (large circle). The approximate time of each colonization event is given either in million years (numbers close to the arrowheads) or is defined as *Recent* (single species that have dispersed onto another landmass). From Voelker (1999) with permission.

tion dynamics. These models include *dispersal models* that focus on the movement alone and *reproduction-dispersal models* that incorporate dispersal with local population growth to model the spatial spread of populations. The rich literature on this subject was recently reviewed and summarized by Turchin (1998, p. 73), who concluded with the following prognosis: "the theory is well developed, and appears to be ready to incorporate whatever quantitative information about movement we can provide." Surprisingly, biogeographers have generally overlooked this progress, despite its clear relevance and potential contribution to the field. A brief overview of some important dispersal and expansion models follows; details can be found in Okubo (1980), Durrett and Levin (1994), Mollison (1995), Shigesada and Kawasaki (1997), and Turchin (1998).

1. Dispersal Models

Dispersal models include phenomenological (empirical curve fitting), diffusion, individual-based, and mechanistic (transport) models. The basic objective of these models is to estimate the probability distribution of dispersal distances, which is often given in a standardized functional form called the *dispersal kernel* (Nathan and Muller-Landau, 2000). A classic dispersal model used in biogeography is the stepping-stone model of MacArthur and Wilson (1967), which assumes that a disseminule moves in a constant direction from one island to another and has a constant probability m per unit time of landing (cessation of movement). Mathematically, these assumptions lead to a widely used phenomenological model—the negative exponential function—in which the probability to disperse to distance x decreases with x at a rate equal to $\ln(m)$. Other widely used phenomenological dispersal models follow the Gaussian distribution and the inverse power law. The Gaussian models usually underestimate observed dispersal data, which almost universally exhibit a leptokurtic distribution (Section II.A). The inverse power law has the problem of singularity (infinite density) at zero and hence cannot be used as a dispersal kernel.

Another set of dispersal models that has been developed recently and is likely to prove beneficial to dispersal biogeography is that of mechanistic models, which simulate the process by a set of mathematical expressions that mimic the relationships and joint effects of the principal operative factors. For example, mechanistic models of seed dispersal by wind typically incorporate factors such as the height of release, the seed free-fall velocity, and wind velocities (Nathan *et al.*, 2000). While complicated to develop and parameterize,

these models have the decisive advantages of actually predicting, rather than merely fitting, changes in disseminule densities with distance, and of potential generalization to other systems. The task of applying and testing models of long-distance dispersal is inherently more complicated than for short-distance dispersal. However, recent technical developments with miniature transmitters, radar, and other remote sensing techniques, as well as molecular markers, are promising a breakthrough in the near future. Important attempts to model the mechanism of long-distance dispersal has already been made for the wind dispersal of aerial plankton (Westbrook and Isard, 1999) and tree seeds (Greene and Johnson, 1995), based on the knowledge of atmospheric dynamics. Similarly, knowledge of sea currents and the winds that drive them can be used to predict the trajectories of passively dispersed marine biota (Jokiel and Martinelli, 1992). Further, dispersal of seeds by birds may to a certain degree be predictable with knowledge of gut retention time and direction and velocity of bird flight (Sun *et al.*, 1997).

From the biogeographic perspective, special attention should be paid to the tail of the distribution, since long-distance dispersal is disproportionately important in range expansions. Mathematically, it has been shown that kernels differing significantly in their tails yield dramatically different rates of population spread, even if they have the same mean. Recent recognition of the need to focus on long-distance dispersal has led to the development of several alternative "fat-tailed" phenomenological and mechanistic dispersal models, as emphasized in Shigesada and Kawasaki (1997), Turchin (1998), Clark *et al.* (1999), and Nathan and Muller-Landau (2000).

Turchin (1998) summarized the relationships between dispersal models and the types of population spread, focusing on the tail of the dispersal kernel. If the tail of the dispersal kernel drops off exponentially or faster with distance (for example, Gaussian tails that are predicted by simple diffusion), then we should observe linear spatial spread in which the range is expanded at a constant rate during each time step. If the tail drops off slower than the exponential tail, as in the so-called "fat-tailed" kernels, the type of spatial spread depends on whether the variance of the kernel is finite or infinite. In both cases the rate of spatial spread accelerates (i.e., increases with successive time steps), but the former reaches an asymptote of a constant expansion rate, while the later continues to increase. Thus, fat-tailed dispersal kernels lead to jump range expansion, while exponential and thinner tails lead to gradual range expansion (Section II.C). Since data on long-distance dispersal is rare, the assumptions of the characteristics of the distribution's tail cannot yet be tested; at the same time it is clear that extrapolating beyond the observed range is questionable.

2. Reproduction-Dispersal Models

The dispersal models that have been discussed are useful tools to understand the general relationships between dispersal and spatial spread. Reproduction-dispersal models, however, are more realistic since they consider the effects of birth and death during the process of range expansion. The simplest reproduction-dispersal models are reaction-diffusion models; and the simplest of those, introduced to biologists by Skellam (1951), model dispersal as random diffusion and local population growth as exponential increase,

$$\frac{dN}{dt} = rN + D \left[\frac{\partial^2 N}{\partial x^2} \right]$$

where N is the population size, t is time, and x is the distance from the source; the two parameters are D, the diffusion coefficient, and r, the intrinsic rate of population growth. Earlier, Ronald Fisher (1937) used a nonlinear variant of this equation in which the unbounded exponential growth (rN) is replaced by asymptotic logistic growth ($rN(1 - N/K)$), where K is the equilibrium population size (or *carrying capacity*). As shown by Fisher, this model predicts, as does Skellam's model, a constant asymptotic front velocity equal to $2\sqrt{rD}$. As discussed in the previous section, this type of gradual range expansion does not incorporate the effect of long-distance dispersal events. Tests of reproduction-dispersal models against observed expansion pattern, starting from Skellam's (1951) pioneering study, have frequently showed that the simple reaction-diffusion approach cannot provide a general description for range expansion. For example, David Andow, Peter Kareiva, Simon Levin, and Akira Okubo (1993) found approximately linear rates of spread in two case studies and accelerating rates of spread in two others. Even for the cases that show linear rates, such as that of muskrats (Section IV.A.5), there was significant variation in space and time. Expansion rates of muskrats were lower in the first 2 to 3 years, possibly because of Allee effects (Section III.B.3), and the average rates range from 0.9 to nearly 80 km/yr. Thus, reaction-diffusion models are best used as a primary null analysis of expansion patterns, followed by more complex approaches, including stratified diffusion, interacting particle systems, and spatial contact models. These models incorporate

both short- and long-distance dispersal and thus are able to predict patterns of jump range expansion as well.

V. BIOGEOGRAPHIC AND EVOLUTIONARY CONSEQUENCES OF DISPERSAL

Dispersal, evolution (adaptation and speciation), and extinction are the fundamental processes in biogeography: together they determine the geographical distribution of organisms. Dispersal biogeography focuses on the effects of dispersal on evolution and on speciation, both within a lineage and for other lineages. The following section outlines some principal relationships; more details are given in related articles on speciation and extinction in this encyclopedia.

A. Speciation

Speciation is the process by which new independently evolving lineages (species) arise from a single ancestral lineage. In general, efficient dispersal between populations (patches) increases the rate of gene flow between them and hence tends to reduce genetic differentiation and to restrain speciation. In contrast, inefficient dispersal between existing populations leads to *geographic isolation* and tends to increase genetic differentiation and to promote speciation. This type of speciation, called *allopatric speciation*, is considered to be the most frequent one. Jump range expansion occurs with very low probability (Section II.C.2) and is thus unlikely to be followed by additional dispersal; this isolation thereby causes rare dispersal events (in this context also called *founder events*) to be relevant to allopatric speciation. The adaptive radiation of Darwin's finches in the Galápagos Islands (Grant, 1999) is one of the many examples of allopatric speciation that has resulted from dispersal events. However, as discussed in Section I, a dispersal barrier may divide a previously continuous distribution, hence allopatric speciation may also result from a *vicariant event*.

A second way in which dispersal affects speciation requires the opposite conditions. The breakdown of a dispersal barrier, and the following biotic interchange (Section IV.A.3), brings organisms to dissimilar environments and thus opens new chances for diversification. The colonization of North American mammals during the Great American Interchange (Section IV.A.3) was followed by extensive diversification in South America. In the same manner dispersal events cause speciation: they affect the genetic structure within the species range, sometimes to the point that the populations are evolving independently, and can be considered as distinct species. Because there is a continuum of genetic structure and evolutionary independence, the final categorization of one or more species is never completely clear. Theoretical models and molecular studies have shown clear effects of long-distance dispersal on the spatial structure of the genetic variation within a species during the postglacial expansion of European and North American biota (Section IV.A.4.b). As predicted by models, jump range expansion well ahead of the "normal" advancing wave generates large patches of low genetic variation, due to the so-called *founder effects*; consequently, genetic diversity decreases from south to north.

B. Extinction

Limited dispersal is associated not only with higher rates of speciation (Section V.A) but generally also with higher rates of extinction. This phenomenon has been observed at various spatial scales. For example, the fossil record show that gastropods with planktonic larval dispersal, typically involved with long-distance dispersal (Box 1), tend to have larger geographical ranges and to persist for longer periods of time than do those with nonplanktonic larvae, which are dispersed locally. A similar phenomenon is observed at regional scales, where populations with higher rates of immigration persist longer due to increased population size and higher levels of genetic variability, which counteracts the deleterious effects of inbreeding. This so-called *rescue effect* has been observed in a variety of organisms. The very similar notion of *source-sink dynamics* attributes a longer persistence time to populations in unfavorable *sink habitats* (where death rates exceed birth rates) due to a rescue effect from populations in favorable *source habitats* (where birth rates exceed death rates). Another relevant term, *mass effect*, emphasizes the contribution of rescue effects to the breeding stock. The regional framework of populations subjected to local extinction and recolonization is treated in *metapopulation models* in which dispersal plays a critical role (Hanski, 1999) (see *Metapopulations*).

While dispersal tends to reduce the probability of extinction of the same species, dispersal, as immigration, may have the opposite effect on other species inhabiting the recipient site. Around 1950, the brown tree snake (*Boiga irregularis*) of Australia and Southeast Asia was accidentally introduced to the Pacific island of Guam (Rodda *et al.*, 1999). This fairly large (up to

3 m) snake, which eats bird eggs from nests either on the ground or in trees, caused the extinction of most of the native birds and the decline of almost all other native terrestrial vertebrates. Despite this and other spectacular examples of massive extinction caused by colonization of a given species, most colonization events appear to have minor consequences. The consequences of colonization of one species to the persistence on other species depend on the species' biology and on postdispersal processes; however, dispersal is the ultimate cause of such extinctions, since it determines whether the prerequisite species is present. Though the particular dispersal mechanisms should not matter to the severity of such incidents, it is noteworthy that in most documented cases dispersal was related to human activity.

VI. APPLIED DISPERSAL BIOGEOGRAPHY

Biogeography in general, and dispersal biogeography in particular, has an important role in managing the world's biodiversity: the processes currently threatening the world's biota typically act on large scales, and the movement (or lack thereof) of organisms is a key process. Dispersal is strongly involved with issues such as invasions, introductions, fragmentation, climatic changes, design of nature reserves, and control of gene flow; it is hard to imagine any issue of applied biogeography that is not related to this process. Thorough discussions of these practical aspects of biogeography can be found in Williamson (1996), Brown and Lomolino (1998), and Spellerberg and Sawyer (1999); this section outlines some major principles of applied dispersal biogeography.

Applied dispersal biogeography deals with all aspects of dispersal that are involved with the management of the earth's biota and aims to predict the causes and the consequences of dispersal events. The mathematical models described above (Section IV.C) are efficient tools on small scales. Although our ability to predict long-distance dispersal has significantly improved recently, much still remains to be done, especially the testing of these predictions. Long-distance dispersal has a substantial stochastic component, which will always remain somewhat unpredictable. However, since most long-distance dispersal events with harmful consequences are related to human activity, our predictive ability of anthropogenic arrivals may be better than for natural arrivals; even unintentional dispersal as that of the brown tree snake to Guam (Section V.B, Box 1) is to some degree predictable and controllable. As emphasized earlier, the variation among organisms in dispersal ability is enormous (Box 1); responses to dispersal barriers and dispersal routes also vary considerably (Section III.B.2). Mathematical models, especially mechanistic ones (Section IV.C.1), can be used to predict dispersal of particular species using knowledge of the species-specific dispersal traits (Box 1) and the effect of environmental factors on dispersal (Sections III.B.2 and III.C).

The invasion of muskrats to Europe (Section IV.A.5) illustrates these points. The muskrat's arrival was related to human activity and the animal's escape was controllable; expansion rates are predictable through simple mechanistic models based on dispersal and reproduction (Section IV.C.2); the variation in expansion rates was related to habitat-related factors (Sections III.C and IV.A.5) and was therefore also predictable. The large number of escapes and the initial slow rates of expansion demonstrate the importance of propagule pressure (Section IV.A.5) and probably of Allee effects (Section III.B.3) as well. Attempts to control this expansion—mostly extensive trapping campaigns—have failed in continental Europe, but have succeeded in Britain and Ireland. Mark Williamsom (1996) suggested that the relatively early response in the British Isles, before the muskrats spread too far and built overly large populations, along with presumably inferior climatic matching that would have led to lower reproduction rates, are possible reasons.

VII. SYNTHESIS

The importance of dispersal in determining the geographic distribution of organisms has been emphasized by Darwin and other dispersalists. These views have been seriously questioned by their opponents, who argue that vicariant events are the principal determinant of distributions. Clearly, both processes occur and their effects vary between species, in space, and in time. The term *dispersal biogeography* is currently restricted to the arguments of extreme dispersalists concerning long-distance dispersal and disjunct distributions. The broader and more fruitful usage of this term adopted here would be the study of the relationship between dispersal the geographical distribution of organisms.

Dispersal is a widespread phenomenon occurring in nearly all organisms. There is a high variability of dispersal ability in organisms, due to differences in their physiological, morphological, and behavioral traits and responses to environmental conditions. Dispersal is a key biological process at the ecological, evolutionary, genetic, and biogeographic levels, which differ in spatial and temporal scales. Long-distance dispersal, disproportionately important at the biogeographic level, is typically rare and very difficult to measure; it contains a large stochastic component but in many cases is a result of a specific mechanism and hence can be predicted. If followed by successful establishment, long-distance dispersal can lead to abrupt long-distance colonization jumps. Short-distance dispersal is overwhelmingly important at the ecological level; at the biogeographic level, multiple short-distance dispersal events over a relatively long time, if followed by establishment, can lead to gradual range expansion. Patterns of speciation and extinction are strongly affected by dispersal, demonstrating the broad biogeographic consequences of this process, which itself is also affected by geography.

The greatest methodological challenge for dispersal biogeography today is to provide direct evidence, especially of long-distance dispersal. Indirect evidence is based on recruitment and therefore incorporates the obscuring effects of postdispersal processes. Yet there are various ways to infer from indirect evidence to dispersal biogeography, and the case studies described earlier have provided many valuable insights for this endeavor. These case studies include colonization of virgin habitats such as oceanic volcanic islands, observations of the biotic consequences of barrier removal and of natural and human-related range expansions, analyses of current distributions, and modeling of spatial spread. Models are extremely important for providing a better understanding of dispersal, therefore enabling prediction; they thus constitute a promising tool for implementing the principals of dispersal biogeography for maintaining biodiversity.

Acknowledgments

I am pleased to acknowledge the financial support of the National Science Foundation, the Andrew Mellon Foundation, and the Schonbrunn Foundation of the Hebrew University of Jerusalem. I am grateful to Simon Levin, Mark Williamson, Kai Chan, Helene Muller-Landau, Nechama Ben-Elihau, Gad Perry, Yossi Heller, and Eitan Tchernov for helpful comments that considerably improved the article.

See Also the Following Articles

BIOGEOGRAPHY, OVERVIEW • CONTINENTAL DRIFT, ECOLOGICAL EFFECTS OF • ISLAND BIOGEOGRAPHY • LIFE

HISTORY, EVOLUTION OF • METAPOPULATIONS •
MIGRATION • PLANT INVASIONS • SPECIATION, PROCESS
OF • SPECIATION, THEORIES OF • VICARIANCE
BIOGEOGRAPHY

Bibliography

Andow, D. A., Kareiva, P. M., Levin, S. A., and Okubo, A. (1993). *Spread of Invading Organisms: Patterns of Spread*. In *Evolution of Insect Pests: Patterns of Variation* (K. C. Kim and B. A. McPheron, Eds.), pp. 219–242. Wiley, New York.

Bremer, K. (1992). Ancestral areas: A cladistic reinterpretation of the center of origin concept. *Systematic Biology* **41**, 436–445.

Brown, J. H., and Lomolino, M. V. (1998). *Biogeography*, 2nd ed. Sinauer, Sunderland, MA.

Browne, J. (1983). *The Secular Ark: Studies in the History of Biogeography*. Yale University Press, New Haven.

Carlquist, S. (1974). *Island Biology*. Columbia University Press, New York.

Clark, J. S., Silman, M., Kern, R., Macklin, E., and HilleRisLambers, J. (1999). Seed dispersal near and far: Patterns across temperate and tropical forests. *Ecology* **80**, 1475–1494.

Cody, M. L., and Overton, J. M. (1996). Short-term evolution of reduced dispersal in island plant populations. *Journal of Ecology* **84**, 53–61.

Darwin, C. (1859). *The Origin of Species by Means of Natural Selection*. John Murray, London.

Dingle, H. (1996). *Migration: The Biology of Life on the Move*. Oxford University Press, Oxford.

Durrett, R., and Levin, S. A. (1994). Stochastic spatial models: A user's guide to ecological applications. *Philosophical Transactions of the Royal Society of London Series B Biological Sciences* **343**, 329–350.

Fisher, R. A. (1937). The wave of advance of advantageous genes. *Annals of Eugenics (London)* **7**, 355–369.

Hall, R., and Holloway, J. D. (Eds.) (1998). *Biogeography and Geological Evolution of SE Asia*. Backhuys, Leiden.

Hengeveld, R. (1989). *Dynamics of Biological Invasions*. Chapman & Hall, London.

Hewitt, G. M. (1999). Post-glacial re-colonization of European biota. *Biological Journal of the Linnean Society* **68**, 87–112.

Johnson, M. L., and Gaines, M. S. (1990). Evolution of dispersal: Theoretical models and empirical tests using birds and mammals. *Annual Review of Ecology and Systematics* **21**, 449–480.

Kadmon, R. (1995). Nested species subsets and geographic isolation: A case study. *Ecology* **76**, 458–465.

Larick, R., and Ciochon, R. L. (1996). The African emergence and early Asian dispersals of the genus *Homo*. *American Scientist* **84**, 538–551.

MacArthur, R. H., and Wilson, E. O. (1967). *The Theory of Island Biogeography*. Princeton University Press, Princeton.

Mollison, D. (Ed.) (1995). *Epidemic Models: Their Structure and Relation to Data*. Cambridge University Press, Cambridge.

Morrone, J. J., and Crisci, J. V. (1995). Historical biogeography: Introduction to methods. *Annual Review of Ecology and Systematics* **26**, 373–401.

Myres, A. A., and Giller, P. S. (Eds.) (1988). *Analytical Biogeography: An Integrated Approach to the Study of Animal and Plant Distributions*. Chapman and Hall, London.

Nathan, R., and Muller-Landau, H. C. (2000). Spatial patterns of seed dispersal, their determinants and consequences for recruitment. *Trends in Ecology and Evolution* (in press).

Nelson, G., and Platnick, N. (1981). *Systematics and Biogeography: Cladistics and Vicariance*. Columbia University Press, New York.

Nelson, G., and Rosen, D. E. (Eds.) (1981). *Vicariance Biogeography: A Critique*. Columbia University Press, New York.

Okubo, A. (1980). *Diffusion and Ecological Problems: Mathematical Models*. Springer, New York.

Pielou, E. C. (1979). *Biogeography*. Wiley & Sons, New York.

van der Pijl, L. (1982). *Principles of Dispersal in Higher Plants*, 3rd ed. Springer, Berlin.

Por, F. D. (1978). *Lessepsian Migration: The Influx of Red Sea Biota into the Mediterranean by Way of the Suez Canal*. Springer, Berlin.

Ridley, H. N. (1930). *The Dispersal of Plants throughout the World*. Reeve, Ashford.

Ronquist, F. (1997). Dispersal-vicariance analysis: A new approach to the quantification of historical biogeography. *Systematic Biology* **46**, 195–203.

Shigesada, N., and Kawasaki, K. (1997). *Biological Invasions: Theory and Practice*. Oxford University Press, New York.

Simpson, G. G. (1965). *The Geography of Evolution*. Chilton, Philadelphia.

Skellam, J. G. (1951). Random dispersal in theoretical populations. *Biometrika* **38**, 196–218.

Spellerberg, I. F., and Sawyer, J. W. D. (1999). *An Introduction to Applied Biogeography*. Cambridge University Press, Cambridge.

Tattersall, I. (1997). Out of Africa again . . . and again? *Scientific American*. **276**, 60–67.

Thornton, I. (1996). *Krakatau: The Destruction and Reassembly of an Island Ecosystem*. Harvard University Press, Cambridge, MA.

Turchin, P. (1998). *Quantitative Analysis of Movement*. Sinauer, Sunderland, MA.

Voelker, G. (1999). Dispersal, vicariance, and clocks: Historical biogeography and speciation in a cosmopolitan passerine genus (*Anthus: Motacillidae*). *Evolution* **53**, 1536–1552.

Wagner, W. L., and Funk, V. A. (Eds.) (1995). *Hawaiian Biogeography: Evolution on a Hot Spot Archipelago*. Smithsonian Institution Press, Washington, DC.

Wiley, E. O. (1981). *Phylogenetics: The Theory and Practice of Phylogenetic Systematics*. Wiley & Sons, New York.

Williamson, M. (1996). *Biological Invasions*. Chapman & Hall, London.

DISTURBANCE, MECHANISMS OF

Frank W. Davis* and Max Moritz[†]
*University of California at Santa Barbara and [†]California Polytechnic State University

I. Disturbance Ecology
II. Disturbance and Biodiversity
III. Mechanical Disturbances
IV. Physico-chemical Disturbances
V. Disturbance by Herbivores and Predators
VI. Interactions among Disturbance Mechanisms
VII. Humans as Agents of Disturbance

GLOSSARY

disturbance Relatively discrete event in time that disrupts ecosystem, community, or population structure and changes resources, substrate availability, or the physical environment.
disturbance regime The collective spatial, temporal, physical, and ecological characteristics of a disturbance process operating in an area.
resilience Measure of a system's ability to recover to its original state when subjected to disturbance.
resistance A measure of system's ability to remain unchanged when subjected to a disturbance.

DISTURBANCES in ecological systems promote characteristic patterns of environmental heterogeneity and regulate ecosystem processes, population dynamics, species interactions, and species diversity. Ecological disturbances are created by mechanical forces, extreme physico-chemical conditions, and biological consumers and pathogens. Each of these three classes of mechanisms has somewhat distinctive characteristics in the way it operates, the kinds of organisms affected, and the biological and physical legacies that remain. This article reviews general principles of disturbance ecology and describes and illustrates different disturbance mechanisms and their interactions, concluding with a brief discussion of the role of humans in altering disturbance regimes.

I. DISTURBANCE ECOLOGY

Ecological communities are subjected to sharp environmental perturbations, such as the passage of a fire, a storm wave, an avalanche, or a large animal, that suddenly reduce standing biomass. Such events also alter ecosystem properties such as energy balance, nutrient fluxes, substrate texture, and chemistry. Generally referred to as "disturbances," these events are distinguished as a special form of environmental variability because they are relatively discrete in both time and space and cause unusual mortality or tissue loss in affected populations. Disturbances promote environmental heterogeneity and free up limiting resources, such as space, light, and nutrients, thereby triggering successional processes of community recovery.

Widespread agents of disturbance in terrestrial ecosystems include fire, wind, extreme temperature, desiccation, gravity (as a force on water, ice, rocks, and soil), and organisms. With the exception of fire, one can find analogs in aquatic systems, in which the main agents are heat, solutes, currents, desiccation, waves, ice, sediments, and organisms. Disturbance agents operate by exerting mechanical force, altering physico-chemical conditions, or through biological consumption and disease. For example, practically any physical mass can mechanically disturb ecosystems at some scale provided it has sufficient velocity to dislodge organisms or kill tissues. In contrast, disturbance by fire involves chemical combustion of biomass. Disturbance of terrestrial communities by prolonged inundation is largely a chemical disturbance resulting from oxygen depletion. Freshets in estuarine and marine environments are also chemical disturbances. These mechanical and physico-chemical disturbances contrast with biological disturbances such as herbivory or predation, which involve tissue removal and digestion by individual mobile consumers.

During the first half of the twentieth century, ecologists paid more attention to post-disturbance processes of species recovery and community succession than to specific disturbance mechanisms. Disturbance was generally treated as a temporary setback to communities that otherwise would tend to develop toward a relatively steady state or "climax community" whose structure and composition were determined by climate and other physical factors and were regulated by endogenous biological interactions. Recently, disturbance has been recognized as intrinsic to and ongoing in virtually all ecological systems. Closer attention has been paid to disturbance processes and their role in promoting characteristic scales of spatial and temporal environmental heterogeneity and in regulating ecosystem processes, population dynamics, species interactions, and species diversity (Paine and Levin, 1981; Sousa, 1984; Pickett and White, 1985). Disturbance ecology has become a familiar term in the lexicon of ecology and a recognized area of emphasis in ecological research.

In practice, it may not be easy to distinguish disturbances from other environmental variation. Most agents of disturbance operate over a continuum, and environmental perturbation is sudden and severe only relative to some set of reference conditions and from the perspective of the affected organisms. The mound of soil produced by a burrowing gopher is a significant disturbance to the underlying herbaceous plants and soil animals but likely to be of little consequence to a large tree a few meters away. A windstorm that topples trees in an open savanna may have no immediate effect on small herbaceous plants located a short distance outside of the canopy. Thus, it is important to bear in mind that disturbance is a relativistic concept and that distur-

bances can span a very broad range of spatial and temporal scales. Not surprisingly, the term disturbance has been applied somewhat indiscriminately in ecology. The most general definition of a disturbance—any process that causes a sudden decrease in standing live biomass and frees up ecological resources (Sousa, 1984)—is perhaps the most unambiguous.

A disturbance differs from a stress in that the latter is a more chronic condition inhibiting the growth or normal functioning of an organism (e.g., a lack of key nutrients or physical abrasion). A disturbance is termed a catastrophe if it causes extraordinary ecological impact.

Variables commonly used to describe a single disturbance event include timing, extent, and magnitude, where magnitude encompasses both intensity (e.g., energy per area per time) and severity (biological impact). These and other stochastic variables, such as event frequency or recurrence interval between events, have statistical properties that serve to define a disturbance regime. More broadly, a disturbance regime is the collective spatial, temporal, physical, and ecological characteristics of a disturbance process operating in an area. Predictability, which can be defined as the inverse of the variance in disturbance frequency, size, and magnitude (Christensen, 1988), is also an important consideration. In general, predictability increases as the spatio-temporal scale of analysis is expanded from local (the typical size of a disturbance event) to regional (the entire area over which the disturbance regime is manifested).

Because the magnitude of disturbance is defined relative to its ecological impact, it is practically tautological that disturbance regimes are dominated by events of relatively low magnitude and high frequency, whereas higher magnitude events are increasingly rare. However, the impact of a disturbance may not increase linearly with size, frequency, or duration. Romme et al. (1998) distinguish three classes of disturbance response: (i) threshold response, (ii) scale-independent response, and (iii) continuous response. Individuals and communities manifest threshold responses when there are discrete limits in their ability to resist a large disturbance (e.g., the wind speed at which a tree is uprooted). Disturbances can have their greatest ecological impact when one or more events follow close on the heels of another, preventing or disrupting normal community recovery (Paine et al., 1998).

Some mechanisms of disturbance, such as earthquakes or storm waves, are exogenous to the biological communities being impacted, whereas others such as treefall or fire could be considered endogenous. In the former, there is little or no feedback between the state of the ecosystem and the likelihood of a disturbance event so that the disturbance regime depends mainly on location and environmental context. In the latter, the likelihood of a disturbance depends on the state of the ecosystem as well as location. Although the categories of endogenous and exogenous disturbances are somewhat artificial, it is useful to examine the relative strength of coupling between disturbance processes and the biota. In many cases, disturbance processes and their effects are tightly coupled to the biological properties of individual organisms and communities. This coupling may promote the formation of specific scales of ecological pattern and reinforce certain ecological and evolutionary processes (Levin, 1992).

In thinking about patterns and processes, it is also useful to distinguish spatially propagating from nonpropagating disturbances. Disturbances such as fire and flood spread from neighboring areas, and the spatial pattern of "susceptible" areas or organisms may have a constraining effect on disturbance dynamics, thereby linking spread to previous disturbance events.

II. DISTURBANCE AND BIODIVERSITY

The "intermediate disturbance hypothesis" (IDH; Connell, 1978) predicts that maximum levels of biodiversity should be observed under some intermediate disturbance frequency because few species are able to tolerate very intense disturbance regimes, and few are able to compete successfully in habitats that experience little or no disturbance. The IDH also implies that maximum diversity should be found at some intermediate span of time since the last disturbance. The IDH has been expanded to incorporate intermediate levels of disturbance intensity and extent, and it has been tested and supported in a wide variety of ecosystems.

The best experimental examples of the IDH come from the study of sessile species competing for space or some space-associated resource. Counter-examples come largely from the study of mobile consumers (e.g., freshwater invertebrates), for which rapid immigration may override local disturbance effects. Defining "intermediate" in the context of specific organisms and choosing the scale at which to measure diversity are also important issues in assessing the validity of the IDH, about which there is ongoing debate. In general, the IDH applies to small-scale disturbances and to plants and sessile filter feeders. The relationship between disturbance and diversity is more complicated at larger scales and may not apply when interactions among multiple trophic levels are considered.

At spatial scales much larger than the characteristic size of single disturbance events, disturbance regimes generate disturbance mosaics that maintain beta diversity in landscapes or regions by promoting co-existence of dispersal-limited competing species or prey species and by maintaining environmental heterogeneity and multiple seral stages (Pickett and White, 1985).

If they recur with sufficiently high frequency (e.g., on average at least once per generation), ecological disturbances can be a strong selective force operating on species' morphology, physiology, and/or behavior. Not surprisingly, many organisms are adapted to and/or depend on specific kinds of disturbances and disturbance regimes. Grime (1979) proposed that herbaceous plant life histories can be ordered along three fundamental axes: stress tolerance, competition, and disturbance. Somewhat analogously, animal species are often called r-strategists or k-strategists depending on whether they have high intrinsic rates of reproduction and tend to be favored by disturbances or whether they have lower reproductive rates but exert competitive dominance in the absence of disturbance. Because local abundance of many species is increased or maintained by disturbance, the long absence of specific kinds of disturbances (e.g., fire or floods) may have large negative impacts on biodiversity.

The life history strategies of some organisms may promote specific disturbance regimes (e.g., some fire-adapted shrub species possess canopy structure and foliar chemistry that promote fire spread). Organisms that recover quickly after a disturbance are said to be resilient to that disturbance, as opposed to those that show little response to disturbance and are considered resistant. These concepts are also applied to ecological communities, and the relationship between community diversity and community stability and resilience has long preoccupied ecologists (Holling, 1973). The subject has received renewed attention due to increasing concern over human-caused species extinctions and community impoverishment associated with habitat fragmentation. To date, the hypothesis that community stability to disturbance increases with community richness has met with mixed results in modeling and empirical studies, in part due to differences in spatial and temporal scale and in how stability is measured.

III. MECHANICAL DISTURBANCES

Most aquatic and terrestrial communities are constantly subjected to mechanical forces that reduce standing biomass at some scale. Avalanches, landslides, and debris flows remove soil, shear off and uproot plants, and bury plants and animals. High winds, snow, and ice uproot trees and break off branches and leaves that then bury other plants and animals. Ice in the nearshore zone plucks and scours littoral communities. Floodwaters topple riparian vegetation and scour streambeds. Waves dislodge encrusting intertidal organisms and break down branching corals. Large animals break twigs and compact the soil as they move. Humans clear or clip vegetation and plow soils. Burrowing animals excavate plants and animals as they dig. In all of these cases, biomass is dislocated but not immediately chemically transformed or consumed. Most important, space occupied by living tissues is evacuated as those tissues are moved someplace else.

Thus, mechanical disturbances leave conspicuous openings and persistent biological legacies, such as large woody debris embedded in landslides, downed trees next to treefall gaps, floating wracks of detached marine plants and animals, or rubble piles of coral. Materials are often transported to a location where they act to create another disturbance. Thus, the legacy of biological debris from mechanical disturbances not only affects the rate and pattern of community recovery but also may influence the timing and location of subsequent disturbance events.

The spatial distribution of mechanical forces in the environment is highly nonrandom. Features such as substrate, topography, and standing biomass create persistent biophysical pathways in which specific mechanical disturbances are concentrated, thereby compounding existing physical environmental heterogeneity. For example, the location of avalanche paths is strongly associated with topography, geology, and lithologic structure. The impacts of severe wind storms on forest communities varies systematically with topography and soil characteristics as well as vegetation composition and structure (which depend, in part, on disturbance history). Wave energy is concentrated on jutting shorelines. Animal burrowing is concentrated in specific soils and sediments. In other words, many mechanical disturbance processes can have somewhat predictable spatial distributions.

Most mechanical disturbances operate indiscriminately (i.e., they do not target specific organisms, as would selective herbivory, for example) and their impact depends on the ability of organisms to withstand the mechanical force. Thus, mechanical disturbances often winnow biological communities. For example, snow avalanches apply a bending stress on trees in their path that is a function of the snow's density and the avalanche's mass and drag. Susceptibility of woody species to breakage (vs bending) increases with tree size, leading to complex interactions between avalanche recurrence frequency and forest structure and composi-

tion in avalanche paths. Similarly, masses of attached barnacles and mussels in rocky intertidal environments become increasingly susceptible to detachment by waves as organism size and density increase. The same mechanical agent may create different kinds of disturbances depending on the size of the organism. In forests, for instance, woody stems may be most susceptible to breakage by wind when they are very small or very large, whereas intermediate-sized stems may be the most susceptible to uprooting by wind (Everham and Brokaw, 1996). When impacts of mechanical disturbances depend on the size or density of organisms, this produces feedbacks that influence population dynamics and create pronounced spatial patchiness at scales well below that of the disturbing process (Levin, 1992).

IV. PHYSICO-CHEMICAL DISTURBANCES

Physico-chemical disturbances, such as extreme cold, prolonged inundation of terrestrial organisms, hypoxic episodes in aquatic systems, freshets in coastal waters, and releases of toxic compounds (e.g., biological exotoxins such as those produced by the dinoflagellate *Pfiesteria piscicida*), reduce biomass by imposing lethal physiological stress. Physico-chemical disturbances do not involve mechanical force (except perhaps in the case of extreme cold when tissue damage is due to intracellular ice formation), and most physical and chemical disturbances leave dead tissues and organisms *in situ*. This contrasts with mechanical disturbances, in which space and resources are freed up through displacement of biological material. There are few studies comparing recovery processes under these two situations, but one might expect differences simply due to the presence or absence of residual detritus that could affect light, nutrient levels, and substrate quality (Reiners, 1983).

Like mechanical disturbances, Physico-chemical disturbances have organism-specific impacts because of species-, age-, and size-specific physiological tolerances to environmental extremes. Thus, like mechanical disturbances, physico-chemical disturbances operate as an environmental filter of biological assemblages.

As mentioned previously, much of the emphasis in disturbance studies has been on sessile organisms such as plants and benthic invertebrates. Some large-scale physico-chemical disturbances may be somewhat distinctive in that they can kill large numbers of highly mobile organisms. For example, episodes of anoxia can cause massive fish kills in estuaries and lakes. Severe cold weather can selectively kill large numbers of birds. In these instances, the suddenness and magnitude of the events prevent animals from escaping, a phenomenon which can also be observed during extreme wildfires.

Fire is obviously a special case of a physico-chemical disturbance because biomass is both transformed and removed in the combustion process. Fire is also unquestionably one of the most pervasive and obvious disturbances in terrestrial ecosystems, and an enormous scientific literature has been devoted to the subject of fire events and fire regimes. It is discussed here to illustrate in more detail the concepts of disturbance mechanisms and disturbance regimes.

The occurrence of fire requires an ignition, fuel combustion, and spread, each of which in turn depends on many physical and biological factors, notably climate, weather, fuel structure and chemistry, and topography. The most important effects of fire are partial or complete combustion of aboveground vegetation, mineralization and deposition of plant tissues as ash and charcoal, extreme heating of the local atmosphere and topsoil, and killing of selected plants, seeds, spores, and animals both above- and belowground.

Figure 1 presents a simple conceptualization of factors that influence fire dynamics at three distinctive spatial and temporal scales in Mediterranean-climate shrublands. The fire fundamentals triangle (Fig. 1a)

captures essential elements of the actual combustion process, which occurs at minute scales and depends on oxygen, heat, and fuel. The timing, size, and severity of a wildfire (or controlled fire) event, which may last hours to weeks and extend over hectares to thousands of square kilometers, depends on local weather, topography, and fuel bed. Finally, the fire regime of a landscape depends on complex interactions among climate, vegetation, and ignition factors operating over decades to centuries.

Fire creates strong biological patterns over a broad range of spatial scales. At the scale of individual plants, biomass combustion, soil heating, and nutrient removal can vary widely, imposing very fine-grained variation in mortality and postfire conditions that is manifested in equally fine-grained patterns of biological recovery. A coarser pattern can be created at the scale of entire populations (vegetation stands). Fire severity and vegetation removal can vary with topography, soil, vegetation, and wind conditions, and distance from source populations for seeds and animal colonists can influence community recovery. At still broader scales, landscapes are composed of fire mosaics, a patchwork of stands with differing fire histories and biotic communities. This landscape mosaic is one of the factors aiding in the long-term persistence of species with different fire tolerances. The spatial patterns from individual plant distributions to landscape mosaics all affect

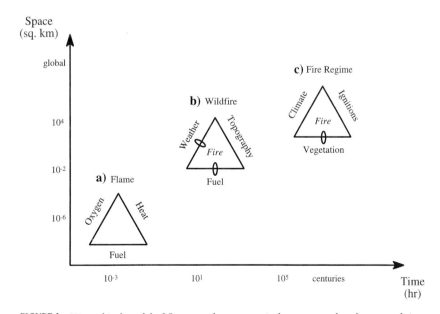

FIGURE 1   Hierarchical model of fire controls at successively coarser scales of space and time. Connections between fire and individual controls indicate potential feedback loops (after Moritz, 1999, with permission).

the location and severity of future fires and their ecological impacts.

Fires are obvious, discrete events, and research on wildfire history has provided many useful models and methods for characterizing temporal and spatial properties of disturbance regimes (Johnson and Gutsell, 1994). Temporal properties such as time-since-fire (survivorship) and fire interval (mortality) are described using probability distribution functions, notably the Weibell and negative exponential distributions. Fire mosaics and their dynamics have been described using spatial Markov chain models, percolation theory, and spatial autocorrelation statistics. These quantitative characterizations are useful for describing changes in fire regime over time, for comparing different areas, and for estimating wildfire risk.

V. DISTURBANCE BY HERBIVORES AND PREDATORS

In many ways, consumers (and pathogens) have the same effects as other agents of disturbance in that biomass is removed and new opportunities for recruitment are created. There are some distinctive features, however, of disturbances by herbivores and carnivores. Most obviously, they operate at the relatively fine scale of individual consumers, and they leave a very different biological legacy than mechanical and physico-chemical disturbances because biomass is concentrated and converted into heat, new tissues, and discrete waste products. The waste products are frequently deposited away from the disturbed area. For example, frass from defoliating caterpillars in tree canopies falls to the soil surface. Shade-seeking ungulates in African savannas deposit a disproportionate amount of urine and feces under trees. Elk migrations lead to a net transport of nitrogen from summer ranges to winter ranges.

Foraging is also a far more selective form of disturbance than other disturbance mechanisms. Organisms (including humans) preferentially occupy specific sites and microsites and focus on specific plant resources, creating patchiness at many scales. Most plants are unpalatable to most herbivores, and selective herbivory by dominant grazers can exert strong, directional effects on community composition, structure, and ecosystem processes. These effects may be amplified by strong feedbacks between consumption, food quantity, and food quality. For example, moose in boreal ecosystems selectively browse nitrogen-rich deciduous species and avoid coniferous species. Over time this promotes nitro-

gen sequestration in conifer litter and accelerates succession of deciduous to coniferous plant communities (Hobbs, 1996). Disease spread from infected to susceptible hosts and the development of resistance in infected individuals is somewhat analogous in terms of the strong spatio-temporal feedbacks between the disturbance process and the state of the system.

Many consumers act as mechanical, chemical, and biological agents of disturbance. Perhaps the most sensational example is that of beavers, which selectively fell trees, dam streams, and inundate floodplains. Similarly, burrowing animals such as pocket gophers can operate as "physical ecosystem engineers" creating pervasive changes in the abiotic and biotic environment by redistributing soil and nutrients, altering soil hydrology and canopy light regimes, and altering vegetation composition and dynamics.

VI. INTERACTIONS AMONG DISTURBANCE MECHANISMS

Disturbances can interact in complex, nonlinear ways across space and time. One disturbance may promote another, such as occurs when dry-season fire in Mediterranean-climate shrublands promotes flooding and debris flows during the ensuing wet season. Conversely, one disturbance mechanism may reduce the likelihood of another. For example, grazing reduces the likelihood of fires in grasslands and the likelihood of crown fires in some forest ecosystems. Nearshore communities that are heavily impacted by wave disturbance may also experience lowered rates of predation. It is not difficult to imagine many such interactions operating across a wide range of spatial and temporal scales.

Disturbances often play an important role in invasion and spread of exotic species. In Hawaii, for instance, invasion of the nitrogen-fixing shrub, *Myrica faya*, was facilitated when native vegetation was thinned by heavy ash deposition from a volcanic eruption. There is a growing list of examples in which disturbance promotes the spread of an invasive species and then that species initiates ecosystem changes that introduce new disturbance processes. For example, in the western United States cattle grazing in shrub steppe promoted the spread of the exotic grass, *Bromus tectorum*, and the presence of a dry grass layer then increased wildfire size and frequency in that system, in which fires were historically uncommon.

As mentioned in Section I, compounded disturbances, or disturbances of communities already stressed by abiotic or biotic forces, can have large and persistent effects on ecological communities if they occur in such rapid succession that they disrupt normal recovery processes. Repeated burning over a series of 2 or more years is a common means of converting shrublands to grasslands in semi-arid ecosystems. In a particularly dramatic example, Hughes (1994) describes the decline of corals in Jamaica due to the combined effects of overfishing, two major hurricanes, and disease that caused mass mortality in sea urchins. These impacts appear to have collectively precipitated a massive phase shift from coral-dominated to algal-dominated reefs.

VII. HUMANS AS AGENTS OF DISTURBANCE

For many millennia humans have been manipulating ecosystem processes with varying impacts on ecosystem function and species distribution. Human activities that create disturbances at spatiotemporal scales to which organisms are adapted generally do not effect long-term changes in community composition or structure. For example, shifting agriculture in the humid tropical forests, like hurricanes, creates small, temporary forest clearings and results in landscape mosaics within which most species can persist. In contrast, commercial forest cutting and cropland and pasture development are large and/or persistent disturbances that can produce regional extinction of species and require much longer periods for forest recovery (Attiwill, 1994).

Another widespread human impact has been to regulate or arrest disturbances such as fires, floods, or animal activities that create habitats on which some species depend. Even when human ecosystem manipulations are, by design, modeled after a particular disturbance, it is difficult to incorporate all disturbance-related changes in the environment. For example, forest harvesting can occur at the spatiotemporal scale of a natural fire regime, but the post-fire nutrient pulse and legacy of dead snags are no longer factors in the recovering ecosystem. Streams may be regulated to mimic periodic flooding, but dams often inhibit fish migrations that would happen at that time (and agencies will understandably stop short of re-initiating very high-magnitude flood events).

As humans fragment terrestrial habitats through land-use conversion and aquatic systems through impoundments, diversions, dredging, trawling, and construction, permanent fragmentation of habitats is superposed on dynamic disturbance mosaics, causing changes in disturbance regimes. Such fragmentation can retard disturbance spread (e.g., the spread of fire or disease), alter disturbance regimes in remnant patches, and create edge environments with novel disturbance regimes. Local human activities can have large cumulative effects. For example, the building of roads alters the natural cycles of erosion and landslide initiation, whereas agricultural land use often leads to the concentrated nutrient inputs and changes storm flows in receiving water bodies. In fact, most human activities have the capacity for altering historical disturbance regimes in some way.

At the broadest scale, humans are changing global climate patterns, and nearly all mechanisms of disturbance and disturbance effects are ultimately tied to climate processes. Changes in disturbance regimes during global climate change are inextricably tied to other ecological responses to changing climate, affecting whether species can persist in their current ranges or can invade new areas of more suitable environmental conditions. The tight coupling between biotic processes and structures and most disturbance patterns and processes—a recurrent theme in this article—creates complex feedbacks that could amplify or moderate the effects of global climate change on disturbance regimes, species, and ecosystems.

See Also the Following Articles

DISEASES, CONSERVATION AND • ECOSYSTEM FUNCTION, PRINCIPLES OF • FIRES, ECOLOGICAL EFFECTS OF • GRAZING, EFFECTS OF • HERBACEOUS VEGETATION, SPECIES RICHNESS IN • HUMAN EFFECTS ON ECOSYSTEMS, OVERVIEW

Bibliography

Attiwill, P. M. (1994). The disturbance of forest ecosystems: The ecological basis for conservative management. *Forest Ecol. Manage.* **63**, 247–300.

Christensen, N. L. (1988). Succession and natural disturbance: Paradigms, problems, and preservation of natural ecosystems. In *Ecosystem Management for Parks and Wilderness* (J. K. Agee and D. R. Johnson, Eds.). Univ. of Washington Press, Seattle.

Connell, J. H. (1978). Diversity in tropical rainforests and coral reefs. *Science* **199**, 1302–1310.

Everham, E. M., and Brokaw, N. L. V. (1996). Forest damage and recovery from catastrophic wind. *Bot. Rev.* **62**(2), 113–185.

Grime, J. P. (1979). Plant Strategies and Vegetation Processes. Wiley, New York.

Hobbs, N. T. (1996). Modification of ecosystems by ungulates. *J. Wildlife Manage.* **60**, 695.

Holling, C. S. (1973). Resilience and stability of ecological systems. *Annu. Rev. Ecol. Syst.* **4**, 1–23.

Hughes, T. P. (1994). Catastrophes, phase shifts, and large scale degradation of a Caribbean coral reef. *Science* **265**, 1547–1551.

Johnson, E. A., and Gutsell, S. L. (1994). Fire frequency models, methods, and interpretations. *Adv. Ecol. Res.* **25**, 239–287.

Levin, S. A. (1992). The problem of pattern and scale in ecology. *Ecology.* **73**, 1943.

Moritz, M. A. (1999). Controls on disturbance regime dynamics: fire in the Los Padres National Forest. Ph.D. dissertation, Department of Geography, University of California, Santa Barbara. 163 pp.

Paine, R. T., and Levin, S. A. (1981). Intertidal landscapes: Disturbance and the dynamics of pattern. *Ecol. Monogr.* **51**, 145–178.

Paine, R. T., Tegner, M. J., and Johnson, E. A. (1998). Compounded perturbations yield ecological surprises. *Ecosystems* **1**, 535–545.

Pickett, S. T. A., and White, P. S. (Eds.) (1985). *The Ecology of Natural Disturbance and Patch Dynamics.* Academic Press, New York.

Reiners, W. A. (1983). Disturbance and basic properties of ecosystem energetics. In *Disturbance and Ecosystems: Components of Response* (H. A. Mooney and M. Godron, Eds.). Springer-Verlag, New York.

Romme, W. H., Everham, E. H., Frelich, L. E., Moritz, M. A., and Sparks, R. E. (1998). Are large, infrequent disturbances qualitatively different from small, frequent disturbances? *Ecosystems* **1**, 524–534.

Sousa, W. P. (1984). The role of disturbance in natural communities. *Annu. Rev. Ecol. Syst.* **15**, 353–391.

DIVERSITY, COMMUNITY/ REGIONAL LEVEL

Howard V. Cornell
University of Delaware

I. Two Perspectives on the Control of Community/
   Regional Diversity
II. Testing for Saturated Communities
III. Matters of Spatial Scale
IV. Synthesis of Community and Regional Perspectives

GLOSSARY

between-habitat diversity Degree of dissimilarity in species composition between two habitats. The greater the number of species shared by two habitats, the lower the between-habitat diversity.

metacommunity An assemblage of species in a patchy landscape, each species comprising a metapopulation in that its distribution shifts among habitat patches because the distribution is determined by colonization and extinction dynamics.

metapopulation Group of subpopulations distributed among habitat patches in a landscape, each subpopulation being subject to periodic extinction and subsequent recolonization from other occupied patches.

richness Simplest measure of diversity, represented by a count of the number of species in some ecologically meaningful unit (community, metacommunity, region) without regard to variation in the number of individuals per species.

saturation Upper limit to species richness within a community, set by species interactions, and indepen-

dent of the pool of colonists to which that community is accessible.

species pool Set of species that are able to colonize a habitat of interest.

type I community An assemblage of species where the richness is a constant proportion of the number of species occurring in a larger geographic unit in which the community is embedded.

type II community An assemblage of species where the richness is independent of the number of species in a larger geographic unit in which the community is embedded.

AN ECOLOGICAL COMMUNITY IS A COLLECTION OF SPECIES LIVING CONTEMPORANEOUSLY IN THE SAME PLACE and comprising populations of individuals that are spatially interspersed and among which direct and indirect interactions can potentially take place. Communities are nested within larger biogeographical regions that comprise many different habitats and that have unique evolutionary histories. Species can disperse among suitable habitats in the region and thus define the species pool for communities occupying those habitats. Species interactions, habitat differences, history, and dispersal have all been hypothesized to influence species diversity in communities, and thus community and regional species diversity are inextricably linked. This article presents two perspectives on the control of community and regional diversity, the nature of the linkage between community and regional diversity, and methods for examining this linkage. Guidelines for synthesizing community and regional perspectives are also considered.

I. TWO PERSPECTIVES ON THE CONTROL OF COMMUNITY/REGIONAL DIVERSITY

A. Local Control of Species Richness

1. The Volterra–Gause Perspective

One of the most obvious and fundamental facts of ecology is that all species do not occur everywhere. Instead, each species has a unique distribution determined by its ecological context and evolutionary history. Nevertheless, species distributions overlap to varying degrees so that some places support more species than others do. Such variations in species richness have been a central issue in ecological and biogeographical studies at least since the turn of the century. Until recently, explanations focused mainly on processes operating within ecological communities over relatively short periods of time. This perspective arose out of a desire to add scientific rigor to ecological studies by focusing on mechanisms and ignoring history. A mechanistic approach made it possible to model ecological systems with simple equations and to test these models and their assumptions with carefully collected data.

One of the earliest truly influential models in community ecology was developed by Vito Volterra (and independently by Alfred Lotka) in the mid-1920's. This model is a set of simple differential equations describing competition between two species in terms of reductions in per capita population growth rates by interactions among individuals. When the model is analyzed at equilibrium, the outcome of competition (coexistence or competitive exclusion) is easily predicted from the intensity of interactions within species relative to those among species. If populations limit their own growth more than the growth of other populations, the two species will coexist. If they limit other populations more than themselves, one will exclude the other. The model and its assumptions were later tested by G. F. Gause in laboratory systems comprising several species of protozoa. The tests lent strong support to the model, and as a result the so-called Volterra–Gause perspective has become a metaphor for the way in which ecologists think about control of species richness.

According to this perspective, species richness in a biogeographical region may indeed be determined in part by history. However, history has little relevance to the upper limits on local richness within communities. Instead local richness is limited primarily by species interactions (predation, parasitism, herbivory, competition, mutualism) in conjunction with abiotic conditions (water, nutrient, and energy availability, chemistry, temperature, disturbance). In other words, the outcome of interactions is determined in large part by the abiotic conditions in which they occur. For example, lower concentrations of a particular nutrient or lower water availability might favor one plant species in competition with another, but higher values of these factors might result in a reversal of competitive dominance. The number of species in the biogeographical region provides a species pool from which the local communities are drawn, and all species have the potential for colonizing all habitats in the region. Since habitats vary in their abiotic conditions and resident assemblages, only those species capable of coexisting in a given milieu will establish successfully. Differences in species richness thus result from differences in the abiotic template and species composition among habitats. In a recent series of publications from the early 1990s, James Drake has shown that the colonization order of species into a habitat can also have some effect, and communities can exhibit multiple stable points representing different levels of species richness.

A crucial element of this perspective that derives directly from the Volterra model is that communities exist at a dynamic equilibrium. For example, in the case of competition, population levels of the resident species are limited by resource availability and physical conditions. Thus, if sufficient time elapses and there are no perturbing disturbances to the community, the number of species will saturate at a level determined by the number of limiting resources and physical factors. At saturation, species are sufficiently similar in their resource use that finer subdivision of resources becomes increasingly difficult and new colonizations into the community must be accompanied by extinctions of current residents. If habitats throughout the region are all identical and saturated with species, then the total number of species in the region is limited to the total number coexisting within each identical habitat. That is, community and regional richness are identical and regional richness is limited by the same factors that limit local richness.

Since all habitats are not identical, species within the region have the option of partitioning habitats as well as resources and thus the same set of species will not occur everywhere in the region. Nevertheless, local processes are still limiting both local and regional richness, but limits on regional richness now result from interplay between species interactions and spatial habitat variation. The greater the number of habitats, the greater the between-habitat diversity and the greater the number of species in the region relative to those within a single community. Historical processes such as massive extinctions might reduce the number of species in the region for a time, and subsequent evolution, speciation, and dispersal might generate a new biota that differs from the previous one. However, the essential feature of this viewpoint is that given sufficient time, the imprint of history on species richness will fade, communities will fill up and become saturated, and processes operating locally will also limit richness regionally.

Even strict adherence to the Volterra–Gause viewpoint still allows for other factors to affect local and regional richness. All else being equal, larger localities and regions will support more species than smaller ones as predicted by the species-area relationship. In addition, some species within a region might be ecologically equivalent and thus might be able to substitute for one another within a given community type. Chance might determine what species occupies what particular community, in which case ecologically equivalent species could increase regional richness without affecting local richness. Finally, spatial interspersion of habitat types might allow species adapted to one habitat type to persist in an adjacent, less favorable habitat type by means of propagule flow. This so-called source–sink effect proposed by Ron Pulliam in 1988 would increase local richness in the less favorable habitat above that which would be self-maintaining in the absence of immigration. Although these factors may have important impacts on species richness, they do not alter the premise that the primary direction of control on species richness is from the local to the regional level.

An ecological community at equilibrium and not subject to vagaries of history evokes a powerful image that continues to influence modern ideas about species diversity. Robert MacArthur's (1972) book represents the pinnacle of this viewpoint, although he recognized that the equilibrium assumption and local control of richness did not apply to all communities. There are clear recent demonstrations that species interactions have strong influences on community structure and might exclude species from the community (see Lawton, 1999, for a brief review). In short, the Volterra–Gause perspective has been a useful guide to a limited understanding of the mechanisms by which particular communities are structured. However, this understanding has led to few if any general patterns or rules that apply to most systems (Lawton, 1999).

2. Species–Energy Theory

Another view of local control of species richness comes from the ecosystem perspective, and has been termed "species–energy theory." The theory was developed mainly by David Wright, David Currie, and Brian Maurer (1993). Species–energy theory posits that local and, ultimately, regional richness are limited by available energy within the local environment. Available energy limits local richness by placing ceilings on the population densities of individual species. As energy decreases, densities decrease, random extinctions increase, and local richness is reduced. Available energy can be scaled up to whole regions by multiplying the energy per unit area by the total area of the region. Thus, population densities and species richness over the whole region are also controlled by energy availability.

Although the theory in this simplified version posits local control of local and regional richness, it differs from the Volterra–Gause viewpoint in that species interactions are of secondary importance in limiting richness. Energy can have direct and independent effects on population levels to increase richness rather than mediating interactions among species. Interactions can be included in the model, but they are not necessary for limits on richness to occur. Species–energy theory is very similar to island biogeography theory, which proposes that species richness is a balance between colonization and extinction rates in a local habitat. Smaller habitats have higher per species extinction rates and thus lower species richness as predicted by the species–area relationship. The main difference between the two theories is that species–energy theory substitutes energy availability for area as the predictor of species richness.

A recent literature review by David Wright and others demonstrated that several proxy variables for energy availability (precipitation, potential or actual evapotranspiration, productivity) are indeed strong predictors of species richness for various taxonomic groups (Table I). However, it is not clear whether energy is the direct cause of richness variation, or whether it simply covaries with other possible causes such as area effects, habitat heterogeneity, species interactions, and historical effects. It is also not clear whether the assumptions of the theory are always met in particular studies. For example, it is assumed that energy limitation should operate at the local scale where individuals of each species seek sufficient energy to persist in the commu-

TABLE I

Energy-Related and Other Factors Which Correlate with Species Richness

Factor cited	Correlations with species richness	
	Significant	Not significant[a]
Energy-related factors		
Mean annual heat and/or humidity variables	29	0
Productivity (sometimes est. as standing crop)	9	0
Nutrient, food availability	3	0
Seasonality	4	0
Other factors		
Habitat complexity, subdivision, microrelief	9	3
Disturbance	5	5(2)
Environmental chemistry (pH, cations)	3	1(1)
Isolation, peninsula effects	2	1(1)
Diversity of food, prey	6	1
Regional richness	2	0
Time, historical factors	1	2(2)
Competition, predation	0	1
Other	3	1

Note: Factors investigated as the principle determinant of species richness in 82 correlations from 53 studies.

[a] Numbers in parentheses indicate the number of correlations that were omitted from further analysis because no correlation coefficient was reported.

(From Wright *et al.* in Ricklefs, R. E. and Schluter, D. (Eds.) (1993). Species Diversity in Ecological Communities: Historical and Geographical Perspectives. University of Chicago Press with permission. © 1993 by University of Chicago Press)

nity. However, it is at the scale of the local community that the correlation between energy and richness is weakest and most variable; the correlation is strongest at larger spatial scales (104–108 km²). It is also assumed that average population size and species richness should increase together with available energy, but this is not always the case. In particular, David Tilman and Pacala have pointed out that plant systems often show an inverse relationship between per species population sizes and energy as predicted by the −3/2 power law, yet species richness commonly increases with energy in the expected way. Despite these limitations, energy can be a useful variable for estimating richness at larger scales because covariation is high regardless of the underlying causes, and because energy data (or proxies) are available for most geographic areas.

At more local scales, species interactions among resident populations may come into play, altering any potential increases in richness with energy availability. Such interactions might partially explain the more variable relationship between energy and richness within communities. Increased energy can lead to increased

biomass of the total community, which in turn can lead to increases or decreases in average population density depending on the intensity of interindividual interactions and the size plasticity of the taxa involved. But it is the changes in the relative allocation of energy among species that are critical for predicting changes in richness. If allocation changes as energy availability increases, for example, caused by shifts in competitive superiority or ability to avoid predation, then the population densities of some species may increase while other species may be driven extinct. Richness can thus increase or decrease with increased energy depending on the circumstances. It is at this point that the species–energy theory converges with the Volterra–Gause viewpoint since the environment is now mediating interactions among species.

3. The Diversity–Productivity Relationship

The idea that richness can change with energy availability in unpredictable ways has had a long history under the guise of the diversity–productivity relationship. Productivity is a measure of energy availability to the community and has been shown to affect local richness in contradictory ways; richness either increases, decreases, or shows no relationship with productivity depending on the assemblage. Recently it has been suggested that these contradictory relationships may just be incomplete segments of an overall hump-shaped, unimodal relationship over a broader range of productivity. In other words, richness peaks at intermediate productivity levels and declines toward the extremes. Different taxa can show richness peaks at different levels of productivity, resulting in unimodal curves that are displaced along the productivity spectrum. No ecological pattern is universal and, accordingly, contradictory relationships between richness and productivity need not always imply a truncated hump. Indeed, evidence for this possibility is currently limited at best (Waide et al., 1999). Monotonic relationships may be just as frequent as unimodal ones. Nevertheless, the unimodal curve is a useful synthetic construct and it is sufficiently frequent to invite explanation. Moreover, the same explanations for each arm of the hump may apply equally well to monotonically increasing or decreasing relationships over a particular range of productivity.

It is not clear why richness shows this strange relationship with productivity, and the pattern remains one of the most important unsolved mysteries in community ecology. The hypothesis that productivity differences are directly responsible for the unimodal relationship assumes that productivity is uniformly distributed among sites. In other words, low-, medium-, and high-productivity environments occur with equal frequency. If medium-productivity sites are more frequent than low- or high-productivity sites, then the unimodal curve might just be a reflection of simple species–area effects. There is some evidence to suggest that this is not the case (Rosenzweig, 1995), but the possibility has not been extensively tested. At the very least, diversity–productivity correlations should be corrected for unequal area effects before a causative link between these variables is seriously considered.

Unimodal curves (as well as monotonically increasing or decreasing ones) can occur at different spatial scales (local, landscape, regional, continental), but their frequency changes as spatial scale increases. Monotonically increasing patterns are more common at larger scales and unimodal patterns are more common at local scales (Fig. 1). Moreover, the mechanisms responsible for patterns at each scale probably differ (Waide et al., 1999). Since the within-community, among-community, and local scales are where ecological processes become important, theories that consider changes in species interactions and local environmen-

tal conditions along productivity gradients are the most relevant.

The decline to the left of the hump may derive from increased environmental stress or the direct effects of energy reduction on population levels as proposed by species–energy theory. Other mechanisms, such as increases in disturbance rate, poor recovery from disturbance due to slower population growth rates, declines in territoriality, and environmental heterogeneity have also been suggested to play a role. The decline to the right of the hump may involve some form of interaction mediation among species, such as increased intensity of competition and predation or increased population dynamic instability within food webs at high productivities. Such declines would then provide strong support for the idea that species interactions limit species richness in accordance with the Volterra–Gause viewpoint. However, it is important to point out that richness declines in highly productive areas may also result from regional effects. For example if productive habitats are relatively rare, they are likely to be surrounded by larger, less productive areas and thus isolated from a large source pool of species adapted to productive conditions. As previously mentioned, area effects may also contribute to low richness. Regional effects must therefore always be taken into consideration when analyzing diversity–productivity relationships.

Tests that distinguish among these various theories are difficult to perform and have thus far been limited. Comparative approaches, although they require less disruption of the community, might not be able to distinguish among alternative theories unless natural "experiments" in which selected mechanisms are not operating can be found. Experimentation has better control of confounding variables, but care must be taken to provide enough time for all processes that contribute to changes in richness to ensure that they be allowed to respond to the manipulation. For example, in a 1999 review, Laura Gough and co-workers found that fertilization experiments consistently resulted in increases in inorganic nutrient concentrations, which in turn intensified species interactions in herbaceous plant communities. Fertilization increased productivity and reduced species richness across a broad range of community type, due primarily to increased dominance by one or a few species. These experiments certainly demonstrate the potential of interaction mediation to limit community richness. However, the relevance of such studies to richness declines in natural communities at highly productive sites is questionable. Fertilization reduces diversity even at poorly productive sites, where theory predicts that diversity should increase.

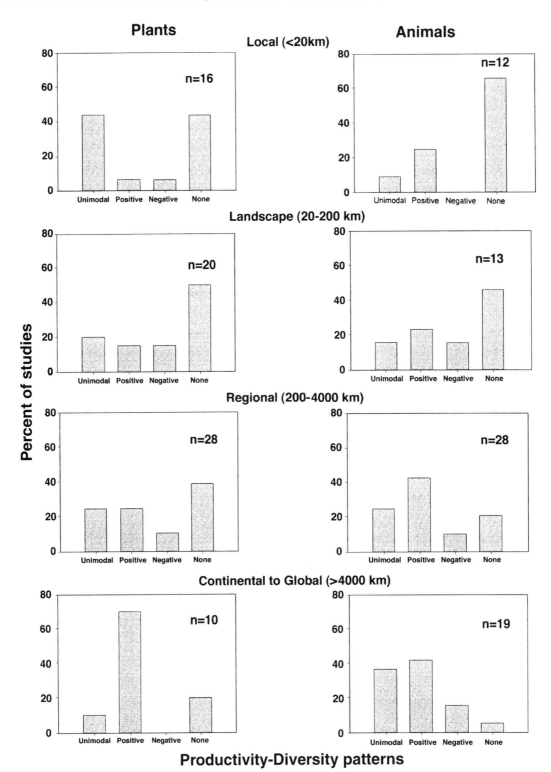

FIGURE 1   Breakdown of published plant and animal studies showing a unimodal, positive, negative, or no relationship between species diversity and productivity at four spatial scales. The number of studies in each histogram (*n*) is shown. (From R. B. Waide, M. R. Willig, C. F. Steiner, G. G. Mittelbach, L. Gough, S. I. Dodson, G. P. Juday, and R. Parmenter. (1999). The relationship between productivity and species richness. *Ann. Rev. Ecol. Systematics,* **30,** 257–300, with permission.)

This anomalous result may be a consequence of the short-term nature and small spatial scale of such experiments. Species adapted to the higher productivities of the fertilized sites might not be able to reach them if they are isolated in space or time. More appropriate experiments should therefore manipulate the species pool to simultaneously test for the effects of productivity and the dispersal effects on richness. If pool limitation is the cause of the anomalous results, then experimentally boosted recruitment to low-productivity sites by an appropriate species pool should generate higher species richness with higher fertilization rates. Naturally, the species selected for the boosted recruitment should be well-adapted to the increased nutrient levels at the fertilized sites. High-productivity sites, on the other hand, should remain depressed in diversity owing to interspecific interactions. Some combination of comparative and experimental approaches will probably be required to solve the mystery of the diversity–productivity relationship. Regardless of the methods used, it is important that it be solved, since it represents one of the clearest possibilities that local richness is locally controlled. It also provides a single framework that subsumes both species–energy theory and Volterra–Gause processes.

4. Theory Doesn't Always Predict Local Control

Without question, the Volterra–Gause viewpoint has stimulated much research on the factors that control species richness. Nevertheless, the fundamental premise that local processes control local and ultimately regional richness may be incorrect for many systems. Even the restrictive Volterra–Gause models do not always predict limits on species richness due to local mechanisms. After appropriate modification to incorporate some realistic features of environments and species, such as spatial heterogeneity, environmental fluctuations, nonlinear population growth responses to environmental change, and trade-offs among species in their ability to compete in different environments, the models often predict that there are no upper limits on local richness (Tilman and Pacala, in Ricklefs and Schluter, 1993). Moreover, the Volterra–Gause assumption that population levels are at equilibrium, may not commonly be met in natural communities. In a classic example, periodic disturbance can drive populations below equilibrium, where the outcomes of interspecific interactions are never fully realized. More species can thus invade than would be allowed by equilibrium population levels.

If real communities have any of these features, it is not likely that local richness will reach saturation as predicted by the simplest models. Evidence from real communities suggests that lack of saturation is indeed common. Lack of saturation raises the possibility that factors external to the community may ultimately determine local richness. As a consequence, the factors that determine regional richness must also be sought elsewhere. In the Volterra–Gause view, regional richness is limited by a combination of the number of species within saturated habitats and the number of habitats within the region. If communities within habitats are not saturated, control on regional richness cannot be exerted at the within-habitat scale. Moreover, the number of habitats within a region is not fixed and may be more an effect of the number and kinds of species in the region than a cause of richness limitation at the between-habitat scale (Rosenzweig, 1995). In other words, the more species there are in the region, the more finely habitats can be partitioned and the more distinct habitat types that can be identified. How, then, do differences in richness among regions arise? And what relevance do these differences have for local richness? An alternative to the Volterra–Gause viewpoint that addresses these questions is thus needed.

B. Regional Control of Richness

1. Regional Processes

The alternative view rests on an idea that actually predates the Volterra–Gause view, namely, that regional richness is set to a large degree by historical biogeographical factors. The importance of history is implicit in the traditional definition of a biogeographical region that relies in part on the presence of a distinctive native biota. The presence of large groups of indigenous species implies that the region has undergone a period of isolation and independent evolution. Thus, the numbers and kinds of species that occur across the region are determined to some degree by its unique historical development. These different histories might result in widely divergent species richnesses among some components of the biota, which in turn may be reflected in local communities. Under this scenario, local richness is not limited by species interactions as predicted by simple Volterra–Gause models, but becomes some function of the number of species in the region that are able to colonize and persist in a given habitat. The richness of the community embedded within the region thus retains some of the region's historical signature.

The key prediction arising out of this scenario that distinguishes it from the Volterra–Gause model is that localities in richer regions can support more species

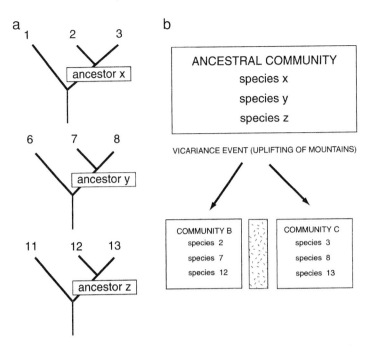

FIGURE 2 Phylogenetic and biogeographical patterns that suggest association by descent. Species sets (2, 7, 12) and (3, 6, 13) comprise communities that were found on opposite sides of a geographic barrier. The phylogeny (A) indicates that species sets (2, 3), (7, 8), and (12, 13) each had a common ancestor. These two patterns suggest that both communities descended from a common ancestral community that exited before the barrier was formed. (B) The species in each extant community are thus associated because their ancestors were associated. (From Brooks and McLennan, in Ricklefs, and Schluter. (1993). Copyright © 1993 by University of Chicago Press.)

than those in poorer ones even though the localities are environmentally similar. This is simply because species co-occurrences within localities are determined to some extent by the same factors that determine regional co-occurrence. In their 1991 book *Phylogeny, Ecology, and Behavior*, Brooks and McLennan pointed out that regional co-occurrence can come about via two processes: association by descent and association by colonization. In association by descent, two species can co-occur because their ancestors co-occurred. In other words, the community was assembled from ancestors in the distant past, and the species present today evolved *in situ* from members of the ancestral community, most likely by geographic isolation (Fig. 2). The community is thus rich or poor because the region in which it is embedded was historically rich or poor. In association by colonization, two species can co-occur because one evolved elsewhere and migrated into the region. Communities are open to colonization because (1) interactions are not limiting richness, (2) the colonizing species uses niche space that is not occupied by the *in*

situ residents, or (3) after contact with the residents, evolution of niche partitioning occurs. In association by colonization, the community is rich or poor because its region has high or low barriers to colonization from elsewhere. It is likely that real communities are mixtures of colonists and *in situ* descendants, and Brooks and McLennan have pioneered the development of phylogenetic methods to distinguish between these two groups.

The influence of higher colonization rates into a region on regional and thus local richness is obvious, but the role of history is less so. Why are some regions historically richer than others? The answer to this question must involve region-specific differences in historical speciation rates, extinction rates, or both. Some processes might act to increase the speciation rate relative to the extinction rate. For example, some taxa may have developed key innovations that allowed them to enter a new adaptive zone and undergo an adaptive radiation. The region may have been simultaneously dissected by colonization barriers, disrupting gene flow among many populations and resulting in a burst of speciation.

Latham and Ricklefs present a possible example in the evolution of frost tolerance and proliferation into the temperate zone by forest trees. Asian temperate forests have three times more tree species than North American forests and six times more than European forests. Because Asia may be particularly diverse topographically and has no geographic barrier between the tropical and temperate zones, such proliferation may have occurred more frequently than in North America and Europe, where such barriers do exist (Ricklefs and Schluter, 1993). Alternatively, some taxa may undergo non-adaptive bursts of speciation resulting from inherent biological attributes, including genetic peculiarities (e.g., chromosomal anomalies or polyploidy), mate recognition and selection, or other factors. A dramatic example is the anomalous diversity of the Hawaiian *Drosophila*. The proliferation of this genus in Hawaii far exceeds that in any other region of equivalent area and has been attributed to a strong tendency for sexual selection in the Hawaiian taxa (see Cornell and Lawton, 1992). Adaptive and nonadaptive radiations might be distinguishable by differences in the relationship between speciation rate and richness. In adaptive radiation, speciation rate should at first exceed extinction rates and then come into balance with extinction as regional richness increases and the adaptive zone fills up. Nonadaptive radiations do not require ecological diversification, and thus speciation and extinction rates should be unaffected by regional richness. Adaptive and nonadaptive radiations are not mutually exclusive and both could be operating in the same clade.

Other processes might act to increase extinction rates relative to speciation rates in some regions. Increased probability of extinction is believed to be linked to low average population size, which in turn is correlated with geographic distribution. Thus any event that reduces a species' range, such as habitat destruction or displacement by an invader, is likely to decrease its time to extinction. Returning to temperate forest trees, glaciation in conjunction with mountain barriers in southern Europe might have squeezed the geographic ranges of many temperate tree species below a critical threshold, driving them to extinction. The absence of mountain barriers in Asia and North America may have allowed temperate species to survive in southern refuges, accounting in part for the higher tree species richness in those regions.

The relative importance of colonization, speciation, and extinction rates in setting regional richness will vary depending on the taxon and the history of the region. Since regional and local richness are linked, an understanding of community assembly will require a new focus on these macroevolutionary processes. Fortunately, new phylogenetic methods involving sister group comparisons and other innovations have been developed by Mitter, Farrell, and others, which can distinguish among various macroevolutionary causes for regional and local richness variation. Such methods have given history and biogeography the consideration they deserve in some modern investigations of community structure (e.g., McPeek and Brown, 2000), but more work needs to be done. Community structure is undoubtedly influenced by contemporaneous processes as well, and distinguishing historical from present-day effects such as energy, local environment, and species interactions will be one of the biggest future challenges in this area.

2. Metacommunity Processes

Once species become part of the biota of a particular region, they can disperse among habitats, and as a result communities can be quite dynamic in space and time. The species within the region can thus be viewed as a species pool for a given local habitat. The species pool will be considerably larger than the number of species in any habitat for at least three reasons: (1) some species in the region will not be able to disperse to all sites because of colonization barriers or isolation-by-distance; (2) all habitats in the region are not favorable for all species; and (3) species can be driven locally extinct by species interactions, demographic and environmental stochasticity, disturbances, or other unfavorable changes in local conditions. It is of some interest to explore situations where species cut off by highly effective colonization barriers are not considered part of the pool for habitats from which they are isolated. The dynamic links between local and regional richness should thus be examined at a spatial scale where all species can reach all habitats over ecological time. Moreover, we should focus on a single habitat type, since looking at more than one habitat will confound regional effects with local environmental variation. This can generate artifactual relationships between local and regional richness (Srivastava, 1999). The metacommunity within the landscape, in its simplest form, fulfills both of these criteria and thus represents an appropriate theoretical context in which to examine these links. The term metacommunity was first used by Gilpin and Hanski in 1991. The metacommunity can be thought of as a pool of species in a fragmented landscape of environmentally identical sites, each species behaving as a metapopulation in that its distribution among sites in the landscape is determined by colonization and extinction. In this scenario, regional processes still determine the potential species pool, but now the effects of the pool on local richness are filtered through the metacommunity.

The simplest metacommunity model derives from Richard Levins' original 1969 equation for a single-species metapopulation:

$$dp_i/dt = c_i p_i (1-p_i) - m_i p_i$$

Tilman and Lehman used this equation to construct a metacommunity model by writing independent equations for as many species as occur in the metacommunity (Tilman and Kareiva, 1997). The value of p_i is the proportion of local sites in the landscape occupied by the ith species, and c_i and m_i are the site-specific colonization and extinction rates, respectively. Thus, when $dp_i/dt = 0$, each p_i is set by a balance between colonization and extinction rates per site. The model assumes that all sites are identical. This assumption permits the examination of links between local richness and pool size without the confounding effects of local environmental variation. If p is defined as the mean of p_i's for all species, then p becomes the average proportion of local sites occupied by a species. Alternatively, it is the average proportion of the metacommunity found at a particular site. Thus, local richness = p(landscape richness)

What this expression tells us is that if p is a constant then local richness is a linear function of landscape richness, a pattern that has been called Type I or proportional sampling (Fig. 3). Proportional sampling predicts that matched habitats embedded in richer landscapes will always have richer communities, but that local richness will be less than landscape richness by a constant proportion. The slope of the relationship, p, will be determined by the size of the pool balanced by the rate of species extinction at individual sites. In this simple model, species-specific colonization and extinction rates are assumed to be independent of pool size and the number of species at a site. Thus the model predicts the relationship between local and landscape richness in the absence of species interactions. Under these conditions, communities will be unsaturated and open to colonization from the species pool.

At least one recent study by Hugueny and Cornell (2000) has shown that the proportional sampling pattern in a real species assemblage can be predicted from colonization and extinction rates. Metacommunity models can be made more realistic by adding rescue effects, by adding provisions for variable population

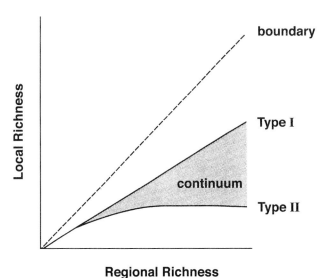

FIGURE 3   The shape of the relationship between local and regional richness in Type I and Type II communities. The linear relationship in Type I communities suggests that they are strongly influenced by the size of the regional species pool. Type II communities are asymptotic, suggesting that they are saturated and that further increases in the regional pool will not affect local richness. Real communities can also occupy the gray zone. (From Cornell and Lawton, 1992.)

sizes among patches, by making habitats differentially suitable to different species, and by incorporating source–sink effects and habitat specialism. Such modifications may alter the details of the Type I curve, but they will not alter the conclusion that pool effects strongly influence community richness. Thus, for the purposes of conceptual distinction between saturated and unsaturated communities, the Levins model suffices.

The Levins model not only predicts the relationship between landscape and local richness; it also shows that the landscape filter constrains the species pool. Most obviously, landscapes that are smaller will have smaller species pools as predicted by the species–area relationship. More subtly, landscapes of a given area that are more highly fragmented will have smaller species pools because they suffer higher landscape-scale extinction rates. Fragmentation also reduces the average proportion of the pool (p) found at a local site by lowering per site colonization rates and increasing per site extinction. Thus any differences in landscape characteristics must be considered when estimating the species pool and evaluating the proportional sampling relationship.

If competitive interactions are added to the model, and it is assumed that competition leads to extinction rather than character displacement and speciation, then increased competition can result in increased extinction rates and/or decreased colonization rates per species. As the species pool increases, competition could become sufficiently intense that colonization by a new species into the community would either fail or result in the rapid extinction of other species. Under this condition, further increases in the pool do not result in increases in local richness and the community becomes saturated with species. At saturation, the relationship between local and landscape richness levels off and becomes independent of the rate of colonization and the size of the species pool. This leveling relationship has been designated Type II (Cornell and Lawton, 1992). Species interactions are thus limiting local richness consistent with the Volterra–Gause perspective. Other dynamic models developed by Caswell and Cohen (in Ricklefs and Schluter, 1993) and Morton and Law (1997), which allow for dispersal, species interactions, and local extinction, also predict leveling off and saturation.

However, the spatial character of many dynamic models allows for realistic complexities that ameliorate the impact of species interactions on local richness. For example, if competition occurs only among close neighbors, and inferior competitors are better dispersers, then a potentially unlimited number of species can coexist locally (Lehman and Tilman, in Tilman and Kareiva, 1997). Other spatial models incorporating disturbance, limited dispersal, spatially local competition, and habitat heterogeneity also predict coexistence at high levels of richness in the face of competitive interactions (Cornell and Karlson, and Lehman and Tilman, in Tilman and Kareiva, 1997). Thus Type I relationships can still occur in the presence of intense competition.

It is important to point out that just because meta-community and other models predict proportional sampling or saturation under certain conditions, it does not mean that the Type I or Type II relationships need be explained exclusively by mechanisms operating at the landscape scale. The models simply represent theoretically plausible ways in which the pattern can be generated. Factors operating at other spatial scales can also produce Type I and Type II relationships. For example, suppose the scale of the analysis is expanded to the entire region. One reason to do this is that regions are easier to define than landscapes and thus more data on regional richness have been collected. Type I and Type II relationships are thus more conventionally represented as plots between local and regional richness (see Fig. 3). Under these circumstances, Type I relationships can result simply from dispersal limitation. That is, when regions of different species richness are compared, a constant proportion of the regional biota is

unable to reach the habitat type of interest. Type II relationships, on the other hand, can be generated by historically stable communities in which all niche space is occupied and into which invasion is limited. In such situations, the community need not be particularly dynamic, need undergo little turnover in ecological time, and its species composition may bear the strong stamp of its historical development. Nevertheless, a Type I relationship would still be an indication that hard limits on local richness have not been reached even in the richest regions, whereas a Type II relationship would suggest saturation. The Type I and Type II relationships should thus be viewed as empirical patterns with multiple causes, but also as evidence that local richness is or is not limited by local factors.

II. TESTING FOR SATURATED COMMUNITIES

A. The Method

The preceding theoretical exploration suggests that one can test for saturation by examining the relationship between local richness and some reasonable estimator of the species pool. Regional richness has been the most commonly used estimator for the reasons stated earlier. To test for saturation, matched habitats are compared in different geographical regions of similar size that support different numbers of species. Habitats in richer regions are presumably exposed to richer species pools. Choosing regions of similar size should reduce any differential effects of colonization barriers that might generate differential dispersal limitation among regions (see Cornell and Lawton, 1992; Caley and Schluter, 1997). If the communities are unsaturated, local richness should increase steadily with increases in regional richness. The shape of the relationship can be linear (Type I) or curvilinear, but it cannot level off. Curvilinearity would suggest some resistance to invasion and thus a tendency toward saturation in richer regions, but that the limit to local richness has not been reached. If communities in the richest regions are saturated, then the curve will reach an asymptote and local richness becomes independent of regional richness (Type II).

The method assumes that matched habitats and species in different regions are similar in ways that are critical to species coexistence. This means that if there is the potential for saturation at some level of local richness determined by local environmental conditions and the characteristics of individual species, that level will be the same from region to region. However, even if matched habitats are environmentally identical, some communities might exhibit multiple stable points and saturate at different levels, for example, due to differences in colonization order as shown by Drake. If the community is near saturation, multiple stable points might increase the variance around the putative asymptote but should not affect the average tendency for local richness to level off.

B. The Tests

Two recent surveys by Cornell and Karlson (1997) and Srivastava (1999) have evaluated the frequency of Type I and Type II relationships in a variety of species assemblages. The surveys were combined by Lawton (1999) to yield 38 papers. Some of the data were not analyzed strictly according to the protocol presented here, and in these cases interpretation of the pattern was conservative and based on common sense. By conservative estimate, there were at least 23 examples of Type I relationships [wood-boring beetles, tiger beetles, fig wasps, lizards, corals, plants, helminth parasites on introduced fishes, 6 examples of fish, 4 examples of plant-feeding insects, 3 examples of parasitoids, 2 examples of birds, and a variety of vertebrate taxa (birds, fish, mammals, reptiles) analyzed separately and in combination]. By contrast, there were 15 examples of curvilinear relationships, and not all of these were Type II (saturating). The examples are classified as saturating (strong), weakly curved (weak), or uncertain (u) as follows: marine crustaceans (u), deep-sea gastropods (weak), tiger bettles (u), 2 examples of helminth parasites on fish (strong), helminth parasites on amphibians (strong), 2 examples of fish (u), 2 examples of birds (weak), 2 examples of mammals (strong), and 3 examples of higher plants (1 strong, 2 weak).

C. Caveats

Several methodological concerns about the saturation test have been raised by various authors. The most relevant of these to the present discussion are that various artifacts such as nonhomogeneous variance structures, insufficient sampling, and overestimates of the true species pool in richer regions sometimes exaggerate the curvilinearity of the relationship between local and regional richness. To the extent that such artifacts are present in some studies, true Type I relationships might appear curvilinear, and curvilinear but unsaturated relationships might appear to be Type II. The count of

unsaturated patterns in the survey is thus a conservative and robust result, at least with respect to the presence of such artifacts.

D. Saturated or Unsaturated?

Given these caveats, at least it can be said that unsaturated patterns are common and widespread in ecological communities. Indeed, unsaturated patterns have been reported about twice as frequently as saturated ones. The survey thus confirms that local communities are often open to regional influences and have not reached any upper limit set by species interactions as predicted by the Volterra–Gause model. Those communities that have reached an asymptote might be saturated but have to be examined closely to rule out artifacts that might be responsible for the Type II shape. A Type II curve is not sufficient in and of itself to conclude that species interactions are limiting local richness. For example, helminth parasite communities on fish and amphibians appear to be saturated, but at least one expert opinion is that they are probably not (see Lawton, 1999).

The survey also supports the theoretical position that strong interspecific interactions are not always sufficient to generate a curvilinear or Type II relationship. Taxa that are known to experience strong interspecific interactions such as fish and birds show Type I or weakly curvilinear relationships that are characteristic of unsaturated communities. Although it is not certain that these particular assemblages were experiencing the strong interactions that are supposedly characteristic of these taxa, the results at least suggest that evidence for species interactions by itself might not permit generalizations about the effects of these interactions on broad patterns of community structure.

E. Experimental Support

Additional evidence from experimental introductions supports the survey results that communities are often unsaturated. Experimental manipulations of the species pool are only just beginning, but they are particularly useful because there is better control of confounding variables. In separate studies by Tilman (1997) and Robinson, Quinn, and Stanton (1995), seeds of native species were added to experimental plots containing different numbers of plant species and establishment was monitored. In the Robinson et al. study, California poppy was the test colonizer, and in the Tilman study, anywhere from 0 to 54 species were added to plots.

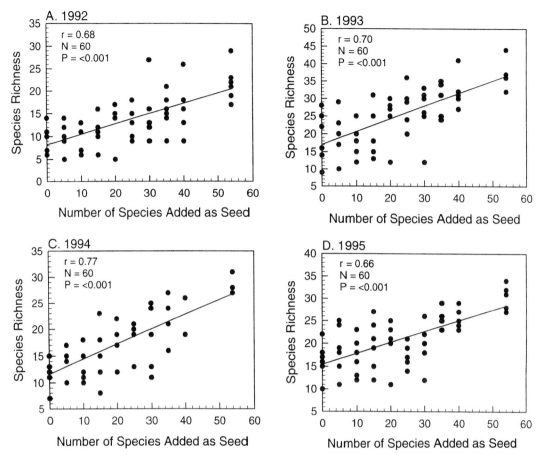

FIGURE 4   Species richness of vascular plants in 0.25-m² prairie plots (1-m² plots in 1995) after the experimental addition of seeds in 1992. Seed additions ranged from 0 to 54 species and the plots were followed for 4 years (1992–1995). The 0 species addition (data on the Y-axis) shows changes in species richness in plots not subject to manipulation. Species richness increased steadily over the entire range of species additions and the increases persisted for at least 4 years. The results suggest that these plant communities are highly recruitment limited and not saturated with species. [From D. Tilman, (1997). Community invasibility, recruitment limitation, and grassland biodiversity. *Ecology* 78, 81.]

Neither study showed evidence of saturation; seeds became established and persisted in many plots. Unsaturation was dramatic in the Tilman study, as the number of species in the experimental plots continued to increase over the full range of species additions and the increases persisted for 4 years (Fig. 4). In other words, an asymptote characteristic of the Type II relationship was never reached. However, plots that were richer in resident species had lower establishment rates, suggesting some resistance to invasion.

The results of these experiments must be interpreted with caution, as they represent short-term responses to pulsed increases in the species pool. There are almost certainly time lags in the response of the community to species additions, and long-term monitoring will be

required to confirm that the "invaders" have become established without driving "residents" extinct over ecological time. Nevertheless, pool manipulations offer a powerful probe for saturation and need to be expanded to other taxa and habitats to test the generality of these results.

III. MATTERS OF SPATIAL SCALE

A central issue in the exploration of saturation in communities is the spatial scale at which local richness should be measured. Ideally, a locality should be ecologically defined. That is, a scale should be chosen within which all species of interest are interspersed and can potentially interact and within which the local environment is homogeneous. Since different species interact at different scales, and environments that are homogeneous for one species may be heterogeneous for others, the ideal locality will probably have indistinct boundaries. It may circumscribe the number of herbivore species associated with a host plant species, the number of insectivorous birds found in a patch of woods, or the number of flowering plant species associated with a serpentine outcrop. Local samples must be standardized in some way so that they are comparable among regions. Typically, the number of individuals or the unit area assessed is uniform, or local assemblages are sampled until the species/sampling curve levels off.

If regional richness is sufficiently high to engender saturation, it is not patently clear at which scale it should occur. Communities are hierarchically structured assemblages of organisms that may be saturated at one scale and not other scales because the intensity of dynamic coupling among species varies with the unit of study. Most comparisons of local and regional richness offer few clues about the relationship between hierarchical structure and saturation because local richness is evaluated at only one scale. Clearly, a more methodological exploration of hierarchical structure that involves multiscale sampling is required. A few studies have made limited attempts at a multi-scale analysis. Caley and Schluter (1997) measured local richness from distributional data of 10 different taxa at two local scales (1% and 10% of regional area) in various regions. All taxa fell along a Type I trajectory at both scales (Fig. 5). In a 1996 paper Kennedy and Guégan showed that parasite communities reached a hard upper limit of four species in individual eels whereas parasite richness in the eel population as a whole continued to increase steadily with regional richness. Finally, data on coral assemblages examined by Karlson and Cornell (2000) showed that local richness was less sensitive to regional richness in 1-m² quadrats than in 10-m line transects. In the latter example, the 1-m² quadrat is so small relative to the size of some coral colonies that local richness, particularly in very rich regions, could be limited simply by the number of individuals that can be packed into this small sample. In all three examples, only two local scales were examined, but for two of these examples, this was sufficient to show differential responses of local richness to changes in regional richness. Clearly, multiscale analyses will permit stronger inferences regarding saturation, or lack thereof, in communities. They will also be useful in defining the spatial arena within which local processes, which might in some cases limit local richness, should be sought.

IV. SYNTHESIS OF COMMUNITY AND REGIONAL PERSPECTIVES

A. A Thought Experiment

Since processes operating at scales larger than the local habitat (e.g., landscape or region) almost surely affect local richness, research emphasis must shift to these larger-scale processes. Only then can patterns of community assembly be completely understood. However, this shift does not imply that local processes should be ignored. Local- and larger-scale processes may both affect local richness and their relative influence must be determined. The advantages of this multi-scale perspective can be illustrated with a thought experiment.

Suppose we have two climatically similar regions with different regional richness due to historical factors. The Asian and European temperate forests discussed previously provide a good example. Suppose further that physically identical habitats of two types (e.g., sandy versus loamy soils) were transplanted into these regions and were allowed to achieve a steady state that showed no influence of initial colonization history. Once the steady state was reached, by how much would per locality tree species richness vary among habitat types and regions? The transplanted habitats would make it possible to partition the contribution to per-locality richness among processes operating at local and regional spatial scales. The design would also make it possible to evaluate interactions between regional and local factors and to examine mediation of species interactions by changes in the local environment.

B. Manipulations of Productivity and the Pool

Although the preceding experiment is possible in principle, it is likely to be difficult to carry out. Also, it is difficult to be sure that climatic conditions that may affect local processes are truly identical among regions. An alternative to habitat transplantation is a two-factorial manipulation of the species pool and local environmental conditions in the same region. For example, suppose some high-productivity lakes in a particular region had unusually low fish species richness. Is the low richness due to interaction mediation or pool limitation? Manipulations could be designed in which the entire fish species pool could be added to lakes of different productivity. In addition, productivity itself could be manipulated by adding fertilizer to particular lakes. Care would have to be taken that the fish were adapted

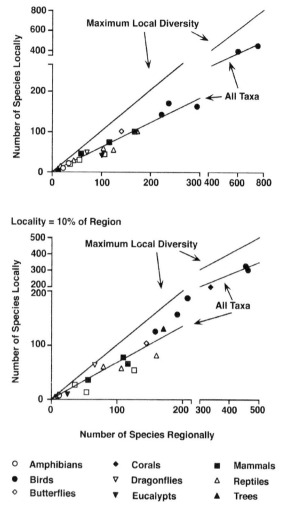

FIGURE 5   Local versus regional richness relationships at two local scales (1% and 10% of regional area) for 23 different taxa that have been combined for this analysis. The relationship is Type I at both scales, with the main difference being that the slope of the curve is roughly 10% lower at the smaller scale, consistent with the species–area relationship. The size of the locality has no effect on the tendency for saturation over this range of spatial scales. (From Caley and Schluter, 1997.)

to live at the prevailing productivity levels. A full two-factorial design would make it possible to sort out pool effects, productivity effects, and pool X productivity interactions on local richness (see Section I). If regional control of community structure is important, then pool effects or pool X productivity interactions should have important effects on local richness.

C. A Comparative Approach

In situations where experiments are intractable or when it is desirable to examine longer-term effects, the prob-

lem can be addressed using a statistical comparative approach. For example, Schluter and Ricklefs (in Ricklefs and Schluter, 1993) have suggested using multifactorial methods to partition the variance in local richness into components that can be explained by habitat differences within regions and species richness differences among regions. To the extent that local environments constrain local richness, species richness in similar habitats from different regions should converge relative to some hypothetical ancestral state of the community (Fig. 6). Detecting convergence is thus equivalent to measuring the variance component due to habi-

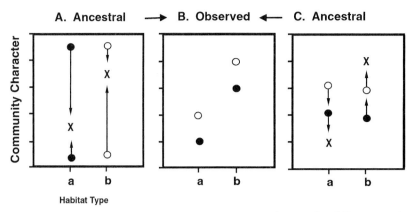

FIGURE 6 Hypothetical changes in communities over time showing that differences in the community due to habitat characteristics can be used to estimate community convergence. The community character in this case is species richness. (A) X indicates the richness value toward which the community will converge because of habitat characteristics. Open circles represent some unusually rich ancestral state of the communities in both habitats because, for example, they are located in a rich region and have been subject to a massive colonization event. Solid circles represent an alternative, depauperate ancestral state, because, for example, the communities are located in a species-poor region and have been subject to a disturbance. (B) Over time, communities will converge on X and the variance components due to habitat and region can be used to estimate the relative contribution of habitat and regional effects on local richness. (C) A slightly different case, but the inferences are the same. [From D. Schluter. (1986). Tests for similarity and convergence of finch communities. *Ecology* 67, 1073.]

tat differences. Important regional effects on local richness will be detectable in the variance component due to differences in regional richness. Schluter and Ricklefs calculated variance components for 36 different studies and found strong convergence in many cases, consistent with significant local effects on local richness. Yet there were strong regional differences in richness within similar habitats as well. Using regression methods to study coral reef communities, Karlson and Cornell (1998) similarly found both local and regional effects to be important.

D. Conclusions

The results from experimental and statistical analyses confirm that both local and regional processes have important effects on local richness. Thus, not only must experimental work on local communities be continued, but investigations need to be expanded to encompass larger-scale processes as well. The Volterra–Gause image of an ecological system at equilibrium and independent of historical processes has undeniable appeal, and has engendered much productive research on community assembly. Nevertheless, integration of historical, biogeographical, and metacommunity processes with local interactions and environmental variation is essential for a complete understanding of richness variation in species assemblages.

See Also the Following Articles

BIOGEOGRAPHY, OVERVIEW • DIVERSITY, MOLECULAR LEVEL • DIVERSITY, ORGANISM LEVEL • ENERGY FLOW AND ECOSYSTEMS • LANDSCAPE DIVERSITY • TEMPERATE FORESTS

Bibliography

Caley, M. J., and Schluter, D. (1997). The relationship between local and regional diversity. *Ecology* 78, 70.

Cornell, H. V., and Lawton, J. H. (1992). Species interactions, local and regional processes, and the limits to richness in ecological communities. *J. Animal Ecol.* 61, 1.

Hugneny, B., and H. V. Cornell, (2000). Predicting the relationship between local and regional species richness from a patch occupancy dynamics model. *J. Animal Ecol.* 69, 194.

Karlson, R. H., and Cornell, H. V. (1998). Scale-dependent variation in local vs. regional effects on coral species richness. *Ecol. Monogr.* 68, 259.

Karlson, R. H., and Cornell, H. V. (2000). Species richness of coral assemblages: Detecting regional influences at local spatial scales. *Ecology*, in preparation.

Lawton, J. H. (1999). Are there general laws in ecology? *Oikos* 84, 177.

MacArthur, R. H. (1972). *Geographical Ecology: Patterns in the Distribution of Species.* Harper and Row, New York.

McPeek, M. A., and Brown, J. M. (2000). Building a regional species pool: Diversification of the *Enallagma* damselflies in eastern North American waters. *Ecology,* **81,** 904.

Morton, R. D., and R. Lau, (1997). Regional species pools and the assembly of local ecological communities. *J. Theoretical Biol.* **187,** 321–331.

Ricklefs, R. E., and Schluter, D. eds. (1993). *Species Diversity in Ecological Communities: Historical and Geographical Perspectives.* University of Chicago Press, Chicago.

Rosenzweig, M. L. (1995). *Species Diversity in Space and Time.* Cambridge University Press, Cambridge, United Kingdom.

Srivastava, D. S. (1999). Using local–regional richness plots to test for species saturation: Pitfalls and potentials. *J. Animal Ecol.* **68,** 1.

Tilman, D., and Kareiva, P. eds. (1997). *Spatial Ecology: The Role of Space in Population Dynamics and Interspecific Interactions.* Princeton University Press, Princeton, NJ.

DIVERSITY, MOLECULAR LEVEL

Carlos A. Machado* and Marcos A. Antezana[†]

*Rutgers University and [†]University of Chicago

GLOSSARY

background selection Selection that takes place at loci other than that of interest. Such events can strongly increase the intensity of genetic drift and hinder the action of selection at a locus of interest. These effects are normally modeled by reducing the effective population size applicable to the locus of interest.

divergence A fixed genetic difference between two species or populations, or the process of evolving such a difference.

effective population size The size that a random mating population must have to experience the same intensity of genetic drift as a population of interest. The size can vary across genes due to changes in background selection across the genome.

functional constraint A limit to the kinds of nucleo-

tides or amino acids which can appear in a gene or protein without compromising function; often called "selective" constraints.

genetic drift Random changes in gene frequency that occur when a finite number of zygotes or propagules is formed by random sampling gametes or genotypes from the previous generation to create the next generation.

mutation A change in the nucleotide sequence of a DNA or RNA molecule.

nearly neutral mutation A mutation whose population genetical dynamics are influenced by both genetic drift and selection.

neutral mutation A mutation whose population genetical dynamics are influenced only by genetic drift.

nonsynonymous mutation A nucleotide change in a coding region that changes an encoded amino acid.

polymorphism A genetic difference between two or more individuals of the same species; to exclude new mutations from the definition, the underlying mutations are required to segregate above an arbitrarily set minimal frequency.

replacement Divergence due to a change of an amino acid in a protein, or the process to evolve such a change.

selection Differential reproduction within a set of genes, individuals, groups, etc. caused by differential performance in fitness determining tasks. The appropriate unit of selection (gene, individual, or group) is determined by the level at which the causation of the fitness differences must be described (interactions or performances at the level of genes, individuals, groups, etc.).

substitution Divergence due to a change of base in a DNA sequence, or the process to evolve such a change.

synonymous mutation A nucleotide change in a coding region that does not change the encoded amino acids.

THIS ARTICLE DEALS WITH MOLECULAR DIVERSITY at the DNA and protein levels and will concentrate on diversity that is genetically heritable. Therefore, molecular diversity shown by other macromolecules and/or that is not heritable will not be discussed. However, this does not imply that the latter diversity is biologically insignificant. Indeed, much of molecular diversity is of this kind, with secondary metabolite profiles in plants and antibody diversity generated via mutation and genetic rearrangements in somatic cells of vertebrates being two outstanding examples. We discuss molecular diversity patterning and causation primarily at the level of single genes and genomic regions. Therefore, important aspects of molecular diversity that concern higher level genomic structuring and evolution will not be treated systematically (e.g., gene duplication, multigene family evolution, concerted evolution, genome size evolution, and mammalian isochore structure).

The article will first deal with DNA and protein diversity within populations and among populations of the same species (molecular polymorphism) and then with that caused by genetic differentiation between species and higher taxa (molecular divergence). Such order is justified because visualizing diversity in this way helps in understanding the evolutionary and population genetical forces and processes that generate it.

The information presented here is almost exclusively drawn from knowledge gathered by molecular evolutionary biologists and molecular population geneticists. Molecular evolutionary research aims at characterizing the degree and forms of molecular divergence, at clarifying how such differentiation takes place, and—by taking advantage of the former—at reconstructing the history of DNA sequences and proteins and the genealogical relationships among extant and sometimes extinct organisms. Molecular population genetics studies the form and extent of molecular polymorphism and tries to identify the forces that pattern it.

I. INTRODUCTION

Phenotypic differences between organisms are due to environmental and genetic differences, and the latter are the result of genetic divergence or polymorphism. When due to divergence, genetic differences often can tell us about the physical and geographical environment in which organisms evolved. When due to polymorphism, they can tell us about the ecological arena that is shaping the genetic constitution of a species (gene pool). Realizing that polymorphism is a common and striking phenomenon in the living world is not only important when trying to make sense of molecular diversity. Indeed, the transition from typological (essentialistic) to "variational-populational" thinking was a major contribution of the Darwinian revolution to biology.

Measuring genetic diversity within and between species is a long-standing interest of biologists. However, this only became possible in a systematic way in the 1960s when novel techniques from molecular biology (protein sequencing and enzyme electrophoresis) began to be applied to the study of evolutionary and population genetical questions. These techniques opened the possibility to study variation at the molecular level (allozyme variation and amino acid sequence divergence) and gave birth to the fields of molecular population genetics and molecular evolution. The availability of protein sequences from various taxa triggered sophisticated studies of the forces driving protein evolution and led to the first reconstructions based on protein sequence data of historical relationships among organisms (phylogeny). However, protein sequencing could not be used for quantitatively oriented studies of polymorphism because of its laboriousness. Electrophoretic studies of variation were carried out in hundreds of taxa and confirmed less conclusive previous observations that hinted at the presence of large amounts of genetic variation within populations. Electrophoresis, however, often failed to detect some amino acid variation so that it could not deliver data of sufficient quality to study the nature of the evolutionary forces maintaining variation and the importance of such variation for the process of adaptation.

The advent of DNA sequencing in the 1980s eliminated resolution problems and made it possible to describe exhaustively the genetic variation segregating at single gene loci. Before DNA studies this was not possible, and in fact investigators developed and tested models that predicted patterns for electrophoretic variation segregating at many loci because only such data could be produced in statistically analyzable counts. The shift to DNA studies allowed researchers to quantify conclusively not only amino acid polymorphism but also coding region variation that does not change the encoded amino acids (synonymous variation) and variation in noncoding regions, both of which were found to occur in copious amounts. This increased drastically the amount of data obtainable from a single locus and thus the number and the statistical quality of contrasts among classes of variation subject to arguably different functional constraints. New and possibly ultimate imagery was developed to describe the processes shaping DNA-level variation within and between populations and new, often more powerful modeling approaches were developed and/or given a more solid empirical grounding (e.g., coalescence and maximum likelihood estimation of genetic distances). However, despite all the technical and conceptual advances in describing and quantifying molecular variation, we still know very little about the genetic basis of adaptation and about the relationship between genetic and phenotypic differences among organisms.

II. THE GENERATION AND MAINTENANCE OF GENETIC DIVERSITY

There are three main sources of genetic variation in natural populations: mutation, recombination, and migration. Mutation is the only mechanism that can generate variation in a genetically homogeneous population since recombination and migration only shuffle and shuttle around preexisting variation, respectively. Mutation is therefore the original source of all genetic variation.

In order to understand the evolutionary and population genetical forces patterning genetic variation, it is useful to consider variation at two main levels: within and between species. The first level constitutes the polymorphism component of genetic diversity, which can also be considered at finer levels if gene flow is structured within the species under study ("geographic" structure). Within sexual populations this type of diversity is more or less freely combined each generation to form individual genotypes. The second level constitutes the divergence component of genetic diversity, which can also be considered at multiple levels depending on how closely related are the species under study. This type of diversity consists of fixed differences between species and is only exchanged in extraordinary circumstances (horizontal gene transfer and introgression).

The patterns of genetic variation within species arise from the combined action of two main sets of processes: (i) a molecular-level origination process in which a mutation in a DNA sequence arises and (ii) a population-level "fate" process that determines whether the mutation is quickly eliminated from the population or

enters the pool of intraspecific variation becoming a polymorphism. The fate process also includes the possibility that the mutation becomes fixed in the population, i.e., part of the divergence component of diversity. If a new mutation is passed on to the next generation of individuals, its final fate depends on the stochastic error produced when creating a finite number of zygotes or propagules by sampling gametes or genotypes from the previous generation (random genetic drift), on demographic forces such as gene flow and changes in population size, as well as on the deterministic action of natural selection if the mutation is deleterious or advantageous.

III. DIVERSITY AS POLYMORPHISM

What is the quantity and the quality of genetic variation found in natural populations? What are the evolutionary forces responsible for the observed patterns of variation in natural populations? What is the meaning of that variation? What is the relationship between adaptive changes, phenotypic changes, and extant genetic variation? These are some of the questions asked by evolutionary geneticists. The following sections provide examples and summarize the current answers to the first two questions but do not elaborate on the last two questions given the unconclusiveness of the current knowledge on the subject.

A. Empirical Observations

Fruit flies (*Drosophila* sp.) are by far the best studied group of organisms with respect to their patterns of heritable molecular diversity. Two randomly chosen homologous sequences of *Drosophila melanogaster* show on average four nucleotide differences per kilobase (kb). A comparison of two full haploid genomes (i.e., including the X chromosome) would give a total of 7.2×10^5 nucleotide differences over the 180-Mb of a typical haploid genome on average, of this species. *Drosophila simulans,* a closely related species, has twice as much variation: eight differences per kilobase and an average of 1 million nucleotide differences between any two haploid genomes (the *D. simulans* haploid size is 130 Mb).

Studies of other taxa are limited to few genes or genomic regions but nevertheless allow interesting contrasts. Humans, for instance, have less variation than fruit flies (one difference per kilobase), but two average human haploid genomes could differ at more nucleotide sites (3.4 Mb) because the human haploid genome size is much larger (3400 Mb). On the other hand, the few nuclear loci studied to date in the common and pygmy chimps (*Pan troglodytes* and *P. paniscus*) show about three times more nucleotide variation than do the homologous loci in humans. In plants, one of the genes encoding the enzyme alcohol dehydrogenase (*Adh1*) shows levels of polymorphism that range from 1.8 differences per kilobase in wild barley (*Hordeum vulgare*) to approximately 18 differences per kilobase in maize (*Zea maize*). In contrast, the homologous gene in *D. melanogaster* shows an average of 6.5 differences per kilobase. These quantities describe more or less the typical range of DNA polymorphism in natural populations.

The question about the quality of the variation (i.e., about the functional consequences of the observed nucleotide polymorphisms) is normally approached by extrapolating from a priori knowledge about the genetic code and from the conservation of certain nucleotides and nucleotide motifs in noncoding regions. For instance, amino acid changes are counted separately from synonymous changes, and changes violating the phylogenetically or functionally determined canonical motifs of regulatory sequences are distinguished from those that do not. With respect to identifying selective pressures affecting different variants, the frequency pattern of mutations (are variants rare or common?) can be very informative. For instance, an indication of deleteriousness is the occurrence in different samples of a certain kind of mutation only at low frequencies (i.e., in only one or few of the sampled individuals).

Drosophila data show that in noncoding regions (flanking regions and introns) there is more variation than in coding regions (2.5 and 2.4 times more variation in *D. melanogaster* and *D. simulans,* respectively) and that synonymous polymorphism is higher than noncoding polymorphism (1.2 and 1.6 times higher, respectively), and higher than amino acid polymorphism (2.7 and 7.6 times higher, respectively). Most amino acid variation tends to be found at low frequencies and so are silent changes away from the major (preferred) synonymous codon within a codon family (i.e., toward minor, unpreferred codons) in genes with biased usage of synonymous codons.

These observations about patterns of variation can be very informative when asking questions such as the following about the nature and sometimes the magnitude of the evolutionary forces generating the patterns: Why do some species have more variation than others? Is it due to differences in the rate of origination of mutations? To differences in the size of their populations? To recent evolutionary, demographic, and ecological events? To the species' distributions over geographic space? To their breeding system? To the genomic context in which the mutations occur (e.g., the intensity of recombination and of deleterious mutation in and around the studied locus)? The latter possibility has been shown to apply in *Drosophila* by two recent studies that found a positive correlation between levels of variation and the rate of recombination.

B. Molecular Biological Considerations

The mutation process is known to be affected by physiological and life history characteristics of the organism which affect the rate and accuracy of DNA replication (generation time, number of divisions before gametogenesis, sex, and age) and by factors that increase damage to DNA (metabolic rate, age, transcription rate, and the frequency and degree of packing and unpacking of chromosomes for expression). DNA damage is caused by exogenous and endogenous agents. Exogenous agents include UV light, ionizing radiation from X-rays, heat, alkylating agents, and hydroxil radicals originated by the interaction of ionizing radiation with water. Endogenous agents include oxidant by-products of normal metabolism. Changes generated by endogenous agents cause the highest fraction of DNA damage. In human cells, oxidants damage between 10^4 and 10^5 nucleotide sites every day. Such damage would be extremely deleterious were it not for the presence of enzymes involved in finding and repairing it. Of the five major types of DNA mutations—substitutions, deletions, insertions, inversions, and duplications—nucleotide substitutions and single base pair deletions or insertions (all called point mutations) account for 85% of the mutation events.

With the exception of macromolecular events such as chromosomal rearrangements, most mutations are the result of errors made by enzymes involved in DNA replication and repair. Replication errors are often due to the incorporation of noncomplementary nucleotides by the DNA polymerase and incorrect subsequent proofreading by the same enzyme. The latter process, however, is normally extremely effective at lowering the error rate by a million-fold relative to the intrinsic misincorporation rate of the polymerase. Table I shows the range of mutation rates observed in several organisms.

C. Cytological and Ontogenetic Considerations

In viruses, prokaryotes, and unicellular eukaryotes—with the notable exception of protists such as *Parame-*

TABLE I
Mutation Rates per Genome Replication[a]

Organism	Mutation rate per base pair	Mutation rate per genome
Bacteriophage Qβ (RNA)	1.5×10^{-3}	6.5
Influenza A virus (RNA)	$>7.3 \times 10^{-3}$	1.0
Bacteriophage M13 (DNA)	7.2×10^{-7}	0.0046
Bacteriophage T4 (DNA)	2.4×10^{-8}	0.0040
Escherichia coli	5.4×10^{-10}	0.0025
Neurospora crassa	7.2×10^{-11}	0.0030
Caenorhabditis elegans	2.3×10^{-10}	0.018
Drosophila melanogaster	3.4×10^{-10}	0.058
Mouse	1.8×10^{-10}	0.49
Human	5.0×10^{-11}	0.16

[a] After Drake et al. (1998).

cium that have a macro- and a micronucleus—all mutations that are not strongly deleterious are heritable and become part of the polymorphism component of diversity affected by population biological and transmission genetical forces such as genetic exchange, recombination, genetic drift, and natural selection. In metazoans, most mutations arise in somatic cells and are not heritable, although they may very well have an impact on the fitness of the individual carrying them. Mutations that arise in the cell lineages leading to propagules or gametes are heritable, however. This applies in general to all multicellular organisms. In organisms with a germline (a terminal differentiation of specialized gamete-producing cells), differences in the timing of germline sequestration and in the number of cell divisions before gametogenesis can be studied carefully and have been shown to have a major impact on the number of new mutations found in gametes. A good example is the observation that most of the evolution at the molecular level in mammals is "male driven": In X- and Y-linked homologous genes the male to female ratio of nucleotide substitution is close to 6 in humans and 2 in rodents, which corresponds closely to the ratio of male to female germline divisions per organismal generation. This means that in humans most mutations participating in molecular divergence arise in the male germline.

D. Factors Influencing the Fate of Mutations

Patterns of molecular diversity other than those created by a continuous flow of short-lived mutations which remain unique to the individuals in which they arise can be very striking. For instance, some of the loci encoding the antigen-presenting major histocompatibility complex (MHC) in humans show extremely high levels of amino acid and synonymous polymorphism as well as many alleles segregating at intermediate frequencies. On the other hand, loci in the human Y chromosome show little polymorphism and mainly low-frequency mutations.

In general, patterns of polymorphism are determined by mutation pressure, genetic drift, and natural selection. The effectiveness of the action of the last two forces depends on the effective population size (see Glossary) and on the fitness effect of the mutation involved. In particular, the transmission genetical sampling effect at the basis of genetic drift phenomena increases in strength with decreasing population size and this in turn hinders the deterministic action of selection.

The patterns of variation of selectively neutral mutations are totally determined by genetic drift, population size, and the rate of mutation to neutral variants. For selected mutations both drift and selection can affect the patterns. In the case of strongly deleterious mutations, the observed pattern is one of mutations unique to each individual and none segregating at high frequencies because such mutations remain in the population for only very few generations. When a steady state is reached between the number of new mutations being produced by mutation and those being removed by selection, one speaks of a mutation–selection balance. There should not be any observable polymorphism patterns for strongly favorable mutations because such mutations are normally fixed soon after being introduced in the population (if they are not immediately lost to drift) and practically do not participate in polymorphism. However, favorable mutations can be maintained at intermediate frequencies, e.g., when the heterozygote has higher fitness. In this case, one speaks of a balanced polymorphism (as in the MHC case previously mentioned). For weakly selected mutations, genetic drift and selection are important. Note that the previous considerations are mainly applicable to recombining populations because in asexual populations and in cases with very weak selection and low recombination intensity, situations can arise that require considering patterns and forces at the level of the whole genome or of long stretches of the genome.

E. Models of Molecular Evolution

The main models that make predictions about the patterns of polymorphism and divergence are the neutral theory proposed in the late 1960s by M. Kimura, the nearly neutral theory proposed in the early 1970s by T. Ohta, and a series of models of natural selection developed by J. Gillespie during the past two decades. The first two models emphasize the importance of genetic drift and mutation, whereas the latter group of models stress the role of natural selection. All models agree that the vast majority of newly arisen mutations are deleterious, but they differ greatly in their assumptions about the relative proportions of neutral and favorable mutations that participate in polymorphism and divergence.

1. Neutral Theory

Kimura's neutral theory postulates that both the fixation of genetic differences between populations and the patterning of genetic variation within populations are the result of random changes in the frequency of neutral alleles driven by genetic drift. This model assumes that most new mutations are either deleterious or neutral and that advantageous mutations arise quite infrequently and are quickly fixed or lost to drift so that they are not important for polymorphism. Most of the theory's predictions for polymorphism patterns assume mutation–drift equilibrium, i.e., that the rate of allele loss to drift is balanced by the production of new alleles by mutation. The predictions therefore apply only when the population size and the mutation rate have been constant for a long period of time.

The theory predicts that the amount of polymorphism within a population is proportional to the effective population size (N) and the mutation rate (μ). Nucleotide polymorphism is normally expressed in terms of the parameter $\theta\ (=4N\mu)$, which can be interpreted as the average number of differences per nucleotide site between two alleles randomly chosen from the population. This divergence is the result of nucleotide mutations that have accumulated since the two alleles last shared a common ancestor. Theory has shown that the expected time back to this ancestor for an autosomal locus in a diploid is $2N$ generations so that when each allele has been evolving independently for $2N$ generations, the expected number of nucleotide differences between them is $2 \times 2N$ times the mutation rate μ, i.e., $4N\mu$.

The neutral theory explains the fact that different genes and regions of the genome show different levels of polymorphism by assuming that the neutral mutation rate varies among such regions and genes because they are under different functional constraints. Recent simulation work has led to a restressing of the realization that the effective population size can be reduced by

selection against deleterious mutations at linked and unlinked sites ("background selection"). This effect can compel neutral polymorphism to change drastically across regions that differ in recombination intensity and deleterious mutation production.

2. Nearly Neutral Theory

Ohta's model, also known as the slightly deleterious allele model and considered by many to be the "dominant paradigm for molecular evolution," was proposed in the early 1970s in order to account for two observations that could not be explained by the neutral model: the generation time effect for DNA divergence and the lower than expected levels of protein variation in natural populations. The model deviates from the neutral theory mainly because it assumes (i) that most observed molecular variation is slightly deleterious, (ii) that mutation rates are correlated with the generation time (g), and (iii) that the rate of molecular divergence for slightly deleterious alleles is inversely proportional to N. The nearly neutral theory is biologically realistic because it proposes plausible correlations of population genetical parameters such as g and N with the origination and fate processes of molecular evolution, and because it avoids the unrealistic assumption of complete neutrality for most variation participating in polymorphism and divergence.

Slightly deleterious alleles, also called weakly selected alleles, are defined as those with selective coefficients close to the neutral boundary ($s \approx 1/N$) and are seen as major contributors to polymorphism and divergence in natural populations. The neutral theory stresses that the fate of neutral variants depends only on random drift, whereas the slightly deleterious model builds on the fact that the fate of slightly deleterious variants depends on the joint action of both drift and selection. Which of these two processes is more important depends on the effective size of the population because the efficacy of selection in removing slightly deleterious variants and fixing advantageous ones is higher in larger populations. In smaller populations slightly deleterious mutations can increase in frequency and become fixed by genetic drift, but in large populations selection prevails and eliminates them quickly. Therefore, the nearly neutral model predicts that if most of the observed variation in natural populations is slightly deleterious, polymorphism patterns should be dominated by rare variants.

The lack of correlation between genetic variation and population size (especially in large populations) was explained by this model with the argument that when divergence and variation involve mainly slightly deleterious alleles, the amount of diversity in a population should be proportional to N but nevertheless reach a plateau when N is large enough for selection to impede that slightly deleterious variants increase in frequency. A main problem with this interpretation is that in organisms with very large population sizes one would not expect any variation if most polymorphism is due to slightly deleterious variants (at least no amino acid polymorphism), which is clearly not the case.

3. Selection Models

Gillespie proposed and developed a series of selection models that are less terse in their assumptions than the neutral and nearly neutral models but that can account for most of the empirical observations about molecular polymorphism and divergence. The models assume that balancing and episodic selection dominate the patterning of divergence and polymorphism at the molecular level. They predict that fluctuations of the environment can lead to the maintenance of polymorphism by balancing selection and, in the right circumstances (when the selective advantage of an allele is greater than the variance of the fitness effects of the alleles at that locus), to the fixation of favorable alleles. The models assume fitness effects that are underdominant, overdominant, or that change in a random environment (SAS-CFF model) or in a temporally fluctuating environment (TIM model). The addition of environmental variability is biologically realistic since the existence of fluctuations in the fitness value of segregating alleles due to changes in the environment over ecological timescales is quite plausible. These models predict patterns of polymorphism similar to those predicted by the neutral models, which makes hypothesis testing very problematic.

F. Statistical Tests on Molecular Polymorphism

Of all the models of molecular evolution, only the neutral theory makes simple quantitative predictions about expected levels and patterning of genetic polymorphism. This fact makes the neutral model a very popular null hypothesis against which to test the importance of other evolutionary forces in shaping patterns of polymorphism. There are two basic families of statistical tests for neutrality: (i) models that test neutrality using polymorphism data from a single locus and (ii) models that test neutrality using polymorphism and divergence data.

The data required for the single locus tests are homologous DNA sequences collected from individuals of a single species. This family of tests compares three estimates of the neutral parameter θ, which equals $4N\mu$ for an autosomal locus in a diploid. The three estimates are straightforwardly calculated from (i) π, the average number of pairwise differences between the sampled alleles which should be equal to θ; (ii) S, the number of polymorphic sites in the sample; and (iii) η, the number of mutations present in only one of the sampled alleles (singletons). Under neutrality the values of these estimates should be the same. The tests therefore ascertain whether departures from zero of the differences between any two of these estimates are statistically significant for the sample at hand.

Tajima, for instance, proposed comparing the values of π and θ_s (θ estimated by S) using his statistic D, which is the difference between π and θ_s, normalized by dividing the statistic by a factor proportional to the variance of the difference. Negative values of D are expected in the presence of purifying selection or following a purging of variation from the region by selection driving the quick fixation (1) of a new mutation or (2) a rapid increase in the frequency of an allele previously present at a lower frequency (selective "sweep"). Positive values of D are expected with balancing selection and population subdivision.

Although the patterns predicted by the neutral theory are always used as null hypothesis, the patterns produced by the two main kinds of selection regimes and/or events (sweeps and balanced polymorphism) have also been described thoroughly, and it has sometimes been striking how well they match the data. For instance, selective sweeps tend to reduce neutral variation in regions linked to the selected site. This "hitchhiking" effect on neutral variation (in this case, non-hitchhiking because most variation does not manage to get a "ride") is more pronounced in zones with low recombination. Balancing selection increases the levels of variation over the neutral expectation in the sites adjacent to the balanced site by making the times to the common ancestor at these linked sites more similar to that of the balanced site (i.e., to the time since the establishment of the polymorphism) than to the neutral time at unlinked sites. If the polymorphism is older than $2N$ generations, one should observe an increase in neutral variation in the region of the balanced site relative to the neutral expectation. Another phenomenon that can affect the level, but much less so the pattern, of neutral variation is "background selection." This effect and regular sweeps may account for the low levels of polymorphism observed in regions with low recombination.

IV. DIVERSITY AS PHYLOGENETIC DIVERGENCE

How could genes, proteins, and genomes become so different, and what limited how different they could become? Are there heterogeneities in the rates of divergence among different proteins, genes, and genomic regions? Do molecules from different organisms evolve at different rates? What are the mechanisms of nucleotide and amino acid substitution? Can we infer the genealogical relationships among different organisms from molecular data? Can we use molecular divergence data to date evolutionary events? These are some of the questions addressed by researchers in the field of molecular evolution. The following sections provide examples and summarize the current answers to the first three questions, i.e., those most relevant to understand the evolutionary dynamics of molecular diversity.

A. Some Empirical Observations

One of the main patterns of divergence that have been described by protein and DNA sequencing studies is that rate of amino acid replacement in different kinds of proteins and in different organisms are extremely variable (Table II). Histone H4, a protein involved in the packaging of DNA in chromosomes, is a striking case of an extremely conserved protein. Comparisons of H4 histones from plants and mammals show that only two amino acid replacements, over the 100 sites compared, have taken place since the two groups last shared a common ancestor about 1 billion years ago. On the other hand, viral envelope proteins, which are under constant challenge by the host immune system, change at extremely high rates—between 1 million to 10,000 times faster than the fastest mammalian protein (interferon-γ).

Table II also shows that rates of substitution at synonymous sites are higher than rates at nonsynonymous sites. This is expected because amino acid changes often disrupt the function of the protein, whereas synonymous changes are less likely to have comparable consequences. This explanation is corroborated by the observation that noncoding regions and pseudogenes (deactivated genes) evolve with rates similar to those of synonymous changes. However, in several proteins amino acid replacements occur more frequently than

TABLE II

Rates of Amino Acid and Silent Substitution (in Events per Site per Year)[a]

Organism	Protein	Nonsynonymous rate	Synonymous rate
Human-rodent	Histone 4	0.00×10^{-9}	3.94×10^{-9}
	Insulin 4	0.20×10^{-9}	3.03×10^{-9}
	Hemoglobin α chain	0.56×10^{-9}	4.38×10^{-9}
	Interleukin-1	1.50×10^{-9}	3.27×10^{-9}
	Ig κ chain	2.03×10^{-9}	5.56×10^{-9}
	Relaxin	2.59×10^{-9}	6.39×10^{-9}
	γ-Interferon	3.06×10^{-9}	5.50×10^{-9}
Drosophila	Adh	0.90×10^{-9}	9.50×10^{-9}
	Esterase 6	3.03×10^{-9}	21.52×10^{-9}
Plants			
Grasses	Adh1	0.32×10^{-9}	7.00×10^{-9}
Grasses	Adh2	0.89×10^{-9}	5.99×10^{-9}
Palms	AdhA	0.41×10^{-9}	2.61×10^{-9}
RNA viruses			
Influenza A	Hemagglutinin	3.59×10^{-3}	13.10×10^{-3}
HIV-1	gag	1.70×10^{-3}	9.70×10^{-3}
DNA viruses			
Hepatitis B	P	1.45×10^{-5}	4.57×10^{-5}

[a] *Drosophila* rates are from comparisons between the *obscura* and the *melanogaster* groups (after Li, 1997).

synonymous substitutions, e.g., in the antigen recognition site of the HLA loci in humans, in two fertilization proteins of abalone, and in proteins involved in male reproduction (Acp26Aa) or associated with male sterility (OdsH) in *Drosophila*. In the latter locus, for instance, more amino acid changes have occurred in less than 1 million years than during the previous 700 million years. The rapid evolution of this gene is likely due to a change in function.

Similarly, different parts of a protein (e.g., protein domains) can diverge more readily than others. A classic example is the heme pocket and the surface regions in hemoglobins α and β, which show a 10-fold difference in their rates of amino acid divergence. A similar situation occurs in the mitochondrial cytochrome oxidase II, whose transmembrane domain evolves at a much faster rate than the rest of the protein.

Another major pattern evidenced by the data is that different taxa can evolve at quite different rates (lineage effects). For instance, a survey of substitution rates in 54 mammalian genes shows that rodents tend to have higher rates of synonymous divergence than primates, but their nonsynonymous rates are not significantly different. Similarly, sharks show a seven-fold reduction in rates of silent mtDNA evolution with respect to primates. Finally, endosymbiotic bacteria associated with aphids show higher rates of amino acid replacement than do closely related free-living strains.

B. The Substitution Process

The fixation of new mutations gives rise to the divergence component of genetic diversity. This process describes how a new allele replaces the previously most common allele in the population, thereby becoming the common ancestor of all alleles at that locus. The dynamics of the substitution process are determined (i) by the rate of origination of new mutations in the population, and (ii) by the probability of fixation of such mutations. The main evolutionary forces affecting this process are mutation, genetic drift, and selection, with their relative importance differing for neutral, weakly selected, and strongly selected mutations.

Neutral autosomal mutants are produced in a diploid population with rate $2N\mu$ per generation, i.e., the number of new mutations produced each generation is equal to the product of the number of gene copies in the population that can mutate, times the mutation rate to neutral alleles per gene per generation μ. The same result applies for selected mutations if we let μ become the rate of mutation to strongly selected or weakly selected alleles.

The probability of fixation of a mutation depends on the selective advantage or disadvantage of the mutation, on the population size, and on the initial frequency of the mutation. For a newly arisen autosomal mutation the initial frequency is $1/2N$ in a diploid population, which for the neutral case equals the probability of fixation of the mutation (since any of the $2N$ alleles present at any given moment can become the common ancestor of all alleles in the population some time later). For a nonrecessive mutation with selective coefficient s close to $1/N$, the probability of fixation is about $2s/(1 - e^{-4Ns})$. If the value of s is positive (advantageous mutant) and the population size is large such that $Ns \gg 1$, the fixation probability is twice the selective advantage of the mutant ($\approx 2s$). Even in this case, however, genetic drift is the main determinant of the fate of new mutations (e.g., 80% of mutations with $s = 0.1$ are lost to drift). For a deleterious mutant (negative s) the fixation probability is lower than that in the neutral case and is practically zero for strongly deleterious mutations.

The rate of substitution equals the average number of new mutant alleles entering the population in each generation times the probability that one of them will fix. For a neutral allele, this equals $(1/2N) \times (2N\mu) = \mu$, i.e., the rate of substitution of neutral alleles equals the mutation rate. For selected mutations the rate of substitution equals $4Ns\mu$, and is therefore dependent on the mutation rate to favorable selected mutations, the population size, and the selection coefficient of the mutant. It is easy to see that for the neutral rate of substitution to remain constant across different populations it suffices that the neutral mutation rate does not change. For selected mutations, however, the substitution rate could vary much more since its constancy requires that three parameters do not change or that each varies in ways that balance changes in the others.

C. Models of Molecular Evolution

From their conception, most theoretical models of molecular evolution dealt with both divergence and polymorphism from a population genetically explicit perspective. In the following sections, we present the predictions for patterns of divergence for the most relevant models (for polymorphism, see Section III).

1. Neutral Theory

Kimura showed that genetic divergence at neutral loci should increase over time due to the fixation of neutral mutations by random drift at a rate that equals the neutral mutation rate μ. This predicted constancy in the rate of substitution matched the empirical observations showing constant rates of amino acid substitutions per year for different proteins in a wide range of vertebrate lineages that led Zuckerkandl and Pauling to propose the molecular clock hypothesis in the early 1960s. The match is often viewed as positive evidence for Kimura's neutral theory, but this is a misunderstanding since selection models can also produce clock-like divergence.

The observed differences in rates of evolution among different genes and the heterogeneity of divergence rates in different regions of a gene are explained by the theory as being the result of differences in the fraction of all mutants that are neutral for each given gene or region (i.e., different functional constraints).

2. Nearly Neutral Theory

Ohta's model assumes that most molecular diversity is slightly deleterious, and predicts that the rate of production and fixation of neutral mutations is negatively correlated with organismal generation time (generation time effect), and that the rate of divergence for slightly deleterious changes is inversely proportional to the population size. The negative correlation between generation time and mutation rate is expected when mutations occur mainly during DNA replication and the number of germinal cell replications per generation is the same across different organisms. The germinal cells of short-lived organisms should indeed experience more DNA replications per year than those of long-lived organisms so that more neutral mutations can be produced in short-lived organisms. This assumption corresponded well with observations from divergence studies from the early 1970s based on DNA hybridization that showed a strong generation time effect for noncoding DNA evolution. The generation time effect can be obscured in comparisons among distantly related organisms when they differ in developmental mode and germline ontogeny (see Section III).

An important contradiction that this theory tried to resolve was that protein evolution appears to be clock-like, whereas noncoding DNA evolution shows a generation time effect. The neutral model explained the former by asserting that most amino acid replacements are neutral. The nearly neutral model asserted that protein divergence is mainly due to the fixation of slightly deleterious alleles, so it had to resort to postulating that generation time is inversely proportional to population size in order to account for the constancy of amino acid replacement rates. It was argued that small organisms have higher mutation rates per year due to their shorter generation times but that due to their larger population sizes, they fix slightly deleterious amino acid changes less readily. On the other hand, organisms with longer generation times have lower mutation rates but smaller population sizes so that they fix slightly deleterious mutations more easily. Therefore, the fact that in actual populations generation time correlates negatively with population size lets the rate of amino acid divergence appear clock-like. Although this negative correlation initially failed to be corroborated, recent studies have shown it to hold across different taxonomic levels.

A major problem with this theory concerns divergence: slightly deleterious variants cannot accumulate indefinitely because they would decrease the fitness of the population, thereby giving new favorable mutations a major role in the process. Additionally, in large populations no divergence should occur because in them selection should be efficient enough to eliminate most slightly deleterious alleles.

3. Selection Models

Divergence data led Gillespie to propose his selection models. He showed that there is excessive variation in amino acid divergence rates in different mammalian proteins using the index of dispersion $R(t)$, which is the ratio of the variance in substitution rate to the mean substitution rate. The index ranged between 2.5 and 7.5, whereas the neutral theory predicted it to be approximately 1.0 (as expected for a simple Poisson process). These high values of $R(t)$, which were not confirmed in a recent *Drosophila* study, led Gillespie to propose that the process of divergence is characterized by bursts of substitutional activity during which alleles previously segregating at high frequencies become rapidly fixed by selection, followed by quiet periods with little substitution and long-lived intermediate frequency polymorphisms. This still allows for a molecular clock if the periods of bursting and stasis succeed each other more or less regularly. One speaks in this case of an episodic molecular clock. Nevertheless, neither these models nor the neutral and nearly neutral models can completely account for the excessive variance of the rate of amino acid replacement in mammals. Improvements of these models will include increasing the time frame over which the fitness effects of mutations change or fluctuate. However, the complexity of the models has already reached the point that they can fit almost any observation, which compromises their value for hypothesis testing. At this point, one will have to assess the plausibility of their assumptions rather than the accuracy of their predictions.

V. COMBINED PREDICTIONS FOR DIVERGENCE AND POLYMORPHISM

Two tests based on the mutual compatibility of divergence and polymorphism patterns have been developed and used with much success to test the neutral theory. When they reject neutrality, outside information can help in further identifying the forces that might have caused the rejection. The Hudson–Kreitman–Aguade (HKA) and the McDonald–Kreitman tests focus on the positive (1) correlation between the (2) amounts of polymorphism and divergence expected under neutrality (given the linear dependence of both patterns on the mutation rate).

The HKA test compares the correlation between polymorphism and divergence at two or more loci. Its rationale is that if loci have different neutral mutation rates, their rates of divergence and levels of polymorphism should be equally affected. Selection is therefore inferred when the ratio of polymorphism to divergence varies too much over loci, a conclusion that requires sophisticated corrections to deal with the fact that, due to stochastic reasons, different loci can have different levels of polymorphism if they are not totally linked.

The McDonald–Kreitman test uses a similar contrast but concentrates on "loci" that are interspersed with each other over the sequence, thereby avoiding the need for the previously discussed corrections and drastically increasing the statistical power. This test can be applied to contrast silent and replacement variation, or unpreferred vs preferred silent changes, but it has mainly been used to compare the former within and between species under the assumption that these two kinds of variation are homogeneously distributed along the coding region studied.

Rejections of neutrality with both tests can be due to departures at the divergence or at the polymorphism level. For instance, an excess of amino acid polymorphism relative to amino acid divergence should be observed if variation is slightly deleterious. However, this "excess" could very well be due to a deficiency in silent polymorphism and/or an excess in silent divergence. If the ratio of amino acid replacement to silent substitution is significantly greater than the ratio of replacement to silent polymorphism, one could infer that natural selection has been fixing amino acid mutants beyond the neutral expectation, but again this could be due to a deficiency in amino acid polymorphism and/or in silent divergence. This is why outside information must be used to identify the causes of a rejection of neutrality by these tests.

VI. OMITTED DIVERSITY PATTERNS

The patterns of molecular diversity presented previously were originally studied in order to test hypotheses about abstract evolutionary and population genetical dynamics. For this reason, they are not necessarily representative of the phenomenon of molecular diversity in a more comprehensive molecular biological, natural historical, and organismic–evolutionary sense. This explains, for instance, the lack of information about the evolutionary dynamics of molecular diversity with developmental consequences. Although biologists are well aware that this kind of diversity may be extremely important in an organismic–evolutionary sense, the current lack of knowledge about it makes it very unsuitable for evolutionary and population genetic research that focuses on hypothesis testing. The concentration on variation suitable for hypothesis testing also explains this article's lack of an extensive presentation of many important molecular biological and biochemical constraints on diversity patterns. For instance, the dinucleotide motif T|A ("|" is the codon boundary) is very underrepresented in most coding regions of *Drosophila*, yeast, and humans, but no research has been done on how it constraints molecular diversity patterns. However, this constraint is quite likely to affect variation patterns in more coding regions than the high codon usage biases shown by most highly translated genes. It is an important task to increase our knowledge of molecular diversity with respect to molecular biological constraints and in the natural historical and organismic–evolutionary sense.

VII. PROBLEMS OF THE MODELS PRESENTED

Most of the neutrality- and selection-based modeling of polymorphism and divergence is based on single-site approaches, and it was not until the 1980s that linkage effects began to be considered. In models with selection, single-site results which assume that no further concomitant mutant is segregating are the standard, and the few efforts to deal with many concomitant mutations assume multiplicative fitness effects of mutations, i.e., that fitness is $(1 - s)^i$ when one carries i mutations.

The latter assumption is made out of convenience and lack of clear alternatives but is not justified a priori nor has it been justified a posteriori. It is indeed not clear how selection acts on, for example, synonymous or insertion/deletion variation, and one cannot exclude that it acts at the level of the total individual genotype via a possibly complex epistatic relationship between the number of mutations and fitness. Similarly, it is very likely that the assumption that silent variation is weakly selected ($s \approx 1/N$) is incorrect because in fact too many species with too disparate values of N nevertheless manage to harbor intermediate levels of variation for synonymous changes and to avoid fixing the major synonymous codon or having codon usages solely determined by mutation pressures. All of this could not be possible under the slightly deleterious model if N changes by more than a factor of 3 up or down from the value $1/s$. This point is also stressed by the fact that genes on the fourth chromosome of *Drosophila* show clear signs of biased usage of synonymous codons despite the fact that the extremely low levels of polymorphism in this chromosome indicate that its local N should be reduced well beyond 10-fold, making the maintenance of such bias not possible. The $s \approx 1/N$ assumption is thus clearly untenable, which does not necessarily imply that more realistic models will not allow nearly neutral dynamics to take place when genes are close, e.g., to an optimum codon usage pattern. Exciting new patterns of polymorphism and divergence will have to be mined in order to address this and many other important questions—an endeavor that will fundamentally bring the evolutionary and population genetical understanding of molecular diversity well beyond its current state.

VIII. THE VALUE OF MOLECULAR DIVERSITY AND CONSERVATION BIOLOGY

Diversity on Earth is endangered, and so is molecular diversity at both the polymorphism and the divergence levels. Much has been written about the value of the divergence component of diversity and about the need to preserve the biological information contained in adaptations evolved over millions of years in species and locally adapted gene pools all over the world. Conservation biologists are aware that the polymorphism component of genetic diversity may be important for the survival of a species and therefore try to preserve it. However, that they do it and the way in which they do it is important not only for evolutionary research but also for the other biological sciences. In fact, the polymorphism patterns of many species with large effective population sizes have been shaped during millions of generations by the action of very small selection intensities. Such patterns may take too long to recreate by expanding a population and allowing selection to act for a sufficient amount of time and/or due to effects too weak to study with standard laboratory techniques. The information in these patterns can indirectly guide researchers to subtle but important molecular biological, cytological, and physiological effects of mutations but will be lost if such populations disappear or their numbers are reduced drastically. Therefore, conservation biologists should also strive to conserve large "unendangered" populations.

See Also the Following Articles

DIVERSITY, COMMUNITY/REGIONAL LEVEL • DIVERSITY, ORGANISM LEVEL • DIVERSITY, TAXONOMIC VERSUS FUNCTIONAL • PHENOTYPE, A HISTORICAL PERSPECTIVE • POPULATION GENETICS

Bibliography

Drake, J. W., Charlesworth, B., Charlesworth, D., and Crow, J. (1998). Rates of spontaneous mutation. *Genetics* **148**, 1667–1686.

Gillespie, J. H. (1991). *The Causes of Molecular Evolution*. Oxford Univ. Press, New York.

Hartl, D. L., and Clark, A. G. (1997). *Principles of Population Genetics*. Sinauer, Sunderland, MA.

Kimura, M. (1983). *The Neutral Theory of Molecular Evolution*. Cambridge Univ. Press, New York.

Li, W.-H. (1997). *Molecular Evolution*. Sinauer, Sunderland, MA.

Ohta, T. (1992). The nearly neutral theory of molecular evolution. *Annu. Rev. Ecol. Syst.* **23**, 263–286.

DIVERSITY, ORGANISM LEVEL

Daniel R. Brooks
University of Toronto

GLOSSARY

entropy An expression or measurement of the energy available for use by a system, including living systems; often described as the extent to which the system tends toward a state of disorder or randomness.

extremophile An organism that requires, or grows optimally in, extreme environmental conditions; e.g., extremes of temperature, pressure, or acidity.

heterotroph Literally, feeder on others; an organism dependent upon organic material from an external source to provide carbon for growth.

organism The smallest entity of life that can function as a whole, distinct from others of the same type; a single living being.

photoautotroph An organism that uses inorganic material as a source of carbon for growth, and light as an energy source; e.g., plants.

ORGANISMS ARE THE SMALLEST functional wholes of life, entities that impose themselves on their sur-roundings and are causally affected by them. Organisms are localized in space and time, connected with their surroundings, including other organisms, through ecological and behavioral interactions and with their evolutionary past through genealogical inheritance. Composed of smaller parts, they are themselves parts of larger wholes, such as demes, populations, species, guilds, communities, and ecosystems. These diverse connections provide opportunities and challenges for any organism-focused efforts to understand, manage, and preserve biodiversity.

I. THE NATURE OF THE ORGANISM

A. The Basic Units of Selection

Long before the advent of modern evolutionary thinking in the nineteenth century, organisms were recognized as a primary focus of observation and explanation of biological systems. Darwin extended this tradition by reserving a central role for organisms and organismal diversity in his theory of evolution by natural selection. He believed that evolutionary change resulted from the interaction of two factors, which he called "the nature of the organism" and the "nature of the conditions." Of these two, Darwin (1872) proposed that the nature of the organism

... seems to be much more the important; for nearly similar variations sometimes arise under, as far as we can judge, dissimilar conditions; and, on the other hand, dissimilar variations arise under conditions which appear to be nearly uniform. (p. 32)

Darwin thought that organisms were historically and developmentally cohesive wholes, and therefore it was in the nature of the organism to produce offspring that were all highly similar (but not identical) to each other and to their parents and other ancestors. He also postulated that reproduction produced variation without regard for environmental conditions and therefore it was in the nature of the organism to produce these offspring in numbers far exceeding the resources available for their support. When this inherent overproduction produced variety in critical characters, natural selection would preserve the versions that were functionally superior in that particular environmental context (adaptations). Whenever an environment changes, those organisms that already had the adaptations necessary to survive would do so, whereas those lacking appropriate adaptations would not. Selection did not create the adaptations—it only determined which ones, if any, would be favored for survival. The production of organismal diversity thus required that organisms be at once autonomous from, and sensitive to, the environment. Darwin's perspective contrasted sharply with Lamarck's proposal that adaptation was an immediate and directed response by organisms to their surroundings. Lamarck also believed that the nature of the organism was important in the production of diversity, but only because all organisms have the same ability to change according to their needs. Therefore, whereas Darwin postulated that the nature of the organism included autonomous, self-regulating properties, Lamarck believed that the nature of the organism was to be directly and completely connected to the environment.

The distinction between Lamarckian adaptationism and Darwinian selectionism became increasingly blurred in the second half of the twentieth century as biologists focused more attention on parts of organisms and less on organisms as wholes. This reductionist movement, driven first by the successes of population genetics and later by the development of molecular methods, may have been an unconscious response to developmental biologists' discoveries of the complexities underlying the transition from DNA sequences to complete organisms—complexities that threatened to swamp simplistic theories based on the mantra of one gene–one trait–one selection vector. Whatever the reasons, losing the perspective on whole organisms led to a loss of Darwin's panoramic view of biological diver-

sity. In the last quarter of the twentieth century there have been many efforts to reemphasize the nature of the organism in evolutionary biology (Brooks and Wiley, 1988; Brooks et al., 1989; Brooks and McLennan, 1990; Depew and Weber, 1995; Kampis, 1991; Kauffman, 1993; Lewontin, 1982, 1983; Maynard Smith and Szathmary, 1995; Odling-Schmee et al., 1996; Raff, 1996; Salthe, 1993; Van de Vijver et al., 1998). These perspectives reassert that orderliness and organization in biological systems result from the interaction of selection processes with the inherent nature of the organism.

B. Organisms as Energy Flow Systems

Lotka (1913, 1925) was among the first twentieth-century authors to discuss biological systems, including organisms, in terms of energy flows and energy partitioning. He recognized that biological systems persist in space and time by transforming energy from one state to another in ways that generate and maintain organized structure. Maurer and Brooks (1991) recognized two classes of such energy transformations. Heat-generating transformations involve a net loss of energy from the system, usually in the form of heat. Conservative transformations involve changing free energy into stored states (e.g., structure). Because all conservative transformations in biological systems are coupled with heat-generating transformations, there is a heavy energetic cost to maintaining structure. Lotka (1913) suggested that the interplay between flow and partitioning of energy in biological systems acts to slow the rate at which energy stored by conservative transformations is degraded by heat-generating transformations.

The development of nonequilibrium thermodynamics in the past 50 years (Prigogine and Wiame, 1946; Prigogine, 1980) has allowed us to generalize Lotka's view of the nature of the organism. Living systems are nonequilibrium thermodynamic systems; they exchange matter and energy irreversibly with their surroundings and they maintain themselves in far from equilibrium conditions. The basic features of nonequilibrium systems can be summarized heuristically as

$$dS = d_e S + d_i S, \quad d_i S > 0$$

Total entropy changes (dS) can be subdivided into two components: $d_e S$, which measures exchanges between the system and its surroundings (changes in the surroundings), and $d_i S$, which measures production by irreversible processes internal to the system (changes within the system). We can also call this heuristic equation the "cost of living" for organisms because all organisms must take in high-grade energy and matter and dissipate lower grade energy to their surroundings in order to survive. Energy degraded in the uptake of raw materials from the surroundings into the system is dissipated into the surroundings ($d_e S$). These exchanges are accompanied by a great deal of waste; hence, $d_e S$ is very large compared with $d_i S$. Nonetheless, it is in the manifestations of internal production that we perceive organismal diversity.

Internal production ($d_i S$) includes (i) dissipation from the system, called the external dissipation function (γ_α, or heat-generating transformations). Heat-generating processes occur when energy and entropy flow in opposite directions, with entropy production tending to move the system toward disordered states; and (ii) dissipation within the system, called the bound dissipation function (γ_μ, or conservative transformations). Conservative transformations are characterized by energy and entropy flowing in the same direction, with entropy production being retained within the system and tending to move the system towards more structured states. In biological systems, γ_μ can be further subdivided into allocations for accumulating biomass (γ_μ^b) and allocations for accumulating genealogical information (γ_μ^i). Heuristically (Brooks and Wiley, 1988),

$$d_i S = \gamma_\alpha + \gamma_\mu^b + \gamma_\mu^i$$

Organisms thus have a dualistic nature. As open thermodynamic systems, they must simultaneously interact with their surroundings and perform critical functions internally. They maintain themselves in a viable state by exchanging matter and energy irreversibly with their surroundings, taking in relatively high-grade energy and using it to perform useful work within themselves. This requires sensing of, and causal engagement with, the surroundings mediated by a physical distinction ("phase separation"; Prigogine, 1980) between the organism and its surroundings. That is, there must be an "inside" and an "outside" of the organism, delineated by a physical boundary. For all organisms, this boundary is provided by cell membranes, which are simultaneously physical barriers between the inside and outside of the organism and highly selective mechanisms for modulating the exchange of matter and energy between the organism and its surroundings. For multicellular organisms, this barrier is a complex of cell membranes.

C. Organisms as Information Systems

Elucidation of the genetic code made it possible to begin thinking of organisms as information systems.

Information systems consist of a source of signals, a channel through which the signals are transmitted, and a receiver to translate the portion of the signals that made it through the channel into information. Gatlin (1972) argued that the genetic system is the source, reproduction and ontogeny are the channel, and the environment is the receiver. Genetic possibilities thus become phenotypic signals as a result of reproduction and ontogeny and become meaningful biological information as a result of causal interactions between the phenotype and the environment. Brooks and McLennan (1990) noted, however, that the environment is not a receiver in a physical sense because its only causal interaction with biological information is the possible elimination of some of it; it does not measure or interpret the information. Rather, the environment acts as interference in the channel, or as a "delete button" monitoring "incoming messages," eliminating relatively less fit organisms.

If biological information is a material part of biological systems, however, it is possible for biological systems to be their own sources and receivers. Current standard evolutionary theory stresses external causality for orderliness in biological systems, so the receiver has been construed as part of the surroundings, i.e., localized in space. Although it is true that biological systems are localized in space, they are also localized in time. In other words, the receiver can be a "time" (Brooks and McLennan, 1990). The source is a genetic system at time t_0, the channel is reproduction and ontogeny, and the receiver is the same genetic system at any given time $t_{1...n}$; thus, the receiver is temporally distinct from the source. If an information source precedes its receiver in time, it can produce the system that acts as receiver, and that system can then become a source. This perspective has been used by information theorists designing self-correcting computer programs, which can enhance their own abilities to store and transmit information efficiently. The same holds true for biological systems: DNA has significant self-repair capabilities.

Biological systems are physical information systems, a type of nonequilibrium thermodynamic system, open to exchanges of matter and energy but maintaining a closed information system internally which functions to reproduce the system—to perpetuate lineages through time. They are able to impose themselves and their functions on their surroundings and thus are self-stabilizing and self-organizing. They produce organized complexity cheaply ($d_i S$ is small compared to $d_e S$, and the portion of $d_i S$ allocated for the information system is small, in part because a small number of chemical templates are used to generate many organisms), variably (because even chemical templates are subject to the statistical mechanical vagaries of the second law of thermodynamics), and functionally (because organisms must exchange matter and energy with their surroundings in order to maintain themselves), but without regard for details of the surroundings (because the information system is embodied in relatively autonomous internal chemical production, d_iS, of the system). As the source and receiver of organized information, they can be the embodiment of the organizing principles for that information. Biological systems thus transmit information through, not to, their surroundings. This supports Darwin's view that it is the (autonomous) nature of the organism that creates the necessary conditions for selection processes, especially natural (environmental) and sexual selection, to occur.

Treating biological systems as physical information systems provides a causal basis for the origin of selection processes consistent with their well-documented causal consequences. Selection processes originate as a result of the necessity that biological systems obtain matter and energy from their surroundings coupled with the relative autonomy of their information systems, which permits production of organisms regardless of the details of their surroundings. Without the constraints provided by this autonomy, there would be no selection; simultaneously, however, constraints provide systems with macroscopic properties that limit the ways in which and the extent to which the system will respond to selection. This means that biological systems should have their own macroscopic, or macroevolutionary, properties that are not reducible to microscopic dynamics. Some of the most important implications of reemphasizing the nature of the organism in the context of thinking of evolution as an informational process involve the manner in which evolutionary theory views major events in the history of life (Maynard Smith and Szathmary, 1995).

D. Putting It All Together

Biological systems, beginning with organisms, are functional wholes with respect to the way they engage their surroundings and with respect to their internal organization. A major component of internal organization is functional integration through the interdependence of parts, and this is most evident in the dynamics of ontogeny (Salthe, 1993; Raff, 1996) and in physiological processes. Evolutionary changes in biological systems do not occur all at the same time; thus, when such changes occur, only part of the system changes. All changes, whether point mutations on a chromosome, genomic doubling, or alteration of part of a complex mating

ritual, must integrate with the rest of the system, which has not changed, in such a way that viable organisms result. The functional necessity of developmental integration creates stability domains within bioinformational phase space (Brooks and Wiley, 1988; Kauffman, 1993; Niklas, 1999). Orderliness and organization in biological systems result from the interaction of selection processes with three aspects of the nature of the organism: (i) historical uniqueness, (ii) cohesive properties, and (iii) hierarchical organization.

Historical uniqueness manifests itself in historical contingency but, more important, in temporal irreversibility. Important organismal processes such as reproduction, development, aging, and death are inherently irreversible. Spontaneous irreversible behavior of this sort always involves growth and increasing complexity, and physical manifestations of at least some of the systems' history (Schanck and Wimsatt, 1988).

Cohesive properties range from membrane-bound nucleic acids to cell–cell adhesion and recognition, sexual reproduction and specific mate recognition systems, and common history of inheritance. Cohesion is especially important to evolutionary explanations; it is the "glue" of functional integration and hierarchical organization that are so characteristic of biological systems (Wake and Roth, 1989). Many biological processes that demonstrate irreversible behavior manifest such changes as a result of interactions among cohesive factors.

Hierarchical organization plays an important role in organismal diversity. Hierarchies provide stability, reinforce boundaries between organisms and their surroundings, allow increasing amounts of complexity without losing organizational coherence, and provide a way in which causation and control can be tied together (Salthe, 1993). Eldredge and Salthe (1984; see also Eldredge, 1985, 1986; Salthe, 1993) emphasized two major forms of hierarchically organized biological structure. The ecological hierarchy (interactors of Hull, 1988) is manifested by patterns of energy flow in ecosystems, and the genealogical hierarchy (replicators of Hull, 1988) is manifested by patterns of ancestral relationships among organisms and species. Environmental and genealogical phenomena are good starting points for investigating hierarchical interactions because they are intimately connected in biology. Prebiotic environmental conditions established the boundary conditions within which life could originate. Conversely, genealogical processes that characterize life are autonomous enough from environmental conditions to be capable of overrunning available resources and of changing the environmental conditions substantially. The longer life exists on this planet, the more it shapes the environment of the planet. Today, much of the environment relevant to biological systems consists of other organisms (Maynard Smith, 1976; Brooks and Wiley, 1988). Organisms imposing themselves on their environments (creating niches) contributes greatly to the far from equilibrium nature of the environment in which evolution occurs (Odling-Schmee *et al.,* 1996).

The intimate relationship between the two hierarchies can be illustrated with a sports metaphor: The ecological hierarchy establishes the dimensions of the playing field, whereas the genealogical hierarchy establishes the rules of the game being played. In other words, biological systems obey rules of self-organization transmitted genealogically (historically) and played out within environmentally defined boundaries. To complicate matters, however, the self-organizing rules of the game by which living systems evolve can produce changes in the dimensions of the playing field. To extend the metaphor, the game may redefine the dimensions of the playing field and may be subsequently constrained by those self-imposed changes. For example, the evolution of photosynthetic prokaryotes from anaerobic ancestors resulted in increased oxygen content in the atmosphere. This increase then altered the diversity and changed the distribution of anaerobic organisms, limiting them to relatively rare environments and paving the way for the evolution of a new array of species. Now that we have a consistent view on the nature of the organism in a physical sense that also accounts for their Darwinian nature, we can discuss observed organismal diversity.

II. DIVERSITY IN THE REAL WORLD: A PRECIS OF CURRENT ORGANISMAL DIVERSITY

Organismal diversity is evolved diversity. This means that no matter how distantly related two organisms are, they still share at least some features in common, such as cell membranes made of lipid-bilayers and a common genetic code, linking them to a single phylogeny of life on this planet. Highlights of organismal diversity can thus be best recounted in a precis of organismal diversification—phylogeny. However, phylogeny is not a linear historical timeline. It is a hierarchically branching pattern of diversification and spread, producing organisms that are historically unique mosaics of ancient and recent traits, both functional and structural. Therefore, it is not possible to give even a precis of organismal

diversity without twists and turns. The perspective on the major transitions in evolution presented by Maynard Smith and Szathmary (1995), however, provides some consistent reference points. Maynard Smith and Szathmary suggested that the major transitions in evolution occurred whenever life functions were compartmentalized and modularized in such a way that the reproductive flow of biological information was enhanced. This can occur either by enhancing the survivorship of organisms in particular environments, thus increasing the likelihood that they will reproduce, or by enhancing the mechanisms by which information is stored and transmitted, regardless of the particular environment in which it evolves. Maynard Smith and Szathmary highlighted what they considered to be the nine most important evolutionary transitions, all of which were examples of enhanced efficiency of information storage and transmission.

One observation that is relevant is that the smallest and the largest organisms known to have existed in the past exist today. Members of *Thermoplasma* are thermophilic and acidophilic prokaryotes living in coal deposits that possess only 1100 kilobase pairs of nucleotides in their genomes, whereas blue whales are oceanic monsters outstripping even the largest of the dinosaurs.

The oldest lineages of organisms on the planet are various prokaryotes, including bacteria and archaeans. These organisms are single celled, lack organelles or membrane-bound genomes, and exhibit only asexual or simple parasexual reproduction. Their cells form only rods, spheres, or spirals, and yet they have inhabited this planet continuously for almost 3 billion years and represent by far the largest number of differentiated lineages on the planet today, including the largest number of unnamed taxa. Representatives of the prokaryotes include chemoautotrophs (thought to have been the lifestyle adopted by the first living things), chemoheterotrophs, photoheterotrophs, and photoautotrophs. Photoautotrophic prokaryotes, primarily Cyanobacteria, were the first to demonstrate that organisms could change the environment of the planet, substantially affecting standing biodiversity. They did this by releasing large amounts of molecular oxygen into a reducing atmosphere, changing it in such a way that most prokaryotes became extinct or restricted to marginal habitats. This release of oxygen, as well as associated nitrogen fixation, by prokaryotic photoautotrophs created the conditions necessary for other life forms to evolve, so it is safe to say that prokaryote evolution played a major role in setting the stage within which all subsequent biological diversification occurred. Prokaryotes today include the so-called Archaea, or extremophiles, which may represent a collection of distantly related remnants of the diversity affected so much by the evolution of those photoautotrophs. Among the extremophiles are members of the halophilic *Halobacterium,* which grow best at pH 11.5, and the acidophilic *Sulfolobus,* growing best at pH 2 or 3 but able to survive at pH 0.9. There are methanogens, producing most of the planet's methane, including a host of symbionts living in the rumens of cattle and their relatives in which they produce digestible carbohydrates for their hosts as well as methane. Another methanogen, *Methanopyrus,* which lives near ocean-bottom thermal vents, grows optimally at 98°C but is able to survive at 110°C and "dies of cold" at 84°C. *Sulfolobus* is also a hyperthermophile, living optimally at 70–75°C and dying at temperatures lower than 55°C. The extreme thermophile *Thermus aquaticus,* isolated from hot springs in Yellowstone National Park, was the original source of high-temperature DNA polymerase, the enzyme used to catalyze the polymerase chain reaction (PCR) technology that has become indispensable to molecular biology. Most other prokaryotes are bacteria, which include some of mankind's most valued and most loathed organisms. Bacteria living in our digestive tracts are essential for proper nutrition, and *Escherichia coli* is the workhorse of molecular biology and biotechnology, including the pharmaceutical industry. On the other hand, diseases such as syphilis, anthrax, diphtheria, tuberculosis, smallpox, tetanus, cholera, and plague and a variety of agents of food poisoning, including *Salmonella* and botulism, are all caused by bacteria. These same disease-causing prokaryotes have also shown a tremendous degree of adaptability in the face of substantial human efforts to eradicate them.

From within this great prokaryotic diversity evolved the first eukaryotes. Most species of eukaryotes, including the oldest surviving lineages, comprise single-celled organisms differing from prokaryotes by well-organized compartmentalization of functions within the cell. Cell energetics are handled by organelles called mitochondria and chloroplasts, which are thought to have originated as endosymbiotic prokaryotes. Digestion and excretion are handled by specialized vesicles within the cell that facilitate the movement of matter and energy in and out of the cell. Synthesis of amino acid building blocks for structural components takes place in specialized organelles called ribosomes. Finally, the storage, production, and transmission of information are compartmentalized inside a membrane-bound nucleus. Powered locomotion, first evolved by prokaryotes in a rudimentary manner, becomes highly effective in unicellular eukaryotes as a result of sophisticated cytoskeletal integration.

Most of the heterotrophic unicellular eukaryotes are grouped together in a nonmonophyletic collection of lineages called the Protista. Within this evolutionarily heterogeneous and diverse collection of lineages are the origins of the first sophisticated sexuality as well as alternation of sexual and asexual generations (diverse examples are found in ciliates, apicomplexans such as malaria, and various algae), the first sophisticated modifications of the cell membrane in ways that modulate organism–environment interactions (diatoms produce complex and beautifully sculpted exoskeletons), and the beginnings of true multi-cellularity, manifested by colonial aggregations of unicells that exhibit some degree of functional differentiation, such as the fruiting bodies of slime molds.

The protists are major underpinnings of the heterotrophic portions of the biosphere. Within the heterotrophic realm, protists represent saprobes and detritivores, predators, and a variety of symbionts, ranging from commensal ciliates living in the tests of sea urchins and the rectums of frogs to mutualist flagellates living in termite guts, providing digestible carbohydrates for the insect hosts in exchange for their own food and a safe haven, and parasites of global socioeconomic significance, including the causal agents of amebic dysentery, malaria, giardiasis, African sleeping sickness, Leishmaniasis, and Chaga's disease. Most show enhanced reproductive output and complex life cycles, but *Giardia* and *Entameba* show some secondary losses of structures.

The photoautotrophic protists, generally called Algae, comprise four distinct lineages: dinoflagellates, red algae, diatoms, and the green algae. The dinoflagellates are best known as the causal agents of red tide. Massive blooms of certain dinoflagellates produce large amounts of waste products that act as nerve toxins capable of killing large numbers of fish or of concentrating in shellfish and killing those who eat them. Red algae also include some basal species that are parasitic in more derived members of the same clades. Diatoms are ubiquitous phytoplankton that represent the primary source of photosynthesis (and oxygen production) in the ocean, although many are freshwater inhabitants. They are characterized by having bipartite siliceous shells (tests), exhibiting species-specific forms and beautiful surface sculpting. Because they are siliceous, these tests are highly resistant to decomposition. As a result, fossil deposits of diatoms are excellent indicators of past environmental conditions. Large deposits produce "diatomaceous earth," which has many industrial uses including insulation, filtration, and metal polishing. Diatomaceous earth has also been proposed as a "nonpolluting" insecticide. Spreading diatomaceous earth on fields may well clog the tracheae of pest insects but will also clog the tracheae of beneficial insects and may also affect the breathing structures of other animals, including those humans who apply it or work in fields on which it has been applied.

True multicellular lineages appear to have evolved four times within the protists: the brown algae and plants, which are photoautotrophs, and the fungi and animals, which are heterotrophs. According to Maynard Smith and Szathmary, multicellularity represents an advance over unicellularity because in multicellular organisms the partitioning of functions can occur at the level of entire cells, tissues, organs, or organ systems rather than having to take place within each cell. Complex reproductive systems can evolve because multicellular organisms can produce gametes without having to become gametes. Given the large numbers of unicellular eukaryote lineages and the small number of multicellular eukaryote lineages, we conclude that this is difficult to achieve, despite the evolutionary payoffs. Perhaps this is because it is difficult to coordinate a population of cells. There are no multicellular prokaryotes (there are bacteria that form strings, strands, and mats, but these are simply aggregations of individual organisms), perhaps because prokaryotes show no functional partitioning at the cellular level. If so, true multicellularity could only have arisen after unicellular eukaryotes evolved. In a complementary vein, there is only a single known example of secondary loss of the multicellular state; this occurred in the Microsporida, a group of obligate parasites of fishes and annelids which were thought to be an apicomplexan protist group until it was shown that they are actually highly modified cnidarians.

Each of the four cases of true multicellularity represents a very different evolutionary path. The least successful of these are the brown algae, sister group of the diatoms. Up to 60 m long in the case of *Macrocystis*, brown algae can be found on ocean shores with pounding surf, in relatively quiet near-shore meadows, or floating in the open ocean. The Sargasso Sea, for example, is an enormous mass of floating *Sargassum*. Although relatively species poor, brown algae can be abundant. They exhibit differentiated tissues and even organs, including holdfasts and filaments, which are analogs of roots and leaves that provide homes to a diverse array of animal life. Some have gas-filled bladders which may contain up to 5% carbon monoxide—enough to kill a human. Cellular cohesion in brown algae is provided by alginic acid in the cell walls. This is a gummy polymer of sugar acids that cements cells and filaments together and provides the glue for the holdfasts. Alginic acid is also an important commercial emulsifier in ice cream, cosmetics, and other products.

Larger pieces of brown algae, especially some kelp species, are considered delicious and healthy food sources for humans.

The remaining three groups of multicellular organisms are relatively closely related to each other, albeit they diverged a very long time ago, and to two groups of protists—the photoautotrophic green algae and the heterotrophic choanoflagellates.

The second group of multicellular photoautotrophs are the plants, sister group of the green algae, or at least of some green algae. Unlike the brown algae, which originated in the sea and have remained there throughout their evolutionary history, plants are the descendants of an aquatic lineage that invaded and diversified on land. Two closely related living relatives of the true plants are *Volvox* and *Micrasterias* (sea lettuce), which form aggregates of single-celled organisms showing some degree of functional differentiation, including the production of large stationary eggs, which are highly differentiated from the very small, flagellated, and highly mobile male gametes, a condition also found in true plants. Cellular cohesion in true plants is provided by plasmodesmata, which are cytoplasmic strands connecting adjacent cells. Many plants exhibit a high degree of adaptability or permissiveness with respect to genome-level evolutionary modifications. Polyploidization is common among many groups, including both autopolyploidization and allopolyploidization as a result of interspecific hybridization.

The land represented an excellent substrate for photoautotrophs because there is more consistent sunlight on land than in the sea, where sunlight of sufficient quality for photosynthesis to occur does not penetrate deeper than approximately 50 m. However, being on land also posed problems, one of which was structural: Air provides little support compared to water. In the absence of inherent structural support, plants could only sprawl on the land—they could never rise above it. Cell walls made rigid by cellulose, a carbohydrate, provided the necessary structural support. Subsequent evolution of vascular tissue in a moss-like ancestor and then a sophisticated tubular transport system (tracheids) in the ancestor of the sister group of the mosses (those plants called, not surprisingly, Tracheophytes) enabled water and nutrients to be transported against gravity to considerable heights. Rigid cell walls and the tracheid system provided opportunities for tremendous vertical growth, represented by the convergent evolution of tree forms in many groups (including giants up to 110 m tall) and a tremendous flow rate of water and nutrients. A single maple tree only 15 m tall may support more than 175,000 leaves with a total surface area of more than 675 m^2, transporting up to 220 liters of water per hour from the soil throughout the tree and eventually into the atmosphere through evaporation from the leaves.

Terrestrial life that is truly independent of aquatic environments must also be associated with suitable reproductive modes. When ancestral plants moved onto land, they brought with them the large stationary eggs characteristic of their closest relatives. In all land plants, when these eggs are fertilized, the resulting embryos are sporophytes contained within a protective covering produced by the parent. Those ancestral plants also brought with them aquatic flagellated sperm, maintaining a dependency on aqueous surroundings, at least for reproduction. Only when nonaquatic male gametes evolved in early angiosperms did land plants become completely terrestrial.

Once firmly established on land, major events in plant diversification were associated with increasing compartmentalization and modularization of the organism, all integrated and coordinated chemically and without a centralized nervous system. They elaborated functionally differentiated nonreproductive body parts, especially roots and leaves, that became specialized habitat for other terrestrial dwellers, creating many novel opportunities for diversification by these others and coevolutionary feedback that affected their own evolution as well. This included the evolution of parasitic plants such as Indian pipe, which has lost its chloroplasts and depends on the host plant for photosynthesis. Their root systems helped plants modify the soil in ways that enhanced the lives of the plants themselves but also affected the lives of plants and other types of organisms living in the soil. Some plants today produce toxins in their roots that diffuse in the soil, inhibiting the growth of other plants. Others, including common garden plants such as marigolds, secrete substances from their roots that deter plant-consuming soil nematodes and thus can be planted as a nonpolluting nematicide border in gardens. Aboveground vertical growth in tracheophytes led to a proliferation of types of habits—plants, shrubs, bushes, vines, and trees. This vertical structuring allowed complex communities of plants to form, within which the various plants created their own ecological interactions with other plants by shading each other. Plants also differ in the relative amount of biomass allocated to below- and aboveground structure. This led to complex ecological interactions between root systems of different plants. In the tropics, for example, most plant biomass is aboveground, whereas in temperate areas in which there is more humus in the soil, a much larger proportion is belowground. In fact, a given plant may participate simultaneously in one set of aboveground and another set of belowground ecological interactions, just with other plants, depending on what species the plant belongs to and where it happens to be growing.

Plant organismal diversity is also associated with major modifications of the reproductive system, each of which has many variations. Primary among these was the evolution of seeds, which are embryos protected by a covering of nutrient tissue provided by the parent. These multicellular protective coats, containing cellular contributions from as many as three generations, are a source of protection and nutrition for the embryos which may allow them to withstand substantial periods of environmental harshness and still be capable of germinating and producing a plant. The earliest seed plants produced seeds exposed directly to the environment, called gymnosperms (naked seeds). Angiosperms (hidden seeds) evolved when the ovaries, containing the seeds, expanded into fruits, hiding the seeds and providing more protection. The evolution of double fertilization at more or less the same point in time, producing triploid endosperm, provided even more nutrition for the embryo. Angiosperm diversification produced some amazing organisms. The seeds of some plants need to be burned by forest fires in order to germinate, and others must pass through the intestinal tracts of herbivorous vertebrates before they germinate. Plant seeds today exhibit an immense array of structural modifications that enhance their dispersal from the parental plant. Those that pass through the intestinal tracts of vertebrate herbivores may be dispersed for many kilometers from their parental plants by the herbivore. Other seed modifications take advantage of wind—from the fluffy dandelions to the robust winged samaras of maples. Another major reproductive innovation was the evolution of flowers, which are specialized and self-contained reproductive modules. Flowers represented an enormous potential for reproductive cohesion and diversification.

The evolution of plants was followed closely by the evolution of herbivores, especially among the nematodes and insects. This in turn was followed by the evolution of a variety of antiherbivore adaptations. Defenses against herbivory include both diminution and elaboration of the part of the plant being eaten, the evolution of surface features such as thorns, and storage of toxic metabolic by-products in tissues that are targets of herbivory. Likewise, the diversification of flowering parts is also associated strongly with the diversification of particular groups of animal pollinators, again most notably insects. In fact, it appears that the evolutionary diversification of flowering plants and of phytophagous insects has been facilitated by their close ecological associations.

Plants exist in many seemingly inhospitable habitats, particularly desert and boreal regions, where the evolution of resistant seeds as well as specialized reproductive, root, and stem and leaf systems allow them to survive and thrive. Cacti and other succulents represent convergent evolution of water-retaining and evaporation-resistant body forms that can survive in desert conditions. Many plants are able to flourish in nearly marine salinity conditions, although none is truly marine. Mangroves, for example, are able to excrete excess salt from the water they absorb in the tropical coastal estuaries in which they live. Plants also thrive in seasonally fluctuating environments. This is helped by the evolution of resistant seeds but is also enhanced by the ability to synchronize growth and reproduction with permissive times of the year. For example, the ability to shed leaves deciduously apparently arose as an adaptive response to seasonally dry climates but today functions in seasonally cold climates as well.

Tapping an almost unlimited energy source (sunlight), coupled with their sophisticated means of compartmentalization, has made plants extremely important to humans and to the biosphere. Terrestrial plants improve air quality by fixing carbon. Some also help improve the ability of soil to sustain life by fixing nitrogen mediated through mutualistic relationships with root-dwelling bacteria; the best known of these mutualists produce the root nodules called peanuts. Plants that have adopted life in freshwater habitats are often quite sensitive to changes in water quality and are thus useful as indicators of pollution. Those having extensive aquatic root systems may also be excellent biofiltration systems; water hyacinths, for example, can actually improve water quality. The cultivation of plants is thought to have been a major event in human evolution. Ironically, the cultivated production of large quantities of edible plant biomass is thought to have provided the conditions for the growth of human population and technology that now threaten our existence by threatening the biosphere. Different plants allocate substantial amounts of nutrients into root systems, especially tuberous ones, into stems and leaves, and into reproductive structures such as nuts, seeds, and fruits. Virtually any part of a plant may be highly edible for humans or their domestic animals. Simultaneously, substances functioning to deter herbivory or competition from other plants may also serve as sources of useful materials for humans, ranging from spices to pharmaceuticals. The American yew tree sequesters a compound called taxol, which has significant effects on certain forms of cancer;

a close relative, the European yew, produces a similar compound called taxotene, which also has anticancer properties.

Fungi represent one of the two major groups of multicellular heterotrophs. Among living groups, fungi are apparently the sister group of choanoflagellates and animals. Cellular cohesion in fungi is provided mostly by default: The vegetative body of a fungus, called a mycelium, comprises many hyphae, which are filamentous multicellular extrusions maintaining cohesion among the cells by forming only incomplete cellular divisions, or septa, between them.

To many people, the presence of fungi indicates that something has gone wrong—one's basement is too damp or one's bread is too old. To others, fungi are something quite different: They are the major recyclers and decomposers of the biosphere, found most commonly in association with the refuse of the biosphere, including that of our own civilization. This is because most fungi are very adaptable saprobes, highly tolerant of great ranges in temperature and osmotic, especially hyperosmotic, conditions. A considerable number of fungi are parasitic and a few are even predators, attacking small animals such as soil nematodes. Fungi are very flexible reproductively, reproducing asexually or sexually with mating types not distinguishable as male or female and exhibiting an enormous diversity of variations on the general eukaryote theme of alternation of generations. Fungi are the primary source of cellulose and lignin breakdown in the environment. These features make them excellent agents for bioremediation. They can become established in highly degraded habitats and actually improve them, creating conditions which can support yet more life. Tropical forest soils, for example, lack much humus but have enormous mats of fungal hyphae forming mycorrhyzal mutualisms on and in plant root systems, which maintain vital nutrients and structural cohesion in the small amount of soil that is present. Fossil evidence suggests that mycorrhyzal associations are at least 300 million years old.

There are four major groups of fungi. The Chytridiomycota, or water molds, are thought to be the sister group of the rest of the fungi. Chytrids have been implicated in recent massive die-offs of amphibians living above 1000 m elevation throughout the world. Many chytrids are known to be parasitic, with some exhibiting complex life cycles involving two different hosts. A few have been found that are predators of nematodes. Members of *Arthrobotrys*, *Dactylaria*, and *Dactyella* produce three-celled loops that entrap nematodes as they wend their way through the soil; once trapped, the nematode cuticle is invaded by hyphae and ingested. The fungi most commonly known as "molds" are members of the Zygomycota. These fungi are also arguably the basis of the modern pharmaceutical industry—the mold that tells you your bread is too old is likely *Penicillium*, the original source of the world's first broad-spectrum antibiotic, penicillin. Two other members of the genus, *P. camembertii* and *P. roqueforti,* are required to produce fine cheeses. Some members of *Aspergillus* are essential to the production of soy sauce and sake, whereas others can cover edible products such as peanuts with carcinogenic aflatoxins. Probably the best known and respected fungi are members of the Basidiomycota. This group includes the edible fungi, called mushrooms, morels, or truffles, in ascending price. This group also includes rusts and smuts, which can cause serious diseases in agricultural plants. Basidiomycotes are also a major source of legal and illegal pharmaceuticals and some of history's most infamous deadly toxins. The sister group of the basidiomycotes are the members of the Ascomycota, which comprise the yeasts and sac fungi. Without members of this group, we would have no bread, yogurt, or many alcoholic beverages, including wine and beer.

On their own, fungi are ubiquitous and numerous throughout the world. In addition, some fungi form mutualistic associations with algae, called lichens, which are perhaps the champion extremophiles of the multicellular world. Extremes of heat and cold and dry and wet do not seem to deter lichens, who often flaunt their hardiness by living on rocks, slowly but inexorably digesting them into dust.

The second group of multicellular heterotrophic eukaryotes are the Metazoa, or true animals. Metazoans are the sister group of the choanoflagellates. This group of multicellular organisms maintains cellular cohesion by means of extracellular matrices of cell-adhesion proteins. Unlike plants, which invaded the land early and diversified there, most animal diversification occurred in the ocean, and various groups have invaded freshwater and terrestrial habitats convergently. Even today, most metazoan organismal diversity is in the sea. A great deal is known about many metazoans, probably because many of them are large enough to be seen with the naked eye and active enough to engage visually oriented organisms such as ourselves. Most metazoans, however, are microscopic. Highly complex and sophisticated adult organisms, such as some rotifers, may be as small as 50 μm.

The array of metazoan body plans provides clear evidence of modularization and compartmentalization. Metazoan bodies range from aggregates of specialized cells in sponges and placozoans to defined body forms composed of specialized tissues in cnidarians and ctenophores, with organ-level organization beginning with platyhelminths, and complex organ systems in the majority of the other metazoans. Body cavities emerged very early in metazoan evolution as effective ways of physically compartmentalizing organ-level modules and laid the evolutionary groundwork for specialized compartments within body cavities, called segments, metamers, and tagmata. Metazoan bodies may be asymmetrical as in sponges, radially symmetrical as in cnidarians, bilaterally symmetrical as in the majority of metazoans, or even biradially symmetrical as in ctenophores and echinoderms. Developmental patterns among metazoans tend to be highly conserved within major groups, although secondary loss of structures is not uncommon; the platyhelminthes appear to be descended from an ancestor that lost its body cavity, and a large group of parasitic platyhelminths, dominated by the true tapeworms, has secondarily lost its digestive tract as well.

Many metazoans are conspicuous because they tend to move around a lot. Modes of locomotion are varied, but most are variations on swimming, creeping, or crawling. Sophisticated locomotion in metazoans is highly correlated with enhanced cephalization and neural integration Cephalopod mollusks have water-jet propulsion provided by specialized siphons. Complex appendages for swimming, walking, and running evolved in the arthropods and the chordates. It is within these two groups that organisms capable of true powered flight emerged.

All metazoans are heterotrophs, but they exhibit the entire range of heterotrophic functions—they are saprobes, detritivores, filter and suspension feeders, predators, herbivores, commensals, mutualists, parasitoids, and parasites. Herbivory is uncommon among marine metazoans but is a major attribute of the structure of terrestrial ecosystems. Modern herbivore diversity is concentrated in the insects, the amniote vertebrates, and the nematodes. Interestingly, no metazoans ever evolved the ability to digest cellulose. Thus, many metazoan herbivores dine on plant cells and tissues that contain nutrients in a form other than cellulose. Those that actually gain nutritional benefit from cellulose are hosts to a variety of prokaryotic and eukaryotic microbes which possess the enzymes necessary to break cellulose down into less complex carbohydrates, which can be absorbed by the metazoan host. Although this form of herbivory seems to be very successful when it evolves, it does not appear to have evolved often, presumably because it is difficult to acquire the proper symbiotic microbes, provide a suitable home for them in the intestine, and provide a means of transmitting them to offspring.

Herbivory has a downside, however. As indicated previously, many plants allocate toxic compounds to the structures eaten by herbivores. The herbivores that have survived such evolutionary changes in their food have evolved ways of avoiding, sequestering, or detoxifying those substances. Metazoans not only graze on plants but also many graze on other metazoans. Parasitism has evolved in virtually every group of metazoans except the chordates. The phylum Acanthocephala is entirely parasitic, whereas the vast majority of platyhelminth species are parasites, including the well-known digeneans (trematodes or flukes) and eucestodes (tapeworms). Many nematodes and arthropods are also parasitic.

Reproductive modes are diversified in metazoans. Most metazoan organisms are either males or females, although hermaphroditism is not uncommon in members of some of the oldest lineages. Gonads may be transient specialized cells or tissues or permanent organs and organ systems. Reflecting their ancient origins and diversification in the ocean, fertilization is plesiomorphically external in most groups. Internal fertilization, in which the female provides the aqueous medium within which the sperm can swim to the ova, has evolved multiple times. The ancient eukaryotic theme of alternation of generations is carried forward into metazoan evolution, although it drops out in several lineages convergently. In many cases, the asexual and sexual portions of the life cycle are ecologically partitioned; in most cnidarians these stages are characterized by distinct medusa and polyp forms which have distinct ecologies. However, even for metazoans that do not exhibit alternation of generations, it is common for the larvae and adults to have very different life-styles. Most larval frogs, for example, are herbivores, whereas virtually all adults are insectivores. Among ecdysozoans (nematodes, rotifers, tardigrades, kynorhynchs, gastrotrichs, acanthocephalans, and arthropods), haplodiploidy is a common theme, setting the stage for extreme sexual size dimorphisms and asymmetrical sex ratios biased heavily toward females. Many metazoans are parthenogenetic, forming unisexual female lineages. In some cases, these lineages are entirely self-sustaining, whereas in others, such as the gynogenetic poeciliid fish called the Amazon molly, sperm from males of a parental species are required to trigger embryogenesis, even though no genetic information from the sperm is incorporated into the embryo.

Another major theme of metazoan organismal diversity is temporal synchronization of reproductive cycles. Perhaps the most extreme case is that of the tropical Pacific polychaete called the Palolo worm, in which the entire species reproduces in a single night each year. A more familiar example is that of anadromous salmonid fishes, in which adults return to their natal streams from the ocean once a year, reproduce, and die. In herbivorous insects reproductive synchronization is tied to cycles of growth and reproduction in the different plants they eat.

One of the most arresting features of animals is complex behavior, especially with respect to reproduction. This can be manifested in sophisticated interactions among males, among females, between males and females, in parental care, or even in complex combinations of all possibilities. Complex mating rituals have been documented in tardigrades, in which males are known to apply ritualized strokes to females to encourage them to reproduce. Some leeches are known to carry their offspring on their abdomens for a period of time after birth. Mollusks (especially cephalopods), arthropods (especially insects), and chordates (especially vertebrates) are the best examples of sophisticated behaviors among metazoans. All three groups have evolved complex control and sensory integration involving neurochemical systems and strong cephalization.

One outcome of integrated and sophisticated behaviors is sociality. Maynard Smith and Szathmary (1995) designated the evolution of social systems as one of the nine major transitions in biological evolution. Social systems permit functional compartmentalization at the level of different organisms within a social group. This level of modularization and compartmentalization is best seen in the social insects, notably bees, ants, and termites, whose colonies function very much like superorganisms. The fundamental feature of these societies is genetically stereotyped individual organisms, most of whom are sterile or sexually inactivated female workers and soldiers (analogs of somatic tissues) along with a few reproductively active males and females (the analogs of reproductive organs). In bees and their relatives, colonies consist of many sexually inactivated females called workers and a smaller number of reproductively active males called drones, whose only function is to reproduce with a single queen, the reproductively active female. The other major group of social hymenopterans, ants, differs from bees and their relatives by having males that die shortly after reproducing with a queen and by exhibiting a tremendous range of morphologically distinct, functionally stereotyped workers and soldiers. Ant societies fascinated Darwin, especially their propensities to wage war, make slaves, use tools, construct complex domiciles, and undertake agriculture, farming fungi and milking aphids for nutrients. The other major group of social insects are termites, whose colonies comprise a single queen and king who do all the reproducing. Sexual inactivation of offspring is maintained from one generation to another by pheromonal spread through trophallaxis, which is the mutual feeding of nymphs. Workers and soldiers produce pheromonal secretions whose proportions in the colony act to maintain suitable proportions of each type of individual in succeeding generations.

Some vertebrates, including many primates, also form social systems. These tend to be much less stereotyped than those of the insects, with more flexibility in roles in which individuals perform multiple tasks and all are reproductively competent. Members of vertebrate societies never lose their individuality. Actual reproductive patterns are constrained by behavioral dominance hierarchies, both male and female, that can be quite complex and can be based in large part on kinship. None, other than humans, come close to the frenetic and compartmentalized social activity seen in the colonies of social insects.

The predominant metazoan groups are the chordates, nematodes, arthropods, mollusks, annelids, and echinoderms. The most organismally diverse group is the Ecdysozoa, which includes the nematodes and the arthropods. Within the arthropods, the insects alone represent more than 1 million named species, with many more remaining undescribed. Nematodes may be as species rich as insects, but only 15,000 nematodes species have been named. Feats of relative strength and tolerance of extreme environmental conditions are legendary among the insects; among the ecdysozoa, however, tardigrades may be the most amazing extremophiles of all, being capable of entering a state of cryptobiosis. Under desiccating conditions, tardigrades can experience a loss of total water content from 85% of their body weight to 3%. Furthermore, while in this state, tardigrades can withstand, sometimes for years, extreme temperatures, including cold to near absolute zero, ionizing radiation, almost total lack of oxygen, and up to 1000 atm of pressure. When moisture levels return to normal, normal life functions resume.

Although they represent a relatively small percentage of the total species on this planet, metazoans have a major impact on the biosphere. Early in the evolutionary history of metazoans, reef-building sponges and corals evolved, creating biogenic alterations in ocean current patterns and novel forms of habitat for additional life forms. Benthic metazoans make use of, and help maintain, the ocean bottom while annelids, nematodes, rotifers, and a variety of arthropods perform the same function for terrestrial soil systems. Many metazoans serve as habitat for other metazoans: In the life cycle of an individual tapeworm, a copepod, a mollusk, a ray-finned fish, and an elasmobranch may all serve as habitat. The many metazoans that are filter or suspension feeders may act as bio-accumulators. In some cases, their activities may enhance local environmental quality, but this may be a mixed blessing. Oysters and other edible mollusks, for example, are excellent at extracting bacteria and other microbes from the water, but if those microbes include fecal coliform bacteria from human sewage the commercial value of the mollusks as food is severely compromised. The array of complex ecological interactions between plants and insects has permitted the classification of insects as "beneficial," meaning that they do not compete with us for plant biomass, "pest," meaning that they do compete with us for plant biomass, or "vector" when they transmit diseases to us or our livestock. We use parasitoid insects as agents of biocontrol by infecting pests and vectors with them.

III. SUMMARY

Organismal diversity is not just amazing, it is important. The extent to which parts of today's planetary array of organisms can be reduced without creating a cascade of extinctions is unknown but hotly debated. It is known that virtually every species has some value to humans, either directly or indirectly. Organisms provide us with food, shelter, raw materials, things of beauty, and pharmaceuticals; help us with our work; and maintain the ability of our air, soil, and water to sustain life, including our own. They are valuable to us indirectly as the building blocks of the biosphere within which we originally evolved and which we still require in order to survive.

There is also clear evidence that all organisms are intimately tied together in the structure of the biosphere because they are all simultaneously parts of larger genealogical and ecological wholes. The significance of the duality of organismal diversity is most apparent in the recognition that new types of organisms are derived from preexisting organisms, while at the same time almost all organisms make extensive use of the biodiversity that predated their origins. Newly evolved organisms always have an impact on preexisting ones. Because it is in the nature of the organism to be relatively autonomous from its surroundings, these interactions are not necessarily positive, often taking the form of "conflicts of interest." The evolutionary resolution of these conflicts of interest has produced an increasingly complex biosphere. As Maynard Smith and Szthmary (1995) observed, each major transition in evolution has been associated with the emergence of organisms, and by extension the entire biosphere, with enhanced abilities to produce, maintain, and transmit information cohesively and also associated with the emergence of novel forms of selection resulting from the evolution of those new organisms. In this way, each newly evolved form of organism becomes intimately involved with both local and global ecology, maintaining the biosphere as a relatively isolated system with its own windows of viability (Ulanowicz, 1997).

See Also the Following Articles

ADAPTATION • ARCHAEA, ORIGIN OF • BIODIVERSITY, ORIGIN OF • DARWIN, CHARLES • DIVERSITY, COMMUNITY/ REGIONAL LEVEL • DIVERSITY, MOLECULAR LEVEL • EUKARYOTES, ORIGIN OF • THERMOPHILES

Bibliography

Brooks, D. R., and McLennan, D. A. (1990). Searching for a general theory of biological evolution. *J. Ideas* **1**, 35–46.

Brooks, D. R., and Wiley, E. O. (1988). *Evolution as Entropy: Toward a Unified Theory of Biology*, 2nd ed. Univ. of Chicago Press, Chicago.

Brooks, D. R., Collier, J., Maurer, B. A., Smith, J. D. H., and Wiley, E. O. (1989). Entropy and information in evolving biological systems. *Biol. Philos.* **4**, 407–432.

Darwin, C. (1872). *The Origin of Species*, 6th ed. John Murray, London.

Depew, D., and Weber, B. (1995). *Darwinism Evolving*. Bradford, Cambridge, MA.

Eldredge, N. (1985). *Unfinished Synthesis*. Columbia Univ. Press, New York.

Eldredge, N. (1986). Information, economics and evolution. *Annu. Rev. Ecol. Syst.* **17**, 351–369.

Eldredge, N., and Salthe, S. N. (1984). Hierarchy and evolution. *Oxford Surv. Evol. Biol.* **1**, 182–206.

Gatlin, L. L. (1972). *Information Theory and the Living System*. Columbia Univ. Press, New York.

Hull, D. L. (1988). *Science as a Process*. Univ. of Chicago Press, Chicago.

Kampis, G. (1991). *Self-Modifying Systems in Biology and Cognitive Science: A New Framework for Dynamics, Information and Complexity*. Pergamon, Oxford.

Kauffman, S. A. (1993). *The Origins of Order: Self-Organization and Selection in Evolution*. Oxford Univ. Press, New York.

Lewontin, R. C. (1982). Organism and environment. In *Learning, Development and Culture* (H. C. Plotkin, Ed.). Wiley, New York.

Lewontin, R. C. (1983). Gene, organism, and environment. In *Evolution from Molecules to Men* (D. S. Bendall, Ed.). Cambridge Univ. Press, Cambridge, UK.

Lotka, A. J. (1913). Evolution from the standpoint of physics, the principle of the persistence of stable forms. *Sci. Am. Suppl.* **75**, 345–346, 354, 379.

Lotka, A. J. (1925). *Elements of Physical Biology*. Williams & Wilkins, Baltimore, MD.

Maurer, B. A., and Brooks, D. R. (1991). Energy flow and entropy production in biological systems. *J. Ideas* **2**, 48–53.

Maynard Smith, J. (1976). What determines the rate of evolution? *Am. Nat.* **110**, 331–338.

Maynard Smith, J., and Szathmary, E. (1995). *The Major Transitions in Evolution*. Freeman Spektrum, Oxford.

Niklas, K. J. (1999). Evolutionary walks through a land plant morphospace. *J. Exp. Bot.* **50**, 39–52.

Odling-Schmee, F. J., Laland, K. N., and Feldman, M. W. (1996). Niche construction. *Am. Nat.* **86**, 309–326.

Prigogine, I. (1980). *From Being to Becoming*. Freeman, San Francisco.

Prigogine, I., and Wiame, J. M. (1946). Biologie et thermodynamique des phénomènes irréversibles. *Experientia* **2**, 451–453.

Raff, R. A. (1996). *The Shape of Life: Genes, Development, and the Evolution of Animal Form*. Univ. of Chicago Press, Chicago.

Salthe, S. N. (1993). *Development and Evolution: Complexity and Change in Biology*. MIT Press, Boston.

Schanck, J. C., and Wimsatt, W. C. (1988). Generative retrenchment and evolution. In *PSA 86* (A. Fine and P. K. Machamer, Eds.), Vol. 2, pp. 33–60. Philosophy of Science Association, East Lansing, MI.

Ulanowicz, R. E. (1997). *Ecology: The Ascendent Perspective*. Columbia Univ. Press, New York.

Van de Vijver, G., Salthe, S. N., and Delpos, M. (Eds.) (1998). *Evolutionary Systems: Biological and Epistemological Perspectives on Selection and Self-Organization*. Kluwer, Dordrecht.

Wake, D. B., and Roth, G. (Eds.) (1989). *Complex Organismal Functions: Integration and Evolution in Vertebrates (Dahlem Workshop)*. Wiley, New York.

DIVERSITY, TAXONOMIC VERSUS FUNCTIONAL

John C. Moore
University of Northern Colorado

I. Species, Communities, and Ecosystems
II. Species, Function, and the Ecological Niche
III. Estimating Diversity
IV. Biodiversity and Systems Theory
V. Linking Taxonomic and Functional Diversity
VI. Natural Selection

GLOSSARY

functional diversity The different types of processes in a community that are important to its structure and dynamic stability.
functional group A group of species that utilize similar resources; synonymous with guild.
guild A group of species that utilize similar resources (usually food).
species evenness A measure of the relative abundance of species in an area.
species richness The number of species in an area.
species turnover The change in the composition of species in an area due to the extinction of some species and the replacement by a new species by colonization.
taxonomic diversity The number and the relative abundance of species in a community.

TAXONOMIC DIVERSITY refers to the number and the relative abundance of species in a community. Functional diversity refers to the different types of processes in a community that are important to its structure and dynamic stability. Taxonomic and functional diversity have been presented as being different, and as the title implies, competing approaches to the study of biodiversity. This article will provide specific examples of how taxonomic and functional diversity are interrelated and how information from both perspectives can be used to estimate biodiversity and propose mechanisms that govern biodiversity.

I. SPECIES, COMMUNITIES, AND ECOSYSTEMS

Linnaeus (1735) and Hutchinson (1959) provided a convenient starting point to address biodiversity in terms of taxonomic and functional diversity. Linnaeus formalized the two kingdom scheme of Plantae and Animalia and made the distinction between sessile photosynthetic plants and motile food-ingesting animals. Predating an understanding of photosynthesis or modern evolutionary theory, this work was descriptive and simply attempted to categorize and count organisms rather than posit mechanisms. Function was at the basis of the classification. In his address to the Society of American Naturalists, Hutchinson, like Linnaeus, made a deliberate distinction between primary producers and heterotrophs by asking "why are there so many kinds of animals?" The unique feature of this address was that it studied biodiversity by invoking principles from ecology and evolution, formalizing niche theory, and including energetics. More important, it marks an early attempt to link species diversity to the functioning of an ecosystem.

The choice of taxonomic versus functional diversity illustrates a broader issue in science. In both cases, traditional taxonomies based on individual traits or ones based on the functional attributes of a species in a community or ecosystem are human constructions. Erhlich and Holm (1962) summarized the dilemma nicely:

> There seems to be no theoretical reason why there must be complete congruence among estimates of relationships based on characters from different developmental stages or on characters from different organs systems of the same stage.... In a taxonomy based on ecological requirements, whales will be more closely related to sharks than to bears. Such a relationship is no more or less "true" than the classical one: It is merely based on different attributes.

At some point the question comes down to the type of information that is used, the way the information is used, and why the information is used.

A. Different Approaches to Studying Biodiversity

There are different subdisciplines within ecology: autoecology (individual/population ecology), community ecology, and ecosystem ecology. The focus and approach of each subdiscipline and the type of information used by each have influenced the degree to which taxonomic or functional diversity have been used (Table I).

1. Individual/Population Approach

Autecology is the study of individual organisms. The approach originally focused on the adaptiveness of an organism's physiology to the environment; however, it has been expanded to include the study of the distribution and dynamics of populations. In terms of biodiversity, autecology has embraced taxonomic diversity.

2. The Community Approach

The community approach defines diversity in terms of the number of species and relative abundances of spe-

TABLE I

Information Used by Different Subdisciplines of Ecology

	Subdiscipline		
	Autecology	Community ecology	Ecosystem ecology
Focus	Individuals/ populations	Populations	Forms of matter
	Species	Species/guilds	Functional groupings
Units	Individuals	Individuals	Biomass
Biodiversity	Taxonomic	Taxonomic/ functional	Functional

cies. This information is used to study the effects of the number and types of interactions and the consequences that the interactions have on the distributions and abundances of the species and the dynamic stability of the community. Community ecology tends to focus more on taxonomic diversity than on functional diversity. An ideal description of a community would include all species and species interactions. In practice, descriptions of communities include a mixture of individual species and aggregations of species. The aggregations have been based on functional attributes defined by similarities in resource utilization among species and on taxonomic affinities (e.g., "mammals").

3. The Ecosystem Approach

The ecosystem approach defines diversity in terms of processes and function within the community. The focus is not on species per se but rather on the products of species interactions, usually different forms of carbon or nitrogen. Underlying this approach is the belief that species interact in ways that reinforce stabilizing feedbacks that operate to maintain a persistent steady state. The feedbacks operate through the availability of nutrients in one pool of that critical to another pool.

A typical description of an ecosystem does not include many species (except for the dominant species) but rather aggregates of species that are similar in the ways in which they process matter relative to a key process. For example, the decomposition of pine needles on a forest floor is a complex process that is influenced by the nutrient quality of the pine needles, the temperature and moisture of the forest floor, and the organisms that are present. For the microbes alone, the process might involve more than 100 species of saprobic fungi and an equal number of strains of saprobic bacteria. Depending on the level of resolution desired, the microbes could be aggregated into a single pool of "decomposers" or separated into pools of "fungi" and "bacteria." The justification for the seemingly casual treatment of the individual species is that the focus is not on how many species are present but rather on the general composition of the pool and the rate at which the pool processes matter.

II. SPECIES, FUNCTION, AND THE ECOLOGICAL NICHE

Taxonomic and functional diversity use similar information but in different ways. The species is central to taxonomic diversity. Species are arranged into phylogenetic relationships that are based on how they acquire energy, their cell types, the level of cellular organization, their embryology and life history, their physiology, and their genetic makeup (RNA and/or DNA). Different classification schemes have been proposed that use all or some of these criteria with different weights given to any one of the criteria. Ecosystem processes are central to functional diversity. Functional diversity has been based on ecological characteristics. Species are aggregated into functional units based on how they use resources. Various schemes have been proposed to aggregate species into functional groupings. Species are arranged into functional relationships based on how they acquire energy, where they acquire energy, the rates at which they process energy (physiology), and when they process energy (life history).

Linking taxonomic and functional diversity would require integrating information that describes the makeup of an organism (its taxonomic characteristics) along with the rates at which it processes and transforms matter (its functional characteristics). Niche theory offers a way to link taxonomic and functional diversity.

A. The Ecological Niche

The ecological niche of a species is defined by the resources it requires. The niche of a species can be viewed graphically as a hypervolume, in which each resource represents an independent axis (Fig. 1). The space that each species occupies within this hypervolume is defined by its resource requirements. An analysis of the resource requirements for several species from different taxa revealed that the principal niche axes are food, habitat use, and time.

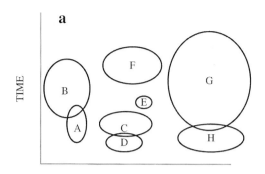

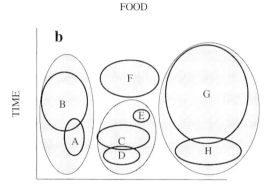

FIGURE 1   Projection of eight species (A–H) onto two of the principal niche axes (food and time). (a) The ovals surrounding each species represent the region of the food and time space that each species occupies. Ovals that overlap with one another represent regions in which the species share (compete for) resources. (b) Ecologists may aggregate species that occupy similar niche space into functional groups (species within the shaded ovals).

B. Functional Groupings of Species

Both the community and the ecosystem perspectives have grouped species into larger units when studying interactions among species or processes mediated by species. The schemes used have ranged from ones of convenience, which are usually based on a broad taxonomic category, to ones that have adopted strict criteria that may include organisms from different taxonomic groupings. For this discussion, the groupings that have been based on the resource requirements of a species and its position in niche space relative to other species makes the most sense, given that function in an ecosystem context is directly related to how species utilize resources. Hence, a functional group is composed of species that utilize similar resources (Fig. 1). There is no standard number of resources (axes in niche space) used to establish functional groups, but many are based on the principal resources of food, habitat, and time. This has resulted in the groupings being given different names. For example, a "guild" is defined as a group of species that feed on similar food in a similar way. Soil ecologists define a "functional group" as a group of species that feed on similar food in a similar way, occupy the same habitat, and possess similar life histories.

III. ESTIMATING DIVERSITY

Procedures to estimate species diversity originated from the fields of information theory and statistics. If presented a code with a set number of symbols (s) of known proportions (p), it is desirable to estimate its information content. Ecologists simply substituted the species (or aggregations of species) for the symbols. Species diversity of a community could be estimated in the same way as the information content of a code. Two aspects of species diversity are the number of species (species richness) and the relative abundance of species in a community (species evenness).

A. Species Richness

Species richness (S) is the number of species within a defined region. The species richness of a region is obtained through sampling or via a census. Because "region" is defined by the observer, species richness has been further categorized into three components to account for changes in spatial scale.

Alpha diversity, sometimes referred to as point diversity, is the species richness that occurs within a given area within a region that is smaller than the entire distribution of the species. Beta diversity is the rate at which species richness increases as one moves in a straight line across a region from one habitat to another habitat. In other words, it is the rate of change in species richness that occurs with a change in spatial scale. Gamma diversity is the species richness within an entire region. As the area being surveyed approaches that of the entire region, alpha diversity approaches gamma diversity and beta diversity approaches zero.

B. Species Evenness

Species evenness takes into account the number of species and the relative abundance of species in a community. Several indices have been proposed. Two of the commonly used measures of evenness are the Shannon index (H) and the Simpson index (D).

The Shannon index (H) is a measure of the informa-

tion content of a community rather than of the particular species that is present. The index is as follows:

$$H = -\sum_{i=1}^{s} p_i \log p_i$$

where s is the number of species in the community (species richness) and p_i is the relative proportion of species i. Hence, if two communities (A and B) with the same number of species (not necessarily the same species) present were compared, and the distributions of p_i were the same, then $H_A = H_B$. On the other hand, if either the number of species or the proportions differed, then $H_A \neq H_B$. In the later case, the community with the greater number of species whose relative abundances were equal would possess the higher diversity.

The Simpson index (D) measures the dominance of a multi-species community and can be thought of as the probability that two individuals selected from a community will be of the same species. The Simpson index was originally proposed as follows:

$$D = -\log \sum_{i=1}^{s} p_i^2$$

where s is the species richness of a community and p_i is the relative proportion of species i. The index can be modified to $1 - D$ to give it the property of increasing as diversity increases (the dominance of a few species decreases).

C. Linking Species Richness and Species Evenness

Species richness and species evenness are special cases of dichotomous-type rarity measures. For dichotomous-type rarity measures for a pool of s species, the rarity of species i, whose function is denoted $R(p_i)$, depends only on its relative abundance (p_i). The rarity measures (Δ) take the following form:

$$\Delta = \sum_{i=1}^{s} p_i R(p_i)$$

The relationships between species richness (S), the Shannon index (H), and the Simpson index (D) are

TABLE II
Rarity Functions and Forms of Diversity Indices

	Rarity function	Diversity index
Species richness (S)	$R(p_i) = (1/p_i) - 1$	$\Delta = s - 1$
Shannon index (H)	$R(p_i) = -\log p_i$	$\Delta = -\Sigma p_i \log p_i$
Simpson index ($1 - D$)	$R(p_i) = 1 - p_i$	$\Delta = \Sigma p_i (1 - p_i)$

presented in Table II. Notice that the indices share three important properties: (i) Δ is maximized when $p_i = 1/s$ for all i species; (ii) given that $p_i = 1/s$, the community with the higher species richness has the greater diversity; and (iii) $\Delta = 0$ if the community possesses a single species ($s = 1$).

IV. BIODIVERSITY AND SYSTEMS THEORY

One of the enduring tenets of ecology in the early part of the twentieth century was that a diverse community was a stable community. The linkage between biodiversity and stability has an intuitive appeal. The more species there are in a community, the more likely they would be tightly networked, and hence the more stable the community would be. Interestingly, function plays an important role in this argument in that in a highly diverse community, there would be alternative species or assemblages of species to perform key tasks to sustain the cycling of matter and the current diversity. These ideas have been challenged and modified. Ecologists agree that biodiversity is important to the function and stability of an ecosystem. The current debates center on the nature of the relationship between biodiversity, function, and stability.

Although useful, the functions used to estimate biodiversity that were presented previously are descriptive measures and imply neither the mechanisms that shape biodiversity nor the importance of biodiversity to maintaining the structure and dynamic stability of a community. If the diversity of one community were higher than that of another, does this increase its likelihood of persisting in time or recovering from a disturbance? The answer depends on the species involved. For example, the colonization of an exotic species increases the species richness of a community but does not necessarily increase its stability—to the contrary, it often destabilizes the system. On the other hand, the addition of a predator can stabilize an ecosystem by tempering the

oscillations in the population densities of its prey through time. Biodiversity is a property of complex dynamic biological systems. The steady state and stability are two concepts important to the study of complex dynamic systems.

A. The Steady State

A steady state is an equilibrial condition in a dynamic process in which the rate of input of a state variable is equal to the rate of output of that variable (Fig. 2). The dynamic process can involve any variable that changes in time. For taxonomic and functional diversity, the variables of interest are the number of species or functional groups, the densities and biomass of each species or functional group, and the mass of key elements within different pools within the ecosystem.

1. The Number of Species

The number of species in a community is at steady state when the rate of colonization equals the rate of extinction (Fig. 3). The classic model of this process, the theory of island biogeography (MacArthur and Wilson, 1967), assumes that the rate of colonization is affected by the distance that the habitat being colonized is from the source of species and by the size of the habitat being colonized. Colonization rates are higher for large habitats that are close to the source of species than for smaller distant habitats. The rate of extinction is affected by the size of the habitat. The rates of extinction for smaller habitats are higher than those for larger habitats.

2. Species Composition

In the model presented previously, although the number of species is constant at steady state, the species composition of the community does not remain constant. The model simply assumes that once the rates of colonization and extinction are equal, then for every new species that enters a habitat, one of the current resident species goes extinct. This point is important when trying to estimate the diversity of a habitat by sampling the habitat over time. If a habitat were sampled long enough, the estimate of the number of species within the habitat would equal the number of species in the habitat serving as the source, even though the source has a higher number of species at its steady state than the habitat being colonized.

3. Population Densities

The population density of an individual species within a habitat is at steady state when the number of births

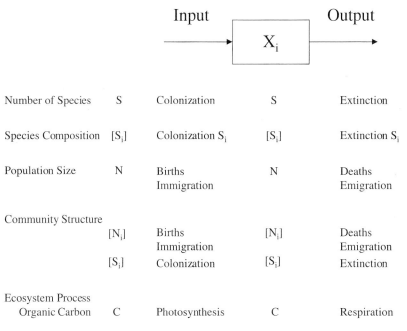

FIGURE 2   Diagram and list of the processes (inputs and outputs) that contribute to the dynamics of a single state variable (X) or several state variables [Xᵢ].

and immigration equal the number of deaths and emigration. When modeling this process, ecologists typically assume that the habitat is closed to immigration and emigration and focus entirely on the rate of birth and the rate of death. More sophisticated approaches that involve multiple habitats will include the immigration and emigration of species among habitats.

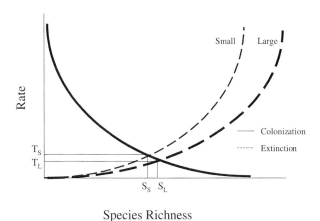

Species Richness

FIGURE 3   The effect of the rates of colonization and extinction on the species richness of large and small islands (S_L and S_S, respectively). T_L and T_S represent the points at which the rates of colonization and extinction are equal.

4. Community Structure

A community is at steady state when the species composition of the community and the densities of the species remain constant. This implies that the species richness and evenness of the community do not change. This condition is rarely met since the species composition and the densities of the species fluctuate through time.

5. Ecosystem Processes

An ecosystem process is at steady state when the rate of input of matter or energy is equal to the rate of output. The input of matter and energy occurs via photosynthesis and from sources external to the ecosystem (e.g., leaves entering a stream). The export of matter and energy occurs through heterotrophic respiration and the physical movement of matter from the ecosystem (e.g., export of leaves from forest to a stream). The production of organic matter from inorganic matter (photosynthesis) and the transformation of organic matter to other organic forms and inorganic matter (heterotrophic respiration, hereafter referred to as respiration) are ecosystem processes that approach a steady state and illustrate the interplay between matter and energy. When the rate of photosynthesis equals the rate of respiration, the system is said to be at steady state

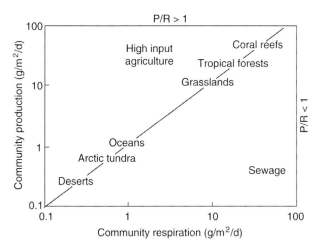

FIGURE 4 The relationship of various ecosystems in terms of their community production (grams per square meter per day), *P*, and community respiration (grams per square meter per day), *R*. The dashed line represents the steady state in terms of *P* and *R* ($P/R = 1$). Note the position of the disturbed ecosystems.

(Fig. 4). In other words, at steady state the ratio of production to respiration for matter and for energy is 1 ($P/R = 1$).

B. Stability

A population, community, or ecosystem is stable if it can persist through time and can recover from minor disturbances. A disturbance is any perturbation to the system that causes any of the species or functional groups to deviate from their steady state. There are many facets to stability and many operational definitions as well. Aspects of stability include local stability, resilience, and oscillation.

1. Local Stability

An ecosystem possesses local stability if it returns to its original steady state following a minor disturbance. The minor disturbance is relative to the sizes of the populations involved and occurs in the neighborhood of the steady state, hence the reference to "local." This concept of stability has been widely used by biologists because it is simple and exact in its interpretation. It has been criticized for the same reasons that it has been accepted. For example, high species diversity has been shown to negatively impact the local stability of model ecosystems. This result was the basis for the hypothesis that dynamic stability constrains biodiversity. However, competing models of ecosystems with different sets of assumptions and criteria for stability, although they do

not disagree that dynamic stability affects biodiversity, yield different results with regard to the nature of the relationship.

The concepts of local stability and the theory of island biogeography may seem at odds with one another, even though they rely on similar mathematics and assumptions and are modeling the same systems. An obvious question is, how can an ecosystem be locally stable if the species composition of the ecosystem is changing? The discrepancy can be clarified by revisiting the processes that are being modeled. When modeling populations within a community, local stability is in reference to the densities or biomass of the populations. The dynamics of the populations are governed by the difference between the rates of birth and death. When modeling the species diversity within the same community, local stability is in reference to the number of species that are present. The dynamics of the number of species is governed by the rates of colonization and extinction. To carry this a step further, the local stability of the carbon or nitrogen cycles within an ecosystem is in reference to the mass of the different forms carbon or nitrogen. The dynamics of the different forms is governed by the functional attributes of the species present and the rates at which the different forms are acted upon.

2. Resilience

Resilience is a measure of a stable system's ability and the time required to recover from a minor disturbance. The principle of resilience can be applied to any state variable present in an ecosystem: population densities, the number of species, or the mass of nutrients. Resilience is a function of the rates of inputs of the state variables, the rate at which the state variables turn over, and the number of state variables.

The return time of a system is a measure of its resilience. More resilient ecosystems have shorter return times than less resilient ones. Return time is estimated as the time between the point when the disturbance altered the ecosystem and the point at which the ecosystem recovers. The persistence of a system is compromised if the frequency of disturbance is too great given the length of return time because a system may not have time to recover from any one incident.

3. Oscillations

The state variables of most ecosystems do not maintain a constant steady-state condition but rather they exhibit oscillations. Oscillations are fluctuations in the magnitude of a state variable through time that can be intrinsic to the dynamics of the system in the absence of a disturbance or can occur during the recovery phase of a system following a disturbance or both. The degree to which a system oscillates can affect its stability, persistence, and diversity. For example, if the amplitude of oscillations are too great or are such that the population size of a species approaches zero, then the likelihood of that species going extinct increases.

C. Modeling Species Interactions

Gardner and Ashby (1970) presented a simple mathematical exercise that illustrated how the number of interacting units (species) within a system and the number of interactions between these units (species interactions) related to the stability of the system. The exercise constructed simple systems of 4, 7, and 10 units assumed to be at a steady state. For a system of a given size, connections were made between the units in a way that ensured that all units would be connected to the system, creating several variations of the system in terms of the connections. Next, the strength of the connections was allowed to vary, and the systems were disturbed by moving the units away from their steady state. Two results emerged from their work. First, as the number of interacting units increased within a system the likelihood of the system returning to a steady state after a small disturbance (stability) decreased. Second, for a given number of interacting units, as the number of connections between the units increased, the stability of the system declined.

There is a clear connection between the simple mathematical exercise presented above and the diversity and stability of ecological communities (May, 1973). The species is substituted for the generic units modeled previously, and the interactions are defined in terms of predation, competition, or mutualism. Like the generic modeling exercise, high biodiversity does not necessarily lead to stability. Additional studies demonstrated that systems composed of species aggregated into blocks of highly interactive species that were connected by weaker interactions were more stable than systems with the same number and types of species and interactions arranged at random. Hence, the way in which the community is constructed affects its diversity and stability.

The types of interactions among species affect the stability of a community. Theory and observation agree in general terms on this point, if not on specifics. For example, mutualistic interactions, whereby the actions of each species benefit the growth and dynamics of one another, and omnivory, in which species derive energy from different trophic levels, have been shown to be destabilizing in simple models, but both are prolific in natural communities. All the dominant terrestrial ecosystems rely on a mutualism between the vascular plants and microbes, and nearly every described ecosystem possesses numerous examples of omnivory.

D. Modeling Species and Processes

Process-oriented models focus on the transformation of elements among various inorganic to organic states through ecosystems. Carbon, nitrogen, and phosphorus are the more common choices of elements to be modeled given their importance to plant and animal growth. Models of species interactions have been connected to models of processes. The general approach to linking these models is to use the structure and formulation adopted when modeling species with the currency used in models of processes. First, species are aggregated into functional groups using one of the schemes described previously, and the interactions among the functional groupings are established. Second, the biomass, expressed in terms of carbon, nitrogen, or phosphorus, of each functional group is estimated. Third, a series of differential equations is developed that establishes which functional groups interact with one another and the nature of the interactions. Parameters within the models include birth and death rates, feeding rates, and the efficiencies by which the functional groups convert consumed matter into new biomass. The models are used to estimate the flow of elements among the functional groups, the flow of elements out of the system, the stability of the ecosystem, and the resilience of the ecosystem.

Three results have emerged from these models. First, the models are sensitive to the changes in the parameters associated with the physiology and life histories of the species within the functional groups. For example, microbes possess high reproductive rates and high rates of energy conversion, and they process matter at a higher rate compared to a functional group dominated by arthropods or nematodes, which possess lower reproductive rates and rates of energy conversion. Second, functional groups are arranged into interactive blocks that form pathways of material flow that originate from a specific source and end with a top predator. The blocked arrangement of functional groups, the distribution of biomass within the ecosystem, and the distribution of the flow of elements through the ecosystem influence the stability and resilience of the system.

V. LINKING TAXONOMIC AND FUNCTIONAL DIVERSITY

The linkage between taxonomic diversity and functional diversity involves many facets. The morphology of an organism determines the type of habitat in which it resides and its ability to colonize new habitats. The physiology of a species influences its adaptiveness to a habitat and the rates and efficiency with which it transforms matter. The life history of an organism influences the rate at which it processes matter (births, deaths, and consumption). Linking taxonomic diversity to functional diversity requires finding measurable attributes that all organisms possess that have meaning to all the ecological subdiscplines, but it also requires flexibility in terms of how the information is used. Three case studies are presented in the following sections. The case studies illustrate how, when studying biodiversity, aspects of both taxonomic and functional diversity are used interchangeably.

A. Diversity of C_3 and C_4 Grasses

The first case study involves a comparison of the distribution of grass species that possess either the C_3 or C_4 photosynthetic pathways in North America. The study illustrates how the information of autecology can be used to find mechanisms to explain patterns in diversity. The study demonstrates how the functional aspects of an organism's physiology influence a species adaptation to a habitat and ultimately the diversity of species with similar physiologies.

Plants that possess C_3 and C_4 photosynthetic pathways differ in terms of their morphology and biochemistry as they relate to the dark reaction of photosynthesis. For C_3 plants, photosynthesis occurs entirely within the outer mesophyll cells, in which CO_2 enters the Calvin–Benson cycle through the plants' stomata. The enzyme (RuBP carboxylase or Rubisco) that incorporates CO_2 into the Calvin–Benson cycle also has an affinity toward O_2. If O_2 were incorporated instead of CO_2 the plant would experience photorespiration—energy is expended, no carbon is fixed, and the plant loses organic carbon. For C_4 plants, there are two types of photosynthetic cells: the outer mesophyll cells and the inner bundle sheath cells. In C_4 plants, CO_2 enters the mesophyll through the stomata and is incorporated into a four-carbon compound before being exported to the bundle sheath cells in which it enters the Calvin–Benson cycle. The additional steps that occur within the C_4 plants prior to the Calvin–Benson cycle cost the

plant additional energy and organic carbon. However, the enzyme (PEP carboxylase) that adds the CO_2 to the four-carbon molecule in the mesophyll cells is more efficient than Rubisco and less likely to initiate the loss of organic carbon to photorespiration. This trade-off places plants with the C_4 pathway at a disadvantage under cool and moist conditions compared to those with the C_3 pathway, but it affords them an advantage in warmer and dry conditions.

Within North America, the northern latitudes are dominated by species of C_3 grasses, whereas the southern latitudes are dominated by species of C_4 grasses. This pattern of diversity has been attributed to the adaptiveness of each pathway to patterns of temperature and light intensity. The lower latitudes possess higher temperature and higher light intensity than the upper latitudes given the orientation of the earth to the sun. Grasses that possess the C_3 pathway have been shown to be more productive than those with the C_4 pathway under conditions of cool temperature and low light intensity. At higher temperatures and light intensity, C_3 plants suffer greater water loss through their stomata and increased photorespiration than do C_4 plants.

B. A Test of the Theory of Island Biogeography

This case study illustrates how information from the community perspective has been used to study biodiversity. The experimental manipulations of several small mangrove islands off the Florida coast by Simberloff and Wilson (1969a,b) and the reanalysis of their work by Heatwole and Levins (1972) offer not only the best validation of the theory of island biogeography but also some insight into how taxonomic diversity and functional diversity are related.

Simberloff and Wilson (1969a,b) conducted a thorough survey of arthropods on four small mangrove islands and then fumigated the islands with methyl bromide to eliminate the arthropods. For 2 years following the disturbance, they sampled for arthropods that recolonized the islands. The results were consistent with those predicted by the model proposed for the theory of island biogeography. The colonization rates were higher and the steady states in species richness were achieved sooner on islands closer to the mainland than on the more distant island. Once a steady state in species richness was achieved, the species composition changed due to the predicted turnover in species.

Heatwole and Levins (1972) provided an explicit link between the recovery process, species diversity, and functional diversity. They aggregated the species into guilds based on the feeding and foraging behaviors of the arthropods (predators, herbivores, decomposers, ants, etc.). One year following the fumigation, the taxonomic diversity of the islands was lower than it was prior to the fumigation (Table III). However, the proportion of species within each of the guilds was the same as it was prior to fumigation. These results suggest that the functional diversity of the islands recovered before the species diversity. In other words, the taxonomic diversity of the communities was in some way dependent on functional characteristics of the community.

C. Invasion of an Exotic Species

The final case study illustrates how elements of taxonomic and functional diversity and information from autecology, community ecology, and ecosystem ecology have been used to study biodiversity. Vitousek and colleagues (1987, 1989) studied the natural history of the invasive plant species *Myrica faya* and the effects that it has had on the nitrogen cycle and plant community development of recent lava flows in Hawaii. Since its introduction to Hawaii from the Canary Islands by Portuguese immigrants at the end of the nineteenth century, *M. faya* had colonized more than 34,365 ha. *Myrica faya* forms a symbiotic association with the soil nitrogen-fixing actinomycete, *Frankia*. The plant possesses two characteristics that separate it from the native nitrogen-fixing plants on the Islands. Unlike the native plant nitrogen fixers (*Acacia koa* and *Sophora chrysophylla*), the seeds of *M. faya* are small, have an edible fleshy fruit, and pass through the digestive systems of birds intact. Moreover, *M. faya* is a more efficient nitrogen fixer than the native species. When combined, these factors make *M. faya* a superior colonizer of recent lava flows, pastures, and forests.

Once established, the *M. faya* and its soil symbiont, *Frankia*, alter the nitrogen status of the soils. For example, soil nitrogen inputs within open canopies of *M.*

faya were four times more than those of similar sites without the exotic plant (23.5 and 5.5 kg per hectare per year, respectively). This added soil nitrogen becomes available to other plant species, facilitating invasions by other exotic species (e.g., strawberry guava, *Psidium catleianum*) to the detriment of native plant species. The functional attributes of a single species and it symbiont affected the biodiversity of a community by altering the course of community development.

VI. NATURAL SELECTION

Species evolve and communities and ecosystems change, but do communities and ecosystems evolve? Does natural selection operate on communities and ecosystems even though they do not possess genes? Natural selection needs to be defined before these questions can be answered. Natural selection is the differential perpetuation of genes in successive generations caused by different degrees of adaptiveness to the environment.

One argument is that ecosystems do evolve because the composition of species within a community changes and natural selection operates on the gene pools of each species. In addition to the changes in the genetic makeup of the species, there are changes in key processes that feed back to reinforce stable configurations or that initiate change. A criticism of this analogy is that natural selection at the ecosystem level for the type of coordination among species that leads to a stable ecosystem is not the same as selection at the organismal level on genes. Selection on genes operates through differential reproductive success of individuals, whereas this may not be true at the ecosystem level since the boundaries of an ecosystem are not so clearly defined.

An alternative view posits that ecosystems do not evolve, but that the species composition of the stable ecosystem is a result of the selection operating on individuals and the coevolution of species interacting with one another. The outcome of selection on individuals

TABLE III

Functional Groups of Arthropods by Number and Relative Proportion (%) on Islands before and 1 Year after Fumigation[a]

	No. and relative proportion (%)							
	Herbivore	Scavenger	Detritus	Wood borer	Ant	Predator	Parasite	Total
Before	55 (34)	7 (4)	13 (8)	8 (5)	32 (20)	36 (22)	12 (7)	163
After	55 (40)	5 (4)	8 (6)	6 (4)	23 (17)	31 (23)	9 (6)	137

[a] Adapted from Heatwole and Levins (1972).

can influence function within a community, which in turn feeds back as a form of selection. Unique to this argument is that the adaptiveness of a species includes the degree to which its dynamics and function contribute to and are compatible with the stability of the ecosystem as a whole.

See Also the Following Articles

GUILDS • HABITAT AND NICHE, CONCEPT OF • ISLAND BIOGEOGRAPHY • MEASUREMENT AND ANALYSIS OF BIODIVERSITY • PLANT INVASIONS • SPECIES DIVERSITY, OVERVIEW • STABILITY, CONCEPT OF • TAXONOMY, METHODS OF

Bibliography

Erhlich, P. R., and Holm, R. W. (1962). Patterns and populations: Basic problems of population biology transcend artificial disciplinary boundaries. *Science* **137**, 652–657.

Gardner, M. R., and Ashby, W. R. (1970). Connectance of large, dynamical (cybernetic) systems: Critical values for stability. *Nature* **228**, 784.

Grassle, J. F., Patil, G. P., Smith, W., and Taille, C. (Eds.) (1979). *Ecological Diversity in Theory and Practice*. International Cooperative, Fairfield, MD.

Heatwole, H., and Levins, R. (1972). Trophic structure, stability and faunal change during colonization. *Ecology* **53**, 531–534.

Hutchinson, G. E. (1959). Homage to Santa Rosalia, or why are there so many kinds of animals. *Am. Nat.* **93**, 145–159.

Leedale, G. F. (1974). IV. How many are the kingdoms of organisms? *Taxon* **23**, 261–270.

Linnaeus, C. (1735). *Systema Naturae Sive Regna Tria Naturae Systematice Proposita per Classes, Ordines, Genera, and Species.* Haak, Lerden.

MacArthur, R. H., and Wilson, E. O. (1967). *The Theory of Island Biogeography,* Monographs in Population Biology, Vol. 1. Princeton Univ. Press, Princeton, NJ.

May, R. M. (1973). *Stability and Complexity in Model Ecosystems,* Monographs in Population Biology, Vol. 6. Princeton Univ. Press, Princeton, NJ.

Pielou, E. C. (1975). *Ecological Diversity.* Wiley, New York.

Rosenzweig, M. L. (1995). *Species Diversity in Space and Time.* Cambridge Univ. Press, Cambridge, UK.

Simberloff, D. S., and Wilson, E. O. (1969a). Experimental zoogeography of islands: The colonization of empty islands. *Ecology* **50**, 267–278.

Simberloff, D. S., and Wilson, E. O. (1969b). Experimental zoogeography of islands: Defaunation and monitoring techniques. *Ecology* **50**, 278–296.

Vitousek, P. M., and Walker, L. R. (1989). Biological invasion by *Myrica faya* in Hawaii: Plant demography, nitrogen fixation, ecosystem effects. *Ecol. Monogr.* **59**, 247–265.

Vitousek, P. M., Walker, L. R., Whiteaker, L. D., Mueller-Dombois, D., and Matson, P. A. (1987). Biological invasion by *Myrica faya* alters ecosystem development in Hawaii. *Science* **238**, 802–804.

DOMESTICATION OF CROP PLANTS

Daniel Zohary
The Hebrew University

ONE OF THE MOST CONSPICUOUS BIOLOGICAL DEVELOPMENTS in the past 10,000 years has been the invention of farming and the explosive evolution of domesticated plants and animals under this new set of conditions. As this chapter will examine, numerous plant species have since been taken from their wild environments and introduced into cultivation—that is, into various human-made systems of maintenance and handling.

I. INTRODUCTION

The original focus of farming was on growing food crops; but this was soon extended to include plants essential for other utilities such as fiber sources, stimulants, medicinal aids, dye crops, forage, and timber production. In addition, the past two or three centuries witnessed another major development, namely a massive introduction of ornamental plant species into cultivation. (Today the number of ornamental species exceeds the number of food crops.) Most recently, plants with medicinal properties are getting new attention. The potentials of scores of such plant species (traditionally collected from the wild) have been evaluated. To ensure a steady supply, and for improving quality and yields, many of them are presently introduced into cultivation. A parallel trend is apparent in forestry. Here, too, dozens of timber producing trees are being quickly domesticated. To date, thousands of plant species, native to the various phytogeographic regions of the World, have been introduced into cultivation. In the majority of these cases, the domestic derivatives have already been drastically altered by this move. So much so that the cultivated forms can no more survive in the wild and have become dependent on humans for their existence. Grain crops, for example, lost the wild type adaptation to disperse their seeds and depend on the harvesting, threshing, and sowing by the farmer. Vegetatively propagated tuber plants (under domestication) are commonly sterile. So are many seedless varieties of fruit crops (e.g., bananas). Moreover, as a result of (a) the ecological shift from the original wild habitats into the greatly different and varied anthropogenic environments and (b) repeated cycles of deliberate selection by humans, many domesticates have greatly diverged under domestication. In most major food crops (e.g., bread wheat, rice, maize, apple, or grapevine) or in many popular ornamentals (e.g., roses, rhododendrons, narcissi, tulips, irises), hundreds or even thousands of distinct cultivated varieties ("cultivars") or ornamental forms are now recognized, fitting the wide range of conditions under which these plants are being maintained and the different and contrasting wishes of the growers who constantly try to improve their plants by selection. All in all, domestication provides us with most impressive examples of rapid evolution in plants. Vast arrays of morphological, anatomical, physiological, and chemical changes have evolved in crops and ornamentals, and this in a very short span of evolutionary time. Moreover, selection under domestication molded the gene pools of the crops (the "cultivated gene pools") to become the backbone for the production of food and other essential utilities for humankind. Today, our very existence depends on this genetic diversity. Thus not only do the crops became dependent on humans; also humans became dependent on crops. Obviously, civilizations and domesticated plants have coevolved; and this coevolution continues today.

II. THE TRADITIONAL AGRICULTURAL SYSTEMS OF THE WORLD

In classical times (about 2500–2000 years ago), and to a large extent until Columbus's time, agriculture was structured quite differently from what we find it today. At present, each country grows a mixture of crops introduced from the various parts of the world. But this fusion is very recent. Traditionally, farming was practiced in largely separated, independent regions, each occupying extensive territories and each possessing a characteristic and largely unique ensemble of native crops. When the Europeans discovered America, they found that the flourishing agricultural civilizations in the New World were based on growing crops that were totally different from what they were familiar with in Eurasia and in Africa. The two hemispheres are indeed very rich in cereals, pulses, tubers, fruits, vegetables, spices, and other types cultivated plants. Yet the list of domesticated plants in each of them is unique. With only one or two exceptions, these two landmasses originally had no crops in common. This strongly indicates that domestication and agriculture in each of them evolved independently.

Also within each of the two hemispheres, one can detect similar patterns. Each landmass seems to contain several historical agricultural regions, previously separated by difficult-to-cross geographic and cultural barriers. Thus the crops of the traditional agriculture system in Southwest Asia, the Mediterranean basin, and temperate Europe were very different from those grown in East Asia. Moreover, the crop composition in each one of these two agricultural systems is very different from the crop assemblage in Africa south of the Sahara. Similarly in the New World, numerous crops grown in South America were not known in Mexico and middle America, and vice versa.

The main food crops in the traditional agricultural regions of the world are given in Appendix 1. A comparison between the various regional lists of crops reveals that each agricultural system usually contains dominant cereals, companion pulses, oil plants, fruit crops, vegetables, spices, tubers and corms, and stimulants. In other words, the composition of crop *kinds* repeats itself. Only the plant species are different in each traditional region (for example, wheats and barley, lentil, and pea in Southwest Asia, the Mediterranean basin, and Europe; rice and soybean in East Asia; maize and *Phaseolus* beans in America; and sorghum and other millets and cowpea in Africa south of the Sahara).

III. THE EMERGENCE OF AGRICULTURE

The shift from hunting and gathering to farming started relatively very late in the history of *Homo sapiens*. The archaeological and botanical evidence already assembled on this development is not even. Some parts of the world (the Mediterranean basin, the Nile Valley, temperate Europe, Southwest Asia, North America) have already been extensively studied. Other parts of the world (East Asia, the Indian subcontinent, South America) are much less explored but at least provide us with some critical evidence on the beginning of agriculture in them. In still others (Africa south of the Sahara, most of the tropical parts of the world) archaeobotanical information is still very poor or almost nonexisting. Even so, the available information clearly shows that farming was independently initiated in several geographically and culturally separated "nuclear areas," both in the Old World and in the New World, between 10,000 and 5,000 years before present (B.P.) uncalibrated radiocarbon time.[1] In each such area cultivation seems to have been independently invented, and native wild plant species—largely unique to each area—were introduced and tested in cultivation. Some evolved to be the "founder crops" that initiated farming in these independent places. Once this new way of life had been successfully established, these domesticates also formed the basis for a territorial expansion of the newly formed agricultural populations and for further development in food production.

A. The First Old World Territory

The oldest and the most extensively studied nuclear area is the "fertile crescent" belt of the Near East. In this rather small geographic territory, a string of early Neolithic farming villages started to appear some 9500 years ago and further established themselves in the subsequent 1500 years. Plant remains retrieved from these early archaeological sites show that eight grain plants growing wild in this Near Eastern "arc" were domesticated at these times. Most common in these preceramic Neolithic Near Eastern sites were remains of three cereals: emmer wheat, einkorn wheat, and barley. They were accompanied by pulses (lentil, pea, chickpea, bitter vetch) and by flax. Soon later, by 8500–8000 B.P., signs of domestication of sheep, goat, cattle, and swine appeared in the sites as well, and the Near Eastern "package" for food production was formed.

Once the package was assembled, and the early Neolithic farming villages were soundly established, farmers started to expand. They did so explosively. By 8000 B.P. this type of agriculture was already introduced into Greece, and by 7000 B.P. it had reached the Danubian basin, the Nile Valley, the Caucasus, and Turkmenistan, and soon afterward it was introduced in the Indus basin. By 6500 to 6000 B.P., grain crop agriculture (and cattle and caprine rearing) was already established all over major parts of temperate Europe—from Ukraine in the east to northern France in the west. More or less at the same time, this new technology also appeared in the middle and western parts of the Mediterranean basin as far as south Spain. All over these vast territories of Southwest Asia, the Nile Valley, the Mediterranean basin, and temperate Europe, agriculture was started

by the introduction of the same Near Eastern crops (wheats and barley, and their companion pulses and flax). Only later were additional plants, native to other parts of this huge region, added (e.g., poppy *Papaver sommiferum*) in the west Mediterranean basin.

Some 3000 years after the start of Neolithic grain agriculture, fruit crop horticulture (based on the invention of vegetative propagation) appeared in this region. As with grain crops, the earliest convincing signs of fruit tree cultivation were found in the Near East. Olive, fig, grapevine, and date palm have been cultivated here at least since Chalcolithic times (5500 B.P.). Horticulture, too, spread quickly. By the late Bronze Age, olive, grapevine, and fig (as well as pomegranate and almond) were already used as principal elements of food production in the Aegean belt; and date palm groves flourished in the warm fringes of the Near East and the eastern Mediterranean basin (including Egypt) and extended eastward as far as the Indus Valley.

From the Bronze Age onward there are sound indications of cultivation of vegetables. Melon, watermelon, onion, garlic, leek, and lettuce were the first vegetable crops grown in Mesopotamia or in Egypt. Definite signs of their cultivation appear by 4500 to 3500 B.P. By 2800 to 2000 B.P., the list of Mediterranean and Southwestern Asian vegetable crops had grown considerably; and beet, turnip, cabbage, radish, carrot, parsnip, celery, parsley, and asparagus had also entered cultivation. More or less at the same period (2400–2000 B.P.), a second group of native fruit trees (those in which cultivation depends on grafting) was also added. Most conspicuous among them are apple, pear, plum, and cherries. Contrary to the earlier crops that were almost all taken into cultivation in the Near East "arc," many of the vegetables and the late fruit trees were probably picked up for domestication, not in the Near Eastern nuclear area but in other parts of the already vast Mediterranean and Southwest Asian system of agriculture. Thus, starting in the Neolithic and ending in classical times, an impressive assemblage of native crops were domesticated and diffused all over Southwest Asia, the Mediterranean basin, and temperate Europe. Most of them remained economically important until today.

B. The Second Old World Territory

A second Asian nuclear area—or more exactly two closely situated, but independent initiations of farming—have been discovered in East Asia. The first was uncovered in the loess soil belt in Honan province and adjacent areas, in the middle part of the Yellow River Valley in North China. The second in the Hupei basin in Central China, along the middle Yangtze River, and also in the wetlands of its delta. The evidence available on the start and on the subsequent development of agriculture in China is still fragmentary compared to that available on the Near East. Yet very ancient farming villages have been discovered both in the northern and in the central parts of this large country. They represent the oldest undisputed signs of plant domestication found yet in East and in South East Asia.

In the Yellow River Valley the oldest farming sites date from 8000 to 7000 B.P. and they contain very large amounts of foxtail millet, indicating that in temperate North China this cereal had been the principal founder crop of the local agriculture. Also in later times foxtail millet continues to be a major food crop in China. More or less in the same time, farming villages seem to have appeared in the middle of the Yangtze River Valley. The earliest sites discovered there date even somewhat older (8500–7000 B.P.) than those in north China. Significantly, in these sites the dominant crop is rice, which later emerges as the most important crop of the vast agricultural economies of East and Southeast Asia. Massive remains of this cereal were unearthed in the Hupei basin, and they continue to appear in the Yangtze River Valley and in later archaeological contexts. These are the oldest remains of rice found yet. But because they occur on the fringe of the wild rice distribution area, some workers suspect that even older rice sites will be discovered in the future further south.

C. The Third Old World Territory

A third territory in the Old World suspected to have had an independent initiation of agriculture is the Savanna belt and its forest margin in Africa south of the Sahara. The traditional agriculture in this region (which stretches from Senegal and Guinea to Sudan) is based, almost entirely, on a rich assemblage of native crops such as sorghum, pearl millet, African rice, fonio, cowpea, and bambara groundnut (Appendix 1). The highlands of Ethiopia border the Savanna belt in the east and also harbor an impressive list of local crops, such as teff, finger millet, coffee, and noog. But here the African plants grow side by side with wheat, barley, and many other Near Eastern elements that reached Ethiopia in the past. All in all, African agriculture is diverse and adapted to a wide range of situations. Apart from the Near Eastern elements in Ethiopia, its rich crop assemblage seems almost fully indigenous. This is borne by the fact that the wild relatives of the crops are restricted in their distribution to this part of the World. Most of them could have been introduced into

cultivation only somewhere in this geographic belt. All this seems to suggest the independent initiation of cultivation. Whether this is indeed the case is yet hard to say. Archaeologically, Africa south of the Sahara is practically unexplored. The few data available on cultivation of pearl millet or other crops are no more than 3000 years old.

D. The American Territory

There is little doubt that farming was independently invented in the Americas. As already noted, in pre-Columbus times the Eastern hemisphere and the Western hemisphere shared almost no crop in common. On basis of this evidence it is hard to imagine agricultural contacts between the two landmasses before that time. Also in America itself archaeological research has revealed definite signs of early cultivation, not in a single area but in several distantly located places.

Relatively early rich finds (the age of which has been recently corrected by direct ^{14}C accelerator mass spectrometry tests to 4700 B.P. onward) were discovered in several caves and rock shelters in central and SW Mexico. They contain numerous remains of what are clearly primitive cultivated maize forms, as well as common bean and squash. Because the wild progenitor race of maize—the principal crop in this triad—is restricted only to this general area, it seems reasonable to regard southwestern Mexico as a nuclear area where maize (and its two accompanying crops) were taken into cultivation. This crop combination also remained the main source for food production in the Aztec days.

Indications of early, possibly independent initiation of cultivation come from South America. Chemical comparisons have shown that the common bean and lima bean were independently domesticated both in middle and in South America. Finds in the central and south Andes show that potato and Quinua cultivation in these elevated areas might be 5000 years old, although dating is yet problematic. Finally, sound evidence has been assembled that independent domestication of plants took place in eastern North America. Here goosefoot *Chenopodium berlandieri,* sunflower, and marsh elders were cultivated at least 4000 to 3500 years ago. This is some 2000 years before maize or any other outsider crop reached this area.

IV. REPRODUCTIVE SYSTEMS IN PLANTS UNDER DOMESTICATION

Cultivated plants differ markedly from wild plants in the way by which they are reproduced and maintained.

In the various families of the flowering plants, cross-pollination is the principle genetic system. Most species build out-breeding populations. Other reproductive systems such as self-pollination and vegetative propagation (including apomixis) occur as well and are even common in some genera and families. Yet compared to sexual reproduction by cross-pollination, their overall weight is small. In contrast, cross-pollination is relatively rare in plants under domestication. Most crops are maintained by one of the following two systems: (a) self-pollination and (b) vegetative propagation. In contrast to cross-pollination, both these reproductive systems are effective in bringing about immediate "fixation" of desired genotypes.

A. Self-pollination

Self-pollination, or more exactly almost full self-pollination, is the principle mating system found in grain crops and in many vegetables. The majority of the 50–60 main grain crops of the world are predominantly self-pollinated. Only a few (such as maize, rye, pearl millet, buckwheat, or scarlet runner bean) are cross-pollinated. Now that the wild progenitors of the majority of the grain crops are already satisfactorily identified, we know that the wild ancestors of the self-pollinated crops are also self-pollinated. In other words, self-pollination in grain crops did not evolve under domestication. It is rather a "preadaptation" of the wild ancestor, which considerably enhances its chance to perform successfully in cultivation. One major advantage of self- over cross-pollination is the fact that self-pollination isolates the crop reproductively from its wild progenitor. It enables the farmer to grow a desired genotype in the same area in which the wild relatives abound without endangering the identity of the cultivar by genetic swamping. A second advantage of self-pollination lies in the genetic structure maintained within the crop. Self-pollination results in splitting the crop's gene-pool into independent homozygous lines. Variation is thus structured in the form of numerous true breeding cultivars. Because they are automatically "fixed" by the pollination system, they can be easily maintained by the farmer, even if they are planted together. In contrast, the preservation of varietal identity in cross-pollinated plants is much more problematic. It requires repeated selection towards the desired norms, constant care to avoid the mixing of types and the prevention of contamination from undesirable plants. It is therefore not surprising that early successes in plant domestication frequently involved selfers. In fact, all the eight Neolithic

Near Eastern founder crops have this system of reproduction.

B. Vegetative Propagation

Vegetative propagation is the second widely adapted means to fix and maintain desired genotypes under domestication. This way of handling prevails in the fruit trees, in tuber and corm crops, and in numerous ornamentals. Here, domestication means first of all changing the reproductive biology of the plants involved by shifting from sexual reproduction (in the wild) to vegetative propagation (under cultivation). As a rule, cultivated varieties of these plants have been maintained as clones by cuttings, rooting of twigs, suckers, or by the more sophisticated technique of grafting (and recently also by meristem tissue culture propagation). This is in sharp contrast with their wild progenitors, which reproduce from seeds. Their wild growing populations are usually variable, maintain themselves through sexual reproduction, and are distinctly cross-pollinated. Consequently, seedlings raised from any mother plant segregates widely in numerous traits, including the size, shape, and palatability of the fruits or the tubers.

In the hands of the grower, vegetative propagation has been a powerful device to prevent genetic segregation and to "fix" desired genotypes. By discarding sexual reproduction and inventing clonal propagation, the farmer was able to select, in a single act, exceptional individuals with desirable traits from among a large numbers of variable, inferior plants. Moreover, the farmer could duplicate the chosen types to obtain genetically identical saplings. This is no small achievement. Because the plants are cross-pollinated and widely heterozygous, most progeny obtained from them (even from "superior" plants) are economically worthless. The change from seed planting to vegetative propagation has been a practical solution to assure a dependable supply of desired genotypes. In most fruit trees—and in many tuber crops and ornamentals—it made domestication possible. The history of the Mediterranean fruit trees illustrates this clearly. Olive, grapevine, fig, and date palm that can be relatively simply cloned (by twig rooting or by suckers) were the first to enter cultivation (already by 5500 B.P.). Other native fruit trees that do not lend themselves to easy rooting (e.g., apple, pear, plum, cherry) were domesticated much later (about 2400–2000 B.P.). Their incorporation into horticulture became possible only after the invention of a new cloning tool, namely scion grafting.

V. CONSCIOUS VERSUS UNCONSCIOUS SELECTION

Two main types of selection operate (and complement each other) in plants under domestication:

1. Conscious selection deliberately applied by the growers for traits of interest to them.
2. Unconscious selection brought about by the fact that the plants concerned were picked up and isolated from their original wild environment and placed in a new (and usually very different) human-made environment. This shift in the ecology led automatically to drastic changes in selection pressures. In response to the introduction of the plants into the anthropogenic environment, numerous adaptations vital for survival in the wild lost their fitness. New sets of traits were automatically selected for to fit the new conditions, resulting in the buildup of characteristic domestic adaptations and the appearance of "domestication syndromes"—each fitting the specific agricultural system provided by the domesticators.

Although the pressures and the effects of conscious selection in shaping domesticated plants are familiar to both plant breeders and crop plant evolutionists, unconscious selection has been less frequently evaluated. Because this type of selection shapes so many of the principal traits that characterize crops, several developments caused by the shift in the ecology are discussed in some detail in the following sections.

VI. METHODS OF MAINTENANCE OF CROPS AND THEIR IMPACT

As already noted, two main methods of maintenance are employed by the cultivators to maintain their crops: (a) planting of seeds and (b) vegetative propagation. The choice between these two agronomic practices is also the choice between two contrasting patterns of selection and modes of evolution under domestication.

With very few exceptions of apomixis (such as nucellar seed in citrus fruits and mango) planting of seeds can be equated with sexual reproduction. Cultivated plants maintained by seeds (the bulk of the grain crops, numerous vegetables and forest trees, some ornamentals) therefore undergo a recombination-and-selection cycle every planting. In other words, such crops had, under domestication, hundreds or even thousands of generations of selection. They have been continually molded either as (a) clusters of inbred lines (in predominantly self-pollinated crops) or as (b) distinct cultivated races (in cross-pollinated plants). In numerous sexually reproducing crops, the results of such repeated selection are indeed striking. Under domestication these crops have diverged considerably from their wild progenitors. They are now distinguished from them by numerous morphological, developmental, physiological, and chemical traits.

Crops maintained by vegetative propagation (most fruit trees and tuber crops, few vegetables, many ornamentals) have had an entirely different history of selection. Cultivars in these crops are not true races but only clonal replications of exceptional individuals, which are as a rule highly heterozygous. They had been picked up by the cultivator from among numerous sexually produced variable individuals and "fixed" by cloning. In terms of selection, the development of a vegetatively propagated cultivar is largely a single-step operation. With the exception of rare somatic mutations, selection is completed the moment the clone is picked up. In traditional fruit tree horticulture, the turnover of clones has been low. Appreciated genotypes were maintained for long periods of time. Thus, clonal crops underwent in cultivation only a few recombination-and-selection cycles. In sharp contrast to sexually reproducing cultivated plants, their cultivars do not represent true breeding races but only clones, which, as a rule, are highly heterozygous and segregate widely when progeny tested. Significantly, the majority of such segregating progeny are not only economically inferior or even worthless, but they often regress toward the means found in wild populations, showing striking resemblance to the wild forms.

VII. THE PURPOSE FOR WHICH THE PLANT IS BEING GROWN

In different crops, different parts of the plant body are being used. Some are raised for their vegetative parts (tubers, leaves, stems, etc.). In others the reproductive parts (inflorescences, flowers, fruits, seeds) constitute the used products. Also the choice of the desired plant parts introduces automatically contrasting selection pressures, particularly in regard to the reproductive system of the crops involved.

When crops are *grown for their seeds* (or at least when they are reproduced by seed), they are kept (like their wild relatives) under constant stabilizing selection that keeps their reproductive system intact. Grain crops provide us with the most rigid cases of such normalizing selection. Yields in these crops depend decisively pon the streamlined development of flowers and fruits, normal chromosome pairing in meiosis, and full fertility. Deviants are weeded out automatically and the reproductive system is kept in balance. It is no wonder that among cultivated plants, grain crops are the most conservative in this regard. They are characterized by strictly balanced chromosome systems and show very little chromosome "problems" under domestication. With very few exceptions (such as the formation of hexaploid bread wheat), the chromosome complements in cultivars of grain crops are identical to those found in their wild progenitors.

Drastic changes in seed fertility (as well as in the chromosome system) can be tolerated when the plants are *grown for their fruits* but are *maintained by vegetative propagation.* Crops in this group (the bulk of the fruit trees) do not depend on seed fertility for their maintenance. Yet they have to keep the basic reproductive growth functions to assure the development of fruits. Moreover, growers usually prefer types in which the size and the number of stones or pips had been reduced, or seedless fleshy fruits. Several solutions for reducing seed fertility without harming fruit set evolve automatically under cultivation. They include the establishment of polyploid clones, some of them meiotically unbalanced (e.g., triploidy in some pears and in bananas), or the incorporation of mutations inducing parthenocarpy—that is, the induction of fruit development without fertilization and without seed set (e.g., bananas, fig, some pears).

Crops *grown for their vegetative parts* and *maintained by vegetative propagation* exhibit the most drastic disruption of their reproductive system and the most variable and bizarre chromosomal situations among cultivated plants. Because such crops are clonally propagated, the conscious selection exerted on them by the growers to increase the output of the desired vegetative part is rarely counterbalanced by normalizing selection to retain sexual reproductive functions. Tropical root and tuber crops provide us with outstanding examples for this development under domestication. Cultivated clones of cassava, yam, sweet potato, or garlic frequently show drastic reduction in flowering. In some, flowering ceases altogether, or almost altogether. When flowers develop they are frequently sterile. Also chromosomally many of these crops are highly polymorphic and frequently contain clones with different levels of polyploidy or variable aneuploid chromosome numbers. Clones may contain 3x, 5x, or even higher

meiotically unbalanced chromosome complements. Thus in the yams, *Dioscorea alata* shows all chromosome levels between 3x to 8x; whereas in *D. esculenta,* 4x, 6x, 9x, and 10x forms are known. Sugarcanes confront us with even more complex chromosome pictures. Cultivated clones in this crop are all highly polyploid and frequently aneuploid. Modern cultivars range from 2n = 100 to 2n = 125 chromosomes. Older cultivars vary from 2n = 80 to 2n = 124.

VIII. THE IMPACT OF SOWING AND REAPING

Traditional grain agriculture is based on *sowing* the seeds of the crop in a tilled field, *reaping* the reproductive parts of the mature plants, and *threshing* out the grains. The introduction of grain plants into such a farming practice initiated, automatically, selection toward the following changes, setting them apart from their wild counterparts.

1. Most conspicuous is the unconscious selection for mutants in which the mature seed is retained on the mother plant (i.e., for the breakdown of the wild mode of seed dissemination). In cereals this implies a shift from shattering spikes or panicles (in wild forms) to nonshattering ones (in plants under domestication). In pulses it usually means the prevention of the pod from dehiscing and from shedding the seeds. In wild wheat, wild barley, or wild maize, the seed dispersal unit comprises a single internode of the ear. Full disarticulation of the ear, at each segment, is thus an essential element for wild type dispersal, and plants in wild populations are constantly selected for quick shattering of their mature spikes. In contrast, the introduction of planting and harvesting brings about automatic, unconscious selection in exactly the opposite direction. Under the new system a sizable proportion of seed produced by brittle plants will shatter and would not be included in the harvest, whereas all grains produced by nonshattering mutants "wait on the stalks" to be reaped by the grower. Under cultivation, nonshattering individuals have therefore a much better chance to contribute their seed to the subsequent sowing. To summarize, nonbrittle mutants that were totally disadvantaged under wild conditions became highly successful under the new system. Thus, when wild cereals are introduced into the system of sowing and reaping, one should expect selection for nonbrittle forms whether or not the cultivator is aware of this trait. Furthermore, the incorpora-

tion of nonbrittle mutants makes the crops fully dependent on humans, as nonshattering plants lose their seed dispersal ability and can no longer survive under wild conditions. For this reason, the presence of nonshattering remains of grain crops in archaeological contexts provide critical indication that these plants were under cultivation. Finally, both theoretical considerations and experimental evidence seem to suggest that at least in wheats and barley, the establishment of nonshattering mutants under the system of sowing and reaping was very likely a fast process. The shift could have been accomplished in the course of several scores of generations.

2. A second major outcome of introducing wild grain plants into this regime of cultivation is the breakdown of the wild mode of their seed germination. Most wild plants, especially annuals, depend for their survival on regulation of their germination in space and time. A common and vital adaptation is a delay of germination and its spread over two or more years. Again, under the system of sowing and reaping such inhibition of germination is automatically selected against. Most grain crops have lost their wild-type germination inhibition patterns. Their seeds germinate fully whenever the farmer plants them.

3. Several other traits characterize cultivated grain crops and are part of the domestication syndrome under sowing and reaping conditions. Because of a rather dense stand in the cultivated field there is a stress to develop forms with erect habit and to reduce the growth of side tillers or branches. Because of the mode of harvest there is a selection for synchronous ripening. In cases of biennial or perennial wild stocks, there is an automatic selection for the shift to annuality. Because the seeds stay protected in the granary, thick shells are selected against. Finally because tilling increases soil fertility there is also a pressure to increase the number of fertile flowers in the reproductive parts of the crops.

IX. APPENDIX 1: CROPS OF VARIOUS REGIONS

This appendix lists the native crops of the various traditional agricultural regions of the world (see Section II). For illustrations and short notes on the various food crops, consult Vaughan and Geissler (1997). The important crops of the world and their evolution are surveyed in Smartt and Simmonds (1995).

SOUTHWEST ASIA, MEDITERRANEAN BASIN, AND TEMPERATE EUROPE

Cereals

Emmer and durum-type wheats *Triticum turgidum*
Einkorn wheat *Triticum monococcum*
Bread wheat *Triticum aestivum*
Barley *Hordeum vulgare*
Rye *Secale cereale*
Common oat *Avena sativa*

Pulses

Lentil *Lens culinaris*
Pea *Pisum sativum*
Chickpea *Cicer arientinum*
Faba bean (broad bean) *Vicia faba*

Oil or Fiber

Flax *Linum usitatissimum*
Oil seed turnip *Brassica rapa*

Fruits and Nuts

Olive *Olea europaea*
Grapevine *Vitis vinifera*
Fig *Ficus carica*
Date palm *Phoenix dactylifera*
Apple *Malus pumila*
Pear *Pyrus communis*
European plum *Prunus domestica*
Sweet cherry *Prunus avium*
Almond *Amygdalus communis*
Walnut *Juglans regia*

Vegetables and Spices

Melon *Cucumis melo*
Watermelon *Citrullus lanatus*
Onion *Allium cepa*
Garlic *Allium sativum*
Leek *Allium porrum*
Lettuce *Lactuca sativa*
Beet *Beta vulgaris*
Turnip *Brassica rapa*
Cabbage *Brassica oleracea*
Carrot *Daucus carota*
Celery *Apium graveolens*
Parsley *Petroselinum sativum*
Asparagus *Asparagus officinalis*
Cumin *Cuminum cyminum*
Coriander *Coriandrum sativum*
Fenugreek *Trigonella foenum-graecum*

EAST ASIAN AGRICULTURE

Cereals and Cereal-Like Grain Crops

Rice *Oryza sativa*
Foxtail millet *Setaria italica*
Broomcorn millet *Panicum miliaceum*
Buckwheat *Fagopyrum esculentum*

Pulses

Soybean *Glycyne max*

Oil or Fiber

Hemp *Cannabis sativa*

Vegetables and Spices

Pe Tsai (Peking cabbage) *Brassica pekinensis*
Pak-choi (Chinese cabbage) *Brassica chinensis*
Rakkuyu *Allium chinense*
Chinese chives *Allium tuberosum*
Ginger *Zingiber officinale*
Chinese pickling melon varieties *Cucumis melo*

Fruits and Nuts

Chinese chestnut *Castanea henryi*
Chinese white pear *Pyrus bretschneideri*
Chinese sand pear *Pyrus pyrifolia*
Loquat *Eriobotrya japonica*
Oriental persimmon *Diospyros Kaki*
Litchi *Litchi chinensis*
Apricot *Prunus armeniaca*
Peach *Prunus persica*

Corms and Tubers

Cinese varieties of turnip *Brassica rapa*
Chinese yam *Dioscorea esculenta*
Lotus *Nelumbium speciosum*

Stimulants

Tea *Camelia sinensis*

Pulses

Sword bean *Canavalia gladiata*
Mung bean *Vigna radiata*
Black gram *Vigna mungo*
Rice bean *Vigna calcarata*

Oil or Fiber

Sesame *Sesamum indicum*
Tree cotton *Gossypium arboreum*
Coconut *Cocus mucifera*
Jute *Corchorus capsularis*

Fruits and Nuts

Banana: cultivated derivatives of *Musa accuminata* and
    M. balbisiana
Citrus fruit crops (several species)
Carambula *Averrhoa carambula*
Jackfruit *Artocarpus integrifolia*
Durian *Durio zibethinus*
Mangosteen *Garcinia mangostana*
Rambutan *Nephelium lapaceum*

Vegetables and Spices

Cucumber *Cucumis sativus*
Egg plant *Solanum melongena*
Black pepper *Piper nigrum*
Cardamom *Elettaria cardamomum*
Tumeric *Cucurma longa*
Nutmeg *Myristica fragrans*
Cloves *Syzygium aromaticum*
Ginger *Zingiber officinale*

Corms and Tubers

Taro *Colocasia esculenta*
Greater yam *Dioscorea alata*
East Indian arrowroot *Tacca leontopetaloides*

Sugar

Sugar-cane *Saccharum officinarum*

SOUTHEAST ASIAN AGRICULTURE
(including the Indian subcontinent)

Cereals

Rice *Oryza sativa*
Job's tears *Coix lachryma-jobi*

SUB-SAHARAN AFRICAN
AGRICULTURE

Cereals

Sorghum *Sorghum bicolor*
Pearlmillet *Pennisetum glaucum*

Finger millet *Eleusine coracana*
Fonio *Digitaria exilis*
Teff *Eragrostis tef*
African rice *Oryza glaberrina*

Pulses

Cowpea *Vigna unguiculata*
Bambara groundnut Voandzeia subterranea

Oil or Fiber

Oil palm *Elaeis guineensis*
Noog *Guizotia abyssinica*
Old Word cotton *Gossypium herbaceum*

Vegetables and Spices

Gherkin *Cucumis anguria*

Corms and Tubers

Yam *Dioscorea cayenensis*
Enset *Ensete ventricosa*

Stimulants

Coffee *Coffea arabica*

AMERICAN AGRICULTURE

Cereals and Pseudo-Cereal Crops

Maize *Zea mays*
Amaranth *Amaranthus cruentus*
Amaranth *Amaranthus caudatus*
Quinua *Chenopodium quinoa*
Goosefoot, *Chenopodium berlandieri*
Marsh elder *Iva annua*

Pulses

Common bean *Phaseolus vulgare*
Lima bean *Phaseolus lunatus*
Scarlet runner bean *Phaseolus coccineus*
Tepary bean *Phaseolus acutifolius*
Jack bean *Canavalia ensiformis*
Peanut (groundnut) *Arachis hypogea*

Oil or Fiber

Sunflower *Helianthus annuus*
Island cotton *Gossypium hirsutum*
Upland cotton *Gossypium barbadense*

Fruits and Nuts

Avocado *Persea americana*
Prickly pears *Opuntia ficus-indica*
Sapodilla *Manilkara zapota*
Guava *Psidium guajava*
Passion fruit *Passiflora edulis*
Papyia *Carica papaya*
Pineapple *Ananas comosus*
Cherimoya *Anona cherimola*

Vegetables and Spices

Tomato *Lycopersicum esculentum*
Peppers *Capsicum annuum* and *C. frutescens*
Squashes and pumpkins *Cucurbita pepo, C. maxima,*
  C. moschata, C. mixta and *C. ficifolia*
Vanilla *Vanilla fragrans*

Corms and Tubers

Potato *Solanum tuberosum*
Sweat potato Ipomoea batatas
Oca *Oxalis tuberosa*
Ysanu *Tropaeolum tuberosum*
Ulluco *Ullucus tuberosus*
Cassava (manioc) *Manihot esculenta*
Cush-cush Yam *Dioscorea trifida*

Stimulants

Cacao *Theobrama cacao*
Tobacco *Nicotiana tabacum*
Coca *Erythroxylon coca*

See Also the Following Articles

AGRICULTURE, TRADITIONAL • BREEDING OF ANIMALS • BREEDING OF PLANTS • CROP IMPROVEMENT AND BIODIVERSITY • EDIBLE PLANTS • PLANT SOURCES OF DRUGS AND CHEMICALS

Bibliography

Harlan, J. R. (1992). *Crops and man* (2nd ed.). Madison, WI: American Society of Agronomy.
Smartt, J., and Simmonds, N. W. (1995). *Evolution of crop plants* (2nd ed.). Harlow, UK: Longman.

Smith, B. D. (1995). *The emergence of agriculture*. New York: Scientific American Library.

Smith, N. J. H., Williams, J. T., Plucknett, D. L., and Talbot, J. P. (1992). *Tropical forests and their crops*. Ithaca, NY: Cornell University Press.

Vaugham, J. G., and Geissler, C. A. (1997). *The new Oxford book of food plants: A guide to the fruit, vegetables, herbs and spices of the world*. Oxford: Oxford University Press.

Zohary, D., and Hopf, M. (2000). *Domestication of plants in the Old World: The origin and spread of cultivated plants in West Asia, Europe and the Nile Valley* (3rd ed.). Oxford: Oxford University Press.

ECOLOGICAL FOOTPRINT, CONCEPT OF

William E. Rees
UBC School of Community and Regional Planning

GLOSSARY

carrying capacity Usually defined as the average maximum number of individuals of a given species that can occupy a particular habitat without permanently impairing the productive capacity of that habitat (see also *human load*).

competitive exclusion The displacement of one species from its habitat or ecological niche by another. When humans appropriate other species' "ecological space," it often leads to the local or even global extinction of the nonhuman organism.

ecological deficit An ecological deficit exists when the "load" (see *human load*) imposed by a given human population on its own territory or habitat (e.g., region, country) exceeds the productive capacity of that habitat. In these circumstances, if it wishes to avoid permanent damage to its local ecosystems, the population must use some biophysical goods and services imported from elsewhere (or, alternatively, lower its material standards).

human load The total "human load" imposed on the "environment" by a specified population is the product of population size times average per capita resource consumption and waste production. The concept of 'load' recognizes that human carrying capacity is a function not only of population size but also of aggregate material and energy throughput. Thus, the human carrying capacity of a defined habitat is its maximum sustainability supportable load.

overshoot A population is in overshoot when it exceeds available carrying capacity. It may survive temporarily but will eventually crash as it depletes vital natural capital (resource) stocks. A population in overshoot may permanently impair the long-term productive potential of its habitat, reducing future carrying capacity.

patch disturbance The measurable habitat and ecosystem modification caused by large animals, including humans, as they forage for food or other resources. Patch disturbance is most pronounced near the den site, temporary camp, or other central place within the overall home range of the individual or group.

sustainability gap The global ecological deficit—that is, the difference between any excessive human load on the ecosphere and the long-term carrying (or load-bearing) capacity of the planet.

ECOLOGICAL FOOTPRINT ANALYSIS is a quantitative tool that represents the ecological load imposed on the earth by humans in spatial terms. Thus, the ecological footprint of a defined population is *the total area of land and water ecosystems required to produce the resources that the population consumes, and to assimilate the wastes that the population generates, wherever on earth the land/water are located.* Ecofootprinting can be used to assess the ecosystem area effectively "appropriated" in support of any specified human population or economic activity. We can then compare this to available productive area (i.e., load-bearing capacity). The size and nature of the human ecological footprint is relevant to biodiversity conservation because energy and material resources extracted from nature to serve human purposes are irreversibly unavailable to other species. The larger the human footprint, the less nonhuman biodiversity. This article describes the contemporary context for ecofootprint analysis and the theory on which it is based. It then discusses the implications of the analysis for sustainable development in general and for the future of biodiversity in particular.

I. THE CONCEPTUAL FRAMEWORK: WHY DO WE NEED ECOFOOTPRINT ANALYSIS?

A. Human Activity and Global Ecological Change

There is little disagreement among scientists that humanity has assumed a major role in changing the face of the earth. Up to one-half of the land has been directly transformed by human action; more than half of the planet's accessible fresh water is already being used by people; atmospheric carbon-dioxide has increased by 30% in the industrial era; more atmospheric nitrogen is fixed and injected into terrestrial ecosystems by humans than by all natural terrestrial processes combined; two-thirds of the world's major fisheries are fully- or overexploited; and biodiversity losses are accelerating (Lubchenco, 1998, Vitousek *et al.*, 1997).

What *is* in dispute is whether any of this really matters to socioeconomic sustainability. The more pessimistic analysts see human domination of the earth and the deterioration of nature as ultimately fatal to civilization itself. Various mechanisms are invoked in coming to this conclusion. For example, because of our culture's near total dependence on liquid petroleum and the present lack of affordable substitutes, energy analyst Richard Duncan (1993) argues that the total life expectancy of western technoindustrial society is less than 100 years—counting from the 1930s! Professor W. M. Hern attacks industrial culture's obsession with growth and compares humanity to a self-destructive cancer on the planet. The continuous expansion of the human enterprise apparently "exhibits all four major characteristics of a malignant process: rapid uncontrolled growth; invasion and destruction of adjacent tissues (ecosystems, in this case); metastasis (colonization and urbanization, in this case); and dedifferentiation (loss of distinctiveness in individual components)" (Hern, 1997). According to geographers J. W. Smith and G. Sauer-Thompson, environmental decay coupled with geopolitical stresses induced by resource scarcity, will inevitably cause industrial society to "self-destruct, producing massive ecological damage, social chaos, and megadeath" (Smith and Sauer-Thomson, 1998). From these perspectives, the human enterprise has already overshot global carrying capacity, is eroding future options, and will inevitably crash—there is virtually nothing we can do to prevent the collapse of global civilization.

At the other extreme, we can take comfort in the view of those scientists and economists who believe that human ingenuity will prevail. Their logic is simple. Stimulated by incipient scarcity, rising prices, and the opportunity for profit, scientists and entrepreneurs will find technological substitutes for exhaustible resources and natural processes. As energy economist R. L. Gordon explains, "the tendency to technical progress is viewed as the most critical economic law involved." And because human ingenuity has historically been remarkable at increasing living standards and warding off the pressures of resource depletion, "the immediate need for avoiding depletion is nil" (Gordon, 1994).

Perhaps the most extreme expression of this belief is revealed in the ebullient optimism of the late Julian Simon who wrote: "Technology exists now to produce in virtually inexhaustible quantities just about all the products made by nature" and "We have in our hands now ... the technology to feed, clothe, and supply energy to an ever-growing population for the next seven billion years" (Simon, 1995). Some conventional economists even argue the benefits of continuous population growth on grounds that "additional people can create more resources than they use up, thanks to technological improvements" (Block, 1990). Clearly, from the techno-optimists' perspective, there are no practical limits at all on human carrying capacity—we don't need to do anything to sustain global civilization but stay our present, increasingly market-driven, global development course.

With such diametrically opposing assessments of the human prospect, both starting from present circumstances and held with equal conviction, it is little wonder that ordinary citizens and policy makers alike are confused about global ecological change and what—if anything—to do about it. This chapter assesses the overall sustainability conundrum using ecological footprint analysis and links the general problem to the specific issue of biodiversity conservation. First, however, we examine the origins of the ongoing perceptual conflict.

B. Dueling Paradigms

Many of the conflicting views about "sustainable development" can be traced to differing fundamental beliefs, values, and assumptions about the nature of reality, particularly humankind-environment relationships. Different societies and interest groups have different preanalytic visions (or worldviews or paradigms), which shape how they interpret any set of facts or circumstances.

Remarkably, people are often unaware that they possess a particular worldview—each person acquires his or her culturally specific paradigm unconsciously, simply by living and growing up in a particular sociocultural milieu. We may therefore be unconscious of the subtle ways in which the prevailing worldview shapes our approach to critical issues or that there may be more viable alternatives. When we think that "the world is flat" was once accepted as self-evident truth, it raises the unsettling possibility that much of our present cultural worldview consists largely of shared illusions.

The dominant social paradigm in western (and increasingly global) technoindustrial culture can be characterized as the "expansionist" or "cornucopian" worldview. As we have already seen, its adherents believe that humankind has achieved mastery over the natural world and that technology will be able to compensate for the depletion of natural resources and the loss of life-support services.

The economics of expansionism is the neoclassical (neoliberal) economics that has come to dominate geopolitics in the past 20 years. Neoclassical models generally represent the economy as a self-regulating and self-sustaining mechanical system whose productivity and growth are not seriously constrained by nature. In effect, conventional economics treats the economy as separate from, and more or less independent of, "the environment." People can therefore be treated solely as economic agents.

By contrast, ecologists, environmentalists, and many pessimists profess an ecological or steady-state worldview. From this perspective, the economy, far from being a separate system, is seen as an inextricably integrated, completely contained, and wholly dependent growing subsystem of a nongrowing ecosphere. People are not only economic agents but also ecological agents.

The ecological perspective recognizes that the so-called environmental crisis is really a human ecological crisis. The distinction is by no means trivial. The former term externalizes the problem, effectively blaming it on a defective environment, which then needs to be fixed. By contrast, the latter term traces the problem to its source—the nature and behavior of people themselves—and suggests that it is the latter that need "fixing." This is the starting point for ecological footprint analysis.

C. Human Ecology/Economy and the "Second Law"

If humans are ecological entities then human activity is governed by natural laws. In particular, economic production requires continuous, irreversible energy and material transformations, and these transformations are ultimately governed by the second law of thermodynamics. This fundamental law is ignored by conventional economic models.

In its simplest form, the second law states that any isolated system will tend toward equilibrium; alternately, the "entropy" of any isolated system always increases. Available energy spontaneously dissipates, concentrations disperse, gradients disappear. An isolated system thus becomes increasingly unstructured in a inexorable slide toward thermodynamic equilibrium. This is a state of maximum entropy in which there is no structure and nothing can happen.

The second law was originally formulated for simple isolated systems close to equilibrium. We now recognize, however, that even complex open systems are subject to the forces of entropic decay. *Any* differentiated far-from-equilibrium system has a natural tendency to erode and unravel.

But not all complex systems do disintegrate in this way. Many biophysical systems, from individual fetuses to the entire ecosphere, actually gain in organizational complexity and mass over time (i.e., they *increase* their distance from equilibrium). This seemingly paradoxical behavior can readily be reconciled with the second law. Biophysical systems exist in loose, nested hierarchies, each component system being contained by the next level up and itself comprising a chain of linked subsystems at lower levels. Living systems can therefore import available energy and material (essergy) from their host environments and use it to maintain their internal integrity against entropic decay and to grow. They also export the resultant waste (entropy) back into their hosts. In effect, modern formulations of the second law posit that all highly ordered systems develop and grow (increase their internal order) at the cost of destroying order at higher levels in the systems hierarchy. Because such systems maintain themselves by continuously degrading and dissipating available energy and matter, they are called "dissipative structures."

Now, the human economy is clearly one such highly ordered, complex, dissipative structure. Both our biological and our industrial metabolisms require enormous inputs of high-grade energy and material resources from the rest of the ecosphere for their maintenance and growth. Indeed, from the ecological perspective, humans are strictly secondary producers— all production by the human enterprise, from the increase in population to the accumulation of manufactured capital, requires the consumption of a much larger quantity of energy and material *first produced by nature* (although we can sometimes increase the amount of incoming solar energy captured by photosythesis through irrigation, or by building terraces for agriculture, for example).

Moreover, recall that within the global hierarchy, the economy is a dependent subsystem of the ecosphere. In effect, the expanding human enterprise is positioned to consume the ecosphere from within. Beyond certain limits, continuous population and material economic growth is, therefore, inherently self-destructive. This reality suggests a necessary second law condition for sustainability: material consumption and waste production by the economy must be no greater than the resource production and waste assimilation capacity of the ecosphere.

D. *Homo Sapiens*: The Archetypal Patch Disturbance Species

Evidence to support the ecological view has accumulated throughout history. Even preagricultural humans had significant effects on local ecosystems' structure and function and on the biodiversity of their habitats.

This was the inevitable consequence of the second law combined with two additional facts of human biology: human beings are large animals with correspondingly large individual energy and material requirements (i.e., we are big consumers), and humans are social beings who live in extended groups. Consequently, whenever human hunter-gatherers invaded a previously "stable" ecosystem, their material demands would significantly alter established energy and material pathways, benefiting some nonhuman species and harming others. People invariably perturb or disturb the systems of which they are a part.

Perhaps the most dramatic evidence is the permanent systemic changes that occur when humans first invade and settle a new habitat. Consider biodiversity loss. The recent paleoecological, anthropological, and archeological literature tells a convincing story of the extinctions of large mammals and birds that accompanied first contact and settlement of their habitats by human beings (Diamond, 1992; Flannery, 1994; Ponting, 1991). It seems that everywhere on earth that paleontologists have studied and that humans first reached within the past fifty thousand years, human arrival approximately coincided with massive prehistoric extinctions. In North America, South America, and Australia, about 72, 80, and 86%, respectively, of large mammal genera ultimately became extinct after human arrival. Scientists estimate that with only Stone Age technology, the Polynesians exterminated more than 2000 bird species, about 15% of the world total.

All this is to emphasize that humans are, *by nature*, a patch-disturbance species, a distinction we share with other large mammals ranging from beavers to elephants. Large animals, due to their size, longevity, and food and habitat requirements, tend to have substantial physical and systemic impacts on the ecosystems that sustain them. A patch-disturbance species may thus be defined as *any organism that, usually by central place foraging, degrades a small "central place" greatly and disturbs a much larger area away from the central core to a lesser extent.*

There is, of course, a major difference between human patch disturbance and that of other species. Because human knowledge and technology are uniquely cumulative, human patch disturbance has been intensifying since the Neolithic. It received a major boost with agriculture and became the dominant force in the ecosphere with the use of fossil fuels and the industrial revolution. (Cheap, plentiful fossil fuels have enabled humans to accelerate the exploitation of everything else.) Today, human patch disturbance is evident on a global scale in the form of such persistent negative trends as greenhouse gas accumulation, increasing climatic variability, ozone depletion, landscape destruction, and accelerating biodiversity loss. Ultimately, the erosion of systems integrity may lead to the irreversible loss of basic life support functions.

Whatever joy one might take from the human symphony, the fact that economic growth now threatens critical biophysical systems sounds a discordant note. Global change implies that economic activity is pressing against the limits of human carrying capacity at the expense of other species. Regrettably, the concept of human carrying capacity has long been rejected by mainstream analysts, but for reasons explained below, it is the organizing principle of ecological footprinting.

II. REVISITING HUMAN CARRYING CAPACITY: THE ROOTS OF ECOFOOTPRINT ANALYSIS

Ecologists who study nonhuman species generally define carrying capacity in terms of the numbers of deer, elk, or other species that a particular habitat type can support indefinitely per unit area. However, because of such factors as climatic variability and community succession, instantaneous carrying capacity is constantly changing. Carrying capacity is therefore not a particularly rigorous concept, even when applied to nonhuman species with simple stable demands on their ecosystems. It is all the more problematic when applied to humans whose demands on the environment are anything but simple and stable.

Indeed, many economists and other cornucopians argue that interregional trade alone is virtually enough to cancel concerns about human carrying capacity. Any region or country that can trade services or surpluses of resource A for needed supplies of resource B need not be limited in numbers or economic growth by the theoretical carrying capacity of its home territory. Furthermore, if trade fails, we can always rely on human ingenuity and technology to increase carrying capacity. These are seemingly powerful arguments, and they explain how such densely populated countries as the Netherlands and the United Kingdom can remain economically vigorous despite having long exceeded their domestic carrying capacities.

But let's examine the assumptions here. The idea that trade acts to increase local carrying capacity treats each trading region as an isolated open system. Looked at in the aggregate the situation becomes cloudier. The world as a whole is a materially closed system. Resources imported to and consumed in region A are no longer available for consumption in the exporting region B, and visa versa. Hence, the exchange may result in a one-time increase in the population of each region taken separately, but it also increases global consumption and may deplete local stocks of natural capital, thereby limiting future options. Thus, although total human load increases, *there is no unambiguous increase in total load-bearing capacity.*

Indeed, in some circumstances unfettered trade can lead to a permanent *loss* of carrying capacity. Global trade exposes pockets of scarce resources everywhere to the largest possible market. This subjects them to ever-greater exploitation pressure, often to the point of depletion or collapse. (Such is the history of trade in fisheries products, for example.) In short, instead of increasing load-bearing capacity, trade often just shuffles it around. This ensures that all countries, their economies happily expanding through trade, hit the (now shrinking) limits to growth simultaneously. Stratospheric ozone depletion is a case in point.

What about human ingenuity? The general argument here is that cumulative knowledge and technological skills enable humans to squeeze more out the environment than nature-in-the-raw can provide. This is achieved mainly by continuously increasing resource productivity and by creating substitutes for any resources that *are* eventually depleted. In these ways, we are able to increase the capacity of any given habitat to support humans and their activities without practical limit.

Let's take these points one at a time. The so-called green revolution, for example, is frequently cited as a miracle of human technological prowess. High-input production agriculture has, in fact, greatly increased the short-term productivity (carrying capacity) of arable land. This has seemingly banished the threat of Malthusian famine and certainly facilitated the modern population explosion. However, this miracle was achieved largely by supplementing renewable sun, soil, and rain with nonrenewable fossil fuel, fertilizer, pesticides, and depletable groundwater at a great cost in soil degradation, biodiversity loss, widespread pollution, and falling water tables. The net effect is a swollen human population, likely in overshoot, and increasingly dependent for survival on nonrenewable artificial inputs, even as pollution mounts and natural long term load-bearing capacity is steadily eroded.

Another potential fix is our demonstrated ability to greatly increase resource productivity, to do more with less. In theory, more efficient technologies should enable a constant energy and material supply to support a given population at a higher material standard, or a higher population at the same material standard, thereby seeming to increase carrying capacity.

Unfortunately, real-world practice has so far betrayed theoretical promise. The steady gains in material efficiency in the post World War II period have been accompanied by both an exploding population *and* steadily increasing consumption. A recent study of resource flows in a selection of the world's most technologically advanced and efficient countries found that although there has been some reduction in the ratio of resource inputs per unit gross domestic product since 1975, there has also been "in most, a gradual rise in *per capita* natural resource use." The study therefore concludes that "meaningful dematerialization, in the sense of an absolute reduction in natural resource use, is not yet taking place" (WRI, 1997, p. 2). In fact, efficiency gains may actually work *against* conservation by lowering prices and raising incomes. This gives people more money to spend on cheaper goods and services so total throughput—human load—keeps expanding.

Finally, overconfidence in resource substitution is also problematic. It is true that human ingenuity has been markedly successful at finding replacements for simple material inputs to production. For example, optical fiber is a product of mind, not of nature, and has displaced much of the demand for copper wire. However, as ecological economist Herman Daly (1991) has emphasized, there are many more circumstances in which technology and nature are complements, not substitutes. More fish boats cannot long substitute for fewer fish; more saws and carpenters are no substitute for less lumber. Indeed, natural capital (e.g., timber, steel, aluminum) is actually a prerequisite for manufactured capital (e.g., fish boat) since the latter is made from the former. More generally, we should emphasize that the growing need for technology-based substitutes (e.g., fertilizer) is a warning signal that natural load-bearing factors (e.g., productive soil) are being permanently eroded.

In summary, confidence that expanding trade and technical wizardry have infinitely expanded human carrying capacity and banished the Malthusian specter forever is premature. Carrying capacity is still very much applicable to humans, and given present trends it is likely to become a major preoccupation of global development policy in coming decades. However, as this discussion will show, we need new approaches to the analysis of carrying capacity, methods that account for uniquely human cultural attributes, including trade and technology.

III. THE LOGIC AND STRUCTURE OF ECOLOGICAL FOOTPRINT ANALYSIS

A. Basic Concepts

Ecological footprint analysis is one such analytic method. Ecofootprinting starts from the premise that, conscious of it or not, modern human beings are integral components of the ecosystems that support them and therefore still very much dependent on "the land" (Rees, 1996; Wackernagel and Rees 1996).

The method also explicitly recognizes (a) that whether one consumes locally-produced products or trade goods, the land connection remains intact, however far removed from the point of consumption some of that land may be, and (b) that no matter how sophisticated our technology, the production/consumption process requires some land- and water-based ecosystems services. Ecofootprint analysis thus incorporates the trade and technology factors simply by inverting the standard carrying capacity ratio: rather than asking what population can be supported by a given area, ecofootprinting estimates how much area is needed to support a given population, regardless of the location of the land or the efficiency of relevant technologies.

Ecofootprinting builds on traditional trophic ecology by constructing what is, in effect, an elaborate "food-web" for the study population. This requires quantifying the material and energy flows supporting the population and identifying corresponding significant sources of resources and sinks for wastes. As we have noted, the human food-web differs significantly from those of other species. In addition to the material and energy required to satisfy the metabolic requirements of our bodies, a human food-web must also account for the needs of our industrial metabolism.

Ecofootprinting is further based on the fact that many material and energy flows (resource consumption and waste production) can be converted into land and water-area equivalents. These are the ecosystem areas required to produce the biophysical goods and services used by the study population. Thus, *the ecological footprint of a specified population is the area of land and water ecosystems required to produce the resources consumed and to assimilate the wastes generated by that population on a continuous basis, wherever on earth the land/water may be located.* A complete ecofootprint analysis would therefore include both the area the population "appropriates" through commodity trade and the area it needs to provide its share of certain free land-

and water-based services of nature (e.g., the carbon sink function).

The area of a given population's ecofootprint actually depends on four factors: the size of the population, the people's average material standard of living, the productivity of the land/water base, and the technological efficiency of resource harvesting, processing, and use. Regardless of how these factors interact, ecofootprinting represents critical natural capital requirements of the study population in terms of corresponding productive land and water area. We can also think of the ecological footprint as representing the extended patch (productive habitat) occupied *ecologically* by the study population.

It is critical to recognize that ecofootprints represent ecologically exclusive areas. The productive capacity used by one human population is not available for use by another. Although two or more human groups may share in the output of the same exporting region, the total ecosystem area appropriated is the sum of the areas required by the individual populations. In the final analysis, all human populations are in competition for the available load-bearing capacity of the earth.

B. The Method in Brief

Population ecological footprints are generally based on final demand for goods and materially intensive services by the referent population. Thus, the first step in calculating the ecological footprint of a study population is to compile, item by item, the total annualized consumption of each significant commodity or consumer good used by that population. Data are obtained from national production and trade statistics and other sources such as United Nations statistical publications. For accuracy, consumption data should be trade-corrected. Thus the population's consumption of wheat can be represented as follows:

$$\text{consumption}_{\text{wheat}} = \text{production}_{\text{wheat}} + \text{imports}_{\text{wheat}} - \text{exports}_{\text{wheat}}$$

The second step is to convert consumption of each item into the land area required to produce that item by dividing total consumption by land productivity or yield. This actually gives us the ecological footprint of the individual item. In general:

$$a_i = c_i / y_i$$

where: a_i is the ecofootprint of item i in hectares, c_i is total consumption of item i in kilograms, and y_i is the yield of item i in kilograms per hectare. Thus, for wheat:

$$a_{\text{wheat}} = c_{\text{wheat}} / y_{\text{wheat}} = \text{kg}_{\text{wheat}} / (\text{kg}_{\text{wheat}} \times \text{ha}_{\text{wheat}}^{-1})$$

Next, the aggregate ecological footprint of the population, (F_p), is determined by summing the footprints for the n individual items:

$$F_p = \sum_{i=1}^{n} a_i$$

Finally, the per capita ecological footprint, f_c, is obtained by dividing the total population footprint by population size, N:

$$f_c = F_p / N$$

For some wastes (such as carbon dioxide emissions) or nutrients (such as phosphates and nitrates), it is also possible to calculate the land (or aquatic) ecosystem area required for sustainable assimilation and recycling. In these cases, the assimilation rate per hectare and year is substituted for y (yield) noted earlier. Box 1 illustrates a basic ecofootprint calculation (aggregate grain consumption in Canada).

We calculate ecofootprint estimates using standard spreadsheet software. Spreadsheets for recent ecofootprint calculations feature up to 132 rows (consumption items) and 15 columns (yield, production, imports, exports, apparent consumption, etc.), with the last column giving the ecofootprint components attributable to each consumption item (see Wackernagel *et al.*, 1999).

Double-counting is avoided whenever identified. For example, if in the analysis of a particular city's ecological

---

Box 1

Sample Calculation: The Ecological Footprint of Grain Consumption in Canada (1993 data)

Total population (N): 28,817,000

Average yield (y): 2744 kg $\times$ ha^{-1} (2.74 tonnes $\times$ ha^{-1})

Domestic production (p): 51,416,000 tonnes (t)

Imports (i): 499,100 t

Exports (e): 26,428,400 t

Domestic consumption ($c = p + i - e$): 25,486,700 t

Ecofootprint of grain consumption by Canadians ($a = c \times y^{-1}$): 9,288,200 ha arable land.

Per capita ecofootprint attributable to grain consumption ($f = a \times N^{-1}$): .32 ha arable land.

footprint one wishes to account for food-related nutrient assimilation but finds that the city's domestic wastes are composted and spread on adjacent agricultural or commercial forest land, then only the agricultural and forest products footprint components should be compiled in the total footprint analysis. Similarly, some consumer products such as leather goods are the by-product of another industry (in this case, beef production). In such cases, one would generally count only the primary land requirements (the grazing and grain lands required for feeding cattle). (For some purposes, of course, it may be of interest to apportion the land among the various products associated with the primary use.)

To estimate typical population ecological footprints (e.g., for whole regions or countries), we usually start with world average land productivities or yields. This greatly simplifies calculations because we do not have to trace all the sources of trade goods and waste sinks or determine the productivity and assimilative capacities of the corresponding production/assimilation areas. As important, using a common base yield facilitates comparison among countries and comparisons of individual countries with global totals. This shortcut is also becoming increasingly realistic for many countries that are heavily dependent on trade flows of commodities from various sources. (Note that the global ecofootprint estimate produced by this method would be the same as the global total estimate that would result if actual yields were used for all items and countries.)

For some kinds of analyses, of course, it might be necessary or useful to base the ecofootprint calculation on actual land/water yields where sources are known and data are available. Also, to compare a country's or region's ecofootprint with the productive or load-bearing capacity of its own domestic territory, we must know the actual productivity of various domestic land categories (cropland, pasture land, forests, carbon sinks, etc.) in order to calculate that country's ecological deficit (or surplus).

For example, if a given country's cropland is three times as productive as world average cropland, we would multiply the area of that country's cropland by a yield factor of three before comparing it to (i.e., subtracting it from) the calculated cropland footprint, as each domestic hectare is the yield equivalent of three global average hectares. This gives a more accurate estimate of the de facto ecological deficit (or surplus) in cropland. Note, however, that if the higher productivity of the domestic land is the result of intensive fertilizer and pesticide use, the apparent reduction in the ecological deficit due to higher cropland productivity will be

at least partially canceled out by the increase in the ecological footprint attributable to the use of agricultural chemicals (fossil fuels are an important feed-stock in the manufacture of these inputs).

In general, we err on the side of caution in making ecofootprint estimates. For example, if there is dispute over, or several estimates of, land productivity, we use the higher estimate (reduces footprint size). Most ecological footprint calculations are therefore likely to under- rather than overestimate.

IV. HUMAN ECOLOGICAL FOOTPRINTS: IMPLICATIONS FOR GLOBAL SUSTAINABILITY

So, how big are our ecological footprints? Most citizens of industrialized high-income countries feel so distanced from nature that it has simply never occurred to them to ask, "How much of the earth's surface is dedicated to supporting just me in the style to which I have become accustomed?" They are generally shocked to learn that average residents of North America, Europe, Japan, and Australia require the biophysical output of 5 to 10 hectares (12–25 acres) of biophysically productive land and water each to support their consumer lifestyles (Rees, 1996; Wackernagel and Rees, 1996; Wackernagel *et al.*, 1999). Such large ecofootprints are also sported by members of wealthy urban elites throughout the developing world.

A. The Functional Footprints of Cities

These findings should alter our perceptions about many things. To begin, they should change how we think about cities and urban land. Almost 30 years ago, American ecologist Eugene Odum wrote, "Great cities are planned and grow without any regard for the fact that they are parasites on the countryside which must somehow supply food, water, air, and degrade huge quantities of waste" (Odum, 1971). Ecofootprinting enables us to quantify the extent of this urban "parasitism." For example, in 1996 Canada's largest city, metropolitan Toronto, had a population of approximately 2,385,000 living in a area of 630 km². Assuming they are typical Canadians, Toronto's citizens had an average ecological footprint of about 7.7 hectares (19 acres). Thus the ecological footprint of metro Toronto was 183,650 km², or about 290 times the size of its then political area.

Most of the city's supportive ecosystems are located at great distance from the people they sustain; indeed, they are scattered all over the planet.

This situation is characteristic of high-income cities. In a particularly comprehensive analysis, Folke *et al.* (1997) estimated that the 29 largest cities of Baltic Europe appropriate for their resource consumption and waste assimilation an area of forest, agricultural, marine, and wetland ecosystems 565 to 1130 times larger than the areas of the cities themselves. A study for the International Institute for Economy and Development in London shows that the biophysical demands of that city alone appropriate an area scattered around the world equivalent to all the ecologically productive land in the United Kingdom.

Such findings have important implications for both urban development and rural sustainability in the 21st century. Some people interpret the global migration of people from the countryside to the city as implying that modern humans are abandoning the countryside and becoming less dependent on the land. This is illusion. The reality is that productive croplands, pasture lands, and forests everywhere are being used more intensely than ever to sustain the world's burgeoning urban populations. The human occupants of cities may think of the latter as their principal habitat, but cities per se represent only a tiny fraction of the total human ecosystem.

There is, of course, a certain mutualism between the city and the countryside. Cities need the resources of rural areas and rural areas benefit from urban markets and technology transfers from cities. However, while rural areas could survive without cities, the dependence of cities on rural environments is absolute. In short, there can be no urban sustainability without rural sustainability. In this light, we might even want to reconsider what we think of as *urban*-dominated land—in a whole-systems ecological sense, the great plains of North America, one of the world's major breadbaskets, are an essential component of the increasingly urban global human ecosystem.

According to United Nations' projections, there will be 27 cities with populations exceeding 10 million by 2015 (up from only 1 in 1950). Forty-four more cities will have populations of 5 to 10 million by the same year. This should trigger at least a cautionary alarm. The world's megacities—all cities for that matter—are dependent on a vastly larger area of productive lands outside their boundaries and political control. But just how secure can any urban population be if the lands represented by its ecological footprint are under threat from ecological change or geopolitical uncertainty?

B. Ecofootprints, Global Development, and Social Equity

It is not just cities that overshoot the productive capacity of ecosystems within their political domain. Most high-income countries have an ecological footprint several times larger than their national territories. In effect, they are running massive ecological deficits with the rest of the world. As a result of continuous population growth, rising material demand, and expanding trade, wealthy consumers in particular have ecological footprints that treat the entire planet as their common patch.

In fact, several ecofootprint studies suggest that humanity is now running a *global* ecological deficit of up to 30% and is accumulating a nonrepayable ecological debt. Even at current average levels of economic production and consumption, the human load already exceeds the long-term carrying capacity of the earth. The direct empirical evidence (e.g., ozone depletion, climate change, fisheries collapse) is becoming the stuff of daily newspaper headlines.

This means that contrary to the assumptions of prevailing international development models, so-called first world material lifestyles are not sustainably extendible to the entire world population using prevailing technology. Although wealthy consumers require the goods and life support services of 5 to 10 hectares of productive land and water there are only about two hectares of such land per person on earth. Meanwhile, the great majority of humankind is understandably not satisfied with its present material lot and is determined to improve it. (Despite the recent industrial revolution in China, the ecological footprint of the average Chinese is less than 1.5 hectares and Bangladeshis get by on only half an hectare.) The problem is, that to support all six billion members of the human family at the average North American material standard would require about three additional earthlike planets (and this does not account for the needs of the three billion more people to come by the middle of the 21st century).

If we are at or beyond the human load-bearing capacity of the planet, then the only way additional population or material growth can be sustained, without ravaging biodiversity and ultimately destroying the ecological basis of human life, is through massive cutbacks in resource consumption by the presently wealthy. In theory this could be achieved through a tightly managed new efficiency revolution (doing *much* more with less) in combination with an absolute reduction in material demand (i.e., a shift to simpler lifestyles).

Other carrying capacity studies have reached similar

conclusions. Using data on current waste discharge rates and the assimilative capacity of biophysical systems, the Sustainable Europe Campaign has estimated that sustainability requires an approximately 50% reduction in the material throughput of the global economy in coming decades. However, in order to create the necessary environmental space for developing countries to claim their fair share of global carrying capacity, developed countries must reduce their consumption of various nonrenewable resources by between 88 and 94% (Carley and Spapens, 1999).

This might seem to be an extreme conclusion and impossible goal. However, the United Nations Environment Program's "Global Environment Outlook 2000" report also argues the need for a tenfold reduction in resource consumption by high-income countries. Even the Geneva-based world Business Council for Sustainable Development acknowledged as early as 1993 that "industrialized world reductions in material throughput, energy use, and environmental degradation of over 90% will be required by 2040 to meet the needs of a growing world population fairly within the planet's ecological means" (BCSD, 1993).

If so-called postindustrial society fails to meet this challenge, the anticipated doubling and doubling again of the human material demand in the next half-century will be disastrous for conservation efforts. As explained next, the expanding human ecological footprint inevitably reduces the ecological space available for other species.

V. HUMAN ECOLOGICAL FOOTPRINTS AND THE BIODIVERSITY CRISIS

Human beings are uniquely successful among large mammals in mastering the variety of environments on earth and in expanding both numerically and spatially over the globe. Our capacity to extend continuously our ecological footprint is attributable to several species-specific qualities of which perhaps three stand out.

First, humans have a remarkably variable diet—we have wide-ranging omnivorous tastes and if we cannot consume something directly (such as grass), we domesticate an animal that will and then eat the animal. This means that humans may well possess the broadest food niche of any vertebrate on the planet. Second, humans are as behaviorally adaptable as they are catholic in their diets. Together, these two factors make virtually any terrestrial ecosystem on earth accessible to *Homo sapiens*. No other species has managed successfully to

colonize and dominate virtually all major terrestrial ecosystem types, from grasslands and forests to deserts and tundra, on all the world's significant land masses. Finally, we are creatures of language, culture, and cumulative learning. Technological advance has enabled humans continuously to increase the intensity of resource exploitation and to extend their ecological footprint into virtually all remaining productive habitats on the planet. For example, thanks to electronic fish-finding devices and ruthlessly effective fishing gear, humans have actually become the functionally dominant marine mammal. By the mid-1990s, the world's fishers were appropriating the equivalent of 25 to 35% of the product of net photosynthesis from the 10% of the oceans that produces 96% of the catchable fish.

We previously noted that human ascendance has historically been achieved at the expense of other species. People's material demands, culminating in today's global consumer society, have steadily appropriated an increasingly disproportionate share of the planet's finite productive and sink capacities. The recent human dominance of ocean ecosystems provides several examples of the resulting biodiversity loss.

To begin, some modern fish-harvesting techniques have devastating impacts on the physical and biological structure of marine habitats, thereby reducing the productivity, diversity, and abundance of both target and no-target species. Various surveys reveal that fishers now drag heavy trawls over all the oceans' continental shelves at least once every two years and some areas are hit several times in a season. In heavily fished regions of the North Sea, trawlers make up to seven passes annually with nets and chains weighing up to five tons. The total area affected by trawls is 150 times larger than the area of forest clear cut globally each year.

Many fish-stock depletions and the destruction of benthic communities are the direct result of the increasing intensity of commercial fishing. However, other systemic problems stem from the fact that expanding the human marine ecofootprint inevitably diverts an ever greater proportion of the finite energy budget of the sea away from other marine predators, including various fish, birds, and mammals.

A particularly telling example of the resultant domino effect can be traced to a tripling of the commercial catch of Alaska pollock since 1986. The reduced availability of pollock apparently precipitated a decline in populations of Steller's sea lions and seals, which feed on pollock. By the late 1990s, sea lion populations in the Gulf of Alaska had fallen by 80 to 90%, in turn depriving local orca (killer whale) populations of their normal primary food source. With the collapse of sea

ECOLOGICAL FOOTPRINT, CONCEPT OF   239

lions and seals, the orcas turned to eating sea otters—smaller, less fatty animals—whose populations have also now also fallen by 80 to 90% since 1990.

But the systemic shifts and biodiversity losses do not end there. Sea otters seem to be a keystone species in the shallow marine kelp forests that constitute their preferred habitat. Among the otters' major prey are sea urchins, which, in turn, graze on kelp. As the otter population declined there was an explosion of sea urchins, leading to the destruction of the kelp beds by overgrazing. So it is that human overfishing for pollock precipitated the collapse of one of the most productive inshore-marine ecosystems in the North Pacific, as so-called sea urchin barrens spread along the coast of Alaska and the adjacent Aleutian Island chain. (Dramatic as this example might seem in its own right, it is only just the final chapter of a longer book. Direct human overharvesting of sea otters for their pelts, beginning in the 19th century, had already long since decimated the kelp forests, which previously characterized much of the Pacific coast of North America from California north.)

Some popular accounts of this phenomenon have emphasized the role of orcas in switching prey from seals to otters. Certainly this was an important link in the chain of events, but the structural change mainly responsible for precipitating the collapse of the near-shore ecosystem was the displacement by humans of sea lions, seals, and ultimately orcas from their ecological niches among the top carnivores in the offshore ecosystem. As humans replaced seals in the food-web, the killer whales were *forced* to switch to sea otters, a much inferior food source.

This example underscores how, whether conscious of it or not, *Homo sapiens* have become the ecologically dominant marine mammal. But the main point to take away is that uncontrolled increases in human appropriations of bioproductivity *inevitably* result in the rerouting of energy and material flows through the affected ecosystems—marine or terrestrial—with consequent and potentially permanent changes in ecosystem structure and function.

We now recognize that several mechanisms are at work as humans extend their ecological footprints. People

1. Passively displace other species from their food niches or appropriate their habitats. (Commercial fishing displaced sea lions, seals, and orcas on the northwest Alaska coast; agriculture pushed bison from the Great Plains of North America; clearing for crops and grazing has destroyed thousands of species in former tropical forests.)
2. Actively eliminate nonhuman competitors—other species that compete with us for "our" food. (We shoot wolves that hunt ungulates, and seals that eat commercially valuable fish; we poison insects that would devour our crops.)
3. Deplete both self-producing and nonrenewable "natural capital" stocks. (Humans overexploit many wild prey populations from rhinos to fish; destroy whole ecosystems such as forests; and deplete vital assets such as groundwater, soils, and fossil fuels.)

These processes are all consumption-related. The first two are forms of competitive exclusion. Technological "man" is simply more effective than other organisms at appropriating nature's bounty for his or her own use. Because energy and material flows used by people are irreversibly unavailable for other species, the latter decline, even to extinction, at least locally.

The third mechanism, stock depletion, is the product of many things, including confidence in technological substitution, blind ignorance, material greed, sheer desperation, and the relentless working of the so-called common property problem on an overcrowded planet. Sometimes it is even the result of willful disregard on the part of those who give no moral standing to other creatures or who simply don't care about the state or fate of the world.

Economic globalization, the sanctioning of greed, the rise of consumerism, and the spread of energy-intensive technologies have intensified all these processes. Overharvesting and habitat destruction are driving what some analysts refer to as "the sixth extinction." According to Biologist E. O. Wilson, the earth is now experiencing the greatest extinction episode since the natural catastrophes at the end of the Paleozoic and Mesozoic eras. The current species extirpation rate is 100 to 1000 times prehuman levels inferred from the fossil and paleontological record. A quarter of bird species have already been extinguished by people and fully a quarter of the 4400 mammal species living today are on a path of decline that, if not reversed, is also likely to end in extinction.

The contemporary human ecofootprint is also characterized by massive increases in waste production. Not surprisingly, the resultant pollution is adding to the toll on biodiversity. A poignant example is the phenomenon of "coral bleaching," which has been in the news since the mid-1980s. This effect is produced when the animal component of coral organisms—corals are a mutualistic union of animal polyps and plant (algae) cells—expel their algal cohabitants as a result of thermal or other stress. The loss of algae leaves the coral pale and washed out (bleached) and unable to grow or reproduce. In the exceptionally warm spring and summer of 1998, coral bleaching associated with a one Celsius degree increase in ocean temperature extended across the tropics, including, for the first time, the Indian Ocean. Reports indicate that 70 to 90% of corals in the Indian Ocean basin appear to have been affected.

Some scientists have referred to the bleaching and die-back of corals as an unprecedented ecological disaster. Coral reefs are among the most diverse and productive ecosystems on the planet. Numerous fisheries, particularly locally important food fisheries, are sustained by the productivity of corals. What economic and social costs would be incurred if the decline of corals and associated reef habitats continues unabated? How will we measure the ultimate consequences of the sheer loss of biodiversity? Most important, what does the apparent spread of the problem say about the state of the ecosphere in general and the marine environment in particular?

These are important questions, but the main point is that to the extent that coral bleaching is caused by human-induced global warming, it is the indirect result of greenhouse gas accumulation in the atmosphere. (The earth's carbon sinks are overflowing.) The attendant biodiversity loss is therefore the distal product of the industrial world's addictive dependence on fossil fuels and our ever-expanding industrial metabolism. Regrettably, if the climate change/global warming hypothesis proves to be correct, coral bleaching will not soon be reversed by any political action feasible under the prevailing global development model.

All these examples underscore the inverse relationship between growth of the human ecological footprint and biodiversity (the competitive exclusion principle at work). Unfortunately, at present, ecofootprint analysis does not directly track biodiversity losses. However, the negative correlation is most directly associated with the conversion of wildlands to agriculture, intensive (including plantation) forestry, and various forms of habitat fragmentation. Many of these conversions are measurable. It may therefore eventually be possible, using the historical record and principles of biogeography, to link some index of biodiversity loss to the growth of population ecofootprints.

Whether or not this relationship can be rigorously defined, the basic point stands. To the extent that further human population and material growth requires the clearing of forests and the intensification of agriculture, and is accompanied by urban sprawl and pollution, it necessarily means less ecological space to sustain nonhuman species and ecosystems. Biodiversity loss is thus virtually certain to accelerate under the prevailing global development scenario.

VI. ASSESSING THE ECOLOGICAL FOOTPRINT APPROACH

A. Conceptual and Methodological Strengths

Ecofootprint analysis brings several methodological strengths to bear in assessing the state of the world. First, it is compatible with various concepts developed by other analysts to address human ecological problems. For example, the ecological footprint concept

- incorporates and extends George Borgstrom's 1960s notion of "ghost acreage," referring to the extra-territorial food lands required to support densely populated regions and countries.
- corresponds closely to Paul Ehrlich's and John Holdren's 1970s definition of human impact on the environment: $I = PAT$, where I is impact, P is population, A is affluence, and T is technology. The population ecological footprint (F_p) corresponds to impact (I) in the latter formulation and is itself a function of population size and consumption. However, because consumption is a function of income (affluence) and the state of technology, I, like F_p, is a function of population and consumption. From this perspective, F_p is an area-based analogue of I.
- provides one measure of human load defined by Catton (1980) as a function of population size and average individual impact (consumption again). The larger the footprint, the greater the load. (Catton defined *carrying capacity* as the "maximum persistently sustainable load.")
- is conceptually related to the embodied energy (emergy) analyses of ecologist Howard Odum and recognizes the importance of the second law of thermodynamics to human affairs as stressed by ecological economists Nicholas Georgescu-Roegen and Herman Daly. Indeed, the area represented by the ecological footprint can be conceived as the photosynthetic surface (solar collector) needed to replace the free energy (essergy or negentropy) dissipated by humans and their industrial metabolism.

However, for all its technical compatibility, perhaps the major strength of the eco-footprint concept is conceptual simplicity and intuitive appeal. First, the 'footprint' metaphor seems particularly effective in communicating the idea that we each have an impact on the Earth for which we are responsible through consumption choices. Second, ecofootprinting consolidates real data on a variety of energy and material flows into a single concrete variable, land area. Land is a particularly powerful indicator because it too is readily understood by ordinary people.

Because the supply of land is finite at any scale, ecofootprint analysis provides a measure of human demand (load) for comparison with supply (load bearing capacity) at all geographic levels from local to global. Such comparisons suggests several secondary indices of (un)sustainability that can be used to stimulate discussion of sustainability options or in establhing measurable policy targets. For example, if we are running a local or regional ecological deficit, the question becomes: Is this a major problem and, if so, what must be done to reduce it? (Note that humanity's total ecological deficit, the sustainability gap, may ultimately be more important than any fiscal deficit yet is totally ignored in most national accounting systems and international developmental planning.)

B. Methodological Limitations

Although it has considerable conceptual and operational strength, ecofootprint analysis does not provide a complete picture of the human ecological dilemma. Some of the methodological limitations are as follows:

1. The Question of Scope

Some critics have argued that the term *ecological footprint* is misleading because the method does not capture the full range of ecologically significant impacts on the ecosphere.

It is true that ecofootprint calculations are not all-encompassing. They do not account for the cause or effects of ozone depletion or endocrine (hormone) mimicry, to cite just two examples. At present, ecofootprinting is best suited for those human load factors that are readily convertible into a land or water ecosystems equivalent. In fact, calculations do not even tell the whole land and water story. Various minor consumption categories that have yet to be included and analysts are only beginning to examine the spatial implications of waste discharges other than carbon dioxide and essential nutrients.

To be sure, it is unlikely ecofootprint analysis will ever cope satisfactorily with various toxic waste discharges into the ambient environment. The biological effects of chronic low-level chemical contamination (e.g., carcinogens), for example, do not readily translate into an appropriated land area. However, this may not be as problematic as it seems. Environmental and ecological economists already agree that society should have zero tolerance for highly toxic chemical wastes and radioactive substances for which the ecosphere has no measurable assimilative capacity. Such substances should simply be banned or phased out (as in the case of chlorinated pesticides and ozone-depleting CFCs).

In any event, no single index can be expected to represent all ecological impacts and this limitation does not invalidate what ecofootprinting does reveal. Rather, it suggests that ecofootprint calculations are almost certainly *underestimates* of actual ecosystem appropriations and that methodological improvements and extensions may well result in considerably larger ecofootprints. As it is, ecofootprinting already provides unambiguous results and suggests clear policy directions. Increasing the accuracy and scope of the analyses may add to our sense of urgency but will not likely shift the direction of needed policy change.

2. Is Ecofootprinting Too Simplistic?

Some commentators have argued that the ecofootprint concept oversimplifies both nature and society and has little predictive value. Both the ecosphere and the economy are dynamic systems, but ecofootprint analysis produces static estimates of a single aggregate variable.

Again, it is true that footprint analysis is not dynamic modeling and cannot directly project the effects of technological change or social adaptation. However, simulation and prediction were never the intent. Ecofootprinting provides a simple ecological camera—each analysis is a snapshot of current demands on nature, a portrait of how things stand *right now* under prevailing technology and social values.

This in itself is important information. Current "snapshots" indicate that humanity has already significantly exceeded carrying capacity using prevailing technology and that some people contribute significantly more to this ecological overshoot than do others. This is sufficient to suggest approximately how much we must reduce our consumption, improve our technology, redistribute wealth, or change our behavior to achieve sustainability.

Of course, once an initial baseline has been established, subsequent ecofootprint assessments can contribute to a time-series study—repeated snapshots over years or decades—to monitor progress toward, for example, reducing national ecological deficits or closing the global sustainability gap. (After all, even a motion picture is a series of snapshots.) Such sequential analyses are fully capable of accounting for the *actual* effects of new technologies, or changes in consumer behavior or cultural values. We can also assess the *potential* impacts of alternative technologies or settlement patterns on the size of a population's ecological footprint in static studies. This involves using assumed levels of consumption or assumed technologies (based on real-world data) in the standard calculations to address such questions as, how would a partial shift from fossil fuels to biomass fuels or solar energy affect a population's ecofootprint?

To summarize this point, ecological footprint analysis is not intended to provide a dynamic window on the future but rather a snapshot in time. As such it can both help to assess current reality and to test alternative "what-if" scenarios on the road to sustainability.

On a related matter, some critics have stressed the impossibility of mapping the state of complex multidimensional processes or systems with a single indicator and that ecofootprint analysis ignores the implications of complexity theory. This comment, while true, misses two points. First, ecofootprinting is not intended to provide a complete picture of any complex system. It generates only one indicator of the state of humanity's engagement with the rest of nature. To the extent that this index is reliable and has policy relevance to sustainability, it should be used in conjunction with economic, social, or any other indicators that bear on the issues at hand.

Second, although science has certainly gained important insights from complex systems theory—the behavior of dynamical, self-organizing, adaptive systems often confounds standard forecasts and derails simplistic approaches to ecosystems management—the fact remains that many phenomena do have simple explanations. In these circumstances, Occam's razor still applies: there is no need to derive a complex explanation where a simple one suffices.

For example, one can hardly imagine a better example of a complex human-dominated system than the world's great cities. Cities are a marvel of biophysical, economic, political, social, and cultural subsystems that somehow come together to function as a complex, dynamic, integrated, self-organizing whole. Nevertheless, if any city as currently conceived were enclosed in an impermeable glass dome whose base coincided with its political or geographic boundaries, it would rapidly disintegrate.

The explanation is simple—the city would be starved of resources and suffocated on its own wastes at the same time. This is because most of the ecosystems and biophysical processes supporting the city lie outside our hypothetical bell jar. Ecofootprint analysis measures the city's (region's, country's, etc.) dependence on nature's services and estimates the ecosystem area needed to produce them. There is certainly room to debate the methods used to make these estimates. However, we don't need to know anything about city governance or the complex relationships among its physical and socioeconomic subsystems to predict the collapse of our city-in-a-jar or to perform an ecofootprint analysis and assess its implications.

3. The Carbon Assimilation (Fossil Energy) Footprint Is Not a Real Land Use

For many high-income industrialized countries, the fossil energy component may make up half of the total ecofootprint as presently calculated. Because science cannot account fully for the global carbon budget and there is dispute over the relative roles of terrestrial and marine ecosystems in carbon sequestration, some critics see energy footprint calculations as unreliable or hypothetical. Others don't regard carbon assimilation as a real land use.

This last criticism reflects our cultural bias that land *use* requires physical occupancy or the consumption of some tangible product of the land. We are unaccustomed to thinking of the provision of other (particularly nonmarket) biophysical services of nature, no matter how essential to sustainability, as effective "uses" of landscape. In terms of thermodynamic law and mass balance, it is this cultural bias that needs adjustment, not ecofootprint theory. All energy and material use is "throughput," implying the generation of wastes, and to the extent that the assimilation and recycling of those wastes (including carbon dioxide) requires an exclusive dedication of land or water, it is legitimate component of ecofootprint calculations.

Returning to the first criticism of energy footprinting, while the details of the carbon budget may be in dispute, there is no question that carbon dioxide levels are increasing in the atmosphere and that these increases represent about half of current carbon emissions from fossil fuel and biomass combustion. This implies that available land and water carbon sinks are insufficient to sequester all anthropogenic carbon dioxide at current rates of emission. In ecofootprint terms, our current global energy footprint is excessive. We are running a global carbon sink deficit in exactly the same sense that many countries run a food-land deficit.

In short, the notion of carbon sink land is not hypothetical even if the real thing is in short supply. Energy footprint calculations are therefore designed to estimate the area of dedicated carbon-sink forests that would be necessary to assimilate a study population's carbon emissions based on the average estimated carbon sequestration rates of the world's growing forests. In fact, various electric utilities are now planting dedicated carbon sink forests to offset the carbon emissions of their fossil fuel-burning generating plants. Similarly, some countries are contemplating developing carbon sink forests as an alternative to reducing their carbon emissions and to be part of a future system of tradable carbon emission rights. (For example, planting a carbon sink forest would release an equivalent quantity of emission rights for sale on the market.) Note that a dedicated carbon sink forest is an exclusive land use insofar as it cannot be harvested for any purpose such as pulp and paper production that would quickly release sequestered carbon back into the atmosphere.

Finally, we should recognize that any alternative to fossil fuels will also generate an energy ecofootprint. For example, the ethanol equivalent of fossil fuel, generated from biomass, would require a fuel-crop growing area considerably larger than the energy footprint based on carbon sinks. Even solar-based alternatives require that a large land area be appropriated for collector surfaces (though this land need not be biologically productive). The point is that all energy sources and related sinks have a corresponding land equivalent and the present method of calculating an energy footprint may actually be fairly conservative.

VII. EPILOGUE

Whatever its weaknesses, the ecological footprint concept is strong enough to have captured both the scientific and popular imaginations. It is now being applied as an assessment tool by nongovernmental organizations and government agencies in towns, cities, and countries around the world. A small industry of consulting firms has sprung up to develop and commercialize various products based on the ecofootprint concept.

Ecofootprinting seems to concretize the *human* ecological crisis in a way other energy and material flows studies do not. Knowing our ecological footprint not only makes us conscious of our personal contribution to global change, but also makes us responsible for doing something about it. Pluralistic democracies can work only if the public are well informed about key policy issues and have a keen sense of their capacity to make a difference in improving the human prospect and those of other species. To the extent that ecological footprint analysis helps to generate a more ecologically literate and politically active populace, it will have achieved its most important objective.

See Also the Following Articles

CARRYING CAPACITY, CONCEPT OF • ENERGY FLOW AND ECOSYSTEMS • ENERGY USE, HUMAN • HUMAN IMPACT ON BIODIVERSITY, OVERVIEW • LOSS OF BIODIVERSITY, OVERVIEW • SUSTAINABILITY, CONCEPT AND PRACTICE OF • URBAN/SUBURBAN ECOLOGY

Bibliography

BCSD (1993). *Getting eco-efficient*. Report of the BCSD First Antwerp Eco-Efficiency Workshop, November 1993. Geneva: Business Council for Sustainable Development.

Block, W. (1990). Environmental problems, private property rights solutions. In *Economics and the environment: A reconciliation* (W. Block, Ed.), pp. 281–332. Vancouver, B.C.: The Fraser Institute.

Carley, M., and P. Spapens (1999). *Sharing the world: Sustainable living and global equity in the 21st century*. London: Earthscan Publications.

Catton, W. R. (1980). *Overshoot: The ecological basis of revolutionary change*. Urbana and Chicago: University of Illinois Press.

Daly, H. (1991). *Steady-state economics*. Washington, D.C.: Island Press.

Diamond, J. (1992). *The third chimpanzee*. New York: HarperCollins.

Duncan, R. C. (1993). The life-expectancy of industrial civilization: The decline to global equilibrium. *Population and Environment* 14, 325–357.

Flannery, T. F. (1994). *The future eaters: An ecological history of the Australasian lands and peoples*. Chatsworth, NSW: Reed Books.

Folke, C., A. Jansson, J. Larsson, and R. Costanza (1997). Ecosystem appropriation by cities. *Ambio* 26, 167–172.

Gordon, R. L. (1994). Energy, exhaustion, environmentalism, and etatism. *The Energy Journal* 15, 1–16.

Hern, W. M. (1997). Is human culture oncogenic for uncontrolled population growth and ecological destruction? *Human Evolution* 12, 97–105.

Lubchenco, J. (1998). Entering the century of the environment: A new social contract for science. *Science* 297, 491–497.

Odum, E. P. (1971). *Fundamentals of Ecology* (3rd ed.). Philadelphia: Saunders.

Ponting, C. (1991). *A green history of the world*. London: Sinclair-Stevenson.

Rees, W. E. (1996). Revisiting carrying capacity: Area-based indicators of sustainability. *Population and Environment* 17, 195–215.

Simon, J. (1995). The state of humanity: Steadily improving. *Cato Policy Report* 17, 5. Washington, D.C.: The Cato Institute.

Smith, J. W., and Sauer-Thompson, G. (1998). Civilization's wake: Ecology, economics and the roots of environmental destruction and neglect. *Population and Environment* 19, 541–575.

Vitousek, P. M., H. A. Mooney, J. Lubchenco, and J. M. Melillo. (1997). Human domination of earth's ecosystems. *Science* 277, 494–499.

Wackernagel, M., L. Onisto, P. Bello, A. C. Linares, I. S. L. Falfán, J. M. Garcia, A. I. S. Guerrero, and M. G. S. Guerrero. (1999). National natural capital accounting with the ecological footprint concept. *Ecological Economics* 29, 375–390.

Wackernagel, M., and W. E. Rees (1996). *Our ecological footprint: Reducing human impact on the earth.* Gabriola Island, BC, and Philadelphia, PA: New Society.

WRI (1997). *Resource flows: The material basis of industrial economies.* Washington, D.C.: World Resources Institute (for World Resources Institute, Wuppertal Institute [Germany], Netherlands Ministry of Housing, and National Institute for Environmental Studies [Japan]).

ECOLOGICAL GENETICS

Beate Nürnberger
Ludwig-Maximilians-Universität München

GLOSSARY

allozyme Enzyme locus that produces more than one electrophoretic variant.

deme Smallest population unit in population genetic models.

frequency dependence Phenomenon that the dynamics of a given ecological or genetic type depend on its frequency in a population.

genetic marker Genetically variable locus that produces distinct variants when analyzed by standard methods such as electrophoresis. It may be based on protein or DNA and may or may not be selectively neutral.

metapopulation Collection of populations that are connected through dispersal and are typically extinction-prone. A key feature is the distinction between local (transient) and regional (persistent) dynamics in one or several interacting species.

phenotypic plasticity Property of a given genotype to express different phenotypes as a function of ecological conditions.

polymorphism Existence of more than one discrete intraspecific type such as distinct phenotypes or genotypes at one locus or several loci.

population structure Subdivision of a given population according to features such as location, age, size, or social status.

ECOLOGICAL GENETICS is an interdisciplinary field of study with the aim to understand adaptation in response to natural selection at the intraspecific level. It applies the principles of neo-Darwinian theory to directly observable microevolutionary processes.

I. INTRODUCTION

Broadly speaking, there are two main objectives at the heart of ecological genetics research: (a) the analysis of the nature and strength of specific selection pressures within a given ecological setting and (b) the genetic basis of the adaptive response. Consequently, ecological genetics draws heavily on two bodies of theory. On the one hand, ecological theory yields predictions about the kinds of adaptations that a given selection pressures might produce. On the other hand, the theory of population genetics provides specific models about allele frequency changes under different selective regimes and in a variety of population structures.

Hutchinson (1957) succinctly captured the focus of this research program in his famous metaphor of the ecological theater in which the evolutionary play is enacted. In this tradition, E. B. Ford (1964) published the first treatise on ecological genetics. Weary about the possible artifacts of evolutionary experiments that are conducted in the laboratory only, he advocated the study of natural populations as an indispensable complement. To him natural populations provided the ultimately relevant setting in which to investigate natural selection and adaptation. Within the past 30 years, the scope of ecological genetics has become even broader thanks to both theoretical and methodological advances.

The evolutionary analysis of genetically complex phenotypic traits rests on the theory of quantitative genetics. The incorporation of this theory into the conceptual framework of ecological genetics has greatly enlarged the range of phenotypic traits under study. Whereas early studies had mostly focused on the dynamics of visible polymorphisms with a relatively simple genetic basis, adaptations in continuously varying characters such as body size, disease resistance, or courtship coloration have since become amenable to ecological genetic analysis. Moreover, the development of a whole spectrum of genetic markers with varying evolutionary dynamics has provided tools for the study of population structure as well as for the spread of selectively favored genes. More recently, genetic markers have been used for the genetic dissection of complex phenotypic traits.

The ecological setting no longer functions as a mere backdrop to the evolutionary process, as Ford saw it, but provides a research focus in its own right. Ecological theory makes testable predictions about the kinds of strategies that are expected to evolve under a given set of environmental conditions. For example, life history theory defines the optimal schedules of growth, reproduction, and intrinsic mortality. Similarly, behavioral ecology makes predictions about such features as optimal breeding systems or diet breadth. For a broad range of traits, ecological theory establishes the link between a given adapted phenotype and the particular selective agent that prompted its formation.

Ubiquitous environmental heterogeneity produces phenotypic differentiation on a variety of spatial scales across a species range. For any one species, there might be broad differentiation with respect to one trait and at the same time a patchwork of local variants in another. Empirical studies have indeed demonstrated a large amount of ecological variation at the intraspecific level, which covers the range from life history traits over behavior and physiology to morphology. Quite often, these differences are hidden from the casual observer and become apparent only through detailed study that extends over more than one generation of the organism in question. Superficial morphological uniformity over large areas therefore does not imply ecological equivalence. The phenotypic changes may be small, and they need not accumulate to change the appearance of the species in the long term. In fact, they might be partly or entirely reversed when the local conditions change. Yet at any one time they can determine the ecological role that the organisms of a particular population play. I will use the term 'ecological variants' to emphasize the context-dependence of these phenotypes. Ecological genetics is beginning to tell us not just about the effects of complex ecological settings on the process of evolution, but also about the role of genetic variants in shaping local ecological interactions.

II. GENETIC VARIATION IN ECOLOGICAL TRAITS

A. Genetic Variation within and between Populations

All phenotypic evolution depends on within-population genetic variance. It is this variance that natural selection can mold to produce either distinct locally coexisting variants or differentiation among populations. There is abundant evidence that this substrate of evolution exists for many phenotypic traits. There is genetic variation within populations of the swallowtail butterflies *Papilio zelicaon* and *P. oregonius* for oviposition preference on a number of host plants. Individuals within two populations of the wood frog, *Rana sylvatica*, differ genetically in their larval development time. Aphids collected from a single field vary in their resistance to a parasitoid wasp, and among the wasps collected from the same field there is genetic variation in their ability to parasitize the local aphids. The sand cricket, *Gryllus firmus*, shows large within population variation for the duration of the male mating call. Several different genetic types that confer resistance to a fungal pathogen exist within populations of the Australian flax, *Linum marginale*. This list could be extended to include many more examples, all of which underscore the fact that the raw material for evolution exists for a broad range of ecological traits.

The extent to which this variation has been transformed into genetic variation between populations is equally impressive. Garter snakes (*Thamnophis elegans*) in coastal populations of California where slugs are common have a much greater innate preference for eating slugs than those from populations further inland. Female guppies vary in their preference for male color patterns even among populations within the same stream. Sticklebacks (*Gasterosteus*) differ geographically in traits such as courtship behavior and cannibalistic tendency. One population of the milkweed bug, *Oncopeltus fasciatus*, shows strong migratory behavior whereas another more southerly population is nonmigratory. Populations of the spider *Agelenopsis aperta* have diverged genetically in number of traits including territorial behavior, predator avoidance, and diet breadth. The wood frog populations just mentioned differ from each other in larval development time and size at metamorphosis.

B. Is the Observed Phenotypic Variation Heritable?

In all of the foregoing examples, the genetic basis of the observed variation has been demonstrated through careful experimentation. The easiest way to demonstrate between population genetic differences is to rear field-collected eggs or seeds from different localities under uniform conditions. Differences between the population means in the trait of interest suggest a genetic component. However, the effect could be purely environmental. More resources in one location could lead to better provisioned offspring that reach a larger body size, develop more rapidly, or have greater disease resistance. Such nongenetic effects can be eliminated or at least minimized if one rears two generations in the laboratory and measures the trait of interest only on the second generation.

Controlled breeding schemes are used in order to determine the within-population genetic variation in sexually reproducing species. If the trait in question has a simple genetic basis and gives rise to a small number of distinct phenotypes, it might be possible to infer the number of loci involved from the Mendelian segregations within crosses. However, such a detailed genetic dissection is impossible in the majority of cases in which phenotypic variation is continuous and a function of both polygenic inheritance and environmental influences. In that case, the resemblance between parents and offspring can be used to estimate the so-called heritability of the trait. Heritability is defined as the proportion of phenotypic variance that is due to herita-

ble genetic variation (see Box 1). Care needs to be taken when one extrapolates from laboratory estimates to the situation in the field. Because heritability is a proportion of total phenotypic variance, V_P, it can be influenced by changes in any of the components of V_P. For example, uniform rearing conditions reduce the environmental

---

Box 1

Heritability

Quantitative genetic theory is built on a model in which a given phenotype is a combination of genetic factors at a large number of loci, each of small effect, and a random component that reflects the environmental influence. There are several types of genetic effects. An additive component represents the sum of allelic effects. Per locus and allele, the allelic effect can be thought of as the average phenotypic effect that the allele would have if it were placed into a large number of genetic backgrounds within the population. In addition, there may be effects due to dominance. If the joint effect of two alleles in a heterozygote is not equal to the sum of the allelic effects, then the difference between additive and observed effect gives the dominance deviation. Similarly, if the effect of an allele depends on which alleles are present at one or more other locus, one talks of epistasis or of an interaction effect.

The phenotypic variance V_P in the population can be broken up into components in an analogous manner:

$$V_P = V_A + V_D + V_I + V_E,$$

where V_A, V_D, V_I, and V_E are the additive, dominance, interaction, and environmental component, respectively. Because sexually reproducing individuals pass on genes but not genotypes to their offspring, all nonadditive effects that result from particular gene combinations in the parent are not heritable. The resemblance between parents and offspring depends only the additive genetic variance. It is this resemblance that determines the evolutionary dynamics of a trait. Hence, heritability, h^2, is defined as the proportion of phenotypic variance that is due to additive genetic variance in the population:

$$h^2 = V_A/V_P.$$

variance component and therefore cause an increase in heritability. Nevertheless, laboratory estimates generally give a good indication whether or not the observed variation has a genetic basis.

An alternative method to demonstrate genetic variation within population in a given trait is to perform an artificial selection experiment. If the trait responds to selection, then the base population must have harbored genetic variation at the underlying loci. Experiments of this type are generally successful. Genetic variation for phenotypic traits appears to be ubiquitous, yet not all of it is necessarily available for adaptive change in nature (discussed later in this chapter)

The famous long-term studies of Darwin's finches on the Galapagos Islands have provided not just field-based estimates of heritabilities but also proof of short-term phenotypic change in response to natural selection. For example, the medium ground finch *Geospiza fortis* on the island Daphne Major shows heritable variation in body weight, wing length, bill size, and tarsus length. These variables are largely correlated, because they are all components of body size. After a period of exceptionally heavy rains, small soft seeds were abundant for the following two years. This food supply differed markedly from the predominantly large and hard seeds of the preceding period of drought. After the rains, the finch size distribution shifted within two years toward smaller birds that showed relatively better survival. Moreover, there was evidence of direct selection in favor of reduced bill width over and above the overall selection on body size.

C. The Alternative: Phenotypic Plasticity

Although genetic differentiation in ecological traits is common, by far not all phenotypic differences between populations need to have a genetic basis, even if they are obviously adaptive. In fact, a surprising amount of intraspecific phenotypic variation lacks a genetic basis. Such variation is referred to as phenotypic plasticity if a given genotype predictably produces different phenotypes under different environmental conditions. This environment-dependent phenotypic repertoire of a genotype is called its norm of reaction. Phenotypic plasticity is thus distinct from random environmental variance due to chance effects that impinge on the expression of traits within a given environment.

In the Texan desert, tadpoles of the spadefoot toad *Scaphiopus couchii* adjust their development time to variation in pond duration. They are able to accelerate their development in rapidly drying ponds and thus increase their chances to evade death from dessication.

However, this apparently adaptive behavior comes at the price of a smaller size at metamorphosis and less well-developed hind limbs, factors which are likely to reduce juvenile survival. Nevertheless, this flexible larval strategy appears adaptive in an environment in which gravid females are seemingly unable to predict pond duration. A large group of examples of phenotypic plasticity involves predator avoidance. The snail *Physella virgata* delays sexual maturity in the presence as opposed to the absence of a predatory crayfish. Small freshwater cladocerans of the genus *Daphnia* develop a variety of defensive structures such as helmets, spines, and neckteeth when they receive chemical cues from predators. Some plants produce defensive chemicals once their leaves have been damaged by herbivores. The sun and shade leaves of many plants are another well-known example of phenotypic plasticity.

In heterogeneous environments, phenotypic plasticity is clearly a valuable strategy if it allows organisms to produce the locally optimal phenotype. The evolution of such adaptive plasticity depends on intricate control mechanisms: relevant environmental cues need to be obtained and processed such that the appropriate developmental switches are triggered in time. The ideal norm of reaction matches the local optima in all environments that an organism is likely to encounter. However, this might be difficult to achieve in reality. Environmental cues might provide only imperfect information about future conditions such that the phenotypic expression shows a time lag relative to the optimum. In addition, adaptive phenotypic expression across environments can be constrained by the underlying genetic machinery. For example, a mutation that increases egg size by a constant amount in two environments might move the population mean closer to the optimum in one environment but could cause a shift away from the optimum in the other. What would be preferable instead is a mutation that has an environment-specific beneficial effect on egg size. Moreover, the norm of reaction is more likely to be optimized for those environmental conditions that are frequently encountered rather than those that occur only rarely. The norm of reaction as a whole can be viewed as a trait in its own right. It has a genetic basis and is subject to the selection pressures that particular patterns of environmental heterogeneity produce. As such it is of concern to the ecological geneticist.

D. Are the Ecological Variants Adaptive?

The question of whether the observed phenotypic variation is adaptive arises similarly for genetic variants and phenotypic plasticity. There are a number of different methods to demonstrate adaptation. For example, one can try to observe the fitness consequences of phenotypic variants directly. In the butterfly *Colias meadii*, male fitness is determined partly by flight performance. Males that are capable of prolonged flight and effective courtship display have higher mating success than those that fly less well. A significant proportion of the variation in flight ability and mating success is explained by the males' genotype at two particular enzyme loci. Both phosphoglucose isomerase (PGI) and phospoglucomutase (PGM) play key roles in the glucose metabolism and thus in the energy supply of the flight muscles. At each of these loci, heterozygous males had better mating success than homozygous males. It appears that heterozygotes have a superior balance of enzyme kinetics and stability.

Adaptive differentiation in Trinidadian guppies, *Poecilia reticulata,* has been subjected to a particularly detailed analysis. There is marked small-scale differentiation among populations in the intensity of the males' color, despite the fact that all females tend to prefer more brightly colored males. The observed differences were shown to represent a balance between sexual selection and varying predation pressures in different habitats. Differences between populations in schooling behavior matched a priori expectations based on predation levels. The action of natural selection could be directly observed in guppies that were transplanted from a high to a low predation site in Trinidad in 1976. Within a few years, the males had become brighter in coloration. In addition there was genetic divergence in life history traits that continued to intensify over 11 years. A second experiment in a different river in Trinidad prompted similarly dramatic phenotypic evolution.

III. ECOLOGICAL VARIANTS IN SPATIALLY STRUCTURED POPULATIONS

A. Scale

Given abundant genetic variation within populations and large-scale environmental heterogeneity, one might expect widely separated populations within a species range to evolve different ecological characteristics. It is much less obvious how closely organisms can track small-scale environmental heterogeneity. Intuitively, one might expect that there is a minimal scale of spatial variation below which local adaptation is impossible. This idea is closely allied to the concept of environmental grain. If environmental conditions change on a smaller scale than that on which individual organisms operate, then the environment is fine grained and perceived in an averaged form. On the other hand, the environment is coarse grained if organisms are exposed to only one of several environmental conditions within their lifetimes. The relative scales of spatial variation in selection pressures and of individual dispersal should thus be critical to the question of local adaptation in heterogeneous environments. Furthermore, the fate of new favorable mutants is influenced by stochastic forces, which can slow the rate of adaptation. The strength of this effect depends in turn on the population structure of the organisms.

B. Selection versus Gene Flow

Consider first the case in which a large, randomly mating population exchanges migrants with the remainder of the species gene pool such that a proportion m of the population is replaced with immigrants every generation. In this population, an allele a at a given locus has arisen with a selective advantage s over the wild-type allele A. This so-called selection coefficient s applies to individuals with the heterozygous genotype Aa relative to the wild-type homozygote AA. The other homozygote aa will initially be too rare to affect the dynamics. For every offspring produced by an AA individual, a heterozygote will on average produce $1 + s$ offspring. Under these conditions, a can become established in the population as long as $s > m$; it will be lost due to the influx of AA immigrants otherwise. In the latter case, migration outweighs selection.

This model is simple, because the only structuring element is the boundary of the population which coincides with the area in which the allele a is favored. A much fuller analysis of spatial dynamics is possible if one assumes a uniform distribution of organisms in space. No populations are delineated. Instead the subset of organisms with which any one individual is likely to interact is determined by the mean dispersal distance, denoted as σ. More specifically, σ represents the standard deviation of the distribution of distances between the organisms' places of birth and reproduction. The dispersal direction is assumed to be random. Thus individuals interact on a local scale that is determined by their movement range, but given enough time genes can in principle flow through the entire gene pool without encountering any physical barriers.

Assume further that the allele a has a selective advantage s inside an "environmental pocket" of specified size and a disadvantage $-s$ outside this region. The minimum size of an environmental pocket in which a could become established is determined by the critical scale $l_c = \sigma / \sqrt{s}$. As intuition suggests, this scale increases with the dispersal range of the organisms and decreases with the strength of selection. In two dimensions, the minimum size of the environmental pocket has to be greater than $1.66\, l_c$. Consider for example a snail species with a mean dispersal distance of $\sigma = 30$ m. If a locally favored allele has a fitness advantage of 1%, then the minimum diameter of the pocket is roughly 500 m. In order to halve that diameter, selection would have to be five times stronger. It follows that the environmental grain associated with local adaptation is often much larger than the movement range of the organisms, even in the case of fairly strong selection.

Moreover, the establishment of the locally favored allele does not imply that it will reach fixation throughout the pocket. It might do so only in a small central portion of the range. At the edges, a cline in allele frequency should form whose steepness again depends on l_c: weak selection or a wide dispersal range would create a shallow cline, which allows the alternate allele to occur at appreciable frequencies inside the pocket. The flux of alleles across its edge further implies that the mean fitness of organisms on either side of it will be reduced relative to areas where the locally favored allele is fixed. The distribution of genotypes thus cannot match the local optimum perfectly, unless the selection pressure is very strong indeed. This conclusion depends on the assumption of a sharp ecotone, where the selection coefficients reverse their sign. The effect on mean fitness would be less pronounced if there was a more gradual environmental change.

The same modeling framework can be used to study the dynamics of quantitative traits in heterogeneous environments. Instead of the change in allele frequency, one considers the change in mean phenotype across an ecotone. Phenotypic variance around that mean is created at any one location by a moderate to large number of segregating loci and by environmental effects on the trait. Natural selection tends to shift the phenotypic mean toward the local optimum. The strength of selection is expressed as the rate of decline in fitness on either side of that optimum. The response of a quantitative trait to spatially varying selection is again governed by the balance between this strength of selection and σ. In addition, the effectiveness of selection depends on the heritability of the trait: a low heritability limits the influence that natural selection can have on the spatial distribution of phenotypes.

The swamping of local adaptation has been documented in a number of empirical studies. Blue tits

(*Parus caeruleus*) tend to lay their eggs shortly after the bud burst of trees such that the peak abundance of caterpillars on fresh foliage coincides with the greatest food demand of the nestlings. Around the Mediterranean coast of France, bud burst occurs later in evergreen forest than in deciduous forest. Blue tit populations in deciduous forest on the mainland, where this forest type predominates, lay their eggs at the "correct" time. On Corsica, where mostly evergreen forest occurs, egg laying happens later, again in accordance with the local food supply. However, mainland populations in small patches of evergreen forest lay their eggs too early, which entails reduced fecundity. Genetic data indicate that this apparent maladaptation is to due to gene flow from predominant deciduous forest.

Similarly, a small riparian population of the spider *Agelenopsis aperta* has diverged from a second population in dry grassland. In contrast to the grassland site, the riparian population experiences less thermal stress, has a better supply of food, but is subject to stronger predation pressure. There are a number of genetic differences between the two populations. The riparian spiders defend smaller territories and are less likely to escalate when an intruder appears. Their generally more fearful behavior also results in a lower prey capture rate. Despite the qualitative agreement with the local selection regime, the spiders still deviate significantly from the optimal strategy as predicted by detailed theoretical models. They still defend excessively large territories, and they show too much aggressive behavior and too little selectivity in their diet. However, a second and much more isolated riparian population does match the model predictions. Direct and indirect measures of gene flow show that the first riparian site cannot reach the local optimum due to massive influx of genes from the surrounding grassland.

On the other hand, evidence for very local adaptation due to strong selection comes from the classic studies of metal contaminated mine sites. Metal tolerant plant genotypes on the mine are separated from nontolerant genotypes on normal soil by only a few meters. Similarly, in the marine snail *Littorina saxatilis*, a cline in allele frequency exists at the aspartate aminotransferase (AAT) locus over only 7 to 10 m at the transition from the submerged surf zone to the splash zone of a rocky shore. After a toxic algal bloom in one year killed all snails in the surf zone, recolonization from the splash zone initially obliterated the allele frequency cline. But within two years, the original allelic distribution was reestablished. The selection coefficient necessary to produce the observed differentiation over such a small spatial scale was estimated to be 0.4.

C. Selection versus Genetic Drift

Chance effects are an integral part of the passage of genes from one generation to the next. Some individuals will always leave more offspring than others even if all have identical fitnesses. If by chance all bearers of a given allele leave more than the average number of offspring, the frequency of that allele in the population increases even though it may not confer a selective advantage. This process of random change in allele frequency is called genetic drift. It interferes with the establishment of favorable mutants and may limit adaptation by reducing the amount of genetic variance in phenotypic traits.

The standard model of drift in a finite population is due to Fisher and Wright (see Box 2). The key feature is the random variance in reproductive success, which poses the risk of imminent chance loss to any rare allele. Even a favorable mutant may disappear from one generation to the next, if it exists in only a few copies and if all its bearers happen to leave no offspring. The probability that an advantageous mutant becomes established in a Wright-Fisher population is $2s$, where s is the selection coefficient of the heterozygote. Thus most mutants that confer a fitness advantage of, say, a few percent will be lost from the population due to chance before they reach appreciable frequencies. In large populations, the deterministic effect of selection begins to dominate once an allele has reached a threshold copy number of roughly $1/s$. In very small populations, drift influences the fate even of common selected alleles. In fact, an allele will be effectively neutral irrespective of its frequency as long as $s < 1/2N$, where N is the population size.

Although many local populations may be small, few of these will be completely isolated. It is therefore more realistic to consider the fate of a favorable mutant in a collection of local populations or demes that are connected through gene flow. Imagine a large number of Wright-Fisher populations that exchange individuals at a rate m through a migrant pool. Allele frequencies in each group of immigrants represent the average for the total population. Gene flow is thus an entirely deterministic force in this model. Moreover, there is no spatial structure, because all demes are equally connected. Taken together these assumptions make up the so-called island model of population genetics. Remarkably, population subdivision here has no effect on the fixation probability of a favorable allele: fixation is equally likely in a set of n demes of size k as in an unstructured population of size nk. The reason for this is that the probability of chance loss of a rare mutant depends on its copy number and not on the size of the population in which it arose.

Although population subdivision per se does not affect fixation probabilities, the addition of local population dynamics does. There is copious empirical evidence in many organisms for pronounced fluctuations in local abundance including the occurrence of local extinctions and (re-)colonizations. The idea that species can persist regionally despite locally transient populations is central to metapopulation theory. As long as changes in abundance are spatially uncorrelated and migration is sufficiently common to ensure the colonization of empty habitat, these local numerical fluctuations tend to average out on the larger scale of the metapopulation. In the population genetic context, fluctuations in population sizes increase the random variance in per capita reproductive success. The more demes within a metapopulation vary in allele frequency, the more likely it is that alleles are differentially affected by local changes in abundance. Drift is thus a stronger force than in the island model with constant local population sizes. Consequently, newly arisen beneficial mutations are yet more likely to be lost again while they are still rare. However, once they have spread to a few demes, their subsequent fate should be determined by selection.

At present, it is difficult to assess the importance of

Box 2

Wright-Fisher Model

Assume an isolated sexual, diploid population of constant size N with nonoverlapping generations. To create the next generation, all individuals contribute evenly to an effectively infinite gamete pool. From this pool $2N$ gametes are drawn that unite at random and represent the adults of the next generation. For a locus with two alleles, this sampling process implies that the allele frequency in generation $t+1$ follows a binomial distribution centered on the frequency in generation t. Thus, there will be successive random changes in frequency through time. Eventually, this process will lead to the fixation of one or the other allele in the population. The probability that a given allele will go to fixation increases with its initial frequency. The average "stepsize" of the random changes per generation increases with decreasing population size. Thus, the time to fixation decreases with population size.

random drift for the establishment of new ecological variants, because we do not know how much adaptation is limited by the supply of novel mutants. Artificial selection experiments suggest that there is typically sufficient standing genetic variance in natural populations to allow for a response to selection. If intraspecific phenotypic evolution depends mostly on shifts in the frequency of existing alleles, then the main role of random drift might be to slow the process of adaptation rather than to inhibit it altogether. This may be of little consequence, if there is only spatial variation. However, environmental heterogeneity typically has a temporal as well as a spatial component. If the local optimum changes through time, then a delay in the adaptive response due to drift might keep the population away from the optimum for most of the time.

IV. INFERENCES FROM THE SPATIAL DISTRIBUTION OF NEUTRAL MARKERS

A. Direct and Indirect Measures of Gene Flow

Gene flow clearly plays a central role in the dynamics of ecological variants in heterogeneous environments. Its role has been the matter of considerable controversy in the past. Does wide ranging gene flow impose limits on intraspecific differentiation or is gene flow on the contrary so limited that populations of a species behave as nearly independent evolutionary units? Based on the available evidence, the latter view is closer to the truth. Nevertheless, it is difficult to predict the potential for local adaptation and differentiation for any given species. Reliable estimates of gene flow are thus highly desirable. They might be obtained by monitoring the movement of marked individuals. However, such direct estimates typically give an underestimate of gene flow. Long distance dispersers will often be missed, yet they play an important role in the spread of genes. Mark-recapture studies provide a snapshot of dispersal whereas a longer-term average is required for evolutionary inferences. On the other hand, not all observed movement necessarily leads to gene flow. For example, immigrants might be less successful than residents in the competition for territories. Much effort has therefore been devoted to the developments of indirect measures of gene flow that can provide a suitably averaged estimate. Most of these methods are based on the spatial distribution of neutral genetic markers.

Consider again the island model: a large number of demes, each of constant size N, are connected by gene flow at a rate m via a common pool of migrants. The overall frequency of a certain selectively neutral allele in the population as a whole, and consequently in the migrant pool, is $\bar{p}$. Genetic drift within demes produces variance in p across demes as a function of local population size. Without gene flow, this process would inevitably lead to the random fixation of one or the other allele in each deme. With gene flow there is an equilibrium amount of differentiation: the divergence among demes due to drift is balanced by the homogenizing effect of gene flow such that the variance in p, V_p, is constant. For a locus with two alleles, Sewall Wright defined the standardized variance in allele frequency as $F_{st} = V_p/\bar{p}$ $(1-\bar{p})$ and showed for the island model that

$$F_{st} \approx \frac{1}{1 + 4\,Nm}.$$

The degree of differentiation among demes in the island model thus depends on the number of migrants per deme per generation. Consider a given combination of N and m. An increase in N reduces random drift such that a smaller migration rate suffices to keep V_p constant. The two forces exactly balance each other. The analysis of the island model shows that only a small amount of gene flow is enough to maintain neutral genetic variability within demes. The threshold number of migrants below which there is a tendency for demes to fix by chance one or the other allele is $Nm = 0.5$, which is equivalent to one migrant every other generation or an F_{st}-value of 0.33.

B. Interpreting F_{st}

At equilibrium, F_{st} expresses the balance between gene flow and drift. As such it can be estimated from the distribution of neutral alleles in any set of populations. However, in principle no inference can be drawn about the relative importance of the two component processes. A given F_{st} estimate may stem from a set of large populations that are connected by weak gene flow or from small populations that exchange migrants at a correspondingly higher rate. Even the interpretation of Nm as the number of migrants is problematic, because real populations differ in important respects from the island model.

Consider populations that fluctuate in size around a certain mean abundance $\bar{N}$. Random drift is disproportionately affected by the smallest population sizes. Therefore, allele frequencies are subject to stronger stochastic forces in this fluctuating population than in a population of constant size $\bar{N}$. In many species, the variance in reproductive success is much larger in males than in females: few males may sire most of the offspring while others do not contribute to the next generation at all. The pool of paternally transmitted alleles is therefore much smaller than the census number of males would suggest. Again, drift forces are stronger as a result and correspond to those in a much smaller ideal population (cf. Box 2). The effective population size N_e of a natural population is equal to the size of an ideal population that experiences the same amount of genetic drift. In the previous two examples and in natural populations in general, N_e will be much smaller than the census population size. It follows that Fst can only be translated into the rather abstract quantity $N_e m$. F_{st} describes a pattern of differentiation. Without additional information about the population structure at hand, it is difficult to infer a process from it.

Further complications arise when F_{st} estimates are low due to recent common descent of the populations rather than high levels of gene flow. Imagine a species that rapidly colonizes an area of patchy habitat from a large pool such that all local populations start out with nearly identical allele frequencies. If gene flow is subsequently reduced to a low level, differentiation among demes will increase with time until the migration-drift equilibrium is reached. During this time period, estimates of F_{st} reflect past history as well as current processes. The relevant time scale in generations is given in the island model by $1/m$ for $N_e m < 1$ and by N_e otherwise. These results imply that F_{st} equilibrates faster than any other measure of genetic differentiation.

C. Extrapolation to Selected Variants

Finally, it is difficult to extrapolate from the dynamics of neutral genes to those that govern selected variants. The fate of a novel mutant is in the first instance determined by the balance between selection and drift within a deme, which depends on s and N, whereas its spread to other demes, once it has reached appreciable frequencies, depends on m. Thus, predictions about the fate of such a mutant in structured populations require knowledge of s, N, and m separately. Different population structures that give rise to the same F_{st} could differ with respect to adaptive processes.

There is a large body of empirical literature on indirect measures of population structure that covers a broad range of taxa. These studies indicate that the balance between gene flow and genetic drift allows in most cases for no more than moderate differentiation in allele frequencies at neutral loci. Most F_{st} values lie between 0 and 0.1; only very few exceed Wright's threshold of 0.33, which implies that most demes tend to be near fixation at a given locus for some neutral allele. The typically low estimates of F_{st} indicate that

local populations are unlikely to be so small and isolated that adaptive processes are severely compromised. The populations are either large and should therefore generate a reasonable supply of new mutations on their own or they are so well connected that they receive a steady stream of new alleles from elsewhere. Even if genetic markers indicate limited within-population variation, there may still be considerable genetic variance in phenotypic traits. In the latter, genetic variation can usually arise at a large number of loci, whereas the single locus mutation rate strictly determines variation in markers.

Limited differentiation at neutral loci does not imply uniformity in adaptive traits. This is illustrated by a number of studies. For example, Scots pine, *Pinus sylvestris*, has a circumpolar distribution occurring from Western Europe across Siberia to northern North America. Allozyme studies have detected only very slight genetic structure across this vast distribution range ($F_{st} = 0.02$). However, direct assessments of the fitness of transplanted versus locally grown trees indicate important differentiation in frost hardiness, the onset of flowering as well as a number of other ecologically important traits. Similarly, the endemic perennial *Primula scotica* has a fixed genotype at 14 out of 15 allozyme loci. Yet common garden experiments reveal significant heritabilities in 6 out of 11 morphological characters. The number of seed capsules, a fitness-related trait, has a heritability of 0.25. Moreover, roughly one-third of all phenotypic variation is due to differences between populations. Clearly, the remaining 15 populations of this species in the far north of the British Isles have not lost their ability to respond to environmental change.

V. THE GENETIC BASIS OF ECOLOGICAL VARIANTS

A. Simple versus Complex Traits

Every phenotypic trait results from a complex interplay of many gene products that influence its mature expression over the course of ontogeny. A mutation at any one of the underlying loci can in principle alter a trait. Yet only those loci that actually are polymorphic at any one time contribute to the trait's heritable variation. It is the genetic basis of phenotypic variation and not of the phenotype per se that is relevant in the context of evolutionary analysis. Traditionally, traits have been referred to as simple if their heritable variation could be attributed to a small number of loci on the basis of crossing experiments. In contrast, a complex trait is typically based on many polymorphic loci as well as on random environmental variation. Its evolutionary dynamics are best analyzed statistically in the framework of quantitative genetics. However, recent developments suggest that the dichotomy of simple versus complex traits is somewhat misleading. Whether few or many loci determine phenotypic variation appears to be more due to the nature of selection than to be an inherent property of the trait. Moreover, modern genetic techniques demonstrate that loci of large effect are segregating for traits that were thus far believed to fit the quantitative genetic model, which assumes a large number of loci, each of which has a small effect on the phenotype.

B. Are Major Mutations Important for Adaptation?

Mutations of large effect have until recently been thought to play only a limited role in phenotypic evolution. After all, when a large and random change is applied to an integrated organism, it is much more likely to cause harm than good. According to this view, such major deleterious mutations are efficiently removed from the gene pool by natural selection, and most adaptive evolution occurs through the accumulation of mutations of small effect. Recent theoretical analyses (Orr, 1998), however, have cast a new light on the genetic basis of adaptation. The contribution of large mutations depends on their frequency of occurrence, their chance of being advantageous, and finally on their chance of fixation given that they are advantageous. The first two criteria indeed limit the importance of large mutations: they are presumably rare and are relatively unlikely to improve fitness. Nevertheless, the small set of large advantageous mutations will be important for evolution if selection is strong and if the current mean phenotype in the population differs from the optimum by a large amount. As adaptation moves the mean toward the optimum, mutations of decreasing effects should become fixed. One should thus expect that especially adaptation to novel conditions involves both mutations or large and of small effect, which tend to become fixed early and late in the process, respectively.

C. The Strength of Selection Shapes Genetic Architecture

The foregoing discussion suggests that the genetic basis of ecological variants is intimately linked to the nature of selection. In fact, the classic case studies of ecological genetics such as the banding patterns of snail shells and the color morphs of butterflies have often been considered to be special cases because of the apparent simplicity of the traits. Although they may well be endpoints of a spectrum, it appears that this spectrum is defined not by some measure of trait complexity but by the strength of selection involved. It is also not surprising that many of these adaptations such as the cryptic dark coloration of moths on sooty tree trunks (industrial melanism) are responses to strong anthropogenic selection pressures.

The application of insecticide is an example for the most extreme selection pressure imaginable: the aim is to kill 100% of the insect population. All cases of insecticide resistance that have been analyzed in detail are based on one or two loci of major effect which are able to transform a susceptible population into a resistant one. They need to be either present in the base population before the application of the pesticide or arise soon thereafter. There are some indications that mutations that allow for adaptation to extreme anthropogenic selection pressures might be in short supply. For example, in the blowfly *Lucilia cuprina*, resistance to each of two different insecticides produced resistances at the same two loci in replicate selection lines. Under less stringent artificial selection, polygenic resistances were observed involving different loci in different lines. These results imply that there were many avenues to evolve weak resistance, but there was only one genetic way to achieve resistance to a given insecticide applied at field strength. Nevertheless, the rapid evolution of resistances in insects and weeds argues against a strong limitation on available mutations.

Minor loci that contributed to a polygenic response were similarly observed in plants that had evolved heavy metal tolerance: there was heritable variation within distinct resistance classes, each of which corresponds to a major mutation. Similarly, in the case of industrial melanism in *Biston betularia*, a trait that is largely attributable to a single dominant allele, one finds additional alleles of relatively smaller effect. These observations underscore the point that the genetic variation in any of these traits is based on a spectrum of allelic effects. Which of these alleles contributes to a given adaptive response should depend on the selection pressures and the difference between the population mean and the optimum.

In comparison to the application of insecticides, artificial selection experiments are more benign because the experimenter needs to keep the selected lines going. Here too, alleles of large effect contribute a large part of the selection response, but typically the response is polygenic. The critical question is how often natural selection on ecological variants is so strong that major mutations are fixed. It seems likely that microevolutionary processes that are of interest to the ecological geneticist will often fall into that category simply because only the rapid response to strong selection is easily observable. Moreover, all of the preceding examples involved so-called directional selection in which the population mean is adjusted to match a new phenotypic optimum. If the population mean is at the optimum, then selection is stabilizing and mutations that cause large deviations from the mean are strongly selected against. The relative importance of stabilizing versus directional selection in natural populations depends on whether or not populations are typically in a state of selective equilibrium. This is at present an unanswered question of considerable interest.

D. QTL Mapping

Even those traits that so far have appeared to have a complex genetic basis have recently become amenable to detailed genetic dissection. The critical methodological advance has been the development of a nearly unlimited supply of genetic markers. These can be used to identify relatively small genomic regions that carry one or more loci affecting a given trait. These regions are referred to as quantitative trait loci or QTLs. The approach requires two or more populations that have diverged in some character of interest and ideally have fixed alternate alleles at each marker locus. Through controlled crosses, one initially establishes a so-called linkage map that represents the relative location of the markers to each other. With the help of further crosses, one searches for statistical associations between the expression of a given trait and a marker locus. Such an association is equivalent to finding a QTL for that trait in the vicinity of the marker. Studies of this type have uncovered a surprising number of loci of large effects that segregate in natural populations. Thus major loci may explain a large proportion of phenotypic variation even in traits that show continuous variation and conform phenotypically to the model of quantitative genetics.

E. Ecological Consequences of Genetic Structure

Whereas the genetic basis of a selection response might be in part determined by the selective regime under which it arose, a given genetic structure can subsequently influence ecological processes. A famous example in this context is the analysis of the mimetic colorations of *Heliconius* butterflies. Both *H. melpomene* and *H. erato* exist in a number of distinct color morphs or races that occur in nonoverlapping regions of their respective species range. Both taxa are unpalatable to avian predators. In any one location, birds learn to avoid the local wing pattern that is shared between co-occurring *Heliconius* species form a so-called mimicry ring. The maintenance of local color morphs is due to frequency dependent selection: the locally frequent wing morph is well protected against predation because birds have learned to avoid it, whereas any rare morph that might appear suffers predation from naïve birds that have yet to learn about its distastefulness.

Through the analysis of numerous hybrid crosses between the local races, the total phenotypic variation of the different color morphs within each of the two species could be attributed to 22 and 17 loci respectively. Some of these have large effects on the formation of pattern components such as wing bands and spots, whereas others modify some detail of an existing pattern. There are complex dominance relationships as well as interactions among loci, such that the effect of a given allele at a particular locus depends on the individual genetic background in which the allele is placed (epistasis, cf. Box 1). For any one pair of geographically adjacent mimetic forms within a species, one finds genetic differences at no more than four major loci plus some small number of modifier loci. Nevertheless, this number of loci is sufficient in principle to produce a large number of recombinants upon interbreeding in areas where the racial distribution ranges adjoin. These animals with recombined wing patterns should suffer greatly increased predation, as any of the local birds would not recognize them as distasteful.

In contrast, there is another mimetic system that involves the swallowtail butterfly *Papilio memnon*. This butterfly is edible, but females gain protection from predation because they mimic other distasteful species. In this case, one finds several different female morphs within populations, which are due to allelic variation at a so-called supergene consisting of several tightly linked loci. As a consequence, only a limited number of phenotypes is produced, each of which may be mimicking different models. The high fitness of these morphs is in part a consequence of the underlying genetic structure. The existence of several unlinked loci as in the case of *Heliconius* would preclude intrapopulational variation in color morphs, because too many of the possible recombinants would not be adaptive.

VI. LIMITATIONS TO THE EVOLUTION OF NEW ECOLOGICAL VARIANTS: TRADEOFFS

A. Antagonistic Pleiotropy

Although the broad spectrum of genetically determined ecological variation within species underscores the efficacy of natural selection, organisms are not infinitely malleable. Even if a trait displays heritable genetic variation, it may not be able to evolve toward a new phenotypic optimum, because the necessary genetic changes could be deleterious with respect to some other trait. If a given set of polymorphic loci affects more than one phenotypic trait, these loci are said to have pleiotropic effects and the traits in question are genetically correlated. If an adaptive change in one trait causes a fitness reduction in a correlated trait, one speaks of antagonistic pleiotropy. In that case, the realized set of phenotypes in a given natural population can at best represent the optimal tradeoff between the conflicting selection pressures on correlated traits.

Direct evidence for such antagonistic pleiotropy comes from a number of laboratory experiments in which a given trait is altered by artificial directional selection for a number of generations. When the selection regime is then relaxed, one observes in the following generations a return of the trait mean toward the original value. This reversal reflects natural selection in the opposite direction due to some correlated trait.

B. Life History Tradeoffs

Strict negative genetic correlations for fitness are at the heart of life history theory. The schedules of growth, reproduction, and mortality together determine the life history of a species. Although one can imagine an arbitrarily fit organism that maximizes its fitness through independent optimization of each life history component, actual life histories are constrained through a web of genetic correlations. The most obvious examples are those that have to do with the allocation of energy. For any one bout of reproduction, for example, a given amount of energy can be allocated to few large or many small offspring. Within-population genetic variation in offspring size and number implies polymorphism at one or more allocation loci. More often, life history tradeoffs involve time delays. For example, deer hinds (*Cervus elaphus*) with fawn enter the winter season with smaller

fat deposits and suffer a higher mortality than those that did not have a fawn that summer. Similarly, female collared flycatchers (*Ficedula albicollis*) that first reproduce at the age of 1 have lower subsequent reproductive success than those that delay first reproduction until they are 2 years old. Aside from tradeoffs between current and future reproduction, those between early reproduction and life span and between reproduction and growth have been demonstrated. In most cases, their physiological and genetic basis remains to be discovered.

Genetic variation in life history strategies allows for their adaptive adjustment. For example, an environment in which adults suffer high mortality from predation selects for earlier reproduction. This strategy brings a net gain in fitness, even if there is a correlated increase in intrinsic adult mortality. The latter is of no consequence in this environment. Due to their intimate relationship with fitness and strong dependence on ecological conditions, one expects and does in fact find abundant intraspecific variation in life history strategies. Moreover, by its very nature local life history adaptation creates genotypes that trade off their fitness in one habitat against that in another one and thus creates ecological specialization.

C. Threshold Traits

A remarkable illustration of a tradeoff is the wing dimorphism in a variety of different insect species. Even within populations, one finds individuals with large functional wings as well as those that are flightless. This is an example of a so-called threshold trait, in which a polygenic basis produces two distinct phenotypes through a developmental switch. Winged individuals are clearly better able to disperse when local conditions deteriorate, yet the production of flight muscles diverts energy away from other organismal functions. The relative proportion of the two types tends to correlate with the local ecological conditions. In the cricket *Gryllus firmus*, the production of a functional flight apparatus entails reduced fecundity in females and the production of a less attractive, shorter mating call in males. The tradeoff, however, goes beyond this simple dichotomy in energy allocation between the two morphs and involves an intricate genetic correlation: in families with a high proportion of long-winged offspring, even flightless sons produce a relatively shorter call than is seen in flightless male offspring from mostly short-winged families.

D. Are Genetic Correlations Constant?

Genetic correlations need not be unbreakable. Like heritabilities, they reflect the genetic architecture in a given population at a given time. For example, only in one of two populations of *Rana sylvatica* was there a strong and positive genetic correlation between development time and size at metamorphosis, while the two traits were uncorrelated in a second location. With time, new mutations can change an existing genetic link between two traits. In the sheep blowfly, *Lucilia cuprina*, the newly evolved resistance to an insecticide initially entailed a fitness cost in the absence of the toxin. This cost was subsequently eliminated through the appearance of additional mutations. These so-called modifiers restored the fitness of the resistant genotype to the same level as that of susceptible wild type. Nevertheless, genetic correlations undoubtedly constrain phenotypic evolution especially over the relatively short time scales that are of interest to the ecological geneticist. They are a likely explanation for the fact that plant genotypes that confer resistance to herbivores typically do not reach fixation in natural populations. Even though genetic correlation may often not be strong enough to rule out an adaptive response, they can certainly slow the process so that a given population might not be able to track spatio-temporal environmental heterogeneity efficiently.

VII. ECOLOGICAL GENETICS OF SPECIES INTERACTIONS

The joint analysis of population genetics and population dynamics has long been on the agenda of ecological geneticists. In fact, the integration of these two large bodies of theory seems like a natural focus in a field of research that is by definition positioned at the interface between ecology and genetics. Yet progress has been hampered by the conceptual and practical complexity that any natural study system poses. Most progress has been made in the fields of host-pathogen and host-parasite interactions.

If several host genotypes interact with several pathogen genotypes, then the question arises how such genetic diversity can be maintained. Imagine a system in which a host population harbors a resistance alleles for every virulence allele in the pathogen population, but each host individual carries only one such allele. The host is resistant against the pathogen strain with matching allele, but it will be infected by all other pathogen strains. Frequency dependence in such a system can in principle lead to a stable equilibrium of multiple genotypes, but the approach to equilibrium in a single population is typically marked by pronounced fluctuations in allele frequency such that most variants could be lost in the process. In a single population of constant size, the system only yields abundant polymorphism when there is a large fitness cost to resistance and virulence. However, if one allows for numerical and genetic dynamics and moreover adds spatial structure, coexistence is possible even under much reduced costs. Genetic types that have been lost locally can be reintroduced through gene flow. Even if the disease dies out in a local population because only resistant host genotypes are left, a colonization event elsewhere may help to reset the system. Stochasticity in local encounters breaks the violent oscillations of frequency-dependent dynamics.

Metapopulation dynamics of this sort have been extensively documented in the interaction of the plant *Silene alba* and its pathogen *Ustilago violacea*. Heritable resistance levels within a host population ranged from 0 to 100%. However, little is known in this system about pathogen virulence types. A lack of close matching in the occurrence of host and pathogen strains was observed in the Australian flax *Linum marginale* and its fungal pathogen *Melampsora lini*. The local co-occurrence of genotypes of host and pathogen in a set of interconnected populations gave little indication that the latter tracked the frequencies of their hosts. Instead, the pattern appears rather haphazard. Gene flow together with local extinctions and colonizations are likely causes of this pattern.

One of the ecological consequences of diverse host and pathogen genotypes is that the chance of epidemics is small and leads to no more than local outbreaks. The metapopulation dynamics are also likely to affect the evolution of resistance and virulence. Consequently, the joint numerical and genetic dynamics are relevant to the process of coevolution. The so-called geographic mosaic theory of coevolution postulates that the evolutionary interactions among closely interacting species vary qualitatively from one population to another depending on the transient local mix of genotypes. In a few places at any one time, these local interactions may be strong and persistent enough to produce truly coevolutionary dynamics. This idea integrates within-population genetic and numerical dynamics with metapopulation dynamics and spatially varying selection. Complex dynamics of this type clearly are commonplace in nature. The development of this theory promises new insights into microevolutionary processes of species interactions.

See Also the Following Articles

DIFFERENTIATION • GENETIC DIVERSITY • INSECTICIDE RESISTANCE • PHENOTYPE, A HISTORICAL PERSPECTIVE

Bibliography

Avise, J. C. (1994). *Molecular markers, natural history, and evolution.* London: Chapman and Hall.

Endler, J. A. (1986). *Natural selection in the wild.* Princeton: Princeton University Press.

Falconer, D. S., and T. F. C. Mackay. (1996). *Introduction to quantitative genetics.* (4th ed.). Harlow: Longman.

Ford, E. B. (1964). *Ecological genetics.* London: Methuen & Co.

Foster, S. A., and J. A. Endler (Eds.) (1999). *Geographic variation in behaviour: perspectives on evolutionary mechanisms.* Oxford: Oxford University Press.

Fritz, R. S., and E. L. Simms (Eds.). (1992). *Plant resistance to herbivores and pathogens.* Chicago: University of Chicago Press.

Grant, B. R., and B. R. Grant (1989). *Evolutionary dynamics of a natural population.* Chicago: Chicago University Press.

Hanski, I., and M. E. Gilpin (Eds.). (1997). *Metapopulation dynamics: Ecology, genetics and evolution.* New York: Academic Press.

Hartl, D. L., and A. G. Clark. (1997). *Principles of population genetics.* (3rd ed.). Sunderland: Sinauer Associates.

Hutchinson, G. E. (1957) Concluding remarks. *Cold Spring Harbour Symp. Quant. Biol.* **22**: 415.

Orr, H. A. (1998). The population genetics of adaptation: The distribution of factors fixed during adaptive evolution. *Evolution* **52**: 935.

Real, L. A. (Ed.). (1994). *Ecological genetics.* Princeton: Princeton University Press.

Rose, M. L., and G. V. Lauder (Eds.). (1996). *Adaptation.* San Diego, CA: Academic Press.

Thompson, J. N. (1994). *The coevolutionary process.* Chicago: University of Chicago Press.

ECOLOGY, CONCEPTS AND THEORIES IN

Peter Kareiva* and Michelle Marvier†
*National Marine Fisheries Service and †Santa Clara University

ECOLOGY IS THE STUDY OF THE DISTRIBUTION AND ABUNDANCE OF ORGANISMS. Ecologists attempt to understand the factors that both promote and limit biodiversity. In practice, the science of ecology blends observation of natural systems, manipulation of environmental and biotic factors in the field, more highly controlled laboratory or microcosm studies, and both qualitative and mathematical theory. The major concepts and theories of ecology concern the dynamics of populations as well as the patterns and processes of communities and ecosystems.

GLOSSARY

coevolution Process of reciprocal genetic changes in interacting species that result from the mutual selection pressures that strongly interacting species can exert on one another.

keystone species Species whose presence fundamentally alters the structure or function of an ecological community or ecosystem.

metapopulation Collection of subpopulations connected by dispersal. Subpopulations within a metapopulation may undergo local extinction and subsequent recolonization.

threshold theorem of epidemiology If the density of susceptible individuals in a population falls below some critical threshold, then a disease may not be able to maintain itself in the population.

I. HISTORICAL DEVELOPMENT OF ECOLOGY'S CENTRAL CONCEPTS AND THEORIES

Ecological theory has three major origins: (1) attempts to manage fisheries, pests, or wildlife, (2) explorations of patterns in nature—especially spatial pattern and the apparent order of species assemblages, and (3) investigations of "the balance of nature" (and how that balance can be disturbed). To some extent, ecological theories developed separately in the sub-disciplines of plant and animal ecology. For example, quantitative plant ecologists tended to focus on patterns of plant dispersion and relationships between individual plants and their environments, whereas animal ecologists focused more on population dynamics. But the most important early ecological models came from neither plant nor animal ecologists; instead these pioneering models were largely contributed by scientists from other fields, such as physics, mathematics, and human demography.

The acceptance of mathematical approaches into mainstream ecology has been slow (and some might argue that mathematical theory is still not accepted by many ecologists). The reluctance of ecologists to embrace mathematical models reflects the different ways that empiricists and theoreticians view the world. In particular, whereas field ecologists are typically concerned with the uniqueness of their study systems and pay attention to the role of history and "details," mathematical ecologists often neglect the role of history and gloss over details to arrive at more general results. Fortunately, there has been increasing reconciliation between the theoretical and empirical approaches to ecology, largely because modern ecological theory has matured enough to allow examination of how details such as age structure, spatial heterogeneity, or environmental variation alter the generalities derived from more simple models. In addition, recent ecological theory relies less on "equilibrium analyses," which tended to downplay historical effects.

Whereas many branches of natural science are grounded in some core body of theory and first principles, this is not the case for ecology. Instead, ecological theory is remarkably varied and discordant, with natural selection as the only fundamental "law." Moreover, although one could argue that little in ecology makes sense without the perspective of natural selection, there is a great deal of useful ecological theory that never makes any contact with evolutionary principles. Thus, even though all ecologists accept evolutionary theory, different subfields of ecology do not share a common view of the most important theoretical questions (such as ecosystem analysis versus population biology). However, in spite of this notable lack of a single unified and ascendant ecological theory, current attention to conservation and resource management has stimulated the practical integration of models that range from evolving gene pools to ecosystem nutrient cycling.

II. POPULATION DYNAMICS

Births and deaths are the bottom line of ecology—this is because variation in birth and death rates determines the distribution, density, and dynamics of populations. However, most population models avoid tabulating particular sequences of births and deaths and instead keep track of average rates of change. In most cases, models

of population change are not precise forecasting tools, but rather are conceptual frameworks with which to understand when a population is expected to stay constant as opposed to fluctuate wildly. Models are also valuable for asking how external factors such as harvesting and habitat destruction might alter a population's long-term fate. Two fundamental features of population dynamics are inescapable:

1. Because of resource limitation, per capita rates of change for populations decline with increasing density.
2. If there is any time lag in the response of demographic rates to declining per capita resources, the population may overshoot their resources, then crash to low levels, then rebound, only to repeat the process again.

Formal mathematical representations of these principles have led to elegant models of nonlinear population dynamics. "Nonlinear" refers to the fact that the rate of change in populations varies nonlinearly with changes in population density. As a consequence, under a wide variety of circumstances populations are prone to fluctuate, sometimes so erratically that a graph of numbers through time would appear random, and such that any small changes in initial population size become amplified through time (an attribute called chaotic dynamics). This theoretical result is now widely appreciated by ecologists, but its illumination in the 1970s represented a major advance in the field. First, it emphasizes that populations may fluctuate without any environmental or climatic variation. Second, the phenomenon of chaos illustrates how complicated and unpredictable population dynamics can arise in the simplest systems, if nonlinear feedbacks are strong enough to cause the overshooting of equilibria followed by pronounced population crashes. Prior to the development of a theory of chaotic and nonlinear dynamics there was a tendency to always look to the external environment as the cause of population fluctuations, and to think that complicated patterns of fluctuation must have complicated explanations.

Applications of population theory have largely focused on three types of species: rare species that are of conservation concern, species that reach extraordinarily high population densities ("outbreaks") that cause severe damage to crops or forests, and finally species that are harvested by humans, especially many fish species.

To assess threats to rare species, ecologists have developed a suite of analytical tools collectively known as population viability analysis (or PVA). PVAs are commonly used to estimate the probability of extinction under a range of alternative management strategies or to quantify the minimum number of individuals that will ensure survival of the species. PVA models may incorporate the effects of environmental and demographic stochasticity, catastrophes, and spatial structure and movement (see Section V). One approach that has been quite successful entails detailed demographic analyses that keep track of the birth and death rates for groups of similar individuals. When the groups are defined by age of the organisms, a life table can be developed. However, demographic matrices are flexible enough to incorporate groups based on size, stage, or sex, as well as age. The principal advantage of demographic matrices is that one can ask how sensitive the growth rate for the entire population is to changes in the survival or fecundity of a particular class of individuals. This approach is being used increasingly to compare the efficacy of alternative management scenarios. Analyses of this nature have revealed that adult survival is generally a very sensitive demographic parameter for long-lived species, whereas fecundity is often the most sensitive parameter for shorter-lived species. These findings have shaped a number of conservation decisions—for example, a demographic matrix applied to loggerhead sea turtles made it clear that even substantial improvements in the survival of baby hatchling turtles would provide little benefit to the turtle population, whereas elevating the survival of adults could dramatically help this threatened long-lived species.

For outbreaking or pest species, population ecologists have asked what factors lead to population outbreaks and whether certain life-history traits are shared among those species with a propensity to outbreak. The goal of this research is to determine how best to control these species and prevent future outbreaks from occurring. One quantitative focus concerning outbreaking species has been on the role of time lags. Time lags can occur when, for example, a population does not respond immediately to changes in its own density, perhaps because the individuals store up energy reserves. Alternatively, a population may lag in its response to one of its natural enemies, such as a predator or a disease. Any such time lag, especially when combined with a high potential for population growth, can lead to dramatic oscillations in population abundance.

Finally, for harvested species there has been a focus on determining the maximum sustainable yield (MSY) that can be harvested from a population without placing it at risk of extinction. Clearly most natural populations do not grow without bound, but rather are regulated around some equilibrium density, called the carrying capacity of the environment. This assumption of density-dependent population growth (a reduction in the per capita rate of population growth as population density increases) is central to the calculation of MSY. In particular, theory predicts that the largest number of individuals can be sustainably harvested when a population is maintained at approximately half its carrying capacity. Although density dependence is a central tenet of population ecology and particularly of fisheries management, actually detecting density dependence from a time series of population abundance estimates remains difficult, and discussions about the appropriate methodology have been contentious. A large part of the difficulty in detecting pattern is the random variation inherent in estimates of population size. While part of the variation is due to observer error, a large fraction is often due to random fluctuations in the environment that change either the population growth rate or the equilibrium population size from one year to the next. It is straightforward to investigate in theory the interplay of density dependence and these external sources of variation. The bigger challenge lies in developing management models for real-world data sets in which the apportionment of variation among observer error and different effects of environmental variation is unknown. Fisheries biologists have made the greatest progress in this area, but the theory is still incomplete; it remains especially challenging to connect models of population dynamics to economic models that integrate the biological and economic aspects of resource management.

III. THEORIES OF SPECIES INTERACTIONS

The traditional core of mathematical ecological theory involves dynamical models of pairwise interactions between a predator and its prey, or between two competing species. Both of these types of interaction lead easily to the extinction of species—either because a predator imprudently consumes its entire food base or because a superior competitor usurps so much of the resources that other species cannot make a living. Not surprisingly then, the focus of species-interaction theory has been to ask what processes mitigate the tendency toward extinction in these antagonistic interactions.

Although the elimination of an inferior competitor seems inherent to competitive interactions, in nature we regularly find many species coexisting. Early on, theoretical ecologists suggested that coexistence generally occurs only when the populations of the two species are limited by different resources. In other words, coexisting species must occupy distinct ecological niches. The concept of the ecological niche motivated numerous attempts to quantify just how different two species must be in order to coexist. Measures of various traits related to resource use, such as beak size, body size, and habitat preferences, were often invoked as evidence for "niche partitioning" (the use of different resources by potentially competing species). Careful attention to such data indicated that this evidence was often not as clear-cut as initially assumed, and research was devoted to alternative explanations for the coexistence of competitors. A second generation of competition models made it clear that competing species may readily coexist without any niche partitioning, as long as some form of disturbance interrupts the exclusion of the inferior competitor. Such disturbance can take the form of predation, or abiotic factors such as storms or droughts. Implicit in disturbance-mediated coexistence is the concept of a trade-off between the ability to be a superior competitor and the ability to recover from a disturbance.

As with competition models, early models of predators and their prey suggested that stable coexistence would require certain special conditions. In the absence of these conditions, predators tend to drive their prey to extinction, and then, having exhausted their food supply, go extinct themselves. One answer to the dilemma of predator–prey coexistence was provided by a classic series of laboratory experiments using predatory mites and their mite prey that live on oranges. These experiments demonstrated that, in a simple "landscape" (a tray of oranges), the mite predator was able to quickly find and consume all of its prey, leading to the extinction of both species. However, more complicated laboratory microcosms that impeded the predator's movement could prolong the coexistence of the predatory and prey mites by as much as an order of magnitude— slower predator movement allowed the prey to stay "one orange ahead" of predators and enjoy a temporary refuge from attack. Subsequent mathematical models have made it clear that this is a general principle: predator–prey interactions can be stabilized by any attribute that affords prey some refuge from attack. The refuge can take the form of patches of habitat, or age or size classes of individuals that enjoy low attack rates. With spatial refugia, there is still a tendency for predator–prey systems to fluctuate locally, but different patches will fluctuate out of phase so that the entire collection of patches never goes through a phase in which prey are everywhere absent.

In the last three decades predator–prey theory has been modified to model plant–herbivore and host–disease (or parasite) interactions. These models are similar to classic predator–prey theory in structure and in terms of the possible dynamics associated with the interactions. However, there are also important differences. First, the effects of herbivory are not as obvious as those of predation—individual plants are usually not killed outright by herbivory, and some plant species may actually perform better after being partially consumed. Also, studies of herbivory have typically had more of an evolutionary emphasis in focusing on how herbivores act as selective pressures on plant traits, especially plant structural and chemical defenses.

Host–disease interactions also have been a rich subject for mathematical modeling, with much of the initial theory originating from work on the epidemiology of human diseases. Epidemiological models reveal that disease spread depends on the transmission rate, virulence, latency, and infectious period of the disease, as well as the density of susceptible hosts. More specifically, a disease will be unable to persist if the density of susceptible hosts falls below a threshold value. Immunization programs represent an application of this result, sometimes called the "threshold theorem of epidemiology." Immunization can lead to the elimination of a disease, because if a sufficiently large percentage of the population is vaccinated, an infected individual will, on average, fail to transmit the disease to a susceptible host before the infected host either recovers or dies. Epidemiological models further reveal that diseases can potentially maintain host populations at low levels or cause pronounced cycles in the host population.

Although much theoretical work has focused on interactions within species pairs, species do not, of course, occur only in pairs in nature. Most natural communities include scores or hundreds of different species, simultaneously interacting with one another to various degrees. Recently, increased attention has been paid to relatively complex interactions occurring among multiple species. The food web or, more generally, interaction web has proven to be a useful way to visualize the interactions among large groups of species with vertical links representing trophic interactions and cross-links representing competitive or mutualistic interactions (Fig. 1). Enormous effort has been focused on understanding the relationship between the patterns of connections among species and the traits of the community itself. However, interactions vary in strength, and it is clearly as important to measure the relative strength of an interaction as to merely document its presence.

The study of multiple species interacting simultane-

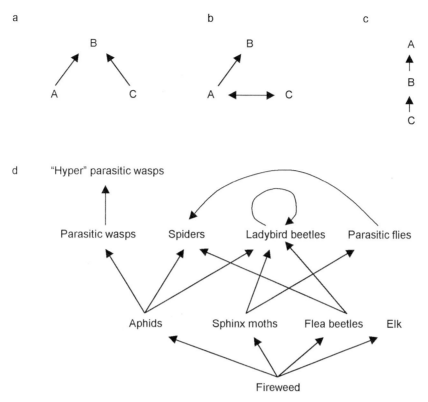

FIGURE 1 Interaction webs demonstrating three types of indirect interaction: (a) apparent competition; (b) predator-mediated competition; and (c) trophic cascade, as well as (d) a more realistic, species-rich interaction web depicting the major interactors associated with fireweed at Mt. St. Helens, in Washington.

ously has led to the recognition that indirect effects can be important determinants of community structure, and several distinct types of indirect effects have been distinguished. For example, apparent competition occurs when two species appear to be competing because, as the population of one species grows, the abundance of the other declines. However, these changes in abundance are not caused by direct competition, but rather by a third species that consumes each prey species in proportion to its relative abundance (see Fig. 1a). A second type of indirect interaction occurs when a consumer species indirectly affects a species via its competitor. For example, a consumer that limits a competitively dominant species may indirectly benefit a competitively inferior species (see Fig. 1b). A third type of indirect effect thought to be common and important is a trophic cascade. An example of a trophic cascade is when a carnivore species limits the abundance of herbivores, resulting in luxuriant plant growth (see Fig. 1c). The presence of a trophic cascade is demonstrated when removal or exclusion of the carnivore results in a dramatic increase in the herbivore population and the sub-

sequent decimation of plant biomass. Field studies are increasingly documenting the ubiquity and importance of indirect interactions. However, the difficulties encountered when studying the simplified food webs depicted in Figs. 1a–1c are greatly exacerbated when trying to understand the interactions occurring in more realistic, species-rich food webs (see Fig. 1d). The significance of this is that simple ecological experiments that focus on just one or two species may fail to reveal the important determinants of community structure.

In sum, theories of species interaction have largely focused on the problem of coexistence or, as it has also been called, the paradox of diversity. These theories address how so many species manage to coexist in nature, whereas simple lab experiments and mathematical models so often predict extinction. A major shift in thinking about this problem was the recognition that natural populations are subjected to environmental variability and other disturbances that can prevent the process of competitive exclusion. In addition, the complexities of multispecies interactions, including indirect interactions, can promote the coexistence of species.

Although the earliest models of species interactions seem abstract and disconnected from reality, ecology has a rich tradition of using experiments to test models. For example, the Russian biologist G. F. Gause pioneered the testing of species-interaction models using laboratory cultures of protozoans and yeast in which he could follow population dynamics on the time-scale of weeks, and thus could rapidly see whether or not a species could persist in his experimental systems. When examining competition between *Paramecium* species, Gause often observed that some species could drive others to extinction, a process termed competitive exclusion (and a result predicted in theory by the differential equations that depict continuous reproduction of two protozoans in competition for the same resource). Currently, laboratory microcosms involving protozoan communities are receiving fresh attention—nowadays these experiments are aimed at testing models of multi-species interactions and interaction webs.

IV. THEORIES OF BIODIVERSITY

Biodiversity is the variety of all living things, and most commonly refers to the number of different species in a particular area. Ecologists have documented striking patterns in the number of species occurring in different regions of the planet. For example, for most groups of species, there is a strong gradient of declining diversity from the equator to the poles, and the vast majority of species are therefore concentrated in tropical and subtropical regions.

Ecologists have proposed a variety of mechanisms that may be responsible for generating global patterns of biodiversity. For instance, areas with more varied landscapes and more topographic relief tend to support more diverse species assemblages than more uniform areas. Thus, spatial heterogeneity is likely to provide more ecological niches for species to occupy and therefore allow more species to coexist. As discussed earlier, classic ecological theory predicts that competitive interactions among species should lead to competitive exclusion and therefore reduced biodiversity. Several of the proposed mechanisms for global diversity patterns focus on forces that counter the process of competitive exclusion. Accordingly, it has been hypothesized that tropical regions harbor many species because their benign climate favors diseases and predators—these predators and diseases increase species diversity by attacking competitively superior species, thereby allowing weak competitors an escape from competitive exclusion. Disturbance is thought to similarly preclude competitive

exclusion, although disturbance that is too intense or that occurs too frequently may result in reduced biodiversity. The idea that diversity should be reduced when disturbance is either too rare or too frequent has been termed the intermediate disturbance hypothesis.

Another robust pattern of biodiversity is the observation that islands generally have fewer species than mainlands, and that larger islands tend to have more species than smaller islands. R. H. MacArthur and E. O. Wilson proposed the equilibrium theory of island biogeography to account for this pattern. According to this theory, there is a balance between colonizations of new species on islands and the subsequent extinctions of established species. On larger islands, colonizations are relatively more frequent and extinctions are less frequent, resulting in a higher equilibrium number of species on larger islands.

Human activities such as the direct harvesting of species, introductions of alien species, habitat destruction, and various forms of habitat degradation have greatly accelerated the loss of biodiversity. Consequently, current extinction rates are estimated be 100 to 1000 times higher than pre-human extinction rates. This extinction crisis has spurred a great deal of scientific interest in the biological *and* ecological functions of biodiversity. Organisms are responsible for a variety of ecosystem functions, including maintenance of the gaseous composition of the atmosphere, regulation of the global climate, generation and maintenance of soils, and recycling of nutrients and waste products. However, it is not obvious whether these biological services require many species or only a handful of species. Much current research is focused on understanding how much (if any) diversity we can afford to lose yet still maintain the necessary ecosystem functions (see Section VII).

Some species play obviously important roles in ecosystems—the addition or deletion of these "keystone" species leads to dramatic changes in ecosystem functions such as productivity or nutrient uptake. However, most species probably do not exert such important effects on ecosystems. In other words, it may be possible to lose a number of species from an ecosystem with little overall impact on ecosystem function. This could be the case if several species that perform approximately the same function are present in the original ecosystem. The situation where multiple species play a similar role has been termed "species redundancy." If species redundancy is a common phenomenon, ecosystem function should be largely independent of species diversity, so long as major functional types are represented. Thus, when one species is lost from an ecosystem, some other species with a similar function may become abundant and compensate for the lost species, leaving the ecosystem as a whole relatively unaffected. Indeed, ecosystem processes often do remain stable despite large fluctuations in the abundance of the various species involved. The term species redundancy may seem to imply that all species are not necessary for an ecosystem to function properly. However, species redundancy may in fact be an essential feature for the long-term health of ecosystems, and the importance of any particular species may only become evident during the occurrence of rare, drastic events.

V. SPATIAL PATTERNS

Spatial patterns and habitat patchiness or heterogeneity have spawned a host of concepts that try to both describe these spatial patterns and explain their impacts and sources. The two most important general results of spatial theory in ecology are:

1. The fact that species interact in spatially extensive and patchy worlds creates opportunities for persistence and coexistence of species that would otherwise not exist.
2. In uniform environments, the interplay of dispersal and nonlinear population dynamics can produce complex spatial patterning in population densities, and this patterning can itself promote coexistence in predator–prey or competitive interactions.

More generally, spatial theory emphasizes that it may be impossible to understand ecological processes and interactions if the focus and scale of a study is too small. This is an important caution, given the fact that many experimental ecology studies are conducted at scales of less than 1 m2.

One of the most important practical applications of spatial theory stems from the use of metapopulation models in conservation biology. Metapopulation theory describes a species as a collection of populations (or patches) in which there is turnover due to local extinction and subsequent recolonization by dispersing individuals, and in which the fate of a species can only be understood by tracking the fate of a collection of populations or patches, as opposed to a single local population or patch. Metapopulation models are of considerable importance in conservation applications because they identify critical thresholds for habitat destruction or for the creation of barriers to plant and animal dispersal, thresholds that, once crossed, imply

doom for a species (because colonization becomes too infrequent to counterbalance local extinctions). It is obvious to anyone who has looked out an airplane window that one of humankind's major impacts on the world is to destroy habitats and to create ever more fragmented landscapes dissected by highways, agricultural fields, and commercial or residential developments that surely hinder the dispersal of species. If this trend proceeds too far, metapopulation theory predicts the slow but inexorable loss of species. Spatially detailed models of human-modified landscapes are increasingly being used to identify strategies for land use and human development that will minimize the undesirable effects of habitat loss and habitat fragmentation. The two most common strategies involve clustering of habitats and the protection of dispersal corridors.

At a more theoretical level, spatially explicit models of species interactions indicate that the interactions themselves can result in pattern formation. For example, high rates of predation combined with limited prey dispersal can result in a dense patch of prey surrounded by a zone that is relatively devoid of prey due to intense predation. This "ring" of intense predation can limit the spread of prey populations. It is important to realize that these spatial patterns can emerge without any underlying environmental heterogeneity. Spatial patterning can allow more species to coexist because the species occupy different portions of the pattern. However, the opportunity for spatial segregation of species into different characteristic portions of spatial patterns requires sufficiently large habitats—if habitats are carved into patches that are too small, patterns cannot be generated, and opportunities for coexistence will be lost. Thus, theories of pattern formation reveal yet another mechanism by which biodiversity can be maintained, and conversely a route by which biodiversity can be degraded.

VI. SCALING LAWS

Because ecological processes operate across a huge range of scales, there have been numerous attempts in ecology to identify scaling laws—laws that allow us to study phenomena on one scale and predict those same phenomena at different scales. Just as important as identifying rules for extrapolating across scales is the idea that there are critical spatial scales at which scaling laws break down and processes fundamentally change in character.

An especially good example of successful extrapolation across scales concerns the prediction of the rate at which exotic species invade new areas and expand their ranges. Simple invasion models extrapolate short-term observations of dispersal (over the scale of one year or less and involving the movements of single individuals) to the broad expanse of an entire species range after decades and over thousands of kilometers. For organisms as diverse as muskrats and cabbage butterflies, invasion models have successfully used small-scale and short-term data to predict long-term rates of range expansion.

To quantitatively examine the legitimacy of extrapolations across scales, it is often useful to graph the variance of a measured variable as a function of the spatial scale at which it was measured. For example, population density could be measured in anything from 1-m^2 quadrats to hectare plots, and the variance in density (per unit area) among plots is expected to decline with increasing plot size. The rate of this decline is a measure of spatial correlation. However, marked changes in the slope relating variance in population density to measurement scale indicate some fundamental shift in biological processes—such as a shift from patterns generated by the dispersal range of a plant's seeds to patterns generated by large-scale climatic heterogeneity. Such nonlinear relationships between variance and scale indicate circumstances where it may not be appropriate to extrapolate across scales.

Most ecologists continue to conduct research at one and only one scale. Although theories regarding scaling laws are still in their infancy, they plainly draw attention to the fact that any extrapolation across scales is tenuous, and can only be justified if the processes have been studied or measured at more than one scale.

VII. ECOSYSTEM THEORIES

Ecosystem theories focus on patterns of energy flow, nutrient cycling, and the factors that control or limit the rates of these processes and the stability or predictability of these processes through time. To assess the flow of energy and nutrients, it is useful to recognize that organisms fill different trophic roles: autotrophs (usually plants) acquire energy, usually from the sun, and nutrients from nonliving materials, heterotrophs derive energy and nutrients by consuming living matter, and decomposers transform once-living materials back to their inorganic forms. Both nutrients and energy flow from autotrophs to heterotrophs and decomposers, and the connections among these groups, constitute a food chain or food web.

The trophic interactions among species affect the

rates of ecosystem functions, including biogeochemical processes, the flow of nutrients, water, and atmospheric gases, and the processing of energy through ecosystems. Species are the key working parts of ecosystems, and therefore biodiversity must be related in some way to ecosystem function. Certainly ecosystems would not function if all species were lost, but it is unclear just how many species are necessary for an ecosystem to function properly. The current extinction crisis has prompted ecologists to assess the relationship between biodiversity and various ecosystem properties. Evidence regarding the importance of biodiversity for ecosystem function largely comes from comparing ecosystems that naturally differ in the number of species present. More recently, ecologists have undertaken more controlled manipulative experiments in which the number of species has been directly varied. Several studies have demonstrated that various measures of ecosystem function, such as production of biomass and nutrient uptake, increase as the number of species present increases. However, some studies also report no effect or even negative relationships between biodiversity and ecosystem processes.

Although some evidence suggests that biodiversity increases or improves the overall functioning of ecosystems, the underlying mechanisms remain unclear. For example, a positive relationship between species diversity and productivity could result simply because including more species increases the chance of including particularly productive, or fast-growing, species. Alternatively, a diverse group of species may use the available resources more efficiently, because each species has a slightly different strategy, resulting in higher overall growth.

A second commonly purported benefit of biodiversity is that more diverse ecosystems may be more stable compared to species-poor ecosystems. Stability can be defined at the community level as fewer invasions and fewer extinctions, meaning that a more stable community will contain a more stable composition of species. However, stability can also be defined at the population level as reduced fluctuations in population size, meaning that a more stable population will contain a more constant number of individuals.

The idea that biodiversity confers stability to ecosystems has enjoyed a long and controversial history. Ecologists initially hypothesized that diverse ecosystems were more stable than those with fewer species because, it was reasoned, with more species in an ecosystem, there are more paths through which energy and nutrients can flow. Consequently, in diverse ecosystems each species should be less affected by changes in the abundance of other species, leading to higher overall stability. However, mathematical models exploring the relationship between biodiversity and population stability indicated that higher species diversity generally reduces the stability of population dynamics of individual species. The conflict between these modeling results and the original intuitions of ecologists remained unresolved for many years.

Recent experimental manipulations of species diversity have helped to resolve this long-standing controversy. In particular, several manipulations of plant and microbe diversity have been performed to examine the stability of various measures of ecosystem function over time. Generally, such studies have shown that, although the abundance of individual species fluctuates more dramatically in high-diversity ecosystems, the total abundance or productivity of all species combined is actually more stable. These empirical results are consistent with both views of diversity and stability. First, the fact that individual species fluctuate more in diverse systems supports the results of mathematical models regarding diversity and stability. On the other hand, the observation that aggregate attributes such as productivity become more stable with increased diversity supports the intuition that multiple pathways for energy (high diversity) foster stability.

Although the relationship between biodiversity and ecosystem stability is becoming increasingly clear, the mechanisms generating this pattern are not. In particular, diverse groups of species may be more stable because complementary species compensate for one another. Alternatively, variation in aggregate measures such as total productivity may increase with richness due to averaging of random fluctuations in the growth of each species. On the basis of simple probability theory, it is expected that as more independently varying species are summed together, the more stable will be the sum of their abundances. The strength of this averaging effect depends on correlations among the species' fluctuations, but a positive relationship between biodiversity and the stability of aggregate measures of ecosystem function should almost always be expected, simply due to averaging.

VIII. EVOLUTIONARY THEORY IN ECOLOGY

Evolutionary theory has entered ecology in two major areas: the evolution of species attributes and how they might vary in a predictable manner, and the coevolution of interacting species, such as the evolution of virulence and resistance in host–pathogen interactions.

There is often a remarkable matching between the traits of organisms and the local conditions in which they live. Ecologists and population geneticists alike have worked to identify the factors contributing to or preventing the adaptation of populations to various aspects of their local environments, including food sources, predators, and microclimatic conditions. Reciprocal transplant and common garden experiments have been the major tools used to test for local adaptation. In particular, there is interest in the balance between selection for locally adapted traits and the immigration of individuals with nonadapted traits into the population. This balance between differentiating selection and the homogenizing effects of migration can create a difficult conundrum for conservation biologists—for a rare species that exists in several small populations, it might be better to pool all the individuals together to bolster the size and viability of a single population. Alternatively, the subpopulations might be so genetically distinct and locally adapted to their environments that pooling the individuals might actually harm the species. Thus, one of the more difficult challenges in conservation is to balance the short-term viability of a species with preservation of its long-term evolutionary potential.

As a species evolves to become locally adapted to other species in its environment, it can exert selection pressure on the co-occurring species, leading to changes in the traits of these species with which it interacts. The reciprocal evolution of two species in response to one another is called coevolution. The term coevolution was originally coined to describe the presence of toxic compounds found in plants and the ability of butterfly species to detoxify these compounds. Indeed, many of the best-known examples of coevolution involve reciprocal changes between adversaries, a sort of ever-escalating "arms race" as might occur between a predator and its prey or between a parasite and its host.

Plant breeders try to directly manipulate the coevolutionary process that occurs between crop plants and the herbivores and pathogens that attack these plants. Selective breeding is used to maximize the defenses naturally present in crop species and, more recently, genetic engineering has allowed breeders to co-opt defenses of very distantly related organisms and introduce completely novel defense traits to crops. Unfortunately, when highly defended crops are planted over the enormous spatial scales of modern farms, herbivore and pathogen species tend to quickly evolve resistance to the plant defenses. Thus, an important application of evolutionary ecology involves determining how we can best deploy plant defensive traits such that evolution on the part of plant pathogens and herbivores does not make the plant defenses obsolete. Theory suggests that alternating plantings of crops with distinct defensive traits may reduce the ability of herbivores to evolve resistance. In addition, incorporating combinations of defensive traits into single individuals should substantially reduce the likelihood that herbivores will become resistant. Thus, theory in evolutionary ecology has contributed concrete strategies that could greatly enhance the "longevity" of plant defenses, with enormous economic consequences.

IX. SYNTHESIS

Of the numerous specific concepts and theories discussed in this article, some general ideas emerge that relate pattern to process and dynamics to structure. First, complicated spatial or temporal patterns in population densities do NOT require environmental explanations or complicated underlying dynamics—rather, we should expect patterning, cycling, and complexity simply because of the feedbacks inherent in ecological systems. This is of more than theoretical interest, because it implies that any management of ecological systems must heed the implications of nonlinear dynamics. In other words, we must anticipate that rates of change in population growth or in number of species may change dramatically and suddenly, not just because of random fluctuations in the environment, but also as a result of deterministic processes. Second, several ecological models, including epidemiological and metapopulation models, point to critical thresholds across which systems change qualitatively (e.g., a disease can no longer persist, or a population dwindles toward extinction because colonizations fail to balance local extinctions). Third, opportunities for coexistence of species arise whenever the world in which organisms live becomes less uniform—spatial heterogeneity creates refugia, disturbances interrupt the process of competitive exclusion, and so on. In contrast, human actions tend to enforce regularity and control on the natural world, and ecological theory suggests that this homogenization of nature is contributing to the erosion of our planet's biodiversity. The consequence of this reduction in biodiversity remains an open theoretical question, with the greatest challenges to ecological theory being the development of models that contend with rich webs of species interactions in complex ecosystems. The future of ecological theory lies in blending human activities with natural processes, so that we can better understand how to encourage a sustainable biosphere while accommodating human interests and demands.

See Also the Following Articles

COEVOLUTION • COMPLEXITY VERSUS DIVERSITY • KEYSTONE SPECIES • POPULATION DYNAMICS • POPULATION VIABILITY ANALYSIS (PVA) • RESOURCE PARTITIONING • SPECIES INTERACTIONS

Bibliography

Gotelli, N. J. (1995). *A Primer of Ecology*. Sinauer Associates, Sunderland, Massachusetts.

Huston, M. A. (1994). *Biological Diversity: The Coexistence of Species on Changing Landscapes*. Cambridge University Press, Cambridge, United Kingdom.

Krebs, C. J. (1994). *Ecology: The Experimental Analysis of Distribution and Abundance*, 4th ed. Addison-Wesley, Menlo Park, California.

Levin, S. A. (1999). *Fragile Domain*. Perseus Books, Reading, Massachusetts.

Ricklefs, R. E., and D. Schluter. (1993). *Species Diversity in Ecological Communities: Historical and Geographical Perspectives*. University of Chicago Press, Chicago.

Roughgarden, J., R. May, and S. A., Levin. (1989). *Perspectives in Theoretical Ecology*. Princeton University Press, Princeton, New Jersey.

Tilman, D., and P. Kareiva. (1997). *Spatial Ecology: The Role of Space in Population Dynamics and Interspecific Interactions*. Princeton University Press, Princeton, New Jersey.

ECOLOGY OF AGRICULTURE

Alison G. Power
Cornell University

GLOSSARY

polyculture Growing two or more crops in the same field at the same time.
resilience The rate at which an ecosystem returns to original conditions after a perturbation.
resistance The degree to which an ecosystem changes in response to a perturbation.
transgenic crop Crop whose genome contains a gene(s) from a distinct species that has been inserted by genetic engineering.

THE POPULARITY OF AGRICULTURE as a focus of ecological study has waxed and waned during the past century. Agricultural systems have been touted as model systems for testing ecological theory because of their relatively well-characterized components, simple food webs, and manipulability. In the past, they have often served as effective model systems for plant ecol-

ogy, plant–insect interactions, and predator–prey theory. In recent years, the growth of sustainable agriculture has resulted in increasing interest in understanding ecological processes in agriculture so that they can be manipulated more effectively for enhancing agricultural productivity and reducing negative environmental impacts of agricultural activities. A variety of types of ecological theory have been applied to agricultural systems, and thus the subdiscipline of "agroecology" encompasses a very broad range of approaches, including physiological ecology, population biology, community ecology, ecosystem ecology, and landscape ecology. All these approaches offer significant insights into the ecology of agricultural systems and provide useful guidelines for improving the functioning of these systems. Because the role of diversity in ecological systems has been a consistent element in much agroecological research, this article focuses on diversity in agricultural systems, with particular attention to areas of agroecology that appear to have the most potential to offer agricultural and environmental enhancement.

I. PLANNED AND UNPLANNED DIVERSITY IN AGROECOSYSTEMS

The biodiversity in agroecosystems can be classified as either "planned" or unplanned diversity. Planned diversity includes the spatial and temporal arrangement of domesticated plants and animals that farmers purposely include in the system. It may also include beneficial organisms that are deliberately added to the agroecosystem, such as biological control agents or plant-associated nitrogen-fixing bacteria. Unplanned diversity includes all the other associated organisms that persist in the system after it has been converted to agriculture or colonize it from the surrounding landscape. This component of diversity is likely to include a variety of herbivores, predators, parasites, and microorganisms that make up the majority of species in any ecosystem, even a simplified one such as an agroecosystem. It has become increasingly clear that the two components of diversity are significantly linked: As planned diversity increases, the diversity of the associated biota also increases. Although the evidence for the plant-based food web is strongest, this relationship also appears to exist for the detritus-based food web.

Given this relationship between planned and unplanned diversity, how do both types of diversity influence agricultural productivity and other ecological processes that occur in and around agroecosystems? The impact of planned diversity is relatively well understood, but much remains to be elucidated about the role of unplanned diversity in agroecosystem function.

II. IMPACT OF PLANNED DIVERSITY ON ECOLOGICAL PROCESSES

Agroecosystems vary dramatically in their complexity and degree of planned diversity, and planned diversity may include genetic diversity, species diversity, structural diversity, and functional diversity (Fig. 1). The input-intensive monoculture systems that dominate commercial agriculture worldwide have low planned

diversity and relatively simplified community assemblages. At the other extreme, traditional agricultural systems and home gardens of the tropics typically have extremely high planned diversity and complex assemblages of associated biota. This range of planned diversity in existing systems has provided rich opportunities for exploring the impact of planned diversity on a suite of ecological processes, including primary production, pest regulation, decomposition, and nutrient cycling.

A. Planned Diversity, Productivity, and Stability

The idea that diversity should impart stability to ecological systems, including agricultural systems, dates back at least to the 1950s and has stimulated much ecological theory, some empirical research, and extensive debate. Until recently, the best evidence for the idea that diverse systems should be more stable than simple ones came from agricultural ecosystems rather than natural ecosystems. Many agricultural studies have demonstrated increased productivity (i.e., significant yield increases) in diverse cropping systems compared to monocultures. However, the majority of these studies address systems of relatively low plant species richness, comparing monocultures to polycultures of two or three species. We expect that the relationship between productivity and plant species richness should resemble the asymptotic curve in Fig. 2, as it appears to do so for natural grasslands (Tilman *et al.*, 1996), but we have little data to indicate where the asymptote might be. Researchers have speculated that the increase in productivity with increasing species richness in agroecosystems may level out at approximately 10 plant species, although this

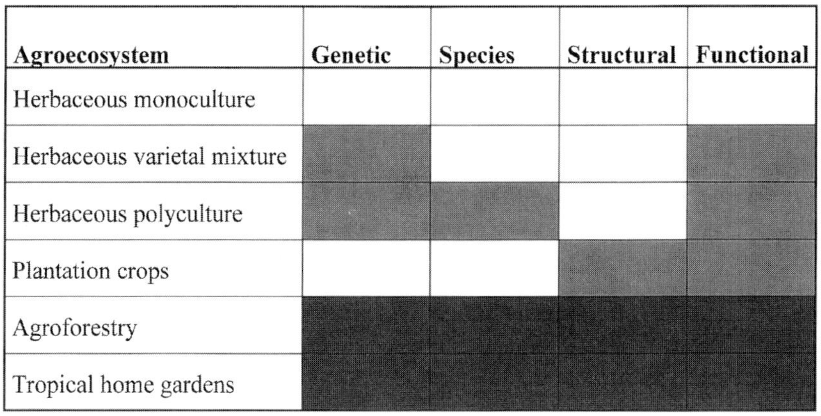

Agroecosystem	Genetic	Species	Structural	Functional
Herbaceous monoculture				
Herbaceous varietal mixture	▨			▨
Herbaceous polyculture	▨	▨		
Plantation crops			▨	▨
Agroforestry	■	■	■	■
Tropical home gardens	■	■	■	■

FIGURE 1   Degree of different types of planned diversity found in some typical agroecosystems. Intensity of shading indicates degree of diversity.

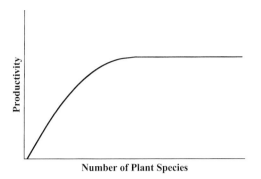

FIGURE 2 Predicted relationship between productivity and plant species richness.

number is likely to vary according to the type of agricultural system and environmental conditions.

The tendency of diverse cropping systems to have higher productivity than would be expected on the basis of the productivity of their component crops grown in monoculture is often called "overyielding." Overyielding may result from a variety of mechanisms, such as more efficient use of resources (light, water, and nutrients) or reduced pest damage, and there have been numerous experimental studies examining these mechanisms. When interspecific competition for a limiting factor is less than intraspecific competition for that factor, overyielding is predicted. Facilitation occurs when one crop modifies the environment in a way that benefits a second crop, for example, by lowering the population of a critical herbivore or releasing nutrients that can be taken up by the second crop. Facilitation may result in overyielding even where direct competition between crops is substantial (Vandermeer, 1989).

There is evidence that diverse cropping systems exhibit greater yield stability and higher productivity, suggesting that resistance to environmental perturbation may be higher in diverse systems. Yield stability has been measured in at least three ways: by calculating coefficients of variation of yield, by computing regressions of yield against an environmental index, and by estimating the probability of crop failure. Polycultures exhibit greater yield stability according to all criteria: Polycultures tend to have lower coefficients of variation than crops grown as separate monocultures, the response of polycultures to environmental change tends to be as stable or more stable than the most stable component crop grown in monoculture, and polycultures tend to have a much lower probability of crop failure than the component crops grown in monoculture. The probability of crop failure is an estimate of risk and lower probabilities result from both the higher

yields of polycultures and the putative yield stability. Overall, studies indicate that diverse cropping systems are more stable than monocultures, both agronomically and economically.

Several mechanisms may lead to greater yield stability in diverse systems. When one crop performs poorly because of drought or pest epidemic, for example, the other crop(s) can compensate, using the space and resources made available. Such compensation is obviously not possible if the crops are grown separately. If the yield advantages of polycultures are greater under stress conditions, then yield stability is higher. This polyculture advantage has been demonstrated for crops under nutrient stress and drought stress. Moreover, where polycultures lead to reduced pest attack, as they often do, then greater yield stability may result from the dampening of pest outbreaks and disease epidemics.

These processes may operate simultaneously in diverse systems and are consistent with the idea that species richness buffers productivity under conditions of environmental variability and that diversity imparts resistance to perturbation. The ability of polycultures to compensate for losses might also be considered to represent resilience. Whereas systems with greater resistance will be less impacted by a perturbation, systems with greater resilience return rapidly and reliably to original conditions.

As a result of this evidence that diverse cropping systems often have greater productivity and stability, there is a persistent conviction among many agroecologists that species-rich natural ecosystems should provide us with models for the design of sustainable agricultural systems. Jackson (1985) and colleagues at the Land Institute in Salina, Kansas, have promoted the idea that agriculture in the U.S. plains should mimic natural prairies by emphasizing polycultures of herbaceous perennials. Others have noted the structural and functional similarity between diverse home gardens in the tropics and humid tropical forests and have suggested that agricultural systems for the tropics should be designed as analogs of natural tropical forests.

Ecological research on the role of diversity in natural grassland systems by Tilman and colleagues (1996) indicates that diverse natural communities may be more productive than simple systems, at least partially as a result of increased nitrogen use efficiency in more diverse systems. These studies also suggest that more diverse plant communities are more resistant to disturbance and more resilient in the face of environmental perturbations such as drought. That is, the productivity of diverse communities appears to decline less during a drought and to return more quickly to predrought levels than is the case for species-poor communities. It remains to be seen, however, whether the "natural systems mimicry" approach can really lead to useful improvements in the productivity and sustainability of agroecosystems.

B. Planned Diversity and Pest Regulation

The planned diversity of the agroecosystem has important effects on herbivorous insects and the microbial community attacking crops. Traditional agricultural systems often include substantial planned genetic and species diversity, and genetically diverse grain crops are used in many parts of the world to control pathogens. In contrast, the low planned diversity of most commercial monocultural systems often results in large crop losses from a pest complex that is less diverse but more abundant than that in more diverse systems. The trend for higher pest densities in monocultures compared to diverse cropping systems is especially strong for specialist insect herbivores with a narrow host range. As planned diversity increases, population densities of these specialist herbivores decrease.

In a classic paper, Root (1973) offered two hypotheses to explain higher densities of herbivorous insects in simple, monocultural systems: the resource concentration hypothesis and the enemies hypothesis. The first hypothesis predicts that herbivores, particularly specialists, in pure, dense host plant stands will be more likely to find their hosts and more likely to survive and reproduce. In contrast, herbivores in less dense or more host plant-diverse stands should be less likely to find their hosts and more likely to lose them. Although the details of herbivore–host interactions vary considerably, subsequent experimental tests of this hypothesis have generally supported it with respect to the effects of diversity, especially for specialist herbivores. Host-finding behavior and insect movement, both colonization and emigration, appear to play important roles in the response of herbivorous insects to agroecosystems. Densities of specialists may be lower in diverse systems because (i) they have difficulty locating hosts due to interference with olfactory or visual cues or (ii) they leave hosts more often due to lower plant quality and then have difficulty relocating them. These behaviors are significantly affected by the chemical, nutritional, and structural diversity that accompanies planned plant species diversity.

The enemies hypothesis predicts that diverse systems should have higher densities of herbivore natural enemies (predators and parasites) because they provide more resources for these natural enemies, such as alternate prey or hosts, nectar, pollen, and refugia. This

hypothesis has also largely been supported in experimental studies. Compared to monocultures, diverse systems are likely to have higher predation rates, higher parasitism rates, and higher ratios of natural enemies to herbivores, all of which may contribute to lower pest densities. Natural enemies comprise one component of the unplanned diversity that increases as planned diversity increases in agroecosystems.

Microbial pathogens also respond to the planned diversity of agroecosystem, but their response is more variable than that of herbivorous insects. Crop diversification can modify the microclimatic conditions that play an important role in the development and severity of plant disease. Pathogen growth and reproduction may be either encouraged or inhibited in more diverse cropping systems, depending on the particular requirements of the organism. The effects of diversity depend on a variety of dispersal processes, infection efficiency, and the rate of disease progress. The majority of viruses, however, are transmitted by insects, and these tend to be found at lower incidence in diverse systems due to the effects of plant species diversity on their insect vectors.

The genetic diversity of crops can dramatically reduce pathogen impacts on crop productivity. Mixtures of genotypes of a single species, such as multiline cultivars and varietal mixtures, have been used effectively to retard the spread and evolution of fungal pathogens in small grains and other crops. There is evidence that they may also have lower densities of insect herbivores and a lower incidence of plant viruses. Typically, these mixtures include both resistant and susceptible crop genotypes, although they may be mixtures of several different resistant genotypes. The reduction in pathogen spread is greater than would be expected on the basis of the proportion of resistant genotypes in the mixtures, and therefore it appears to be due to the effects of diversity per se on the ability of pathogens to disperse.

High planned diversity that includes genetic diversity, species diversity, and structural diversity thus has a strong influence on populations of herbivorous insects and plant pathogens. The intentional manipulation of planned diversity for the purpose of pest control is common in traditional agricultural systems, and it offers real potential for increasing the productivity of contemporary agricultural systems without the negative environmental impacts of pesticides.

C. Planned Diversity and Soil Processes

Planned diversity may also have significant impacts on the soil community. Plant pathogens and their antagonists in the soil are well-known to respond to crop diversity. Long-term, continuous monoculture can result in dramatic shifts in the competitive balance among microbial species and increase the aggressiveness of plant pathogens. Conversely, populations of plant parasitic fungi, bacteria, and nematodes may decline when monocultures of their host plants are replaced by diverse cropping systems. Rhizosphere microbial communities, including a variety of bacteria and fungi, respond to plant species composition of the cropping system, plant phenology, plant nutrient status, and even plant genotype. Plant mutualists such as mycorrhizal fungi or nitrogen-fixing bacteria may be strongly determined by crop composition and phenology. Some ecological processes such as nitrogen cycling may be substantially controlled by crop diversity and plant composition since the inclusion of legumes can increase rates of biological nitrogen fixation. Due to the effects of litter diversity, decomposition rates may also be quite responsive to planned diversity.

Despite an increasing awareness of the effects of planned crop diversity on the soil community, our ability to predict the impacts of manipulating planned diversity is limited because our understanding of the complex interactions among soil organisms and between plants and soil microbes is still rather weak. Greater attention to the ecological dynamics of these soil communities under different cropping systems will aid in determining the extent to which we can manage these communities to enhance agricultural productivity using planned diversity.

III. UNPLANNED DIVERSITY AND ECOLOGICAL SERVICES

The unplanned diversity that accompanies planned diversity in agricultural systems can provide many ecological services to agriculture. This means that the conservation of biodiversity can offer significant benefits to agriculture. Uncultivated species, including wild relatives of crops that occur in and around the agroecosystem, are an important source of germplasm for developing new crops and cultivars. Increasing planned crop diversity can augment the resources available to pollinators and to natural enemies such as parasitic wasps, resulting in higher populations of these beneficial organisms. Increasing planned diversity may also foster beneficial soil organisms and the conservation of functional processes, such as decomposition and nutrient cycling.

Overall, unplanned diversity within the agroecosystem affects plant and soil processes that can in turn affect crop productivity. For example, many traditional cacao and coffee systems in the tropics resemble forest ecosystems since they are well shaded and humid, with a diversity of shade trees, a diverse herbaceous understory, and a thick leaf litter layer. Thus, these traditional agroecosystems have high planned species diversity, high genetic diversity, and high structural diversity. Recent studies have shown that shaded cacao and coffee plantations support higher levels of biodiversity than full-sun plantations grown without shade trees, i.e., monocultures (Perfecto et al., 1996). The shaded systems may be traditional "rustic" systems in which the understory of native forests is cleared and the crop is planted under the natural forest overstory. They may also be improved systems in which a few carefully chosen tree species are planted interspersed among the crop. In either case, these shaded systems have high planned diversity accompanied by high unplanned diversity.

Shaded coffee and cacao have been demonstrated to be good habitats for migrant and resident forest birds in the neotropics, and these birds may be important consumers of pest insects. They also contain relatively high diversities of insects, lizards, and other fauna, some of which contribute to pest regulation. Pollination also appears to be enhanced in shaded systems. Since many of the commonly used shade trees are leguminous, the trees reduce the amount of nitrogen inputs required. In addition, the shade provides significant weed control and, at least in cacao, reduces the incidence of several important diseases. The high levels of unplanned diversity thus contribute significantly to agricultural productivity. Moreover, by serving as habitat for a relatively high diversity of vertebrate and invertebrate animals, such agroecosystems may enhance the conservation value of a agricultural landscape.

A. Unplanned Diversity and Pest Regulation

One of the ecological services provided by the unplanned diversity in agroecosystems is the regulation of herbivorous insects and microbial pathogens by competitors, predators, and parasites. As discussed previously, high planned diversity often results in higher densities of predators and parasites. Even in agroecosystems of low planned diversity, the diversity of natural enemies may be quite high as long as pesticides are not used. For example, paddy rice monocultures managed without pesticides can have surprisingly high diversities of herbivorous insects, predators, and parasitoids compared to similar monocultures in which pesticides are used. Recent pest management programs in Southeast

Asian paddy rice have taken deliberate advantage of this diversity and drastically reduced pesticide inputs without sacrificing yields. In traditionally managed rice fields, the predators are likely to include fish and amphibians, which contribute to pest regulation and also provide additional nutritional resources for farm families.

B. Unplanned Diversity and Soil Processes

Despite some significant advances in our understanding of soil processes, we know relatively little about the soil biota and their impact on agricultural productivity. In natural ecosystems, decomposition and soil nutrient cycling are regulated by a diverse community of invertebrates and microorganisms, such as termites, earthworms, nematodes, fungi, and bacteria. The composition, abundance, and activity levels of the soil biota in agricultural systems are markedly different from those in surrounding natural ecosystems. For example, the diversity and abundance of soil insects and earthworms in tropical agroecosystems are typically significantly reduced compared to those of a wide range of undisturbed tropical ecosystems. In cases in which abundance remains high in agricultural systems, the soil communities are often dominated by a single or small number of species highly adapted to the modified environment.

The changes in the soil community under agriculture result from a variety of perturbations to the soil environment. The initial conversion from undisturbed ecosystems typically involves the removal and/or burning of plant biomass followed by tillage—activities which have drastic impacts on soil structure and soil chemistry. The physical changes at the soil surface amplify diurnal and seasonal fluctuations of temperature and moisture. In addition, organic inputs to the soil are significantly reduced as a result of plant biomass removal, and the chemical composition of organic inputs is altered. These extreme modifications of the soil environment can result in the elimination of some soil organisms and, at a minimum, are likely to change the competitive balance among species. To the extent that agricultural systems minimize these perturbations (e.g., by reducing tillage or burning), the impacts on the soil community may be less severe.

Despite these well-known effects of agricultural conversion on the soil community, the link between loss of soil biodiversity and various ecological processes has not been extensively studied. Decomposition and nutrient mineralization, for example, are controlled by the activities of a diverse community of organisms. It is not clear that the loss of some species will result in

significant changes to these functions, but the degree of functional redundancy among different species is still controversial. On the other hand, some processes such as nitrogen fixation are carried out by very specific organisms, whose loss might substantially affect nitrogen cycling. Given the importance of soil biota for decomposition and nutrient cycling, it is essential that the link between this component of unplanned diversity and function be explored more fully.

IV. IMPACTS OF AGRICULTURE IN THE LANDSCAPE

Ecological concepts have often been utilized to address the consequences of agricultural perturbations for natural ecosystems in agricultural landscapes. For example, the impacts of nutrient inputs and pesticides for downstream aquatic and terrestrial ecosystems have been well studied. Much of the pesticide applied to crops fails to reach the target pests and instead moves into adjacent ecosystems, where it can have significant impacts on the diversity and abundance of nontarget species. Studies have shown that pesticides have strong effects on trophic dynamics and a range of ecosystem processes in complex natural ecosystems, both aquatic and terrestrial. Similarly, movement of nitrogen from agroecosystems to natural ecosystems through leaching or deposition may lead to significant impacts at several trophic levels. Increased nitrogen loading may reduce plant species richness while increasing the biomass of dominant species. In response to these changes, both pathogens and herbivorous insects are likely to exhibit lower species richness but higher population abundance, just as they do in agricultural systems. Changes in pathogen and herbivore populations may in turn influence the productivity and fitness of their host plants. Even in the absence of changes in herbivore population pressure, research has shown that the effects of herbivory on plant growth and reproduction may be more severe at high levels of nutrient availability. As with the complex effects of pesticides, these impacts of inadvertent fertilization are likely to impact trophic dynamics and ecosystem processes.

A. Ecological Risks of Genetically Engineered Crops

In general, the impacts of agricultural technologies such as pesticides and chemical fertilizers were characterized many years or decades after the introduction of the technology and well after their use was widespread.

In contrast, ecologists began to call attention to the ecological consequences of genetically engineered crops before these crops were widely available commercially. In the past few years, however, these transgenic crops have rapidly become common in agricultural systems throughout the world, but the ecological impacts of these crops are still poorly understood. Many ecologists throughout the world are beginning to investigate these impacts in order to develop a realistic assessment of the potential risks to the ecology of both natural and agricultural ecosystems. Despite the relatively small number of studies undertaken to date, this research is having an important influence on how transgenic crops are being deployed.

Ecologists are investigating many different types of risks, including changes in the ecological competency of the crop, transgene movement from crops to wild relatives, the evolution of resistant pests, risks to non-target organisms, recombination between genetically engineered plants expressing virus resistance and natural viruses leading to new viral genotypes, and changes in crop management strategies and cropping systems. These different types of risk are receiving differing degrees of attention from ecologists.

There is evidence for changes in the ecological competency of a few crops. Through genetic engineering, the crop may acquire particular traits that allow it to survive better in natural ecosystems or as a volunteer in agricultural systems. Such traits might increase seed survivorship, increase crop resistance to pests or diseases, or increase the competitiveness of the crop relative to wild plants. This process could lead to invasion of natural ecosystems and ecological disruption. In addition, genetically engineered traits such as herbicide resistance can make volunteer crops persistent weeds in subsequent cropping systems. This problem has already been detected with canola (rapeseed) that has transgenic herbicide resistance.

Genetically engineered traits may also move from crops to wild relatives of crops through natural processes of gene flow and hybridization. Many recent studies have examined the degree of hybridization between crops and wild relatives. Successful hybridization under natural conditions has been demonstrated for many crop–wild relative associations, and some studies have documented the persistence of crop genes in populations of wild relatives over several generations. Traits such as pest and disease resistance or herbicide tolerance may provide a selective advantage to wild relatives of crops and allow them to become weeds in agroecosystems or to increase in abundance in natural ecosystems and thereby disrupt natural patterns of plant distribution.

Increasingly, ecologists are exploring the possible effects of these transgenic crops on nontarget organisms, including natural enemies and soil biota, and evidence for detrimental effects is accumulating. For example, lacewing predators fed on prey that consumed transgenic crops expressing *Bacillus thuriingiensis* (*Bt*) toxins (bacterial insecticides) have higher mortality than those fed on prey that did not consume *Bt* toxins. Similarly, parasitic wasps that develop in *Bt*-treated host insects have higher mortality. Through the incorporation of crop residues into the soil, *Bt* toxins can accumulate in the soil and cause changes in the abundance and species composition of soil microorganisms, thus affecting decomposition and nutrient cycling.

There is much less work on the ecological implications of viral recombination, but recombination has been demonstrated several times between replicating natural viruses in host plants and transgenes that have been inserted into host genomes. It has been demonstrated that this recombination can lead to changes in viral competitiveness and pathogenicity to hosts. Recombination may also alter transmission characteristics and allow transmission by new species of vectors, which would augment the opportunities for virus establishment in new host species. Thus, the potential for ecological impacts of new, recombinant viruses arising from transgenic crops is significant and warrants more attention from ecologists.

V. FUTURE DIRECTIONS

Although the study of agricultural ecosystems has provided many insights into the role of diversity in agroecosystem function and the ecological impacts of agriculture, the subdiscipline of "agroecology" is still in its infancy. Two areas that urgently warrant more attention from ecologists are (i) the influence of planned and unplanned agroecosystem diversity on soil processes such as decomposition and nutrient cycling and (ii) the ecological consequences of widespread adoption of genetically engineered crops. A better understanding of the ecological processes operating in both areas could result in significant benefits to both agriculture and environmental protection.

See Also the Following Articles

AGRICULTURE, INDUSTRIALIZED • AGRICULTURE, SUSTAINABLE • BREEDING OF PLANTS • PARASITISM • PLANT CONSERVATION, OVERVIEW • PLANT-SOIL INTERACTIONS

Bibliography

Altieri, M. A. (1995). *Agroecology: The Science of Sustainable Agriculture*. Westview, Boulder, CO.

Collins, W. W. and Qualset, C. O. (Eds.) (1999). *Biodiversity in Agroecosystems*. CRC Press, Boca Raton, FL.

Gliessman, S. R. (1998). *Agroecology. Ecological Processes in Sustainable Agriculture*. Ann Arbor Press, Chelsea, MI.

Jackson, L. E. (Ed.) (1997). *Ecology in Agriculture*. Academic Press, New York.

Jackson, W. (1985). *New Roots for Agriculture*. Univ. of Nebraska Press, Lincoln.

Perfecto, I. Rice, R. A. Greenberg, R. and van der Voort, M. E. (1996). Shade coffee: A disappearing refuge for biodiversity. *BioScience* **46**, 598–608.

Root, R. B. (1973). Organization of a plant–arthropod association in simple and diverse habitats: The fauna of collards (*Brassica oleraceae*). *Ecol. Monogr.* **43**, 95–124.

Tilman, D. Wedin, D. and Knops, J. (1996). Productivity and sustainability influenced by biodiversity in grassland ecosystems. *Nature* **379**, 718–720.

Vandermeer, J. (1989). *The Ecology of Intercropping*. Cambridge Univ. Press, Cambridge, UK.

ECONOMIC GROWTH AND THE ENVIRONMENT

Karl-Göran Mäler
Beijer International Institute of Ecological Economics

THIS ARTICLE DISCUSSES three questions: What is economic growth? Will economic growth deteriorate or improve the environment? and What is the role of environmental resources for economic growth?

I. INTRODUCTION

The title of this article has many meanings. Economic analysis of *environment and growth* can mean many different things. In this article, the focus is on the following issues:

- What is economic growth?
- Will economic growth deteriorate or improve the environment?
- What is the role of environmental resources for economic growth?

The first question is typically answered by growth in gross domestic product (GDP) or GDP per capita. However, seldom is it explained why this is an interesting definition of growth. Of course, behind this notion of growth is the notion that GDP or gross national product (GNP) is a measure of well-being in society. It is therefore necessary to discuss what we mean by well-being and possible indices for such a concept. This will, of course, immediately bring us to a discussion of green accounting. There is a huge literature on this subject, but in this article I will not try to survey those contributions but instead examine one particular formulation of the conceptual basis of green accounting in order to determine the difficulties in creating such an index.[1]

The second question has during the past 10 years been discussed in terms of so-called Kuznets curves—that is, empirically estimated relations between environmental quality and GDP per capita. The general finding for many pollutants is that a country with a very low income does not have much pollution but when the scale of the economy grows, for example, as measured by GDP per capita, emissions of these pollutants will increase. However, when the income per capita is high enough, the economy will reach a turning point and pollution will decrease with further increases in per capita income. This has been taken by the World Bank to mean that growth implies a win–win strategy. By growing, a country may in the long run obtain both increased material well-being and a better environment. Of course, this kind of analysis has been criticized, from conceptual points of views and also from econometric and empirical points of view. Once again, the literature

---

[1] The discussion of these issues is based on Dasgupta and Mäler (2000) and Mäler (1999).

258

is huge, and no attempt to summarize this literature will be done. However, some central issues will be discussed.

Finally, the role of environmental resources as inputs for the growth process has very seldom been studied. When Solow (1957) studied economic growth in the United States, he found that most of the growth could not be explained by the two traditional factors of production (capital and labor), and he attributed the "unexplained" growth or the residual to technical progress. Later investigations by Dale Jorgenson, Zwi Grilliches, and others[2] reduced the residual considerably (e.g., by using better measurements of capital and by incorporating human capital), but it still remains quite high. This means that economists basically cannot account for all the productivity growth Western economies experienced during the past 100 years. To solve this problem, endogenous growth theory developed during the 1980s and the beginning of 1990s. The main idea is to incorporate the externalities that research and development activities generate because the benefits from these cannot be captured completely by the companies that carry out these activities. These externalities will therefore contribute to growth without being accounted for in the usual way that we analyze productivity growth. However, there are other externalities that also should be accounted for, such as networking externalities or social capital (Dasgupta, 1999). In this context, there is one particular externality that is of interest— environmental or resource externality. Because environmental resources or natural resources in general often lack well-defined property rights, they are not transacted on markets and therefore not accounted for in the standard national accounts (SNAs.) This omission of natural capital in growth accounting studies may explain at least part of the residual.

II. NET NATIONAL PRODUCT

Aggregate measures such as GNP or GDP were not constructed to measure welfare or well-being. They were aimed at measuring total monetary income in an economy and the total supply of goods produced in man-made facilities to provide economic planners with a database suitable for economic policies. However, despite warnings from many economists, GNP and GDP became used as indices of welfare. During approximately the past 25 years, economists have tried to define welfare and construct indices measuring welfare, and

there is now a substantial literature on the subject. Instead of reviewing the literature, I will discuss some issues not very much discussed. Furthermore, most of the literature is based on the assumption that the economy is following an optimal growth path, which is rarely a realistic assumption. Instead, we will assume that we are able to make forecasts arbitrarily into the future.[3]

The issues that will be discussed in this section are

- The notion of social well-being
- Social well-being and population growth
- The search for a linear index—net national product (NNP) as a marginal social cost–benefit criterion for elementary projects
- The search for a linear index—NNP as a measure of sustainable development
- Sustainable development and technical progress

A. Social Well-Being

In order to discuss social well-being, it is useful to imagine a social planner who is in charge of planning an economy from now (date 0) to infinity. Koopmans has convincingly shown (1960, 1972a, b) that the rational social planner should evaluate different policies with regard to how they affect the present value of the stream of future utilities, where the utility function is strictly concave and is a function of all the factors that influence the current well-being. In other words, the social welfare function is written

$$W = \int_0^\infty e^{-\delta\tau} u(c(\tau))\, d\tau \tag{1}$$

where $c(\tau)$ is the rate of consumption at moment τ. In principle, $c(\tau)$ is a vector, the components of which represent all the factors that affect well-being at time τ. However, it will not change any conclusions if we restrict ourselves to the interpretation that c is the aggregate consumption good.

One of Koopmans' condition is that $\delta > 0$; that is, the discount rate must be strictly positive. The discount rate is necessary in order to be able to define a social well ordering over all possible future paths of utilities. However, this is the rate for discounting utilities. The

[2] Jorgenson (1995a, b) provides masterful surveys of these issues.

[3] The basic reason behind the assumption of an optimal growth path in much of the literature is that predictions become very simple. It will be shown that nothing essential will change when we switch to arbitrary allocation mechanisms.

rate for discounting consumption will in general be different and depends on the growth rate of consumption. The formula is

$$r = \delta + \eta g \qquad (2)$$

where r is the consumption discount rate, δ is the utility discount rate, ν is the elasticity of marginal utility of consumption, and g is the growth rate of consumption. It is clear that if δ is small and the growth rate is sufficiently negative, the social planner will want to discount future costs and benefits measured with consumption as a numeraire (i.e., the way we usually measure them in cost–benefit analysis) with a negative interest rate.

The strict concavity of the utility function guarantees an attitude in favor of equalization between generations. Holding everything else unchanged, a transfer from a rich generation to a poor generation will increase social welfare.

1. Mathematical Preludes

No in-depth mathematics will be used in this section and an elementary knowledge of calculus should be enough for the derivations made here. However, reading this section is not necessary for an understanding of later sections.

At time t, the economy has inherited assets from the past. Denote the vector of these assets K_t. Assume that the social planner can predict the future flow of consumption, given K_t. In order to do this, the social planner must know how these assets are going to be used (now and in the future) for consumption and for accumulation of some of the assets and perhaps depletion of others (in particular, stocks of nonrenewable resources). Such predictions can be summarized by an *allocation mechanism*, which will be called α:

$$c(\tau) = \alpha(\tau, t, K_t), \, \forall \, \tau > t \qquad (3)$$

Thus, the mapping α is the information the social planner uses for mapping forecasts.

Note that α is a function of the initial stocks but also of calendar time t. The reason for this is that there may be autonomous changing over time. If there is exogenous technical progress, for example, future predicted consumption will of course depend on this and α must therefore depend on calendar time. Other reasons for autonomous changes in the allocation mechanism may be trends in trade and forecasted changes in government policies. I will discuss these issues later.

Given the allocation mechanism, the social welfare

beginning at time t will be a function of the capital stocks at t, the allocation mechanism, and the calendar time. Thus,

$$W = V(t, \alpha, K_t) \qquad (4)$$

V gives the social welfare achievable from the initial stocks K_t with the allocation mechanism α.

A marginal change in one of the stocks at time t will change the welfare, and this change is basically the marginal value of that stock; that is,

$$p_i(t) = \frac{\partial V(t, \alpha, K_t)}{\partial K_i} \qquad (5)$$

where K_{it} is the stock of asset i at time t. In most of the literature on green accounting, the accounting prices are those that support an optimal growth path. The prices defined here do not. However, they reflect the true values of the initial stocks at time t.

It is practical to define the Hamiltonian H at time t as

$$H_t = u(c(t)) + \sum_{i=1}^{N} p_{it} \frac{dK_{it}}{dt} \qquad (6)$$

One can now show that[4]

$$\frac{dp_i}{dt} = \delta p_i - \frac{\partial H_t}{\partial K_i}, \, i = 1, 2, \ldots, N \qquad (7)$$

Although these equations are formally identical to the equations governing the costate variables in Pontryagin's maximum principle, they do not reflect any optimization except that the prices are the true measures of the marginal social values of the initial stocks.

B. Population and Welfare

So far, I have neglected population changes in the discussion of social welfare. However, we know that humans have a finite life and new generations will be born. How do we take this into account when we try to formulate a welfare function for a society?

The simplest way is perhaps to assume that population changes exogenously in a predictable manner. Then, at each moment in time we can represent the welfare of the cohort by the utility of an individual with average per capita consumption multiplied by the number of individuals. However, this is seriously flawed

---

[4] See Mäler (1999) for a proof.

in that fertility behavior is not exogenous. We now know that human reproduction behavior is governed by many social and economic factors. There are many studies showing that reproduction depends on the need for support during retirement, on the need for labor, on the need for risk diversification, and of course because children can be regarded as consumption goods—that is, they give immediate welfare to the parents. Dasgupta (1993, 1997) provides discussions of these issues and references to empirical material. Thus, we should treat fertility as endogenous, something that depends on the economic system in all its characteristics. However, this is exactly what the allocation mechanism introduced in the previous section represents. Thus, the population at any particular future point of time is determined by α.

It is now easy to assume that the welfare function can be written

$$\int_0^\infty e^{-\delta t} L(\alpha(0, K_0)) u(c(t))\, dt \qquad (8)$$

However, this formulation does not take into account the fundamental difference between a potential human being and an actual human. What is the value of having one more individual to that individual? There is no reasonable answer to this question [see Dasgupta (1997) for an extremely interesting discussion of these issues]. So far, we do not have an ethical base for assessing welfare to endogenous population changes.

C. A Marginal Cost–Benefit Rule

Define NNP with utility as the numeraire as the linearized Hamiltonian

$$\text{NNP} = u'(c)c + \sum_{i=1}^N p_{it} \frac{dK_{it}}{dt} \qquad (9)$$

If, instead, we choose consumption at time 0 as the numeraire, $\widetilde{\text{NNP}}$ is defined by

$$\widetilde{\text{NNP}} = c + \sum_{i=1}^N \tilde{p}_{it} \frac{dK_{it}}{dt} \qquad (10)$$

where $\tilde{p}_{it}$ is $p_{it}/u'(c(0))$.

Note that NNP or $\widetilde{\text{NNP}}$ is a linear index in all quantities, which makes it an interesting measure. A nonlinear measure such as the Hamiltonian is hopeless in terms of empirical applications. We cannot estimate all the consumers' surpluses needed to get an empirical estimate of the Hamiltonian.

Second, note that c is basically a proxy for all variables that affect current utility. Therefore, all these variables must be included in a generalized consumption set. Among these variables is the disutility of labor, which leads us to the conclusion that the wage bill, as long as the wage rate reflects the disutility of work, should be deducted. Immediate environmental damages should also be deducted.[5] In all these cases the damages should be valued at a rate that reflects the marginal disutility of the environmental damage with consumption as the numeraire.

Third, note that the investment part of the NNP must include net changes in all assets that will affect future well-being. This means of course that the current narrow asset boundary in SNAs must be substantially extended.

Also note that the asset prices p_i must correctly reflect social the scarcity values captured by Eq. (7). These prices will not always coincide with market prices. However, surprisingly often, market prices will capture the social scarcity values of assets that are transacted on markets. However, for all the ecological assets, the values must be estimated by modeling the ecological systems and their connectedness. There is a growing literature on these procedures.[6]

It should be noted that the NNP measure so defined deviates radically from the United Nations recommendations. The NNP measure we have defined has one important property, namely, that if there is a small perturbation (so that price changes are small) of the allocation mechanism during a small time interval, and if this small perturbation increases the social well-being, then it will also increase NNP and vice versa.[7] This is what makes NNP interesting—a linear marginal cost—benefit criterion with which we can evaluate such perturbations (which henceforth will be called policy reforms).

Here, we consider discreet time, and the following example will indicate the use of the theorem stated previously. Assume that there is one allocation mechanism α and we consider a perturbation or policy reform $\Delta\alpha$ that for the first period prescribes the same allocation as α did for the second period. We compute the change in NNP when we make this policy reform and this change is positive. This can be interpreted as saying that the second period gives higher social welfare than the first period. By following the allocation mechanism

---

[5] Mäler (1991) discusses these issues in detail.
[6] Freeman (1992) provides a good introduction.
[7] For a proof, see Mäler (1999) or Dasgupta and Mäler (2000).

α, welfare increases over time. However, note that this depends on the assumption that the prices do not change much.

This theorem can now easily be generalized to situations in which we take atemporal equity into account, and even population changes can be accommodated. The trick is to find the appropriate shadow prices.

D. The Hamiltonian as Constant-Equivalent Utility

In the previous two sections it was shown that NNP can be used as an index for conducting social cost–benefit analysis of policy reforms. However, the theoretical literature on green NNP has been directed toward a quite different end (Weitzman, 1998). It has argued that NNP measures "constant-equivalent consumption."

It can be shown that the present value of a constant stream of utility equal to the Hamiltonian at time zero is the maximum feasible constant utility stream that an economy can afford. This was first shown by Weitzman in the case in which the utility function is linear. In this case, NNP and the Hamiltonian coincide, and NNP gives the maximum constant equivalent consumption.

However, a linear utility function is ethically flawed: It is insensitive to distributional issues. Furthermore, a large body of evidence concerning household saving behavior is at odds with linear utility functions.

E. Social Well-Being and the Concept of Sustainability

The World Commission (1987) defined "sustainable development" as an economic program in which, loosely speaking, the well-being of future generations is not jeopardized. There are many possible interpretations of this. Consider the following:

a. An economic development is sustainable if $dU_t/dt > 0$, where $U_0 > \lim U_t$ as $t \to -0$.

b. An economic development is sustainable if $dU_t/dt > 0$.

c. An economic development is sustainable if $dV/dt > 0$, where

$$V_t = \int_0^\infty e^{\delta t} U(C_t)\, dt$$

It is clear that a. lacks ethical foundation. For example, it may be desirable to reduce U in the short run

in order to accumulate assets so that the flow of U is still higher in the future. In this sense, b. offers greater flexibility in ethical reasoning: It permits initial sacrifices in the current standard of living, U (a burden assumed by the generation engaged in the reasoning), but requires that no future generation should have to experience a decline in its standard of living.

In contrast to b., the focus of c. as a notion of sustainable development is social well-being, V. The criterion permits the first generation to make initial sacrifices in V (relative to the past) but requires that social well-being should never decline in the future. Note that although b. implies c., c. does not imply b. In other words, c. is more general. In what follows, we adopt c. as our notion of sustainable development and develop criteria for judging whether a given economic program represents sustainable development.

One can now easily show that

$$\frac{dV}{dt} = \frac{\partial V}{\partial t} + \sum p_i \frac{dK_i}{dt}$$

The first term on the right-hand side is the autonomous change in V, reflecting exogenous technical progress or exogenous changes in terms of trade. We will discuss this term in the last section. For now, we assume that there are no autonomous changes. The previous equation simply indicates that the rate of change in welfare is equal to the value of the net investment in the economy. Thus, an economic program represents sustainable development if, and only if, the program net investment in the economy's capital assets is always nonnegative.

The result has intuitive appeal. It says that social welfare is higher today than it was yesterday if the economy is wealthier today. Here, an economy's "wealth" is interpreted as the accounting value of all its capital assets, and wealth comparisons are made at constant prices. In a famous article, Samuelson (1961) argued in connection with national income accounting that welfare comparisons should deal with "wealth-like" entities. Proposition 4 formalizes this insight.

Note, however, that what we have obtained is an equivalence result: The theorem cannot on its own tell us if sustainable development is feasible. Whether the economy is capable of growing wealthier indefinitely depends on, among other things, the extent to which different assets are substitutable in production.

This idea of examining changes in wealth dates back to Samuelson (1961). The connection to sustainable development was first suggested by Pearce and Atkinson (1993), although they did not derive the theorem rigorously and they did not interpret it as an equivalence result. The idea was then investigated by the World Bank, which published many attempts to measure the value of net investment or genuine savings [see *Expanding the Measure of Wealth* (1997) and Kunte and Clemens (1998) for further details.]

Note that we cannot in general use NNP to determine whether the economy is on a sustainable path. Only if the prices are stationary is an increasing NNP equivalent to saying that the economy is on a sustainable path.

In conclusion, NNP can be used for cost benefit analysis in situations in which prices do not change very much. In particular, in such a situation it may give some guidance as to whether social welfare is increasing or not. Note that there is no conflict between growth in the NNP and the concern for the environment. These concerns have been incorporated into the linear index. If the economy is on a path in which environment is degraded, this will be captured by the NNP, and it can be expected that NNP will decrease if the degradation is sufficiently large.

NNP is not useful for determining whether the economy is on a sustainable path because the relative prices will change along the path. Here, concepts related to wealth become more interesting. In particular, the value of the net investments in a time period gives a correct criterion for judging whether the economy is on a sustainable path. However, it is only a necessary condition. The sufficient criterion is that in each future period, the value of net investments must be positive. A priori, we do not know whether this is feasible.

III. ENVIRONMENTAL KUZNETS CURVES

Since the early 1990s, economists and econometricians have been industrious in estimating empirical relations between GDP and environmental quality indicators. Typically, they have used cross-sectional data—country data for a particular year. For some quality indicators, they have found a relation that looks like an inverse U (Fig. 1).

The interpretation is simple. Very poor countries with low industrialization are not damaging the environment much and as a result the environmental quality indicator is good (in this case, a low sulfur emission). When the country starts the growth process, the environmental damages will increase. When income per

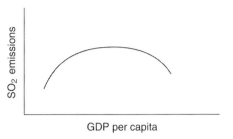

FIGURE 1　An estimated empirical relation between GDP and environmental quality indicators.

capita increases, the demand for a cleaner environment will increase, environmental legislation will be enacted, and the environment will improve. Because the curve is similar (but inverse) to the curves Kuznets (*Economic Growth and Income Inequality*, 1955) estimated for the relationship between income inequality and GDP per capita, they have been known as the environmental Kuznets curves. For an overview and examples of environmental Kuznets curves, see *Environment and Development*, **2** (4), October 1997.

The breakthrough for this kind of analysis came with the World Development Report (1992), which was devoted to the environment. The existence of these curves was taken to be an indication of possible win–win situations. By stimulating growth, one would also in the long run stimulate environmental regulations. It is clear, however, that the analysis is quite muddy complex.

First, environmental Kuznets curves have only been established for a few environmental indicators. Second, at least earlier, the specifications of the relation between the quality measure and GDP per capita were quadratics, which almost certainly forced the curves to look like an inverse U. Third, the studies did not include environmental spillovers between countries. Fourth, they did not examine possible relations between indices of freedom and the environment. It has been shown that whenever people have had the right to voice concern about environmental degradation, the environment has improved, irrespective of the per capita income. Locally, in many cultures people have had a direct influence on, for example, local water pollution, which has improved even in very poor countries. It is therefore possible that the environmental Kuznets curves measure not the effect of income growth but growth in freedom.

However, most important is that the environmental Kuznets curves—if they exist in any meaningful way—

are not optimal in any sense of the concept. Typically, a country would do much better to develop institutional and regulatory reforms for better environmental management than just trust that economic growth will automatically solve environmental problems. In this sense, the belief in environmental Kuznets curves and the associated belief in win–win situations are very dangerous.

IV. TECHNOLOGICAL CHANGE AND GROWTH ACCOUNTING

Productivity growth as discussed previously is productivity growth in GDP in which the flow of services that are not transacted on markets are excluded (with of course some important exceptions, e.g., government output). Here, we are interested in *productivity growth in NNP* as defined previously. There is no reason why productivity growth in GDP would correspond to a similar growth of NNP.[8] Furthermore, if, as Dale Jorgensen (1995a, Chap. 1) pointed out, changes in energy prices affected the observed productivity growth rate substantially, it should be clear that the utilization of other natural resources should also affect the growth rate of the total production. Thus, we should analyze the following issues:

• Growth in NNP
• The role of natural capital in the growth of NNP

The following calculations are very speculative, but I hope that some day someone will find it worthwhile to study the empirical consequences from the analysis. I use the simplest possible model to demonstrate the main ideas (which probably will be seen to be perhaps too obvious).

Let there be one aggregate good x which is produced with man-made capital K_x and with one input representing natural capital services, denoted y. Thus,

$$x = f(K_x, y) \qquad (11)$$

Assume that the production function f is homogeneous to degree 1.

[8] In fact, Repetto *et al.* (1989) showed that by including depreciation of oil reserves, deforestation, and soil erosion in Indonesia, the growth rate was reduced to approximately half the official growth rate in GDP.

The total output is used for consumption c, for investment in man-made capital $I = K_x$, and for investment in man-made capital for exploiting the natural resources $I_y = K_y$. Let the production function for y be

$$y = g(K_y) \qquad (12)$$

Thus, y is an intermediary good in the system and x will correspond to GDP:

$$x = c + K_x + K_y = c + K \qquad (13)$$

Assume that capital is freely mobile between the two uses so that we can reasonably talk about the total capital stock $K = K_x + K_y$. Then we can define the productivity growth in GDP as

$$\theta = \frac{1}{x}\frac{dx}{dt} - \frac{1}{K}\frac{dK}{dt} \qquad (14)$$

Now, let the dynamics of the natural or ecological capital K_E be

$$\frac{dK_E}{dt} = \phi(K_E) - y \qquad (15)$$

Assume that there is an institutional failure in the economy so that there are no assigned property rights to natural capital and therefore no market transactions in K_E. This is of course the reason why investments in natural capital are not included in GDP.

NNP according to the definition in equation is

$$\text{NNP} = c + p_E\frac{dK_E}{dt} + \frac{dK}{dt} = x + p_E\frac{dK_E}{dt} \qquad (16)$$

It is clear from Eq. (16) that a positive productivity growth for GDP can be offset by depreciation of the stock of natural capital so that the productivity growth of NNP is zero or even negative. In fact, if we assume that $\dot{K} = 0$, we have

$$\frac{1}{\text{NNP}}\frac{d\text{NNP}}{dt} - \frac{1}{K_E}\frac{dK_E}{dt} = \frac{\text{GDP}}{\text{NNP}}\theta + \frac{\dot{K}_E}{K_E}\left\{\frac{p_E K_E}{\text{NNP}}\frac{\ddot{K}_E}{\dot{K}_E} - 1\right\} \qquad (17)$$

The last bracket in Eq. (17) will surely be negative becaus of the overexploitation of the resource as a result of the institutional failure. Thus, this bracket will compensate for the positive productivity growth of GDP. In this case, because we know that the production functions are stationary, we know for sure that the productivity growth rate of NNP (if the production functions have constant returns to scale in K_x, K_y, and K_E) must be zero. We can therefore conclude that even if measured GDP growth rates show a positive productivity growth, this does not necessarily imply that NNP will show growth irrespective of factor inputs.

This result also follows from the observation that conventional SNA regards the inputs of resources as intermediary goods (if there are costs in producing them and it neglects them completely if they are supplied free of any costs to the production processes), and the estimation of an aggregate production function with only labor and capital as inputs, necessarily will be biased because resources are the basis for material production. Thus, the total factor productivity of GDP should be measured as

$$\frac{1}{x}\frac{dx}{dt} - w_K\frac{1}{K}\frac{dK}{dt} - w_\nu\frac{1}{\nu}\frac{d\nu}{dt} \qquad (18)$$

where w_K is the value share of man-made capital and w_ν is the value share of the resource input. This will necessarily be zero if the production function is linearly homogeneous and stationary. Thus, it may be that the neglect of resource inputs in the production functions is an important explanatory variable of the measured positive factor productivity growth. However, the only way to find out whether this is important is to carry out new empirical measurements. In conclusion, it does not seem reasonable to assume autonomous productivity growth in models designed to create a new accounting system for welfare measurements. Therefore, the argument for a nonstationary value function is not necessarily true.

See Also the Following Articles

Bibliography

Dasgupta, P. (1993). The population problem. *J. Econ. Literature*, 1879–1902.
Dasgupta, P. (1997). The population problem. *J. Ecol. Econ.*
Dasgupta, P. (1999). Social capital. *Economic growth and income inequality (1955). Am. Econ. Rev.* **45**, 1–28.
Dasgupta, P., and Mäler, K.-G. (2000). Net National Product, Wealth and Social Well-being. *Environ. Dev. Econ.* **5** (1 & 2), 69–94.
Expanding the Measure of Wealth. (1997). World Bank, Washington, D.C.
Freeman, A., III, (1992). *The Measurement of Environmental and Resource Values*. Washington, D.C.: Resources for the Future.
Jorgenson, D. (1995a). *Productivity*, Vol. 1. MIT Press, Cambridge.
Jorgenson, D. W. (1995b). *Productivity*, Vol. 2. MIT Press, Cambridge.
Koopmans, T. (1960). Stationary ordinal utility and impatience. *Econometrica* **28**, 287–309.
Koopmans, T. (1972a). Representation of preference orderings with independent components of consumption. In *Decision and Organization* (C. B. McGuire and R. Radner, Eds). North-Holland, Amsterdam.
Koopmans, T. (1972b). Representation of preferences ordering over time. In *Decision and Organization* (C. B. McGuire R. Radners, Eds.) North-Holland, Amsterdam.
Kunte, A. K. H. J. D., and Clemens, M. (1998), Estimating national wealth: Methodology and results, Technical report. World Bank, Environment Department, Washington, D.C.
Mäler, K. (1991). National accounting and environmental resources. *Environ. Resour. Econ.*, 1–5.
Mäler, K.-G. (1999). National income, welfare measurements and local prices.
Pearce, D., and Atkinson, G. (1993). Capital theory and the measurement of sustainable development: An indicator of weak sustainability. *Ecol. Econ.* 8(2), 103–108.
Repetto, R., Magrath, W. W. M. B. C., and Rossini, F. (1989). *Wasting Assets: Natural Resources and the National Income Accounts*. World Resources Institute, Washington, D.C.
Samuelson, P. (1961). The evaluation of social income: Capital formation and wealth. *In The Theory of Capital* (F. Lutz D. Hague, Eds). Macmillan, London.
Solow, R. (1957). Technological progress and productivity change. *Rev. Econ. Stat.* **39**, 312–320.
World Development Report (1992). Oxford Univ. Press, New York.

ECONOMIC VALUE OF BIODIVERSITY, MEASUREMENTS OF

Robert Mendelsohn
Yale University

GLOSSARY

biodiversity Number of species on the planet.
local diversity Number of species or ecosystems in a small geographic area.
market value Benefit of wildlife or species as an input into the economy.
quality-of-life value Benefit of wildlife or species to people's well-being or happiness.
willingness-to-pay Expression by people of strength of preference in dollar terms.

BIODIVERSITY IS USED IN MANY DIFFERENT WAYS including the number of species on the planet, the number of species in one place, the number of individuals of desired species, the genetic diversity of a given species, and the ecological diversity of ecosystems. Each of these meanings implies a set of different values. This article reviews how we measure these different values and provides an overview of the values that have been measured to date. There are two broad types of biodiversity values. Market values reflect how biodiversity serves as an input to economic activities such as agricul-

ture, construction, and medicine. Quality-of-life values reflect ways that biodiversity makes our lives richer simply by increasing our appreciation of life itself.

I. INTRODUCTION

Before one can describe the values that society places on biodiversity, it is first necessary to define what biodiversity means to people. Unfortunately, the term has come to mean many different things, which only complicates valuation. A narrow definition is that biodiversity represents the number of species on the planet. Hence the loss of a species diminishes biodiversity. This definition clearly links biodiversity with endangered species. Biodiversity has also been used to compare one place to another. Local diversity measures this type of biodiversity: places with more biodiversity have more species per hectare. This definition recognizes the value of complex ecosystems, as well as the value of a variety of ecosystems across a landscape. A third definition of biodiversity reflects the value of all living things: a place with more songbirds, more deer, more butterflies, and more trees is a "better" place. In this case, biodiversity is a pseudonym for the overall population. The term biodiversity means all of these things to society, making a precise measurement of the value of biodiversity difficult.

There is no single value of biodiversity since there is no single definition of what the word actually means. Biodiversity values are really only clear as one moves from the general term to specific questions. Rather than determining a single value of biodiversity, the literature on valuation has attempted to measure how important specific aspects of nature are to people. For example, what would people pay to prevent a specific species from going extinct? What would people sacrifice to keep more complex ecosystems such as mixed-species forests in place? How much income would people give up to have a variety of ecosystems across the landscape rather than having only one or two? What would people pay to increase the population of particular creatures in their local woods?

Biodiversity values have often been further divided into two types: market and quality-of-life values. The market values reflect the interaction between nature and the economy, whereas the quality-of-life values involve interactions directly between nature and people. Methods to measure market values are well developed and we consequently know more about market values. Quality-of-life values are more difficult to measure. There are many of such values we have just begun to understand, much less measure. Although these quality-of-life benefits are technically difficult to measure, they cannot be dismissed because they are likely to be the primary motivation for protecting species and wildlife populations.

The remainder of this article is divided into a review of biodiversity values for market and quality-of-life services. In each case, we begin discussing the methods and then review existing measurements. For both market and quality-of-life, we examine species diversity, local diversity, and natural populations. The article concludes with an overview of what has and has not been measured, discusses research priorities, and briefly hints at policy implications.

II. VALUE TO THE ECONOMY

Biodiversity is one of many inputs into the economy. From medicine to timber, biodiversity provides raw materials for many economic processes. Although these inputs are not always traded in markets, they nonetheless fuel market activities. By valuing the resulting market activities, one can deduce the value that each input contributed. Of course, one must be careful to value all inputs to each economic process. One cannot assign the market value of outputs to just one of the many inputs required to make a good. For example, plant biodiversity may be an input to pharmaceutical research but it is only one of many inputs, including laboratories, technicians, and clinical research. In this section, each of the multiple definitions of biodiversity is examined to determine what value they contribute to the economy.

A. Diversity

The diversity of species on the earth is of economic value because the diversity offers many alternative services. For example, the diversity of plants has been used to find cures for diseases. Plants, in their constant struggle for survival, have developed numerous defenses to limit consumption by predators. Some of these chemicals have important medical qualities. Many existing pharmaceuticals have come from plants. Some 47 drugs have come from tropical forest plants alone (Soejarto and Farnsworth, 1989). Given that there are 250,000 plant species in the world, there is a strong possibility that additional undiscovered drugs lie hidden in this diversity. Assuming that one out of a million samples contains a hidden drug, there may be as many as 325 useful drugs still hidden in tropical plants alone (Balick and Mendelsohn, 1995).

What is the value of these hidden drugs? The drugs themselves are worth billions of dollars. A successful drug is worth about $94 million to a drug company (OTA, 1993). Such a drug is worth about $450 million to society when one includes the resulting health benefits, taxes, and revenues beyond the patent period (Balick and Mendelsohn, 1995). Multiplying both figures by the number of hidden drugs suggests that private companies would pay $30 billion and society would pay $145 billion for these drugs. This is the value of the drugs, but not the value of the biological component since many other inputs went into creating each drug. For example, finding the samples, identifying active chemicals, and developing a drug that can be marketed costs over $100 million per drug. Pharmaceutical companies should not be willing to pay much for samples at all since the cost of developing new drugs from these samples is likely to be greater than the expected revenues. In contrast, society would still find the drugs valuable since society obtains more benefits from the drugs than companies can get. Society would be willing to pay about $40 billion for these samples. When spread over 125,000 plants, this amounts to about $320,000 per plant. Allocating this value to each of the 3 million hectares of tropical forests suggests that each hectare is worth an average of about $13. Of course, once the plants on a particular hectare are investigated, they will no longer have any more value in this regard. This raises serious questions about how to collect these revenues since many landowners will compete to be the one and only collector of each plant.

Diversity may also be of value to crops and trees because it yields a broad gene pool. Although this gene pool may not be needed today, it could serve as a resource against future pathogens and problems. Because farmers and foresters rely on improved varieties, many scientists are concerned that the crop gene pool has become too narrow. Despite the success of modern farms and forests, these scientists argue that farmers need more diversity and thus it is imperative that we protect plants in the wild that are related to modern crops. Similar concerns are raised to protect old varieties that have fallen out of favor. It is possible that noncommercial plants related to valuable crops could be valuable in the future in order to create a broader gene pool to be used against future pathogens. However, it is hard to quantify the market value of this potential since most modern crop varieties appear to be doing well in relying on a more select gene pool that is already well protected. It would be useful to quantify the additional value of sustaining marginal crop varieties against some future date when they might be needed. The known cost of sustaining varieties that have little value today would have to be weighed against the uncertain benefit in the future when they might be needed.

B. Local Diversity

Local diversity has often been championed as a defense against the ravages of nature. Whether one is considering hurricanes, ice storms, and other severe weather or whether one is concerned about insects, rust, or viral attack, ecologists argue that more diverse systems are more robust. For example, by planting a diversity of crops, a farmer can protect his or her income from weather, insects, and pathogens. Even if some crops are destroyed, the other crops will likely survive, thus assuring at least some income. Modern farms tend to plant a wide set of crops to protect against uncertainty from varying market prices as well as the adversities facing all farmers. Similar schemes can protect forests as well. By growing trees from a wide range of species and in a number of age groups, foresters can reduce the likelihood of catastrophic events.

Some efforts to combine forests and farms have also been practicing local diversity. Windbreaks across flat areas have been used as a way to protect crops from high winds. Mixed agroforestry methods have been used in the tropics to provide a portfolio of outputs (Padoch *et al.,* 1985) as well as to sustain soils (Montagnini and Mendelsohn, 1997). These land use planning practices have been successful in reduc-ing some of the damages that are more easily afflicted to a homogeneous landscape.

C. Population

Sometimes plants and animals are a direct input to the economy. In this case, biodiversity represents the size of relevant populations. The most significant biological population to humans lies in the agricultural sector. Approximately 1.9 billion tons of grain, 0.6 billion tons of tubers, and 1.5 billion tons of sugar crops are grown each year. In addition, approximately 181 million tons of meat are harvested annually. Agriculture accounts for approximately 6% of the world's gross domestic product, or about $1.2 trillion overall. This number reflects the marginal value of food. Of course, people would pay a lot more for food rather than be without it entirely. Assuming that people would pay twice as much for food if only half of the food was available, the all-or-nothing value of food would be $2.4 trillion.

Another economic sector that depends on raw materials from nature is forestry. The timber industry harvests trees from more or less natural settings and converts them into lumber and paper products. For the last 200 years, the industry has relied heavily on cutting natural stands of trees. Originally, this was done unsustainably as the industry harvested the vast stands of old growth around the world. As these stands have become rare, much of the industry has shifted to growing trees on plantations. The worldwide industry is approaching sustainability as regeneration now almost balances harvest rates. Once this balance is reached and maintained, trees can be harvested indefinitely from a given land base. The timber industry currently harvests 1600 million cubic meters every year (IPCC, 1996). Over half comes from temperate forests (890 million m^3), another fourth from boreal forests (430 million m^3), and the remainder from tropical forests (280 million m^3). In contrast, most of the world's land in forest is currently tropical (27.6 million km^2) and boreal (13.8 million km^2), with only 5.4 million km^2 in the temperate region (IPCC, 1996). The timber industry is most dependent on the productive lands in the temperate and subtropical parts of the globe.

This industry produces an annual output worth $418 billion (IPCC, 1996). However, a large fraction of this estimate is value added from production. The value of the trees themselves is more difficult to estimate. Assuming that the value of harvesting an additional tree, the stumpage value, is about $20/m^3, the annual market value of the 1600 million m^3 would be about $32 billion per year. This is the marginal value of the harvest.

The most abundant physical product from the forest is not industrial wood but rather fuelwood and charcoal. Each year approximately 1750 million m³ of wood is used for energy. The vast majority of this use occurs in tropical regions, which rely on 1370 m³ for energy. Temperate forests produce 20% of the wood energy (370 million m³) and boreal forests produce very little (10 million m³) (IPCC, 1996). Although the use of wood for energy consumes more forest biomass than any other use, it is not the most valuable activity. Wood is often used by communities that cannot afford more expensive sources of energy (such as fossil fuels). The average price of a cubic meter of wood for biomass is probably in the neighborhood of about $5/m³. The global market value of fuelwood is around $9 billion annually.

In addition to these foregoing products from trees, there is a host of nontimber forest products (NTFPs) that people enjoy around the world. The more diverse the forest, the more NTFPs that are present. Yet even in relatively simple northern forests, people enjoy wild rice, berries, mushrooms, boughs, maple sugar, and wood resins. In the tropical zone, forests produce natural medicines, fodder, fruits, latex, rattan, roofing material, dyes, and other wood resins. It is difficult to quantify the global value of these products since most are traded only locally and so escape official enumeration. However, local studies of selected tropical forests suggest that these NTFPs are often as valuable as timber (Peters *et al.*, 1989; Balick and Mendelsohn, 1992; Grimes *et al.*, 1994; Houghton and Mendelsohn, 1996; Muniz-Miret *et al.*, 1996). As a rough approximation, it is likely that the value of NTFPs is equal to the value of timber from tropical forests, or about $6 billion each year.

III. QUALITY-OF-LIFE

Although biodiversity makes a large contribution to the world's economy, it is likely that the most important services that humans receive from nature are to our quality-of-life. Many aspects of biodiversity are not used by the economy and so provide no market value. Nonetheless, if we were to lose these species and populations, people would suffer a great loss. People clearly enjoy nature and obtain a great deal of pleasure from it.

A. Diversity

The earth contains an amazing diversity of plants and animals. Approximately 1.5 million living species have been identified. Almost 1.25 million of these species are arthropods, or insects. Scientists estimate that there are as many as 250,000 flowering plants, 80,000 fungi, and 100,000 mollusks. Only 30,000 species are vertebrates. Although there have been some major mass extinctions in the geological record, the number of higher-order species that have perished in industrialized nations is few. In fact, the vast preponderance of species lost today are the assumed losses of insects from the clearing of tropical forests. Scientists predict that each of these hectares likely contains a certain number of unique insects that are lost when the trees are removed. This is a predicted loss, since no one has actually identified the insects in these forests that have now disappeared.

Surveys of public opinion clearly indicate widespread and strong interest for species protection. Despite frequent attacks from land developers, business groups, and some industries, public support for endangered species and other wildlife continues to be strong. However, public preferences for which species deserve protection are complex, for U.S. society is not equally concerned with all species. People express the strongest preference for flagship or charismatic species (Kellert, 1984). These include North American animals such as the grizzly bear, black-footed ferret, Florida panther, and bald eagle. International examples include lions, elephants, tigers, rhinoceroses, seals, monkeys, and apes. In general, people appear to be willing to pay the most for these charismatic animals. For example, using a survey on the willingness to pay, Brookshire *et al.* (1983) found that people would pay between $10 and $26 to protect grizzly bears and Boyle and Bishop (1986) found that eagles were worth between $10 and $31 per person. Despite the broad enthusiasm for species protection, it is not clear that people are willing to sacrifice very much to protect lower-order animals and plants. For example, Boyle and Bishop (1986) found that people were only willing to pay between $4 and $5 to protect the striped shiner, a small fish. Valuation studies have not specifically determined a value for a broad class of lower-order animals. It is most likely the case that people would pay for their protection as part of a larger plan to protect entire ecosystems. However, the valuation of entire ecosystems has not yet been done.

B. Local Diversity

Scientists have determined that certain ecosystems have more species per hectare than others. For example, species diversity generally rises as one approaches the equator from either pole. The highest number of endemic species per hectare tends to be in tropical locations. It is possible that people have preferences concerning whether they would like to have more species in the ecosystems that they live in. Unfortunately this has not been tested, so it is difficult to know. It is clear that certain "hot spot" locations with high numbers of endemic species have become desirable travel destinations. Madagascar, the Philippines, and the Galapagos Islands are all sites where ecotourists flock to see the high numbers of endemic species.

Local diversity can also mean that there are many ecosystems within a small region. Sometimes terrain can create small microsites with unique climates. For example, mountainous regions near the coast can have a wide variety of ecosystems in a relatively small area. This diversity gives people many landscapes to enjoy locally. Although most of this diversity is a natural phenomenon, it can be affected by land use patterns. For example, if people drain wetlands, the only land remaining may be an upland ecosystem. By draining the wetlands, people have simplified the landscape. The many recent laws to protect wetlands and other ecosystems suggest that this diversity has considerable value to people. Curiously, little valuation research has been done on the value of heterogeneous landscapes.

C. Populations

There is no question that people value local wildlife populations. A survey conducted by the U.S. Fish and Wildlife Service (1991) found that approximately one-half of Americans participated in some wildlife-related recreation. Of these, 76 million people enjoyed nonconsumptive activities such as bird watching, 35 million people fished, and 14 million hunted. Throughout America, people appear to derive great pleasure from having wildlife around them. Even with such common creatures as squirrels, sparrows, and pigeons, people are entertained by chance encounters with animals.

Economists have assigned values to wildlife interactions by using survey methods. For example, the techniques used to study endangered species protection (Brookshire et al., 1983; Boyle and Bishop, 1986) have also been applied to places and activities such as the Maine woods, duck hunting, and goose hunting.

Economists also measure values by observing the travel preferences of people. Because many people interact with wildlife by taking trips to outdoor sites, their choice of sites reveals how much they want to see different things. For example, with the simple travel cost method, visitation rates are compared with distances.

The farther someone is willing to travel to see a site, the more valuable the site is. The relationship between visitation rates and distance reveals how much people are willing to pay to see the site. By calculating the area under this visitation function minus actual travel costs, one can determine the value of the site for a person. The total value of the site would be the sum of the values for all the visitors (Clawson and Knetsch, 1966; Brown et al., 1983). This model has also been extended to examine a set of sites (Burt and Brewer, 1987; Cichetti et al., 1992).

A weakness of the simple travel cost method, however, is that it values sites only in their entirety. One of the questions facing land managers is the valuation of changing site characteristics. For example, a manager might consider harvesting some old growth at a site or increasing deer or elk populations. To place a value on site characteristics, one must examine the set of sites that people choose from. Two techniques have been developed to value site characteristics: discrete choice and hedonic travel cost. The discrete choice method looks at all the available sites and determines a quality index that can explain why people choose particular places more often (Morey, 1981). The hedonic travel cost method examines the attributes of the places people choose and tries to estimate the marginal cost or value that people assign to each attribute. The hedonic travel cost then goes further and tries to estimate a demand function for each attribute (Brown and Mendelsohn, 1984).

The literature on valuing site attributes indicates that natural site qualities are important to people. For example, steelhead trout anglers care a lot about the number of fish in a stream (Brown and Mendelsohn, 1984). In Washington state, they would pay an average of $5 per trip for just 1 more fish per day. Hunters care a lot about the number of deer per acre. Urban hunters would pay an extra $4 for one more deer per square mile in Pennsylvania (Mendelsohn, 1984). Hikers care about old-growth forests, clear-cuts, and the size of open areas in wilderness hikes in Washington (Englin and Mendelsohn, 1991). Their consumer surplus per trip is about $70 for old growth, $14 to avoid clear-cuts, and $67 for large rock and ice areas. This literature also revealed that what people would pay for certain characteristics declined as they experienced more of them. Although people would pay $88 to avoid a 20% reduction in the steelhead population at an average site, they would pay very little for more steelhead at a site where there is already a large population. The quality-of-life value of wildlife populations consequently appears to depend on their abundance.

See Also the Following Articles

AESTHETIC FACTORS • ECONOMIC GROWTH AND THE ENVIRONMENT • ENDANGERED MAMMALS • MARKET ECONOMY AND BIODIVERSITY • PLANT SOURCES OF DRUGS AND CHEMICALS • TIMBER INDUSTRY • TOURISM, ROLE OF

Bibliography

Balick, M., and R. Mendelsohn. (1992). The economic value of traditional medicine from tropical rain forests. *Conservation Biol.* 6, 128–130.

Balick, M., and R. Mendelsohn. (1995). The value of undiscovered pharmaceuticals in tropical forests. *Econ. Botany* 49, 223–228.

Boyle, K., and R. Bishop. (1986). The economic value of endangered species in wildlife. *Trans. Fifty First N. Am. Wildlife and Natural Resources Council.*

Brookshire, D., L. Eubanks, and A. Randall. (1983). Estimating option prices and existence values for wildlife resources. *Land Econ.* 59, 1–15.

Brown, G., and R. Mendelsohn. (1984). The hedonic travel cost method. *Rev. Econ. Statist.* 66, 427–433.

Brown, W., C. Sorhus, B. Chou-Yang, and J. Richards. (1983). Using individual observations to estimate recreation demand functions. *J. Am. Agric. Econ.* 65, 154–157.

Burt, O., and D. Brewer. (1987). Estimation of net social benefits from outdoor recreation. *Econometrica* 39, 813–828.

Cichetti, C., A. Fisher, and V. K. Smith. (1992). An econometric evaluation of a generalized consumer surplus measure: The Mineral King controversy. *Econometrica* 44, 1259–1276.

Clawson, M., and J. Knetsch. (1966). *Economics of Outdoor Recreation.* Johns Hopkins University Press, Baltimore.

Clayton, C., and R. Mendelsohn. (1993). The value of watchable wildlife: A case study of McNeil River. *J. Environ. Management* 39, 101–106.

Englin, J., and R. Mendelsohn. (1991). A hedonic travel cost analysis for valuation of multiple components of site quality: The recreation value of forest management. *J. Environ. Econ. Management* 21, 275–290.

Grimes, A., B. Bennett, P. Jahnige, S. Loomis, M. Burnham, K. On-
thank, W. Palacios, C. Ceron, D. Neill, M. Balick, and R. Mendelsohn. (1994). The value of tropical forests: A study of nontimber forest products in the primary forest of the upper Napo Province, Ecuador. *Ambio* 23, 405–410.

Houghton, K., and R. Mendelsohn. (1996). An economic analysis of multiple-use forestry in Nepal. *Ambio* 25, 156–159.

International Panel on Climate Change (IPCC). (1996). *Climate Change (1995): Impacts, Adaptations, and Mitigation of Climate Change: Scientific–Technical Analysis.* Cambridge University Press, Cambridge, United Kingdom.

Maille, P., and R. Mendelsohn. (1993). Valuing ecotourism in Madagascar. *J. Environ. Management* 38, 213–218.

Montagnini, F., and R. Mendelsohn. (1997). Managing forest fallows: Improving the economics of swidden agriculture. *Ambio* 26, 118–123.

Morey, E. (1981). The demand for site-specific recreational activities: A characteristics approach. *J. Environ. Econ. Management* 8, 345–371.

Muniz-Miret, N., R. Vamos, M. Hiraoka, R. Mendelsohn, and F. Montagnini. (1996). The economic value of managing the acai palm. *Forest Ecol. Management* 87, 163–173.

Office of Technology Assessment (OTA). (1993). *Pharmaceutical R&D: Costs, Risks and Rewards.* U.S. Government Printing Office, Washington D.C.

Padoch, C., J. Chota Inuma, W. De Jong, and J. Unruh. (1985). Amazonian agroforestry: A market-oriented system in Peru. *Agroforestry Syst.* 3, 47–58.

Peters, C., A. Gentry, and R. Mendelsohn. (1989). Valuation of an Amazonian rain forest. *Nature* 339, 655–656.

Soejarto, D., and N. Farnsworth. (1989). Tropical rain forests: Potential source of new drugs? *Perspect. Biol. Med.* 32, 244–256.

Sohngen, B., R. Mendelsohn, and R. Sedjo. (1999). Forest conservation, management, and global timber markets. *Am. J. Agric. Econ.* 81, 1–13.

Tobias, D., and R. Mendelsohn. (1991). Valuing ecotourism at a tropical rain forest reserve. *Ambio* 20, 91–93.

U.S. Fish and Wildlife Service. (1991). *National Survey of Fishing, Hunting, and Wildlife-Related Recreation.* U.S. Government Printing Office, Washington, D.C.

Vaughan, W., and C. Russell. (1982). Valuing a fishing day: An application of a systematic varying parameter model. *Land Econ.* 58, 450–463.

ECONOMIC VALUE OF BIODIVERSITY, OVERVIEW

Partha Dasgupta
University of Cambridge and Beijer International Institute of Ecological Economics

GLOSSARY

accounting price In this context, a measure of the economic value of a resource in social terms, described as the increase in social well-being that would be enjoyed if one unit more of the resource were made available without cost, and expressed as the difference between the market price of the resource and the tax or subsidy that ought to be imposed on it.

bifurcation point A transitional boundary in the state of a dynamic system (such as an ecosystem), at which the stability or basic nature of the system changes due to alterations in its fundamental characteristics.

intrinsic value A measurement or description of the values inherent in a resource by virtue of its existence (e.g., a living organism), as opposed to its value for present or future use.

option value A variation on the concept of USE VALUE (see below); a measurement or description of the potential value of a resource for possible future use.

net social benefit A measure of the social benefit of a project, obtained by identifying the projects positives (favorable outputs) and its negatives (inputs, unfavorable outputs), then multiplying these by the accounting price of each commondity involved. The net result of comparing these positive and negative values will indicate the social benefit of the project.

separatrix Locus of points on the state space of a dynamical system separating different basins of attraction.

shadow price Another term for ACCOUNTING PRICE.

substitution possibility The possibility that the depletion of a given resource can be compensated for by one or more of various forms of substitution; e.g., by substituting a more plentiful natural resource for a scarce one, or by substituting a manufactured product for a natural one.

ANY ASSESSMENT OF THE VALUE of biodiversity should begin with an account of why we need to value it and the reasons market values would not be expected to suffice for the purpose. The first three sections discuss these matters in the wider context of valuing natural resources (biodiversity is but a special case). A transitional section (Section IV) shows how values can be translated into prescriptions for economic policy. Sections V–VII discuss the special problems that arise in valuing biodiversity and the techniques that are available for coping with these problems.

I. THE SUBJECT

Resource economists view the natural environment through the lens of population ecology. Since the latter's focus is the dynamics of interacting populations, it has proven useful in resource economics to regard the functioning of the rest of the ecosystem as exogenously given. Well-known illustrations of this viewpoint include the use of the logistic function to chart the time path of the population of a single species of fish, the study of predator–prey interactions by means of variants of the Lotka–Volterra equations, and the estimation of growth in biomass of a species of trees at a given site. A prominent concern in resource economics has been to determine the rates at which a single resource would be harvested in different institutional settings. Thus, not only have socially optimum harvest rates been analyzed (Spence, 1974; Clark, 1976; Dasgupta and Heal, 1979; Hartwick, 1993) but also economists have determined harvest rates when harvesters have free access to the resource (Gordon, 1954; Dasgupta and Heal, 1979; Dasgupta, 1982). We now have an understanding of the effect on harvest rates of harvesting costs, the rates at which harvesters discount future costs and benefits, the productivity of the resource *in situ*, the "worth" of the harvest to harvesters, and the property rights regime in which the harvesting is done.

Environmental economists, in contrast, base their studies on ecosystem ecology. Here, the focus is on objects such as energy at different trophic levels and its rate of flow among them, the distribution and flows of biochemical substances in soils and bodies of water, and the flows of gases and particulates in the atmosphere. The motivation is to study the biotic and abiotic processes which underlie various ecosystem functions. Economic studies of global warming, eutrophication of lakes, the management of rangelands, purification of water in watersheds, and the pollution of estuaries are examples of such endeavor (Mäler, 1974; Costanza, 1991; Nordhaus, 1994; Perrings and Walker, 1995; Chichilnisky and Heal, 1998). Such studies have provided valuable insights into the effects on ecosystems of the character of economic activities, as driven by technology, costs and revenues, discount rates, and the property rights regime that governs the ecosystem.

Formally, it is useful to interpret differences between resource and environmental economics in terms of the state variables that are taken to characterize complex systems (state variables are frequently called stock variables). In resource economics, state variables are "quantities" (biomass units for forests, cow dung, and crop residue; herd size, expressed in numbers, for animal populations; acre-feet for aquifers; and so on). In environmental economics state variables are frequently "quality" indices, such as those for air, soil, or water (although quality, for example, salinity, is typically inferred from quantity indices), with the intention to draw attention to ecosystem services. State variables are summary statistics, reflecting different forms of aggregation. Therein lies their virtue: They enable the analyst to study complex systems by means of a few strategically chosen variables.[1]

The viewpoint just discussed—that of distinguishing resource and environmental economics in terms of the state variables that summarize complex systems—has enabled economists to integrate problems of resource management with those of environmental pollution (Dasgupta, 1982). Insights from one field of study have been used for gaining an understanding of the others. The viewpoint also reminds us that environmental and resource economics is the study of renewable natural resource systems, when subject to human predation. Since it would be convenient to refer to environmental and resource economics by an overarching name, I do so in this article with the term ecological economics.[2]

The economics of biodiversity, still in its infancy [Perrings *et al.* (1994, 1995) are pioneering collections of studies on the subject], is all of the above, but it includes an additional complication in that it recognizes that aggregate statistics can mislead—for example, to devise schemes to manage a resource without understanding its function within the ecosystem of which it is a part can be a road to disaster because the ecosystem's biodiversity can be a source of its productivity, including, for example, the system's stability (Tilman and Downing, 1994; Tilman, 1997). In this article, I provide an account of valuation problems in ecological economics; they are, naturally, valuation problems as well in the economics of biodiversity. I then discuss some of the special problems that arise in the latter and techniques available for resolving them.

II. MARKET FAILURE

If you were to browse among such leading Western journals in ecological economics as the *Journal of Envi-*

---

[1] This two-fold classification, based on the character of state variables, is implicit in contemporary ecology (Ehrlich *et al.*, 1977; Roughgarden *et al.*, 1989).

[2] I am able to usurp the term from the literature because it appears to have no fixed meaning: "Ecological economics" seems to mean different things to different people.

ronmental Economics and Management, you would discover that a central concern in the field has been to devise ways by which it would be possible to ascertain the "value" of natural resources and the services they provide. A question that would inevitably occur to you is, why? Why should there be a special need to determine the worth of such resources? Why not rely on their market prices? More generally, we may ask why it would not do to rely on markets to guide decisions bearing on the natural resource base, be they global or local, in the way we do for so many other goods and services. In other words, we may ask why markets are not an adequate set of institutions for protecting the "environment."

The answer is that for many natural resources, markets simply do not exist. In some cases they do not exist because the costs of negotiation and monitoring are too high. One class of examples is provided by economic activities which are affected by ecological interactions involving long geographical distances (e.g., the effects of uplands deforestation on downstream activities hundreds of miles away) and another by large temporal distances (e.g., the effect of carbon emission on climate in the distant future, in a world in which forward markets are nonexistent because future generations are not present today to negotiate with us). Then there are cases (e.g., the atmosphere, aquifers, and the open seas) in which the nature of the physical situation (viz., the migratory nature of the resource) makes private property rights impractical and therefore keeps markets from existing; in other cases, ill-specified or unprotected property rights prevent their existence or make markets function wrongly even when they do exist. In other words, environmental problems are often caused by market failure (but see Section III). Indeed, the phenomenon of externalities (i.e., exchanges among people which occur without their consent) looms large in what has traditionally been called environmental economics.[3]

Problems arising from an absence of forward markets for "transactions" between the present generation and those in the distant future are no doubt ameliorated by the fact that we care about our children's well-being and know that they, in turn, will care for theirs in an intergenerational sequence. This means, by recursion, that even if we do not care directly about the well-being of our distant descendants, we do care about them

indirectly. However, there is a distinct possibility that our implicit concern for the distant future via such recursion is inadequate due, for example, to institutional failure in other spheres of economic activity. This is why economists have argued that market rates of interest do not reflect socially desirable discount rates (Lind, 1982; Arrow *et al.*, 1996; Portney and Weyant, 1999). In other words, market failure involves not only misallocation of resources in the present but also misallocation across time.

In each of these cases, the market prices of goods and services fail to reflect their social worth; typically, they are less than their social worth. In economics, the social worth of goods and services is called accounting price (or shadow price). The accounting price of a resource is the increase in social well-being which would be enjoyed if a unit more of the resource were made available costlessly. Therefore, a resource's accounting price is the difference between its market price and the tax (or subsidy) that should be imposed on it. Needless to say, accounting prices reflect social objectives, ecological and technological constraints, and the extent to which resources are available.

It should be noted that externalities do not create market distortions; they are a form of market distortion. The presence of externalities leads to a wedge between market prices and accounting prices. Generally, laissez-faire economies are not much good at producing publicly observable signals of environmental scarcities. To illustrate, if there were free access to a resource base, the market price of the resource *in situ* would be zero. However, because it is in limited supply, its accounting price would be positive. Therefore, there is a directional bias in environmental externalities: Market failure typically results in an excessive use of the natural resource base, not an insufficient one.

One way to improve matters would be to impose regulations on resource users, for example, restrictions on effluent discharges, quotas on fish harvests, and bans on logging. Another would be to introduce a system of taxes, often called Pigovian taxes [in honor of Pigou (1920), who first discussed the difference between private and social costs in the context of environmental pollution]. Pollution charges, charges on the amount of fish harvested, and stumpage fees are examples. The idea underlying Pigouvian taxes is to bring market prices (inclusive of taxes) in line with accounting prices. Each of the two schemes, quotas and taxes, has its advantages and disadvantages; some of the differences between the two become salient once we recognize not only that ecological processes are stochastic but also that resource users and government agencies do not

---

[3] The early literature on ecological economics identified market failure as the underlying cause of environmental problems (Pigou, 1920; Lindahl, 1958; Arrow, 1971; Meade, 1973; Mäler, 1974; Baumol and Oates, 1975; Dasgupta and Heal, 1979).

have the same information about local ecology (e.g., the cost of waste disposal) (Meade, 1973; Weitzman, 1974; Baumol and Oates, 1975; Dasgupta, 1982). I will not discuss details here, but three points are worth noting. First, the two schemes are distributionally not equivalent: Under a quota, resource rents are captured by harvesters and polluters, whereas under a tax system they are collected by the tax authority. Second, the imposition of Pigouvian taxes provides greater incentives to resource users to explore resource-saving technological improvements. This is because if the users are taxed, they pay more for the resource than they would have if they had been issued quotas instead. Third, environmental taxes, when properly designed, remove market distortions. In addition, there is a presumption that tax revenues, thus collected, would enable the government to reduce distortionary taxes (e.g., taxes on earned income). There is thus a presumption that Pigovian taxes yield a "double dividend" (Bovenberg and van der Ploeg, 1994; Goulder, 1995; Bovenberg and Goulder, 1996; but see Bohm, 1996), a rhetorical phrase that has been much used in recent years to persuade governments to impose "green" taxes. Matters of public finance have been a recurrent theme in ecological economics (Baumol and Oates, 1975; Cropper and Oates, 1992; Carraro and Siniscalco, 1996).[4]

III. INSTITUTIONAL FAILURES AND POVERTY: GLOBAL VS LOCAL ENVIRONMENTAL PROBLEMS

Thus far, market failure. Recently, however, certain patterns of environmental deterioration have been traced to government failure. For example, Binswanger (1991) argued that, in Brazil, the exemption from taxation of virtually all agricultural income (allied to the fact that logging is regarded as proof of land occupancy) has provided strong incentives to the rich to acquire forest land and to then deforest it. He argued that the subsidy the government has thereby provided to the private sector has been so large that a reduction in deforestation (via a removal of subsidies) is in Brazil's

interests and not merely in the interest of the rest of the world. This has implications for international negotiations. The current consensus appears to be that, as a country, Brazil has much to lose from reducing the rate of deforestation in which it is engaged. If this were true, there would be a case for the rest of the world to subsidize Brazil, providing compensation for losses it would sustain if it were to restrain itself. However, as Binswanger's account suggests, it is not clear if the consensus is correct.[5] Elsewhere (and by extension), one would imagine that the massive agricultural subsidies in the European Union considerably influence agricultural practices in ways that inflict substantial damage on the environment.

However, it is important to note that the causes of environmental problems are not limited to market and government failure; problems also occur because such microinstitutions as the household can function badly. In poor communities, for example, men typically have the bulk of the political voice. We should then expect public investment in, for example, resource regeneration to be guided by male preferences, not female needs. On matters of afforestation in the drylands, for instance, we should expect women to favor planting for fuelwood and men for fruit trees because it is the women and children who collect fuelwood, whereas men control cash income (and fruit can be sold in the market). This explains why, even as the sources of fuelwood continue to recede, fruit trees are often planted (Dasgupta, 1993).

That political instability (at the extreme, civil war) is a direct cause of resource degradation is obvious. What is not obvious is that it is a hidden cause as well. Political instability creates uncertainty in property rights. In its presence, people are reluctant to make the investments that are necessary for environmental protection and improvement: The expected returns on such forms of investment are low. In a study comprising 120 countries, Deacon (1994) offered statistical evidence of a positive link between political instability and forest depletion.

Together, these examples reflect the environmental consequences of institutional failure. They have a wide reach, and in recent years they have often been discussed within the context of the thesis that environmental degradation, such as eroding soil, receding forests, and vanishing water supplies, is a cause of accentuated poverty among the rural poor in poor countries. There is truth in this. However, there is also accumulated

---

[4] A hybrid policy instrument, which involves the government issuing a fixed number of transferable licenses, combines some of the features of quotas and Pigouvian taxes. For example, the scheme resembles quotas in that resource rents are not captured by government, and it resembles Pigouvian taxes in that at the margin license holders pay the accounting price of the resource for its use. See Tietenberg (1980, 1990) for a discussion of transferable licenses, both in theory and in practice.

[5] In a broader discussion of the conversion of forests into ranches in the Amazon basin, Schneider (1995) demonstrated that the construction of roads through the forests has also been a potent force.

evidence that poverty can be a cause of environmental degradation (Dasgupta and Mäler, 1991, 1995; Dasgupta, 1993a, 1999; Ehrlich et al., 1995). This reverse causality occurs because some natural resources (e.g., ponds and rivers) are essential for survival in normal times, whereas others (e.g., forest products) are also a source of supplementary income in times of acute economic stress. In changing circumstances (e.g., economic development in urban centers), social norms which previously had maintained long-term economic relationships among members of a community tend to break down. Some (e.g., the able-bodied and mobile) gain, whereas others (e.g., women, the old, and the very young) lose and become poorer. In extreme cases, the breakdown of social norms also means that local resources which earlier were subject to communitarian regulations become "open access," with all the attended consequences.

These links between rural poverty and the state of the local natural resource base in poor countries offer a possible pathway along which poverty, resource degradation, and even high fertility feed on one another in a synergistic manner over time (Dasgupta, 1993a, 1995, 2000). Recent experiences in sub-Saharan Africa and Pakistan are not inconsistent with this (Cleaver and Schreiber, 1994; Filmer and Pritchett, 1996). Indeed, an erosion of the local natural resource base can make certain categories of people destitutes even while the economy's gross national product increases. The belief that entire populations can always be relied on to make the shift from resource-based, subsistence existence to a high-income, industrial one is belied both by evidence and by theory.

These two causes of resource degradation (i.e., institutional failure and poverty) pull in different directions and are together not unrelated to an intellectual tension between the concerns people share about global warming and acid rains, which sweep across regions, nations, and continents, and about those matters (e.g., the decline in firewood or water sources) which are specific to the needs and concerns of the poor in as small a group as a village community. Environmental problems present themselves differently to different people. In part, it is a reflection of the tension previously discussed and is a source of misunderstanding of people's attitudes. Some people, for example, identify environmental problems with poverty and unprecedented population growth in the South, whereas others identify them with wealth and unprecedented expenditure patterns in the North (I use the geographical terms in their current geopolitical sense). Even though debates between the two groups often become shrill, each vision

is in part correct. There is no single environmental problem, and therefore no single valuation problem; rather, there is a large collection of them (Dasgupta and Mäler, 1995; Reardon and Vosti, 1995; Vincent et al., 1997). Thus, growth in industrial wastes and resource use have been allied to increased economic activity; and in the former Socialist block neither preventive nor curative measures have kept pace with the production of waste. Moreover, the scale of the human enterprise, both by virtue of unprecedented increases in the size of the world's population and by the extent of economic activity, has so stretched the capabilities of ecosystems that humankind can today rightly be characterized as Earth's dominant species (Vitousek et al., 1997). These observations loom large not only in ecological economics but also in the more general writings of environmentalists and in the professional writings of ecologists in the West.

On the other hand, economic growth has brought with it improvements in the quality of many natural resources. The large-scale availability of potable water and the increased protection of human populations against both water- and air-borne diseases in industrial countries have in great measure occurred in the wake of growth in national income these countries have enjoyed during approximately the past 200 years. Moreover, the physical environment inside the home has improved beyond measure with economic growth. For example, cooking in Southeast Asia continues to be a central route to respiratory illnesses among women. Such positive links between economic growth and environmental quality often go unnoted by environmentalists in the North. I believe that this lacuna is another reflection of the fact that it is easy to overlook the enormous heterogeneity of Earth's natural resource base, ranging from the atmosphere, oceans, and landscapes to water holes, grazing fields, and sources of fuelwood. Both this heterogeneity and the diversity of the human condition throughout the world need constantly to be kept in mind in discussions of the value of biodiversity in different locations.

IV. VALUING RESOURCES AND EVALUATING PROJECTS

Since institutional failures abound in our dealings with Earth, the commercial profitability of economic activities, for example, of investment projects (projects for short), is frequently not an adequate measure of their social worth. Therefore, recourse should be taken to social cost–benefit analysis, the purpose of which is to estimate the impact of projects on human well-being, now and in the future. Note that, if undertaken, a project would be a perturbation to the economy. Therefore, a project consisting of the construction of a dam, for example, would be a perturbation to an economy without the dam. The economic forecast sans the project can be thought of as the status quo.

Analyzing the consequences of a project would involve estimating the need for labor, intermediate products, raw materials, and output as well as predicting the ecological effects of the project. These consequences need to be specified for each future period (see Section VII for a formalization). Since there is never sufficient knowledge to make precise estimates of the consequences, project evaluators should quantify estimates of the uncertainties, preferably in terms of probabilities. This means that, in general, project designers should model the integrated ecological and economic system. However, in practice this is infrequently done.[6]

In order to derive a good estimate of a project's social benefits and costs, one should in principle value every commodity involved in it. The procedure devised by economists is to select some readily measurable bundle of goods ordinarily consumed and define the value of any other commodity as the amount of the bundle society would be willing to give up for it. This is a workable way for estimating the commodity's accounting price. The net social benefit of a project in any given period of its life is obtained by multiplying the project's inputs and outputs in that period by their corresponding accounting prices and adding them (outputs of "goods" are taken to be positive, and output of "bads" and inputs are taken to be negative). Using a suitable discount rate (often called the social discount rate; Arrow et al., 1996; Portney and Weyant, 1999), the net social benefits yielded by a project in each period are added. Projects which yield a positive present discounted value of net social benefits are recommended for acceptance, whereas those yielding a negative present discounted value of net social benefits are rejected. The theory of social cost–benefit analysis has been developed by economists during the past 50 years and is now, to all intents and purposes, complete (Dasgupta et al., 1972; Little and Mirrlees, 1974; Dasgupta and Mäler, 2000).[7]

A prior exercise (i.e., prior to conducting social cost–benefit analysis) is to estimate accounting prices. A great deal of work in ecological economics has been directed at discovering methods for estimating accounting prices of natural resources.

It must be remembered that the kinds of resources I am considering here are sometimes of direct use in consumption (as with fisheries), sometimes of indirect use as inputs in production (as with plankton, which serves as food for fish), and sometimes both (as with drinking and irrigation water). The value may be utilitarian (e.g., as a source of food or as a keystone species), it may be aesthetic (e.g., a shoreline), or it may be intrinsic; indeed, it may be all these things.

Economists have devised various methods for estimating accounting prices. As would be expected, the prices of some natural resources are easier to estimate than those of others. There are standard techniques for determining the accounting prices of irrigation water, fisheries, timber, and agricultural soil (Anderson, 1987; Repetto et al., 1989; Solorzano et al., 1991; Vincent et al., 1997). They involve estimating the resource's use value. For example, the value of a piece of agricultural land, such as agricultural land, would be the present discounted value of the flow of net profits it is expected to generate from cultivation minus the environmental damage caused by the pesticides and herbicides to be used. Such an approach can also be used for estimating losses associated with water-logging and overgrazing. Reductions in air- or water-borne pollution can be valued in terms of improvements in health (e.g., reductions in the number of days people would be expected to be ill; World Bank, 1992). Other techniques have been devised for valuing "amenities" such as places of scenic beauty.[8]

Methods have also been devised for giving expression to the "precautionary principle," which assumes a particularly subtle form when applied to resources such as genetic material in tropical forests. The subtlety arises from a combination of uncertainty in the future use values of these resources and irreversibility when they are lost. The twin presence of uncertainty and irrevers-

---

[6] However, there are signs of change. See, for example, Perrings et al. (1994, 1995), Vincent et al. (1997), Chopra and Kadekodi (1999), and various issues of the journal Environment and Development Economics.

[7] Daily et al. (1999) provide a nontechnical account of the role of social cost–benefit analysis in environmental management.

[8] One popular method involves asking people hypothetical questions concerning their willingness to pay for preserving the amenity (this is called the "contingent valuation method"); another involves estimating from sample surveys the distribution of costs that visitors from different locations have incurred to view the site (this is called the "travel cost method"). A third involves inferring how much people are willing to pay for enjoying the amenity (e.g., clean air) from the commercial value of land at sites which offer the amenity (this is called the "hedonic price" of land). See Mäler and Wyzga (1976), Mitchell and Carson (1989), and Freeman (1992).

ibility implies that preservation of the stock has a value in addition to its current use value, namely, the value of extending society's set of future options. Future options have an additional worth because, with the passage of time, more information is expected to be forthcoming about the resource's use value. This additional worth is often called an option value (Arrow and Fisher, 1974; Henry, 1974; Fisher and Hanemann, 1986). The accounting price of a resource, at the very least, is the sum of its direct use value and its option value.

These techniques enable us to estimate the use value of a given resource. The resource's accounting price may well exceed this. Why? Because there may be additional values "embodied" in a resource. An additional consideration is applicable to living resources: their intrinsic worth as living resources. It would be absurd to suppose that the value of a blue whale is embodied entirely in its flesh and oil, or that the value of game in Kenyan safari parks is simply the present discounted value of the flow of tourists' willingness to pay to view them. The idea of intrinsic worth of living things is inherent not only in traditional religious systems of ethics but also in modern ethical theories. Therefore, the question is not so much whether living creatures have intrinsic worth but rather regards the ways of assessing this worth. Because it is almost impossible to get a quantitative handle on intrinsic worth, the correct thing to do is to take note of it, keep an eye on it, and call attention to it in public debate if the stock is threatened with destruction.

We may conclude that the social worth of natural resources can be divided into three parts: their use value, their option value, and their intrinsic value. The components appear in different proportions, depending on the resource. For example, oil and natural gas would not be thought to have intrinsic value, nor perhaps an option value, but they have use value. On the other hand, primates would be thought to be intrinsically valuable.

It must also be emphasized that the purpose of estimating environmental accounting prices is not to value the entire environment; rather, it is to evaluate the benefits and costs associated with changes made to the environment due to human activities. Prices, whether actual or accounting, have significance only when there are potential exchanges from which choices have to be made (e.g., when one has to choose among alternative investment projects). Thus, the statement that a particular act of investment can be expected to degrade the environment by, for example, $1 million annually has meaning because it says, among other things, that if the investment were not to be undertaken, humanity

would enjoy an additional $1 million of annual benefits in the form of environmental services. The statement also has operational significance: The estimate could (and should) be used for calculating the rate of return attributable to the investment in question.

Contrast such an estimate of the value of an incremental change in the natural resource base with the one which says that, worldwide, the flow of environmental services is currently worth, in total, $33 trillion annually (Costanza *et al.*, 1997). The former is meaningful because it presumes that humanity will survive the incremental change and be there to experience and assess the change. The reason the latter should cause us to balk is that if environmental services were to cease, life would not exist. However, who would be there to receive $33 trillion of annual benefits if humanity wished to exchange its very existence for them? This is a case in which the value of an entire something has no meaning and therefore is of no use, even though the value of incremental changes to that same something not only has meaning but also has use.

V. BIODIVERSITY: NECESSITY OR LUXURY?

Biodiversity has been neglected in economics. Scratch an economist, and you are likely to find someone who regards the natural resource base as an "amenity." Thus, it is commonplace that, to quote a recent editorial in London's *Independent* (December 4, 1999), "[economic] growth is good for the environment because countries need to put poverty behind them in order to care" or that, to quote the *Economist* (December 4, 1999, p. 17), "trade improves the environment, because it raises incomes, and the richer people are, the more willing they are to devote resources to cleaning up their living space."

I quote these views not to question that poverty should be a phenomenon of the past, nor to suggest that arbitrary restrictions on trade do not cause much harm, but rather to show that natural resources are widely viewed as luxuries. This is, of course, an incorrect view: The natural resource base is not a luxury. Producing, as it does a multitude of ecosystem services, the natural resource base is a necessity.[9] Indeed, the

---

[9] As stressed elsewhere in the encyclopedia, these services include maintaining a genetic library, preserving and regenerating soil, recycling nutrients, controlling floods, filtering pollutants, assimilating waste, pollinating crops, operating the hydrological cycle, and maintaining the gaseous composition of the atmosphere.

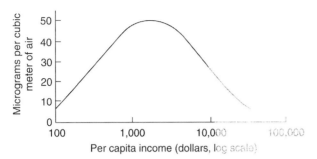

FIGURE 1 Urban concentrations of sulfur dioxide (reproduced with permission from World Bank, 1992, Fig. 4).

ecological economics to be found in the North does not exactly resonate among those who worry about degradation of the natural resource base in the South and the additional hardship this brings to the many among the poorest of people whose lives depend directly on that base.[10] Needless to say, it does not resonate among ecologists, qua ecologists, either.

This perspective of viewing natural resources as luxuries is expressed in the World Bank (1992) report, in which it was suggested that there is an empirical relationship between gross domestic product (GDP) per head and concentrations of industrial pollutants. Based on the historical experience of the Organization for Economic Cooperation and Development countries, it was argued in the document that, when GDP per head is low concentrations of atmospheric pollutants [e.g., sulfur dioxide (SO_2)] increase as GDP per head increases, but when GDP per head is high concentrations decrease as GDP per head increases further. In other words, it was found that the functional relationship between GDP per head and concentrations of industrial pollutants has an inverted U shape (Fig. 1). Among economists, this relationship has been christened the "environmental Kuznets curve".[11]

Panayotou (1992) reported the inverted U shape in cross-country data on GDP per head and deforestation and emissions of SO_2, nitrogen oxides (NO_x), and particulate matters. Sweden, for example, was found to lie on the downward part of the curve. Indeed, time series on timber stocks and sulfur and nitrogen emissions in Sweden, during the 1980s, are consistent with this: Timber stocks have increased, and the emission rates of sulfur and nitrogen oxides have declined.

Note that, excluding forests (but see Section VI), the previously mentioned findings concern mobile pollutants. Their mobility hides the fact that earlier emissions have had to find somewhere to lodge: Since matter is conserved, inputs and outputs of material must balance (Ayres and Kneese, 1969; d'Arge et al., 1970; Mäler, 1974). The point is that my emissions would affect not only me but also you if you lived downwind or downstream. However, even if we did not consider this point, the logic underlying the environmental Kuznets curve is that resource degradation is reversible: Degrade all you want now, you can always recover the stock later because Earth can be relied on to rejuvenate it. The science of biodiversity has shown this presumption to be false. The presence of ecological thresholds implies that damage to ecosystems can be irreversible. As an overarching metaphor for "trade-offs" between manufactured wealth and resource degradation, the environmental Kuznets curve has to be rejected.[12]

VI. SUBSTITUTION POSSIBILITIES[13]

The belief that constraints arising from resource depletion can be overcome as countries become wealthier in terms of their manufactured and human capital assets is frequently based on a subtler idea than the one that underlies the environmental Kuznets curve. The belief is based on possibilities of substitution.

Resource constraints facing an economy can be eased by four types of substitution. First, there can be substitution of one thing for another in consumption (nylon and rayon cloth substituting for cotton and wool, pulses substituting for meat, etc.). Second, manufactured capital can substitute for labor and natural resources in production (the wheel and double-glazing are two extreme examples). Third, novel production techniques can substitute for old ones. For example, the replacement of the piston with the steam turbine (i.e., converting from reciprocating to rotary motion) was introduced into power plants and ships more than 100 years ago. The innovation was an enormous energy saver in engines. Fourth, and most important, natural resources

---

[10] I have discussed these concerns in greater detail elsewhere (Dasgupta, 1982, 1993a, 1996, 1997).

[11] This is a misnomer. The original Kuznets curve, which was an inverted U, related income inequality to real national income per head on the basis of historical cross-country evidence.

[12] For more extensive discussions of the environmental Kuznets curve, see Arrow et al. (1995) and the responses their article elicited in symposia built round the article in Ecological Economics 15 (1995), Ecological Applications 6 (1996), and Environment and Development Economics 1 (1996); also see the special issue of Environment and Development 2(4) (1997).

[13] This and the following section are based on Dasgupta and Mäler (2000) and Dasgupta et al. (2000a,b).

can substitute for one another. This involves the belief that, as each resource (e.g., each species) is depleted, there are similar substitutes lying in wait, either at the same site or elsewhere. If this were true, then even as constraints increasingly tighten on any one resource base, humanity would be able move to other resource bases, either at the same site or elsewhere. The enormous additions to the sources of industrial energy that have been realized (successively, human and animal power, wind, timber, coal, oil, and natural gas and, recently, nuclear) are a prime historical illustration of this possibility.[14]

Humans have been "substituting" one thing for another since time immemorial. Even the conversion of forests into agricultural land in England in the Middle Ages was a form of substitution: Large ecosystems were transformed to produce more food. However, the pace and scale of substitution in recent centuries have been unprecedented. Landes (1998) argued that substitution created the industrial revolution in England in the eighteenth century. The extraordinary economic progress experienced in Western Europe and North America since then (during the past two centuries GDP per head in Western Europe has increased approximately 20-fold), and recently in east Asia, has also been a consequence of substitution. Spatial dispersion of ecosystems has enabled this to happen. The ecological transformation of rural England in the Middle Ages presumably reduced the nation's biodiversity, but it increased income without any direct effect on global productivity.

However, that was in the past, and we must deal with the present. A question currently much debated is whether it is possible for the scale of human activity to be increased substantially beyond what it is today without placing undue stress on the major ecosystems that remain. In any event, the cost of substituting manufactured capital for natural resources can be high. Low-cost substitutes could turn out to be not so low-cost if accounting prices were used in the costing, not market prices. Even when accounting prices are not used, degrading natural capital and substituting it with manufactured capital can be uneconomic. Chichilnisky and Heal (1998) compared the costs of restoring the ecological functioning of the Catskill watershed ecosystem in New York State to the costs of replacing the natural water purification and filtration services the ecosystem has provided in the past by building a water-purification plant costing $8 billion. They showed the overwhelming economic advantages of preservation compared to construction: Independent of the other services the Catskill watershed provides and ignoring the annual running costs of $300 million for a filtration plant, the capital costs alone showed a more than sixfold advantage for investing in the natural capital base.

Degradation of a natural resource base (e.g., destruction of native populations of flora and fauna) not only affects the volume and quality of ecosystem services the base provides but also challenges the system's resilience, which is its capacity to absorb disturbances, or perturbations, without undergoing fundamental changes in its functional characteristics. The way to interpret an ecosystem's loss of resilience is to view it as having moved to a new stability domain, which is another way of saying that the system, having crossed a threshold, has been captured by a different attractor (Levin et al., 1998; Levin, 1999; Brock et al., 1999). Sudden changes in the character of shallow lakes (e.g., from clear to eutrophied water), owing to increases in the input of nutrients (Scheffer, 1997; Carpenter et al., 1998), and the transformation of grasslands into shrublands, consequent upon nonadaptive cattle management practices (Perrings and Walker, 1995), are examples. Human societies have on occasion been unable to avoid suffering from unexpected changes in their local ecosystems. Fishermen on Lake Victoria and the nomads in the now shrublands of southern Africa are examples from recent years.

Biodiversity appears to be a key to ecosystem resilience. However, even today it is a popular belief that the utilitarian value of biodiversity is based mainly in the potential uses of genetic material (e.g., for pharmaceutical purposes) or, in other words, that its social worth is almost wholly an option value. Preservation of biodiversity is seen as a way to hold a diverse portfolio of assets with uncertain payoff. However, as other contributions to this encyclopedia make clear, biodiversity, appropriately conceived, is essential for the maintenance of a wide variety of services on which humans depend for survival. This has the important corollary that to invoke the idea of substitutability among natural resources in order to play down the use value of biodiversity, as people frequently do (Simon, 1981, 1994), is a wrong intellectual move. The point is that if an ecosystem's biodiversity is necessary for it to be able to continue providing us with its services, the importance of that same biodiversity cannot be downplayed by the mere hope that for every species there are substitute

---

[14] However, these shifts have not been without unanticipated collective costs. Global warming, associated with the burning of fossil fuels (an "externality"), did not feature in economic computations in earlier decades. See Dasgupta (1993b) for a less coarse partition of substitution possibilities than that provided by the four-way classification.

species lying in wait within that same ecosystem. In other words, there is an inconsistency in this line of reasoning. Recall the famous analogy in Ehrlich and Ehrlich (1981) relating species in an ecosystem to rivets in an airplane. One by one, perhaps, species may disappear and not be missed. Eventually, however, the cumulative effect of loss of biodiversity will lead to the crash of ecosystem functioning, just as the cumulative loss of redundant rivets will lead to the crash of an airplane.

VII. ECOSYSTEM BIFURCATIONS AND DISCONTINUOUS VALUE FUNCTIONS

How do discontinuities in the social worth of ecosystems affect valuation exercises and social cost–benefit analysis? To answer this, it helps to formalize.

Consider an ecosystem describable by N state variables, indexed by i and j ($i, j = 1, 2, \ldots, N$). For concreteness, we may think of each state variable as reflecting the population size of a particular species. (As noted in Section I, problems of environmental pollution can be formulated in a similar manner.) Denote time by t (≥ 0) and let S_{it} be the population size of i at t. Time is taken to be a continuous variable. We imagine, therefore, that the dynamics of the ecosystem can be described by a system of (nonlinear) differential equations. For expositional ease, we assume that the system is deterministic.

Let the net reproduction rate of i at t be F_{it}. Since the ecosystem is coupled, F_{it} is a function of the stocks at t. This I write as $F_{it}(S_{1t}, S_{2t}, \ldots, S_{Nt})$, for $i = 1, 2, \ldots, N$. I assume that ecologists have estimated these functions. Next, assume that the ecosystem dynamics are autonomous. This means that F_{it} is not an explicit function of t. Therefore, I drop the subscript t from F_{it} and write the function as $F_i(S_{1t}, S_{2t}, \ldots, S_{Nt})$. In all the applications of this framework with which I am familiar, F_i is taken to be a differentiable function. Let us assume this is true.

The analysis begins at $t = 0$ (the "present"). Denote by X_{it}, the rate at which species i is harvested at time t. We now imagine that economists have studied the human–ecosystem interactions in question. They have inquired into the structure of property rights, demand conditions, government policies, and so forth. On the basis of this they have concluded that harvests are based on an implicit policy, in that they are time autonomous and are functions solely of stocks. Therefore, we may write $X_{it} = X_i(S_{1t}, S_{2t}, \ldots, S_{Nt})$. Assume that X_i is piecewise continuous and possesses right and left par-

tial derivatives everywhere. This is a technical assumption and a good one. For example, optimal policy functions for those ecosystem management problems that have been studied have been found to possess this property (Skiba, 1978; Brock et al., 1999). Moreover, actual harvest rates have frequently been known to be approximately constant over time. Both sets of examples satisfy the assumption.[15] No doubt some of the X_{it}'s would be zero. For example, it could be that only one species in the ecosystem is ever harvested (e.g., because it is the only one that has economic worth). We should think of X_{it} as a forecast. It should be stressed that $X_i(S_{1t}, S_{2t}, \ldots, S_{Nt})$ is not necessarily a socially optimal harvest policy function. It can be an actual policy function within an imperfect institution (e.g., the ecosystem could be one to which there is free access).

The rate of increase of S_{it} is the difference between F_i and X_{it}. Therefore, given the economists' forecast for X_{it}, mathematicians would be able to forecast S_{it} by solving the "coupled" system of differential equations:

$$dS_{it}/dt = F_i(S_{1t}, S_{2t}, \ldots, S_{Nt}) - X_i(S_{1t}, S_{2t}, \ldots, S_{Nt}), \text{ for all } i \tag{1}$$

For simplicity of exposition, assume that the social worth of the ecosystem is autonomous in time. We may then express that worth by a scalar V. Since V would be a function of the stocks, we may write it as $V(S_{1t}, S_{2t}, \ldots, S_{Nt})$. V is the value at t of the entire ecosystem. It is the maximum amount society should be "willing to pay" at t for the ecosystem's survival if the stocks of the N resources were $S_{1t}, S_{2t}, \ldots,$ and S_{Nt}, respectively. Any alternative use of the site (e.g., conversion into an urban center) would have to be worth at least V if the alternative were to be acceptable. The form of V would depend on the availability of substitutes for those species that are harvested. Again, to keep the mathematical notation from becoming too complex, I assume that there are no substitutes available at low cost from outside the ecosystem (e.g., because the community doing the harvesting is not near other sources of livelihood). Using Eq. (1), it is possible to use the forecast on harvest rates to determine forecasts on stocks. This in turn makes it possible to forecast the time path of V.

---

[15] Actual harvest rates frequently display time trends, for example, because population and income grow. Time trends in X_{it} would render the system of Eq. (1) below nonautonomous. In the text, I restrict the discussion to autonomous systems because I understand the mathematics of autonomous better than that of nonautonomus ones. However, experience with simple nonautonomous systems suggests that the arguments I offer later in the text apply to them as well.

At one level the valuation problem is now "solved": $V(S_{1t}, S_{2t}, \ldots, S_{Nt})$ would be the value of the ecosystem at t. It would be the social worth of the ecosystem at t. The problem is that V is typically a nonlinear function, which means that it is very difficult to estimate. The task of valuing ecosystems would be made much easier if recourse were taken to estimating accounting prices. The advantage would be that since accounting prices reflect the social worth of marginal units of the various populations, we could use such prices to construct a linear index of the ecosystem's value.

Assume that V is differentiable everywhere. Let P_{it} be the accounting price of i at t. From the discussion in Section IV and from our assumption that no substitute resources are near the human community in question, we know that

$$P_{it} \equiv \partial V / \partial S_{it}, \text{ for all } i \text{ and all } t \text{ } [16] \qquad (2)$$

At time t the value of species i would be $P_{it}S_{it}$. It follows that the value of the ecosystem would be $\Sigma_i P_{it}S_{it}$. Note that this is a linear function of stocks, with the weights being accounting prices.

In Section IV it was noted that a "project" can be thought of as a perturbation of the forecast X_{it}. Therefore, a project can be denoted as $(\Delta X_{1t}, \Delta X_{2t}, \ldots, \Delta X_{Nt})$, for $t \geqq 0$. (Δ denotes an operator signifying "small difference.")[17] Some of the ΔX_{it}'s would be zero. Nevertheless, the project would be expected to perturb future stocks of all the resources since this is what a strongly coupled ecosystem would be expected to display.

Let r be the social rate of discount and let C_{it} be the unit cost of harvesting i at t.[18] It follows that the present discounted value of the flow of net social benefits from the project is

$$\int_0^\infty e^{-rt} [\Sigma_i (P_{it} - C_{it}) \Delta X_{it}] \, dt \qquad (3)$$

If Eq. (3) is positive, the project should be accepted; if it is negative, the project should be rejected.

It can be argued that projects, as I have defined them here, are merely "small" perturbations, whereas redirecting economic activity so as to avoid damaging

an ecosystem irrevocably could involve drastic change. However, it should be noted that one way to conceptualize a "large" perturbation is to regard it as the sum of many small perturbations. A large perturbation (i.e., a large project) could then be evaluated by repeated use of Eq. (3).

However, if this route is not adopted, social cost–benefit analysis of large projects requires the project evaluator to estimate the large changes in V consequent on the adoption of large changes in economic policy. Accounting prices, reflecting the social worth of marginal units of the various resources (Eq. 2), would then not suffice: The evaluator would need to integrate over the marginal units so as to estimate "consumer surpluses," to use a term familiar in economics.

So far so good. However, there is a problem with the account: It is unreasonable to assume that V is differentiable, even continuous, everywhere. Ecosystems are nonlinear. So, even if it were reasonable to suppose that V is differentiable everywhere else, it would be wrong to suppose that it is even continuous at separatrixes and bifurcation points.[19]

However, if the X_i's are not optimal, V can be discontinuous at points on separatrixes and bifurcation points. This causes problems because accounting prices cannot be defined at such points. Now, we will study the implications of this for biodiversity valuation and social cost–benefit analysis.

Experience with nonlinear models of ecosystems tells us that, under the assumptions we have made, there could be at most a countable number of bifurcation points and separatrixes. This is fortunate because it means that points on the stock space that are "troublesome" are nongeneric. Therefore, let us assume this is so. In Fig. 2, the matter is illustrated in the context of an ecosystem comprising a single species. Figure 2 depicts the case in which the separatrix is a single point, S^*, reflecting a threshold. For example, it could be that, under the harvesting policy $X(S)$, the species would become extinct if its population were below S^* but would be harvested in a sustainable manner if the population were in excess of S^*. Therefore, stocks to the

---

[16] If substitutes were available, P_{it} would be the minimum of $\partial V/\partial S_{it}$ and the accounting price of the substitute. I want to avoid such complications here.

[17] Note that for all i and all t, $\Delta X_{it} = \Sigma_j [\partial X_i/\partial S_{it}][dS_{jt}/dt]\Delta t$.

[18] C_{it} could depend on stock sizes at t. For example, the unit cost of fishing depends not only on the technology available for fishing and the price of fishing equipment but also on the stock in the fishery: The larger the stock, the smaller the unit cost.

[19] The location of bifurcation points depends on, among other things, the X_i's. Therefore, they are endogenous to the economic–ecological system. In an important early contribution, Skiba (1978) showed via an example that if harvest functions are optimal, V is continuous even at points where harvests are discontinuous. This means that accounting prices are not uniquely specified at such points. However, V can be shown to possess right- and left-partial derivatives there. Therefore, accounting prices can be used for evaluation purposes, even though they are not uniquely given at every point on the space of resource stocks.

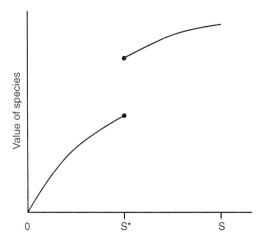

FIGURE 2 Nonlinear model depicting value of species in relation to function of the stock, where the ecosystem comprises a single species.

right and left of S^* represent different basins of attraction. Figure 2 reasonably depicts a case in which the value of the species, $V(S)$, is an increasing function of the stock. It is assumed to be continuous (indeed, differentiable) everywhere except at S^*, where it jumps (an irrevocably dying population being a lot less valuable than a sustainable one). Of course, the location of S^* depends on $X(S)$: Change the harvesting policy slightly, and S^* will shift slightly. The influence of $X(S)$ on S^* has to be estimated if ecologists and economists are to offer policy advice.

Now, excepting by fluke, the stock at $t = 0$ would be different from S^*. Therefore, let us assume it is different. If the project is sufficiently small, the account of social cost–benefit analysis given previously remains valid: The system would not cross into a different basin of attraction. However, a good theory should be extendable to fluke cases. Moreover, actual projects are frequently not "small" so that acceptance of a project or its rejection could mean that the ecosystem is eventually in one basin of attraction rather than in another. How do we extend the theory to handle the possibility that the ecosystem crosses into a different basin of attraction? In particular, is the repeated use of Eq. (3) a feasible means of evaluating projects?

It is also clear where the problem lies if we try using Eq. (3). The problem lies in that an accounting price cannot be defined at S^*. This means that a project which involves the stock passing through S^* cannot be evaluated by means of a linear index of social profitability. The height of the jump would have to be estimated and put to use in social cost–benefit analysis. Ecologists and economists would have to combine their expertise

to locate S^* and identify the functional form of $V(S)$, both on the right and on the left of S^*. Estimating the height of the jump involves measuring consumer surpluses, a point noted earlier. In short, at least one small project in the series of small projects which add up to the large project in question would not be assessable by means of Eq. (3). This causes difficulties.

However, there is a way to avoid the problem. We have been studying deterministic systems. Introducing uncertainty about the location of S^* can help matters by smoothing the value function. To see how, imagine that $V(S)$ represents the expected value of the resource's social worth at S. If the location of S^* were a smooth probability distribution, $V(S)$ would be a continuous, even a differentiable, function. In this case, an accounting price of the resource would be definable at all S (with S^* being a smooth random variable). A linear index of the social profitability of projects could then be constructed. The methods of social cost–benefit analysis outlined earlier would remain valid.

It is not often that introducing realism simplifies analysis. Valuing biodiversity would seem to be an exception.

See Also the Following Articles

BIODIVERSITY AS A COMMODITY • ECONOMIC GROWTH AND THE ENVIRONMENT • ECONOMIC VALUE OF BIODIVERSITY, MEASUREMENT OF • HUMAN EFFECTS ON ECOSYSTEMS, OVERVIEW • MARKET ECONOMY AND BIODIVERSITY

Bibliography

Anderson, D. (1987). *The Economics of Afforestation*. Johns Hopkins Univ. Press, Baltimore.

d'Arge, R., Ayres, R., and Kneese, A. V. (1970). *Economics and the Environment: A Materials Balance Approach*. Johns Hopkins Univ. Press, Baltimore.

Arrow, K. J. (1971). Political and economic evaluation of social effects of externalities. In *Frontiers of Quantitative Economics* (M. Intriligator, Ed.). North-Holland, Amsterdam.

Arrow, K. J., and Fisher, A. (1974). Preservation, uncertainty and irreversibility. *Q. J. Econ.* 88, 312.

Arrow, K., Bolin, B., Costanza, R., Dasgupta, P., Folke, C., Holling, C. H., Jansson, B.-O., Levin, S. A., Mäler, K.-G., Perrings, C., and Pimentel, D. (1995). Economic growth, carrying capacity, and the environment. *Science* 268, 520. (Reprinted in *Environ. Dev. Econ.* 1, 104, 1996).

Arrow, K. J., Cline, W. R., Mäler, K.-G., Munasinghe, M., Squitieri, R., and Stiglitz, J. E. (1996). Intertemporal equity, discounting, and economic efficiency. In *Climate Change 1995: Economic and Social Dimensions of Climate Change, Contribution of Working Group III to the Second Assessment Report of IPCC, the Intergovernmental Panel on Climate Change* (J. P. Bruce, H. Lee, and E. F. Haites, Eds.). Cambridge Univ. Press, Cambridge, UK.

Ayres, R., and Kneese, A. V. (1969). Production, consumption and externalities. *Am. Econ. Rev.* **59**, 282.

Baumol, W. M., and Oates, W. (1975). *The Theory of Environmental Policy: Externalities, Public Outlays and the Quality of Life.* Prentice Hall, Englewood Cliffs, NJ.

Binswanger, H. (1991). Brazilian policies that encourage deforestation in the Amazon. *World Dev.* **19**, 821.

Bohm, P. (1996). Environmental taxation and the double-dividend: Fact or fallacy?, GEC Working Paper Series No. 96-01. University College, London.

Bovenberg, A. L., and Goulder, L. (1996). Optimal environmental taxation in the presence of other taxes: General equilibrium analyses. *Am. Econ. Rev.* **86**, 766.

Bovenberg, A. L., and van der Ploeg, F. (1994). Environmental policy, public finance, and the labour market in a second-best world. *J. Public Econ.* **55**, 349.

Brock, W. A., Mäler, K.-G., and Perrings, C. (1999). The economic analysis of nonlinear dynamic systems. Discussion paper, Beijer International Institute of Ecological Economics, Stockholm.

Carpenter, S. R., Ludwig, D,. and Brock, W. A. (1998). Management of eutrophication for lakes subject to potentially irreversible change. Discussion paper, Beijer International Institute of Ecological Economics, Stockholm.

Carraro, C., and Siniscalco, D. (Eds.) (1996). *Environmental Fiscal Reform and Unemployment.* Kluwer, Dordrecht.

Chichilnisky, G., and Heal, G. M. (1998). Economic returns from the biosphere. *Nature* **391**, 629.

Chopra, K., and Kadekodi, G. (1999). *Operationalising Sustainable Development: Economic–Ecological Modelling for Developing Countries.* Sage, New Delhi.

Clark, C. W. (1976). *Mathematical Bioeconomics: The Optimal Management of Renewable Resources.* Wiley, New York.

Cleaver, K. M., and Schreiber, G. A. (1994). Reversing the Spiral: The Population, Agriculture, and Environment Nexus in sub-Saharan Africa. World Bank, Washington, D.C.

Costanza, R. R. (Ed.) (1991). *Ecological Economics: The Science and Management of Sustainability.* Columbia Univ. Press, New York.

Costanza, R. R., d'Arge, R., de Groot, R., Farber, S., Grasso, M., Hannon, B., Limburg, K., Naeem, S., O'Neill, R. V., Paruelo, J., Raskin, R. G., Sutton, P., and van den Belt, M. (1997). The value of the world's ecosystem services and natural capital. *Nature* **387**, 253.

Cropper, M., and Oates, W. (1992). Environmental economics: A survey. *J. Econ. Literature* **30**, 675.

Daily, G., Söderqvist, T., Aniyar, S., Arrow, K. J., Dasgupta, P., Ehrlich, P., Folke, C., Jansson, A, Jansson, B.-O., Kautsky, N., Levin, S., Lubchenco, J., Mäler, K.-G., Simpson, D., Starrett, D., Tilman, D., and Walker, B. (1999). Valuing resources. Discussion paper, Beijer International Institute of Ecological Economics, Stockholm.

Dasgupta, P. (1982). *The Control of Resources.* Harvard Univ. Press, Cambridge, MA.

Dasgupta, P. (1993a). *An Inquiry into Well-Being and Destitution.* Clarendon, Oxford.

Dasgupta, P. (1993b). Natural resources in the age of substitutability. In *Handbook of Natural Resource and Energy Economics* (A. V. Kneese and J. L. Sweeney, Eds.). North-Holland, Amsterdam.

Dasgupta, P. (1995). Population, poverty, and the local environment. *Sci. Am.* **272**(2), 40.

Dasgupta, P. (1999). Economic progress and the idea of social capital. In *Social Capital: A Multifaceted Perspective* (P. Dasgupta and I. Serageldin, Eds.). World Bank, Washington, D.C.

Dasgupta, P. (2000). Reproductive externalities and fertility behaviour. *Eur. Econ. Rev.* (*Papers Proc.*), in press.

Dasgupta, P., and Heal, G. M. (1979). *Economic Theory and Exhaustible Resources.* Cambridge Univ. Press, Cambridge, UK.

Dasgupta, P., and Mäler, K.-G. (1991). The environment and emerging development issues. Proceedings of the Annual World Bank Conference on Development Economics, 1990 (Supplement to the World Bank Economic Review and the World Bank Research Observer), p. 101. World Bank, Washington, D.C.

Dasgupta, P., and Mäler, K.-G. (1995). Poverty, institutions, and the environmental resource-base. In *Handbook of Development Economics* (J. Behrman and T. N. Srinivasan, Eds.), Vol. IIIA. North-Holland, Amsterdam.

Dasgupta, P., and Mäler, K.-G. (2000). Net national product, wealth and social well-being. *Environ. Dev. Econ.* **5**, 69.

Dasgupta, P., Marglin, S., and Sen, A. (1972). *Guidelines for Project Evaluation.* United Nations, New York.

Dasgupta, P., Levin, S., and Lubchenco, J. (2000a). Economic pathways to ecological sustainability: Challenges for the new millennium. *Bioscience* **50**, 337.

Dasgupta, P., Levin, S., Lubchenco, J., and Mäler, K.-G. (2000b). Ecosystem services and economic substitutability. Discussion paper, Beijer International Institute of Ecological Economics, Stockholm.

Deacon, R. T. (1994). Deforestation and the rule of law in a cross section of countries. *Land Econ.* **70**, 414.

Ehrlich, P. R., and Ehrlich, A. H. (1981). *Extinction: The Causes and Consequences of the Disappearance of Species.* Random House, New York.

Ehrlich, P. R., Ehrlich, A. H., and Holdren, J. P. (1977). *EcoScience: Population, Resources, Environment.* Freeman, San Francisco.

Ehrlich, P. R., Ehrlich, A. H., and Daily, G. (1995). *The Stork and the Plow: The Equity Answer to the Human Dilemma.* Putnam's Sons, New York.

Filmer, D., and Pritchett, L. (1996). Environmental degradation and the demand for children. Research Project on Social and Environmental Consequences of Growth-Oriented Policies, Working Paper No. 2. World Bank, Washington, D.C.

Fisher, A., and Hanemann, M. (1986). Option value and the extinction of species. *Adv. Appl. Microecon.* **4**, 169.

Freeman, A. M., III (1992). *The Measurement of Environmental and Resource Values: Theory and Methods.* Resources for the Future, Washington, D.C.

Gordon, H. S. (1954). The economic theory of common-property resources. *J. Political Econ.* **62**, 124.

Goulder, L. (1995). Environmental taxation and the double dividend: A reader's guide. *Int. Tax Public Finance* **2**, 157.

Hartwick, J. (1993). Forestry economics, deforestation, and national accounting. In *Toward Improved Accounting for the Environment* (E. Lutz, Ed.). World Bank, Washington, D.C.

Henry, C. (1974). Investment decisions under uncertainty: The irreversibility effect. *Am. Econ. Rev.* **64**, 1006.

Landes, D. (1998). *The Wealth and Poverty of Nations.* Norton, New York.

Levin, S. (1999). *Fragile Dominion: Complexity and the Commons.* Perseus, Reading, MA.

Levin, S., Barrett, S., Aniyar, S., Baumol, W., Bliss, C., Bolin, B., Dasgupta, P., Ehrlich, P., Folke, C., Gren, I.-M., Holling, C. S., Jansson, A., Jansson, B.-O., Mäler, K.-G., Martin, D., Perrings, C., and Shenshinski, E. (1998). Resilience in natural and socioeconomic systems. *Environ. Dev. Econ.* **3**, 222.

Lind, R. C. (Ed.) (1982). *Discounting for Time and Risk in Energy Planning*. Johns Hopkins Univ. Press, Baltimore.

Lindahl, E. R. (1958). Some controversial questions in the theory of taxation. In *Classics in the Theory of Public Finance* (R. A. Musgrave and A. T. Peacock, Eds.). Macmillan, London. (Original work published in Swedish in 1928).

Little, I. M. D., and Mirrlees, J. A. (1974). *Project Appraisal and Planning for Developing Countries*. Heinemann, London.

Mäler, K.-G. (1974). *Environmental Economics: A Theoretical Enquiry*. Johns Hopkins Univ. Press, Baltimore.

Mäler, K.-G., and Wyzga, R. E. (1976). *Economic Measurement of Environmental Damage*. Organization for Economic Cooperation and Development, Paris.

Meade, J. E. (1973). *The Theory of Externalities*. Institute Universitaire de Hautes Etudes Internationales, Geneva.

Mitchell, R. C., and Carson, R. T. (1989). *Using Surveys to Value Public Goods: The Contingent Valuation Method*. Resources for the Future, Washington, D.C.

Nordhaus, W. D. (1994). *Managing the Global Commons: The Economics of Climate Change*. MIT Press, Cambridge, MA.

Panayotou, T. (1992). Environmental Kuznets curves: Empirical tests and policy implications. Discussion paper, Harvard Institute for International Development, Harvard University, Cambridge, MA.

Perrings, C., and Walker, B. W. (1995). Biodiversity loss and the economics of discontinuous change in semi-arid rangelands. In *Biodiversity Loss: Economic and Ecological Issues* (C. Perrings *et al.*, Eds). Cambridge Univ. Press, Cambridge, UK.

Perrings, C., Mäler, K.-G., Folke, C., Holling, C. S., and Jansson, B.-O. (Eds.) (1994). *Biodiversity Conservation: Problems and Policies*. Kluwer, Dordrecht.

Perrings, C., Mäler, K.-G., Folke, C., Holling, C. S., and Jansson, B.-O. (Eds.) (1995). *Biodiversity Loss: Economic and Ecological Issues*. Cambridge Univ. Press, Cambridge, UK.

Pigou, A. C. (1920). *The Economics of Welfare*. Macmillan, London.

Portney, P. R., and Weyant, J. P. (Eds.) (1999). *Discounting and Intergenerational Equity*. Resources for the Future, Washington, D.C.

Reardon, T., and Vosti, S. A. (1995). Links between rural poverty and the environment in developing countries: Asset categories and investment poverty. *World Dev.* **23**, 1495.

Repetto, R., *et al.* (1989). *Wasting Assets: Natural Resources and the National Income Accounts*. World Resources Institute, Washington, D.C.

Roughgarden, J., May, R. M., and Levin, S. A. (Eds.) (1989). *Perspectives in Ecological Theory*. Princeton Univ. Press, Princeton, NJ.

Scheffer, M. (1997). *The Ecology of Shallow Lakes*. Chapman & Hall, New York.

Schneider, R. R. (1995). Government and the economy on the Amazon frontier, World Bank Environment Paper No. 11. World Bank, Washington, D.C.

Simon, J. L. (1981). *The Ultimate Resource*. Princeton Univ. Press, Princeton, NJ.

Simon, J. L. (1994). Post-debate statement. In *Scarcity or Abundance? A Debate on the Environment* (N. Myers and J. L. Simon, Eds.). Norton, New York.

Skiba, A. K. (1978). Optimal growth with a convex-concave production function. *Econometrica* **46**, 527.

Solorzano, R., *et al.* (1991). *Accounts Overdue: Natural Resource Depreciation in Costa Rica*. World Resources Institute, Washington, D.C.

Spence, A. M. (1974). Blue whales and optimal control theory. In *Systems Approaches and Environmental Problems* (H. Göttinger, Ed.). Vandenhoek & Ruprecht, Göttingen.

Tietenberg, T. (1980). Transferable discharge permits and the control of stationary source air pollution: A survey and synthesis. *Land Econ.* **56**, 391.

Tietenberg, T. (1990). Economic instruments for environmental regulation. *Oxford Rev. Econ. Policy* **6**, 17.

Tilman, D. (1997). Biodiversity and ecosystem functioning. In *Nature's Services: Societal Dependence on Natural Ecosystems* (G. Daily, Ed.). Island Press, Washington, D.C.

Tilman, D., and Downing, J. A. (1994). Biodiversity and stability in grasslands. *Nature* **367**, 363.

Vincent, J., Ali, R. M., and Associates (1997). *Environment and Development in a Resource-Rich Economy: Malaysia under the New Economic*. Harvard Institute for International Development, Cambridge, MA.

Vitousek, P. M., Mooney, H. A., Lubchenco, J., and Melillo, J. M. (1997). Human domination of Earth's ecosystem. *Science* **277**, 494.

Weitzman, M. L. (1974). Prices vs quantities. *Rev. Econ. Stud.* **41**, 477.

World Bank (1992). *World Development Report*. Oxford Univ. Press, New York.

ECOSYSTEM, CONCEPT OF

Eugene P. Odum
University of Georgia

I. Introduction
II. The Ecosystem Concept
III. The Human-Dominated Techno-ecosystem

GLOSSARY

autotroph Literally, a self-feeder; an organism that is able to utilize inorganic carbon (carbon dioxide) as the sole carbon source for growth; for example, green plants and certain bacteria.

black box Entity that can be examined at the system level without specifying its internal contents.

heterotroph Literally, a feeder on others; an organism that is dependent on organic material from an external source to provide carbon for growth; for example, vertebrates.

industrialized agriculture Modern form of agriculture that differs from traditional agriculture in the use of elaborate and expensive machinery, the control of pests with toxic chemicals rather than biocontrols, fertilization by synthetic rather than organic products, excessive consumption of water, and farm ownership and management by corporations rather than individuals.

input environment Collective term for all energy and materials moving into a given system.

mega-city Modern city with a large, expanding population, characterized by high consumption levels of energy, water, and food from sources outside the city.

output environment Collective term for all energy and materials moving out of a given system.

techno-ecosystem Technology-based ecosystem in the contemporary world that is fundamentally distinct from natural ecosystems in the use of energy sources other than sunlight (fossil fuels, nuclear power), an urbanized concentration of human population, and the generation of substantial amounts of air and water pollutants and waste materials.

LIVING ORGANISMS AND THEIR NONLIVING (ABIOTIC) ENVIRONMENT are inseparably interrelated and interact with each other. An ecological system, or ecosystem, is any unit (a biosystem) that includes all the organisms (the biotic community) in a given area interacting with the physical environment so that a flow of energy leads to clearly defined biotic structures and cycling of materials between living and nonliving parts. An ecosystem is more than a geographical unit (or ecoregion); it is a functional system unit with inputs and outputs, and with boundaries that can be either natural or arbitrary.

I. INTRODUCTION

The ecosystem is the first unit in the molecule to ecosphere hierarchy (as shown in Fig. 1) that is complete,

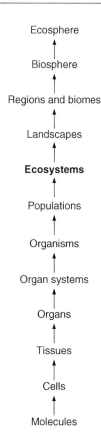

Ecosphere
↑
Biosphere
↑
Regions and biomes
↑
Landscapes
↑
Ecosystems
↑
Populations
↑
Organisms
↑
Organ systems
↑
Organs
↑
Tissues
↑
Cells
↑
Molecules

FIGURE 1 Hierarchical organization of living systems. The ecosystem is the first level that is complete.

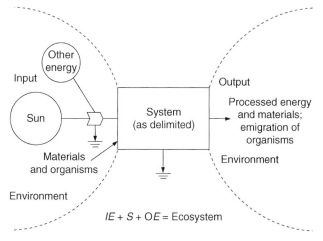

$IE + S + OE =$ Ecosystem

FIGURE 2 This ecosystem model emphasizes the external environment, which must be considered an integral part of the ecosystem concept. See Fig. 3 for a model emphasizing internal structures and processes.

that is, it has all the components, biological and physical, necessary for survival. Accordingly, it is the basic unit around which to organize both theory and practice in ecology. Furthermore, as the shortcomings of the "piecemeal" short-term technologic and economic approaches to dealing with complex problems become ever more evident with each passing year, management at this level, that is, *ecosystem management,* emerges as the challenge for the future.

Since ecosystems are functionally open systems, consideration of both inputs and outputs is an important part of the concept, as shown in Fig. 2. A diversity of species and genetic forms, together with a variety of functions and niches, are essential properties of natural ecosystems. Ecosystem diversity provides redundancy in times of environmental uncertainty.

II. THE ECOSYSTEM CONCEPT

The term "ecosystem" was first proposed in 1935 by the British ecologist A. G. Tansley, but of course the

concept is by no means so recent. Allusions to the idea of the unity of organisms and the environment (as well as the oneness of humans and nature) can be found as far back in written history as one might care to look. Not until the late 1800s did formal statements begin to appear, interestingly enough, in a parallel manner in the American, European, and Russian ecological literature. Thus, in 1877 Karl Mobius wrote (in German) about the community of organisms in an oyster reef as a "biocoenosis," and in 1887 S. A. Forbes, an American, wrote his classic essay on the lake as a "microcosm." The pioneering Russian V. V. Dokuchaev and his chief disciple, G. F. Morozov (who specialized in forest ecology), emphasized the concept of the "biocoenosis," a term later expanded by Russian ecologists to "geobiocoenosis" (Sukachev, 1959).

In addition to biologists, physical scientists and social scientists began to consider the idea that both nature and human societies function as systems. In 1925, the physical chemist A. J. Lotka wrote in a book entitled *Elements of Physical Biology* that the organic and inorganic worlds function as a single system to such an extent that it is impossible to understand either part without understanding the whole. It is significant that a biologist (Tansley) and a physical scientist (Lotka) independently and at about the same time came up with the idea of the ecological system. Because Tansley coined the word "ecosystem" and it caught on, he gets most of the credit, which should be shared with Lotka.

In the 1930s, social scientists developed the holistic concept of regionalism, especially Howard W. Odum, who used social indicators to compare the southern region of the United States with other regions (Odum, 1936; Odum and Moore, 1938). More recently, Machlis *et al.* (1997) and Force and Maddie (1997) have promoted the idea of the human ecosystem that combines biological ecology and social theories as a basis for practical ecosystem management. Accordingly, the concept of the ecosystem now brings together organisms, the physical environment, and humans.

As shown in Fig. 2, a graphic model of an ecosystem can consist of a box that we can label the system, which represents the area we are interested in, and two large funnels that we can label *input environment* and *output environment*. The boundary for the system can be arbitrary (whatever is convenient or of interest), delineating an area such as a block of forest or a section of beach, or it can be natural, such as the shore of a lake where the whole lake is to be the system, or ridges as boundaries of a watershed.

Energy is a necessary input. The sun is the ultimate energy source for the biosphere and directly supports most natural ecosystems within the biosphere. But there

are other energy sources that may be important for many ecosystems, for example, wind, rain, water flow, or fuel (the major source for urban-industrial society). Energy also flows out of the system in the form of heat and in other transformed or processed forms such as organic matter (e.g., food and waste products) and pollutants. Water, air, and nutrients necessary for life, along with all kinds of other materials, constantly enter and leave the ecosystem. And, of course, organisms and their propagules (seeds and other reproductive stages) enter (immigrate) or leave (emigrate).

In Fig. 2 the system part of the ecosystem is shown as a "black box," which is defined by modelers as a unit whose general role or function can be evaluated without specifying its internal contents. Figure 3 is a graphic model of the solar-powered natural ecosystem showing internal system components and functions. The interactions of the three basic components, namely, (1) the community, (2) the flow of energy, and (3) the cycling of materials, are diagrammed in this simplified compartment model. Energy flow is one-way; some of the incoming solar energy is transformed and upgraded in

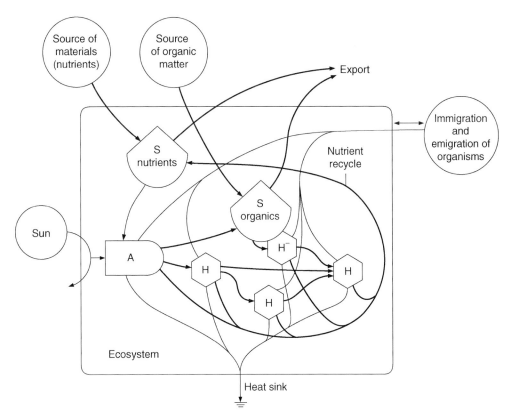

FIGURE 3 A functional diagram of a natural ecosystem, with emphasis on internal dynamics involving energy flow, material cycles, and storage (S), as well as food webs of autotrophs (A) and heterotrophs (H).

quality (i.e., converted into organic matter, a more up-graded form of energy than sunlight) by the community, but most of it is degraded and passes through and out of the system as low-quality heat energy (heat sink). Energy can be stored, then "fed back," or exported, as shown in the diagram, but it cannot be reused. In contrast with energy, materials, including the nutrients necessary for life (carbon, nitrogen, phosphorus, and so on), and water can be used over and over again. The efficiency of recycling and the magnitude of imports and exports of nutrients vary widely with the type of ecosystem.

Each "box" in Fig. 3 is given a distinctive shape that indicates its general function. Circles are renewable energy sources, bullets are autotrophs, hexagons are heterotrophs, and the round-bottomed shapes are storages (in this case of nutrients and organics). The community is depicted as a "food web" of autotrophs and heterotrophs linked together with appropriate energy flows, nutrient cycles, and storages.

Both graphic models (Figs. 2 and 3) emphasize that a conceptually complete ecosystem includes inputs and outputs along with the system as delimited, that is, an ecosystem = IE + S + OE (input environment + system + output environment). This scheme solves the problem of where to draw lines around an entity that one wishes to consider, because it does not matter very much how the box portion of the ecosystem is delimited. Often, natural boundaries, such as a lakeshore or forest edge, or political ones, such as city limits, make convenient boundaries, but limits can just as well be arbitrary so long as they can be accurately designated in a geometric sense. The box is not all there is to the ecosystem, because if the box were an impervious container, its living contents (lake or city) would die.

It is important to emphasize that it is the *diversity of ecosystem functions* including microbial recycling, inputs and outputs as well as habitats and human land uses that need to be maintained, not just the diversity of species or biodiversity in the narrow sense.

III. THE HUMAN-DOMINATED TECHNO-ECOSYSTEM

Current urban-industrial society not only impacts natural life-support ecosystems in negative and sometimes positive ways, but has created entirely new arrangements that we can call *techno-ecosystems* that

are competitive with and parasitic on natural ecosystems. These human-made systems involve new, powerful energy sources, technology, money, and fuel-powered cities that have little or no parallel in nature. It is imperative that techno-ecosystems interface with natural life-support ecosystems in a more positive or mutualistic manner than is now the case if our rapidly growing urban-industrial society is to survive in a finite world.

Before the industrial revolution, humans were a part of—rather than apart from—natural ecosystems. In the ecosystem model of Fig. 3, humans functioned as top predators and omnivores (the terminal H box in the food web). Early agriculture, as is the case with traditional or preindustrial agriculture as still widely practiced in many parts of the world, was compatible with natural systems and often enriched the landscape in addition to providing food. But with the increasing use of fossil fuels and atomic fission—energy sources many times more powerful than sunlight—together with the mushrooming growth of cities and increasing use of money-based market economics as the basis for decision making, the model of Fig. 3 is no longer adequate. We need to create a new model for this techno-ecosystem, a term suggested by pioneer landscape ecologist Zev Naveh (1982).

Figure 4 is our graphic model for these new (in terms of human history) fuel-powered systems. It includes the four components listed earlier: powerful energy sources, technology, money, and cities. The model shows the inputs of the new fuel energy sources and natural resources, and the increasing outputs of air, water, and solid waste pollution that are very much larger and more toxic than anything that comes out of natural ecosystems. In Fig. 5 we add to the techno-ecosystem model some natural ecosystems that provide life-supporting goods and services (breathing, drinking, and eating!) and that maintain homeorhetic (i.e., pulsing) global balances in the atmosphere, soils, freshwater, and oceans. Note that money circulates as a two-way flow between society and human-made systems, but not natural systems, thereby creating a vast market failure when society fails to pay for ecosystem services.

A modern city,[1] of course, is the major component of the fabricated techno-ecosystem. It is a very energetic hot spot that requires a large area of low-energy natural

---

[1] The term "city" is used synonymously with the geographers' term "standard metropolitan district (SMD)," which includes industrial areas and residential suburbs that often extend far beyond official city limits.

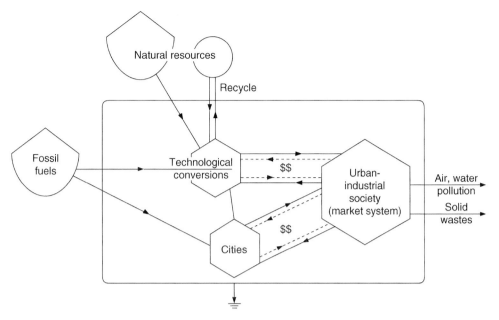

FIGURE 4   A human-dominated techno-ecosystem.

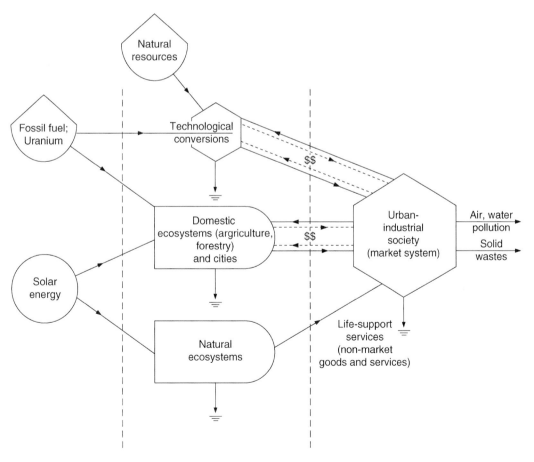

FIGURE 5   A human-dominated techno-ecosystem and natural ecosystem.

and seminatural countryside to maintain it. Current cities clean and recycle no air and or water (to the point of being redrinkable), grow little or no food, and generate a huge waste stream that impacts wide areas of downstream rural landscapes and oceans. The city does export money that pays for some natural resources, and the city provides many desirable cultural institutions, such as museums and symphonies, that are not available in rural areas.

In summary, cities are essentially parasites on the low-energy countryside. To call a city a parasite is not to belittle it, but to be realistic. In undisturbed nature, parasites and hosts tend to coevolve for coexistence; otherwise, if the parasite takes too much from its host, both die if the parasite has only one host. Currently humans have only one habitable host—the earth.

Especially threatening to the global life-support eco-systems is the explosive growth of *mega-cities* in the less-developed nations, caused in part by the increasing dominance of another techno-ecosystem, that of *industrialized agriculture,* with its often excessive consumption of water and use of toxic and enriching chemicals. These systems produce more food products per unit of space, but in turn are prodigious polluters and by their economic might drive small farmers out of business worldwide, forcing them into cities that are unable to assimilate them. This current situation illustrates what engineer and former president of MIT Paul Gray (1989) has written: "A paradox of our time is the mixed blessing of almost every technological development." In other words, technology has its destructive as well as beneficial side. To bring the natural and technical ecosystems into a mutualist relationship will be society's greatest challenge in the twenty-first century.

See Also the Following Articles

ECOSYSTEM SERVICES, CONCEPT OF • ENERGY FLOW AND ECOSYSTEMS • HUMAN EFFECTS ON ECOSYSTEMS, OVERVIEW

Bibliography

Force, J. E., and G. E. Maddie. (1997). The human ecosystem. *Society and Natural Resources* **10**, 369–382.

Gray, P. E. (1989). *The paradox of technological development.* In *Technology and the Environment,* pp. 192–205. National Academy Press, Washington, D.C.

Lotka, A. J. (1925). *Elements of Physical Biology.* Williams and Wilkins, Baltimore.

Naveh, Z. (1982). Landscape ecology as an emerging branch of human ecosystem science. *Adv. Ecol. Res.* **12**, 189–237.

Odum, H. W. (1936). *Southern Regions of the United States.* University of North Carolina Press, Chapel Hill.

Odum, H. W. and H. E. Moore (1938). *American Regionalism.* Henry Holt, New York.

Machlis, G. E., J. E. Force, and W. R. Birch (1997). The human ecosystem as an organizing concept in ecosystem management. *Society and Natural Resources* **10**, 347–367.

Sukachev, V. N. (1959). The correlation between the concepts "forest ecosystem" and "forest biogeocoenose" and their importance for classification of forests. *IX International Botanical Congress* **2**, 387.

Tansley, A. G. (1935). The use and abuse of vegetational concept terms. *Ecology* **16**, 284–307.

ECOSYSTEM FUNCTION MEASUREMENT, AQUATIC AND MARINE COMMUNITIES

John T. Lehman
University of Michigan

GLOSSARY

aerobic In the presence of oxygen.

allochthonous Imported from outside the ecosystem.

biogeochemical Consisting of biological and abiotic transformations.

chemoautotrophy Use of energy-yielding chemical reactions as an energy source for synthesis of organic matter from inorganic precursors.

cohort A group of individuals of the same age that can be identified within a population.

compensation depth Depth where photosynthesis and respiration are in balance.

euphotic zone Water depth with sufficient light for photosynthesis.

photoautotrophy Use of light as an energy source for synthesis of organic matter from inorganic precursors.

plankton The community of organisms suspended in the water.

production Newly formed biomass of a population or

trophic level, including the organic matter eliminated during the period of observation.

trophic level Position in a food chain, defined by the number of energy transfer steps to that level.

trophogenic region Region where net production of organic matter occurs by photoautotrophy or chemoautotrophy.

tropholytic region Region where respiration and decomposition of organic matter proceed in the absence of primary production.

INVESTIGATIONS OF ECOSYSTEM FUNCTION in fresh and saltwater have been dominated by two major themes. The themes originate in two different ways of thinking about biological production. Biological processes can be regarded fundamentally as conduits for the flow of mass and energy across trophic levels, or as the mechanisms by which individuals and populations maintain their existence. These processes include photosynthesis, mineral nutrient uptake, gathering of food, and extracting nutrition from it, as well as life history patterns, behavior, and survival strategies.

I. CONCEPTUAL FRAMEWORK

Investigation of biological mass flux and energy transfer relies on the first law of thermodynamics. Inquiries rooted in this theme require basic accounting for inputs and outputs balanced on carbon mass, nitrogen mass, or the chemically bound potential energy present in organic matter. The different bases for accounting are interlinked by the common stoichiometry of all protoplasm. As Alfred Redfield and his colleagues began to point out in the 1930s, there are consistent ratios among carbon, nitrogen, and phosphorus in living matter drawn from lakes and oceans. The ratios are not as rigid as the elemental composition of a crystalline mineral, but they are reliable within limits. They stem from the fact that organisms exist as biochemical aggregates of proteins, lipids, carbohydrates, and nucleic acids. The basic biochemical building blocks are universal, so the scope for differences among species is limited. If a class of organisms dominates the organic nitrogen pool within an ecosystem, it likely dominates in terms of organic carbon or calories as well. Thus the different measurement bases converge to common results.

The energy and mass-based approaches offer consistency of accounting, but they are intrinsically abiotic. They are indifferent to the forces that govern self-propagating genetic entities, for which the law of natural selection is as important as the laws of thermodynamics. Survival, persistence, and production of viable offspring are properties that define successful species, whether or not they dominate ecosystem biomass and material flux. An associated theme in measuring ecosystem function focuses on the properties of genetic entities rather than on the properties of conservation laws. This theme is keyed to the processes that add or subtract individuals within an ecosystem, processes such as birth, death, or migration, as well as the explicit match between individual age and the passage of time. Viewed through a lens that traces heritable lines of descent, mass and energy transformations are incidental consequences. Transformations of energy and material are the byproducts of life struggles whose object is indefinite persistence.

In 1961, G. E. Hutchinson coined the phrase "paradox of the plankton" to define the challenge to explanation presented by the species diversity of plankton communities. Nets towed from several hundred meters depth to the surface in the Pacific Ocean north of Hawaii, for example, routinely collect more than 300 species of zooplankton, and the richness of phytoplankton species in the surface waters is equally great. The oceanic pelagic region is the oldest continual habitat on the planet and its denizens are the product of ceaseless natural experiments. Near coastal regions and in lakes and rivers, the permanent plankton are joined by the larvae of benthic organisms. To Hutchinson, the con-

temporaneous existence of so many distinct genetic lines in habitats that lack obvious structural complexity begged important questions about coexistence, interactions, competition, and resource use. Faced with a bewildering array of potential species interactions, investigations of the marine pelagic ecosystems have emphasized mass and energy flow rather than individual species dynamics. Compounding the problem of overwhelming diversity is the fact that marine plankton are notoriously difficult to census accurately, owing to the physical movements of water masses, to such an extent that some workers believe that time series data that are essential to population studies are nearly impossible to gather.

In an effort to span the breach between measurements of organic composition in units of carbon or nitrogen and the realities of the ways matter is packaged as individual organisms, there has been great interest in finding rules of general validity and broad predictive strength that blend the approaches. Special attention has been paid to relationships based on individual body size or biomass. These size-based or allometric models of ecosystem processes hold the promise of predicting metabolic rates and trophic interactions from simple measurements. Extensive tabulations have become available for physiological processes like respiration, motility, body growth, and feeding rates versus individual size. This activity has spurred the drive toward models of ecosystem dynamics that treat the size structure of organisms present in a system rather than the phyletic composition of the communities.

II. BIOLOGICAL COMPARISONS BETWEEN FRESHWATER AND MARINE ECOSYSTEMS

Freshwater ecosystems exhibit reduced species richness compared with marine systems. Many important invertebrate groups (e.g., Echinodermata, Ctenophora, Chaetognatha) have failed to colonize freshwater habitats, although in some cases their roles have been assumed by successful radiation of aquatic insects, particularly the Diptera, in lakes and streams. Much of the difference may be owed to the greater depth, antiquity, and continuity of oceanic plankton environments. The role of age alone is problematic, however. The great ancient lakes of the planet such as Baikal, Tanganyika, and Malawi, with basin ages measured in hundreds of thousands to a few millions of years, exhibit endemic species radiation of some groups but not all. The zooplankton faunas of the ancient lakes are notably undiversified, despite rich endemism among some fish, mollusks, or amphipods.

Not only are there differences in overall species richness, but there are some differences in latitudinal trends as well. In general, the diversity of marine plankton is greater at low latitudes than at high latitudes. The latitudinal trend in species richness for freshwater plankton is the opposite. Tropical lakes have abbreviated zooplankton faunas compared with temperate sites. The tropical lakes are depauperate of large-bodied Cladocera and copepods, but they are poor in small-bodied pelagic rotifers as well.

III. PRIMARY PRODUCTION

A basic unit of measurement, and also a fundamental basis for comparisons among aquatic ecosystems, is the rate of primary biological production. Several methods are used, but they rely on the overall chemical reaction of carbon fixation in photosynthesis:

$$12\,H_2A + 6\,CO_2 \xrightarrow{h\nu} C_6H_{12}O_6 + 12\,A + 6\,H_2O \quad (1)$$

where A represents an element that serves as electron donor for the photochemical oxidation-reduction reaction. In photoautotrophic reactions of algae and bacteria in lakes and oceans, oxygen, sulfur, or reduced organic compounds are used as the electron donors. Methods for measuring primary production differ depending on whether the focus is on the transformation of inorganic carbon into organic matter, the resulting release of oxidized product, or changes in the internal cellular redox system or photochemical state. The different methods are not equivalent, and the stoichiometry of carbon fixed to oxidation product released is not strictly 1 to 2 as indicated in Equation 1. Equation 1 is a simplification of the true synthetic reactions involved with cellular growth. Synthesis products are not only hexose sugars, but include all sorts of carbohydrates, as well as proteins, lipids, and nucleic acids. The photosynthetic quotient (PQ) of oxidized product released to carbon fixed varies according to the dominant synthesis products.

In practical application, measurement of primary production relies on one of two approaches. Either measurements are made on subsets of the natural community, which are enclosed and isolated for an experimental time duration, or the measurements are made directly on the natural community without isolation.

Both approaches have advantages and potential complications.

A. Oxygen-Based Methods

To avoid the complexity of diffusion or advection of dissolved substances and gases in or out of a water parcel, most measurements of primary production are conducted in enclosures. During the 1920s, oceanographers introduced one such method that is still in wide use: the light and dark bottle oxygen method. The method relies on making three measurements of oxygen concentration: (a) the initial concentration of dissolved oxygen in a water sample at the start of the experiment, (b) the final concentration of oxygen in a water sample that was enclosed in a transparent bottle called the Light bottle, and (c) the final concentration of oxygen in a water sample enclosed in an opaque bottle called the Dark bottle. From these three experimental measurements it is possible to deduce rates of respiration (R), gross primary production (GPP), and net primary production (NPP):

$$R = ([O_2]_{Initial} - [O_2]_{Dark})/\Delta t \quad (2)$$
$$GPP = ([O_2]_{Light} - [O_2]_{Dark})/\Delta t \quad (3)$$
$$NPP = ([O_2]_{Light} - [O_2]_{Initial})/\Delta t \quad (4)$$

Where Δt is the duration of the incubation. The experimental bottles are either suspended at specified depths in the water column at the sampling site for duration of the experimental incubation, or they are placed in light and temperature conditions that simulate submarine conditions. The duration of experimental incubation Δt is invariably a compromise between the desire to obtain measurable changes in oxygen concentration and the desire to minimize artifacts resulting from prolonged exposure, such as growth of microorganisms on bottle surfaces or development of supersaturated oxygen concentrations inside the bottles. Typical incubation durations range from 4 to 24 hr.

B. Carbon-Based Methods

Methods that record fixation of inorganic carbon into organic matter rely either on measuring the removal of inorganic carbon or the appearance of new carbon in the organic matter. The pool of inorganic carbon available for photosynthesis in aqueous solution includes not only dissolved aqueous carbon dioxide, but also the reaction products of carbon dioxide with water molecules.

Carbon dioxide reacts with water to form carbonic acid:

$$CO_2 + H_2O = H_2CO_3 \tag{5}$$

Carbonic acid is a weak acid that dissociates to form its conjugate bases bicarbonate and carbonate:

$$H_2CO_3 = HCO_3^- + H^+ \tag{6}$$
$$HCO_3^- = CO_3^{2-} + H^+ \tag{7}$$

At equilibrium conditions the concentration ratio of carbon dioxide to carbonic acid is about 600 to 1. Proportions of carbon dioxide to bicarbonate to carbonate vary widely, but at the environmental pH of the ocean and many lakes, bicarbonate concentrations exceed all other molecular carbon species. The pool of dissolved inorganic carbon (DIC) available to photosynthesis is defined as

$$DIC = CO_2 + HCO_3^- + CO_3^{2-} \tag{8}$$

The three dominant chemical species that comprise DIC exist in known stoichiometric relationships to each other as functions of pH, temperature, and ionic strength of the aqueous solution. There are several ways to measure or calculate DIC. An aqueous sample may be acidified, transforming all DIC to carbon dioxide, and the resulting gas can be measured. Alternatively, pH and CO_2 may be measured and then the other species are calculated. Or the concentrations of the conjugate bases can be measured by acid titration, and from their total plus measured pH, the DIC is calculated.

C. C-14 Method

An alternative to measuring oxygen by light and dark bottle method is a method that uses radiocarbon as ^{14}C-bicarbonate as a tracer. A small but known amount of radiocarbon is introduced to a water sample and the bottle is incubated for a time period in the same manner as with the oxygen method. At termination of the incubation period, the contents of the experimental bottles are filtered. Radiocarbon retained on the filters is considered to represent inorganic carbon fixed into particulate organic matter during the experiment. Any fixed carbon that leaked into the water from algal cells can also be assessed by acidifying the filtrate under vacuum to drive residual inorganic carbon out of solution. The radioactivity of residual organic release products can then be measured. The beta particles emitted from dis-

integrating C-14 nuclei are detected and counted by various methods, including gas or planchet Geiger-Muller (ionization) detector, liquid scintillation, or track autoradiography.

Calculation of total carbon fixation by the C-14 method requires knowledge of the specific activity of the inorganic carbon pool and of the isotopic fractionation differences between C-12 and C-14. Specific activity is the ratio of added radioisotope activity to the mass of stable isotope that the radioisotope is intended to trace. In this method the radiocarbon traces the pool of DIC. Uptake of inorganic carbon by photosynthetic organisms involves the kinetic process of diffusion and active transport across cell membranes. In such processes there are differences in the mobility of isotopes of the same element owing to differences in molecular mass. Empirical study has demonstrated that the ratio of C-12 uptake to C-14 uptake is approximately 1.06. Accordingly, the biological fixation of inorganic carbon is calculated as

$$\begin{aligned} DIC\ fixed = 1.06 &\times (DIC\ in\ the\ sample) \\ &\times (radiocarbon\ fixed)/ \\ &(total\ inorganic\ radiocarbon\ added) \end{aligned} \tag{9}$$

Both light and dark bottles are used in estimation of primary production by the C-14 method, but interpretation of results from the dark bottles are quite different from interpretations for the oxygen method. In the case of C-14, the dark bottle does not measure respiration, but rather it records the rate of nonphotosynthetic carboxylation reactions by algae and bacteria. In simultaneous comparisons between C-14 and oxygen methods, it appears that C-14 measures something intermediate between net and gross primary production. The reason for this is that some but not all of the carbon fixed during the experiment is preferentially respired during the incubation period.

D. Fluorescence Methods

Indirect methods for assessing primary production rely on the fact that photosystem pigments alternate between a ground state in which they are able to absorb excitation photons of characteristic wavelengths and an energized state in which they either transfer the excitation energy to a chemical reaction or fluoresce a photon at longer wavelength and return to their ground state. The proportion of pigments in one state or the other varies with the rate of primary production expressed per unit mass of pigment. The usual reference pigment is chlorophyll *a*, the antenna pigment present in all aerobic photoautotrophs.

Application of indirect approaches requires calibration against direct measurements of primary production. Their attraction is their potential to estimate primary production in natural communities without isolation.

E. Interpretation and Analysis of Primary Production Results

Rates of primary production vary with depth according to light intensity, temperature, algal biomass, and physiological state of the algae. The photosynthetically active spectrum includes light from 400 to 700 nm. Near surface, light intensities are often sufficient to saturate the antenna pigments with photons. At high light, rates of photosynthesis are limited by the rates of biochemical dark reactions, specifically the carboxylation reactions, rather than by interception of light photons. Because carboxylation is enzyme dependent, the maximum rates of carbon fixation are temperature dependent and they vary with nutritional status as well.

Light intensities diminish with depth. At low light, the rate of primary production is limited by the rate at which photons are intercepted, and carbon fixation varies directly with light intensity. Temperature variations are less important at low light because in photochemical reactions limited by interception of photons, the role of temperature-dependent enzyme kinetics is minimized.

Respiration rates vary with temperature. Near the surface, as a daily average, gross primary production often exceeds respiration and net primary production is positive. At greater depth, however, gross primary production declines and a depth is reached at which net primary production equals zero. This depth is termed the compensation depth. For purposes of predicting the physiological state and growth response of the algae, it is important to recognize that organisms in aqueous suspension do not always maintain fixed vertical positions in the water. What is most important is the integral difference between the rates of gross primary production and of respiration during 24 hours:

$$\int_0^{24} \int_0^{z_{mix}} GPP(z, t) - R(z, t) \, dz \, dt \qquad (10)$$

where z_{mix} is the maximum depth of water involved in vertical mixing during 24 hours. When the integral defined by Equation 10 is positive, net growth will occur. Seasonal variations in photoperiod, water trans-

parency, temperature, and species composition influence the components of Equation 10, and so the critical mixing depth for net positive algal growth varies likewise.

IV. SECONDARY PRODUCTION

The principle of mass and energy balance used in primary production studies can be extended to heterotrophs. Among herbivores, predators, and detritivores, material and energy are considered commodities that continually turn over through biochemical means and by replacement of individuals within a population. Secondary production is the total growth increments over a given time interval as experienced by all individuals within a population that were alive at the beginning of the interval, whether or not they survive the complete interval. Unlike primary production, there is no distinction between gross and net production; secondary production (P) is inherently a net balance of several factors:

$$P = I - Eg - Ex - R - M - G \qquad (11)$$

where I is ingestion, Eg is egestion of unassimilated food, Ex is excretion, R is respiration, M is molts or exuvia, and G is gametes and reproductive products such as egg yolk and spermatophore capsules.

The efficiency of energy or material transfer during secondary production is expressed by several quotients. Assimilation efficiency is the ratio of assimilation (I − Eg) to ingestion (I). Net growth efficiency is the ratio of growth (P) to assimilation. Gross growth efficiency is the ratio of growth (P) to ingestion (I). Ecological efficiency is sometimes defined for predator-prey trophic interactions. Ecological efficiency is the ratio of predator production to prey production.

A. Bioenergetic Analysis

Bioenergetic analysis places its frame of reference on individual organisms of specified size, weight, or age. When Equation 11 is applied at the individual level, the result (P) is incremental growth. The units of accounting can be total body mass, content of carbon, nitrogen, or phosphorus, or calories. Analysis is facilitated if each of the terms on the right side of Equation 11 can be expressed with reference to body mass. For example, ingestion rate may be expressed as a function of food types and availability, temperature, and body mass of the consumer. Respiration may be expressed as a function of temperature, activity level, and body mass. In this way, the incremental growth of individual organisms is linked with organism size and a set of environmental conditions. The approach can then be extended to the population level by summing calculated growth increments across the empirical size distribution of the population.

B. Biomass Accrual

The standing stock biomass of a population is the sum of the weights of all individuals in the population. Changes in biomass occur owing to changes either in number (N), or mean weight (W):

$$\Delta \text{Biomass} = \Delta(\text{N} \times \text{W}) = \text{N} \, \Delta\text{W} + \text{W} \, \Delta\text{N} \qquad (12)$$

Equation 12 suggests a means to calculate secondary production for populations consisting of distinct cohorts. Presence of distinct cohorts is often the case among long-lived, seasonally reproducing populations. In such cases, a survivorship chart or life table for each cohort (decreasing N versus time) defines ΔN. A parallel chart or table reporting the average individual weights of cohort members versus age or time defines ΔW.

When cohorts cannot be distinguished, as happens with continually reproducing populations, secondary production can be calculated by summing the growth increments (Δw_i) of individual life history stages:

$$P = \text{N}_1 \, \Delta w_1/T_1 + \text{N}_2 \, \Delta w_2/T_2 + \text{N}_3 \, \Delta w_3/T_3 \\ + \cdots + \text{N}_k \, \Delta w_k/T_k \qquad (13)$$

where the subscripts denote individual stages (e.g., instars) or size classes and where T_i represents the mean duration of each stage. Abundance at each stage (N_i) is considered to be the mean stage-specific abundances during the period of interest.

C. Birth and Death Rate Analysis

Numerical changes in populations can be represented as a simple function of population size (N):

$$dN/dt = r \times N \qquad (14)$$

where r is the net intrinsic population growth rate (time^{-1}) resulting from additions and removals:

$$r = \text{birth} - \text{death} + \text{immigration} - \text{emigration} \qquad (15)$$

In cases where changes owing to migration are not an issue, net population growth amounts to the balance between birthrate (b) and death rate (d):

$$r = b - d \qquad (16)$$

Over short time intervals an assumption is made that birth and death rates are stationary, so that the simple integral solution to Equation 14 is

$$N(t + \Delta t) = N(t)e^{r\Delta t} \qquad (17)$$

and net intrinsic population growth rate r can be calculated from population sizes at the beginning and end of any time interval Δt:

$$r = \ln[N(t + \Delta t)/(N(t)]/\Delta t \qquad (18)$$

To decompose the net rate r into its individual components, an independent estimate is needed for either birth or death rate.

D. Egg Ratio Method

The egg ratio method, pioneered by W. T. Edmondson, calculates birthrate from demographic properties of populations in nature. The method is particularly successful in application to parthenogenic populations, especially to species in which females produce and brood clutches of eggs, such as among rotifers and Cladocera.

The egg ratio, E, is defined as the ratio of eggs and embryos to the sum of all postembryonic stages (juveniles and adults) in the population. From egg ratio and duration of development for the embryonic stage (D), the best estimate for population birthrate (b) is

$$b = \ln(E + 1)/D \qquad (19)$$

Derivation of Equation 19 requires assumption of a stable age structure for the population, and further that all embryonic and postembryonic stages experience a common death rate (d), such that d can be calculated from r and b by difference:

$$d = b - r \qquad (20)$$

These assumptions are most nearly applicable to short-lived species with abbreviated juvenile stages. Accuracy of the birthrate estimate is improved if the egg ratio is based on a census of egg stages that are in final phases of development and if the development time is correspondingly restricted in range. Most of the populations to which this analysis is applied are ectotherms, so egg development times are strong reciprocal functions of water temperature.

E. Instar Analysis

For populations that exhibit distinct morphological stages, such as size classes or the developmental instars of arthropods, it is possible to estimate stage-specific mortality. This method requires knowledge of the durations (T_i) of each stage including their variation with time and water temperature. The numerical change in each stage is defined as

$$dN_i/dt = N_{i-1}/T_{i-1} - N_i/T_i - d_iN_i \qquad (21)$$

The first term on the right side of Equation 21 represents recruitment of new individuals into the stage or age class from the preceding stage. The second term is recruitment out of the stage, and the third term is mortality within the stage. This approach assumes a uniform age distribution within each stage. It further assumes that the calculation time step is small compared to the stage durations T.

Instar analysis may be combined with birthrate estimation by the egg ratio method to estimate recruitment into the youngest instar category. Addition of mass or energy information about each instar or stage permits further calculation of secondary production according to Equation 13.

V. BIOGEOCHEMISTRY AND NUTRIENT CYCLING

Transformations of matter and energy in aquatic ecosystems are dependent on, and in turn influence, the availability of all the elemental constituents of protoplasm including both major components like carbon and nitrogen as well as trace constituents like iron and manganese. Even elements and compounds that are not essential constituents of biological material are affected by ecosystem processes. For example, redox reactions and pH changes caused by biological reactions can influence the solubility and reactivity of metals like mercury and cadmium, which can become toxic.

Alfred Redfield is credited with the conceptual theory of "biochemical circulation" within ocean ecosystems, and the theory is equally applicable to fresh waters. According to this view, processes of synthesis and decomposition may be separated in space and time, but they remain linked. Organic matter is generated at the ocean surface where radiant energy from sunlight permits positive net primary production. Carbon is assimilated from inorganic sources into organic form, together with nitrogen, phosphorus, silicon, iron, and all the other constituents of protoplasm. Over time, a portion of this synthesized organic matter will be transported actively or passively to deeper depths in perpetual darkness where photosynthesis is impossible. Once there, the allochthonous organic matter is used in secondary production, but its ultimate fate is to be recycled to inorganic mineral form or to be buried in sediments for long periods of geological time. Organic matter that is not transferred to the depths is recycled *in situ*, owing to the production inefficiencies of multiple secondary production pathways including herbivory, predation, and detritivory.

Physical processes of upwelling and thermohaline circulation permit some mineral nutrients dissolved in deep water to be returned to the surface for another round of synthesis reactions. Over geological time scales even the elements buried in the ocean sediments can become available again through weathering of the ancient deposits. This alternation between the inorganic state of an element and a state in which it is associated with organic processes at both biological and geological time scales is the basis for the term biogeochemistry.

For elements such as carbon and nitrogen, organic transformations entail changes in oxidation state and allied chemical properties. For others such as phosphorus and silicon, transformations affect not oxidation state but rather solubility.

Elemental nutrients that are essential for the production of organic matter can be traced through ecosystem transformations owing to the principle of conservation of matter. Quantitative illumination of the pathways has led to production of descriptive "cycles" for each element. Different aquatic ecosystems exhibit characteristic magnitudes of different pathways. Even water itself is involved in a hydrological cycle, in which water alternates among liquid, gas, and solid phases, and among ocean, atmosphere, and terrestrial environments. Water evaporates from the sea surface to the atmosphere, from where some of it falls as rain on the continents.

Surface and groundwater on the continents react with carbon dioxide from the air and from respiration of soil biota. The resulting solution thus includes carbonic acid. In some regions the carbonic acid is supplemented by strong acids introduced by anthropogenic activities. For example, combustion of fossil fuels introduces oxides of nitrogen and sulfur into the atmosphere, because the elements were components of the ancient proto-plasm that underwent diagenesis to form the fuel sources. After reacting with atmospheric water vapor, the oxides enter aqueous solution as nitric and sulfuric acids. The resulting mix of weak and strong acid is the primary agent of chemical weathering, a process that dissolves the rocks of the continents. Chemical weathering accounts for most of the ions in river water.

Elements that enter the ocean by runoff from the continents or as wind-blown dust are subject to further reactions and removal processes. Biological processes are prominent among these. Microbiological processes reduce sulfate to sulfide, followed by loss of hydrogen sulfide gas or chemical precipitation of metal sulfides. Inorganic carbon is transformed into particulate organic matter or is precipitated from solution with calcium as the skeletons of marine organisms. Silica is precipitated likewise, as cell wall covering or skeletal material for a variety of plants and animals.

Inputs of new material and removal processes are in close balance, for the bulk chemical composition of the ocean appears to be in steady state. The relationship between material flux and pool sizes in biogeochemical cycles is expressed through the concept of residence time, defined as

$$\text{Residence time (yr)} = \frac{\text{Amount of a given element in the ocean}}{\text{Rate of addition or removal per year}} \quad (22)$$

Elements that exhibit great biological reactivity, such as the nutrient elements N, P, and Si, have short residence times and are usually present at low concentrations. Elements with lower significance to biological reactions like Na and Cl have long residence times, and correspondingly account for most of the dissolved salt in sea water.

A. New and Regenerated Production

In the late 1960s, R. C. Dugdale and J. J. Goering introduced a conceptual model of biological production in the surface ocean euphotic zone. The model represented the euphotic zone as a compartment or box with inputs and outputs of nitrogen. The inputs were delivered via upwelling, nitrogen-fixation of diazo-nitrogen, and river runoff from the continents. Outputs were the sinking material consisting of dead plankton and fecal pellets. In addition to the inputs and outputs, they described internal recycling processes within the euphotic zone, through which nitrogen was regenerated by zooplankton and bacteria.

To maintain steady state, the outputs from the euphotic zone must equal the inputs, irrespective of the magnitudes of regeneration. Dugdale and Goering reasoned that primary biological production that was sustained by the inputs, or "new" nitrogen, should be called new production, and that the primary production sustained by regenerated nitrogen should be called regenerated production. "New" nitrogen typically arrives in the form of nitrate, a form that is thermodynamically favored at the redox potential of surface sea water, and is the form to which organic matter is decomposed under aerobic conditions. "Regenerated" nitrogen is typically ammonium, which is at the same reduced oxidation state as the primary amino nitrogen that comprises proteins. Most aquatic animals are ammonotelic in their excretion metabolism. Thus, the forms of nitrogen used to support primary production of organic matter are a clue to the processes that dominate production.

The concepts were further refined by R. W. Eppley and B. J. Peterson into a modern ecosystem paradigm of biological oceanography. New production is sustained by imported nutrients. It is considered to be equal to the sum of sinking flux of organic matter plus transfers of organic matter to higher trophic levels, which may be exported from the ecosystem by, for example, fisheries yields. By empirical evidence, ecosystems dominated by "new" production exhibit higher primary production overall and are more capable of sustaining exploitation such as harvest of fisheries stocks. Ecosystems dominated by "regenerated" production cannot endure prolonged exploitation of their sustaining elements.

The concept of new production and its linkage with export production has further implications for global biogeochemical cycles and specifically for carbon dioxide levels in the global atmosphere. Organic matter that is exported from the surface ocean to the deep sea represents a sink for DIC from the surface ocean. Deficits of DIC are replenished in part by dissolution of carbon dioxide from the atmosphere. As a consequence of this linkage, the fraction of global oceanic primary production that is new production represents an upper limit to the amount of carbon that can be exported and sequestered in the deep ocean.

Techniques for measuring the rates of nutrient regeneration differ depending on whether the measurements are conducted in trophogenic regions or in tropholytic regions. In tropholytic regions, primary production is minimal and thus regeneration can be assessed by measuring the accumulation of mineral nutrients directly. In trophogenic regions, synthesis reactions dominate over regeneration, and regenerated nutrients are quickly reassimilated. Techniques for measuring nutrient recycling under such conditions rely on assessment of uptake and regeneration simultaneously, such as by isotope dilution. In isotope dilution experiments, a deliberate tracer such as ^{15}N-ammonium is introduced to a closed system in which ^{14}N is the predominant natural isotope. Over time, the ratios of ^{15}N to ^{14}N in the ammonium pool become diluted as regenerated ^{14}N-ammonium is added, even while uptake of both isotope forms proceeds.

VI. STABLE ISOTOPES

Stable isotopes of natural elements are ubiquitous. They have often been collected and used as deliberate tracers, such as in studies of new and regenerated production, but they can be measured and used at natural occurrence as well. Owing to the proportions in which different nuclear species survived the first instants of the universe, and the proportions that have been retained in the planetary composition of the Earth, many common elements exhibit more than one stable form. Hydrogen, carbon, nitrogen, oxygen, and sulfur, for example, all have two or more nuclear configurations that differ in numbers of neutrons and hence differ in atomic mass. Oxygen-16, O-17, and O-18 are stable (not radioactive), but they exist in different proportions in the environment. The different forms can be distinguished by mass spectrometry. The preferred form of analysis is by isotope ratio mass spectrometer, which compares the stable isotope ratios in a sample with ratios in a standard.

Measurement of stable isotope ratios in natural materials has been a technique used by geochemists for decades to help interpret complex processes. The relative proportions of the stable isotopes vary from one medium or substance to another. Mechanisms that can change the proportions of two stable isotopes of the same element are called fractionation processes. Fractionations result from differing mobilities and reactivities by the same chemical element owing to differences in atomic mass. Both kinetic and thermodynamic factors affect the magnitudes of fractionation. Five general rules apply:

1. Elements with high atomic mass generally exhibit smaller fractionation among their isotopes than do lightweight elements.
2. If multiple isotopes of an element exist, fractionation is greatest between isotopes that have the largest atomic mass difference.

3. In unidirectional reactions, the lighter isotope becomes enriched in the endpoint.
4. In reactions that reach equilibrium, the lighter isotope is enriched in the reactant compound or phase that has the weaker bond strength.
5. As temperature increases, fractionation decreases asymptotically to zero.

Changes in the fractionation ratios with temperature have been the basis for using stable isotopes in geochemical studies as "paleothermometers".

Atmosphere, terrestrial, and marine environments are heavily dominated by single isotope forms of common elements, and the alternate isotopes are usually very rare. For example, in the atmosphere, O-16, O-17, and O-18 co-occur in overall ratios by atoms as $1:0.0004:0.002$. Similarly, C-12 and C-13 occur in atmospheric carbon dioxide in the mean ratio $1:0.011$. Given such disparities of magnitude, isotope ratio differences between samples and standard are calculated in δ notation with units of parts per thousand, or per mil ($^0/_{00}$). For example,

$$\delta-^{18}O = [(^{18}O/^{16}O)_{sample}/(^{18}O/^{16}O)_{standard} - 1] \times 1000 \quad (23)$$

Isotopes are useful in ecosystem studies because physical reactions and enzyme reactions cause reproducible discrimination between isotopes. Reaction differences lead to differences in the composition of biological material, skeletal products, and inorganic nutrient pools.

Stable isotopes have been used to infer and trace the origins and pathways of materials in aquatic ecosystems. They are useful when different sources are isotopically unique and when the sources change in predictable ways. Successful applications require that the magnitudes of all isotope fractionations that result from physical, chemical, and biological processes be known.

Isotopes of carbon and nitrogen have been used to trace food-web relations in freshwater and marine environments. Phytoplankton discriminate against C-13 during photosynthetic carbon fixation, just as they discriminate against C-14. The magnitude for discrimination against C-13 is less than for C-14, or about -20 to -30 $^0/_{00}$ with respect to the source DIC. Organic matter that originates from terrestrial sources sometimes has very different carbon isotope ratios than material that is of aquatic origin. Terrestrial plants fractionate atmospheric CO_2 differentially during photosynthesis according to whether they use C3, C4, or CAM pathways for carbon fixation. Similarly, organic matter generated microbiologically from methane as a substrate typically exhibits unique carbon isotope ratios. Stable isotopes may thus be used to identify chemoautotrophic pathways as well as photoautotrophic ones.

Carbon isotopic ratios in animals resemble those of their diets within about 1 $^0/_{00}$. Slight enrichment of C-13 in an animal versus its diet is presumed to be caused by preferential loss of C-12 during respiration or by preferential assimilation of C-13 from the food. Synthesis of lipid rather than protein can also influence the isotope ratios because there are reported fractionation differences between biochemical constituents. As long as the isotope ratios in different potential food sources are well defined and consistently different, the relatively conservative levels of fractionation for carbon by trophic level can indicate whether an animal is eating its food from one source, another source, or a definable mixture of the two.

Nitrogen isotope ratios also reflect the composition of an animal's diet, but the animal is usually enriched with N-15 by about +3 $^0/_{00}$ compared with its food. This increased retention of N-15 over N-14 seems to be favored by thermodynamic factors associated with transamination reactions, and by kinetic factors leading to preferential loss of N-14 as excreted ammonium and urea. Reproducible differences in δ–^{15}N with trophic level provide a diagnostic measure of relative trophic position among organisms that depend on the same primary source of organic matter. Carnivores are isotopically heavier than herbivores, which in turn are isotopically heavier than the algae; omnivores are isotopically intermediate between herbivores and carnivores.

Interpretation of nitrogen isotope dynamics can be complicated by temporal and spatial variations in isotope fractionation and nutrient source materials. In laboratory experiments when nutrients are not limiting, isotope fractionations associated with assimilation of nitrate and of ammonium can be large. However, if a nutrient element becomes limiting such that all of it is consumed, there can be no isotopic fractionation. In nature, seasonal alternation between times of nutrient excess and nutrient limitation is common, with the result that isotope ratios can vary greatly in the source material and in the primary producers. Such variations provide useful information about nutrient dynamics, but simultaneously introduce complication to interpretations of trophic relations particularly for long lived organisms.

See Also the Following Articles

BIOGEOCHEMICAL CYCLES • ECOSYSTEM FUNCTION MEASUREMENT, PLANT COMMUNITIES • MARINE ECOSYSTEMS

Bibliography

Downing, J. A., and F. H. Rigler (Eds.). (1984). *A manual on methods for the assessment of secondary productivity in fresh waters.* (2nd ed.). Oxford: Blackwell.
Edmondson, W. T. (1974). Secondary production. *Mitt. Int. Ver. Limnol.* 20: 229–272.
Griffiths, H. (Ed.). (1998). *Stable isotopes.* Oxford: Bios Scientific Publishers.
Lajtha, K., and R. H. Michener (Eds.). (1994). *Stable isotopes in ecology and environmental science.* Oxford: Blackwell.
Lampert, W., and U. Sommer. (1997). *Limnoecology: The ecology of lakes and streams.* Oxford: Oxford University Press.
Vollenweider, R. A. (Ed.). (1974). *A manual on methods for measuring primary production in aquatic environments.* (2d ed.). London: Blackwell.

ECOSYSTEM FUNCTION MEASUREMENT, TERRESTRIAL COMMUNITIES

Sandra Díaz
Universidad Nacional de Córdoba—CONICET

GLOSSARY

biodiversity Number and composition of genotypes, species, functional types, and/or landscape units present in a given system.

ecosystem Conceptual view of an assemblage of organisms and of physical and chemical components in their immediate environment, and the flow of materials and energy between them.

ecosystem functioning Flow of energy and materials through the arrangement of biotic and abiotic components of an ecosystem.

ecosystem stability Capacity of an ecosystem to persist in the same state. It has two components. Ecosystem resistance is the ability to stay in the same state in the face of perturbation. Ecosystem resilience is the ability to return to its former state following a perturbation.

resource dynamics Inputs, outputs, and internal cycling of key resources, such as carbon, water, and mineral nutrients, in an ecosystem.

trophic transfer The amount of biomass and/or energy which is transferred from the primary producers to the herbivore-based food chain rather than directly from plants to detritivores.

THIS ARTICLE DEALS WITH the functioning of terrestrial ecosystems with emphasis on primary producers and their relationship to biodiversity. Primary producers, in particular vascular plants, are the major sustainers of terrestrial life. They determine the amount of energy and materials available to terrestrial food webs. Biodiversity in this context involves not only the number and relative abundance of species (taxonomic diversity) but also the kind and relative abundance of traits they possess (functional diversity). These traits influence, and in turn are influenced by, major ecosystem processes. Ecosystem resource dynamics (carbon, nutrients, and water) and ecosystem stability in the face of disturbance (resistance and resilience) are discussed, including conceptual issues and major theoretical and methodological approaches to their study.

I. WHY MEASURE ECOSYSTEM FUNCTIONING?

Ecosystem functioning involves processes such as primary production, trophic transfer from plants to animals, nutrient cycling, water dynamics, and heat transfer. Traditionally, the term has referred to functioning in equilibrium, namely, the amount (how much), the rate (how fast), and sometimes the seasonal variations of those processes. Some authors believe that the responses of ecosystem processes to perturbation, resilience and resistance, should also be incorporated into the term. This is the approach taken in this article (Fig. 1). Ecosystem processes provide essential goods and services to humankind, such as food; fiber; fodder; fuel; water provision, control, and detoxification; amelioration of weather; soil formation; retention of carbon that otherwise would be released into the atmosphere contributing to climate change; and medicinal, recreational, and cultural resources. The impact of changes in the quantity and quality of services provided by various terrestrial ecosystems to humankind has been gaining increased attention, and in some cases their costs are very high compared with their traditional market value. The measurement of how much seed, green biomass, and animal biomass an ecosystem can produce

or how much water and nutrients can be retained *in situ* are key inputs for an evaluation of ecosystem services to humankind. Concepts, methods, and data retrieval on ecosystem functioning have progressed enormously in the past few decades, and today an impressive body of quantitative information on flows and stocks of materials and energy through a wide spectrum of ecosystems is available. Much less is known, however, of the effects of perturbations on these processes. Physiological processes underlying biomass production, water balance, and nutrient cycling are well understood. However, there are still many theoretical and methodological difficulties and information gaps in the prediction of processes at coarser scales of time (decades or centuries) and space (ecosystems, landscapes, and biomes). In order to determine how the biosphere will respond to the changes in climate, atmospheric composition, and land use projected for the future, accurate predictions at these levels are vital.

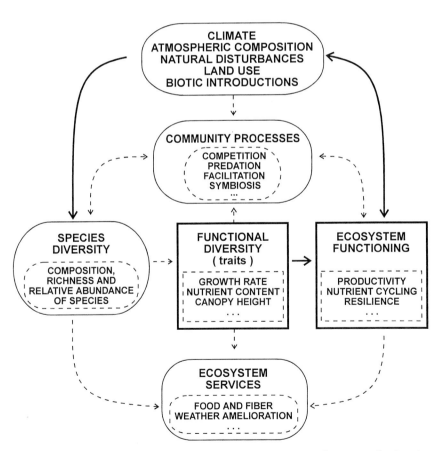

FIGURE 1 Conceptual diagram of the relationships between ecosystem functioning, biodiversity, ecosystem services, and external structuring factors. Rectangular boxes and solid arrows represent the main focus of this article (partially based on Chapin *et al.*, 1998).

II. ECOSYSTEM RESOURCE DYNAMICS

Three main processes will be analyzed here, which are linked to the acquisition and loss of chief resources for terrestrial ecosystems: light, carbon, water, and mineral nutrients. These processes are biomass production (primary productivity and trophic transfer), nutrient (especially nitrogen) cycling, and water dynamics. There is a strong association between water flow, carbon assimilation, and nutrient uptake by a plant. Since carbon dioxide (CO_2) input and water output occur through open stomata, and nutrient and water uptake occur through root hairs and fine roots, they are usually linked. Nutrient availability controls the increase in plant mass, and incorporation of mineral nutrients into biomass is not possible without carbon assimilation. The derived processes at the ecosystem level, namely, biomass production, water balance, and nutrient cycling, are therefore closely linked and basically regulated by net radiation, temperature and precipitation, soil composition and structure, and the plants that are present. Any consideration of carbon, nutrient, or water balance in isolation is probably artificial. The examples presented here are described in different sections for the sake of clarity, but they highlight the strong interdependence of these processes.

A. Approaches to Analysis and Measurement of Ecosystem Functioning

1. Whole-Ecosystem Approaches and Approaches Based on Community Structure and Composition

The measurement of ecosystem functioning focuses on the sizes of major pools of resources, such as water, carbon, and mineral nutrients, and on the rates of flows connecting them. The emphasis on one of these two aspects has varied with time and authors. Not until the 1970s did compartmental models become an important tool in "systems ecology" (illustrations of this view can be found in the classic textbook on ecology by E. P. Odum, 1971). In these models, boxes represent pools or stocks (usually major trophic levels), and arrows signify the flows between them. This represents a first and very useful approximation. However, pools are often treated as "black boxes" in which only the general size, and not the internal composition, is important. This often makes this approach too coarse for the management of real situations and of very little predictive value for novel situations. In most cases, the whole pool is not altered, but instead subtle intrapool changes occur

until a threshold is reached or substances are chemically altered or sequestered. This is because organisms which are members of the same pool tend to respond in an individualistic way and have slightly different properties. As a consequence, other approaches have focused on the internal composition of pools (what taxa or functional types are present within the boxes and in what abundance) and how changes in that composition can alter flow rates between pools. These approaches are strongly based on community structure and composition and will be addressed in Section IV.

2. Measurement of Short-Term Resource Dynamics and Long-Term, Large-Scale Ecosystem Processes

Some methods of ecosystem analysis emphasize the measurement of short-term process rates (e.g., photosynthesis, nutrient uptake, and evapotranspiration), whereas others emphasize the measurement of the size of pools accumulated as a result of these processes, with or without consideration of their internal components (e.g., biomass production, nitrogen stocks in different compartments, and water use efficiency integrated over a whole season) (Table I). The choice depends mostly on the objectives pursued and on the scale of observation selected. However, as a general rule, methods based on short-term flows provide more mechanistic understanding, but they do not provide a clear picture of what happens at the time and space scales most relevant to ecosystem functioning (meters to thousands of kilometers and months to decades or centuries). Measurements of flow rates (F measurements in Table I) tend to be very precise but highly variable. They give a snapshot of ecosystem functioning, but they do not necessarily reflect processes over a growing season. For example, short-term (hourly or daily) variation in gas exchange per leaf area is not directly reflected in annual biomass production. Plant growth and vegetation productivity cannot be directly equated with leaf photosynthetic rates. The productivity of plant communities is determined more by the amount of photosynthetic tissue (which is in turn controlled by carbon allocation) than by photosynthetic rate. There is often only a slight relationship between these two parameters. Gas exchange measures the instantaneous plant performance, whereas growth necessarily integrates along time and reflects plant allocation patterns. At the community level, additional regulations occur; therefore, caution is necessary in generalizing the ecological meaning of leaf physiological processes for production. For example, a 10- to 40-fold difference between species in maximum rates of photosynthesis at the leaf level is reduced

TABLE I

Examples of Ecosystem-Level Measurements with Emphasis on Short-Term Flow Rates (F) and on Longer Term Pool Sizes (P)[a]

Parameter measured	Usual units	Techniques	Emphasis
Biomass production			
Photosynthesis	μmol CO_2 cm^{-2} s^{-1} or mg CO_2 m^{-2} hr^{-1}	Gas exchange systems (CO_2, O_2), or simultaneous H_2O and CO_2 exchange	F
Standing biomass	g m^{-2} or t ha^{-1}	Harvest at the end of growing season	P
Productivity	g m^{-2} year^{-1} or t ha^{-1} year^{-1}	Sequential harvest; equations including climatic and soil factors; satellite imagery	P
Cover or frequency	% ground area (cover); % total interceptions (frequency)	Estimation through nondestructive measurement, which can be converted into units of biomass using allometric equations developed on similar neighboring individuals that are destructively harvested	P
Trophic transfer			
Consumer biomass	g or kJ m^{-2}; t ha^{-1}	Harvest (with or without replacement); equations including NPP	P
CO_2 efflux from respiration (estimates microfaunal biomass)	μmol CO_2 m^{-2} or mm^{-3} s^{-1}	Gas exchange measurements	F
Secondary productivity	g or kJ m^{-2} year^{-1}; t ha^{-1} year^{-1}; No. animals m^{-2} year^{-1}	Sequential harvest (with or without replacement); animal countings; equations including NPP	P
Consumption	g or kJ m^{-2} year^{-1}; t ha^{-1} year^{-1}; % plant tissue; % leaf surface	Plant harvest in herbivore-free and grazed situations; estimation of proportion of plant tissue removed; equations including NPP	P
Nutrient cycling			
Decomposition through biomass loss	% initial dw	Sequential weighing of litter samples	P
Decomposition through CO_2 efflux from soil or jars	μmol CO_2 m^{-2} s^{-1} or μg CO_2 g dw substrate^{-1} hr^{-1}	Gas exchange system (soil-only control samples or isotope labeling sometimes used in order to distinguish from root respiration)	F
Nutrient status	mg nutrient g dw^{-1} or % nutrient dw	Harvest followed by chemical analysis; estimation through remote sensing	P
Nutrient use efficiency in litter (amount of biomass produced per unit of nutrient expended and lost; it reflects initial leaf status and efficiency of nutrient resorption)	kg dw mol N^{-1} in litter	Harvest followed by chemical analysis	P
Nutrient uptake	mg nutrient g dw^{-1} year^{-1}	Change in nutrient pool in live biomass over time interval; estimated through sequential harvest followed by chemical analysis of samples	P
Nutrient uptake potential	μg radioisotope g^{-1} hr^{-1}; pg radioisotope mg^{-1} min^{-1}	Soil labeling with radioisotopes such as ^{32}P, ^{33}P, and ^{42}K followed by monitoring of isotope appearance in plant tissue	F
N transformations in the ecosystem (e.g., "tightness" or "openness" of nitrogen cycle)	δ ^{15}N (‰)	Ratio of stable isotopes ^{15}N to ^{14}N in plant material compared with standard (air) using mass spectrometer	P
Nitrogen in soil available to plants	NO_3^- + NH_4^+ − total nitrogen (μg g^{-1}; %)	Soil sampling followed by chemical analysis; ion exchange resin procedure	P

continues

Continued

Parameter measured	Usual units	Techniques	Emphasis
Mineralization rate	μg N g substrate dw^{-1} day^{-1}	Tracing of ^{15}N released from control samples and from samples treated with chloroform fumigation to kill microbial cells; ion exchange resin procedures	F
Nutrient retention or loss	mg nutrient L^{-1} leachate; μg nutrient g^{-1} soil or plant sample; ^{15}N recovered (mg; %)	Measured as nutrient concentration in leachates (the higher the concentration in leachates, the lower the retention by the ecosystem) or as ^{15}N recovered in plants and soil organic matter	P
Water balance			
Water content	mg H_2O g dw^{-1} or % H_2O dw	Harvest followed by measurement of fresh and dry weights	P
Water use efficiency (C assimilated per unit water transpired, integrated over a long period, e.g., a whole season or longer)	μmol CO_2 mmol H_2O^{-1}; δ ^{13}C ‰	Ratio of stable isotopes ^{13}C to ^{12}C in plant material compared with standard (PeeDee Belemnite)	P
Evapotranspiration	mg H_2O m^{-2} min^{-1}	H_2O exchange measurements	F
Water potential (the more negative the water potential, the more negative the balance between absorption and transpiration)	MPa	Pressure chamber measurements (pressure needed to expel a drop of sap out of the xylem)	F

[a] More details can be found in Pearcy *et al.* (1989).

to only a factor of two to four at the primary production level between different forest types.

Therefore, for integral ecosystem responses, methods which focus on changes of pools (P measurements in Table I) are usually more meaningful and/or convenient. This point is illustrated by Fig. 2 and by some examples of the use of different methods for assessing biomass production at different scales (see Section II,B). An example is provided in Fig. 3. A long-term study of the responses of tundra-dominant plants to different treatments showed that processes that are readily integrated at annual time steps (shoot growth, mortality, and allocation) were more useful than instantaneous physiological measurements in predicting decadal vegetation changes. Strong treatment effects on photosynthetic rate (Fig. 3a) and nutrient uptake were poorly related to longer term changes in production and nutrient concentration (Figs. 3b and 3c). This was probably due to the operation of buffering mechanisms (e.g., allocation, nutrient relations, altered phenology, interspecific interactions, and positive and negative feedback mechanisms) which compensate for immediate physiological responses to the environment.

B. Biomass Production

Net primary production (NPP) is strictly defined as the difference between the energy fixed by autotrophs and

their respiration, and it is most commonly equated to increments in biomass per unit of land surface and time. Because the increment in biomass over a given time depends on the rate at which new biomass is produced and also on the initial amount of carbon-assimilating photosynthetic tissue, stands with a large standing biomass often show higher NPP than stands with lower biomass. Therefore, another useful concept is that of relative productivity rate, or the time needed by a vegetation stand to produce its standing biomass. For example, the estimated relative productivity rate for a dry tropical forest can be many years, whereas in an annual grassland it is less than 1 year.

The fate of assimilated carbon—that is, whether it is allocated to increase the pools of aboveground or belowground biomass, root exudates, litter, soil organic matter, grazers, symbionts, or parasites—varies strongly between ecosystems, depending on prevailing climatic conditions, disturbance regimes, and allocation patterns of dominant plant functional types (Fig. 4).

At the regional scale, net primary production can be largely accounted for by climatic factors. For example, precipitation, potential evapotranspiration, and radiation are enough to account for the aboveground net primary production (ANPP) of North American forests, deserts, and grasslands. In regions of the United States with up to 1400 mm of annual rainfall, annual precipita-

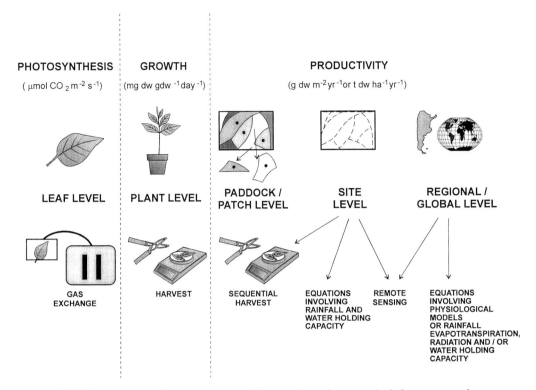

FIGURE 2   Carbon assimilation processes at different scales and some methods for measuring them.

tion is enough to account for 90% of the variability in ANPP of grasslands (Fig. 5a). At higher precipitation, ANPP depends more on other factors, and equations based on annual rainfall lose part of their predictive power. At the site level, variability in production seems to be accounted for by annual precipitation and soil water-holding capacity (whc; Fig. 5b). Soil whc can have a positive or negative effect depending on the precipitation value. In dry regions, major losses of soil water occur via bare soil evaporation. However, where sandy soils occur, bare soil evaporation is lower than in loamy soils because water penetrates deeper into the soil. For the same reason, surface runoff is also lower in sandy soils than in loamy soils. In more humid regions, substantial water losses occur via deep percolation, which is reduced in soils with high whc. This is known as the inverse texture hypothesis, proposed by I. Noy-Meir in 1973.

At finer scales of analysis (e.g., paddocks and vegetation patches), more variables are needed to account for ANPP. Species composition and land-use regime become important factors, although drivers at a coarser scale are still in operation and constrain responses (e.g., irrespective of management or species composition, annual precipitation will set an upper boundary to ANPP). For instance, in Argentine montane and pampean natu-

ral grasslands, ANPP decreased between 50% and more than 300% when subjected to moderate to heavy grazing. Species composition is crucial at this level; for example, ANPP tends to be higher in legume-dominated pastures than in grass-dominated ones because legume growth is much less limited by soil nitrogen availability due to their capacity for symbiotic nitrogen fixing.

Biomass production from local to global scales can also be estimated by remote sensing. The normalized difference vegetation index, derived from the reflectance in the red and infrared bands measured by the metereological satellites NOAA/AVHRR (National Oceanic and Atmospheric Administration/Advanced Very High Resolution Radiometer), shows strong correlation with vegetation processes such as photosynthesis and primary productivity and has been widely used to assess primary production (Fig. 6).

C. Trophic Transfer

Trophic transfer can be defined as the amount of biomass and/or energy which is transferred from the primary producers to the herbivore-based food chain rather than directly from plants to detritivores. In this article, only the transfer from plants to herbivores will be analyzed. Two concepts are relevant: consumption,

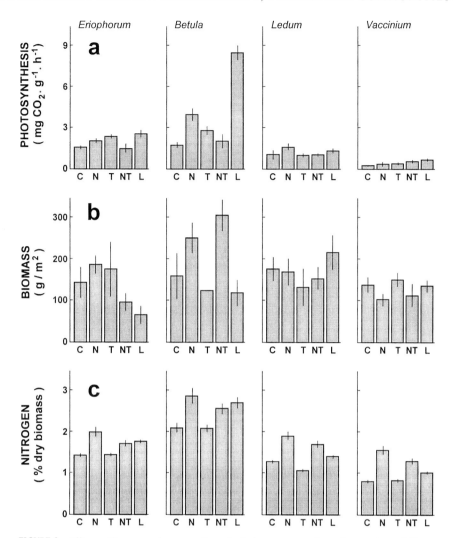

FIGURE 3 Effects of 3-year environmental manipulations on tussock tundra species at different scales: (a) photosynthetic rate at 20°C and in full sun, (b) total peak-season biomass (excluding roots), and (c) leaf nitrogen content. Treatments are control (C), nutrient addition (N), temperature increase (T), nutrient × temperature (NT), and light attenuation (L) (reproduced with permission from Chapin and Shaver, 1996).

or the amount of plant biomass consumed by herbivores, and net secondary productivity (NSP), or the amount of biomass/energy at the herbivore level which is available to carnivores. Both depend on the amount of available ANPP, the quality of plant biomass, the kind of metabolism of the herbivores, and the linkages between herbivores and plants.

1. Effects of Nutrient Availability, Plant Production, and Plant Quality on Herbivore Performance

Plant nutrient quality can directly affect animal populations. Plants growing in low-nutrient sites tend to have elevated concentrations of carbon-based secondary compounds that deter consumption by making a higher proportion of plant biomass unavailable. In addition, low foliar concentrations of nutrients can result in either decreased or increased consumption by herbivores. In the second case, herbivores (most commonly insects) consume higher quantities of biomass in order to meet nutritional requirements, with or without reduced fitness as the final result. This effect, sometimes called the "nutrient dilution effect", is particularly true in the case of generalist herbivores, whose opportunities for co-evolutionary adjustment to the chemistry of a particular plant species are slight.

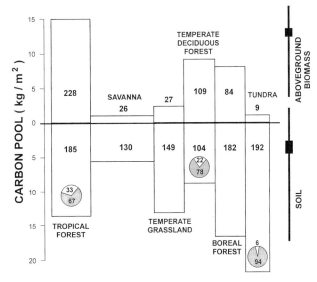

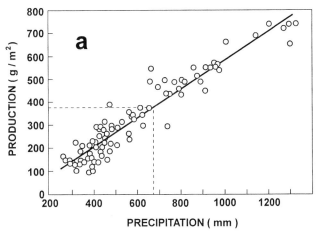

FIGURE 4 Carbon pools in major ecosystem types. Soil stocks include biomass, soil organic mass, and litter. Pie diagrams indicate percentage of soil carbon in belowground biomass (gray) and in soil organic mass (white) [modified from Anderson (1991) Physiological Plant Pathology, and Larcher, Fig. 2.81 (1995) © Springer-Verlag, with permission.].

Secondary chemicals in foliage seem to be more important as regulators of the types of herbivores prevailing in different ecosystems and their dietary habits, than as regulators of consumption at the ecosystem level. There is evidence that herbivore biomass, consumption, and productivity are closely correlated with plant productivity across a wide range of ecosystems, such as deserts, savannas, agricultural grasslands, tropical forests, and salt marshes (Fig. 7). Respiratory costs per unit production at the consumer trophic level are higher for homeotherms (such as ungulates) than for heterotherms (such as insects). In forests, in which most herbivores are heterotherms, most production is allocated to wood, whereas in grasslands, in which homeotherms predominate, a much higher production is allocated to green tissue. This explains why net foliage productivity predicts consumption considerably better than net aboveground biomass.

Herbivore biomass and consumption increase as a power of net aboveground primary production, whereas net secondary production increases linearly (Fig. 7). This indicates that highly productive ecosystems sustain a larger level of herbivory per unit of net aboveground primary production than unproductive systems. This larger level of herbivory, however, is accompanied by a lower secondary production per unit of consumption. Because the foliage unconsumed by herbivores will flow into decomposer food webs, the

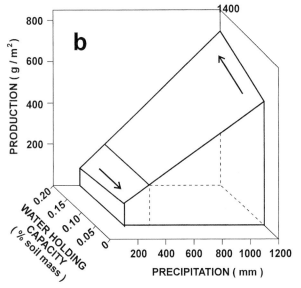

FIGURE 5 Regional- and site-level controls over aboveground net primary production (ANPP) of U.S. grasslands. (a) Annual precipitation (APPT) is the main factor at the regional level, with ANPP = 0.6 (APPT − 56) (r^2 = 0.90), where 0.6 represents the average water use efficiency of the community, and 56 mm/year is the "ineffective precipitation" (precipitation volume which is not enough to result in production). Addition of temperature and potential evapotranspiration did not improve the model. (b) Annual precipitation and soil water-holding capacity (whc) are the main factors at the site level, with ANPP = 32 + 0.45 APPT − 352 whc + 0.95 whc APPT; r^2 = 0.67) (reproduced with permission from Sala et al., 1988).

relationship shown in Fig. 7 between consumption by herbivores and net aboveground primary production indicates that the relative importance of the direct flow to detritus decreases as ecosystem productivity increases. However, the relative amount of litter which is actually consumed is small, even in the most productive systems.

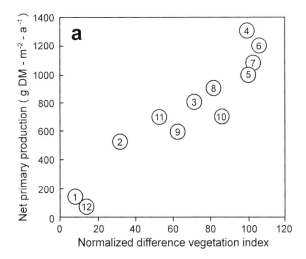

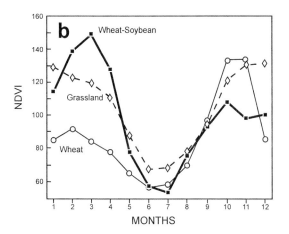

FIGURE 6 Use of normalized difference vegetation index (NDVI) in the estimation of annual and seasonal patterns of primary production. (a) Relationship between the net primary production and NDVI of different vegetation types: 1, tundra; 2, tundra–taiga ecotone; 3, boreal coniferous belt; 4, humid temperate coniferous forest; 5, transition from coniferous to deciduous broad-leaved forests; 6, deciduous forests; 7, oak–pine mixed forests; 8, pine forests; 9, grassland; 10, agricultural land; 11, bushland; 12, desert (reproduced with permission from Physiological Plant Pathology, Larcher, Fig. 2.77, 1995, © Springer-Verlag). (b) Seasonal changes in NDVI for a native grassland, a wheat field, and double-cropping wheat–soybean in the Argentine Pampas (reproduced with permission from Sala and Paruelo, 1997).

2. Effects of Herbivores on Nutrient Cycling and Primary Productivity

Animal activity can affect nutrient cycling directly and indirectly. The amount of ANPP consumed by herbivores can vary from less than 10% in tropical rain forests to more than 50% in meadows (Larcher, 1995). Herbivores can short-circuit the decomposer pathway since urine and feces are much easier to decompose than

plant litter. They can also redistribute nutrients and create patches (ungulates, rodents, and ants), and they can promote secondary compound production in plants, which may further deter grazers and may retard decomposition. Mammals, ants, and termites can play important roles in spatial distribution of nutrients at different scales, with impacts on the dynamics of the whole food web.

In many systems primary productivity increases with light grazing, then decreases, and finally decreases more or less sharply as grazing becomes severe. This was at the heart of the herbivore optimization curve hypothesis proposed by S. J. McNaughton in the 1970s. Herbivore optimization models are still under debate, and to determine whether total (aboveground and belowground) biomass actually increases under herbivory is operationally very difficult. However, supporting evidence has been found in a wide range of terrestrial ecosystems, including not only grasslands and wild and domestic ungulates but also forbs and geese, trees and mosses, and crops and birds. It is generally accepted that the highest production rates of plants occur when grazing occurs but it is not too high.

3. Bottom-up and Top-down Controls of Food Chains

Bottom-up or resource control in food chains emphasizes the importance of resource availability for primary producers and the subsequent energy and nutrient flow through a series of trophic levels. Organisms at each trophic level are food limited. The rationale of the opposite, or top-down, view is that organisms at the top of the food chain are food limited but those at lower trophic levels are alternatively predator and food limited. Both mechanisms are recognized to occur in nature, with their importance varying from place to place. S. D. Fretwell and L. Oksanen proposed that the importance of top-down control increases with primary productivity. This idea has recently been empirically demonstrated for invertebrate herbivory and predation on limestone grasslands through a series of experiments summarized in Fig. 8 involving pesticide treatments, transplant of turves, and the use of bioassays (lettuce discs and blowfly maggots in order to evaluate the degree of herbivory and predation, respectively).

D. Nutrient Cycling

1. Nutrient Capture, Retention, and Release by Plants

A key, albeit controversial, concept in relation to nutrient cycling is that of resource use efficiency, or the

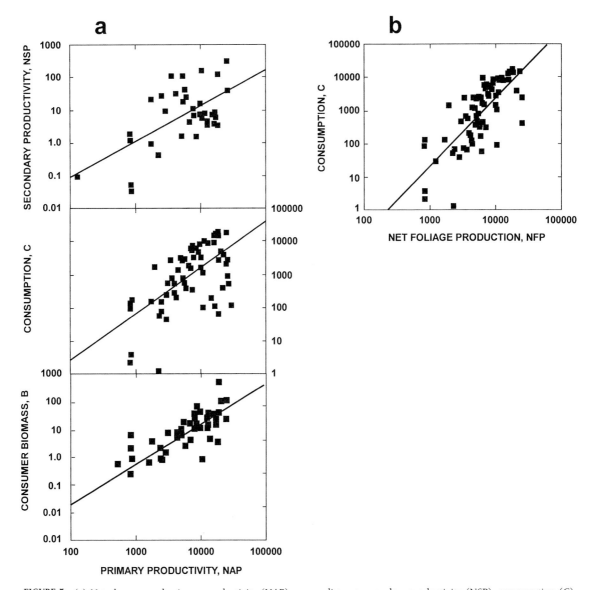

FIGURE 7   (a) Net aboveground primary productivity (NAP) can predict net secondary productivity (NSP), consumption (C), and consumer biomass (B). Log NSP = 1.10(log NAP) − 327, r^2 = 0.364; log C = 1.35(log NAP) − 2.32, r^2 = 0.367; log B = 1.52(log NAP) − 4.79, r^2 = 0.367. (b) Net foliage primary productivity (NFP) is a better predictor of consumption. Log C = 2.04(log NFP) − 4.80, r^2 = 0.594. Units are kJ m^{-2} year^{-1} except for biomass, which is kJ m^{-2}. All relationships are significant at $p < 0.0001$ (modified with permission from *Nature*, McNaughton *et al.*, Copyright 1989 Macmillan Magazines Limited).

relationship between a limiting resource (light, nitrogen, and water) and a biological process (photosynthesis and primary production). Nutrient use efficiency was originally defined in the early 1980s by P. Vitousek as the total NPP (above- plus belowground) per unit nutrient absorbed annually. In practice, it has been usually measured as the ratio of dry mass to nutrient content in litter (Table I), which is a good index of the nutrient economy in a stand as a whole and is based on information reasonably easy to obtain. Recently, a distinction has been proposed by J. Pastor and S. D. Bridgham between nutrient use efficiency (production per unit nutrient uptake) and nutrient response efficiency (production per unit nutrient available).

2. Major Controls over Nutrient Cycling

The main factors underlying variations in nutrient cycling in different ecosystems are climatic factors, soil

a

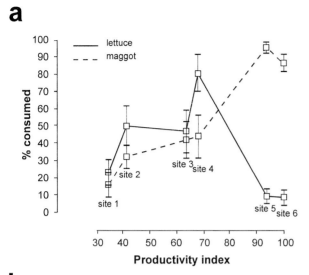

b

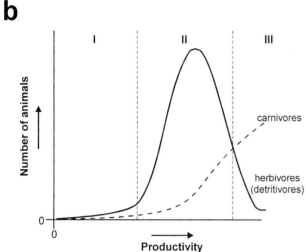

FIGURE 8 (a) Estimation of herbivore and carnivore activity, measured as a percentage of lettuce discs and maggots consumed, respectively, and (b) illustration of Fretwell–Oksanen theory of interaction between trophic dynamics and primary productivity in British limestone grasslands with low (I), intermediate (II), and high (III) productivity. At very low productivity, the vegetation did not experience a detectable amount of herbivory. Vegetation of intermediate productivity supported a high level of herbivory and responded strongly to the removal of herbivores. Productivity in this situation is believed to be insufficient to sustain a high intensity of "top-down" control. In the case of highly productive vegetation, carnivory was strong and the intensity of the top-down control by carnivores on plant mass removal by herbivores was maximum (reproduced with permission from Fraser and Grime, 1997).

fertility, time from major disturbance events, and species composition. Nitrogen content is a strong control of productivity, carbon exchange, and composition on many ecosystems. In some cases, other macronutrients

also exert strong controls (e.g., phosphorus and calcium in some tropical rain forests and grasslands, respectively).

3. Open vs Tight Nutrient Cycles

A concept strongly linked with nutrient use is the degree of "tightness" or "openness" of nutrient cycling in different ecosystems. This refers to the relative importance of within-system nutrient cycling vs external inputs and outputs. Tropical forests are considered systems with tighter nutrient cycling than temperate forests because within-system nutrient recycling is more important than influx into or effluxes out of the system.

4. The Effects of Rainfall

Concentrations of available nutrients in the soil are relatively (although not absolutely) high in semiarid sites and decrease with increasing precipitation. Soil organic carbon tends to follow the opposite pattern, increasing with increasing rainfall. While total soil nutrient content may also increase, carbon : nutrient ratios in soil increase with higher mean annual precipitation. This suggests that, as rainfall increases, rate of carbon accumulation in soils is higher than that of total nutrient accumulation due to differences in mineralization.

Changes in nutrient cycling "tightness" with climatic factors are well illustrated by an analysis of soil and foliar nutrients in a rainfall gradient in Hawaii (Fig. 9). As rainfall increases with altitude, there is a shift from relatively high nutrient availability to relatively high carbon gain by producers, indicated by a decrease of leaf mass and leaf nitrogen concentration and an increase in lignin concentration with altitude (Fig. 9b). A progressively depleted ^{15}N signature in both soils and vegetation in the wetter sites (Fig. 9a) suggests that N cycling shifts from more open at the drier sites (larger turnover) to tighter (smaller losses) as precipitation increases.

5. The Effects of Soil Type, Land Use, and Vegetation Structure

A comparison among different types of Amazonian rain forest ecosystem, differing in soil properties and topographic positions (Table II), illustrates how patterns of nutrient allocation depend on soil chemical properties and flooding regimes. In turn, soil properties influence nutrient supply, and flooding regimes affect nutrient uptake ability. Mixed and *guaco* forests, located in higher topographical positions, show relatively high nitrogen contents in both soil and vegetation. In the tall *caatinga* forest, the proportion of total nitrogen in living biomass is much higher than in the mixed forest

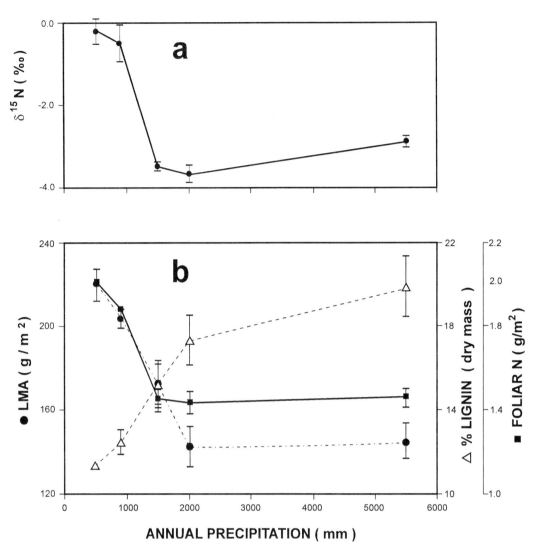

FIGURE 9 Influence of precipitation on nutrient cycling. Nitrogen cycle becomes increasingly closed with increased precipitation in Hawaii. (a) δ ^{15}N average values of leaves of seven species. (b) Leaf mass per area (LMA), lignin concentration, and foliar N concentration of the dominant *Metrosideros polymorpha* Gaud (modified with permission from Austin and Vitousek, 1998).

and *guaco* forests, and decomposition is slower, associated with waterlogging and nitrogen limitation. Nitrogen circulates in larger amounts in mixed and *guaco* forest, which show more "open" nitrogen cycling than tall *caatinga* forest.

Changes in vegetation structure due to land use can lead to different nutrient relations in sites under similar climatic and original soil conditions. A comparison between a tropical dry deciduous forest and savannas derived from the same forest and now maintained by grazing and fire in northern India (Table III) shows that nutrient cycling is faster in the vegetation, litter, and soil of the savanna. The forest shows a higher nutrient use efficiency: Whereas the biomass and nutrient content are much higher, the annual net production and nutrient uptake are similar to those of the savanna. In the savanna there is smaller permanent nutrient storage, and significant nutrient leakage from the system, reflected in lower soil content, indicating more open nutrient cycling.

E. Decomposition: From Nutrient Organic to Inorganic Forms Available to Plants

The process of decomposition, or disintegration of plant and animal residues by the soil detrital food web, is a

TABLE II

Patterns in Nutrient Flows and Stocks in Three Different Amazonian Forests under Similar Macroclimate[a]

	Forest		
	Mixed	*Guaco*	Tall *caatinga*
Soil	Concretional oxisol	Yellow ultisols	Tropaquods
Geomorphological features	Rolling hills, laterite cap, never flooded	Clayey hillsides, never flooded, sometimes water saturated	Sandy valley fills, flood-prone
Total biomass (t ha^{-1})	310	465	400
Leaf fall (t ha^{-1} year^{-1})	6.5	7	5.2
Disappearance constant (k; year^{-1})	3.68–0.34	—	0.93–0.62
Biomass : nitrogen ratio (g : g)	~110	~120	~260
Litter N content (kg ha^{-1})	137	69.5	132
Soil N content (at 10 cm depth) (kg ha^{-1})	1474	2490	716

[a] Data from Medina and Cuevas (1989).

key step in nutrient cycling since it makes nutrients already present in the system available for new plant growth. Some plants can incorporate organic forms of nitrogen with or without the intervention of mycorrhizal symbionts. However, these cases seem mostly restricted to very nutrient-poor systems. In most ecosystems, the majority of primary production is not consumed by herbivores but passes directly to detritus. A smaller fraction of primary production is incorporated into herbivores and carnivores which becomes detritus when these organisms die. The ultimate end product of organic matter breakdown is inorganic forms of carbon (CO_2) and nutrients (nitrates, ammonium, and phosphates). Decomposition depends on soil environment (water potential, temperature, and aeration), microsite characteristics (slope, texture, drainage, aspect, and cover type), substrate quality, and composition of decomposer community (size and specific composition of animal and microbe assemblages and synergistic or antagonistic relationships among them).

The control of decomposition by macroclimatic parameters is very strong. Whereas mean NPP increases by a factor of approximately 20 from tundra to tropical rain forests, mean residence times of dead organic mat-

TABLE III

Nutrient Cycling in Two Tropical Dry Deciduous Vegetation Types under the Same Climatic and Soil Conditions, With Different Land-Use History in Northern India[a]

	Dry deciduous forest			Savanna		
Production (t ha^{-1} year^{-1})	15			11		
Total biomass (t ha^{-1})	95			67		
	N	P	K	N	P	K
Nutrient uptake per unit energy captured (mg 1000 kcal^{-1})	3000	225	1720	3290	320	2740
Vegetation nutrient content (kg ha^{-1})	680	53	451	87	8	74
Litter nutrient content (kg ha^{-1})	37	3	11	14	1	11
Soil nutrient content (0–30 cm depth) (kg ha^{-1})	2906	126	377	2386	134	160
Total litter fall (kg ha^{-1} year^{-1})	80	6	38	82	8	77
Total nutrient release (leaf + root decomposition; kg ha^{-1} year^{-1})	104	8	58	124	11	107
Nutrient retention in vegetation (kg ha^{-1} year^{-1})	62	4	38	20	2	17

[a] Data from Singh (1989).

TABLE IV

Relationships between Leaf Litter Quality (C : N) and Decomposition Rate (% dry mass loss) in a Wide Variety of Plant Species[a]

Reference	Plant material	r	p	n
1	British shrubs and trees	−0.780	<0.01	12
2	Mediterranean shrubs and trees	−0.720	<0.05	8
3	South American temperate and subtropical graminoids, forbs, succulents, shrubs, and trees	−0.520	<0.001	52

[a] Reference: 1, Cornelissen (1996); 2, Gallardo and Merino (1993); 3. Pérez-Harguindeguy et al. (1999). r, Spearman's correlation coefficient; p, significance level; n, number of species involved.

ter decrease by a factor of approximately 200 primarily because soil temperatures limit decomposition more than air temperatures limit production.

At a more local scale, litter quality plays an extremely important role in determining decomposition rates. The relative velocity of decomposition of litter from different plant species tends to remain constant under different environmental conditions and to be strongly correlated with the palatability of living leaves. This suggests that the same compounds which determine palatability aboveground control litter decomposition by soil microbiota. The most widely used indexes to describe litter quality are the carbon : nitrogen and lignin : nitrogen ratios, with higher ratios being associated with lower decomposition rates (Table IV).

1. Mineralization and Immobilization of Nutrients in the Soil

The release of organically bound nutrients into the inorganic form available to most plants is called mineralization. Because decomposer soil organisms require these nutrients to be incorporated into their bodies, decomposition can only proceed if there is adequate nutrient supply to the decomposers. Otherwise, they can act as a net sink for available nutrients in the system rather than as a net source, and decomposition is then said to be nutrient limited. The use of nutrients by decomposers of nutrient-deficient substrates is termed "immobilization." Mineralization and immobilization tend to occur simultaneously in most systems; therefore, it is important to determine which of them predominates (whether net mineralization or immobilization is occurring in a system at a given time). For example, the decrease in soil fertility immediately after the addition of cereal straw to the soil is due to immobilization, with carbon : nitrogen ratios of 100 to 150 : 1. In arable soils, a carbon : nitrogen ratio of 20 : 1 is considered the threshold between net immobilization and mineralization.

2. Litter Decompositon and Composition of Detrital Food Webs

The composition of the decomposer communities, which include macrofauna (earthworms and arthropods that carry out initial comminution, mixing, and dispersion of litter and microbial propagules), mesofauna (springtails, mites, and enchytraeid worms), and microbiota (fungi, bacteria, protozoa, nematodes, and actinomycetes), also determines the local decomposition rate. Grazing on the bacteria, actinomycetes, and fungi is important (often indispensable) for net mineralization to occur. Several experiments manipulating densities of bacteria, fungi, and their predators (e.g., protozoa and nematodes) and macrofauna (e.g., millipedes) have shown that the composition of the detrital food web is a strong factor determining the balance between net mineralization and immobilization. Belowground grazing on bacteria and fungi and aboveground grazing on plants have both been shown to increase microbial activity and mineralization.

F. Water Dynamics

Water flow through the plant compartment of the soil–plant–atmosphere continuum is regulated at the plant–air interface, where the transition of liquid water to vapor occurs and the steepest gradient of water potential exists. The shoots are exposed to the low water potential of the air and a flow of water through the plant is set in motion. In this way, and depending on xylem structure and cavitation, the plant component of the system bridges the steep water potential gradient between soil and air.

Vapor loss from land and surfaces is governed by leaf area, by stomatal and aerodynamic conductances of plant canopies, and by the contribution of evaporation from soils, and it affects numerous terrestrial processes ranging from the biogeochemical cycling of elements to the development of regional climate. The aerodynamic conditions above and in the canopy, determined by the density and the architecture of a vegetation stand, strongly influence transpiration. These can also affect climate, and therefore water availability, at the regional level. For example, deforestation, by reducing roughness and increasing albedo, tends to result in higher temperatures and decreased precipitation.

Most of the terrestrial evaporative water loss passes through the stomata pores of plant leaves. Given certain meteorological conditions, the strongest biotic determinants of transpiration from vegetation are stomatal opening (usually expressed as γ = diffusive conductance of leaves for water vapor) and the amount of leaf area per land area (leaf area index; LAI). Maximum canopy conductance increases with LAI up to a point. Below a LAI of about 3, wet soil evaporation contributes significantly to total ecosystem evaporation. Examples of vegetation types with LAI below 3 are most deserts and tundra ecosystems, some temperate grasslands, and some semiarid scrubs dominated by members of the genera *Eucalyptus* or *Acacia*.

Maximum diffusive conductance of canopy plus soil and maximum photosynthetic rate tend to be correlated, with the correlation being better for herbaceous plants; woody plants tend to assimilate less carbon per unit of transpirational water loss. Evapotranspiration and carbon assimilation are so coupled that water use efficiency at the ecosystem level (how much biomass can be built up per unit of water transpired) can be approximately estimated as NPP/precipitation.

Because of the link between water loss and CO_2 absorption at the leaf level, and because it is a very important factor in soil processes, water availability can also indirectly control ecosystems by affecting cycling of carbon and nitrogen (see Sections II,D,3 and II,D,4). There is evidence suggesting that the nutritional status of a canopy type determines the capacity for exchanging gases with the atmosphere. Maximum evaporation conductances of vegetation plus soil and carbon assimilation rates have been shown to be determined by plant nutrition across very different ecosystem types (Fig. 10). However, these patterns apply for maximum rates, which do not always reflect seasonal carbon or water balances. For example, coniferous forests are less sensitive than grasslands to soil drought. They appear to close their stomata at lower soil water content than do herbaceous plants. In addition, woody vegetations are generally deeper rooted and thus access a larger soil and water volume than do herbaceous species. Apparently, the main variable that determines seasonal carbon balance is the total length of the growing season, which usually depends on the functional rooting depth of the plants. Figure 10 suggests

that the carbon and water balance will be strongly dependent on maximum stomatal conductance and its relation to nutrition as well as the length of the growing season, which is dependent on rooting depth. The constraints during the growing season, rather than those during the unfavorable season, seem to determine the success of different plant life-forms in different regions of the world. In 1982, E.-D. Schulze proposed that optimization at the whole-plant level of leaf longevity and carbon allocation into photosynthetic, nonphotosynthetic, and above- and below-ground tissue seems to underlie the dominance of different life-forms in different climates.

III. ECOSYSTEM FUNCTIONING UNDER DISTURBANCE: RESISTANCE AND RESILIENCE

Ecosystem stability is often divided into two components: resistance and resilience. Ecosystem resistance is the ability of a system to avoid change—the capacity to stay in the same state in the face of perturbation (e.g., fire, unusual frost or drought, plowing, eutrophication, and pollutant input). Ecosystem resilience is the rate at which a system returns to its former state after being displaced from it by a perturbation. Not all aspects of ecosystem functioning are equally resistant or resilient. The most commonly measured ones (which are not necessarily the most important or sensitive ones) are species composition and biomass. Usually, a standard or ad hoc index is constructed that relates vegetation structure or composition before the perturbation to that after the perturbation. A higher dissimilarity means the ecosystem has a lower resistance. The longer amount of time needed to achieve maximum similarity between the predisturbance and postdisturbance situation, the lower the resilience. Some authors have used indicators other than community composition, such as nutrient loss rate, to estimate stability; the higher the relative nutrient loss rate following disturbance, the lower the resistance, and the longer the time to restore "normal" nutrient loss rate, the lower the resilience.

Evidence has accumulated for the idea that both components of ecosystem stability are primarily determined by key traits of the dominant plant species. Highly productive communities, dominated by fast-growing plants, tend to have high resilience and low resistance, with the opposite being true for communities dominated by slow-growing plants. Productivity and seed production (especially persistent seeds) favor resilience, whereas preferential allocation to storage and

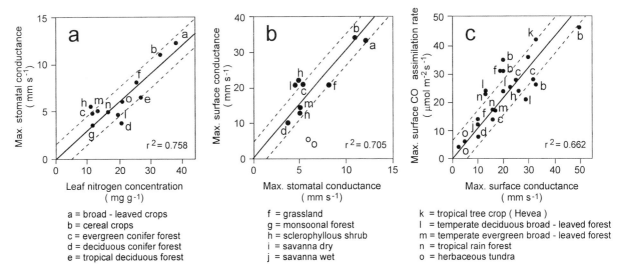

a = broad - leaved crops
b = cereal crops
c = evergreen conifer forest
d = deciduous conifer forest
e = tropical deciduous forest

f = grassland
g = monsoonal forest
h = sclerophyllous shrub
i = savanna dry
j = savanna wet

k = tropical tree crop (Hevea)
l = temperate deciduous broad - leaved forest
m = temperate evergreen broad - leaved forest
n = tropical rain forest
o = herbaceous tundra

FIGURE 10   Relationships between water dynamics, carbon assimilation, and plant nutrition in different vegetation ecosystems. (a) Maximum stomatal conductance vs nitrogen concentration (solid regression line through the origin: $y = 0.3012\ x$; dashed lines, SE of $y = 1.358$). (b) Maximum surface conductance vs maximum stomatal conductance ($y = 2.996\ x$; SE of $y = 4.495$). (c) Maximum surface CO_2 assimilation rate vs maximum surface conductance ($y = 1.048\ x$; SE of $y = 6.445$) (reproduced with permission from the *Annual Review of Ecology and Systematics*, Vol. 25, © 1994 by Annual Reviews, www.AnnualReviews.org).

defense favors resistance. These ideas were formalized in the early 1980s on the basis of studies of old-field successional communities (Fig. 11). Recently, new experimental support has been provided for the idea that vegetation resistance and resilience in the face of extreme events are a function of the nutrient stress tolerance of the component species. Figure 12 clearly illustrates that resistance to extreme events increases, and resilience decreases, with increasing nutrient stress tolerance in herbaceous communities.

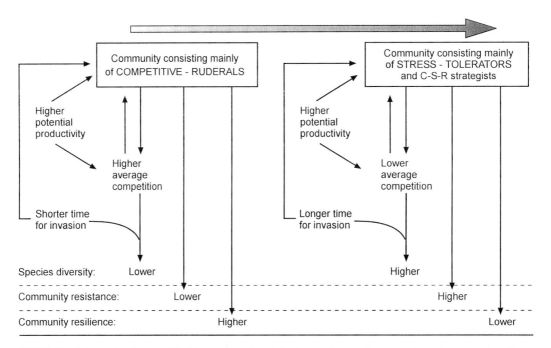

FIGURE 11   Relationship between life history strategies of dominant plants and community resistance and resilience (reproduced with permission from Kluwer Academic Publishers, Lepš *et al.*, Fig. 5 © 1982 with kind permission from Kluwer Academic Publishers.).

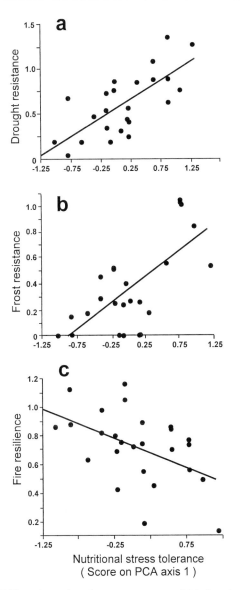

FIGURE 12 Relationships between estimates of (a) drought resistance, (b) frost resistance, (c) fire resilience and nutrient stress tolerance (expressed as scores on a PCA axis) for 26 British herbaceous species. Experimental treatments of fire, drought, and frosts were applied to turves of natural herbaceous vegetation. The resistance of each species was measured as the ratio of the biomass of that species in the treated turves to its biomass in the control turves immediately after the application of the treatments. The resilience (capacity for recovery) of each species was measured as the ratio of the biomass of that species in the treated turves to its biomass in the control turves 8 weeks and 1 year after applying the treatments (modified from MacGillivray et al., 1995, with permission of Blackwell Science Ltd.).

This immobilization can substantially control the nutrient losses as long as a sufficient amount of plant residue is left on the site to sustain the microorganisms that immobilize the nutrients in their biomass. Therefore, soil organic mass is a major determinant of ecosystem resistance. On the other hand, systems with tight nutrient cycles have increased return time to equilibrium (lower resilience). When substantial amounts of biomass are removed from these systems, causing high nutrient loss, recovery may be very slow because there is little throughflow of nutrients coming from outside the system compared with the nutrient lost. As a general rule, higher nutrient mean residence time in an ecosystem increases its resistance and decreases its resilience.

IV. ECOSYSTEM FUNCTIONING AND FUNCTIONAL DIVERSITY

A. Taxonomic and Functional Diversity

The most common way of assessing biodiversity is the measurement of the number and relative abundance of species per area. Taxonomic diversity per se is an important parameter for conservation, but it provides little information on the magnitude and rate of ecosystem processes. On the other hand, as discussed in Section II,A, measurements of pools and flow rates at the whole ecosystem level provide little information about a system's conservation value or about its likely response in the face of a perturbation. An intermediate approach is the one based on plant functional types. Functional types are sets of plants sharing similar responses to environmental conditions and similar effects on major ecosystem processes. This approach is as old as ecology itself, but it has gained renewed interest in the past few years because it bridges the gap between individual species and whole-ecosystem functioning, explains particular values of pools and flow rates, and improves predictions of how ecosystems can be modified by the introduction of new abiotic (e.g., fertilization) or biotic factors (e.g., invasion by alien species). The practical value of distinguishing discrete "types" is obvious. However, it is important to bear in mind that many real plants represent transitions between, rather than typical examples of, different functional types.

B. Ecosystem Functioning and Biodiversity

Although the role of species richness per se remains controversial, and the evidence is sparse, the impor-

The size and the turnover rate of the detrital compartment also have implications for ecosystem stability. Ecosystems with high carbon : nutrient ratios in the soil are effective at immobilizing large amounts of nutrients.

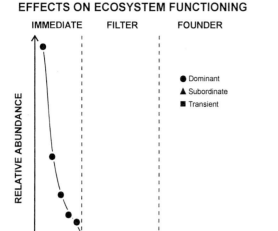

FIGURE 13 Proposed ecosystem effects of species in different positions along an idealized dominance–diversity curve. Transients are those species unable to regenerate and persist *in situ*; their sources are soil seed bank and seed rain from the surrounding landscape. Immediate effect, control over major processes of resource dynamics; filter effect, positive or negative influence on the regeneration of dominants following major perturbations; founder effect, reservoir of potential colonizing dominant and subordinate species in the event of ecosystem reassembly (adapted from Grime, 1998 with permission from Blackwell Science Ltd.).

tance of the role of functional types, particularly the dominants, is incontestable (Fig. 1). Until the early 1980s, studies on the relationship between diversity and ecosystem functioning usually emphasized the impacts of ecosystem processes on biodiversity. Recently, the question has been reversed, and there is much more emphasis on how diversity influences ecosystem functioning. Biodiversity has moved from the *y* to the *x* axis. Dominant species or functional types, which account for most of the standing biomass and energy flow, tend to be the most important (and sometimes the only) components of biodiversity accounting for ecosystem functioning at the local level. This, and the fact that they tend to represent a small proportion of the local species assemblage, provides strong evidence that further studies on the role of biodiversity on ecosystem functioning should focus on them. However, the likely (but far less known) role of subordinates and even rare species in maintaining long-term ecosystem functioning, especially in the face of disturbances, has also been stressed (Fig. 13).

Another interesting aspect of the relationship between ecosystem function and biodiversity is the fact that, despite sharp turnover in species or functional type composition, some ecosystem-level parameters often remain constant along environmental gradients. Individual species or functional types follow an optimum response along any environmental gradient (often reflected in growth). However, if the assemblage is rich enough, those optima are not detectable when ecosystem-level parameters are analyzed along gradients because suboptimum species are replaced by other members of the assemblage as environmental conditions change.

C. Links between Plant Functional Traits and Ecosystem Functioning

The existence of suites of vegetative traits consistently associated across taxa and ecosystems (e.g., plants that "go for it" and plants that "sit and wait") has been repeatedly documented and is increasingly accepted. Central to the plant functional type approach is the idea that life history, allocation, and phenological, physiological, architectural, and reproductive traits of dominant plant species appear to be associated in a limited number of combinations and strongly determine vegetation structure and ecosystem functioning (Fig. 14). By being primary components of ecosystem-level fluxes of matter and energy, dominant plant functional types are indirect determinants of the biogeochemical cycles of carbon, water, and nutrients. Individual examples of strong links between plant traits and ecosystem processes have accumulated in the past few decades (Table V).

Major ecosystem processes appear more strongly and directly linked to recurrent suites of vegetative traits than to regeneration traits (Fig. 14). Plant regeneration traits appear to be under the control of selective forces different from those that operate on resource dynamics during the established phase. *In situ* plant regeneration by seed is an important aspect of resilience and also strongly influences migration across the landscape in the face of climatic changes. Seed production, germination, dispersion, and establishment are therefore key aspects in determining ecosystem functioning in the face of major changes of climate and land use, although their role has proved much more difficult to document than that of adult-phase traits.

D. Positive-Feedback Switches

The effect of the biotic component on ecosystem dynamics often takes the form of "switches" or positive-feedback processes in which members of a community

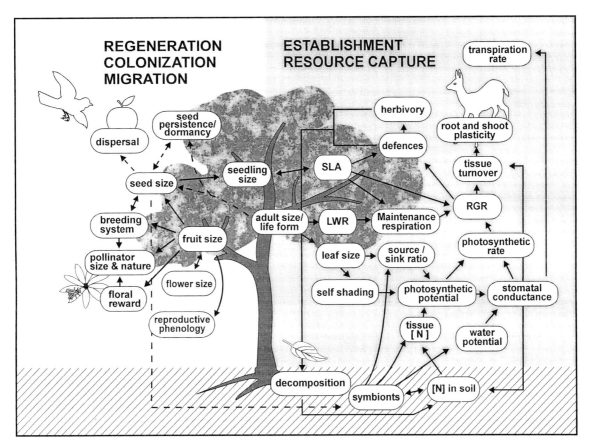

FIGURE 14   Some of the most common associations among vegetative traits (right), among regeneration traits (left), between vegetative and regenerative traits, and between traits and major ecosystem processes. SLA, specific leaf area; LWR, leaf weight ratio; RGR, relative growth rate.

modify their environment, making it more suitable for themselves. This term was coined in 1992 by J. B. Wilson and A. D. Q. Agnew, who provided an extensive list of examples, some of which are shown in Fig. 15.

V. MEASUREMENT, ANALYSIS, AND PREDICTION OF ECOSYSTEM FUNCTIONING: MAJOR PROTOCOLS AND OBSTACLES

A. Monitoring, Field Manipulations, and Synthesized Ecosystems: A Gradient of Questions and Methods

The understanding of how ecosystems function can be achieved by a whole gradient of protocols, from simply documenting what is happening in real systems to synthesizing model ecosystems from scratch. These ex-

treme approaches represent a trade-off between realism and precision and between the documentation of general patterns and that of the specific mechanisms underlying them. The monitoring of real ecosystems often needs to be performed by "soft" approaches. It is important at the initial heuristic stage and when testing whether results found in simpler experiments can be reasonably applied to the real world (synthesis stage). Experiments on synthesized ecosystems, on the other hand, are ideal for the stage of testing specific hypothesis on mechanisms which may account for the patterns observed (analytic stage). Field manipulations lay in the middle of this gradient (Fig. 16).

1. Monitoring Real Ecosystems: Approaches Based on Hard and Soft Traits

Processes in natural ecosystems can be recorded by applying the usual methods (hard approaches) of measuring pools and flows described previously (see Sec-

TABLE V

Examples of Individual Plant Traits which Strongly Influence Processes of the
Community/Ecosystems in which They Are Dominant[a]

Individual trait	Ecosystem/community process
Relative growth rate	Productivity, resilience, trophic transfer
Leaf turnover rate	Nutrient cycling, detritivore diversity and biomass
Nutrient content	Nutrient cycling, trophic transfer, detritivore diversity and biomass
Biomass	Trophic transfer, carbon sequestration, flammability
Life span	Resistance
Canopy structure	Aerodynamic conductance, interception, water relations, runoff, roughness/albedo, temperature buffering, soil stability, consumer biodiversity
Secondary growth	Carbon sequestration, trophic transfer, nutrient cycling
Ramification	Structural complexity, consumer biodiversity, resistance (particularly drought), temperature buffering
Root architecture	Water uptake, soil stability
Reserve organs	Resilience
Pollination mode	Expansion over landscape
Persistent seed bank	Resilience
Seed number	Expansion over landscape
Dispersal mode	Expansion over landscape
The presence of root symbionts	Diversity, nutrient cycling, carbon sequestration, rate of succession

[a] Modified from Díaz et al. (1999).

tions A,1 and A,2). They are sometimes termed hard approaches because they are quantitative and usually represent a direct measurement of a process. Alternatively, ecosystem functioning can be inferred from the presence and abundance of plant traits which are easily measured but at the same time have clear implications at the ecosystem level. These are called soft traits because they are usually indirect indicators of ecosystem processes. Some examples of soft traits are given in Table V. The soft approaches have the advantage that they need a minimum investment in financial and technological resources and can be utilized to characterize extensive areas or high numbers of systems in a short period of time. The result is usually a comparative estimation of ecosystem functioning (Fig. 17), which then needs to be calibrated against a hard approach. This is achieved by testing how well the soft traits correlate with hard traits, which have more direct and well-documented relationships with ecosystem processes (e.g., the soft trait leaf tensile strength correlates well with

the hard trait decomposition rate, and the soft traits seed mass and shape correlate well with the hard trait seed persistence in the soil; Fig. 18).

2. Experimental Manipulations Involving "Natural" vs Synthesized Ecosystems

The major disadvantage of field experiments is that the degree of control of independent and external variables is low compared to manipulating them. The amount of "noise" (unwanted variance) is usually high, and enough replicates in order to reduce it are usually unavailable or involve prohibitive costs. On the other hand, synthesized ecosystems, such as microcosms, never come close to the realism of experiments involving manipulations of real ecosystems, they are always a badly simplified version of nature, and they have severe size limitations (e.g., for ecosystems based on woody vegetation). Considering the realism–precision trade-off illustrated in Fig. 16, and taking into account

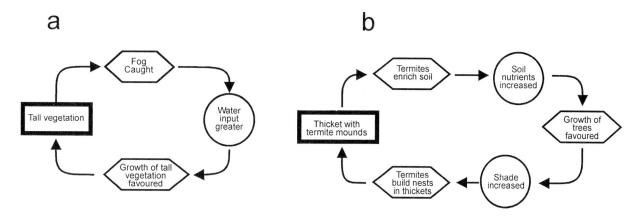

FIGURE 15   Switches of (a) fog precipitation on a hillside in the montane tropics, (b) termite mounds in tropical savanna, and (c) water (sediment entrapment)/salt with salt pans on a salt marsh (reproduced with permission from Wilson and Agnew, 1992).

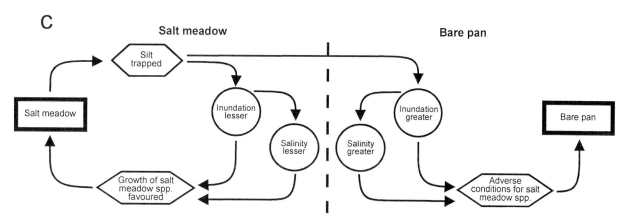

FIGURE 16   General characteristics of different approaches to the study of ecosystems.

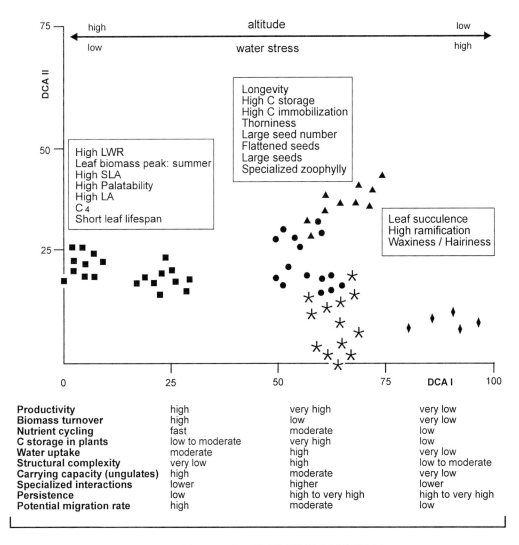

FIGURE 17   Characterization of different ecosystems along a climatic gradient by means of multivariate analysis (DCA) of "soft" traits. ■, montane grasslands; ✷, montane woodlands; ●, xerophytic woodlands and woodland–shrubland communities; ▲, open xerophytic shrublands; ◆, halophytic vegetation on poorly drained soils. Plant traits and expected community/ecosystem processes associated with different sectors of the ordination plane are displayed in boxes and at the bottom, respectively (reproduction of Fig. 3 in Díaz and Cabido, *J. Veg. Science* **8**, 463–474, 1997 with permission).

that some processes scale up poorly from single systems to complex ecosystems (see Section II,A,2), field experiments are indispensable. However, microcosms work has produced highly relevant insights into ecosystem functioning, including relationships between trophic levels, community roles of symbionts, feedbacks involved in soil fertility and climate manipulation, and some of the most controversial and inspiring experiments on the role of biodiversity in determining ecosystem functioning. Field manipulations, such as the application of treatments to real vegetation plots, or the

selective removal of certain components of the community have some limitations. These include the difficulty of separating the effects of removal of one component from the effect of the disturbance caused during the removal process and the difficulty of minimizing spatial heterogeneity that usually masks treatment effects. However, they are an excellent intermediate step between laboratory experiments and the long-term monitoring of undisturbed systems. Additionally, they have produced important evidence of the difficulties and potentialities of scaling up from individual physiology to

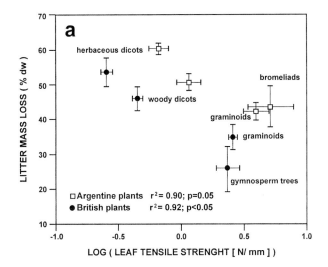

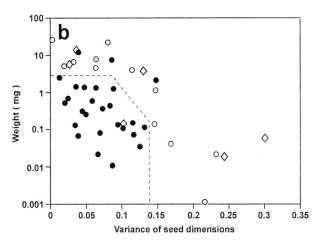

FIGURE 18 Examples of calibration of soft against hard vegetative and regeneration traits. (a) Percentage leaf litter mass loss of different species assemblages in terms of life-form and higher taxonomy as a function of tensile strength of fresh leaves (reproduced with permission from Cornelissen *et al.*, 1999). (b) Relationship between seed weight and variance of seed dimensions in 44 British species: ●, species with seeds which persist in the soil for at least 5 years; ○, species with seeds which persist in the soil for <5 years; ◇, species whose seed bank type could not be determined. The dashed line encloses the region within which all seeds examined are long lived (reproduced from Thompson *et al.*, 1993, with permission of Blackwell Science Ltd.).

ecosystem functioning (see Fig. 3), and they are likely to continue to provide fruitful insight in ecosystem ecology.

Acknowledgments

I am grateful to Diana Abal-Solís for drawing the figures and to two reviewers for fruitful comments.

See Also the Following Articles

ECOSYSTEM, CONCEPT OF • ECOSYSTEM FUNCTION MEASUREMENT, AQUATIC AND MARINE COMMUNITIES • ECOSYSTEM FUNCTION, PRINCIPLES OF • MEASUREMENT AND ANALYSIS OF BIODIVERSITY • TERRESTRIAL ECOSYSTEMS

Bibliography

Anderson, J. M. (1991). The effects of climate change on decomposition processes in grassland and coniferous forests. *Ecol. Appl.* 3, 326–347.

Austin, A. T., and Vitousek, P. M. (1998). Nutrient dynamics on a precipitation gradient in Hawai'i. *Oecologia* 113, 519–529.

Begon, M., Harper, J. L., and Townsend, C. (1990). *Ecology: Individuals, Populations, and Communities.* Blackwell, Oxford.

Chapin, F. S., III, and Shaver, G. R. (1996). Physiological and growth responses of Arctic plants to a field experiment simulating climate change. *Ecology* 77, 822–840.

Chapin, F. S., Walker, B. H., Hobbs, R. J., Hooper, D. U., Lawton, J. H., Sala, O. E., and Tilman, D. (1997). Biotic control of the functioning of ecosystems. *Science* 277, 500–503.

Cornelissen, J. H. C. (1996). An experimental comparison of leaf decomposition rates in a wide range of temperate plant species and types. *J. Ecol.* 84, 573–582.

Cornelissen, J. H. C., Pérez-Harguindeguy, N., Díaz, S., Grime, J. P., Marzano, B., Cabido, M., Vendramini, F., and Cerabolini, B. (1999). Leaf structure and defence control litter decomposition rate across species, life forms and continents. *New Phytol.* 143, 191–200.

DeAngelis, D. L. (1992). *Dynamics of Nutrient Cycling and Food Webs.* Chapman & Hall, London.

Díaz, S., and Cabido, M. (1997). Plant functional types and ecosystem function in relation to global change. *J. Veg. Sci.* 8, 463–474.

Díaz, S., Cabido, M., and Casanoves, F. (1999). Functional implications of trait–environmet linkages in plant communities. In *Ecological Assembly Rules* (E. Weiher and P. A. Keddy, Eds.), pp. 338–362. Cambridge Univ. Press, Cambridge, UK.

Ehleringer, J. R., and Field, C. B. (1993). *Scaling Physiological Processes—Leaf to Globe.* Academic Press, San Diego.

Fraser, L. H., and Grime, J. P. (1997). Primary productivity and trophic dynamics investigated in a North Derbyshire, UK, dale. *Oikos* 80, 499–508.

Gallardo, A., and Merino, J. (1993). Leaf decomposition in two Mediterranean ecosystems of southwest Spain: Influence of substrate quality. *Ecology* 74, 152–161.

Grime, J. P. (1998). Benefits of plant diversity to ecosystems: Immediate, filter and founder effects. *J. Ecol.* 86, 901–910.

Herms, D. A., and Mattson, W. J. (1992). The dilemma of plants: To grow or defend. *Q. Rev. Biol.* 3, 283–335.

Lambers, H., Poorter, H., and Van Vuuren, M. M. I. (1998). *Inherent Variation in Plant Growth—Physiological Mechanisms and Ecological Consequences.* Backhuys, Leiden.

Lambers, H., Chapin, F. S., III, and Pons, T. (1999). *Plant Physiological Ecology.* Springer, Berlin.

Larcher, W. (1995). *Physiological Plant Ecology.* Springer, Berlin.

Lepš, J., Osbornová-Kosinová, J., and Rejmánek, M. (1982). Community stability, complexity and species life-history strategies. *Vegetatio* 50, 53–63.

MacGillivray, C. W., Grime, J. P., and the ISP Team (1995). Testing predictions of the resistance and resilience of vegetation subjected to extreme events. *Funct. Ecol.* **9**, 640–649.

Marschner, H. (1986). *Mineral Nutrition of Higher Plants.* Academic Press, London.

McNaughton, S. J., Oesterheld, M., Frank, D. A., and Williams, K. J. (1989). Ecosystem-level patterns of primary productivity and herbivory in terrestrial habitats. *Nature* **341**, 142–144.

Medina, E., and Cuevas, E. (1989). Patterns of nutrient accumulation and release in Amazonian forests of the upper Rio Negro Basin. In *Mineral Nutrients in Tropical Forests and Savanna Ecosystems* (J. Proctor, Ed.), pp. 217–240. Blackwell, Oxford.

Mooney, H. A., Medina, E., and Schindler, D. W. (1991). *Ecosystem Experiments*, SCOPE Series No. 45. Wiley, Chichester, UK.

Mooney, H. A., Cushman, J. H., Medina, E., Sala, O. E., and Schulze, E.-D. (1996). *Functional Roles of Biodiversity.* Wiley, Chichester, UK.

Pearcy, R. W., Ehleringer, J., Mooney, H. A., and Rundel, P. W. (1989). *Plant Physiological Ecology—Field Methods and Instrumentation.* Chapman & Hall, London.

Pérez-Harguindeguy, N., Díaz, S., Cornelissen, J. H. C., Vendramini, F., Cabido, M., and Castellanos, A. (1999). Chemistry and toughness predict leaf litter decomposition rates over a wide spectrum of functional types and taxa in central Argentina. *Plant and Soil* **218**, 21–30.

Sala, O. E., and Paruelo, J. M. (1997). Ecosystem services in grasslands. In *Nature's Services: Societal Dependence of Natural Ecosystems* (G. C. Daily, Ed.), pp. 237–252. Island Press, Washington, DC.

Sala, O. E., Parton, W. J., Joyce, L. A., and Levenroth, W. K. (1988). Primary production of the central grassland region of the United States. *Ecology* **69**, 40–45.

Schulze, E.-D., and Caldwell, M. M. (1995). *Ecophysiology of Photosynthesis.* Springer, Berlin.

Schulze, E.-D., and Mooney, H. A. (1994). *Biodiversity and Ecosystem Function.* Springer, Berlin.

Schulze, E.-D., Kelliher, F. M., Körner, C., Lloyd, J., and Levning, R. (1994). Relationships among maximum stomatal conductance, ecosystem surface conductance, carbon assimilation rate and plant nitrogen nutrition: A global ecology scaling exercise. *Annu. Rev. Ecol. Syst.* **25**, 629–660.

Singh, K. P. (1989). Mineral nutrient dynamics in tropical dry deciduous forest and savanna ecosystems in India. In *Mineral Nutrients in Tropical Forests and Savanna Ecosystems* (J. Proctor, Ed.), pp. 153–168. Blackwell, Oxford.

Smith, T. M., Shugart, H. H., and Woodward, F. I. (1999). *Plant Functional Types.* Cambridge Univ. Press, Cambridge, UK.

Thompson, K., Band, S. R., and Hodgson, J. G. (1993). Seed size and shape predict persistence in soil. *Funct. Ecol.* **7**, 236–241.

Vitousek, P. M., and Sanford, R. L., Jr., (1986). Nutrient cycling in moist tropical forest. *Annu. Rev. Ecol. Syst.* **17**, 137–167.

Wilson, J. B., and Agnew, A. D. Q. (1992). "Positive-feedback switches in plant communities." *Adv. Ecol. Res.* **23**, 263–335.

ECOSYSTEM FUNCTION, PRINCIPLES OF

Ross A. Virginia* and Diana H. Wall†

*Dartmouth College; †Colorado State University

GLOSSARY

ecosystem All the individuals, species, and populations in a spatially defined area and the interactions among them and with the abiotic environment.

ecosystem functioning The sum total of processes such as the cycling of matter, energy, and nutrients operating at the ecosystem level.

functional group A group of species that perform similar roles in an ecosystem process.

nutrient cycle (or biogeochemical cycle) The repeated pathway of mineral elements, such as carbon, nitrogen, phosphorus, and water, from the environment through organisms and back into the environment.

succession The predictable change in species that occupy an area over time caused by a change in biotic or abiotic factors benefiting some species but at the expense of others.

ECOSYSTEMS ARE COMPOSED OF COMMUNITIES of organisms that interact with one another and the abiotic environment. The interactions of organisms and their environment are represented in processes that are called ecosystem functions. The capture of solar energy (photosynthesis), the cycling of nutrients, and the stability of ecosystem functioning are influenced by biodiversity. An understanding of how biodiversity and ecosystem functioning are related is necessary for determining how to sustain human populations in the future.

I. DEVELOPMENT OF THE ECOSYSTEM CONCEPT

The concept of the ecosystem as a functioning unit in the natural world is a relatively recent one. The term ecosystem was coined by the British ecologist Tansley in 1935 and has since become a common word in science and with the public. An ecosystem encompasses all the organisms of a given area and their relationships with one another and the physical or abiotic environment. The ecosystem contains the linkages and dynamic interactions between life and the environment, many of which are essential to society. A focus on the ecosystem as the unit of study represents a shift from studying the ecology and behavior of individual organisms and species (natural history) to the study of processes and how they influence or are influenced by organisms and their interactions with the environment.

Dividing the complexity of nature into convenient units of study is required for scientific investigation but can present problems. Ecological systems can be organized in a hierarchy of increasing levels of organization and complexity: individual, population, species, community, ecosystem, landscape, and biome. The size (scale) of an ecosystem is defined by the purposes of the study. Ecosystems may have distinct boundaries as in the case of a lake or a watershed. More often, the boundaries of one ecosystem (a forest) may grade gradually into another (a meadow) across an intermediate area called an ecotone. The ecotone is often a zone of higher diversity because it may be a suitable habitat for species from each of the adjoining ecosystems. At one extreme of scale, the earth is sometimes treated as an ecosystem. At the other extreme, the complex symbiotic community of organisms inhabiting the gut of a termite has all the functional properties of an ecosystem. The definition and delineation of an ecosystem has practical importance because ecosystems are increasingly seen as a functional unit for resource and conservation management purposes. It has become evident that the management of lands for sustained levels of ecosystem services and natural resources requires an understanding of how ecosystems function, how they respond to disturbance, and how the role of biodiversity is regulating their function and stability.

II. ECOSYSTEM FUNCTIONING AND ECOSYSTEM SERVICES

Society depends on the functioning of ecosystems for many essential ecosystem services on which we place economic and aesthetic value (Daily, 1997). Ecosystem functioning results from the collective activities of organisms and their life processes (production, consumption, and excretion) and the effects of these activities on the condition of the environment. These functions (services when they provide utility to humans) include production of food, fuel, and fiber, the cycling and purification of water, and the maintenance of organisms that have a role in ecosystem functioning or that provide products for human use (Table I). Humans are rapidly changing the earth's ecosystems and their services by altering land use or by harvesting biological resources (forest cutting and fisheries) (Vitousek *et al.*, 1997). Approximately 40% of the earth's primary production is diverted to human use. One consequence of these economic activities is an abrupt increase in the rate of change in biological diversity leading to species extinction, replacement of high-biodiversity ecosystems with less diverse managed systems, and invasions of natural ecosystems by exotic species. This pattern of ecosystem

TABLE I

Examples of the Biological and Physical Processes or Interactions That Contribute to Important Ecosystems Functions

Process	Ecosystem function
Photosynthesis	Primary production
Plant nutrient uptake	
Microbial respiration	Decomposition
Soil and sediment food web dynamics	
Nitrification	Nitrogen cycling
Denitrification	
Nitrogen fixation	
Plant transpiration	Hydrologic cycle
Root activity	
Mineral weathering	Soil formation
Soil bioturbation	
Vegetation succession	
Predator–prey interactions	Biological control

change has raised serious concern that the functioning and stability of our global ecosystem are threatened by the loss of biodiversity.

A. What Do Ecosystem Scientists Study?

Ecosystems share certain characteristics and functions that allow scientists to study ecosystem types (e.g., deciduous forest, temperate grassland, arctic tundra, coral reef, and deep-ocean hydrothermal vents) that vary greatly in structure, biodiversity, and spatial extent. For example, all ecosystems require inputs of energy (usually solar) and a supply of the mineral elements (nutrients) essential for life. These inputs support many ecological processes operating at multiple scales. For example, sunlight, carbon dioxide, and water are inputs for the process of photosynthesis, which can be measured and studied at the scale of individual cells, a leaf, the plant canopy, or an entire ecosystem. Photosynthesis acting with other processes such as mineral uptake by roots combine to create an ecosystem function—primary productivity.

Scientists can discover basic principles about the behavior of ecosystems by studying the functions that very different ecosystems, such as the polar desert of Antarctica and the rangelands of the southwestern United States, share in common (Virginia and Wall, 1999). The movement of energy and materials within and between ecosystems and the role of organisms in mediating these processes are the parameters used by

TABLE II

Examples of Ecosystem Services That Would Be
Affected by a Decline in Ecosystem Function[a]

Pest control
Insect pollination
Fisheries
Climate regulation
Soil retention
Flood control
Soil formation and maintenance of soil fertility
Cycling of matter
Composition of the atmosphere
Maintenance of genetic diversity

[a] Based on Daily (1997).

ecosystem scientists to compare the functioning of eco-systems and their responses to disturbance. Some of the important processes and functions central to the integrity and sustained activity of an ecosystem are summarized in Table II. Ecosystem scientists study the rate at which ecosystems remove carbon from the atmosphere by photosynthesis, store it in the soil as organic matter, and then return the stored carbon to the atmosphere during decomposition. They study how nitrogen is cycled through ecosystems to sustain continued plant productivity. Our knowledge of how carbon and nitrogen move in the ecosystem helps us to understand when an ecosystem has been seriously altered by humans, for example, by adding nitrogen in the form of air pollution (acid rain) and fertilizers.

Many basic principles provide insight into the functioning of ecosystems and their response to human use and disturbance. Here, we will consider some of the essential functions of ecosystems and examine the principles that govern their operation, with an emphasis on the role of organisms (biodiversity) in determining ecosystem functioning.

III. IMPORTANT ECOSYSTEM FUNCTIONS

A. Ecosystem Productivity

A central process of most ecosystems is photosynthesis, the capture of solar radiation and its conversion to stored chemical forms (biomass). Plants require sunlight, water, and essential nutrients for the processes of photosynthesis. Photosynthesis is coupled with other plant processes that result in plant growth, i.e., the

accumulation of biomass. Primary productivity, the change in plant biomass per unit area and time, is an important index of ecosystem function. Primary productivity (often referred to as ecosystem productivity) has been related to plant species diversity as well as the diversity of organisms (soil biota) that influence the availability of limiting resources. Humans depend on ecosystem productivity as the basis of our agriculture and forestry and fisheries. Thus, factors that alter ecosystem productivity (e.g., climate change and biodiversity loss) affect us directly.

Ecosystems with high rates of primary productivity have favorable amounts of the resources required for plant growth and optimal climate. These systems also tend to have higher diversity (Table III). The highest rates of terrestrial ecosystem productivity are seen in the tropics, where temperature and moisture are favorable for plant growth throughout the year. In contrast, water-limited hot and cold deserts have much lower productivity, averaging less than 10% of that of tropical systems.

1. Limits to Ecosystem Productivity

A basic principle invoked to explain variation among ecosystems in their productivity is Liebig's Law of Minimum. Justus Liebig formulated this concept during pioneering studies of the mineral nutrition of plants in the early 1800s. He found that addition of a single "limiting element" to a soil would increase plant growth. Once this element was in sufficient supply, another mineral element would have to be supplied in increased amounts to stimulate additional increases in plant growth. From these observations, he proposed that a limiting factor was responsible for limiting the growth or reproduction of an organism or population. This

TABLE III

Typical Values for the Net Primary Productivity
of Major Ecosystems[a]

Ecosystem type	Net primary production (g C/m²/year)	Relative species diversity
Tropical rain forest	900	Highest
Temperate forest	540	Intermediate
Grassland	315	Intermediate
Desert	32	Low
Extreme desert	1.5	Lowest

[a] Ecosystem productivity and biodiversity are often positively related.

factor might be a chemical factor (a growth-stimulating nutrient such as nitrogen), a physical factor such as moisture, or a biological factor such as the presence of a competing species. Thus, any change in a limiting factor is expected to have large effects on ecosystem functioning.

There are many examples in which a change in a limiting factor alters ecosystem function. The large increase in the amount of nitrogen cycling in the environment from fertilizers and fossil fuel should have significant effects on rates of ecosystem functions since nitrogen frequently is the primary limiting element for plant growth in terrestrial ecosystems. Humans have doubled the rate of nitrogen inputs to ecosystems with increases in carbon storage and declines in biodiversity (Vitousek *et al.*, 1997). In fact, the forests of the northeastern United States may have reached "saturation" in their ability to absorb and retain anthropogenic inputs of nitrogen.

Are plant species diversity and primary production related? Ecologists are accumulating evidence from experiments in controlled growth facilities and in the field that ecosystem primary productivity increases with increasing plant species diversity. The theoretical basis for the expectation that productivity and diversity should be related derives from an understanding of how limiting resources (water and nutrients) are distributed in ecosystems and an appreciation for the diversity of physiological or "functional" traits that organisms have evolved to capture and utilize these resources for growth. Differences between plant species in rooting depth, phenology (seasonality of growth), photosynthetic rates, and other physiological traits allow multispecies communities to more fully utilize the available resources.

The ability of diverse plant communities to obtain higher productivity than low-diversity systems is demonstrated in traditional (low-input) agriculture in which polycultures (multiple-species plantings) often have higher yields than single-species plantings (monocultures) (Gliessman, 1998). For example, corn (*Zea mays*) yields at comparable densities are higher when corn is grown in the presence of nitrogen-fixing beans (*Vicia* spp.). The bean crop forms a symbiotic association with bacteria that "fix" atmospheric nitrogen (N_2) to other inorganic forms (ammonia and NH_3) useable by plants. The nitrogen fixed by the bean crop improves the overall supply of this limiting element in the soil and increases the growth of the interplanted corn. The functioning provided by the diverse corn–bean–nitrogen-fixing bacteria association is often replaced in intensive agriculture by applying inorganic nitrogen fertilizers. With external inputs (fertilizers) the corn monoculture can produce higher yields than can the polyculture. Substituting an industrial source of nitrogen for a biological source has environmental costs resulting from the production and combustion of fossil fuels used to produce fertilizers. In addition, overapplication of fertilizers is a major source of water pollution in surface and groundwaters.

There are similar examples of diversity influencing productivity in natural ecosystems. In a California grassland ecosystem, Hooper and Vitousek (1997) manipulated the number of plant functional groups in a community (early vs late-season forbs, perennial grasses, and nitrogen-fixing plants) in combinations of one to four groups in a given plot. They found that the number of plant functional groups was not the main factor that determined productivity. Rather, certain functional characteristics of individual species within functional groups contributed more to ecosystem productivity than overall diversity of the plot. This study points to the complexity of trying to simply relate species diversity to function. As a general principle, ecologists recognize that some species play particularly important roles in regulating important ecosystem functions such as productivity and nutrient cycling.

B. Keystone Species

Certain species, termed keystone species, have a disproportionate influence (relative to their biomass) on ecosystem functioning. The loss of a keystone species will produce a cascade of effects on the diversity and function of the remainder of the ecosystem (Bond, 1993). Consequently, since keystone species can control ecosystem diversity and associated ecosystem functions, they and the habitats they live in often receive high priority in conservation management plans. There are many well-documented studies of keystone species and how they interact with ecosystem functioning, e.g., the North Pacific sea otter preys on sea urchins, which consume kelp. In the absence of the keystone predator, sea urchin populations increase and create areas devoid of kelp and, consequently, the myriad of fish and other species that depend on the kelp forest (Fig. 1). This is an example of a food web—the representation of trophic (feeding) relationships between species in an ecosystem.

There are many examples of keystone species in terrestrial ecosystems. A large change in African elephant numbers has dramatic effects on the diversity

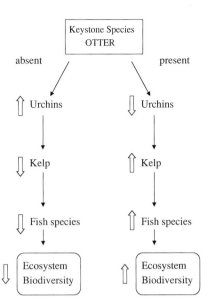

FIGURE 1 The influence of a keystone species on the biodiversity of an entire ecosystem can be large. Arrows indicate an increase or decrease in population size or species diversity in response to the presence or absence of the keystone species. The removal of the Pacific sea otter from California coastal ecosystems leads to the loss of the kelp community and many fish species.

and structure of the vegetation types (savanna woodlands and forests) they consume, altering ecosystem productivity, soil nutrient cycles, and plant community diversity. The much smaller tsetse fly shares the elephant's habitat and also has the attributes of a keystone species. The tsetse fly is the vector for the human disease sleeping sickness (African trypanosomiasis). This biting fly also influences the behavior of large herbivores that tend to avoid heavily infested areas. Consequently, herbivore-related impacts on plant communities and associated ecosystem functions are altered in tsetse-occupied ecosystems. This small insect may control the biodiversity of large tracts of Africa through another mechanism. Diverse native ecosystems have been "protected" from agricultural development and species loss because humans avoid regions where the tsetse and therefore sleeping sickness are endemic.

C. Nutrient Cycling

The sustained functioning of any ecosystem requires a minimum number of species to develop the intricate relationships between producers, consumers, and decomposers that regulate the flow of energy and nutrients. The productivity of all ecosystems is dependent on the cycling of essential elements. The movement and biological transformations of organic matter and nutrients are mediated by biota, especially those found in soil and sediments (Wall and Virginia, 1999). Therefore, changes in the biodiversity of ecosystems can alter biogeochemical processes.

1. Succession

Scientists study the process of ecological succession (ecosystem change with time, often in response to disturbance) in part to untangle relationships between biodiversity and function. Although not all ecosystems follow a predictable pathway as they develop in time, examples of succession highlight the linkage between organisms and diversity and ecosystem function. They include the recovery of a forest after harvest or following damage by a hurricane, the reestablishment of grassland following fire, and the old-field succession of natural vegetation reclaiming abandoned agricultural land. During succession, ecosystems change in generally predictable ways as they accumulate species, increase in biomass, and gain structural complexity. Odum (1969) proposed a model of ecological succession (development) that relates ecosystem diversity, structure, and functioning as ecosystems redevelop and "mature" following disturbance (Table IV). Odum's model related the stability (constancy) of function and the conservation of nutrients to increasing diversity—themes that are at the center of biodiversity and ecosystem research today.

TABLE IV

A Model of Ecological Succession Showing Relative Changes in Energy Flow, Nutrient Cycling, and Diversity over Time[a]

Ecosystem trait	Ecosystem status	
	Developing	Mature
Energetics		
Net primary production	High	Low
Food chains	Linear	Web-like
Communities		
Species diversity	Low	High
Nutrient cycling		
Mineral cycles	Open	Closed
Nutrient conservation	Poor	Good
System dynamics		
Stability	Poor	Good

[a] Based on Odum (1969).

The relationships represented in Odum's (1969) model between ecosystem function and diversity are elucidated in the Hubbard Brook watershed experiment (Likens and Bormann, 1995). One of the first long-term ecosystem studies, the Hubbard Brook project began in 1963 in the White Mountains of New Hampshire. The study was designed to understand the process of forest recovery following harvest with a focus on ecosystem functions related to production, nutrient cycling, and nutrient loss. Measurements of the mature intact deciduous forest showed that less than 0.1% of the nitrogen contained in living forest biomass and dead organic matter in the soil and litter left the site in stream flow. A nutrient cycle in which outputs are low and internal recycling of nutrients is high (the loop from soil to vegetation and back to soil) is called a closed nutrient cycle.

After the unperturbed patterns of growth and nutrient cycling were known, an entire Hubbard Brook watershed was clear-cut. What followed was a dramatic change in ecosystem functioning. Stream flow increased by approximately 40% because water use by plants had been nearly eliminated by the forest harvest. The previously "closed" nutrient cycle of this forest became "open." After clear-cut the concentrations of nitrogen (nitrate) in the stream water draining the watershed increased approximately 60-fold. Concentrations of elements that are important to the biology of the ecosystem leaked into the streams and were exported from the ecosystem. Elements not essential to plant growth or required in very small amounts (e.g., sodium) were not lost to the same degree, indicating their cycling was not regulated by biotic activity of the forest. Odum (1969) predicted that nutrient losses would decline with increasing plant biomass and function. After the Hubbard Brook forest was allowed to regrow (undergo succession), nutrients resumed being absorbed by plants and nutrient losses to streams declined to near baseline levels. The Hubbard Brook ecosystem experiment informed forest management practices by providing a better understanding of how forest removal and regrowth affect the retention of soil nutrients and therefore the long-term productivity and diversity of the ecosystem.

D. Ecosystem Stability

Ecosystems are dynamic. They experience change in species composition and function in response to variations in climate and an array of disturbances. Fire, flood, drought, frost, and biological events such as the out-

break of pathogens and pests can "stress" ecosystems and alter their condition. Ecosystems vary widely in their responses to disturbance. The ability of an ecosystem to withstand stress without a loss of function (resistance) or to recover rapidly from disturbance (resilience) is an important ecosystem trait. Some ecosystems, such as tropical forests, appear very stable (high resistance and resilience) and their functioning is little affected by variations in factors external to the system (e.g., weather). Ecosystems with high resilience are buffered against perturbation. Many ecosystems, however, show large decreases in productivity and biodiversity when disturbed. These ecosystems are "fragile" and have low resistance.

The relationship between ecosystem stability and diversity has been the subject of many field studies and theoretical tests using mathematical modeling. Ecologists have hypothesized that ecosystems with high biodiversity are more resistant (will experience less change) in response to a given level of disturbance and will also exhibit resilience—a high rate of recovery to predisturbance functioning (Folke et al., 1996).

Does diversity influence the stability of ecosystem functioning? There is experimental evidence that it can do so (Chapin et al., 1997). Several mechanisms have been proposed and tested to varying degrees to examine this relationship (Chapin et al., 1997). Higher species diversity means that the trophic structure (feeding relationships among species) of the ecosystem is more complex, providing alternate pathways for energy flow within and between trophic levels (producers, consumers, and decomposers). Alternative pathways for energy transfers within the ecosystem could increase resistance to disturbance (species loss). Naeem and Li (1997) tested the hypothesis that redundancy (multiple species with similar functions in a food web) would stabilize ecosystem functioning by creating experimental microcosms with a varying number of species in each functional group. The simple systems contained producers (algae), decomposers (bacteria), and a primary and secondary consumer trophic level (protists)—the trophic structure of a typical aquatic ecosystem. Nutrient levels, light, and the number of species per trophic level were manipulated, and the biomass and density of the producers and decomposers were measured as an indicator of ecosystem functioning. As the number of species in a trophic level increased, the biomass and density of replicate communities were more consistent. Thus, the communities with more species were more predictable in function (biomass production) and had higher reliability, i.e., the probability that an ecosystem will provide a given level of performance over a specified period of time.

Higher species diversity may ensure functioning by reducing the risk of invasion by species that have the capacity to alter the structure or function of the ecosystem. An example is the higher resistance of species-rich natural systems to pest outbreaks compared to low-diversity agricultural ecosystems growing under the same environmental conditions. The spatial arrangement of individuals in an ecosystem can affect their risk to disease, predation, or consumption. In higher diversity systems the mean distance between individuals of the same species is on average greater than that of low-diversity systems. The wider spacing of individuals acts to slow the movement of pathogenic organisms, which should limit the occurrence of pest outbreaks that alter the performance of the ecosystem. These and other observations lead to the general expectation that diversity increases the resistance of ecosystems to disturbance.

The benefits of biodiversity to ecosystem functioning should be multiple since the processes of production and nutrient cycling are coupled by the biological interactions of organisms. The response of a Minnesota grassland to a severe drought (disturbance) illustrates this principle (Tilman *et al.*, 1996). In 1987 and 1988, a drought decreased productivity of the grassland. The species diversity of experimental plots prior to the drought explained the degree of productivity loss. Diverse plots experienced about a 50% decline in productivity, whereas productivity in the least diverse plots declined by more than 90%. The greater resistance of the higher diversity plots resulted from compensatory increases in productivity shown by drought-resistant species. The more diverse plots also had lower concentrations of nitrate in the rooting zone, indicating a more efficient use of this limiting resource.

This experiment demonstrates that species diversity has an effect on productivity and nutrient cycling and that declining species diversity influences these functions. However, we lack an understanding of the mechanisms producing these patterns of ecosystem response to disturbance and biodiversity change. Increasing diversity may increase the chance that a single drought-adapted and productive species will be present in the community, ensuring relatively high productivity. Alternatively, higher diversity may provide for a more efficient utilization of limiting resources, as suggested by the lower soil nitrate in more diverse plots. Before the basic relationships between biodiversity and ecosystem functioning can be more fully formalized, we need more

detailed information on the critical levels (thresholds) of diversity associated with specific ecosystem functions and how environmental conditions operating over time alter their relationship (Folke *et al.*, 1996).

IV. CONCLUSIONS

Humans have become major agents of environmental change and influence the biodiversity and structure of ecosystems in many ways. Air pollution, clearing of natural systems for agriculture, forestry and urban development, the spread of exotic species, changes in the composition of the atmosphere, and other anthropogenic influences are altering ecosystem functioning. By changing ecosystem biodiversity and altering the processes that biota mediate, we significantly decrease the ability of ecosystems to provide services and resources for our use. The management of ecosystems for sustained levels of services and the restoration of damaged ecosystems will require greater knowledge about the role that species play in ecosystems functions related to production and nutrient cycling. Although we cannot know with certainty the roles of most species in ecosystems, it is prudent to assume that all biodiversity is essential to ecosystem function and stability and should be valued and protected.

See Also the Following Articles

ECOSYSTEM, CONCEPT OF • ECOSYSTEM SERVICES, CONCEPT OF • ENERGY FLOW AND ECOSYSTEMS • KEYSTONE SPECIES • NITROGEN CYCLE

Bibliography

Bond, W. J. (1993). Keystone species. In *Biodiversity and Ecosystem Function* (E.-D. Schulze and H. A. Mooney, Eds.), pp. 237–253. Springer-Verlag, New York.

Chapin, F. S., III, Sala, O. E., Burke, I. C., Grime, J. P., Hooper, D. C., Laurenroth, W. K., Lombard, A., Mooney, H. A., Mosier, A. R., Naeem, S., Pacala, S. W., Roy, J., Steffen, W. L., and Tilman, D. (1998). Ecosystem consequences of changing biodiversity. *BioScience* 48, 45–52.

Daily, G. C. (Ed.) (1997). *Nature's Services. Societal Dependence on Natural Ecosystems.* Island Press, Washington, D.C.

Folke, C., Hollings, C. S., and Perrings, C. (1996). Biological diversity, ecosystems, and the human scale. *Ecol. Appl.* 6, 1018–1024.

Gleissman, S. R. (1998). *Agroecology: Ecological Processes in Sustainable Agriculture.* Sleeping Bear Press, Chelsea, MI.

Hooper, D. U., and Vitousek, P. M. (1997). The effects of plant composition and diversity on ecosystem processes. *Science* 277, 1302–1305.

Likens, G. E., and Bormann, F. H. (1995). *Biogeochemistry of a Forested Ecosystem,* 2nd ed. Springer-Verlag, New York.

Naeem, S., and Li, S. (1997). Biodiversity enhances ecosystem reliability. *Nature* **390**, 507–509.

Odum, E. P. (1969). The strategy of ecosystem development. *Science* **164**, 262–270.

Pimm, S. L. (1984). The complexity and stability of ecosystems. *Nature* **370**, 321–326.

Tilman, D., Wedin, D., and Knops, J. (1996). Productivity and sustainability influenced by biodiversity in grassland ecosystems. *Nature* **379**, 718–720.

Virginia, R. A., and Wall, D. H. (1999). How soils structure communities in the Antarctic Dry Valleys. *BioScience* **49**, 973–983.

Vitousek, P. M., Aber, J. D., Howarth, R. W., Likens, G. E., Matson, P. A., Schindler, D. W., Schlesinger, W. H., and Tilman, D. G. (1997). Human alteration of the global nitrogen cycle: Sources and consequences. *Ecol. Appl.* **7**, 737–750.

Wall, D. H., and Virginia, R. A. (2000). The world beneath our feet: Soil biodiversity and ecosystem functioning. In *Nature and Human Society: The Quest for a Sustainable World* (P. Raven and T. A. Williams, Eds.), pp. 225–241. National Academy of Sciences Press, Washington, DC.

ECOSYSTEM SERVICES, CONCEPT OF

Gretchen Daily* and Shamik Dasgupta[†]

*Stanford University and †University College London

GLOSSARY

ecosystem services The wide array of conditions and processes through which ecosystems, and their biodiversity, confer benefits on humanity; these include the production of goods, life-support functions, life-fulfilling conditions, and preservation of options.

marginal value Economic value of the next incremental unit of something. In this context, marginal values are those associated with managing the next small unit of an ecosystem in a particular way (e.g., preserving, rather than clearing, the next unit of forest).

ECOSYSTEM SERVICES ARE ESSENTIAL TO HUMAN EXISTENCE, and yet their supply is seriously threatened by the intensification of human impacts on the environment. This article provides an overview of issues concerning the identification, biophysical and economic characterization, and safeguarding of ecosystem services.

I. OVERVIEW

Human societies derive many benefits from natural ecosystems. These include the production of a diversity of ecosystem goods, or extractive benefits, such as seafood, timber, biomass fuels, and precursors to many industrial and pharmaceutical products. The harvest and trade of these goods represent an important and familiar part of the economy. Ecosystem services also include non-extractive benefits—fundamental life-support processes including pollination, water purification, renewal of soil fertility, and climate regulation. Ecosystem services include life-fulfilling functions, encompassing aesthetic beauty and the cultural, intellectual, and spiritual values derived from nature. Finally, preservation of the option to use these (or new) services in the future is also an important service in itself.

One way to appreciate the nature and value of ecosystem services is to imagine trying to set up a happy existence on the moon. Assume for the sake of argument that the moon miraculously already had some of the basic conditions for supporting human life, such as an atmosphere and climate similar to those on Earth. After packing one's prized possessions, the big question would be, Which of Earth's millions of species would be required to sustain the lunar colony?

Tackling the problem systematically, one could first choose from among all the species exploited directly for food, drink, spice, fiber and timber, pharmaceuticals, industrial products (such as waxes, lac, rubber, and oils), and so on. Even being selective, this list could amount to hundreds, or even several thousand, species. The space-ship would be filling up before even beginning to add the species crucial to supporting those at the top of one's list. Which are these unsung heroes? No one knows which—nor even approximately how many—species are required to sustain human life. This means that rather than listing species directly, you would have to list instead the life-support functions required by your lunar colony; then you could guess at the types and numbers of species required to perform each. At a bare minimum, other companions on the spaceship would have to include species capable of supplying a whole suite of ecosystem services that Earthlings take for granted. Table I provides a classification of important ecosystem services.

Armed with this preliminary list of services, one could begin to determine which types and numbers of species are required to perform each. This is no simple task. Consider soil fertility. Soil organisms play important and often unique roles in the circulation of matter in every ecosystem on Earth; they are crucial to the chemical conversion and physical transfer of essential nutrients to higher plants, and all larger organisms, including humans, depend on them. The abundance of soil organisms is tremendous: under a square yard of pasture in Denmark, for instance, the soil was found to be inhabited by roughly 50,000 small earthworms and their relatives, 50,000 insects and mites, and nearly 12 million roundworms. And that is not all. A single gram (a pinch) of soil has yielded an estimated 30,000 protozoa, 50,000 algae, 400,000 fungi, and billions of individual bacteria (Overgaard-Nielsen, 1955). Which to bring to the moon? Most of these species have never been subjected to even cursory inspection. Yet the sobering fact of the matter is, as Ed Wilson put it: they don't need us, but we need them.

In the early 1990s, the first Biosphere 2 "mission" carried out this thought experiment to the greatest degree possible on Earth. Eight people were enclosed in a 3.15-acre closed ecosystem, featuring agricultural land plus a wide array of natural habitats (desert, savanna, tropical forest, wetland, and even a miniature ocean). The aim was to demonstrate a (mostly) closed system that could supply people with their material needs for two years. Yet in spite of an investment of over $200 million in the design, construction, and operation of this model Earth (including $1 million in annual energy inputs), it proved impossible to do so and the experiment was shut down early in failure. Numerous unpleasant and unexpected problems arose, including a fall in atmospheric oxygen concentration to 14% (the level normally found at an elevation of 17,500 feet);

TABLE I

A Classification of Ecosystem Services with Illustrative Examples

Production of Goods

Food
　Terrestrial animal and plant products
　Forage
　Seafood
　Spices

Pharmaceuticals
　Medicinal products
　Precursors to synthetic pharmaceuticals

Durable materials
　Natural fiber
　Timber

Energy
　Biomass fuels
　Low-sediment water for hydropower

Industrial products
　Waxes, oils, fragrances, dyes, latex, rubber, etc.
　Precursors to many synthetic products

Genetic resources
　Which enhance the production of many of these goods

Regeneration Processes

Cycling and filtration processes
　Detoxification and decomposition of wastes
　Generation and renewal of soil fertility
　Purification of air
　Purification of water

Translocation processes
　Dispersal of seeds necessary for revegetation
　Pollination of crops and natural vegetation

Stabilizing Processes
　Coastal and river channel stability
　Compensation of one species for another under varying
　　conditions
　Control of the majority of potential pest species
　Moderation of weather extremes (such as of temperature and
　　wind)
　Partial stabilization of climate
　Regulation of hydrological cycles (mitigation of floods and
　　droughts)

Life-Fulfilling Functions
　Aesthetic beauty
　Cultural, intellectual, and spiritual inspiration
　Existence value
　Scientific discovery
　Serenity

Preservation of Options
　Maintenance of the ecological components and systems needed
　　for future supply of these goods and services and others
　　awaiting discovery

high spikes in carbon dioxide concentrations; nitrous oxide concentrations high enough to impair brain function; rapid species extinctions (including 19 of 25 vertebrate species and all pollinators, thereby dooming most of the plant species to eventual extinction as well); overgrowth of aggressive vines and algal mats; and, to top it all off, population explosions of crazy ants, cockroaches, and katydids. Even heroic personal efforts on the part of the Biospherians did not suffice to make the system viable and sustainable for humans nor many nonhuman species, illustrating the tremendous expense and difficulty of replicating many basic ecosystem services (Cohen and Tilman, 1996).

Ecosystem services are generated by a complex of natural cycles, driven by solar energy, that constitute the workings of the biosphere—the thin layer near Earth's surface that contains all known life. The cycles operate on very different scales. Biogeochemical cycles, such as the movement of the element carbon through the living and physical environment, are truly global and reach from the top of the atmosphere to deep into soils and ocean-bottom sediments. Life cycles of bacteria, in contrast, may be completed in an area much smaller than the period at the end of this sentence. The cycles also operate at very different rates. The biogeochemical cycling of carbon, for instance, occurs at a rate that is orders of magnitude faster than that of phosphorus, just as the life cycles of microorganisms may be many orders of magnitude faster than those of trees.

All of these cycles are ancient, the product of billions of years of evolution, and have existed in forms very similar to those seen today for at least hundreds of millions of years. They are absolutely pervasive, but unnoticed by most human beings going about their daily lives. Who, for example, gives a thought to the part of the carbon cycle that connects him or her to the plants in the garden outside, to plankton in the Indian Ocean, or to Julius Caesar? Noticed or not, human beings depend utterly on the continuation of natural cycles for their very existence. If the life cycles of predators that naturally control most potential pests of crops were interrupted, it is unlikely that pesticides could satisfactorily take their place. If the life cycles of pollinators of plants of economic importance ceased, society would face serious social and economic consequences. If the carbon cycle were badly disrupted, rapid climatic change could threaten the existence of civilization. In general, human beings lack both the knowledge and the ability to substitute for the functions performed by these and other cycles.

For millennia, humanity has drawn benefits from

these cycles without causing global disruption. Yet because most of these benefits are not traded in markets, they carry no price tags that could alert society to changes in their supply or to deterioration of the underlying ecological systems that generate them. Escalating impacts of human activities on forests, wetlands, and other natural ecosystems imperil the delivery of ecosystem services. The primary threats are land use changes that cause losses in biodiversity as well as disruption of carbon, nitrogen, and other biogeochemical cycles; human-caused invasions of exotic species; releases of toxic substances; possible rapid climate change; and depletion of stratospheric ozone. Because threats to these systems are increasing, there is a critical need for identification and monitoring of ecosystem services both locally and globally, and for the incorporation of their value into decision-making processes.

Based on available scientific evidence, it is certain that:

- Ecosystem services are essential to civilization.
- Ecosystem services operate on such a grand scale and in such intricate and little-explored ways that most could not be replaced by technology.
- Human activities are already impairing the flow of ecosystem services on a large scale.
- If current trends continue, humanity will dramatically alter virtually all of Earth's remaining natural ecosystems within a few decades.

In addressing these points in more detail, we shall first consider briefly the history of concern for ecosystem services. Then we shall explore the functioning of a small set of services in biophysical terms; the economic characterization of ecosystem services is discussed elsewhere in the Encyclopedia. Finally, we will examine the utility of the Ecosystem Services Framework, in the abstract and then in operation, reviewing recent developments in the safeguarding of ecosystem services.

II. HISTORY

Interestingly, the nature and value of Earth's life-support systems have been illuminated primarily through their disruption and loss. Thus, for instance, deforestation has demonstrated the critical role of forests in the hydrological cycle—in particular, in mitigating floods, droughts, the erosive forces of wind and rain, and silting of dams and irrigation canals. Release of toxic substances, whether accidental or deliberate, has revealed the nature and value of physical and chemical processes, governed in part by a diversity of microorganisms, that disperse and break down hazardous materials. Thinning of the stratospheric ozone layer sharpened awareness of the value of its service in screening out harmful ultraviolet radiation.

A cognizance of ecosystem services, expressed in terms of their loss, dates back at least to Plato around 400 B.C. and probably much earlier:

> What now remains of the formerly rich land is like the skeleton of a sick man with all the fat and soft earth having wasted away and only the bare framework remaining. Formerly, many of the mountains were arable. The plains that were full of rich soil are now marshes. Hills that were once covered with forests and produced abundant pasture now produce only food for bees. Once the land was enriched by yearly rains, which were not lost, as they are now, by flowing from the bare land into the sea. The soil was deep, it absorbed and kept the water ... , and the water that soaked into the hills fed springs and running streams everywhere. Now the abandoned shrines at spots where formerly there were springs attest that our description of the land is true.
>
> —PLATO
>
> (Quoted in Daily, G. C. 1997. *Nature's Services*, p. 6.)

Mooney and Ehrlich (1997) trace modern concern for ecosystem services to George Perkins Marsh, a lawyer, politician, and scholar. Indeed, his 1864 book *Man and Nature* describes a wide array of services, again, often expressed in terms of their loss. Remarking on the terrain of the former Roman Empire, he notes that it "is either deserted by civilized man and surrendered to hopeless desolation, or at least greatly reduced in both productiveness and population" (p. 9). He goes on to describe the reduction of hydrological services: "Vast forests have disappeared from mountain spurs and ridges, the vegetable earth ... [is] washed away; meadows, once fertilized by irrigation, are waste and unproductive, because ... the springs that fed them dried up; rivers famous in history and song have shrunk to humble brooklets" (p. 9). Marsh also made connections between deforestation and climate: "With the disappearance of the forest, all is changed. At one season, the earth parts with its warmth by radiation to an open sky—receives, at another, an immoderate heat from the unobstructed rays of the sun. Hence the climate becomes excessive, and the soil is alternately parched

by the fervors of summer, and seared by the rigors of winter. Bleak winds sweep unresisted over its surface, drift away the snow that sheltered it from the frost, and dry up its scanty moisture" (p. 186). Finally, he even wrote of decomposition services: "The carnivorous, and often the herbivorous insects render an important service to man by consuming dead and decaying animal and vegetable matter, the decomposition of which would otherwise fill the air with effluvia noxious to health" (p. 95).

Following World War II, other eloquent writers on the environment emerged, including Fairfield Osborn (*Our Plundered Planet*, 1948), William Vogt (*Road to Survival*, 1948), and Aldo Leopold (*A Sand County Almanac and Sketches from Here and There*, 1949). Each discusses ecosystem services without using the term explicitly. In *The Population Bomb* (1968), Paul Ehrlich describes anthropogenic disruption of ecosystems and the societal consequences of doing so, addressing the need to maintain important aspects of ecosystem functioning. Along these lines, the *Study of Critical Environmental Problems* (1970) presents a list of key "environmental services" that would decline with a decline in "ecosystem function," including many in Table I. This list was expanded upon by Holdren and Ehrlich (1974). By the early 1980s, efforts were initiated to investigate two questions: the extent to which ecosystem function (and the delivery of services) depends on biodiversity, and the extent to which technological substitutes could replace ecosystem services. The first question is addressed elsewhere in this Encyclopedia, and the second question was tackled by Ehrlich and Mooney (1983). Work on these topics proliferated and, in 1997, a collective effort was made to synthesize the wealth of scientific information that had accumulated on the functioning of ecosystem services, with a preliminary exploration of their economic value, and of key issues meriting further work (Daily, 1997).

III. BIOPHYSICAL CHARACTERIZATION

Food production is arguably humanity's most essential activity. It is also the most important proximate cause of biodiversity loss worldwide, involving major direct and indirect impacts, including: (i) conversion of natural habitat to agricultural use, (ii) facilitation of biotic invasion through habitat alteration and trade, and (iii) application of chemical fertilizers and pesticides. Global demand for food is expected to double over the period 1990–2030. In Asia and Africa, food needs are projected to increase by a factor of 2.3 and 5, respectively, with a sevenfold increase or more in some countries. It is difficult to forecast how these needs will be met. Whatever course is taken, it is reasonable to expect that growing human pressures will continue to alter ecosystems, probably very dramatically.

At the same time, food production is highly dependent on ecosystem services. In what ways? Let us consider four types of services that are key to agriculture: climate stability, services supplied by soil, pollination, and pest control. In each case, an overview of the functioning of the service will be given, along with a general assessment of its importance.

A. Climate Stability

Earth's climate has fluctuated tremendously since humanity came into being, but it has been relatively stable since the invention of agriculture around 10,000 years ago. At the peak of the last ice age 20,000 years ago, for example, much of Europe and North America were covered by mile-thick ice sheets. Even relatively recently, from 1550 to 1850, Europe was significantly cooler during a period known as the Little Ice Age. Many of these changes in climate are thought to be caused by alterations in Earth's orbital rotation or in the energy output of the sun, or by sudden perturbations (such as violent volcanic eruptions and asteroid impacts) or more gradual tectonic events (such as the development of the Himalayas). Remarkably, climate has been buffered enough through all of these changes to sustain life for at least 3.5 billion years—although climate change is thought to have induced the past mass extinctions. Human activities now threaten to induce rapid climatic change, with wide-reaching consequences for society.

Climate plays a major role in the evolution and distribution of life over the planet. Most scientists would also agree that life itself is a principal factor in the regulation of global climate, helping to dampen the effects of episodic perturbations through negative feedback mechanisms that offset climate oscillations with changes in greenhouse gas concentrations. For instance, natural ecosystems may have helped to stabilize climate by removing CO_2 from the atmosphere as the sun grew brighter over millions of years. Life may also exert a destabilizing (positive) feedback with climate change, particularly during interglacial–ice age transitions. One such mechanism is the fertilization of phytoplankton that is thought to occur when climatic cooling leads to sea level reductions, which expose continental shelves and enhance nutrient runoff to the oceans. Since the bulk of many phytoplankton species is a calcium car-

bonate shell, increasing productivity would remove carbon dioxide from the oceans (and the atmosphere), further cooling the planet. Warming trends may also be enhanced by life, such as through the stimulation of decomposition rates of dead organic matter, which may release CO_2 to the atmosphere. The relative influence of stabilizing and destabilizing feedbacks remains uncertain; what is clear is that climate and natural ecosystems are tightly coupled, and maintaining the stability of that coupled system is an important ecosystem service.

Ecosystems also help moderate regional and local weather. For instance, transpiration of plants in the morning contributes moisture to the atmosphere that then falls in thunderstorms in the afternoon, damping both moisture loss and surface temperature rise. In the Amazon, for example, 50% of the mean annual precipitation is recycled via evapotranspiration by the forest itself. Amazon deforestation could so dramatically reduce precipitation that the forest might be unable to reestablish itself following large-scale destruction. Temperature extremes are also moderated by forests, which provide shade and surface cooling and also act as insulators, blocking searing winds and trapping warmth by acting as a local greenhouse agent.

B. Services Supplied by Soil

The ecosystem services supplied by soil are so tightly interrelated as to make any discrete classification arbitrary. We shall briefly touch on six services here: (i) buffering and moderation of the hydrological cycle; (ii) physical support of plants; (iii) retention and delivery of nutrients to plants; (iv) disposal of wastes and dead organic matter; (v) renewal of soil fertility; and (vi) regulation of major element cycles.

An enormous amount of water, about 119,000 km^3, falls annually on Earth's land surface—enough to cover the land to an average depth of 1 meter. Much of this water is soaked up by soils and is gradually meted out to plant roots and into aquifers and surface streams. Without soil, rainfall would rush off the land in flash floods. Plant foliage, roots, and residues shield the soil from the full, destructive force of raindrops, holding it in place. Rain on denuded landscapes compacts the surface and turns soil rapidly to mud (especially if it has been loosened by tillage), which clogs surface cavities in the soil, reduces infiltration, increases runoff, and further enhances clogging in a positive feedback. Detached soil particles are splashed downslope and carried off by running water.

Soil also shelters seeds and provides physical support

as they sprout and mature into adult plants. The energetic costs to plants of packaging and storing seeds and of anchoring plant roots would be enormous without soil. Human-engineered hydroponic systems grow plants in the absence of soil and their cost provides a lower-bound index to the value of this service. The physical support trays and stands used in such operations amount to about U.S. $55,000 per hectare.

Third, soil retains and delivers nutrients to plants. Tiny soil particles, primarily bits of humus and clays, carry a surface electrical charge that is generally negative. This property holds positively charged nutrients (cations) near the surface, in proximity to plant roots, allowing them to be taken up gradually. Otherwise, they would quickly be leached away. Soil also acts as a buffer in the application of fertilizers, holding the fertilizer ions on soil exchange sites until required by plants.

Fourth, soil plays a paramount role in the decomposition of dead organic matter and wastes, in the process rendering harmless many potential human pathogens. People generate a tremendous amount of waste, including household garbage, industrial waste, crop and forestry residues, and sewage from their own populations and their billions of domesticated animals. A rough approximation of the amount of dead organic matter and waste (mostly agricultural residues) processed each year is 130 billion metric tons, about 30% of which is associated with human activities. Fortunately, there is a diverse array of decomposing organisms—ranging from vultures to tiny bacteria—that earn their living by extracting energy from the large, complex organic molecules found in many types of waste.

The simple inorganic chemicals that result from natural decomposition are eventually returned to plants as nutrients. Thus the decomposition of wastes and the recycling of nutrients, the fifth service, are two aspects of the same process. The fertility of soils—that is, their ability to supply nutrients to plants—is largely the result of the activities of diverse species of bacteria, fungi, algae, crustacea, mites, termites, springtails, millipedes, and worms, all of which, as groups, play important roles. Some bacteria are responsible for "fixing" nitrogen (a key element in proteins), drawing it out of the atmosphere and converting it to forms usable by plants and, ultimately, human beings. Certain types of fungi are essential to supplying nutrients to many kinds of trees. Earthworms and ants act as "mechanical blenders" by breaking up and mixing plant and microbial material and other matter.

Finally, soils are a key factor in regulating Earth's major element cycles (e.g., of carbon, nitrogen, and sulfur). The amount of carbon and nitrogen stored in soils dwarfs that in vegetation, for example, by factors of ≈ 1.8 and ≈ 18, respectively. The importance to society of maintaining an approximate steady state in the stocks and fluxes of major elements can be most easily appreciated by considering the consequences of their recent disruption. Alterations in the carbon and nitrogen cycles can be costly, long term, and in many cases irreversible on a time scale of interest to society. The former are leading to a buildup of greenhouse gases in the atmosphere; the latter cause acid precipitation, eutrophication, and contamination of groundwater and surface drinking water sources by high nitrate-nitrogen levels.

C. Pollination

Animal pollination is required for the successful reproduction of most flowering plants, including both wild plants and about 70% of the agricultural crop species that feed the world. Over 100,000 different animal species (including bats, bees, beetles, birds, butterflies, and flies) are known to provide "free" pollination services, which maintain croplands, backyard gardens, meadows, forests, and rangelands. The availability of these pollinators depends on the existence of a wide variety of habitat types needed for their feeding, successful breeding, and completion of their life cycles.

Considering the proportions in which we consume different crops, about one-third of human food comes from animal-pollinated plants. Critical grain crops, such as rice and wheat, are not animal-pollinated. Without natural pollination services, yields of other important crops would decline precipitously and many wild plant species would become extinct. In the United States alone, billions of dollars of crops would be lost (annually) without the services of wild, native pollinators—those sustained by natural habitats adjacent to farmlands. Pollination by honeybees, originally imported from Europe, is extremely important as well, but they are presently in decline, making pollinators from natural ecosystems all the more important. Management of the honeybee in the New World is currently threatened by the movement of, and hybridization with, the aggressive African strain of honeybee that was accidentally released in Brazil in 1956. Diseases of honeybee colonies are also causing a marked decline in the number of managed colonies. Meanwhile, the diversity of pollinators available to both wild and domesticated plants is diminishing: more than 60 genera of pollinators include species now considered to be threatened, endangered, or extinct.

D. Natural Pest Control Services

Humanity's competitors for food, timber, cotton, and other fiber are called pests; these include numerous herbivorous insects, rodents, fungi, snails, nematodes, and viruses. These pests destroy an estimated 25% (up to as much as 50%) of crops, before and after harvest. In addition, numerous weeds compete directly with crops for water, light, and soil nutrients, further limiting yields.

Chemical pesticides, and the strategies with which they are applied, can have harmful unintended consequences. First, resistance is now found in over 500 insect and mite pests, over 100 weeds, and in about 150 plant pathogens. This means that achieving a fairly constant level of pest control requires applying higher and higher doses of pesticides—or continuously developing new, more powerful chemicals to combat evolving pest defenses. Second, populations of natural pest enemies are decimated by heavy pesticide use. Not having the same evolutionary experience with plant chemicals that the pests themselves have had, and typically having much smaller population sizes (being higher on food chains), natural predators are often more susceptible to synthetic poisons than are the pests. Destruction of predator populations leads to explosions of their prey and the "promotion" of species previously not classified as pests to pest status. In California in the 1970s, for instance, 24 of the 25 most important agricultural pests had been "created" by the overuse of pesticides. Third, exposure to pesticides and herbicides may pose serious health risks to humans and many other organisms; the recently discovered declines in human sperm counts may be attributable in part to such exposure.

Fortunately, an estimated 99% of potential crop pests are controlled by natural enemies, including many birds, spiders, parasitic wasps and flies, ladybugs, fungi, viral diseases, and numerous other types of organisms. These natural biological control agents save farmers and society billions of dollars annually by protecting crops and reducing the need for chemical control.

IV. THE ECOSYSTEM SERVICES FRAMEWORK

All of these services are presently under threat. When human activities approach or exceed the environment's capacity to sustain them, growth in those activities is rarely brought to an immediate halt. Rather, the people so engaged suddenly find themselves confronted with a set of trade-offs in the allocation of resources to competing uses and users. These trade-offs are becoming increasingly vexing and difficult to resolve, from both ethical and practical perspectives. They involve our most important ideals (such as ensuring a prosperous future for our children), our oldest tensions (such as between individual and societal interests), and sometimes our bloodiest tendencies.

At the local level, allocation of land or water to competing activities often involves a zero-sum game. This is apparent in the widespread loss of water and land from native habitat to farms, and increasingly to urban and industrial uses. On what basis should such allocations be decided? How can individual preferences for alternative allocations be aggregated fairly? How can the costs and benefits of alternative schemes be distributed fairly? And how can the parties with the most at stake—future generations—be represented at the bargaining table? At the international level, these questions are writ large. Consider efforts to allocate among nations permits to produce chlorofluorocarbons, to harvest certain marine fish stocks, or to use the global carbon dioxide sink. How these questions are decided will profoundly influence the willingness of nations and individual actors to make and comply with agreements.

The ecosystem services approach provides a conceptual framework for helping to resolve these trade-offs. The framework recognizes natural ecosystems and their biodiversity as capital assets that, if properly managed, will yield a stream of life-support goods and services over time. Relative to physical, human, or financial capital, renewable natural capital (embodied in ecosystems) is poorly understood, typically undervalued, scarcely monitored, and—in many important cases—undergoing rapid depletion. Up until now, there has been little incentive to measure or manage natural capital: it has been treated as essentially inexhaustible.

Measuring natural capital involves assessing both its stock and its importance to society. Neither is simple to do, but the latter is particularly difficult: valuation in economic or other (e.g., cultural, spiritual) terms involves resolving fundamental philosophical issues (such as the underlying bases for value), the establishment of context, and the defining of objectives and preferences, all of which are inherently subjective. Even after doing so, one is faced with formidable technical difficulties with interpreting information about the world and transforming it into a quantitative measure of value. Just as the full value of a human being cannot be captured in economic terms (on the basis of his or her wage-earning power, or the economic value of his or her constituent materials), there exists no absolute value of ecosystem services waiting to be discovered and revealed to the world by a member of the intellectual community. Other entries in the Encyclopedia discuss these issues and describe alternative empirical valuation techniques, their applicability to different types of ecosystem services, and the advantages and limitations of their use.

As a whole, ecosystem services have infinite value because human life could not be sustained without them. The evaluation of the trade-offs currently facing society, however, requires that we estimate the marginal value of ecosystem services (the value yielded by an additional unit of the service, all else held constant) to determine the costs of losing—or the benefits of preserving—a given amount or quality of services. The information needed to estimate marginal values is also difficult to obtain and is presently unavailable for many aspects of the services. Nonetheless, even imperfect measures of their value, if understood as such, may prove superior to ignoring ecosystem services altogether, as is generally done in decision making today.

V. SAFEGUARDING

How can the Ecosystem Services Framework be made operational to ensure the safeguarding of key services? Here success hinges on the institutional framework in which valuation is done and incorporated into policy. Valuation is not a solution in itself—it is merely one way of organizing information in the much larger politics of decision making. The actual safeguarding of ecosystem services will require the establishment (or strengthening) of institutions that reward such action. Institutions must be tailored to local ecological, economic, and cultural circumstances and provide a payoff for conservation in a relatively short and certain time frame to be effective (see Heal, in press).

The management by New York City of its drinking water quality offers a model example of how ecosystem services can be characterized (both biophysically and economically), monitored, and safeguarded. The city's water originates in the Catskill Mountains, about a hundred miles to the north of the city, and was once considered so pure and salubrious that it was bottled and sold throughout the Northeast. In recent years, the natural purification system of the Catskills has been overwhelmed by sewage and agricultural runoff, resulting in a reduction in water quality below minimum drinking standards. The United States Environmental Protection Agency (EPA) required that the city remedy this problem.

The city administration first considered replacing the natural system with a filtration plant, but found that it would cost an estimated $6–8 billion in capital plus $300 million in annual operations. These high costs prompted investigation of an alternative solution, namely, restoring and safeguarding the natural purification services of the Catskills. This would involve the purchase of land in and around the watershed to protect it, as well as subsidizing several changes on privately owned land: the upgrading of sewage treatment plants, the improvement of farming practices, and "environmentally sound" economic development. The total cost of this option was estimated at about $1.5 billion.

Thus, New York City faced a choice of investing $6–8 billion in physical capital or $1.5 billion in natural capital. In spite of some political opposition, the latter option clearly appeared to be the more attractive, so the city floated an Environmental Bond Issue to fund its implementation. This financial mechanism captures the important economic and public health values of a natural asset (the watershed) and distributes them to those assuming stewardship responsibilities for the asset and its services.

The Catskills supply many other valuable services, such as flood control, carbon sequestration (locking up, in forests and the soils beneath them, carbon dioxide, an important greenhouse gas), biodiversity conservation, and—perhaps above all—beauty, serenity, and spiritual inspiration. Moreover, these services benefit others besides water consumers in New York City. It would be absurd to try to express the full ecosystem service value of the Catskills in a dollar figure. In this case, fortunately, there was no reason to try: even a lower-bound estimate of the value of the natural asset was sufficient to induce adoption of a conservation policy.

The challenge is to extend this model to other geographic locations and to other services. In other places, water quality is certainly a growing concern: the United Nations Environment Programme reports that most diseases in the less-developed world are caused by contaminated water, and that 50% of people in these countries suffer from one or more water-related diseases. An estimated 10–15% of the U.S. population currently consumes water from systems that violate EPA contaminant standards. The EPA forecasts that communities will have to spend $140 billion over the next 20 years to maintain drinking water quality at minimum required standards.

To evaluate the potential for extending the New York City model, Walter Reid posed this question: How much land could be protected with an economic justification, using water quality as a major goal? A first-order approximation suggests that the area would be significant: 10% of U.S. land area and 14% of global land area under current population sizes. Reid further points out that hydroelectricity generation is another important good whose supply, like that of drinking water, economically justifies watershed protection in many instances. Costa Rica, for example, derives 99% of its electricity from hydroelectric plants; most of the water generating this electricity flows from protected areas. In 1995, Costa Rica established a small tax on water and electricity use, the revenues from which are now recycled into managing the protected areas to limit erosion and sedimentation and to maintain high water flows. Numerous ancillary benefits accrue in this scheme, including protection of valuable flood control, biodiversity conservation, and ecotourism services.

To extend this model to other places and other services, appropriate educational, financial, and legal institutions, tailored to cultural and economic circumstances, will be required. Without these, statements from ecologists and economists that ecosystems are important and valuable assets will accomplish little. Promising new institutions for safeguarding ecosystem services have emerged in a wide array of cultures and economies (e.g., Australia, Madagascar, the United States, Vietnam); at a variety of scales, from local to international; and among governments, nongovernmental organizations, and private sectors. The services safeguarded by these emerging institutions include pollination; pest control; water supply for drinking, for irrigation, and for hydropower generation; maintenance of soil fertility; sustainable harvesting of tropical timber; provision of aesthetic beauty; and even decomposition (of orange peels produced by Del Oro, an orange juice company in Costa Rica, which are carefully distributed in a reserve area).

VI. CONCLUSIONS

An enormous payoff could result from further research in the characterization (biophysical and economic) of ecosystem services and in the development of institutions for their safeguarding. A series of basic questions, spanning a wide array of disciplines, beg addressing in this area. With emphasis here on biophysical characterization, they include:

- Which ecosystems supply what services? What is the scale of delivery, transport, and consumption of the services?
- What are the relationships between the quantity or

quality of services and the condition (e.g., relatively pristine versus heavily modified) or areal extent of the ecosystem supplying them? Where do critical thresholds lie?
- To what extent do the services depend on biodiversity?
- To what extent, and over what time scale, are the services amenable to repair?
- How interdependent are the services? How does safeguarding or damaging one influence the functioning of others?
- What indicators could be used to monitor accurately and efficiently the changes in the supply or quality of ecosystem services?
- How effectively, and on how large a scale, can existing or foreseeable human technology substitute for ecosystem services?
- What are the main sources of uncertainty regarding ecosystem services, and how important are they? How can the uncertainty best be quantified and incorporated into policy?
- How can economic principles and tools best be brought to bear on the management of natural capital?
- Given that many values of ecosystems lie mostly in the future (and will always lie mostly in the future), how should future benefits be valued, in economic, cultural, or other terms?
- What financial, legal, and other social institutions are needed to safeguard critical ecosystem services? How can their development be catalyzed?

While a great deal is known about the functioning of ecosystems and the supply of services in general, abstract terms, there is a paucity of information on particular, local ecosystems and economies. Very little is known about marginal values (the net benefit or cost associated with protecting or destroying the next unit of an ecosystem) or about the nonlinearities in ecosystem responses to human impact. Often this information is not acquired until after it is too late to reverse harm done (e.g., after heavy flooding). And we still have relatively little experience with institutional mechanisms for safeguarding ecosystem services: these will vary with the different ecological and economic characteristics of the services (such as whether they are public or private goods, the time period and spatial scale over which benefits are realized, and so on).

Further development of case studies addressing these issues would be most helpful. Such work would define the envelope of opportunities and limitations in applying this conceptual framework; it would illuminate how general are the findings from specific localities; and it would serve as a guide to policy development. In the New York City case, for instance, officials are purchasing land and changing agricultural and municipal practices in the hopes of restoring the natural water purification services of the Catskills—all with quite limited scientific information. In this particular case, and generally, success in the policy arena hinges on whether the scientific underpinnings of policies are sound.

See Also the Following Articles

BIODIVERSITY AS A COMMODITY • ECONOMIC VALUE OF BIODIVERSITY, OVERVIEW • ECOSYSTEM, CONCEPT OF • ECOSYSTEM FUNCTION, PRINCIPLES OF • HISTORICAL AWARENESS OF BIODIVERSITY

Bibliography

Chichilnisky, G., and Heal, G. (1998). Economic returns from the biosphere. *Nature* **391**, 629–630.

Cohen, J. E., Tilman, D. (1996). Biosphere 2 and biodiversity: The lessons so far. *Science* **274**, 1150–1151.

Daily, G. C. (ed.). (1997). *Nature's Services: Societal Dependence on Natural Ecosystems*. Island Press, Washington, D.C.

Daily, G. C., Alexander, S., Ehrlich, P. R., Goulder, L. H., Lubchenco, J., Matson, P. A., Mooney, H. A., Postel, S., Schneider, S. H., Tilman, D., and Woodwell, G. M. (1997). Ecosystem services: Benefits supplied to human societies by natural ecosystems. *Issues in Ecology* **2**, 1–16.

Daily, G. C., Soederqvist, T., Arrow, K., Dasgupta, P., Ehrlich, P., Folke, C., Jansson, A.-M., Jansson, B.-O., Levin, S., Lubchenco, J., Mäler, K.-G., Starrett, D., Tilman, D., and Walker, B. (In review). *The value of nature and the nature of value*.

Ehrlich, P. R., and Mooney, H. A. (1983). Extinction, substitution, and ecosystem services. *BioScience* **33**, 248–254.

Heal, G. (In press). *Nature and the Marketplace*. Island Press, Washington, D.C.

Holdren, J. P., and Ehrlich, P. R. (1974). Human population and the global environment. *Amer. Sci.* **62**, 282–292.

Jansson, A. M., Hammer, M., Folke, C., and Costanza, R. (eds.). (1994). *Investing in Natural Capital: The Ecological Economics Approach to Sustainability*. Island Press, Washington, D.C.

Mooney, H. A., and Ehrlich, P. R. (1997). Ecosystem services: A fragmentary history. In *Nature's Services: Societal Dependence on Natural Ecosystems* (G. C. Daily, ed.), pp. 11–19. Island Press, Washington, D.C.

Overgaard-Nielsen, C. (1955). Studies on Enchytraeidae 2: Field studies. *Natura Jutlandica* **4**, 55–58.

Reid, W. A. (In press). A business plan for ecosystem services: Extending the New York City watershed model to other geographic regions and other ecosystem services.

Vitousek, P. M., Mooney, H. A., Lubchenco, J., and Melillo, J. M. (1997). Human domination of Earth's ecosystems. *Science* **277**, 494–499.

ECOTOXICOLOGY

J. M. Lynch, A. Wiseman, and F. A. A. M. De Leij
University of Surrey

GLOSSARY

bio-accumulation The accumulation of toxicants or other chemicals by successive organisms in the food chain.

biological integration Organization of life functions in distinct, self-regulating units (cells, tissues, organs, organisms, populations, communities, and ecosystems).

biosphere The region of the earth where life exists, including parts of the lithosphere, hydrosphere, and atmosphere

ecosystem A self-regulating assembly of communities of animals, plants, and microorganisms interacting with one another and with their nonliving environment.

endocrine disruptor Chemical that interferes with the chemical communication (mainly performed by hormones) within an organism.

remediation Removal of toxicants from a contaminated environment using chemical, physical or biological means.

resolution The magnitude of a biological response due to a chemical insult.

specificity The ability to relate a chemical effect to a distinct biological function or organism.

xenobiotics Toxic substances not naturally produced within organisms.

ECOTOXICOLOGY CAN BE DEFINED as "the study of harmful effects of chemicals upon ecosystems." This implies that ecotoxicology is not concerned with the detection of chemicals per se, but that it deals with biological effects of (toxic) chemicals that contaminate (or have contaminated) the environment. Biological effects induced by chemicals can be anything from a molecular effect (e.g., genetic change) in an individual organism to effects on the biosphere as a whole.

I. INTRODUCTION

The manifest principles of ecotoxicology involve the application of the principles of toxicology to the environment: focused on human activity leading to the release of molecules such as polynuclear aromatic hydrocarbons (PAH), polychlorinated biphenyls (PCBs), xenoestrogens (XEs), chlorofluorocarbons (CFCs), pesticides, heavy metals, radionucleotides, greenhouse gasses (CO_2, CH_4), sulfur dioxide (SO_2) and oxides of nitrogen (NO_x) into the environment.

Biomonitoring is undertaken sometimes as the "prelude to remediation." Remediation aims to remove potential toxic substances from a contaminated site, thus restoring ecosystem function as far as is reasonably possible following the removal of the ecotoxicity by biological, physical, or chemical means.

Molecular toxicology is the branch of toxicology (the study of the effects of manifestly poisonous substances on individuals) that adopts the biochemical approach to the understanding of the detrimental threats to life, often recognized by a range of morbid conditions in a variety of fauna and flora. Such pathology can be understood at a molecular level in terms of biomolecular damage undergone by such macromolecules as DNA (deoxyribonucleic acid: the genetic determinant mainly residing in the nucleus of the living cell), RNA (ribonucleic acid: responsible for transfer of the genetic message inherent in DNA structure sequence to the cytoplasm of the cell), and diverse proteins constructed from a choice of sequence of residues of 20 amino acids (obtained mainly from dietary protein hydrolysis during digestion).

It is the increased understanding of chemistry and its associated biochemistry that allows a meaningful prediction of toxicity of chemicals in all forms of life (ranging from bacteria and higher fungi to plants and animals including *Homo sapiens*). Although no meaningful distinction can be made between natural and manmade chemicals released into the environment, detoxifying enzymes such as the ubiquitous oxygen-requiring enzyme family, cytochromes P-450, present in most forms of life, are useful biomarkers for constitutive or acquired tolerance to chemicals such as polycyclic aromatic hydrocarbons (PAHs), polycyclic halogenated biphenyls (PCBs), and xenoestrogens (XE). In the course of millions of years of biological evolution of species, survivors now display a range of such biochemical defenses against toxicants including the more recent pollutants released through industrial activity in processing and manufacturing industries.

Particular concern has been displayed recently for the contamination of the environment with endocrine disruptors (environmental hormones). These mimic the biochemistry of natural estrogens (such as 17β-estradiol) and these have been associated with feminization of males, leading to displayed hermaphroditism in a few cases, in polluted environments. Some of these endocrine disruptors are derived from the release of industrial chemicals and are therefore referred to as xenoestrogens (foreign-compound estrogens). This is using the accepted nomenclature of xenobiotics as "foreign" or "manmade" chemicals, as contrasted with natural chemicals of which the environment, the diet, and all life is composed.

Several important concepts have been incorporated into the principles of ecotoxicology. For example, "hazardous chemicals" have the potential for causing harm to a variety of organisms, but the assessment of the risk to the biosphere posed by these chemicals must be based on consideration of the exposure (dose) to individuals, populations and communities as well as direct toxicity to the different species involved. Environmental exposure is not only determined by the quantity of a chemical released but depends also on the characteristics of a particular environment and the distribution of the biota in that environment. Soil that contains large quantities of clay minerals and humic substances might lead to immobilization of hazardous chemicals through absorption, while chemicals released in aquatic environments might be dispersed rapidly. In both cases, toxicity to the biota living in these environments might be limited as exposure (bioavailability) is small. Also the distribution of different species in the environment is not random, but populations of a particular species are concentrated at particular sites (niche). For example, in soil some species are found in the litter that lays on the soil surface (epigeic and hemigeic species), while others frequent the deeper mineral soil layers (eudaphic species). It is clear that epigeic and hemigeic species are more likely to be affected by toxic chemicals that enter the soil environment from above than eudaphic species. Besides distribution, exposure will be determined by feeding patterns. Organisms that feed on prey that is likely to have accumulated the substance will be exposed to larger quantities of the toxicant than those that feed on substances that are relatively uncontaminated. Especially recalcitrant, low water soluble chemicals such as PCBs and organochlorines (DDT, for example) can be biomagnified in the environment to lethal concentrations higher up the food chain. Furthermore, toxicity of each hazardous chemical should be related to the metabolism of each major species present, as affected by the toxicant. Therefore, signs of toxicity need to be sought by a battery of methods before meaningful conclusions can be drawn on the environmental risk of a potential chemical released in the environment; environmental pollutants at a specified dose may cause changes in finely balanced ecosystems due to differential toxicity to different species and differential exposure of different species that inhabit the environment.

Much of ecotoxicology relates to the toxic effects of

chemicals in natural ecosystems, but it must be remembered that man is an integral part of these ecosystems and effects on the other biotic components of the ecosystems of which he is part can affect him directly or indirectly. For example, in relation to bioaccumulation of toxic chemicals, substances like PCBs, chlorinated organopesticides, and heavy metals, are likely to accumulate (and cause toxic effects) in the human body because of the simple fact that man is high up in the food chain. Indirectly, man is dependent on the biosphere for the maintenance of the climate, purification of water and air, and the provision of a sustained food supply. Disruption of biological activity that impairs the capacity of the living world to fulfill these functions will ultimately affect humankind's well-being and survival. In this respect, therefore, human toxicology and ecotoxicology are of necessity closely interlinked.

II. APPLICATION OF DIFFERENT SCIENTIFIC DISCIPLINES TO ECOTOXICOLOGY

A. Chemistry

Many different xenotoxicants of many chemical structures are now ubiquitous in aquatic, terrestrial, and atmospheric environments. Most, however, are present at extremely low concentration. For example, there may be 60,000 molecular-species of "hormone-disrupters" such as the xenoestrogens that may be detected (albeit by remarkably sensitive detector systems that can respond to even a few hundred molecules for a positive identification). Nevertheless, most of these xenoestrogens are extremely weak molecular-mimics of human estrogens, and many of these can, paradoxically, stop the estrogenic response of particular target tissues.

Proposed chemical approaches to ecotoxicology are often analytical in nature and seek to quantify the effect of each chemical species that can be identified and assayed so as to define the concentration-dependence of any hazardous substance and its consequent environmental or human risk. Quick tests for biohazards are essential, such as that achieved in the famous Ames Test for mutagenicity (and related carcinogenicity) in selected bacteria, for example, by the polyaromatic hydrocarbon, benzo(a)pyrene. Such mutagenicity (novel growth-behavior) due to genetic (DNA) damage may manifest as cancer in mammalian organs (because of loss of growth-control in particular tissues, due to subtle alterations in regulation of cellular growth by particular genes).

B. Biochemistry

Biochemistry studies the chemical processes that take place in living organisms. The main pertinent principle of biochemistry that can be applied to ecotoxicology is the modern concept of enzyme (biocatalytic proteins) regulation by molecular-intermediates of tissue metabolism in cells. Many xenobiotics (foreign chemicals unusual in nature) can interfere with the finely balanced biochemical reactions of living cells by perturbation of the web of molecular interactions necessary for life.

Additional problems are due to the molecular damage to DNA (deoxyribonucleic acid) and its related RNA (ribonucleic acid) along with cellular membranes (phospholipids that are readily destroyed by reactive oxygen species (ROS), generated by reactions of these chemicals with atmospheric molecular-oxygen). The ecotoxicity of ROS may, however, be ameliorated by antioxidants in the environment (and by dietary antioxidants such as vitamins C, E, A, and D as well as phytoestrogens in the diet).

C. Microbiology

The activity of microorganisms is at the basis for most functions on which life on this planet depends. Due to their almost limitless metabolic capacities, microbial communities are responsible for the transformation and recycling of organic and inorganic molecules in the environment. This activity results in the maintenance of nutrient cycles, the maintenance of soil fertility, and the detoxification of toxic substances in the environment. Furthermore, many microorganisms are intimately associated with every conceivable higher life form on this earth. They perform functions without which higher organisms could not function. Examples of this kind are organelles that have a microbial origin, including mitochondria, which are responsible for the energy generation in eukaryotic organisms (organisms with a true nucleus), and chlorophyll that allows algae and higher plants to convert carbon dioxide into sugars via photosynthesis. Other symbiotic interactions between microorganisms and plants that are of crucial importance for plant growth and nutrition are dinitrogen fixing microorganisms and mycorrhizae, which have close associations with the roots of many plant species. In fact plant roots have adapted specially to accommodate for these microorganisms. Furthermore, microorganisms associated with the gut of animals provide essential vitamins that animals cannot synthesize themselves. On the negative side, microorganisms are responsible for causing a wide variety of diseases in plants and animals. Although this is negative from the viewpoint of the organism (or population) that is affected, disease causing organisms are important in ensuring that excessive population growth of particular populations is curtailed, thus ensuring the maintenance of an environment that contains a rich diversity of species. The interdependence between microorganisms and higher life forms means that toxic effects that affect either group of organisms can have important consequences for the other.

D. Biotechnology

Biotechnology is an applied science that aims to harness different life forms for the benefit of man. Agriculture, antibiotic production, and bioremediation are examples of such applications. Recent advances in molecular biology have given biotechnologists new tools to change the nature of life by genetic modification. This technology has resulted in the creation of organisms that would very unlikely have arisen in nature via normal mating and exchange of genetic material. It is now possible for humans to combine the genes of widely different organisms into one organism, giving this recombinant organism the means to express novel characteristics. Such "novel characteristics" include resistance to diseases and pests, resistance to pesticides, or the means to produce products that are of commercial or medical value. Whereas these organisms are of potential value to society, it is necessary to evaluate their environmental impact before they can be safely released into the environment. Therefore, with the advance of biological science comes the responsibility for biotechnologists to ensure that the integrity of the environment is maintained by ensuring that the biological functions that allow life on this planet to thrive are not impaired. Interestingly, recombinant technology has also opened the way to construct organisms that can be used to monitor the environmental impact of toxicants more sensitively. An important interface of ecotoxicology with biotechnology arises in the use of such genetically modified organisms as biomonitoring tools. However, the detection (and assay) of ecotoxicants can be achieved by a variety of methods. An interesting example is the detection of carbon monoxide. This can be achieved using a caged canary (that responds sensitively to carbon monoxide in coal mines) or using an enzyme-based detector electrode that functions as a biosensor for this gas. A detailed account of biotechnological techniques for biomonitoring and bioremediation in relation to ecotoxicology can be found in the book by Lynch and Wiseman (1998).

E. Ecology

Ecology is the study of the interactions between organisms and their environment that determine the distribution and abundance of organisms. These interactions can be studied at different levels of biological integration starting from individuals, which are part of populations, which are part of communities, which are part of ecosystems, which are part of the earth's biosphere. To gain a living from nature, humans have to understand the interactions that determine the abundance and distributions of organisms on which human life depends. This is not only important for the harvest of species from natural ecosystems by hunting, fishing, and or gathering of plant products, but also where the aim is to modify the natural environment to yield predominantly products that are of economical value as is the case in agriculture and forestry. Not only do we need to understand how we can most efficiently use the environment, but also increasingly we need to be aware of the impact of human activities on ecosystem function in general. Pollution of the environment by human activities (manufacturing, mining, waste disposal, transport, energy use, pesticide use, etc.) can have a large impact on ecosystem function via effects at all levels of biological integration. Especially since the scale of human activities has increased so dramatically over the past century there is a real danger that human activity interferes significantly with the basic biological processes of sustained life, including our own. Therefore, ecological knowledge applied to ecotoxicology aims to evaluate the consequences of human activity to ecosystem function, especially in relation to pollution caused by human activity.

F. Toxicology

Important principles of toxicology that relate to ecotoxicology include the concepts of extent of exposure, persistence, and distribution of chemicals in the environment. Subsequently predictions can be made on the toxicity of such chemicals to individual organisms or populations. The starting point of such analysis is often the chemical structure of each toxin, which will allow some prediction of the behavior of the chemical in the environment to be made. One of the best examples of this kind is the environmental impact assessment of the insecticide DDT. DDT is almost completely insoluble in water but readily soluble in fat. Furthermore, DDT and its degradation product DDE are highly persistent. Its low solubility means that it is easily dispersed in aquatic environments (in the case of DDT it is found at low concentrations all over the globe) and only accumulates in places with a high fat content (i.e., living organisms). Once it has entered living organisms, it persists and accumulates in the fat tissues of organisms that are higher up the food chain resulting in toxicity. For this reason, populations of both fish eating birds and bird eating raptors such as peregrines and hawks were badly affected by DDT, even in cases where they inhabited pristine environments.

Another aspect of toxicology is risk analysis. This quantitative topic is most easily studied when death rates within a population can be quoted. Whereas the rates can be determined experimentally for animals, plants, and microorganisms, the determination of the effects on the human population is inevitably more difficult to determine, and one of the few situations where effects are well known is the mutagenic effects of radionucleotides, resulting from contamination caused by accidents at nuclear power installations or nuclear warfare.

III. ENTRY, MOVEMENT, AND FATE OF POLLUTANTS IN ECOSYSTEMS

Organic and metallic (including radionucleide) toxicants that enter the environment are regarded as pollutants of air, soil, and water. Their movement and transport is through air and water depending on their volatility and solubility, respectively. Soil and the detritus of sediments provide important solid supports for adsorption to regulate the movement and flow of the pollutants through terrestrial ecosystems, as well as influencing the localization and persistence of the pollutants in the environment. In this respect clay and humus particles that are electrostatically charged have an important role to play. Some organisms will take up the pollutant and concentrate them in their cells, a phenomenon known as bioaccumulation or biomagnification. This is particularly serious where the toxicant enters the food chain leading to progressive accumulation of toxic molecules higher up the chain, including humans. This is often the case with molecules that are relatively biologically inert such as heavy metals, PCBs, and organochlorine insecticides. Accumulation of such molecules in different species can lead to toxicity expressed as reduced growth, reduced fecundity, changes in behavior, susceptibility to diseases, or increased mortality. However, on the positive side, where there is storage of the pollutant in the cell, as is normally the situation with (heavy) metals, it is possible that the

pollutant might be harvested and therefore removed from the environment. This especially applies to plants that hyper accumulate metals in their tissues. The application of this process for the cleanup of contaminated land is known of phytoremediation. Also, many pollutants are biologically degradable and can therefore be detoxified. Both plants and mammals have a range of enzymes that are involved in detoxification of molecules that are potentially harmful to them (cytochromes P-450, for example), whereas the almost limitless metabolic capacity of a wide variety of microorganisms allows degradation of pollutants, especially hydrocarbons, into nontoxic molecules such as water and carbon dioxide. The stimulation of microbial degradation and metabolism of pollutants is known as bioremediation and is currently used to clean up contaminated land while microbial degradation of pollutants present in sewage and water is used in a variety of water purification systems. The fate of pollutants in the environment is therefore not only dependent on the molecular characteristics of the pollutant but also on a range of biotic and abiotic factors.

IV. BIOMONITORING

The ultimate goal of biomonitoring is to use biological effects resulting from chemical exposure for making predictions and deductions about the quality of the environment for life in general and human life in particular. Biomonitoring activities can be conducted on various levels of biological integration, over a period of time. Several overlapping terms are commonly used in this process. It is worth noting that some of the tools are, strictly speaking, alternatives for chemical/physical methods to measure bioavailability of chemicals in the environment. In other words, they do not provide information on ecological effects per se.

- The term *bioreporter* is used for a molecular tool, often genetically modified cells, to transform the presence of a chemical into an easy measurable signal (e.g., luminescence).
- The measurement of biochemical and physiological variables in individuals or in their excretion products, providing information on exposure or damage, is indicated by the term *biomarker*.
- A *biosensor* or *bioprobe* is a physical device that allows the detection of a chemical as an electrical signal derived from a biocatalyst, such as an enzyme or an antibody.
- A *bioassay* is a toxicological test system in which

TABLE I

Some Tests of Chemicals That Affect Soil Function

Retention capacity of soil and endangering of groundwater	Living space for plant production	Living space for soil communities
Growth (algae)	Dehydrogenase (*Bacillus cereus*)	Biomass (microorganisms)
Immobilization (*Daphnia*)	Nodulation (*Rhizobium*)	Enzyme activity (microorganisms)
Light emission (*Photobacterium phosphoreum*)	Shoot or root growth (various plants)	Nitrification (bacteria)
Mortality (fish, nematodes)	Yield (various crops)	Mortality (earthworms)
Mutagenicity (Ames, Umu)		Reproduction (collembola, earthworms)
		Respiration (microorganisms)

the activity of a chemical is measured as an adverse effect on a population of a specific test species.

- A *bioindicator* is used to detect the effects of chemicals at the individual, population, community, or even ecosystem level.

Important in this respect is that when we move up to the higher levels of biological integration (populations, communities, ecosystems), effects are increasingly difficult to link to a specific chemical, even though the effects measured are of more importance. Furthermore, it takes an increasingly long time for higher levels of integration to respond to a chemical insult. For example, DDT took years to bioaccumulate up the food chain resulting in negative effects at the population/community level. On the other hand, chemicals that are apparently nontoxic at the lower levels of biological integration (carbon dioxide, chlorofluorocarbons (CFCs), sulfur dioxide, etc.) can have profound effects on all levels of biological integration. These chemicals might not directly affect biological processes but affect the chemical balance of the earth's biosphere (air, water, and soil). For example, greenhouse gases (carbon dioxide, methane, and water vapor) trap the sun's long wave radiation augmenting global warming, which could lead to profound changes of the world's ecosystems. CFCs contribute to the destruction of the earth's protective ozone layer, leading to an increase in harmful UV radiation that reaches the earth, resulting in increased lethal mutation rates and cancer. Sulfur dioxide and NO_x emissions lead to acid rain and acidification of soil and water, which in turn leads to the release of toxic quantities of metal ions in soil and water that kill both vegetation and animal life. An interesting example of this kind is the acidification of lakes in Scandinavian countries as a result of acid rain. The resulting lower pH of the lake water means that the water can contain

a higher concentration of aluminum ions. However, the pH around the gills of fish is slightly alkaline, resulting in the precipitation of aluminum on the gills. This results in the impairment of gill function and suffocation of the fish. Even though Scandinavian lakes affected in this way look perfectly clean, they are devoid of fish.

A variety of specific tests have been employed to investigate the effects of chemicals on the environment. Generally the biotic effects used for biomonitoring deploy biota from the environment and expose them to different concentrations of the test chemical. Mortality of fishes, for example, has been used to assess toxicity of chemicals that are used in the soil environment (Table I). In the soil environments a variety of effects can be investigated as to whether the retention capacity of soil would affect the toxicity of a chemical leading to leaching into groundwater or the living space for plant production and soil communities. In this respect, the assays relate to the effects on populations, gene products (enzymes and metabolites), or impact (plant bioassays) (Fig. 1). However, the exciting development of

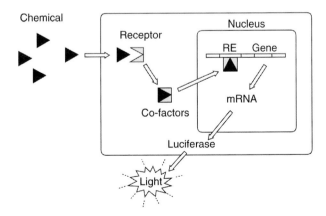

FIGURE 1   Receptor/reporter gene assays; RE: responsive element.

recent years is the recognition that these effects are all controlled by genes and another method of biomonitoring is to use gene (DNA) probes (Fig. 2). This option is very much in its early stages and the concept is to link a gene, which responds to a specific chemical or general effect, to a promoter DNA sequence and a reporter DNA sequence that either encodes for an enzymatic effect or any other specific signal. It is therefore possible to use genes, which respond to mutagenic substances with a light signal that is proportional to its mutagenic potential in the environment. Whereas this approach offers much promise, the reality is that it is complementary to the other methods currently deployed and should be used in conjunction with them.

A particular useful activity, which has been monitored in aquatic environments, is cytochrome P450. (Table II). This is the terminal component of the mixed-function oxygenase, which catalyses expoxidation, hydroxylation, dealkylation, and desulfurisation, all critical in detoxification of toxic molecules in invertebrates and vertebrates (including humans). To date, more than 750 cytochrome P450 genes have been identified (Nelson, 1998). The cytochrome P450/A1 (CYPAI) isoform is a biotransformation enzyme involved in a wide range of xenobiotic metabolisms (chemicals foreign to the natural environment) including metabolism of polyaromatic hydrocarbons (PAHs). Usually CYPIAI has been used as a biomarker in fish and mollusks. The linking of this enzymatic function with its genetic control offers a great deal of potential for the biomonitoring of toxicology of both marine and freshwater ecosystems.

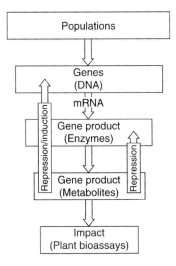

FIGURE 2 Levels of perturbation measurement.

TABLE II

Hepatic Cytochrome P450 1A (CYPIA) as a Biomarker of Organic Pollution in Fish

Fish species
Atlantic tomcod (*Microgadus tomcod*), cod (*Gadus morbua*), Largemouth bass (*Micropterus salmoides*), Rainbow trout (*Oncorbynchus mykiss*), Whitefish (*Coregornus lavaratus*)
Point sources of pollution
Chemical plants, incineration plants, industrial complexes, landfill sites, oil industry, oil spills, natural oil seeps, pulp mills, sewage
Diffuse areas of pollution
Bays, bights, estuaries, fjords, gulfs, harbors, lakes, offshore areas, river systems.

V. BIOLOGICAL PROCESSES

The question can be raised why it is necessary to analyze complicated biological processes if the chemical in question can be measured directly using chemical or physical methodologies. There are at least three situations where biomonitoring provides answers that cannot be arrived at using direct chemical measurements. The first use for biomonitoring is in situations where one wants to trace the chemical history of an environment. For example, the pH history of lakes can be deduced from the percentage acidophilous and alkalophilous species within the diatom communities preserved in sediment cores (Renberg and Hellberg, 1982). Similarly, pollen records preserved in soil provide information on past vegetation from which the past environmental conditions can be deduced. The second use of biomonitoring is a situation where the activity of a chemical is short-lived but the effects are persistent. For example, pesticides that disperse rapidly or are broken down rapidly cannot be measured easily, but their biological effects might persist for long periods. The insecticide deltamethrin is active only for a few hours after spraying, while effects on sensitive ground spiders and beetle fauna remain visible for weeks after spraying (Everts, *et al.*, 1989; Jagers op Akkerhuis, 1993). The third reason, and probably the most important one, for using biomonitoring is that effects of chemicals are dependent on the interactions that take place within the environment itself after their release. These interactions can be of a physical/chemical nature where the chemical is immobilized on to soil particles and is therefore not bioavailable. For example, the toxicity of metals, such as copper and zinc, in soil is determined by their bioavailability in soil. Both copper and zinc ions bind to soil minerals (mainly clay particles) and humic substances and as such are unavailable to the biota present in soil. Soils contain large quantities of these potential binding sites and therefore toxic effects of heavy metals are often buffered, even at relatively high concentrations. Because of this soil-buffering capacity, it usually takes some time of sustained metal inputs before metal ions become bioavailable at concentrations that are toxic. Both the mineral and organic fractions of soil interact with metal ions, providing negatively charged surfaces on to which cations can bind. Such adsorption displaces the cation that was previously counter-balancing the colloid's negative charge; hence the ability of colloids to adsorb ions in this way is known as the soil's cation exchange capacity (CEC). Heavy metals may also specifically adsorb on to hydrated oxides of aluminium, iron, and manganese by the formation of partly covalent bonds and can be chelated by solid-phase humic substances or by other organic ligands. The adsorption of metal ions is also affected by the soil pH. Acidification decreases the CEC of humic substances when carboxyl and ammonium groups will become protonated, reducing their capacity to bind metal ions. The increase in hydrogen ions will also increase the competition for negative binding sites, leading to the displacement of metal ions. Therefore, the size distribution of mineral particles, the amount of organic matter and the soil pH will to a large degree determine the metal buffering capacity of the soil (Gupta, 1991).

Besides physical/chemical interactions, chemicals interact with the biota present in the soil. For example, sustained use of specific pesticides leads to the development of microbial communities that degrade the pesticide, reducing their half-life time and therefore their environmental impact. On the other hand, chemicals such as PCBs and organochlorides are likely to accumulate up the food chain resulting in high concentrations of residues distant from the point of release, thus increasing their environmental impact. As biological systems are not just collections of independent organisms but are strongly dependent on each other for their survival, it follows that chemical effects on one species can have important consequences for species that depend on the affected species. An example of this type of indirect effect is the decline of the gray partridge (*Perdix perdix*) in West Sussex, England. It could be shown that 48% of chick mortality was explained by the density

of preferred insects. Given that reduced availability of insects is the key to the partridge decline, it was postulated that this might be due to increased pesticide usage, especially since there was a strong correlation between the partridge decline and increased herbicide use. The effect of the herbicides was shown to be indirect (they are relatively nontoxic to partridges), removing the host plants on which the insect larvae fed that were the food source of the partridge chicks, resulting in less insects and therefore less food for the partridge chicks. In conclusion, because biological effects are dependent on the behavior (movement, persistence, bioavailability) of the chemical in the environment, the sensitivity of biota and the role these biota play in ecosystem function, biomonitoring is often the only available way for monitoring chemical effects.

VI. EARLY WARNING SYSTEMS

Life as a whole can be seen as an hierarchical systems with distinct levels of biological integration. Going from a low level of organization to a high level of organization we can recognize cells that make up tissues, that make up organs, that are part of an organism; these organisms are arranged in populations that form part of communities, that are ultimately part of ecosystems. In general, it can be said that biological systems are at each level of organization buffered to resist change in overall performance by adjusting the components that make up that level of integration. The result of these buffering effects is that effects at lower levels of integration are dampened so that they do not affect higher levels of integration. Only when the buffering capacity at a certain level of integration is exceeded will the next level of integration be affected. This implies that toxic effects can be measured at the lower levels of biological integration before they result in effects at higher levels of integration. Because the aim of biomonitoring is "to provide a means by which (impending) changes in ecosystem function can be detected," compensatory effects at lower levels of integration can provide an early warning system of adverse chemical effects. A good example of such an early warning system is the thinning of bird eggshells as a result of DDE (one of the metabolites of DDT). DDE affects Ca deposition in eggshells (physiological change). However, only when egg shell thinning is in excess of 16 to 18% will this lead to a decrease in breeding success of the birds affected. Eggshell thinning is therefore an early warning system for population change. Not all bird populations are likely to be affected as DDE accumulates up the food chain. Only birds that are at the top of the food chain (raptors and fish-eating birds) will be affected in the first instance. However it can be assumed that if DDT use had continued, the decline in raptors and fish-eating birds would eventually have resulted in community changes.

What is important in this context is to recognize which factors determine the buffering capacity at each level of biological integration. These factors can be used profitably in biomonitoring. It is, however, important to recognize that if we are interested in impending changes at the community level, it is probably most sensible to look for changes that are occurring at the level below (i.e. the population level). Similarly, impending population effects are most profitably investigated using effects on individuals. However, changes at the organism level will be too sensitive for making reliable predictions at community level and above. This simple fact is often overlooked in biomonitoring where current molecular technology has allowed precise investigation at the lowest levels of integration. These techniques allow the development of sensing systems that pick up changes that occur at the molecular and physiological level. Whereas these approaches are valuable for medical purposes in which we are interested in warning systems that help prevent problems at the individual level, they are inappropriate for monitoring ecosystem changes, in that they are far too sensitive. In choosing the most appropriate monitoring tool, one has to evaluate its relevance, reliability, robustness, responsiveness, and reproducibility. These factors, known as the 5 Rs, will differ according to level of biological integration and the composition of biological units that make up that level. It is unlikely that an environmental monitoring system will be "a freeze dried, talking bug on a stick" (van Straalen, 1998). As interactions that govern processes at each level of biological integration become apparent, new and more appropriate biomonitoring tools will become available. The simple fact that only a fraction of the species that make up "biological life" are classified shows that we have still a long way to go.

VII. SPECIFICITY AND RESOLUTION

Within each level of biological integration the biological units that make up the level of integration display different degrees of specificity and sensitivity to different forms of (chemical) stress. With the term *specificity* it is meant that a biological effect can be related to a specific stimulus. For example, inhibition of the enzyme ALA dehydrogenase is only induced by lead, while induction of an immune response might be triggered by a wide variety of causes. The term *resolution* is used to indicate the ability of a bioindicator to respond to small environmental changes. For example, a specialized predator at the top of a food chain is a more sensitive indicator of bioaccumulation of persistent pesticides than omnivores with a varied diet. Although it seems that sensitive indicators with a great resolving power hold the greatest promise for biomonitoring, these properties are likely to result in overcaution and false alarms. In general, it can be said that indicators at lower levels of biological integration are more sensitive than characteristics that relate to higher levels of biological integration. Therefore, indicator species are more sensitive than diversity indices, for example. What makes a "good" bioindicator will depend on the information that is required. Looking at a "soil ecosystem" for example, the biological components that make up the system consist of plants, vertebrates, invertebrates (insects, earthworms, nematodes, mites and springtails), protozoa, bacteria, and fungi, all of which perform crucial roles in the maintenance of soil ecosystem function. When the system is exposed to a potentially toxic chemical the ability of each population to cope with this exposure will depend on genotypic plasticity within the population, phenotypic plasticity, mutation rate, metabolic capacity, and level of exposure. Genotypic plasticity will be high in microorganisms that multiply fast and can acquire resistance genes via mutations or by acquisition of plasmids carrying resistance genes. As a result, adaptation of microbial communities to chemical stress is likely to be high. Furthermore, because of the almost limitless metabolic diversity displayed by microorganisms, populations that contribute to the detoxification of a toxic chemical might increase rapidly (comparable with the activation of the immune system as a response to foreign molecules/organisms that enter a higher organism). Other organisms (plants, vertebrates, and invertebrates) that have a large phenotypic plasticity might respond to a chemical insult by avoiding those areas that are most contaminated. Furthermore, spatial distribution of the biota in the soil environment and their mobility will both determine exposure and therefore toxicity. Only when there is a clear understanding of "normal" distribution and diversity of the biota in a given ecosystem is it possible to relate biological patterns to the effects of a toxicant in a meaningful way.

Because of the biological complexity of ecosystems, efforts have been made to classify organisms not according to species but according to similarities in life strategy. Biological life can be roughly classified by the way organisms spend resources on their offspring. Organisms that invest in large numbers of offspring (giving each limited resources) are termed r-strategists, while those species that invest in few offspring (each of which is provided with ample resources) are termed K-strategists. Looking at the spectrum of life, there is a general trend that large organisms are K-strategists, while small ones are r-strategists. Although this is true to a large extent, within a certain body class size there are organisms that have adopted r- or K-type survival strategies. For example, mice are typical r-strategists (they have many large litters each year), while bats produce every two years a single offspring, making them typical K-strategists. Not surprising, K-strategists are characteristic for stable, climax environments while r-strategists are characteristic for unstable pioneer environments. In terms of "stress," K-strategists are adapted to cope with biological stress (competition, predation, etc.), while r-strategists are adapted to cope with physical/chemical stress but are not good at coping with biological stress. In general, it can therefore be stated that r-strategists are insensitive to chemical stress, while K-strategists are sensitive. When an ecosystem is challenged by toxins (pesticides, for example), the r-strategists (the pest organisms) are the first to develop pesticide resistance, while the K-strategists (nonpest species) will disappear. The reason for this might be found in the fact that large numbers of offspring provide large genetic variability leading to rapid selection of the best-adapted individuals. Chemical/physical stress on an ecosystem will therefore lead to a shift in the balance between r- and K-strategists in favor of r-strategists. This principle has been used successfully to monitor environmental pollution and disturbance using colonizer/persister (c-p) indices applied to nematodes (Bongers, 1990) or perturbations of rhizophere microbial communities (De Leij *et al.*, 1993).

Another useful way of looking at communities and ecosystems is by classifying the organisms that inhabit that ecosystem by their "trophic function." Using this approach, organisms can be classified (aboveground) as primary producers (plants and algae), herbivores, predators, and top predators, while organisms that feed on a variety of food sources are classified as omnivores. In general, food chains have no more than four trophic levels because less and less energy is available higher up the food chain. Furthermore, dependence of predators and top predators on exclusive feeding sources make them vulnerable to perturbations lower down the food chain. Using the concept of trophic levels, it can be argued that specialized feeders are more sensitive to environmental stress (when this stress affects their food source) than organisms that are less specialized. Not

surprisingly, when recalcitrant chemicals enter the food chain it is the specialized feeders (for example, peregrines and sparrow hawks that feed exclusively on birds) that are the first to be affected by toxins. Beside the fact that they are exposed to relatively high levels of a pollutant, these organisms are often very skillful specialized hunters. Any impairment of these skills due to a toxin will result in starvation. Therefore, these species can be sensitive bioindicators of toxins, even at sublethal concentrations.

Another important ecological concept that can be used to simplify ecosystems in a meaningful way is the concept of keystone species. The idea is that even though there is dependency between all species that are involved in a certain food web, some species are more important than others. It is important in this context to make a distinction between dominant species that derive their importance to the ecosystem in terms of biomass or energy flow and keystone species. Keystone species are often relatively rare but have a major impact because of their key regulatory function in the system as a whole. This is because these species are involved in more links or interact with parts of the ecosystem that maintain living conditions for a wide variety of other species. In other words, the species diversity of an ecosystem is determined in large part by a few keystone species on which many others depend. A detrimental effect on these species will result in the collapse of the whole ecosystem (or at least a large part thereof). Several different categories of keystone species have been recognized, such as keystone predators, keystone prey, and keystone habitat modifiers. One of the best-studied examples of the importance of keystone predators is the starfish (*Pisaster*) and predatory whelks (*Nucella*) in pools along the rocky inter tidal zone. Removal of these predators led to the disappearance of 80% of the prey species and the nearly complete dominance of the prey community by mussels (*Mytilus*). Sea otters (*Enhydra lutis*) have also been labeled as keystone species as they limit the density of sea urchins, which, in turn, eat kelp and other macro algae that provide the habitat for a large variety of species. Without sea otters the sea urchin population explodes resulting in the destruction of the kelp forests leading to the loss of habitat for those species that depend for their survival on the kelp forests for reproduction, shelter, and food. The way keystone species are affected by toxins is an important consideration in ecotoxicology, because they represent the Achilles' heel of an ecosystem. Not surprisingly, keystone species such as earthworms and honeybees are often used in toxicity testing because they are important for ecosystem function in general.

Clearly a negative effect on ecosystem function is often due to alterations of complex interactions within the ecosystem itself. In most cases, toxic effects are buffered due to the compensating capacity of an ecosystem. However, occasionally toxic effects can affect species that are responsible for essential functions in that ecosystem. Further insight in the way that different species contribute to ecosystem function is an essential requirement for the development of the most appropriate biomonitoring tools for assessing the ecotoxicological effects of chemicals.

VIII. CONCLUSION

Currently there is growing concern internationally about environmental toxicants affecting biodiversity and human health. Biodiversity has come to the fore by the United Nations meetings, the first of which was in Rio de Janeiro in 1992. The medical profession is increasingly concerned about a range of conditions that might be induced by chemicals in the environment. This is resulting in moves to increase legislation. The study of ecotoxicology seems therefore certain to grow in importance, as well as there being a stimulus to prevent problems occurring in the first place via the development of clean technology and stimulation of sustainable land use. On the other hand, ecotoxicology can help to identify existing pollution problems, which in many cases can be cleaned up using a variety of remediation strategies.

See Also the Following Articles

ACID RAIN AND DEPOSITION • AIR POLLUTION • ECOLOGICAL FOOTPRINT, CONCEPT OF • ENVIRONMENTAL IMPACT, CONCEPT AND MEASUREMENT OF • GREENHOUSE EFFECT • KEYSTONE SPECIES • SOIL BIOTA, SYSTEMS AND PROCESSES

Bibliography

Alloway, B. (1995). Heavy Metals in Soil. Blackie Academic and Professional, Glasgow.
Bongers, T. (1990). The maturity index: an ecological measure of environmental disturbance based on nematode species composition. Oecologia 83, 14–19.
Carson, R. (1965). Silent Spring. Penguin Books, Harmondsworth, England
De Leij, F. A. A. M., Whipps, J. M., and Lynch, J. M. (1993). The use of colony development for the characterisation of bacterial communities in soil and on roots. Microbial Ecology 27, 81–97.
Everts, J. W., Aukema, B., Hengeveld, R, and Koeman, J. H. (1989). Side effects of pesticides on ground dwelling predatory arthropods in arable ecosystems. Environmental Pollution 59, 203–225.
Gupta, S. K. (1991). Metabolisable metal in antropogenic contaminated soils and its ecological significance. In Impact of Heavy Metals in the Environment (J. P. Vernet, Ed.), pp. 299–310. Elsevier, Amsterdam.
Jagers op Akkerhuis, G. A. J. M., and Van der Voet, H. (1992). A dose-effect relationship for the effect of deltamethrin on a linyphiid spider population in winter wheat. Archives of Environmental Contamination Toxicology 22, 114–121.
Lovelock, J. E. (1979). Gaia: A New Look at Life on Earth. Oxford University Press, Oxford.
Lynch, J. M., and Wiseman, A. (Eds.) (1998). Environmental Biomonitoring: The Biotechnology Ecotoxicology Interface. Cambridge University Press, Cambridge.
McCarthy, J. F., and Shugart, L. R. (1990). Biomarkers of Environmental Contamination. Lewis Publications, Boca Roton.
Nelson, D. R. (1998). Metazoan cytochrome P450. Comparative Biochemistry and Physiology 121 Part C, 15–22.
Potts, G. R. (1986). The Partridge. Collins, London.
Renberg, I., and Hellberg, T. (1982). The pH history of lakes in Soutwestern Sweden, as calculated from the subfossil diatom flora of sediments. Ambio 11, 30–33.
Van Straalen, N. M. (1998). Evaluation of bioindicator systems derived from soil arthropod communities. Applied Soil Ecology 9, 429–437.
Walker, C. H., Hopkin, S. P., Silby, R. M., and Peakall, D. B. (1996). Principles of Ecotoxicology. Taylor and Francis, London.
Zelikoff, J. T., Schepers, J., and Lynch, J. M. (Eds.). (1997). Ecotoxicology: Responses, Biomarkers and Risk Assessment. Fair Haven, NJ: SOS Publications.

EDIBLE PLANTS

Eduardo H. Rapoport and Barbara S. Drausal
Universidad Nacional del Comahue and CONICET, Argentina

GLOSSARY

cultivar Cultivated variety or genetic strain of a domesticated food plant.
domesticate Plant that has been selected by humans and adapted for use as a food crop, nutrient, fiber, or other purpose.
ethnobotany Study of the variety, natural history, and characteristics of the plants used by human cultures.

THE VARIETY OF PLANTS NORMALLY CONSUMED BY HUMANS represents a tiny proportion of what nature supplies. Of the estimated 270,000 plant species recognized in the world, about 12,500 species from over 3000 genera are considered edible to humans.

I. FOOD HABITS

Animals, being terrestrial or marine, have limitations with respect to the variety of species they consume. They depend on what nature offers in the place where each animal lives. Since the geographic ranges of species are relatively widespread, the individuals of a given animal species may vary their food resources in different sites of the range, according to their opportunities. For this reason, the variety of plants ingested by a given species is always greater than the variety of plants ingested by an individual or by an entire population.

Domestic as well as wild herbivores may eat a wide variety of food items, but they have clear preferences for particular plant species. Some ungulates are capable of selecting the most nutritious individual plants among those of the same species. They select forage mainly by smell, and secondarily by taste. Volatile substances in the plants may either inhibit or attract foragers, and are largely responsible for their palatability. According to Klein (1970), the nutritive value of plants available to wild ruminants depends on the stage of maturity of the vegetation, with highest nutritive quality coinciding with the initiation of growth, as well as with soil type and climate. Rapid growth in plants is correlated with high nutritive quality. Similar conclusions were reported by Gardarsson and Moss (1970) in a study of food selection by the Icelandic ptarmigan (*Lagopus mutus*). This bird consumed leaves and flowers of 8, 11, 10, and 11 species, respectively, in summer, autumn, winter, and spring.

Similarly, the European hare (*Lepus europaeus*) in northwestern Patagonia makes use of 17 to 21 species of plants in a given season out of a total of 28 species that it consumes over the year. Table I provides a short sample of the variety of food consumed by different

TABLE I

The Number of Food Plants Consumed Annually by Different Mammal Species

Mammal species	Common name	No. of plant species	Source
Alouatta fusca	Brown howler monkey, SE Brazil	52	Galetti *et al.* (1994)
Ateles spp.	Spider monkey, Panama	14	Milton (1981)
Bos taurus	Cattle, NW Patagonia	23	Relva (1998)
Capra hircus	Goat, Mendoza, Argentina	76	Dalmasso *et al.* (1995)
Cebus apella	Capuchin monkey, SE Brazil	73	Galetti and Pedroni (1994)
Cervus elaphus	European red deer, NW Patagonia	34	Relva (1998)
Chiropotes satanas	Monkey, Venezuela	29	Kinsey and Norconk (1993)
Ctenomys mendocinus	Tuco-tuco, Argentina	28	Madoeri (1993)
Lagidium viscacia	Vizcacha de la sierra, Argentina	21	Galende and Grigera (1998)
Lama guanicoe	Guanaco, Mendoza, Argentina	47	Candia and Dalmasso (1995)
Lepus europaeus	European hare, NW Patagonia	28	Galende and Grigera (1998)
Ovis aries	Soay sheep, St. Kilka Island, U.K.	~12	Gwynne and Boyd (1970)
Pithecia pithecia	Monkey, Venezuela	25	Kinsey and Norconk (1993)

herbivorous and omnivorous mammals. The case of the Soay sheep was included to show how a domestic herbivore may restrict its diet in a species-poor environment. Of course, the range of foods ingested by other animals may vary widely. Monophagous insects restrict themselves to only one plant species, soil amoebae (*Acanthamoeba*) normally ingest five species of microscopic algae (Heal and Felton, 1970), and polyphagous, pest arthropods may feed on more than 300 species of crops and wild plants. The number of food plants eaten by humans is not far from the figures shown in Table I.

Domestic and wild ungulates may show a copious range of food species but they focus on a few, preferred plants. A number of less palatable species are used only in time of food scarcity. Studies performed in western Argentina by M. B. Kufner and S. Monge showed that the rodent *Lagostomus maximus* increases the variety of its food sources in degraded habitats and during droughts.

A. Standards of Consumption

On an individual basis, people use a small number of plant items per day, perhaps between 10 and 20 species or products. A normal diet includes common vegetables, fruits, seeds (in the form of flour or oil), roots, sugar, beverages (beer, wine, colas), condiments, teas, and herbals. The number of plant species normally used by an individual over a year, however, is about 100, although this is limited by the number of edible plant items commonly offered by popular markets and supermarkets. Taking into account all the variety of greens,

vegetables, fruits, grains, nuts, and condiments in an exceptionally well-provisioned supermarket, the figure (including different varieties and brands) may rise to 600 according to Duke (1992).

The stomach contents of two mummies that were found well-preserved in Danish bogs provided interesting information about the gastronomic habits of people during the Iron Age. Their last meals contained 66 different plant taxa (Godwin, 1960; King, 1966), many of which are nowadays considered as cosmopolitan weeds. Before the invention of agriculture, in the Paleolithic Age, humans were hunter-gatherers, and probably had a better knowledge of the variety of edible wild plants than modern people. This knowledge, however, has slowly been lost since the Neolithic and, in present times, is still lost after one or two generations of acculturation in aboriginal communities (Plotkin, 1993). The process of "civilization" goes hand in hand with the loss of knowledge, as well as with the abandonment of traditional crop varieties and the habit of gathering wild plants. But at the same time, new cultivars, coming from distant countries, are constantly increasing the variety of foods. A recent case is that of the kiwi fruit (*Actinidia deliciosa*), which originated in China and then was renamed, cultivated, and popularized in New Zealand. The roots of the ahipa (*Pachyrhizus ahipa*) and jícama (*P. erosus*), of Central American origin, have become increasingly popular in the United States and Southeast Asia, especially the latter species, which has a similar texture and flavor to the bamboo shoots and is used by Asian food restaurants in Western countries (National Research Council, 1989; Brücher, 1989). Rice

(*Oryza sativa*), of Asian origin, is at present the most popular staple in warm countries of South America, while the South American potato (*Solanum tuberosum*) has become "mandatory" in the European cuisine.

II. THE DIVERSITY OF FOOD PLANTS

No one has compiled a complete record of edible plants for the entire world. The Food and Agriculture Organization (FAO), part of the United Nations, publishes an annual report of the production of the commercially most important foods. This list includes about a hundred species of plants. In *The Oxford Book of Food Plants* (Nicholson et al., 1969), the number increases to 389 species distributed among 81 plant families. These are both locally and widely known cultivated plants. Duke (1992) estimated that North American Indians ate 1112 plant species. This figure is set at 1886 species according to Moerman (1998). More than 3000 edible species are carefully listed and commented on in the voluminous book 'Cornucopia', compiled by Facciola (1990), but in its preface Noel D. Vietmeyer suggests that there are about 20,000 edible species across the world. Probably, the most complete inventory is Kunkel's (1984) book, which lists roughly 12,560 species from 3100 genera belonging to about 400 families of flowering plants and ferns. This list, however, is being constantly enriched by the contribution of many ethnobotanical studies.

The proportion of edible plant species in slightly disturbed communities is variable. In the Sonoran Desert it is about 15%. Ona Indians from Tierra del Fuego made use of at least 6% of this island flora, whereas the Chácobo Indians, in the Bolivian Amazon, use 21% of their surrounding flora. Medium to highly disturbed communities may contain similar or higher proportions of edible species for human consumption. For example, in western Uruguay the proportion is 17%, in southwestern Córdoba province (Argentina) it is 19%, in the outskirts of Havana (Cuba) it is 33%, in swidden (slash-and-burn) fields of northern India it is 43%, and in experimental fields in Saskatchewan (Canada) it may reach 61% of all wild plants. Yet studies like these do not necessarily reveal the actual possibilities offered by nature, but rather the knowledge of informants and/or the perspicacity, experience, and field-work time employed by the investigators. According to the estimate that about 10% of any flora represents food resources, then 10%, or 27,000, of the 270,000 species of plants already recognized by world botanists should be edible.

Since historical times, because of written testimony,

Europe has conserved people's knowledge of gastronomic matters. From the botanical point of view, the United Kingdom is probably the best-known country in the world. If we compare the floristic list compiled by Martin in 1976 with Kunkel's list of food plants, and discard the exotic species, hybrids, and other subspecific taxa, as well as plants used only during famine times, we can verify that out of 1503 species considered, 350 are edible. In other words, 23% of the British flora is edible. Thus, we have two estimates of the possible richness of edible vascular plants—10 and 23%—and they represent between 27,000 and 62,000 species, respectively, based on the 270,000 known at present. Because the description of the world flora has not been completed yet, the final list of comestibles will probably increase in the future.

By comparison, less than 2% of the Central American flora is eatable, based on the list prepared by Duke (1992). Possible explanations for this remarkable difference are: (1) greater taxonomic ignorance and/or less exploration of the natural resources—because the flora of Central America is much richer than the flora of Britain, humans may have concentrated on fewer, more abundant and profitable plants, and disregarded the less useful ones; and (2) widespread loss of cultural heritage and environmental knowledge following the conquest and colonization by European countries.

A. The Most Prolific Taxa

A first, rough estimation at higher taxonomic ranks indicates that the proportions of edible species are quite similar to the proportions of "common" (edible and nonedible species) species present in the plant kingdom. The right-hand column in Table II is based on a random sample of 1790 food plants appearing in Kunkel (1984).

If the property of being eatable or palatable were

TABLE II

A Comparison between Common and Edible Species in Higher Taxonomic Groups (Figures Represent Percentages of Their Respective Totals)

Taxonomic group	Common species ($n = 270,000$)	Edible species ($n = 1790$)
Pterydophyta	3.9	1.9
Gymnospermae	0.3	1.0
Dicotiledoneae	69.9	75.5
Monocotiledoneae	25.9	21.6

TABLE III

A Ranking of the Abundance of Genera per Plant Family

Common species	Edible species
Compositae (Asteraceae)	Compositae
Orchidaceae	Leguminosae
Leguminosae (Fabaceae, Mimosaceae)	Gramineae
Gramineae (Poaceae, Bambusaceae)	Cactaceae
Rubiaceae	Umbelliferae
Cruciferae (Brassicaceae)	Palmae
Umbelliferae (Apiaceae)	Rubiaceae
Euphorbiaceae	Labiatae
Liliaceae	Euphorbiaceae
Asclepiadaceae	Cruciferae
Acanthaceae	Rosaceae
Labiatae (Lamiaceae)	Myrtaceae
Palmae (Arecaceae)	Araceae
Scrophulariaceae	Apocynaceae
Rutaceae	Moraceae

randomly distributed among the different taxa, then it would be predictable that the most numerous families of plants would contain a higher number of edible species (Table III). This relation seems to be valid since many of the most prolific families are also among the most productive in edible plants. This is the case with the Compositae (Asteraceae), Leguminosae (Fabaceae), Gramineae (Poaceae), Euphorbiaceae, and Rubiaceae, which occupy the top positions in the ranking of both common species and edible species. The most numerous of all plant families, the Orchidaceae, however, has few edibles. Among the most suggestive cases is the number of genera in the Cactaceae family, which appears in 4th place for edibles but is 31st in the rank of common plants.

There is no clear relation among the ratios of number of species per genus that allows us to differentiate edible from common species. Among families with the lowest ratios are the Asclepiadaceae (8.0), Cruciferae (8.6), and Rutaceae (10.0). Families that are richer in species per genus are the Begoniaceae (255.0), Aizoaceae (208.3), and Eriocaulaceae (92.3), yet they show no evidence of having experienced a process of selection, that is, of "proclivity" or "rejection" by humans, that would have led them to speciate toward palatability or distastefulness.

Of the 389 more frequently cultivated species considered by Nicholson *et al.* (1969), the ranking goes as follows: Rosaceae (13.3% of total species), Leguminosae (8.5%), Gramineae (6.4%), Compositae (5.9%), Umbelliferae (5.4%), and Cruciferae (5.1%). They are followed by Palmae, Cucurbitaceae, Rutaceae, Alliaceae, Chenopodiaceae, Dioscoreaceae, Annonaceae, Ericaceae, Grossulariaceae, and 66 less prolific families. Kunkel (1984) states that the Rosaceae is the richest family among food plants. The analysis of a sample of 6222 items from his list confirms this assessment. Rosaceae appears at the top of the list, comprising 5.8% of the cases, insofar as Leguminosae are split into Fabaceae (4th place), Mimosaceae (6th place), and Caesalpinaceae (82nd place). On the contrary, if the latter three families are considered as a unit, the Leguminosae stand in first place, comprising 6.7% of the sample. The rank of the first 30 families is shown in Table IV.

B. Edible Parts

Some plant genera are extremely abundant in edible species and may show particular tendencies toward a given kind of food (Table V). For example, all of the 205 species of *Rubus* appearing in Kunkel's list provide edible fruits. Among them, there are three species whose leaves are also used as tea. Similarly, the 80 or more

TABLE IV

A Ranking of the Abundance of Edible Species in the 30 Most Prolific Families[a]

1. Rosaceae	9. Rubiaceae	17. Polygonaceae	25. Caesalpiniaceae
2. Compositae (Asteraceae)	10. Myrtaceae	18. Palmae (Arecaceae)	26. Annonaceae
3. Dioscoreaceae	11. Solanaceae	19. Ericaceae	27. Zingiberaceae
4. Fabaceae	12. Cactaceae	20. Rutaceae	28. Vitaceae
5. Liliaceae	13. Gramineae (Poaceae)	21. Umbelliferae (Apiaceae)	29. Araceae
6. Mimosaceae	14. Fagaceae	22. Sapotaceae	30. Cucurbitaceae
7. Moraceae	15. Euphorbiaceae	23. Guttiferae (Clusiaceae)	
8. Ebenaceae	16. Cruciferae (Brassicaceae)	24. Asclepiadaceae	

[a] The richest family of common edible and nonedible species, the Orchidaceae, appears in 45th place.

TABLE V

The Most Prolific Genera of Food Plants

Genus	Family	No. of food species	Total No. of species	% food species
1. *Rubus*	Rosaceae	205	2500–3000	7–8
2. *Ficus*	Moraceae	137	700	20
3. *Dioscorea*	Dioscoreaceae	110	600	18
4. *Solanum*	Solanaceae	100	1400–1700	6–7
5. *Acacia*	Leguminosae	80	800	10
6. *Eugenia*	Myrtaceae	79	800	10
7. *Diospyros*	Ebenaceae	69	200	35
8. *Garcinia*	Guttiferae	68	400	17
9. *Quercus*	Fagaceae	67	470–1000	7–14
10. *Vaccinium*	Ericaceae	66	300–400	17–22
11. *Passiflora*	Passifloraceae	58	500	12
12. *Opuntia*	Cactaceae	52	250	21

species of *Prunus* provide edible fruits, as also occurs with *Rosa* spp. and *Ribes* spp. The majority of *Piper* species are used as black or white pepper or as a spice for curries. *Rumex* provides 44 species with leaves used as vegetables and 3 species with edible roots. Of the 100 edible species of *Solanum*, 59 are used only for their fruits, 20 species only for their tubers, 14 for both fruits and leaves, 6 species only for their leaves, and a single species exclusively for its seeds.

An analysis of the data compiled by Duke (1972) for Central America indicates that 85% of the plant species are used for a single purpose, 10% have two uses, 4% have three uses, and 1% have four uses (leaves, flowers, fruits, and seeds). A similar analysis of the 350 edible species of the native British flora yields slightly different proportions: 78%, 19%, <3%, and <1% for one, two, three, and four purposes, respectively. One of the exceptional cases of variability of uses is that of

TABLE VI

An Estimate of Food Usage (as Percentages of Regional Totals) from Different Sources

	World flora (Facciola, 1990)	World flora (Kunkel, 1984)[a]	Panama Isthmus (Duke, 1972)	United Kingdom (Martin, 1976)	Andes (NRC, 1989)	Cambodia (Ito, 1969)	Botswana (Campbell, 1986)
Leaves[b]	21.8	28.9	20.6	46.8	7.6	22.8	16.2
Fruits	19.6	30.5	37.4	9.5	61.4	29.9	33.3
Seeds	13.9	13.2	17.5	10.1	10.6	17.0	8.1
Condiments, flavorings	11.5	6.2	3.2	6.9	6.1	6.5	0.0
Tea, herbals	9.5	2.9	1.6	7.6	0.0	3.1	4.0
Beverages	7.0	0.8	6.0	0.0	2.3	1.7	4.0
Flowers, capers	6.7	4.0	4.1	4.8	0.0	9.2	1.0
Roots[c]	5.6	8.7	6.3	12.6	12.1	7.5	26.3
Saps[d]	4.2	3.9	2.2	0.6	0.0	2.0	7.1
Barks	0.2	0.9	1.0	1.1	0.0	0.3	0.0
No. of species	3000	1790	182	350	98	186	77
No. of items (multiple uses)	6311	2442	336	476	132	294	99

[a] A more detailed analysis of food usages appears in Table VIII.

[b] Including stems, sprouts, and meristems.

[c] Including bulbs and rhizomes.

[d] Including gums, latex, sugars, and masticatories.

TABLE VII

The Distribution of Food Plants by Growth Form in Different Geographic Regions[a]

Growth form	Total No. of spp.	Percentages (from totals)	Nearctic (N. America)	Neotropical (C. & S. America)	Indomalayan	Australian	Ethiopian (Africa S. of the Sahara)	Palearctic (Eurasia and N. Africa)
Trees	504	25.9	14.3	24.9	33.9	45.1	35.4	9.6
Shrubs	452	23.2	31.1	31.6	20.0	17.7	16.5	20.0
Herbs	787	40.4	51.2	23.7	33.9	32.4	38.8	66.5
Vines	203	10.4	3.4	19.8	12.1	4.9	9.3	3.9
Totals	1946	~100.0	12.2	22.3	25.4	5.2	16.6	18.2

[a] The sample includes 1790 species. Because, a number of species are shared between two or more regions, the total number of items classified increases to 1946. The third column, as well as the bottom row, are percentages of the total number of species considered. The remaining columns are percentages calculated from their respective regional subtotals.

hops (*Humulus lupulus*). Its leaves, roots, flowers, and bark may be used as food and as condiment. The pumpkin (*Cucurbita moschata*) provides fruits, seeds, flowers, young leaves, and shoots for human consumption.

It is clear from Table VI that some kinds of food are scarcely used, such as bark, flowers, sap, and liquorice. In contrast, leaves (including stems, sprouts, seedlings, and shoots) and fruits seem to be the most preferred food.

III. GEOGRAPHIC PATTERNS OF FOOD PLANTS

On the basis of a sample of 1790 species from Kunkel's list of food plants, it appears that edible herbs are more numerous than trees, shrubs, and vines. "Vines" include all the climbing, creeping, and epiphytic plants. Table VII shows that in decreasing order of species richness, the Indomalayan regions appear first, followed by the Neotropical, Palearctic, Ethiopian, Nearctic, and Australian (or Australasian) regions. By means of a chi-square test, at a significance level $P < 0.05$, the Nearctic and Australian regions show greater, and the Palearctic lower, proportions of food trees than expected. The proportion of edible shrubs does not vary significantly among regions, although at a $P < 0.1$ the Neotropics seem to have a higher proportion than the rest of the world. The Palearctic region shows a higher and the Neotropics a lower proportion of herbs, while the latter region is richer in vines ($P < 0.05$). In contrast, the Northern Hemisphere (Palearctic and Nearctic regions) shows a significantly lower proportion of vines than the Southern Hemisphere.

According to biogeographic regions (Table VIII), the Palearctic shows a significantly greater proportion of species that provide edible leaves, stems, and sprouts than the other regions. The Australian region is characterized by a greater proportion of edible seeds, and the Neotropics by its abundance of fruit species. Of course, these differences were derived from the sample analyzed by us. For instance, the appearance of zero values for beverages from the Australian and Palearctic regions does not mean that there are no species of this kind in their floras; they were simply not registered in our sample of 1790 species. On the contrary, the high proportion of edible fruits in the Neotropical region is repeatedly verified in ethnobotanical studies performed in different countries of Central and South America. These contrasts may be attributed to differential characteristics of seed dispersal evolution within their plant communities, as well as to prevailing cultural trends in the use of natural resources. The possibility of ethnobotanists showing biased attention toward particular kinds of food should not be discarded.

IV. ARE EDIBLE WEEDS, WEEDS?

The habit of gathering wild food plants has not been totally lost. E. R. Spencer maintained that "any plant is a weed if it insists upon growing where the husbandman wants another plant to grow. It is a plant out of place in the eye of man; in the nice eye of nature it is very much in place." Many of our dearest crops, however, have originated from weeds. In the course of time, humans learned how to profit from them. Oats (*Avena sativa*), foxtail millet (*Setaria italica* derived from *S. viridis*), chicory (*Cichorium intybus*), pak choi (*Brassica rapa*), spinach beet (*Beta vulgaris*), and many more

TABLE VIII

A Sample of 1790 Species used in 2442 ways, Classified by Types of Food in Different Regions (Figures Are Percentages of Regional Totals)

	Nearctic	Neotropical	Indomalayan	Australian	Ethiopian	Palearctic
Leaves[a]	24.3	14.3	30.0	25.0	27.3	48.0
Fruits	28.3	52.5	31.9	14.1	27.3	14.4
Seeds	12.8	11.7	13.7	28.9	13.3	10.0
Condiments, flavorings	3.0	3.9	8.2	2.3	6.4	8.7
Tea, herbals	4.9	2.3	2.2	4.7	2.2	3.6
Beverages	1.0	2.8	0.3	0.0	0.2	0.0
Flowers, capers	3.3	2.7	4.8	1.6	5.4	4.2
Roots[b]	15.8	7.6	5.8	8.6	9.1	9.1
Saps[c]	4.6	2.0	2.0	13.3	8.1	1.7
Barks	2.0	0.4	1.1	1.6	0.7	0.2
No. of items	304	487	643	128	407	471

[a] Including stems, sprouts or shoots, and meristems.

[b] Including bulbs and rhizomes.

[c] Including gums, latex, sugars, and masticatories.

species appeared originally as invaders in ancient cultivated fields. Furthermore, several cosmopolitan "weeds," such as dandelion (*Taraxacum officinale*), garden rocket (*Eruca vesicaria*), common purslane (*Portulaca oleracea*), and shepherd's purse (*Capsella bursa-pastoris*), are nowadays cultivated for the specialty food market.

After analyzing a set of 22,521 species of plants inhabiting natural ecosystems of North and South America, we concluded that 11.3% were edible. In another sample of 1264 species from seminatural communities in the temperate Neotropical region, edibles were 20.3% of the total. But if only weeds are considered (2455 widely spread species), the proportion mounts to 35.6%. This demonstrates that the degree of environmental disturbance correlates with the proportion of weeds, which in turn increases the proportion of edible food resources. This should not be surprising given that 16 of the world's 18 most aggressive weeds provide parts for human consumption. Because weeds are so numerous (more than 10,000 species catalogued) and so abundant (averaging 1.3 and 2.1 tons/ha in a temperate and a tropical area of northwestern Patagonia and eastern Mexico, respectively; Díaz-Betancourt *et al.*, 1999; Rapoport *et al.*, 1998), they stand ambiguously as both enemies and potential benefactors of humanity.

The prospects for the future of food plant diversity appear to be auspicious. Further research should be conducted to manage and profit from these varied and abundant natural resources. At present, more than 15,000 species of food plants are recorded and this figure is constantly growing. Yet most of the world is fed with about 20 crops. As has been pointed out by Facciola (1990), 8000 cultivars of apples have been developed by humans, but only a handful are available in supermarkets.

See Also the Following Articles

AGRICULTURE, TRADITIONAL • DOMESTICATION OF CROP PLANTS • PLANT BIODIVERSITY, OVERVIEW

Bibliography

Brücher, H. (1989). *Useful Plants of Neotropical Origin and Their Wild Relatives*. Springer-Verlag, New York.

Campbell, A. (1986). The use of wild food plants, and drought in Botswana. *J. Arid Environ.* **11**, 81–91.

Díaz-Betancourt, M., Ghermandi, L., Ladio, A., López-Moreno, I. R., Raffaele, E., and Rapoport, E. H. (1999). Weeds as a source for human consumption. A comparison between tropical and temperate Latin America. *Rev. Biol. Tropical,* **47**, 329–338.

Duke, J. A. (1972). *Isthmian Ethnobotanical Dictionary*. Fulton, Maryland.

Duke, J. A. (1992). *Handbook of Edible Weeds*. CRC Press, Boca Raton, Florida.

Facciola, S. (1990). *Cornucopia. A Sourcebook of Edible Plants*. Kampong Publications, Vista, California. [A second, revised edition was published in 1998.]

Gardarsson, A., and Moss, R. (1970). Selection of food by Icelandic Ptarmingan in relation to its availability and nutritive value. In *Animal Population in Relation to Their Food Resources*. (A. Watson, ed.), pp. 47–71. Blackwell, Oxford.

Godwin, H. (1960). The history of weed in Britain. In *The Biology of Weeds* (J. L. Harper, ed.), pp. 1–10. Blackwell, Oxford.

Heal, O. W., and Fulton, M. J. (1970). Soil amoeba: Their food and their reaction to microflora exudates. In *Animal Population in Relation to Their Food Resources* (A. Watson, ed.), pp. 145–162. Blackwell, Oxford.

Ito, S. (1969). Economical plants and their distribution in Cambodia. *J. Agric. Sci., Tokyo Univ. of Agriculture* 13, 163–218.

King, L. W. (1966). *Weeds of the World. Biology and Control.* Leonard Hill, London.

Klein, D. K. (1970). Food selection by North American deer and their response to overutilization or preferred plant species. In *Animal Population in Relation to Their Food Resources* (A. Watson, ed.), pp. 25–46. Blackwell, Oxford.

Kunkel, G. (1984). *Plants for Human Consumption.* Koeltz Scientific Books, Koenigsten, Germany.

Martin, W. K. (1976). *The Concise British Flora in Colour.* Ebury Press and Michael Joseph, London.

Moerman, D. E. (1998). *Native American Ethnobotany.* Timber Press, Portland, Oregon.

National Research Council. (1989). *Lost Crops of the Incas.* National Academy Press, Washington, D.C.

Nicholson, B. E., Harrison, S. G., Masefield, G. B., and Wallis, M. (1969). *The Oxford Book of Food Plants.* Oxford University Press, London. [A second, updated edition appeared as J. G. Vaughan and C. A. Geissler. (1997). *The New Oxford Book of Food Plants: A Guide to the Fruit, Vegetables, Herbs, and Spices of the World.*]

Plotkin, M. J. (1993). *Tales of a Shaman's Apprentice.* Penguin Books, New York.

Rapoport, E. H., Ladio, A., Raffaele, E., Ghermandi, L., and Sanz, E. H. (1998). Malezas comestibles. *Ciencia Hoy* 9(49), 30–43.

EDUCATION AND BIODIVERSITY

Shirley M. Malcom
The American Association for the Advancement of Science

GLOSSARY

formal education Education that takes place in a school environment.
informal education Education that takes place in non-school settings, such as museums, zoos, parks, or through the media.
out of school experiences Things that individuals experience that may support or reinforce learning. These might include participation in youth-serving organizations or hobbies such as bird watching.

THIS CHAPTER REFLECTS a view from the science education and scientific communities of what all high school graduates would know and understand about biodiversity. This is a statement of learning goals and does not necessarily represent the actual understandings that an average student would attain through study

of science through grade 12 in the typical American high school.

I. INTRODUCTION

The Convention on Biological Diversity articulates a case for nations of the world to come together to undertake activities to improve conservation of biodiversity and sustainable use of biological resources. As of June 1997 more than 170 nations had ratified the convention. In addition to calls for better management, more research, and study and international and regional cooperation, there was also recognition of the role of education, public participation, public information, and the development of a cadre of professionals to support the goals of the convention.

A recent report to President Clinton from the President's Committee of Advisors on Science and Technology (PCAST) makes recommendations to strengthen "the understanding and management of biological resources" (PCAST, 1998). Among the recommendations in *Teaming with Life: Investing in Science to Understand and Use America's Living Capital* are calls for increased opportunities for formal and informal education centered on biodiversity and ecosystems, for interactions between scientists and students, and for continuing professional education for K–12 teachers.

This chapter will outline aspects of formal and informal education focused on biodiversity. Specifically, it will outline learning goals around biodiversity for K–12 education, recommended school-based experiences that can lead to an attainment of these goals, and evidence as to the extent to which these understandings are being achieved by students. The chapter will review school-based environmental education initiatives and describe how these compare to and differ from education about biodiversity. The chapter will then continue with a discussion of informal learning opportunities in "places of science" as well as those available through youth-serving organizations, tourism, and field experiences. The section on general education will conclude with information on public interest in and awareness of biodiversity.

A brief section will discuss biodiversity and tertiary education: biodiversity themes and courses as part of liberal education as well as issues in the education of professionals and specialists who work in biodiversity research, management, and conservation.

II. LEARNING ABOUT BIODIVERSITY: K–12 EDUCATION

How do students learn about biodiversity? What specific concepts must they learn and what ideas must they acquire to support that understanding? What class work, materials, curriculum, set of courses, and experiences would provide an adequate background so that they come to an understanding of this concept?

In 1985 the American Association for the Advancement of Science (AAAS) began Project 2061, a long-term initiative to reform K–12 education in natural and social sciences, mathematics, and technology. *Science for All Americans*, published in 1989, was conceived as a statement of learning goals for science, mathematics, and technology education, defining what all students should know and be able to do by the time they conclude secondary education. *Science for All Americans* includes learning goals related to biodiversity (see Box 1). *Benchmarks for Science Literacy (Benchmarks)*, published by AAAS in 1993, and the *National Science Education Standards (Standards)*, published by the National Research Council in 1996, describe the ideas that students must grasp at different ages (grades or developmental levels) in order to understand science concepts, including the overall concept of biodiversity.

A. Grades K–2

For young children in early primary education, *Benchmarks* recommends that students learn the following:

Box 1

Biodiversity: What Would a Science Literate Adult Understand?

Diversity of Life

Millions of different types of individual organisms inhabit the earth at any one time—some very similar to each other, some very different. Biologists classify organisms into a hierarchy of groups and subgroups on the basis of similarities and differences in their structure and behavior. One of the most general distinctions among organisms is between plants, which get their energy directly from sunlight, and animals, which consume the energy-rich foods initially synthesized by plants. But not all organisms are clearly one or the other. For example, there are single-celled organisms without organized nuclei (bacteria) that are classified as a distinct group.

Animals and plants have a great variety of body plans, with different overall structures and arrangements of internal parts to perform the basic operations of making or finding food, deriving energy and materials from it, synthesizing new materials, and reproducing. When scientists classify organisms, they consider details of anatomy to be more relevant than behavior or general appearance. For example, because of such features as milk-producing glands and brain structure, whales and bats are classified as being more nearly alike than are whales and fish or bats and birds. At different degrees of relatedness, dogs are classified with fish as having backbones, with cows as having hair, and with cats as being meat eaters.

For sexually reproducing organisms, a species comprises all organisms that can mate with one another to produce fertile offspring. The definition of species is not precise, however; at the boundaries it may be difficult to decide on the exact classification of a particular organism. Indeed, classification systems are not part of nature. Rather, they are frameworks created by biologists for describing the vast diversity of organisms, suggesting relationships among living things, and framing research questions.

The variety of the earth's life forms is apparent not only from the study of anatomical and behavioral similarities and differences among organisms but also from the study of similarities and differences among their molecules. The most complex molecules built up in living organisms are chains of smaller molecules. The various kinds of small molecules are much the same in all life forms, but the specific sequences of components that make up the very complex molecules are characteristic of a given species. For example, DNA molecules are long chains linking just four kinds of smaller molecules, whose precise sequence encodes genetic information. The closeness or remoteness of the relationship between organisms can be inferred from the extent to which their DNA sequences are similar. The relatedness of organisms inferred from similarity in their molecular structure closely matches the classification based on anatomical similarities.

The preservation of a diversity of species is important to human beings. We depend on two food webs to obtain the energy and materials necessary for life. One starts with microscopic ocean plants and seaweed and includes animals that feed on them and animals that feed on those animals. The other one begins with land plants and includes animals that feed on them, and so forth. The elaborate interdependencies among species stabilize these food webs. Minor disruptions in a particular location tend to lead to changes that eventually restore the system. But large disturbances of living populations or their environments may result in irreversible changes in the food webs. Maintaining diversity increases the likelihood that some varieties will have characteristics suitable to survival under changed conditions. As noted in *Science for All Americans*, Project 2061, from the American Association for the Advancement of Science:

> Our planet's essential goods and services depend on the variety and variability of genes, species, populations and ecosystems. Biological resources feed and clothe us and provide housing, medicines and spiritual nourishment. The natural ecosystems of forests, savannahs, pastures and rangelands, deserts, tundras, rivers, lakes and seas contain most of the Earth's biodiversity. Farmers' fields and gardens are also of great importance as repositories, while gene banks, botanical gardens, zoos and other germ-plasm repositories make a small but significant contribution. The current decline in biodiversity is largely the result of human activity and represents a serious threat to human development. (1989, pp. 60–61).

- Some animals and plants are alike in the way they look and in the things they do, and others are very different from one another.
- Plants and animals have features that help them live in different environments.
- Stories sometimes give plants and animals attributes they really do not have.

Recommended learning activities include providing students the opportunity to observe a variety of plants and animals in the classroom; on the school grounds; in the community; at home in parks, streams, and gardens; and at the zoo. The observations would prompt students to pursue questions about how the organisms live, where they are found, or how they interact with other organisms.

B. Grades 3–5

For children in later primary education, *Benchmarks* recommends that students learn the following:

- A great variety of kinds of living things can be sorted into groups in many ways using various features to decide which things belong to which group.
- Features used for grouping depend on the purpose of the grouping.

Recommended learning activities include providing students the opportunity to learn about an increasing variety of living organisms and offering them a chance to invent schemes for classification. Students would be encouraged to develop different classification schemes (without being introduced to the Linnean classification system) and shown how their usefulness varies depending on the purpose of the classification. The purpose of the work would be to help students develop a deeper understanding about the relatedness of organisms.

C. Grades 6–8

For children in upper primary and lower secondary education *Benchmarks* argues that science should provide students with opportunities to enrich their growing knowledge of the diversity of life on the planet and to begin to connect that knowledge to what they are learning in geography. That is, whenever students study a particular region in the world, they should learn about the plants and animals found there and how they are like or unlike those found elsewhere.

Food patterns of development and external and internal structures would all be used to illustrate interrelationships, interdependence, similarities, and differences. Students would be introduced to the features that biologists use in classification systems and would be taught why these classifications are made.

Benchmarks suggests that students in this group should know the following:

- One of the most general distinctions among organisms is between plants, which use sunlight to make their own food, and animals, which consume energy-rich foods. Some kinds of organisms, many of them microscopic, cannot be neatly classified as either plants or animals.
- Animals and plants have a great variety of body plans and internal structures that contribute to their being able to make or find food and reproduce.
- Similarities among organisms are found in internal anatomical features, which can be used to infer the degree of relatedness among organisms. In classifying organisms, biologists consider details of internal and external structures to be more important than behavior or general appearance.
- For sexually reproducing organisms, a species comprises all organisms that can mate with one another to produce fertile offspring.
- All organisms, including the human species, are part of and depend on two main interconnected global food webs. One includes microscopic ocean plants, the animals that feed on them, and finally the animals that feed on those animals. The other web includes land plants, the animals that feed on them, and so forth. The cycles continue indefinitely because organisms decompose after death to return food material to the environment.

D. Grades 9–12

For students at the secondary level, curricular objectives lead to understanding diversity within and among species by looking at "same and different" features at a molecular level. Students would learn the following:

- The variation of organisms within a species increases the likelihood that at least some members of the species will survive under changed environmental conditions, and a great diversity of species increases the chance that at least some living things

will survive in the face of large changes in the environment.
- The degrees of kinship between organisms or species can be estimated from the similarity of their DNA sequences, which often closely matches their classification based on anatomical similarities.

Understanding built up over this period of study would lead students to comprehend the diversity of ecosystems, diversity of species, and the genetic diversity within species.

While *Benchmarks* sets out a recommended sequence of learning goals to help students come to an understanding of biodiversity as a complex idea, it is not clear that most students have access to the education, ideas, concepts, and learning experiences needed to achieve such understandings. Therefore it would be necessary to explore what students are taught or expected to learn over time during their schooling.

E. Biodiversity and School Science

Formal education in science is an important contributor to students' fundamental understandings about science. While self-directed study—books, articles, the Internet, museum visits, and field experiences—augment science learning for many students, the quality of the curriculum, textbooks, and other instructional materials, the preparation of teachers, the school-mediated experiences provided to students both inside and outside of the classroom all interact to shape what students take away from school science. Understandings of biodiversity would be based on accumulated experiences and knowledge. These would include the early school focus on "natural history" and "nature study" and development of an "intuitive" understanding of biological diversity and the relationships among living organisms (National Research Council, 1990). Students' out-of-school experiences, where such are available, would reinforce school learning. The curriculum focus shifts in lower and upper secondary levels (grades 6–12) to more formal, taxonomic instruction.

This pattern of topic coverage for life sciences concepts is present in the curriculum of other countries around the world and was prevalent among the majority of the 50 educational systems of countries that participated in the Third Mathematics and Science Study (or TIMSS). Data were collected in 1993 and results presented in 1996.

In the United States, according to statistics provided by the U. S. Department of Education, biology is the most frequently taken high school science course, with over 93% of 1994 graduates of public high schools reporting having completed such a class. Over 97% of graduates from nonpublic high schools reported completing biology. Analysis of content core is not available to determine the extent to which the ideas critical to a student's understanding of biodiversity are actually taught. It is also not clear if earlier foundational ideas are provided to students as a part of instruction at primary and lower secondary levels.

In the United States there has been considerable discussion about the adequacy, pacing and structure of curriculum and, especially, of textbooks in middle and high school biology. Researchers have criticized their "encyclopedic" nature, with too many ideas covered too superficially, and too much focus on vocabulary at the expense of big ideas. A recent study that evaluated the science textbooks used in middle grades against three concepts from *Benchmarks* and *Standards*, one each from the earth, physical, and life sciences, led AAAS Project 2061 to conclude that all nine titles examined were inadequate to help students achieve understanding of the ideas that they were attempting to explain.

Efforts to actually measure what students know about science are undertaken through the regularly administered National Assessment of Educational Progress and recently have been internationally benchmarked through TIMSS. At lower and upper secondary levels, the results suggest that U. S. students gain little real understanding of the big ideas of science, including life sciences, in spite of the fact that they are most likely to have had formal classes in this area than in other areas of science. It has largely been in the study of biology that the basic concepts of biodiversity have been advanced; diversity within species, across species and ecosystems; relatedness of species; interdependence among living things; threats to this diversity due to population pressures; loss of habitat; changes in species over time; natural extinctions; and human induced extinctions. While biodiversity and biodiversity education are more recent concepts, nature study, conservation, and environmental education have older roots, in a more general emphasis on environmental improvement and appreciation. The following section will explore some of the history of environmental education and consider how it might relate to biodiversity education.

III. ENVIRONMENTAL EDUCATION

In 1977 the world's first intergovernmental conference on environmental education was organized by the United Nations Education, Science and Cultural Organization (UNESCO) in cooperation with the United Nations Environment Programme and held in Tbilisi,

Georgia (Soviet Union). The Tbilisi Declaration, adopted at the conference, stressed the importance of environmental education in the preservation and improvement of the world's environment. Agenda 21, Chapter 36, outlines issues related to formal education, public awareness and training to promote sustainable development and emphasizes environmental and development education as an essential aspect of all education. There were also calls for linking to environmental education in recommendations from the World Conference on Education for All (held in Jomtien, Thailand in 1990) that urged a move toward universal access to basic education for girls and boys.

The recommendations include calls for better informal education, promotion of environmentally sound leisure and tourism activities, programs to involve young people and children, as well as respect for and support of efforts to promote dissemination of traditional and socially learned knowledge through mechanisms based in local cultures.

Environmental education was called "nature study" when it got its start in the 1920s with Junior Audubon Clubs teaching children to appreciate nature. According to Karen Schmidt in a December 13, 1996, article in *Science*, the movement was transformed into conservation education in the 1930s when the Dust Bowl environmental tragedy led to incorporation of ideas into some schools about the management of natural resources.

With the initiation of Earth Day in 1970 and passage of the National Environmental Education Act of 1970, teachers received supplemental training in environmental education. Many states enacted their own environmental education laws, and schools began incorporating these topics into science classes. On the 20th anniversary of Earth Day, President Bush signed the National Environmental Education Act of 1990, which created an Office of Environmental Education at the Environmental Protection Agency and supported curriculum development and teacher training in the states.

In the United States concern has been expressed about the place of environmental education, as currently configured into the curriculum, and especially about the quality of programs sometimes offered in lower secondary schools. This has included concerns about an advocacy orientation in instruction, about the need for and balance in materials used and instruction provided. The North American Association for Environmental Education (NAAEE), a professional group that includes college faculty and K–12 educators among its members, developed "Guidelines for Excellence" to help guide teachers and others in the selection of quality, balanced materials.

Teaming with Life makes a strong recommendation that environmental education have a stronger base in science, using scientifically grounded curricula. Innovative programs such as GLOBE (see Box 2) that depend on student-scientist partnerships and collection of real data may point the way to science education based around environmental and biodiversity concerns.

In 1996 a national conference held in Washington, D.C., spotlighted projects that depended on student-scientist partnerships, including a number that focused on biodiversity as learning themes. Included among these were "Classroom Feederwatch," a program of authentic research for grades 5–8 developed by Cornell Laboratory of Ornithology and TERC, with funding from the National Science Foundation. In this program, students (and their teachers) learn to identify birds, to ask scientific questions and to design experiments to answer them. Students analyze and display data to answer their questions and collect and share data in a research database used by professional ornithologists in their studies of bird populations. Students thus become part of a larger team of those contributing to biodiversity studies. (For more information see http:/birdsource. cornell.edu/cfw/watiscf.htm.)

IV. LEARNING IN THE INFORMAL EDUCATION SECTOR

A. Overview

A wide range of informal education experiences are available for adults and children to extend their knowledge about biodiversity. These include organizations that incorporate biodiversity education and exhibition within their missions, such as the following:

- Zoos
- Botanical gardens
- Aquariums
- Museums
- National parks

Depending on their size, these "places of science" might also incorporate research, collections, conservation, or other functions important to biodiversity. Signage and docent-led and audio tours provide additional information to visitors. Classes, lectures, and workshops (including those for teachers) are also often provided to

---

Box 2

Global Learning and Observation to Benefit the Environment (GLOBE)

GLOBE, a hands-on, school-based international environmental science program, was introduced by U.S. Vice President Al Gore in April 1994 and began operation on Earth Day in April 1995. In 1999 there were more than 6000 participating schools in more than 70 countries. GLOBE brings together students, teachers, and scientists from around the world to enhance environmental awareness of individuals worldwide, increase scientific understanding of the earth, and support improved student achievement in science and mathematics.

Students make environmental observations, or take environmental measurements near their school site, report their observations via the Internet, receive and use GLOBE images created from the combined data, and study the environment by relating their observations to larger environmental topics. GLOBE educational materials were developed by environmental educators and curriculum development specialists working with scientists. Materials are used in schools under the guidance of teachers who have received training using GLOBE materials. Teachers include GLOBE activities as appropriate within their local curriculum. GLOBE materials are translated into the six United Nations' languages (Arabic, Chinese, English, French, Russian, and Spanish) and are also available in Czech, Estonian, German, Greek, Hebrew, and Japanese. GLOBE international partners sign bilateral agreements with the United States and manage participation of schools in their countries. In the United States, GLOBE is administered by an interagency team that includes the National Oceanic and Atmospheric Administration, the National Aeronautics and Space Administration, the National Science Foundation, the Environmental Protection Agency, the Departments of State and Education, and others.

Observations range from basic weather parameters (temperature, atmospheric pressure, and precipitation) to measurements such as water chemistry, biodiversity, and biomass assessment. The focus to date has been more heavily directed toward physical systems measurements. Potential is great to increase the biodiversity and life systems aspects of the project.

extend the learning experience. Increasingly, materials and web sites incorporate aspects of a visit, bringing resources to audiences at a distance to make some part of the visitor experience remotely available.

Informal education also includes more intensive immersion experiences such as where it is incorporated into visits to natural preserves such as parks (including those within national park systems) and forests, or that provided through ecotourism.

Interpretive programs using volunteers, staff, and written and video materials provide enhanced learning experiences by bringing the science, the issues, and the concerns into sharp focus as a part of the overall environmental experience. Programs such as Earthwatch have biodiversity-focused visits that involve the participant in the research as data collector. Other informal learning opportunities are available through television and IMAX programs, web sites, and books.

Youth-serving groups provide a broad range of activities and experiences that can support education around concepts of biodiversity (see Box 3). Many may involve long-term projects of environmental monitoring, animal and plant breeding, habitat restoration, and other activities undertaken individually or in groups.

Box 3

Selected Youth Serving Groups Providing Informal Science/Environmental Education

American Camping Association (ACA)

Founded in 1910 ACA provides an accrediting mechanism for camps. Of the more than 2200 accredited camps listed, 284 provided nature/environmental study as part of the camp experience.

Boy Scouts of America (BSA)

Enrollment: 5.6 million
Program: Incorporated in 1910 BSA provides programs for boys that include outdoor skills, nature study, and conservation activities through an elaborated badge structure and group activities. BSA is a charter member of the World Scout Conference.

World Scout Conference includes 145 member associations representing more than 25 million scouts.

Boys and Girls Clubs of America

Enrollment: 3 million youth; served in 1,006 local organizations
Program: Activities include outdoor and environment education (The Ultimate Journey).

Camp Fire Boys and Girls

Enrollment: 667,000 young people (birth–age 21)
Program: Camping and environmental education programs offering children an appreciation and commitment to the natural environment.

Girl Scouts of the U.S.A. (GSUSA)

Enrollment: 3.5 million
Program: Activities include out-of-doors, nature study. Badge structure that includes environmental issues. GSUSA is member of the World Association of Girl Guides and Girl Scouts.

World Association of Girl Guides and Girl Scouts undertakes world projects including building world citizenship. Environment is one of the themes of this program.

Girls Incorporated

Enrollment: 350,000 girls (ages 6–18) at more than 1000 sites nationwide
Program: Activities include experiences in mathematics and science education through Operation SMART

National 4-H Clubs

Enrollment: 6.0 million in more than 76,500 clubs
Program: Part of the U.S. Department of Agriculture's Cooperative Extension Service established in 1914. Project areas include agricultural and natural sciences, and technology; Cornell Nest Box network.

B. Zoos and Other "Places of Science"

Animal parks were established by and for rulers. Maier and Page, in their volume Zoo: The Modern Ark (1990), describe how animals were kept by royalty for entertainment and as a show of wealth. The third dynasty ruler of the Sumerian city of Ur had a park that dated around 2300 B.C. A millennium later as civilization spread in the Near East and Asia, rulers and pharaohs exchanged "exotic" animals for their zoos. Emperor Wu Wang of the Chou dynasty laid out a zoological garden called the Park of Intelligence. Animal collections were found around the globe in early civilizations such as in Egypt some 3500 years ago.

Alexander the Great, perhaps influenced by Aristotle's private menagerie, installed what was perhaps the first public zoo in Alexandria, Egypt.

With the coming of the "Dark Ages" of Europe, monasteries became the keepers of menageries and game parks. When Cortes arrived in the Aztec capital of Tenochtitlan, he found a large zoo behind Emperor Montezuma's palace. Zoos in India were established by Akbar toward the end of the 16th century. He, like the Aztecs, employed people specially trained to care for and medically tend to animals.

The zoo at Vienna was reinvigorated by Maria Theresa and her husband as the Imperial Menagerie at Schönbrun for the convenience and entertainment of the nobility. The zoo remains today as likely the oldest in continuous operation, dating from the 1750s.

Democratization of Europe and establishment of urban centers that accompanied industrialization led to the "modern zoo" as a repository of exotic specimens of life that were to be studied as a way of understanding "flora and fauna" of the world. Public monies (rather than private patrons or royal largesse) were available to begin systematic scholarly study. Maier and Page date the modern zoo to 1826 when the Zoological Society of London founded the zoological gardens at Regent's Park for the purpose of understanding the natural history of the animals inhabiting the reaches of the British Empire.

Since zoos as public institutions had to raise funds and attract money (independent of their research and conservation goals), they had to become popular attractions. Zoo organizers also had to learn to manage space and figure out and meet animals' requirements, such as for social interaction. Zoos' role in conservation became educational as they raised visitor awareness about endangered species and loss of habitat. Where larger zoos also developed significant breeding herds, they established breeding farms. In San Diego, for example, this "wild animal park" has become an additional attraction.

The National Zoological Park (National Zoo), associated with the Smithsonian Institution, established a "biopark," Amazonia, to emphasize the relationships among soil, plant, invertebrate, and other animal forms and the need to preserve the habitats of the world. Zoos, aquariums, and game parks are being seen as tools to affect public attitudes regarding the variety of life on earth.

As these "places of science" intentionally blend education and entertainment they are increasingly adding materials from museum collections and incorporating interactive exhibits from science—technology centers to reinforce conservation messages, concern about loss of species numbers, and diversity and loss of habitat.

1. Botanical Gardens

In 1989 the World Resources Institute estimated that 150 million persons visited some 1500 botanical gardens around the world. In addition to visits and guided tours, gardens offered continuing education for adults, workshops and hands-on experiences for children and families, and professional education courses and seminars for K–12 teachers. The New York Botanical Garden and Missouri Botanical Garden are examples of two of 21 member gardens of the American Association of Botanical Gardens and Arboreta offering graduate studies programs, usually in collaboration with universities in their area.

2. Museums

Through collections, education programs, exhibitions, and graduate-level research, museums have been very active in promoting biodiversity in both the formal and informal sectors. The American Museum of Natural History in New York (AMNH) provides an interesting example of an institution with current involvement in all these areas:

- Exhibition. The 11,000-square-foot Hall of Biodiversity is the newest permanent exhibit of AMNH and uses collections, interactive technologies, and an immersive environmental replica of a portion of the rain forest of the Central African Republic—complete with sound, smell, movement, and running water—to provide a unique visitor experience.
- Graduate and continuing education. The Center for Biodiversity and Conservation collaborates within and outside the museum in the development of courses and programs. AMNH is home to the oldest and largest doctoral and postdoctoral training

program of any scientific museum in the world, collaborating with Yale, Columbia, Cornell, and City University of New York.

- Education. The National Center for Science Literacy, Education and Technology supported by the National Aeronautics and Space Administration (NASA) has developed a number of projects related to the theme of biodiversity, including *Biodiversity Counts: A Student Inventory Project*, a program for middle school students across the United States to inventory plant and animal life in their communities and to share their findings through publications and on-line field journals.

C. Biodiversity "Experiences" and Resources

Overview

For most adults, biodiversity education will take place in the informal sector as they read books; visit zoos, museums, and national parks; listen to lectures; and watch programs on the increasing number of science- and nature-based cable channels, public television, or the increased coverage of science on the news or news magazines. Others will visit the World Wide Web, where an increasing number of excellent sites developed by universities, museums, federal agencies, and nonprofit organizations provide high-quality information. Several examples of resources for biodiversity education available to the adult public follow.

1. Earthwatch

Earthwatch Institute is an international nonprofit organization founded in 1971 that supports scientific field research worldwide. Volunteers participate in actual field research, assisting scientists in gathering data. Since its beginnings it has "mobilized 150 projects around the globe, resulting in the discovery of 2000 species, the establishment of 12 national parks, and the founding of eight museums." The Earthwatch web site lists active projects that volunteers can join in seven topical areas including Endangered Ecosystems and Biodiversity. The site also includes virtual field trips. Visit http://www.earthwatch.org.

2. National Biological Information Infrastructure (NBII)

The NBII attempts to organize the disparate sources of information available through agencies, departments, museums, and other organizations, providing a source of links to sites. A "Biodiversity, Systematics and Collec-tions" section connects to other web sites, many of which have education or "for kids" sections. Many federal agencies such as the U.S. Geological Survey, the Environmental Protection Agency, NASA, and others have relevant sites. Visit http://www.nbii.gov/biodiversity.

3. Nonprofit Environmental and Biodiversity Groups

A number of organizations produce materials to support education about environmental and biodiversity issues. These include groups such as the World Wildlife Fund, the Sierra Club, and the Audubon Society. These groups develop a wide variety of public information and educational materials.

While most mainstream advocacy groups are conscious of the need to "get the science right" and to present balanced viewpoints concerns are sometimes expressed about school use of materials that emanate from an "advocacy position." Guidelines have been developed by NAAEE to assist educators in assessing the scientific accuracy of such materials.

V. PUBLIC AWARENESS

How much does the public understand about environmental issues in general and biodiversity in particular, and what are the attitudes toward these issues? Surveys from a number of sources indicate that there is strong public interest in and support for issues related to the environment. The National Science Board's *Science and Engineering Indicators* (1998) suggested strong interest and "informedness" of the public around environment and health topics, especially when compared with other science and technology areas, and interest and support were stronger among women than men. The National Environmental Report card, an attitudinal and knowledge survey of American adults conducted by the National Environmental Education and Training Foundation and Roper Starch Worldwide, concluded that there was "an alarming lack of knowledge about some of our most critical environmental problems." With regard to biodiversity, however, 73% of adults surveyed correctly responded about the direct relationship between species loss and habitat destruction.

In 1995 the Consultative Group on Biological Diversity initiated the Biodiversity Project, a public education effort by a nonprofit grant makers' forum to "assess public opinion on biodiversity, to develop collaborative strategies to increase public awareness and engagement; and lay the groundwork to implement those strategies."

In June 1998 a "summit" on biodiversity and environmental education was convened at the American Museum of Natural History as a Biodiversity Educators Summit. The summit was supported by evidence from focus groups and surveys and co-convened by The Project, AMNH, and World Wildlife Fund. The 1996 Biodiversity Poll, conducted by the public opinion research firms Belden & Russonello and R/S/M and reported at the Summit, revealed the following about the environment and biodiversity:

- People care about the environment, but it isn't in the top tier of public concerns.
- Of environmental concerns, the public considers the most serious problems to be toxic waste, destruction of the rain forest, loss of places in nature, and air and water quality.
- Extinction is a concern, but it is not high on the list.
- People understand that nature is connected and interdependent, but most people do not recognize or use the word *biodiversity*. Only 2 in 10 said they had heard about the "loss of biological diversity."
- The public understands that species are declining and that human activity is largely responsible. But the public does not understand much about specific reasons or about the seriousness of the rate of loss.
- Public support for biodiversity conservation (once biodiversity is explained) is wide—87%. But this support is shallow.
- Countervailing pressures (values) can peel support away from biodiversity protection. These include concerns about jobs, individual property rights, comfort and convenience, and preservation of "unattractive" species. However, 51% of Americans agree that the world would suffer if such "unattractive species" (e.g., mosquitoes) are eliminated.

The reasons Americans think biodiversity should be conserved included personal and family issues (79%), responsibility to future generations (71%), and spiritual concerns of stewardship (67%) (i.e., Nature is God's work).

Another survey of biological scientists, science educators, and the general public conducted by the AMNH in April 1998 revealed that most scientists believe we are in the middle of a mass extinction largely caused by human activity. While 70% of the scientists surveyed rated loss of biodiversity as major and urgent, the general public was generally unaware of species loss and the threats this posed. Even science teachers, who are

aware of the biodiversity crisis, did not believe that there was mass extinction, and only 38% of teachers rated themselves as being very familiar with the concept of biodiversity. Both 1996 and 1998 polls revealed a large gap between scientists' perception and the public's awareness and concern about biodiversity, this in spite of the public's perceived attentiveness and informedness about environmental issues in general. (For more information, see http://www.biodiversityproject.org/eesummit.htm.)

The gap may relate to the following:

- A general lack of attention to biodiversity and its consequences by the media
- The way that the public message about biodiversity is conveyed
- The biological understandings that members of the public bring to the discussion

Interestingly, the spottiness of adult knowledge and concepts (some individual ideas understood, but not the overall concepts nor the consequences that flow from them) tracks with observations made about the K–12 student understanding of life sciences ideas. For most members of the public, high school is the last time for a formal course in the life sciences. After that, they generally rely on the informal sector for additional information and updates to their knowledge.

VI. COLLEGES AND UNIVERSITIES

According to *Science and Engineering Indicators*, persons taking college-level science courses are more likely to be informed about, supportive of, and interested in science and technology topics. College level courses in biology influence knowledge and attitudes of the public. *Beyond Biology 101*, produced by the Howard Hughes Medical Institute, describes efforts to transform college-level courses in biology for both majors and nonmajors, moving away from vocabulary-driven courses to ones that are more integrative and that include meaningful laboratory and field experiences. A number of programs are described at http://www.hhmi.org/BeyondBio101. These include an innovative program in human biology developed and in place since the 1960s at Stanford University. The program, taught by faculty from biology, education, anthropology, psychology, and other disciplines, focuses on the relationship between human biology and human behavior including human interactions with environment. It is a major course of interdisciplinary study and provides introductions to integrative biology for nonmajors.

The program for majors involves students in a range of activities designed to get them to think like scientists. Biodiversity and human impacts on the environment are explicit foci of instruction, and in the late 1980s and 1990s the college program was "translated" into a middle grades life sciences project (HumBio). Biodiversity concepts and activities are explicitly included among the curriculum materials in the ecology theme.

While most four-year institutions offer majors in biology or related subspecialties (including ecology), biodiversity was found as an area of major concentration at only a few U.S. colleges and universities. A search of the site, Academic Programs in Conservation Biology (http://www.conbio.rice.edu/programs), yielded programs at Columbia University, Illionis State University, San Diego State University, State University of New York at Albany, University of California at Riverside, University of Massachusetts at Amherst, University of Southwestern Louisiana, University of Wisconsin at Madison, and Yale University. Searches for "sustainable development" added to this list programs at Cornell University, Harvard University, University of Georgia, University of Maryland, University of Miami, and University of Minnesota. Distance learning classes were also available in these fields and listed at http://eelink.net.

Other countries that have evidence of strong interest in collegiate and university studies in biodiversity include Canada and Australia. The United Nations University, Institute for Natural Resources in Africa (UNU/INRA) has a number of relevant programs in the following categories:

- Soil and water management: Soil fertility restoration and maintenance.
- Conservation of biodiversity. Genetic improvement and increased utilization of Africa's indigenous food crops and useful plants
- Conservation and management of mineral resources.

Education and training are key areas of interest to UNU/INRA. In cooperation with other agencies of the UN, the program develops curriculum and contributes to training in areas such as ecological economics, natural resource economics and environmental accounting, germplasm and biodiversity conservation, wildlife management, and taxonomy. Gender and Natural Resources is a major cross-cutting theme in the work of the institute.

VII. SYSTEMATICS RESEARCH AND TRAINING

While the demand for expertise in conservation, biodiversity, and systematics has been increasing, concern is being expressed about the human resources, especially in developing countries, available to manage and inform natural resources utilization around the globe. A 1995 workshop on Priorities in Systematics Research and Training organized by the United Kingdom Systematics Forum and held at the Linnean Society of London raised issues about the adequacy of support for systematic biology, the declining interest in systematics among students, and the decline in the teaching of systematics in many universities.

A search for university departments worldwide that provide training in systematics and taxonomy revealed 24 institutions, 10 in the United States. This may be deceptive, however, in that a number of institutions provide graduate training in partnership with research-oriented museums and botanical gardens. New emphases such as work in molecular systematics may exist in cellular, molecular, microbiology or biochemistry programs. BIO NET-INTERNATIONAL is a global network of people and institutions that develop biosystematics capacity in developing countries. Training at all levels (in service, short courses, distance courses, and joint graduate programs) is a major focus of the network's activities.

VIII. CONCLUSION

A combination of school-based learning and out of school experiences combine to provide young people with knowledge of and attitudes about biodiversity that they then take into adulthood. For those who pursue higher education, college-level courses are available in some institutions that integrate biodiversity education into larger biological, environmental, or human impacts courses. Other adults must depend on the informal education sector, with experiences provided by a variety of different institutions and media. Biodiversity education may also, in some cultures, rely on community transmission of locally held knowledge of plants and animals of a region. Whatever the process for developing understanding, education and public awareness have been seen as crucial precursors to building support for biodiversity.

Additional Information

American Museum of Natural History. Please visit *http://www.amnh.-org/index.html* or http://www.amnh.org/education/ for additional information.

Center for Biodiversity and Conservation. Please visit http://research.amnh.org/biodiversity/ or contact the Center for Biodiversity and Conservation, American Museum of Natural History, Central Park West at 79th Street, New York, NY 10024; telephone: (212) 769 5742; fax (212) 769 5292; or via e-mail at *biodiversity@-amnh.org*.

Consulative Group on Biological Diversity. Please contact Lincoln and Torney, Presidio Building 1014, P.O. Box 29361, San Francisco, CA 94129-0361; telephone: (415) 561-6575; fax (415) 561-6490; e-mail *cgbdcgbd.org*; or visit *http://www.biodiversityproject.org/more.htm#cgbd* for additional information.

NAAEE. Please contact NAAEE Headquarters, 1825 Connecticut Avenue, NW, Suite 800, Washington, DC 20009-5708; telephone: (202) 884-8912; fax: (202) 884-8455; or visit *http://www.naaee.org/html/staff.html*.

NAEP. Please contact Bob Clemons, National Center for Education Statistics, 555 New Jersey Avenue, NW, Washington, DC 20208; telephone: (202) 219–1690; or visit *http://nces.ed.gov/nationsreportcard/site/contact.asp*.

National Environmental Education and Training Foundation. Please visit http://www.neetf.org/.

National Center for Science Literacy, Education and Technology. Please visit http://www.amnhonline.org/nationalcenter/ for additional information.

TERC. Please contact TERC, 2067 Massachusetts Avenue, Cambridge, MA 02140; telephone: (617) 547-0430; fax: (617) 349-3535; or visit *http://www.terc.edu/*.

TIMSS. Please contact the U.S. TIMSS National Research Center, Michigan State University, College of Education, 455 Erickson Hall, East Lansing, MI 48824-1034; telephone: (517) 353-7755; fax: (517) 432-1727; or visit http://ustimss.msu.edu/.

UNESCO. Please visit *http://www.unesco.org/*.

United Nations University/Institute for Natural Resources in Africa. Please visit http://www.unu.edu/inra/research.htm.

See Also the Following Articles

BIODIVERSITY, DEFINITION OF • CONSERVATION EFFORTS, CONTEMPORARY • GOVERNMENT LEGISLATION AND REGULATION • HISTORICAL AWARENESS OF BIODIVERSITY • HUMAN IMPACT ON BIODIVERSITY, OVERVIEW

Bibliography

Committee on High School Biology Education, National Research Council. (1990). *Fulfilling the Promise: Biology Education in the Nation's Schools*. National Academy Press, Washington, D.C.

Maier, F., Page, J., and Durrell, G. *Zoo: The Modern Ark*. (1990). Facts on File, New York.

National Research Council. (1996). *National Science Education Standards*. National Academy Press, Washington, D.C.

National Science Board. (1998). *Science and Engineering Indicators, 1998*. National Science Foundation, Arlington, VA.

President's Committee of Advisers on Science and Technology (PCAST) Panel on Biodiversity and Ecosystems. (1998). *Teaming with Life: Investing in Science to Understand and Use America's Living Capital*. PCAST, Washington, D.C.

Project 2061, American Association for the Advancement of Science. (1993). *Benchmarks for Science Literacy*. Oxford University Press, New York.

Project 2061, American Association for the Advancement of Science. (1989). *Science for All Americans*. Oxford University Press, New York.

ENDANGERED BIRDS

N. J. Collar
BirdLife International

I. Introduction
II. The Identification of Endangered Birds
III. Trends and Factors in the Endangerment of Birds
IV. Approaches to the Conservation of Threatened Birds

GLOSSARY

biological species concept Concept of a species as a population or series of populations that are reproductively isolated from other groups, as well as the degree of morphological similarity.
endangerment Condition in which a species is at risk of extinction.
phylogenetic species concept Concept of a species in which species-level identity is determined by members sharing distinct characteristics.

ENDANGERED BIRDS ARE DEFINED CHIEFLY AT THE SPECIES LEVEL (although the definition of "species" remains contentious), and at the global and national levels. New global criteria apply thresholds on decline rate, population size, and range size to identify endangerment. Some 11% of the world's avifauna are at risk, but altogether 20% of species give cause for concern. Most endangered birds (70%) have popula-

tions of less than 10,000 mature individuals. The Philippines, Indonesia, Brazil, and Colombia possess the most significant proportions of endangered birds. Tropical forest loss is the greatest threat, but there are many other reasons for elevated vulnerability (through range restriction, occurrence on islands, use of restricted habitat, etc.). Remedial actions include detailed research and documentation, site and habitat protection, and intensive multifaceted management programs.

I. INTRODUCTION

Endangerment is the condition in which a species or subspecies of animal or plant is at risk of extinction. The nature of the impending extinction may be local, national, or global. There is no vocabulary to differentiate between these conditions; as a consequence, there is often confusion over appropriate priority levels. A trend to use *extirpation* to indicate nonglobal extinction brings its own problems, since the active verb *extirpate* is synonymous with *eradicate;* there is also the problem that disappearing subspecies of species may be both extinct *and* extirpated (extinct as taxonomically distinct forms, extirpated as representatives of a higher taxonomic unit). This article concentrates primarily on bird species at risk of global extinction, but nationally and locally endangered forms, including subspecies, are also considered; so *extirpated* is here used to mean locally extinct. The terms *endangered, threatened,* and *at risk* are commonly used interchangeably; *endangered* is prevalent in U.S. usage, but IUCN/The World Conservation Union status categories use *threatened* as a generic term and *Endangered* (with a capital E) for a particular status. In this article, *endangered* is preferred, except in contexts involving the IUCN threat categories.

II. THE IDENTIFICATION OF ENDANGERED BIRDS

A. Taxonomy and the Identification of Target Units

In general, decision makers look to science in two ways for help in establishing priorities in species conservation: the first is the degree of a taxon's risk of extinction, and the second is the degree of its evolutionary differentiation. Many people are surprised that these aspects, particularly the second, remain hard to assess. Ever since the passing of the United States Endangered Species Act, the term "endangered species" has become widely used in popular and indeed jocular English parlance; so it seems extraordinary that debate still rages over what a species is. Nonetheless, its definition is a pervasive problem, and the profile of this issue will increase as more morphologically distinct, local forms, currently considered subspecies, come under pressure from human development activities.

When considering closely related taxa in which there is no geographical contact, and therefore no test for reproductive isolation (the key criterion under the prevalent Biological Species Concept or BSC), taxonomists are compelled to rely on the degree of their morphological similarity to judge, subjectively, whether they are conspecific or not. Among birds there are thousands of such cases, in part because flight has allowed them to colonize so many offshore and oceanic islands, where they have evolved features that distinguish them at some level from continental or other island stock.

Some influential museum-based ornithologists have recently advocated a narrow phylogenetic species concept (PSC), which bestows species-level identity on any population whose members fully share distinct characters, irrespective of hybrid zones. Many of these populations currently have taxonomic standing as subspecies, and proponents of the PSC expect its adoption to cause a doubling of the number of avian species, to about 20,000 from the roughly 9500 in current usage. However, BSC supporters contend that, rather than clarifying the status of disjunct populations, the PSC shifts the difficulty to a yet more complex and subjective level,

where very minor differences (sometimes biochemical only, and certainly not used to recognize even subspecies) may or may not be regarded as sufficiently distinct and consistent to admit species-level status, so that no final number of avian species can be even approximated.

This debate is important, not least because biological diversity is directly at stake. Some years ago the BirdLife Red Data Book program eliminated subspecies from its concern, on the basis that there were simply too many to document and that species had to take priority. This decision, made with the knowledge that many threatened subspecies will benefit from site management for threatened species with which they are sympatric, still tends to expose some subspecies, especially those with small ranges and therefore with relatively constrained populations, to the vagaries of deteriorating global conditions. Unless a country is as wealthy as the United States, whose legislation embraces subspecies and indeed populations, or has little or no other biological heritage in which to invest (for birds, Barbados is one such), the chances of intervention on behalf of these forms are relatively low.

Nevertheless, subspecies are far less stable as taxonomic entities than are species, and are easily erected and just as easily subsumed (prior to 1950 at least two museum ornithologists, Oberholser and Koelz, each established over 100 subspecies in the course of a single paper!). Because of this, they are generally unwelcome in law, which requires widespread long-term agreement on taxonomy to be able to function. Consequently, species rather than subspecies are the units of concern in most national and international legislation.

Partly perhaps as a consequence of this, many ornithologists seem willing to countenance the steady, consistent "unlumping" of many forms, on the basis of multiple character differences extending beyond morphology to voice, behavior, and even habitat. For conservation purposes this process needs to be expedited to ensure that specific identity is not bestowed too late for intervention. However, the elevation of many weakly distinct forms to species level may actually make the identification of conservation targets *harder,* for if resources are insufficient to support all the resulting endangered species, many deserving cases (in terms of their evolutionary distinctiveness) might be lost amid the competing claims of virtual look-alikes.

B. Scale and the Identification of Target Units

Birds can be endangered at the global, broad regional (e.g., subcontinental), national, narrow regional (e.g., provincial), and local levels. It is entirely legitimate for countries and specialist interests to seek to "red-list" (i.e., list as endangered) bird taxa at these various levels, although as the scale becomes smaller, the various legitimate causes may, theoretically at least, begin to conflict with one another. However, decreasing scale is widely accepted as correlating with decreasing priority, so conflicts of interest are unusual.

Global endangerment is the most important priority level, since total extinction is a far worse risk than any other form. In the past 35 years or so, global endangerment of birds has been registered principally through the Red Data Book program of the International Council for Bird Preservation (now BirdLife International). This program, accepted by IUCN/The World Conservation Union as the official source of globally threatened bird listings, has developed into a long-term, continent-oriented project to create detailed profiles of every endangered species (Collar and Stuart, 1985; Collar *et al.,* 1992; Collar *et al.,* 1999), based on the rationale that all information relevant to a species's conservation should be included. This in turn has led to the need for abbreviated global listings (Collar *et al.,* 1994).

The species that have found their way into these full and abbreviated Red Data Books are normally found on national "red lists," but not always at the expected priority level, owing to the various algorithms that non-global assessment tends to involve. Thus an apparently logical ranking system in the 1984 *South African Red Data Book: Birds* resulted in widespread and common species such as Egyptian Vulture (*Neophron percnopterus*) and House Martin (*Delichon urbica*)—both found in Europe, the latter in huge numbers—coming out higher (2/102 and 6/102, respectively) than the top-ranking globally threatened South African endemic Rudd's Lark (*Heteromirafra ruddii*)(22/102).

Much more frequently a species is declared nationally endangered without being globally endangered, although the issues at stake may sometimes be so momentous that global endangerment is often assumed. This is the case with the Houbara (*Chlamydotis undulata*), judged by many to be unsustainably exploited by Gulf State hunters in most of its range and yet on Central Asian evidence still outside the IUCN criteria thresholds. Similarly, the Spotted Owl (*Strix occidentalis*) is an endangered species under U.S. law, but it still misses the IUCN criteria because its populations north and south of its U.S. range render it unlikely to die out within a relatively short time frame.

However, clearly the most important red list after the global list must be at the national level, and somewhat surprisingly this emphasis has been increased by the recently inaugurated Convention on Biological Diversity. Although the Convention has a supposedly global remit and overview, it has devolved responsibility for actions in defense of biodiversity to national agencies as framed by *national* perceptions. Parties to the Convention thus somewhat unfortunately run the risk of focusing only on elements of their natural patrimony that they regard as relevant, at the expense of species identified through international perspectives.

C. Criteria for the Identification of Target Units

Under protocols still being developed by IUCN/The World Conservation Union, the global red-listing of a taxon may occur only if its conservation status is judged to satisfy at least one of a set of universal quantitative criteria (IUCN Species Survival Commission, 1994). Since a species can only become extinct by decreases in population and range size, these criteria set thresholds on these parameters as well as on decline rate by which to measure eligibility for and degree of threatened status. In IUCN terminology, the word *threatened* means what has hitherto been called *endangered,* whereas "Endangered" denotes a specific conservation status.

In crude form, the criteria stipulate that, to qualify as threatened, a species must possess a total population (A) declining at a rate (projected or past) of 20% over 10 years or three generations, or (B) within a range of less than 20,000 km^2 and declining, or (C) of less than 10,000 mature individuals and declining, or (D1) of less than 1000 mature individuals, or (D2) within a range of less than 100 km^2. Species meeting any one of these criteria qualify as threatened with the category Vulnerable; nested thresholds qualify species for the categories Endangered and Critically Endangered. Subspecies can be subjected to the same criteria, but this may result in the curious circumstance where all races of a species qualify as at risk (e.g., all five races of an island species with declining populations of under 10,000) but the species itself does not (total population still above 40,000).

The general experience with birds, almost certainly as with all animals and plants, is that populations tend to be significantly underestimated: for example, one observer on the New Caledonian island of Uvea judged there to be 70–90 Uvea (Horned) Parakeets (*Eunymphicus (cornutus) uveensis*) in 1993, but more intensive fieldwork later that year yielded a formal estimate of 617 ± 379. Similarly, a CAMP (Conservation Assessment and Management Plan) run by the Captive Breeding Specialist Group of IUCN in 1992 suggested a total population of some 5000 Tanimbar Corellas (*Cacatua goffini*), whereas analysis of quantified data from field-work that same year produced an estimate of 300,000–400,000 birds. Consequently, lists of threatened species cannot be expected to remain stable: while some will be added over time as their situation deteriorates or their taxonomy is revised, others will be removed as their true status is revealed.

In 1988, 1030 bird species were identified as at risk of extinction. In 1994, this figure, based on the new IUCN criteria, rose to 1111. It was not, however, the case that 81 species were added to the 1988 complement. In fact only 816 species were common to both lists. The 214 disappearances from and 295 additions to the 1994 list largely resulted from new "pioneering" knowledge (involving new areas or new identification insights) rather than from "monitoring" updates that disclosed a clear trend. Exploration is thus still the strongest biological data source, and "rare" species (for which see Kunin and Gaston, 1998) sometimes prove to be relatively common in some part of their range, or in some previously uninvestigated habitat. Consequently, attempts to predict future extinction rates using changes in red lists have been premature: the changes in question are not real-world events (Crosby *et al.*, 1996).

A degree of red list stability derives from some species being destined to remain endangered in perpetuity. This is because of their irremediably small ranges or populations (a circumstance that has made the criteria unpopular in some quarters, since no active threat need exist to trigger the listing). These are species—avian examples including the Lava Gull (*Larus fuliginosus*), with 300–400 pairs maximum, and the Tinian Monarch (*Monarcha takatsukasae*), on an island of less than 100 km² despite its estimated 40,000 individuals—for which the price of survival is eternal vigilance.

Appropriate criteria for use at the national level are still under development by IUCN; meanwhile, a good model is that of Avery *et al.* (in Coulson and Crockford, 1995).

III. TRENDS AND FACTORS IN THE ENDANGERMENT OF BIRDS

A. Globally Threatened Birds in 1994: Numbers, Criteria, and Extinction Rate Predictions

The 1111 bird species judged to be at risk of extinction in 1994 represented 11% of the world's avifauna. More-over, a further 66 (1%) species were then listed as Data Deficient and 875 (9%) as Near-Threatened, so that altogether over 20% of all bird species were identified as being of some global conservation concern. The majority of threatened species were classified as Vulnerable (704, 63%), with 235 (21%) Endangered and 168 (15%) Critically Endangered.

The commonest criterion triggered by threatened birds was C, which combines small population (<10,000) with significant decline. As many as 764 species (approaching 70%) of all threatened birds were judged (or, under the precautionary principle, thought likely) to fulfill this criterion; thus 8% of all bird species are known or suspected to have dangerously low populations. The other four criteria proved to be rather evenly distributed: A (rapid decline) and B (small range with significant decline) were triggered by around 400 species each, with somewhat smaller numbers triggering D1 (very small population) and D2 (very small range).

The new IUCN criteria attach hypothetical probabilities of extinction to the different categories of threat, on which basis 400 species may be expected to become extinct, without remedial action, in the coming century; however, it will take 1750 years for 90% of the 1111 listed threatened species to disappear. Comparison with the 1988 listing allowed three different listing recruitment rates to be tested on two models, indicating that with current trends between 400 and 1200 species of bird may die out within the next 100 years, with a time to extinction for half the planet's avifauna (ca. 4850 species) of 800–2800 years (Crosby *et al.*, 1996). Although these figures are less pessimistic than other recent estimates derived from less robust data, human pressures on the environment will only increase in the foreseeable future, so these extinction rates will probably prove much too conservative.

B. Globally Threatened Birds: Regions, Countries, Habitats

The majority of threatened bird species occur in Asia and the New World, with relatively few in Africa. The top ten countries for the highest numbers of threatened species are Indonesia (104), Brazil (103), Philippines and China (both 86), India (71), Colombia (62), Peru (60), Ecuador (50), United States (46), and Vietnam (45). Asian countries predominate in this list; the United States ranks high because of its Pacific territories. The highest African countries, ranking 21st and 22nd, are Tanzania (30) and Madagascar (28).

Priority countries might be selected on this basis,

but several further filters can be applied. Because species in the higher categories of threat are likely to become extinct sooner, a reranking involving just those 403 species that are either Critically Endangered or Endangered yields a top ten of Brazil (47), Philippines (45), Colombia (31), United States (25), Indonesia and Mexico (20 each), Peru (18), and Vietnam, Ecuador, and Argentina (16 each). Using this category, the emphasis shifts dramatically to the New World.

A further filter involves only those species from the preceding analysis that are nationally endemic, thereby indicating the degree of "ultimate responsibility" that falls to these countries as their most urgent bird conservation tasks. The Philippines (40) emerge far ahead of Brazil (32), itself far ahead of Colombia (24), United States (17), Mexico (13), Indonesia and New Zealand (12 each), Australia (11), Madagascar (10), and Peru (9). Although it was fairly obvious from the first analysis, given their small land area, that the Philippines would be a priority area, this refined analysis offers startling evidence of the critical importance of the country in terms of avian biodiversity and its impending loss.

The catastrophic erasure of forests from the planet in the course of the twentieth century means that most threatened birds are (mostly tropical) forest dwellers; the only surprising thing is that the figure is as low as 65%. Wetland species account for 9%, scrubland for another 9%, and grassland for 6%. The relatively low forest representation can be explained in part by the fact that the largest tracts of forest, in Amazonia, the Congo basin, and Borneo, are—despite the destruction visited upon them—still too extensive, and the species they contain too widespread, to have resulted in more than a handful of listings from these areas.

C. Causes of Endangerment in Birds

Endangered birds suffer from a range of different threats. Some 52% of them are affected by habitat loss and degradation (although this figure is almost certainly higher; indeed, low-level and hence unreported habitat loss could probably safely be indicated for the other 48%); this theme is explored in the following paragraphs. The next most important threat is simply restriction of range or population, involving 23% of all threatened birds. Hunting afflicts 8%, introduced species 6%, and trade 3%.

Until recently the notion that habitat loss was important in the demise of the Passenger Pigeon (*Ectopistes migratorius*) had not been entertained, but Bucher (1992) showed how the species was a specialist on seeds produced in masting events whose scale and geographic

location varied from year to year, that is, that were patchy in both space and time. Human settlement of the east and center of the North American continent fragmented the native forests to the point where the lapse in both time and distance between masting events simply became too great. Despite the settlers' prodigious slaughter of birds, which has always been blamed for their disappearance, it appears that it may have been their axes, not their guns, that caused the loss of their quarry; the last wild birds very possibly *starved* to death.

An equally celebrated North American species, perhaps still extant, is the Eskimo Curlew (*Numenius borealis*), which, like the Passenger Pigeon, used to be hunted in phenomenal numbers in the nineteenth century and likewise never recovered after the slaughter finally abated. It seems likely, however, that the loss of its Argentine grassland wintering grounds and North American prairie spring stopover sites was to blame. Loss of stopover habitat is beginning to supplant hunting (never a convincing case) as the best explanation for the virtual disappearance of the closely related Slender-billed Curlew (*N. tenuirostris*), which breeds in western Siberia and migrates southwest to the Mediterranean basin, almost certainly using the once extensive east–west Russian steppes along the way.

Other notable examples of this space/time vulnerability include the Thick-billed Parrot (*Rhynchopsitta pachyrhyncha*), Purple-winged Ground-dove (*Claravis godefrida*), Andean (*Phoenicopterus andinus*), and Puna Flamingos (*P. jamesi*), Lesser Florican (*Sypheotides indica*), and Resplendent Quetzal (*Pharomachrus mocinno*). The parrot shows the same trait as the Passenger Pigeon, since it is dependent on pine seed, a notoriously unpredictable resource. The species is nomadic, but as its native pine forests in Mexico's Sierra Madre are further fragmented, there is a serious danger that a cone-crop failure will leave the last populations "stranded" too far from food for any to survive. The ground-dove specializes on the seeds of forest bamboo in southeast Brazil. So much forest has been destroyed within its range, and bamboo seedset is so temporally patchy, that the species has become one of the rarest in the country.

The flamingos move between lakes in search of appropriate conditions (which shift over time) and are therefore exposed to the possibility that human damage to even a small number of sites may one day leave the species with nowhere to go. The florican selects different grassland sites from year to year in western India, depending on the effects of local rainfall. Grassland is under enormous human pressure in India, and conserving tracts that may be empty of birds for several years at a time is not a simple proposition. Post-breeding quetzals are now known to make complex short-distance movements to several different areas, so that many more tracts of forest than one or two may be needed to ensure the long-term survival of viable populations of this species.

In Australia, many birds have been affected by human alteration of the natural fire regime. The Paradise Parrot (*Psephotus pulcherrimus*) almost certainly became extinct (it may conceivably survive somewhere) owing to new burning patterns that suited livestock but not the grasses on whose seeds the parrot subsisted. Similar problems afflict the food supply of the Golden-shouldered Parrot (*P. chrysopterygius*) and the habitat of the Noisy Scrub-bird (*Atrichornis clamosus*).

The reason why restriction of range ranks so highly in the list of threats is because any cause of decline is likely to affect the entire species too quickly for human intervention to help. Such species are often restricted to islands, and when those islands are oceanic, the birds have usually evolved in the absence of continental pressures from mammalian predators. As a consequence they are behaviorally and physically adapted in ways that leave them highly vulnerable when continental predators become established within their ranges, through either the direct or indirect agency of humans.

These behavioral adaptations are not degenerative. The Dodo (*Raphus cucullatus*) was given its name from the Portuguese slang for "stupid," *doido*. But island animals that are entirely tame, or that nest in what to human eyes are ludicrously undefended places, or that have lost the function of their wings are not evolutionary failures. On the contrary, these seemingly disadvantageous attributes are the result of continuing evolutionary pressures. Wings cost energy to carry and maintain, and in the absence of predators they offer no return on the investment in such energy. Shyness costs its possessors dearly if less shy creatures have more time to exploit whatever resource is at stake. Nesting in inaccessible places is needlessly expensive if there is no risk in nesting on the ground in the open. Moreover, the relatively stable conditions on tropical islands tend to promote marked K-selected traits (e.g., slow reproductive rates). So the very things that render island birds so vulnerable to aggressive, fast-breeding, continental animals have actually been selected *for* through narrower, often intraspecific competition.

Most avian extinctions since 1600 have been on islands (King, in Moors, 1985), and a significant number of threatened birds today are island species. The impact of rats has been and remains massive (see Atkinson, in Moors, 1985): Magenta Petrel (*Pterodroma magentae*), Zino's Petrel (*P. madeira*), Tuamotu Sandpiper (*Proso-*

bonia cancellata), Polynesian Ground-dove (*Gallicolumba erythroptera*), Seychelles Paradise-flycatcher (*Terpsiphone corvina*), and Rarotonga Monarch (*Pomarea dimidiata*) are a few of the Critically Endangered birds whose fate is directly linked to the invasion of their islands by rats. In some cases where cats have also been introduced, it is not clear which predator is the greater culprit, and these affected species often live on larger islands where very little can be done to help. The New Caledonian Rail (*Gallirallus lafresnayus*), Cuba's Zapata Rail (*Cyanolimnas cerverai*), the Samoan Moorhen (*Gallinula sylvestris*), and virtually the entire endemic avifauna of Hawaii (the little of it that survives, but notably the honeycreepers, Drepanididae) are good examples of this uncertainty and impotence. Cats alone are responsible for the plight of some species, such as Townsend's Shearwater (*Puffinus auricularis*), Socorro Dove (*Zenaida graysoni*), and the Marquesan Ground-dove (*Gallicolumba rubescens*). Mongooses, which were often released to devour rats or control snakes, are major threats to species such as Hawaiian Duck (*Anas wyvilliana*) and St Lucia's Semper's Warbler (*Leucopeza semperi*).

Ancient lakes, like oceanic islands, often harbor endemic faunas and floras that are highly susceptible to exotic introductions. Among the birds, the grebe family Podicepitidae has been particularly hard hit, with the Alaotra Grebe (*Tachybaptus rufolavatus*) of Madagascar and Junín Grebe (*Podiceps taczanowskii*) of Peru close to extinction; the Atitlán Grebe (*Podylimbus gigas*) of Guatemala and Colombian Grebe (*Podiceps andinus*) have already vanished.

Linear water bodies may similarly expose certain species to extinction risks from a single event or series of events: Scaly-sided Merganser (*Mergus squamatus*), Brazilian Merganser (*M. octosetaceus*), Wrybill (*Anarhynchus frontalis*), Rufous-throated Dipper (*Cinclus schulzi*), and Luzon Water-redstart (*Rhyacornis bicolor*) are all vulnerable in this way. Moreover, species that concentrate in a small area for even part of their life cycle may be abnormally exposed to danger. For example, the entire world population of Ascension Frigatebird (*Fregata aquila*) breeds on a single stack smaller than a municipal parking lot. Even birds that are briefly drawn to individual fruiting trees are liable to suffer: as many as 40 Visayan Wrinkled Hornbills (*Aceros waldeni*) were shot in a single tree over the course of a single day in October, 1997, an event that quite possibly killed 50% of the population.

The foregoing examples represent "spatial" threats. Other threats are better characterized as "temporal," involving a seemingly innocent event whose consequences cannot be remedied by the time they become apparent. New Zealand's Kaka (*Nestor meridionalis*) and Yellowhead (*Mohoua ochrocephala*) face enormous difficulties now that introduced wasps compete with them for honeydew, an extremely important foodstuff (breeding success in the Kaka is directly correlated with honeydew intake in the previous autumn). The White-headed Duck (*Oxyura leucocephala*) faces long-term extinction through hybridization with its New World counterpart, the Ruddy Duck (*O. jamaicensis*), which became feral in Britain in the 1960s and is now spreading into its range. In both cases the costs and logistics of eradication are too great to consider.

Sometimes threats come not from exotics but from natives expanding their ranges, often owing to human modifications of habitat. The spread of cowbirds (*Molothrus*) through the Americas and the Caribbean is particularly worrying. Birds such as Kirtland's Warbler (*Dendroica kirtlandii*) and Black-capped Vireo (*Vireo atricapillus*) need constant-effort programs to reduce cowbird brood-parasitism to tolerable levels. Similarly the spread through the Caribbean of the Pearly-eyed Thrasher (*Margarops fuscatus*), a nest-hole competitor, has been viewed with alarm by the Puerto Rican Amazon (*Amazona vittata*) recovery teams.

Analysis of Neotropical data (Collar *et al.*, 1997) shows that the avian families with significantly high numbers of threatened species suffer from particular threats in combination. Thus the parrots (Psittacidae) have the greatest proportion of threatened birds in any family (28%), closely followed by the curassows and guans (Cracidae, 26%), and then the tinamous (Tinamidae, 15%, still almost double the 8% rate of endangerment in the New World avifauna as a whole). All three families are sensitive to habitat loss; but the parrots also experience intensive trapping for trade purposes, and the cracids and tinamous are no less intensively exploited for food. Possession of a distinct economic value within a beleaguered habitat type, particularly one with some (bio-)geographic restriction, confers a strong likelihood of endangerment.

However, there are many endangered birds whose rarity has abidingly obscure causes. Among these are the White-winged Duck (*Cairina scutulata*, Southeast Asia), Giant Ibis (*Pseudibis gigantea*, Indochina), Himalayan Mountain-quail (*Ophrysia superciliosa*, India), Bornean Peacock-pheasant (*Polyplectron schleiermacheri*, Borneo), Negros Fruit-dove (*Ptilinopus arcanus*, Negros, Philippines), Blue-headed Lorikeet (*Charmosyna toxopei*, Buru, Indonesia), New Caledonian Lorikeet (*C. diadema*), Forest Owlet (*Athene blewitti*, India), Liberian Greenbul (*Phyllastrephus leucolepis*), Cone-billed Tana-

ger (*Conothraupis mesoleuca*, Brazil), Cherry-throated Tanager (*Nemosia rourei*, Brazil), Ibadan Malimbe (*Malimbus ibadanensis*, southern Nigeria), and Isabela Oriole (*Oriolus isabellae*, Luzon, Philippines). All are bafflingly rarer—some are known only by a single museum specimen—than might be inferred from the habitat apparently available to them. Unidentified factors must afflict them all, and clearly *sometimes* natural causes may be in play, particularly unseen ones such as diseases and infestations (see May, in Coulson and Crockford, 1995), which are known to afflict the Iphis Monarch (*Pomarea iphis*) and Gouldian Finch (*Erythrura gouldiae*). The safest assumption in cases of inexplicable rarity, at least on continents, is that habitat degradation or loss is in some way involved.

IV. APPROACHES TO THE CONSERVATION OF THREATENED BIRDS

A. Research and Synthesis

The primary conservation need of an endangered species is information. There is a common behavioral trait among academics of ignoring literature much older than 10 years, presumably partly reflecting the assumption that the data contained in such literature are incorporated into more recent work. Often, in fact, the information has been ignored; sometimes, however, it has been used but, on proper reconsideration, proves to be faulty. There is therefore considerable virtue in seeking out, assembling, and critically evaluating all information relating to an endangered species before deciding on the most appropriate remedial or merely investigative action. Despite the need for speed in cases of species at risk, precipitate intervention can waste hundreds of thousands of dollars and even prejudice attitudes against a species when its real needs are finally recognized. For example, captive breeding management for the Philippine Eagle (*Pithecophaga jefferyi*), a costly long-term, but to date unsuccessful program, appears to have come into being in response to somewhat overcautious estimates of the bird's population size.

Recent BirdLife Red Data Books have sought to present detailed syntheses of relevant data, extending to translations from languages with which biologists may not be familiar, so that a clear picture of a situation can rapidly be considered and the options objectively assessed. However, there are many other examples of the careful construction of evidence in endangered species management. One of the most notable is the exhaustive 400-page review of the history, plight, and management of the Puerto Rican Amazon (*Amazona vittata*) assembled by members of the team that spent some 20 years piecing together the facts (Snyder *et al.*, 1987).

The importance of individual study of endangered birds cannot be overstated, although there can be difficulties (over logistics, permissions, the generation of statistically useful data) and dangers (sometimes to the student, sometimes to the species) that militate strongly against such work. For larger species, radio-tracking is becoming increasingly valuable as a means of recovering large quantities of information from a relatively small investment of effort: species as different as Black-faced Spoonbill (*Platalea minor*) and Madagascar Serpent-eagle (*Eutriorchis astur*) have yielded data on movements and daily behavior patterns that are crucial to their long-term management.

B. Site and Habitat Conservation

The conservation of sites at which endangered birds occur, and of the habitat they are known to occupy, is the primary management technique for ensuring their survival. The key tool is the protected area, which for larger sites is usually reflected in law as a national park, indicating the clear public interest of setting aside a major proportion of a country for noneconomic reasons. Smaller areas are often designated as nature or biological reserves, and are frequently considered more as refuges or scientific laboratories for research purposes than as sites with a broader public service; public access can be more difficult than in national parks. However, the *size* of many national parks is important for conserving viable populations of larger, low-density species. Naturally it is appropriate to seek to save these species where they are sympatric, thereby maximizing the efficiency of the expense. Evaluations such as those by Wege and Long (1995), Stotz *et al.* (1996), and Stattersfield *et al.* (1998) provide clear rationales for the targeting of conservation resources in such a way as to secure not just individual species but the key representatives of biogeographic regions.

In general, endangered birds on continents require larger-scale habitat conservation, whereas those on islands need intensive multi-faceted management (aspects of which are treated in Section III,D). The cardinal element in successful site conservation is local support, backed of course by national government. In its work since 1983 to conserve the montane forests of western Cameroon (to which 25 species of bird are endemic, 9 of them threatened), BirdLife International has imple-

mented a major ICDP (integrated conservation and development project) at Mt. Kilum-Ijim to aid local human communities around the mountains, and this has led to a widespread appreciation of the biological value of the forest without generating the kind of hostility that goes with attempting to create strict exclusion zones. The same kind of program operates at Arabuko-Sokoke Forest in Kenya, home to six endangered birds. Other conservation organizations have been using the same fundamental formula—that local people must be made part of the solution, not demonized as part of the problem.

Campaigns to promote interest in and support for species conservation programs are vital elements of those programs. These may take the form of extension work related to ICDP promotion of sustainable use of local resources, or more direct appeals to people to appreciate the unique value of the wildlife in their neighborhood.

C. Trade Controls and International Legislation

Although trade is not a strong factor in the endangerment of birds in general, it is important for a few groups of species, most notably the parrots. The major international instrument for the control of trade is the Washington Convention, universally known as CITES (Convention on International Trade in Endangered Species), which has been in operation since the 1970s. Animals and plants may be registered in three ways: on Appendix I, which essentially prohibits all movement of the species in question; on Appendix II, which prohibits all commercial trade except under license; and on Appendix III, which allows a particular nation to prohibit trade across its borders irrespective of a species' status elsewhere.

Because of the immense volume of traffic in parrots, and the problem of identification in so diverse a family (around 350 species), all but three species were placed on Appendix II in 1981. This move had the intention if not the effect of giving protection to the more endangered parrots—naturally their rarity increased their desirability among bird-fanciers—which, if listed alone on the Appendix, could easily have been traded indiscriminately under other names without customs officials necessarily being able to identify them. Appendix II species have quotas set by exporting countries, in theory based on data that show the exploitation to be sustainable. Where the evidence suggests that it is not, movement to Appendix I is supposed to bestow immunity.

Curiously—but as a measure of the power that some trade interests can exert—listing on Appendix I can actually stimulate trade (a) while the species is still at the proposal stage for upgrading but also (b) following listing, on the basis of its enhanced rarity value. This happened to the Hyacinth Macaw (*Anodorhynchus hyacinthinus*), whose rarity in the wild today is primarily a consequence of a trapping blitz in the 1980s. In general, CITES trade controls have some effect, and allow for very useful monitoring of changes in trading fashions over time. However, for truly prized species such as Lear's Macaw (*A. leari*) and Spix's Macaw (*Cyanopsitta spixii*), both from Brazil, there is little that any control system can do to eliminate smuggling: the financial inducements are simply too strong. The same tends to be true in Indonesia for exquisite songsters like the Straw-headed Bulbul (*Pycnonotus zeylanicus*) and good-lookers like the Bali Starling (*Leucopsar rothschildi*).

Other important international instruments for endangered birds are the Ramsar Convention (for important wetlands), Bonn Convention (for migratory species), and Berne Convention (for European species). The Convention on Biological Diversity, already mentioned, ought to be the cornerstone of endangered species conservation across the planet, but considerable effort by nongovernmental organizations is needed to ensure the inclusion of global priorities in the national conservation strategies that each party to the convention is obliged to produce and implement.

D. Management Techniques for "Critically Endangered" Birds

1. Habitat Restoration

Major conservation projects almost invariably involve some habitat restoration work. However, for species at the brink of extinction through habitat loss, the emphasis falls more directly on the rapid replanting of food-plants and land areas. Lear's Macaw appears to be constrained by the availability of licurí palms (*Syagrus coronata*), many stands of which show no sign of regeneration owing to cattle-grazing: programs have long been planned to establish many new groves within the species' range. The Cebu Flowerpecker (*Dicaeum quadricolor*) survives in an area of heavily degraded forest that consists of a mere 3 km²: efforts are under way to reforest adjacent areas within the next 30 years. There is now an emerging discipline of restoration ecology, and it is likely that a great deal of conservation energy in the twenty-first century will be channeled into reconfiguring habitats that were ruined during the twentieth.

2. Control and/or Restriction of Aliens and Natives

Eradication of exotic predators and pests from islands has been pioneered in New Zealand, where several small offshore islands have been rid of various mammals in order both to preserve resident breeding species and to translocate stricken native species from the main islands (Clout and Craig, in Coulson and Crockford, 1995). This painstakingly systematic restoration of islands—mainly involving the elimination of cats, rats, and mustelids—has been essential to the survival of birds such as the Black Petrel (*Procellaria parkinsoni*) and Kakapo (*Strigops habroptilus*), and the expertise generated is now being exported to other islands in the Pacific and Indian Oceans.

Researchers on Mauritius recently discovered why the native, Critically Endangered Pink Pigeon (*Nesoenas mayeri*) and Mauritius Fody (*Foudia rubra*) have managed to persist in the face of rat predation. These birds now nest almost exclusively in a grove of exotic *Cryptomeria japonica*, whose bark produces a sticky gum and whose leaves consist of spiny needles, both of which discourage rats from getting at nests. Curiously, then, in rare cases it appears that exotic vegetation can help rather than hinder native species in their struggle against exotic predators; in this instance the planting of more *Cryptomeria* is clearly called for.

It is not always the case that exotic predators are the problem. In New Zealand, the native Weka (*Gallirallus australis*) has proved to be a significant influence in depressing the numbers of Little Spotted Kiwi (*Apteryx owenii*) marooned on Kapiti Island, of Cook's Petrel (*Pterodroma cookii*) on Codfish Island (until removed), and of Chatham Oystercatcher (*Haematopus chathamensis*) on various Chatham islands. On Bermuda, White-tailed Tropicbirds (*Phaethon lepturus*) outcompete Cahows (*Pterodroma cahow*) for nests and have to be controlled. As already mentioned, brood-parasitism by cowbirds requires intensive local control efforts to prevent the suppression of breeding success in Kirtland's Warbler and Black-capped Vireo.

3. Captive Breeding

The role of captive breeding in endangered birds has long been controversial. The prevailing view among conservation biologists is that captive breeding is not a major management tool for endangered birds, and that indeed it can positively distract attention and resources from serious problems that affect the species. The Philippine Eagle has already been mentioned; similarly, cranes and parrots have in the past been launched into *ex situ* programs in the mistaken belief that major benefits will result (both species imprint heavily, and parrots, many of which learn survival techniques through observation of their parents, are particularly disadvantaged for return to the wild). Captive breeding has many other drawbacks, notably the particular dangers of disease transmission to wild birds from captive stock (Snyder *et al.*, 1996).

Nevertheless, captive breeding has achieved several outstanding successes. The Northern Bald Ibis (*Geronticus eremita*) and California Condor (*Gymnogyps californianus*) both flourish in captivity while efforts to improve environmental conditions in the wild continue. The Lord Howe Rail (*Gallirallus sylvestris*) and Guam Rail (*G. owstoni*), members of a notoriously vulnerable family (since so many rails reached islands in the past and proceeded to lose the power of flight), probably only persist thanks to *ex situ* regimes. The Socorro Dove (*Zenaida graysoni*) survives only because a few Californian bird-fanciers took and bred specimens at a time before its native island was overrun by exotic predators. In 1998 the Bali Starling (*Leucopsar rothschildi*) was within eight individuals of becoming extinct in the wild owing to relentless poaching for trade, but zoos throughout the world have bred the species such that many hundreds and perhaps thousands survive for possible reintroduction when poaching has finally been brought under control.

There is also a role for veterinarians in intensive *in situ* management programs, through their expertise in providing appropriate advice on toxicity of nestbox materials, control of nest parasites, composition and secure provision of supplementary foods, disease screening, and minimization in cross-fostering and translocation exercises.

4. Reintroduction and Translocation

The extirpation of a species at a discrete site may have been caused by a short-term or remediable factor, so that restocking with individuals from elsewhere can be undertaken. This is usually relevant or worthwhile only when the species is globally endangered, but whatever the urgency, the endeavor requires careful planning (Black, 1991). A series of feasibility assessments should determine the site's continuing ecological suitability (the original constraining factors must no longer operate), the threats it faces, the availability of appropriate stock, the socioeconomic implications, and local, national, and international awareness needs. The birds must be in optimal condition, and if captive-bred they should have been reared in disease-free conditions and in such a way as to be behaviorally and genetically fit

for independence in nature. Monitoring of the project is crucial, and it should be documented for circulation to other biologists contemplating such efforts.

Bird species that have benefited from well-designed reintroduction programs include the Nene or Hawaiian Goose (*Branta sandvicensis*), California Condor, Peregrine (*Falco peregrinus*), and Chatham Islands Snipe (*Coenocorypha pusilla*). Those that have not include Cheer Pheasant (*Catreus wallichii*) in Margalla Hills, Pakistan (fox predation), Shore Plover on Mangere Island, New Zealand (the birds flew back to South East Island), and Thick-billed Parrots in the United States (the wild-caught birds flew back to Mexico; the captive-bred ones flew nowhere at all).

Translocation to previously unoccupied sites has become a greatly valued technique, if only as a temporary measure while efforts are concentrated on habitat restoration or predator eradication at the native site. Beneficiaries of this approach include Niuafoou Megapode (*Megapodius pritchardii*), *Vini* lorikeets, Kakapo, Guam Rail following captive breeding, Seychelles Magpie-robin (*Copsychus sechellarum*), and Seychelles Warbler (*Acrocephalus seychellensis*). The work done in preparation for the Seychelles Warbler translocations showed that target islands had many times the insect abundance of the host island, Cousin. This was clearly related to the absence of predation pressure on the insect fauna, and—unsurprisingly but still very strikingly—when birds were released onto the target islands they began breeding almost immediately, and very rapidly expanded their numbers. On Cousin the birds had been so packed that they bred only very slowly, with one offspring tending to stay on territory and help at the nest, waiting for a parent to die (Komdeur, 1997). This observation underscores the value of the food resource base in managing endangered birds, and leads to the next point.

5. Supplementary Feeding

The value of increasing food availability for species whose populations need rapid growth seems to have been perceived only relatively recently, perhaps beginning with work on Peregrines (see Temple, 1978). The technique has been adopted for endangered birds on Mauritius, principally with the Mauritius Kestrel (*Falco punctatus*), for which extra food clearly enhanced reproductive output (Jones *et al.*, in Coulson and Crockford, 1995). New Zealand workers supporting efforts on Mauritius have found the same with the Kakapo: normally the species breeds only in response to major masting events, which may occur once every five years, but with dietary supplements it appears to be capable of breeding every year. On Hokkaido, Red-crowned Cranes (*Grus japonensis*) have recovered from near-extirpation through a combination of better protection and extensive food provision. Clearly, as a relatively short-term measure supplementary feeding can be crucial in producing an unnaturally rapid reproductive output, a very desirable effect in critically low populations where the genetic value of every individual needs to be maximized.

6. Nest-Site Provision or Enhancement

Where nest sites are limiting (often the case with hole-nesting species), the provision of nestboxes or the enhancement of natural cavities may be appropriate. Puerto Rican Amazons ignored the former but benefited from the latter. Mauritius Kestrels greatly benefited from the creation of nesting ledges. The provision of grilles at entrances has helped prevent White-tailed Tropicbirds from appropriating Cahow burrows. In the Galápagos, a new, more secure colony of Dark-rumped Petrels (*P. phaeopygia*) was established by digging nest burrows and playing calls among them at night.

7. Cross-Fostering and Cross-Breeding

Many large raptors and most cranes lay two eggs but normally rear only one young, so biologists attempting to increase productivity of endangered forms of such birds have long sought to make use of the expendable second eggs by taking them for hatching and captive breeding. In the case of the Whooping Crane (*Grus americana*), an attempt was made to establish a second population by placing such eggs in the nests of Sandhill Cranes (*G. canadensis*), but this did not result in a breeding population, possibly as a result of imprinting. The most famous instance of cross-fostering involved the Black Robin (*Petroica traversi*), whose population fell to five individuals in 1980 but recovered by the placing of some eggs in the nests of Chatham Island Tit (*P. macrocephala*).

Cross-breeding with another subspecies was regarded as the last hope of the Dusky Seaside Sparrow (*Ammospiza maritimus nigrescens*), but the last stock died out before the endeavor could begin. However, it has been used on the endemic Norfolk Island race *undulata* of the New Zealand Boobook (Morepork, *Ninox novaeseelandiae*), which by 1986 had been reduced to a single female. After nestboxes were erected to overcome an immediate shortage, two male nominate male boobooks were introduced to the island; one of them paired with the female and in due course produced a string of hybrid offspring. The taxonomic (and legal) implications of this technique may not have been con-sidered, but it clearly represents a means of preserving genetic diversity.

8. The Role of the Concerned Citizen

Finally, it is worth stressing that much of the advocacy for endangered birds, and many of the insights into their plight and salvation, comes from national and international conservation organizations. One of the most valuable things that any sympathetic individual can do to help endangered birds is simply to join or support such organizations. Their conservation effectiveness depends not only on the financial security derived from a broad membership base, but also on being recognized as the representative mouthpieces of a constituency consisting of millions of concerned citizens.

See Also the Following Articles

BIRDS, BIODIVERSITY OF • CONSERVATION EFFORTS, CONTEMPORARY • ENDANGERED MAMMALS • ENDANGERED REPTILES AND AMPHIBIANS • EXTINCTIONS, MODERN INSTANCES OF

Bibliography

Black, J. M. (1991). Reintroduction and restocking: Guidelines for bird recovery programmes. *Bird Conserv. Int.* 1, 329–334.

Bucher, E. H. (1992). The causes of extinction of the Passenger Pigeon. In *Current Ornithology 9* (D. M. Power, ed.), pp. 1–36. Plenum Press, New York.

Collar, N. J., and Stuart, S. N. (1985). *Threatened Birds of Africa and Related Islands: The ICBP/IUCN Red Data Book*. International Council for Bird Preservation/International Union for Conservation of Nature and Natural Resources, Cambridge, United Kingdom.

Collar, N. J., Gonzaga, L. P., Krabbe, N., Madroño Nieto, A., Naranjo, L. G., Parker, T. A., and Wege, D. C. (1992). *Threatened Birds of the Americas: The ICBP/IUCN Red Data Book*. International Council for Bird Preservation, Cambridge, United Kingdom.

Collar, N. J., Crosby, M. J., and Stattersfield, A. J. (1994). *Birds to Watch 2: The World List of Threatened Birds*, BirdLife Conservation Series 4. BirdLife International, Cambridge, United Kingdom.

Collar, N. J., Wege, D. C., and Long, A. J. (1997). Patterns and causes of endangerment in the New World avifauna. In *Studies in Neotropical Ornithology Honoring Ted Parker* (J. V. Remsen, ed.), Ornithological Monograph 48, pp. 237–260. American Ornithologists' Union, Washington, D.C.

Collar, N. J., Tabaranza, B. R., Mallari, N. A. D., Villasper, J. M., Lowen, J. C., Tobias, J. A., Long, A. J., and Crosby, M. J. (1999). *Threatened Birds of the Philippines*. Bookmark, Inc., Manila.

Coulson, J., and Crockford, N. J. (eds.). (1995). Bird conservation: The science and the action. *Ibis* 137, Suppl. 1.

Crosby, M. J., Stattersfield, A. J., Collar, N. J., and Bibby, C. J. (1996). Predicting avian extinction rates. *Biodiversity Lett.* 2, 182–185.

IUCN Species Survival Commission. (1994). IUCN Red List categories, as approved by the 40th meeting of the IUCN Council, Gland,

Switzerland, 30 November 1994. IUCN/The World Conservation Union, Gland.

Kunin, W. E., and Gaston, K. J. (1998). *The Biology of Rarity*. Chapman & Hall, London.

Moors, P. J. (ed.). (1985). *Conservation of Island Birds*. International Council for Bird Preservation, Cambridge, United Kingdom.

Snyder, N. F. R., Wiley, J. W., and Kepler, C. B. (1987). *The Parrots of Luquillo*. Western Foundation of Vertebrate Zoology, Los Angeles.

Snyder, N. F. R., Derrickson, S. R., Beissinger, S. R., Wiley, J. W., Smith, T. B., Toone, W. B., and Miller, B. (1996). Limitations of captive breeding in endangered species recovery. *Conserv. Biol.* **10**, 338–348.

Stattersfield, A. J., Crosby, M. J., Long, A. J., and Wege, D. C. (1998). *Endemic Bird Areas of the World: Priorities for Biodiversity Conservation*, Conservation Series 7. BirdLife International, Cambridge, United Kingdom.

Stotz, D. F., Fitzpatrick, J. W., Parker, T. A., and Moskovits, D. K. (1996). *Neotropical Birds: Ecology and Conservation*. University of Chicago Press, Chicago.

Temple, S. A. (ed.). (1978). *Endangered Birds: Management Techniques for Preserving Threatened Species*. University of Wisconsin Press, Madison.

Wege, D. C., and Long, A. J. (1995). *Key Areas for Threatened Birds in the Neotropics*. BirdLife International, Cambridge, United Kingdom.

ENDANGERED ECOSYSTEMS

Raymond C. Nias
World Wide Fund for Nature Australia

I. Introduction
II. Major Ecosystems at Risk
III. Conservation of Ecosystems

GLOSSARY

biological diversity The variety of genes, species, and ecosystems in the living world.

biome Group of ecosystems with similar characteristics, usually a major vegetation type such as tropical rain forest or grassland.

community All of the populations of organisms that live within a given area.

conservation status Relative likelihood of extinction of a species or community.

deforestation Physical removal of trees and consequent reduction in forest cover.

ecoregion Geographic region of similar ecological processes and ecosystems.

ecosystem Community of species and their environment.

endangered Very likely to become extinct within the near future.

endemism Degree to which species are unique to a specific location.

THE SCALE OF HUMAN IMPACT ON THE NATURAL ENVIRONMENT has resulted in the decline and loss of many ecosystems. All of the world's major biomes are to some degree adversely affected by human activities, with forests, scrub and grasslands, freshwater and coastal ecosystems being the most threatened. Specific ecosystems at risk include dry tropical forests, especially on islands, temperate forests, Mediterranean shrublands, temperate grasslands, and coral reefs. Overall, the large marine and oceanic ecosystems, polar regions, and dry deserts are generally less threatened.

I. INTRODUCTION

In terms of area, there seems little doubt that the recent expansion of mechanized agriculture has been the main factor in a rapid conversion of suitable ecosystems into agricultural production. Large areas of forest, woodland, and grassland have been destroyed following the expansion of agriculture, starting in southwestern Asia, China, the Mediterranean, and Europe.

Many temperate ecosystems, such as forests, woodlands, grasslands, and wetlands, are endangered as a result. Similar trends are now obvious in the tropical regions of the world, exacerbated by poor forestry practices and the replacement of forests with cash crops. Even where natural ecosystems have not been subject to deliberate modification, human impacts are detectable as a result of exploitation, pollution, and the presence of foreign species.

Ecosystems can be considered endangered when their extent, species composition, or the natural processes that sustain life become sufficiently disrupted or degraded. The difficulty in describing the exact extent and status of an ecosystem, however, makes the assessment of their conservation status much more difficult than it is for a species of organism. In particular, the conservation status of an ecosystem depends greatly on the scale at which the ecosystem is considered. From the major global biomes such as tropical forests, or deserts, to the regional and local scale, various threats operate that affect the likely persistence of that ecosystem, or its components. The ecoregional approach offers good potential for classifying, assessing, and managing ecosystems at the regional scale—combining environmental and geographic attributes to identify specific regions of high biodiversity.

Efforts to conserve ecosystems are being made throughout the world, but these are often hampered by conflicting land-use issues and a general lack of understanding of ecosystem processes. At the global level, international conventions and agreements have established a legal framework for ecosystem conservation. Several countries have begun the task of classifying ecosystems and determining threats. The conservation of natural ecosystems will require coordinated action at a variety of levels, from international to local, and will require a fundamental change in the patterns of resource exploitation and consumption.

A. Definition of an Ecosystem

Nearly sixty years after the development of the concept of an ecosystem, it is now apparent that many have been lost through human activities, or so significantly altered that they may be considered endangered. Our understanding of ecosystems has changed profoundly, with much greater recognition now given to their dynamic and complex nature and the potential for human activities to change their biological and physical components.

The international Convention on Biological Diversity, developed prior to the United Nations Conference on Environment and Development held in 1992, defines an ecosystem as "a dynamic complex of plant, animal and micro-organism communities and their non-living environment interacting as a functional unit." By itself, however, the term "ecosystem" does not connote any specific dimensions and the boundaries of ecosystems

are often indistinct. Even in aquatic ecosystems, where the presence of water helps to identify the lateral boundaries of lakes and rivers, the extent of an aquatic ecosystem will fluctuate with changing water levels, floods, and tides. In the marine environment, ecosystems may be defined by particular attributes of water temperature or chemistry, or by the characteristics of the substrate, in addition to specific plant communities (e.g., seagrass beds or kelp forests). Other ecosystems may be defined by specific geographic or special physical characteristics, such as caves or sea-mounts.

The uncertainty in defining the exact extent of an ecosystem has hampered attempts to quantify their conservation status. Ecosystems can be increasingly finely described and there is no consensus on which level is the most appropriate for conservation. At the broadest level, ecosystems cross international boundaries and are subject to a wide range of threats and human influences. These major ecosystem types can be sub-divided, as they vary enormously in their biotic composition, environment, and functioning, often as a result of major geographic features such as mountain ranges, bays, and river catchments. At a fine level, an ecosystem may comprise a small group of underwater cave systems, sand dunes, an isolated island and fringing coral reef, or a mossy high-altitude swamp that covers no more than a few hectares.

The concept of an endangered ecosystem is relatively new compared to the widely accepted concept of an endangered species. The World Conservation Union (IUCN) has maintained an internationally recognized system of "Red Data Books" (e.g., Walter and Gillett, 1998) for defining and listing threatened species for many years, but no universally applied system has been developed for ecosystems. Similarly, there is no recognized system for measuring the relative conservation status of an ecosystem and it is often not clear in what sense the whole of the ecosystem can be seen to be threatened, especially if many of the resident and visible animals and plants remain reasonably abundant. While it is clear in most cases at what point a species has become extinct, it is much less clear at what point a specific ecosystem ceases to exist.

A paucity of information on many ecosystems, as well as the lack of a standard approach, has hampered attempts at national or global assessments of conservation status. The European Environment Agency assessment (Stanners and Bourdeau, 1995), for example, found that incomplete and fragmented data made it impossible to prepare a complete and systematic assessment of Europe's ecosystems. Information about nature conservation at the international level was found to be focused largely on protected areas only and frequently did not cover parameters such as habitat types, human pressures, or type of management that might be important in determining their status.

B. Overviews of Endangered Ecosystems

1. General Reviews

Despite the various problems and general lack of knowledge, there have been several attempts to provide an overview of the conservation status of the world's ecosystems. The most notable of these overviews have been compiled by the World Conservation Monitoring Centre (e.g., WCMC, 1992) and the United Nations Environment Programme (UNEP) (e.g., Heywood, 1995). Regional assessments of varying thoroughness have been completed for Europe (Stanners and Bourdeau, 1995), United States (Noss and Peters, 1995), and the Asia and Pacific regions (UNESCAP/Asian Development Bank, 1995).

Reviews of particular ecosystem types have been completed by a number of agencies. The United Nations Food and Agriculture Organisation, for example, produces a regular update of forest ecosystems (e.g., FAO, 1997), and some coverage of grasslands and other drier ecosystems is provided by UNEP (e.g., Middleton and Thomas, 1997). Marine ecosystems subject to review include mangroves and coral reefs (World Resources Institute, 1996) and oceans (IUCN, 1991). A number of regional freshwater ecosystem reviews have been completed (e.g., Dugan, 1993), and wetland conservation databases are maintained by Wetlands International, World Resources Institute, and other agencies.

2. Endangered Ecosystems in the United States

As one of the most extensively cleared and developed of the world's large countries, the United States has experienced a dramatic decline in the extent of many ecosystems in the past few hundred years. Using the decline in extent of major vegetation associations, Grossman et al. (1994) described some 371 globally rare terrestrial and wetland plant communities in the United States. Another 482 communities required further definition before they could be mapped and ranked with certainty. Noss and Peters (1995) used this system to identify hundreds of threatened ecosystems. They based their assessment on four factors—if the ecosystems have been greatly reduced since European settlement, if they are currently very small in extent, if they have a number of threatened species, and/or if the con-

tinued threat to their existence is high. The greatest reduction in extent has occurred in the prairies and other grasslands, savannas (such as the oak savannas in the Midwest), and some forest and wetland ecosystems. Twenty-seven ecosystem types have lost more than 98% of their original extent, including spruce-fir forest in the southwest Appalachians, pine rockland habitat in south Florida, wet and mesic coastal prairies in Louisiana, sedge meadows in Wisconsin, and Palouse prairies in the Pacific Northwest (Table I).

3. Endangered Ecosystems of Australia

In Australia, several studies have also attempted to identify endangered ecosystems, particularly in relation to forests subject to commercial logging operations. The most extensive of these inventories has been carried out as part of the Regional Forest Agreement (RFA) process, which is a national effort to create a system of protected areas within regional forest management plans across Australia's major commercial forest areas. The RFA process defined an endangered forest ecosystem as one "where its distribution has contracted to less than 10% of its former range or the total area has contracted to less than 10% of its former area, or where

TABLE I

Endangered Ecosystems of the United States[a]

South Florida landscape
Southern Appalachian spruce–fir forest
Longleaf pine forest and savanna
Eastern grassland, savanna, and barrens
Northwestern grassland and savanna
California native grassland
Coastal communities in lower 48 states and Hawaii
Southwestern riparian forest
Southern California coastal sage scrub
Hawaiian dry forest
Large streams and rivers in lower 48 states and Hawaii
Cave and karst systems
Tallgrass prairie
California riparian forest and wetlands
Florida scrub
Ancient eastern deciduous forest
Ancient forest of Pacific Northwest
Ancient red and white pine forest, Great Lakes states
Ancient ponderosa pine forest
Midwestern wetland
Southern forested wetland

[a] After Noss and Peters (1995).

90% of its area is in small patches which are subject to threatening processes and unlikely to persist." These studies have resulted in the identification and subsequent protection of many endangered forest ecosystems. Although these efforts have not resolved all outstanding issues relating to forest conservation, or adequately protected all forest types, it has been clear that a rigorous definition and assessment process is a prerequisite for successful conservation of these ecosystems.

A major study has also been undertaken to identify threatened ecosystems in the south-west corner of Western Australia—an area of extremely high plant diversity. Based on a modification of the Conservation Status categories adopted by the IUCN/World Conservation Union (e.g., Walter and Gillett, 1998), the study is one of the first attempts to carry out a scientifically rigorous and repeatable assessment of the status of ecosystems, referred to in the report as "ecological communities." Criteria are established for each of the IUCN categories: Presumed Extinct (modified to Presumed Totally Destroyed), Critically Endangered, Endangered, Vulnerable, Data Deficient, and Lower Risk. As with the species categories, each ecosystem is assessed against a set of modified criteria relating to the degree to which the original extent of the ecosystem has been reduced, the number of occurrences of the ecosystem and the extent to which it is restricted or isolated, and the degree to which the ecosystem has been modified and could be rehabilitated or restored. A total of 110 ecological communities were identified as being possibly threatened. Sixteen communities were found to be Critically Endangered, 7 as Endangered, 10 as Vulnerable, and 5 as Data Deficient. The most severely threatened ecosystems tended to be forest, woodland, and shrubland communities restricted to specific soil types, and the wetlands associated with these communities. As with the Regional Forest Agreement process, this approach allows for a much more objective assessment and, importantly, allows conservation targets to be established as part of the management of the ecosystem.

Efforts to define and identify endangered ecosystems have also been advanced by the adoption of national endangered species conservation legislation in Australia (the Endangered Species Protection Act of 1992), which includes a legal definition for endangered ecosystems. Under this legislation, specific ecosystems can be identified and listed on appropriate schedules and guidelines have been established for the assessment of status. Importantly, the legislation recognizes that ecosystems are defined not only by their boundaries and extent, but also by the identity and number of species (structure)

and the interactions that link species and their environment (processes).

Box 1

Guidelines for identification of an endangered ecological community under Australia's Endangered Species Protection Act of 1992.

1. Community may already be extinct.
2. Community is subject to current and continuing threats likely to lead to extinction as demonstrated by one or more of:
   (a) Marked decrease in geographic distribution.
   (b) Marked alteration of community structure.
   (c) Loss or decline of native species that are believed to play a major role in the community.
   (d) Restricted geographic distribution such that the community could be lost rapidly by the action of a threatening process.
   (e) Community processes being altered to the extent that a marked alteration of community structure will occur.

C. The Ecoregion Approach

In recognition of the difficulty involved in describing and measuring the status of ecosystems, an approach based on the combination of environmentally and geographically similar features into "ecoregions" has been developed. Olson and Dinerstein (1998) used this approach to identify some 200 of the world's most significant ecoregions by assessing the number of species and levels of endemism they contained. The advantage of such an approach is that it allows a reasonably precise area to be defined, some level of risk to be determined in relation to the various biological components, and a reasonable basis for comparison, at least for terrestrial ecosystems. Given the dynamic nature of ecosystems, and the fact that they are as much a result of interaction between species as they are of a particular locality, it probably makes more sense to use the ecoregion concept as a basis for comparing conservation status. Much of the overview of endangered ecosystems in this article is therefore based on the ecoregion concept.

II. MAJOR ECOSYSTEMS AT RISK

A. Forests

1. General Status

It is variously estimated that between 1700 and 1980, the total area of all land remaining in a natural state decreased by more than one-third, from about 95% to about 65%, mostly as a result of expanding human populations and agriculture. In densely populated subregions, such as the western part of Europe and southern China, this level of conversion has been much greater, with less than 30% now remaining. The large-scale loss of ecosystems resulting from human influence can be seen most readily with the conversion of fertile temperate forests for cropland and pasture. Considerable decline in forest area had already taken place by 1700, especially in the Mediterranean Basin and the Indus Valley (areas occupied by ancient Egyptian, Indian, Greek, and Roman civilizations) and in northern and northwestern China. Large declines also occurred in northwestern Europe during the Middle Ages. Primary or old-growth forest (more than 200 years old) is now only a small part of the world's total forest area. Primary forest cover has been greatly reduced in most industrial countries and is rapidly decreasing in less-developed ones.

In contrast to the Mediterranean region, development in North America did not lead to the almost complete loss of forest. Over some 150 years, an initial rapid conversion of forested land was followed by a slowing down in clearance rates, and eventually by a stabilization. In Australia, some 47% of all forests were cleared over the last 200 years, but mostly during the period of rapid economic expansion after the end of the Second World War. A similar pattern of forest loss is now occurring in the developing (mostly tropical) world. These patterns are again fueled by expanding human populations and agriculture, with the added impetus of industrial crops such as rubber and oil palm.

2. Major Threats to Forests

Deforestation is a serious problem throughout the world's major forest areas, although its causes and magnitude vary by region. In Africa, for example, the major cause of deforestation is forest clearance for agriculture (particularly commercial farming and to some extent shifting cultivation) and the harvesting of fuelwood. Commercial logging is limited, but settlement and agriculture around roads built to transport timber have resulted in additional clearing of forest areas.

The total area of tropical forests in Africa in 1990 was estimated to be approximately 530 million hectares (ha), compared with 569 million ha in 1980, a loss of nearly 7%. The remaining forests in humid West Africa are disappearing at the rate of about 2% a year, and exceeding 5% in the extreme case of Côte d'Ivoire. The relic blocks of forests left at Gola in Sierra Leone, at Sapo in Liberia, and at Tai in Côte d'Ivoire are now of global importance as the last significant remains of the structurally complex and species-rich forests of the upper Guinea zone. Some areas, such as Fouta Djallon, Mount Nimba, and Loma at the head of major watersheds in western Africa (the Niger, Senegal, and Gambia Rivers), harbor exceptional biodiversity.

Similarly in Latin America, deforestation is due mainly to the expansion of agriculture. This process is driven by a number of factors, the most important being the displacement of peasant farmers from traditional farming areas, large-scale settlement programs such as in Rondônia in Brazil, and the deliberate promotion of livestock-raising activities in the Amazon through financial incentives. In addition, commercial logging, collection of firewood for household use, and road construction have all led to the unsustainable exploitation of the region's forests. Deforestation has been most severe on the Pacific coast of Central America, where less than 2% of the original 550,000 square kilometers remains, and within the Atlantic forests of Brazil, which now has only 4% of its original 1 million km² as pristine forests. The rapid loss of highly diverse native forests is of particular concern, because they are often replaced by species-poor pastures or by monocultures of exotic timber species. Most of the endangered tropical plants in Brazil (65%) are found in this highly endangered tropical forest ecosystem.

In the Asia and Pacific regions, three countries—Australia, Indonesia, and China—account for 52% of the remaining forest cover. As a result of industrialization, agricultural expansion, and poor forestry practices in many countries, deforestation is one of the major environmental issues in the region. Various agencies, such as the FAO, report that deforestation in the region increased from 2 million ha per year during 1976–1981 to 3.9 million ha per year in 1981–1990, with Bangladesh, Pakistan, the Philippines, and Thailand experiencing the fastest rate of forest clearance. The Pacific subregion has the lowest rate of deforestation (around 130,000 ha per year), most of which occurs in Papua

New Guinea. Rapid population growth has contributed to depletion of forests not only through land clearing for cultivation but also through over-harvesting of forests for fuelwood, roundwood, and fodder. At the current rate of harvesting, the remaining timber reserves in Asia may not last for more than 40 years.

Although the extent of forest cover in Europe may be significant and stable, or even increasing in parts, many of the elements of the forest ecosystem have been severely modified or lost and for this reason are considered to be threatened. Mature natural forest ecosystems are very rare. In northern and central Europe this rarity has been caused primarily by intensive logging, resulting in a significant decrease in the integrity and natural condition of forests. Forests have been modified significantly to increase their commercial value at the expense of natural ecosystem complexity and structure. They are more likely to contain uniform, relatively even-aged stands, with little dead timber, and with a significantly reduced complement of animal species. In addition, there has been extensive use of nonnative species such as Sitka spruce (*Picea sitchensis*) in the north and eucalyptus (*Eucalyptus* spp.) in the south.

The Mediterranean forests have lost much of their natural condition as a result of excessive logging, fire, and overgrazing by livestock, especially goats. Forests are especially threatened at the boundaries of their natural distribution, such as the forest–steppe regions of southern Ukraine and Russia or in the Mediterranean Basin. Most of central Europe's alluvial forests along all major river courses have virtually disappeared. Airborne pollution (mostly in the form of acid deposition and photochemical smog) is also a major cause of damage to forest ecosystems in Europe, particularly in central and eastern Europe. First reported in the Black and Bavarian forests of Germany in the early 1970s, damage from air pollution may now be detected in a quarter of all European trees.

One of the growing threats to particular forest ecosystems stems from the vast increases in tourism to remote and spectacular regions such the Himalayas. The fragile high-altitude ecosystems of the Annapurna Himal, Nepal's most densely populated and ethnically diverse region, have been badly affected by a boom in tourist numbers. To provide trekkers with hot food and showers, the lodges in a single village are obliged to cut one hectare of virgin rhododendron forest each year. Throughout Nepal, about 400,000 ha of forest are cleared each year—an annual loss of 3% of its forests. Between 1975 and 1980, 15% of Nepal's forests were destroyed. Moreover, each hectare of logged forest re-

sults in the erosion of anywhere from 30 to 75 metric tons of soil annually.

3. Endangered Forest Ecosystems

Table II shows examples of some of the most endangered and biologically diverse forest ecoregions in the world. It is clear from this table that many tropical moist forests are endangered, particularly forests in western Africa, the lowland forests of Southeast Asia, and forests on islands such as Madagascar and the Philippines. Rapid development of these regions for agriculture and large-scale commercial logging have been primarily responsible for this extensive forest loss. Large-scale fire now poses a major problem in some of the previously contiguous forests of Indonesia and Malaysia, especially during periods of prolonged drought.

Similar problems face the seasonally dry forests, especially the monsoonal forests and conifer forests in the tropical and subtropical regions of the world, especially because these are often more restricted in their original distribution. The forests of Madagascar are of particular concern, as are the dry tropical forests of Central and South America, both of which are biologically distinct and now much reduced. The remaining dry forests of Madagascar represent some of the richest dry forest ecosystems in the world in terms of their natural diversity. Around 165 million years ago, Madagascar drifted away from Africa, which allowed its native species to evolve in isolation from the mainland continent. Among the many unique species are the lemurs—members of the primate family found only on Madagascar and the neighboring Comoros Islands—including the recently discovered mouse lemur (*Microcebes myoxinus*) and golden-crowned sifaka (*Propithecus tattersalli*), both of which appear to be restricted to this ecosystem.

Temperate forests have also been extensively cleared for hundreds of years and few large intact examples remain. Forest loss has been most severe in western Europe, and most examples of this forest type are now restricted to central and eastern Europe. Clearing has also proceeded apace in the temperate forests of China and around the Himalayan region in the past few decades. Most examples of temperate rain forests, including the southern temperate beech forests (*Nothofagus* spp.) of Chile and New Zealand, are now much reduced in extent.

The Caucasus ecoregion harbors some of the most diverse and distinctive temperate conifer and broadleaf forests in Eurasia, as well as rich woodlands, steppe, and grasslands. One of the world's seven temperate rain forests formerly occurred here, but this habitat has been

TABLE II

Endangered Forest Ecoregions[a]

Region	Tropical and subtropical moist broadleaf forests	Tropical and subtropical dry and monsoon broadleaf forests	Conifer and temperate broadleaf forests
Africa	Madagascar moist forests Guinean moist forests (West Africa) Eastern Arc montane forests East African coastal forests Albertine Rift highland forests (central Africa) Seychelles and Mascarene Islands forests Gulf of Guinea island forests Macronesian (Atlantic islands) forests Congolian coastal forests (West Africa)	Madagascar dry forests Maputaland–Pondoland dry forests	
Asia	Western Ghats moist forests (India) Sri Lankan moist forests Peninsular Malaysia lowland and montane forests Sumatran–Nicobar Islands forests Northern Borneo–Palawan moist forests Philippines moist forests Southeast China subtropical forests Hainan Island forests (China) Nansei Shoto Archipelago forests (Japan)		Western Himalayan temperate forests Central China temperate forests Eastern Himalayan broadleaf and conifer forests
Australasia and Pacific	New Caledonia moist forests New Zealand subtropical forests Lord Howe and Norfolk Island forests (Australia) Hawaiian moist forests South Pacific island forests	New Caledonia dry forests Hawaiian dry forests	
Europe and Russia			Southern European montane forests
Latin America and Caribbean	Brazilian Atlantic forests Northern Andean montane forests Coastal Venezuela montane forests Greater Antilles moist forests Varzea flooded forests	Bolivian lowland dry forests Tumbesian and North Inter-Andean Valleys dry forests Southern Mexican dry forests	Mexican pine–oak forests Greater Antilles pine forests Valdivian temperate rain forests (Chile)
North America			Klamath–Siskiyou coniferous forests (U.S.A.) Appalachian and mixed mesophytic forests (U.S.A.) Pacific temperate rain forests (U.S.A. and Canada) Sierra Nevada conifer forests Southeastern conifer and broadleaf forests (U.S.A.)

[a] After Olson and Dinerstein (1998).

virtually destroyed. Lowland forests are dominated by oak, chestnut, and lime, while higher regions are covered by beech, fir, and spruce. Although the ecoregion covers a relatively small area, its varied range of landscapes and climates, and its geographic position at the

edge of Europe and Asia, has given the ecoregion a high level of biological diversity, with many European and Asian species represented. In the Caucasus Mountains alone, more than 6000 species of vascular plants have been recorded (of which 20% are endemic), as well as

8300 species of ferns and 1200 types of lichens and mosses.

The endangered temperate forests found in southern China are also among the world's most biodiverse temperate regions. Subtropical evergreen broadleaf forest is the dominant vegetation of this ecoregion, which also includes alpine ecosystems, steppe grasslands, and wetlands. Alpine vegetation of the Qiinghai–Tibet plateau predominates in the west and the higher regions of the Hengduan Mountains of the southwest. Deciduous broad-leaved forests of the warm temperate zone occur north of the province. Though the biological richness of China and Sichuan is well recognized, international attention has largely focused on a single species that has become a symbol for conservation efforts worldwide—the giant panda (*Ailuropoda melanoleuca*). Sichuan province is home to the vast majority of the world's wild pandas. Loss of vegetation cover is one of the main threats to wildlife in the province and half of the panda's habitat (10,000 km²) was lost between 1974 and 1989 alone. Logging and small-scale agriculture are the main reasons for this decline. Hunting in wildlife reserves and human disturbance from people collecting medicinal plants are additional concerns.

B. Scrub, Heath, and Grasslands

1. General Status

Despite their often uniform appearance, low-stature vegetation communities of the type referred to as scrubs, heaths, and grasslands are complex and varied, with many regional variations and subtle differences resulting from soil type and climate. They are usually present where soil nutrients, water availability, or climate are not conducive to the growth of forests. They are well suited, however, to grazing by domestic animals and for conversion to cultivation, especially cereal crops. As a result of their suitability for agricultural development, temperate scrubs, heaths, and grasslands are among the most modified of the world's major biomes.

Large areas of current and former temperate scrub and grasslands occur throughout continental Europe and Asia and North America. Significant areas also occur in the Southern Hemisphere, in southern Australia, southern Africa, and temperate South America. In northern polar regions, these communities extend into the Arctic Circle, forming a vast patchwork of semi-frozen and permanently frozen swamps and low vegetation. This latter region appears to be relatively intact, although the future impacts of global warming might considerably alter the distribution and extent of these ecosystems.

European communities of this type range from the Arctic tundra to the Mediterranean shrublands, including many different types of woody shrub communities (including maquis and dwarf shrub heath), tall herb stands, and many types of grassland. They occur at all altitudes and can range from very sparse to very dense cover, and from a few centimeters to 2 or 3 m in height. Human activities have modified most European scrub and grassland plant communities other than those found on mountaintops or on very poor soils. In some cases, particularly where wetland drainage or forest clearance for timber and farming has taken place, new scrub and grassland ecosystems have been created on lands that were previously wetland or forest.

Closer to the tropics, vast areas of tropical savannas and grasslands can still be found in Africa, Australia, and South America. Large parts of these savannas are used for pastoralism, and the impact of large numbers of grazing animals, often poor soils, changed fire regimes, and human settlements have modified and degraded many areas. As native vegetation cover declines, soil erosion proceeds and turns these semi-arid ecosystems into deserts.

In both hot and cold regions, lack of water typically creates heathlike low scrub, sparse grassland, and desert communities (with characteristic species such as cacti and other succulents), many with a high level of endemism and species richness. Large areas have been cleared for cultivation in North America, Australia, and Europe in regions where rainfall is reasonably predictable.

2. Endangered Scrub, Heath, and Grassland Ecosystems

Table III shows some of the most endangered and biologically diverse ecoregions of this type in the world. Many temperate ecosystems have been lost or so extensively modified that it is difficult to determine their original nature. In Australia, for example, less than 1% of the original temperate lowland grasslands remain, much of this restricted to roadsides, railway lines, cemeteries, and lightly grazed unimproved pastures. Some of the world's most floristically diverse regions, such as the mixed shrublands of the Mediterranean, South Africa, and southwestern Australia are now highly fragmented and suffering from impacts such as dryland salinity and more frequent or intense fires.

Given their suitability for grazing animals and agriculture, it is not surprising that large areas of grasslands have been converted from their natural state. In eastern

TABLE III

Endangered Scrub, Grassland, and Savanna Ecoregions[a]

Region	Grasslands, savannas, and shrublands	Tropical montane grasslands and savannas	Deserts and xeric shrublands	Mediterranean shrublands and woodlands
Africa	Angolan escarpment woodlands Zambezian flooded savannas	Ethiopian highlands Zambezian montane savannas and woodlands South African montane grasslands and shrublands	Namib and Karoo Deserts and shrublands Madagascar spiny desert	Fynbos (South Africa)
Asia	Terai-Duar savannas and grasslands (north Indian subcontinent) Red Sea fog woodlands		Central Asian deserts	
Australasia and Pacific				Southwest woodlands and heaths (Australia)
Europe and Russia				Mediterranean shrublands
Latin America and Caribbean			Atacama Desert	Chilean matorral
North America	Tallgrass prairies (U.S.A.) Everglades flooded grasslands			California chaparral and woodlands

[a] After Olson and Dinerstein (1998).

Europe the steppe ecosystems have become extremely rare and large areas of undisturbed steppes are now found only in the south, where they border semidesert regions (north and west of the Caspian Sea, i.e., the Volga Delta and Terek region). The principal cause for the loss of scrub and grassland habitats over the last 50 years has been the advent of widespread agricultural mechanization, which has resulted in the cultivation of large areas of natural and semi-natural grassland. Remaining grasslands are isolated and highly fragmented, and usually restricted to steep slopes and ground with thin soils. Increased use of fertilizers and biocides has further threatened many species. The vast grassy plains of North America have been drastically reduced. The tallgrass prairies of the American Midwest and Great Plains are a unique set of plant communities that once covered vast areas of the continent and supported huge herds of grazing animals such as the bison. Originally thought to be poor agricultural lands, because of the scarcity of trees, the prairie ecosystem was discovered to be immensely fertile, and during the period from the early 1800s to 1930, about 90% of this ecosystem was converted to farmland.

One of the most biologically diverse and endangered of all the world's ecosystems are the fynbos shrublands of the southwestern and southern Cape of South Africa.

Although relatively small in extent, the 470 km^2 of the Cape Peninsula, including Table Mountain, are home to 2256 different plant species, including one of the highest concentrations of endangered plant species. The fynbos is under serious threat from increasing urban expansion (especially in the Cape Flats area) and invasion by exotic weed species, including a number of Australian acacias. Similar shrubland communities can be found along the coastlines of the Mediterranean, southwestern Australia, southern California, and Chile, and in each case they are considered to be threatened.

C. Wetlands

1. General Status

The term "wetland" covers a variety of inland, coastal, and marine habitats that are temporarily or permanently under water. The convention for wetlands of international importance (the Ramsar Convention) defines wetlands as "areas of marsh, fen, peatland or water whether natural or artificial, permanent or temporary, with water that is static or flowing, fresh, brackish or salt, including areas of marine water the depth of which at low tide does not exceed six metres." With such a wide definition and range of ecosystems classified as wetlands, and the dynamic nature of floods, tides, and rainfall, it is difficult to accurately gauge the total extent of wetland ecosystems or to precisely define their location. Nevertheless, there have been a number of attempts to classify the world's wetlands and to provide an overview of their extent and status at the regional level (e.g., Scott and Poole, 1989; WCMC, 1992).

In 1971, the Ramsar Convention was signed in the small Iranian town of Ramsar to protect wetlands and promote their "wise use." Acceptance of the convention has since grown and most countries are now signatories. The existence of the International Wetlands Convention (as it is also known) has resulted in the development of a catalog of many of the world's most significant wetlands. In addition, the Convention allows the recording of "sites where changes in ecological character have occurred, are occurring or are likely to occur"— the Montreaux Record. By 1998, 61 sites were listed on the Montreaux Record, including the Florida Everglades (United States), Chilka Lake (India), and the 400,000-ha Shadegan Marshes and mudflats of Khor al Amaya and Khor Musa in Iran.

2. European Wetlands

Given the long agricultural and industrial history of Europe and the Mediterranean region, it is not surprising that most of its wetlands have been significantly modified or lost. Only in the extreme north do large wetland ecosystems remain relatively intact. Of the more than 300 European and Mediterranean wetlands of international importance designated in the Ramsar Convention, 80% are considered to be threatened to some extent. Of major concern has been the continued loss and degradation of freshwater habitats in the Mediterranean Basin, including many inland lakes and coastal lagoons and estuaries. In Spain alone, more than 60% of all inland freshwater wetlands have disappeared during the last 25 years.

The major threats to wetland ecosystems in this region result from human activities, most notably the diversion of water supplies for industry and agriculture, drainage for urban expansion and grazing, and more recently pollution from industrial sources. The use of lake and river water for industrial purposes often affects water quantity (when abstracting water) and water quality (when reintroducing it as wastewater, sometimes polluted or of higher temperature). Many industries, such as pulp and paper production and mining, wash large quantities of particulate matter into lakes and rivers.

The harmful increase in nutrient levels (eutrophication) occurs widely where the catchments of lakes and

other wetlands are intensively farmed or densely populated. Fish-farming operations and sewage effluent discharge may also result in the eutrophication of water bodies. As a result of increased nutrient levels (most commonly nitrogen- and phosphate-based fertilizers or animal and human wastes), there is an increase in phytoplankton production and microbial decay with consequent deoxygenation of the water and production of toxins.

The Aral Sea of southwestern Asia is one of the most threatened wetland ecosystems in the world, and will require major international effort and cooperation to restore it to a reasonable state of health. The Aral Sea is fed by the Amu Dar'ya and Syr Dar'ya Rivers, flowing from the mountains of Tian Shan and Pamirs. These waters are the main source of water for Uzbekistan, Kyrgyzstan, Tajikistan, Turkmenistan, and the southern part of Kazakstan. Traditionally, about half of the water was used for irrigation and the other half flowed into the Aral Sea. Intensive development of cotton and other irrigated crops over the last few decades has reduced the inflow of these rivers to about 3% of former levels.

Since 1960 the Aral Sea has lost two-thirds of its volume and its salinity level is now approaching that of seawater. Almost all the native organisms have died out. The 3.3 million ha of exposed seabed have become a source of wind-borne salt and agricultural residues that now spread over surrounding areas. The five republics developed a water distribution agreement in 1992 and an Aral Sea Programme in 1995, which aims to stabilize the region and improve the management of the waters in the basin.

In some cases, the source of wetland pollution may be distant, with contaminants being transported through the air in the form of industrial emissions. One of the best-known examples of this threat to wetland ecosystems was the acidification of European rivers and lakes, especially in the southern parts of Scandinavia. During the 1950s and 1960s, at the peak of sulfur deposition, the rate of acidification was several hundred times that of the natural process and thousands of Scandinavian lakes became too acidic to support their original fish populations. Despite a decline in atmospheric pollution over the last two decades and partial mitigation through liming activities, acidification continues to have detrimental effects on these lake ecosystems.

Marshes, bogs, and fens are a group of vegetated wetlands that were once widespread in Europe. Since early medieval times, however, these habitats have been drained or mined for peat. Though some extensive areas of peatlands and similar ecosystems still exist in the northern (boreal) and eastern regions of Europe, extensive areas of the original peatlands of western Europe (including Scandinavia) have disappeared. The vast marshy wetlands of Finland, for example, have declined by about half, primarily as a result of natural drying, drainage for agriculture, and peat extraction.

3. North American Wetlands

With about 18% of the world's fresh surface water, the Great Lakes system on the Canada–U.S. border is one of the most extensive freshwater ecosystems. Spanning more than 1200 km from east to west, the Great Lakes—Superior, Michigan, Huron, Erie, and Ontario—together with the St. Lawrence River system, are surrounded by large human populations and dense concentrations of industry.

Beginning in the 1950s, concerns arose about the increasing eutrophication of the lakes, loss of wetlands and other habitats, the impact of exotic species on native fish stocks and aquatic ecosystems, and environmental contamination by persistent toxic chemicals. Major efforts to manage the lake system have since done a great deal to restore environmental quality. Nutrient levels have been managed successfully and there has been a decline in the concentrations of some toxic chemicals such as persistent organic pollutants derived from pesticides and industrial processes. The flora and fauna of Lake Superior have recovered and are again in good condition. The system remains in a fragile balance, however, and concentrations of many toxic chemicals remain high. Aquatic habitats and wetlands are still in a poor state overall, and the aquatic ecosystems of Lakes Michigan, Ontario, and eastern Erie remain highly modified.

Equally famous are the Everglades of the U.S. Southeast—the largest example of a rare flooded grassland ecosystem. For the past 30 years, south Florida has experienced an accelerated loss of this large and unusual wetland complex. From the Kissimmee River, a series of marshy oxbow lakes flow into Lake Okeechobee, one of America's largest freshwater lakes. Okeechobee in turn feeds slowly into the Everglades itself, known as the fabled "River of Grass." To the south, the shallow estuary of Florida Bay once contained a very rich marine life, and an extensive coral reef system thrived off the Florida Keys. Excessive consumption of water for agriculture and residential populations has led to a decline in wading bird populations by some 90%, and the Everglades have shrunk to a tiny fraction of what they were a half-century ago. Seagrass die-offs and massive algal blooms have contaminated much of the previously clear waters of Florida Bay with silt, which now threatens to smother the coral reefs. An introduced tree species of the genus *Melaleuca* is now widespread and has displaced some of the native flora. Altogether there are now some 55 federally listed endangered or threatened species in the area.

4. Wetlands of the Neotropics

Large freshwater ecosystems are also found throughout South America, the largest and most famous being those within the Amazon River Basin. The rivers of this region are being polluted by a number of different sources, but the most important ones are industrial and urban wastewaters from large industrial cities, wastewater from mining industries, and agricultural runoff. As a result, many of the region's water resources are now chemically and biologically contaminated. Several rivers in Colombia, including the Medellín and the Bogotá, are almost totally devoid of life as a result of reductions in dissolved oxygen. Other rivers, especially in areas subjected to mining operations, are heavily contaminated with toxic wastes and overburden.

Large quantities of agricultural contaminants are disposed of in streams flowing into the Caribbean Sea, where there is clear evidence of elevated levels of phosphorus, nitrates, potassium, pesticides such as DDT, and organic effluent. Pollution and saltwater intrusion also threatens groundwater supplies across the wider Caribbean. In Venezuela, for example, the overuse of aquifers has resulted in a reduced flow of freshwater to nearby reef areas.

The impoundment of river water to supply massive hydroelectric schemes in the region has transformed a number of large river systems into a series of chained lakes. In the upper and middle Paraná system in Brazil, Paraguay, and Argentina, for example, these large projects have resulted in major environmental impacts, including the restriction of fish movements, invasion by floating weeds, and eutrophication of reservoirs. One of the world's most significant wetland ecosystems, the Pantanal, is threatened by the development of navigable channels as part of the vast Hidrovia Project in the Paraná–Paraguay fluvial system.

5. Wetlands of Asia

Some of the most thorough reviews of threatened wetlands have been carried out by nongovernmental organizations in East Asia, such as Wetlands International (e.g., Scott and Poole, 1989). Over half of the wetlands of international importance in Asia show a moderate to high degree of threat, with the largest number of threatened sites occurring in China, India, Indonesia, Malaysia, Pakistan, the Philippines, and Sri Lanka.

Hunting and human disturbance are significant threats to wetlands in Asia, as are human settlements, drainage for agriculture, pollution, fishing, and logging.

The rivers, lakes, and other freshwater habitats of the lowland Sunda region (western Indonesia, Malaysia, and Brunei) contain the most significant freshwater diversity of any Asian wetland ecosystem. Forest loss is by far the most serious threat to these freshwater ecosystems as it disrupts the flow of nutrients, leads to water temperature rise because of decreased shading, and lowers concentrations of dissolved oxygen, resulting in conditions that are unsuitable for many fishes and freshwater invertebrates. Road construction, logging, and shifting agriculture, particularly in mountainous regions, result in soil erosion and thus increased sediment loads in rivers, which further reduces water quality. Hunting also represents a major threat to some aquatic species in this ecoregion, such as crocodiles and freshwater turtles. Water pollution is a constant and increasing pressure on many water systems. Part of this problem stems from increased motor traffic on the larger rivers, but chemical runoff from agriculture and the dumping of untreated waste into lakes and rivers are also major concerns. Tourism will also contribute to pollution if it is allowed to develop in an uncoordinated manner and scenic areas such as Lake Toba in northern Sumatra are particularly vulnerable to pollution.

D. Coastal and Marine Ecosystems

1. General Status

Given that up to 60% of the world's human population lives along the coast, it seems inevitable that many highly diverse and productive coastal ecosystems should be particularly threatened. Early human settlements were common along coastlines as they offered a combination of marine and land-based sources of food, navigable rivers, and access to maritime trading routes. With the advent of industrial development, coasts also offered a superficially simple solution to the growing human populations and their wastes, namely, dumping in large coastal rivers. With few areas of the globe now out of reach by modern transportation, even the most remote coastlines are now subject to tourism impacts as travelers seek ever newer opportunities to enjoy unspoiled beaches.

The impacts of human activities on marine and coastal areas fall into three main categories: (1) habitat destruction and degradation through direct activities such as reclamation, coastal constructions, dredging, and damaging fishing techniques; (2) deterioration in quality of habitat and of the water through pollution; and (3) damage to biological communities and living resources through nonsustainable levels of harvesting. The World Resources Institute (1996) reported that more than one-third of the world's coastlines are under a high degree of threat from development-related activities. More than 70% of the European coastline is considered to be under a high degree of threat.

2. Major Threats to Coastal and Marine Ecosystems

Threats to coastal and marine ecosystems arise from a wide variety of human impacts and vary considerably in different parts of the world. In the temperate and industrialized regions, areas such as the Wadden Zee along the Netherlands coast show the effects of many centuries of human impact. Much of the biological richness of the Wadden Zee derives from its mudflats (waddens), which appear to be decreasing as a result of the construction of dykes and barriers that affect not only the immediate vicinity but also the flow and direction of water currents and sedimentation patterns. Pollution and eutrophication, caused by synthetic organic substances, heavy metals, oil, and a surplus of nutrients, have been identified as a major threat to the area's wildlife populations. The use of tributyl tin (TBT) as an antifouling agent in paint has resulted in a sharp decline of two species of whelk. Shorebirds are threatened by oil spillages in the Wadden Zee, and pollutants such as chlorinated hydrocarbons have caused reduced breeding success among common tern (*Sterna hirundo*) populations. High levels of pollutants may also interfere with the immune systems of certain species, and in 1988 a large number of harbor seals died from a viral outbreak. Many of the region's natural resources have been overexploited, and as early as the end of the nineteenth century native oysters began to disappear. Natural mussel beds have also dramatically declined, and in large parts of the Wadden Zee mechanized cockle fisheries continue to disturb the mud surface.

In comparison to the temperate regions, coastlines in much of the tropics are only beginning to be affected by industrial development. More typically, the major threats are rapidly expanding human settlements, the release of untreated sewage, and depletion of natural resources such as mangroves or fish. Coastal marine ecosystems in eastern Africa are biologically diverse, but they are also increasingly threatened by expansion of coastal populations and development. Corals form an almost continuous fringing reef along the entire coastline of eastern Africa. The western coasts of the islands of Zanzibar, Pemba, and Mafia are characterized by patch reefs, whereas well-developed fringing reefs predominate on the eastern sides. One of the most biologically diverse areas—the Bazaruto Archipelago in Mozambique—supports a range of marine habitats, including deep-sea areas, coral reefs, rocky intertidal areas, sandy beaches, tidal sand flats, seagrass meadows, and mangrove communities. Marine turtles and many species of marine mammals are found in these coastal waters, including the highly threatened dugong (*Dugong dugon*). The mangroves are spawning and nursery habitats for many fish and crustacean species, and provide an invaluable ecological service by filtering riverine sediments that can damage nearby coral reefs. Environmentally harmful fishing practices, such as dynamite fishing, as well as overfishing, pollution, and uncontrolled tourism development, threaten the survival of these reefs and other coastal ecosystems throughout the region.

3. Endangered Coastal and Marine Ecosystems

a. Sand Dunes

Sand dunes are unique ecosystems occurring at the margins of the land and sea, founded upon the sands that are washed up by the movements of the sea. Sand dunes may stabilize and become forested or covered in shrubs and grasses. Alternatively the dunes may move landward, or be dispersed by subsequent tides. Over time, substantial areas, such as Fraser Island off the eastern coast of Australia, can be formed by this process. In Europe there are less than 428,000 ha of dunelands remaining on Atlantic coasts and widespread afforestation with exotic species has changed their character. The situation is more acute along the southern European coastline, where sandy beaches are the main attraction for many of the 100 million tourists who visit the Mediterranean each year. Construction of coastal towns and resorts has resulted in an estimated 71% loss in dune areas since 1900.

b. Mangroves

Under the right conditions, extensive mangrove forests can be found throughout the tropics in shallow bays and inlets and along rivers. These biologically rich communities have been subjected to enormous pressure from human settlement. By 1996, it was estimated that half of the world's mangroves had been destroyed. The main reasons for this loss in the past have been the felling of firewood or the production of charcoal. More recently, the rapidly expanding shrimp aquaculture industry represents the greatest threat to the

world's remaining mangroves. The clearing of mangrove forests to make way for shrimp ponds is responsible for about half of all the mangrove loss in recent years, and has greatly reduced the extent of mangroves in Southeast Asia, South Asia, and Latin America. Thailand has lost more than half of its mangrove forests since 1960 as a result of the booming shrimp aquaculture industry. In Ecuador, estimates of mangrove loss range from 20% to nearly one-half of Ecuador's original 362,000 ha of mangrove forest. Typically, the shrimp ponds are abandoned after a few years and new areas have to be cleared.

c. Coral Reefs

Although coral reefs occupy a small fraction (less than 0.2%) of the total area of the world's oceans, they are among the most biologically diverse of all ecosystems. Formed by the accumulation of calcium carbonate deposited by certain corals and algae over thousand of years, coral reefs cover some 600,000 km^2 and are the largest structures formed by living creatures. The destruction of coral reefs is caused by many human activities, ranging from coastal development and destructive fishing practices to overexploitation of resources, marine pollution, and runoff from agricultural activities and deforestation. A number of reefs of particular biological interest are also under the most serious threat, including almost all the reefs of the Philippines and coral communities in coastal Indonesia, Tanzania, the Comoros, and the Lesser Antilles in the Caribbean.

An analysis of the conservation status of the world's coral reefs (Bryant *et al.*, 1998) found that 58% were potentially threatened by human activity. Coral reefs in Southeast Asia are the most threatened, where an estimated 80% are at risk as a result of coastal development and fishing-related pressures. The study also concluded that coral reef ecosystems are very poorly conserved, with at least 40 countries lacking any marine protected areas for the conservation of coral reefs. A widespread phenomenon—coral bleaching—became evident during the extreme El Niño weather pattern of 1982–1983. Corals became stressed by high temperatures and expelled the symbiotic zooxanthellae that provide their nutrients, leading to coral death. Large areas of coral bleaching were recorded during this period, as well as in the subsequent El Niño event in the mid-1990s. There is considerable speculation that the severity of these events may be linked to global warming caused by the accumulation of carbon dioxide and other "greenhouse" gases in the atmosphere.

The Nansei Shoto Islands, a chain of 200 islands off southwestern Japan, contain some of the most extensive and biologically diverse coral reefs in the western Pacific. The relative isolation of the islands' marine and terrestrial ecosystems has produced unusual patterns of endemism. On Shiraho reef there are at least 120 species of coral, including the oldest and biggest community of blue coral (*Heliopora coerulea*) in the Northern Hemisphere and extensive colonies of the massive porous coral (*Porites australiensis*), Madracis coral (genus *Madracis*), and staghorn coral (family Acroporidae), as well as more than 300 species of fish. These coral reef ecosystems have become seriously threatened since 1972, with up to 90% lost as a result of silt runoff from construction, farming, and logging activities.

d. Island Ecosystems

In many ways islands are among the most vulnerable of ecosystems. Islands are particularly susceptible to invasion by exotic species or rapid depletion of their resources following human colonization. Some island ecosystems have evolved in long isolation from nearby landmasses and may contain many unique plant and animal species. They are also the breeding ground for many marine species, such as turtles, seals, and seabirds, that may be vulnerable to predators on the larger landmasses. In some cases, birds have become flightless as a result of the absence of predators, making them especially vulnerable should humans or exotic species arrive. It is not surprising then that most animal and plant extinctions that have occurred in historic times have been island species (WCMC, 1992). Islands that are of particular concern for both their significant biological richness and vulnerability to human impact are those of New Caledonia, Hawaii, Madagascar, and the Galápagos.

The Galápagos Islands represent one of the most outstanding examples of the evolutionary processes that influence isolated island ecosystems and their susceptibility to human impacts. Across the many individual islands can be found arid lowlands of open cactus forest, transitional subtropical forest, moist dense forest at higher elevations, and treeless upland areas covered with ferns and grasses. Habitat destruction and degradation from grazing by introduced goats, sheep, and cattle, as well as uncontrolled fires, are a major threat. Introduced pest species such as rats and cats prey on native species. Poaching of rare and threatened species is increasing, as is the overexploitation of many marine species such as sea cucumbers (*Ischitopus fuscus*) and sharks. Tourism

and settlement from the mainland pose additional pressures on an already stressed ecosystem.

4. Oceans

Although the boundaries between marine ecosystems do not appear as distinct as those on land, marine ecosystems may be identified as distinct and moving water masses, often with identifiable and characteristic planktonic assemblages. The sea bottom (benthic systems) can be defined on the basis of sediment type, again associated with characteristic faunal and floral groups, though in some areas other physical and chemical features such as turbidity, light, or salinity help define the ecosystem.

Long considered a virtually infinite resource, the oceans are beginning to show signs of detrimental human impact, including algal blooms, disease outbreaks, and dramatic changes in species composition. In contrast to most terrestrial ecosystems, human impacts on open-water systems and deeper benthic systems are often not specific to any particular ecosystem. Almost no marine ecosystem can be considered immune to the effects of pollutants owing to their dispersal in the marine environment or through the food web.

Endangered marine ecosystems are most obvious in those parts of the world subject to extensive industrialization, such as the North Atlantic Ocean and the Baltic, Caspian, Mediterranean, and other seas adjacent to Europe. Other ocean systems that would appear to be at a high risk include the Sea of Japan, South China Sea, Caribbean Sea, and Gulf of Mexico.

Studies of European marine systems show that land-based pollution and diffuse sources of pollution are the major causes in a general deterioration in water quality, from polychlorinated biphenyls (PCBs) in the Baltic to oil and pesticides in the Caspian. Organic wastes, including sewage, by-products from industries (pulp and paper mills or tanneries), and fertilizers contained in runoff from agriculture, are common throughout European marine ecosystems. Three-quarters of the region's pollution originates from only three countries—France, Spain, and Italy—and causes the contamination of seafood and eutrophication of enclosed bays. Throughout the region, 80% of municipal sewage is discharged untreated from coastal cities, and up to 600,000 million tons of crude oil are released annually from oil-related operations and shipping. Agricultural runoff in the form of pesticides, including persistent organochlorinated compounds, contaminate the food chain, and nitrate and phosphate run-off cause eutrophication and increased episodes of algal growth, including the poisonous "red tides." Chemical pollution, bottom-trawling fishing practices, and the introduction of exotic species have contributed to the depletion of indigenous marine life, such as *Posidonia* (seagrass) meadows.

Increasingly, overfishing is being implicated as a threat to marine ecosystems, particularly in areas such as the North Sea, which are subject to intensive exploitation. Depletion of commercial fish populations may cause changes in the species composition and adversely affect populations of birds, seals, and cetaceans that feed on these species. In addition to depletion of fish stocks, some fishing techniques have a direct impact on nontarget populations. Trawling or dragging of fishing gear, for example, can have devastating local impacts on benthic ecosystems such as the rare Norwegian coral communities (*Paragorgia arborea*) or the diverse communities associated with the horse mussel (*Modiolus modiolus*) at Strangford Loch in Scotland. In many parts of the world, other fishing techniques have also severely damaged marine ecosystems. Drift-netting is particularly notorious for the high by-catch of species such as sea turtles, cetaceans, and seals, whereas long-line fishing has been responsible for the deaths of tens of thousands of albatross and other seabirds in southern oceans. The trawling of seamounts has the potential to destroy these unique ecosystems and their numerous unusual and restricted species.

III. CONSERVATION OF ECOSYSTEMS

There is an urgent need to improve our knowledge of the distribution and conditions of the world's ecosystems. Current national and international initiatives for developing habitat classifications and evaluation methodologies need further coordination. The information that is available points to the inescapable fact that the extent and quality of remaining natural ecosystems are in decline and in certain cases this decline is accelerating. Many ecosystems are at the point of disappearing, or being irreparably modified. Table IV gives a summary of threats to major ecosystem types. It is clear that the major threats arise as the result of (1) the need to continually increase

TABLE IV

Threats to the World's Major Biomes

Biome	Main threats
Tropical moist forests	Clearing for agriculture and plantations, large-scale forestry operations, development of roads, towns, and other urban infrastructure, fire.
Temperate forests	Clearing for agriculture and plantations, large-scale forestry operations, development of roads, towns, and other urban infrastructure, fire, introduced exotic species, air pollution.
Boreal forests	Large-scale forestry operations, fire, development of roads, towns, and other urban infrastructure.
Tropical woodlands and savannas	Clearing for agriculture and plantations, irrigated crops, pastoral development and grazing by domestic stock, feral animals (e.g., goats), development of roads, towns, and other urban infrastructure, fire, introduced exotic species.
Temperate woodlands	Clearing for agriculture and plantations, large-scale forestry operations, development of roads, towns, and other urban infrastructure, fire, introduced exotic species, air pollution.
Scrubs and grasslands	Clearing for agriculture and plantations, irrigated crops, pastoral development and grazing by domestic stock, feral animals (e.g., goats), development of roads, towns, and other urban infrastructure, fire, introduced exotic species.
Deserts	Large-scale irrigation developments, changes in water regimes (e.g., water extraction for irrigation), changes in fire regimes, introduced plant and animal species, urban encroachment.
Tundra and ice sheets	Pollution (land and marine), climate change.
Freshwater wetlands	Water extraction, pollution (especially eutrophication, persistent organic pollutants, and heavy metals), drainage and altered flows, large-scale dams and weirs, urban development, introduced species (especially nonnative fish, invertebrates, and water plants).
Coasts and shallow seas	Direct impacts from dredging and reclamation, coastal development and urban infrastructure (including cities, ports, and harbors), marine and land-based sources of pollution, introduced marine organisms, climate change (especially for coral reefs and coastal wetlands), overfishing and destructive fishing practices (e.g., dynamite and cyanide fishing, bottom trawling).
Oceans	Marine pollution (including oil pollution and dumping of toxic chemicals), overfishing and destructive fishing practices.

human food production, (2) the production of waste from human settlements and industry, (3) the overexploitation of natural resources such as timber and fish, and (4) the accidental or deliberate release of exotic organisms into the environment.

The conservation of ecosystems depends largely on the management of human-induced impacts. Factors that threaten ecosystems can be seen as falling into three categories: factors that reduce the extent of an ecosystem, factors that alter the species composition of an ecosystem, and factors that disturb ecosystem processes.

The clearing of land for agriculture, as well as the draining of wetlands, often results in the almost complete removal of the original ecosystem. Trees and larger plants are felled and removed, and the remaining vegetation is often burned. Such clearing, or deforestation, may be exacerbated by the commercial exploitation of forests and other factors. Wetlands are drained by cutting channels or by damming the supply of water upstream. The dry land that is left behind may be cropped or used for housing and other developments. The almost complete loss of many ecosystems has been a feature of agricultural development in the temperate regions of the world.

Evidence of widespread ecosystem disturbances can be seen from the beginnings of modern agriculture in Europe, the Middle East, and North Africa. During this period, large areas of forest were cleared and many wetlands were drained as agriculture and human settlement expanded. Similar patterns can be observed in China and South Asia. During the 1800s and early 1900s, large areas of North America and Australia also became subject to widespread agricultural development and, in the past 50 years, this expansion has spread rapidly into tropical regions. Deforestation in the humid tropics is now of major concern and many tropical forest and wetland ecosystems in central and west Africa, Latin America, and Southeast Asia are endangered. In many cases, the extent of removal is so complete that the original ecosystem must be considered completely destroyed or, at best, confined to small unviable remnants. In such circumstances there are few options for conserving the remnants, and areas that have been set aside, such as nature reserves, may provide the only tenuously viable examples of the original ecosystems.

Apart from complete destruction, most of the world's remaining natural ecosystems show some effects of human intervention. Commonly, these are factors that alter the distribution or abundance of species within an ecosystem. Although species populations constantly change, disruptions to these natural cycles can have widespread effects. Certain species, for example, are referred to as "keystone" species because of their vital role in maintaining the populations of other species (e.g., they may pollinate flowers or be large predators). When such changes affect so-called "keystone" species, then larger impacts may be observed in the rest of the ecosystem. Extensive hunting, the harvesting of timber, grazing of livestock, commercial exploitation of fisheries, or other natural resources often affect populations of keystone species and thus alter natural ecosystems, sometimes to the point where they can no longer sustain their original plants and animals. The deliberate or accidental spread of exotic animals and plants also disrupts the ecosystem as these new predators or competitors flourish at the expense of the original species. The pollution of land, water, and air may affect ecosystems by introducing additional nutrients or toxic substances into food chains, causing deaths or reduced reproductive success and consequent changes in species composition. The cumulative impact of these factors can often be observed at some distance or point in time from the original impact. The modification of a river high in the mountains, for example, may eventually affect the water quantity and quality reaching a coastal marine ecosystem, changing that ecosystem and in turn affecting the viability of coastal communities. In such cases it may be possible to manage these uses sustainably, and even to restore and rehabilitate many of the attributes of the original ecosystem.

In extreme cases, the species composition of an ecosystem or its natural processes may be so disturbed from the original that they do not appear to be capable of recovery. Such disturbance may occur, for example, with the loss of nutrients, soils, or water supply. A good example of changes in ecosystems bought about by disruptions in natural processes can be seen in the case of fire. Many ecosystem types (particularly some open forest, shrubland, and grassland communities) are adapted to frequent fires and their species may exhibit a range of responses to fire that ensure their survival and continuation. On the other hand, ecosystems such as temperate and tropical rain forests may be extremely susceptible to fires and the subsequent loss of nutrients that follows heavy rains and leaching.

The effects of climate change will pose new threats to natural ecosystems, particularly those at the edge of their climatic tolerance. Species composition might change rapidly as specialized species are replaced by more generalist species able to take advantage of changing conditions. Of course change is a natural part of evolution and ecosystems should not be considered as static and unchanging in time. As continents drift over millennia, climates change, and species evolve and disappear, ecosystems are constantly being modified. However, it is the pace at which ecosystem change is now occurring that threatens the capacity of these natural systems to respond and adapt.

One of the earliest responses to the loss of ecosystems was the development of national parks and other protected areas. Originally established largely as a means of providing outdoor recreation for increasingly urbanized societies, protected areas now fulfill many important functions. One of the most important functions of a protected area system is to safeguard a representative sample of major ecosystems from development, and most countries now have some form of protection for achieving this goal. It has become clear, however, that protected areas cannot provide sufficient ecosystem protection in isolation from other measures. In particular, as ecosystems become fragmented, many of the ecological processes that sustain these areas degrade. Pollution, soil erosion, the invasion of exotic species, and changes in the frequency or intensity of fires can all have an impact on the viability of all but the largest protected area. Ecosystem conservation needs to take into account all of these factors, and this is one reason why planning models such as the ecoregion approach are being adopted. Ecoregional approaches operate in a series of different variations in catchment management plans, regional resource management plans, and some of the large marine protected areas (including the Great Barrier Reef Marine Park of Australia).

The fundamental basis of this bioregional planning approach is to define a region that contains a relatively identifiable set of ecosystems and human impacts that can be managed through common structures (such as a planning authority or a provincial government). The factors that influence the natural environment must be understood, as well as the interaction between the resources available and the economic and social infrastructure. The key attributes of the environment that must be protected need to be identified, limits on the total levels of resource extraction (such as fishing or water use) need to be established, and then equitable rules for the distribution of these resources need to be negotiated. In addition, external factors that operate in the region should be identified and agreements reached with external agencies and governments to mitigate these impacts.

The maintenance, protection, and restoration of degraded ecosystems will depend to a significant extent on the capacity and willingness of people and their governments to reduce and control the impact of human development. There are few proven methods to achieve this outcome, and indeed the trend of humanity over the past millennium has been in the opposite direction—increasing populations, greater levels of per capita resource consumption, and the rapid transformation of natural ecosystems to provide the economic resources necessary for this expansion.

Legal protection and management are clearly necessary in many cases to halt the rapid loss of ecosystem types. Although the first stated goal of the U.S. Endangered Species Act is "to provide the means whereby the ecosystems upon which endangered species and threatened species depend may be conserved," it is clear that even this relatively powerful legislation has only limited ability to achieve the aim of ecosystem conservation. The situation is obviously much more difficult in many poor and developing countries. In many cases the intent of environmental law vastly exceeds the willingness or capacity of the country to effectively implement its provisions. In such cases, international agencies, organizations, and funding bodies are required to help build this capacity and to provide training and institutional strengthening. International treaties and conventions, such as the Convention on Biological Diversity, established as a result of the 1992 Earth Summit, provide an important framework and guidelines for such international cooperation.

The major pressures on ecosystems and their biodiversity arise directly and indirectly from the amount of resources being consumed by a growing human population. Many social and economic factors contribute to this impact. Overall, however, it is the demand for food, water, housing, and material goods, and the necessary economic infrastructure required to produce these goods, that places stress on natural ecosystems. Eventually the expansion of human impact on the environment will need to be modified dramatically if more than a small collection of protected ecosystems are to survive in the future. Controlling our accelerating consumption of natural resources will require fundamental changes in the pattern of human development. Central among these changes must be a shift in social expectations and economic forces toward the conservation of resources, the responsible valuing of ecological processes, and the removal of incentives to resource overexploitation and ecosystem destruction. The achievement of these changes would realize the principles of "sustainable development" as devised in the late 1980s (e.g., World Commission on Environment and Development, 1987). The pursuit of these principles are considered by many to be essential if the current loss of ecosystems is to be halted.

See Also the Following Articles

DEFORESTATION • DIVERSITY, COMMUNITY/REGIONAL LEVEL • ECOSYSTEM, CONCEPT OF • ECOSYSTEM SERVICES, CONCEPT OF • ENDEMISM

Bibliography

Bryant, D., Burke, L., McManus, J., and Spalding, M. (1998). *Reefs at Risk*. World Resources Institute, Washington, D.C.

Dugan, P. (1993). *Wetlands in Danger: A World Conservation Atlas*. Oxford University Press, New York.

Food and Agriculture Organisation (FAO). (1997). *State of the World's Forests*. United Nations FAO, Rome.

Grossman, D. H., Goodin, K. L., and Reuss, C. L. (1994). *Rare Plant Communities of the Conterminous United States: An Initial Survey*. The Nature Conservancy, Washington, D.C.

Heywood, V. H. (1995). *Global Biodiversity Assessment*. United Nations Environment Programme/Cambridge University Press, Cambridge, United Kingdom.

IUCN (World Conservation Union). (1991). *Oceans: A Mitchell Beazley World Conservation Atlas*. Mitchell Beazley, London.

Middleton, N. J., and Thomas, D. S. G. (1997). *World Atlas of Desertification*. United Nations Environment Programme/Edward Arnold, New York.

Noss, R. F., and Peters, R. L. (1995). *Endangered Ecosystems. A Status Report on America's Vanishing Habitat and Wildlife*. Defenders of Wildlife, Washington, D.C.

Olson, D. M., and Dinerstein, E. (1998). The Global 200: A representation approach to conserving the earth's most valuable ecoregions. *Conservation Biol.* **12**, 502–515.

Scott, D. A., and Poole, C. M. (1989). *A Status Overview of Asian Wetlands*, No. 53. Asian Wetland Bureau, Kuala Lumpur, Malaysia.

Stanners, D., and Bourdeau, P. (1995). *Europe's Environment: The Dobris Assessment*. European Environment Agency, Copenhagen.

UNESCAP/Asian Development Bank. (1995). *State of the Environment in Asia and the Pacific*. UNESCAP, Bangkok.

Walter, K. S., and Gillett, H. J. (eds.). (1998). *1997 IUCN Red List of Threatened Plants*. The World Conservation Union, Gland, Switzerland/Cambridge, United Kingdom.

World Commission on Environment and Development. (1987). *Our Common Future*. Oxford University Press, Oxford, United Kingdom.

World Conservation Monitoring Centre. (1992). *Global Biodiversity: Status of the Earth's Living Resources*. Chapman and Hall, London.

World Resources Institute. (1996). *World Resources. A Guide to the Global Environment. 1996–97*. Oxford University Press, New York.

ENDANGERED FRESHWATER INVERTEBRATES

David L. Strayer
Institute of Ecosystem Studies

GLOSSARY

alien species Species that has been moved and established outside of its native range as a result of human activities; also called exotic species, introduced species, nonindigenous species.

aquifer Geological formation that contains and allows movement of groundwater.

endangered species Species that is at substantial risk of extinction as a result of human activities.

endemic species Species that occurs only over a limited geographical range.

eutrophication Process of increasing the productivity of an ecosystem by enriching it with nutrients.

groundwater Water that occurs in saturated soils and geological formations beneath Earth's surface.

THE EARTH'S FRESHWATERS CONTAIN A BEWILDERING DIVERSITY OF INVERTEBRATE LIFE. More than 70,000 species have already been described, and tens of thousands of species remain to be discovered and described by scientists. This diversity is not spread evenly over the surface of the globe, but is concentrated in local "hot spots," usually geologically ancient lakes, streams, or groundwaters. These hot spots often contain dozens to hundreds of species of freshwater invertebrates that are found nowhere else in the world. Because freshwaters are such an important resource for people, and have been used intensively for water supply, power, irrigation, fisheries, navigation, waste disposal, and as sites for cities, environmental conditions in many of the world's freshwaters have been altered greatly from their original states. Especially where hot spots of diversity coincide with areas of intensive human development, many freshwater invertebrates have disappeared from their native habitats. Some invertebrate species have already become extinct, and thousands of others are in danger of disappearing from the earth. Careful management of freshwaters, especially in hot spots of high biological diversity, is needed to prevent catastrophic extinctions of freshwater invertebrates in the future.

I. INTRODUCTION

A. Freshwater Invertebrates of the World

Over 70,000 species of freshwater invertebrates have been described, representing about 570 families and 16 phyla. The world's freshwater invertebrate fauna is actually much larger than this; probably 10,000–100,000 species await discovery, and new genera and families are discovered regularly. We are particularly ignorant about what lives in groundwaters, what lives outside of Europe and parts of North America, and small, soft-bodied invertebrates. Especially widespread and species-rich groups of freshwater invertebrates include insects, crustaceans, mollusks, mites, nematodes, and rotifers. Invertebrates live in nearly all kinds of freshwater habitats; lakes, rivers, brooks, ephemeral ponds, wetlands, caves, alluvial groundwaters, and even hot springs each contain a rich and characteristic invertebrate community. Among important freshwater habitats, perhaps only the deepest groundwaters usually lack invertebrates. A typical lake or stream contains a few hundred species of invertebrates representing several dozen families and 8–12 phyla. By comparison with the better-known vertebrates, freshwater invertebrates possess a wide range of biological traits. Life spans range from days to more than a century. Many invertebrates reproduce sexually, others reproduce asexually by budding or parthenogenesis, and still others change their sexuality or mode of reproduction depending on environmental conditions. Some invertebrates produce eggs or other reproductive bodies that remain viable for years to centuries. Freshwater invertebrates include herbivores, bacteriovores, fungivores, predators, and parasites, and exhibit a wide range of specialized morphologies and behaviors to aid in food gathering. Some even use symbiotic algae to photosynthesize!

B. What Is "Endangered"?

Various terms such as "endangered," "threatened," "imperiled," and "at risk" have been used to describe species that are in danger of extinction through human activities. Conservation organizations and governments typically have tried to develop a graded series of carefully defined terms, running from species only remotely threatened with extinction to those on the verge of extinction. Table I shows an example of such a system. In this article, "endangered" is used loosely to mean a species or population that is at substantial risk of becoming extinct over the next few decades as a result of human activities.

C. Causes of Endangerment

The specific causes of endangerment of freshwater invertebrates are highly varied from case to case. It would be impractical (and probably not very illuminating) to discuss all the known cases of endangerment of freshwater invertebrates. Further, because information about the world's freshwater invertebrates is still so incomplete, a catalog of known cases of endangerment many be misleading. Instead, the focus here is on the problem of endangerment in a more general way, and selected case studies are used to illustrate major points.

Endangerment is a product of three factors: the preexisting vulnerability of a species, the pressure of human activities, and the sensitivity of the species to specific human activities (Fig. 1). Thus, a species may become endangered if it already was vulnerable to extinction prior to human involvement, if human activi-

TABLE I

The Nature Conservancy's System for Ranking the Global Conservation Status of Species

GX	Presumed Extinct: believed to be extinct throughout its range. Not located despite intensive searches and virtually no likelihood that it will be rediscovered.
GH	Possibly Extinct: known only from historical occurrences. Still some hope of rediscovery.
G1	Critically Imperiled: critically imperiled globally because of extreme rarity or because of some factor(s) making it especially vulnerable to extinction. Typically 5 or fewer occurrences or very few remaining individuals (<1000).
G2	Imperiled: imperiled globally because of extreme rarity or because of some factor(s) making it especially vulnerable to extinction. Typically 6 to 20 occurrences or few remaining individuals (1000 to 3000).
G3	Vulnerable: vulnerable globally either because very rare and local throughout its range, found only in a restricted range (even if abundant at some locations), or because of other factors making it vulnerable to extinction. Typically 21 to 100 occurrences or between 3000 and 10,000 individuals.
G4	Apparently Secure: uncommon but not rare, and usually widespread. Possibly cause for long-term concern. Typically more than 100 occurrences globally or more than 10,000 individuals.
G5	Secure: common, typically widespread and abundant.

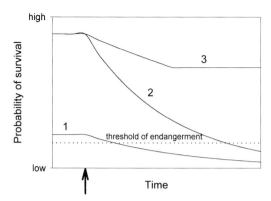

FIGURE 1 Diagram illustrating general causes of species endangerment. The three solid lines represent different species, and the dashed line shows the probability of long-term species survival below which we regard a species as endangered. The arrow shows the point at which human impacts began. Species 1 and 2 are now endangered, but for different reasons. Species 1 became endangered because it was vulnerable to extinction before human impacts, and was somewhat sensitive to human activities. Species 2 and 3 were not very vulnerable to extinction prior to human impacts. Species 2 was either highly sensitive to human activities or lived in an area where human activities were intensive, whereas species 3 was either not very sensitive to human activities or lived in an area where human activities were weak.

ties heavily affect most of the regions or habitats that it occupies, or if it is especially sensitive to a particular human activity. Conversely, a species is likely to avoid endangerment only if it evades all three of these conditions. Each of these factors will be discussed in more detail.

II. VULNERABILITY OF FRESHWATER INVERTEBRATES

A. Small Ranges

Many species of freshwater invertebrates had small ranges even before human intervention. Species with small ranges are called narrowly endemic species. For example, half of the 281 North American pearly mussel species were found in only one to three states, even before human intervention. Such species may have had elevated probabilities of extinction through natural catastrophes, and certainly are especially vulnerable to human activities (Fig. 2). Small natural ranges often arise though a small number of understandable processes. Because these processes are focused in certain regions and on species with characteristic biological traits, narrowly endemic species often are clustered together into small regions and concentrated in certain taxonomic or ecological groups.

1. Causes of Small Ranges

A primary cause of small ranges in freshwater invertebrates is the limited dispersal abilities of these animals. All freshwater habitats are islands in a sea of terrestrial habitats, and are more or less isolated from other similar habitats. Although streams are connected into drainage networks, the streams of one drainage network are isolated from those in other drainage networks. The dispersal abilities of freshwater invertebrates, and thus the perceived isolation of freshwater habitats, vary widely.

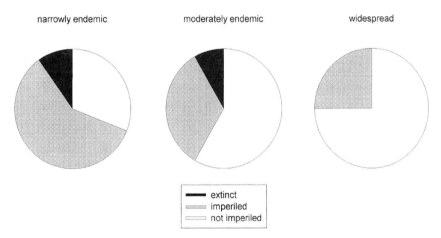

FIGURE 2 Current conservation status of North American pearly mussels (Unionoida) as a function of their native range sizes. Narrowly endemic species were found in one to three states and provinces, moderately endemic species in four to six states and provinces, and widespread species in more than six states and provinces. Conservation status from Williams *et al.* (1993).

For animals like dragonflies, whose long-lived aerial adults are strong fliers, or ectoproct bryozoans, whose tough resting stages ("statoblasts") are readily dispersed by migratory waterfowl, the separation of freshwaters probably does not present an important barrier to dispersal or gene flow. For other animals, such as fragile groundwater crustaceans that are poor swimmers, avoid the light, and lack tough dispersal stages, adjacent streams or aquifers may be nearly as remote as distant continents, and even small barriers may prevent migration and gene flow. For instance, the present-day distribution of microparasellid isopods nearly follows the pattern of marine beaches from over 20 million years ago (Fig. 3), where these species presumably arose and from which they subsequently apparently have been unable to disperse.

The isolation of freshwater habitats may produce small ranges in two ways. First, endemic species with small ranges may evolve in place following infrequent crossing of dispersal barriers, resulting in a group of more or less closely related species whose ranges are separated by barriers to dispersal. Second, a formerly widespread species may be eliminated from most of its former range, for instance by a changing climate or the arrival of a competitor, stranded in small refuges, and be unable subsequently to disperse out of the refuges. This second mechanism may become especially important for the freshwater biota if humans cause large changes in regional or global climate, especially because habitat alterations and pollution have eliminated many of the natural dispersal corridors between freshwater habitats.

Finally, a species may have a small range because is requires an unusual habitat, which is itself rare. For example, the thermosbaenacean crustacean *Thermosbaena mirabilis* was described from ancient Roman warm baths and is known from only a few thermal springs in Tunisia. Its small range presumably derives from its unusual habitat requirements as well as its limited dispersal abilities.

2. "Hot Spots" of High Endemism

Because of processes of speciation, extinction, and dispersal do not occur uniformly over the earth, species richness and endemism vary greatly across the world's freshwaters. Some bodies of water contain more than 1000 invertebrate species, many of them unique to that single body of water. At the other extreme, some bodies

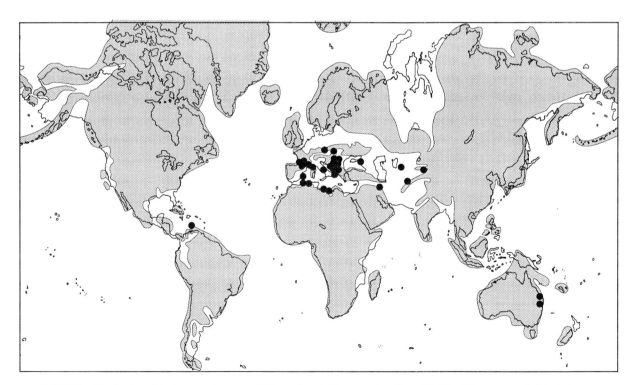

FIGURE 3 Distribution of freshwater microparasellid isopods and Oligocene shorelines (24–37 million years ago). Stippled areas were land during the Oligocene and solid circles show places where freshwater microparasellids have been found.

of water support fewer than 100 invertebrate species, all of them widely distributed. We might expect sites of high richness and endemism to be habitats of great age, habitats where dispersal is limited, either by geographic isolation or by characteristics of the habitat, or habitats that harbor animal groups that are prone to speciate. Thus, many ancient lakes (Baikal in Siberia, Tanganyika and Malawi in Africa) and river systems (the Tennessee in the United States, the Mekong in Southeast Asia) that have not been recently disturbed by glaciation, marine submergence, or desiccation support unique assemblages of invertebrates. Many aquifers seem to contain a high proportion of endemic species, probably because both the characteristics of aquifers (slow water flow, tortuous passageways within aquifers, and barriers between aquifers) and the characteristics of their inhabitants (e.g., fragile bodies, strong thigmotaxis) discourage long-range dispersal. Conversely, glacial lakes and temporary ponds rarely support locally endemic species. Although we know many hot spots of freshwater invertebrate diversity, it probably is not yet possible to produce a reliable global map that shows all major hot spots.

3. Species with High Endemism

Groups of animals vary in their tendency to form new species with small ranges. As already suggested, narrow dispersal probably allows the development of local species, whereas broad dispersal probably provides so much gene flow across populations that speciation is unlikely to occur. Other traits that have been suggested to encourage local speciation include a requirement for outcrossing (as opposed to selfing hermaphroditism or parthenogenesis), production of large young, live-bearing (as opposed to egg-laying), and narrow habitat requirements. The importance of each of these (and other) factors is unclear, but it is clear that groups of freshwater invertebrates do differ widely in their degree of endemism.

Figure 4 shows two examples. Dragonflies, most of which are strong fliers and easily cross drainage divides, are much less likely to have small ranges than pearly mussels, which do not readily cross drainage divides. Even the same group of animals may have dramatically different degrees of endemism depending on the habitat occupied. Thus, groundwater cyclopoid copepods have very much smaller ranges than their relatives in surface waters. This difference presumably arises because groundwater animals have distinctive behaviors and especially because dispersal between aquifers is more difficult than dispersal between lakes or streams.

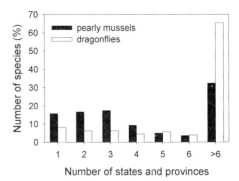
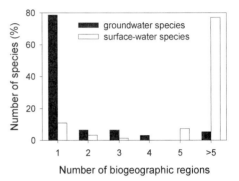

FIGURE 4 Different kinds of animals have different characteristic range sizes. The upper panel shows the number of states or provinces occupied by species of North American dragonflies and pearly mussels. The lower panel shows the number of biogeographic regions occupied by groundwater and surface-water cyclopoid copepods in Europe.

4. Traits of Narrowly Endemic Species

Narrowly endemic species may possess traits (other than small range size) that influence their vulnerability to endangerment. For example, the limited dispersal abilities and specialized habitat requirements of many narrowly endemic species may make them particularly sensitive to and slow to recover from catastrophes, whether natural or human-caused. Further, some environments that contain endemic species may encourage the development of traits that influence species vulnerability. Thus, in the food-poor groundwater environment, many animals have sparse populations, delayed maturity, and low reproductive rates, all of which probably add to their sensitivity to human impacts.

B. Sparse Populations

Species may be vulnerable to endangerment because their populations are sparse. Because population densities of freshwater invertebrates are much less well known than their geographic ranges, relatively little is

Box 1

Rivers in the Southeast United States

The U.S. Southeast (extending roughly from the Ohio River south to the coastal plain of Alabama and Georgia, plus the highlands of Arkansas, Missouri, and Oklahoma) contains ancient river systems with an extraordinarily rich biota. This region was not covered by Pleistocene glaciers, nor was it covered by the sea or desiccated for hundreds of millions of years, so river systems like the Tennessee, Cumberland, and Alabama (and their associated aquifers) are very old. The rivers and groundwaters of the Southeast are examples of biologically rich ecosystems that have suffered badly from human activities.

The freshwater invertebrate fauna of the Southeast contains hundreds of species of mollusks, crustaceans, insects, mites, and other animals (Fig. 7) that are found nowhere else in the world. Dozens of genera and two families of invertebrates [Parvidrilidae (Oligochaeta) and Neoplanorbidae (Gastropoda), the latter now probably extinct] are known only from this region. Many of these species have small ranges within the Southeast and may have occurred in only one stream. Despite a long history of scientific study in the Southeast, new species and genera of freshwater invertebrates are discovered regularly in the region.

As is the case for many river systems, the streams and rivers of the Southeast have been profoundly affected by human activities. Impoundments have been especially damaging to the invertebrate fauna. All of the large rivers in the region have been extensively impounded for flood control, hydroelectric power, and navigation, to the point that some of the large rivers have been converted into a continuous series of reservoirs. These reservoir systems differ from natural rivers in their hydrology, temperature, chemistry, sediments, and so on, and often are unsuitable for the native riverine biota. Other physical alterations of stream channels, such as channelization, dredging, diking, and instream gravel mining, also have severely damaged the freshwater biota in parts of this region. Further, as in most developed countries, these river systems have been badly polluted by toxins, nutrients, and sediments from industries, farms, and cities. Coal is mined in parts of the Southeast, which brings acid mine drainage and fine sediments into streams. Finally, invasive species, particularly the zebra mussel and biotic exchanges through the Tennessee–Tombigbee Waterway, may affect the southeastern freshwater fauna.

As a result of these massive changes to southeastern rivers, much of the freshwater invertebrate fauna is extinct or imperiled. Among the mollusks, the only group for which reasonably complete data are available, about 60 species and 4 genera from the Southeast are now extinct. Literally hundreds of additional mollusk species, representing over half of the native fauna, are threatened or endangered. Hundreds of southeastern crayfish and aquatic insects are likewise rare or endangered, and additional species of small, poorly known animals like copepods, isopods, amphipods, and oligochaetes are doubtlessly extinct or at risk of extinction.

The southeastern fauna is now receiving some protection from the U.S. Endangered Species Act and parallel state laws. Nevertheless, unless the continuing damaging effects of human activities like impoundments are reversed or remediated, it is difficult to be optimistic about the long-term prospects for the southeastern freshwater biota.

Sources: Benz and Collins, 1997; Lydeard and Mayden, 1995.

known about the occurrence of sparse populations of freshwater invertebrates. Likewise, relatively little is known about the causes of population sparseness, or about how population density per se affects the probability of species extinction. Generally, large-bodied animals have lower population densities than small-bodied animals, and predators have lower population densities than their prey, although many exceptions exist to these generalizations. Further, population densities often are lower near the edge of a species range than in its center, and may be lower in unproductive habitats than in more productive habitats. Thus, we might expect to find sparse populations especially in large-bodied invertebrates and in unproductive habitats like deep lakes and groundwaters.

III. PRESSURE OF HUMAN ACTIVITIES

Human activities endanger freshwater invertebrates in many different ways. Five broad classes of activities

Mound Springs of the Great Artesian Basin in Australia

Much of the arid interior of eastern Australia is underlain by a large aquifer called the Great Artesian Basin. Freshwater and brackish-water springs occur along the margins of this aquifer in Queensland, New South Wales, and South Australia. These springs range in size from small, moist seeps to large (>100 liters/sec), flowing springs, and some of them have built up large (>10 m high) hills of sand and mineral deposits, and so are locally called "mound springs." Although perhaps connected to one another in the past, when the Australian climate was wetter, the springs are now separated from one another by a few meters to many kilometers of desert, and are not connected by streams or rivers.

Like many springs in arid regions, the springs of the Great Artesian Basin support animals that live nowhere else in the world. Many of these species are found in only one or a few neighboring springs. Only the fish and the snails of these Australian springs have received serious study. About 25 species and 3 genera of snails have so far been found to be endemic to the springs (Fig. 8). All of the endemic snails belong to the Hydrobiidae, a widespread family that has produced flocks of endemic species in springs, caves, and groundwaters in the Balkans, the arid Southwest of the United States and Mexico, and elsewhere. Although these snails may be very abundant in the Australian springs (>1,000,000/m²), some species are restricted to one or a few springs, and all are highly vulnerable to human impacts. The endemic fauna is thought to have originated by speciation in the more or less isolated springs, perhaps after a more widespread fauna was stranded in the springs by an increasingly arid Australian climate in the Pleistocene.

The chief threat to the spring fauna is from development of wells in the Great Artesian Basin. Because this is an arid region, there is great demand for water for humans and livestock. When new wells are brought into production, the groundwater level drops, causing springs to dry up. Additional threats include the trampling of springs by livestock, which has badly degraded many springs, conversion of springs into pools by excavation or damming, and introduction of alien species (so far, the mosquitofish, *Gambusia*

affinis, seems to be the only potentially damaging alien in the springs). Over the past two decades, the springs of the Great Artesian Basin have come to be recognized as important habitats for conservation, and steps are being taken to limit at least local impacts from grazing and habitat alterations. Nevertheless, many springs, especially in New South Wales, have dried up as a result of groundwater extraction, and many have been badly altered by livestock or people. It seems likely that at least some of the unique invertebrates of the Great Artesian Basin have gone extinct, and the remaining fauna is at risk of loss.

In arid regions around the world, extreme isolation of aquatic habitats has promoted speciation and development of endemic invertebrate faunas. As in the Great Artesian Basin, water in arid regions is a critically important resource that has been exploited heavily by people. Consequently, freshwater invertebrates of arid regions around the world are endangered by forces similar to those at work in this basin.

Sources: Knott and Jasinska, 1998; Ponder, 1986, 1995; Ponder and Clark, 1990; Ponder *et al.*, 1989.

that are especially important in altering fresh waters and endangering their inhabitants are considered here.

A. Habitat Destruction and Degradation

Humans have massively altered the physical characteristics of many freshwaters, usually without consideration for consequent effects on the biota. These physical alterations probably are the chief cause of endangerment of freshwater invertebrates. Dams have been especially damaging (e.g., Fig. 5). Above the dam, running-water habitats are converted into an artificial pool that is usually unsuited to the native invertebrate species. Below the dam, the water temperature and flows of water and sediment often are so altered that downstream reaches support a highly artificial biota as well. Finally, the dam itself is a barrier that blocks normal migrations and movements of the riverine biota. Thus, even a single dam may endanger the riverine biota for hundreds of kilometers, and many river systems are now dismembered by dozens or even hundreds of dams.

In many rivers and streams, floodplains and other

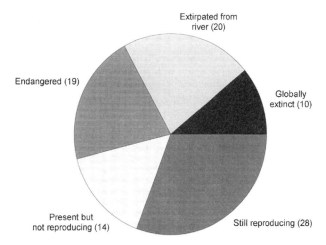

FIGURE 5   Status of the 91 species of freshwater mussel species that formerly occurred in the Tennessee River (United States), which now consists largely of a series of reservoirs. (From Benz and Collins, 1997.)

shallow-water marginal habitats have been destroyed by dredging, channelization, or filling, or separated from the main channel by dikes and levees. Further, humans often simplify shoreline habitats in both lakes and streams by straightening or filling shorelines and removing trees along and in the water. Because marginal habitats often are important for the feeding and spawning of the freshwater biota, the loss or isolation of these habitats may have grave consequences for the native biota. When water is removed from a river for irrigation or held up in a reservoir for hydroelectric generation or flood control, downstream reaches may dry up or lose critically important floods. Several of the world's major rivers (e.g., the Colorado and the Ganges) no longer flow to the sea during dry periods. Likewise, the drawdown of many of the world's aquifers from overuse of groundwater presumably has major effects on the groundwater biota, although these effects have scarcely been studied. Finally, mining of underwater deposits (for gravel or gold, for instance) may have devastating effects at the site of mining, as well as far downstream through sediment transport and far upstream though headcutting of the streambed.

B. Pollution

Water pollution is another widespread activity that has had severe effects on freshwater invertebrates. Rivers and lakes often have been used for waste disposal. These wastes include sewage and other organic matter, the

decomposition of which may reduce concentrations of dissolved oxygen to levels too low to support most species of invertebrates. Other wastes include substances (e.g., mercury used in gold mining, acid precipitation from power plants and automobiles) that are directly toxic to freshwater invertebrates. Particularly in industrialized regions, long reaches of streams and rivers have been nearly sterilized of invertebrates as a result of severe, chronic pollution (e.g., Fig. 6).

In addition to pollution caused by deliberate waste disposal, pollution may arise from a wide range of human activities in the watershed. Thus, conversion of forests or native grasslands to agricultural fields or development typically greatly increases loadings of sediments, nutrients, and toxins that are washed in from the altered watershed. These "non-point-source" pollutants are more difficult to track down and control than point loadings of pollutants from factories, yet may have equally serious effects on freshwater ecosystems. While pollution has come under partial control in many developed parts of the world, residual pollution from past releases and inadvertent spills still damage the freshwater biota. Spectacular recent examples include a large spill of pesticides into the River Rhine following a fire at a Sandoz chemical plant, which killed fish and invertebrates for hundreds of kilometers, and the overturning of a truck that spilled a rubber accelerant into the Clinch River, Virginia, which killed most aquatic animals in a 10-km reach, including hundreds of endangered mussels. This was the largest "take" of endangered species in the United States since the Endangered Species Act was passed in 1973. Of course, in many less-developed parts of the world, water pollution is still poorly controlled.

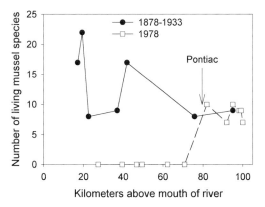

FIGURE 6   Destruction of the freshwater mussel fauna in the Clinton River, Michigan (United States), in the mid-twentieth century by pollution from the city of Pontiac.

Box 3

Lake Baikal, Russia

Lake Baikal, part of a rift system in southeastern Siberia, is the oldest (>25 million years old) and deepest (>1600 m deep) lake in the world, and covers 31,500 km². It is the only great rift lake that is oxygenated to its bottom, allowing colonization of the entire lake by a wide range of invertebrates. So far, more than 1400 species of invertebrates have been found in the lake, about 60% of which live nowhere else in the world. Many genera and four families of invertebrates [Lubomirskiidae (Porifera), Baicalarciidae (Turbellaria), Baicaliidae (Gastropoda), Benedictiidae (Gastropoda)] are endemic to Baikal. Probably the most remarkable group of endemic species in Baikal is the huge flock (46 genera and >250 species) of endemic gammarid amphipod crustaceans (Fig. 9). These amphipods, which constitute more than one-third of all gammarid species in the world, have diversified into a wide range of morphologies and behaviors and occupy a range of ecological niches, including planktonic and benthic herbivores, detritivores, predators, and semiparasites on sponges. New species of invertebrates are found regularly in Baikal, so it is clear that the true diversity in the lake is even higher than these figures imply. A distinctive feature of the Baikal fauna is that, though nearshore areas contain a mixture of endemic and widespread species, the open water and abyssal sediments are inhabited chiefly by Baikalian endemics.

In many ancient lakes (e.g., Victoria in Africa, Biwa in Japan, Lanao in the Philippines), pollution, habitat destruction, overfishing, and introductions of alien species have extinguished many endemic species. Baikal has been protected by its remoteness and vast size, and its fauna has so far been relatively unaffected by human activities. Nevertheless, industrial and domestic waste and siltation arising from deforestation of the catchment have polluted nearshore areas. Although lakewide water quality seems not to have suffered yet, pollution is a concern in Baikal because the long residence time of water in the lake means that contaminants entering Baikal may remain in the lake for a very long time. In addition, alien species (e.g., including several fish and the aquatic plant *Elodea canadensis*) have the potential to affect the nearshore fauna. Lake Baikal is a remarkable example of an ancient lake with a richly endemic invertebrate fauna that may yet be preserved through careful management.

Sources: Kozhov, 1963; Kozhova and Izmest'eva, 1998; Martens *et al.*, 1994.

C. Direct Harvest

Some freshwater invertebrates are harvested for human use, which may contribute to their endangerment. Freshwater mussels have been fished for their shells, pearls, and meat since prehistoric times. Especially in the nineteenth and twentieth centuries, harvest rates became so high that many populations were locally depleted in Europe and North America. For example, over 13 million kg of shells from living unionids were taken from Illinois streams and rivers in a single year during peak harvests. Another heavily harvested freshwater invertebrate is the medicinal leech *Hirudo medicinalis*, which was collected in large numbers in Europe when blood-letting was widely practiced in the eighteenth and nineteenth centuries. As a result of this collecting and widespread pollution and habitat destruction, this animal is now threatened. Harvests of Australian crayfish from the wild are now hundreds of tons per year and have contributed to the endangerment of some species. Other invertebrates are collected for bait or the pet trade, which may contribute to local depletion of populations. Invertebrates sometimes are protected by harvest regulations (e.g., closed seasons, size regulations, bag limits), but such regulations may be inadequately conceived and poorly enforced. Fortunately, for economic reasons, harvests usually (but not always) concentrate on common species rather than rare ones.

D. Alien Species

Humans often move species outside of their native ranges. Such introductions may be deliberate, such as the stocking of trout throughout much of the temperate world, or inadvertent, such as the widespread movement of species in ships' ballast water. Whatever the cause, these alien species often have strong ecological impacts, and sometimes are responsible for the endangerment of freshwater invertebrates. A spectacular example is the loss of native unionid mussel populations throughout much of northeastern North America as a result of competition with the introduced zebra mussel (*Dreissena polymorpha*). It is projected that the zebra mussel will be the final blow that will drive several species of unionids into global extinction. Alien species may also serve as important predators of freshwater invertebrates, as in the case of the brown trout (*Salmo trutta*), which when introduced to Tasmania apparently preyed on and reduced the range of the unusual and endemic anaspidacean crustacean *Anaspides tasmaniae*. Because the effects of alien species tend to be cumulative and difficult to reverse, this is a difficult and growing problem in invertebrate conservation.

E. Global Climate Change

Humans have changed the chemistry of the earth's atmosphere so much that significant changes in global climate are expected in the twenty-first century. At this point, it is difficult to make precise predictions about how these changes will affect specific bodies of freshwater. In many bodies of water the changes may be varied and large, involving such diverse characteristics as temperature, hydrology, water level, rising sea level, stratification, the nature and severity of disturbances, increases in damaging ultraviolet light, water chemistry, riparian vegetation, and food quality. In regions where freshwater becomes scarcer while human demands for water continue to grow, human destruction and degradation of freshwater habitats probably will become more severe. Even though we cannot yet specify the details of global climate change, it is almost certain that this change will endanger or extinguish many freshwater invertebrate species. A rapid, large change in climate will make habitats unsuitable for some of their native species. To survive, such a species will have to disperse to a body of water with suitable ecological conditions. As we have seen, though, the dispersal rates of many freshwater invertebrates are slow, almost surely too slow to keep up with the pace of climate change that current models predict. Further, human modifications to waterways (e.g., impoundments) probably have made long-distance dispersal more difficult.

IV. NUMBER AND DISTRIBUTION OF ENDANGERED FRESHWATER INVERTEBRATES

How many of the world's freshwater invertebrates are endangered, and where do they live? The most comprehensive list of endangered animals available is the IUCN Red List (compiled by the World Conservation Union),

which includes 1151 species of freshwater invertebrates (Table II). Although an enormous amount of work by experts went into compiling the IUCN list, it is clearly incomplete. The list is dominated by large, conspicuous, and attractive animals (mollusks, decapods, dragonflies, and damselflies). There is no reason to believe that smaller and less conspicuous animals are less endangered, but there is simply insufficient information on the status and trends of their populations to identify many endangered species. Likewise, almost 80% of IUCN-listed species are from North and Central America, Australia, or Europe, which probably reflects the geographical distribution of conservation biologists as much as the actual distribution of endangered freshwater invertebrates. Another way to assess global en-

TABLE II

Numbers of Species of Freshwater Invertebrates Included on the 1996 IUCN Red List of Threatened Animals, by Taxonomic Group and Continent

Turbellaria (flatworms)	1
Hirudinoidea (leeches)	1
Gastropoda (snails, limpets)	375
Bivalvia (clams, mussels)	192
Amphipoda (scuds)	73
Syncarida (anaspidaceans, bathynellaceans)	4
Cladocera (water fleas)	8
Anostraca (fairy shrimps)	24
Conchostraca (clam shrimps)	4
Notostraca (tadpole shrimps)	1
Copepoda (copepods)	78
Decapoda (crayfish, crabs, prawns)	169
Isopoda (sow bugs)	34
Ostracoda (seed shrimps)	9
Ephemeroptera (mayflies)	3
Plecoptera (stoneflies)	3
Odonata (dragonflies, damselflies)	143
Coleoptera (beetles)	22
Trichoptera (caddisflies)	3
Diptera (true flies)	4
North and Central America	601
South America	21
Africa	143
Europe	147
Asia	62
Australia	163
Oceanic islands	14
Total	1151

dangerment of freshwater invertebrates is to examine where damaging human activities coincide with areas of high endemism (Table III). Areas where freshwater invertebrates are especially likely to be endangered include river systems throughout much of the unglaciated world, groundwaters and springs in arid and semiarid regions, and many industrialized areas.

Except in or near areas covered by Pleistocene ice, recently emerged from the sea, or desiccated, river systems often support species that are endemic to that drainage basin. Most river systems have been very highly modified through impoundment and other physical modifications, water withdrawals, and pollution. Thus, we could project that most old river systems probably contain endemic invertebrates, and that many of these invertebrates probably are endangered as a result of human activities (see Box 1 on the U.S. Southeast, for example). Only species that are good dispersers and thus live in many drainage basins or are habitat generalists and thus occur in nonriverine environments are likely to escape endangerment. Human impacts on rivers are almost global, and even river systems that have not yet been heavily impounded and modified (e.g., some basins in Southeast Asia and central Africa) are facing impoundment and other large modifications in the coming decades.

A second area where we might expect to find many endangered invertebrates are groundwaters and springs in arid and semiarid regions (e.g., northern Africa, the American Great Plains and Southwest, and Australia).

These environments typically support highly endemic invertebrate faunas that are probably very sensitive to human impacts. Throughout much of the world, humans living in arid and semiarid regions are pumping water out of aquifers faster than it can be replenished, resulting in large, rapid drops in the water table, which in turn dries up springs and aquifers. We know that desert spring communities are endangered (see Box 2 on Australian springs), and it is possible that aridland aquifers are experiencing large but unseen losses in biodiversity.

In addition to these current threats to freshwater invertebrates, we can expect increasing problems in any regions of rapid human population growth or economic development (e.g., China and Southeast Asia) from the wide range of impacts that typically accompany human populations. Further, any rapid changes in climate probably will cause endangerment and extinction of freshwater invertebrates. These climate changes are projected to be most severe in middle to high latitudes. Because of glaciation, endemism of freshwater invertebrates is higher in midlatitudes than at high latitudes, so impacts of climate change may be most severe at midlatitudes, especially where the freshwater fauna and environments already have been damaged by human activities. Again, the poorly dispersing species will probably be most severely affected.

As to the question of how many freshwater invertebrates actually are endangered (or extinct) globally, there is no certain answer at present. Only for a few

TABLE III

Expected Global Patterns of Endangerment of Freshwater Invertebrates as a Function of Human Activities

Activity	Geographic distribution	Groups of animals affected
Impoundment	Global, especially North America, China, India, and arid regions	Many, especially migratory species or those that depend on flooding or turbid water
Physical alterations (diking, channelization, shoreline modification)	Global, especially highly developed regions	Many, especially those that depend on marginal habitats (shallows, floodplains)
Water withdrawal	Global in arid and semiarid regions	Many, perhaps especially species living in groundwaters and springs
Toxic pollution	Global, especially industrialized regions	Most species
Eutrophication	Global, especially densely populated or farmed regions	Species of lake profundal sediments or plant beds
Harvest	Locally important throughout the world	Bivalves, decapods, sometimes others
Alien species	Global in surface waters, perhaps rare in groundwaters	Many
Climate changes	Global, especially in high latitudes	Many, perhaps especially species that disperse poorly

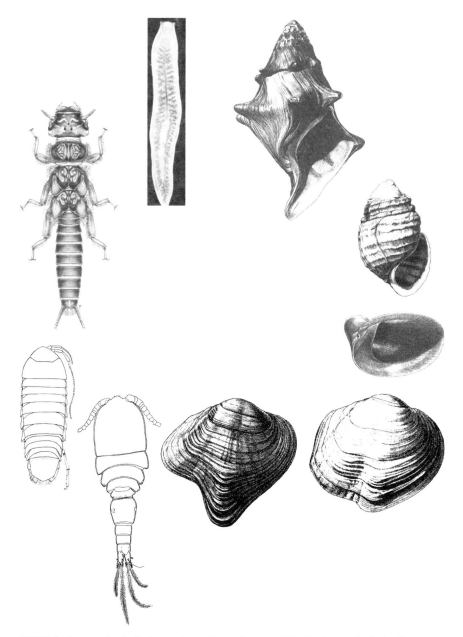

FIGURE 7 Some endemic freshwater invertebrates from the U.S. Southeast. Clockwise from upper left (with length in parentheses): the flatworm *Sphalloplana holsingeri* (14 mm), the snails *Io fluviatilis* (73 mm), *Gyrotoma alabamensis** (25 mm), and *Amphigyra alabamensis** (2 mm wide), female (50 mm wide) and male (38 mm wide) of the pearly mussel *Epioblasma lewisii,** the copepod *Rheocyclops carolinianus* (0.4 mm), the isopod *Antrolana lira* (20 mm), and the stonefly *Beloneuria georgiana* (22 mm). Species marked with an asterisk are thought to be extinct. [From Bowman, T. E. (1964). *Antrolana lira*, a new genus and species of troglobitic cirolanid isopod from Madison Cave, Virginia. *Int J Speleol* **1**, 229–244; Burch, J. B. (1975). *Freshwater Unionacean Clams (Mollusca: Pelecypoda) of North America.* Malacological Publications, Hamburg, Michigan; Burch, J. B. (1989). *North American Freshwater Snails.* Malacological Publications, Hamburg, Michigan; Kenk, R. (1977). Freshwater triclads (Turbellaria) of North America. IX. The genus *Sphalloplana. Smithsonian Contributions to Zoology* **246**, 1–38; Reid, J. W., *et al.* (1999). *Rheocyclops,* a new genus of copepods from the southeastern and central United States (Copepoda: Cyclopoida: Cyclopidae). *J. Crustacean Biol.* **19**, 384–396; Stewart, K. W., and B. P. Stark (artist Jean A. Stanger). (1988). *Nymphs of North American Stonefly Genera (Plecoptera)*, Thomas Say Foundation Vol. 12, Entomological Society of America. All figures are reprinted with permission.]

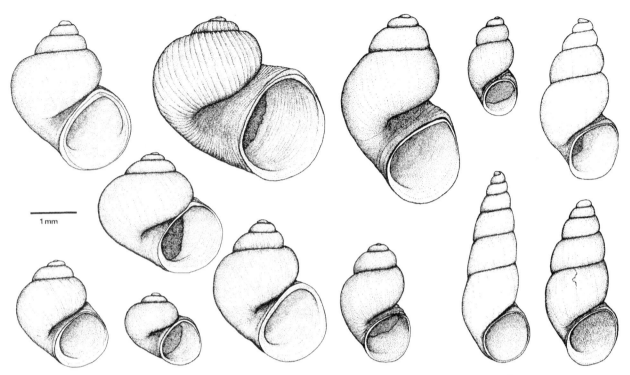

FIGURE 8 Endemic hydrobiid snails (*Jardinella* spp.) from springs in Australia's Great Artesian Basin. Shells are 1.5–3.5 mm high. (From Ponder and Clark, 1990, with permission.)

conspicuous and well-studied invertebrates (e.g., unionoid mussels, odonates) are the IUCN estimates likely to be an accurate reflection of actual endangerments. A reasonable guess might be that 3000–10,000 of the world's freshwater invertebrate species are extinct or endangered as a result of human activities.

V. PROTECTION OF ENDANGERED FRESHWATER INVERTEBRATES

Although freshwater invertebrates are protected by international, national, and local regulations, this protection often is inadequate, for several reasons. First, lists of protected species usually underlist invertebrates and include only the largest, most conspicuous species. For instance, in the United States, only 111 domestic species of freshwater invertebrates are protected under the Endangered Species Act (compared with 318 species of vertebrates), all but 25 of them mollusks. Thus, many species of endangered freshwater invertebrates, especially small and inconspicuous animals, are not being protected by existing regulations. Second, simply being

listed as a protected species may not provide enough help to endangered invertebrates. In many countries, resources for managing endangered species are insufficient, and attention naturally goes to the larger, more charismatic species. In a recent year in the United States, half of all money spent on endangered species was directed to just seven species, all of them vertebrates. As a result, so little money and attention may be spent on invertebrates that plans for recovery of invertebrate species often are general and not pursued aggressively. Third, current approaches to species protection may be inadequate to protect species over the long term. Often, legal protection focuses on trying to prevent further losses from the remaining small populations of an endangered species, without adequate attention to removing the threats that endangered the species in the first place. Consequently, legal protection may slow the rate at which an invertebrate species approaches extinction without reversing its downward trajectory.

How might we more effectively reduce extinction rates of freshwater invertebrates? First, we need to devise ways to protect species without formally listing them. Human activities will endanger or extinguish many species of freshwater invertebrates before we ever

FIGURE 9 Endemic gammarid amphipods from Lake Baikal, showing some of the wide diversity of body forms. Body lengths are 1.5–6 mm. [From Salemaa, H., and R. Kamaltynov. (1994). The chromosome number of endemic Amphipoda and Isopoda—An evolutionary paradox in the ancient lakes Ohrid and Baikal. *Arch. Hydrobiol. Ergeb. Limnol.* **44**, 247–256, with permission.]

gather enough information on their status to satisfy requirements for legal listing, and before we have enough information on their biology to develop effective species-specific plans to protect them. One way to do this is to take advantage of the fact that many species of endemic freshwater invertebrates co-occur in hot spots by protecting such hot spots from the most damaging of human activities (e.g., dams, excessive water withdrawals, toxic pollution). Such a program of hot spot protection will require better identification, recognition, and protection of hot spots. Nonetheless, it may require less research and provide more effective protection to the world's freshwater invertebrates than existing species-based programs. Of course, not all endangered species occur in hot spots, so species-based research and protection will necessarily have to accompany any program of hot spot protection. Further, conservation biologists and policymakers need to be more aggressive in identifying and removing threats that endanger species rather than just trying to protect the few populations that have somehow escaped threats. This will require creative thinking about how to preserve or restore essential features of habitat without making unrealistic demands on humans.

A second class of possible solutions could be focused on alleviating the dispersal limitations that are so acute for many freshwater invertebrates by actively establishing new populations of endangered species. This class of solutions is motivated by two main concerns. First, simple protection of existing populations of endangered species may fail to assure long-term survival because natural or human-caused catastrophes (e.g., the Sandoz spill) or normal population fluctuations ultimately may drive many isolated populations into extinction. Second, it seems likely that global climate change may occur faster than the abilities of some freshwater invertebrates to disperse into suitable habitats. To preserve species under these conditions it may be necessary to deliberately establish populations of endangered species in new locations where suitable habitat exists. Several serious problems attend this approach. First, we currently cannot reliably identify "suitable habitat" for most freshwater invertebrates. Second, we do not have good protocols for reintroductions for most species. Third, species introduced outside their native ranges may have unpredictable and undesirable effects on ecosystems and other species. Finally, many biologists feel that it is unethical to introduce species outside of their known historical ranges (of course, for many invertebrate species, the known historical range is much smaller than the actual and unknowable historical range).

Despite these problems, it may be necessary to confront the problem of species reestablishments, especially if climate change in the twenty-first century is substantial. Clearly, we will need much better information on how fast species are able to disperse in response to a changing climate (to identify which species, if any, will perish without intervention), practical information on how to establish populations of freshwater invertebrates, and reliable models to predict whether the species we move will have undesirable effects in their new homes.

See Also the Following Articles

ENDANGERED MARINE INVERTEBRATES • ENDANGERED TERRESTRIAL INVERTEBRATES • ENDEMISM • EUTROPHICATION/OLIGOTROPHICATION • HOTSPOTS • INTRODUCED SPECIES • INVERTEBRATES, FRESHWATER, OVERVIEW • LAKE AND POND ECOSYSTEMS

Bibliography

Benz, G. W., and D. E. Collins (eds.). (1997). *Aquatic Fauna in Peril: The Southeastern Perspective*, Southeast Aquatic Research Institute Special Publication 1. Lenz Design and Communications, Decatur, Georgia.

Knott, B., and E. J. Jasinska. (1998). Mound springs of Australia. In *Studies in Crenobiology: The Biology of Springs and Springbrooks* (L. Botosaneanu, ed.), pp. 23–28. Backhuys, Leiden, Netherlands.

Kozhov, M. M. (1963). *Lake Baikal and Its Life*. Dr. W. Junk, The Hague.

Kozhova, O. M., and L. R. Izmest'eva (eds.). (1998). *Lake Baikal. Evolution and Biodiversity*. Backhuys, Leiden, Netherlands.

Lydeard, C., and R. L. Mayden. (1995). A diverse and endangered aquatic ecosystem of the Southeast United States. *Conservation Biol.* 9, 800–805.

Martens, K. B., Goddeeris, and G. Coulter (eds.). (1994). Speciation in ancient lakes. In *Advances in Limnology*, Vol. 44. E. Schweizerbart'sche Verlagsbuchhandlung, Stuttgart.

New, T. R. (1995). *An Introduction to Invertebrate Conservation Biology*. Oxford University Press, Oxford, United Kingdom.

Ponder, W. F. (1986). Mound springs of the Great Artesian Basin. In *Limnology in Australia* (P. De Deckker and W. D. Williams, eds.), pp. 403–420. CSIRO, East Melbourne, Australia.

Ponder, W. F. (1995). Mound spring snails of the Australian Great Artesian Basin. In *The Conservation Biology of Molluscs* (E. A. Kay, ed.), pp. 13–18, Occasional Paper of the IUCN Species Survival Commission 9. IUCN, Gland, Switzerland.

Ponder, W. F., and G. A. Clark. (1990). A radiation of hydrobiid snails in threatened artesian springs in western Queensland. *Records of the Australian Museum* 42, 301–363.

Ponder, W. F., R. Herschler, and B. Jenkins. (1989). An endemic radiation of hydrobiid snails from artesian springs in northern South Australia: Their taxonomy, physiology, distribution and anatomy. *Malacologia* 31, 1–140.

Williams, J. D., M. L. Warren, K. S. Cummings, J. L. Harris, and R. J. Neves. (1993). Conservation status of the freshwater mussels of the United States. *Fisheries* 18(9), 6–22.

ENDANGERED MAMMALS

Peter Zahler
Wildlife Conservation Society; University of Massachusetts, Amherst

GLOSSARY

anthropogenic Caused by humans.
depauperate Impoverished, as in a region with low taxonomic diversity.
endemic Native and restricted to a specific geographic region.
fossorial Adapted for living or digging underground.
stochastic event An event that occurs by chance; a random event.
taxon Any taxonomic group, at any level, that is considered distinct enough from other groups to be considered a separate unit; plural taxa.
therapsids An order of reptiles that existed during the late Paleozoic and early Mesozoic and from which mammals are believed to have evolved.

MAMMALS CONSTITUTE ONLY about 10% of all vertebrates and less than 0.3% of all known species of organisms, yet endangered mammals (mammals facing an imminent threat of extinction) have received a disproportionate amount of attention and conservation efforts. To a large extent this is the result of the higher visibility and appeal of mammals compared to members of other taxa. This stems in part from our own relationship to this group, both evolutionarily and in the roles that other mammals play in a wide range of human activities from subsistence hunting to sport and commercial activities to domestication for food, work, and companionship.

I. MAMMALIAN DIVERSITY

Mammals can be found from the arctic to the tropics, on every continent, on many of the smaller islands, and in all of the oceans. There are wholly aquatic mammals (Cetacea), semiaquatic mammals, terrestrial mammals, arboreal mammals, fossorial mammals, and even aerial mammals (Chiroptera). Mammals fall into three main groups: the monotremes, who lay shell-covered eggs and have a number of physiological structures in common with reptiles, including a single urogenital opening; the marsupials, noted for a suite of physiological traits including an incomplete placenta, a female bifid reproductive tract, and usually an abdominal pouch for the almost-embryonic young; and the eutherians, who have a chorioallantoic placenta and give birth to relatively precocial young. Mammals may live for over a hundred years, as suspected in some cetaceans, or as short as one year, as in the "annual" males of the marsupial *Antechinus*. Mammalian social systems vary from solitary individuals through a wide range of social organizations to the termite-like sociality of the naked mole rat (*Heterocephalus glaber*). Mammals can have individual home ranges as small as a few square meters to systems where groups of tens of thousands migrate together (e.g., wildebeest, caribou) and migrations may cover thousands of kilometers (e.g., gray whales). Mammals have evolved to feed on virtually anything that might qualify as edible, from invertebrates, to other mammals, to fungi, grasses, leaves, bark, and even conifer needles. The variety of niches utilized by mammals and their behavioral and physiological adaptations are enormous—it is difficult to think of a niche not already occupied by a mammal, from aquatic grazer (the Sirenia) to aerial piscivore (*Noctilio* and two other genera of bats).

Not only can mammals be found in most places on earth, many species can have dramatic effects on their ecosystems. Classic examples include the alteration of drainage systems and vegetation patterns caused by beaver dams; the alteration of bush and forest to grassland caused by elephant foraging; the alteration of forest pattern and structure caused by sciurid seed predation, hoarding, and burial; and the alteration of grassland pattern and structure caused by grazing from both large ungulates such as American bison (*Bison bison*) and a suite of African ungulates, and small rodents such as prairie dogs (*Cynomys*) and viscachas (*Lagostomus maximus*). Many of these effects have been shown to increase biodiversity of both plants and animals within the region of modification.

Despite the high visibility and appeal of mammals and their significance to ecosystems, we still know astonishingly little about most species. This is especially true of the smaller, more cryptic, and nocturnal groups, such as the rodents, insectivores, and bats, which make up the bulk of mammalian species diversity. Medellin and Soberon (1999) point out that 459 new species of mammals were described between 1983 and 1993 (partly due to laboratory genetic investigations), with over half weighing under 100 g. For many of these and other species, there is not even basic information on population size, geographic range, or even whether the species is still extant. This also holds true for larger, more conspicuous species. For example, there is almost

no information regarding population numbers, specific locations, or even existence for the dhole (*Cuon alpinus*), a large, pack-living canid, throughout most of its putative range in Asia.

II. A BRIEF HISTORY

A. Early Mammalian History

The history of mammals has been traced back approximately 250 million years to offshoots of the therapsid reptiles in the Triassic. According to limited fossil evidence, the first major radiation of mammals occurred in the Jurassic, about 190 million years ago. Although the diversity of mammals was increasing at this time, with marsupials and placentals in evidence by the early Cretaceous, these early mammals tended to have small and fairly uniform body sizes. A number of hypotheses have been generated to explain both the increase in mammalian diversity and the small body size. The breakup and then reattachment of the continents allowed for isolation, speciation, and then faunal exchange, and the development of angiosperm plants also undoubtedly resulted in coevolutionary adaptations and speciation by a wide range of mammals. Small body size perhaps was influenced by the dominance of reptiles, which may have forced mammals into a secretive, nocturnal lifestyle where a small body size would be an advantage. An explosive mammalian radiation occurred in the Paleocene, again partly from repeated separations, isolation, and reintroductions as continents continued their movements, and perhaps further spurred by the loss of a wide range of large reptilian competitors during the great extinction event of 65 million years ago.

B. The Pleistocene Extinctions

Between 30,000 and 10,000 years ago a number of sudden and major die-offs occurred, almost entirely among large terrestrial mammals. Approximately 11,000 years ago in North America, a megafaunal extinction event claimed up to 33 genera of large mammals within a span of only 1000 to 2000 years, including mastodons, mammoths, ground sloths, camels, horses, and various large predators such as saber-toothed cats and dire wolves. A similar extinction event involving 46 genera of mammals occurred in South America between 15,000 and 10,000 years ago. In Australia, 13 genera of large mammals disappeared about 30,000 to 20,000 years ago. Two hypotheses have been forwarded to explain these extinction events. The first is climate change, as in the end of the Ice Age in North America. The second hypothesis relates to the arrival and subsequent hunting of large herbivores by humans, as it appears that all three extinction events occurred at approximately the same time that humans colonized and spread across these continents.

C. Recent Mammalian Extinctions

Many island mammal fauna suffered extinction events similar to those that occurred on the continents during the Pleistocene, although most island extinctions occurred in more recent times. On islands in the Mediterranean, 13 endemic genera vanished, including small goats and even dwarf elephants; most disappeared about 4000 years ago. A dwarf form of the mammoth disappeared from Wrangell Island at about the same time. In Madagascar, eight genera of lemurs, two genera of pygmy hippos, and what is now considered a distinct order, the Bibymalagasia (comprising two species that are variously aligned with the aardvarks and the ungulates), became extinct about 1000 years ago. In each of these cases human activity, either hunting, habitat alteration, or a combination, is thought to be the main cause. Island extinction events have continued into modern times. In the Caribbean Islands, 5 insectivores, 12 rodents, a raccoon, and a seal have all become extinct since 1600—almost a quarter of all recorded mammal extinctions during that period (Fig. 1).

Although the total number of mammalian extinctions in recent times does not appear to be large, the current rate of extinction, about .01% per year, is anywhere from 100 to 1000 times greater than what would be predicted from the fossil record. This rate will very likely increase in the near future. The status of 4355 species of mammals was reviewed in the 1996 IUCN Red List of Threatened Animals. Of those reviewed, approximately 40% were considered at risk, from critically endangered to near threatened. This percentage is undoubtedly a low estimate, for as mentioned the status of many mammal species is still unknown.

III. CAUSES AND THREATS

Many of the dangers that threaten other taxa also threaten mammals. These dangers include habitat loss, exploitation, disease, and exotic introductions, but there are some differences in the manner and level of these threats for mammals as compared to other taxa. For example, human persecution and harvesting pres-

sure is probably greater for mammals than for most other taxa, with the probable exception of fish. True mammalian specialists are rare compared to taxa such as insects, where many species are entirely dependent on only one species of plant and thus are vulnerable to habitat changes that alter community structure. However, many mammals are relatively large, and their habitat needs in terms of area are greater than for most other taxa, putting them at high risk from habitat loss and fragmentation.

A number of historical mammalian extinctions have occurred due to human exploitation or introductions rather than habitat loss. Of 18 mammal species that have become extinct since 1600 and whose cause of extinction is known, eight became extinct due to direct human persecution, eight became extinct as the result of introduced predators and competitors, and only two became extinct because of habitat destruction. However, it is apparent that many mammalian species are now under a growing threat from habitat changes. The World Conservation Monitoring Centre (1992) estimated that habitat modification or loss is the major danger for more than three-quarters of all threatened mammals, and this number is likely to grow as the human population continues to increase and expand.

A. Mammalian Physiology and its Relationship to Threats

Mammals constitute a class of organisms that span an enormous size range of approximately eight orders of magnitude, from 2 g in some bats and shrews to 190,000 kg in the blue whale. However, the majority of mammals tend toward large sizes compared to other taxa, and they tend to have a correspondingly long maturation and slow reproductive rate. This puts them at risk from exploitation and other threats, as breeding rates may not be able to keep up with losses in a population. An extreme example is the blue whale (*Balaenoptera musculus*), whose population was lowered to 1% of its original numbers in less than a century and has not yet shown any appreciable increase, despite complete protection for more than 30 years. Large body size often means large habitat needs, which also puts mammals at risk. Estimates for self-sustaining populations of carnivores such as brown bears (*Ursus arctos*) and African wild dogs (*Lycaon pictus*) range from 1000 to more than 10,000 km² of suitable habitat, which means that many protected areas are too small to sustain long-term populations of these species.

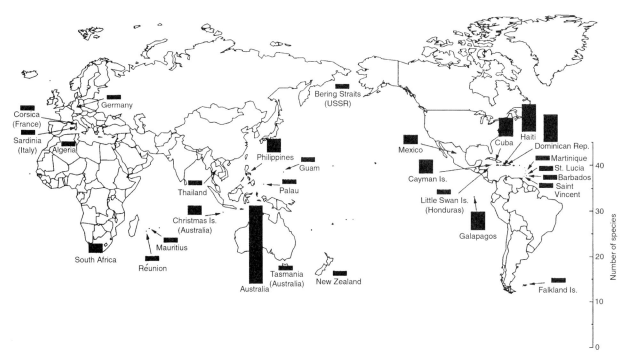

FIGURE 1 Known mammal extinctions since 1600.

B. Habitat Loss, Degradation, and Encroachment

Habitat loss and fragmentation is the single greatest threat to biodiversity worldwide, and this certainly holds true for mammals today. Conversion of habitats by humans into other land uses can fragment and separate mammal populations and increase the likelihood of local population extinctions and eventual species extinction. Rapid deforestation of tropical areas is a growing threat to a number of mammalian species, including many large, wide-ranging, or specialist species of primates, cats, and forest ungulates, as well as numerous small species with restricted ranges such as rodents, insectivores, and marsupials. Most of these species cannot adapt to a highly fragmented or altered landscape, and the few that do adapt often come into conflict with humans by feeding on crops or livestock.

The example of the giant panda (*Ailuropoda melanoleuca*) shows some of the complexities related to fragmentation and habitat loss. Pandas feed primarily on bamboo that may live for decades but then tends to flower, seed, and die en masse within certain areas. When this happens pandas must switch to other bamboo species, often having to move to new locations to find these alternative food sources. The increase in human population within the panda's range in China has limited most populations of pandas to very small islands of habitat. A recent seeding and die-off of three species of bamboo resulted in the starvation of over 10% of the world's remaining wild panda population. Widely separated and very small populations of pandas may not be viable over the long term, even without the problems faced from the fluctuations in their food source.

Because mammals are often relatively poor dispersers, the creation of corridors linking habitats has been suggested as a way to help some species, especially large or wide-ranging (including nomadic or migratory) ones. But for many mammals the necessary size and structure of corridors is unknown, and few management plans have yet to put this idea into practice.

C. Genetic Loss

It is generally assumed that a population's long-term survival is at least partly dependent on sufficient genetic variation for individual fitness and population adaptability. Loss of genetic diversity and reduced fitness from inbreeding depression and the chance fixation of detrimental alleles has been presumed to reduce adaptive potential and increase the probability of extinction.

However, this has been notoriously difficult to quantify and has not yet been proven to have directly caused a decline in a mammal's population in the wild. Other events, such as disease or predation, often complicate analysis, especially as lowered fitness may increase an individual's susceptibility to these factors. There are some examples for which reduced genetic fitness may well be a cause for declines in mammal populations. The genetic bottleneck that occurred in cheetahs (*Acinonyx jubatus*) in the Pleistocene has led to extremely low genetic variability among populations in sub-Saharan Africa and is believed to be the cause of sperm aberrations and low sperm counts. Isolation and reduction in numbers leading to inbreeding of Florida panthers (*Puma concolor coryi*) is considered to be the cause of low sperm counts, malformed spermatozoa, and cryptorchidism (undescended testicles) as well as an increase in heart defects within this population.

Another potential genetic threat to endangered mammals is hybridization. Documented cases of hybridizations involving endangered species include red wolves (*Canis rufus*) with coyotes (*C. latrans*), Ethiopian wolves (*C. simensis*) with domestic dogs (*C. familiaris*), and Asian wild asses (*Equus hemionus*) with domestic horses (*E. caballus*). Hybridization with domesticated forms or relatives is also seen as a threat to wild yaks (*Bos grunniens*), Bactrian camels (*Camelus bactrianus*), Mongolian wild horses (*Equus ferus przewalskii*), and European wild cats (*Felis silvestris*), among others.

Another genetic issue is the loss of native species and their replacement by common generalist species characteristic of human landscapes. This can result in a local numerical increase in species diversity but a worldwide loss of genetic diversity.

D. Livestock and Disease

Domestic livestock and their attendant parasites and diseases have had dramatic effects on some wild mammal populations. Land clearing and overgrazing from domestic cattle, sheep, and goats have altered and fragmented landscapes, increased desertifaction, and reduced forage for other herbivores. The African rinderpest epidemics of the late 1800s, introduced by cattle, devastated a wide range of wild ungulate populations across the continent. More recently, African horse sickness decimated Indian wild asses, canine distemper killed one-quarter of the lions (*Panthera leo*) in the Serengeti National Park, and all eight packs of wild dogs (*Lycaon pictus*) involved in a study in the Serengeti region disappeared in 1991 with rabies as the suspected or confirmed cause. Disease can spread in the opposite direction, with equally problematic results for wildlife; for example, bison have been shot when they wandered outside of protected areas in Yellowstone National Park in an attempt to control the spread of brucellosis to cattle.

Another factor related to livestock and disease has been the erection of fences to separate wildlife and domestic animals. The results have sometimes been dramatic, such as the case in Botswana, where enormous fences were built in an attempt to control the transmission of disease. These fences blocked the natural migration routes of wild ungulates during the dry season and resulted in the deaths of tens of thousands of red hartebeest (*Alcelaphus bucelaphus caama*), blue wildebeest (*Connochaetes taurinus*), and other antelope.

E. Exploitation

Exploitation of mammals has occurred for many different reasons. Subsistence hunting is a growing pressure due to the explosion of the human population, especially in developing countries where much of the populations' dietary protein may come from wildlife. Along with the increase in the human population has come an increase in the number of accurate firearms, including automatic weapons, that are now available to subsistence hunters. Subsistence hunting especially threatens ungulates, such as various deer and tapir species in Asia and South America—a third of all deer species and all four tapir species are now considered threatened or near threatened. Subsistence hunting also threatens aquatic mammals whose ranges have already been limited from other causes or who, like sirenians and river dolphins, can primarily be found in coastal or riverine areas. Steller's sea cow (*Hydrodamalis gigas*), the largest and only cold-water sirenian, was hunted to extinction by hungry sailors and other visitors to the north Pacific within 27 years of the first scientific description of this monotypic genus.

Commercial meat hunting has also threatened some species of mammals. The saiga antelope (*Saiga tatarica*), a migratory species of the central Asian steppes that once numbered in the millions, was hunted almost to extinction for its meat, hides, and horns, with the population dropping to a low of about 1000 individuals in the early part of the 1900s.

Until recently the fur trade was an important part of many countries' economies and cultures. In North America this trade led to the depletion of many fur-bearing species by the early to middle 1800s, including beaver, marten, fisher, otter, and a number of species of seals. Fur hunting for sea otter (*Enhydra lutris*) caused the population of this species to drop from an estimated 300,000 to under 2,000 by 1911, and hunting for skins was the main cause of depletion in numbers for the now-extinct Caribbean monk seal (*Monachus tropicalis*). More recently the market for fur resulted in heavy losses for most wild cat species, including large cats such as tigers and jaguars and the smaller spotted cats of South America and Asia. However, it should be noted that other mammal species, especially wide-ranging, abundant, and fast-breeding species such as the muskrat, raccoon, and coyote, have maintained their numbers despite heavy trapping efforts that continue to this day.

A second result of the fur trade was the introduction of fur-bearing animals outside of their ranges, with often unintended results. The accidental release of American mink (*Mustela vison*) into Europe now threatens the European mink (*M. lutreola*), as the American species breeds earlier and appears to be outcompeting its European counterpart.

Commercial hunting is not limited to fur. The hunting of whales for oil as well as baleen led to severe depletions of populations of many species in both hemispheres, with some populations driven to extinction and others lowered to below 5% of their original estimated numbers. Although bans have been in place for most species of whales for some time, many populations have yet to show signs of recovery.

Exploitation for body parts used in traditional medicine has also led to the decline in a number of mammal species. Almost every part of a tiger (*Panthera tigris*) is used for medicinal purposes, including the feet, fat, bones, blood, testes, penis, bile, whiskers, claws, and tail. The Asiatic black bear (*Ursus thibetanus*) is under great pressure from poaching for its gallbladder, which is used in traditional remedies, and the poaching of rhinos for their horns, which are used both for knife handles and traditional medicines, is the major cause of their precipitous decline. Although habitat loss, trophy hunting, and persecution played an important role in the initial decrease in tigers, bears, and rhinos, poaching for the medicinal market is now the greatest threat to their continued survival.

F. Persecution

Persecution usually takes the form of purposeful attempts to control or extirpate a species. Examples include persecution of potential livestock predators such as wolves and tigers, potential livestock disease threats such as antelope species in Africa and bison in Yellowstone National Park, potential livestock competitors

such as prairie dogs in the United States, and, in the case of the American bison in the 1800s, an effort to control and eradicate the indigenous human culture that once depended on this large ungulate. In some cases, control efforts have driven nontarget species to endangered status, such as poison control efforts aimed at coyotes that eradicated the swift fox (*Vulpes velox*) throughout much of its range, and the prairie dog eradication efforts that eliminated the rodents over 98% of their range and drove the black-footed ferret, a predatory specialist on prairie dogs, to the edge of extinction.

G. Interference

Interference can occur in a number of ways. As mentioned, fences have interfered with migratory ungulates in Africa and elsewhere. Dams have negatively affected Ganges and Indus river dolphins (*Platanista gangetica and P. minor*) by fragmenting populations, increasing siltation, and reducing fish prey. For the critically endangered North Atlantic right whale (*Eubalaena glacialis*) injuries from boat collisions and entanglement in fishing and lobster gear have accounted for as much as a third of all mortalities in a year, while for manatees in Florida boat collisions have accounted for up to a quarter of all mortality in a year. Death from net entanglement is thought to exceed reproductive rate in Hector's dolphin (*Cephalorhynchus hectori*), a small coastal species from New Zealand. Spinner dolphin (*Stenella*) populations were significantly reduced due to drowning by tuna seiners who targeted the species because of their association with tuna schools. All three species of spinner are now listed as conservation dependent by IUCN. In Florida, the greatest cause of mortality for the endangered subspecies of key deer (*Odocoileus virginianus clavium*) and Florida panther is from being struck by cars. Even tourism may pose a threat. For example, tourist vehicle chases in desert environments have led to death from exhaustion for threatened African wild asses (*Equus asinus*) and addax (*Addax nasomaculatus*) in northern Africa. The expansion of the whale watching industry has led to concerns that the increasing number of boats, motor noise, and chases may negatively effect whale behavior, including disturbing migratory patterns, breeding efforts, and even separating mothers and young.

War is an extreme example of interference, and it can have an equally dramatic effect on already-rare mammal populations. Although there are arguments that historical tribal warfare may have created source buffer zones for large mammal populations (Martin and Szuter, 1999), today's warfare results in numerous negative effects including habitat destruction from defoliation and bombing efforts and the increase in killing of wildlife due to improved hunting efficiency from the availability of modern firearms. This combination is thought to have had an effect on most wild mammals in Vietnam, including the threatened Douc langur (*Pygathrix nemaeus*). Soldiers and guards may also negatively affect wildlife. For example, the disappearance of most of the remaining population of Marco Polo sheep (*Ovis ammon polii*) along the China-Pakistan border is attributed to target practice and trophy hunting from armed border forces. War may also cause local increases in subsistence hunting and habitat destruction from desperate indigenous and displaced people, especially in and around formerly protected areas. This appears to be the case in Ethiopia and Somalia, where rare antelope and wild ass are heavily hunted, and in western Africa, where mountain gorillas (*Gorilla gorilla beringei*) are under mounting pressure from human refugees seeking an escape from fighting as well as food and firewood in protected forests.

H. Pollution

Most documented cases of pollution affecting wildlife relate to taxa other than mammals, such as DDT and peregrine falcons or acidification and brook trout. There are cases where pollution has had an effect on populations of threatened or endangered mammals. For example, bats in southwestern North America suffered from DDT spraying, and a number of European bats have been affected by wood preservatives used in buildings that serve as roosts. Most documented mammal-pollution relationships, however, have involved aquatic systems. Perhaps the most obvious example was the death of thousands of sea otters from the effects of the Exxon *Valdez* oil spill in Alaska. Less obvious cases include the St. Lawrence population of beluga whales (*Delphinapterus leucas*) that have been shown to have high levels of organochlorines that have been implicated in cancers, ulcers, infections, and decreased fertility compared to open ocean belugas. Decreased pup production has been noted in seals fed fish from polluted waters in Europe, and mercury released from mining is thought to pose a serious threat to Amazon river dolphins (*Inia*) and manatees (*Trichechus inunguis*).

Pollution may also have a more subtle effect by increasing a populations' vulnerability to disease by depressing the immune system. Seals have suffered outbreaks of disease in the Mediterranean, northeast Atlantic, and in Lake Baikal. Immune depression has been shown to exist in harbor seals (*Phoca vitulina*) from the Baltic. Pollution is thought to have led to infections from a calcivirus in California sea lions (*Zalophus californianus*) that caused abortion and premature parturition. Dall porpoises (*Phocoenoides dalli*) have been found to have an inverse correlation between serum testosterone level and DDE concentrations in their blubber. Pollution may also have more indirect effects, such as increased mortality of fish prey for river dolphins in the Ganges and Indus.

I. Introductions

Introductions of exotics, both purposeful and accidental, have had very strong impacts on some native species of mammals. These impacts may involve predation, competition, and the concurrent introduction of diseases or parasites. Most documented introduction problems have involved mammals as the pest species and have taken place on islands. Cases include the introduction of goats, sheep, and other herbivores to islands where endemic plants lack defenses against herbivory, and the introduction of omnivores such as rats, pigs, and macaques or carnivores such as mongooses and stoats to islands where endemic animals lack defenses against predators. The introduction of rabbits (*Oryctolagus cuniculus*) to Laysan appears to have driven three species of birds to extinction from grazing that destroyed avian food sources, both plants and insects dependent on the plants. The results from the explosive population growth of rabbits introduced into Australia is well documented, although predation by nonnative red foxes probably has had a greater negative effect on indigenous marsupials. Both pet and feral cats and dogs also pose a threat to wildlife. Feral cats in Australia are estimated to kill more than 400 million native mammals, birds, and reptiles each year, while in Argentina predation by dogs is the biggest threat to the endangered pink fairy armadillo (*Chlamyphorus truncatus*). Introduced aoudad (*Ammotragus lervia*) and feral burros in the southwest United States are considered to be potential competitors and disease vectors for the desert bighorn sheep (*Ovis canadensis nelsoni*). Introduced deer species have had major effects on vegetation in New Zealand, and in South America they may be serious competitors of the endangered Andean huemel (*Hippocamelus bisculus*).

J. Multiple Threats

For many species, a series or combination of the factors listed have led to precipitous population declines and endangerment. For the tiger, trophy hunting, eradication programs, market exploitation for fur and traditional medicines, and loss of habitat all played a role in the extinction of three races and a decline from perhaps over 100,000 tigers worldwide in 1920 to as few as 5000 today.

Even for mammals that have recovered from previous threats, new factors may arise to threaten the species. The saiga antelope recovered from commercial hunting once a ban and then regulations were put in place, rebounding from a low of only a thousand to more than 2 million in just 50 years. However, overgrazing, desertification, migratory barriers, and poaching for medicinal products now threaten the species, and the population has dropped to around 1 million animals in four widely separated regions.

IV. LOCATIONS FOR ENDANGERED MAMMALS

A. Geographic Analysis of Endangered Mammals

As with most other taxa, mammalian species diversity tends to increase with decreasing latitude. The top eleven countries for mammalian diversity are Mexico (450), Indonesia (436), United States (428), Zaire (415), Brazil and China (394), Columbia and Kenya (359), Peru (344), Tanzania (322), and India (316). Only the United States is not found entirely or primarily within the tropical zone. The number of threatened mammals is shown in Table I, with Indonesia leading the list with 128 and China and India tied for second with 75 each. Most of the listed countries also have some of the highest human densities in the world. However, some of these countries, such as Australia, have high numbers of threatened species yet relatively low human densities. In most of these cases a second factor, endemism, can explain much of the data.

B. Endemism: The Case of Australian and Island Mammals

Oceanic islands tend to have depauperate mammalian faunas, primarily because terrestrial mammals make poor oceanic dispersers. Many islands also do not have enough habitat to maintain viable populations of terrestrial mammal species. However, on islands that terrestrial mammals were able to colonize and that were large enough to sustain populations, isolation often led to speciation. Some examples of island endemism are truly

spectacular, and often these locations also contain some of the most threatened species in the world. Indonesia, an archipelago consisting of more than 13,000 islands, has the second-greatest species richness in the world and also has the greatest number of endemic mammal species with 201. Indonesia also has the greatest number of threatened mammals with 128 (29% of its mammalian fauna). In Australia, out of a total of 252 species of mammals, 198 species are considered endemic. Twenty species are recorded as extinct since 1600 and, of the remaining species, 58 (23%) are considered threatened with extinction. Madagascar is the fourth largest island in the world and has been separated from mainland Africa since the Cretaceous. Because of its long isolation, all but one of the terrestrial species are endemic forms (the exception is a pig that may have been introduced). There are five families and 15 genera of living endemic primates (the lemurs), an endemic family of insectivores (the Tenrecidae), three endemic subfamilies of the carnivore family Viverridae, at least eight endemic genera of rodents, and a family of bats (the Myzopodidae). Of the endemic species, 46 (44%) are currently threatened, many from the recent loss of 90% of the island's original forest cover. Papua New Guinea has 214 species of mammals, of which 57 are endemic; 57 species are also considered threatened. In Cuba almost half (15 out of 31) of the mammal species are endemic, and 9 of these are considered threatened. As mentioned, many of these islands have already suffered extinction spasms: of a total of 88 mammal species recorded as extinct since 1600, 53 are from islands, with 20 more extinctions occurring in Australia.

V. WHICH MAMMAL TAXA ARE MOST THREATENED?

A. IUCN Red Data Book

Lists such as the Red Data Books are useful in drawing attention to the plight of certain species and in developing conservation and management efforts, including the creation of protected areas. The 1996 IUCN Red List of Threatened Animals attempted to assess the conservation status of every mammal species in the world. Each species was placed within one of eight categories. The first three are categories of threat: critically endangered, endangered, and vulnerable. The next three are lower risk categories: conservation dependent, near threatened, and least concern. The last two categories are data deficient and not evaluated. As well as these categories, two more exist: extinct and extinct in the

TABLE I

Countries with the Most Species of Threatened (CR, EN, VU) Mammals

	Number of threatened mammals	Total number of mammals in country	Percent threatened	Total number of endemics
Indonesia	128	436	29	201
China	75	394	19	77
India	75	316	24	44
Brazil	71	394	18	96
Mexico	64	450	14	140
Australia	58	252	23	198
Papua New Guinea	57	214	27	57
Philippines	49	153	32	97
Madagascar	46	105	44	77
Peru	46	344	13	45
Kenya	43	359	12	21
Malaysia	42	286	15	27
Zaire	38	415	9	28
Viet Nam	38	213	18	7
Ethiopia	35	255	14	31
USA	35	428	8	101
Colombia	35	359	10	28
Thailand	34	265	13	7
South Africa	33	247	13	27
Tanzania	33	322	10	14

From IUCN Red List of Threatened Animals (IUCN Gland, Switzerland, and Cambridge, UK, 1996) with permission.

wild. Out of a total of 4355 species of mammals whose status was reviewed by IUCN, 25% were considered threatened. A total of 4% were considered critically endangered, 7% endangered, and 14% were considered vulnerable to extinction. Another 16% were considered to be conservation dependent or near threatened (Fig. 2). This compares with 11% of birds threatened (2%, 2%, and 7%), with 9% near threatened. A possible explanation for why 25% of mammals are under direct threat as compared to only 11% of birds is that mammals are not as efficient dispersers. Birds in general are more likely to be found over a wide area, and as habitats become fragmented or disappear some populations are likely to survive, and individual birds can also move to new locations more easily than mammals.

The recent development of programs such as national Red Data Books that cover individual countries or the Natural Heritage Data Centers that cover North and most of Central America has led to a new spatial scaling of endangerment. These local lists often list many more species and subspecies of mammals than appear in the IUCN Red Data Book. This is both from inclusion of species that are locally but not globally endangered and endemic subspecies that have yet to receive international attention. The inclusion in a national or regional list of an otherwise-common species whose range barely reaches the borders of a country may at first seem questionable. However, there is the likelihood that it is a genetically distinct subpopulation adapted to living in what would otherwise be considered marginal habitat for the species.

B. Endangered Mammals by Taxa

An analysis of endangered mammals by taxa, using the IUCN Red List of Threatened Animals (1996) shows that among the larger orders of mammals primates are the most endangered, with 41% of the species under threat (critically endangered, endangered, or vulnerable). All primate species are either on Appendix I or II of the Convention on International Trade in Endangered Species of Wild Fauna and Flora (CITES). Primate threats are many and varied, but the principal threat is habitat destruction. Because most primates have arboreal habits, tropical forest loss and fragmentation have had strong negative effects. Captures for the pet trade,

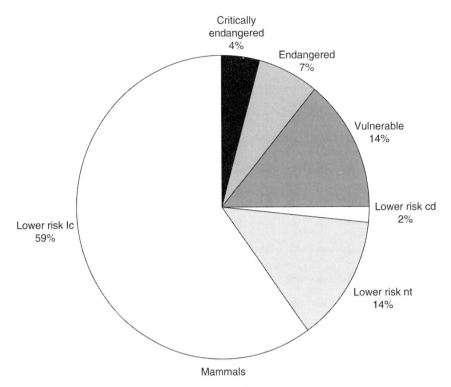

FIGURE 2 IUCN red list of threatened mammals (1996). cd, conservation dependent; lc, least conserved; nt, near threatened (used with permission).

zoos, and medical research had a major impact on a number of species until recent legislation limited wild captures. Other species have been affected by subsistence hunting or accidental take from snares set for other species.

Insectivores are the next most threatened order, with almost 41% of species under threat. Many of these species are poor dispersers found in one or a few populations that are especially vulnerable to habitat alteration and fragmentation. Artiodactyls, or even-toed hoofed mammals, have 31% of the species under threat. Roughly 28% of all marsupial species are threatened, with 33% of species from neotropical marsupial orders and 26% from Australasian orders threatened. Approximately 25% of Chiroptera are threatened. Although much attention has been given to threatened members of the order Carnivora, "only" about one-quarter (24%) of the species are threatened. Rodents have about 16% of species listed as threatened, but because of the relatively large number of rodent species, this order has the highest number of species under threat with a total of 330. Among smaller orders, almost 58% of Perissodactyls (odd-toed hoofed mammals, such as horses, tapirs and rhinos) are threatened, with 4 of the 18 species critically

endangered. Almost half (7 out of 15) of the species of Macroscelidea (elephant shrews) are considered threatened. Four out of five Sirenia (manatees and dugongs) are listed as vulnerable, whereas both Proboscidea (elephants) are listed as endangered.

VI. CONSERVATION

A. Conservation Legislation

An enormous number of treaties and laws govern protection for wildlife or their habitat. These measures include prohibitions or controls on taking, collection, possession, and trade of specific species, control of exotic species, and the protection of habitat, either directly through the creation of protected areas or through indirect measures such as tax incentives, permits, and zoning.

Three levels of conservation legislation exist. The first is worldwide treaties, usually involving conservation of habitat as in RAMSAR's wetland protection, pollution issues, commercial exploitation and trade, or combinations of these issues, as in the Convention on

Biological Diversity of 1992. The second is regional treaties, such as the ASEAN Agreement covering Southeast Asia or the Berne Convention for Europe. The last involves individual species or species groups, such as the creation of the International Whaling Commission or the 1973 Agreement on the Conservation of Polar Bears.

CITES (Convention on International Trade in Endangered Species of Wild Fauna and Flora), is an example of a worldwide treaty that is especially important for conservation of endangered mammals. CITES was established in 1973 to deal with the enormous level of international trade in wildlife, with estimated earnings as high as US$20 billion. CITES has three appendices, with Appendix I listing species threatened with extinction, Appendix II listing species that may become threatened with extinction unless trade is regulated, and Appendix III listing species protected under national law. Unfortunately, a treaty is only as good as the on-ground enforcement of its provisions, and while CITES has gone a long way toward stemming the tide of illegal trade in wildlife, black market commercial exploitation continues to threaten many species throughout the world.

B. Species Survival Commissions

The International Union for the Conservation of Nature and Natural Resources (IUCN) is an international union of states, agencies, and organizations dedicated to the conservation of biodiversity, primarily through monitoring and advisory roles. IUCN now has more than 100 specialist groups within the Species Survival Commission (SSC), each focused on a particular taxa. More than 30 of the specialist groups work on mammals, ranging from entire orders such as the Rodent Specialist Group to individual species such as the Asian Elephant Specialist Group. SSC specialist groups consist of volunteer scientists and other experts who provide information on their taxa and even lobby governments for conservation and research efforts. One method of providing this information is through Status Surveys and Conservation Action Plans that describe trends, threats, and conservation options, which most specialist groups have now published for their particular taxa.

C. Protected Areas and Preserve Sizes

With human population increasing and unmodified habitat rapidly dwindling, habitat protection and maintenance is certainly the single most important conservation method for the preservation of biodiversity, includ-

ing mammals. Large mammals often live at low densities and over large individual areas. A few protected meadows with the appropriate plant species may be enough habitat to maintain a population of butterflies, but an enormous tract of wilderness is necessary to sustain a viable population of tigers or rhinos. Sadly, many large mammals, such as rhinos and tigers, are now entirely or mostly found within park boundaries, and it is uncertain whether populations are large enough to maintain themselves in the face of stochastic events and anthropogenic changes. Mammals within protected areas often come into conflict with people sharing the land or living on the edges of the preserves, and poaching may continue to cause declines in mammal populations within protected areas for a variety of social or economic reasons. A new focus on off-reserve or nonprotected area conservation of threatened mammals faces daunting challenges from multiple use, social, cultural, and economic growth issues. However, off-reserve conservation may be the only hope for some mammals, as the money and political will to continue creating protected areas is limited.

D. Captive Breeding and Reintroduction

Captive breeding has become a well-accepted way of managing and increasing populations of critically endangered species. New techniques have led to breakthroughs in captive breeding with some species, including the use of extra-specific surrogate mothers and embryo manipulation, including transfer, cryo-preservation, and microsurgical division. However, captive breeding has not worked in every case. For example, the critically endangered Sumatran rhino (*Dicerorhinus sumatrensis*) has not been bred in captivity despite a number of attempts, and a third of the 27 captive animals have died.

A number of captive breeding programs have led to reintroduction efforts for mammals that were extinct in the wild, including wisent (*Bison bonasus*) in Europe, Arabian oryx (*Oryx leucoryx*) in Oman, and Père David's deer (*Elaphurus davidianus*) in China. Other reintroduction programs have attempted to supplement decreasing wild populations; for example, 25% of wild golden lion tamarins (*Leontopithecus rosalia*) are now from captive-bred stock. Perhaps the best-known case of captive breeding and reintroduction of a mammal involves the black-footed ferret (*Mustela nigripes*). This small carnivore was once common across the North American plains, where it specialized in hunting large, colonial ground squirrels called prairie dogs (*Cynomys*). Prairie dogs were considered to be pests and competitors with cattle, so an intensive eradication program eventually reduced historic prairie dog range by over 95%. The black-footed ferret subsequently declined and was thought to be extinct in the wild in the 1970s, although a small number of ferrets still existed in captivity. This captive colony suffered from physical problems, perhaps related to inbreeding as well as disease, and the colony died out in 1979, leading to fears that the species had truly gone extinct. However, in 1981 a new colony of ferrets was discovered in Wyoming. The first six ferrets to be captured for breeding died of canine distemper. The remaining wild ferrets were captured and successful captive breeding resulted in their numbers reaching about 300, with some released back to the wild. However, the program has not been without criticism: politically motivated decisions and arguments between state, federal, and private organizations have caused numerous problems and may have even jeopardized the success of the project at times. Finding suitable prairie dog colonies (both in terms of finding colonies of adequate size and the political difficulty of maintaining large numbers of what many consider to be a "pest" species) for continued reintroduction is yet another roadblock to ferret recovery.

Another successful and controversial example has been the reintroduction of gray wolves (*Canis lupus*) into the Rocky Mountain region of the United States, beginning in 1995 and including most recently Yellowstone National Park. The wolf population in Yellowstone, numbering more than 100, has already had an obvious effect on the ecosystem of the area, causing a decrease in some species (i.e., coyotes, through competition and killing) and an increase in other species (due at least in part to an increase in carrion from wolf kills). Wolves also appear to be altering the behavior of their ungulate prey, which in turn is having other effects on the system. This introduction has not been without controversy, partly stemming from wolf predation on livestock. During a 3-year period wolves in the Rockies killed more than 250 sheep and cattle, but a compensation fund for livestock lost to wolves has helped mitigate some of the antiwolf feelings from ranchers. Despite these problems, the success of the program has led to discussions regarding possible reintroduction of wolves into the northeastern United States.

Other predator reintroduction attempts have not fared so well, however. Recent efforts to reintroduce the Mexican subspecies of the wolf to the southwestern United States resulted in the shooting death of 5 of the first 11 released wolves. Attempts to reintroduce lynx (*Lynx lynx*) into parts of the Swiss Alps and the Adiron- dacks of New York have failed. Another lynx introduction at the southern limit of the species' range in the Colorado Rocky Mountains has also had problems, with 4 of the first 13 animals starving to death. The success of many reintroductions has also been threatened by legal challenges from individuals or organizations opposed to land use restrictions or worried about potential direct conflicts with humans or livestock.

Reintroductions do not always involve captive breeding. A number of African antelope and Eurasian Caprinae have been successfully translocated from still-viable populations into areas where they had been eradicated. One issue regarding this technique is whether there was a distinct local genotype. This is a serious problem if translocations are being considered as a method to bolster an existing but declining or no-longer-viable population, in which case mixing genotypes may result in the swamping of possible adaptive differences. There have also been introductions into areas that were not part of the original range of a species, with occasional unintended results of competition with indigenous species and habitat destruction. In one example, 11 Bahamian hutias (*Geocapromys ingrahami*), an endangered rodent, were introduced to a nearby island in a successful attempt to establish a second population of the species. However, it appeared that the hutia had not lived on this second island at least within historic time, and after 16 years the population had expanded to the point where seven plant species had vanished from the island from overgrazing.

Reintroductions are a complex undertaking, involving a solid understanding of the biology of the species, from feeding and breeding to individual and population genetics. The animals must be trained to survive in the wild, from finding food to socializing and, in the case of arboreal primates, even learning to move through the trees properly. The location for reintroduction should be part of the original range, be large enough to sustain a viable population, contain the correct habitat specifications, and no longer have whatever pressures resulted in the original loss of the species in the wild. Introduced animals must be closely monitored, and there must strong political and local support and extended funding for the project. Thus captive breeding and reintroduction may only be appropriate in specific cases where this kind of information and support is available.

E. Economic Incentives

Because reserves only cover approximately 5% of the earth's surface, and captive breeding and reintroductions are only feasible for certain well-studied and well-funded species, other methods will be necessary to ensure the survival of a number of mammal species. One method is economic utilization, which applies the fact that mammals have always been heavily exploited. It also has the important potential advantage of paying for itself. Because much of the present extinction crisis of mammals has an economic incentive, this idea has won wide if cautious acceptance, although controversy still exists on issues such as lack of data on population numbers and dynamics, whether overuse is inevitable, and concerns over animal welfare.

Sports and trophy hunting has been linked with historic declines in a number of mammal species, often in conjunction with other factors such as habitat loss. However, today sports and trophy hunting is being considered as a management tool. For some species, especially big game animals, it is suggested that conservation can "pay for itself" when fees from hunters are redirected back into management. This can be most effective for polygynous species where males are not a limiting resource and are also considered desirable as trophies. Licensing and management for white-tailed deer in North America has resulted in a population explosion of that species to the point where it is now necessary to actively control deer as pests in many locations. Trophy hunting fees may also be directed back to local communities to help replace losses from subsistence hunting or nuisance animals and encourage conservation awareness and behavior. Wild goat trophy hunting programs in the mountains of northern Pakistan have resulted in locally effective conservation efforts on behalf of populations of ibex (*Capra ibex*) and markhor (*Capra falconeri*), although the long-term success of these projects is still in doubt. In Zimbabwe, the Communal Areas Management Programme for Indigenous Resources (CAMPFIRE) has had success with trophy hunting as an economic incentive for local people to manage their wildlife in a sustainable manner. However, sports hunting as a management tool is only possible when populations of the target species are large enough to support regular culling, monitoring of the target species is constant, regulations are capable of being altered if necessary, and where the political situation enables funds generated from hunting fees to be returned either directly to continued conservation efforts or to local people to encourage their support. From the biological side, potential problems associated with the removal of dominant males include skewed sex and age ratios, lowered genetic variability, altered behavior of other group members and even the destruction of group cohesion in some targeted species, such as big-

horn sheep (*Ovis canadensis*) and elephants (*Loxodonta africana*).

Tourism is another method of raising funds so that the organism under protection pays for itself. This has worked well for a few charismatic species, such as mountain gorillas, but the vast majority of threatened mammal species are not capable of generating a level of public interest that will result in tourism revenues. However, it has been argued that certain charismatic species may act as "umbrellas" to create protection that will cover other species.

F. Umbrella Species: The Case for and against Charismatic Megafauna

The umbrella principle of conservation involves the protection of a large or wide-ranging species in the hopes that this will result in the protection of a number of other, small species. This concept has numerous advantages: it is easier to mobilize public interest toward large and charismatic species, it is easier to get funding for conservation work, and efforts to conserve large species are likely to result in conservation of large numbers of smaller species, due to the large habitat needs of the large species overlapping or including those of smaller ones.

The tiger is often used as an example of a species where conservation success will likely mean protection for a number of other species. However, the 15 reserves set up for tigers in India have not even adequately protected tigers, and poaching and encroachment threatens the ability of these reserves to protect many other species found within them. Another problem associated with the concept of mammalian umbrella species is that areas of species richness or endemism for other taxa, such as invertebrates, may not be within the designated core area of protection for a large mammal. A third issue of concern is that management for a single species, such as antipoaching efforts for tigers, may not satisfactorily protect other species whose main threats may involve other factors such as fire, fragmentation, or the spread of exotic species.

VII. PRESENT AND FUTURE TRENDS

Although mammals get a disproportionate amount of attention and funding for conservation efforts, the future for many species still looks grim. As of 1996, one-quarter of all mammal species were considered to be threatened. For a number of these mammals, a long-term commitment will be needed scientifically, financially, and politically if they are to survive. However, species-based conservation efforts for mammals can have enormous costs. It has been estimated that conservation of the northern subspecies of the white rhino (*Ceratotherium simum cottoni*) runs about US$10,000 a year per rhino, while the cost over seven years for reintroduction of the golden lion tamarin was estimated at more than US$1.5 million. The cost of creating and maintaining a preserve that will protect populations of many mammal species (as well as other taxa) is only a fraction of that needed for species-based management.

Unfortunately, for many endangered mammals there is a desperate need for immediate and focused management efforts. There also is a tremendous need for more protected areas and better protection within the reserves themselves, and a need for directing conservation efforts toward nonprotected land, whether it be through zoning of public and private lands or the use of private ranches or communal lands. For this to succeed, efforts must also be directed toward educating local people about conservation issues, minimizing wild mammal-people conflicts and developing compensation schemes, and managing the threat of disease. In all cases, research and monitoring of wildlife populations and threats is critical, especially in situations where sustainable harvesting is being attempted.

See Also the Following Articles

CAPTIVE BREEDING AND REINTRODUCTION • ENDANGERED BIRDS • ENDANGERED REPTILES AND AMPHIBIANS • EXTINCTIONS, MODERN INSTANCES OF • MAMMALS, BIODIVERSITY OF • MAMMALS, CONSERVATION EFFORTS FOR • MARINE MAMMALS, EXTINCTIONS OF

Bibliography

Beacham, W., and Beetz, K. H. (1998). *Beacham's guide to international endangered species*. Osprey: Beacham Publishing Corp.

Burton, J. A., and Pearson, B. (1987). *The Collins guide to rare mammals of the world*. Lexington: The Stephen Greene Press.

IUCN. (1996). 1996 IUCN Red List of Threatened Animals. J. Baillie and B. Groombridge (Eds.). Gland and Cambridge: IUCN.

Martin, P. S., and Szuter, C. R. (1999). War zones and game sinks in Lewis and Clark's West. *Conservation Biology*, 13(1): 36–45.

Medellin, R. A., and Soberon, J. (1999). Predictions of mammal diversity on four land masses. *Conservation Biology*, 13(1): 143–149.

Nowak, R. M. (1999). *Walker's Mammals of the World*. Vols. 1 and 2 (6th ed.). Baltimore and London: The Johns Hopkins University Press.

World Conservation Monitoring Centre. (1992). *Global biodiversity: Status of the earth's living resources*. London: Chapman and Hall.

ENDANGERED MARINE INVERTEBRATES

James T. Carlton
Williams College—Mystic Seaport

GLOSSARY

critically endangered A species facing an extremely high risk of extinction in the wild in the immediate future.

endangered A species not critically endangered but facing a very high risk of extinction in the wild in the near future.

euryhaline Able to live over a wide range of salinities, from brackish to fully marine waters.

planktotrophic Larval forms of invertebrates that feed on plankton to survive and grow.

stenohaline Able to live only in fully marine waters.

threatened Vulnerable, endangered, and critically endangered species.

vulnerable A species not critically endangered or endangered but facing a high risk of extinction in the wild in the medium-term future.

ALTHOUGH ONLY A FEW MARINE INVERTEBRATE SPECIES are known to be endangered, there may be hundreds to thousands of species actually at risk of extinction. The ability to determine the conservation status of such species faces considerable challenges. Habitat destruction, fisheries activities, and the introduction of exotic species are among the primary causes leading to the endangerment of marine invertebrates.

I. ENDANGERED MARINE ORGANISMS—CONSIDERATIONS AND DEFINITIONS

More marine organisms are threatened by increasing human pressures in the 21st century than ever before. Of the many enduring challenges in marine conservation science, one of the greatest is securing an objective understanding of extinction risk of individual species. With increasing attention paid in the world's oceans to the broad concept of "threatened and endangered species"—attention cascading from similar concerns in terrestrial ecosystems—the question of perceived versus actual vulnerability of many marine animals and plants is a pressing one for scientists and conservationists.

In the following treatment one group of marine organisms—the invertebrates—is considered relative to endangerment and potential extinction. In turn, given their broad range of reproductive strategies, habitats, and susceptibility to human-induced extirpation, invertebrates may serve as model systems to understand endangerment in the broader array of marine life, including algae (seaweeds), seagrasses, vertebrates, bacteria, protists, and fungi.

What constitutes endangerment in marine invertebrates? The IUCN Red List (Baillie and Groombridge, 1996) considers three categories: *critically endangered, endangered, and vulnerable* species, all grouped under the broad aegis of *threatened.* Assignment of a species to one of these three categories is based on quantitative assessments (Table I). Information required includes temporal data (for example, rate of population decline, population fluctuations), spatial data (for example, extent of occurrence, number of populations), and population data (for example, size of population, number of

adults). A species is *critically endangered* when "it is facing an extremely high risk of extinction in the wild in the immediate future"; a species is *endangered* when "it is not critically endangered but is facing a very high risk of extinction in the wild in the near future"; and a species is *vulnerable* when "it is not critically endangered or endangered but is facing a high risk of extinction in the wild in the medium-term future."

There are other scales of endangerment as well, and these are parallel to scales of extinction. *Local endangerment* occurs when a species faces extinction in a small area or habitat. *Regional endangerment* occurs when a species faces extinction in a broad geographic region. *Global endangerment* occurs when a species faces extinction everywhere. *Functional endangerment* occurs when

TABLE I

IUCN Categories and Criteria of Extinction Risk

Y	= Year
G	= Generation
o/m	= Order of magnitude
subpops	= Subpopulations

	Critically endangered	Endangered	Vulnerable
A. Declining Populations			
Population decline rate at least	80% in 10Y or 3G	50% in 10Y or 3G	20% in 10Y or 3G
Using either			
1. population reduction observed, estimated, inferred, or suspected in the past			
or			
2. population decline projected or suspected in the future based on			
a. direct observation			
b. an index of abundance appropriate for the taxon			
c. a decline in area of occupancy, extent of occurrence, or quality of habitat			
d. actual or potential levels of exploitation			
e. the effects of introduced species, hybridization, pathogens, pollutants, competitors or parasites			
B. Small Distribution and Decline or Fluctuation			
Either			
extent of occurrence	<100 km^2	<5000 km^2	<20,000 km^2
or			
area of occupancy	<10 km^2	<500 km^2	<2,000 km^2
and two of the following three:			
1. either severely fragmented (isolated subpopulations) or known to exist at a number of locations	1	5	10
2. continuing decline in any of the following:	Any rate	Any rate	Any rate
a. extent of occurrence			
b. area of occupancy			

continues

Continued

	Critically endangered	Endangered	Vulnerable
c. area, extent, or quality of habitat			
d. number of locations or subpopulations			
e. number of adults			
3. fluctuating in any of the following:	>1 o/m	>1 o/m	>1 o/m
a. extent of occurrence			
b. area of occupancy			
c. number of locations or subpopulations			
d. number of adults			
C. Small Population Size and Decline			
Number of adults:	<250	<2,500	<10,000
and one of the following two:			
1. rapid decline rate	25% in 3Y or 1G	20% in 15Y or 2G	10% in 10Y or 3G
2. continuing decline	Any rate	Any rate	Any rate
and either			
a. fragmented	All subpops 50	All subpops 250	All subpops 1,000
or			
b. all individuals in a single subpopulation			
D. Very Small or Restricted			
Either			
1. number of adults	<50	<250	<1,000
or			
2. population is susceptible	N/A	N/A	Area of occupancy <100 km² or number of locations <5
E. Quantitative Analysis			
Indicating the probability of extinction in the wild to be at least	50% in 10Y or 3G	20% in 20Y or 5G	10% in 100Y

Modified from Baillie and Groombridge (1996).

a species faces loss of its role in structuring a community (that is, influencing the diversity, distribution, or abundance of other species) or mediating energy flow in an ecosystem.

In practice, these categories may be expressed as a dichotomy between (1) species that have been demonstrably reduced to one or a few populations (whether adjacent or widespread) everywhere and thus face global extinction, and (2) species that have been demonstrably reduced in some locations but still maintain populations over a large area and thus face "only" local or regional extinction.

A final category of endangerment involves international transport and trade. Under Appendices I and II of the 1997 Convention on International Trade in Endangered Species of Wild Fauna and Flora (CITES),

in terms of marine invertebrates, all species of giant clams (family Tridacnidae), the queen (pink) conch snail *Strombus gigas*, all corals (Scleractinia), as well as other Anthozoa (Coenothecalia, Stolonifera, and Antipatharia) and certain hydrozoan corals (Milleporina and Stylasterina) are listed. Appendix I species are considered "the most endangered" and include "all species threatened with extinction which are or may be affected by trade." Appendix II includes "other species at serious risk . . . which although not necessarily currently threatened with extinction may become so unless trade is subject to strict regulation."

These CITES *trade protected* species may thus, in popular or legal writing, become *endangered* species, although many would not be so listed (as either critically endangered, endangered, or vulnerable) under

IUCN quantitative guidelines. As an example the North Atlantic coral *Lophelia pertusa*, which ranges to 1500 m depth from Scandinavia to the Mediterranean (and which can occur on the Norwegian shelf in deep water (270–310 m) in the form of coral reefs over 13 km in length, 10 to 35 m in height and up to 300 m in width) was reported in the popular press in 1999 as "endangered," citing CITES listing of all corals, after being found on North Sea oil platforms. Awarding species protected status may provide certain benefits, such as regulated fisheries, regulated trade and transport, habitat protection, and so forth, even if the species is not close to extinction per se. Alternatively, very broad application of the concept of vulnerability to species that are not immediately threatened may decrease sensitivity to the issue of potential extinction and thus increase the difficulty in affording truly endangered species adequate protection.

II. WHY IS IT HARD TO KNOW IF A MARINE INVERTEBRATE IS ENDANGERED?

Knowledge of marine biodiversity is poor and getting poorer. The names, distribution, and ecology of most marine organisms are not well known. There are fewer professional systematists and biogeographers with every passing generation. It thus difficult to assess the status of now uncommon and rare species—those that are becoming or have become endangered. And just as in terrestrial ecosystems, many species in the oceans may go extinct or become endangered before they are described. In familiar and relatively easily accessible marine environments in Pacific, Atlantic, and Gulf of Mexico seas, often more than half of the species encountered in certain groups are not known to science (Table II). Indeed, most of the species in the oceans are undescribed, even though the majority of the largest common species in shallow waters have been given Latin names. More generally, the last half of the 20th century, in parallel with *increased interest* in the scale of modern human-mediated change in the oceans, has been marked by *decreased documentation* of changes in the status, range, and natural history of marine invertebrates. Carlton *et al.* (1991) reported that a common species of intertidal snail along the New England coast became extinct about 1930, but the species remains listed as "common" in many modern seashell books. Gathering the precise data required under the IUCN guidelines (Table I) thus may often pose considerable challenges.

As a result, Carlton *et al.* (1999) were able to document only four species of marine snails that have become extinct in historical time. Roberts and Hawkins (1999) identified no further candidates among marine invertebrates for clear evidence of extinction. Turgeon *et al.* (1998) list no North American marine mollusks as endangered or threatened. Does this situation mean that marine invertebrates are generally "safe" or "immune" from extinction, or does this reflect our level of

TABLE II

The Extent of Undescribed Marine Invertebrate Diversity

Location and general habitat	Invertebrate group	Common name	Number of undescribed species out of total collected
Gulf of Mexico: Shallow shelf sediments	Crustacea: Copepoda Harpacticoida	Copepods	Up to 27 of 29
New Guinea: Shallow lagoon	Mollusca: Gastropoda: Opisthobranchia	Sea slugs	310 of 564
Philippines: Shallow waters of one island	Mollusca: Gastropoda Opisthobranchia	Sea slugs	135 of 320
Georges Bank: Shallow shelf sediments	Annelida: Polychaeta	Worms	124 of 372
Hawaiian Islands: Coral reef sediment on one island	Annelida: Polychaeta	Worms	112 of 158
Great Barrier Reef: Shallow waters of two islands	Platyhelminthes: Turbellaria: Polycladida	Flatworms	123 of 134

Modified from National Research Council (1995).

knowledge of the conservation status of marine life? Modern authors have concluded that the latter is now the case, and that a "tip of the iceberg" understanding of marine endangerment now exists.

III. WHAT CAUSES THE ENDANGERMENT OF MARINE INVERTEBRATES?

Both natural and human-mediated changes to the environment can lead to the endangerment and extinction of marine organisms. Examples of natural changes include volcanic eruptions destroying endemic island marine communities, tropical storm activity (such as hurricanes and monsoons) causing destruction of reef communities, and other severe weather systems, such as sustained winter freezing impacting temperate shore communities. While many cases of these natural impacts have been documented, none have been linked to the extinction or endangerment of any marine invertebrate species (although such may have occurred).

An example of the ocean-wide demise of a marine invertebrate by what may have been a natural series of events is the case of the eelgrass limpet *Lottia alveus alveus*. This snail, which occurred from Labrador to New York, lived solely on the blades of the eelgrass *Zostera marina*. Between 1930 and 1933, 90% of the eelgrass in this region died due to a disease caused by the slime mold *Labyrinthula zosterae*. The last known population of the limpet occurred on Mt. Desert Island, in the Gulf of Maine, where in 1929 it occurred by the thousands. *Lottia* was a stenohaline species; however, its host, the eelgrass (with a broader physiological repertoire than the limpet) survived in brackish-water refugia below the salt tolerance of the slime mold. A distinct subspecies—or sibling species—of *Lottia alveus* survives in the North Pacific Ocean, from Alaska to British Columbia. It may be noted that the possibility remains that *Labyrinthula* was introduced to North America in some manner.

In an important lesson in understanding the relative susceptibility of marine organisms to extinction, another North American Atlantic eelgrass specialist, the sea slug *Elysia catulus*, did not become extinct, because it, like the eelgrass, is euryhaline. Disease events that impact the host of specialized species could, in theory, lead to species endangerment as well, and it may be that for a period of time *Elysia* would have been regarded as a threatened species until the reexpansion and recolonization of eelgrass over the following decades.

Human-induced pressures on marine communities include fishing, chemical pollution, destruction of habitat, invasions of nonindigenous (exotic) species, and atmospheric chemical alterations leading to global climate change. We know a great deal about many of these pressures. However, our record of changes in the distribution and abundance of marine life in the 19th and 20th centuries is inverse to our record of the rapid increase of these anthropogenic activities.

Overall, habitat destruction, fisheries (both overexploitation and the secondary impacts of fishing) and the invasion of nonindigenous (exotic) species are held to be the primary causes of endangerment to marine organisms. The global extinction of three species of marine mollusks, the southern California rocky shore limpet *Collisella edmitchelli*, the Chinese periwinkle *Littoraria flammea*, and the southern California mudflat hornsnail *Cerithidea fuscata*, have all been linked to habitat destruction. While overfishing has not yet been linked to the endangerment of a marine invertebrate, vast declines in marine invertebrate populations around the world due to fishing and fishing activities are increasingly recorded. The North Atlantic whelk *Buccinum undatum* is now absent in areas of the southern and central North Sea and the Dutch Wadden Sea where it was abundant in the 1970s. Overfishing, combined with a disease of unknown origin, have led to such vast declines in commercial sponges in the Mediterranean Sea that some workers consider certain species to be on the "brink of extinction." Here again *regional extinction* needs to be distinguished from *global extinction*, although unrelenting fisheries within an enclosed basin could lead these extinction states to synonymy.

Trawling on the sea floor (dragging nets, chains, bags, scoops, and a plethora of similar devices on the bottom) represents a combination of both habitat destruction and fisheries and has extended the reach of human perturbation well onto the continental shelves of the world, far from the marine habitats that have been classically viewed as those able to be destroyed.

While trawling is particularly a concern of the 20th and 21st centuries, the roots are deep, as they are with many human-induced pressures on the oceans. As early as 1376 the Commons petitioned the King of England "that the great and long iron of the wondyrchoun [a net] runs so heavily and hardly over the ground when fishing that it destroys the flowers of the land below water there" (from the Records of the Parliament in the Reign of Edward III). It seems plausible, given some six and more centuries of trawling in selected regions of the world, that certain invertebrate species would have been heavily impacted. However, no reports of endangerment or extinction are in hand, a situation resulting, as discussed earlier, from a lack of knowledge of the distribution of marine invertebrates on the sea floor (and noting again the number of undescribed species found in shelf habitats [Table II]).

Underscoring the probability that marine benthic invertebrates may be endangered by consistent habitat destruction (and emphasizing the difficulty of recognizing such species) is the discovery that a large fish, the barndoor skate *Raja laevis*, formerly common and widespread in the Northwest Atlantic Ocean, virtually disappeared in the 1990s from benthic trawls, when it used to be common as trawl bycatch. Given that the demise of *Raja*, a 1.5 m long fish, was not recognized until 1999, the chances of detecting the loss of much smaller organisms such as most invertebrates is all the more unlikely.

IV. WHAT WE KNOW: EXAMPLES OF ENDANGERED MARINE INVERTEBRATES

The IUCN "Red List" lists 13 marine invertebrate species as high-profile threatened species: one species is considered endangered, two critically endangered, and ten vulnerable. Evaluating whether these species (10 mollusks, 1 crab, 1 sea anemone, and 1 seafan) are globally threatened is often a challenge, as no published data are cited for any listing. Often species are suggested for *global listing* based on *local or regional extinctions* of populations or based on the rate at which local or regional populations are discovered to be in decline. While such observations may be a harbinger for what may eventually become global concerns, the data may be lacking to understand the full extent of endangerment. As examples, the status of 4 of these 13 species is as follows:

- The tiny Northeastern Pacific crab *Parapinnixa affinis* is listed by IUCN as endangered. It lives in bays in shallow, perhaps largely intertidal, mud bottoms, as a commensal in the tubes of terebellid polychaete worms. It was first collected in 1895 in what was to become Los Angeles–Long Beach Harbors, and as late as 1932 it was collected at several locations between Los Angeles and San Diego. No further material from southern California has surfaced. Given the extraordinary modifications of these embayments in the 20th century, with little or no semblance of original habitat often remaining, the demise of this (and many other) spe-

cies would not be unexpected. However, it may remain common in similar habitats of Baja California, the Gulf of California, and further south, sites that remain largely unexplored. *P. affinis* was "rediscovered" in one such location, in Tortugas Bay, southern Baja California, in 1987. It thus may be an example of *local* and *regional extinction.*

- The salt marsh sea anemone *Nematostella vectensis* is listed by IUCN as "vulnerable," also a high-profile endangerment category. This species occurs, often in large numbers, in salt marsh pools in the Northwest Atlantic Ocean from Nova Scotia to the Gulf of Mexico and in the Northeast Pacific Ocean from Washington to California. It also occurs in England, its type locality (but not apparently otherwise in Europe); given its widespread distribution in North America, it may be introduced in the Eastern Atlantic Ocean. In England it is considered rare (based on localized populations and declining habitat, although where found it may be abundant) and is given protected status. The *habitat* of this anemone has, throughout portions of its range, been fundamentally altered (if not completely removed), and thus the *abundance* of this anemone has inevitably been reduced, but it remains common in many areas. It too would appear to be an example of *local extinction.*

- The giant clam *Tridacna derasa* is listed (along with other species of *Tridacna*) as vulnerable. These clams characterized many shallow-water reef environments up to the mid-20th century, before commercial removal caused their widespread disappearance and population declines. There is thus no doubt that both the range and abundance of a number of *Tridacna* species have been severely modified, and indeed certain populations are extinct. However, *T. derasa*, as an example, remains widespread throughout a vast area of Australasia and the Indo-Pacific and is widely used in mariculture.

- The sea fan *Eunicella verrucosa* is listed as vulnerable. It lives at sublittoral depths of 10 m and deeper on sheltered rocky bottoms. The species is widely distributed from northern Europe to Africa, and it remains common in a number of sites.

All of these species have suffered at the hands of habitat destruction (*Parapinnixa, Nematostella*) or human exploitation (*Tridacna, Eunicella*). Because of knowledge lacunae and provincial perspectives, they illustrate the difficulty in ranking threat status and in attempting to assess whether *local and regional extinctions* imply impending *global extinction.* There are many species, however, that may be certainly endangered. We discuss examples here (Tables III and IV).

The invertebrates—and all species—that live in marine habitats most influenced by human activity have sustained the most damage. It is in these habitats that endangered species are especially to be sought. These include salt marshes, mangroves, estuaries, lagoons, beaches, dunes, the supralittoral (maritime or strand line) zone, and seagrass beds. Especially vulnerable are habitats within regions that are now fragmented remnants of their previous extent. It is estimated that approximately half of the world's salt marshes and mangroves have been destroyed and along with them, of course, an untold number of species. Seagrasses— particularly the estuarine species, such as the eelgrass *Zostera*, turtle grass *Thalassia*, widgeon grass *Ruppia*, shoal and star grasses *Halodule*, manatee grass *Syringodium*, and others that support unique species of snails, chitons, and other invertebrates precisely adapted to their blades—serve as particularly compelling harbingers of threat to coastal marine biodiversity.

Several examples are available of endangered species living where expanding human populations impinge on these fragile shoreline habitats. Two small sea slugs,

the anaspidean sea hare *Phyllaplysia smaragda* and the sacoglossan slug *Stiliger vossi*, known only from Florida lagoons, have been missing for many years. *Phyllaplysia* was first described in 1977 and last found in 1981. It was only known from the Indian River Lagoon of eastern Florida. It ate epiphytic algae growing on the manatee grass *Syringodium*, a plant subsequently obliterated from the sea slug's type locality (although the plant remains widespread elsewhere in southern waters, including Florida and the Caribbean, and in many of which locations *Phyllaplysia* has been sought unsuccessfully). *Stiliger*, another herbivore, was first and last collected in 1960 in Biscayne Bay, southeast Florida, despite years of subsequent searching. While subtropical Caribbean lagoons and bays remain to be thoroughly explored, there appears to be little doubt that these slugs have been and remain at risk because of living in habitats proximal to rapidly expanding human populations.

The false limpet *Siphonaria compressa* is represented by what is probably one last remaining population in the Langebaan Lagoon on the west coast of South Africa. It is listed by IUCN as critically endangered. Little is known of its life history (although it may have nonplanktonic development), population dynamics, or environmental tolerances. It is restricted, however, to living on the eelgrass *Zostera capensis*, the width of whose narrow blades it matches (thus the name of the limpet). In this regard, it is an ecological equivalent of the northern hemisphere true limpet *Lottia alveus*, which similarly was, in the North Atlantic Ocean, restricted to the eelgrass *Zostera marina*. Although Langebaan Lagoon is part of a national park and is a site identified under the Ramsar Convention on Wetlands of International Importance, regional harbor development leading to pollution, the introduction of nonindigenous species, severe weather events, disease, or other phenomena could threaten this last stand of *Siphonaria*.

An unusual example of an endangered taxon in a coastal habitat is the small (1.0 cm) talitrid amphipod crustacean (beach hopper) *Transorchestia enigmatica*. It is an intertidal species, eating decaying plant material, known only from a single beach approximately 0.8 km in length on the shores of an estuarine lagoon (Lake Merritt) in the middle of a densely urbanized region (Oakland) of San Francisco Bay. It is closely related to the South Pacific Ocean species *Transorchestia chiliensis*, which is known from New Zealand and Chile. *Transorchestia enigmatica* is presumed to have been carried into San Francisco Bay, perhaps in the 19th century, in the dry (beach) ballast of ships from the southern hemisphere. However, no populations of *T.*

TABLE III

Systematic Classification of Examples of
Endangered and Extinct Marine Invertebrates

Phylum Mollusca	
Class Gastropoda	
Order Patellogastropoda	
Lottia alveus alveus	Eelgrass limpet
Order Archaeogastropoda	
Haliotis sorenseni	White abalone
Order Anaspidea	
Phyllaplysia smaragda	Emerald seahare
Order Sacoglossa	
Stiliger vossi	Sea slug
Order Basommatophora	
Siphonaria compressa	False limpet
Class Bivalvia	
Order Mytiloida	
Mytilus trossulus	Blue mussel
Phylum Crustacea	
Class Malacostraca	
Order Amphipoda	
Transorchestia enigmatica	Beach hopper

TABLE IV

Which Marine Invertebrates May Be at Risk of Being Endangered?

Risk group	Examples
Species with restricted distributions	
Species restricted to lagoons, estuaries, isolated marshes and mangroves, remnant seagrass beds, and other limited or now-restricted coastal habitats	Seahare *Phyllaplysia smaragda*
	Seaslug *Stiliger vossi*
	Limpet *Siphonaria compressa*
	Amphipod *Transorchestia enigmatica*
	Many other species in coastal habitats, including *estuarine seagrass stenotypic endemics.*
Species with limited distributions under heavy extraction pressure or under pressure from related extraction activities (such as bottom trawling)	White abalone *Haliotis sorenseni*; selected benthic invertebrates in areas of heavy trawling; endemic coral reef species.
Short-range endemics with nonplanktotrophic development, especially those under extraction pressure, fisheries activities, or at risk from coastal development and pollution	Many invertebrates species in numerous phyla, including short-range cone snails (*Conus* spp.) listed by IUCN; endemic coral reef species (all invertebrates); species restricted to islands, offshore banks, and sea mounts.
Species restricted to extinguishable habitat	
Species restricted to an extinguishable (but not endangered) living host, which could itself be extirpated by some means (such as a provincial-wide or ocean-wide disease)	Eelgrass limpet *Lottia alveus* (extinct in the Atlantic Ocean)
Species impacted by exotics	
Species impacted by the introduction of exotic (nonindigenous) species	Blue mussel *Mytilus trossulus* in southern California (a *regional* and *local extinction*)
Species associated with other endangered species	
Commensals, symbionts, and parasites of endangered marine vertebrates, invertebrates, and plants	*Species associated with*
	* Mammal: *Phocoena sinus* (Gulf of California vaquita)
	* Fish: *Syngnathus affinis* (Texas pipefish)
	* Fish: *Raja laevis* (barndoor skate)
	* Bird: *Oceanodroma macrodactyla* (Guadalupe storm petrel)*
	* Other endangered vertebrates
	* Endangered invertebrates
	* Endangered plants

* If not already extinct.

chiliensis in either Chile or New Zealand are identical morphologically to those found in California; the population in San Francisco Bay represents a unique sibling morphotype.

Among various possibilities to explain this phenomenon are (1) that the morphological distinctiveness of this population has originated from reduced genetic heterozygosity (due to founder effect) followed by genetic drift and mutation in allopatry, (2) *T. enigmatica* could represent a morphotype that remains to be discovered in Chile or New Zealand, and (3) *T. enigmatica* may have become extinct in the South Pacific Ocean, given that it almost certainly originated in the early harbors of Chile or New Zealand, which themselves have undergone extensive modifications and lost much of their own original littoral habitats.

Although occasional specimens of this shore-dwelling amphipod are found after storms around the lake margin or in a small canal connecting the lake to San Francisco Bay, the only reproducing population is on the one beach noted. The construction of a seawall, extensive gardening activity on the bluff above the beach, or other physical perturbation could render *T. enigmatica* extinct.

Thus the seahare *Phyllaplysia smaragda*, the sea slug *Stiliger vossi*, the false limpet *Siphonaria compressa*, and the beach hopper *Transorchestia enigmatica*, as well as the sea anemone *Nematostella vectensis* and the crab *Parapinnixa affinis* discussed earlier, were or are all associated with estuarine and lagoonal habitats. A great many species in similar habitats may now be at equal or greater risk of extinction.

Another risk group are those species with relatively limited distributions (even if such may be an ocean basin!) under heavy, inexorable extraction pressure. Under this category we find one of the most compelling examples of endangerment among invertebrates in the oceans.

The white abalone, *Haliotis sorenseni*, is a large gastropod (up to 22 cm long) which formerly ranged from Point Conception in southern California to the islands of central Baja California, on low-relief rocky reefs in sublittoral waters to depths of about 65 m. Its modern postfisheries range, especially its northern limits, is not known. A number of species of abalone, a widely sought after mollusk, occur along the Pacific coast of North America and have since the 1850s attracted sport and commercial fisheries both for the shell and the meat. Marked depletion of certain species on the California coast was noted by the turn of the 20th century and regulatory laws were passed in 1903 to set size limits. Keep noted in 1904 that the "persistent warfare" against the red abalone *Haliotis rufescens* had rendered large specimens rare compared to about 1884. Edwards (1913) reported that the green abalone *Haliotis fulgens* and the pink abalone *Haliotis corrugata*, which about 1893 occurred piled four and five deep on rocks near Avalon, Santa Catalina Island (southern California), were by 1913 entirely gone.

The white abalone *H. sorenseni* was not described as a distinct species until 1940. It is a deeper water species that entered the commercial fishery only later, after shallow water abalone were exhausted. Thus there are no comparable early records of depletion. However, in remarkable parallel with the reports of Edwards on the green and pink abalones are the much later findings of Davis *et al.* (1998) who consider white abalone to be on the brink of extinction. Davis and colleagues searched more than 107,000 square m of white abalone habitat on the California Channel Islands, the species' historical center of abundance. By the early 1990s at depths of 25 to 42 m (where in the 1970s mean densities were 2000 to 10,000 abalone per hectare) mean white abalone densities had decreased to 1.6 ± 0.5 per hectare. In 1996 and 1997, at depths of 27 to 67 m, densities had further dropped to extremely low numbers, 1.0 ± 0.4 per hectare.

Commercial harvests virtually ceased after the 1970s, but no fishery-independent assessment of white abalone populations was made until 1992 to 1993; in the meantime the fishery remained open until 1996, based on a minimum harvest size of 15.3 cm and a closed season during spawning. All surviving white abalone known in the wild are adult and dying of old age. Davis *et*

al. suggest that while spontaneous recovery is highly unlikely, even in the absence of any further harvesting, a mariculture program may be able to save the species from global extinction.

White abalone produce planktotrophic larvae. However, as the last known recruitment event occurred in the late 1960s or 1970s, and with population densities now far below 1 per square m, successful fertilization may now be impossible, and complete reproductive failure of the species appears to be at hand. Unless unknown populations exist in refugia in deeper water (abalones are rarely if ever seen below 67 m) or populations of sufficient size exist to the south in Mexican waters, the total size of the white abalone population may now be fewer than 1000 individuals.

The white abalone offers the following lessons about the pathway to endangerment: a combination of restricted habitat (a narrow sublittoral zone), the ability to discover and remove most individuals by advanced diving technology, a severe disruption of the life history of the species such that sporadic recruitment (that likely relied on sufficient adult densities) was interrupted, and the lack of a fishery management framework have all led to the demise of this once common open ocean species. No juvenile abalone have been found in many years. An intensive, focused fishery on the white abalone began only about 1965 but vanished 35 years later, perhaps with the mollusk itself.

There are several other guilds of at-risk marine organisms as well. Mentioned earlier were species restricted to an extinguishable (but not necessarily itself endangered) living host, which could be impacted by a disease or other agent. The demise of the eelgrass limpet *Lottia* falls under this category. Endangered species would also include, of course, those invertebrates that are commensals, parasites, and symbionts of other endangered marine organisms (Table IV). Finally, the invasion of exotic species may lead to at least local or regional extinctions (Geller, 1999). The native mussel *Mytilus trossulus*, once abundant in southern California both historically and in native American shell middens, was rendered virtually regionally extinct by the invasion in the 20th century of the Mediterranean look-alike mussel *Mytilus galloprovincialis*.

V. WHAT WE DO NOT KNOW: THE POTENTIAL SCALE OF ENDANGERMENT

Given the extent of the modification, destruction, devastation, and extirpation that has characterized coastal marine environments over the past 1000 and more years, hundreds or thousands of species of marine invertebrates may in fact be endangered, as opposed to the few cases reviewed here or listed under IUCN. For example, it has been estimated that given that a minimum of 5% of the world's coral reefs have been degraded to a nonreef state, and based on estimates of coral reef biodiversity per unit area, as few as 1,000 species of coral reef invertebrates may have already become extinct in modern times (although, as with many tropical forest ecosystems, we have no records of these species, many of which may not have been described before going extinct). It follows that many thousands of coral reef invertebrates may be endangered as well, given that coral reef destruction continues largely unabated in many parts of the world.

More generally, the demise of the white abalone offers a striking counterpoint to the ancient assumption of the invulnerability of marine life to exhaustion by human extraction. Lamarck in 1809 wrote,

> Animals living in the waters, especially the sea waters . . . are protected from the destruction of their species by man. Their multiplication is so rapid and their means of evading pursuit or traps are so great, that there is no likelihood of his being able to destroy the entire species of any of these animals.

These words were echoed by Byron in 1818 in a similar fashion,

> Man marks the earth with ruin, his control
> Stops with the shore.

These turn-of-the-19th century views of the oceans were clearly incorrect and at the turn of the 21st century the reverse now appears to be true.

See Also the Following Articles

ENDANGERED FRESHWATER INVERTEBRATES •
ENDANGERED TERRESTRIAL INVERTEBRATES •
INVERTEBRATES, MARINE, OVERVIEW •
MARINE ECOSYSTEMS

Bibliography

Baillie, J., and Groombridge, B. (1996). *1996 IUCN Red List of Threatened Animals.* IUCN, Gland, Switzerland.

Byron, G. G. (1818). Apostrophe to the ocean. In *Childe Harold's Pilgrimage.* John Murray, London.

Carlton, J. T. (1993). Neoextinctions of marine invertebrates. *American Zoologist* **33**, 499–509.

Carlton, J. T., Geller, J. B., Reaka-Kudla, M. L., and Norse, E. A. (1999). Historical extinction in the sea. *Annual Review of Ecology and Systematics* **30**, 515–538.

Carlton, J. T., Vermeij, G. J., Lindberg, D. R., Carlton, D. A., and Dudley, E. C. (1991). The first historical extinction of a marine invertebrate in an ocean basin: The demise of the eelgrass limpet *Lottia alveus. Biological Bulletin* **180**, 72–80.

Casey, J. M., and Myers, R. A. (1998). Near extinction of a large, widely distributed fish. *Science* **281**, 690–692.

Clark, K. B. (1994). Ascoglossan (= Sacoglossa) molluscs in the Florida Keys: Rare marine invertebrates at special risk. *Bulletin Marine Science* **54**, 900–916.

Davis, G. E., Haaker, P. L., and Richards, D. V. (1998). The perilous condition of white abalone *Haliotis sorenseni*, Bartsch, 1940. *Journal of Shellfish Research* **17**, 871–875.

Edwards, C. L. (1913). The abalones of California. *Popular Science Monthly*, June 1913, 532–550.

Gaino, E., and Pronzato, R. (1992). Disease and overfishing drive commercial sponges to the brink of extinction in the Mediterranean basin. Bollettino dei Musei e degli Istituti Biologici, Universita di Genovà, 56–57: 218–224.

Geller, J. B. (1999). Decline of a native mussel masked by sibling species invasion. *Conservation Biology* **13**, 661–664.

Herbert, D. G. (1999). *Siphonaria compressa*, South Africa's most endangered marine mollusc. *South African Journal of Science* **95**, 77–79.

Keep, J. (1904). *West American Shells.* The Whittaker and Ray Company Publishers, San Francisco.

Lamarck, J.-B. (1809). *Philosophie Zoologique.* Dentu l'Auteur, Paris.

National Research Council (1995). *Understanding Marine Biodiversity: A Research Agenda for the Nation.* National Academy Press, Washington, DC.

Roberts, C. M., and Hawkins, J. P. (1999). Extinction risk in the sea. *Trends in Ecology and Evolution* **14**, 241–246.

Turgeon, D. D., *et al.* (1998). *Common and scientific names of aquatic invertebrates from the United States and Canada: Mollusks*, 2nd ed. American Fisheries Society Special Publication 26. Bethesda, Maryland.

Watling, L., and Norse, E. A. (1998). Special section: Effects of mobile fishing gear on marine benthos. *Conservation Biology* **12**, 1178–1179 (followed by 7 papers on the topic, pp. 1180–1240).

ENDANGERED PLANTS

Thomas J. Stohlgren
Colorado State University

GLOSSARY

critical habitat Habitat of a threatened or endangered species that is itself threatened by destruction, disturbance, modification, or human activity, potentially resulting in a reduction in the numbers, distribution, or reasonable expansion or recovery of that species.

endangered species Those species in danger of extinction throughout all or a significant portion of their range.

endemism Condition in which a species' distribution is restricted to a given geographic region.

rare species Species with small world populations that are not presently listed as endangered or vulnerable, but are at risk because of their small population size.

threatened species Those species that are likely to become endangered in the foreseeable future throughout all or a significant portion of their range.

THE GROWING PUBLIC CONCERN about endangered plants is well founded. It is estimated that there are between 235,000 (Raven *et al.*, 1986) and 270,000 (IUCN, 1998) species of vascular plants (including ferns and fern allies, gymnosperms, and flowering plants) in the world today; two-thirds of these species are found in the tropics (Raven *et al.*, 1986). The World Conservation Union (IUCN, 1998) currently lists 33,798 species (12.5% of the world's flora) as threatened, and 6,522 species of plants (2.4% of the world's flora) as endangered. Threatened plants are found among 369 plant families in 200 countries. The U.S. Fish and Wildlife Service reports 540 endangered species and 130 threatened species of flowering plants in the United States (U.S.) as of November 30, 1998.

I. INTRODUCTION

Extinction has always played an important role in nature: 95 to 99% of all species that ever existed are now extinct. It is the current rapid rate of extinction that has many plant ecologists worried. Worldwide, as many as 654 species of plants have gone extinct since A.D. 1600 (Heywood, 1995). In this time period, about 110 plant species may have gone extinct in Hawaii according to records of the U.S. Fish and Wildlife Service and The Nature Conservancy. Estimates of plant extinctions in the continental United States range from below 25 (IUCN, 1998) to above 90 (Davis *et al.*, 1986; Flather *et al.*, 1994). The discrepancy in estimates is due to incomplete systematic surveys, limited monitoring activities, and poor information on viable seed banks. Estimates of extinctions in other countries include 71 plant species in Australia, 53 in South Africa, 47 in Mauritius, and 23 in Cuba. Three endemic species of ebony (*Diospyros* spp.) went extinct in the Mascarene Islands before they were described. Nuttall's mudwort (*Micranthemum micranthemoides*) had been recorded in Delaware, Maryland, New Jersey, New York, Pennsylvania, Virginia, and the District of Columbia, but, despite extensive searches it has not been seen since 1941. The single individual of *Dicliptera dodsonii*, which clings to life in western Ecuador (Gentry, 1986), may be the next victim of accelerated extinction.

Estimates of future plant species extinctions vary widely. Raven *et al.* (1986) estimated that about 40,000 tropical plant species may go extinct in the wild within the next several decades. The New York Botanical Garden suggests that about 700 of the more than 2000 species of threatened and endangered plant species in the United States may be extinct in the next 10 years. The Center for Plant Conservation reported that 680 U.S. plant species were critically endangered, with approximately 253 species estimated to become extinct in 5 years and 427 species to become extinct in 10 years (based on unpublished data in 1988). Today, more than 10 years later, there is little evidence of these mass extinctions, but the concern may still be well founded in several areas.

Hawaii provides a good example of endangered plants. Up to 47% of the Hawaiian flora may meet the criteria for official listing by the U.S. Fish and Wildlife Service, with as many as 16% immediately threatened with extinction (see the Missouri Botanical Garden website listed in the Bibliography). The genus *Hibiscadelphus* in Hawaii includes only six species (with two extinct species) with a total of 14 live individuals, and one species with only one individual (Cody, 1986). The Hawaii Endangered Plant Task Force, which includes many federal, state, and non-government partners, now tracks 597 threatened, endangered, and rare plant species.

There is a growing backlog of candidates for listing as endangered species. Between July, 1976, and August, 1992, about 21 species per year were added to the U.S. endangered species list. In the second half of that 16-year period, 33 species per year were added to the list (Flather *et al.*, 1994). More than 200 plant species are being petitioned for future listing in the United States.

It has been said that "all species are rare somewhere" (Orians, 1997), because most plant species have larger populations in well-suited habitats and smaller populations in marginal or sub-optimal habitats. Many threatened vascular plants in Finland were found to be rare largely because the species were in marginal rather than suitable habitat (Cropper, 1993). Along with the increasing threats of rapid environmental change, habitat loss, contaminants, and invasive exotic species, we may face an uphill battle in protecting our treasured botanical resources in these sub-optimal habitats.

Maintaining biodiversity requires a considerable understanding of rarity and the processes and stresses that promote the endangerment of certain plant species. This article reviews the kinds of rarity, patterns of endangerment, causes of endangerment, consequences of rarity, and legal mandates to protect endangered plant species. It then provides selected examples of endangered plants and concludes with the management and social implications of protecting endangered plants.

II. KINDS OF RARITY

There are many published definitions of rarity. In an extreme example, DuMond (1973) stated nine criteria for rarity, including species that are: (1) found out of its expected context; (2) particularly subject to extinction or severe reduction in total population size by human activities that have already caused a significant population reduction; (3) found only in a very specific habitat of limited occurrence; (4) thought to be a relict of a no-longer extant vegetation association; (5) an indicator of a unique extant vegetation association; (6) recognized as an example of a wide, disjunction pattern; (7) at its natural distribution limits within the area in question; (8) known to be introduced and has become naturalized only on a very small scale; or (9) does not consistently occur as a member of any particular natural plant community. Gaston (1997) described rarity in another way, noting that rare species can be delimited on the basis of one, two, or at most a few of the following variables: abundance, range size, habitat specificity (habitat occupancy), temporal persistence (e.g., taxon age), threat (probability of, or time to, extinction), gene flow, genetic diversity, endemism, and taxonomic distinctness. Rare species in the IUCN Red Book (1998) are described in terms of population size rather than an assessment of extinction risk, which is reserved for threatened and endangered status. The Nature Conservancy and associated network of Natural Heritage Programs rate species endangerment based on five levels of global rarity (i.e., G1, G2, G3, etc.), national rarity (N1, N2, etc.), and state rarity (S1, S2, etc.).

The kinds of rarity that concern conservation biologists relate specifically to the potential vulnerability to extinction. Commonly recognized kinds of rarity (Rabinowitz, 1981; Cropper, 1993) include species with:

- Small populations, large geographic range, and occurrences in several habitats. For example, American chaffseed (*Schwalbea americana*) had a historical distribution from Mississippi to Massachusetts. It resides now in 20 relatively small populations in five states, with the greatest populations in South Carolina. It can survive in several vegetation types as long as there is enough light in the understory. Fire suppression, which allows for succession and canopy closure, threatens several *Schwalbea* populations. Another example is *Hypochoeris maculata*, which has small populations over a wide range in Great Britain.
- Large populations, large geographic range, but occurrences in specific habitats. For example, several grassland species of *Calochortus* (lilies) in California have large populations following wildfires. They were thought to be far more widespread prior to a century of fire suppression and encroachment of shrubs into grassland habitats. Sparse populations in a large geographic range but in specific habitats. For example, *Psilotum nudum* grows only on rocky outcrops and always in sparse populations, but it occurs in Australia, New Zealand, and Easter Island (Cropper, 1993).
- Small geographic range, but locally abundant in specific habitats. For example, the 48 endemic species of Bignonaceae in Amazonia have a total home range of only a few thousand square kilometers, but some of these species are locally abundant in restricted habitats (Gentry, 1986). A specific example in Scotland is *Primula scotica*, which has a tiny range with a few large populations.
- Small geographic range, and sparse in specific habitats. This may be the most critical type of rarity, for small populations restricted to small and specific habitats are highly vulnerable to extinction. For example, the scouring of a new stone quarry in South Africa ended life in the wild for *Moraea loubseri*, a small, sparsely population iris (Stermer, 1995).
- Small populations of concern in any region despite populations elsewhere. For example, many floristically poor northern European countries typically have 200 or more plant species listed as threatened or endangered. However, many of these species in-

vaded broadly since the last Ice Age, and they are buffered from extinction with widely scattered populations in several countries (Davis *et al.*, 1986).
- Small populations of new species. For example, a newly discovered species of *Calyptranthes* from El Yunque, Puerto Rico, has a population of four trees, and *Auerodendron pauciflorum* from Quebradillas and *Calyptranthes luquillensis* from the Luquillo Mountains have only five individuals each (Cody, 1986).

Generally, species may be ranked for protection based on overall rarity, magnitude of potential threats to populations, and the immediacy of the threats. Thus, a monotypic genus with high-magnitude and imminent threats might receive a higher priority ranking than a subspecies with moderate or non-imminent potential threats.

III. ENDANGERMENT PATTERNS

Slightly over 90% of threatened plants are single-country endemics. Species with restricted ranges face the greatest threat of extinction. It follows that endemic plant species on islands appear to be the most highly endangered. About 95% of the plant species on the Canary Islands are endemic, and 50% are considered endangered (Cody, 1986). *Senecio heritieri*, for example, is restricted to a small area of rocky slopes on the south coast of the island of Tenerife, one of the Canary Islands. On Crete, two-thirds of the 155 endemic species are endangered (Cody, 1986). Compare these numbers to those of the entire United States, where there are about 20,000 plant species of which 2,050 species are rare and threatened.

About half the plant species in Mediterranean climate areas (parts of California, South Africa, Australia, and the Mediterranean basin) are narrow endemics that dominate the threatened and endangered species lists in their various countries (Fig. 1) (Davis *et al.*, 1986). California, for example, contains 669 endemic species of the 2,050 species on the U.S. rare and threatened list (Davis *et al.*, 1986). Chile has 50 forest tree species; 47 of these species are endemic and 38 species are listed as endangered, vulnerable, or rare.

Over half of the rare and endangered plants of the continental United States grow within the borders of 12 western states (Arizona, California, Colorado, Idaho, Montana, Nevada, New Mexico, Texas, Utah, Oregon, Washington, and Wyoming; Fig. 2) (Flather *et al.*, 1994). However, among U.S. states Hawaii is the hot

FIGURE 1 Distribution of rare plants in the world. Data from the IUCN Red List for 1997 (IUCN, 1998).

FIGURE 2 Distribution of rare plants in the continental United States. The "top 10" continental states for rare plants include six western states, and California clearly dominates. [From C. H. Flather *et al.*, *BioScience* **48**(5), pp. 365–376. ©1998 American Institute of Biological Sciences.]

spot of endangerment: about 950 native plant species of Hawaii are extinct, endangered, or threatened (Raven *et al.*, 1986).

Endangerment patterns vary by habitat type. Flather *et al.* (1994) reported that 44% of threatened and endangered plants in the continental United States are found in rangelands. Twenty-five percent of endangered plant species were associated with barren land or rocky habitats. However, wetlands, which comprise only 5% of the land base, contain 15% of the listed species.

Endangerment patterns also vary by region and land use characteristics. In northern California, southern Nevada/Sonoran Basin, and the eastern Gulf Coast, >50% of the listed species are associated with wetland habitats. U.S. Department of Defense lands, which comprise only 3.4% of federally administered lands, contain 26% of the listed species. The high number of rare plants in the U.S. Southwest when combined with rapid development and land use change, leads to special problems for the western states that wish to conserve their floral heritage.

Worldwide, Australia and New Zealand have attracted much attention. Official listings show 1,931 endangered plant species in Australia and 22 species in New Zealand (Stermer, 1995). Australia has already lost 117 plant species to extinction in modern times (Davis *et al.*, 1986), and has 3,329 plant species that are considered rare or threatened (6.2% of the flora). An estimated 993 species are at risk of extinction in the next 50 years (Cropper, 1993).

Some plant families and genera are more prone to extinction than others. Twenty vascular plant families have at least 50% of their species threatened (e.g., Brunelliaceae, 91.9%; Zamiaceae, 88.9%; Araucariaceae, 78.9%; Taxaceae, 75%; Limnanthaceae, 72.7%). In Australia, species-rich genera such as *Acacia, Eucalyptus*, and *Grevillia* contain a proportionally greater number of rare species than do species-poor genera. Gymnosperms, containing relatively ancient species, may be less adapted to today's rapidly changing environment (IUCN, 1998). Cacti in the United States are particularly prone to extinction: 72 of 268 native species are very rare (Benson, 1982). In California, shrubs and subshrubs of the species-rich genera *Arctostaphylos, Ceanothus, Eriogonum*, and *Ribes* have a higher proportion of threatened and endangered plants relative to their proportion in the flora (Cody, 1986).

Curiously, some rare plant habitats may have human origins. Some moorlands, or blanket bogs, in western Europe were created around 7,700 years ago by human activities of burning and grazing. These heathlands and grasslands today support many endangered plant species.

IV. CAUSES OF ENDANGERMENT

There are several natural causes of rarity that lead to endangerment. Ancient taxa, like cycads, are thought to be prone to extinction as they cling to increasingly shrinking relict habitats. New taxa, resulting from recent speciation, are thought to be susceptible to extinction because they have not had time to spread in distribution to reduce risk. Rare species generally lack an ability to rapidly colonize areas and they are often poor competitors. Many rare species are edaphically restricted to specific soils or geology. However, based on recent rates of extinction, natural causes of rarity (individually or combined) are no match for human-related causes of endangerment.

Species usually become endangered because of multiple, human-related causes. The overwhelming cause of plant species endangerment is habitat loss, directly linked to agriculture, forestry, and urbanization. Remaining small populations have a higher risk of extirpation (local extinction) or extinction. Coinciding with habitat loss is reduced habitat quality related to invasive species, grazing, and other land-use changes (Flather *et al.*, 1994). There are many examples of habitat loss to choose from. In the wetter forested Sierra Madres of Mexico, Guatemalan fir (*Abies guatemalensis*) or Pinabete trees are considered endangered from over-exploitation of old-growth forests, land-use change, and livestock grazing (Burton, 1991). *Persea theobromifola*, once an important timber species in western Ecuador, has been reduced to fewer than 12 trees at Rio Palenque (Gentry, 1986). In the United States Burton (1991) reported that only 1% of the 1,036,000 km^2 (400,000 mi^2) of tallgrass prairie remains, now existing in isolated patches and small nature reserves. Over half of all continental U.S. wetland and aquatic habitats have been destroyed. Fire suppression and succession, invasive exotic plants, and habitat fragmentation will make it increasingly difficult for rare species to persist.

Causes of rarity vary by region. Though habitat loss was generalized as an overall problem, agricultural development was specifically noted in the southern Appalachia region, whereas urban development and forest clearing were specifically noted in Florida. In the Gulf Coast areas, shoreline modification and development were specifically noted in endangerment patterns (Flather *et al.*, 1994).

Commercial exploitation is also a problem. Cacti in the southwestern United States are a prime example. Legislation regulating the collection of cacti in Arizona and California date from 1929. Yet, in a single year, over half a million small cacti were illegally collected for sale in the United States, Japan, and Europe (Burton, 1991). One of the two colonies of the 2.5-cm-diameter Nellie Cory cactus (*Coryphantha minima*) was completely eliminated by thieves in the 1960s. The rarest species demand the highest prices from collectors. Rare orchids are plucked one by one by various " collectors." Some threatened orchids in India, such as *Paphiopedilum druri, Dendrobium pauciflorum, D. nobile*, and *Diplomeris hirsuta*, are used for medicinal purposes. As such, they are facing increasing pressure from growing and aging human populations.

Natural environmental change can both cause and maintain rarity. Cropper (1993) reported that some rare species, such as *Leptorhychos gatesii* in Australia, are observed only following natural fire. A riparian herb, *Collomia rawsoniana*, required periodic high and low streamflows to persist. Likewise, disruption of the natural hydrology of Lake Okeechobee, the largest lake in Florida, has led to the endangered listing of *Cucurbita okeechobeensis*. Land management decisions can obviously upset required disturbance patterns. Fire suppression activities and flood control will likely increase the rarity of such species.

The introduction of exotic grazers such as goats, pigs, rabbits, and sheep often leads to devastated native floras. For example, 47 of the 49 native plant species on St. Helena Island in the south Atlantic Ocean are now rare or threatened. The island was once dominated by beautiful forests of St. Helena ebony (*Trochetia melanoxylon*). Goats were introduced on the island in 1513, and the goat population skyrocketed by 1588. Goats ate ebony seedlings and humans used mature trees as fuel, so by 1810 the forests were destroyed and the ebony was extinct. About 260 naturalized exotic plant species were also introduced to the island (Davis *et al.*, 1986).

Introduced rabbits have greatly threatened *Acacia carnie* in arid western New South Wales, Australia (Cropper, 1993), and nearly decimated *Dudleya traskiae* (ironically called the Santa Barbara live-forever) on Santa Barbara Island off California (Benseler, 1987). Hawaii' s beautiful Haleakala silversword (*Argyroxiphium sandwicense* ssp. *macrocephalum*), with a flower stalk up to 2 m tall, almost went extinct in the 1920s due to vandalism and grazing by introduced cattle and goats. Like many endangered plant species, several additional threats must be held at bay. A major effort in

Hawaii is under way to protect the habitat of the Haleakala silversword from invasive exotic plants (*Verbascum thapsus*, mullein; and *Pennisetum setaceum*, fountain grass) and the Argentine ant (*Iridomyrmex humilis*). The Argentine ant poses a significant threat to native pollinators of the silversword.

Competition from invasive exotic plant species may be a major contributor in the future to native plant endangerment. Large, nearly pure stands of exotic purple loosestrife (*Lythrum salicaria*) have directly influenced the endangered small spikerush (*Eleocharis parvula*) in New York and Long's bulrush (*Scirpus longii*) in Massachusetts.

V. CONSEQUENCES OF RARITY

The most obvious consequence of rarity is extinction. Charles Darwin, like many naturalists of his time and since, recognized that rarity often preceded extinction. Small populations of sessile organisms are vulnerable to catastrophes. Landslides, fire, flooding, hurricanes, and other disturbances can simply wipe out populations. Small populations of vascular plants are also vulnerable to breeding problems from higher variability in breeding success caused by inbreeding. Inbreeding has been documented in several localized endemics such as *Limnanthes bakeri*, a vernal pool species in Mendocino County, California. In contrast, a common congener *Limnanthes douglassii* reproduces almost exclusively by cross-fertilization. *Stephanomeria malheurensis*, a plant species confined to one small hilltop in Burns, Oregon, is auto-fertile, whereas wide-spread congeners are not. Reproductive failure is not uncommon in sparse species. Lower genetic variation in small populations may also make them more vulnerable to rapid environmental change.

Not all small populations march rapidly towards extinction. Many rare plant populations can persist for centuries and millennia. Sparse populations can often avoid pathogens and herbivory. *Pinus ponderosa*, which may have been restricted to small refugia in Arizona and New Mexico at the end of the last Ice Age, has become the most widespread pine in the western United States; it now occurs from Mexico to Canada, and from California to Nebraska. Other small populations may speciate (form new species). For example, two species of *Ranunculus* in alpine areas of the North Island of New Zealand (*R. verticillata* and *R. insignis*) have given rise to *R. nivicola*.

VI. LEGAL MANDATES TO PROTECT ENDANGERED PLANTS

A. Policy and Legal Mandates throughout the World

The International Plant Protection Convention held in Rome in 1951 set forth recommendations for the protection and promotion of plant life throughout the world. Since then, the International Union for Conservation of Nature and Natural Resources (IUCN; now known as the World Conservation Union) has taken center stage in the protection of the world's flora. The IUCN Plant Red Book strongly defines "endangered species" as a species in danger of extinction and whose survival is unlikely if the causal factors (e.g., over-exploitation, extensive habitat destruction) continue operating, "including taxa whose numbers have been reduced to a critical level or whose habitats have been so drastically reduced that they are deemed in immediate danger of extinction" (IUCN, 1998). It also identifies and tracks "vulnerable species" as those "believed likely to move into the endangered category in the near future if the causal factors continue operating." Finally, it tracks "rare species"—"taxa with small world populations that are not presently endangered or vulnerable, but are at risk." These IUCN classifications are determined by scientists and government officials around the world and classified species are not necessarily afforded legal protection after designation (Stermer, 1995).

Several countries have policies or legislation that protect endangered plant species. Following the Convention on International Trade in Endangered Species (CITES) of Wild Flora in 1973, 113 countries have agreed not to trade certain threatened species. Many countries augment these agreements with additional legislation. For example, the Mauritius National Plant Protection Legislation (the Plants Act of 1976) and the Forest and Reserves Act (1983) legally protects endangered species and habitats in the territories of Mauritius. Enforcement of endangered species laws and policies in many countries is generally considered to be well-intentioned but weak.

B. The United States Endangered Species Act

The intent of the U.S. Endangered Species Act of 1973 (16 USC1531-1543) is to prevent further decline and help restore endangered and threatened species and the habitats upon which such species depend and to "provide a means whereby the ecosystems upon which endangered species and threatened species depend may be conserved" (Greenwalt and Gehringer, 1975). Thus, the Act recognizes the inseparable link between protecting a species, its habitat, and the surrounding ecosystem. The Act also provides broad-ranging protection for all species threatened with extinction in the "foreseeable future."

The Act makes the "taking" of endangered species anywhere within the United States a federal offense, requires federal agencies to use their existing authorities to conserve listed species, prohibits federal agencies from taking actions that may jeopardize a species' existence, provides a formal structure for listing endangered species, and provides a means for citizens to bring suit against any federal agency for failure to meet its obligations under the Act (Flather *et al.*, 1994).

Ayensu and DeFilipps (1978) noted that a species may be rare at the edge of its range, but not endangered or threatened as a whole. In determining national endangered, threatened, and extinction status, the total range and abundance of the species must be considered. However, states may further protect a species threatened with extirpation. Still, the cost of protecting individual species and habitats against multiple stresses is high, and the reality is that enforcement on public and private lands is generally weak.

VII. SELECTED EXAMPLES OF ENDANGERED PLANTS

Coleus forskohlii (Willdenow)

Coleus forskohlii, a 40-cm-tall, rare herb, is found in the Yunnan Province of China, Bhutan, India, Nepal, Sri Lanka, and Africa at about 2300 m on steep slopes (Fig. 3). Compounds extracted from the roots have long been a Hindu and Ayurvedic traditional medicine. Rampant collection has increased the rarity of this species. A German pharmaceutical company holds at least six U.S. patents for use of the plant as treatments for high blood pressure, cardiovascular disease, colic, respiratory problems, insomnia, painful urination, and convulsions. This is sometimes referred to as "biopiracy," when the intellectual property of indigenous peoples is appropriated and used by foreign companies to develop and patent commercial products.

Fitzroya cupressoides

Fitzroya cupressoides is an ancient tree species in southern Chile (Fig. 4). Only 5% of the world's temperate

Coleus forskohlii root

The root of *Coleus forskohlii* is the only source of the biologically active compound Forskohlin. This ancient medicinal herb has been re-discovered by modern pharmaceutical researchers who have patented uses of the plant. *Coleus forskohlii* grows in the Yunnan Province of China, and in Bhutan, India, Nepal, Sri Lanka and Africa.

FIGURE 3 *Coleus forskohlii.*

forests are in the Southern Hemisphere, and one-third of the threatened temperate forests occur in Chile. The highest biodiversity in any temperate forest is also found in Chile. These forests are remarkably productive, with some of the world's largest concentrations of biomass. Old-growth *Fitzroya cupressoides* trees often reach 4 to 5 m in diameter and may live for 4000 years. One-third of Chile's forests were burned or cleared by 1955. Some *Fitzroya* populations are protected in nature reserves, but most of the forests containing this species are privately owned.

Sarracenia rubra ssp. alabamensis

The Alabama canebrake pitcher-plant is a carnivorous plant with maroon flowers on 0.6-m stalks originating from rhizomes (underground roots; Fig. 5). It is found in sandy and gravelly bogs, seeps, springs, and swamps, and flowers from late April to early June. *Sarracenia*

rubra ssp. *alabamensis* is restricted to only 12 localized sites in a three-county area in central Alabama. Four of the 12 sites have 70 to 300 plants each, and half the populations have 2 to 20 plants each. Much of the original habitat (16 other sites) has been modified or destroyed by agriculture and construction of farm ponds in boggy areas. Fire exclusion, gravel mining, and invasive plants pose additional threats. Several populations have also been lost or degraded by plant collectors.

Encephalartos longifolius

South African cycads are ancient gymnosperms and the most primitive living seed-bearing plants on Earth (Fig. 6). They flourished 50 to 60 million years ago and provided forage for dinosaurs before then. In South Africa, all 40 cycad species are endangered, and some species are extinct in the wild. The thick-trunked plants

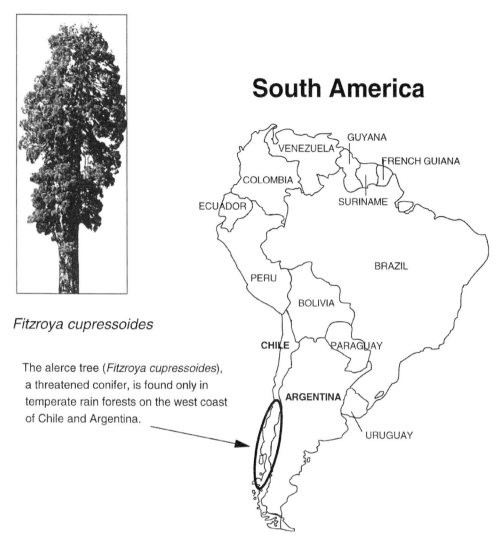

South America

Fitzroya cupressoides

The alerce tree (*Fitzroya cupressoides*),
a threatened conifer, is found only in
temperate rain forests on the west coast
of Chile and Argentina.

FIGURE 4 *Fitzroya cupressoides.* (Photograph courtesy of Dr. T. T. Veblen.)

with rigid spiked leaves grow very slowly; some species take 100 years to grow 1 m. Habitat loss and competition from invasive exotic plant species are often cited as causes of rarity, but illegal collecting is also a major problem. South Africa has some of the world's strictest laws controlling cycad theft, but the thefts continue. Some cycads now have implanted microchips for identification and tracking purposes in an effort to curb poaching.

Pterostylis truncata

Brittle greenwood is a ground-dwelling orchid that emerges in the fall with large, squat flowers (Fig. 7). It is a clonal plant that can regenerate vegetatively, and its flowers can be pollinated by a small flies. *Pterostylis truncata* is found in only three locations in south-central New South Wales, Australia. One of the populations is threatened by introduced European rabbits (*Oryctolagus cuniculus*), feral goats (*Capra hircus*), and eastern gray kangaroos (*Macropus giganteus*). An additional recent threat comes from the highly invasive weed *Chrysanthemoides monifilera*. The second population is on private land, where it is potentially threatened by kangaroos and rabbits. The third population occurs in a forest reserve; here it is threatened from trampling by orchid enthusiasts, weed invasions, and bird predation.

VIII. IMPLICATIONS

Reducing the rate of habitat loss around the world will be difficult. Between 1980 and 1990, species-rich tropi-

The Alabama canebrake pitcher-plant *(Sarracenia rubra ssp. alabamensis)* is an endangered, carnivorous plant. It is restricted to only 12 sites in central Alabama. Many factors have contributed to the listing of this plant as endangered, including habitat destruction, fire suppression, and the introduction of non-native plants.

Sarracenia rubra ssp. alabamensis

FIGURE 5 *Sarracenia rubra* ssp. *alabamensis.*

cal forests were cleared at the rate of 6.3 million ha/yr (15.4 million acres/yr), or 0.8% of the forest per year (Heywood, 1995). Deforestation causes habitat loss, habitat fragmentation, and edge effects at the boundaries, so the effects are greater than indicated by the actual deforested area. Large areas will be lost to urbanization and agriculture. Many dry grasslands in Germany have been converted to range and arable land in recent decades, making it more difficult to protect the

nation's 164 threatened plant species (Heywood, 1995). Thus, saving critical habitats throughout the world is seen as increasingly important and extremely urgent.

A species-by-species approach to rare plant conservation is expensive and difficult. However, the unstated assumption of the habitat preservation approach is that species are inseparably linked to habitats, which in turn are stable and predictable. Yet because habitats may be neither stable nor predictable (Flather *et al.*, 1994),

Encephalartos species with male cones.

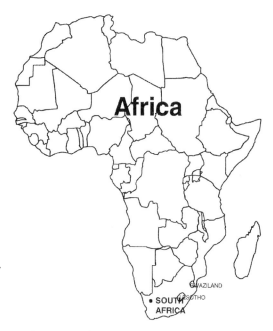

Encephalartos is the second largest genus in the Cycad genera. All Encephalartos are endemic to the African continent, with most in xeric habitats. Natural populations of Encephalartos suffer intense pressure from plant collectors.

FIGURE 6   *Encephalartos longifolius.* (Photograph courtesy of Dr. D. A. Steingraeber.)

both species and habitats must be monitored to protect endangered plants and animals. In Australia, management techniques to protect critical habitats and species include: burning or slashing overgrown vegetation, removal of weeds, removal of grazing animals, hand pollination of selected species, propagation and seed storage, reintroduction into restored habitats, and quarantine to reduce the threat of introduced pathogens (Cropper, 1993).

Less than 10% of named plant species have been analyzed for medicinal or nutritional properties (Stermer, 1995). Nonetheless, 25 to 40% of all drug prescriptions in the United States contain plant ingredients (Durant and Saito, 1985), and many of these ingredients cannot be synthetically made. About 80% of people in developing countries use traditional medicines. As more and more plant species go extinct, so may our chances to find the next heart medicine (e.g., *Digitalis;* foxglove) or treatment for childhood leukemia (from *Catharanthus roseus;* the rosy periwinkle).

To improve the effectiveness of biodiversity conservation, increased emphasis is needed on systematic sur-

veys and monitoring. Many species may be classified as rare owing to poor surveys, as evidenced by the rate of species discoveries. Since 1970, for example, one botanist has discovered 55 new plant species in Utah, and over 80 species have been named since 1970 (Durant and Saito, 1985). Botanists searching the well-studied Rocky Mountain National Park in Colorado added over 100 species to the Park's plant checklist between 1987 and 1992 (Stohlgren *et al.*, 1997). On average, one or two native plant species are added to the flora of New York each year. The Nature Conservancy and Natural Heritage Programs have found that plant surveys often show that many plant species are more common than previously believed. For example, several populations of rare orchids have been found in Australia after intensive searches. Systematic surveys of plants and rare habitats are badly needed.

The preservation of intact ecosystems may be the most promising way to protect endangered plants. It is now widely understood that maintaining natural disturbance regimes, such as fire and flooding, is important for many rare species (Cropper, 1993), and that preserv-

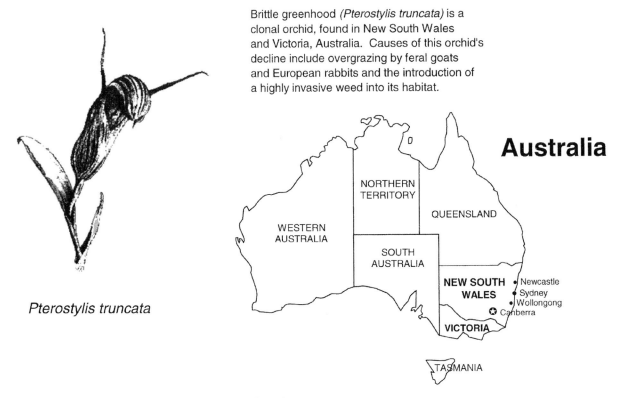

Brittle greenhood *(Pterostylis truncata)* is a
clonal orchid, found in New South Wales
and Victoria, Australia. Causes of this orchid's
decline include overgrazing by feral goats
and European rabbits and the introduction of
a highly invasive weed into its habitat.

Pterostylis truncata

FIGURE 7   *Pterostylis truncata.*

ing the habitat around rare plant locations is vitally
important. However, subtle assaults to endangered species may include the use of fertilizers, herbicides, and
insecticides that may reduce plant pollinators, invasive
plant species that may out-compete the endangered
plants, and introduced foreign diseases, insects, or
domesticated/feral animals that may further endanger
plants. Where habitat loss and degradation have been
significant, habitat restoration efforts and species reintroductions are needed.

Cultivation, the last resort for endangered plants,
may become necessary in some cases to preserve genetic
variation. Franklinia *(Franklinia alatamaha)* is a familiar example. Franklinia is a small tree that was restricted
naturally to the Altamaha River basin in southeastern
Georgia. Now extirpated in the wild, the species survives as an ornamental throughout the eastern United
States. In Australia, 1053 of 3329 rare and threatened
plant species are found only in botanical gardens, and
515 plant species are held in only one collection (Cropper, 1993). Several botanical gardens in the United
States have active programs to study, collect, and grow
rare and endangered plants from all over the world.
However, duplicate and more complete collections are
needed in most countries, and such a system of collec-

tions should not be viewed as a substitute for preserving
native habitats and ecosystems.

Some plants have appeared to bounce back from
extinction. Mountain golden heather *(Hudsonia montana)* was described by Thomas Nuttall in 1818 in North
Carolina, and was thought to be extinct in the 1960's.
About 2000 individuals were found in the Blue Ridge
Mountains around 1990 (Burton, 1991). Running buffalo clover *(Trifolium stoloniferum)* was also believed to
be extinct, but it was rediscovered in 1983 in West
Virginia and later discovered in four other states. Also in
1983, *Lomatium peckianum* was rediscovered in Oregon
after "disappearing" for over 50 years. In 1985, one
species of clover *(Trifolium microcephalum)* that had
not been seen since the Lewis and Clark expedition
in 1805–1806 was rediscovered. Such stories are not
uncommon. Eleven plant species in New York have
been "rediscovered" in the past 10 years. However, ex-
pecting miraculous rediscoveries for many long-lost
species is probably unrealistic.

This article has focused on endangered vascular
plants. However, many non-vascular plants are similarly threatened. For example, >50% of the mushrooms
in Europe are listed as endangered or threatened in at
least one country. Species that were once common,

such as *Hydnum repandum*, have been extirpated in some countries. Over-collection of the lichen *Gymnoderma lineare* in the Great Smoky Mountains National Park has led to its recommendation for listing. Air pollution and acid deposition are known to adversely affect many non-vascular plant species. The assessment of endangered plants cannot be done without assessing the interacting species (e.g., pollinators and symbiotic fungi) and the ecological processes that affect complexes of rare and common species.

Understanding the causes and consequences of rarity also requires a comprehensive knowledge of biology, evolutionary and recent history, and species demography. Often, the causes of population decline may be elusive. For example, *Torreya taxifolia*, the Florida torreya, is a narrowly restricted endemic conifer that has suffered catastrophic declines since the 1950s. Pathologists and ecologists have studied the problem relentlessly and found no obvious cause for the decline. There are now fewer than 1500 trees in the wild, with no solution in sight. In many cases, careful field and laboratory experiments may be necessary to isolate the causes of rarity.

It is equally important to increase public awareness about the ecology of rarity and the economic and social consequences of losing our endangered plants. In the end, protecting endangered species and biodiversity is a question of ethics and values. Endangered plants are best viewed as valuable resources and inherently valuable species with which we share the Earth.

Acknowledgements

April Owen and Lisa Schell assisted with the literature review and graphics. Curtis Flather provided information and Fig. 2. Geneva Chong, April Owen, and Lisa Schell provided helpful comments on an earlier version of the manuscript. To all I am grateful.

See Also the Following Articles

ENDANGERED ECOSYSTEMS • ENDEMISM • EXTINCTION, RATES OF • PLANT BIODIVERSITY, OVERVIEW • PLANT INVASIONS

Bibliography

Ayensu, E. S., and DeFilipps, R. A. (1978). *Endangered and Threatened Plants of the United States.* Smithsonian Institution/World Wildlife Fund, Washington, D.C.

Benseler, R. W. (1987). Conservation ethics, animals, and rare plant protection. In *Conservation and Management of Rare and Endangered Plants* (T. S. Elias, ed.), pp. 623–626. California Native Plant Society, Sacramento.

Benson, Lyman D. (1982). *The Cacti of the United States and Canada.* Stanford University Press, Stanford, California.

Burton, J. A. (ed.). (1991). *The Atlas of Endangered Species.* Macmillan Publishing, New York.

Cody, M. L. (1986). Diversity, rarity, and conservation in Mediterranean-climate regions. In *Conservation Biology, the Science of Scarcity and Diversity* (M. E. Soulé, ed.), pp. 122–152. Sinauer Associates, Sunderland, Massachusetts.

Cropper, S. C. (1993). *Management of Endangered Plants.* CSIRO Publications, East Melbourne, Australia.

Davis, S. D., Droop, S. D. M., Gregerson, P., Henson, L., Leon, C. J., Villa-Lobos, J. L., Synge, H., and Zantovska, J. (1986). *Plants in Danger: What Do We Know?* International Union for Conservation of Nature and Natural Resources, Gland, Switzerland.

DuMond, D. M. (1973). A guide for the selection of rare, unique and endangered plants. *Castanea: The Journal of the Southern Appalachian Botanical Club* **38**(4), 387–395.

Durant, M., and Saito, M. (1985). The hazardous life of our rarest plants. *Audubon* **87**(4), 50–61.

Flather, C. H., Joyce, L. A., and Bloomgarden, C. A. (1994). *Species Endangerment Patterns in the United States*, General Technical Report RM-241. USDA Forest Service, Rocky Mountain Forest and Range Experiment Station, Fort Collins, Colorado.

Gaston, K. J. (1997). What is rarity? In *The Biology of Rarity: Causes and Consequences of Rare–Common Differences* (W. E. Kunin and K. J. Gaston, eds.), pp. 31–47. Chapman and Hall, London.

Gentry, A. H. (1986). Endemism in tropical versus temperate plant communities. In *Conservation Biology, the Science of Scarcity and Diversity* (M. E. Soulé, ed.), pp. 153–181. Sinauer Associates, Sunderland, Massachusetts.

Greenwalt, L. A., and Gehringer, J. W. (1975). Endangered and threatened species: Notice on critical habitat areas. *Federal Register* **40**(78), 17764–17765.

Heywood, V. H. (ed.). (1995). *Global Biodiversity Assessment.* United Nations Environment Program/Cambridge University Press, Cambridge, United Kingdom.

IUCN. (1998). *1997 IUCN Red List of Threatened Plants.* (K. S. Walter and H. J. Gillett, eds.). World Conservation Union (IUCN), Morges, Switzerland.

Lucas, G., and Synge, H. (1978). *Red Book Data.* IUCN, Morges, Switzerland.

Missouri Botanical Garden. Online, June 2000, http://www.mobot.org

Orians, G. H. (1997). Evolved consequences of rarity. In *The Biology of Rarity: Causes and Consequences of Rare–Common Differences* (W. E. Kunin and K. J. Gaston. eds.), pp. 190–208. Chapman and Hall, London.

Rabinowitz, D. (1981). Seven forms of rarity. In *The Biological Aspects of Rare Plant Conservation* (H. Synge, ed.), pp. 205–217. John Wiley & Sons, New York.

Rapoport, E. H. (1982). *Areography: Geographical Strategies of Species.* Pergamon Press, New York.

Raven, P. H., Evert, R. F., and Eichhorn, S. E. (1986). *Biology of Plants.* Worth Publishers, New York.

Stermer, D. (1995). *Vanishing Flora: Endangered Plants around the World.* Harry N. Abrams, New York.

Stohlgren, T. J., Chong, G. W., Kalkhan, M. A., and Schell, L. D. (1997). Rapid assessment of plant diversity patterns: A methodology for landscapes. *Ecol. Monitoring Assessment* **48**, 25–43.

U.S. Endangered Species Act (1973). Title 16–Conservation. Chapter 35–Endangered Species.

ENDANGERED REPTILES AND AMPHIBIANS

Tim Halliday
The Open University

GLOSSARY

amphibian Member of a class of vertebrates (the Amphibia), comprising frogs and toads (order Anura), newts and salamanders (order Caudata), and caecilians (order Gymnophiona), which typically return to water to breed and pass through an aquatic larval stage with gills. Amphibians have a moist skin without scales, which is permeable to water and gases.

reptile Member of a class of vertebrates (the Reptilia), comprising turtles and tortoises (order Testudinata), lizards, snakes, and worm-lizards (order Squamata), the tuatara (order Rhynchocephalia), and crocodiles and alligators (order Crocodylia), which typically lay eggs with a leathery, impermeable shell. Reptiles have a dry, horny skin with scales, plates, or scutes.

THE EARTH CURRENTLY CONTAINS approximately 7150 species of reptiles and 4600 species of amphibians. These data are underestimates because new species are still being discovered, partly as a result of exploration of previously unknown parts of the world and partly because new genetic techniques have revealed that what were thought to be single species are in fact several species. Many reptile and amphibian species are currently listed as threatened or endangered, and several have recently become extinct. Current estimates of the number of species at risk of extinction do not reflect the true extent of the threat faced by these two groups. Both groups are threatened by a variety of environmental factors.

I. REPTILES AND AMPHIBIANS

The reptiles and amphibians are two groups of vertebrate animals that are commonly grouped together, and the study of the two groups is called herpetology. In reality, however, the biological differences between reptiles and amphibians are more numerous and extensive than are their similarities. Amphibians are descended from fishes and, like fishes, both amphibians and reptiles are ectothermic, meaning that the heat that they require to maintain physiological processes is derived externally, directly or indirectly from the sun. Reptiles are the ancestors of the birds and the mammals, which are both endothermic, meaning that body heat is primarily generated within their bodies. Their dependence on external heat sources limits both amphibians and reptiles to tropical and temperate regions of the earth and both groups are especially numerous in the tropics;

80% of the world's amphibian and reptile species live in tropical habitats.

There are two major differences between amphibians and reptiles. First, amphibians have a thin, permeable skin that severely limits their ability to retain water within the body. In contrast, reptiles are covered in impermeable scales and are able to retain water very effectively. Second, the eggs of amphibians are bounded by only a thin membrane and can only survive if they are kept wet. The eggs of reptiles are covered by a hard shell, providing a sealed environment for the developing embryo, which does not depend on external sources of water. These two differences mean that, whereas reptiles have successfully colonized very dry habitats, including deserts, amphibians are restricted to those habitats in which water is available for all or most of the year.

Neither reptiles nor amphibians are as well-known, in terms of documentation of species, as either the birds or the mammals. Neither group has been studied as intensively as birds and mammals, both are rather secretive in their habits, and both are very numerous in parts of the world, notably tropical forests, that have not been fully explored. Therefore, new species of amphibians and reptiles are being discovered and described at a much higher rate that is the case for birds or mammals. The number of recognized species of amphibians is approximately 4600 and of reptiles approximately 7150, but both these figures are continuously being revised upwards as new species are described.

There are three major groups of amphibians: the frogs and toads (approximately 4500 species), the salamanders and newts (approximately 415 species), and the caecilians (approximately 165 species). Reptiles are divided into four major groups: the lizards, snakes, and worm-lizards (approximately 6850 species); the turtles and tortoises (approximately 260 species); the alligators and crocodiles (approximately 22 species); and the tuataras (1 species).

II. THE CONSERVATION STATUS OF REPTILES AND AMPHIBIANS

An accurate estimate of the number of reptile and amphibian species that are currently threatened with extinction, or which have recently become extinct, is not currently available. Table I presents two recent estimates of the numbers of endangered and threatened reptiles and amphibians. The disparity between the two sets of data, and in the total numbers of species that they each assume, reflects the current incompleteness of our knowledge of these two groups and of their conservation status. Currently, major efforts are being made, for example, by the Species Survival Commission of the International Union for the Conservation of Nature (IUCN), to correct this situation by compiling accurate lists of reptile and amphibian species and making assessments of their current status.

TABLE I

Estimated Numbers of Threatened and Endangered Reptiles and Amphibians

| Taxonomic group | Number of species | | | | Approximate number of described species | |
| | Threatened | | Endangered | | | |
	A	B	A	B	A	B
Turtles and tortoises	78	8	11	29		
Sphenodon	1			1		
Lizards	43	14	9	23		
Snakes	33	7	7	9		
Crocodiles and alligators	15	2	11	19		
REPTILES	170	31	38	81	4771	6500
Salamanders and newts	25	3	2	6		
Frogs and toads	32	2	6	12		
AMPHIBIANS	57	5	8	18	4014	4500

A, IUCN Red List, 1990 (World Conservation Monitoring Centre 1992); B, US Fish & Wildlife Service, 1994 (Pough et al., 1998).

III. THREATS TO REPTILES AND AMPHIBIANS

Many environmental factors have been identified that are detrimental to the continued survival of reptile and amphibian species. Some of these are common to the two groups of animals; others are more relevant to one group than the other.

A. Habitat Destruction

The single most important factor that adversely affects reptiles and amphibians is habitat destruction and modification. Many habitats that are essential for the two groups are shrinking or disappearing at an accelerating rate as a result of human population growth and economic development. Tropical forests provide some of the most species-rich habitats in the world and are particularly vulnerable to destruction by humans, with the land being turned over to agriculture or to provide residential areas for people. In 1991, the United Nations Food and Agriculture Organization reported that the world's tropical forests were being destroyed at a rate that was 50% faster than a decade previously. Tropical forests support a high diversity of reptile and amphibian species, many of them still to be described. At the current rate of deforestation, within 30 years all extensive areas of tropical forest, together with their reptile and amphibian fauna, will have disappeared.

Amphibians are particularly dependent on freshwater habitats, especially for breeding. The World Wildlife Fund (1998) published a report which suggests that freshwater habitats are the most seriously threatened on Earth. Between 1970 and 1995, the diversity of freshwater species decreased by 50%, a faster rate of decline than has been detected in any other component of the global ecosystem. Although there is legal protection for some of the world's larger lakes and rivers, or at least parts of them, largely because of their importance in providing fish to support the human population, the smaller streams, ponds, and swamps that are essential habitat for amphibians are generally not protected. The draining of wetlands to make way for agriculture and housing has had a particularly serious negative effect on many amphibian populations.

In much of western Europe, traditional methods of agriculture previously provided good habitat for amphibians in the form of small woodlands, hedgerows, and numerous ponds created to water livestock. In the past 50 years, however, agricultural practices have changed and all these landscape features have been destroyed over very large areas. In parts of Britain, for example, the number of ponds suitable for amphibians to breed in has declined by 90% in the past 50 years.

Habitat destruction in a given area is not always total; often, small pockets of forest, heathland, or wetland are set aside for conservation purposes. This results in the fragmentation of previous areas of habitat and there is increasing evidence that habitat fragmentation is a serious threat to the continued survival of species that it was assumed were afforded some degree of protection. Fragmentation of habitat leads to the isolation of small populations of reptile or amphibian species. Inbreeding in such populations reduces their genetic diversity, and their isolation prevents interchange of individuals and thus of genetic variation with other populations. As a result, isolated populations tend to decline slowly and eventually die out, even though they are protected. Many reptiles and amphibians have only limited powers of dispersal, and even a road built through an area of otherwise suitable habitat will reduce dispersal.

B. Climate Change

There is increasing evidence that the earth's climate is undergoing major changes as a result of human activities such as the destruction of forests. Most notably, average temperatures are steadily increasing in many parts of the world and there are major, long-term changes in rainfall patterns. The long-term effects of such changes on reptiles and amphibians are largely a matter of speculation, but there is evidence that they have had an impact on some species. For reptiles, temperature is crucial because many species rely on basking in the sun to raise their body temperature to a level at which they can carry out essential activities such as feeding and reproduction. For amphibians, rainfall patterns are critical because they determine whether, and for how long, the small streams and ponds in which they breed contain water.

The apparent extinction of the golden toad (*Bufo periglenes*), along with several other frog species, in the montane forest of Costa Rica appears to be due, at least in part, to climate change (Pounds *et al.*, 1999). A critical factor in the habitat of this species is low-lying cloud, which provides the water that maintains water flow in the small streams in which many of the native frogs breed. The extent of such cloud has decreased in recent years, leading to extensive reductions in the amount of available water and thus a general reduction in stream habitat. In Britain, winters are becoming increasingly less severe and there is evidence that several native amphibians are now breeding earlier in the spring than they were 20 years ago.

C. Ultraviolet Radiation

The steady erosion of the ozone layer in the earth's stratosphere has led to an increase in the amount of ultraviolet (UV) radiation reaching the earth's surface. Such radiation, especially UV-B, is harmful to living organisms and there is accumulating evidence that it may be a factor in the decline of some amphibian species, especially those living at high altitudes, where incident levels of UV-B are highest. Increased UV-B causes the genetic material DNA to mutate, leading to the abnormal development and eventual death of embryos. Frog and toad species that lay their eggs close to the surface of water, which filters out UV-B, may be particularly susceptible to increased UV-B. Although there is considerable experimental evidence that ambient levels of UV-B radiation are harmful to the early life stages of amphibians (Blaustein et al., 1995), it is not clear to what extent elevated UV-B has been a factor in the decline of amphibian populations in nature (Alford and Richards, 1999).

Amphibian species vary in their susceptibility to elevated UV-B. Some species, such as the Pacific tree frog (Hyla regilla), produces high levels of the enzyme photolyase, which repairs DNA damage of the kind caused by UV-B. This species is one of those that has not declined in the Pacific Northwest of North America, unlike species with lower photolyase levels, such as the Cascades frog (Rana cascadae) and the western toad (Bufo boreas) (Blaustein et al., 1994). If elevated UV-B does affect natural populations of amphibians, it is likely that it acts synergistically with other adverse environmental factors, such as acidification and pathogens. There is experimental evidence that both these factors have a more detrimental effect on the survival of amphibian embryos when they are combined with elevated UV-B.

D. Pollution

Environmental pollution has caused the decline or extinction of some local populations of amphibians, and it is likely that it also has a widespread harmful effect. For example, declines of amphibians in Yosemite National Park in California appear to be due to chemical pollution that has drifted on the wind from agricultural areas many miles away. Amphibians may provide sensitive biological indicators of pollution because their highly permeable skin rapidly absorbs toxic substances.

Examples of pollutants include fertilizers, herbicides, pesticides, heavy metals, and poisoning resulting from logging and mining operations.

A major form of pollution that can affect very large areas is atmospheric acid deposition or acid rain. Much of the rain that falls in regions downwind from major industrial areas, such as in the eastern United States, Scandinavia, and western Europe, is markedly acidic, with a pH of approximately 4.5; unpolluted rain has a pH of approximately 5.6. Acid rain lowers the pH of natural water bodies below the tolerance level of many species of amphibians. The toxic effects of low pH on amphibian development are well-documented. Acidic conditions reduce the mobility of sperm and may cause them to disintegrate, with a consequent reduction in the fertilization success of eggs. Eggs that are fertilized may develop abnormally; if they hatch, they produce deformed tadpoles that soon die. Acid precipitation has been implicated in the declines of tiger salamanders (Ambystoma tigrinum) in the Rocky Mountains of Colorado and of the natterjack toad (Bufo calamita) in lowland heaths in Britain.

Pollution in the form of solid plastic waste is harmful to some sea turtles that may eat it. Green turtles (Chelonia mydas) eat plastic bags while feeding on aquatic plant food and leatherback turtles (Dermochelys coriacea) mistake plastic bags for their jellyfish prey. Half of the sea turtles examined in some localities have plastic debris in their intestines. This may interfere with their digestion, respiration, and buoyancy, and some plastics are toxic.

Some chemical compounds of human origin are readily absorbed by animals and interfere with their endocrine systems; these are known as endocrine disrupters and include estrogen mimics, which are compounds that disrupt the reproductive development of both sexes (Stebbins and Cohen, 1995). Males may become feminized to varying degrees, for example, suffering lowered sperm counts. Reptiles are especially susceptible to such effects because of their sex-determination mechanisms which, in many species, are very dependent on environmental factors. Of particular concern are polychlorinated biphenyls (PCBs), industrial chemicals such as those used in fire retardants and adhesives, which persist for a very long time and accumulate in the environment. Some PCBs have a molecular structure so similar to estrogen that they mimic its effects when they enter an animal's body. PCBs can turn male red-eared slider turtles (Trachemys scripta) into females, drastically reducing the reproductive success of populations.

Many pesticides, such as DDT, are not only poisonous to a wide range of animals but also, at low concentrations, have endocrine-disrupting effects. Populations of alligators in Florida have been adversely affected by such compounds because of their feminizing effects on males. Many of these compounds, including PCBs, are highly volatile and so are transported very widely in the environment from their point of origin. The long-term effects of widespread chemical pollution of this kind are poorly understood but are very alarming not just for amphibians and reptiles but also for animals of all kinds, including humans.

Another kind of pollution which may affect amphibians in agricultural habitats results from nitrates derived from fertilizers. Recent research has shown that frog tadpoles reared in water containing nitrates at levels low enough to be considered safe for human consumption suffered physical abnormalities, paralysis, and death (Macro and Blaustein, 1999). Nitrate fertilizers are widely used throughout the world and may constitute a serious threat to many amphibian populations.

E. Disease

Diseases are a natural cause of morbidity and mortality among animals, but very little is known about their role as a determinant of population size among reptiles and amphibians. In the past few years, however, major outbreaks of disease among amphibians in many parts of the world have occurred, and some of these have had a major impact on the populations of many species. During the 1990s, mass mortalities occurred among populations of the common frog (*Rana temporaria*) in southeast Britain. These were caused by ranaviruses similar to pathogens that have caused mass mortalities among tiger salamanders (*Ambystoma tigrinum*) at several localities in the United States and Canada during the past 5 years (Daszak *et al.*, 1999). Recently, a disease called chytridiomycosis has been identified in Australia and Central America, where it has wiped out many populations of frogs, and among captive amphibians in the United States (Daszak *et al.*, 1999). The disease is caused by a chytrid fungus, which is an organism that invades the skin of amphibians. Whether it kills them by blocking water and oxygen transport across the skin or by producing a lethal toxin is not known. These sudden, major outbreaks of the same, previously unknown disease in widely distant parts of the world raises several questions that are urgently being addressed: Is the chytrid fungus a new organism that has only just evolved? Is it a long-established organism that has recently found its way to these localities and, if so, how? One possibility is that the fungus is not new and has

long been distributed worldwide, and that what has changed is the susceptibility of amphibians to it. It has been suggested that, as a result of one or more kinds of environmental stress, the immune system of amphibians has been compromised so that they are no longer able to survive infection by the fungus.

F. Commercial Exploitation

Many people in many parts of the world have long eaten amphibians and reptiles because they are a good and readily available source of protein. Until recently, such exploitation rarely had a serious impact on natural populations because it was localized, seasonal, and of low intensity. Unfortunately, modern commercialization of amphibians and reptiles for the world's luxury food market is generally done with little regard for the long-term protection of natural populations. Most of the frogs that are killed for human consumption are not a vital component of the diet of local people in the developing world but are a luxury item in the diet of people in developed countries.

The scale of trade in frog legs has long been substantial, but it is beginning to be controlled and reduced. In 1976, 2.5 million kg of frog legs was imported into the United States, mostly from India and Japan. In France, the annual consumption is estimated to be 3000 or 4000 tons, imported mostly from Bangladesh and Indonesia. Until recently, 200 million pairs of legs were exported annually from Asia to the United States, Europe, and Australia; however, since 1987, India has banned this trade because declines in natural frog populations had caused dramatic increases in the densities of insect pests.

There is also a substantial trade in reptile meat. Between 1979 and 1987, the hunting of alligators in Louisiana yielded 45,000 kg of alligator meat each year. Many turtles are heavily exploited as a food source, both by indigenous people and to supply luxury food markets in developed countries. The green iguana (*Iguana iguana*) has declined severely as the result of being hunted for food. Eaten by humans in Central America for centuries, destruction of its habitat and the expansion of the human population have led to it being hunted much more intensely in the past 30 years. In the late 1960s, as many as 150,000 iguanas were eaten each year in Nicaragua alone. Not only is iguana flesh regarded as a great delicacy but also their fat and their eggs are used for a variety of medicinal purposes, including a cure for impotence.

The highly durable skin of reptiles makes it an excellent alternative to leather, and reptile skins have long been used for making footwear, purses, belts, and many other accessories. Many species of pythons, boas, varanid lizards, and crocodiles are declining because they are hunted for their skins. In 1981, the United States imported 304,189 pairs of shoes made from boa skin and 176,204 pairs of python skin shoes. In the absence of commercial breeding programs for these species, this trade was entirely based on the exploitation of wild populations. Today, trade in reptile skin is increasingly regulated and sustained by breeding animals in captivity, but there remains a substantial illegal market in skins derived from natural populations.

Products derived from reptiles and amphibians are constituents of many traditional medicines and are used in many other ways. Broths made from snakes and tortoises are believed to combat many diseases; the fat of monitor lizards (*Varanus* spp.) is used to combat skin infections; sea turtle eggs are used as aphrodisiacs; turtle oil is used to make perfumes and lubricants; and the skin secretions of poison-dart frogs have been used in hunting by Central and South American Indians, and those of many toad species are used in many parts of the world as hallucinogens. Whether any of these forms of exploitation has had a harmful effect on natural populations is doubtful, but the future exploitation of amphibians and reptiles could either seriously threaten their continued existence or ensure their survival, depending on how it is managed. There is a lengthening list of compounds, found in the skin of frogs, that have considerable potential as medicines. For example, poison-dart frogs of the genus *Epipedobates* produce a unique alkaloid called epibatidine which is a much more effective painkiller than morphine and that appears not to be addictive.

G. Introduced Species

As the human population has expanded and people have colonized new parts of the world, they have taken with them, deliberately or accidentally, a variety of organisms that have been harmful to indigenous wildlife. Populations of many reptile species that are endemic to islands have been devastated by the introduction of alien rats, cats, dogs, and other animals that feed on them or their eggs, and introduced pigs and goats have destroyed the ground cover that supports their food supply and which they need to escape from predators. The tuataras (*Sphenodon*) have become extinct on the two major islands and on many of the smaller islands of New Zealand and are now confined to a very few small islands that remain free of alien immigrants. Introduced rats have caused the extinction of several frog

species of the genus *Leiopelma* on New Zealand; only three such species still exist, reduced to very small populations on remote, rat-free islands.

The eggs and larvae of amphibians make easy prey for fishes and other freshwater predators, and many amphibian species are dependent on water bodies that are free of fish to breed successfully. In many parts of the world, exotic fishes have been introduced by humans to the detriment of native amphibians. In California, trout have been introduced, to provide sport fishing, to high-altitude lakes and have caused the decline of several amphibian species; in many parts of the world, mosquito fish (*Gambusia affinis*) have been introduced to control mosquitos and other insect pests, with equally disastrous results for native amphibians.

Some reptiles and amphibians have been implicated in the declines of other species where they have been introduced to parts of the world in which they did not previously occur. The most notable example is the marine toad (*Bufo marinus*), a native of Central America, which was deliberately introduced to Hawaii, Australia, and many other places to control insect pests in sugarcane plantations. In Australia, where it is known as the cane toad, it has spread relentlessly outwards from the Queensland coast to the detriment of much of Australia's native fauna, including many frog species. The American bullfrog (*Rana catesbeiana*), a native of eastern North America, has been introduced to many parts of the world to be farmed to supply the trade in frog legs, and it has had a harmful effect on many native frog species both because it is a greatly superior competitor and because it is a predator of smaller frogs. Brown tree snakes (*Boiga irregularis*), accidentally introduced onto the island of Guam, have seriously reduced or extirpated not only populations of endemic birds but also several native reptiles.

H. Species at High Risk of Extinction

The fact that introduced species, such as cane toads and the American bullfrog, can multiply and spread to the point at which they become a serious threat to other species indicates that population declines among amphibians are not due to a single factor that affects all species. Although many species are adversely affected by current environmental conditions, at least some can thrive. This raises the question of whether there are particular characteristics of certain reptiles and amphibians that make them particularly susceptible to environmental changes.

There are many reptile and amphibian species, notably many turtles, that live for a long time but have delayed sexual maturity and low fecundity. Such life history characteristics limit the capacity of a population that has been reduced, for example, by human exploitation, to build up its numbers. Some species have very low reproductive rates because they do not breed every year. This is a feature of many amphibian species that live in near-desert habitats in which sufficient rain to fill their breeding ponds falls only occasionally.

Many amphibians move only small distances during the course of their lives. Such species have very poor dispersal abilities, with the result that, if their habitat is destroyed or modified, they do not have the option of colonizing suitable habitat elsewhere. There are many species that have very specialized habitat requirements and very restricted distributions. Such species are vulnerable to any very small change in their environment. Examples of such species are the golden toad (*B. periglenes*) of Costa Rica and two species of gastric-brooding frog (*Rheobatrachus*), restricted to a very few streams in southeast Queensland, Australia. All three species appear to have become extinct in the past 10 years.

Animals of all kinds that live on small oceanic islands are especially vulnerable to extinction, and reptiles and amphibians are no exception. Island species are often very vulnerable to introduced predators, such as rats, cats, dogs, and mongooses, having evolved in an environment in which such predators were previously absent. Small islands are also highly susceptible to habitat destruction and excessive hunting. Among endangered reptiles that live on islands are giant tortoises such as those found on Aldabra and the Galapagos.

Some reptiles and amphibians are vulnerable to exploitation by humans and to natural predation because they gather to breed in large numbers. A marine turtle, Kemp's ridley (*Lepidochelys kempi*), was reduced to near extinction in the 1970s by the commercial exploitation of breeding females and eggs, gathered at their coastal breeding sites in Mexico and Texas. This species is now internationally protected and its numbers are recovering.

Migratory species, such as some sea turtles, are very vulnerable both because their conservation requires that they be protected in two, often distant, parts of the world and because they have to make long journeys across possibly hazardous environments. The green turtle (*Chelonia mydas*) is protected in Australia but is exploited in Indonesia. Many amphibians migrate to specific breeding sites that are used year after year. High mortality can occur when these sites are near roads. In some places in the United States and Canada, roads that cross frog or salamander migration paths are closed in the spring, and in several places in Europe special tunnels have been built to allow frogs and toads to pass safely under roads.

IV. DECLINING AMPHIBIAN POPULATIONS

In the Monteverde cloud forest reserve in Costa Rica, the formation of large mating aggregations of the spectacularly colorful golden toad (*B. perigienes*) used to be an annual event. In 1989, however, this species failed to appear and it has not been seen since (Pounds and Crump, 1994). The golden toad has become the icon of the declining amphibian phenomenon, but it is only one of many species throughout the world that have disappeared because of this deeply disturbing process. The Monteverde reserve was established to protect biodiversity and yet, since 1990, 40% of its native frog and toad species have disappeared (Pounds *et al.*, 1997). Similar catastrophic declines of amphibian populations in supposedly pristine, protected habitats have occurred in Queensland, Australia, the Atlantic forests of Brazil, and the Pacific Northwest of North America. In a section of California's Sierra Nevada that includes Yosemite National Park, 5 of 7 amphibian species have seriously declined in recent years. In protected remnants of tropical rain forest in eastern Queensland, 14 species of stream-dwelling frogs have drastically declined or totally disappeared. These events carry a disturbing message for conservationists: As one scientist said at a recent workshop on amphibian declines, "locking up nature just isn't working."

Although it is clear that the majority of the world's amphibians that have declined or become extinct in the past 50 years have done so as a result of habitat change or destruction, these declines in protected areas suggest that some other kind of process is adversely affecting amphibian populations. Because of their dual life history, spent partly in water and partly on land, and because none of their life stages (egg, larva, or adult) have the kind of protective covering possessed by animals such as insects, reptiles, birds, and mammals, it has been argued that amphibians are especially sensitive to environmental insults such as chemical pollution (Stebbins and Cohen, 1995). According to this argument, amphibian declines may be the prelude to an environmental catastrophe that could affect many forms of life. Like the coal miner's canary, frogs may be providing an early warning to all biodiversity.

It is becoming increasingly clear that this kind of argument is not applicable to all amphibians. Indeed, a feature of all the declines that have occurred recently in supposedly protected habitats, such as Monteverde, is that although some species have declined or vanished, others have been quite unaffected. It is not clear, however, what character or characters differentiate amphibian species that have declined in pristine areas from those that have not. Amphibians are a very diverse group of animals in terms of their habits, their life histories, and their physiology. For example, although some species do have a highly permeable skin and are extremely sensitive to pollutants, for others the skin is highly effective as a protective covering. The American bullfrog (*R. catesbeiana*), for example, seems to be remarkably resistant to the effects of pesticides.

The current intense interest in amphibian population declines began in 1989 at the first World Congress of Herpetology. In 1991, the Declining Amphibian Populations Task Force (DAPTF) was set up under the aegis of the Species Survival Commission of IUCN, and by 1993 more than 500 populations of amphibians on five continents had been listed as declining or of serious conservation concern (Vial and Saylor, 1993). The DAPTF is continuing its work and plans to publish major overviews of the current status of amphibians throughout the world in 2001. Currently, it is clear from the information already gathered by the DAPTF that (i) amphibian population declines are widespread throughout the world and (ii) that they are not due to a single, global cause. Rather, they are the result of many factors, with more than one factor usually being implicated in any particular instance.

A. A Possible Link with Deformities?

Since a group of children found some severely deformed frogs in Minnesota in 1995, there has been much media attention devoted to this phenomenon in the United States, often making a link with amphibian declines there and elsewhere. These deformities reflect abnormalities in development and include missing eyes, digits, or entire limbs as well as extra limbs, sometimes growing from unexpected parts of the body. There are several reasons for being cautious about a possible link between deformities and the global decline phenomenon, however. First, such deformities are not new but rather have been reported periodically for more than 200 years, especially in the mid-northern United States. Second, although reports have become more frequent recently, there is no reason to believe that deformed frogs have become more common; there are simply more people looking for them. Third, developmental

abnormalities are caused by a variety of natural causes, including parasites and injury following attack by predators, and by anthropogenic factors, such as chemical pollution and increased UV-B radiation.

Deformed amphibians are a cause for concern because they are often symptomatic of local environmental degradation. Before we can conclude anything of wider significance from them, however, we have to remember that they are both naturally occurring and typical of amphibians and that they do not represent a new phenomenon. Only additional research into the role of various pollutants and of UV-B into the development of deformities can reveal to what extent they are another facet of processes that are adversely affecting amphibians on a global scale.

See Also the Following Articles

AMPHIBIANS, BIODIVERSITY OF • ENDANGERED BIRDS • ENDANGERED MAMMALS • REPTILES, BIODIVERSITY OF

Bibliography

Alford, R. A., and Richards, S. J. (1999). Global amphibian declines: A problem in applied ecology. *Annu. Rev. Ecol. Syst.* **30**, 133–165.

Blaustein, A. R., Hoffman, P. D., Hokit, D. G., Kiesecker, J. M., Walls, S. D., and Hays, J. B. (1994). UV repair and resistance to solar UV-B in amphibian eggs: A link to population declines? *Proc. Natl. Acad. Sci. USA* **91**, 1791–1795.

Blaustein, A. R., Edmund, B., Kiesecker, J. M., Beatty, J. J., and Hokit, D. G. (1995). Ambient ultraviolet radiation causes mortality in salamander eggs. *Ecol. Appl.* **5**, 740–743.

Daszak, P., Berger, L., Cunningham, A. A., Hyatt, A. D., Green, D. E., and Speare, R. (1999). Emerging infectious diseases and amphibian population declines. *Emerging Infect. Dis.* **5**, 735–748.

Marco, A., and Blaustein, A. R. (1999). Sensitivity to nitrate and nitrite in pond-breeding amphibians from the Pacific Northwest, USA. *Environ. Toxicol. Chem.* **18**, 2836–2839.

Pounds, J. A., and Crump, M. L. (1994). Amphibian declines and climate disturbance: The case of the golden toad and the harlequin frog. *Conserv. Biol.* **8**, 72–85.

Pounds, J. A., Fogden, M. P. L., and Campbell, J. H. (1999). Biological response to climate change on a tropical mountain. *Nature* **398**, 611–615.

Pough, F. H., Andrews, R. M., Cadle, J. E., Crump, M. L., Savitsky, A. H., and Wells, K. D. (1998). *Herpetology.* Prentice Hall, Englewood Cliffs, NJ.

Stebbins, R. C., and Cohen, N. W. (1995). *A Natural History of Amphibians.* Princeton Univ. Press. Princeton, NJ.

Vial, J. L., and Saylor, L. (1993). The status of amphibian populations: A compilation and analysis, IUCN/SSC Declining Amphibian Populations Task Force, Working Document No. 1. The World Conservation Union, Gland, Switzerland.

World Conservation Monitoring Centre. (1992). *Global Biodiversity. Status of the Earth's Living Resources.* Chapman & Hall, London.

World Wildlife Fund (WWF). (1998). *Living Planet Report 1998.* WWF International, Gland, Switzerland.

ENDANGERED TERRESTRIAL INVERTEBRATES

Mark Deyrup
Archbold Biological Station

GLOSSARY

alien invertebrates Invertebrates intentionally or accidentally imported by humans into new geographic areas.

coevolution Long-term evolutionary adaptation of species to each other (e.g., mutually beneficial relationships between bees and flowering plants).

endangered species Species that are likely to become extinct in the near future because of normal human activities. Examples of such activities are land clearing for agriculture or housing and accidental importation of invasive species through commerce.

endemic species Species confined to areas where they evolved (e.g., flightless crickets in the Hawaiian Islands).

habitat specialists Species found only in a specific habitat (e.g., species found only in forests of Sequoia trees).

terrestrial animals Animals that live on land for their entire lives (e.g., spiders) as opposed to animals that live in water for their entire lives (e.g., lobsters). There are also many amphibious animals that spend part of their lives in both places (e.g., dragonflies).

terrestrial invertebrates Animals that are not vertebrates (such as fish, reptiles, amphibians, birds, and mammals) that live on land for their entire lives. Examples are all insects except for those with aquatic larvae; nonaquatic mites and nematodes; and all spiders, millipedes, centipedes, and scorpions.

SIMPLE STATEMENTS CAN RAISE the most complicated questions. There are two simple facts about endangered terrestrial invertebrates. The first is that there are great numbers of these endangered animals—many thousands of species. The second is that nobody knows enough to make a comprehensive list of these species. Without such a list, how is it possible to say whether this list would be long or short and whether it would really contain many thousands of species? What is the evidence that there are huge numbers of endangered terrestrial invertebrates that are missing from lists of endangered species? Why aren't biologists moving more quickly to add the missing names to the list? Are terrestrial invertebrates particularly vulnerable to ongoing rapid changes caused by our own species? If there are such large numbers, what does this mean for the forests, fields, deserts, and other habitats that are threatened with the loss of these species? What does it mean for our own species?

I. EVIDENCE OF MULTITUDES OF ENDANGERED TERRESTRIAL INVERTEBRATES

The belief that there are very large numbers of endangered terrestrial invertebrates is based on a combination of knowledge and logic. The great majority of all animals are terrestrial invertebrates, in which are included at least three-fourths of the insects, most mites, terrestrial mollusks, a large percentage of the nematodes, and all the spiders, millipedes, and centipedes. The last three groups total more than all the vertebrates. If the factors that endanger other animals also affect terrestrial invertebrates, the number of endangered terrestrial invertebrates must be large indeed. These factors are primarily habitat destruction and the introduction of nonnative species into new areas, and it is known that they affect terrestrial invertebrates.

Hawaii, showcase of beleaguered biota, is a revealing example of these factors at work on terrestrial invertebrates. Of the total number of animals that live on the Hawaiian Islands and in the surrounding sea, about three-fourths are terrestrial invertebrates, 99% of which are found only on the Hawaiian Islands. The islands are so strongly affected by habitat destruction that whole habitat types, containing all their habitat-specific invertebrates, are considered endangered. Invasions of alien plants and animals are so severe that whole groups of invertebrates have been devastated, such as the Hawaiian land snails, which were originally believed to number approximately 1000 species but now number approximately 500, with most of the remaining species endangered. The Hawaiian Islands may provide the best opportunity to document relatively easily the scale of the threat to terrestrial invertebrates in especially vulnerable places, such as oceanic islands (Fig. 1).

II. AWARENESS AND UNDERSTANDING OF ENDANGERED TERRESTRIAL INVERTEBRATES

Why is it taking so long to recognize and document endangered terrestrial invertebrates? There seem to be both logistical and strategic considerations involved.

FIGURE 1 The Hawaiian tree snail *Achatinella mustelina* is 1 of 41 species in its genus; 22 of these species are extinct, and the remaining 19 are severely threatened. These elegant, specialized, and slow-growing snails have been clobbered by a series of disasters. There was a shell-collecting frenzy of the 1800s, which also decimated populations of Florida tree snails. This was followed by the destruction of most of the forest habitat in which *Achatinella* species live. Finally, in 1955, a predatory snail species was introduced to combat an introduced species of African snail that is destructive to Hawaiian agriculture. The killer snail quickly eliminated several species of *Achatinella*, and several other species survive only as captive colonies.

A. Logistical Problems

A major logistical problem is an insufficiency of experts. There are so many species of terrestrial invertebrates that it is easy to find groups with hundreds or thousands of species that can only be identified by two or three people. Who can identify the dark-winged fungus gnats, Indonesian bark beetles, the egg parasites of tropical sac spiders, the mites associated with millipedes, or any one of innumerable large groups of small organisms? The few people who do study these groups are generally overwhelmed with species that are undescribed and with species whose biology is almost completely unknown. Recognizing any of these species as endangered requires documentation of distribution and abundance—levels of information that will probably not be available for most terrestrial invertebrates for approximately another century (Fig. 2).

A second logistical problem is that specialized knowledge is needed not only for identifying most terrestrial invertebrates but also for finding and counting them. Species that are rare in museum collections are not necessarily rare in nature; they are frequently species that are difficult to find because their habits are unknown. Certain flightless pygmy mole crickets, for example, are very common in ancient dune areas in Florida, but until recently there were no specimens in museums because nobody knew to look for these insects just under the surface of the sand just after a heavy rain. Many grasshoppers have species-specific songs which biologists must learn to study the abundance of the

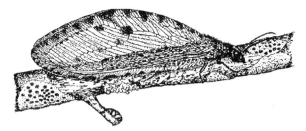

FIGURE 2 With regard to terrestrial invertebrates, there may be few people with the specialized knowledge to identify even large and easily characterized species. Many species could become rare or extinct without triggering alarms. The giant lacewing, *Polystoechotes punctatus*, apparently a common and widespread insect a century ago, is now so rare that few entomologists have seen a living individual. Nobody is writing distraught letters asking about the disappearance of this animal, and it does not appear on any list of endangered species. Nobody is doing any methodical surveys for this species; nobody knows how one would go about making such a survey because the ecology of the insect is poorly known. This species, which is relatively large (7-cm wingspread), distinctive, and of special interest because it is so primitive, has gone unrecognized as an endangered species. The great number of smaller and less conspicuous invertebrate species have that have become rare or extinct have even less of a chance of attracting attention.

FIGURE 3 The red widow spider (*Latrodectus bishoppi*) is rare in the sense that it has a very small geographic range (a few areas of the Florida peninsula) and a strict habitat requirement (Florida scrub). Within its remnant patches of habitat, the red widow is sometimes very common, whereas at other times it is extremely rare. This is probably due to predation by natural enemies such as spider wasps and egg sac parasites. If one took a single survey in an area, one might conclude that this species was endangered or not at all endangered, depending on the population level at the time of the survey. Is this species endangered? Nobody knows. It depends on whether there are factors that could wipe out populations when they are at their low point, since the habitat for the species is now in smaller patches that are farther apart.

adult insects; this has recently become a tool for studying endangered grasshoppers. Estimating the abundance of a species also requires specific knowledge of life cycles and population dynamics. Many terrestrial invertebrates have a dormant period in the egg or pupal stage, during which they are almost impossible to find. Invertebrates often produce large numbers of offspring per female, and the percentage of survival of these offspring may be strongly dependent on variable factors, such as weather or the population levels of certain predators. This means that populations may go through frequent fluctuations of abundance and rarity, neither of which may be good predictors of the long-term survival of a population (Fig. 3).

Given all these logistical problems, however, there are still ways of quickly identifying large numbers of genuinely endangered invertebrates. There are whole groups of invertebrates that include many habitat specialists and have poor dispersal abilities, and therefore they are likely to be endangered if their habitat is reduced to small fragments. Examples of these unfortunate animals are the land snails, millipedes, flightless beetles, flightless grasshoppers and crickets, and several groups of primitive flies. It would be relatively easy to compile a long list of endangered invertebrates by surveying any old and distinctive habitat that is rapidly disappearing. There are some ecological groups of invertebrates that tend to have specific, coevolved relationships with other organisms and are endangered by anything that threatens their host. Examples of these excessively specialized invertebrates are most leaf-mining and gall-making insects and mites, seed predator insects, insects and mites associated with plant genera that contain only one species, pollen feeders that visit one or a few species of flowers, and the fleas, lice, feather mites, and den inhabitants of vertebrates. Since most species of endangered plants and vertebrate animals are likely to have at least one host-specific invertebrate, it would be relatively easy to compile a long list of endangered invertebrates by studying the invertebrate associates of larger and better known endangered species.

In summary, there are currently insurmountable difficulties that prevent the recognition of more than a small percentage of the probable number of endangered terrestrial invertebrates. Even this small percentage of easily recognized species, however, is not receiving the intensive attention that one might expect, which suggests that there are additional problems with recognizing endangered invertebrates.

B. Strategic Considerations

There are two species of bird lice (*Franciscoloa thompsoni* and *Neopsitticonirmus emersoni*) that live only on the threatened Philippine cockatoo. These are examples of endangered invertebrates that might go unrecognized for strategic reasons. First, lice have no general appeal, and to say that this particular parrot is infested with unique and interesting species of lice is most unlikely to further spur efforts to save the bird. Simultaneously, there seems to be no strategic need to recognize the endangered species of lice because their survival depends on that of their host, whose status is already recognized. In the long run, however, it is important to understand that each species of organism represents a complex of ecological relationships, and the presence of a host-specific parasite is indicative of ancient sequences of adaptation which add to the significance of the host species. Moreover, host animals may be molded by their parasites in ways that are just beginning to be acknowledged by biologists. Therefore, for example, mutual grooming behavior may have evolved through the presence of body parasites on places that are difficult to reach, and this mutual grooming may have a major role in social bonding between individuals. Some biologists also believe that strenuous courtship performances and extravagant male ornamentation may evolve to demonstrate vigor and resilience in the face of the parasite load that is borne by almost all wild animals (Fig. 4).

At a more general level of strategy, conservationists are caught in the same contradiction that afflicts all biological educators. On the one hand, simple messages are the most effective. On the other hand, life is irreducibly complex. The story of the Florida scrub jay, for example, is complicated enough without considering all the grasshoppers, beetles, millipedes, and other invertebrates that are also restricted to Florida scrub habitat. The choice is often made to focus on a few species whose protection through habitat preservation will automatically protect a large number of other species,

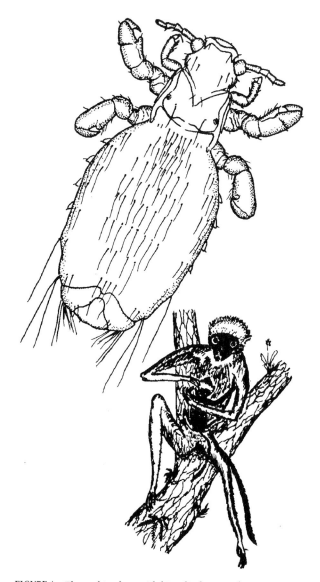

FIGURE 4 The sucking louse, *Phthirpediculus propitheci*, is a parasite on the body of the endangered sifaka lemur, *Propithecus diadema*. This louse, which is indubitably endangered because of the status of its host, seems an ideal candidate for Gilbert and Sullivan's "little list of society offenders who might well be underground, and never would be missed!" Maybe not, however, because sifakas, like most other primates, do much of their socializing during episodes of mutual grooming, and the removal of the stimulus for this interaction could change the intimate and supportive behavior of the species.

such as terrestrial invertebrates. It may seem that this is an unscientific approach, one that panders to a perceived bias against small, crawling creatures. In reality, it may be more a question of available information—both scientific information and information that is part of our human heritage. The biology of the Florida scrub jay is known in great detail, including the complex of factors that threaten its existence, whereas the biology of the Florida scrub millipede is almost completely unknown. Likewise, almost everybody has an intuitive understanding of the significance of habitat loss to the Florida scrub jay because this bird, like our own species, is relatively intelligent, home-loving, hierarchical, territorial, and lives in nuclear families. Accurate and detailed communication about the world of the Florida scrub millipede is not currently possible. If one were to make the choice between studying a few endangered species in detail or studying many endangered species more superficially (a choice that few biologists actually make), it might make sense to study a few species in-depth (Fig. 5).

In the long run, however, the study of a great variety of endangered species provides a wealth of fine-grain

information on management and biogeography that is omitted when pandas or tigers are used as "umbrella species" whose protection helps to protect thousands of species that share the same habitat. A portion of the earth's endangered biota is contained in sites that do not support any endangered vertebrate species. The endangered blue butterflies of the genus *Maculinea* often occur in small sites in which there are no endangered vertebrates. Certain small islands near the larger islands of New Zealand lack remaining endangered vertebrates but have retained some endangered invertebrates. Although some endangered invertebrates quickly disappear from habitat fragments, others persist, together with endangered plants, long after the habitat-specific vertebrates have been extirpated. Management of endangered invertebrates may require consideration of microhabitat features that are less likely to be crucial to an endangered vertebrate, such as a species of crane or a species of antelope, because these larger animals have more general requirements. *Maculinea* butterflies provide a good example. Their caterpillars not only require certain food plants but also must spend part of their lives in the nests of particular species of *Myrmica* ants, which also have habitat requirements. Rotting tree trunks are a more generic example of a microhabitat; they are the only home for an enormous number of invertebrate species. The loss of this habitat type through salvage logging might have a small effect on vertebrate species but a great effect on invertebrates. Invasions of alien invertebrates are generally more of a threat to endangered invertebrates than to endangered vertebrates because these invaders are likely to be direct predators or competitors of native invertebrates.

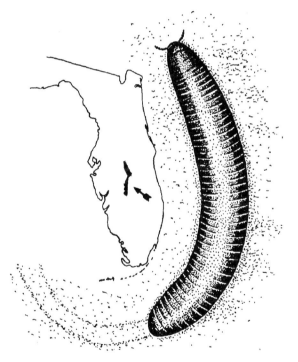

FIGURE 5 The Florida scrub millipede, *Floridobolus penneri*, is a large gray species that burrows in the deep sand of Florida's Lake Wales Ridge. It emerges at night to feed on dead scrub oak leaves and to disperse. This species is an example of an unrecognized endangered invertebrate that receives some protection only because it occurs on several sites on which the endangered Florida scrub jay also occurs. The Florida scrub millipede is absent, however, from most of the range of the Florida scrub jay and is present in some small habitat fragments in which the jay is absent; therefore, the bird is not a truly effective "umbrella" for the invertebrate. Perhaps this millipede also has some special microhabitat requirements.

III. THE SPECIAL PERILS OF SPECIALISTS

The very factors that have made the terrestrial invertebrates such a huge, successful group have guaranteed that large numbers of their species will become endangered as humans change and destroy natural habitats. The small size of terrestrial invertebrates has allowed them to specialize on miniature resources. Huge numbers of species, for example, feed on only one part of a single species or genus of plant. Many other species are internal parasites in a narrow range of insect hosts. The great advantage to this specialization is that it has allowed spectacular efficiency in finding and exploiting resources. The small bark beetle, *Cactopinus hubbardi*, which raises its larvae in the injured tissue lining the cavities made by woodpeckers in saguaro cactus, only needs to deal with the nutrients and defensive compounds of one kind of plant. It is probably able to zero in on its breeding sites, which are unlikely to be numerous in any one area, by following an odor plume emanating from the injured cactus.

Since terrestrial invertebrates are able to rest in a dormant state with minimal energy expenditure, they can evolve life cycles synchronized with the availability of their resources, and this also increases the ability to specialize. The hundreds of species of solitary bees in arid habitats of Mexico and the southwestern United States can synchronize their emergence with the seasonal changes or the periodic rains that stimulate blooming in their plant hosts. This, combined with coevolution between the mouthparts of bees and the architecture of flowers, has led to many species-specific bee and flower relationships. The evolution of many specializations is driven by the benefits of greater efficiency and less competition accrued by specialists. When humans disturb natural habitats, however, extreme specialization is a liability for many inhabitants because even temporary loss of a resource can eliminate the species that depended on that resource (Fig. 6).

The small size of terrestrial invertebrates allows them to maintain thriving populations of habitat specialists in a small area, such as an isolated mountaintop with some alpine habitat that was colonized at the end of a glacial period. Evolutionary biologists believe that speciation is most likely to occur in such isolated populations peripheral to much larger populations. The isolated subspecies (really species in the making) of butterflies known as arctics (*Oenis*) and alpines (*Erebia*) show that this process can occur over a relatively short time. In areas in which there has been long-term isolation of habitat fragments, it is often possible to find isolated populations that have diverged so much that they are clearly distinct species that could never merge, even if they were brought back together. Some groups of flightless Orthoptera provide good examples of this process of speciation at work. Many of these distinctive forms and species of terrestrial invertebrates could persist indefinitely in the small areas of habitat where they now occur, but these species and forms can be considered endangered because they could easily be eliminated by an episode of habitat destruction that would be small by current standards (Fig. 7).

Terrestrial invertebrates as a group are not at risk because there are many species that are generalists or are widely distributed. A significant proportion of the total diversity of invertebrates, however, is composed

FIGURE 6 The Schaus swallowtail (*Papilio aristodemus ponceanus*), whose caterpillar feeds on young leaves of torchwood and wild lime, is a good example of how specialization contributes to both the evolution of species and the vulnerability of species. The young and tender leaves and shoots of plants have concentrated nutrients and are easy to chew, so they are usually greatly preferred by leaf-eating insects such as caterpillars. Plants, however, usually protect these young tissues with chemicals. Much of the diversity of butterflies and moths is based on species-specific adaptations: timing of the life cycle to make the best use of growth spurts of plants and specialized detoxification systems for particular plant poisons. The new growth of plants is concentrated at the top of the plant where there is the most light. Pollutants and pesticides, such as aerial sprays for adult mosquitos, land mostly of the tops of the plants, where they are most likely to be consumed by caterpillars on young leaves. The Schaus swallowtail seems to quickly disappear in areas sprayed for mosquitos.

of species that are highly specialized or restricted to one or a few small patches of habitat.

IV. THE SIGNIFICANCE OF ENDANGERED TERRESTRIAL INVERTEBRATES

A sparrow falls, Thou art mindful; A spider is gone, art Thou vexed?

Why should anybody be concerned about endangered terrestrial invertebrates? Some people view this as a question of ethics: They believe that *Homo sapiens* is not the only species with a right to exist, or they believe that it is wrong to rob all future generations of the rich biological heritage that was passed down to us. It is true that there have been previous waves of extinction

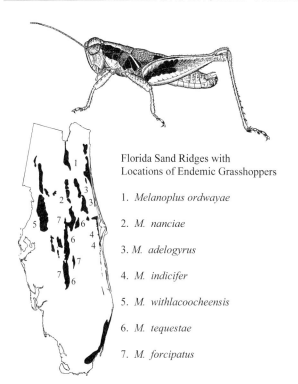

Florida Sand Ridges with
Locations of Endemic Grasshoppers

1. *Melanoplus ordwayae*

2. *M. nanciae*

3. *M. adelogyrus*

4. *M. indicifer*

5. *M. withlacoocheensis*

6. *M. tequestae*

7. *M. forcipatus*

FIGURE 7 In Florida a series of isolated ancient dune fields have their own species of grasshoppers that are unable to live in the surrounding lowlands. Since these upland sites escape the flooding that is common in lower areas, they are well suited for housing and agriculture, and these grasshoppers have much reduced habitat and at least three appear to be endangered. Distinctive endemic species such as these grasshoppers also serve as biogeographic indicators, showing that the areas they inhabit have been isolated for a long time and might have additional endemic plants and animals. The number of examples of endangered Florida invertebrates appearing in this article is due to my familiarity with these species; there are many examples of endangered invertebrates almost everywhere.

of invertebrates caused by such events as ice ages or perhaps the impact of asteroids. Ethics, however, denies to us the innocence of an ice sheet or shooting star. A foundation stone of ethics is that we must endeavor to understand and take responsibility for the consequences of our actions. Since our species is strongly guided by ethics, ethical considerations may be as real and important as the dictates of materialism. The origin and nature of ethical attitudes toward other species have been considered by several scientists, especially Edward Wilson. Even from a materialistic standpoint, however, it makes sense to be concerned about endangered invertebrates, for a variety of reasons.

First, endangered terrestrial invertebrates may be viewed as repositories of information. Currently, we have neither the time nor the skill to interpret this information. Some endangered invertebrates might pro-

duce useful chemicals. They might have innovative defenses against fungi or bacteria. They might have peculiar genetic systems or developmental pathways that are easy to analyze. Endangered species might inspire new areas of microengineering. They might display a variety of physiological mechanisms to deal with extreme environmental conditions. They may present more detailed evidence of evolutionary trends. They may be convenient indicators of environmental change. They might have novel types of mutualistic relationships with other animals or with plants. They may be remnant populations of formerly abundant species with important roles in restored ecosystems. Introduced pests might be controlled by invertebrates that are rare or endangered in their homeland.

In addition to these potential material benefits, nonmaterial attributes of endangered invertebrates could have their own materialistic spin. Humans are willing to spend prodigious effort and money on nonmaterial things, such as entertainment, aesthetics, or the opportunity to make new discoveries. Many endangered species of terrestrial invertebrates might achieve their highest value as examples of the beauty and intricacy of life; considering the psychological needs of our species, this is an offering that can never come too often or in too many guises. Already, the conservation of several species of butterflies is pushed by the market value of aesthetically pleasing specimens or of live specimens for butterfly houses that charge admission.

With regard to terrestrial invertebrates, we are currently in an inspiring but frustrating state of ignorance in which any useful quality that we can imagine may well be represented in the group, but our imaginations are clearly insufficient to the task.

Second, the things that we care about are often dependent on small, unconsidered details: the linchpin that holds the wheel on the axle or the knot at the end of the thread anchoring the stitching. From this truth, it is easy to postulate that there are ecological systems that are tenuously held together by endangered invertebrates. Good examples might be found in old, isolated systems with a very limited number of species, such as a cave or an oceanic island. The situations in which a single species has this linchpin role are likely to be few and exceptional. Most large-scale ecological systems, such as a large area of rain forest in New Guinea or desert in Arizona, must have impressive built-in versatility or they would not have survived the many natural changes affecting the planet before the advent of our species. On the other hand, biological systems of all kinds can only withstand a certain degree of stress. It is difficult to recognize the approach of a breaking point, much less identify the relatively small events that could serve as triggers.

Third, although a single rare and endangered species of invertebrate is unlikely to determine the fate of a large ecological system, rare invertebrates may be important in the aggregate (Fig. 8). In the remaining natural habitats of the world, especially those in warm and warm-temperate climates, most of the total diversity of invertebrates is composed of relatively rare species. Many of these rare species could quickly become endangered species if their habitat changes in some major way. This change could be general habitat disturbance, or it could be habitat fragmentation. It could also be caused by extensive contamination by pesticides to control agricultural pests. Aerial application of pesticides to control biting insects is a special problem because a reduction in the incidence of human disease can be used to excuse even the most destructive of practices. Most insect-borne diseases, however, are cycling in

heavily settled areas; they seldom sweep out of large tracts of natural habitat. In the future, invertebrates might be endangered by genetically altered plants or pathogens that carry self-replicating pesticides from agricultural areas into other habitats. One of the greatest threats to rare invertebrates today is invasion by exotic invertebrates. All these kinds of threats, which often occur in combinations, have a single general effect on ecosystems: They replace many specialized invertebrate species with a few generalists. The cumulative effect of this loss of invertebrate diversity may be to threaten populations of large, noticeable organisms, such as certain large vertebrates or plants. At a more basic level, however, much of the efficiency and precision of energy flow through ecosystems relies on the rarer and more specialized species of invertebrates, just as the efficiency and precision in our communication relies on the availability of a huge number of rarely used and specialized words.

V. NO SECOND CHANCES

Each species of endangered invertebrate is an old and irreplaceable entity. This seems like an obvious point, but one so often hears of "new" species of invertebrates that it is easy to unconsciously absorb the impression that invertebrates are evolving at a prodigious rate. These new species are always old species that are newly discovered. It is true that in special circumstances invertebrate populations can diverge relatively rapidly, evolving behavioral and physiological differences specific to the different conditions in different sites. Most invertebrate species, however, are distinguished by morphological specializations that probably took a long time to evolve. This can be seen by examining the fauna of recently formed islands, such as the Bahamas. The Bahamas were submerged by the sea about 135,000 years ago and emerged again approximately 100,000 years ago. Although 100,000 years is a long time by human standards, it is evidently short by evolutionary standards—insufficient for the evolution of a major group of Bahamian species of terrestrial invertebrates or terrestrial plants.

Recently, systematists (specialists who study the evolutionary relatedness of animals and plants) have put forward the argument that special attention should be given to endangered species that are the last remnants of evolutionary lines that largely disappeared millions of years ago. It is noted that these relicts may have unusual kinds of adaptations and also provide glimpses

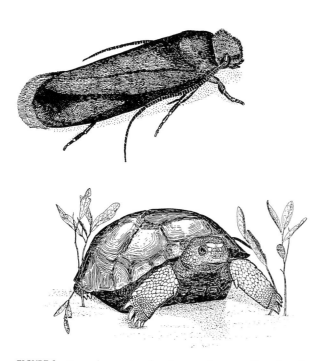

FIGURE 8 General ecosystem functions, such as recycling of nutrients, are most efficiently done by many specialists rather than by a few generalists. The plates that cover the shells of tortoises are made of keratin—a protein that, because of its strong disulfide bonds, few scavengers can digest. However, there is a moth, *Ceratophaga vicinella*, whose caterpillar appears to have a specialized diet of the shells of dead gopher tortoises. There are innumerable specialized invertebrates that feed on particular kinds of wood, bark, dung, and carrion. No single one of these species is likely to be important in an ecosystem, but as a group they have a large impact.

of ancient life on our planet. Certain endangered terrestrial invertebrates would be high on the list of species that would benefit from special consideration of relicts: Onychophora, giant mites, redwood Thysanura, and many primitive lineages known from small areas of the tropics or south temperate areas. Although there is some logic to priority for the most ancient and conservative of the endangered terrestrial invertebrates, unless systematists are willing to work closely with ecologists, the identification of relict lineages is not very useful. Moreover, although ancient relicts may be important for historical reasons and because they embody peculiar adaptations, they may have less relevance to modern systems than species that evolved from widespread lineages only a few million years ago. For example, useful natural enemies of pests of solanaceous crops (such as potatoes, peppers, and tomatoes) are most likely to be found among the large numbers of wasps and flies that attack insects on wild Solanaceae in some remnant habitat in the Andes Mountains.

In summary, terrestrial invertebrate species, once lost, cannot "reevolve," and the generation of new species that might restore lost diversity occurs on a timescale too long to be relevant to our own species. In other words, every year there are fewer species of terrestrial invertebrates than ever before in human history; every year there are more species of terrestrial invertebrates than there ever will be again, even should human civilization endure 100,000 years.

See Also the Following Articles

ENDANGERED FRESHWATER INVERTEBRATES • ENDANGERED MARINE INVERTEBRATES • INVERTEBRATES, TERRESTRIAL, OVERVIEW • TERRESTRIAL ECOSYSTEMS

Bibliography

Brown, K. S., Jr. (1997). Diversity, disturbance and sustainable use of Neotropical forests: Insects as indicators for conservation monitoring. *J. Insect Conserv.* **1**, 25–42.

Collins, N. M., and Thomas, J. A. (Eds.) (1991). *The Conservation of Insects and Their Habitats.* Academic Press, San Diego.

Deyrup, M., and Franz, R. (Eds.) (1994). *Rare and Endangered Biota of Florida. Vol. IV. Invertebrates.* Univ. Press of Florida, Gainesville.

Elmes, G. W., Thomas, J. A., Wardlaw, J. C., Hochberg, M. E., Clarke, R. T., and Simcox, D. J. (1998). The ecology of *Myrmica* ants in relation to the conservation of Maculinea butterflies. *J. Insect Conserv.* **1**, 25–42.

Gagné, W. C. (1988). Conservation priorities in Hawaiian natural systems: Increased public awareness and conservation action are required. *Bioscience* **38**, 264–271.

Grove, S. J., and Stork, N. E. (1999). The conservation of saproxylic insects in tropical forests: A research agenda. *J. Insect Conserv.* **3**, 67–74.

Howarth, F. G., Sohmer, S. H., and Duckworth, W. D. (1988). Hawaiian natural history and conservation efforts: What's left is worth saving. *Bioscience* **38**, 232–237.

Matyot, P. (1998). The orthopteroids of the Seychelles: A threatened island fauna. *J. Insect Conserv.* **2**, 235–246.

McKnight, B. N. (Ed.) (1993). *Biological Pollution: The Control and Impact of Invasive Exotic Species.* Indiana Academy of Science, Indianapolis.

Middleton, S., and Liittschwager, D. (1994). *Witness: Endangered Species of North America.* Chronicle, San Francisco.

New, T. R. (1991). *Butterfly Conservation.* Oxford Univ. Press, Oxford.

Porter, S. D., and Savignano, D. A. (1990). Invasion of polygyne fire ants decimates native ants and disrupts arthropod community. *Ecology* **71**, 2095–2106.

Riede, K. (1998). Acoustic monitoring of Orthoptera and its potential for conservation. *J. Insect Conserv.* **2**, 217–223.

Swengel, A. B., and Swengel, S. R. (1997). Co-occurrence of prairie and barrens butterflies: Applications to ecosystem conservation. *J. Insect Conserv.* **1**, 131–144.

Tepedino, V. (1997). Wild bees and floral jewels. *Wings* **20**, 8–11.

Wilson, E. O. (1988). *Biodiversity.* National Academy Press, Washington, D.C.

Wilson, E. O. (1992). *The Diversity of Life.* Belknap Press of Harvard Univ. Press, Cambridge, MA.

ENDEMISM

R. M. Cowling
University of Cape Town

GLOSSARY

biotope Region that is distinguished by particular environmental conditions (climate, soil, altitude, etc.) and therefore a characteristic assemblage of organisms.

stenotopic Referring to taxa with restricted habitat requirements (i.e., confined to a single biotope) and hence restricted distributions.

ENDEMIC TAXA ARE THOSE RESTRICTED TO A SPECIFIED GEOGRAPHICAL AREA. Therefore, the concept is a relative one; the patterns, correlates, and causes of endemism will vary according to the size and location of the geographical area, as well as the taxonomy and phylogenetic relatedness of the assemblage under consideration. At a global scale, all taxa are endemic and there is relatively little to say on the topic. Most research has focused on species that are endemic to relatively small areas. In this sense, endemism is best viewed as a form of rarity, that is, range-restricted rarity. This article presents biogeographical, evolutionary, ecological, and conservation perspectives on endemism and discusses generalizations regarding the patterns, correlates, and causes of species-level endemism in relatively small areas.

I. CATEGORIES

Endemics may be categorized according to their spatial distribution, inferred evolutionary age, affinities, and local abundance.

A. Spatial Distribution

Endemics are loosely and commonly categorized in four contexts of spatial distribution: site or restricted area; biotope; biogeographical region; and political area.

B. Evolutionary Age and Affinity

Categorization of endemics according to evolutionary age and affinity is summarized in Box 1. These schemes have been widely used by botanists but rarely by zoologists. Problems associated with the schemes are that (1) age is regarded as a categorical rather than continuous variable; (2) the establishment of relationships among

Box 1

Categorization of Endemics According to Evolutionary Age and Affinities

A. Engler's scheme, published in 1882.

1. Neoendemics: comprising clusters of closely related species and subspecies that have evolved relatively recently.
2. Palaeoendemics: comprising phylogenetically high-ranking taxa, usually monotypic sections, subgenera, or genera that may be regarded as evolutionary relics.

R. C. Favarger and J. Constandriopoulos's scheme, published in 1961. This scheme uses cytological data to provide a more rigorous basis for assessing the age and affinities of endemics.

1. Palaeoendemics: ancient isolated taxa with a high ploidy level, whose diploid ancestors are extinct or unknown.
2. Schizoendemics: vicariant species of equal ploidy level, resulting from either gradual or rapid divergence.
3. Patroendemics: restricted diploid species that have spawned younger, widespread polyploid species.
4. Apoendemics: polyploid endemics that are derived from widespread species of a lower ploidy level.

Schizo-, patro-, and apoendemics are further subdivisions of Engler's neoendemics.

taxa lacks rigor; and (3) many palaeoendemics are diploid.

Phylogenetic methods, which consider the distribution of characters among taxa in a cladistic context, provide a rigorous categorization of endemics in terms of relative age and propinquity of descent. In this context, low-ranking taxa correspond to neoendemics and high-ranking taxa to palaeoendemics (cf. Box 1). An absolute estimate of the age of endemics can be given when congruent phylogenetic relationships correlate with identifiable historical events.

C. Local Abundance

The classical, biogeographical perspective on endemism has tended not to consider the local abundance of spe-

cies. However, in the more recently developed ecological and conservation perspectives, in which endemism is conceived as a category of rarity, population abundance is invariably explicitly considered. Thus, geographical range size as a categorical variable (wide/narrow) has been used as one of the factors in defining seven forms of rarity recognized for plants. Endemics (narrow range) may belong to any four categories of rarity according to biotope specificity (broad/restricted) and local population size (somewhere large/everywhere small).

II. PERSPECTIVES

The concept of endemism has a long history in biology, dating back to A. P. De Candolle's treatise, published in 1820. Most research on the topic has been in the field of descriptive biogeography, where distribution patterns of taxa have been used to define centers of endemism at various spatial scales. This approach provides a static perspective of endemism.

Over the past few decades, historical biogeographers have evaluated areas of endemism for monophyletic lineages in a phylogenetic context. This approach provides a dynamic perspective of endemism, especially when endemic taxa show congruent phylogenetic relationships that can be correlated with historical events.

Evolutionary biologists, studying both fossil and extant biotas, have explored the role of range restriction as a cause and consequence of speciation. Recently, several statistical techniques have been employed—collectively termed the comparative method—to exploit the phylogenetic relationships among species to extract independent information on the evolutionary correlates of endemism. These techniques acknowledge that related species may have similar range sizes, that is, range size cannot be assumed to be independent among species. However, in at least some cases, variance in range sizes seems to be partitioned mostly at the species level.

Community ecologists have conceptualized endemism as one of several forms of rarity, namely, range-restricted rarity, and have explored its role as an explanatory variable for taxon-specific ecological traits, such as local population size, body size, reproductive fitness, and dispersal distance. Increasingly, they are using comparative methods to correct for phylogenetic relatedness among biotas. However, for every cause–effect relationship documented, there are numerous exceptions.

Conservation biologists view range-restricted rarity

as an attribute that predisposes a taxon to extinction. They seek to understand the abiotic and biotic correlates of this form of rarity as a basis for management guidelines that will reduce rates of extinction. A distinction is often, although not always, made between naturally rare species that may have some adaptation to rarity and those that have previously been widespread and are now restricted. Conservation planners often use patterns of endemism to identify reserve systems that are representative of a region's biodiversity. Many reserve selection algorithms have been formulated to select sites that have a unique or endemic complement of species.

III. MEASUREMENT

In quantifying patterns of endemism, the units of measurement (spatial scale and taxonomic entity), the mode of reporting of the data (percentages or counts), and a number of biases all influence the interpretation of the results. Of great importance is the relative nature of endemism: evaluation is always dependent on the spatial context and biological assemblage under consideration. This section provides a clarification of the problems and approaches associated with the measurement of endemism.

A. Units of Measurement

A variety of methods have been used to measure the range sizes of taxa. A useful distinction can be made between measures that attempt to estimate the extent of occurrence of a taxon—the distance between the outermost limits of a species' occurrence—and the area of occupancy—the area over which the species is actually found. The latter measure is particularly relevant for ecological studies that seek correlations between range size and environmental tolerances, as well as for conservation planning research; extent of occurrence is widely used in biogeographic studies.

Measures of endemism invariably seek to identify a subset of taxa within an assemblage that can be classified as having a lower than average range size. Within the biotas of larger-scale regions—biogeographic areas or countries—many researchers have recognized "local" endemics as a distinct category. However, the range size, or extent of occurrence, for defining this category is often arbitrarily set, varying between 50,000 km² (for Neotropical birds and plants, as well as for birds globally) to 2000 km² or less (for plants in the Cape

Floristic Region). Endemics with extremely small range sizes—<5 km²—are regarded as point endemics. An approach that is increasingly being used is to evaluate endemism as a continuous variable, calculated as the sum of the inverse range sizes of all taxa in each quadrat (cell grid or map unit).

From both the biogeographical and ecological perspectives, patterns of endemism are best studied in relation to ecologically homogeneous, biogeographical regions. However, conservation planners often use political regions or property boundaries when evaluating endemism, since these may be the most effective decision-making unit for the preservation of endemics.

The taxonomic or phylogenetic scales employed also influence patterns of endemism. Centers of endemism identified on the basis of patterns among low-ranking taxa (sub-species or closely related species) often differ from those where the units are high-ranking members of the same lineage. Similarly, the spatial scale for defining endemism will vary among different taxa of the same rank.

B. Percentage versus Counts

Endemism may be expressed as a percentage of all extant taxa present, or as the absolute number of endemics in an area. Depicting plant endemics in biogeographic regions as percentages or counts, and using area and latitude as explanatory variables, results in different patterns with different significance (Fig. 1). Some species-poor areas, such as oceanic islands and arid regions, although low in actual numbers of endemics, may support a high percentage of endemic taxa. Others areas, such as Madagascar, the Cape Floristic Region, and parts of the Neotropics combine high richness and high endemism for some taxa. Ideally, both measures of endemism should be considered when explaining patterns, but seldom are.

C. Biases

Endemism is influenced by taxonomic interpretation, sampling error, and human perceptions of rarity. Of particular importance is the fact that limited geographical exploration, as well as variation in the application of taxonomic concepts, introduces biases in the identification of endemics and the significance of their status. Pseudoendemics are widespread species incorrectly classified as endemics, whereas nonapparent endemics are endemic species that are incorrectly classified as widespread. The fact that widespread species are usually more thoroughly researched than those with smaller

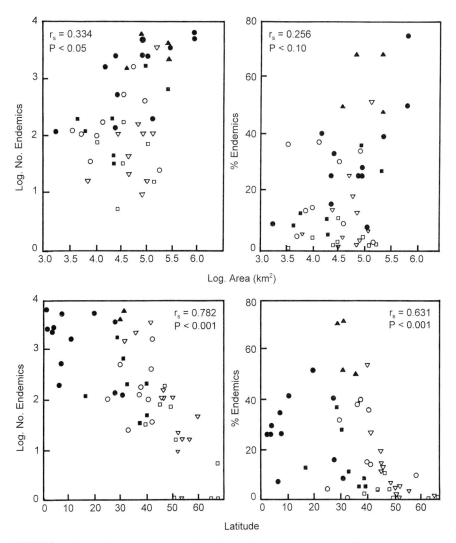

FIGURE 1   Relationships (Spearman's rank correlation) between two measures of endemism (percent-ages and counts) and area and latitude for plants in 52 biogeographical units in tropical and subtropical forests and savanna (●), temperate forest and woodland (○), Mediterranean-climate shrubland and woodland (▲), warm desert and steppe (■), cold desert and steppe (□), and boreal forest and tundra (▽) on continental landmasses across the globe. (Reprinted with permission from Cowling and Samways, 1995. Endemism and biodiversity. Cambridge University Press.)

range sizes introduces biases in studies that explore the correlates of range size.

IV. PATTERNS

There are very clear global and regional patterns of endemism for a wide range of taxa: endemics are not randomly distributed across the globe. However, these patterns are constrained by poor taxonomic knowledge and distributional data in key areas (e.g., the tropics) and for some taxa (e.g., most invertebrate groups).

A. Latitudinal Gradients

The incidence of endemism for whole assemblages in biogeographic zones increases with decreasing latitude (see Fig. 1). Range sizes, as measured by latitudinal extent, increase for a wide range of organisms above a latitude of approximately 40°–50°N, but the same patterns are not evident in the Southern Hemisphere. There are many patterns that are not consistent with the generalization—termed Rapoport's Rule—that range sizes of taxa decrease with decreasing latitude, as a consequence of greater ecological specialization in less seasonal environments. For example, very high endemism for terrestrial taxa is recorded in the mid-latitudes of the Southern Hemisphere, particularly in and adjacent to Mediterranean-climate regions. Marine teleost fishes have smaller range sizes at higher than at lower latitudes, and endemism for marine algae peaks in mid-latitude areas. These patterns are probably more a product of speciation and extinction processes than contemporary ecological conditions. Thus, widespread glaciation during the Pleistocene at high latitudes in the Northern Hemisphere resulted in the extinction of less tolerant terrestrial taxa. In mid-latitude Mediterranean-climate regions that escaped glaciation, rates of speciation, at least for plants, have overwhelmed extinction rates, resulting in an accumulation of habitat-specialist, range-restricted species.

B. Centers

Many centers of endemism—areas of higher than average concentrations of range-restricted taxa—have been recognized globally and regionally, principally for higher plants and large-bodied terrestrial vertebrate faunas. Generally, many groups of organisms show a concentration of centers at lower latitudes (Fig. 2). Following from the previous section, it is no surprise that the high-latitude areas of the Northern Hemisphere support few centers. However, this is not always the case in the Southern Hemisphere, where large numbers of range-restricted taxa occur in middle- to high-latitude landmasses that were never glaciated during the Pleistocene.

C. Congruence

Overlapping or congruent areas of endemism for different taxa have been used extensively by biogeographers to reconstruct historical events. Patterns of congruence of endemism are also important for identifying reserve systems that maximize the preservation of different biotas.

Although strong patterns of congruence have been recognized for some taxa at the global scale—for example, swallowtail butterflies and tiger beetles, amphibians, birds, and mammals—higher plant centers often do not coincide with faunal centers. Nonetheless, on the basis of congruent patterns of endemic species diversity for mammals, reptiles, amphibians, and higher plants, it has been possible to identify 17 "megadiversity countries" (Table I), that is, political units of high conservation value.

At a finer scale, patterns are highly variable among different taxa and in different regions, and no generalizations have emerged. This lack of strong congruence underlines the fact that endemism is an expression of many different causes, both ecological and historical.

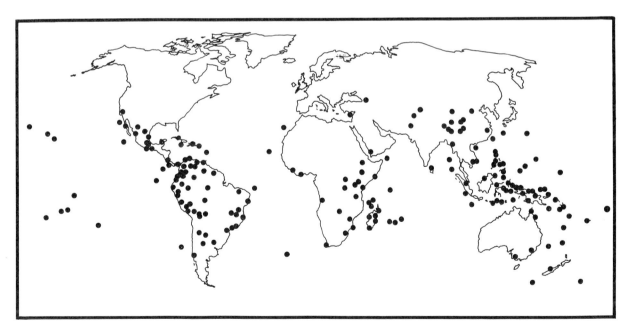

FIGURE 2   The distribution of Endemic Bird Areas of the globe, as recognized by Birdlife International. These centers are identified on the basis of the distributions of 2609 bird species that have had in historical times a global breeding range of less than 50,000 km². (Reprinted with permission from Bibby *et al.*, 1992.)

TABLE I

Vertebrate and Higher Plant Endemism in the World's 17 Megadiversity Countries

Country	Area (km² × 10³)	Mammals	Birds	Reptiles	Amphibians	Plants
Brazil	8512.0	131 (4)[a]	>191 (3)	172 (5)	294 (2)	ca. 17,500
Indonesia	1916.6	201 (2)	397 (1)	150 (6)	100 (11)	ca. 16,000
Colombia	1141.7	28	>142 (5)	97 (11)	367 (1)	ca. 16,000
Mexico	1972.5	140 (3)	125 (6)	368 (2)	169 (6)	ca. 12,500
Australia	7686.8	210 (1)	355 (2)	616 (1)	169 (5)	14,458
Madagascar	587.0	77 (8)	103 (8)	274 (3)	176 (3)	ca. 9200
China	9561.0	77 (7)	99 (9)	133 (7)	175 (4)	ca. 10,000
Philippines	300.8	116 (5)	183 (4)	131 (8)	44	ca. 5000
India	3287.8	44 (12)	52 (12)	187 (4)	110 (10)	ca. 7500
Peru	1285.2	46 (11)	109 (7)	98 (10)	>89 (12)	5356
PNG Papua New Guinea	475.4	57 (9)	85 (10)	79	134 (8)	ca. 13,000
Ecuador	283.6	21	37	114 (9)	138 (7)	ca. 4500
United States	9372.1	101 (6)	71 (11)	90	126 (9)	4036
Venezuela	912.1	11	45	57	76	ca. 6000
Malaysia	329.7	27	11	68	57	ca. 7250
South Africa	1221.0	27	7	76	36	16,500
Democratic Republic of Congo	2344.0	28	23	33	53	3200

[a] Figures in parentheses are rankings for the number of endemic species among the top 12 countries.

V. CORRELATES AND CAUSES

Range size or degree of endemism shows some clear relationships with a wide array of abiotic and biotic factors. These correlations are very useful in conservation biology since they may be used to identify factors that predispose endemic species to extinction. However, correlates may be either a cause or a consequence of endemism. To identify the causes of endemism in an evolutionary context, comparative methods that exploit phylogenetic relationships must be employed.

The causes of endemism are complex and numerous, and include intolerance of widespread habitats, niche specialization, isolation in marginal habitats owing to climate change, phylogenetic predisposition to narrow habitat selection, competition from alien species, and recent speciation of isolates in marginal habitats. Therefore, historical processes, contemporary ecological factors, and inherent biological properties of lineages are involved. In many cases, historical factors may be overriding, resulting in a poor relationship between measures of endemism and explanatory variables reflecting the contemporary environment.

Establishing correlates is a useful step in explaining patterns and causes of endemism. Most pertinent studies have addressed the following question: When compared to more widespread taxa, are endemics, however defined, a random subset of the biota with regard to abiotic and biotic factors? Developing these profiles, however, has been complicated by different definitions of endemism, multiple interactions between different traits, and a failure to consider phylogenetic relatedness.

This section provides a brief review of the abiotic and biotic interspecific correlates of narrow range size, and concludes with an assessment of the role of endemism in speciation.

A. Regional Species Richness

There is often a positive relationship between the incidence of endemism and regional-scale richness. This results from the importance of high habitat-related and geographical compositional turnover (beta and gamma diversity, respectively) in producing regional richness. Habitat specialists (or stenotopic species) and geographical vicariants often have narrow range sizes. However, there are also many cases where patterns of endemism and diversity are largely noncoincident. Examples include plants in the Neotropics, birds in the Andes, dragonflies and terrestrial vertebrates in southern Africa, and the biotas of many oceanic islands and deserts.

B. Area

As a generalization, proportionate and absolute measures of endemism increase with increasing area (see Fig. 1), irrespective of the taxonomic level. However, the relationship between number of endemic species (counts) and area is not as tight as that for the more widely studied species–area relationship. This results from the lack of congruence between endemism and richness in many areas (e.g., arid lands and oceanic islands).

C. Abiotic Environmental Factors

Levels of endemism may vary in a predictable way along gradients of rainfall, temperature, productivity, and habitat heterogeneity. Models that accurately predict levels of endemism on the basis of easily measurable environmental variables have been used for the rapid identification of endemic-rich areas.

For higher plants, levels of endemism increase with increasing productivity, with increasing elevation (reflecting increased habitat heterogeneity and isolation in high-altitude areas), and with higher rainfall in low- and middle-latitude areas, although many exceptions to these patterns exist. In the Mediterranean-climate regions of the Cape and southwestern Australia, there is a negative relationship between local endemism and soil fertility. In the California Floristic Province, palaeo-endemics are clustered in the wettest and driest areas, whereas neoendemics occur in transitional rainfall areas

where rates of speciation are highest. Similar patterns exist for Afrotropical birds and Neotropical butterflies. For a wide range of marine taxa, endemism is more pronounced in exposed and variable nearshore environments than in the more stable distant-shore habitats.

D. Biotope

Geographically isolated areas and biotopes, such as certain islands, mountain peaks, ancient lakes, caves, thermal vents, hot springs, vernal pools, the abyssal zone, and chemically imbalanced substrata, support a disproportionately high number of stenotopic endemics.

Most studies have focused on endemism on islands, mountains, and unusual substrata. Generally, larger continental islands such as Madagascar, New Caledonia, and New Zealand support the greatest number and proportion of endemic taxa, especially of higher plants. Elevational range explains the incidence of plant endemism on the Canary Islands and bird endemism in Indian Ocean archipelagos, suggesting the importance of topographical diversity. Continental islands are typically rich in palaeoendemics, whereas some taxa have undergone extensive and unusual adaptive radiation on oceanic islands such as the Canaries and Hawaii.

Mountains are also often rich in endemics, in both tropical and temperate regions, but not in recently glaciated, high-latitude areas of the Northern Hemisphere. Many desert inselbergs (granitic outcrops) act as mesic refugia that support endemics; this is particularly pronounced for plants in middle Asia. As on islands, endemism on mountains results from both historical (e.g., isolation) and ecological (e.g., heterogeneity) factors.

The restriction of endemic plant species to nutritionally imbalanced substrata, especially when these occur in an islandlike configuration, is widespread in Mediterranean-climate and humid tropical regions. These sites provide both a strong selective force for the evolution of neoendemics and a refuge from competition for palaeoendemics. The restriction of animal taxa to unusual substrata has not been studied in any detail, but is likely to be a response to habitat effects on vegetation structure rather than nutritional peculiarities per se.

E. Biology

Very few studies have addressed the relationships between restricted range size and biological factors such as body size, growth form, life-history traits, population size, and genetic architecture. Of these studies, few have

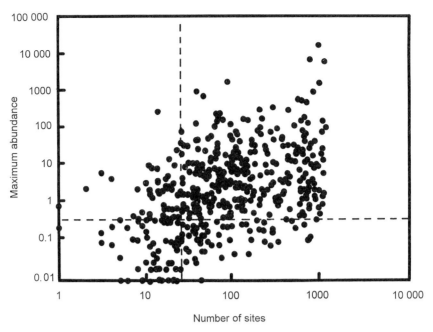

FIGURE 3 Relationship between the maximum local abundance value for North American birds and the number of sites at which each was recorded. (Reprinted with permission from Gaston, 1994.)

considered multiple trait interactions or phylogenetic relatedness.

As a generalization, there is a positive, albeit weak relationship between a species' range size and its local population abundance for a wide range of taxa (Fig. 3). However, not all endemics have low local abundances; indeed, many narrow plant endemics are extremely abundant locally. There are a number of hypotheses to explain the positive relationship between range size and local abundance. These are based principally on artefacts (e.g., sampling effects), resource use, metapopulation dynamics, and spatially independent rates of population growth. This pattern and its causes are currently attracting considerable attention.

There is a broad positive relationship between geographical range size and body size for animal species (Fig. 4). There is also a trend, both within regional floras and specific taxa, for low-stature growth forms to be overrepresented among plant endemics. This is especially true of the South African Mediterranean-climate region, where low shrubs (Fig. 5) dominate the endemic flora. Among Neotropical forest plants, endemics tend to be herbs, shrubs, or epiphytes rather than forest trees, whereas in the rain forests of Sri Lanka, endemics are overrepresented among long-lived, late-successional trees.

Gigantism is a common feature among some plant groups endemic to alpine habitats at low latitudes.

Among animals, gigantism, dwarfism, and flightlessness are widespread among island endemics, as well as among some continental endemics associated with insular biotopes.

The reproductive correlates of endemism have been more extensively studied than other biological attributes. There are a number of pertinent generalizations, although exceptions exist for all of them. Range-

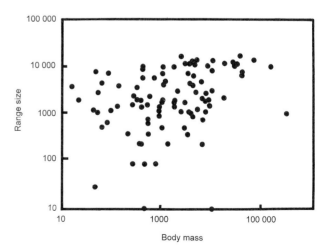

FIGURE 4 Relationship between the geographic range size (km² × 1000) and the body mass (g) of Neotropical forest mammals. (Reprinted with permission from Gaston, 1994.)

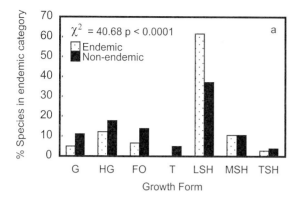

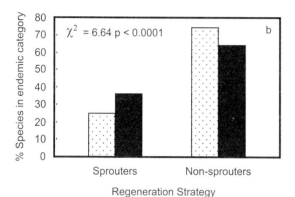

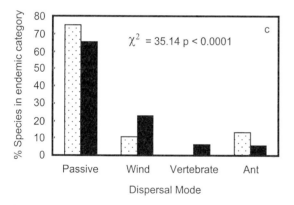

FIGURE 5 Percentage of endemic and nonendemic species in the Langeberg mountain flora (Cape Floristic Region, South Africa) in (a) seven growth form classes (G = geophyte, HG = graminoid, FO = forb, T = tree, LSH = low shrub, MSH = mid high shrub, TSH = tall shrub); (b) two postfire regeneration classes; and (c) four dispersal mode classes. Chi-square analyses were performed on untransformed data. (Reprinted from *Biological Conservation* **72**, D. J. McDonald and R. M. Cowling. Towards a profile of an endemic mountain fynbos flora: Implications for conservation, pp. 1–12. Copyright 1995, with permission from Elsevier Science.)

restricted species differ from common ones in that they:

- tend to be self-compatible or rely on asexual reproduction;
- tend not to be wind-pollinated or have other inefficient forms of pollen transfer;
- invest less in reproduction;
- have poorer dispersal abilities;
- have shorter generation times.

The last two attributes are shown in Fig. 5, where short-distance ant dispersal and fire sensitivity (rapid generation time) are overrepresented among plant endemics in a mountain region of the Cape Floristic Region in South Africa. However, these and other reproductive traits, such as seed size, seed number, and reproductive investment, all interact in complex ways. Furthermore, these traits are not independently distributed among species.

Many studies indicate that plant and animal endemics have lower levels of genetic variation in comparison with widespread congeners. This may be due to several factors, including adaptations to narrow ecological conditions, small population size, and self-incompatibility in plants. However, there are also cases of little difference in genetic diversity between closely related endemic and widespread plant species.

F. Taxonomy and Phylogeny

Many biotas that are endemic to biogeographic regions are not a random phylogenetic assemblage. Some plant families are significantly overrepresented among the endemic floras in many parts of the globe, especially in Mediterranean-climate regions. The same is true of certain dragonfly families in southern Africa. Among plants, Cyperaceae and Poaceae are underrepresented as endemics in many floras throughout the world. In many cases, these patterns can be attributed to taxon-specific biological attributes that predispose a lineage to endemism. Thus, the existence of discernible phylogenetic correlates of endemism implies that range size may be an evolutionarily stable character of a lineage. Hence there is a need for the comparative approach to assess the role of phylogenetic relatedness in explaining patterns. However, it is important to establish the taxonomic level at which these relationships are manifested. For example, for several data sets, the majority of variation in range size is explained at the level of species within genera.

G. Endemism and Speciation

At face value, the relationship between range size and speciation appears to be quite simple: a reduction in range size will always accompany a speciation event, and a species nearing extinction—in an advanced stage of the taxon cycle—will occupy a limited range size. The deeper issue of the extent to which range size is a cause or consequence of speciation is a question of considerable interest.

There has been a long-standing and as yet unresolved debate regarding the causal relationship between range size and speciation. The arguments assume positive relationships between range size, population size, and dispersal ability. One viewpoint suggests that owing to extensive gene flow and reduced extinction rates, widespread taxa should have lower rates of speciation than range-restricted taxa. An alternative hypothesis is that owing to greater genetic variability and a higher frequency of founder effects, species that comprise large and well-dispersed populations that occupy large range sizes are prone to vicariant speciation.

Many studies of fossil and extant lineages suggest that turnover (speciation and extinction) is associated with relatively low local population abundance, poor dispersal, and narrow range size. Clearly, at extremely low values for these variables, extinction rates will overwhelm rates of speciation. Elevated speciation and extinction rates are also associated with increased specialization, reduced body size, and increased generation times; all of these are correlates of narrow endemism. Thus, endemism and its correlates are responsible not only for enhanced rates of speciation, but also rapid rates of extinction. In E. S. Vrba's parlance, these processes are flip sides of the same coin.

The alternative view, that speciation is associated with large, centrally located and wide-ranging populations, and that peripheral isolates are relictual taxa, also has support. Ultimately, aspects of this debate will be resolved by studies that assess range size and its correlates in a phylogenetic context.

VI. CONSERVATION

Because of their restricted geographical range size, high habitat specificity, and generally low population abundance, endemics are more vulnerable to extinction than are widespread and common species, as a result of both deterministic (habitat transformation) and stochastic (small population effects) factors. Therefore, considerable attention has been given to the conservation of local endemics. Attempts have been made to use the correlates of local endemism to devise management plans that will reduce anthropogenic extinctions.

Recent advances in systematic conservation planning have identified priorities for conservation on the basis of complementarity of biotas (representation), but also for the retention of biodiversity in the face of threatening processes. This approach involves the assessment of the irreplaceability of an area—a measure of the likelihood that the area will be needed to achieve a conservation goal—and its vulnerability to biodiversity loss as a result of current or impending threatening processes. Endemic-rich areas inevitably emerge as priorities since they combine high irreplaceability, owing to their unique biota, and high vulnerability, since endemics are prone to extinction. However, some endemics, particularly plants, may be preadapted to persist in small populations and could be effectively preserved in small, fragmented areas.

VII. CONCLUSIONS

There are few generalizations regarding geographical patterns and correlations of endemism. This is understandable, given that definitions of endemism are mostly study-specific, and that endemism is partly a consequence of regional-specific historical events acting on phylogenetically distinct biotas. Furthermore, species with similar range size often have different local abundances that are likely to be manifested in very different biological attributes. Finally, within-region analyses invariably lump together palaeoendemics and neoendemics, groups with different origins and phylogenetic relationships, and often, different biologies. The recent trend to correct for phylogenetic relatedness holds much promise for understanding the ecological and evolutionary correlates of endemism.

The most active fields of research currently are studies on the correlates of range size, particularly local population abundance, body size, and reproductive traits; the role of endemism in reserve selection, especially as a measure of irreplaceability and surrogate measure of vulnerability; and historical reconstructions using congruent areas of endemism in phylogenetic studies. Much less classical biogeographic research is being carried out on the identification of centers of endemism, despite the fact that reliable distribution data are lacking for many areas and taxonomic groups. This lack of data has serious consequences for the identification of endemic-rich areas for conservation purposes.

See Also the Following Articles

BIODIVERSITY-RICH COUNTRIES • BIOGEOGRAPHY, OVERVIEW • DIVERSITY, COMMUNITY/REGIONAL LEVEL • EXTINCTION, CAUSES OF • ISLAND BIOGEOGRAPHY

Bibliography

Anderson, S. (1994). Area and endemism. *Quart. Rev. Biol.* **69**, 451–471.

Bibby, C. J., Crosby, M. J., Heath, M. F., Johnson, T. H., Long, A. J., Stattersfield, A. J., and Thirgood, S. J. (1992). *Putting Biodiversity on the Map: Global Priorities for Conservation.* ICBP, Cambridge.

Cowling, R. M., and Samways, M. J. (1995). Endemism and biodiversity. In *Global Biodiversity Assessment* (V. H. Heywood, ed.), pp. 174–191. Cambridge University Press, Cambridge, United Kingdom.

Gaston, K. J. (1994). *Rarity.* Chapman & Hall, London.

Kruckeberg, A. R., and Rabinowitz, D. (1985). Biological aspects of endemism in higher plants. *Annu. Rev. Ecol. Syst.* **16**, 447–479.

Kunin, W. E., and Gaston, K. J. (eds.). (1997). *The Biology of Rarity. Causes and Consequences of Rare–Common Differences.* Chapman & Hall, London.

Major, J. (1988). Endemism: A botanical perspective. In *Analytical Biogeography. An Integrated Approach to the Study of Animal and Plant Distributions* (A. A. Myers and P. S. Giller, eds.), pp. 117–146. Chapman & Hall, New York.

McDonald, D. J., and Cowling, R. M. (1995). Towards a profile of an endemic mountain fynbos flora: Implications for conservation. *Biol. Conserv.* **72**, 1–12.

Ricklefs, R. E., and Schluter, D. (1993). *Species Diversity in Ecological Communities. Historical and Geographical Perspectives.* University of Chicago Press, Chicago.

ENERGY FLOW AND ECOSYSTEMS

Alan P. Covich
Colorado State University

GLOSSARY

chemoautotrophs Microbes that use inorganic compounds as a source of carbon and energy and function as primary producers.

decomposition The biotic breakdown of dead organic matter (detritus) by bacteria and fungi that releases carbon dioxide and nutrients for recycling.

ecosystems Composed of species assemblages (producers and consumers) that interact with each other and their associated abiotic environment within well-defined natural or conceptual boundaries.

food chains Composed of species that are connected by the flow of energy and material from producers to consumers.

food webs "Flow maps" that depict connections among multiple food chains.

functional groups Aggregations of species that perform similar ecosystem processes, such as grazers, suspension or filter feeders, leaf shredders, predators, and decomposers.

photoautotrophs These, such as green plants and some bacteria, use solar energy and inorganic compounds to synthesize organic matter as primary producers.

primary productivity The rate of synthesis of organic matter by plants (biomass per unit area of habitat per unit time or, in some cases, biomass per unit volume per unit time).

secondary productivity The rate of assimilation and growth by animals (biomass per unit area of habitat per unit time or, in some cases, biomass per unit volume per unit time).

trophic cascades These occur when changes in the presence or absence (or shifts in abundance) of a top predator alter the production at several lower trophic levels; primary and secondary production at lower levels are alternately constrained or unconstrained by the feeding activities of consumers at upper levels.

trophic levels Groups of individuals classified as primary producers or primary or secondary consumers within food webs; individuals feeding both as primary and secondary consumers are omnivores. A single species may be represented on more than one trophic level.

ECOSYSTEMS ARE THERMODYNAMICALLY OPEN, hierarchically organized communities of producers, consumers, and decomposers together with the abiotic factors that influence species growth, reproduction, and dispersal. These abiotic factors include the flow of energy and the circulation of materials together with the geological, hydrological, and atmospheric forces that influence habitat quality, species distributions, and species abundances. Energy flows through many species, and the way in which this flow affects the persistence of ecosystems is influenced by land-use changes, precipitation, soil erosion, and other physical constraints such as geomorphology.

Energy flow through ecosystems is essential for nutrients to cycle through food webs. These food webs are often subwebs of more complex species assemblages and may be only partial descriptions of more complex hierarchies of energy flows. The hierarchy of species' interactions in natural food webs typically results in some important feedback loops and recycling of nutrients and materials within the conceptually defined boundaries of an ecosystem. Many species of producers and consumers are usually interconnected and some may be interdependent. Food webs are diagrams that can function as "flow maps" to document which species interact with other species, either directly or indirectly, as energy flows through the community and determines the movement of nutrients and other materials.

Different species have important functional values, such as for organic matter production (plant and animal growth) or organic matter breakdown (decomposition), oxygen production, nitrogen fixation, and nutrient cycling. These species have intrinsic values as the unique end products of evolution, and native species are likely to have adapted specific ways to respond to local or regional environmental disturbances. Conceptually, the loss of even a single native species, or the introduction of a nonnative species, could alter how the other remaining native species continue to perform different ecosystem functions. Disruptions of ecosystem processes are known to have occurred after certain well-adapted, native species were lost through local extinction following intense (pulse) or prolonged (press) disturbances. However, predicting which species are essential to ecosystem functions has generally remained difficult because information is lacking on many species interactions as well as on life history, adaptations to different disturbances, and dispersal abilities among key species. Ecosystem studies can provide a broad perspective regarding species relationships and recycling of essential nutrients. Species' shifts in patterns of abundance (or local extinctions) following natural and anthropogenic disturbances illustrate how some key species regulate nutrient cycling and other ecosystem

functions. Field testing of many concepts related to understanding the importance of key species in ecosystem functioning is just beginning. Results of these long-term ecosystem studies can provide guidelines for the stewardship of biodiversity.

I. ECOSYSTEM BOUNDARIES: INPUTS, OUTPUTS, AND TRANSFORMATIONS OF ENERGY

An ecosystem approach can be used to address many different questions spanning scales from the global biosphere to small ponds or patches of habitat. As the questions change, so do the boundaries and the complexities of species interactions within and among different compartments or across trophic levels. Ecosystem boundaries are often defined to include natural species assemblages and to analyze inputs and outputs of energy and materials for cross-site comparisons of efficiencies in energy transfers and studies of changing conditions. These analyses often take the form of mathematical models such as computer simulations or individual-based models of species and their specific functions within the biotic assemblage and environmental conditions under study.

One ecosystem may export nutrients and organic matter (stored energy) to other ecosystems so that cross-site linkages often become important. For example, in studies of nutrient cycling in terrestrial ecosystems, the definition of boundaries would likely have some compartments of organic matter production by living plants and their relationships with herbivores and carnivores. Macro- and micronutrient inputs would likely be derived from the atmosphere through dry deposition of particulates—nitrogen gas being taken up by some nitrogen-fixing species of microbes. Other sources of nutrients, especially phosphorus, would come from weathering and erosion of soil and bedrock deposits, with movement among other compartments derived from actions of wind and water. This combination of interactions illustrates the importance of defining clear boundaries for subsystems within the complete ecosystem so that measurements of movements (fluxes) among compartments can be measured accurately. Generally, ecosystems and their boundaries are abstractions that can only be useful and insightful when combined with sufficient knowledge regarding the natural history and general ecology of communities and their physical environment. There is wide recognition that a combination of direct field observation, experimentation, and modeling is essential when conducting ecosystem studies.

Currently, fundamental questions dealing with relationships between energy flow and the species-specific roles of organisms are attracting increased attention. For example, can results of controlled small-scaled (fine-grained) experimental studies of productivity be used to predict responses of other natural assemblages at larger scales (coarse grained) of ecosystem dynamics? Does energy flow through an ecosystem increase, decrease, or remain the same if one species goes locally extinct but the abundances of other "similar species" change rapidly to compensate for the lost species? Under what environmental conditions do species substitute for one another and compensate functionally for the lost species? These and many other questions are beginning to be answered, but studies related to biodiversity and the persistence of species assemblages remain incomplete. A series of symposia during the 1990s dealt with the relationships between biodiversity and ecosystem functions (Fig. 1) and stimulated many new

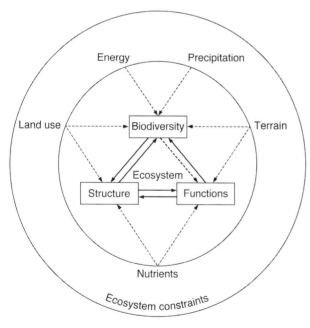

FIGURE 1 Interactions among various factors constrain how ecosystems function. The climatic controls over inputs of solar energy and precipitation are two "forcing functions" which strongly influence how ecosystems operate. Land-use changes are also important in determining how water and nutrients move through the ecosystem and influence biodiversity. Nutrients moving among biotic and abiotic components are a type of "transfer function" that depends on abiotic and biotic factors (modified from Schulze and Mooney, 1994).

ideas (Schulze and Mooney, 1994; Orians et al., 1996; Lawton, 1997; Palmer et al., 1997; Naeem, 1998; Wall, 1999).

A. A Historical Perspective: The Ecosystem Concept

How energy moves from one group of species to another has been an active area of study at least since Charles Darwin and Alfred Russel Wallace first wrote in the 1850s about the interconnections among species. They were intrigued with the general proportions of population abundances that were thought to exist among different groups of large predators and their prey, and they emphasized competition and predation as important factors for regulating species interactions. These early observers lacked a conceptual approach to what was later viewed as ecosystem-level dynamics. Darwin's studies of earthworms and their roles in soil development and his views on the roles of multiple species in the "tangled bank" metaphor stimulated others to consider how these many interactions could be viewed holistically. In 1887, Stephen Alfred Forbes described material cycles within lakes and used a table of predator–prey data to examine which fish species consumed similar or different prey species in Illinois rivers and lakes. Forbes drew his conceptual boundaries and predation matrix to coincide with the shoreline and emphasized putting the many pieces together in a type of system homeostasis in his new metaphor of the lake as a microcosm. This holistic view was taken up in a different way by E. A. Birge in 1915 with his work on heat budgets of lakes in Wisconsin. Birge measured inflows and outflows of energy, and in so doing he set the stage for viewing ecosystems in terms of their general physical attributes rather than numerous component parts. This "black box" approach allowed for observations at larger scales without full analysis of the controlling variables within the process under study.

The British ecologist, Charles Elton, recognized that food chains existed in the context of energy flow and nitrogen cycling from his early work with V. S. Summerhayes on Spitzbergen and Bear Islands. The sum of these chains formed a food web which Elton called "food cycles" and included in a diagram combining nutrient cycling and energy flows over landscapes scales. Elton built on earlier concepts of Victor Shelford and introduced the idea of a "pyramid of numbers" and a "pyramid of energy" that reflected general patterns of community organization. More individual plants and plant material (biomass) can usually be harvested in a plot than herbivores or the relatively rare carnivores.

(Biomass is the weight of all individuals of a species or trophic level found in a square meter of habitat that can be converted into calories of energy.) These concepts of pyramids of biomass and energy led others to formalize ways to analyze these relationships. However, it was not until 1935 that Sir Arthur George Tansley first defined ecosystems as the "basic units of nature." Since then, ecologists have recognized and debated the importance of ecosystems as objects of study (Golley, 1993).

B. The Trophic Dynamic Concept

A major breakthrough came when Raymond Lindeman and G. Evelyn Hutchinson first proposed the concept of trophic levels (Lindeman, 1942; Hutchinson, 1948). Lindeman emphasized that aggregations of individuals rely on similar sources of energy and that this grouping of species provided a conceptual structure and a means for quantification and testing predictions. Lindeman built his concept on Elton's pyramid of numbers by emphasizing the importance of biomass as well as the linkages of "food chains" and "food cycles" as Elton had done. In addition, Lindeman used concepts from studies of biogeochemistry by V. I. Vernadsky. Vernadsky stressed integrating abiotic and biotic relationships among components of the biosphere, atmosphere, hydrosphere, and geosphere (Hagen, 1992).

These ideas on ecological efficiency of energy transfers within the biosphere were also being developed by Lindeman's mentor, G. E. Hutchinson at Yale University, in his long-term studies of Linsley Pond, Connecticut. Lindeman recognized that groups of species both "co-act" and "re-act" in their relationships with each other and with their abiotic environment. He further emphasized that these patterns of productivity change as the biotic and abiotic community develops over time through succession.

Based on his own detailed studies at Cedar Bog Lake, Minnesota (and other studies by Chauncey Juday on Lake Mendota, Wisconsin, who was actively studying lake productivity with E. A. Birge), Lindeman proposed that lake productivity increased during succession as nutrients accumulated in the basin during eutrophication. This natural aging process was thought to reach a climax stage and undergo senescence before developing into a bog forest. Some of his thinking on patterns of change in productivity over time was influenced by earlier studies by G. E. Hutchinson and Anne Wollack on Linsley Pond; Lindeman (1942) noted that their work suggested that "these generalized changes in the rate of production may be expressed as a sigmoid curve showing a rough resemblance to the growth curve of

an organism or of a homogenous population" (p. 409). Such analogies between ontogeny of individuals, populations, communities, and ecosystems were commonly used during that time period. However, as more accurate data on rates of sedimentation became available and isotopic dating improved, the generality did not hold up and multiple patterns of lake eutrophication were later documented.

The concept of trophic dynamics and its focus on transfers of energy between trophic levels was not widely accepted until a new postwar influx of investigators began detailed studies of energetics. Several questions have persisted for decades: How can energy flow regulate the number of trophic levels within an ecosystem? How does one group of consumers regulate the numbers of individuals and energy flow in other trophic levels? These questions and many others were rapidly taken up by ecologists such as G. Evelyn Hutchinson, Eugene Odum, and Howard Odum and their students in the 1940s and into the 1970s.

After considerable debate and continued developments of the concept, many ecologists today rely heavily on models and field experiments using modifications of Lindeman's approach (Hairston and Hairston, 1997). Some still have concerns about fundamental issues regarding how trophic levels are defined relative to the complexities of natural food webs (DeRuiter et al., 1996). As a result, there are various definitions of what constitutes a trophic level. Because most descriptive field-based studies of food webs really study subwebs and therefore are incomplete, a thorough test of predictions regarding food chain length or long-term stability relationships derived from trophic models is usually not feasible.

Conceptually, the predictions of how energy flow regulates trophic dynamics are relatively straightforward. First, some of the initial energy entering the first trophic level is lost by reflectance from the plants, lost as heat, expended in metabolism and evapotranspiration, or lost because of a less than complete coverage of foliage (leaf area) or algal volume. Thus, some warming occurs by energy absorption by the physical habitat (e.g., soil, rock, or water). Then, from constraints imposed by the second law of thermodynamics, energy is lost at each step in the flow of energy from the first trophic level to successively higher levels within food webs. The use of solar energy or chemical energy by primary producers and the consumption of plants and animals at higher trophic levels are relatively inefficient because some energy is lost to metabolism and as heat at each transfer across trophic levels. The total energy flow through the plant trophic level is termed gross primary productivity (GPP). Once the energetic costs of respiration are subtracted from GPP, the remaining energy is called net primary productivity (NPP). This NPP is usually a very small portion of the available solar energy that entered the first trophic level. NPP is the only amount available for transfer to upper trophic levels. In some ecosystems, there is an "energy subsidy" provided by inputs of organic matter from another ecosystem. For example, leaf litter entering a stream, lake, or open cave can be an essential source of stored energy for use by detritivores in a different ecosystem than the forest ecosystem in which it was produced. In all energy transfers between trophic levels, the assimilation of energy is variable but generally of low efficiency. Typically, efficiencies (output:input ratios) are less than 10%, but higher values are known for some food webs. The consistently low values that were first measured in ecological studies in the 1940s and 1950s led to the hypothesis that the number of trophic levels within any food web was determined primarily by the amount of incoming energy and the efficiency of energy transfers.

C. Food Webs and Trophic Levels

Empirical studies demonstrate that most food webs contain fewer than four trophic levels (DeRuiter et al., 1996). However, the number of trophic levels is not a consistent measure because of the complexity of feeding relationships over time and space, the mobility of consumer species, and the movement of food resources across ecosystem boundaries. Many species vary in how they obtain their energy and how efficient they are at different stages of their life histories and under different conditions. Among consumer species, many rapidly growing juveniles or reproductive adults require high-quality, nutrient-rich foods. These same individuals typically feed on lower protein foods when they become nonreproductive adults. Numerous species are omnivores and feed on plants and animals from different trophic levels. Because of these complexities, there has not been complete agreement on how to operationally define trophic levels.

The transformation of inorganic elements into organic matter requires energy to be converted into biomass by species of algae, green plants, and a few types of bacteria. These micro- and macroautotrophs are often represented by many species. The degree of similarity (niche overlap) in their abilities to produce and to store organic matter is important in predicting the consequences of any losses of species. Many species have evolved into persistent assemblages that store carbon and nutrients such as nitrogen and phosphorus. Energy stored in the form of plant-produced organic matter is later passed on directly to grazing species and then indirectly to predators within food webs.

Efficiency of energy transfer from one trophic level to the next higher level is of fundamental importance in understanding conceptually how different ecosystems function. Measures of efficiency, however, are only a part of the explanation for why some ecosystems have longer food chain lengths than others. Relatively "inefficient" food webs with few trophic levels appear to be adapted to certain types of frequent disturbances. Ecologists realize that a single explanation or mechanism is unlikely to account for all the various complexities that exist in determining how ecosystems are organized in terms of energy flow. However, comparisons among well-studied ecosystems and their numbers of trophic levels (food chain length) can provide a useful basis for predicting vulnerability of food webs to major disturbances (Orians et al., 1996) and movement of toxins such as mercury and other heavy metals.

Analysis of guilds and that of functional groups are different approaches used to study shifts in feeding behavior and to complement trophic-level analysis. Guilds are defined as assemblages of species or individual age classes that share a common source of energy at any given time (e.g., nectar-feeding birds and bees and insect-feeding birds and spiders). Functional groups are defined as being similar in their mode of feeding (e.g., filter feeders, shredders, and pursuit predators), but individuals may use a variety of different sources as resource availability shifts. Thus, individuals of a single species may be distributed over several trophic levels and belong to different guilds and functional groups during each individual's life span and reproductive period. As nutritional requirements change and the availabilities of different types of food resources also change, individuals can often adapt to find different available sources of energy. Analyses of similarities in these adaptations and the degree of overlapping functionality are used to understand the degree to which producer and consumer species are interdependent. Numerous complex linkages (e.g., herbivory, predation, decomposition, parasitism, and mutualism) imply that few species are likely to be complete substitutes for other species (Frost et al., 1995; Naeem, 1998; Covich et al., 1999; Crowl et al., 2000). As discussed later, some species may interact positively, negatively, or neutrally in association with other species. Such complex relationships among species, especially under changing environmental conditions, complicate field experiments and make predictive models difficult to test fully.

D. Controls of Energy Flow in Food Webs

A small increase in species richness can have a large effect on how energy flows through food webs. The main features are the number of linkages among species and, especially, the type and strength of those linkages (Paine, 1969). For example, a simple trophic structure would be a linear series of three species (A–C) in a food chain with one species in each trophic level (Fig. 2). Thus, a single species of plant is consumed by a single species of herbivore, which is consumed by a single predator species. Although analysis is relatively definitive in these types of communities with few species, this simple food chain structure may preclude consideration of some questions of general concern, such as resiliency of the assemblage following a disturbance and species loss.

If one more plant species is added (D) to a simple community, then this slightly more species-rich food web provides some important additional dynamics in terms of alternative pathways for energy to flow. Moreover, the herbivore (B) can switch from one food resource to another and this additional complexity increases generality and realism incrementally. With two herbivore species (B and C) the food web is more complex and the predator (A) has a choice of prey resources. Even with the same number of species, much more realism is added by considering the predator (A) to be an omnivore, and even more is added if the predator and one of the herbivores (C) are also cannibalistic. These simple diagrams show how quickly the types and numbers of linkages (connectance) within food webs

can change the dynamics of energy flow even with only a few species. This last example is typical of some low-diversity stream food webs on isolated tropical islands currently under study (Covich et al., 1999; Crowl et al., 2000) and discussed later.

E. Ecosystem Analysis

General rules regarding the relationships between energy flow and the control of food web complexity are currently incomplete. In some habitats a complex relationship apparently does exist among the total annual amount (and seasonal distribution) of energy, the nutrient inputs to ecosystems, and the number of different species in a habitat. In other habitats there is no evidence for a cause-and-effect relationship among the rate of energy flow, species growth and productivity, and the number of species in an ecosystem. Other likely variables include evolutionary time and biogeographical distributions as well as the frequency and intensity of disturbances. The particular species composition of a food web may also alter productivity. Empirical evidence for predicting the importance of species-specific relationships is increasing, but methods for establishing which species regulate ecosystem functions remain controversial. Currently, only a few studies have focused on the species-specific roles to determine which species have unique roles and how these roles shift as environmental conditions change.

Although recent experiments have examined some fundamental relationships, we do not have a full understanding of the effects of varied energy inputs on the

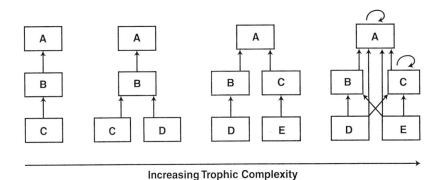

Increasing Trophic Complexity

FIGURE 2 Energy flow through simple food chains and webs. Small increases in species number lead to high trophic complexity as connectance (arrows) and cannibalism (looped arrows) increase. The number and strength of connections among species are more important in regulating the flow of energy through food webs than simply the number of different species in food webs.

richness of consumer species that naturally coexist in ecosystems. For example, in most tropical forests and coral reefs there is generally a high number of species, but the cause and effect of this species diversity are open to different interpretations other than the potential importance of relatively high and continuous inputs of energy (Waide *et al.*, 1999).

At the earth's surface the annual, seasonal, elevational, and latitudinal distribution of solar energy provides varied inputs of energy to deserts, grasslands, forests, wetlands, lakes, rivers, and oceans. Generally, energy always flows through ecosystems but does so at different rates under different global geographical locations and local conditions of slope and aspect. Depending on latitude, ecosystems generally receive a seasonally pulsed or a continuous annual supply of solar energy for primary producers. The rate of energy flow and associated biological productivity are dependent not only on the availability of energy but also on water, on combinations of different macronutrients (nitrogen and phosphorus) and micronutrients (trace elements such as iron, manganese, and silica), and on the presence of an assemblage of interactive plant and animal species. Some natural food webs in extreme environments, such as hot springs, saline lakes, caves, or certain deep-sea thermal vents, have relatively simple linear food chains and have species adapted for specific habitats. For example, dark, closed caves that are deep underground only receive indirect sources of detrital energy from sunlit surfaces aboveground and are typically characterized by a relatively low number of endemic species not found on the surface. Deep-sea thermal vents in the oceans rely solely on chemical energy derived from microbial breakdown of gases such as hydrogen sulfide and are characterized by sulfur bacteria and unique consumer species. These simple ecosystems continue to provide an opportunity to test some fundamental concepts regarding food webs and energy flow relationships.

II. MULTIPLE ENERGY PATHWAYS

It has been evident since Lindeman's work that energy travels along different pathways and includes microbial species and macrospecies in various interconnected relationships. Experimental assemblages are now being widely used to provide some insights into which mechanisms control ecosystem dynamics. There is evidence for biotic control mechanisms (interspecific competition for resources, predation, parasitism, and mutualism), abiotic controls (nutrient limitation and fre-

quent and/or intense disturbances), and combinations of controls in different ecosystems. These studies have also provided important insights regarding two main energy pathways. The distinction between direct, solar-driven, photosynthetically based food webs and indirect transfers of stored energy in the form of detritus (that can be wind driven or washed into habitats) has sorted energy flows into two main classes. The earliest work on ecosystems recognized this bimodal classification and it remains an important organizing framework in linking aspects of terrestrial and aquatic ecology (Polis *et al.*, 1997; Covich *et al.*, 1999; Wall, 1999).

The importance of organic detritus as a means for storage of energy was recognized by studies of Jerry S. Olson in the 1960s at Oak Ridge National Laboratory on the carbon cycle. Analysis of time lags requires an understanding of how rapidly organic detritus accumulates and then breaks down to recycle carbon, nitrogen, phosphorus, and other materials. These insights are critical in current discussions regarding carbon dioxide accumulation in the atmosphere as fossil fuels are burned (i.e., coal, oil, and natural gas taken from storage that accumulated over geological timescales and are now being rapidly cycled back into the atmosphere following combustion). Debates regarding global warming, the greenhouse effect, and where the "missing" carbon is in the present-day ecosystem require a thorough understanding of the entire biosphere and the carbon cycle as it relates to other nutrient cycles.

Recent studies of deep-sea vents and hot mineral springs define a third distinct class of ecosystems that is not solar driven but depends on chemical energy sources used by chemosynthetic microbes. Geologic sources of hydrogen sulfide and other gases provide examples of chemical energy pathways that may well have been the first modes of ecosystem formation by the earliest microbial species on Earth before the evolution of photosynthetic species. Various lines of evidence, such as the banded iron formations in pre-Cambrian rock strata, indicate that the earliest atmosphere lacked oxygen, suggesting that chemoautotrophs dominated the first phases of evolution. Once oxygen-producing photoautotrophs evolved and dominated the oceans and lakes (and later developed terrestrial forms of green plants), their high levels of primary productivity resulted in an accumulation of oxygen in the atmosphere and a decrease in carbon dioxide (possibly through carbon uptake and storage by plants and deposition of sedimentary limestones). This early shift into a photoautotrophically based ecosystem apparently put the chemoautotrophs at a competitive disadvantage in

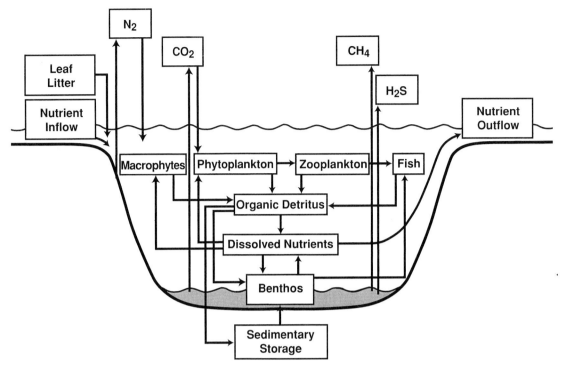

FIGURE 3 A nutrient-rich lake ecosystem and the flow of energy and materials through major compartments. The inputs of dissolved nutrients and particulates (leaf litter and suspended organics) move through the lake ecosystem at a rate determined primarily by the amount of solar energy entering from the surface of the lake, the stored energy in the form of organic matter inputs (for bacteria, fungi, and detritivores), and throughflow of water. Productive lakes can be either sources or sinks of nutrients relative to downstream river and lake ecosystems, depending on their uptake and storage of energy and nutrients. Shallow lakes have a higher surface to volume ratio than deep lakes and are usually more productive because a larger proportion of their volume receives solar energy inputs. Winds mix nutrients from bottom waters and often circulate limited concentrations of dissolved phosphorus that flux from the deeper sediments. Energy transformations from solar input to green plants to herbivores and carnivores are essential for nutrients to move through the food web and be recycled (reproduced with permission from Covich et al., 1999).

an oxygen-rich environment. These remnants of the earliest species of microbes now dominate only in deep-sea vents and hot mineral springs, in which they are still well adapted to compete.

A. Lake Ecosystems

To illustrate the flow of energy through ecosystems it is useful to consider some examples derived from lake studies. These convenient habitats have been used for comparative ecosystem studies because distinct boundaries provide clear definitions of inputs and outputs (Fig. 3). The main boundaries include any inflowing and outflowing rivers as well as the lake surface–atmosphere and the sediment–water interfaces and also shorelines and topographic ridges (that delimit drainage basins). Water temperatures, nutrient inflows and outflows, and mixing and transport processes all influence species distributions and abundances in generally predictable ways.

Solar energy transformed into organic matter through the process of photosynthesis is the main source of energy for most ecosystems, especially in large lakes. Different sizes and types of plants in lakes vary greatly in their rates of productivity. For example, in shallow-water ecosystems solar energy can be used by microphotoautotrophs (attached algae or suspended phytoplankton) and macrophotoautotrophs (pond weeds such as cattails and water lilies). The ratio of the biomass of organisms relative to their rate of production is termed "biomass turnover time." Biomass is generally measured by multiplying the number of individuals in a population by each individual's weight. Turnover time is related to how rapidly organisms increase in biomass during their life span. Rapid turnover is associated with high rates of productivity by small, fast-growing individuals of plants and animals. Small species with rapid turnover usually exploit resources at relatively fine spatial scales.

B. Importance of Depth and Spatial Heterogeneity

Different wavelengths of light energy penetrate into waters of different depths. Shade-tolerant species of plants living in deep water (or in deep shade of the canopy trees in terrestrial ecosystems) have distinct physiological adaptations that allow them to obtain sufficient energy to grow and reproduce even at low light intensities. The range of wavelengths that is used by plants is light visible to the human eye. This range of visible light (between approximately 400 and 700 nm) is termed photosynthetically active radiation. Short wavelength (ultraviolet light) and long wavelength (infrared light) are not used in photosynthesis but are important in regulating floating and emergent aquatic plants (or terrestrial plants in other ecosystems) because ultraviolet radiation degrades organic molecules and infrared increases leaf temperatures. Daytime warming from the sun is essential for some species to survive, especially insects and cold-blooded vertebrates (poikilotherms). Other transfers of solar energy are essential to various lake ecosystem processes (e.g., sediment and nutrient transport by wind and water). Work done by solar-generated winds mixes the lake and thereby influences nutrient cycling. Similarly, wind-driven currents disperse planktonic larvae and aerate deeper waters. Winds also increase salinity through evaporation.

Consumer species are limited by the availability of plant-derived organic materials (or microbially derived organic materials from chemoautotrophs). The well-mixed, brightly lit, open water (pelagic zone) is dominated by small species of zooplankton that are well adapted to feed on suspended phytoplankton. Microbial breakdown of dead organic matter (detritus or seston) and grazing on algae (by numerous invertebrate and vertebrate species) both function to recycle nutrients in the open waters. In other distinct habitats such as the shallow littoral zone near the shoreline, a vegetated zone is dominated by invertebrate species that feed on larger, rooted plants that have a slower turnover rate compared to the small, suspended algae. Some species of fish live in the littoral zone while young and move

in and out of the pelagic open waters as they grow larger. As these predators increase in size, they switch from feeding on invertebrates and small fishes to consuming larger fish and crayfish prey.

III. INTERNAL AND EXTERNAL NUTRIENT CYCLING

Ecosystems are generally "open" with regard to external additions of nutrients from the atmosphere and from catchment basins. Deserts, grasslands, forests, and wetlands as well as most lakes and rivers continue to receive new supplies of nutrients from precipitation (rain and snow), from dust and eroded soils that are carried to the ecosystem by wind and water, and from human-derived fertilizers, sewage effluent, air pollutants, and other agricultural and industrial wastes. Decomposers recycle essential elements for continued productivity by other species. As a result of rapid decomposition, some ecosystems can be relatively "closed" in that nutrients may be rapidly recycled internally so that these nutrients remain within the ecosystem's boundaries and are not transported or lost to other ecosystems. Nutrients can recycle internally from temporary storage in living tissue (biomass), detritus, or storage in soils and sediments.

Different species of plants and consumers within the food web and specific environmental conditions (such as warm temperatures, low concentrations of nutrients, and high pH) influence rates of internal cycling. Land-use practices also have a large effect on vegetation and on how nutrients recycle within a catchment or move into streams and lakes. For example, by measuring the concentrations of nutrients in a stream draining forested catchments, a team of ecologists (led by Gene Likens and Herbert Bormann) identified different pathways for nutrients such as gaseous nitrogen and erosional phosphorus in Hubbard Brook, New Hampshire, from a series of long-term studies on the forest and its components (streams, lakes, and catchments). The research at Hubbard Brook demonstrated the importance of large-scale experiments (such as removal of forest vegetation) to determine how various components are related. The increased outputs of some major nutrients (nitrogen, phosphorus, and calcium) following the experimental manipulation of the forest cover provided information about the role of trees in taking up and storing water and nutrients. By tracking the movement of phosphorus out of the forest and into the stream, Judy Meyer concluded that pulses of stream flow were the primary mechanism governing movement of this relatively scarce element. Later studies led others to recognize the generality of Meyer's observations and to identify the important role that aquatic mosses play in taking up and storing phosphorus in stream ecosystems. These studies at Hubbard Brook and elsewhere later became important in understanding how calcium biogeochemistry and acid deposition altered the forests, soils, and stream chemistry. Comparative watershed studies were conducted by researchers at several sites, such as the Coweeta Hydrologic Laboratory in North Carolina and the H. J. Andrews Experimental Forest in Oregon.

A. Loops, Spirals, and Chains

Different species can be complementary in their roles as they interact in obtaining their energy. For example, benthic species transfer nutrients deposited in the sediments and stored in organic detritus back into overlying waters and thus help move the stored energy up food chains and to other associated food webs. Burrowing and mixing of sediments by benthic species enhance productivity of freshwater ecosystems and lead to interconnected food webs. The rate of movement from one form to another can be facilitated by how species use resources in different ways. Size differences in the resources and the species can be extremely important.

Microbial species play especially effective roles in internal nutrient cycling by using organic detritus as an energy source. Their small size and fast turnover (biomass:productivity ratios) of carbon and nitrogen make them highly important. Bacteria and fungi rapidly break down dead organic matter before it accumulates within terrestrial soil and aquatic sediment (benthic) ecosystems. In many soil, stream, and estuarine ecosystems the amount of energy and nutrients cycled by microbes is relatively larger than that of the photosynthetic pathway. The soil and sediment biota have similar groups of specialized species that shred organic detritus (from leaves, fine roots, dead algae, and other small invertebrate consumers). Fine particulate organic matter (FPOM) is composed of small fragments of detritus and aggregates of dissolved organic matter (DOM) that form from breakdown products and from cell exudates and leachates in aquatic ecosystems (such as groundwaters, streams, lakes, and estuaries). The release of DOM and inorganic nutrients is accelerated by some species of microbes and is a source of energy and nutrients for other microbial species, such as bacteria, fungi, and ciliated protozoans. DOM is also a food resource for

larger invertebrate consumers and provides a parallel set of pathways for energy flow.

Invertebrate detritivores break down dead organic materials to obtain their energy. Many detritivores are dependent on microbes to condition the detritus before it can be consumed by invertebrates or vertebrates. Nutrient cycling allows for continued energy flow through the ecosystem. The continued input and flow of energy through the food web is likewise required for nutrient uptake and recycling. Numerous species of bacteria and fungi form a "microbial loop" that provides for high rates of energy flow by breaking down organic matter. Excretion of wastes by consumers provides one source of nutrients for species in the microbial loop. The microbial loop is especially important in marine and freshwater pelagic (open-water) ecosystems. For example, a crater lake (e.g., Crater Lake, Oregon) with a limited surface area and positioned in a volcanic depression usually has a very small drainage basin compared to large lakes with extensive drainage runoff. A lake with a small surface area will also receive very little atmospheric inputs of nutrients. Primary production remains low and mostly dependent on rapid, internal recycling of nutrients from fine, suspended detritus (dead plankton) as a result of breakdown by microbial species. Nutrients may only slowly accumulate in the sediments of the deep crater lake and not be mixed by currents back into the upper layers of well-lit waters (photic zone). Whenever nutrients are low (as in midsummer in some more productive lakes following the spring bloom of algae) the role of microbes and invertebrate species in cycling nutrients is important to maintaining energy flow in the ecosystem.

Similarly, nutrient concentrations are relatively low in the open ocean, far away from continental sources of nutrient runoff and from coastal upwellings of deeper, nutrient-rich waters. Microbes attach to dead plankton and use this organic detritus as a source of energy. In the process of breaking down this fine detritus, the microbes release nutrients for further growth by phytoplankton (floating algae). Nutrients are also released in the photic zone by leakages from algal cells and by excretion of herbivores. The small size of the detritus particles slows their rate of sinking through the well-lighted photic zone and allows the microbes to release nutrients where phytoplanktonic photosynthesis is not light limited. Larger pieces of detritus fall more rapidly and accumulate in deeper, darker waters where short-term seasonal storage occurs. The nutrients in these deep, dark waters are not available for continued photosynthesis until this entire layer of nutrient-rich water is mixed vertically by wind- and gravity-driven currents back into the photic zone and taken out of temporary, seasonal storage. In very deep lakes, the wind energy is insufficient to mix the entire volume and nutrients accumulate over many years in these deep waters.

There are also some important horizontal linkages in that nutrients and organisms (especially differently sized fishes and wind-driven currents carrying drifting zooplankton) move from the shallow, nearshore waters (littoral zones) of lakes to the open waters (pelagic zones). In these surface waters, nutrients are also recycled by the grazing zooplankton (feeding on phytoplanktonic algae) and by consumption of zooplankton and phytoplankton by fishes. These nutrients are returned to the well-illuminated surface waters, where they are again available for continued algal growth.

This series of vertical and horizontal transformations of organic detritus and primary production of algae in the pelagic zones has some similar analogs in streams, in which the dynamics occur mostly horizontally, but to some extent vertically, along the network of stream channels. The combined vertical and horizontal currents in stream channels form a spiral in the downstream flow of water that carries nutrients and organisms various distances. Dissolved nutrients move in and out of solution as they are briefly taken out of solution by adsorption on sediments and by active uptake by microbes and attached algae (growing along the bottom of the channel), and are later consumed and released back into the water by cell leakage, excretion from grazers, and predatory fishes. This spiral pattern of nutrients being transported downstream while moving in and out of sediments and organisms and then taken up again by other organisms or adsorbed onto sediment particles farther downstream was first examined by ecologists at the Oak Ridge National Laboratory using isotopes to trace pathways and is now widely studied, for example, with stable isotopes of nitrogen (N^{15}). The distance that a particular nutrient element such as nitrogen moves downstream is termed the "spiral length." If the nutrient is taken from solution and tightly held in different species' biomass, then the spiral length is relatively short and the role of microbes and other organisms along the channel is relatively important compared to that for streams where biotic interactions are less significant and during other periods when the spiral length is longer.

Use of functional groups such as shredders and filter feeders (or scrapers, burrowers, and predators) allows for analysis of groups of different species in terms of particular attributes (e.g., how each species processes different sizes of detritus). In both vertically and horizontally structured interactions, the processing rate of

organic detritus by microbes is accelerated by the presence of those invertebrate species that convert large organic fragments into finer fragments. If one species (shredders) uses larger sizes of suspended particles and breaks them down into smaller particles as a result of its feeding, then the downstream supply of FPOM is increased for use by other species (filter feeders). However, if the species that shreds detritus is lost, the filter-feeding species may not function effectively. Different species may form "processing chains" that require particular combinations of linked-species to complete certain ecosystem processes efficiently. These processing chains are one way that increased species diversity can increase efficiency of processing detrital resources, especially if detritus is available at low levels of abundance or if the detritus quality is low. Similar linkages are known to occur among burrowing organisms in soils and sediments where strong interactions are generally important in understanding how different ecosystems function (Wall, 1999).

IV. BIODIVERSITY EFFECTS ON ENERGY FLOW

Productivity and related ecosystem processes such as nutrient cycling and decomposition are generally known to be influenced by particular species. Species attributes, including length of life span, rates of dispersal, and tolerance of frequent and intense disturbances under various environmental conditions, have important effects on ecosystem processes. Key processes (such as nitrogen fixation) performed by only a limited number of species or particular modes of feeding characteristic of only a few species are recognized as important characteristics associated with biodiversity (Palmer et al., 1997; Covich et al., 1999). The importance of a single species in biogeochemical cycling was demonstrated by Peter Vitousek and his students in studies of the invasion of some habitats in Hawaii by the nitrogen-fixing plant *Myrica faya* and by determining the long-term consequences for nitrogen cycling and impacts on the ecosystem.

David Tilman and others conducted several experiments to test the hypothesis that primary productivity in prairie grasslands is related to the number of species of plants grown in the same plot. These experiments and others have the advantage of considerable replication and careful controls but were conducted at relatively small scales. Although the relationship between species richness and energy flow in ecosystems has attracted considerable study, the scale of these experiments and the composition of the species assemblages studied have resulted in different interpretations of general relationships. Results of field tests suggest that the number of species per se is apparently not as important as the particular attributes of different species that relate to their efficiency of nutrient uptake and retention, as well as growth. The larger the number of species included in a study, the more likely some species will be included that are well adapted for the conditions and function effectively. Thus, although ecologists have conducted field tests to determine how different species alter ecosystem processes, there is no complete consensus on how energy flow by itself influences species richness or vice versa (Tilman *et al.*, 1998; Waide *et al.*, 1999). Part of this lack of consensus is a result of using different scales and methods in field studies of many different types of ecosystems (e.g., boreal and tropical forests, grasslands, deserts, deep and shallow lakes, and large and small rivers). Theoretical and conceptual developments are being actively developed and are stimulating additional field testing of these relationships between biodiversity and productivity.

A. Top-Down and Bottom-Up Regulation of Energy Flow

In the 1960s, some of the first experiments designed to identify the importance of different species and key attributes included the removal of predators. Because top predators are relatively few in numbers, their removal can be operationally feasible over some relatively large areas. Robert Paine removed predatory starfish from the intertidal zone and observed a shift in the relative and absolute abundances of some of the molluscan prey species. His results demonstrated that competing species can be held in check by selective predation, especially if the predators consume more individuals of the more numerous prey species. Work by James Estes in the 1970s on the effects of declines in abundances of sea otters and increased abundances of sea urchins and other prey (as well as the shifts of algal regrowth in kelp beds) along the west coast of North America also demonstrated that predators had large effects on entire food webs. These "keystone species" were viewed as important regulators of energy flow in natural food webs in that they had a disproportionate effect despite their relatively small numbers or biomass.

Recognizing the importance of keystone species led some ecologists to use this idea for managing certain ecosystems. Researchers introduced the concept of biomanipulation to study different combinations and abundances of algal species and grazing species in order to regulate nutrient cycling in aquatic ecosystems. In a series of lake studies by several ecologists (especially Joseph Shapiro, J. Hrbacek, John Brooks, and Stanley Dodson), they controlled the presence or absence of top fish predators. From these studies it became apparent that consumers in upper trophic levels could regulate populations of prey, and this regulation in turn had consequences for lower trophic levels as well as nutrient cycling. Manipulating consumer species as a means of removing algae or altering the uptake and storage of nutrients became an area of active study in order to manage lakes and improve water quality. These types of food web studies were also recognized as a means to monitor and to understand movements of toxic compounds such as DDT and mercury in lake ecosystems.

Studies of introduced predators by Paine and Zaret further emphasized that nonnative predatory species disrupted food webs. However, some ecologists argued that native predators rarely altered prey populations or shifted species composition. Steve Carpenter and James Kitchell demonstrated that these species changes did occur in open-water food webs of temperate lakes. Certain fishes could selectively remove large-sized zooplankton grazers (as Brooks and Dodson had earlier demonstrated), resulting in increased biomass of phytoplankton and decreased levels of dissolved nutrients in a trophic cascade (Carpenter and Kitchell, 1993). In other lakes additions of certain predatory fish species had major effects on phytoplankton growth and nutrient cycling through the predators' effects on grazers. Carpenter and Kitchell performed a series of whole-lake experiments to show that fish predation altered zooplankton species composition and size spectra. Shifts in sizes and types of zooplankton herbivores (and other invertebrate predatory species) in turn altered the phytoplankton community and its productivity. These trophic cascades have mostly been observed in aquatic ecosystems (Carpenter and Kitchell, 1993; Power *et al.*, 1996). Recent studies, however, in a tropical forest demonstrate top-down effects through four trophic levels.

Many studies have demonstrated that bottom-up control by nutrients also influences the rate of energy flow in food webs. For many years, additions of nutrients were known to increase phytoplankton biomass and to alter species compositions of algae. Zooplankton abundances and species richness, in turn, often decline when the quality of their algal food resources is reduced. Thus, the distinct pathways of energy flow are altered by the amounts and proportions of nutrients available to algae and bacteria. High phosphorus concentrations lead to phytoplankton communities dominated by blue-green algae (cyanobacteria) because some species of blue-greens can fix nitrogen (incorporate it into their cells) from nitrogen gas in the atmosphere. Their capacity to use high amounts of both phosphorus and nitrogen allows them to outcompete species of green algae and diatoms that cannot use atmospheric sources of nitrogen. Blue-greens can also take up and store nutrients internally beyond their metabolic needs for nutrients and deprive other species of recycled nutrients. Many species of blue-greens have distinct features (slime sheaths, toxins, and large sizes of colonies) that make them relatively unpalatable to many grazers. Some species of blue-greens produce toxic substances. By avoiding grazers, blue-green algal species can grow rapidly in the upper waters of lakes and ponds, thereby shading out the competing species. Blue-green algae often form long, chain-like colonies of cells that are less readily filtered by certain species of large zooplankton. Thus, zooplankton production often declines in waters dominated by blue-greens, and with fewer large zooplankton available as prey the production by fish predators also declines.

In many ecosystems a combination of top-down and bottom-up control can be expected. The roles of specific species in determining how nutrients are used and how rapidly energy flows through different food webs are being studied. The generality of these complex relationships and the predictability of patterns of food web responses to additions of nutrients or the additions or deletions of predators are not completely established. There are some traits that appear to increase the likelihood of certain patterns, and these are under intensive study. For example, as plants and animals die and sink into deep, density-stratified lakes the breakdown of organic matter releases nutrients that can accumulate and dissolved oxygen can be depleted. Deoxygenation sets up chemical conditions in the sediments that release previously bound nutrients that are added to bottom waters. During mid- or late summer nutrients will be scarce in uppermost brightly lit waters (photic zone) in temperature-density stratified lakes because essential nutrients have become increasingly concentrated (during the spring and early summer growing season) in the lowest layers of water below the photic zone. With these accumulated nutrients stored seasonally in deep waters in which light is limited and nutrients cannot be taken up by plants, the roles of grazers and predators are less likely to control nutrient dynamics. Biomanipulation of consumer species may be more effective in shallow, frequently wind-mixed lakes and ponds.

B. Stressful Environments as Testing Grounds: Ecosystem Services and Biodiversity

In some ecosystems the number of producer and consumer species is relatively low because they are adapted to severe environments. Similar situations can occur on isolated islands or in deep caves or hot springs. Only a few species can tolerate the physiological stresses associated with very high or low temperatures or highly variable or extreme salinities, acidities, or nutrient concentrations. Such sites are useful for conducting field experiments because the low number of species can be studied in detail and manipulated over relatively brief periods.

Because of their distance from mainland sources of colonizing species, headwater streams on tropical islands contain a few abundant species of freshwater decapods (shrimp and crabs) and only a few species of other detritivores. Recent studies in streams on Caribbean islands illustrate how species differ in their effectiveness as leaf shredders. Furthermore, these studies show that different species form processing chains and interact to transform suspended organic detritus into benthic biomass that is retained within headwater food webs rather than being washed downstream or accumulating in deep pools.

Whole-pool experiments (Covich *et al.*, 1999; Crowl *et al.*, 2000) in the Luquillo Experimental Forest on the island of Puerto Rico were used to study rates of leaf litter processing. Leaves of *Cecropia schrevenriana* were placed into pools that were cleared of other detritus. To start the experiments, either *Xiphocaris elongata* or *Atya lanipes* were placed into the treatment pools in which other detritivorous shrimp had been removed. Controls were pools with *Cecropia* but with no shrimp. Downstream concentrations of suspended fine, medium, and coarse particulate organic matter, dissolved organic carbon, total dissolved nitrogen, nitrate, and sulfate were measured during 23 days.

Results of the experiments illustrate how differently these two species function as detritivores. Both species of shrimp accelerated leaf breakdown rates relative to controls where only microbial decomposition occurred. *Xiphocaris* shredded *Cecropia* leaves much faster than *Atya*. *Xiphocaris* rapidly shredded the large, intact *Cecropia* leaves and converted them into fine suspended particulates. *Xiphocaris* increased the rate of downstream export of particulate organic matter and concentrations of total dissolved nitrogen and dissolved organic carbon relative to controls. These differences in processing rates also affected downstream distributions of suspended particulate organics and nutrients. *Atya* increased rates of leaf breakdown less than did *Xiphocaris* and apparently filtered out fine organic particulates resulting in less downstream export. *Atya* are especially well adapted to shred conditioned leaves and to scrape microbes from leaf surfaces at low flows. *Atya* are also well adapted with highly modified cheliped fans to filter suspended organic particulates from the water when flow rates are higher than 20 cm/sec. Although both of these shrimp are detritivores, they are not complete substitutes for each other. Their co-location and relative abundances affect rates of detrital processing when they form detrital processing chains.

V. WHY SPECIES MATTER

Recent studies have demonstrated that some ecosystem processes apparently do change as the number of species increases (Tilman *et al.*, 1998). One of the most cited examples is the relationship between increased numbers of species in mixtures of annual plants that are grown in experimental plots. The total area of the plot covered by plant growth (an indirect measure of primary productivity) increased as the number of species increased from 1 to 24. The mechanisms for this relationship are not clear, but progress is being made in interpreting the effects of species composition on these and similar replicated experiments dealing with primary productivity (Hector, 1998).

Linking species to different ecosystem processes (such as productivity and decomposition) highlights the importance of how organisms interact. In some cases, these interactions may facilitate how rapidly processes occur. For example, decomposition of organic matter prevents buildup of organic detritus, which could lead to increased or reduced growth of macrophytes depending on the amounts and types of litter. In other cases the role of different benthic species in breaking down litter can prevent deoxygenation of lakes and streams and maintain supplies of clean water (Covich, 1993; Covich *et al.*, 1999).

VI. FUTURE STUDIES

Given the rapid and accelerated loss of species and the irreversibility of global extinction, it is imperative that experimental and conceptual studies do more to examine multiple levels of ecological organization, from populations and communities to ecosystems and land-scapes, before more species are lost. Unfortunately, we lack sufficient information about how the loss of different species can disrupt natural ecosystem services. These losses represent a type of "warning light" that should draw more attention to analyzing the consequences of losing native species and introducing nonnative species that may disrupt ecosystem dynamics.

Improved techniques using stable isotopes, enhanced computer models, and many other developments have expanded the means of addressing ecosystem questions. However, only recently have ecologists evaluated the importance of species-specific attributes in ecosystem processes. Part of the reason for this slowness is that complete food webs are complex and dynamic. Furthermore, the specific functions of a single species are difficult to isolate from the functions of other species in most natural communities. Probably the most important factor has been the development of different scientific perspectives by population ecologists, community ecologists, and ecosystem ecologists. The recent emphasis on integration across subdisciplines has created a new perspective on the importance of different species' roles in performing different ecosystems "services" or functions, such as nutrient cycling and productivity.

See Also the Following Articles

CARBON CYCLE • ECOSYSTEM, CONCEPT OF • ENERGY USE, HUMAN • FOOD WEBS • LAKE AND POND ECOSYSTEMS • TROPHIC LEVELS

Bibliography

Carpenter, S. R., and Kitchell, J. F. (1993). *The Trophic Cascade in Lakes*. Cambridge Univ. Press, Cambridge, UK.
Covich, A. P. (1993). Water and ecosystems. In *Water in Crisis* (P. H. Gleick, Ed.), pp. 40–55. Oxford Univ. Press, Oxford.
Covich, A. P., Palmer, M. A., and Crowl, T. A. (1999). The role of benthic invertebrate species in freshwater ecosystems. *BioScience* 49, 119.
Crowl, T. A., McDowell, W. H., Covich, A. P., and Johnson, S. L. (2000). Species specific effects of freshwater shrimp on detrital processing and localized nutrient dynamics in a montane tropical rain forest stream. *Ecology*, in press.
DeRuiter, P. C., Neutel, A. M., and Moore, J. C. (1996). Energetics and stability in below-ground food webs. In *Food Webs: Integration of Patterns and Dynamics* (G. A. Polis and K. O. Winemiller, Eds.), pp. 201–210. Chapman & Hall, New York.
Frost, T. M., Carpenter, S. R., Ives, A. R., and Kratz, T. K. (1995). Species compensation and complementarity in ecosystem function. In *Linking Species & Ecosystem* (C. G. Jones and J. H. Lawton, Eds.), pp. 224–239. Chapman & Hall, New York.
Golley, F. B. (1993). *A History of the Ecosystem Concept in Ecology*. Yale Univ. Press, New Haven, CT.

Hagen, J. B. (1992). *An Entangled Bank: The Origins of Ecosystem Ecology.* Rutgers Univ. Press, New Brunswick, NJ.

Hairston, N. G., and Hairston, N. G., Jr. (1997). Does food web complexity eliminate trophic-level dynamics? *Am. Nat.* **149**, 1001.

Hector, A. (1998). The effect of diversity on productivity: Detecting the role of species complementarity. *Oikos* **82**, 597.

Hutchinson, G. E. (1948). Circular causal systems in ecology. *Ann. N. Y. Acad. Sci.* **50**, 221.

Lawton, J. H. (1997). The role of species in ecosystems: Aspects of ecological complexity and biological diversity. In *Biodiversity: An Ecological Perspective* (T. Abe, S. A. Levin, and M. Higashi, Eds.), pp. 215–228. Springer-Verlag, New York.

Lindeman, R. L. (1942). The trophic-dynamic aspect of ecology. *Ecology* **23**, 399.

Naeem, S. (1998). Species redundancy and ecosystem reliability. *Conserv. Biol.* **12**, 39.

Orians, G. H., Dirzo, R., and Cushman, J. H. (Eds.) (1996). *Biodiversity and Ecosystem Processes in Tropical Forests.* Springer-Verlag, New York.

Paine, R. T. (1969). A note on trophic complexity and community stability. *Am. Nat.* **103**, 91.

Palmer, M., Covich, A. P., Findlay, B. J., Gibert, J., Hyde, K. D.,

Johnson, R. K., Kairesalo, T., Lake, S., Lovell, C. R., Naiman, R. J., Ricci, C., Sabater, F., and Strayer, D. (1997). Biodiversity and ecosystem processes in freshwater sediments. *Ambio* **26**, 571.

Polis, G. A., Anderson, W. B., and Holt, R. D. (1997). Toward an integration of landscape and food web ecology—The dynamics of spatially subsidized food webs. *Annu. Rev. Ecol. Syst.* **28**, 289.

Power, M. E., Parker, M. S., and Wootton, J. T. (1996). Disturbance and food chain length. In *Food Webs: Integration of Patterns and Dynamics* (G. A. Polis and K. O. Winemiller, Eds.), pp. 286–297. Chapman & Hall, New York.

Schulze, E.-D., and Mooney, H. A. (Eds.) (1994). *Biodiversity and Ecosystem Function.* Springer-Verlag, Berlin.

Tilman, D., Lehman, C. L., and Bristow, C. E. (1998). Diversity–stability relationships: Statistical inevitability or ecological consequences? *Am. Nat.* **151**, 277.

Waide, R. B., Willig, M. R., Steiner, C. F., Mittelbach, G., Gough, L., Dodson, S. I., Juday, G. P., and Parmenter, R. (1999). The relationship between productivity and species richness. *Annu. Rev. Ecol. Syst.* **30**, 257.

Wall, D. H. (1999). Biodiversity and ecosystem functioning. *BioScience* **49**, 107.

ENERGY USE, HUMAN

Patrick Gonzalez
U.S. Agency for International Development

I. Patterns and Scale of Human Energy Use
II. Implications of the Laws of Thermodynamics
III. Biodiversity Impacts of Industrial Energy
IV. Biodiversity Impacts of Traditional Energy
V. Future Energy Paths

GLOSSARY

energy The capacity to perform work. Potential energy is this capacity stored as position (e.g., in a gravitational or electromagnetic field) or as structure (e.g., chemical or nuclear bonds). Kinetic energy is this capacity as manifested by the motion of matter. The joule (J) is the common SI unit of energy, where 1 J equals the amount of energy required to increase by one Kelvin the temperature of one gram of water. Other units include kilocalories (kcal), kilowatt-hours (kWh), and British thermal units (BTU).

energy, industrial Forms of energy generally transformed in bulk at centralized facilities by means of complex technology. The major forms of industrial energy are oil, coal, natural gas, nuclear, and hydroelectric. In addition to hydroelectric, industrial energy also includes other technologically complex methods of harnessing renewable energy, including photovoltaics, electricity-generating wind turbines, and geothermal turbines.

energy, nonrenewable Forms of energy whose trans-

formation consumes the energy source. The major forms include oil, coal, natural gas, and nuclear.

energy, renewable Forms of energy whose transformation does not consume the ultimate source of the energy, harnessing instead solar radiation, wind, the motion of water, or geologic heat. The major forms of renewable energy are solar, biomass, wind, hydropower, and geothermal. The forms of renewable energy that depend on complex technology are forms of industrial energy. The simpler renewable systems are forms of traditional energy.

energy, traditional Forms of energy generally dispersed in nature, renewable, utilized in small quantities by rural populations, and often not counted in government statistics. The principal forms of traditional energy are firewood, charcoal, crop residues, dung, and small wind and water mills.

energy efficiency A measure of the performance of an energy system. First law efficiency, the most commonly used measure, equals the ratio of desired energy output to the energy input. Second law efficiency equals the ratio of the heat or work usefully transferred by a system to the maximum possible heat or work usefully transferable by any system using the same energy input.

entropy A measure of disorder or randomness at the microscopic level. The entropy of a completely ordered system (e.g., a system at a temperature of absolute zero) is zero.

fossil fuels Forms of stored energy produced by the action of pressure and temperature on organic matter buried over geologic time. The major types of fossil fuels are oil, natural gas, and coal.

law of thermodynamics, first Physical principle that energy is neither created nor destroyed, only converted between different forms. Energy is therefore conserved. In thermodynamic terms, the change in energy of a system equals the difference of the heat absorbed by the system and the work performed by the system on its surroundings.

law of thermodynamics, second Physical principle that any system will tend to change toward a condition of increasing disorder and randomness. In thermodynamic terms, entropy must increase for spontaneous change to occur in an isolated system.

power The rate of energy transformation over time. The watt (W) is the common SI unit of power, where 1 W equals the power expended by the transformation of one joule in one second.

HUMAN ENERGY USE is the extraction, collection, harnessing, and conversion of energy into forms that available technologies can utilize. Our energy use directly alters patterns of biodiversity through changes in land use and through industrial pollution. Indirectly, human energy use is changing global biodiversity through the emission of greenhouse gases that cause global climate change and through other broad environmental effects of industrialization. Whereas the direct effects cause acute damage, the indirect effects generally induce chronic harm. Because human energy use is equivalent to the product of population, per capita economic production, and energy use per unit of economic production, each of these factors can exert an equivalent indirect impact on biodiversity. Several other chapters in the *Encyclopedia of Biodiversity* cover important topics closely related to human energy use. Consequently, this chapter focuses on issues most unique to human energy use. Related entries include Acid Rain and Depositions; Air Pollution; Economic Growth and the Environment; Greenhouse Effect; Pollution, Overview.

I. PATTERNS AND SCALE OF HUMAN ENERGY USE

We use energy both to meet our subsistence needs and to satisfy our wants. In a subsistence society, a farmer's wife will burn wood to cook the day's meals. In an industrial society, a couple will jump in the car on Saturday night to go to a movie. Yet the forms of energy involved in these activities—wood, gasoline, electricity—constitute just the means to desired end-uses—cooking, driving, operating a theater—that ultimately provide unique services—food, transportation, entertainment.

As used by humans, energy falls into two broad categories: industrial and traditional. Industrial energy includes those forms of energy generally transformed in bulk at centralized facilities by means of complex technology. In general, these forms fuel the technology developed in the two-and-a-half centuries that have passed since the Industrial Revolution. The major forms of industrial energy are oil, coal, natural gas, nuclear, and hydroelectric. Industrial energy also includes other technologically complex methods of harnessing solar radiation, wind, and heat, including photovoltaics, electricity-generating wind turbines, and geothermal turbines.

Traditional energy includes those forms generally dispersed in nature and utilized in small quantities by rural people. The principal forms are firewood, charcoal, crop residues, dung, and small wind and water mills. Humans most depended on these forms of energy in the early stages of the development of the species. Because traditional energy sources occur widely and because their transformation does not rely on complex technology, they constitute the most important sources today for rural people in the less industrialized parts of the world. In most cases, a rural household will harvest its own traditional energy sources for its own needs. Because no commercial transaction occurs in these situations, and because most governments do not regulate the use of traditional sources, official statistics do not closely track traditional energy use.

Traditional energy is one form of renewable energy, which includes those forms of energy whose transformation does not consume the ultimate source of the energy. Renewable energy harnesses solar radiation, wind, the motion of water, or geologic heat. The major forms of renewable energy are solar, biomass, wind, hydropower, and geothermal. Conversely, the nonrenewable energy systems consume the very source of the energy, most notably, oil, coal, natural gas, and nuclear fuel.

Besides traditional energy and industrial hydroelectric energy, renewables include a host of recently developed, sometimes technologically complex, methods of harnessing sunlight, wind, water, or heat. These other renewable energy forms include photovoltaics,

TABLE I

1997 Energy Use (TW) by Region and Energy Source. Data from FAO 1997, PCAST 1997, and BP 1998.

	Oil	Natural gas	Coal	Nuclear	Hydroelectric	Traditional	Total	Percent of total
Africa	0.2	0.1	0.1	<0.05	<0.05	0.4	0.8	5%
Asia and Oceania	1.6	0.5	1.5	0.2	0.2	0.8	4.8	33%
Europe	1.1	0.5	0.5	0.3	0.2	0.2	2.8	19%
Latin America	0.4	0.2	<0.05	<0.05	0.2	0.2	1.0	7%
United States and Canada	1.3	0.9	0.8	0.3	0.3	0.2	3.8	26%
Former Soviet Union	0.3	0.6	0.3	0.1	0.1	<0.05	1.4	9%
World	4.8	2.8	3.3	0.9	1.0	1.8	14.6	100%
Percent of total	33%	19%	22%	6%	7%	12%	100%	

electricity-generating wind turbines, geothermal turbines, and other technologies in development. These sources require some of the complex machinery associated with industrial energy, yet depend only upon nondestructive methods of harnessing natural energy sources.

In 1997, the world rate of industrial energy use totaled 12.8 TW (BP 1998). Estimates of the rate of traditional energy use fall in the range of 1.7–1.9 TW (Johansson *et al.* 1993, PCAST 1997, unpublished International Energy Agency data). Of this, firewood and charcoal account for 0.7 to 1.1 TW (FAO 1997, unpublished FAO data). Total world energy use amounted to approximately 14.6 TW, or 14.6 trillion W. As a comparison, this rate of energy use is equivalent to the power drawn continuously by 146 billion light bulbs rated at 100 W. To put this in another perspective, consider that utilities in the United States generally built nuclear plants at a standard rating of 1 GW. So world energy use in 1997 required the equivalent of the continuous output of 14,600 standard nuclear plants.

Table I shows global energy use in 1997 by region and by energy source. The world depends on industrial energy sources for almost 90% of its energy use. Industrial countries, including the United States, Canada, countries of Europe, countries of the former Soviet Union, Japan, China, and India, account for most industrial energy use. Most industrial energy sources are nonrenewable fossil fuels and nuclear. Over a third of industrial energy goes to electricity generation.

Traditional energy comprises only approximately one-tenth of world energy use. Mainly nonindustrial countries in Africa, Asia, and Latin America account for most of the world's traditional energy use. In these countries, firewood and charcoal constitute the primary sources of energy. Indeed, firewood and charcoal pro-

vide more than 70% of the energy used by more than 30 countries in these regions.

The world uses renewable energy sources for only one-fifth of its energy use. The main renewables and their approximate rates of use are firewood and charcoal (0.7–1.1 TW), large hydroelectric (1 TW), agricultural crop residues (50 GW), biomass electric (25 GW), small hydroelectric (20 GW), wind electric (8 GW), geothermal (7 GW), urban waste (1 GW), biomass methane (1 GW), energy crops (500 MW), and photovoltaics (400 MW).

Figure 1 shows the tremendous increase in world energy use over time. In the 20th century alone, energy use has increased by a factor of 12. While total biomass use has remained constant, the world has witnessed an explosion in the use of fossil fuels.

Figure 2 shows the share of the United States in world population, economic production, and industrial energy use in 1997. Although the United States hosts

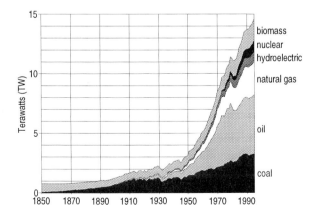

FIGURE 1 World energy use 1850–1995 (data from WEC and IIASA 1995).

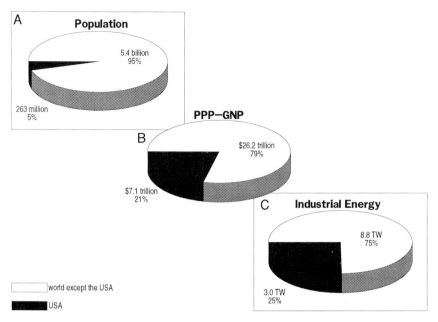

FIGURE 2   Share of the United States in world population, economic production, and industrial energy use in 1995 (data from the World Bank and IEA 1997a, 1997b).(a) Population. (b) Gross national product (GNP) adjusted for purchasing power parity (PPP). (c) Industrial energy use. (This total of 11.8 TW counts only the energy output of hydroelectric generators. Counting the equivalent input if the electricity were produced by nonrenewables, the method used in Table 1, would increase the total to 12.2 TW.)

only 5% of the world's population, it generates 21% of the world's economic production and uses 25% of the world's energy. The average 1995 industrial energy use per person in the United States of 11,200 W cap^{-1} greatly exceeded the world average of 2000 W cap^{-1}, as well as the industrial energy use in other industrial countries, such as the United Kingdom at 5400 W cap^{-1}. On average, each American uses 10 times the amount of energy as each person in the People's Republic of China (1000 W cap^{-1}) and 30 times the amount of energy of each citizen of India (370 W cap^{-1}). Figures 3a and Figure 3b show the 10 countries with the highest and the 10 countries with the lowest industrial energy use per person.

One measure of energy efficiency is energy intensity, the amount of energy used per unit of economic production, generally per dollar of gross national product, adjusted for purchasing power parity. The 1995 industrial energy intensity of the United States, 0.42 W $^{-1}$, exceeded the world average of 0.35 W $^{-1}$. Figures 3c and Figures 3d show the 10 countries with the highest and 10 ten countries with the lowest industrial energy intensity.

Concerning energy end use, detailed data on a global scale are not gathered. In the United States,

however, the Department of Energy does regularly survey energy end use. Americans use approximately 40% of total energy for industrial processes and agriculture. Approximately 35% of energy use goes to cooling, heating, lighting, and maintaining commercial and residential buildings. The remaining 25%, almost all from oil, goes to transportation. Passengers vehicles use half of all transportation energy. The high energy per unit volume and the flexibility of a liquid render petroleum products extremely convenient for powering vehicles.

Globally, a third of energy use goes to electricity generation, mainly from coal, hydroelectric, and nuclear. Power plants release two-thirds of that as waste heat (see the next section). The remaining third mainly goes into the end use of industrial processing with the balance going to cooling, heating, and lighting.

Households generally use the traditional energy sources of firewood and charcoal for the end uses of cooking and heating. Generally, cooking a joule of food requires 2 J of firewood wood or 8 J of wood converted to charcoal. Consequently, rural people use 1 to 2 kg wood cap^{-1} d^{-1} for a rate of energy use of 250 to 500 W cap^{-1}. Actually, a total of only 20 to 40 W cap^{-1} actually enters the cooked food and warmed people.

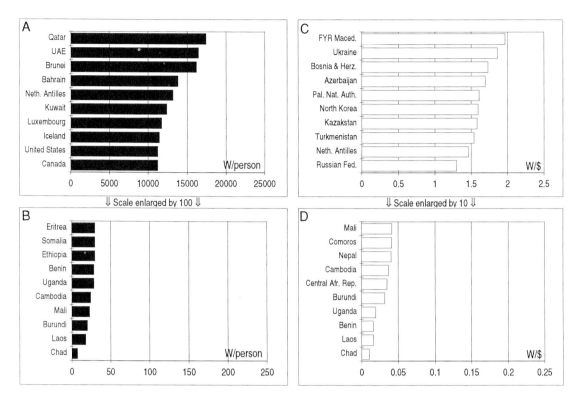

FIGURE 3 Per capita industrial energy use and energy intensity of economic production in 1995 (data from the World Bank and IEA 1997a, 1997b). (a) Ten countries with the highest per capita industrial energy use. (b) Ten countries with the lowest per capita industrial energy use. (c) Ten economies with the highest industrial energy use per dollar of economic output. (d) Ten economies with the lowest industrial energy use per dollar of economic output.

Open fires will diffuse the rest as waste heat (see the next section).

In urban areas of nonindustrial countries, people often rely on charcoal for energy. Even though the conversion of wood to charcoal releases waste heat, the end product has higher energy per unit mass than firewood. This makes charcoal easier to store and transport than firewood. Urban people use 100 to 150 kg charcoal $cap^1 y^{-1}$, requiring 800 to 1200 kg wood $cap^1 y^{-1}$. The ultimate end-use energy requirement is 30 to 45 W cap^{-1}.

II. IMPLICATIONS OF THE LAWS OF THERMODYNAMICS

The first law of thermodynamics states that energy is neither created nor destroyed, only converted between different forms. This is the principle of conservation of energy. The first law means whatever energy a process does not convert into useful forms must still go somewhere. The nonuseful energy does not just disappear.

Humans use the environment as the sink for this waste energy.

The second law of thermodynamics states that any system will tend to change toward a condition of increasing disorder and randomness. This is the principle of increasing entropy. The second law means that no energy transformation can convert 100% of one energy form completely into a useful form. The process will always release amounts of energy wasted in forms that are unrecoverable due to the disorderliness or randomness of the waste energy forms. The fewer energy transformations that a system contains, the fewer chances for random second law energy losses.

For example, the objective of an automobile's internal combustion engine is the conversion of chemical energy in the covalent bonds of hydrocarbons in gasoline to heat energy of an expanding fuel-air mixture in the piston, to kinetic energy of the drive shaft, to kinetic energy of the main axle. No matter how efficient the engine and automobile technology is, the conversion process will always waste energy as heat in the friction of engine parts, sound in the banging of vehicle components, heat in the friction of tires on the road, kinetic energy of the wind displaced by the vehicle, and countless other unrecoverable losses.

As another example, the objective of a coal-fired electric power plant is the transformation of the chemical energy in the coal to heat energy in the boiler, to heat energy in steam, to kinetic energy of a turbine fan, to electromagnetic energy in the generator coil. Along the way, the conversion processes lose energy as the light and sound of the boiler fire, the vibration of turbine parts, the heat of power plant components, and, most significantly, the waste heat carried by the power plant cooling water.

Theoretically, the maximum efficiency across a heat gradient is the Carnot efficiency:

$$\eta = 1 - \frac{\text{Temperature of heat sink}}{\text{Temperature of transformation}}$$

with temperatures in Kelvin.

For a coal-fired power plant, materials limit boiler temperatures to 1000 to 1200 K. At an ambient environmental temperature of 293 K, the maximum efficiency will be 70 to 75%. Typically, coal plants only achieve 30 to 35%, releasing two-thirds of the total as waste heat.

Table II gives various formulations of the first and second laws of thermodynamics. The inevitability of entropy losses makes the colloquial interpretation of the second law "You can't even break even."

III. BIODIVERSITY IMPACTS OF INDUSTRIAL ENERGY

A. Oil

The major impacts of oil on biodiversity derive from a fuel and use cycle that ranges over vast areas of terrestrial and marine habitat. Exploration, drilling, crude oil transport, refining, and utilization in vehicles change land use and introduce industrial pollution to land and sea.

Petroleum, or oil, consists of a complex mixture of hydrocarbons formed over geologic time from organic matter compressed under anoxic conditions. The most important chemical constituents are alkanes such as octane and methane and aromatics such as benzene and toluene.

The majority of oil deposits derive from aquatic plants and bacteria deposited in inland seas and coastal basins during the Cretaceous Period 100 million years ago. In the early stages of formation, bacteria initiated the anoxic reduction of the organic matter. Over time, pressure and temperature replaced microbial activity as the main agent of transformation. Eventually, these forces drove off most of the water, oxygen, and nitrogen from the condensate, leaving carbon and hydrogen compounds. Dispersed between sediment granules, the oil eventually migrated to low pressure geologic traps at depths of 1 to 7 km. Today, oil fields occur at an average depth of 1.5 km. On average, the stoichiometric composition of crude oil is $CH_{1.5}$, with a very small amount of sulfur.

Petroleum exploration entails geologic surveys over extensive areas often with low human populations and relatively undisturbed natural communities. Exploratory surveys generate vehicle traffic and temporary dwellings that bring localized disturbances, but the most serious impacts occur with seismic detection. This method involves controlled detonations along lines or at points so that seismometers can extrapolate the layout of subsurface formations. These activities destroy areas of vegetation, disturb certain animals, especially ground-nesting birds, and fragment habitat. If such activities disturb animal behavior during breeding times, the impact can last over many growth periods.

Edwin L. Drake drilled the world's first commercial oil well in Titusville, Pennsylvania, in 1859. All oil wells require access roads, and high-volume wells require buildings and electric and water lines. This infrastructure destroys vegetation and takes land away from animal habitat. The more extensive an exploited oil field, the wider the habitat impacts extend. Infrastructure at the Prudhoe Bay field, opened for drilling in 1968, now extends over 1700 km² of Arctic tundra. This has noticeably displaced calving of *Rangifer tarandus* (caribou) from the field. Likewise, proposed exploitation of Area 1002 in the Arctic National Wildlife Refuge would disrupt the migration routes of the porcupine caribou herd to its calving grounds.

TABLE II

Formulations of the First and Second Laws of Thermodynamics

	First law	Second law
Universal	The total energy in the universe is constant.	All physical processes proceed such that the entropy of the universe increases.
Concise	Energy is conserved.	Entropy increases.
Colloquial	You can't get something for nothing.	You can't even break even.

Drilling operations produce water and air pollution. Serious water pollution comes from the vast amount of used drilling muds, which are lubricating substances pumped down to the drilling bit to carry away rock cuttings, to keep the bit from overheating, and to protect the drilling shaft from surrounding rock. Drilling muds consist of water mixed with light molecular weight oils. Used muds contain bits of metal from drilling components and any trace metals mobilized out of the drilled rock.

All stages of the oil production system from drilling operations to end use spill oil into surface and ground waters. Globally, oil spills into surface waters total more than 3 million tons each year. Half of these spills come from oil production, 40% come from nonpoint urban runoff, and the remaining come from natural seeps. At the start, spills occur at well blowouts when equipment fails to contain naturally high fluid pressures in oil-bearing strata. Spills also occur along the significant lengths of pipeline from the wellhead to tank farms to supertanker ports to refineries to gas stations. Pipes, valves, and tanks leak from fatigue and from human error.

These and other problems roused concern when oil companies first proposed construction of the Trans-Alaska pipeline to carry crude oil 1300 km from the North Slope to the Gulf of Alaska. When it eventually started operations on June 20, 1977, the Trans-Alaska pipeline integrated a set of environmental protection features. To prevent thawing of permafrost areas, brackets elevate 700 km of pipeline to heights of 3 m. Heat pipes at the bracket legs dissipate heat generated by the friction of oil passing through the pipe.

The elevated sections serve as underpasses for caribou. Over buried sections in certain permafrost areas, construction engineers designed refrigerated overpasses for caribou. Bridges carry the pipeline over 800 streams. Zigzags along the pipeline translate longitudinal movement of pipes expanding under heat to lateral movement, reducing the risk of leakage. Oil companies revegetated areas denuded by construction activities.

An access road now open to the public parallels the entire length of the pipeline. This road has opened up a strip of habitat to human contact, possibly changing behaviors among caribou and other mammals.

On March 24, 1989, the supertanker *Exxon Valdez* ran aground on Bligh Reef in Prince William Sound in the Gulf of Alaska, ruptured, and poured out 41 million liters of crude oil, the largest oil spill ever in U.S. waters. The spill caused acute damage to birds, marine mammals, and intertidal communities. The spill also caused chronic damage to fish species and intertidal and subtidal communities. The progression of the spill demonstrates a pattern repeated in smaller spills that occur frequently in the world's shipping lanes.

Oil floats on top of water. Gravity and wind will spread a floating slick out to a thickness of 0.5 to 10 μm. Patches 0.1 to 5 mm thick can cover just 10% of the total slick area yet contain 90% of the total slick volume. Some oil dissolves and emulsifies into the water column, forming emulsions containing 80% H_2O. Oil generally will not sink to depths below 20 m. In the *Exxon Valdez* spill, recovery teams deployed an array of countermeasures that included booms, skimmers, sorbents, pumps, burning, and surfactants for chemical dispersion.

Exposure to sunlight initiates photolysis of hydrocarbons into lighter molecular weight compounds. Heterotrophic bacteria will also oxidize hydrocarbons to smaller compounds, CO_2, and water. The lightest hydrocarbons, as well as aromatic compounds such as benzene, volatilize. Loss of the lighter fraction leaves the remaining slick more viscous over time. This thick oil forms tar balls and pancake-like forms. Oil from the *Exxon Valdez* eventually spread across hundreds of kilometers of beaches, penetrating deeply into cobbled stretches and mussel beds. Today, oil still persists beneath the surface layer of rocks in many areas.

Three years after the *Exxon Valdez* spill, photolysis degraded 70% of the original oil. Bacteria then eliminated photolysis products amounting to 50% of the original crude oil; the other 20% evaporated. Work crews recovered 14% of the spill. Thirteen percent of the original oil sank into subtidal sediments. Beaches absorbed 2%, leaving 1% still suspended in the water column.

The spill occurred in early spring, just before the young of many species emerged to rejuvenate marine animal populations. *Clupea pallasi* (Pacific herring) were spawning inshore. Millions of *Oncorhynchus gorbuscha* (pink salmon) fry were soon to be washed from gravel spawning beds into the spring plankton bloom offshore. *Phoca vitulina* (harbor seal) and *Enhydra lutris* (sea otter) pups were testing the frigid waters. Seabirds were beginning to converge on breeding colonies in the gulf. Consequently, the oil devastated populations of birds, marine mammals, and fish.

Thousands of birds can die in even moderate spills, but the *Exxon Valdez* spill eventually killed more than a quarter of a million birds of more than 90 species, the greatest demonstrated mortality of birds from any oil spill. Workers physically recovered 36,000 carcasses. Of these, 8000 were *Brachyramphus marmoratus* (marbled murrelets) and 150 were *Haliaeetus leucocephalus* (bald eagles). The spill killed individuals of two *Fratercula* spp. (puffins) and four *Gavla* spp. (loons). For the following three years, fewer breeding *Uria aalge* (common murres) showed up at spring colonies.

Oil coats feathers, matting and waterlogging them. The water repellency, buoyancy, and insulating properties of plumage derive from a precise, orderly arrangement of feather barbules and barbicelles. Contact with oil disrupts these arrangements. Soaked birds can die of hypothermia and drowning. Those that survive risk chronic exposure to toxic organic compounds through ingestion, inhalation of fumes, or absorption. Moreover, eggs are highly sensitive to contact with oil.

Many of the aromatics in petroleum, including benzene, toluene, xylene, and phenols, are lethal to animals on contact and carcinogenic under chronic exposure. Moreover, polycyclic aromatic hydrocarbons bond to lipophilic sites, an affinity that magnifies these compounds up the food chain. Even when not deadly, sublethal disruption of physiology or behavior activities can reduce resistance to infection and cause generalized stress.

In the wake of the *Exxon Valdez* spill, 300 of the 2200 harbor seals in Prince William Sound died, as well as 3500 to 5500 out of 10,000 sea otters. Like birds, oil obliterates the insulating properties of marine mammal pelage, leaving them to die of hypothermia. In addition, oil can clog the nostrils of seals, causing them to suffocate. Whales, insulated not by hair, but by layers of oily blubber, resist the effects of oil, although a well blowout off Santa Barbara, California, in 1969 led to the death of gray whales. In Prince William Sound, harbor seals may experience chronic problems because oil is accumulating in their bile and fatty tissues. Since the spill, sea otters have continued to experience elevated mortality.

An unfortunate coincidence has resulted in the geographic juxtaposition of important commercial fisheries and high yield offshore oil fields on the continental shelves. Not only do oil spills invariably cause fish kills, but chronic effects also reduce fish fitness years after initial exposures.

Oil at the air-water interface acts as a physical barrier interfering with gas exchange. In fact, oil has been a traditional line of defense used for mosquito larvae control. Under a thick slick, fish larvae can suffocate.

Fish eggs, which often float at the sea surface, and fish larvae, which are often distributed in the upper water column, both occur in the areas of highest oil concentration. Hydrocarbons and aromatics damage eggs on contact. Oil concentrations will quickly exceed the LC_{50} of 1 to 10 ppm for fish larvae. The early life stages of intertidally spawning fish are especially susceptible.

Exposure to toxics from the *Exxon Valdez* spill has caused chronic problems in *Clupea pallasi* (Pacific herring), *Oncorhynchus clarki* (cutthroat trout), *Oncorhynchus gorbuscha* (pink salmon), and *Salvelinus malma* (Dolly Varden). Fish species have shown elevated egg, larvae, and adult mortality, larval deformities, and poor adult growth rates, even in situations of constant food supply. Fish tissues in some species contain elevated concentrations of toxics.

Over time, tidal action spreads and coats the shore of the intertidal zone in a band of oil. This oil ring smothers intertidal invertebrates, crustaceans, mussels, barnacles, limpets, and algae. Oil coatings will asphyxiate filter feeders. Recovery crews use hot water washes to clean oil coated shores, an effective method, yet destructive to intertidal organisms.

Oil spills also damage phytoplankton and other marine plants. Oil absorbs photosynthetically active radiation, so direct coating hinders plant growth and increases plant tissue temperatures. Aromatics may disrupt the orderly arrangement of grana in chloroplasts. An increase in ruderals characterizes the changes in plant species diversity. Blue-green algae blooms will increase eutrophic conditions.

Oil refining focuses on the catalytic cracking of carbon-carbon bonds of long-chain alkanes for the production of lower molecular weight hydrocarbons. Refineries try to recover every possibly useful organic compound, from the light products methane, benzene, toluene, and kerosene, to medium-weight products like gasoline and diesel fuel, to heavy tars and asphalt. These processes, as well as sulfur recovery, inevitably generate water pollution.

Most constituents of petroleum and refined oil products volatilize easily. Consequently, each step of the petroleum fuel cycle generates air pollution. Methane, ethane, benzene, toluene, and other compounds will evaporate from crude oil exposed to air. The major emissions from oil refineries include CH_4, CO, CO_2, H_2S, NO_x, and SO_2.

This section has concentrated on the impacts from the core stages of the petroleum fuel cycle: exploration, extraction, transport, and refining. Nevertheless, manufacture of the infrastructure and materials needed for these end uses generates industrial pollution and requires land. Moreover, armed conflicts caused, in part, by efforts to control access to oil fields and refineries take human life and directly disrupt ecosystems.

The combustion of refined oil products for transportation, heating, and other end uses generates perhaps the gravest by-product of the entire fuel cycle—carbon dioxide, the principal greenhouse gas (see Greenhouse Effect).

B. Natural Gas

Natural gas is a mixture of light hydrocarbons that exists at a gaseous state at standard temperature and pressure. Methane (CH_4) is the main constituent, but the presence of higher molecular weight alkanes, including ethane, propane, and butane, changes the average stoichiometric composition for natural gas with the water vapor removed to 0.79 $CH_{3.62}$. Formed by the same processes that formed oil, natural gas is often found at the top of oil deposits. The most voluminous natural gas reservoirs occur in Cretaceous strata. The land use changes brought by the exploration and extraction of natural gas produce the same biodiversity impacts as described for oil.

In the nineteenth century, companies had not yet erected natural gas pipelines or processing facilities. Moreover, industry had not yet developed much technology for using natural gas. Because companies found natural gas uneconomical to exploit, they just burned it off to reduce the risk of fire and explosion. The entire history of natural gas production has flared the equivalent of 8 years worth of U.S. energy use. Today, U.S. companies generally flare only small amounts at refineries, but companies from other countries flare enough that the total amount flared amounts to 5% of global natural gas production.

Gas companies generally pump natural gas straight from the well to a processing plant, eliminating the need for storage facilities at the wellhead and thus reducing the potential for leakage. Gas companies generally divide natural gas into three fractions: natural gas liquids (NGL), liquefied petroleum gas (LPG), and liquefied natural gas (LNG). NGL consists of the higher molecular weight fraction of natural gas that often settles out by gravity. Processing of natural gas from oil wells produces liquefied petroleum gas (LPG). Finally, pressurization of natural gas produces liquefied natural gas (LNG), a product that is expensive because of the special containers required for transport.

The major end-uses of natural gas, cooking and heating, burn the fuel directly with no further transformation. Electricity generation from natural gas uses a gas turbine, which directly uses the hot gas products of combustion to turn the turbine fan, eliminating the intermediate step of steam generation used in oil and coal-fired plants. Cogeneration plants increase the energy efficiency of gas turbine systems by utilizing the waste heat of gas turbines for space heating or industrial processes.

The extraction and combustion of natural gas pollute much less than the extraction and combustion of oil. Because it exists in the gaseous state for much of the fuel cycle, natural gas exploitation does not produce significant amounts of water pollution. However, methane itself is a greenhouse gas, and its combustion produces the main greenhouse gas, CO_2.

C. Coal

Coal consists of hard carbonaceous material formed by the compression and transformation of terrestrial plant matter rich in cellulose buried at the bottom of ancient freshwater swamps and bogs. The richest coal-bearing strata date from the Cretaceous period 100 to 200 million years ago and from the Permian period 250 million years ago. Similar to the process of petroleum formation, the deposited plant matter undergoes incomplete decay in anoxic conditions.

In geologic time, the pressure of overlying rock and the temperatures generated therein drive off oxygen and hydrogen, leaving thick seams of reduced carbonaceous rock containing much more organic than mineral matter. The average stoichiometric equation of coal is $0.75CH_{0.8}$, but elemental sulfur also contaminates most coal deposits. The four major types of coal, in order of decreasing carbon content and increasing sulfur, are anthracite, bituminous, subbituminous, and lignite. Bituminous coal is the most physically abundant type worldwide. Peat, the partially oxidized, moist, organic soil that forms in marshes and bogs, is the very early precursor to coal. In certain areas, people burn peat for heating, cooking, and light.

The coal fuel cycle extends from extraction at the mine to conversion at a power plant to distribution across the electric grid to end uses in lighting, heating, and all the uses of electricity.

Coal mines generally fall into three types: deep, open pit, and strip. Deep mines extend down to a depth of around 1 km. Open pit mines reach down to 300 m. Strip mining generally removes the upper 30 m of land surface. Coal mines consume land, not just for areas actually excavated and areas used to dump unwanted extracted rock, but also for the support infrastructure of buildings, roads, and rail lines.

Deep and open pit mines remove huge amounts of rock, termed overburden, lying over the coal. The land over deep mines will sink, a process termed subsidence, drastically changing the topography, hydrology, and microclimatic profile of a landscape. This can destroy vegetation and alter important animal habitat characteristics. Underground coal fires in abandoned mines and refuse banks will not only exacerbate subsidence, but they will also release CO_2 and other air pollutants.

Miners dump the huge amounts of unwanted extracted rock, termed mine tailings, in abandoned parts of active mines or on the surface. Pyrite (FeS_2) usually comprises a signicant fraction of the tailings. The reaction of water and pyrite produces sulfuric acid (H_2SO_4). In addition to being poisonous to plant and animal life, sulfuric acid mobilizes other toxic substances. The leaching of acids, trace metals, dissolved solids, and toxic organics produces a brew known as acid mine drainage that can devastate surface waters. Selenium and cadmium often occur in high concentrations in tailings, so acid mine drainage can initiate the bioaccumulation and bioconcentration of these trace metals in the surviving sections of the food chain.

Surface mining consumes vast tracts of land. Heavy machinery remove the upper layer of a landscape to expose relatively shallow coal seams, completely destroying the mined area. Although coal companies generally fill back the overburden into the mined area and replant it, strip-mined land never recovers its original characteristics. Replanting even creates opportunities for ruderals to expand where perennial plant species may have dominated. Rodents and other animals that adapt readily to human disturbance also take advantage of reclaimed areas.

Coal mines often need to impound surface streams to satisfy the significant water needs of mine operations. These needs include water cannon drilling, transport by slurry, fugitive dust spraying, coal washing, and size sorting.

Mines crush and screen coal for uniform sizing, then wash and dry the coal for open air storage. The fugitive emissions from these processes consist of particulates that coat any exposed surface, blocking photosynthetically active radiation from plants, contaminating food and water sources for animals, and acidifying affected soil. Leaching of toxic substances from coal storage piles can also add to the pollution of surface waters. Rail transport provides the most cost-effective means of moving the bulky commodity of coal. Fugitive emissions from unit trains increase the particulate load in rail corridors. To save money on rail transport, many utilities will site electric power plants next to the mine then wire out the electricity. In certain regions, this shifts the pollutant load from urban areas to less polluted rural areas.

Most coal worldwide goes to electricity generation.

A conventional power plant burns coal in a boiler to boil water that circulates through a closed loop system of pipes. The steam from the boiler enters a steam turbine to turn huge fans that power an electric generator that converts kinetic energy to electric energy. As a principal of physics, the movement of a conductor across a magnetic field creates electric current in the conductor. In the coal-fired power plant electric generator, the conductor consists of stationary coils of wire surrounding a magnet on a shaft rotated by the turbine fan. Much of the steam that moves through the fan transfers its heat energy to the kinetic energy of the fan, causing the steam to condense back to water. A condenser will then allow heat to transfer from any steam that continues past the turbine to an external supply of cold water. The water in the internal loop from the condenser returns back to the boiler to enter the steam cycle again.

Coal combustion releases CO, CO_2, SO_2, NO_x, particulates, fly ash, arsenic, cadmium, chromium, mercury, and selenium. Nearly 40% of anthropogenic CO_2 emissions come from burning coal, whereas coal burning produces 80% of human SO_2 emissions. Consequently, greenhouse gases and acid precipitation may constitute the agents of coal's most extensive environmental effects.

The slag remaining from coal burned in the boiler contains high amounts of trace metals, especially cadmium and mercury. In addition, the sludge from flue gas desulfurization units, the pollution control devices known as scrubbers, contains trace metals and toxic organics. The disposal of this sludge presents problems for land use and water quality.

Internal steam turbine water is the working fluid circulating from the boiler to the turbine to the condenser and back to the boiler. Cooling water is the medium that draws heat from the internal steam turbine water. In most conventional coal-fired power plants, the internal steam turbine water remains separate from power plant cooling water. A typical condenser consists of copper coils, carrying cooling water, that pass through larger structures carrying the internal steam turbine water. As a physical principle, heat passes from the steam turbine water through the walls of the copper coils into the cooling water.

A 1 GW coal-fired power plant typically requires 4 million m^3 per day for all operations, mostly for cooling. These water needs dictate the necessity to locate a plant next to a natural water body. Power plants mainly use fresh water because of the corrosive effects of salt water. Water withdrawals change the hydrology of a watershed, changing water levels, surface area of mudflats, surface area of wetlands, and other important habitat characteristics that can strand hydrophilic plant species such as *Salix* spp. (willows) and harm fish and shorebird populations. Impingement on intake screens kills significant numbers of fish and other aquatic species. Organisms that get through the screens undergo entrainment through the condenser, causing even greater mortality. The stress that any surviving organisms undergo reduces their fitness considerably.

All power plants, including coal, oil, and nuclear, generate three-quarters of the waste heat dumped into U.S. surface waters and into the atmosphere above the United States. Once-through systems dump the waste heat directly into local waters. Cooling towers dump waste heat into the atmosphere, condensing steam from the air. Cooling ponds provide a buffer for releasing some of the heat from cooling water into the atmosphere, reducing the temperature of the cooling water before it enters surface waters.

Thermal discharges into freshwater and coastal zones cause a host of negative effects on aquatic species:

1. Direct lethality to fish and crustaceans at water temperatures $\geq 35°C$.
2. Decrease in dissolved oxygen.
3. Increase in metabolic rates and nutrition needs for fish and changes in nutrition requirements for other taxa.
4. Displacement of diatoms by green and blue-green algae.
5. Inhibition of vertical migratory behavior by zooplankton.
6. Thermal plume blockage of migratory fish movement.
7. Avoidance of warm areas by migratory waterfowl.
8. Early emergence of aquatic insect adult life stages into inhospitable environmental conditions.
9. Copper contamination from condenser coils.

Long-range transmission of electricity occurs across high-voltage lines strung on metal towers up to 30 m tall. The 115 kV network in the United States stretches across 200,000 km and occupies 2 million ha. The clear cutting of corridors 30 to 60 m wide for transmission easements directly changes the vegetation and plant life in cut areas. Periodic clearing maintains and intensifies the original changes. The areas that remain favor ruderals and animal species that adapt readily to human disturbance, such as *Odocoileus virginianus* (white tailed deer). Herbicides used for periodic clearing can hurt insect and bird species. Transmission line corridors fragment habitat and increase the area of habitat suscep-

tible to edge effects. The cleared areas can also block migrating land animals.

Short-range electricity transmission occurs across low-voltage lines strung on wood, metal, or concrete poles generally 5 m tall. Harvesting wood poles can produce all the potential biodiversity impacts of commercial logging, monospecic plantations, and milling. In many countries, utilities treat the wood with creosote to guard against the action of insects and weather. Creosote, a by-product of crude oil refining, contains significant amounts of toxic organics that can leach and contaminate surface waters.

The material and energy needs for building the massive infrastructure of the coal fuel cycle produce wide-ranging environmental effects. Because most coal goes to electricity generation, the end uses of coal produce the environmental effects associated with climate control, lighting, commercial production machinery, residential appliances, and other electric devices.

D. Nuclear Fission

Nuclear fission is the splitting of high molecular weight elements to release energy held among protons and neutrons in the nucleus of the atom. Uranium and plutonium are the elements that provide the most effective yield from fission at current levels of technology. A fission reaction produces energy in the form of light, heat, motion of the split pieces, and radiation. Radiation consists of kinetic energy of small molecules and atomic particles and electromagnetic energy of photons traveling at certain frequencies. When radiation passes through living tissue, the particles or photons impart their energy to atoms and molecules in the tissue, disrupting molecular and atomic structures.

The fission products themselves will continue to emit radiation until they reach a stable atomic state. Strontium-90 and Cesium-137 have half-lives of decades, but Plutonium-239 decays with a half-life of 25,000 y and Iodine-129 will halve in mass only after 17 million years.

Nuclear fission plants require highly processed uranium fuel. A 1 GW fission plant requires 150,000 Mt U_3O_8-containing ore to fabricate enough fuel for one year. Milling this removes 150 Mt U_3O_8. In order to concentrate Uranium-235, a conversion plant converts U_3O_8 to 188 Mt of UF_6 gas. Differential diffusion of the UF_6 separates 31 Mt UF_6 enriched in Uranium-235. A fuel fabrication plant then produces 30 Mt of UO_2 pellets.

The mining and milling of uranium ore creates most of the same environmental problems already described for deep mining for coal and for coal processing. Conversion, enrichment, and fuel fabrication require fluorine gas, which is lethal on contact to animals, damages vegetation, and reacts to form toxic by-products.

According to the International Atomic Energy Agency (IAEA), at the end of 1998, 434 nuclear fission plants were operating in 33 countries around the world; 104 nuclear fission plants were operational in the United States. These plants possessed a combined rated capacity of 349 GW and generated 2300 TWy of electricity, 16% of the world total. Plant operating experience reached 9000 plant years.

Nuclear plants generate electricity in a steam cycle very close to that employed in coal plants, except that nuclear fission provides heat to the boiler. The higher operating temperatures require more cooling water than a coal-fired plant of the same electric generation capacity. A 1 GW nuclear fusion plant requires 6 million m^3 of cooling water each day, so the effects of water intake and thermal discharge described in the previous coal section are all more serious for nuclear plants.

Because nuclear plants involve combustion only in construction and in support vehicles for operations, they produce few air emissions. Nuclear plants do, however, produce long-lived radioactive wastes. Low-level wastes include reactor containment water, worker clothing, exposed tools, and plant fixtures irradiated for limited periods of time. High-level wastes consist of the spent fuel and the fuel rods in which they are encased.

Permanent disposal of these wastes in a manner that isolates their radiation from the living world has proven an intractable task. In 1999 the U.S. Department of Energy finally opened the Waste Isolation Pilot Plant for low level wastes in the Carlsbad Cavern system of New Mexico. The department has also been working on a repository for high-level wastes deep under Yucca Mountain, Nevada.

The greatest single release of nuclear radiation came from the Chernobyl Unit 4 accident on April 26, 1986, in the Republic of Ukraine in the then–Soviet Union. Operator error combined with design drawbacks of the RBMK graphite moderated reactor resulted in a virtually instantaneous catastrophic increase of thermal power and in a steam explosion. The explosion destroyed the reactor, releasing over 3% of the reactor fuel and up to 60% of the volatile products in the reactor core, mainly Iodine-131, Cesium-134, and Cesium-137. The accident deposited radioactive fallout over the entire northern hemisphere.

Twenty-eight people died from acute radiation doses, while more than 6500 may contract fatal cancers

through the year 2080. The Soviet government evacuated all people from a zone of 30 km radius and constructed a cement sarcophagus to contain the remains of the reactor core.

Lethal radiation killed many conifers and small mammals within 10 km of the accident in the first few weeks, but populations have since mostly recovered. By 1996, radioactive decay had diminished the amount of radioactive materials in the immediate area to 1% of their original amount, mainly as Cesium-137 in topsoil. Trees have accumulated Cesium-137 in growth rings. Grass, mushrooms, and berries also continue to incorporate the isotope, perpetuating a source of exposure for species that feed on contaminated plants. Aquatic ecosystems have generally tolerated the radioactivity concentrating in sediments, although fish may be accumulating radionuclides.

Ecologists have still not determined the long-term genetic effects of the fallout from Chernobyl. Apparently, the accident did not eliminate any plant or animal species, except where cleanup activities involved soil removal. Indeed, as a result of the evacuation, some plant and animal populations have thrived.

From 1961 to 1976, ecologists, led by George M. Woodwell, examined the chronic effects of irradiating a forest at Brookhaven, New York. Gamma radiation from Cesium-137 caused sensitive species to die, allowing resistant species and ruderals to invade. Species richness in 2 m square plots fell by half.

E. Hydroelectric

Hydroelectric systems harness the potential energy represented by an elevated mass. The potential energy of water at elevation will convert to increased kinetic energy of the water when the water runs to a lower elevation. A dam concentrates the difference in elevation, termed hydraulic head, in a spillway equipped with a turbine and electric generator. The electricity produced immediately enters the electric grid. In this manner, a hydroelectric plant will generate electricity with few direct air emissions and little thermal discharge. The principal effects of hydroelectric plants come from the total physical and hydrologic alteration and partial inundation of a watershed. Besides the forced removal of people and inundation of homes, hydroelectric plants also cause significant ecological changes.

More than 40,000 large dams now straddle rivers around the world, creating reservoirs that inundate more than 400,000 km^2. The Akasambo Dam on the Volta River in Ghana created the largest impoundment in the world, covering 8500 km^2. The Three Gorges

Dam under construction in the People's Republic of China will be the hydroelectric plant with the highest generation capacity in the world, 18.2 GW. The project, under construction in the period 1993–2009, will flood 1100 km² along 600 km of the world's third longest river, the Yangtze, and displace 1.2 million people. The dam will require 26 million m³ of concrete.

The inundation of formerly dry land submerges vegetation and immediately decreases the area of animal habitat. Lost forests represent ecosystem services and biomass wasted to decomposition. In Brazil, a country that depends on hydroelectric for 20% of its industrial energy, the land requirement for hydroelectric reservoirs averages 450 km² GW⁻¹ with a range of 17 to 10,000 km² GW⁻¹.

A dam blocks nutrient-rich sediment that a river system otherwise would have deposited in floodplains, wetlands, and at the outlet delta. Not only does the sediment buildup fill in a reservoir and eventually impair electricity generation, but the blocked sediment also represents a source of organic carbon and other nutrients wasted at the bottom of the reservoir. At the delta, bay and estuary topography changes, mudflat areas decrease, and nutrient-rich upwellings can decline. The Aswan High Dam blocks 98% of the 120 million tons of sediment that the Nile had carried to the sea each year, formerly depositing 10 million tons on the floodplain and delta. Consequently, soil depth has thinned and agricultural production has declined in the Nile Valley. Blockage of sediment and fresh water by the Akasombo Dam in Ghana has caused the decline of clam populations in the Volta estuary, and populations of *Sphyraena barracuda* (barracuda) offshore in the Gulf of Guinea. In addition, coasts become more susceptible to erosion.

Utilities start and stop the flow of water based on electricity and operational requirements. One operational objective is to smooth out natural extremes in the flood regime. This will usually change the meandering response and other channeling processes of a river. Ever since the Glen Canyon Dam removed spring floods in the near downstream section of the Colorado River, sandbar erosion has increased because the river does not flow fast or deep enough to move the amount of silt required for extensive sandbar formation. The resulting disappearance of some riparian tree species has led to the decline of *Empidonax traillii* (Southwestern willow flycatcher) and other birds. To mitigate the problem, the U.S. Department of the Interior staged a controlled flood in 1996.

For some dammed rivers, the flow of water unburdened by silt can deepen the riverbed. This lowers the water table of surrounding land. Also, the depletion of riverbed gravel can harm any species of fish, insect, mollusk, or crustacean that requires gravel bottoms to spawn. Many insect, amphibian, and fish species also use gravel areas for habitat or for protection.

The depth of a reservoir will often keep water at a temperature lower than that in the native river. For example, the Glen Canyon Dam changed the water temperature in the near downstream section of the Colorado River from a range of 0 to 27°C to a relatively constant 8°C. This has been a major factor in the extinction of *Ptychocheitus lucius* (Colorado squawfish), *Gila robusta* (roundtail chub), and *Gila elegans* (bonytail chub) and in the endangerment of five other fish species. Whereas the release water is clear, reservoir water often becomes slightly eutrophic and turbid. This degraded water quality can harm certain species.

The impacts of dams on anadromous fish relate to the migratory behavior and timing of the life cycles of these unique species. Dams render hazardous the downstream migration of young fish and block the upstream migration of adults. Moreover, salinity and temperature adaptations occur on a precise schedule, making long delays lethal. Disoriented and fatigued fish more easily fall prey to predation. Despite the deployment of extraordinary means in contemporary times to facilitate fish migration, including fish ladders, elevators, and trap and haul trucking, dams have eliminated anadromous species from many rivers. Runs of *Salmo salar* (Atlantic Salmon) and *Alosa sapidissima* (American shad) have disappeared from many rivers in the Northeast United States. In the Columbia River Basin in the Northwest United States, overfishing, pesticide runoff, and hydroelectric plants have endangered populations of *Oncorhynchus nerka* (Snake River sockeye salmon) and *Oncorhynchus tshawytscha* (Snake River chinook salmon). The physical barrier formed by a dam can even divide populations of aquatic species, altering patterns of gene flow and genetic variation.

Exotic fish species adapted to human disturbance and introduced into reservoirs for sport fishing will often outcompete native species. In the 1350 km² reservoir straddling the Brazil-Paraguay border behind the Itaipu Dam—with a generation capacity of 12.6 GW that currently ranks it the highest in the world—a nonnative species, *Plagioscyon squamosissimus* (curvina) has become the second most numerous species.

PCBs released from circuit breakers and oil leaking from machinery constitute the worst direct industrial pollution from dams. These and toxic organics build up in sediments and magnify up through the food chain. Impingement of aquatic organisms on intake screens and entrainment through turbines kills many individuals and causes stress and injury in survivors.

The material and energy needs for building the massive infrastructure of the hydroelectric energy cycle produces wide-ranging environmental effects. The end uses of hydroelectricity will produce the environmental effects associated with climate control, lighting, commercial production machinery, residential appliances, and other electric devices. Because smelting aluminum from bauxite ore requires a large amount of electricity, aluminum smelting comprises an end use closely tied to the hydroelectric option. The air emissions from smelters include CO, CO_2, particulates, NO_X, and trace metals. Major water emissions include trace metals and sulfates.

F. Renewable Energy Technologies

Renewable energy includes those forms of energy whose transformation does not necessarily consume the ultimate source of the energy, harnessing instead solar radiation, wind, the motion of water, or geologic heat. This section covers renewable energy technologies, including solar heating, solar thermal electric, solar photovoltaics, electric wind turbines, biomass-to-electricity conversion, biomass-to-alcohol fuels, and geothermal electric. These are renewable energy systems that depend on complex technology, so they are forms of industrial energy. The following section on traditional energy covers the simpler forms of renewable energy—firewood and charcoal.

Renewable energy forms share the physical characteristics of site specificity, variable availability, diffuse flow, and low or no fuel costs. Except for biomass-to-electricity conversion, renewable energy technologies do not involve combustion, so they do not directly produce much air pollution. The major environmental impacts derive from the fabrication, installation, and maintenance of renewable energy devices.

Solar energy systems fall into the categories of passive and active. Passive solar technologies consist of architectural forms that more effectively follow the diurnal and seasonal patterns of sunlight for the efficient heating and cooling of a building. Passive systems use the natural phenomena of radiation and convection. On the other hand, active systems use moving devices to achieve heat transfer. The simplest active systems use pipes or other collectors to heat water for residential or commercial use. For the most part, solar heating is environmentally benign.

Solar thermal uses arrays of reflective collectors to focus sunlight on a water boiler for the turbine production of electricity. These systems require significant amounts of land for the parabolic or trough collectors. Because solar is generally economically feasible only in hot sunny areas, sites are generally arid and water is scarce. Water withdrawals for the turbine and for washing the collectors can damage aquatic ecosystems. The bright arrays can also harm birds.

Photovoltaics are solid-state devices in which photons stimulate the emission of electrons and semiconductor materials channel the electrons for collection. In this way, photovoltaics directly convert sunlight to electricity with no moving parts, except for devices that move photovoltaics to track the sun, and no water, except for water to occasionally wash photovoltaic surfaces.

The fabrication of photovoltaic (PV) cells produces noxious environmental impacts. The first step is mining the quartz that constitutes the base material of a PV cell, so this produces many of the impacts on aquatic and terrestrial biodiversity described for coal mining. Then the production of metallurgical grade silicon requires the refining of quartz to 99% purity at 3000°C in an electric arc furnace. The production of semiconductor grade silicon occurs through a fluidized bed reaction of the silicon with hydrochloric acid. Then the production of semiconductor grade polycrystalline silicon occurs by electrically heating at 1000°C the semiconductor-grade silicon for vapor deposition on a silicone substrate. Remelting the polycrystalline silicon produces a form that can grow into crystals. These crystals are sawed into wafers 0.5 mm thick, wired, and encapsulated in glass 3 mm thick.

Trace metals are used to dope the semiconductor for the principal types of photovoltaic cells, including gallium arsenide, copper indium diselenide, cadmium telluride, indium phosphide, and cadmium sulfide. The trace metals, together with chlorinated organic solvents and phosgene gas, produce hazardous air, water, and solid wastes that can be lethal on contact or carcinogenic in small doses.

The conversion of biomass into electricity involves burning specially grown wood or crops in low pressure boilers to power steam turbines or the gasification of organic matter into methane to power gas turbines. The United States currently possesses a biomass electricity-generating capacity of 7.6 GW. Biomass-for-energy plants in the United States often employ cogeneration to provide process heat for an adjacent industrial facility. The principal species used include short rotation trees *Populus* spp. (poplars, aspen, cottonwoods), *Platanus* spp. (sycamore), and *Acer saccharinum* (silver maple) grown at densities of 1600 to 5000 trees ha^{-1}. Herbaceous energy crops include *Panicum virgatum* (switchgrass) and *Andropogon gerardii* (big bluestem). Brazil generates electricity and cogenerates heat from the organic wastes, or bagasse, left from the processing of sugarcane and orange juice.

The environmental impact of an energy crop depends on the previous land use as well as the cultivation techniques of both the new crop and any previous old crops. An energy crop can generate negative effects on biodiversity if it is grown in monoculture, if it is grown using pesticides, and if vegetation is clear-cut to prepare for the energy crop. Still, if previous land-uses were less environmentally sound than the energy crop, then the energy crop constitutes a mitigating practice. Previous sections detail the negative biodiversity effects of electric turbines and condensers.

The conversion of biomass into alcohol fuels also requires the dedicated growing of energy crops. Fermentation of cellulose and other complex carbohydrates produces ethanol, which certain engines can burn straight or mixed with gasoline. The United States now produces 4 billion liters of ethanol annually, mainly from corn. Brazil produces enough ethanol from sugarcane to provide for 10% of the country's energy use.

Otto Cycle engines burn neat ethanol, a mixture of 96% ethanol and 4% water. Modified conventional automobile engines can burn gasohol, a mixture of 78% gasoline and 22% ethanol. At one time, half of the automobiles in Brazil ran on gasohol, the other half on ethanol. Because the combustion of ethanol mainly produces CO_2 and water, with much smaller amounts of hydrocarbons and NO_x than gasoline combustion, ethanol used for transportation mitigates the most harmful direct effects of petroleum.

For centuries, society has captured wind for moving sailing ships, pumping water, and milling grain. Contemporary wind turbines also power electric generators. Rated at 100 to 300 kW per wind turbine, the steel machines reach heights of 10 m with pinwheel diameters up to 7 m. Arranged in arrays of up to hundreds of turbines, wind "farms" occupy considerable land areas. The greatest arrays cover unique areas in the Altamont Pass in the San Francisco Bay Area in California, the Tehachapi Pass in Southern California, in the Netherlands, and Denmark. Wind farms fragment terrestrial habitats and access road networks cause soil erosion. Spinning turbines can also kill birds.

Geothermal energy captures the heat of hot geologic formations, generating more than 7 GW of electricity worldwide. Geothermal plants sink pipes down to either capture deep hot water or to inject water for it to boil on contact with hot rocks. This process mobilizes trace metals contained in certain geologic strata and releases H_2S gas associated with geothermal deposits.

IV. BIODIVERSITY IMPACTS OF TRADITIONAL ENERGY

The most important sources of traditional energy are firewood and charcoal, which is produced from firewood. Local people harvest firewood either by coppicing, or cutting at the base, moderately sized shrubs, by lopping branches off mature trees, or only rarely by felling whole trees and splitting the logs.

In semiarid areas of Africa, women prefer the straight, moderately sized branches that only coppiced shrubs produce. Each year, women and, sometimes, their husbands or fathers go out in the dry season and cut at the base shrubs mainly in the family *Combretaceae*, carry the branches back to the village, and let them dry out. Just before the first rains, men and women cut a store of firewood for the rainy season. This serves, first, to avoid cutting wood that is wet and difficult to burn and, second, to get a time-consuming and strenuous chore out of the way before the exhausting and rushed rainy season. Coppiced shrubs will resprout in the rainy season and, in a year, regrow a full set of branches. When shrubs become scarce, women begin to pull down branches from adult trees, sometimes using long-handled hooks. This harms the growth potential of a tree by removing shoot apical meristem tissues and only provides difficult, thorny branches. When branches are exhausted, women fall back on noxious, dead stalks of spurges, family *Euphorbiaciae*. The last resort is animal dung. Only rarely will people cut down an adult tree for their own firewood needs. Men cut down trees for firewood for community events, large baptisms, weddings, or funerals, but even then, men prefer trees that have already died because these yield dry, more combustible wood.

Although women carry firewood for rural use, rural people load beasts of burden and carts to transport wood for sale in urban areas. So a town or city can produce land use changes far beyond its borders.

The low energy density of wood makes its transport onerous relative to the energy gained. Conversion of firewood to charcoal creates a product with double the energy per unit mass, but emits as waste heat up to two-thirds of the energy contained in the original wood. Charcoal makers cut down live and dead trees, particularly prizing sturdy tree trunks. In the field, they pile the wood, cover it with soil to form a kiln 1 to 3 m in height and 1 to 3 m in diameter, and ignite a slow burn. Over 3 to 6 days, the wood converts to charcoal by partially anaerobic pyrolysis.

Firewood harvesting can superimpose locally severe changes in biological diversity over wider alterations caused by long-term climate change. Global warming, the El Niño Southern Oscillation, and desertification all lead to systemic ecological changes at a regional scale. Embedded within these regions, firewood harvesting can reduce vegetative cover in less extensive areas where people depend on wood for their energy use.

The impacts of desertification in the West African Sahel clearly illustrate this complex situation. In Senegal, anthropogenic and climate factors caused a decline in forest species richness of one-third in the last half of the twentieth century (Gonzalez 1997).

Rainfall in the Sahel has shown a persistent downward trend in the past four decades, with the rainfall average of all years since 1919 falling at Louga, Senegal (15″37′ N, 16″14′ W) from 470 mm in 1953 to 400 mm in 1993. Serious droughts have hit in the periods 1910–1914, 1942–1949, and 1968–1973. An increase in human population has coincided with the decline in rainfall. The population of Senegal doubled in the period 1945–1988, growing at a rate of 0.025 y^{-1}.

In Northwest Senegal, the average forest species richness of areas of 400 ha fell from 64 ± 2 species ca. 1945 to 43 ± 2 species in 1993. Moreover, densities of trees of height ≥3 m declined from 10 ± 0.3 trees ha^{-1} in 1954 to 7.8 ± 0.3 trees ha^{-1} in 1989. Both the fall in species richness of 33 ± 5% and the decrease in tree densities of 23 ± 5% translate to a rate of −0.8% per year.

In West Africa, rainfall increases and evapotranspiration decreases toward the equator, creating a gradient that differentiates species into three broad bands of increasingly mesic vegetation: the vegetation zones of the Sahel, the Sudan, and Guinea. In Senegal, arid Sahel species (e.g., Family *Mimosaceae*) expanded in the north, tracking a concomitant retraction of mesic Sudan (e.g., Family *Caesalpiniaceae*) and Guinean species (e.g., Family *Bombacaciae*) to the south. Vegetation zones shifted southwest 25 to 30 km in the period ca. 1945–1993 (Fig. 4), a rate of 500 to 600 m y^{-1}, foreshadowing the magnitude of projected shifts driven by CO_2-induced climate change. The historical change acted through a higher mortality among mesic species, leaving drought-resistant species to dominate the remaining tree cover. The most notable species that have experienced local extinctions include *Dalbergia melanoyxlon* (Senegal ebony), *Prosopis africana* (ironwood),

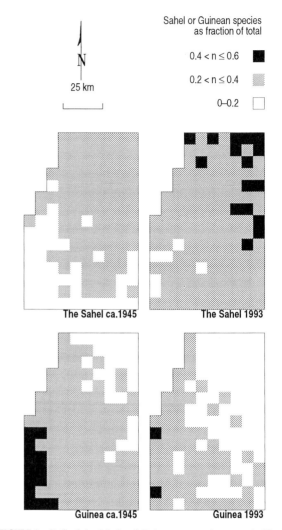

FIGURE 4 Shift of the Sahel and Guinean vegetation zones in Northwest Senegal from ca. 1945 to 1993 (Gonzalez 1997).

Sterculia setigera (mbep), and *Tamarindus indica* (tamarind).

Out of 215 ecological and socioeconomic variables, multivariate statistical analyses identifies rainfall and temperature as the most significant factors explaining the distribution and densities of trees and shrubs in Northwest Senegal. Rainfall and temperature override local anthropogenic factors.

Examination of dead trees along the coast supports a predominance of climatic over local anthropogenic factors. The sparsely populated coast offers a view of the state of the countryside before cultivation. Natural stands of *Euphorbia balsamifera* still occur there. In contrast, elsewhere in the Senegal Sahel, farmers have cut all natural stands of this species and replanted it along field boundaries. In the collective memory of local people, vast areas along the coast have not been cultivated. Dead trees still stand along the coast, but they show no ax marks or any sign that humans directly caused their death.

On a subcontinental scale, however, human activities may have caused the decline in rainfall. Deforestation of tropical rain forests in the Congo vegetation zone from the Republic of Guinea to Côte d'Ivoire may have reduced the evapotranspiration inputs essential to the maintenance of the Southwest Monsoon. Reduced rainfall over an extended period would reduce the vegetation cover in the Guinea zone. This in turn would decrease rainfall and vegetation in the Sudan, eventually reducing rainfall and vegetation in the Sahel. Thus, human activities in the distant rain forests may initiate a concatenation of climatic links that ultimately touch the Sahel.

Nevertheless, population growth has undoubtedly placed increasingly inordinate pressures on the area's vegetative cover. In Northwest Senegal, rural firewood use exceeds firewood production from shrubs over 90% of the land area, affecting 95% of the rural population. The rural population density of 45 people km^2 exceeded the 1993 carrying capacity of firewood from shrubs of 13 people km^2 (range 1–21 people km^2). The rural population density has exceeded carrying capacity since 1956.

The standing biomass of trees across the research area decreased from 14 t ha^{-1} in 1956 to 12 t ha^{-1} in 1993, matching a cumulative firewood deficit in the same period of 2 t ha^{-1}. The reduction in standing biomass released carbon into the atmosphere at a rate of 60 kg C cap^{-1} y^{-1}, somewhat less than the 100 kg C cap^{-1} y^{1} released from the burning of fossil fuels, mainly by the urban industrial and transport sectors.

Not only do the quantitative uses of firewood and charcoal exceed the area's wood production, but the fall in species richness has also reduced people's options qualitatively. For example, rural women depend on two particular shrub species for firewood because of the size of the branches, high wood density, and ease of collection. Beyond that, few fallback species remain. The fraction of women that reported shrub species as most prevalent in firewood use fell from 87% ca. 1945 to 50% in 1993. With respect to traditional medicine, 25 useful species have diminished significantly. Furthermore, eight species that provided fruit, leaves, and gum in past droughts have disappeared from as much as 53% of their range. If a grave famine hit the area in its current condition,

people would not be able to find the emergency foods that saved others in past episodes.

In the Sahel, the natural regeneration of local species could halt the declines in biodiversity and forest biomass. Natural regeneration is a traditional practice in which farmers and herders protect and promote the growth of young native trees. Traditionally, local people protect small trees that have germinated naturally or resprouted from roots, prune them to promote growth of the apical meristem, and, if necessary, set a stake to straighten the small tree.

Natural regeneration has expanded *Acacia albida* from an original restricted range along rivers in Southern Africa over thousands of km^2 up through the Sahel and the Sudan. In Senegal, the Sereer have protected dense parks of *Acacia albida* and *Adansonia digitata* in wide areas south of the research area. On the Mossi Plateau in Burkina Faso, farmers have similarly protected expanses of *Butyrospermum parkii* for the valuable oil from the tree's seeds. Across the Sahel, leather workers protect *Acacia nilotica adansonii* for the tannin enriched bark.

Natural regeneration requires no external inputs. It concerns species well known and appreciated by villagers. It focuses on young trees that have demonstrated their hardiness by surviving with no human caretaker, no watering, and no special treatment. Furthermore, natural regeneration not only augments the supply of wood, poles, fruit, medicine, and other products, it puts trees where farmers and herders really need them: in fields to maintain soil fertility and in pastures to provide forage.

Although photosynthetic activity in semiarid lands is an inefficient conversion of the total available solar radiation, the inefficiency of human tools renders end uses even more inefficient in the final conversion into heat and light. Table III shows this energy chain from sunshine to wood end use in the West African Sahel.

TABLE III

Energy Chain from Sunshine to Wood End Use in the West African Sahel (W ha^{-1})

Insolation at ground	2,400,000
NPP	1,720
Total wood production	118
Human wood energy use	213
Imported fossil fuels	93
Food consumption	53
Human wood energy end-use	13

Gonzalez 1997.

Therefore, another practice that can serve to conserve vegetative cover in rural areas dependent on firewood is the use of improved efficiency cook stoves. In many areas, women customarily cook with a kettle over an open fire. International development agencies have worked to develop and introduce stoves such as the *ban ak suuf* in Senegal, a horseshoe-shaped hearth constructed from clay and cow dung that provides a more enclosed combustion space to more effectively channel heat to the cooking vessel. The *lorena* in Guatemala is another earthen stove. The *jiko* in Kenya and *sakkanal* in Senegal are enclosed metal or ceramic charcoal stoves that more effectively contain heat than do traditional open charcoal burners.

V. FUTURE ENERGY PATHS

Human energy use directly alters patterns of biodiversity through changes in land use and through industrial pollution. Indirectly, human energy use is changing global biodiversity through the emission of greenhouse gases that cause global climate change and through other broad environmental effects of industrialization. Not only does the direct processing of energy generate environmental impacts, but the end uses that convenient energy forms make possible produce impacts locally and globally.

Table IV summarizes the major environmental impacts of human energy use on biodiversity. Almost every source requires land (Table V), a requirement that leads to habitat fragmentation and destruction. Globally, the climate change caused by CO_2 emissions constitutes the major impact of fossil fuels, but nonfossil fuel sources also produce air and water pollution. No energy transformation system operates without negative environmental effects, yet renewable sources generally restrict harmful effects to the capital formation stage and do not produce much ongoing pollution.

Holdren and Ehrlich (1974) proposed that environmental impact is equivalent to the multiplicative effect

TABLE IV

Major Sources of Biodiversity Impacts from Human Energy Use

	Oil	Natural gas	Coal	Nuclear fission	Hydroelectric	Renewable technologies	Wood
Habitat destruction and fragmentation	Exploration, access roads, pipelines	Exploration	Mining, electricity transmission lines	Mining, electricity transmission lines	Flooding vast areas, changes to hydrology of rivers	Land requirement for collectors	Unsustainable harvesting can eliminate or fragment habitat
Water pollution	Oil spills, drilling muds		Acid leachate from tailings, water removal for processing and cooling water	Acid leachate from tailings, water removal for processing	Thermal changes	Toxics from photovoltaic production	
Effects on aquatic organisms	Oil spills		Entrainment, impingement, thermal pollution	Entrainment, impingement, thermal pollution	Complete alteration of habitat, barriers to migration, entrainment, impingement		
Air pollution	CO_2, toxic organic compounds from refining	CO_2, flaring, volatilization of CH_4	CO_2, SO_2	Radiation, toxic halogenated compounds in fuel processing			CO_2
Soil	Oil spills		Tailings	Radioactive waste		Toxic solid wastes from photovoltaic production	Erosion possible with unsustainable harvesting
Major end-uses	Automobiles	Cooking, heating	Electricity	Electricity	Electricity, smelters	Electricity	Cooking, heating

TABLE V

Land Requirements and Major Air Emissions
for Electric Generation

	Land req. (ha MW^{-1})	(t GW^{-1} h^{-1})		
		CO$_2$	NO$_x$	SO$_2$
Geothermal	0.1–0.3	57		
Natural gas turbine	0.3–0.8	500		
Wind electric	0.4–1.7	7		
Nuclear	0.8–1.0	8	0.03	0.03
Coal	0.8–8.0	1000	3	3
Solar thermal electric	1–4	3.6		
Hydroelectric	2–1000	3		
Photovoltaics	3–7	5	0.008	0.02
Biomass	150–300		0.6	0.2

OTA 1995.

TABLE VI

Estimates of World Energy Resources
at Current Technologies

Nonrenewable stocks	TWy
Petroleum	600
Natural gas (conventional)	400
Coal	5000
Heavy oils, tar sands, unconventional gas	1000
Uranium	3000
Renewable flows	**TW**
Solar electric	52
Biomass	26
Hydroelectric	1.2
Wind electric	1.5

Holdren 1991.

of population, affluence, and technology:

$$\text{Environmental Impact} = \text{Population} \times \frac{\text{Resource use}}{\text{Person}}$$

$$\times \frac{\text{Environmental impact}}{\text{Resource use}}$$

People now call this identity the IPAT equation (impact = population × affluence × technology). Because the environmental impact of human energy use is proportional to the rate of energy use, and energy use is proportional to economic production, then the IPAT equation for energy becomes

$$\text{Environmental impact} \, \alpha \, \text{energy use}$$

$$= \text{Population} \times \frac{\text{GNP}}{\text{Person}} \times \frac{\text{Energy use}}{\text{GNP}}$$

Here, economic production per person indicates the level of material affluence, while energy use per unit of economic production indicates the level of technological efficiency. This highlights the great leverage that both energy conservation and efficiency wield to reduce the environmental impact of energy use. Indeed, improvements in energy efficiency reduced the energy intensity of economic activity in the United States by nearly one-third between 1975 and 1995.

For a 2100 AD global population of 10 million people to stay within the environmental limits of the earth, Holdren (1991) suggests that industrial countries im-

prove their energy efficiency to allow for an increase in economic activity in the nonindustrial countries so that everyone converges on an average use of 3 kW per person. This would increase total world energy use to 30 TW, more than double today's total. Imagine a world with twice as many nuclear power plants, coal mines, automobiles, and other energy infrastructure as today. Because many environmental impacts increase exponentially, the total impact will be much more than double.

Yet, the historical path of industrialization has left the world with only costly and environmentally disruptive energy alternatives. The earliest exploitation of fossil fuels depleted the most convenient oil and gas deposits. This is why fossil fuel production over time follows the bell-shaped Hubbert Curve. Not only has this left the current generation with deposits that are farther in polar and desert regions, deeper underground, and dispersed, but the low-cost energy has also shaped the expectations of people around the world for inexpensive on-demand energy services. Societies even subsidize the provision of convenient energy through infrastructure support to energy industries, tax breaks to oil drillers, preferential treatment to automobile companies, and other schemes.

The depletion of nonrenewable resources (Table VI) and other serious environmental and social constraints hobble most energy options for the future: recoverable oil and gas reserves will last only another 50 years; coal burning releases the principal agent of global warming, CO$_2$; biomass energy requires vast amounts of land; the small number of exploitable sites limits the potential for hydroelectric and wind power; and health and safety concerns prevent expansion of nuclear energy.

As a response to these constraints, governments, industry, and academia have placed enormous effort into the development of technologies such as electric vehicles, hydrogen cars, fuel cells, and nuclear fusion. Still, the future of human energy use may not lie with exotic devices. Instead, renewable energy sources, conservation, and efficiency of end use could form the future energy path of the world. Amory Lovins (1977) called this the "soft path." This would occur if societies set as their goal the provision of services, not just the acquisition of energy stocks and devices. In effect, people don't require light bulbs, they need illumination.

Acknowledgment

The author wishes to thank John P. Holdren, from whose work, both published and unpublished, much of the material in this entry derives.

See Also the Following Articles

DEFORESTATION • DESERTIFICATION • ECOLOGICAL FOOTPRINT, CONCEPT OF • ENERGY FLOW AND ECOSYSTEMS • GREENHOUSE EFFECT • HUMAN EFFECTS ON ECOSYSTEMS, OVERVIEW

Bibliography

British Petroleum Company, p.l.c. (BP). (1998). *BP statistical review of world energy 1997.* London: BP.

Costanza, R., and H. E. Daly. (1992). Natural capital and sustainable development. *Conservation Biology* **6**: 37.

Food and Agriculture Organization (FAO) of the United Nations. (1997). *State of the world's forests 1997.* Rome: FAO.

Georgescu-Roegen, N. (1971). *The entropy law and the economic process.* Cambridge, MA: Harvard University Press.

Gever, J., R. Kaufmann, D. Skole, and C. Vörösmarty. (1986). *Beyond oil.* Cambridge, MA: Ballinger.

Gonzalez, P. (1997). *Dynamics of biodiversity and human carrying capacity in the Senegal Sahel.* Ph.D. dissertation, University of California, Berkeley, CA.

Hall, C. A. S., C. J. Cleveland, and R. Kaufmann. (1992). *Energy and resource quality.* Niwot, CO: University Press of Colorado.

Holdren, J. P. (1991). Population and the energy problem. *Population and Environment* **12**: 231.

Holdren, J. P., and P. R. Ehrlich. (1974). Human population and the global environment. *American Scientist* **62**: 282.

International Energy Agency (IEA). (1997a). *Energy balances of OECD countries, 1994–1995.* Paris: Organisation for Economic Co-operation and Development.

International Energy Agency. (1997b). *Energy statistics and balances of non-OECD countries, 1994–1995.* Paris: Organisation for Economic Co-operation and Development.

Johansson, T. B., H. Kelly, A. K. N. Reddy, and R. H. Williams (Eds.). (1993). *Renewable energy.* Washington, DC: Island Press.

Lovins, A. B. (1977). *Soft energy paths: Toward a durable peace.* Cambridge, MA: Ballinger.

Office of Technology Assessment (OTA). (1995). *Renewing our energy future.* Washington, DC: U.S. Government Printing Office.

Pimentel, D., and M. Pimentel. (1979). *Food, energy, and society.* London: Edward Arnold.

President's Committee of Advisors on Science and Technology (PCAST). (1997). *Report to the President on federal energy research and development for the challenges of the twenty-first century.* Washington, DC: U.S. Office of Science and Technology Policy.

World Energy Council (WEC) and International Institute for Applied Systems Analysis (IIASA). (1995). *Global energy perspectives to 2050 and beyond.* London: WEC.

ENVIRONMENTAL ETHICS

Richard B. Primack* and Philip J. Cafaro†
*Boston University and †Colorado State University

GLOSSARY

anthropocentrism Position that only human beings have moral worth or intrinsic value.

biocentrism Position that all living beings have moral worth or intrinsic value.

deep ecology Activist philosophy that advocates radical personal and political change to protect wild nature.

ecofeminism Liberation philosophy that draws connections between preserving nature and promoting women's rights.

environmental ethics Philosophical discipline that specifies proper human relationships to the natural world.

ethical holism Position that complex aggregates such as species, ecosystems, or human societies have intrinsic value.

instrumental value Value of something relative to human interests or desires.

intrinsic value Value of something independent of its value to people.

rights Justified claims to the protection of one's important interests.

stewardship Responsibility to protect biological diversity, as given by God or accepted from society.

ETHICS IS THE BRANCH OF PHILOSOPHY that seeks knowledge of human flourishing and right conduct toward others, so that we may act upon it. Modern philosophers have tended to limit their ethical concern to human beings, but throughout history people have also attempted to cultivate proper relationships to the rest of nature. Recently philosophers have turned to this topic, largely in response to environmental degradation and the loss of biodiversity, and have created a new discipline: environmental ethics. Environmental ethicists attempt to specify appropriate human relationships to the nonhuman, natural world. In the course of their work they have developed strong ethical arguments for preserving biodiversity. They have also challenged conventional views of happiness and human welfare and the materialistic values at the base of much modern life. While environmental ethics treats the full range of environmental issues, from air pollution to nuclear risk assessment, this article focuses on ethical issues directly related to the preservation of biodiversity.

Encyclopedia of Biodiversity, Volume 2
Copyright © 2001 by Academic Press. All rights of reproduction in any form reserved.

I. DUTIES TO PROTECT BIODIVERSITY: INTRINSIC VALUE ARGUMENTS

A. From Economics to Ethics

Biodiversity has great economic value, both direct and indirect. Direct economic values are provided by wild game, seafood, fuelwood and timber products, and indigenous and high-technology medicines. Indirect economic values accrue from biodiversity's roles in waste disposal, climate regulation, protecting soil and water resources, recreation and ecotourism, and much more. Both directly and indirectly biodiversity safeguards or adds to human wealth, often justifying its protection in purely economic terms.

Thus economic arguments by themselves provide a basis for valuing and protecting species, especially when we expand our concept of economic value from short-term, next-quarter profits to include longer-term and indirect benefits. But economic arguments can also provide grounds for extinguishing species or for saving one species rather than another. In conventional economic terms, low value will be given to species having small populations or limited geographical range, small physical size or unattractive appearance, no immediate use to people, or no relationship to other species of economic importance. Such qualities may characterize a substantial proportion of the world's species, particularly insects and other invertebrates, fungi, nonflowering plants, bacteria, and protists. Halting profitable developments or making costly attempts to preserve these species may not have any obvious economic justification. In some circumstances, economic justification could exist for destroying an endangered species, particularly organisms that cause disease or attack crop plants. Still, many people believe that such destruction is morally wrong even if it is economically profitable.

To say that an object has economic value is to say that it is useful to human beings or that they desire to possess it. Put another way, it has an instrumental value: someone can use it as an instrument for his or her purposes. On the other hand, we recognize that at least some entities, such as human beings, have value regardless of whether anyone else values or uses them. These entities have an intrinsic value: a value that is grounded not in their usefulness to others, but in what they are themselves.

Human beings, we usually think, have both instrumental and intrinsic value. Because we have intrinsic value we possess certain rights that no one can legitimately infringe, even if it is in their self-interest to do

so. Conversely we have certain duties towards other people that specify how we should or should not treat them in various situations. We cannot help but look at the people we interact with each day in terms of their usefulness to us, but if we look at them solely in these terms we disrespect their (and our) humanity and if we treat them solely as a means to our own ends we are likely to behave immorally.

Many conservationists argue that similar duties restrict the morally acceptable treatment of wild nature. They consider it wrong to destroy a rare woodland or cause a species to go extinct, even if this action is in an individual's or corporation's self-interest. Not only have we no right to destroy any species, we have a moral responsibility to actively protect species from extinction as the result of our activities. Opponents counter that this position is illegitimate because only human beings have intrinsic value. Unless our actions affect other people, directly or indirectly, any treatment of the natural world is morally acceptable.

B. Extensionist Arguments for Intrinsic Value

Proponents provide both extensionist and nonextensionist arguments for the intrinsic value of wild nature. Extensionist arguments ask what qualities give intrinsic value to human beings, then assert that some other beings possess these same qualities. Therefore, they conclude, we should recognize the intrinsic value of these other beings and extend proper treatment to them.

One common justification for valuing human beings is our ability to reason. But some of the so-called "higher animals" also seem to possess the rudiments of reason. Chimpanzees and gorillas have been taught sign language involving several hundred words; wolves have an elaborate social life and the ability to coordinate long hunts; dolphins, whales, and other cetaceans send complex signals that we are just beginning to understand. Many argue that because of these factors, we should not hunt these animals for food or sport, use them in research, or in general treat them solely as means to our own ends.

Other philosophers believe that sentience—the ability to perceive the surrounding world and feel pleasure and pain—demands moral consideration. The extensionist argument here states that because pain is bad we should avoid inflicting pain unnecessarily on others who can feel it—even if those others have fur, feathers, scales, or numerous legs. Immoral actions against human beings are wrong, at base, because they cause unnecessary pain; actions that cause other animals unnecessary pain are likewise immoral. Millions of people around the world act on such beliefs by abstaining from eating animals or using products that can only be procured by killing or harming them. Even most meat-eaters accept that inflicting gratuitous pain on animals is wrong, suggesting a widespread belief that where there is sentience there exists some moral responsibility.

One possible problem with the extensionist arguments just discussed, from the point of view of conservationists, is that they only encompass certain organisms. Basing intrinsic value on some rudimentary form of reasoning or complex mental experience would appear to rule out most animals, for example, almost all invertebrate species. Basing moral consideration on the ability to feel pain includes a wider class of animals, but again many simpler animals and all plants, fungi, and single-celled organisms are ruled out.

However, extensionist arguments can be extended further by recognizing that all organisms have a drive to stay alive and reproduce. In the same way that we acknowledge the rights of people to live, have children, and satisfy basic needs, we may extend these rights to individuals of all species. In his autobiography, Albert Schweitzer maintained that "a man is ethical only when life, as such, is sacred to him, that of plants and animals as that of his fellow men, and when he devotes himself helpfully to all life that is in need of help."

This position may lead to a very demanding code of conduct! Believers in strict biocentric equality assert that it is always wrong to kill individuals of any species (because of their intrinsic value) unless we need to do so to survive. Others believe that using nature to provide necessities and some measure of comfort is morally acceptable, but not to provide luxury goods. In this view, cutting down trees for firewood or to build a modest house is morally acceptable, particularly if this is done sustainably, whereas harvesting mahogany trees from rain forests to make expensive furniture for wealthy individuals half a world away is morally unacceptable. Whatever the particular judgments made, recognition of an intrinsically valuable organic world leads to distinctions between essential and inessential human uses and to a more limited use of natural resources.

C. Nonextensionist Arguments for Intrinsic Value

In addition to extensionist arguments, which point up similarities between wild nature and intrinsically valuable humans, there are arguments that find value in nature without referring to such similarities. Some believe it is a mistake to value other beings only for the

ways in which they resemble humans. Writing in *The Monist* in 1992, Robert Elliot suggested that natural organisms may have the following properties that give them intrinsic value: "diversity, stability, complexity, beauty, harmony, creativity, organization, intricacy, elegance, and richness." These are qualities of natural organisms that we can appreciate—and that may call forth responses of personal restraint and active protection.

All species represent unique biological solutions to the problem of survival. They have solved the challenges placed before them by their environments and thrived. Some people see a value in this creativity and time-tested uniqueness. Others appreciate the beauty and elegance of the natural forms created by this process. Still others value the complexity and ingenious structures that science and close observation have revealed. People who know and value this uniqueness, beauty, and complexity feel a special horror at its permanent disappearance. After all, if an individual dies it may be replaced by another individual more or less the same, but take away the last passenger pigeon or giant moa and their like will never return. Nonextensionist arguments thus support preservation of species as well as protection of individual organisms (see Section I,E).

Species interact in complex ways in natural communities. The loss of one species may have far-reaching consequences for other members of the community: other species may become extinct or the entire community may become destabilized. As we learn more about global processes we are finding out that many chemical and physical characteristics of the atmosphere, the climate, and the ocean are linked to biological processes in a self-regulating manner. More diverse biological communities may be better able to deal with environmental disturbances such as drought and global climate change. For these reasons, if we value some species, we should arguably protect all species.

D. Anthropocentric Denials of Intrinsic Value

Skeptics reply that even though some people do value non-human organisms and species, we are not obligated to do so, because only human beings have intrinsic value or genuine rights. Humans have a value beyond all other beings because only we are fully conscious and rational. Unless our actions affect other people, directly or indirectly, any treatment of the natural world is morally acceptable.

Such a viewpoint is anthropocentric—locating value solely in humans—and to many it seems the most obvious common sense, while departures from it seem irrational or overly sentimental. The appellation "tree hugger" expresses this view, suggesting inappropriate sentiments toward trees, leading to inappropriate actions. It also suggests a callous disregard for the interests of people who cut trees for a living, whom we really should care about (consider the bumper sticker "Hug a Logger, Not a Tree"). People who value nature counter that anthropocentrism is selfish and that speciesism—the privileged treatment of one species over another—is no more justified than racism or sexism.

Although the debate between anthropocentrists and biocentrists has tended to incorporate all the uncertainties attending ethical justification in general, some clarification may still result. Charges of irrationalism to the contrary, it is possible to love and value nonhuman nature and act on this view. The challenge to biocentrists is to fashion fulfilling lives that limit their negative environmental impacts and help preserve and celebrate nature. Anthropocentrism also remains a rationally defensible position, which may be consistently acted upon. Anyone who values humanity based on qualities that we share with other species, however, cannot consistently deny intrinsic value to those other species. Furthermore, anthropocentrists who value humanity primarily for our ability to reason may consider the many arguments in favor of lessening pollution and preserving wild nature that appeal to our rational self-interest. Anthropocentrists are more likely than biocentrists to accept some amount of pollution in rivers or the extinction of certain species, but they also acknowledge human needs for clean air and drinking water, the enjoyment we get from fishing, swimming, and canoeing, and the value of biodiversity to science, art, and business. When it comes to particular environmental policies, anthropocentrists and biocentrists may find considerable common ground.

E. Ethical Holism

The preservation of biodiversity seems to demand that the needs of endangered species take precedence over the needs of individual organisms. For example, the U.S. National Park Service killed hundreds of introduced rabbits on Santa Barbara Island off the California coast to protect a few plants of the endangered Santa Barbara live-forever (*Dudleya traskiae*)(Fig. 1). In this case, one endangered species was judged to be more valuable than hundreds of individual animals of a common species. Similarly, conservation biologists would not find it acceptable to destroy the last remnant of a rare biological community even if every species living there could be maintained in captivity; the ecological interactions and

FIGURE 1 Beasts versus the biosphere. Government agencies judged the continued existence of the endangered plant Santa Barbara live-forever (*Dudley traskiae;* the tall plant in the photo) to be more valuable than the common rabbits on its island home. The rabbits, which fed on the plant's fleshy leaves (shown at the bottom right), were killed to stop their destruction of this fragile plant species. (National Park Service photograph.)

evolutionary processes of the community would be lost if the species only lived in captivity. These examples illustrate that most conservationists are holists, finding value in larger groupings, such as species and biological communities. They are thus sometimes willing to sacrifice the interests of individuals to preserve species and communities.

Many writers, especially animal welfare advocates, have difficulty assigning rights to species. Peter Singer and Tom Regan, two prominent philosophical animal welfare advocates, argue that species are not conscious entities and so do not have interests. In their view, to sacrifice the genuine interests of an individual animal, who can suffer or possess rights, to the imagined interests of a species, which cannot, is mistaken. Many animal welfare advocates also reject conservationists' special concern for native species over exotics: an animal is an animal, with a greater or lesser ability to suffer, and this ability should determine our actions, rather than its point of origin.

On both biological and ethical grounds, however, most conservation biologists argue that species, rather than individual organisms, are the appropriate targets of conservation efforts. All individuals eventually die; it is the species that continues, evolves, and sometimes forms new species. In a sense, individuals are temporary representatives of species. Whether or not we allow them rights, species carry great value as the repositories of the accumulated experience and history of millions of previous life-forms through their continuous, evolutionary adaptation to changing environments. The premature extinction of a species due to human activities destroys this natural process and obliterates its history. It can be regarded as a "superkilling," because it kills future generations of the species and eliminates whole lines from the processes of evolution and speciation.

Furthermore, conservationists typically argue that species should be prevented from spreading beyond their natural ranges as a result of direct or indirect human activity. For example, the zebra mussel (*Dreissena polymorpha*) is native to the Caspian Sea, but it has recently become an aggressive invader of North American aquatic habitats (Fig. 2). Arguably this species should be destroyed whenever possible in North America, for two reasons. First, exotics often displace

FIGURE 2 Unwelcome guests. The Zebra mussel (*Dreissena polymorpha*), a native of the Caspian Sea, was accidentally introduced into the Great Lakes and associated rivers in 1988. Since then the species has formed dense populations over a wide and ever-increasing area, outcompeting and choking out native species. In this case, thumb-sized Zebra mussels have almost totally encrusted a crayfish shell. (Photograph courtesy of Ontario Ministry of Natural Resources from the Great Lakes Sea Grant Exotic Species Library.)

native species, sometimes contributing to their extinction. To prevent this loss of biological diversity we destroy individual exotics, on the assumption that the exotic species will continue to thrive in its native habitats. Second, and more controversially, many conservationists deny intrinsic value to individual plants and animals that have spread beyond their natural range due to human activities. Part of what gives a species value is its unique evolutionary history and its ecological roles in native habitats, both of which are tied to particular places and biological communities. When these species instead invade new natural areas and destabilize or radically change their species composition, these justifications of their value no longer hold. Exotic species contribute to biodiversity in their native habitats but often diminish biodiversity when they become established and common in new locales.

II. DUTIES TO PROTECT BIODIVERSITY: OBLIGATIONS HUMAN AND DIVINE

A. Duties to Humans

That human beings have direct moral duties to other species remains controversial; that we have direct moral duties to other human beings is not (although great disagreement remains regarding their scope). Our duties to other human beings may support protection of the environment and preservation of biodiversity, based on their instrumental value to intrinsically valuable people.

We have responsibilities to our neighbors and fellow citizens. Because of this we must arguably minimize damage to the natural environment, since such damage harms not only other species but people as well. Increasingly, connections are being made between environmental pollution and high levels of human disease. For example, massive environmental pollution in the former Soviet Union has greatly increased cancer, birth defects, lead poisoning, and lung disease among the population. Much pollution and environmental degradation is unnecessary and could be minimized with better planning. Often pollution occurs because corporate leaders are unwilling to spend money to prevent it, despite the resulting ill effects on the health, wealth, or happiness of other people. Citizens must recognize the biological and social costs of environmental damage and force corporations to be good neighbors and governments to enact and enforce strong environmental laws, for the common good.

We also have responsibilities to future generations. Economic decision making tends to focus on the short term and it is this economic system that is driving environmental change and degradation. In response, many ethicists have emphasized the importance of intergenerational justice. Humanity's unprecedented numbers and technological power mean that one generation can now radically remake the earth that the next will inhabit, for better or worse. If in our daily living we degrade the natural resources of the earth and cause species to become extinct, future generations will pay the price in terms of a lower quality of life.

This truth gives us arguments not just for keeping ecosystems safe for present human health, but also for preserving biodiversity for future human use, enjoyment, and development. For example, if species and wild places are lost, children will be deprived of one of their most exciting experiences in growing up—the wonder of seeing "new" animals and plants in the wild. These concerns are a powerful motivating force for the members of organizations such as the Sierra Club and The Nature Conservancy, who see themselves as land stewards preserving biodiversity for future generations.

Writing in the *Encyclopedia of Environmental Biology* in 1995, Holmes Rolston predicted: "It is safe to say that in the decades ahead, the quality of life will decline in proportion to the loss of biotic diversity, though it is often thought that one must sacrifice biotic diversity to improve human life." Of course, this contention is debatable. Many argue that job creation and increased wealth are more important to future generations than the preservation of biodiversity. Debate on this issue, while inconclusive, is essential, because it forces us to specify the actual, long-term benefits of development projects and the sorts of societies we want to create for our children. Given our unprecedented power and the tendency to put personal interests above the common good, even inconclusive debate may be a force for creating a better future for all.

At a minimum, our duties to posterity seem to require us to live sustainably—that is, limiting our consumption so as not to degrade essential life-support systems or deplete natural resources that future generations will need. The world's governments have formally recognized this through treaties governing air pollution, whale hunting, and disposal of wastes in the ocean. One important treaty, the Montreal Protocol of 1987, reduced the permitted use of ozone-depleting chlorofluorocarbons, leading to a phase-out of production three years later. The grave dangers these chemicals posed to future generations, with their potential to thin the earth's protective ozone layer and cause millions of

new cases of deadly skin cancer, clearly outweighed the interests of producers and users of CFCs, especially since many of those uses were not essential to human survival (e.g., in luxury items such as hairsprays and air conditioners) and alternatives were available or quickly developed for most uses.

Similar appeals to the good of future generations have so far proved less successful in convincing world governments to reduce fossil fuel use to slow global climate change. This is partly due to strenuous lobbying by oil and automotive corporations and oil-producing nations, whose leaders have implicitly placed short-term profits above the health and well-being of future generations. Partly though it is because the risks of increased fossil fuel use, though grave, are less certain than those associated with continued use of CFCs, while the economic costs of decreasing fossil fuel use are far greater.

Human maturity leads naturally to self-restraint and a respect for others. Many conservationists agree with Arne Naess (1989), who writes that the further maturation of the human species will involve an "identification with all life forms" and "the acknowledgment of the intrinsic value of these forms" in an expanding circle of moral obligations. Moving outward from oneself, the circle would include duties to our family and relatives, our local community, our country, all humanity, mammals (Save the Whales!), all animals (Save the Snail Darter!), all species (Save the Yellow Lady Slipper Orchid!), ecosystems (Save the Rain Forest!), and ultimately the whole earth (Fig. 3). Such an expansion of ethical concern involves new limitations on acceptable actions, but also new opportunities for personal growth and flourishing.

Although some worry that recognizing intrinsic value in nature devalues human beings, many environmental philosophers argue that respect for human life and human diversity is compatible with a respect for nonhuman nature. Citizens of all countries will be more likely to accept their responsibility for protecting biological diversity, they assert, when they have full political rights, a secure livelihood, and an awareness of environmental issues. Some of the most exciting developments in conservation biology involve supporting the economic development of disadvantaged rural people in ways that are linked to the protection of biological diversity. Helping poor women establish sustainable plots of cash crops and achieve a degree of economic independence may reduce the need to overharvest wild species. Working with indigenous people to establish legal title to their land may give them the means to protect the biological communities in which they live.

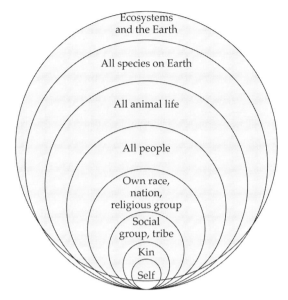

FIGURE 3 Human development. An ethical sequence in which the individual extends concern outward beyond the self to progressively more inclusive levels. (Courtesy of Reed Noss.)

Actions taken to protect species and biological communities should whenever possible benefit people as well.

B. Duties to God: Religious Stewardship

Various preindustrial cultures successfully coexisted with a rich local flora and fauna for hundreds of years, in part because their religions and societal ethics encouraged personal responsibility and thoughtful use of resources. People in these societies respected wild animals and plants even as they harvested them or borrowed their habitat for human purposes. For example, the Cherokee Indians of the southeastern United States spoke special prayers to the deer that they killed, telling the deer spirit that they indeed needed the meat, that they would not waste it, and that they would bury the bones with due solemnity. Traditional peoples often treated rivers, mountains, and other ecosystems as sacred places to be approached with reverence and an appreciation for what they were, rather than for what human beings could make of them.

Many modern religious adherents abhor the destruction of species, because they are God's creation. If God created the world, then presumably the species God created have value. Within the Jewish and Christian traditions, human responsibility for protecting animal species is explicitly described in the Bible as part of a covenant with God. The Book of Genesis describes the creation of the earth's biological diversity as a divine

act, after which "God saw that it was good" and "blessed them." In the story of Noah's Ark, God commanded Noah to save two of all species—not just the ones human beings found useful. God provided detailed instructions for building the ark, an early species rescue project, saying "Keep them alive with you." After the flood subsided, the animals were released to repopulate the earth. This story, versions of which were told throughout the ancient Near East, can be interpreted as an early awareness of the importance of biological diversity.

The prophet Mohammed, founder of Islam, continued this theme of human responsibility, saying "The world is green and beautiful and God has appointed you as His stewards over it. He sees how you acquit yourselves." Many other religious traditions also support the preservation of nature. For example, Hinduism locates divinity in certain animals and recognizes a basic kinship between humans and other beings, including the transmigration of souls from one species to another. A primary ethical concept in Hinduism and other Indian religions, such as Jainism and Buddhism, is *ahimsa*—avoiding unnecessary harm to life. In attempting to live this ideal, many religious people become vegetarians and live as simply as possible.

III. BIODIVERSITY AND HUMAN FLOURISHING

Economic arguments stress that we should preserve biological diversity because it is in our material self-interest. Ethical arguments based on the intrinsic value of wild nature, our duties to other human beings, or our duties to God stress that we should sometimes act altruistically; that is, we should set aside our personal interests to preserve biological diversity. A second kind of ethical argument appeals to a fuller understanding of our self-interest, arguing that preserving biodiversity and developing our knowledge of it will make us better and happier people. The following are the main arguments for preserving biodiversity in our own enlightened self-interest.

A. Health and Wealth

It cannot be repeated too often that biological diversity preserves our basic life-support systems of food production, water supply, oxygen replenishment, waste disposal, soil conservation, and more. People are healthier and more productive in clean, intact environments. We depend on this and should value it. Similarly, biodiversity allows us to create tremendous economic wealth, directly and indirectly. An article published in *Nature* in 1997 by Robert Costanza and others, "The Value of the World's Ecosystem Services and Natural Capital," estimated that the world's ecosystems provide $32 trillion per year of value to people, substantially higher than the $18 trillion per year of goods and services those people produce themselves. In other words, human society could not exist without what the natural world provides us for free, nor could we afford to pay for substitutes even if they existed.

B. Aesthetic and Recreational Enjoyment

Nearly everyone enjoys wildlife and landscapes at an aesthetic level and this is part of a good life. The beauty of a field of wildflowers in Glacier National Park or a migrating warbler on a spring morning in a city park enriches the lives of those who appreciate them. For many people, a high quality of life involves experiencing nature first-hand. Simply reading about species or seeing them in museums, gardens, zoos, or videos will not suffice. Hiking, canoeing, nature photography, and bird watching are physically, intellectually, and emotionally satisfying.

Hundreds of millions of people spend tens of billions of dollars annually in these pursuits, proof enough of their value. As the world becomes more crowded, it becomes ever clearer that these activities are in competition with other human uses. We are creating a world with ever-diminishing opportunities for aesthetic and recreational encounters with wild nature (Fig. 4). If species and ecosystems are not to disappear altogether, they must be consciously preserved.

C. Artistic Expression and Scientific Knowledge

Throughout history, poets, writers, painters, sculptors, and musicians of all cultures have drawn inspiration from wild nature. Nature provides countless forms and symbols for visual artists to render and interpret, while poets have often found their greatest inspiration in either wild nature or pastoral countrysides (Fig. 5). Preserving biological diversity preserves possibilities for all artists. It also allows those of us who appreciate such creative acts access to the sources and experiences that inspired them. A loss in biological diversity could very well limit the creative energies of people in the future and thus restrict the development of human culture. For example, if many species of whales, butterflies, and

FIGURE 4 Reaching out. Most people find interacting with other species to be an educational and uplifting experience. Here people greet a minke whale that is being rescued after it became entangled in a trawler's gill net; the float behind the whale was attached to the net to keep the whale at the surface so it could breathe. Later, rescuers were able to release the whale from the netting. (Photograph by Scott Kraus, New England Aquarium.)

FIGURE 5 Celebrating nature. Rare wildflowers and butterflies are the inspiration for botanical sculptor Patrick O'Hara. In his studio in western Ireland, O'Hara molds, sculpts, and paints delicate porcelain scenes from nature that inspire an appreciation of biodiversity in a worldwide audience. (Photograph courtesy of Patrick O'Hara from *Essentials of Conservation Biology*, p. 138, Fig. 6.3.)

orchids go extinct in the next few decades, whole sets of imagery will be lost to the direct experience of future generations of artists.

Like art, our growing knowledge of nature through science is one of humanity's greatest achievements. This knowledge is facilitated by the preservation of wild nature. Wild areas allow the study of natural ecological interactions. Wild species preserve the record of evolution. Young people are inspired to become scientists through personal contacts with nature, while those who do not pursue science professionally can take a basic scientific knowledge and apply it to an understanding of their own local fields, forests, and streams.

Two of the central mysteries in the world of science are how life originated and how the diversity of life found on earth today came about. Thousands of biologists are working on these questions and are coming ever closer to the answers. Recent discoveries of bacteria deep in the earth's crust and new species of animals in tropical rain forests are important developments in this exciting story. New techniques of molecular biology allow greater insight into the relationships of living species as well as some extinct species known only from fossils. But when species become extinct, important clues are lost and the mysteries become harder to solve.

If *Homo sapiens'* closest living relatives, the great apes, disappear from the wild, we will lose important information regarding human physical and social evolution.

D. Historical Understanding and Religious Inspiration

Knowing nature, both scientifically and through personal experience, is a key to self-knowledge and an understanding of human history. In walking the landscapes that our ancestors walked, we gain insight into how they experienced the world, at a slower pace and without mechanized aids. People often forget just how recently humankind has moved to ultrafast transportation, fully illuminated cities that shut out the night, and other aspects of modern life. Preserving natural areas allows us to develop our historical imaginations.

Many religions have traditions of "wandering in the wilderness" in order to commune with God or with spirits. In the Western tradition, Moses, Isaiah, Jesus, and St. Francis of Assisi all sought out the solitude of wilderness. So did the Chinese sage Lao-tzu, the Japanese Zen poet Basho, and generations of Lakota vision-seekers. Being in nature allows us to clear and focus our minds and sometimes experience the transcendent. When we are surrounded by the artifacts of civilization, our minds stay fully focused on human purposes and our everyday lives. Religion probably would not disappear from a totally tamed human environment, but perhaps it would become diminished for many.

In brief, while the preservation of biodiversity sets limits to some human activities, it is a prerequisite for our continued enjoyment of others. There are good reasons to believe that preserving and exploring biodiversity can make us better, happier people. Many conservationists are convinced that a better understanding of our true self-interest would lead to greater efforts to protect biological diversity.

IV. ENVIRONMENTAL PHILOSOPHIES

A. Deep Ecology

Paul Sears, recognizing that increased knowledge would lead to the questioning of destructive practices common in modern society, and often taken for granted, called ecology a "subversive science." During the twentieth century, ecologists, nature writers, and practicing environmentalists have increasingly articulated an appreciation of nature and spoken of the need for changes in human lifestyles to protect it. "Green" political parties and activist conservation organizations such as

Greenpeace, EarthFirst!, and India's Chipko movement have appeared throughout the world.

One well-developed environmental philosophy that supports this activism is known as deep ecology. Deep ecology builds on two basic premises: biocentric equality and self-realization. Biocentric equality expresses "the intuition . . . that all things in the biosphere have an equal right to live and blossom and to reach their own individual forms of unfolding" (Devall and Sessions, 1985). Humans have a right to live and thrive, but so do the other organisms with whom we share the planet.

Self-realization describes a striving to grow, develop, and find fulfillment. Human beings are assumed to share this goal with the rest of nature. In the case of humans, deep ecologists believe that true self-realization involves spiritual growth and an expansion of knowledge and concern, rather than an increase in material wealth. Individual human self-realization should lead to a concern with all of nature: the preservation and development of the larger wholes to which we belong, including both human and natural communities. Deep ecologists articulate this position in opposition to a "dominant worldview" that makes human concerns paramount and views human happiness in materialistic terms (see the following table). Deep ecology's idealism and its call to action make it an appealing philosophy for people concerned with protecting biodiversity.

B. Ecofeminism

Ecofeminism represents another recent, important development within environmental ethics. Ecofeminists see a connection between the domination of women and the domination of nature. They attempt to specify links between these two forms of domination, and advocate both environmental protection and full, equal rights for women.

Like deep ecologists, ecofeminists argue that how we conceptualize wild and human nature has important ethical consequences. Karen Warren argues that some conceptual frameworks are inherently oppressive, because they split the world up into "value dualisms, i.e., disjunctive pairs in which the disjuncts are seen as oppositional (rather than as complementary) and exclusive (rather than as inclusive), and which place higher value on one disjunct rather than the other" (Armstrong and Botzler, 1998). Thus men are privileged over women, human culture over nature, animals over plants, and reason over emotion. The inferior disjuncts tend to be lumped together: women may be seen as closer to nature, more emotional, or less rational. Such conceptual frameworks legitimize discriminatory treatment of those "others" who fall on the wrong side of these value dualisms.

Ecofeminists launch a three-pronged attack on such a "logic of domination." First, they may deny that certain differences exist, that they are hard and fast differences, or that they are as extreme as they are portrayed. For example, most ecofeminists simply deny that women are less rational or more emotional than men. In like fashion, they may point up similarities between humans and the other animals and downplay our differences. Second, ecofeminists may deny that actual differences make some beings morally superior to others. Human beings are superior reasoners compared to frogs, but that does not mean that we are therefore morally superior, in the sense that our interests should always trump theirs. Third, ecofeminists may deny that moral superiority underwrites domination. Even if humans are morally superior to frogs, it may still be wrong to kill them for food or destroy the last members of a rare frog species to create a new subdivision. Perhaps the proper response to those inferior to us in abilities or moral value is care and restraint.

C. Humanism: Our Default Mode

Like deep ecologists and ecofeminists, conventional ethical philosophers, including Kantians, utilitarians, and theologically grounded ethicists, have begun to

Dominant worldview	Deep ecology
Humans dominant over nature	Humans living in harmony with nature
Natural environment and species are resources for humans	All nature has intrinsic worth, regardless of human needs
A growing human population with a rising standard of living	A stable human population living simply
Earth's resources are unlimited	Earth's resources are limited and must be used carefully
Ever higher technology brings progress and solutions	Appropriate technology must be used with respect for the earth
Emphasis on material progress	Emphasis on spiritual and ethical progress
Strong central government	Local control, organized according to watersheds, bioregions, or other natural units

address environmental issues. Most philosophers who have considered these matters have found strong reasons to support environmental protection. As previously noted, biocentrists and anthropocentrists may agree on a wide range of measures to protect the environment, despite great philosophical differences.

Still, some philosophers and many members of the general public remain unconvinced of any moral imperative to protect biological diversity. Their position may be characterized as humanism, a philosophy committed to the following propositions:

1. Biological diversity exists for humans, has no value apart from humans, and need not exist apart from humans.
2. The transformation of wild nature into natural resources adds value to nature, since nature possesses value only in human use. Indeed, the ever more thorough transformation of wild nature allows increased human numbers to lead longer, happier, and better lives.
3. The creation of just societies filled with flourishing individuals is the highest achievement of which humans are capable. We should judge ourselves based on our technological, scientific, artistic, and social progress—not on whether we preserve nature.

Like the philosophies discussed earlier, humanism may be developed in a variety of ways. A humanist's ideal society may be more or less egalitarian, wealthy, or stable, and more or less racially and culturally diverse. But humanists share a belief in the centrality of humans. They generally applaud increased human numbers, wealth, and technological power and the development of new arts and activities that flourish in highly artificial environments. For these reasons they see little to lament in the loss of biodiversity.

Humanism may be called humanity's default mode, since current trends are moving us more and more in this direction, whether it is desirable or not. We are creating a world with much less biological diversity and this artificial world has come to seem normal, indeed natural, for many people. It is not, however, inevitable. Human beings can reconnect to nature and curb our activities that threaten it. Environmental ethics reminds us that we have many reasons to value and protect earth's remaining biological diversity.

See Also the Following Articles

ECONOMIC VALUE OF BIODIVERSITY, OVERVIEW • ETHICAL ISSUES IN BIODIVERSITY PROTECTION • LITERARY PERSPECTIVES ON BIODIVERSITY • RELIGIOUS TRADITIONS AND BIODIVERSITY • SOCIAL AND CULTURAL FACTORS • STEWARDSHIP, CONCEPT OF • TRADITIONAL CONSERVATION PRACTICES

Bibliography

Armstrong, S., and Botzler, R. (eds.). (1998). *Environmental Ethics: Divergence and Convergence*. McGraw–Hill, New York.

Callicott, J. B. (1994). *Earth's Insights: A Multicultural Survey of Ecological Ethics from the Mediterranean Basin to the Australian Outback*. University of California Press, Berkeley.

DesJardins, J. (ed.). (1999). *Environmental Ethics: Concepts, Policy, Theory*. Mayfield Publishing, Mountain View, California.

Devall, B., and Sessions, G. (1985). *Deep Ecology: Living as if Nature Mattered*. Gibbs Smith Publisher, Salt Lake City, Utah.

Grant, L. (1996). *Juggernaut: Growth on a Finite Planet*. Seven Locks Press, Santa Ana, California.

Hargrove, E. (ed.). (1979–). *Environmental Ethics: An Interdisciplinary Journal*. University of North Texas, Denton.

Kellert, S. (1996). *The Value of Life: Biological Diversity and Human Society*. Island Press/Shearwater Books, Washington, D.C.

Leopold, A. (1987). *A Sand County Almanac: And Sketches Here and There*. Oxford University Press, New York.

Naess, A. (1989). *Ecology, Community, and Lifestyle: Outline of an Ecosophy*. Cambridge University Press, Cambridge, United Kingdom.

O'Neill, J. (1993). *Ecology, Policy and Politics: Human Well-being and the Natural World*. Routledge, London.

Orr, D. (1994). *Ecological Literacy: Education and the Transition to a Postmodern World*. State University of New York Press, Albany.

Rolston, H., III. (1994). *Conserving Natural Value*. Columbia University Press, New York.

Thoreau, H. (1989). *Walden*. Princeton University Press, Princeton, New Jersey.

VanDeVeer, D., and Pierce, C. (eds.). (1998). *The Environmental Ethics and Policy Book: Philosophy, Ecology, Economics*. Wadsworth, Belmont, California.

Warren, K. (ed.). (1996). *Ecological Feminist Philosophies*. Indiana University Press, Bloomington.

ENVIRONMENTAL IMPACT, CONCEPT AND MEASUREMENT OF

Ellen W. Chu* and James R. Karr[†]
*Northwest Environment Watch and [†]University of Washington

GLOSSARY

biological integrity Wholeness of a living system, including the capacity to sustain the full range of organisms and processes having evolved in a region.

biosphere The thin layer of life at the surface of the earth.

biota Living things, in particular, the flora and fauna of a region.

biotic impoverishment Systematic reduction in the earth's capacity to support life.

environment Surroundings; the complex of physical, chemical, and biotic factors acting upon a living system and influencing its form and survival.

health A flourishing condition, well-being; capacity for self-renewal.

impact A forceful contact; a major effect of one thing on another.

ALL LIVING THINGS influence their environment. For the first time in the earth's history, however, the environmental impact of one species, *Homo sapiens*, is the principal agent of global change. Humans' most harmful impact is worldwide degradation of living systems—an impact that threatens humans' own life-support network. The 21st century's greatest challenge for scientists, decision makers, and citizens worldwide will be to understand and control human environmental impact and to protect the health and integrity of the biosphere.

I. NATURAL AND HUMAN DISTURBANCE

On the outermost rocks of scenic Big Sur's windswept headlands live rubbery palm tree look-alikes called sea palms (*Postelsia palmaeformis*). Graceful forests of these brown algae extend along the Pacific coast from central California northward to Alaska, always where the waves hit hardest. For organisms living from the high-tide line to the lower surf zone, wave action is life renewing, bringing in oxygen and nutrients and carrying away reproductive products to colonize other shores. Sea palms are able to colonize rocky open places that other intertidal organisms cannot; they thrive on the violence of the waves.

A few miles north of Big Sur, fields of artichokes,

brussels sprouts, and strawberries stretch to the seaward edge of the coastal plain. These benchlands have been farmed for two centuries by successive waves of European settlers. The area's land vegetation is dominated by nonnative plants brought in with agriculture. Countless streams and rivulets carry seaward a legacy of eroded soil, fertilizers, and pesticides. Many small creeks are choked with algae, which flourish because of excess nitrogen and phosphorus from fertilizers and sewer outfalls; many plants and animals die or are otherwise harmed by the eroded soils and seeping chemicals. Even before farming and European settlement, however, humans were shaping the landscape: for thousands of years, native Costanoan people harvested shellfish from the intertidal zones and set fires to burn off woody shrubs and trees in favor of the grassland they preferred.

Along this coast and others around the world, relentless wave action and human occupation are both driving forces of landscape change; both can be said to "disturb" the environments they influence. Both forces have an environmental impact, but the effects of the natural disturbance differ fundamentally from the effects of human-driven change.

Over millennia, organisms evolve to contend with routine disturbances in their natural environments. Those that do not adapt go extinct. Those that survive are molded by evolution and biogeography to succeed within the natural disturbance regime. Even unusual or seemingly catastrophic events, like tidal waves from earthquakes, are an integral part of the ecological context to which organisms adapt over long time spans. In contrast, human-caused disturbances—including those that, like chemical-laden runoff, are invisible to a casual observer—outpace the capacity of living systems to respond. In evolutionary terms, human effects are recent and outside the experience of most organisms. Over a mere two centuries—barely more than two human lifetimes—human disturbance has profoundly altered the biota that had persisted on the central California coast for hundreds of thousands of years.

Understanding the environmental impacts of human actions is one of modern science's greatest challenges. Understanding the consequences of those impacts, and managing them to protect the well-being of human society and other life on earth, is humanity's greatest challenge.

II. HISTORY OF HUMAN IMPACTS

All organisms change their environment as they live, grow, and reproduce. Some organisms, like beavers or

elephants, change their surroundings so dramatically that they have been called "ecosystem engineers." Beaver dams alter the flow of rivers, increase dissolved oxygen in downstream waters, create wetlands, and modify streamside zones. African elephants convert vast areas of wooded savanna to open grassland by toppling trees as they browse. Change brought about by living things, including the first humans, has been slow and incremental in evolutionary terms. Even ecosystem engineers evolve along with other inhabitants of their ecosystems, developing ways to coexist.

The human evolutionary line began about 7 million years ago in Africa. It took some 5 or 6 million years for protohumans to spread from there to Asia and then Europe. These early humans lived a hunting-gathering life, using a variety of stone tools to feed on easy-to-kill prey; they differed little from other primates trying to survive. But by 13,000 years ago, modern humans had become unlike any other species, having spread to all the continents and many islands across the globe. Still, most human-environment interactions at this stage consisted of threats to human health and well-being that came *from* the environment, including injuries in fights or wars, periodic famine, vector-borne diseases, and accidents (Table I). Then, some 10,000 years ago, people began to domesticate plants and animals. Instead of searching for food, they began to produce food.

Food production changed the course of human and environmental history. Domestication of plants and animals enabled people to adopt a sedentary lifestyle. Populations grew as agriculture developed, because larger sedentary populations both demanded and enabled more food production. With animal domestication, contagious diseases of pets and livestock adapted to new, human hosts. Diseases spread more quickly in crowded conditions; inadequate sanitation compounded the effects. From agriculture, civilization followed, and with it, cities, writing, engineering, and political empires.

In just 10,000 years, these developments led to 6 billion people on Earth, industrial societies, and a global economy founded on complicated technologies. Now the predominant human-environment interaction consists of threats *to* all four of the earth's environmental "spheres":

• *Geosphere (lithosphere), the earth's crust and upper mantle, containing nonrenewable fossil fuels, minerals, and nutrients plants require.* The activities of plants, animals, and microorganisms weather mineral soils and rocks, create organic soils, and alter erosion and sedimentation rates. Humans mine minerals, metals, and

TABLE I

Challenges to Human Health and Welfare over the Past 200,000 Years[a]

Years before 2000 A.D.	Major events	Human population	Challenges[b]
200,000	Appearance of anatomically modern humans	?	Accidents
			Combat
			Vector-borne infection
			Periodic famine
10,000	Emergence of agriculture	8 million	Nutritional deficiencies
	Rise of towns and cities		Contagious diseases
			Local ecological disruptions
150	Industrial revolution	1.3 billion	Chemical toxins
50	Modern affluence	3 billion	Overnutrition
			Global ecological disruption

[a] Modified from McMichael, A. J. (1993). *Planetary Overload: Global Environmental Change and the Health of the Human Species.* Cambridge University Press, Cambridge, UK.

[b] New challenges are listed in the era they began; all continue through the present.

gems; extract fossil fuels including coal, oil, and natural gas; and increase erosion and sedimentation by removing or altering natural plant cover through agriculture, logging, and urbanization.

• *Atmosphere, the thin envelope of gases that encircles the planet.* Living systems maintain the earth's atmosphere, its temperature, and the amount of water it contains by continually generating oxygen through photosynthesis, holding down atmospheric carbon dioxide, and changing the amount and forms of other gases. Humans release toxic chemicals into the air and alter the climate by raising the atmospheric concentration of greenhouse gases such as carbon dioxide and methane through the burning of fossil fuels in motor vehicles, electricity-generating plants, and airplanes.

• *Hydrosphere, the earth's liquid surface and underground water; its polar ice caps, oceanic icebergs, and terrestrial permafrost; and its atmospheric water vapor.* Living systems alter the water cycle by modifying Earth's temperature and the amount of water plants send into the atmosphere in a process called evapotranspiration. Humans build dams, irrigation canals, drinking-water delivery mechanisms, and wastewater treatment systems. They use water to generate electricity; they mine groundwater for farming and drinking; they alter the flows of surface waters for everything from transportation to gold mining; they drain wetlands to gain land area and abate waterborne diseases. Modern human interference in global climate is likely to alter the entire planetary water cycle.

• *Biosphere, the earth's living systems, which occupy a thin layer from the deepest oceans to the top of the highest mountains.* Life on Earth emerged 3.9 billion years ago and has sustained itself through changes in form, diversity, and detail since then. No other known planet supports life. Like other predators, humans affect the populations of their prey, on a small scale millions of years ago and on much larger scales since. Thousands of years ago, humans hunted the large mammals of Australia and the Americas to the point where many became extinct. Humans have domesticated plants through agriculture and massively redistributed vegetation through cutting of forests, burning of grasslands, or land uses creating vast areas of desert. Human actions today have precipitated a spasm of extinctions rivaling five previous mass extinctions caused by astronomical or geological forces; each of these eliminated more than 70% of species then existing.

Yet despite today's world-altering technologies and computer-generated virtual reality, humans are as dependent on their environments as other organisms are. History, not just ecology, is very clear on this point. From the Old Kingdom of Egypt more than 4000 years ago to the culture that created the huge stone monoliths on Easter Island between 500 and 1550 A.D. to the 1930s dust bowl of North America, civilizations or ways of life have prospered and failed by using and (mostly unwittingly) abusing natural resources.

In Old Egypt, the resource was the valley of the Nile,

richly fertilized with sediment at each river flooding, laced with canals and side streams, blessed with a luxuriant delta. Agriculture flourished, populations swelled, until unusually severe droughts brought on the civilization's collapse. On Easter Island, the resource was trees, which gave Polynesians colonizing the island in the fifth century the means to build shelter, canoes for fishing the unproductive surrounding waters, and log rollers for moving the ceremonial stone monuments for which the island is famous. Deforestation caused by a growing human population not only eliminated the humans' source of wood, but also further deprived the already poor soil of nutrients and made it impossible to sustain the agriculture that had sustained the island's civilization. On the dry Great Plains of North America, settlers were convinced that rain would follow the plow, and so they plowed homestead after homestead, only to watch their homesteads' soils literally blow away in the wind.

In these cases, and many others, human civilizations misjudged or damaged the environment such that their ways of life could not continue in those places. In each case, a human culture was operating precisely the way it was supposed to. The Egyptians' culture enabled them to prosper on the Nile's natural bounty in good years and make do by sharing with others in bad years, but it could not survive the unusually severe droughts. The Easter Islanders overpopulated the island and used up its resources. The dust bowl farmers were living out their culture's view of dominating and exploiting the land for all it was worth. The inevitable outcome in all three cases was an environmental catastrophe. Scholars now know that human actions often exacerbate the effects of climatic or other natural cycles to bring down civilizations or ways of life. What humans still label random "acts of God," such as unusually frequent or intense bad weather, is often a stage in the natural course of events, a fluctuation, such as El Niño, in a natural disturbance cycle whose consequences are enlarged by human activities.

Natural, particularly living, systems have always provided the capital to fuel human economies. When populations were small, humans making a living off nature's free wealth caused no more disruption than did other ecosystem engineers. But with 6 billion people occupying or using resources from every place on Earth, humans are overwhelming the ability of other life-forms to make a living. We are depleting the planet's natural wealth. In fact, the scientific consensus at the turn of the millennium is that *Homo sapiens*—a single species—rivals geological and climatic forces in its impact on the rest of the biosphere.

III. BIOTIC IMPOVERISHMENT

The first step in dealing with the present impact of human activity is to correctly identify the nature of humanity's relationship with the environment and how human actions affect that relationship. Many people still see the environment as something people must overcome, or they regard environmental "needs" as something that ought to be balanced against human needs (for example, jobs *versus* the environment). Most people still see humans' relationship with the environment as a one-way street: the environment provides us with goods and services, and our job is to keep the goods and services coming.

Thus, when asked to name humanity's principal environmental challenges, people usually think of running out of nonrenewable raw materials and energy, or about water and air pollution. Our environmental research and development institutions focus on ways technology can help solve each problem, such as fuel cells to provide clean, potentially renewable energy or scrubbers to curb smokestack pollution. Even when people worry about "biodiversity loss," they are primarily concerned with stopping species extinctions rather than with understanding the underlying losses leading up to species extinctions or the broader biological crisis that extinctions signal.

This perspective misses a crucial point: humanity has a two-way relationship with the biosphere. The reason pollution, energy use, extinction, and dozens of other human impacts are important is their larger impact on the biosphere. At this point in the planet's history, one species is compromising the earth's ability to support the living systems that evolved here over millions of years.

The systematic reduction in the earth's capacity to support life, termed biotic impoverishment, is thus the most important human-caused environmental impact. At best, the ethics of this impact are questionable; at worst, we are jeopardizing our own survival.

The connection between extinction and biotic impoverishment is intuitively obvious. By overharvesting fish, overcutting forests, overgrazing grasslands, or paving over land for cities, humans are clearly killing other organisms outright or eliminating their habitats, thereby driving species to extinction. But biotic impoverishment takes many forms besides extinction. It encompasses three categories of human impacts on the biosphere: (a) indirect depletion of living systems through alterations in physical and chemical environ-

TABLE II

The Many Faces of Biotic Impoverishment[a]

Indirect depletion of living systems through alterations in physical and chemical environments

1. Degradation of water (redirected flows, depletion of surface and ground water, wetland drainage, organic enrichment; destruction and alteration of aquatic biota)
2. Soil depletion (destruction of soil structure, erosion, salinization, desertification, acidification, nutrient leaching; destruction and alteration of soil biota)
3. Chemical contamination (land, air, and water pollution from pesticides, herbicides, heavy metals, and toxic synthetic chemicals and atmospheric ozone depletion; kills, extinctions, biodiversity loss, bioaccumulation, hormone disruption, immunological deficiencies, reproductive and developmental anomalies, respiratory disease, intergenerational effects)
4. Altered biogeochemical cycles (alteration of the water cycle, nutrient enrichment, acid rain, fossil fuel emissions, particulate pollution; degradation of land and water biota and outbreaks of pests, pathogens, and red tides)
5. Global climate change (rising greenhouse gases, altered precipitation and airflow patterns, rising temperatures; effects on individual and community health and shifts among and within global ecosystems)

Direct depletion of nonhuman life

1. Overharvest of renewable resources such as fish and timber (depleted populations, extinctions, altered food webs)
2. Habitat fragmentation and loss (extinctions, biotic homogenization, emerging and reemerging pests and pathogens)
3. Biotic homogenization (extinctions and invasions)
4. Genetic engineering (homogenization of crops, antibiotic resistance, potential extinctions and invasions if genes escape, other unknown ecological effects)

Direct degradation of human life

1. Epidemics and emerging diseases (occupational hazards, asthma and other respiratory ills, pandemics, Ebola, AIDS, hantavirus, tuberculosis, Lyme disease, antibiotic resistance, diseases of overnutrition, higher human death rates)
2. Loss of cultural diversity (genocide, ethnic cleansing, loss of cultural and linguistic diversity, loss of knowledge)
3. Reduced quality of life (environmental refugees, malnutrition and starvation, failure to thrive, poverty)
4. Environmental injustice (environmental discrimination and racism; economic exploitation; growing gaps between rich and poor individuals, segments of society, and nations; gender inequities trampling of the environmental and economic rights of future generations)
5. Political instability (resource wars, civil violence especially under intransigent regimes, international terrorism, increased number of environmental refugees)
6. Cumulative effects (environmental surprises, increased frequency of "natural" catastrophes, "boom and bust" cycles, collapse of civilizations)

[a] Modified from Karr, J. R., and Chu, E. W. (1995). Ecological integrity: Reclaiming lost connections. In *Perspectives in Ecological Integrity* (L. Westra and J. Lemons, Eds.), pp. 34–48. Kluwer Academic, Dordrecht, Netherlands.

ments, (b) direct depletion of nonhuman life, and (c) direct degradation of human life (Table II). Identifying and understanding the biological significance of human actions—their effects on living systems, including our own social and economic systems—is the key to developing effective ways to manage our impacts.

A. Indirect Biotic Depletion

Humans affect virtually all the physical and chemical systems life depends on: water, soils, air, and the biogeochemical cycles linking them. Some human-driven physical and chemical changes have no repercussions on the biota; others do, becoming agents of biotic impoverishment.

1. Degradation of Water

Humans probably spend more energy, money, and time trying to control the movement and availability of water than using any other natural resource. In the process, we contaminate water; move water across and out of natural basins; deplete surface and groundwater; shift the timing and amount of flow in rivers, straighten or build dikes to constrain the rivers, and alter natural flood patterns. We change the amount, timing, and chemistry of fresh water reaching coastal regions and dry up wetlands, lakes, and inland seas. Our demands are outrunning supplies of this nonrenewable resource, and the scale of our transformations risks altering the planet's water cycle.

Physical alterations to the planet's waters, combined with massive industrial and residential pollution, have taken a heavy toll on nonhuman aquatic life. Ten percent of the world oceans' coral reefs have died from pollution and other causes. Globally, the number of so-called oceanic dead zones, where there is little or no dissolved oxygen, tripled in the last 30 years of the 20th century. The biota of freshwater systems fares no better. A 4-year survey of the freshwater fishes inhabiting Malaysian rivers in the late 1980s found only 46% of the 266 known Malaysian species. Nearly 40% of North America's freshwater fishes are at risk of extinction; two-thirds of freshwater mussels and crayfishes and one-third of amphibians that depend on aquatic habitats in the United States are rare or imperiled. Wetlands worldwide continue to disappear; the lower 48 United States lost 53% of its wetlands between the 1700s and mid-1980s. Such losses destroy major fish and shellfish nurseries, natural flood and pollution control, and habitat for countless plants and animals.

Humans now use 54% of the accessible water runoff on Earth, a figure that is likely to grow to 70% by 2025. By then, more than a third of the world's population could suffer shortages of fresh water for drinking and irrigation. Groundwater aquifers in many of the world's most important crop-producing regions are being drained faster than they can be replenished. Natural flood regimes, as in the Nile River basin, no longer spread nutrient-rich silt across floodplains to nourish agriculture; indeed, the High Dam at Aswan traps so much silt behind it that the Nile delta, essential to Egypt's modern economy, is falling into the sea. In the past 50 years, the number of large dams on the world's rivers grew more than seven times, to some 38,000 today. The huge Three Gorges Dam across China's Yangtze River promises to create a 690-km serpentine lake behind it and displace 1.2 million Chinese citizens before its scheduled completion in 2009; the impact on the river's fertile delta farming region and downstream estuaries has just begun to be assessed. Whole inland seas, such as the Aral Sea in Uzbekistan, are drying up because the streams feeding them have so little water. In addition to eliminating habitat for resident organisms, the seas' drying is bringing diseases to surrounding human populations. Diseases caused by waterborne pathogens are making a comeback even in industrialized nations.

2. Soil Depletion

Hardly just "dirt," soil is a living underground system that makes it possible for raw elements from air, water, and bedrock to be physically and chemically assembled, disassembled, and reassembled with the aid of living macro- and microorganisms into the thin green shell of life above ground. Accumulated over thousands of years, soil cannot be renewed in any time frame useful to humans alive today, or even to their great-grand-children.

Humans degrade soils when they compact it, erode it, disrupt its organic and inorganic structure, turn it too salty for life, and cause desertification. Urbanization, logging, mining, overgrazing, altering soil moisture, air pollution, fires, chemical pollution, and leaching out of minerals all damage soils. Thanks to removal of vegetative cover, mining, agriculture, and other activities, the world's topsoils are eroded by wind and water ten to hundreds of times faster than they are renewed (at roughly 1 ton per hectare per year). Soils constitute the foundation of human agriculture, yet agriculture, including livestock raising, is the worst culprit in degrading soils. Agricultural practices have eroded or degraded an area equal to 38% of present cropland. Besides topsoil erosion, the damage includes salting and water-logging of poorly managed irrigated lands, compaction by heavy machinery and the hooves of livestock, and pollution from excessive fertilizers, animal wastes, and pesticides.

Living, dead, and decomposing organic matter is the key to soil structure and fertility. Soil depleted of organic matter is less permeable to water and air and thus less able to support either aboveground plants or oxygen-dependent microbes. It's the linkages between soil's inorganic components and the soil biota (naturalist E. O. Wilson's "little things that run the world") that give soil its life-sustaining capacity. A clear-cut forest patch whose soil biota has been damaged beyond recovery can no longer sustain trees, no matter how many are planted; another clear-cut patch whose soil community is still functioning will support new tree growth. Destroying soil biota exerts a whole series of impoverishing biotic effects both below and above ground.

3. Chemical Contamination

In 1962 Rachel Carson's landmark book *Silent Spring* alerted the world to the pervasiveness of synthetic chemicals produced since World War II. Some 75,000 synthetic chemicals are in use today. True to one company's slogan, many of these have brought "better living through chemistry," providing new fabrics and lighter manufacturing materials, antibiotics, and life-saving drugs. But industrial nations have carelessly pumped chemicals into every medium. Chemicals directly poison organisms; they accumulate in physical surroundings and are passed up the food chain.

Many chemicals travel the ocean and atmospheric currents to sites far from their source; sulfur emissions from the U.S. Midwest, for example, fall to earth again as acid rain in Europe, killing forests and so acidifying streams and lakes that they too effectively die. China's burning of soft coal sends air pollution all the way to northwestern North America; the heavy haze hanging over China's chief farming regions may be cutting agricultural production by as much as a third. Chlorofluorocarbons (CFCs), once widely used as refrigerants, have damaged the atmospheric ozone layer, which moderates how much ultraviolet radiation reaches the earth, and opened ozone "holes" over the Arctic and Antarctic. Chemicals ranging from prescription drugs flowing out of sewage plants to pesticides to heavy metals to cancer-causing by-products of countless manufacturing processes now lace the world's water, soil, and air and the bodies of all living things, including humans. Carson's book revealed the real danger of chemical pollution: we have not simply altered the chemistry of water, soil, and air but harmed the biosphere as well.

Although many of the most obviously deadly compounds were banned in the 1970s, they continue to impoverish the biosphere. Polychlorinated biphenyls (PCBs)—stable, nonflammable compounds once used in electrical transformers and many other industrial and household applications—accumulate and persist in soils and the food web. They are found in polar bears and arctic villagers; they are implicated in reproductive disorders, particularly in such animals as marine mammals, whose long lives, thick fat layers where chemicals concentrate, and position as top predators make them especially vulnerable. The agricultural pesticide DDT, sprayed with abandon in the 1940s and 1950s, even directly on children, severely thinned wild birds' eggshells. Populations of birds such as the brown pelican and bald eagle dropped precipitously by the 1970s but began to recover in the last decade of the 20th century.

Chemicals cause cancer, disrupt hormonal systems, provoke asthma, and impair the functioning of immune systems. They have intergenerational effects, such as intellectual impairment in children whose mothers have eaten contaminated fish. What's more, a half century of pesticide and antibiotic overuse has bred resistance to these chemicals among insects, plants, and microbes,

prompting fears of new and reemerging scourges. The litany of chemicals' effects on living things is so long that chemical pollution *is* humans' environmental impact in most people's minds; yet it is just one form of biotic impoverishment.

4. Altered Biogeochemical Cycles

All the substances found in living things, such as water, carbon, nitrogen, phosphorus, and sulfur, cycle through ecosystems in biogeochemical cycles. Human activities alter or have the potential to alter all these cycles. Sometimes the results stem from changing the amount or the precise chemistry of the cycled substance; in other cases, humans alter biogeochemical cycles by changing the biota itself.

Freshwater use, dams, and other engineering feats change the amount and rate of river flow to the oceans and increase evaporation rates, directly affecting the water cycle and indirectly impoverishing aquatic life. Direct human modifications of living systems also perturb the water cycle. In the tall "cloud forests" of redwoods in northern California or Douglas firs in the Pacific Northwest, the trees gather more moisture directly from the clouds than falls as rain. Among the effects of logging in such forests is a depletion of stream flows, which shifts the water cycle. In South Africa, European settlers supplemented the treeless native scrub, or fynbos, with such trees as pines and Australian acacias from similar Mediterranean climates. Because these trees are larger and thirstier than the native scrub, regional water tables have fallen sharply.

Human activity has altered the global nitrogen cycle by greatly increasing the amount of nitrogen fixed from the atmosphere (combined into compounds usable by living things). The increase comes mostly from deliberate addition of nitrogen to soils as fertilizer but also as a by-product of the burning of fossil fuels. Agriculture, livestock raising, and individual yard maintenance chronically add tons of excess nutrients, including nitrogen and phosphorus, to soils and water. The additions are often invisible; their biological impacts are often dramatic. Increased nutrients in coastal waters, for example, trigger blooms of toxic dinoflagellates, the algae that cause red tides, fish kills, and tumors and other diseases in varied sea creatures. When huge blooms of algae die, they fall to the seafloor, where their decomposition so robs the water of oxygen that fish and other marine organisms can no longer live there. With nitrogen concentrations in the Mississippi River two to three times as high as they were 50 years ago, a gigantic dead zone forms in the Gulf of Mexico every summer; it covered a record 20,020 square kilometers in the summer of 1999.

The burning of fossil fuels is radically altering the carbon cycle, primarily by greatly increasing the atmospheric concentration of carbon dioxide. With other greenhouse gases, such as methane and oxides of nitrogen, carbon dioxide helps keep the earth's surface at a livable temperature and drives plant photosynthesis, but since the industrial revolution, atmospheric carbon dioxide concentrations have risen 30% and are now widely thought to be disrupting the planet's climate.

5. Global Climate Change

The Intergovernmental Panel on Climate Change (IPCC), comprising 2500 of the world's preeminent atmospheric scientists, had concluded by 1995 that humans have caused at least part of the atmospheric warming observed globally since the beginning of the 20th century. The concentrations of heat-trapping gases in the atmosphere are at their highest level in more than 200,000 years. The 20th century in the Northern Hemisphere has been described as the warmest of the past millennium; the 1990s ranked as that century's warmest decade.

Higher global temperatures precipitate a whole series of effects, making the study of climate change, and of humans' role in it, complex and controversial. Spring now arrives one week earlier in the Northern Hemisphere. Polar glaciers and ice sheets are receding. The large-scale circulation of global air masses is shifting and, with them, the large-scale cycles in ocean currents, including the periodic warming and cooling in the tropical Pacific Ocean known as El Niño and La Niña. As a result, the distribution, timing, and amount of rain and snow are also changing, making the weather seem more unpredictable than ever. Unusually warm or cold winters, massive hurricanes such as those that devastated North Carolina in late summer 1999, and weather-related damage to human life and property are all predicted to increase with global warming. Weather-related damage totaled $92 billion in 1998, exceeding the total for the entire decade of the 1980s, even after adjusting for inflation.

Rising carbon dioxide concentrations may enhance plant photosynthesis and growth where other nutrients are not limiting. Rising temperatures may shift the ranges of many plants and animals, shifting the composition and distribution of the world's biomes. The resulting disruptions to habitats will have far-reaching implications not only for the displaced plants and animals but also for the goods and services humans depend on from living systems.

B. Direct Depletion of Nonhuman Life

From their beginnings as hunter-gatherers, humans have become highly efficient, machine-aided ecosystem engineers and predators. We transform the land so it produces what we need or want; we harvest the oceans in addition to reaping our own fields; we cover the land, even agricultural land, with sprawling cities. All these activities directly affect the ability of other lifeforms to survive and reproduce. We deplete nonhuman life by eliminating some forms and favoring others; the result is a loss of genetic, population, and species diversity. We are irreversibly homogenizing life on Earth, in effect exercising an "unnatural selection" that is erasing the diversity generated by millions of years of evolution by natural selection. One species is now determining which other species will survive, reproduce, and thereby contribute the raw material for future evolution.

1. Overharvest of Renewable Resources

In the 1930s, so many sardines were scooped from the waters off Monterey's Cannery Row that the population collapsed, taking other sea creatures and human livelihoods with it; the species has still not recovered fully. According to the U.S. National Marine Fisheries Service, nearly 80% of commercially valuable fish of known status were overfished or "fished to their full potential" by 1993. Atlantic commercial fish species at their lowest levels in history include tunas, marlins, cod, and swordfish. Overfishing not only depletes the target species but changes the whole structure of marine food webs.

Marine mammals, including whales, seals, sea lions, manatees, and sea otters, were so badly depleted by human hunters that one species, Steller's sea cow (*Hydrodamalis gigas*), went extinct; many other species almost disappeared. In the 19th century, Russian fur traders wiped out sea otters (*Enhydra lutris*) along the central California coast; with the otters gone, their principal prey, purple sea urchins (*Stronglyocentrotus purpuratus*) overran the offshore forests of giant kelp (*Macrocystis pyrifera*), decimating the kelp fronds and the habitat they provided for countless other marine creatures, including commercially harvested fishes. Thanks to three decades of protection, most marine mammal populations are slowly rebounding—only to face food shortages following disruptions in regional marine food webs due to fishing, changing oceanic conditions, and contamination.

Timber harvest has stripped vegetation from the Amazonian rain forests to mountainsides on all continents, diminishing and fragmenting habitat for innumerable forest and stream organisms, eroding soils, worsening floods, and contributing significantly to global carbon dioxide emissions. In the Northern Hemisphere, only 10% remains of old-growth temperate rain forests. The uniform stands of trees usually replanted after logging do not replace the diversity lost with the native forest, any more than monocultures of corn replace the diversity within native tallgrass prairies.

2. Habitat Fragmentation and Loss

Human land use inevitably damages or destroys habitat for some living thing. But satellite-mounted remote-sensing instruments over the past two decades reveal transformations of a scale unimaginable in centuries past. At one extreme, row-crop agriculture or urban-industrial areas occupy only 10 to 15% of the earth's land surface; pastureland occupies another 6 to 8%. At the other extreme, every alteration of the land alters the physical, chemical, and biological surroundings of anything living there. Estimates of the share of land wholly transformed or degraded by humans fall at around 50%. Landscapes that have not been entirely converted to human use have been cut into fragments. In *Song of the Dodo*, writer David Quammen likens our actions to starting with a fine Persian carpet and then slicing it neatly into 36 equal pieces; we still have the same square footage but not 36 nice Persian rugs, only ragged, nonfunctional fragments.

Our roads, farms, cities, feedlots, and ranches either destroy or fragment the habitats of most large carnivorous mammals. Mining and oil drilling damage the soil, remove vegetation, and pollute marine areas. Grazing compacts soil and sends silt and manure into streams, where they harm stream life. Habitat damage may lead directly to extinction, or it may isolate organisms in ways that make them extremely vulnerable to natural disturbances, climate change, or further human disturbance.

3. Biotic Homogenization

"The one process now going on that will take millions of years to correct," writes E. O. Wilson, "is the loss of genetic and species diversity by the destruction of natural habitats. This is the folly our descendants are least likely to forgive us." Both deliberately and unwittingly, humans are rearranging the earth's living components, reducing diversity and homogenizing biotas around the world. The present, continuing loss of genetic diversity, of populations, and of species vastly exceeds background rates. At the same time, our global economy is transporting species worldwide at unprecedented scales.

The globe is now experiencing its sixth mass extinction, the largest since the fall of the dinosaurs 65 million years ago; present extinction rates are thought to be on the order of 100 to 1000 times those before people dominated the earth. Many of the world's bird species have been driven to extinction by humans, especially on oceanic islands. According to a survey by the World Conservation Union, at least one of every eight plant species is threatened with extinction. Although mammals and birds receive the most attention, massive extinctions of plants, which form the basis of the biosphere's food webs, undermine life-support foundations. Some scientists are calling extinction "the worst biological tragedy," but extinction too is but another symptom of global biotic impoverishment.

Ever since they began to spread over the globe, people have transported other organisms with them, sometimes for food, sometimes for aesthetic reasons, and most often inadvertently. With the mobility of modern societies and today's especially speedy globalization of trade, the introduction of alien species has reached epidemic proportions, causing some scientists to label it "biological pollution." Aliens—zebra mussels (*Dreissena polymorpha*) and tamarisks, or saltcedar (*Tamarix spp.*), in North America; the Red Sea jellyfish *Rhopilema nomadica* and the common aquarium alga *Caulerpa taxifolia* now choking the Mediterranean Sea; and Leidy's comb jelly (*Mnemiopsis leidyi*) of northeastern America in the Black Sea, to name just a few—are present everywhere, and they usually thrive and spread at the expense of native species. On many islands, for example, more than half the plant species are not native, and in many continental areas, the figure reaches 20% or more.

Such alien invasions cause extinctions and, when added to other extinctions and the deliberate monocultures of agricultural crops, worsen biotic homogenization. Introduced species are fast catching up with habitat loss and fragmentation as the major engines of ecological deterioration.

4. Genetic Engineering

Humans have been manipulating their crop plants and domesticated animals for 10,000 years or so—selecting seeds or individuals, breeding, and cross-breeding. The goal was something better, bigger, tastier, hardier, or all of the above; success was sometimes elusive, but the result was crop homogenization. Of the myriad strains of potatoes domesticated by South American cultures, for example, only one was accepted and cultivated when potatoes first made it to Europe. The new crop made it possible to feed more people from an equivalent area of land and initially staved off malnutrition. But the strain succumbed to a fungal potato blight in the 1800s; had more than one strain been cultivated, the tragic Irish potato famines might have been averted.

In the last few decades of the 20th century, people began to manipulate genes directly using the tools of molecular biotechnology, even cloning sheep and cows from adult body cells. U.S. farmers routinely plant their fields with corn whose genetic material incorporates a bacterial gene resistant to certain pathogens. More than 40 genetically altered crops have been approved for sale to U.S. farmers since 1992, with genes borrowed from bacteria, viruses, and insects; worldwide, upward of 40 million hectares are planted with genetically modified crops. Biotechnologists see this new-millennium green revolution as finally solving world hunger. But other scientists worry about unknown human and ecological health risks; these concerns have stirred a deep scientific debate, akin to the debate over pesticides in Rachel Carson's time.

One worrisome practice is plant genetic engineers' technique of attaching the genes they want to introduce into plants to an antibiotic-resistance gene. They can then easily select those plants that have acquired the desired genes by treating them with the antibiotic, which kills any nonresistant plants. Critics worry that the antibiotic-resistant genes could spread to human pathogens and worsen an already growing antibiotic-resistance problem. Another threat comes from allergies humans might have or develop in response to genetically modified foods.

Supporters of genetic engineering believe that genetically altered crops pose few ecological risks. But studies in the late 1990s indicated that pollen from genetically engineered "Bt corn" can kill monarch butterfly caterpillars. Bt is a strain of bacteria that has been used since the 1980s as a pesticidal spray; its genes have also been inserted directly into corn and other crops. Studies in radishes, sorghum, canola, and sunflowers found that genes from an engineered plant can jump to wild relatives through interbreeding. The fear is that a gene conferring insect or herbicide resistance might spread through wild plants, creating invasive superweeds, which could potentially lower crop yields and further disrupt natural ecosystems.

Great economic and environmental benefits are claimed for the products of biotechnology: higher yields, lower synthetic pesticide and herbicide use, increased farmer profits. But so far the benefits have not always materialized, and growing consumer and scientific concern may slow the spread of genetically modified crops, at least in the developed world.

C. Direct Degradation of Human Life

Human biotic impacts are not confined to other species; human cultures themselves have suffered from the cascading indirect and direct impacts humans exert on the rest of nature. Over the past hundred years, human technology has been a two-edged sword with regard to public health. Wonder drugs controlled common pathogens at the same time that natural selection strengthened those pathogens' ability to resist the drugs. Reservoirs in the tropics made water supplies more reliable for humans but also created ideal environments for human parasites. Industrialization exposed human society to a remarkable array of toxic substances.

Though "man's inhumanity to man" has been both fact and the subject of discourse for thousands of years, the discussions have mostly been removed from any environmental context. Few people today regard social ills as "environmental impacts" or humans as part of a "biota." But diminished societal well-being—whether manifest in high death rates or poor quality of life—shares many of its roots with diminished nonhuman life as a form of biotic impoverishment.

1. Epidemics and Emerging Diseases

The interface between the environment and human health is the core of the discipline known as environmental health. Among the environmental challenges to public health are the direct effects of toxic chemicals; occupational health threats, including exposures to hazardous materials on the job; and sanitation and hazardous waste disposal. Exploitation of nonrenewable natural resources—including coal mining (causing black lung disease), rock quarrying or other mining operations (causing silicosis), and petroleum extraction and refining—often chronically impairs workers' health and shortens their lives. Farmworkers around the world suffer long-term ills from high exposures to pesticides and herbicides. Partly because of increased air pollution, asthma rates are rising, particularly in big cities. Synthetic volatile solvents are used in products from shoes to semiconductors, producing lung diseases and toxic wastes. Nuclear weapons production starting in World War II, and the associated contamination, have been linked to a variety of diseases.

Infectious diseases have challenged human populations throughout history, playing a significant role in their evolution and cultural development. The 20th century brought major successes in eradicating such infectious diseases as smallpox, polio, and many waterborne illnesses. But toward the century's end, emerging and reemerging diseases were again reaching pandemic proportions. Human migrations have been the main source of epidemics throughout history, but environmental and social changes resulting from human activities are accelerating pathogen traffic. Infectious diseases thought to be on the wane—including tuberculosis, malaria, cholera, diptheria, leptospirosis, encephalitis, and dengue fever—began a resurgence. In addition, seemingly new scourges—Ebola virus, hantavirus, HIV/AIDS, legionnaires' disease, and Lyme disease—are also spreading, often, it appears, from animal hosts to humans as people encroach further upon previously undisturbed regions.

Especially in highly developed countries such as the United States, diseases of affluence and overconsumption are also taking a toll. Heart disease is the number one cause of death in the United States; overnutrition and obesity due to sedentary, technology-driven lifestyles, particularly among children, is chronic and rising. One estimate put the share of U.S. children considered overweight or obese at one in four.

2. Loss of Cultural Diversity

Although not conventionally regarded as elements of biodiversity, human languages, customs, agricultural systems, technologies, and political systems have evolved out of specific regional environments. Like other organisms' adaptive traits and behaviors, these elements of human culture constitute unique natural histories adapted, like any natural history, to the biogeographical context in which they arose. Yet unique indigenous human cultures, including knowledge of local medicines and geographically specialized economies, are disappearing even more rapidly than the natural systems that nurtured them.

Modern technology, transportation, and trade are moving the world toward a globalized culture, thereby reducing human biological and cultural diversity. Linguists, for example, are predicting that at least half of the modern world's 6000 languages will become extinct in the twenty-first century. Indigenous knowledge and practices are dwindling fast with the spread of Euro-American culture. This loss of human biodiversity is in every way as troubling as the loss of nonhuman biodiversity.

3. Reduced Quality of Life

Human activities that degrade environmental conditions threaten not only the biosphere but also humans' own quality of life. Many sectors of society now recognize this link. In 1990 the United Steelworkers of America endorsed a report saying "We cannot protect steelworker jobs by ignoring environmental problems" and "the greatest threat to our children's future may lie in the destruction of their environment." Independent of race or economic class, declining quality of life is manifest in both the increased incidence of asthma in the United States caused by environmental contaminants and the high disease rates in the former Soviet Bloc after decades of unregulated pollution. Even with explicit legal requirements that industries release information on their toxic emissions, many people throughout the world still lack both information and the decision-making power that would give them any control over the quality of their lives.

Food availability, which depends on environmental conditions, is a basic determinant of quality of life. Yet according to the World Health Organization, roughly half the population in all nations suffers from two forms of poor nutrition: undernutrition and overnutrition. A swollen abdomen is now a symptom shared by malnourished children, who lack calories and protein, and overweight residents of the developed world, who suffer clogged arteries and heart disease because of eating too much food.

Deeply concerned about the degraded environment and resulting quality of life in his homeland, Ogoni activist Ken Saro-Wiwa stated, shortly before he was executed by the Nigerian government in 1995, "The environment is man's first right. Without a safe environment, man cannot exist to claim other rights, be they political, social, or economic." Environmental degradation and desperate living conditions have spawned a new kind of refugee: the number of international environmental refugees exceeded the number of political refugees around the world for the first time in 1999. Environmental refugees flee homelands devastated by flooding from dam building, extraction of mineral resources, desertification, and unjust policies of national and international institutions. Such degradation preempts many fundamental human rights, including the rights to health, livelihood, culture, privacy, and property.

4. Environmental Injustice

Making a living from nature's wealth has consistently opened gaps between haves and have-nots, between those who bear the brunt of environmental damage to their home places and those who do not, and between the rights of people alive now and those of future generations; these disparities too are part of biotic impoverishment. Inequitable access to "man's first right"—a healthy local environment—has come to be known as environmental injustice.

Environmental injustices, such as institutional racism, occur in industrial and nonindustrial nations. Injustice can be overt, as when land-use planning sites landfills, incinerators, and hazardous waste facilities in minority communities, or when environmental agencies levy lower fines for hazardous waste violations that are lower in minority communities than in white communities. Less overt, but no less unjust, is the harm done to one community when unsound environmental practices benefit another, as when clear-cut logging in the highlands of northwestern North America benefits logging communities while damaging the livelihoods of lowland fishing communities touched by debris flows, sedimentation, and downstream flooding.

The plight of the working poor and the disparities between rich and poor are also examples of biotic impoverishment within the human community. According to the United Nations Research Institute for Social Development, the collective wealth of the world's 358 billionaires equaled the combined income of the poorest 2.4 billion people in 1994. In the United States during the last decade of the twentieth century, the incomes of poor and middle-class families stagnated or fell, despite a booming stock market. The Center on Budget and Policy Priorities and the Economic Policy Institute of Washington D.C. reported that, between 1988 and 1998, earnings of the poorest fifth of American families rose less than 1%, while earnings of the richest fifth jumped 15%.

Despite great advances in the welfare of women and children over the past century, poverty still plagues both groups. Children from impoverished communities, even in affluent nations, suffer from the lethargy and impaired physical and intellectual development known as failure to thrive. Poverty forces many children to work the land or in industrial sweatshops; lack of education prevents them from attaining their intellectual potential. This impoverishment in the lives of women and children is as much a symptom of biotic impoverishment as are deforestation, invasive alien organisms, or species extinctions.

The creation of a "third world" to supply raw materials and labor to the dominant European civilization after 1500—and the resulting schism between today's developed and developing nations—is perhaps the grossest example of human and environmental domination leading to continued injustice. Exploitation of poor countries and their citizens by richer, consumer nations (as well as by the governments of developing nations themselves) persists today in agriculture, wild materials

harvesting, and textile and other manufacturing sweatshops. In the mid-1990s, industrial countries consumed 86% of the globe's aluminum, 81% of its paper, 80% of its iron and steel, 75% of its energy, and 61% of its meat; they are thus responsible for most of the environmental degradation associated with producing these goods. Yet most of the actual degradation takes place in poorer, developing countries.

Ironically, local environmental injustice often arises in the name of national development. Locals resist projects that seem socially disruptive or ecologically unsustainable while governments push them as potential boons to the national economy. China's Three Gorges Dam presents one example; another is the government-backed exploitation, fought by Saro-Wiwa, of Ogoniland's oil reserves in southern Nigeria by the Shell Petroleum Development Corporation. After Saro-Wiwa's execution, the Ogoni were left, without a voice, to deal with a scarred and oil-polluted homeland.

Governments of developing nations are starting to demand a share of profits made from biological resources within their borders, including pharmaceuticals from plants and animals or new strains bred from local crop varieties. Nature's "free wealth" is now commonly viewed as a "genetic resource," the raw material for the biotechnology age like fossil fuels or minerals for the industrial age. Still, profits tend to flow to the elite, wealthy classes of both developing and developed nations while indigenous groups remain marginalized.

Little by little, so-called community-based conservation and development initiatives are being mounted by local citizens worldwide to combat these injustices. And one might contend that all efforts to protect the biosphere constitute a fight for the rights of future generations to an environment that can support them.

5. Political Instability

It was only during the last two decades of the 20th century that environmental issues found a place on international diplomatic agendas, as governments began to see threats to political security in an environmentally destabilized world. The 1987 Montreal Protocol on Substances That Deplete the Ozone Layer, which forced countries to curb their use of CFCs and other ozone-destroying chemicals, was an example of governments, nongovernmental organizations, and industry successfully working together to safeguard part of the environmental commons. More often, diplomacy stalled in conflicts over natural resources: arguments over water rights held up Israeli-Palestinian peace agreements; fights over fish erupted between Canada and the United States, Spain, and Portugal.

Scholars, including Norman Myers and Thomas Homer-Dixon, began calling attention to the links among growing populations; environmental scarcity and degradation; environmental injustice; and civil violence, including urban unrest. Myers noted that we cannot dispatch the military to turn back the advancing deserts, rising seas, or the greenhouse effect. Homer-Dixon and others pointed out that nations with high population growth, rising urbanization and unemployment, and depleted resources are becoming increasingly hard to govern except by tyrannical means. Perversely, the laudable goals of democracy and increased human rights can lead to civil riots in such nations as India, Pakistan, Indonesia, and China. Environmental and economic injustices worldwide leave no nation immune to this type of threat.

6. Cumulative Effects

If scientists have learned anything about the factors leading to biotic impoverishment, they have learned that the factors' cumulative effects can take on surprising dimensions. The multiple stresses of global climatic cycles such as El Niño–Southern Oscillation, natural disasters like droughts or floods, biotic impoverishment, and political instability have changed the course of history. Civilizations as far-flung as ancient Egypt, Peru, Easter Island, and the American Southwest prospered and collapsed because of unwise management of their environments. The city of Ubar, built on desert sands, literally disappeared into the sinkhole created by drawing too much water out of its great well. In modern Sahelian Africa, a combination of well digging and improved medical care and sanitation led to a threefold population increase; sedentary ways, heavy taxes imposed by a colonial government, and an impoverished people took the place of a nomadic culture evolved within the desert's realities.

During the last decade of the 20th century, numerous natural disasters befell nations around the world: massive flooding along the Mississippi River; devastating hurricanes in the Caribbean and southeastern United States; catastrophic landslides and floods in Honduras, China, and Venezuela; and destructive earthquakes in Turkey and Taiwan. Neither the rains nor the earthquakes were caused principally by human activity, but the cumulative effects of human land uses and management practices—from dikes separating the Mississippi from its floodplain to hog and poultry farms in North Carolina to clear-cut logging in Honduras to crowded cities in Turkey and Taiwan—made the disasters much worse in loss of human life and property.

IV. ROOT CAUSES OF HUMAN IMPACT

The ultimate cause of humans' massive environmental impact is our reproductive and consumptive behavior, which has given us spectacular success as a species. The very things that enabled humans to thrive in nearly every environment have magnified our impacts on those environments, and the technological and political steps we take to "mitigate" our impacts often worsen them. There are simply too many of us, we take too much from the natural world, and we ask it to absorb too much waste.

A. Fragmented Worldviews, Fragmented Worlds

For most of human history, people remained tied to their natural surroundings. Even as agriculture, writing, and technology advanced, barriers of geography, language, and culture kept humans a diverse lot, each group depending on mostly local and regional resources for survival. Their worldviews, and resulting economies, reflected this dependency. In northwestern North America, for example, a native economy centered on the abundance of salmon began to develop about 3000 years ago. At its core was the concept of the gift and a belief system that treated all parts of the earth—animate and inanimate—as equal members of a community. In this and other ancient gift economies, a gift was not a possession that could be owned; rather, it had to be passed on, creating a cycle of obligatory returns. Individuals or tribes gained prestige through the size of their gifts, not the amount of wealth they accumulated.

This system coevolved with the migratory habits of the salmon, which moved en masse upriver to spawn each year. Because the Indians viewed salmon as equals to themselves, killing salmon represented a gift of food from salmon to people. Fishers were obligated to treat salmon with respect or risk losing this vital gift. The exchange of gifts between salmon and humans—food for respectful treatment—minimized waste and the chance of overharvesting and ensured a continuous supply of food. Further, the perennial trading of gifts among the people effectively redistributed the natural wealth brought each year by fluctuating populations of migrating fish, leveling out the boom-and-bust cycles that usually accompany reliance on an uncertain resource.

In modern times, the gift economy has gone entirely, along with the egalitarian worldview that accompanied it. We now have a redistributive economy tied not to an exchange of gifts with nature but to the exploitation of nature and to the technologies that enhance that exploitation. Nature became a resource for humans rather than an equal to humans. In economic terms, natural resources fell under the heading of "land" in an economic trinity comprising three factors of production: land, labor, and capital. Land and resources, including crops, became commodities, expendable or easily substitutable forms of capital, whose value was determined solely by their value in the human marketplace.

In 1776 Adam Smith published his famous *Inquiry into the Nature and Causes of the Wealth of Nations*, in which he argued that society is merely the sum of its individuals, that the social good is the sum of individual wants, and that markets (the "invisible hand") automatically guide individual behavior to the common good. Crucial to his theories was division of labor and the idea that all the factors of production were freely mobile; his mechanistic views created an economic rationale for no longer regarding individuals as members of a community linked by moral, social, and ecological bonds.

About the same time, fueling and fueled by the beginnings of the industrial revolution, the study of the natural world was transforming itself into modern physics, chemistry, geology, and biology. Before the mid-19th century, those who studied the natural world—early 19th-century German biogeographer Baron Alexander von Humboldt and his disciple Charles Darwin among them—took an integrated view of science and nature, including humans. For von Humboldt, understanding interdependence was the "noblest and most important result" of scientific inquiry. But this integrated natural philosophy was soon supplanted by more atomistic views, which fit better with industrialization.

Mass production of new machines relied on division of labor and interchangeable parts. Like automobiles on an assembly line, natural phenomena too were broken down into their supposed component parts in a reductionism that has dominated science ever since. Rushing to gain in-depth, specialized knowledge, science and society lost sight of the need to tie the knowledge together. Disciplinary specialization replaced integrative scholarship.

Neoclassical economics, which arose around 1870, ushered in the economic worldview that rules today. A good's value was no longer tied to the labor required to make it but derived instead from its scarcity. A good's price was determined only by the interaction of supply and demand. As part of "land," natural resources therefore became part of the human economy, rather than the material foundation that makes the human economy possible. Because of its doctrine of infinite substitutability, neoclassical economics rejects any limits on growth; forgotten are the classical economic thinkers and contemporaries of von Humboldt, including Thomas Malthus and John Stuart Mill, who saw limits to the growth of human population and material well-being.

B. Too Many and Too Much

In October 1999, the global human population reached 6 billion. It doubled from 3 billion to 6 billion in about 40 years—before most post–World War II baby boomers even reached retirement age. From the appearance of *Homo sapiens* about half a million years ago, it took humans until 1804 to reach their first billion, 123 years to double to 2 billion, and 33 years to achieve 3 billion. With fertility rates declining in developed countries, China, and some developing countries where women are gaining education and economic power, and with pandemics like AIDS claiming more lives, the United Nations predicts that world population will reach 8.9 billion by 2050. Lined up one by one, the world's 6 billion people would stretch nearly to the moon; jammed together in a crowd, they would just fit into the state of Texas.

But these 6 billion people consume far more resources than can be produced by a piece of earth the size of Texas. As already noted, humans appropriate about 40% of global plant production, 54% of the earth's freshwater runoff, and enough of the ocean's bounty to deplete or nearly extinguish two-thirds of the world's major marine fisheries. In energy terms, a human's food consumption is about 2500 to 3000 calories a day, about the same as that of a common dolphin. But with all the other energy and materials humans use, the rise in global per capita energy and material consumption has soared even faster than population growth in the past 40 years. Mostly in the form of fossil fuels, North Americans consume six times as much energy as the average human. Now, instead of coevolving with a natural economy, global society is coevolving with fossil fuels and the American Dream.

V. MEASURING ENVIRONMENTAL IMPACTS

For most of the 20th century, environmental indicators tracked primarily two classes of information: counts of administrative activities in support of environmental protection and the supply of products to people. Counting bureaucratic achievements—such as the numbers of environmental laws passed, permits issued, enforcement actions taken, or treatment plants constructed—focuses on actions rather than on the results of those actions. Ultimately such counts provide only limited information about real environmental status and trends.

Resource managers concentrate on resource supply. Water managers, for example, measure "water quantity," allocating water to domestic, industrial, and agricultural uses and rarely reserving any to sustain aquatic life or to protect scenic and recreational values. Foresters, farmers, and fishers count board-feet of timber, bushels of grain, and tons of fish harvested. All these measures keep track of commodity production, not broader environmental quality. Even biologists' counts of threatened and endangered species—which would seem to measure biotic impoverishment directly—still focus narrowly on biological parts, not ecological wholes. Enumerating threatened and endangered species is just like counting any other commodity, reinforcing humanity's view that we know which parts of the biota are important or valuable.

Society needs to use to better effect the environmental measures it has had (Table III). It particularly needs objective measures more directly tied to the condition, or health, of the environment so that people can judge whether their actions are compromising that condition. Such measures should be quantitative yet easy to understand and communicate; they should be cost-effective and applicable in many circumstances. Unlike narrow criteria tracking only commodity or extinction numbers, they should provide reliable signals about status and trends in ecological systems. Ideally, effective indicators should describe the present condition of a place, document change, diagnose its cause, and predict what is likely to happen. They should reveal not only risks from present activities but also potential benefits from alternative management decisions.

Most important, these indicators should, either singly or in combination, give information explicitly about living systems. Measurements of physical or chemical factors can sometimes act as surrogates for direct biological measurements, but only when the connection between those measures and living systems is clearly understood. Too often we make assumptions—when water managers assume that chemically clean water equals a healthy aquatic biota, for example—that turn out to be wrong and fail to protect living systems.

As environmental concerns grow more urgent, several new measurements aim to direct the public's attention to the value of living things. Using the value system best understood by the people and institutions driving the global economic engine, some researchers assign standard monetary amounts to the goods and services humans derive from nature. A calculation by Robert Costanza and his colleagues puts the average total monetary value of natural capital and ecological services in 16 biomes at US$33,000 billion per year—an amount exceeding the gross global product. Another study, by ecologist David Pimentel and his colleagues, calculated separate values for specific biological services, such as soil formation, crop breeding, or pollination; by summing these figures, these researchers estimate the total economic benefits of biodiversity for the United States at $319 billion, for the world at $2928 billion—in other words, 5 and 11% of the gross national and gross global products, respectively. Still another of Pimentel's analyses reports that the approximately 50,000 nonnative species in the United States cause major environmental damage and reparation costs amounting to $137 billion a year. Though this monetary approach does not create a comprehensive indicator of environmental quality, it certainly points out that ecological values ignored by the global economy are very high.

Another accounting approach, pioneered by Canadian landscape planners Mathis Wackernagel and William Rees, translates humans' impact on nature, particularly resource consumption, into a metaphorical "ecological footprint." Researchers estimate the amount of land and water area required by a city, town, nation, or other human community to produce consumed resources and absorb the generated wastes; they then compare the physical area occupied by a city or country with the area required to supply that city or country's needs. The 29 largest cities of Baltic Europe, for example, appropriate areas of forest, agricultural, marine, and wetland ecosystems that are at least 565 to 1130 times larger than the areas of the cities themselves. National ecological footprints range from a high of 10.3 hectares per person for the United States to 0.5 hectares per person for Bangladesh. Thirty-four of 52 nations are now operating under ecological deficits; that is, their material standards are subsidized by exploiting other nations' natural capital.

Raising all 6 billion people on Earth at the start of the 21st century to living standards, and thus ecological footprints, equal to those in the United States would require an additional two planets. Clearly, humans are consuming more resources, and discarding more waste, than the earth's living systems can produce or absorb in a given time period. This gap is the global sustainability gap the world now faces.

TABLE III

Plausible Indicators of Environmental Quality[a]

Indirect depletion of living systems through alterations in physical and chemical environments

1. Degradation of water (chemical contaminant concentrations; river flows, rainfall, runoff)
2. Soil depletion (erosion rates, desertification rates, salt accumulation in soils)
3. Chemical contamination (pollutant and toxic emissions; pollutant and toxic concentrations in air, water, soil, and living organisms)
4. Altered biogeochemical cycles (river flows and lake levels; amount of nutrients going into water bodies, or nutrient loading; nutrient concentrations in water bodies; chlorophyll concentrations reflecting nutrient-triggered algal blooms; oxygen depletion in water bodies; trophic status of lakes; changes in air and soil chemistry; atmospheric greenhouse gas concentrations)
5. Global climate change (atmospheric greenhouse gas concentrations, change in atmospheric temperatures, distribution and intensity of severe storms or droughts)

Direct depletion of nonhuman life

1. Overharvest of renewable resources such as fish and timber (tons of fish harvested, fish population "escapement," hatchery fish released and recovered; board-feet of timber harvested, forest regrowth rates, "stumpage," or standing timber; ecological footprints)
2. Habitat fragmentation and loss (area of forest or wetland habitats remaining, landscape connectivity, rates of habitat destruction)
3. Biotic homogenization (number of extinct, threatened, and endangered taxonomic groups; spread of nonindigenous species; local or regional diversity; damage and reparation costs of invasions or extinctions; major shifts in species distributions)
4. Genetic engineering (diversity among cultivated crop strains, genetic diversity within strains, escape of genetically engineered organisms or traits to wild populations)

Direct degradation of human life

1. Epidemics and emerging diseases (death and disease infection rates, or mortality and morbidity; geographic spread of diseases; recovery rates; frequency and spread of antibiotic and other drug resistance)
2. Loss of cultural diversity (extinction of languages, disappearance of cultures)
3. Reduced quality of life (population size and growth, changes in death rates or average life spans, infant mortality rates, teen pregnancy rates, number of chronically malnourished individuals, starvation rates, literacy rates, rates of stress and other diseases of affluence, length of work week, child or other forced labor, employment shifts by economic sector)
4. Environmental injustice (siting of toxic waste dumps or waste emissions relative to resident communities, economic exploitation of certain groups, worker strikes, wage and income gaps, unemployment rates for different economic sectors)
5. Political instability (frequency of domestic and international strife, environmental terrorism rates, number of environmental refugees, ethnic "cleansing")
6. Cumulative effects (frequency of "rare" "natural" disasters; costs of weather-related property damage; human death tolls; government subsidies of environmentally destructive activities such as fishery overcapitalization, below-cost timber sales, water projects, and agricultural supports; replacement costs for ecological services; pricing that reflects environmental costs; "green" taxes; rise in polycultural practices; number of organic farms)

[a] These indicators have been or could be used to monitor status and trends in environmental quality, including dimensions of biotic impoverishment. Without a full spectrum of indicators, however, and without coupling them to direct measures of biological condition, only a partial or segmented view of environmental quality (the degree of biotic impoverishment) will emerge.

A few indexes have integrated economic, environmental, and human community indicators into a general measure of sustainability. The index of environmental trends for nine industrialized countries incorporates ratings of air, land, and water quality; chemical and waste generation; and energy use since 1970. By its rankings, environmental quality in the United States has gone down by 22% while Denmark has declined by 11%. Social scientists Herman Daly and John Cobb developed an index of sustainable economic welfare that adjusts the United States' gross national product by factoring in environmental good things and factoring out environmental bad things. Public expenditures on education, for example, are weighted as "goods" while costs of pollution cleanup, depletion of natural resources, and treating environmentally related illnesses are counted as "bads." Instead of showing continual growth as gross national product does, the index of sustainable economic welfare has remained nearly unchanged over three decades.

Such approaches offer important insights into the dangers of local and global resource consumption and consumerism and paint a clearer picture than ever before of humans' domination of the earth and the biosphere. But the accounting systems are still human centered. They still do not measure the condition of the biosphere itself. We may know that biodiversity's services are worth huge sums of money and that our hometown's ecological footprint is much bigger than our town, but how do we know whether specific actions damage living systems or that other actions benefit them? How do we know if aggregate human activity is diminishing life on Earth? To answer this question, we need direct measures of the condition of living systems.

More comprehensive than monetary or footprint analyses, biological assessment directly measures the attributes of living systems to determine the condition of a landscape. The very presence of living systems—sea palms on the California coast, salmon in Pacific Northwest waters, monk seals in the Mediterranean Sea—says that the conditions those organisms need to survive are also present. A biota is thus the most direct and integrative indicator of local, regional, or global biological condition. Biological assessments give us a way to evaluate whether our monetary valuations and ecological footprints are telling the truth about human impact on the biosphere. Biological assessments permit a new level of integration because living systems, including human cultures, register the accumulated effects of all forms of degradation caused by human actions.

Direct, comprehensive biological assessment has been done for many aquatic systems; measures are less developed for terrestrial systems. The index of biological integrity (IBI), for example, was developed in 1981 to assess the health of streams in the U.S. Midwest and has since helped scientists, resource managers, and citizen volunteers to understand, protect, and restore rivers worldwide. The index borrows a page from well-known composite indexes of economic performance—the concept of multiple indicators—and applies it to animals and plants in bodies of water. The specific measurements (Table IV) are sensitive to a broad range of human effects in waterways, such as sedimentation, nutrient enrichment, toxic chemicals, physical habitat destruction, and altered flows. The resulting index combines the responses to human actions of both biological parts (such as species) and processes (such as food web dynamics).

Indexes of biological integrity have been developed for a number of aquatic and terrestrial environments; the widest-used indexes for assessing rivers examine fishes and benthic (bottom-dwelling) invertebrates.

TABLE IV

Biological Attributes in Two Indexes of Biological Integrity

Benthic invertebrates	Fish
Total number of taxa	Number of native fish species
Number of mayfly taxa	Number of riffle-benthic insectivore species
Number of stonefly taxa	Number of water-column insectivore species
Number of caddisfly taxa	Number of pool-benthic insectivore species
Number of intolerant taxa	Number of intolerant species
Number of long-lived taxa	Relative abundance of omnivores
Number of clinger taxa	Relative abundance of insectivores
Relative abundance of tolerant taxa	Relative abundance of tolerant taxa
Relative abundance of predators	Relative abundance of top carnivores
Dominance	Relative abundance of diseased or deformed individuals

These groups are abundant and easily sampled, and the species living in virtually any water body represent a diversity of anatomical, ecological, and behavioral adaptations. As humans alter watersheds and water bodies, shifts occur in taxonomic richness (biodiversity), species composition (which species are present), individual health, and feeding and reproductive relationships.

Sampling the inhabitants of a stream can tell us much about a stream and its landscape. Biological diversity is higher upstream of wastewater treatment plants than downstream, for example; at the same location, year-to-year variation is low (Fig. 1). Biological sampling also reveals differences between urban and rural streams. For instance, samples of invertebrates from one of the best streams in rural King County, Washington (United States) contain 27 kinds, or taxa, of invertebrates; similar samples from an urban stream in Seattle contain only 7. The rural stream has 18 taxa of mayflies, stoneflies, and caddisflies, the urban stream only 2 or 3. When these and other metrics are combined in an index based on invertebrates, the resulting benthic IBI (B-IBI) ranks the condition, or health, of a stream numerically (Table V). The B-IBI for the rural stream in King County was 46 (the 10-metric index maximum is 50); that for the urban stream, 12 (the index minimum is 10).

A benthic IBI can also be used to compare sites in different regions. Nearly pristine areas in Wyoming's Grand Teton National Park have near-maximum B-IBIs. Streams with moderate recreation taking place in their

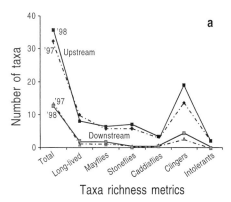

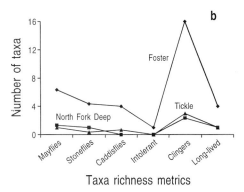

FIGURE 1   (a) Biodiversity is higher at sites upstream of wastewater treatment outfalls than downstream. At Tickle Creek near Portland, Oregon (United States), taxa richness differed little between years but differed dramatically between sites upstream of a wastewater outfall and downstream. (b) Taxa richness also differed between two creeks with wastewater outfalls (Tickle and North Fork Deep) and one creek without an outfall (Foster). All three streams flowed through watersheds with similar land uses.

watersheds have B-IBIs that are not significantly lower than those with no human presence, but places where recreation is heavy are clearly damaged. Urban streams in the nearby town of Jackson are even more degraded, yet not as bad as urban streams in Seattle.

The core message embodied in biological assessment is that preventing harmful environmental impacts goes beyond narrow protection of clean water or clear skies, even beyond protecting single desired species. Certain species may be valuable for commerce or sport, but these species do not exist in isolation. We cannot predict which organisms are vital for the survival of commercial species or species we want for other reasons. Failing to protect phytoplankton, zooplankton, insects, higher plants, bacteria, or fungi ignores the key contributions of these groups to healthy biotic communities.

No matter how important a particular species is to humans, it cannot persist outside the biological context that sustains it. Direct biological assessment objectively measures this context.

VI. RECOGNIZING AND MANAGING ENVIRONMENTAL IMPACTS

Every animal is alert to dangers in its environment. A microscopic protozoan gliding through water responds to light, temperature, and chemicals in its path; it turns or retreats at the first sign of something noxious. A bird looking for food must decide when to pursue prey and when not, because pursuit might expose the bird to predators. The bird might risk pursuit when it is hungry but not when it has young to protect. Animals that assess risks properly and adjust their behavior are more likely to survive; in nature, flawed risk assessment often means death or the end of a genetic line.

Humans too are natural risk assessors. Each person chooses whether to smoke or drink, to drive a car or ride a motorcycle and at what speeds, to fly or take the train, to engage in "extreme" sports or go for a woodland stroll. Each decision is the result of a partially objective, partially subjective internal calculus that weighs benefits and risks against one another.

Risk is a combination of two factors: the numerical probability that an adverse event will occur and the consequences of the adverse event. People may not always have the right signals about these two components, however, and so base their risk calculus on the wrong clues. Urban dwellers in the United States generally feel that it is safer to drive home on a Saturday night than to fly in a jetliner, for example. Even though the numerical odds of an accident are much higher on the highway than in the air, people fear more the consequences of an airliner falling out of the sky.

Human society also strives to reduce its collective exposure to risks, primarily through government agencies responsible for protecting the public's interests. Governments do not hesitate to use military power and international agreements to protect their people from external aggression. They have, albeit more reluctantly, also used their regulatory power to reduce workplace risks and risks associated with consumer products like automobiles. But people and their governments have been much less successful in defining and reducing a broad range of ecological risks, largely because they have denied that the threats are real.

Society has long behaved as if its activities did not

TABLE V
Biological Responses to Different Land Uses

Region	Land use	B-IBI[a]
King County, Washington	Rural	46
	Urban Seattle	12
Grand Teton region, Wyoming	Little or no human activity	48
	Light to moderate recreation	44
	Heavy recreation	32
	Urban Jackson Hole	21
Clackamas County, Oregon[b]	Upstream of wastewater treatment plant	
	Tickle Creek up (1997, 1998)	40, 42
	Foster Creek	34
	Downstream of wastewater treatment plant	
	Tickle Creek down (1997, 1998)	14, 16
	North Fork Deep Creek	10

[a] Benthic index of biological integrity: the highest possible score is 50, the lowest is 10.
[b] See Fig. 1 for graphs of selected B-IBI metrics at these sites.

entail any risks to nonhuman living systems, that is, any ecological risks. The plans generated by economists, technologists, engineers, and even ecologists typically assume that the lost and damaged components of living systems are unimportant or can be repaired or replaced. Widespread ecological degradation has resulted directly from the failure of modern society to properly assess the ecological risks it faces. Like ancient Egypt's or Easter Island's fate, our civilization's future depends on our ability to recognize this deficiency and correct it.

Risk assessment as formally practiced by various government agencies began as a way to evaluate the effects on human health of toxic substances, usually the effects of single substances, such as pollutants or drugs, from single sources, such as a chemical plant. During the 1990s, the focus widened to encompass mixtures of substances and also ecological risks. Ecological risk assessment by the U.S. Environmental Protection Agency asks five questions: Is there a problem? What is the nature of the problem? What are the exposure and ecological effects? (A hazard to which no one or nothing is exposed is not considered to pose any risk.) How can we summarize and explain the problem to "stakeholders" (both at-risk populations and those whose activities would be curtailed)? How can we manage the risks?

Even though these are good questions, ecological risk management has not made any visible headway in stemming biotic impoverishment. Its central failing comes from an inability to correctly answer the second question, What is the nature of the problem? Our pres-

ent political, social, and economic systems simply do not give us the right clues about what is at risk. None of society's most familiar indicators—housing starts, gross national product, index of leading economic indicators, educational testing scores, air quality alerts, even number of threatened and endangered species—measure the consequences, or risks, of losing living systems.

Moreover, ecological risk assessment is still an assessment of the *risks* of business as usual, not an evaluation of potential *benefits* from alternatives to business as usual. When deciding what level of grazing to allow, for example, ecological risk assessors still examine the risks to, or impacts on, existing resources such as soil, water, and vegetation. They do not ask what the benefits of different grazing regimes would be; neither do they consider the benefits of no grazing at all, even though the no-grazing alternative might offer the greatest ecological, and ultimately societal, benefits.

If biotic impoverishment is the problem, then it makes more sense to direct environmental policy toward protecting the integrity of biotic systems. Aldo Leopold, in *A Sand County Almanac*, was the first to invoke the concept of integrity in an ecological sense: "A thing is right when it tends to preserve the integrity, stability, and beauty of the biotic community. It is wrong when it tends to do otherwise." Integrity implies a wholeness or unimpaired condition. In present biological usage, *integrity* refers to the condition at sites with little or no influence from human activity; the organisms there are the products of natural evolutionary and biogeographic processes in the absence of humans. Tying the concept of integrity to an evolutionary framework provides a benchmark against which to evaluate sites that humans have altered.

Directing policy toward protecting biological integrity—as called for in the United States' Clean Water Act, Canada's National Park Act, and water policy directives being established by the European Union—does not, however, mean that humans must cease all activity that disrupts a pristine earthly biota. The demands of feeding, clothing, and housing billions of people mean that few places on Earth will maintain a biota with evolutionary and biogeographic integrity. Rather, measurements founded on the evolutionary idea of integrity allow us to directly assess biotic condition and to compare that condition with what might be expected in a place with little or no human influence. At least then we can make an informed choice: continue with activities that degrade biotic condition or think of an alternative.

Ethical arguments aside, humans' own dependence on living systems makes it in our interest to manage our activities so they do not compromise a place's ability to support those activities in the future; that ability can be called ecological health. Ecological health describes the preferred state of sites heavily used for human purposes: croplands, tree farms, water bodies stocked for fish, and urban areas. Integrity in an evolutionary sense cannot be a goal at these places, but we should avoid practices that so damage places that we cannot continue to use them. Agricultural practices that leave soils salted, lower regional water tables, and erode fertile topsoil faster than it can be renewed destroy prospects for future agriculture; such practices are unhealthy.

In contrast to risk assessment, and more akin to ecological benefits assessment, striving to protect biological integrity is more likely to lead away from technological fixes for environmental problems and toward practices that prevent ecological degradation and encourage restoration. Biological integrity as a policy goal turns our focus away from maximizing the goods and services provided for the human economy and toward ways to manage human affairs within the bounds set by the natural economy. It begins to turn our attention away from "How much stress can landscapes and ecosystems absorb?" to "How can responsible human actions protect and restore ecosystems?" It could even help avoid contentious fiscal allocation issues that arise when massive emergency operations—like rescuing endangered Pacific salmon in the Columbia River—are needed to pull species back from the brink of extinction.

To be sure, managing for biological integrity requires a deep commitment to self-imposed limits on human population size, curbs on consumerism, less-selfish attitudes toward land stewardship, and the realization that the biosphere matters. Instead of calling on human technical and spiritual wellsprings to manage resources, we have to call on them for managing human affairs. We have to find and use appropriate measurements for all the factors contributing to biotic impoverishment, be they climate change, overharvesting, agriculture, or environmental injustice.

Managing our affairs to prevent ecological risks requires integrating ecological systems and human social and political systems. The ecological world is a complex, variable system. We cannot predict with certainty the intensity of the next El Niño drought or rainstorm. We cannot know with assurance whether the cumulative effects in a place will turn the next earthquake into a disaster or an inconvenience. We cannot calculate the absolute risks (or benefits) of pumping (or not pumping) a given tonnage of carbon dioxide into the air. Purely quantitative risk assessment only works in the absence of such uncertainties.

Instead, when managing for ecological risks, people and their governments need to expect the unexpected and develop formal, yet flexible means of coping with environmental surprises. Rather than plunge ahead with projects entailing ecological risks because they *can* be done, decision makers should follow the precautionary principle, which holds that regulators should act to prevent potential environmental harm even in the absence of certainty. It acknowledges the existence of uncertainty rather than denying it, and it includes mechanisms to safeguard against potentially harmful effects.

Though inappropriate ecological risk assessment and management is more often the norm today, modern institutions *can* recognize ecological threats correctly and respond to them in time. The Montreal Protocol is a prime example. Just over a decade after its adoption, satellite measurements in the stratosphere indicated that harmful chlorine pollution was in fact on the decline.

VII. RECLAIMING LOST CONNECTIONS

Early in the 20th century, two sciences of "home maintenance" began to flourish: the young science of ecology (from the Greek *oikos*, meaning home) and a maturing neoclassical economics (also from *oikos*). Ecology arose to document and understand the interactions between organisms and their living and nonliving surroundings—in essence, how organisms make a living in the natural economy. In fact, Ernst Haeckel, who coined the term in the 1860s, defined *ecology* [in an 1870 article] as "the body of knowledge concerning the economy of nature." Neoclassical economics reinforced humans' self-appointed dominion over nature's free wealth and brought unparalleled gains in societal welfare in some places, but it also divorced the human economy from the natural one on which it stands.

Monitored by both those sciences, human actions and their effects have reached scales unprecedented in the history of life. We have altered the earth's physical and chemical environment, changed the planet's water and nutrient cycles, and shifted its climate. We have unleashed the greatest mass extinction in 65 million years and disrupted the structure and function of non-human and human communities worldwide. In trying to make our own living, we have contributed not only to the global loss of individual lives and loss of species but, worse, to the loss of life itself—the removal of *bio* from the biosphere. In the end, life—and earth's capacity to sustain life—is at risk.

Early in the 21st century, we need a new science and art of home maintenance, one that helps us understand and interpret the consequences of human-driven change. We need a new story, a different worldview, to guide our behavior, one that is in harmony with nature's economy. We must seek a balance between our modern industrial economies and our homelands' natural economies. As individuals and as societies, we need to understand the consequences for the present and future biosphere of what we do.

To reclaim our connections to the natural world, those of us insulated within industrial societies have to start small, with connections we can understand—like what plants and animals live in our backyards, which chemicals are used to grow our coffee, which laborers glued the soles on our walking shoes, where the pulp in our newspapers came from. We have to pay attention to how living systems respond to our behavior. In order to live, we have to let live. In his 1995 contribution to *A New Century for Natural Resources Management*,

environmental ethicist Holmes Rolston III writes, "The next millennium is, some say, the epoch of the end of nature. But another hope is that we can launch a millennium of culture in harmony with nature."

See Also the Following Articles

BIOGEOCHEMICAL CYCLES • ECOLOGICAL FOOTPRINT, CONCEPT OF • ECOTOXICOLOGY • ENERGY USE, HUMAN • ENVIRONMENTAL ETHICS • HUMAN EFFECTS ON ECOSYSTEMS, OVERVIEW

Bibliography

American Association for the Advancement of Science. (1997). Human-dominated ecosystems (special section). *Science* 277, 486–529.

Baskin, Y. (1997). *The Work of Nature: How the Diversity of Life Sustains Us.* Island Press, Washington, D.C.

Bright, C. (1998). *Life Out of Bounds: Bioinvasion in a Borderless World.* W. W. Norton, New York.

Bullard, R. D. (1997). *Unequal Protection: Environmental Justice and Communities of Color.* Sierra Club Books, San Francisco.

Diamond, J. (1997). *Guns, Germs, and Steel: The Fates of Human Societies.* W. W. Norton, New York.

Fagan, B. (1999). *Floods, Famines, and Emperors: El Niño and the Fate of Civilizations.* Basic Books, New York.

Homer-Dixon, T. F. (1999). *Environment, Scarcity, and Violence.* Princeton University Press, Princeton, New Jersey.

Karr, J. R., and Chu, E. W. (1999). *Restoring Life in Running Waters: Better Biological Monitoring.* Island Press, Washington, D.C.

McMichael, A. J. (1993). *Planetary Overload: Global Environmental Change and the Health of the Human Species.* Cambridge University Press, Cambridge.

Myers, N. (1993). *Ultimate Security: The Environmental Basis of Political Stability.* W. W. Norton, New York.

Norgaard, R. B. (1994). *Development Betrayed: The End of Progress and a Coevolutionary Revisioning of the Future.* Routledge, London.

Prugh, T., Costanza, R., Cumberland, J. H., Daly, H., Goodland, R., and Norgaard, R. B. (1995). *Natural Capital and Human Economic Survival.* ISEE Press, Solomons, MD.

Wackernagel, M., and Rees, W. E. (1996). *Our Ecological Footprint: Reducing Human Impact on the Earth.* New Society Press, Gabriola Island, British Columbia.

Wilson, E. O. (1992). *The Diversity of Life.* Harvard University Press, Cambridge, MA.

Woodwell, G. M. (1990). *The Earth in Transition: Patterns and Processes of Biotic Impoverishment.* Cambridge University Press, Cambridge.

ESTUARINE ECOSYSTEMS

G. Carleton Ray
University of Virginia

The cold remote islands
And the blue estuaries
Where what breathes, breathes
The restless wind of the inlets
And what drinks, drinks
The incoming tide.

LOUISE BOGAN, "NIGHT"

I. Introduction
II. Definition and Classification
III. Estuarine Biodiversity
IV. Ecological Function of Biodiversity
V. A Case Study: The Chesapeake Bay
VI. Future Challenges

GLOSSARY

biological diversity (biodiversity) The collection of genomes, species, and ecosystems occurring in a geographically defined region (NRC, 1995).

coastal zone Zone whose terrestrial boundary is defined by (a) the inland extent of astronomical tidal influence or (b) the inland limit of penetration of marine aerosols within the atmospheric boundary layer and including both salts and suspended liquids, whichever is greater; the seaward limit is defined by (a) the outer extent of the continental shelf (approximately 200 m depth) or (b) the limits of territorial waters, whichever is greater (Hayden *et al.,* 1984).

estuary Semi-enclosed coastal body of water that has a free connection with the open sea and within which seawater is measurably diluted with freshwater derived from land drainage (Pritchard, 1967).

functional diversity Variety of different responses to environmental change, especially the diverse time and space scales with which organisms react to each other and to the environment (Steele, 1991).

metapopulation An abstraction of the population to a higher level at which individuals frequently move from one place (population) to another, typically across habitat types that are not suitable for their feeding and breeding activities, and often with substantial risk of failing to locate another suitable habitat patch in which to settle (Hanski and Gilpin, 1991).

ESTUARIES ARE AMONG THE MOST IMPORTANT INTERCONNECTIONS between land and sea. They are situated in the coastal zone, which accounts for a disproportionate amount of global ecological functions. For example, the coastal zone (modified from Pernetta and Milliman, 1995):

- occupies only 18% of the surface of the globe, 8% of the ocean surface, and 0.5% of ocean volume;
- but provides for up to 50% of global denitrification, 80% of global organic matter burial, 90% of global

sedimentary mineralization, 75–90% of the global sink of suspended river load and its associated elements/pollutants, and in excess of 50% of present-day global carbonate deposition;

- also supplies approximately a quarter of global primary production, around 14% of global ocean production, and 90% of the world fish catch.

It follows that estuaries, as major pathways of aquatic exchange between land and sea, are major influences on a large proportion of these functions.

I. INTRODUCTION

Some of the steepest environmental gradients on planet Earth occur in the coastal zone, where land, sea, and atmosphere uniquely interact to exchange energy and materials. Also, the dynamic linkages among biological, physical, and chemical systems are exceptionally strong in estuaries, and are characterized by cyclic changes that occur at different frequencies—such as for tides, salinity cycles, freshwater inputs, light, and temperature stratification. Estuaries also bear the brunt of extreme events, such as flooding, storms, hurricanes, and seasonal sea ice. All of these are of importance for organisms, which have evolved suites of adaptive mechanisms to cope.

Estuaries have usually been considered as transitional areas between freshwater and saltwater environments. However, relatively few species are totally confined to estuarine conditions, even though various stages of many species' life cycles are estuary-dependent. This raises questions about whether estuaries can be considered as transitional or as more-or-less autonomous ecosystems in their own right. The distribution of biodiversity provides important information toward the resolution of this apparent dichotomy, which needless to say is essential for conservation and management.

Our present knowledge about estuary-dependent biodiversity is sparse. Fundamental questions remain about species distributions in estuaries, in what ways species are adapted to estuaries, and how some species may affect others by means of structural or functional interrelationships. Furthermore, the diversity of estuaries relative to other ecosystems remains to be clarified. These questions require both ultimate, historical-evolutionary explanations and proximate, functional-ecological explanations.

Despite the location of estuaries in the critical portion of Earth called the "coastal zone," the *Global Bio-diversity Assessment* (Heywood and Watson, 1995) contains no sections specifically devoted to them; the term "estuary" does not even appear in the index! Nevertheless, this volume does characterize biodiversity as comprising three disciplines, which also apply to estuaries: (1) taxonomy: provides the reference system and depicts the pattern or tree of diversity for all organisms; (2) genetics: gives a direct knowledge of the gene variations found within and between species; and (3) ecology: provides knowledge of the varied ecological systems in which taxonomic and genetic diversity are located, and it also provides the functional components. Evolutionary biology brings these together, as it "provides explanations of how biodiversity arose, and the processes, such as speciation and extinction, by which it continues to change."

The third aspect of biodiversity, namely, the functional-ecological aspect, is the focus of this article. In this respect, it is worthwhile to note that, even today, estuarine science continues to be organized along disciplinary lines. Although the study of land–seascape ecology of estuaries remains in its infancy, there are extensive publications on geomorphology, land–sea interactions, coastal zone management, and other disciplines from which to gain an integrated understanding of estuaries. Nevertheless, a comprehensive understanding of the functional biodiversity of estuaries remains a future goal.

Estuaries became topics of intensive concern and research only in the mid-twentieth century. This is ironic, as humans have lived in close proximity to estuaries and have been dependent on them and their biological resources for millennia. Reasons for human proximity to and dependence on estuarine environments are both social and ecological, for estuaries are ecologically diverse and productive, making possible the sustainment of large and sophisticated human societies. Indeed, it is more than coincidental that among the first known city-states were those of the lower reaches of the Tigris and Euphrates Rivers of Mesopotamia.

The distribution of estuaries corresponds to regional and coastal characteristics; that is, they tend to be extensive, large, and numerous where coastal plains are wide and flat, but are relatively small where coastal plains are steep and narrow. Particularly in the former, estuaries and associated lagoons constitute a much higher percentage of the coasts than is generally recognized. In fact, many of the world's largest cities (London, New York, Karachi, Amsterdam, Alexandria, Tokyo, etc.) have been built on or near drained marshes or filled land adjacent to estuaries. In the United States, 80–90% of the Atlantic and Gulf Coasts and 10–20% of the

Pacific Coast consist of estuaries and lagoons (Emery, 1967).

Estuaries are best understood in the context of the coastal zone, definitions of which vary. Ketchum (1972) was among the first to take a functional perspective, that the coastal zone "is the broad interface between land and water where production, consumption, and exchange processes occur at high rates of intensity." NERC (1992), on the other hand, defined the coastal zone as: "An indefinite zone of land and sea that straddles the shoreline; includes all land that is the product of, and/or at risk from (Holocene) marine processes, and extends seaward from the shoreline to water depths of about 30 m." The key element is "marine processes" and, from that point of view, it seems best to adopt Ketchum's broader view. Accordingly, Hayden *et al.* (1984) adopted Ketchum's definition (see Glossary), which makes sense of such interactions as the existence of coastal vegetation under the influence of aerosols, sedimentation induced by freshwater flows and atmosphere–ocean processes, and the coastal distribution of aquatic biota worldwide. With respect to the latter, Nelson (1984) estimated that of about 21,700 described species of fishes, about 8400 (39%) occur in freshwater and 2700 (12%) are oceanic. Nearly half of these fishes (10,600 species, or 49%) are coastal, that is, occur from estuaries to the outer extent of the continental shelf. This proliferation of fish diversity is powerful evidence of the functional importance and the extent of the coastal zone.

Within this coastal zone context, Pritchard's (1967) definition of "estuary" also makes sense (see Glossary). However, other definitions must be acknowledged. For example, Mann (1982) defined an estuary as "a region where river water mixes with, and measurably dilutes, sea-water." Yet this definition could include semi-enclosed seas (e.g., the Baltic), plumes of large rivers, and diluted water off open coasts, making difficult any geographic analysis of estuarine biodiversity or function. Additionally, Pritchard's definition takes account of Pleistocene rises and falls in sea level, as well as of terrestrial processes, such as sedimentation, which clearly affect the distributions of aquatic biota.

Thus, estuaries are best defined functionally in a land–sea context and as important portions of the coastal zone. In this context, estuaries are subject to rapid environmental, structural-functional change, which has major consequences for biodiversity. Hydrological, biological, and sedimentary processes and events may substantially alter or destroy estuaries at many spatial and temporal scales. The estuaries that we now see are the result of the latest major episodes of

sea level fall and rise and, in fact, the age of the present estuaries is only about 1% of the age of the continental shelf (Emery, 1967). It is reasonable to assume that the communities of estuarine biota that exist today are as young and equally subject to change.

Many estuaries around the world have been studied in some detail. The North American bias in this article reflects the considerable body of research that has been conducted on North American estuaries during the past few decades, motivated unfortunately by the depleted, over-enriched, polluted, and over-populated states of many of them, some aspects of which will be examined in the Chesapeake Bay case study in Section V.

II. ESTUARINE CLASSIFICATION

Classification is essential as a comparative reference system, for otherwise data and information cannot be made comparable among estuaries. Various classifications, or typologies, of estuaries have been attempted, but these are mostly physical; no typology is directed specifically to biodiversity, even though the distributions of estuarine species have resulted in various classification schemes.

To my knowledge, the first classification was the so-called "Venice system" (Anonymous, 1959), in which estuaries were divided into salinity zones. This was modified later by Bulger *et al.* (1993) on the basis of species' salinity tolerances. These two schemes align rather closely and may be compared as follows (Anonymous, 1959 = V; Bulger *et al.*, 1993 = B; ppt = parts per thousand):

Limnetic: freshwater, 0.5 ppt (V); freshwater, 4 ppt (B)
Oligohaline: 0.5–5 ppt (V); 2–14 ppt (B)
Mesohaline: 5–18 ppt (V); 11–18 ppt (B)
Polyhaline: 18–30 ppt (V); 16–27 ppt (B)
Euhaline: 30 ppt–full marine (V); 24–ppt marine (B)

The reason for the differences in salinity ranges between the Venice system and Bulger *et al.* is that the former was derived from salinity, whereas the latter was derived analytically from species' salinity tolerances, in which the zones would be expected to overlap. In both cases, however, the compartments are over-simplistic, as estuaries exhibit many characteristics that influence biotic distribution and the distinction of estuarine zones, variably identified as "upper reaches," "upper-middle reaches," "lower reaches," and so forth. Nor do salinity-derived systems distinguish zones according to variations in bottom type, water movement, volume of flow, and other attributes important to the biota.

Another classification concerns basin geomorphology, which is of obvious importance for circulation patterns. Classification on this basis appears in many texts and may be summarized as:

- coastal plain estuary (drowned river valley): Usually confined to areas with a wide coastal plain where seawater has invaded existing rivers because of sea level rise since the Pleistocene Ice Age. Generally the up-estuary limit is where chlorinity is about 0.06% (salinity about 0.1%); above this point there may be a portion of tidal freshwater.
- fjord: Generally U-shaped in cross section, in which the sides are steep and have been glaciated. May be fed by a river, have a deep basin, and a shallow sill may be present near the mouth.
- bar-built: Occurs in flat, low-lying areas, where sand tends to be deposited in bars lying parallel to the coast. Usually shallow and wind-mixed. Can be a composite of drowned river valleys and embayments, and occurs when offshore sand barriers are built between headlands into a chain to enclose the body of water. May be fed by multiple rivers, but the total drainage area is usually not large.
- tectonic: A miscellaneous category including estuaries formed from faults or folding of Earth's crust. Often have an excess of freshwater flow.

The interchange of freshwater and seawater provides yet another classification. The inlet (mouth) must be of sufficient dimension to allow mixing of seawater and freshwater, and the dilution of seawater provides the density gradients that drive characteristic circulation patterns. In terms of this interchange, the general classification is:

- salt wedge: Wherein a layer of relatively fresh water flows out at the surface.
- partially mixed (moderately stratified): Wherein tidal flow, turbulence, and mixing are increased, tending to erase the salt wedge.
- vertically homogeneous: Wherein tidal flow is strong, river runoff is weak, and all stratification is broken down.

Combinations of these typologies are possible; that is, it may be possible to find a stratified or a mixed bar-built estuary, or a fjord with a salt wedge or not. Furthermore, the extents of salinity zones can vary considerably for all categories. Such combinations of structure and hydrologic process result in highly varied conditions in the distributions of, for example, sediment, phytoplankton, submerged aquatic vegetation, and fishes and invertebrates. Additionally, variations in freshwater inputs, circulation, turbulence, and mixing can modify the typology.

A final classification concerns estuarine evolution, such as that of Roy (1984) for estuaries of New South Wales, Australia. There, estuaries are of three successional types: drowned river valleys, barrier estuaries, and saline coastal lakes. All are characterized by infilling during relatively short time spans. This affects their size, configuration, the invasion of mangroves and other aquatic vegetation, and fish communities. Biodiversity maxima are reached in the intermediate stages, because faunal population densities and species diversity increase with ecological complexity. However, as infilling becomes more advanced, the estuary becomes simplified and biological diversity declines. Therefore, estuarine geology, hydrology, and biology form a hierarchical succession.

III. ESTUARINE BIODIVERSITY

From the foregoing discussion, the impression may be gained that estuaries are simply transitional and, therefore, not biologically diverse. Indeed, Sanders (1968) found that estuaries are relatively non-diverse biologically, but also noted: "What is significant is that each environment seems to have its own characteristic rate of species increment." This is to say that salinity, for example, is an important determinant of the distribution of the biota, but also that estuaries exhibit high habitat and land–seascape diversity, a consequence of which is high variability among the biota and a high degree of biotic interaction. Thus, estuarine biotic communities would be expected to be especially varied and complex, contrary to earlier impressions of estuarine biological and ecological simplicity. Additionally, their biota have evolved resiliency to disturbance, both natural and human-caused. This is expressed at species, community, and ecosystem levels, leading to the impression that estuarine species are facultative with respect to estuaries as preferred environments. These characteristics have resulted in a tendency to describe any species that enters estuaries, or those that tolerate brackish waters, as "estuarine," which can be misleading. Nevertheless, some species seem to be restricted to estuarine and near-shore environments, at least at some life-history stage. A notable example concerns temperate oysters, which build extensive reefs in estuaries and lagoons and nowhere else. These reefs provide habitat for dozens of species, representative of almost every animal phylum.

Carriker (1967) noted that estuarine biota have adapted in different ways to estuarine conditions; for example, oligohaline organisms disappear at the head of the estuary; euryhaline species constitute the majority of the estuarine biota, as they can tolerate salinities as low as 5 ppt, as well as full salt water; and stenohaline species do not tolerate salinities of <25 ppt and are found only at the mouths of estuaries or on open seashores. This leaves "true estuarine organisms"—those relatively few species that are restricted to estuaries and that are best represented in the upper and middle reaches. Carriker concentrated mainly on benthic invertebrates, but concluded that an "estuarine biocenose" may be justified as a discrete functional aggregation of interdependent, regularly recurring, dominant, benthic populations that are strongly represented numerically. He acknowledged that much needs to be learned of ecology and life histories to justify this, but that the estuarine biotope appears to be more than "just a simple overlapping of factors (an ecotone) extending from the sea and the land, but is characterized by a unique set of its own factors arising from within the estuary from the materials and forces contributed by its bounding environments" (Carriker, 1967).

Some of the dominant, or "true," macroscopic biota of estuaries that he named are the plants—*Spartina alterniflora, Zostera marina, Ruppia maritima, Cymodocea mamatorium, Rhizophora mangle,* and *Avicennia nitida,* and the invertebrates—*Nereis diversicolor, Balanus improvisus,* xanthid mud crabs, *Uca pugnax, Callinectes sapidus, Mya arenaria, Mytilus edulis, Modiolus demissus,* and *Crassostrea virginica.* Additionally, he noted that characteristic estuarine habitats include tidal marshes, mangrove swamps, seagrasses, oyster reefs, soft clam–clam worm flats, and others. Finally, Carriker stated that: "Little is known of the sum of these effects on community structure, but they do emphasize the need to consider benthic organisms in the context of the total ecosystem rather than as an independent benthic biocenose." This statement, made a third of a century ago, has yet to be fully realized.

Fishes are the best known of aquatic groups in a general sense, mostly due to their commercial value. Therefore, insights into "estuarine dependency" may be best revealed through their study. One reason for this is their mobility in which various life-history stages inhabit quite different environments. Winemiller (1995) reviewed fish ecology and made the following points. First, fishes are by far the most diverse verte-

brates, and they inhabit an incredibly wide range of aquatic habitats from pole to pole. Second, fishes are ecologically diverse, with a wide variety of food habits, behaviors, reproductive habits, physiologies, and morphologies. Third, fishes exhibit a range of life-history strategies that result from trade-offs among various attributes, including clutch and egg size; these strategies can be classified as opportunistic, periodic, and equilibrium, but a range of intermediate strategies also exist. Finally, fishes and their diversity in ecosystems can be used as "indicators" of environmental conditions.

Recently, much attention has been directed toward the early life histories of fishes, as this is closely related to recruitment and, therefore, of much interest to fish ecologists and to fisheries. Houde (1997) provided a review of the selection factors that are of special importance in this regard. Able and Fahay (1998) extended studies on juvenile stages of fishes to "estuarine dependence" and determined that the numbers of permanent estuarine residents is relatively low, at least in part because estuaries exhibit extremes in environmental conditions. Also, the fish diversity of estuaries is augmented by transients, such as freshwater species that occasionally occur in estuaries and marine species that spawn at sea but whose young use estuaries as nurseries. Therefore, the estuarine fish fauna includes both residents and transients and a wide range of sizes, ages, and adaptations. In addition, those species that have successfully invaded estuaries usually inhabit only a small number of broad niches, implying that larger estuaries have larger numbers of species owing to increased habitat and niche complexity.

Able and Fahay found that, of the species for which good information is available, 60% are transients, 28% are residents (uncannily close to the "educated guess" of C. R. Robins and myself that 27% are "obligate" on estuaries; see Section V), 6% are infrequent, and 6% are unclassified. Furthermore, they have suggested the following adaptive groups for juveniles:

Group I. Facultative estuarine breeders: species whose nurseries are either in estuaries or on the inner shelf (e.g., *Centroptristis striata, Brevoortia tyrannus*).

Group II. Seasonal residents: species whose adults migrate into estuaries to spawn in spring or summer (e.g., *Menidia menidia, Mustelus canis*).

Group III. Anadromous species: species whose adults migrate through estuaries in order to spawn in freshwaters (e.g., *Morone saxatilis, Alosa* spp.).

Groups IV–VI. Early users, delayed users, and distant spawners: species that spawn exclusively in the ocean, but the location, timing, and manner of use of estuaries by young-of-the-year juveniles vary (e.g., *Pollachius virens, Prionotus carolinus, Mugil cephalus*).

Group VII. Expatriates: species whose estuarine larvae come from distant spawning (e.g., *Chaetodon ocellatus, Monacanthus hispidus*).

Group VIII. Summer spawners: the largest group, represented by shallow-water spawners whose larvae develop in the immediate vicinity of spawning sites (e.g., *Cyprinodon variegatus, Fundulus heteroclitus*).

Group IX. Winter–spring spawners: a few species that spawn in the winter or spring (e.g., *Pseudopleuronectes americanus*).

Group X. Migrating spawners: species that undergo spawning migrations within the estuary (e.g., *Morone americana*).

Group XI. Species difficult to classify: species for which some populations appear to be estuarine and other populations do not (e.g., *Tautogolabrus adspersus*).

Able and Fahay (1998) caution that, for fishes at least, "estuarine dependence" depends on the resolution of three areas of research: (1) the need to sample well-defined areas thoroughly for habitat evaluation; (2) assessment of the effects of habitat loss; and (3) more detail on temporal and spatial use of habitats where early stages are collected. In short, a coherent understanding of the life-history factors that control the early life histories of fishes remains to be accomplished. The same no doubt holds for invertebrates. For macroscopic plants, the situation is perhaps less uncertain, as their life histories are simpler and assessments are more easily accomplished.

In sum, most truly estuarine species are typically resistant to environmental variations due to the extreme conditions of estuaries, and/or take advantage of favorable situations; consequently, they do not appear to have strong habitat associations. This makes difficult the strict establishment of a definition of "estuarine dependency." Also, the seaward boundary of an "estuary" is often blurred, so that the definition of "dependency" is hampered by lack of comparative, quantitative data from offshore habitats. The easiest distinctions are for those species for which at least one stage is shown to be physiologically or behaviorally obligate, but good natural history and experimental data are required for this. Therefore, the question "What is an estuarine species?" remains elusive. In addition, the oft-made contention that estuaries with similar habitats may support similar species assemblages seems reasonable, but may be misleading if assumptions of estuarine dependency

are based on occurrence rather than in an adaptive-evolutionary sense.

IV. ECOLOGICAL FUNCTION OF BIODIVERSITY

In addition to genome, species, and ecosystem aspects of biodiversity, a fourth category must be considered, namely, "functional diversity" (Steele, 1991; see Glossary), which concerns ecological functions with respect to environmental maintenance and change. Ecological functions within the coastal zone and its estuaries are complex and variable, and they must be understood before we can interpret the composition and patterns of biodiversity. Holligan and Reiners (1992) listed a number of factors that underlie the biological diversity of the coastal zone and its estuaries, first for natural processes:

Exchanges of Materials Riverine and atmospheric export and import, groundwater exchange, and ocean–land material transport operate at various levels, but are presently poorly understood. [Recent information on anadromous fishes is shedding light on organic-matter transport; e.g., Hesslein *et al.* (1991); Bilby *et al.* (1996); Garman and Macko (1998).]

Physico-chemical Properties The coastal zone is a region of high energy exchange due to interactive oceanic and atmospheric forcing associated with topographical discontinuities, density gradients caused by freshwater inflows, and seasonal heat exchanges. Deltas, estuaries, and lagoons are the major sites for transformation and accumulation of organic matter and sediment, and all are highly variable spatially and temporally, so that their average conditions are not good indicators of net fluxes. Estuaries, in particular, are "sites of complex interactions, related to salinity gradients, phase transformation involving particle–water reactions, and to biological processes that cause biogeochemical transformations" (Holligan and Reiners, 1992).

Biological Properties Favorable conditions of light and nutrients in the coastal zone maintain high rates of primary productivity that are several times greater than for the open ocean, and even greater than for certain coastal upwelling areas; some coastal systems, such as salt marshes, mangrove swamps, mudflats, beds of aquatic vegetation, and coral reefs, exhibit even higher productivity. Some areas act as sources, others as sinks, and the nature of the coupling of primary productivity to the bottom or to open waters may determine community structure and function.

Biogeochemical Processes Organic matter is readily reoxidized in coastal waters, but some poorly drained areas may become anaerobic. This is especially apparent in the bottom water of estuaries in summer, when temperatures are high.

Many present-day human activities influence both ecological functions and biological diversity:

Altered Delivery of Freshwater Freshwater impoundment by damming has decreased total discharge into estuaries and coastal seas by about 15% since the 1950s, an amount equivalent to a change in sea level of -0.7 mm/yr. Seasonal flows have also been altered; alteration in the residence time of water in estuaries may have far-reaching effects on chemical processes.

Changes in the Transport and Fate of Suspended Matter Coastal subsidence, sediment starvation and consolidation, and nutrient levels have all been altered by human interventions. Land clearing especially on steep slopes, has increased sedimentation.

Chemical Modification Nutrients, eutrophy, and blooms have become widespread and their frequency seems to be increasing. Contaminants that are of most concern include heavy metals, synthetic organic compounds, radionuclides, and hydrocarbons.

Ecosystem Modification This takes many forms, from physical change, to habitat loss, to depletion of resources. The worst-affected areas are those with high human population densities, such as Southeast Asia, and along temperate coasts that have significant sources of pollutants, such as the Baltic Sea.

Longer-term processes that influence biodiversity are the effects of climate change, especially in response to global warming, should that continue to occur:

Natural Variations in Climate Many climate-change studies describe possible variations in the altered distributions of biota. However, rather subtle changes in climatic conditions can induce large ecological changes that reflect the sensitive nature of marine food chains to climate and to climate-dependent factors such as nutri-

ent levels and salinity. The direct effects of climate are difficult to distinguish from those incurred by humans.

Temperature The largest climate changes are expected in the higher latitudes. Thus, the poleward extension of climate-sensitive species is to be expected in case of global warming. Temperature changes can also affect behavior and physiology (e.g., reproduction, feeding and food availability, predation, migration), so that predictions are destined to be speculative.

Wind Wind strongly influences upwelling and stratification, thus affecting productivity through nutrient and light availability. According to most climate change scenarios, wind intensity is expected to increase.

Extreme Events Short time-scale events are also expected to increase with climate warming, and these may induce dramatic, long-term changes. A single storm lasting <5 days can result in sand transport equivalent to two-thirds of the total for an average year. Tsunamis have had the greatest effects recorded to date.

Changes in Sea Level Presently, sea level is rising faster than the rate during the late Holocene due to a combination of thermal expansion of seawater and melting of ice as the climate warms. Severe impacts of sea level rise on deltas and estuaries are already apparent, partly because they are low-lying, strongly perturbed by humans, and exhibit enhanced erosion and subsidence. Natural communities of plants and animals play a crucial role in determining the response of the coastal zone to changes in sea level.

This array of effects requires the development of research programs to address hypotheses that are relevant to the ecological function of estuarine biodiversity. Among many possibilities, the following seem essential (slightly modified from Solbrig, 1991):

- For species: no aspect of life history has any influence on extinction probability.
- For communities: keystone species are essential for maintaining species richness in communities under all environmental conditions.
- For ecosystems: removal or addition of functional or structural groups that produce changes in temporal or spatial configuration of landscape elements will have no significant effect on ecosystem properties over a range of time and space scales.

These hypotheses can be clarified by means of a case-by-case examination (see the Chesapeake Bay case study). For example, some species seem very alike in their life histories. However, redundancy in species function may mean that diversity and function are somewhat independent of one another. Many species of benthic infauna and epifauna are extremely abundant and ecologically important in estuaries. Many feed on sediments, and those with complete alimentary canals can consolidate organic residues into often long-lived, sculptured pellets. The question is: Many species have similar ecological requirements and, therefore, are species replaceable?

With respect to physical structure, Roy (1984) stated that the ecology of an estuary depends on the geological stage it has reached in its evolutionary progression, and that the rate and direction of natural change provide a yardstick to assess impacts induced by humans. However, as Roy emphasized, factors influencing estuary development include (1) inherited factors, mainly of a geological nature, that control the size and shape of the basin and the nature of the sediment supply, and (2) contemporary factors of a process nature (such as tides, river discharge, waves, etc.) that influence modes of sedimentation, hydrodynamics, and the biota. This prompts the question: To what extent are structure and biodiversity related?

Mann (1982) observed that, in general, estuaries are more productive than adjacent shelf systems, bringing up the question of nutrient flushing. That is, estuaries tend to act as nutrient traps. Many are enriched by pollution; the Hudson is a spectacular example of enrichment of a large shelf area well beyond its mouth. Within 600 km^2 of sea at the apex of the New York Bight, phytoplankton production amounted to about 370 g C/m^2/day, compared with only 100 g C/m^2/yr at the edge of the shelf. Mann and Lazier (1991) also noted that the dynamics of coastal waters, including estuaries, are made complex by: (1) shallowness, resulting in relatively mixed water that may extend to the bottom, and dead biological material that may accumulate to release nutrients that are carried rapidly to surface waters; (2) tidal currents that create turbulent mixing, which has especially marked effects on food particles, fertilization of planktonic eggs, and larval dispersal; and (3) barriers to convection imposed by coastlines, meaning that wind drives surface water away from the coast, and upwelling is the only way for it to be replaced, bringing nutrients to the surface. The question here is: To what extent are enrichment and/or pollution and circulation related to biodiversity?

Turning to larval transport, a variety of organisms have adapted to the seaward flow of low-salinity water and a compensatory landward flow of bottom water in estuaries. Organisms can make vertical migrations to maintain themselves in the estuary, or to enter or leave it seasonally. For example, estuarine larval transport and retention mechanisms are evident on two scales: circulation patterns on a large, regional scale and small-scale, local water motion. There is evidence that oyster larvae (*Crassostrea virginicus*) rise into the water column to be carried upstream, and that this is cued by increasing salinity associated with increasing upstream flow; larvae of the blue crab (*Callinectes sapidus*), on the other hand, occur in maximum numbers in surface waters at the mouth of Chesapeake Bay at night as the salinity falls on the ebb tide (Boicourt, 1982). From this and other evidence, it has been concluded that the crab larvae develop offshore, then reinvade as megalopa larvae or juveniles. Fishes have also been shown to vary their depths, some rising into surface waters during flood to remain in the estuary, and others doing the opposite to be taken out to sea. Thus, many invertebrates and fishes utilize the two-layered estuarine structure for dispersal, and this may not be entirely passive, as has often been assumed. Despite some improved knowledge, Boicourt's conclusion is still pertinent, that the larval transport and retention problem "stands at the state of the art in both physical and biological fields." The question is: Does recruitment depend on return or retention (in the strict sense) as the operative process, and to what extent do larvae determine their own fates?

As another example of the importance of functional diversity, juveniles of the five species of Pacific salmons (*Onchorhynchus* spp.) vary in time spent in estuaries, but for all of them a high proportion of their prey tends to be detritus feeders (Healey, 1982). This means that the configuration of the estuary and the efficiency of entrapment of detrital matter are important for juvenile salmon habitat. Retention of detritus is enhanced by restricted exchange with the ocean and low bed-load transport. Marshes and submerged aquatic vegetation are efficient detritus traps, and these habitats also shelter salmon from predation. Thus, it may be hypothesized that the complex of intertidal marshes, tidal creeks and secondary river channels, lower intertidal and subtidal weed beds, and basin morphology all contribute to the carrying capacity of the estuary for young salmon, and that the appropriate configurations must be conserved if salmon production is to be maintained. The question here concerns how the complexity of the land–seascape enhances biodiversity, and how this may operate differently for closely related species.

From these examples, it is apparent that, insofar as ecosystem functioning is concerned, the addition or deletion of species, structural groups, or essential processes can have profound effects on the capacity of an estuary to maintain its biodiversity. This is especially true for "keystone" species, which have influences out of proportion to their density or biomass. Likewise, the fragmentation and/or simplification of habitats and of land–seascapes may have profound effects on estuaries, since these impacts shift ecological complexity and community structure and function. Furthermore, it is likely that the functional autonomy of estuaries depends on their size and the time intervals of various processes. That is, the degree to which an ecological system may be autonomous depends on the extent to which it is independent of the ecological dynamics outside its domain. Of course, no ecosystem can be completely independent owing to the climatic, ecological, and geological connections among all portions of Earth. However, the larger the domain, the more it may tend to be autonomous during the time spans of investigation. Consideration of autonomy requires one to consider to what extent estuaries are forced functionally by the dynamics of the contributing watershed and adjacent shelf (e.g., tides, currents, flushing, river inputs, storms). Obviously, the elucidation of autonomy for a domain of a given size is not a simple endeavor. However, the simple fact is that under many management regimes, autonomy may be incorrectly assumed.

V. A CASE STUDY: THE CHESAPEAKE BAY

Chesapeake Bay is one of Earth's largest estuaries. Its origin is that of a drowned river valley. This is the case for many estuaries associated with coastal plains, wherein the dominant processes are sedimentary and erosional and whereby the bottom is largely soft sand and mud. Chesapeake Bay's one major hard feature is that of the oyster reef, formed by the eastern Oyster, *Crassostrea virginicus*.

Many scientists have observed the drastic decline of oysters and of oyster reefs during the past hundred years and more, and the associated ecosystem effects. From a structural point of view, oyster reefs represent a unique and dominant biogenic structure of the Bay. Their distribution and ecological importance during the mid-1800s were analyzed by McCormick-Ray (1998). Their loss would be expected to have extensive repercussions on biological, hydrological, erosional, and sedimentary patterns and processes, all of which can have major influences on biological diversity. Indeed,

history has borne out this conclusion. For example, a review by Rothschild *et al.* (1994) stated that "considerable concern is voiced regarding Chesapeake Bay water quality and the effects of disease on oysters" and that "the effects of a diminished oyster population abundance certainly must have changed the 'ecology' of Chesapeake Bay, and these effects must have become evident at the time of maximum stock decline (1884 to 1910)."

To understand the ecosystem effects of the oyster and oyster reefs better, one must begin at the regional scale, wherein the coastal zone is conceived as a nested hierarchical system (Ray *et al.*, 1997). The regional scale is that of biogeographic and physiographic provinces. The mesoscale is represented by major regional subdivisions, such as watersheds, estuaries, coastal islands, lagoons, and coastal–ocean fronts that separate major marine regimes. The smallest scale is that of the interacting mosaics of land–seascapes, for example, wetlands, hard and soft bottoms, and water masses that are distinguished by salinity, temperature, and density. The oyster reef represents this latter scale.

This hierarchy is illustrated in Fig. 1, which indicates top-down "controls" and bottom-up "feedbacks" and which places estuaries in a central role. First, the biogeographic province (and/or "region") is an area whose limits are defined by the relative homogeneity of the biota. For example, the traditionally accepted boundaries for the Virginian Province are Cape Cod, Massachusetts, to Cape Hatteras, North Carolina. These capes are significant points of deflection for major ocean currents, principally the warm, north-flowing Gulf Stream and the cold, south-flowing Labrador Current. At these capes, dramatic changes in coastal characteristics, such as water temperatures and circulation patterns, occur and these physical features play major roles in determining the ranges of the biota. One major feature of the Virginian Province is the presence of very large estuaries, such as the Chesapeake and Delaware Bays.

Species' ranges respond to these large-scale attributes, as well as to species' physiological and behavioral adaptations. Fishes are a case in point. Of the almost 1100 East Coast fish species, 556 species presently occur in the Virginian–Carolinian region (Ray, 1997; Ray *et al.*, 1997). Estuary-dependent species are drawn from this species pool. As discussed earlier, "estuary-dependent" has usually been interpreted very broadly. C. R. Robins and I re-examined this matter and concluded that occurrence and even abundance of fishes in estuaries do not necessarily infer "dependence." Rather, we determined that a species must be truly "obligate" in an evolutionary, adaptive sense for this definition to apply; that is, if estuaries were removed,

Levels of Habitat Description

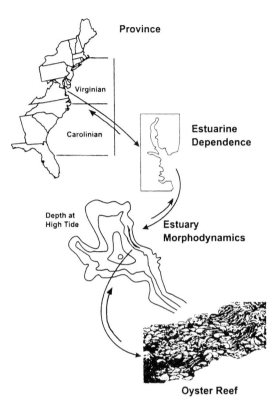

FIGURE 1 A hierarchical model of coastal zone relationships, showing top-down "controls" and bottom-up "feedbacks" of coastal zone interactions, involving levels from biogeographic provinces, to estuaries, to the oyster reef. The biogeographic province provides the species pool from which estuaries may draw "estuary-dependent" representatives. This biota is influenced by the morphometrics of individual estuaries, leading to different species communities among the estuaries in a biogeographic region. The oyster is a "keystone" species both biologically and ecologically, as the reefs it builds influence the morphometrics of the estuaries in which it occurs. Over-harvesting of oysters in the Chesapeake Bay, and elsewhere, has had major effects on estuarine function, structure, and probably biodiversity as well. (From Ray et al., 1997.)

"dependent" species would be at risk of significant depletion, even to the point of local or regional extirpation. According to this definition, we determined that 151 species (27% of 556 species) qualify as "estuary dependent," less than has been assumed in the past, but still a significant part of the total. This figure is remarkably consistent with the results of Able and Fahay (1998: see Section III). A principal components analysis of the ranges of these species resulted in four assemblages. Figure 2 shows these assemblages and demonstrates that so-called "faunal breaks" between provinces must be viewed as gradients, and not as "boundaries" in a rigid sense.

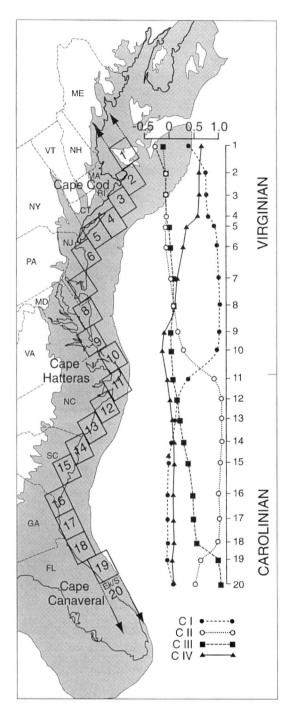

FIGURE 2 A principal components analysis of the ranges of 151 Carolinian and Virginian estuary-dependent species revealed the following assemblages: Component I = Virginian; Component II = Carolinian; Component III = tropical; and Component IV = boreal. These four assemblages overlap, as would be expected. (From Ray et al., 1997.)

This finding brings up the following question: How might changes in estuaries, human-caused or not, influence the composition of these fish assemblages? For insight into an answer, we must examine the dynamics of estuaries themselves. Many factors interact to characterize an estuary. Among these are drainage area, tides and mixing, estuary area, depth, dimension, water column stratification, floods, habitat types, and many others. A principal components analysis (Ray et al., 1997) revealed five components that may influence biological diversity: estuarine dimensions, dominance of marine processes, co-dominance of marine and freshwater processes, fjord-like attributes, and surface area. The interplay of these factors may be used to classify estuaries into the following types: (1) those that are long and wide with extensive catchment areas; (2) large, embayed, well-stratified estuaries with extensive seawater zones; (3) marine-dominated, deep, and well-stratified estuaries; (4) long and narrow, fjord-like estuaries, with large tidal prisms; and (5) estuaries with large surface areas. Chesapeake Bay falls somewhere between the first and second categories.

It seems reasonable, from what we know of the natural histories of the biota, that these estuarine types would be expected to host different communities of species, and further that different disturbance regimes would be expected to affect these estuarine types and their species' communities differently. The conclusion seems obvious that biotic communities will differ among estuaries and that seasonal or weather-related changes in salinity and other factors will be reflected in the variability of biotic patterns. Furthermore, because the great majority of estuarine fishes, in particular, also occur over the continental shelf, fluctuations of estuarine fish communities would also be reflected, up-scale, by shelf-fish communities.

This approach offers a series of environmental top-down "controls" over biodiversity and ecosystem function. But this can not totally explain what might be the consequence of bottom-up environmental alterations. That is, the prediction of biodiversity and faunal dynamics requires that the response of the organism to the environment at different scales and the modifications the organism may make to the environment both be made explicit. For example, Fig. 1 indicates that the decline or removal of a species or a local structure, in this case oyster reefs, will influence the total biological diversity of the system by influencing environmental conditions through environmental feedbacks.

For the Chesapeake Bay, and many other Virginian–Carolinian estuaries, oysters are especially critical because they form reefs, which influence biodiversity at many levels (McCormick-Ray, 1998). The location of these reefs is not accidental. Their formation depends on the geometry of the estuarine basin, tidal stream channels and meanders, and other factors. Furthermore, oyster reefs influence estuarine development, sedimentation, and water clarity, and thus the formation of habitats (e.g., submerged aquatic vegetation, marshes, soft bottoms, and hard bottoms) for a host of organisms. In sum, the eastern oyster appears to be a classic example of a "keystone" species at the level of the ecosystem. Structurally and functionally, individual oysters and the reefs they build strongly influence species diversity and productivity. Additionally, the distribution of oyster reefs may be of fundamental importance to development of the estuarine land–seascape.

Another type of feedback concerns the fact that most species exist as a number of separate populations that mix together as one or more "metapopulations." For example, an estuary-dependent species, such as menhaden (Brevoortia tyrannus), forms populations in individual estuaries, and these populations assemble over the shelf to form one or more metapopulations. Furthermore, these metapopulations join those of other species and become part of the shelf "metacommunity," as illustrated in Figure 3. It follows that fluctuations of any one metapopulation within any one estuary will affect the total "metacommunity" to a greater or lesser extent (Ray, 1997). This form of biodiversity concerns community composition, not necessarily the presence or absence of individual species, and is strongly affected by functional alterations of estuaries. The conclusion is that at the scale of the large, regional ecosystem, each estuary may be conceived in terms of the sum total of estuaries and is responsible, to a greater or lesser degree, for the overall large-scale dynamics of the biogeographic region. This approach fuses concepts of landscape ecology with metapopulation theory.

The concepts presented in the case of the Chesapeake Bay suggest controls and feedbacks among organisms and the environment at several scales, in which one fundamental factor seems clear. East Coast estuaries have been perturbed in many ways, but one of the most dramatic for the Chesapeake Bay has been the depletion of oyster reefs and the practical eradication of their functional ecosystem role. Although data are lacking that would explain beyond doubt what changes have been perpetrated by the oyster's decline in Chesapeake Bay, it seems apparent that, at the very least, the oyster reef's demise has had a marked effect on the distributions of estuarine species, not necessarily because the

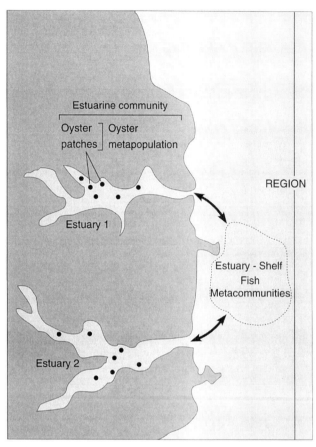

FIGURE 3   The concept of estuarine metapopulations and shelf meta-
communities. Oyster reef metapopulations influence estuarine morphometrics and biodiversity. Consequently, the fish biota of various
estuaries influence the fish metacommunity of the shelf.

reef is required habitat, but because of its functional
importance to the Bay as a whole. It is possible that
these effects may have cascaded up-scale to the adjacent
continental shelf.

VI. FUTURE CHALLENGES

I make three points in conclusion. The first concerns
the need for greatly increased attention to the natural
histories of estuarine and shelf species. The natural
histories of these organisms underlie both theory and
management practice. The minimal requirements for
informed conservation and management are descriptions of species' life histories in the context of their
environmental relationships.

Second, many estuarine organisms range widely and
form metapopulations over the shelf, as components of

estuary–shelf communities. Thus, the minimal scale
for sustainability of biodiversity becomes that of the
biogeographic region. Quantitative, landscape-level descriptions of the regional coastal zone, including estuarine habitats, are a necessary prerequisite for conservation and management.

Third, it has become a truism in ecology that no
one scale adequately describes ecosystem phenomena.
Rather, the interaction among phenomena on different
scales must become the centerpiece of research and
management. This strongly suggests that explanations
for fluctuations in biodiversity, including those within
biotic communities and at regional scales, will continue to be obscure until multiscale ecosystem functions are better understood. Ecosystem management
is the logical outcome of interdisciplinary, multiscale
knowledge. This recognizes that understanding the
ecology and diversity of coastal zone biota depends
in large part on understanding land–sea and estuarine
interactions, and also on the joint application of
metapopulation and land–seascape theory and
methods.

The National Research Council (NRC, 1995) stated
that a major future research objective is "to understand the patterns, processes and consequences of
changing marine biological diversity by focusing on
critical environmental issues and their threshold effects,
and to address these effects at spatial scales from local
to regional." This objective cannot be met absent a
specific consideration of estuaries as major, scale-
dependent pathways of biotic and abiotic interchanges.
Estuarine biodiversity, structure, and function have
been severely modified by humans around the globe.
Nevertheless, many estuaries remain either good candidates for restoration or relatively rich, productive, and
resilient. Documentation of impacts is severely hampered by lack of long-term baseline information, inadequate assessment of biodiversity, lack of trained taxonomists, and difficulty in sampling.

Nevertheless, an extensive estuarine literature is now
available, and it illustrates that control of pollution,
development, excessive natural resource extractions,
and changes in ecosystem function urgently need to
be addressed. Problems may not be eliminated, only
ameliorated, but increased understanding is essential
for the future sustainability of estuaries. Carriker (1967)
put the matter boldly three decades ago: "There is consequently an urgency to study estuaries before unenlight-
ened defacement obliterates them and before it becomes
expedient to investigate them primarily as outdoor pollution laboratories."

See Also the Following Articles

COASTAL BEACH ECOSYSTEMS • FRESHWATER
ECOSYSTEMS • INTERTIDAL ECOSYSTEMS •
LAKE AND POND ECOSYSTEMS • MARINE
ECOSYSTEMS • RIVER ECOSYSTEMS •
WETLANDS ECOSYSTEMS

Bibliography

Able, K. W., and Fahay M. P. (1998). The *First Year in the Life of Fishes in the Middle Atlantic Bight.* Rutgers University Press, New Brunswick, New Jersey.
Anonymous. (1959). Symposium on the classification of brackish waters. April 8–14, 1958, Venice, Italy. *Archivo di Oceanografia e Limnologia,* Vol. II, supplemento.
Bilby, R. E., Fransen, B. R., and Bisson P. A. (1996). Incorporation of nitrogen and carbon from spawning coho salmon into the trophic system of small streams: Evidence from stable isotopes. *Canad. J. Fisheries Aquatic Sci.* 53, 164–173.
Boicourt, W. C. (1982). Estuarine larval retention mechanisms on two scales. In *Estuarine Comparisons* (V. S. Kennedy, ed.), pp. 445–457. Academic Press, New York.
Bulger, A. J., Hayden, B. P., Monaco, M. E., Nelson, D. M., and McCormick-Ray, M. G. (1993). Biology-based estuarine salinity zones derived from a multivariate analysis. *Estuaries* 16(2), 311–322.
Carriker, M. R. (1967). Ecology of estuarine benthic invertebrates: A perspective. In *Estuaries* (G. H. Lauff, ed.), AAAS Sci. Publ. No. 83, pp. 442–487. American Association for the Advancement of Science, Washington, D.C.
Emery, K. O. (1967). Estuaries and lagoons in relation to continental shelves. In *Estuaries* (G. H. Lauff, ed.), AAAS Sci. Publ. No. 83, pp. 9–11. American Association for the Advancement of Science, Washington, D.C.
Garman, G. C., and Macko, S. A. (1998). Contribution of marine-derived organic matter to an Atlantic coast, freshwater, tidal stream by anadromous clupeid fishes. *J. North Amer. Benthol. Soc.* 17(3), 277–285.
Hanski, I., and Gilpin, M. (1991). Metapopulation dynamics: Brief history and conceptual domain. *Biol. J. Linnaean Soc.* 42, 3–16.
Hayden, B. P., Ray, G. C., and Dolan, R. (1984). Classification of coastal and marine environments. *Environ. Conservation* 2(3), 199–207.
Healey, M. C. (1982). Juvenile salmon in estuaries: The life support system. In *Estuarine Comparisons* (V. S. Kennedy, ed.), pp. 315–341. Academic Press, New York.
Hesslein, R. H., Capel, M. J., Fox, D. E., and Hallard, K. A. (1991). Stable isotopes of sulfur, carbon and nitrogen as indicators of trophic level and fish migration in the lower Mackensie River basin, Canada. *Canad. J. Fisheries Aquatic Sci.* 48, 2258–2265.
Heywood, V. H., and Watson, R. T. (eds.). (1995). *Global Biodiversity Assessment.* Published for the United Nations Environment Programme by Cambridge University Press, Cambridge, United Kingdom.
Holligan, P. M., and Reiners, W. A. (1992). Predicting the responses of the coastal zone to global change. *Adv. Ecol. Res.* 22, 211–255.
Houde, E. W. (1997). Patterns and consequences of selective processes in teleost early life histories. In *Early Life History and Recruitment in Fish Populations* (R. C. Chambers and E. A. Trippel, eds.), pp. 173–196. Chapman and Hall, London.
Ketchum, B. H. (ed.). (1972). *The Water's Edge: Critical Problems of the Coastal Zone.* Massachusetts Institute of Technology Press, Cambridge, Massachusetts.
Mann, K. H. (1982). *Ecology of Coastal Waters: A Systems Approach.* University of California Press, Berkeley.
Mann, K. H., and Lazier, J. R. N. (1991). *Dynamics of Marine Systems: Biological–Physical Interactions in the Oceans.* Blackwell Scientific Publications, Boston.
McCormick-Ray, M. G. (1998). Oyster reefs in 1878 seascape pattern—Winslow revisited. *Estuaries* 21(4B), 784–800.
Nelson, J. S. (1984). *Fishes of the World,* 2nd ed. Wiley–Interscience, New York.
National Research Council (NRC). (1995). *Understanding Marine Biodiversity.* National Research Council, Washington, D.C.
Natural Environment Research Council (NERC). (1992). *Land–Ocean Interaction Study (LOIS): Science Plan for a Community Research Project.* Natural Environment Research Council, Swindon, Wiltshire, United Kingdom.
Pernetta, J. C., and Milliman, J. D. (eds.). (1995) *Land–Ocean Interactions in the Coastal Zone: Implementation Plan.* Global Change Report No. 33. International Geosphere–Biosphere Programme of the International Council of Scientific Unions (ICSU), Stockholm.
Pritchard, D. W. (1967). What is an estuary: Physical viewpoint. In *Estuaries* (G. H. Lauff, ed.), AAAS Sci. Publ. No. 83, pp. 3–5. American Association for the Advancement of Science, Washington, D.C.
Ray, G. C. (1997). Do the metapopulation dynamics of estuarine fishes influence the stability of shelf systems? *Bull. Mar. Sci.* 60(3), 1040–1049.
Ray, G. C., Hayden, B. P., McCormick-Ray, M. G., and Smith, T. M. (1997). Land–seascape diversity of the U.S. east coast coastal zone with particular reference to estuaries. In *Marine Biodiversity: Causes and Consequences* (R. F. G. Ormond, J. D. Gage, and M. V. Angel, eds.), pp. 337–371. Cambridge University Press, Cambridge, United Kingdom.
Rothschild, B. J., Ault, J. S., Goulletquer, P., and Heral, M. (1994). Decline of the Chesapeake Bay oyster population: A century of habitat destruction and overfishing. *Mar. Ecol. Progress Ser.* 111, 29–39.
Roy, P. S. (1984). New South Wales estuaries: Their origin and evolution. In *Coastal Geomorphology in Australia* (B. G. Thom, ed.), pp. 99–121. Academic Press, Orlando, Florida.
Sanders, H. L. (1968). Marine benthic diversity: A comparative study. *Amer. Naturalist* 102, 243–282.
Solbrig, O. T. (1991). *From Genes to Ecosystems: A Research Agenda for Biodiversity.* Report of a IUBS–UNESCO Workshop, June 1991, Harvard Forest, Massachusetts. International Union of Biological Sciences, Paris.
Steele, J. H. (1991). Marine functional diversity. *BioScience* 41(7), 470–474.
Winemiller, K. O. (1995). Fish ecology. In *Encyclopedia of Environmental Biology,* Vol. 2, pp. 49–65. Academic Press, Orlando, Florida.

ETHICAL ISSUES IN BIODIVERSITY PROTECTION

Philip J. Cafaro* and Richard B. Primack†
*Colorado State University and †Boston University

GLOSSARY

anthropocentrism Position that only human beings have moral worth or intrinsic value.

ecosystem health Condition of an ecosystem, whether natural, managed, or human-dominated, that is free from influences that would damage or destroy its characteristic structures and functions.

ecosystem integrity Condition of an ecosystem that is largely free from human interference and possesses a species composition and functional organization comparable to those of natural ecosystems in the region.

instrumental value Value of something relative to human interests or desires.

intrinsic value Value of something independent of its value to people.

rights Justified claims that others respect and protect one's important interests.

sustainability Ability of a society or a particular human activity to continue indefinitely without depleting resources or damaging the environment.

wilderness Large area that remains essentially unmanaged and unmodified by human beings.

CONSERVATIONISTS AND CONSERVATION biologists agree on a basic ethical commitment to preserve biodiversity. But ethical disagreements and dilemmas arise in the attempt to realize this commitment. This article examines some important ethical issues faced by those who value and seek to protect biodiversity.

I. CONSERVATIONISTS' ETHICAL CONSENSUS

There is widespread disagreement in modern societies regarding the proper human relationship to the rest of the natural world. Conservationists and conservation biologists, however, mostly agree on the following ethical principles: the diversity of organisms is good; ecological complexity and natural evolution are good; the untimely extinction of populations, species and biological communities is bad; biological diversity has great value both to people and in its own right; and human beings have both strong altruistic and strong self-interested reasons for preserving biodiversity.

Conservationists justify these principles in very different ways. One may value biodiversity because it is God's creation and testifies to God's glory; another because it is the natural culmination of evolution and global forces that have occurred over hundreds of millions of years; a third because of its value to science, perhaps humanity's noblest pursuit; a fourth because of its beauty and the enjoyment that current and future generations may take in it; a fifth because its intrinsic value grounds a strong duty to protect it, regardless of human interests; a sixth because preserving biodiversity helps preserve human life-support systems. Such justifications may be more or less anthropocentric—centered on human interests. However different their ultimate justifications, conservationists tend to accept these ethical principles and work with others to preserve biodiversity.

Difficult ethical and practical issues emerge when we ask two further kinds of questions. First, what do these general principles entail? Given a high valuation of nature and a sense of personal responsibility to help protect it, what exactly is required of us as scientists and citizens? Second, how does our concern to protect biodiversity mesh with our obligations to humans? What if they clash? Such questions come up in numerous practical contexts and must be answered, although we cannot wait for certainty or full consensus in order to act.

II. CONSERVATION GOALS: HEALTH, INTEGRITY, SUSTAINABILITY

For much of the twentieth century, public and private land management goals have been primarily economic. It has become increasingly clear, however, that an exclusive focus on economic productivity leads to environmental degradation: pollution, the extinction of species, and the creation of a progressively simplified landscape. In response, laws have been passed directing government land managers to preserve the health and integrity of ecosystems. A focus on short-term economic returns also tends to undermine the long-term economic productivity of biological systems, such as forests and fisheries. Recognizing this, governments have begun to promote sustainability as an ideal in the use of natural resources. Conservationists hope that health, integrity, and sustainability specify goals concrete enough to guide us in our actions while commending themselves as basic values that a wide variety of people will share.

A. Health

Responding to widespread criticism of forestry practices, Canada's Provincial Forest Ministers agreed in 1992 that "our goal is to maintain and enhance the long-term health of our forest ecosystems, for the benefit of all living things both nationally and globally" (Westra and Lemons, 1995). Government agencies around the world have similarly embraced the notion of land health. Participants in a major symposium on ecological health defined it as follows:

> An ecological system is healthy and free from "distress syndrome" if it is stable and sustainable—that is, if it is active and maintains its organization and autonomy over time and is resilient to stress. (Costanza et al., 1992).

Just as a healthy human being is free from disease and able to perform his or her characteristic functions well, a healthy forest or stream is free from air or water pollution, siltation, or invasions of exotic organisms. Such "stresses" impede or radically alter the systems' natural functions (such as photosynthesis or net primary productivity) and structure (often eliminating sensitive species and simplifying biological communities)(Table I).

Healthy ecosystems provide both natural biodiversity and a wide array of products and services valued by humans. For this reason, economic productivity (as measured in board feet, fish caught, or crops yielded) can sometimes stand as one measure of overall ecosystem health. Rapidly declining fish catches in Lake Victoria, due to the introduction of the Nile perch, signaled a decrease in the system's health. However, economic productivity may sometimes be increased by a radical simplification of an ecosystem, as when diverse, natural forests are converted to single-species, even-aged pine plantations, or when natural prairies are converted to cattle ranches (Fig. 1). Such simplified ecosystems are not necessarily economically productive in the long run, but if they are, then whether they are seen as unhealthy depends on whether preservation of native biodiversity is part of our definition of land health.

If simplified ecosystems are managed well, they may stay healthy for humans and a reduced biota, remain stable and resilient to stress, not pollute surrounding ecosystems (e.g., not dump excessive silt or fertilizer in streams), and continue to perform valuable ecosystem services. Because some relatively intensive land use is necessary for human survival and this can be done better or worse, it makes sense to apply concepts of

TABLE I

Characteristic Response of Ecosystems to Stress

Types of stress	Nutrient pool	Primary productivity	Size distribution	Species diversity	System retrogression
Harvesting of renewable resources					
Aquatic	*	*	−	−	+
Terrestrial	−	−	−	−	+
Pollutant discharges					
Aquatic	+	+	−	−	+
Terrestrial	−	−	−	−	+
Physical restructuring					
Aquatic	*	*	−	−	+
Terrestrial	−	−	−	−	+
Introduction of exotics					
Aquatic	*	*	*	−	+
Terrestrial	*	*	*	*	+
Extreme natural events					
Aquatic	*	*	−	−	+
Terrestrial	−	−	−	−	+

Note: Signs (+ or −) indicate direction of change compared with normal functioning of relatively unstressed systems. An asterisk indicates that a characteristic response was not sufficiently determined. Rapport *et al.* (1985).

land health to simplified ecosystems. At the same time, we must recognize that preserving as much biodiversity as possible in managed systems (croplands, lakes, forests) is desirable and that excessive simplification across a large landscape leads to extinction of species and loss of characteristic natural communities. Ecosystem health must be supplemented with the ideal of ecosystem integrity—at least in some areas (see the next section).

We should also remember an important difference between human health and ecosystem health. Individual human beings grow old, decay, and die, and while we fight this we also accept it. In contrast, we want healthy ecosystems in perpetuity, for their own good and the good of our descendants. Thus ecosystem health must also be supplemented with the ideal of ecological sustainability (see Section II,C).

B. Integrity

Amendments to the U.S. Clean Water Act of 1972 first set ecological integrity as a management goal by calling for the restoration and maintenance of "the chemical, physical, and biological integrity of the nation's waters" (Costanza *et al.*, 1992). Rolston (1994) noted that "both integrity and health are combined fact-value words. Both convey the idea of wholeness and of unbroken functioning." But ecosystem integrity implies a greater measure of freedom from past and current human ma-

nipulation and a closer approximation to natural structure and functioning than does ecosystem health. Karr and Dudley (1981) defined ecological integrity as "the capability of supporting and maintaining a balanced, integrated, adaptive community of organisms having a species composition, diversity, and functional organization comparable to that of natural habitats of the region."

Integrity is a key environmental ideal because it encompasses the full preservation of biological diversity, including individual organisms, species, natural communities, and the ecosystem processes that have created and sustained them. It has taken a long time for this ideal to be acknowledged, even in supposedly fully preserved areas. For example, wolves, mountain lions, and other predators were routinely shot, trapped, and poisoned in U.S. national parks up until the 1960s. It is only in the last ten years that wolves, once extirpated, have been reintroduced into Yellowstone National Park. For most of this century, fires were suppressed throughout the national parks; they were perceived as dangerous to both people and forests. Over time, however, fire ecologists documented the historical role of fires in creating the landscape and preserving fire-dependent species of plants and animals. In response, land managers have introduced controlled burning into parks and allowed some lightning-ignited fires to burn. Improved ecological understanding and an attempt to distinguish between what is good for the ecosystem and what is

A

B

C

FIGURE 1   Healthy and unhealthy rangelands. (A) A natural grassland with numerous native species on the National Bison Range, federally protected land in the state of Montana. (B) Cattle graze on natural grassland. (C) Overgrazed grassland takes on the appearance of a desert and native species are eliminated. (Photographs courtesy of the U.S. Fish and Wildlife Service and the U.S. Forest Service.)

comfortable or familiar for people have shown that wolves and fire belong in the parks.

The fact that ecosystems are not as strongly integrated, clearly bounded, or stable as individual organisms complicates attempts to specify ecological integrity in particular cases. It is sometimes difficult to stipulate

which outside influences represent assaults on integrity rather than mere changes. Recent study suggests that many ecosystems are relatively loose assemblages of species; that these assemblages may be relatively recent and transient creations; and that even left to themselves many natural communities will not necessarily reach a particular, invariant climax state but may instead reach any one of a number of more or less stable states, or no stable state whatsoever. If species assemblages are always changing naturally, why distinguish species that are extirpated or introduced by humans, as when conservationists reintroduce wolves in the Rockies or eradicate wild boars in the Great Smoky Mountains? If there is no one natural end point to succession, why assume that human-created early-successional stages (fields of shrubs after clear-cuts) have less integrity than naturally occurring late-successional stages (old-growth forests)?

Still, ecosystems typically go through characteristic successional stages and support characteristic (if not invariant or exact) species assemblages. Though natural species assemblages change, this usually occurs on a time-scale that allows for much stability, the development of detailed interactions between organisms, and the increase of biological diversity at the landscape level. In contrast, human-induced changes typically decrease biodiversity and always lead to a landscape that is partly our creation. This loss of independence arguably marks a qualitative change in an ecosystem's natural history and a corresponding loss of ecosystem integrity. Applying this complex concept to particular ecosystems thus involves attention to scale, knowledge of an ecosystem's particular history, and comparison with the structures and functioning of similar ecosystems. Maximizing ecological integrity involves balancing the sometimes conflicting goals of freedom from human interference and preservation of historical natural communities.

Despite these complications, the desire to restore degraded ecosystems demands robust conceptions of ecosystem integrity and health. Ecologists attempting to restore degraded pasture-lands to native prairie in Wisconsin, worked-out strip mines to forests in Appalachia, or drained lands back to wetlands in Florida must set specific objectives (Fig. 2). As with land managers seeking to limit their effects on relatively pristine areas, restorationists have taken health and integrity, defined in relation to natural baselines, as goals for restored landscapes.

C. Sustainability

"Live Sustainably!" has become a rallying cry in the environmental movement. Like health and integrity, sustainability is a term that blends facts and values, helping us carry out our environmental convictions. At a minimum, sustainability involves preserving resources for future generations. But this minimum will not satisfy conservation biologists, who insist that part of what must be sustained is the full complement of biological diversity, for its own sake and for the benefit of humans.

Wide differences exist here. For instance, in 1987 the United Nations World Commission on Environment and Development (the Brundtland Commission) defined sustainable development as "development that meets the needs of the present without compromising the ability of future generations to meet their own needs." The focus here is wholly on human beings. In this definition the mass extinction of species is sustainable, provided future generations of people can meet their self-defined needs. R. Noss summarized the shortcomings of such narrow, anthropocentric accounts of sustainability as follows:

> A failure of those who promote sustainability to consider environmental and social limits to growth; an unwillingness to address the unsustainability of the current human population, much less its expected growth; a reluctance to confront the implications of the lifestyles of average citizens of the more affluent societies . . . a failure to recognize the claims of other species to their share of the planet's resources. (Westra and Lemons, 1995)

Contrast the Brundtland definition with Barbier's more generous definition of sustainable development: "*to maximise simultaneously* the biological system goals (genetic diversity, resilience, biological productivity), economic system goals (satisfaction of basic needs, enhancement of equity, increasing useful goods and services), and social system goals (cultural diversity, institutional sustainability, social justice, participation)" (Munda, 1997). Such a definition implies a different conception of human development and a more restrained treatment of the non-human world than has prevailed up to this point in human history. Satisfying the basic needs of all people and creating just and flourishing societies are part of this goal; limitless wealth creation and unbounded consumerism are not.

Conservationists are particularly concerned to keep the definition of sustainability based in ethics and biology and not solely in traditional economics. Economically based definitions often define sustainability in terms of indefinite economic growth, but in the crowded, fossil-fuel-propelled world of the twenty-first century, continued emphasis on economic growth will inevitably undermine ecosystem health and integrity worldwide. We cannot sustain increased consumption

A

B

FIGURE 2 Restoration of ecological integrity. (A) In the late 1930s, members of the Civilian Conservation Corps participated in a University of Wisconsin project to restore the wild species to a Midwestern prairie. (B) The prairie as it looked 50 years later. (Photographs from the University of Wisconsin Arboretum and Archives.)

and increased human populations and protect biodiversity at the same time.

III. MANAGEMENT AND NONMANAGEMENT

Setting aside parks or wilderness areas where human beings are not permitted to use resources has been conservationists' most effective tool in conserving biological diversity. Here the results from conservation biology are clear. Wild areas preserve greater numbers of native species than do areas managed for agriculture or forestry. Large protected areas are more likely than smaller ones to preserve full complements of native species. Some species need large areas of habitat to preserve minimum viable populations. Some species tolerate little human disturbance. Larger areas are less likely to suffer "edge effects" that render otherwise good habitat unusable by certain species.

A dilemma arises, however, because in order to preserve certain wild species or communities, we may have to actively manage wild or semi-wild areas. Rare, threatened plant populations may have to be fenced off from high populations of browsers or new populations created from seeds or cuttings. Small, widely scattered populations of an endangered mammal may have to be gathered together to preserve genetic variety or create viable populations. Controlled burning may have to be carefully introduced into a forest or field to preserve fire-dependent species (Fig. 3).

Most conservationists accept reintroductions of indigenous species into wild areas and other manipulations that have as their goal the preservation or recreation of indigenous flora and fauna. But they sharply distinguish actions that have this goal from actions that support further resource extraction or tourist comfort (road building, tree harvesting, construction of lodges and restaurants), which they continue to oppose in wild areas. Further, they see value in the absence of manipulation itself and tend to view the need for heroic measures in pristine areas as an indication of what has already been lost.

As a practical matter, hands-on management will often be necessary to preserve biodiversity. Even our largest natural areas are increasingly "islands" of diversity surrounded by development. These islands will likely lose species over time, as predicted by conservation biology theory and confirmed by empirical study; more native biodiversity may be preserved by augmenting populations and managing habitat. Other semi-wild areas will continue to be utilized by humans; these areas support wild species that may also need management in order to survive. Some manipulation will be needed, in perpetuity, if we are going to preserve as much biodiversity as possible.

It should be noted, however, that heroic measures and intrusive management often fail to preserve species. Skeptics also warn that a casual acceptance of manipulation may lead to a loss of biodiversity, if all sorts of human purposes are allowed to vie with nature preservation on our park and forest lands. Too often, management of nature becomes a substitute for management of ourselves. Rather than close popular trails or campgrounds in Yellowstone National Park, for example, the U.S. National Park Service exterminates "problem" grizzly bears, despite their status as an endangered species.

Arguably, too, wilderness is itself a part of biodiversity and not just a means to preserve it. Conservation biologists commonly define biodiversity to include species diversity, genetic diversity, and community diversity: the latter is defined as different biological communities and their associations with the physical environment (the ecosystem). But if biodiversity includes biological communities and eco-systems, their disappearance or development into something essentially different constitutes a loss of biodiversity.

So, conservationists must make hard choices to protect biodiversity. For example, some models of global climate change suggest that more than 10% of the plant species in many U.S. states will not be able to survive new climatic conditions in the next few centuries—they will have to migrate northward or die out. If species are in danger of going extinct in the wild because of global climate change, the last remaining individuals may have to be preserved by us; hopefully, wild populations of these species can then be reestablished in new protected areas where the climate is suitable for them. In these cases of ecological triage, we should arguably move from a wild species—not our creation, not domesticated—to a wild species, living and continuing to evolve in the wild. The goal should be a future in which species survive without having to be moved, monitored, or supported by us and in which we rein in the human overpopulation, overdevelopment, and overconsumption that make such biodiversity management necessary. By managing ourselves more wisely, we can limit the need to manage wild nature.

IV. BIODIVERSITY PROTECTION AND ANIMAL RIGHTS

Conservationists typically argue that preserving species and whole biological communities should take precedence over preserving individual animals. In practice, this means that exotic animals may be sacrificed when necessary to preserve native species and overpopulous native or exotic animals may be sacrificed to preserve land health. These positions have provoked arguments with animal rights advocates, who might seem to be conservationists' natural allies.

For example, U.S. government agencies judged the continued existence of the endangered plant Santa Barbara live-forever (*Dudleya traskiae*) to be more valuable than the common rabbits that had been introduced on its island home. The rabbits, which fed on the plant's fleshy leaves, were killed to stop the destruction of this fragile plant species. In another example, state officials sanctioned the culling of hundreds of deer in the Quabbin Preserve in central Massachusetts to prevent over-browsing as a way to allow tree regeneration and protect water quality. Protection of the overall health of the forest and continuation of basic ecosystem functioning (forest regeneration, water purification) outweighed the interests of individual deer.

A

FIGURE 3   Conservation management: intervention versus leave-it-alone. (A) Heathland in
protected areas of Cape Cod, Massachusetts, is regularly burned to maintain open habitat and
protect wildflowers and other rare species. (Photograph by P. Dunwiddie.) (B) This old-
growth stand in the Olympic National Forest in Washington is the result of many years of
keeping human disturbance to an absolute minimum. (Photograph by Thomas Kitchin/Tom
Stack & Associates.)

Such practices seem mere common sense to many conservationists, but many animal rights advocates argue that they are misguided, for two reasons. First, animals (at least some higher animals) have rights or interests based on their consciousness or sentience (their ability to feel pleasure or pain). This means that we cannot sacrifice them casually for other goals, such as a healthy forest. Second, species and forests do not have rights or interests, because these complex aggregates are too loosely organized (see earlier discussion). To sacrifice the real interests of a higher animal to the bogus interests of a species or biological community is thus ethically mistaken.

Conservationists' answer that as we learn more about biodiversity, we realize that its greatest value lies not in individuals but in the wholes that those individuals help constitute. It is the species that persists and evolves rather than the individuals, which come and go. It is the natural community that sustains and generates new forms of life. While an ethic based on individual rights best specifies appropriate human interactions, a less individualistic ethic is needed to properly value the natural world.

This ethical shift seems to be necessary for effective conservation. However, animal rights advocates have given strong arguments for respecting individual higher animals such as dogs, cats, rabbits, and wild horses. Conservationists, who usually value species over individuals, may concede that individual animals should not be sacrificed casually, but only if it truly is necessary to preserve species or the health of biological communities. They might add that individual animals should always be culled humanely, that is, with a minimum of suffering. This would rule out most trapping and much sport hunting, although these activities are routinely defended on conservation grounds.

For example, Quabbin Preserve managers might carefully monitor deer populations and tree regeneration and authorize hunting where appropriate. All deer kills would have to be performed humanely, and in some years no hunts may be authorized. Whereas a traditional wildlife management position would authorize hunting game birds and trapping beavers, mink, and muskrats as long as this was done sustainably, a position responsive to animal welfare might not allow any of these activities, unless they could be justified with reference to the good of the forests or particular species.

Conservation and animal rights ethics both typically lead beyond anthropocentrism—the belief that only human interests matter—and they are usually comple-

mentary. For example, a concern for individual animals leads logically to vegetarianism, which prevents much animal suffering and bloodshed. But since "eating lower on the food chain" is less energy and land-use intensive, it also helps conserve resources and preserve biodiversity. Limiting consumption and distinguishing between needs, comforts, and luxuries are precisely what are needed to carry out both ethics. Nothing prevents individuals from living highly consistent lives that incorporate both ideals. We may act to limit animal suffering and preserve wild nature so that the entire realm of biodiversity can flourish.

V. BIODIVERSITY PROTECTION AND HUMAN RIGHTS

Conservationists often argue that individual organisms, species, and biological communities have intrinsic value—that is, value in addition to their usefulness to human beings—that obliges us to preserve them. This means human interests should sometimes be sacrificed to preserve biodiversity. For example, a developer should not build a housing project that destroys critical habitat for an endangered songbird species—whether or not it is legal to do so. The continued existence of this species is more important than the developer's profits. His less-than-vital interests should be foregone or pursued in ways that do not sacrifice something infinitely more valuable: a natural species. In many cases, however, preserving biodiversity and furthering human interests are complementary goals, especially if we take a broad view of human interests. Present and future generations have strong material, aesthetic, scientific, recreational, and spiritual interests in preserving biodiversity.

A. Owls and People

The coniferous forests of the U.S. Pacific Northwest are some of the most majestic on earth, with thousand-year-old trees towering over crystal clear streams filled with salmon. As commercial timber harvesting escalated after World War II, the big trees fell and species dependent on old-growth forests such as the northern spotted owl (*Strix occidentalis caurina*)(Fig. 4), the marbled murrelet (*Brachyramphus marmoratus*), and various salmon species (*Oncorhynchus* spp.) declined dramatically. In response, environmentalists staged logging blockades and tree sittings during the 1980s and suc-

FIGURE 4   Natives. (A) The northern spotted owl is an indicator species for old-growth forests of the Pacific Northwest, a habitat coveted for its rich timber stands. (B) Demonstrators protest government and corporate policies that destroy forests and threaten species extinctions.

cessfully sued the U.S. Forest Service under the Endangered Species Act, resulting in drastic limitations on tree harvests in the 1990s.

This contentious political battle was often billed as "owls versus people," but such a simplistic formulation was misleading. Reduced harvests resulted in the loss of thousands of logging jobs, but some reduction was inevitable, given unsustainably high harvest levels based on a one-time cutting of thousand-year-old trees. Logging reductions helped preserve and create jobs in fishing and tourism and enhanced watershed maintenance, flood control, and other objectives. A calculation based on the full spectrum of long-term human interests would likely have resulted in substantially decreased timber harvesting. Still, old-growth forests do have great economic value as timber, and preserving more in reserves rather than cutting the big trees and moving to high-yield forest plantations may indeed lead to lower profits for industry and fewer overall jobs.

Are such trade-offs ever justifiable to prevent the extinction of species or important biological communities? Forestry industry executives and loggers' unions said no. The interests of a logger in supporting his family should supersede the interests of animals or trees. Logging restrictions to protect other species are wrong because they lead to real human hardship for loggers and their communities. Environmentalists insisted that more logging jobs and higher corporate profits did not outweigh the permanent loss of species and the old-growth temperate rain forest itself. They also noted that industry and government had made no efforts to preserve logging and mill jobs when technological changes, log exports, or overharvesting had led to job loss in the past (see Fig. 4).

These considerations suggest a solution to the dilemma: harness the economic system to maximize both human interests and biodiversity protection. Whether we focus on intrinsic or instrumental value, the human communities of the Pacific Northwest have an interest in preserving their remaining old-growth forests. Fortunately, the regional economy generates enough wealth to provide laid-off workers with guaranteed health care, generous severance pay, and job retraining for those who want it. The big trees could be preserved and displaced workers supported by their fellow citizens. Such a course would not reconcile all conflicting interests or solve all problems: economic "winners" would have to be taxed to support economic "losers," some loggers would have to give up a valued livelihood, and some businesses would have to forego profits or even fold. But many of the costs of environmental protection would be borne by society as a whole, which benefits from the preservation of biodiversity.

B. Property Rights

When environmentalists won suits to limit logging in the Pacific Northwest, the resulting bans affected both private and public lands. To some observers, the restrictions on private lands were unjustified. In their view, the right to own and control property is essential. Dictating to individuals or corporations what they can or cannot do on their own land is seen as an intolerable abridgment of that freedom. In devising a plan to continue tree harvesting in the region, the U.S. government focused preservation efforts on public lands, allowing maximum freedom to private landowners.

Does private property ownership give owners unlimited control over their land, even if their actions contribute to the extinction of a rare or endangered species? Historically it often has, but morally it does not. Both the intrinsic value of species and duties to our fellow citizens to preserve a common biological heritage argue against such actions, regardless of how much profit must be foregone. Land ownership confers both rights and responsibilities.

What then of a small landowner who needs to cut an old-growth stand in order to send a child to college or keep the land in the family? Clearly as the need becomes greater, the ethical justification increases, even if some loss of biodiversity occurs. Hopefully, though, this small owner is not so poor that he must view his land solely as a money-generator. The ideal is some compromise in which the land is used in a responsible way that also preserves biological diversity.

Finally, it is important to distinguish between individual small owners and the large timber corporations that rival some national governments in the size of their landholdings. Arguments for preserving the freedom of small forest owners do not justify laissez-faire policies for large corporate landowners. Corporate managers' personal freedoms are not at stake in the same way, and their management decisions have vastly greater effects on natural and human communities. No government that cares about preserving its citizens' biological heritage will fail to regulate the environmental impacts of large corporate landowners.

C. Wilderness in Less-Developed Countries

Recently, some writers have argued that biodiversity preservation is a specifically North American or Western preoccupation, whose promotion in less-developed nations amounts to cultural imperialism. According to Ramachandra Guha, for example, wilderness preservation is inappropriate and unnecessary in these countries, whose peoples face more pressing environmental issues centered on meeting basic human needs. He argues that:

> the setting aside of wilderness areas has resulted in a direct transfer of resources from the poor to the rich [in the less-developed nations]. Thus, Project Tiger, a network of parks hailed by the international conservation community as an outstanding success, sharply posits the interests of the tiger against those of poor peasants living in and around the reserve. The designation of tiger reserves was made possible only by the physical displacement of existing villages and their inhabitants; their management requires the continuing exclusion of peasants and livestock . . . transplant[ing] the American system of national parks onto Indian soil. In no case have the needs of the local population been taken into account, and as in many parts of Africa, the designated wildlands are managed primarily for the benefit of rich tourists. (Desjardins, 1999)

There is some truth in Guha's claims. Several of Project Tiger's reserves were built around old hunting preserves from which the poor had long been excluded, while others were set up on former state or communally owned lands and displaced numerous villages, causing real hardship to thousands. Grazing and tree cutting in the core areas of the reserves were prohibited, both to limit human–tiger conflicts and to preserve a more complete flora and fauna.

On the other hand, India's wild tigers were clearly headed for extinction when Project Tiger was undertaken: between 1900 and 1972, the year it was initiated, the tiger population fell from tens of thousands to just 1800 individuals. Strictly limiting tiger hunting had not halted this decline. Habitat conservation was clearly necessary to protect the tigers over the long term and, given the requirements of tigers and their inevitable conflicts with humans, this had to involve some displacement and restrictions on local inhabitants.

Should Project Tiger have been attempted? Guha suggests not: he believes it represented an unjust appropriation of the resources of poor people. At the same time, a focus on tigers went hand in hand with ignoring

"environmental problems that impinge far more directly on the lives of the poor—e.g., fuel, fodder, water shortages, soil erosion, and air and water pollution . . . [which are] far more pressing environmental problems within the Third World" (Desjardins, 1999). Others disagree, arguing that habitat preservation and restrictions on human economic use of this habitat were necessary and appropriate responses to the imminent extinction of the Bengal tiger. This would have been a great cultural loss to the Indian people and an unjust destruction of an intrinsically valuable species. Many conservationists do not accept the extinction of species as a "less pressing" problem than those mentioned by Guha, because they do not accept his exclusive focus on human interests, nor his narrow definition of those interests (Fig. 5).

Recently, wildlife managers in the less-developed countries have been experimenting with ways to give local people a stake in the success of wildlife and park conservation, with encouraging results. In Africa, these efforts have included disbursing a percentage

FIGURE 5 Valuing biodiversity in less-developed countries. Buddhist priests in Thailand offer prayers and blessings to protect communal forests and sacred groves from commercial logging operations.

of park revenues directly to local villages, increasing efforts to hire locals as guides and forest wardens, compensating for crop damage done by wildlife straying outside of parks, and giving local people input into protected areas management. In India, beginning in the late 1980s, the Ranthambhore Foundation and other groups began work to promote sustainable development in the villages on the periphery of the Project Tiger reserves. This has included replanting denuded forests and improving fodder—to decrease pressure to graze and cut firewood within the reserves—as well as efforts to educate local children in natural history and conservation principles.

Justice demands that local people be treated with respect and their interests considered in all programs to protect wildlife and wildlands. Such respect and consideration will not solve all problems, however, particularly in a country as overpopulated as India, whose human population has increased from 280 million to 910 million in this century. In some instances, if we want to preserve wildlife, we will have to sacrifice human interests.

Overall, securing human rights and furthering essential human interests should strengthen efforts to preserve biodiversity. The United Nations General Assembly has affirmed that all human beings have fundamental rights to clean air, clean water, pure food, and a healthy environment generally. The measures needed to secure these human rights would also benefit wildlife and natural systems. Environmentalists have argued that future generations have a right to an undiminished natural heritage. Respect for this right would halt many environmentally destructive development projects. Natural areas are often threatened by remote governments and corporate managers, while their strongest defenders tend to be local inhabitants who know and love these areas. Securing political rights and economic justice for local inhabitants should help preserve these areas.

VI. THE IDEAL OF THE CONSERVATION BIOLOGIST

Four hundred years ago, modern science's great pioneer and prophet Francis Bacon located the primary value of science in its creation of a powerful technology for controlling nature. "The End of our Foundation is the knowledge of Causes, and secret motions of things," Bacon wrote in *The New Atlantis,* "and the enlarging of

FIGURE 6 Role models. Primatologists Dian Fossey (left), Jane Goodall (center), and Birute Galdikas began by studying animal behavior, but eventually devoted themselves to conservation activism.

the bounds of Human Empire, to the effecting of all things possible." Without denying science's utilitarian benefits, conservation biologists attempt to move in a different direction, developing the knowledge needed to preserve nature rather than to change, control, or exploit it.

Both historical and contemporary examples attest that science can be carried out in this spirit. Aldo Leopold and Rachel Carson, Paul Ehrlich and E. O. Wilson, and many others from a wide variety of disciplines and backgrounds have combined contributions to science, literature, education, public policy, and conservation. The challenge for the scientist is to personally cultivate a loving knowledge of nature and to speak about nature in a way that promotes its celebration and protection (Fig. 6).

A. Professional Decisions

If conservation biologists are to help successfully preserve biodiversity, they must take on several active roles in addition to their scientific pursuits. For a start, they must become more effective educators in public forums. Conservation biologists often teach college students and write technical papers addressing environmental issues, but they need to reach a wider range of people by also speaking in villages, elementary and secondary schools, parks, and neighborhood gatherings. They must spend more time writing articles and editorials for newspapers and magazines, as well as speaking on radio, television, and other mass media. Because tenure, promotion, and professional prestige often accrue for technical work and peer-reviewed science but not for more popular and disseminative work, conservation biologists may face difficult choices when deciding whether to devote time to the latter.

Similarly, research projects describing habitat requirements or methods of transplanting an endangered species may have important conservation implications, but little potential to make basic advances in ecological or biological theory. Again, hard choices between professional glory and conservation usefulness may be necessary. As conservation biology becomes more established as a discipline, the system of rewards in place for practical conservation work will hopefully improve. Tenure and promotion committees at universities should acknowledge the value in community outreach and research that addresses local conservation issues.

Conservation biologists must also become politically active. Involvement in the political process allows conservation biologists to influence the passage of new laws to support the preservation of biological diversity or to argue against harmful legislation. Though much of the political process is time-consuming and tedious, it is often the only way to accomplish major conservation goals, such as acquiring new parks and reserves. Conservation biologists need to master the language and methods of the legal process and form effective alliances with environmental lawyers, citizen groups, and politicians. With their detailed knowledge of specific organisms and ecosystems, biologists are well placed to alert the general public to threats to biodiversity. They may also present this natural heritage in an appealing and inspiring way that leads to its preservation.

Such outreach and activism take time away from the pursuit of pure science. But many believe that at this point in human history, we no longer have the luxury to pursue knowledge for its own sake. To learn about biological diversity today is to learn about the threats facing it. These threats demand an active response.

Fortunately, there is room for many types of work within conservation biology. Field and laboratory biologists who perform detailed natural history studies and genetic analyses on endangered species are necessary; so are hands-on managers who put together species

recovery programs and coordinate interagency partnerships to facilitate them; so are teachers and writers who popularize wild nature; and so are political firebrands who alert the public to biodiversity losses and corporate environmental crime.

B. Personal Choices

What about conservation biologists' private lives? Should a conservation biologist own a sports utility vehicle, for example? It gets terrible gas mileage, yet it might greatly facilitate her field work. Should a conservation biologist take a spectacular eco-tourist vacation to Amazonia? The trip uses prodigious resources, yet he might gather information and take slides to educate audiences back home about important environmental issues. Then again, such reasoning may hide our real motivations: to drive in maximum comfort, to travel and enjoy ourselves. There is nothing wrong with comfort and enjoyment per se, but such high-consumption activities take a great toll on world biodiversity. Perhaps a conservation biologist should set a more restrained example.

Integrity is a key human virtue. It means living in conformity with our values and resolutely striving toward worthwhile goals. As spokespeople for biodiversity protection, conservation biologists will make more converts if they practice conservation in their own lives. Above all, this means limiting their personal consumption of resources through owning fewer things, recycling, using public transportation, and taking other appropriate actions. However, it is not necessary to put on a hair shirt or engage in a quest for absolute purity. Showing that a life devoted to conservation can be stimulating and enjoyable is itself an important conservation message.

Should a conservation biologist dive into political debates? Give up time in the lab to meet with a group of concerned citizens? Drive a car to work? Eat meat? Have children? These difficult professional and personal questions must be faced by conservation biologists and other committed conservationists. Responses may legitimately differ, but we should not forget that the fate of world biodiversity depends, in part, on our answers.

See Also the Following Articles

BIODIVERSITY AS A COMMODITY • CONSERVATION BIOLOGY, DISCIPLINE OF • CONSERVATION EFFORTS, CONTEMPORARY • ENVIRONMENTAL ETHICS • HUMAN EFFECTS ON ECOSYSTEMS, OVERVIEW • PROPERTY RIGHTS AND BIODIVERSITY • SUSTAINABILITY, CONCEPT AND PRACTICE OF • WILDLIFE MANAGEMENT

Bibliography

Callicott, J., and Nelson, M. (eds.). (1998). *The Great New Wilderness Debate*. University of Georgia Press, Athens.

Carson, R. (1962). *Silent Spring*. Houghton Mifflin, Boston.

Costanza, R., Norton, B., and Haskell, B. (eds.). (1992). *Ecosystem Health: New Goals for Environmental Management*. Island Press, Washington, D.C.

Daley, H. (1996). *Beyond Growth: The Economics of Sustainable Development*. Beacon Press, Boston.

Desjardins, J. (1999). *Environmental Ethics: Concepts, Policy, Theory*. Mayfield Publishing, Mountain View, California.

Grumbine, R. (1992). *Ghost Bears: Exploring the Biodiversity Crisis*. Island Press, Washington, D. C.

Karr, J., and Dudley, D. (1981). Ecological perspective on water quality goals. *Environmental Management* 5, 55–68.

Lear, L. (1997). *Rachel Carson: Witness for Nature*. Henry Holt, New York.

Milbrath, L. (1989). *Envisioning a Sustainable Society: Learning Our Way Out*. State University of New York Press, Albany.

Munda, G. (1997). Environmental economics, ecological economics, and the concept of sustainable development. *Environ. Values* 6, 213–233.

Rolston, H., III. (1994). *Conserving Natural Value*. Columbia University Press, New York.

Segal, J. (1999). *Graceful Simplicity: Toward a Philosophy and Politics of Simple Living*. Henry Holt, New York.

Sutherland, W. (ed.). (1998). *Conservation Science and Action*. Blackwell Science, Oxford, United Kingdom.

Westra, L., and Lemons, J. (eds.). (1995). *Perspectives on Ecological Integrity*. Kluwer Academic Publishers, Dordrecht, Netherlands.

Wilson, E. O. (1994). *Naturalist*. Island Press, Washington, D.C.

ETHNOBIOLOGY AND ETHNOECOLOGY

Gary J. Martin
The Global Diversity Foundation, Morocco

GLOSSARY

analysis, emic and etic Concepts derived from the linguistic terms "phonetics" (representing speech sounds by precise and unique symbols and by technical descriptions of articulation, as practiced by trained linguists) and "phonemics" (characterization of speech through a minimal number of symbols, typically recognized by the speakers of a language). By extension, etic refers to the external explanation of cultural knowledge and practice (such as the use of Linnean taxonomy or scientific nomenclature to describe local useful plants), whereas emic denotes the internal perspective of local people (e.g., ethnobiological categories and nomenclature).

economic botany As originally conceived, a branch of applied botany that arose during the colonial period to identify and characterize economically important plants and the products derived from them. Currently, it is a scientific endeavor that seeks to document the properties of useful plants through agronomic, archaeological, ecological, ethnobotanical, genetic, historical, phytochemical, and other empirical approaches. It overlaps broadly with ethnobiol-

ogy because both fields have witnessed a similar development in theory and methodology in recent years.

ecosystems, anthropogenic and natural Communities of organisms and their environment formed either through human action or through natural processes. In practice, it is difficult to establish the extent to which an ecosystem is anthropogenic or natural, reflecting the current and historical impact of people on the environment.

ethnobiology A term coined in 1935 that has been defined as the study of the reciprocal interactions between people and the biological organisms in their local environment and, recently, as the study of biological sciences as practiced in the present and the past by local people throughout the world. Many researchers consider that ethnobiology comprises numerous subfields, such as ethnobotany, ethnoecology, ethnoscience, and ethnozoology, but there is no consensus on this point.

ethnobotany and ethnozoology Approaches to studying the reciprocal interactions between people and the plants and animals in their local environment. This definition has been criticized as broad and open-ended, but it captures the common goals of analyzing traditional biological knowledge and assessing human impact on the environment. These approaches include subfields such as paleoethnobotany and paleoethnozoology, which evaluate archeological evidence on the past interactions between people, plants, and animals.

ethnoecology Typically defined as the study of local knowledge and management of ecological interactions. Recently, some researchers have proposed an alternate definition, considering ethnoecology as an emerging field that focuses on local peoples' perception and management of complex and coevolved relationships between the cultural, ecological, and economic components of anthropogenic and natural ecosystems. It is concerned with the interaction between knowledge, practice, and production, and it is oriented toward applied research on conservation and community development.

ethnoscience Arose as a minor subfield of ethnography concerned with recording in great detail local peoples' knowledge of biological organisms and the physical environment. Later, the term came to be used in a more restricted sense by cognitive and linguistic anthropologists to refer to local classificatory systems (as an object of study) and their semantic analysis (as a methodological approach). In France, the term is used to refer to ethnobiological studies in general.

indigenous, local, and traditional Adjectives used by anthropologists, ethnobiologists, and other academics to describe people, practices, and knowledge. Indigenous denotes people (and their cultural practices and knowledge) who claim to be the original or long-term inhabitants of a particular place, in contrast to more recent colonizers. Traditional refers to established lifestyles, practices, and beliefs that guide cultural continuity and innovation—a definition that recognizes that traditions are always in a process of adaptation and change. Local, preferred by many researchers because it is the broadest and least value-laden term, indicates cultures that are found in a specific part of the world. It is commonly used to refer to people, whether long-term residents or recent arrivals (rural or urban), who make a living from the land and are knowledgeable about the biological resources in their environment.

ALTHOUGH THE 100th anniversary of ethnobotany (coined in 1896), the golden anniversary of ethnobiology (first used in 1935), and the silver anniversary of ethnoecology (appearing in 1954) have passed, there is no consensus on the precise definition of these fields. This is explained in part because of their relatively recent origin and the current surge in their theoretical, conceptual, and methodological refinement. Disagreement over definitions is typical of multidisciplinary

fields; in the words of Brent Berlin, ethnobiology "combines the intuitions, skills, and biases of both the anthropologist and the biologist, often in quite unequal mixtures."

In one sense, ethnobotany, ethnobiology, and ethnoecology are new terms for old practices. People have been exploring the usefulness of diverse plants, animals, and ecosystems since the dawn of humanity. Documentation of local people's perception of the environment emerged slowly over thousands of years as scholars from many cultural traditions recorded local ways of classifying and using plants and animals. The onset of European colonization of Africa, Asia, the Pacific, and the New World gave added impetus to the study of local knowledge of tropical and temperate organisms and ecosystems.

Toward the end of the nineteenth century, academics began to use the prefix *ethno-* to refer to the way that local people view the natural world, in contrast to the perspective of natural scientists trained in universities. They coined terms such as "ethnobotany" (first used in print by Harshberger in 1896) and "ethnozoology" to describe these emerging fields of study that crossed the boundaries of natural and social sciences. Interest in traditional environmental knowledge continued apace in the early twentieth century, and in 1935 Castetter coined the term "ethnobiology," setting as its agenda the systematic analysis of data collected by ethnobotanists and ethnozoologists to achieve a deeper understanding of local peoples' knowledge and lifestyles. Economic botany gained importance as a parallel field focused on useful plants and the products derived from them. In 1954, Harold Conklin proposed the term "ethnoecology," originally conceived as a holistic and integrated approach to understanding local ecological knowledge and practice on their own terms, even while drawing on the concepts and methods of diverse scientific disciplines. A focus on classificatory systems and the linguistic and anthropological methods used to analyze them gave high visibility to an approach called ethnoscience.

In the 1980s and 1990s, further development of these various lines of research gave rise to new definitions, innovative theoretical orientations, and sophisticated qualitative and quantitative methodological approaches applied to local knowledge of the environment. In addition, ethnobiology expanded beyond its original geographical borders as the field gained importance in countries such as China, India, and Mexico. There is currently a new synthesis emerging—as yet without consensus—that defines ethnobiology as the study of biological sciences as practiced by local people throughout the world, comprising both empirical knowledge (*savoir*) and technical know-how (*savoir-faire*), and inclusive of subfields such as economic botany, ethnobotany, ethnoecology, and ethnozoology.

I. HISTORICAL DEVELOPMENT OF ETHNOBIOLOGY

Despite the insights provided by archaeological and historical linguistic studies, setting an even approximate date of the emergence of local biological knowledge is a matter of opinion. A detailed understanding of the natural world was key to the independent emergence of plant and animal domestication over a period ranging from 8500 B.C. in southwest Asia to 2500 B.C. in the eastern United States. However, environmental knowledge reaches even further back into history, when hunting and gathering dominated subsistence activities.

Some researchers would place the beginning of human ecological knowledge at the dawn of humanity, approximately 7 million years ago. Early human ancestors, who lived on the African continent 2.5 million years ago, apparently fashioned stone tools for harvesting and processing food, probably allowing them to adapt to new environmental conditions. It is widely assumed that humans have been observing natural phenomena, distinguishing between biological organisms and discovering their uses ever since the emergence of early *Homo sapiens* approximately 500,000 years ago. The archeological record reveals that by 50,000 years ago, Cro-Magnons had developed technologies for construction, fishing, gathering, and hunting that were dependent on a detailed understanding of plants, animals, and other elements of the natural environment.

Just as no one knows exactly when ecological knowledge appeared on the cultural landscape, there is no clue when the original precursors of ethnobiologists came on the scene. The first critical observations of other peoples' ways of perceiving nature are probably as old as culture contact itself. Because these observations went unrecorded, the origin of the study of traditional biological knowledge is lost in history.

In the absence of other evidence, the historical roots of ethnobiology can be seen emerging over a period of several thousand years, when students of natural history from Greek, Roman, Egyptian, Chinese, Indian, Arabic, Native American, European, and other cultures began to record popular beliefs in scholarly texts. Original studies focused on medicinal botany, agriculture, and horticulture—activities that drew heavily on the knowledge of local people.

An overview of these early texts and later works reveals that ethnobiology and natural history have evolved—much like the biological species and ecosystems that are their focus—through a process of punctuated equilibrium. Certain historical periods are marked by an intensive effort to expand empirical knowledge of natural phenomena, often by incorporating local lore, whereas other epochs are characterized by an unquestioned acceptance of published works.

A. Early Scholars in Europe

Academics often trace their intellectual history to the era of Greek philosophers, who lived more than 2300 years before our time. This is when classical botany and zoology were brought into existence by scholars such as Aristotle, who sought to summarize all current knowledge about plants and animals in encyclopedic works. It was as part of this endeavor that scholars in Europe first made a systematic study of what local people knew about the environment. In part, Aristotle and other early naturalists such as Theophrastus—who, as author of *De Historia Plantarum* and other works, is considered the father of botany—rejected many local supernatural beliefs in their quest to understand the natural world. Simultaneously, they drew on common-sense explanations and empirical knowledge of local people when describing the classification and use of plants and animals.

Among the people who followed in the footsteps of these early naturalists is Dioscorides, a military physician born in Asia Minor in the first century A.D. He wrote *De Materia Medica*, a treatise on medicinal plants which was the standard reference of botanists, medical doctors, and other scholars in Europe for 1500 years. Apart from drawing on previous herbals, Dioscorides learned much about herbal remedies by interacting with local people he encountered during his wide-ranging travels with the Roman army in the Mediterranean region. Pliny the Elder, a Roman scholar who was one of Dioscorides' contemporaries, recorded extensive plant lore in his 37-volume encyclopedia called *Historiarum mundi or Natural History*. He devoted 9 volumes to medicinal plants, making frequent reference to traditional practices and knowledge.

B. The Doldrums of the Middle Ages

The documentation of local knowledge that marked the origin of biological thought in Greek and Roman Antiquity was much less evident in the Middle Ages in Europe. The decline of the Roman Empire virtually halted scholarly research on natural history and resulted in the destruction of much existing literature of the epoch. Throughout the Middle Ages, Europeans based their studies of medicinal plants almost entirely on the works of Theophrastus, Pliny, Dioscorides, and other early naturalists. Physicians from across the continent relied heavily on *De Materia Medica*, often trying unsuccessfully to match the local flora to the approximately 600 Mediterranean species described by Dioscorides instead of documenting the popular knowledge of their own region. As anthropologist Scott Atran (1990) summarized,

> After Aristotle, the practice of copying descriptions and illustrations of living kinds from previous sources superseded actual field experience in the schools of late antiquity. Well into the Renaissance, scholastic "naturalists" took it for granted that the local flora and fauna of northern and central Europe could be fully categorized under the Mediterranean plant and animal types found in ancient works. Herbals and bestiaries of the time were far removed from any empirical base.

C. The Golden Age of the Moors

Despite this stagnation on the part of European scholars, general knowledge of medicinal plants was enriched by the flow of information coming from the Arab world, particularly through Spain (Andalucia), Sicily, and North Africa. Although dedicated in part to translating the works of Aristotle, Theophrastus, Dioscorides, and other classical writers, Moorish and other medieval scholars in these regions pursued empirical research that they applied to practical ends in agriculture, astronomy, botany, mathematics, medicine, and other fields. Although there are records of Arabic writings on botany dating to the ninth century, it was particularly in the twelfth and thirteenth centuries that scholars became prolific in recording precise original observations on plant and animal biology, conducting experiments on agricultural crops, and attempting to classify plants systematically. One of the key early scholars was Abû-l-Kheyr Al-Ichbili, apparently the "anonymous botanist of Seville" of the latter half of the twelfth century who wrote the *'Umdat at-tabîb,* a botanical treatise that contains detailed descriptions of the anatomy, habitat, and local names of plants. Maimonides, a Jewish scholar resident in southern Spain during the twelfth century, contributed works such as *Book Explaining Medicinal*

Drugs and *Treatise on Poisons.* The thirteenth century brought the works of Rachid-eddin Ibn Es-Sûri in the Machreq (eastern Arabic region), who is credited with the first herbal containing illustrations, which were made by a collaborating artist from both living and pressed plants. Ibn Es-Sûri traveled widely, describing formerly undocumented plants from Syria, Palestine, and Egypt that he discovered by interacting with local people of the region. His efforts were matched in the Maghreb (western Arabic region) by Abû-l-'Abbas En-Nabati—who documented local plants and their uses during his extensive travels in what is now Spain, North Africa, Syria, and Iraq—and later scholars such as his student Ibn Al-Baytar, author of the *Treatise on Simples or Jami' al-mufradat,* which contains information on the synonymy (including Berber and other local names), description, properties, and uses of approximately 1400 species. Throughout this period, descriptions of specific plants were drawn from multicultural sources, including Berber plant knowledge and the traditional practices of Jewish pharmacists who lived throughout Europe and North Africa, passing their profession from one generation to the next. As the Moors were forced from Andalucia and other parts of Europe, Arab science fell into decline. Later botanical treatises, including the *Hadîqat al-azhâr* of the sixteenth century Moroccan scholar and medical doctor Al-Wazir Al-Ghassani, were largely based on the *'Umdat at-tabîb* and other early works.

D. Scholars from Other Cultural Traditions

A similar pattern of initial empirical discovery and cross-cultural learning mixed with centuries of uncritical acceptance of written works is evident in other cultural traditions. In China, the first scholarly studies of traditional biological knowledge are thought to date at least to the fifth century B.C., approximately 200 years before early Greek philosophers began recording their ideas about botany and medicine. During this epoch, the Chinese philosopher Confucius is said to have encouraged his students to study ancient knowledge, including traditional names of plants and animals.

In the first or second century A.D., about the time that Dioscorides was afoot in the Mediterranean, Chinese scholars published the first of many *materia medica,* referred to generally as *bencao* (from a combined term for tree and grass) in Chinese, which contain information on minerals, plants, and animals used traditionally in the treatment of illness. Medical doctors of the fifth century A.D. revised these early *bencao,* providing a major reclassification of the various types of natural medicines used at the time. The resulting pharmacopoeia went unchanged and unchallenged for many centuries, much as the classic work of Dioscorides in Europe. At the end of the sixteenth century, a Chinese doctor named Li Shizhen began to emulate the practice of empirical observation that Confucius had advocated many centuries before. His major work, the *Bencao Gangmu* or *Compendium of Materia Medica,* contains information on more than 10,000 herbal remedies that he obtained not only by studying ancient texts but also by traveling to the countryside to talk with people.

Ayurveda, a system of medicine which putatively began in India during the sixth century B.C. and spread to Sri Lanka by the third century B.C. and into Tibet by the seventh century A.D., was partially based on traditional knowledge. According to folklore, shepherds and forest dwellers familiar with the types and properties of medicinal plants first discovered the remedies used in this oriental medical practice. Their knowledge was discussed in various literary religious works called *vedas* (from the Sanskrit word for knowledge), which were apparently written in India approximately 3200 years ago, after millennia of oral transmission. Ayurvedic scholars later compiled additional empirical observations in a series of books referred to as the *Nighants,* or Vedic glossaries. During the subsequent period of foreign domination and internal conflict that brought innovation and documentation of local knowledge to a standstill, these standard texts of ayurveda remained unchanged.

Other ancient written sources that document local biological knowledge were in part the product of culture contact and changes in political and economic dominance. In the New World, for example, the Aztecs broadened their own sophisticated knowledge of medicine and agriculture as they sought tribute and learned of new useful plants from the different Mesoamerican cultures they conquered. The Aztecs cultivated many newly discovered species in extensive highland botanical gardens tended by people from various geographical regions of Mesoamerica. The depth and richness of the preconquest indigenous knowledge of the natural world are demonstrated by scholarly works, including the *Badianus Manuscript,* an illustrated herbal written in 1552 by two Aztecs who had been educated by Catholic missionaries. One author, Martin de la Cruz, was an indigenous physician who had acquired his medical knowledge empirically. The Mayas and Incas had similar literate traditions and they doubtlessly recorded some aspects of the ecological knowledge of the various ethnic groups they dominated at the height of their political power and cultural development. Many of these New World written sources of local knowledge were victims of the conquest, destroyed by overzealous missionaries and conquerors who wished to impose European culture, languages, and religion on the people of the New World.

E. The Renaissance and Exploration

During the sixteenth and seventeenth centuries, Renaissance botanists began to emulate the methods that Dioscorides had applied approximately 1500 years previously, bringing an end to the intellectual stagnation that characterized the Middle Ages in Europe. They carefully observed plants in the field and inquired about their local names and uses in Germany, Holland, Italy, and other parts of Europe. This experience served them well when faced with the influx of exotic species from areas of the world discovered and colonized by Europeans during this period. The diversity of biological organisms discovered by explorers stimulated Linnaeus, Darwin, and other natural scientists to formulate many of the concepts that are the building blocks of modern-day systematics and evolutionary studies. Although Linnaeus left some notes and sketches on the use of plants by local people, his greatest contribution to the future field of ethnobiology was the incorporation of notions of folk biology and nomenclature, including the concept of morphological affinity as a criterion for defining taxa, in the scientific classification of plants.

This Renaissance was the golden age of the European herbals. The shift from manuscripts (produced by hand) to wood-cut and metal-engraved herbals published in large numbers allowed new botanical knowledge to be disseminated widely.

The quest to exploit local knowledge and economically important species which went along with colonization inspired adventurers, missionaries, and natural historians to record their observations on traditional biological knowledge in many parts of both the New World and the Old World. As ethnobotanist Richard Ford (1978) described,

A rapid progression of expeditions came to North America to discover and to colonize, and the chronicles of adventure are a record of the utilitarian value of an unfamiliar landscape and the use the indigenous people made of it. Its economic potential certainly had priority to any interest in attitudes about the land. The observations ... provided the first natural history of North America and the bases for the beginning of ethnobotany.

From the sixteenth century onwards, researchers began to focus increasing attention on the biological wealth of tropical countries and the benefits it promised for Europeans. To this end, scholars drew on the knowledge of local people, who continually experiment with cultivated and managed species in anthropogenic ecosystems and wild plants harvested in natural ecosystems. Scholars consulted both written sources, such as Ayurvedic works and Chinese pharmacopoeias, and oral history to produce extensive encyclopedias of useful plants from around the world, ranging from the *Coloquios dos simples e drogas da India*, written by the Portuguese explorer Garcia ab Orta in 1563, to the 12-volume *Hortus Malabaricus* of Van Rheede published in the late 1700s and A *Dictionary of Economic Products from the Malay Peninsula* produced in 2 volumes by Burkhill in 1935.

Another notable example is the herbal of Rumphius, a seventeenth-century natural historian from Germany who spent nearly 50 years in Asia working for the Dutch East-Indies Company. Increasingly released from his administrative duties but still under the employ of the company, he focused his attention on studying useful plants, animals, and minerals in various regions that today constitute Indonesia. He provided descriptions of more than 700 medicinal or toxic plants, published posthumously in the six volumes of the *Herbarium Amboinense*.

F. Creation of New Fields

Although Rumphius and his counterparts approached natural history as a holistic phenomenon, scholarly activity in later centuries began to reveal a fragmentation of research into distinct disciplines, marking the beginning of reductionism in the sciences. Subdivisions of science became more clearly defined, and professional practitioners began to specialize in specific subfields. Theoretical science was increasingly applied to technological innovation, stimulating the growth of educational institutions and commercial enterprises.

For example, pharmacognosy (the study of naturally occurring compounds that can be used medicinally and in other ways) became recognized as a separate field of endeavor in the early nineteenth century. It focused on the identification, preparation, and commercialization of drugs, which mostly came from natural sources, espe-

cially plants. This enterprise, combined with tropical exploration, yielded such novel cures as quinine, which was successfully used to control malaria in Europe and other regions.

As research on traditional biological knowledge expanded in the late nineteenth and twentieth centuries, several lines of research became apparent, reflecting the diversity of researchers who began to appreciate the importance of the subject. The historical development of these various fields is intertwined, and the definition of each is in flux. There may be an emerging consensus—one supported in this article—that ethnobiology is the broadest field, comprising ethnobotany, ethnozoology, ethnoecology, and economic botany as subfields. Other colleagues would posit ethnoecology or even ethnoscience as the most inclusive term. In practice, this debate on terminology is less significant than the realization that these fields broadly overlap, and that the slight differences of opinion on definition are dwarfed by the general agreement on theory, concepts, and methods.

1. Ethnobotany and Related Fields

By the end of the nineteenth century, researchers began to recognize the study of traditional biological knowledge as a separate discipline. John W. Harshberger, a professor of biology, initiated the fashion of using the prefix *ethno* to indicate the study of local people's natural history. In 1896, he used the term ethnobotany in print, and it began to replace names such as "aboriginal botany" and "botanical ethnography" that had been used previously by other authors. In the words of Richard Ford (1978), after a "half century of scientific attention and an even longer history of casual observations," the study of other people's interaction with nature finally had a name and recognition as a distinct line of academic endeavor.

The emergence of ethnobotany, ethnozoology, and related fields coincided with important developments in the natural and social sciences toward the end of the nineteenth century. The diverse elements of natural history, including botany, zoology, pharmacognosy, and other fields, began to mature into distinct disciplines, each with separate methods and goals. Scott Atran (1990) characterized this as the "breakaway of science," a time when natural historians began to leave behind commonsense descriptions of natural phenomena—drawn in part from local peoples' perception and classification of nature—in order to embrace rigorous experimental methods. Social scientists began to focus on separate aspects of human society and culture, with the consequent emergence of fields such as anthropology, linguistics, and sociology.

2. Economic Botany

Nineteenth-century botanists, who focused primarily on the utility of plants and only secondarily on local culture, began to refer to their approach as economic botany. The goal of their research has been to document local uses of plants and to organize the resulting data according to the global system of plant classification. They have produced detailed works on plants employed by local people for food, medicine, textiles, utensils, and many other purposes (Schultes and Raffauf, 1990).

Research on the commercial value and utility of plants expanded as botanists from the United States and Europe explored the New and Old World tropics in search of products that would increase the wealth of developed countries and the well-being of people in general. Today, economic botanists continue to search for marketable products in tropical forests and elsewhere, but they are increasingly interested in how the commercialization of these resources can contribute to resolving the poverty, malnutrition, and diminished social status of local people as well as spurring economic development in developing countries. An increasingly important offshoot of this enterprise is bioprospecting (a term derived from "biodiversity prospecting"), the search for useful and novel products (including chemical constituents of medicinal value) from plants, animals, fungi, and other biological organisms. Other economic botanists are concerned with developing theoretical and methodological approaches to understanding the subsistence and commercial value of plant resources, now and in the past.

3. Ethnoscience

While botanists were establishing economic botany, anthropologists and other social scientists were developing a different perspective. In the tradition of ethnography developed by anthropologist Franz Boas, ethnoscience emerged as a minor subfield dedicated to recording in minute detail local peoples' knowledge of biological organisms and the physical environment. The subfield underwent a further transition in the 1950s and 1960s, when cognitive and linguistic anthropologists began to focus on the empirical categories, social rules, symbolic systems, and modes of behavior that reflect how local people perceive the natural world. These early anthropological studies formed the foundations for a new ethnoscientific approach that advocated rigorous analyses of ethnobiological knowledge, with particular emphasis on systems of ethnobiological classification. Thus, whereas economic botany emerged as a utilitarian practice firmly rooted in commerce and development (and later developed theoretical frameworks), ethnoscience arose as an intellectual endeavor oriented toward a deeper understanding of human culture and cognition.

4. Ethnoecology

Even though he is most associated with the development of the ethnoscientific approach, Conklin is credited with coining the term ethnoecology in 1954. Given the precedent set by terms such as ethnobotany and ethnozoology, it would be natural to assume that ethnoecology would refer to the study of local perceptions of ecological processes, such as nutrient cycling, vegetational succession, or the interactions between plants and animals. An increasing number of researchers propose a different definition, using the term to refer to local peoples' perception and management of the complex and coevolved relationships between cultural, ecological, and economic components of anthropogenic and natural ecosystems. This emerging subfield, much as the broader field of ethnobiology, is concerned with the interaction between knowledge, practice, and production and is oriented toward applied research on conservation and community development. Mexican ecologist Victor Toledo stated that the aim of ethnoecology should be the ecological evaluation of the intellectual and practical activities that people carry out during their appropriation of natural resources.

Although the definition of ethnobiology includes a reference to knowledge and know-how (*savoir* and *savoir-faire*), for ethnoecologists the distinction is between an ethnobiological *corpus,* local peoples' repertories of concepts, perceptions, and symbolic representations of nature, and *praxis,* the art, science, and skill of appropriating nature and biological resources. The interrelationship between knowledge and practice is manifested in production, as people apply their intellectual understanding of nature to the everyday tasks of farming, gathering, and hunting for subsistence and commercial purposes. In order to understand these complex interactions, ethnoecologists seek to elucidate how the management of anthropogenic and natural ecosystems—and the biological organisms they harbor—has arisen through a process of coevolution between the environment, knowledge, technology, social organization, and values of local peoples.

Although attractive conceptually, the development of this conception of ethnoecology has been limited by the lack of a unifying theoretical framework and a practical methodology. This distinguishes it from ethnobiology, which is developing a central organizing theory, an orientation toward hypothesis testing, and an increasingly elaborate set of qualitative and quantitative methods, drawn in part from ethnobotany, ethnozoology, and economic botany.

5. Ethnobiology

Clément (1998) proposes that the starting point for ethnobiology—as the field which integrates related approaches such as economic botany, ethnobotany, ethnozoology, and ethnoscience—is the 1860s, when the first designations for the field began to be used by American and European scientists. In a historical sketch that spans a period of more than 130 years, he discusses the origins, key theories, and methodological approaches of the main trends of ethnobiology. Although no such historical framework of a scientific discipline is without controversy and potential modification, Clément's synopsis is a serious effort to provide a detailed historical analysis of ethnobiology.

Clément divides the development of the discipline into three eras and seven periods (Table I). The preclassical period, from 1860 to 1953, is dedicated to gathering empirical data on the uses of plants and animals from an etic perspective and to the first syntheses that begin to define the scope of the discipline. During the classical period (1954–1980), there is a shift to studies carried out from an emic perspective and a particular focus on ethnobiological classification. An increase in collaborative work between academic specialists and local people and the formation of professional associations of ethnobiology characterize the postclassical period, from 1981 to the present. Later in the period, there is an increased focus on the appropriation and management of plant and animal resources and a concern for application of research results to the resolution of environmental and social problems. This historical review provides an appropriate starting point for considering the current trends in basic and applied ethnobiological research.

II. CURRENT TRENDS IN ETHNOBIOLOGY

Recent studies in ethnobiology can be classified in three general, interrelated areas. Documentation and analysis of uses of plants and animals is the oldest aspect of ethnobiology. A focus on knowledge, which became especially popular in the 1950s and 1960s, characterizes the study of perception and classification of the natural world. In the 1980s and 1990s, attention shifted to local management of biological resources and the environment, often drawing on ecological concepts and methods. These areas are combined in various degrees in the following major trends in ethnobiological research.

A. Cognitive Mapping

Ethnobiologists have documented local peoples' spatial conceptualization, including their ability to locate biological resources, discern landscape features, and identify different types of vegetation. This geographical literacy is linked to an aptitude for assessing the potential productivity of the environment, often through the recognition of plant species that indicate fertility or sterility of soils. This expertise derives in part from local peoples' sophisticated perception of how the various elements of the ecosystem (organisms, soils, climate, topography, etc.) form an interdependent whole. In addition, it is derived from their classification of key geographical landmarks that are labeled by specific toponyms, or geographical place names. Local views of the landscape are being integrated into geographical information systems, which in turn are useful in creating biological resource maps and management plans for specific areas. When united with ethnobiological inventories and studies of local categorization of ecological succession, community mapping allows researchers to test hypotheses on whether the highest number of useful plants and animals come from primary or secondary forest, or anthropogenic versus natural ecosystems.

Local peoples' mastery in selecting plant populations that yield the best fruit, the most potent medicine, or the best materials for construction is attributed in part to their grasp of the landscape. These mental maps also explain their adeptness at selecting the best place to cultivate the earth, create human settlements, or leave natural areas that maintain soil fertility, water purity, and other environmental benefits. Researchers have demonstrated that these skills can degenerate when local people find themselves pressured by economic needs to overexploit the resources available to them or are forced onto lands for which they are maladapted culturally.

The ability of some local people to integrate and recall complex information on the local environment often gives them special proficiency as stewards of community reserves and other protected areas. Their ability to assess the quality of useful biological resources and to locate areas where these organisms are found in

TABLE I

Features of Various Historical Periods and Stages in Ethnobiology[a]

Period	Stage	Dates	Features
Preclassic (1860–1954)	Economic uses	1860–1899	Studies of biological resources and their utility carried out by researchers affiliated with major museums and universities; general lack of appreciation of the sophistication of local knowledge and subsistence systems from an emic perspective
	Information gathering	1900–1931	Greater empirical depth in research but continued emphasis on economic uses of plants and animals; better appreciation of complexity of local knowledge and use of plants and animals, especially as reflected in systematic attempts to record local terminology, myths and beliefs, and knowledge of anatomy and behavior; emergence of comparative studies and standard methods
	The first syntheses	1932–1953	Emergence of ethnobiology as a distinct field of enquiry and appearance of the first syntheses that delimit its scope; increasing distinction between economic botany and ethnobotany, with the latter emphasizing the systematic documentation of local knowledge and management of plants; continued lack of recognition of scientific aspects of traditional biological knowledge
Classic period (1954–1980)	Emic knowledge	1954–1968	Emergence of ethnoscience, leading to a focus on the organization of knowledge systems from the local perspective, with insights from linguistics and empirical anthropological methods; relegation of the study of plant and animal resources to secondary importance; beginning of interest in ethnobiological classification and appreciation of the scientific basis of traditional knowledge
	Classification	1969–1980	Focus on ethnobiological classification, including principles of categorization and nomenclature, and the analysis of correspondence between scientific and local classifications; accumulation of evidence for the scientific basis of local biological knowledge; increasing interest in ethnobiology beyond the United States and Europe, especially in Latin America and the Pacific
Postclassic (1981 to present)	Associations	1981–1992	Production of major empirical works based on close collaboration between academic and local researchers; development of theoretical approaches beyond classification, including gender relations in resource use, cultural significance of plants, and historical reconstruction of ethnobiological knowledge systems; emergence of academic societies and specialized journals of ethnobiology, especially in developing countries
	Resource management	1993 to present	Publication of standard methods manuals, quantitative techniques, and innovative empirical studies; emergence of concern about applying ethnobiology to conservation and development; renewed interest in economic botany, including nutritional and medicinal benefits of plants, but incorporating novel theoretical and methodological approaches and informed participation by local people.

[a] Adapted from Clément (1998).

greatest density makes them indispensable members of research teams that seek to identify priority areas for conservation and management. Foresters and conservation biologists draw on this expertise when deciding how to zone natural areas according to various land-use options. Local peoples' classification of land units also plays a key role in justification for claims of ancestral domain and other forms of ownership of the lands they have long occupied.

B. Resource Management and Valuation

In order to ensure subsistence production and to earn a living, local people draw on their detailed ecological knowledge to manage the diverse microenvironments and biological organisms in their communities. Recent studies have shown that the process of plant domestication through selection of preferred varieties occurs not only in cultivated fields but also in other parts of the anthropogenic landscape. From the cultivation of these domesticated plants to the harvesting of wild useful species, local people engage in many ecological practices that are often energy efficient and sustainable, at least under traditional conditions. They capitalize on the consumptive use value of natural resources when they harvest plants and hunt animals for subsistence purposes and on the productive use value when they barter or sell agricultural and forest products. They also benefit, as does the whole world, from the nonconsumptive value that comes from ensuring the viability of ecosystem function in general, including watersheds, nutrient cycles, climate, soils, and other elements.

By investigating the link between local knowledge, practices, and production, ethnobiologists assess the value of anthropogenic and natural ecosystems and the rationality of resource harvesting decisions made by small-scale farmers and gatherers of forest products. Much of this research is carried out in 1-ha plots, agricultural fields, home gardens, and other measured study sites, resulting in quantitative assessments of sustainability and value. These methods allow researchers to evaluate the hidden costs of tropical forest destruction, the economic benefits derived from both subsistence and commercial use of wild species, and the environmental advantages of maintaining forest cover as a way of buffering local climate and preserving the purity of local air, water, and soil. Through these perspectives, conservation biologists, development specialists, and communities monitor the sustainability of current productive practices, propose new methods of managing fields and forests, and select new biological species that can be domesticated, cultivated, or gathered locally. Studies of local systems of resource management can also enrich the work of ecologists, who seek to restore the diversity and value of forest ecosystems damaged by mismanagement or natural catastrophes. In addition, evidence of long-term management of biological organisms and ecosystems reinforces local peoples' claims for traditional resource rights, including just compensation when novel biological resources are commercialized.

C. Scientific Covalidation

A key activity of ethnobiologists is to understand the rationale behind the way local people interact with the natural environment. In research laboratories, scientists carry out a broad array of analyses that seek to corroborate the efficacy of local uses of plants and animals, ranging from the identification of active compounds in medicinal plants to appraisal of the tensile strength of natural fibers and assessment of the nutrient content of wild foods. In the field, ecologists assess how the yields of agroecosystems compare with those achieved by large-scale monocultures and to what extent traditional methods of wildlands management are sustainable. Ethnobiologists compare ethnobiological categories with scientific taxa, judging the extent to which local biological classifications correspond to biosystematics. This range of activities is aimed at revealing the logic, from a scientific perspective, of the thought and practices of local people. Because much is understood about the classification of specific cultural domains,

such as plants, animals, soil, climatic zones, and vegetation types, there is a call for ethnobiologists to study local peoples' perceptions of ecological interactions, the reciprocal relationships of various elements of the ecosystem.

Scientific covalidation allows researchers to understand how to optimize the value—and also to ensure the safety and efficacy—of plants and animals that are consumed or commercialized locally. In addition, covalidation provides insights on how traditional knowledge can be incorporated in the management of protected areas. These studies reveal the breadth of local people's ecological knowledge and the wealth of resources available in natural areas, highlighting their potential value on the world market and their contribution to local subsistence. In addition, this research plays an important role in convincing protected area managers of the value of including local people in conservation and development projects.

D. Ethnobiological Classification

Continuing a trend initiated by ethnoscientists in the 1950s, ethnobiologists are documenting how local people classify diverse elements of the natural environment. Based on fieldwork in diverse cultures, they describe complex interrelated sets of categories for plants, animals, soils, climates, vegetation, illnesses, food, and other cultural domains and natural phenomena.

Much attention has focused on describing universal similarities in the ways in which local people perceive the natural world. Many generalizations on the categorization, naming, and identification of plants and animals are now widely accepted, whereas others continue to provoke controversy, especially among anthropologists who place emphasis on cultural relativity or the uniqueness of each ethnic group. In particular, there is disagreement over why people are motivated to classify various elements of the environment. Some researchers seek a utilitarian explanation, suggesting that people enhance their ability to fulfill their basic subsistence needs by naming and classifying useful plants, animals, soils, and other natural features. Those who follow an intellectualist line of reasoning argue that there is a universal human tendency to categorize plants and animals according to their overall appearance or symbolic role. These researchers note that local people tend to group organisms with a similar morphology or behavior, regardless of their cultural utility. In a similar vein, some colleagues propose an ecological rationale, noting that some aspects of ethnobiological classification can be elucidated by reference to the role and interaction of plants and animals in diverse ecosystems. Although there is an emerging consensus that classification is motivated by a combination of these factors, there continues to be much debate about which is the most important dimension.

Studies of ethnobiological classification contribute to understanding—in part by reference to the global system of plant systematics—how local people perceive and manage natural resources. When carried out in a participatory way, ethnobiological inventories provide an opportunity for local people and researchers to work together to document the distribution, management, and use of biological resources locally and globally. They produce the baseline data needed to produce bilingual and bicultural manuals that compare and contrast different ways of classifying, managing, and using biological organisms. These ethnobiological manuals contribute to applied programs of conservation and development by highlighting culturally significant species that local people are harvesting from the wild, managing in anthropogenic landscapes, or cultivating in gardens and fields. Often, these species are selected for use in initiatives that promote reforestation, sustainable harvesting of minor forest products, or cultivation of useful plants that contribute to the well-being and income of local people.

E. Knowledge Variation

There are significant differences in the way local people perceive and use biological resources. Ethnobiological studies reveal that some plants and animals are known by a majority of the population, but others are the domain of curers, the elderly, women, or members of another social group. There are significant differences in the biological knowledge of people living in separate communities, belonging to distinct ethnic groups, speaking different languages, or subsisting in diverse ecological zones. These differences can be explained in part by the fact that each person's knowledge is correlated with sociological characteristics such as age, gender, occupation, education, social status, and zone of residence. Perception and management of plants and animals are also affected by each individual's life experiences—for example, if he or she has suffered from a major illness, migrated to other communities, or worked as an apprentice to a plant specialist.

Awareness of these differences is essential when assessing the depth and breadth of traditional knowledge and the diversity of ecological practices in a particular area. It is now recognized that researchers run the risk of obtaining a biased perspective of local biological knowledge if they work with few informants. In contrast, consultation with a representative cross-section of local people can provide a relatively accurate portrayal of perception of the natural world. Ethnobiologists are analyzing patterns of agreement among different individuals, producing a quantitative method of identifying the most culturally significant plant and animal species, soil types, forest zones, and other elements of the landscape. These results ensure accuracy when preparing natural resource management plans and popular manuals produced with local communities. They can also play a role in understanding the knowledge held by specialist user groups and how they are transmitting it from one generation to the next.

F. Mechanisms of Change

Another principle that has emerged in recent studies is that knowledge about the natural environment and ways of managing biological resources are not static. They change as people move from one region to another, as youth reinterpret what they have learned from elders, and as cultures come into contact with each other. Ethnobiologists are analyzing the dynamics of these cultural changes and assessing how they can enrich or impoverish local knowledge systems.

Archaeology, linguistic reconstruction, archival research, and oral history open a window to the past, allowing researchers to discover how ecological knowledge and resource management have evolved during decades and centuries of political, demographic, and economic change. Studies of historical ecology are revealing which species, ethnobiological categories, and ecological practices have persisted, changed, or disappeared over time, including those that have become widespread through borrowing across cultural boundaries. Research that focuses on world events in recent centuries allows historians to assess the impact of colonization and other forms of culture contact on local ways of perceiving and managing the natural world.

Many regions of the world are currently in an intense period of change characterized by not only the destruction of wildlands and loss of biological diversity but also the transformation of traditional biological knowledge. In some regions, young people are not learning what their elders know about the environment, particularly as traditional ways of using biological organisms and managing natural areas fade away. Specialized knowledge held by only a few curers or spiritual leaders is lost when no apprentice is found to carry on traditional medical or religious practices. This loss of knowledge is often linked to increasing contact with national and international cultures and is often exacerbated as local people incorporate formal education, major religions, and migration into their lifestyles.

Ethnobiologists are playing a practical role in assessing the extent of cultural transformation and searching for ways of promoting the survival of local ways of classifying, using, and managing natural resources. These actions are based on the assumption that participation in the joint management of ecosystems and resources stimulates local people to retain and build on empirical knowledge and practices acquired during the tens, hundreds, or thousands of years that they and their ancestors have resided in the region. Studying the evolution of ethnobiological knowledge elucidates mechanisms of cultural resistance, allowing us to understand why local ecological knowledge persists and how we can reinforce the mechanisms of resistance. These studies also highlight the impact that local people have had on the natural environment over time, often providing evidence that they are responsible for maintaining biological diversity and stimulating innovative ideas for conservation in the future.

G. Ritual, Religion, and Symbolism

Plants and animals play an important role in rituals and spiritual practices and are a recurring element in myths, legends, and stories. Some researchers assert that local people, because of their cosmology or understanding of the universe, relate to the natural environment in ways fundamentally different from that found in, for example, European or American culture. A central element in this conception is that traditional lifestyles link people to nature in a way that provides them with a special understanding of nature. A common way for outsiders to gain access to the world of ritual, magic, and religion of local peoples is through apprenticeship with spiritual leaders, often accompanied by the use of psychoactive plants.

Whether or not these generalized notions of differences between global knowledge systems and traditional knowledge are valid in all cases, it is undeniable that spiritual beliefs about the forest and associated taboos on the use of natural resources are important elements of conservation and sustainable use of natural resources in many cultures. When people begin to abandon these beliefs, traditional controls on resource exploitation often disappear, potentially leading to devastation of formerly protected areas such as sacred groves. Ethnobiologists are increasingly aware of the need to record these belief systems and to verify empirically what impact they have on conservation and use of natu-

ral resources. In areas in which traditional cosmology is still a viable element of local culture, conservation biologists are exploring ways of integrating it with the management of protected areas.

H. The Internationalization of Ethnobiology

Although ethnobiology as an academic discipline originated in Europe and the United States, it has now been embraced by researchers in many developing countries who have subsequently adapted the techniques and concepts to their own goals and local conditions. The emergence of professional societies of ethnobotanists in developing countries, ranging from the Indian Society of Ethnobotanists in 1980 to the Asociación Mexicana de Etnobiología in 1993 and the Sociedade Brasileira de Etnobiologia e Etnoecologia in 1997, is evidence of this trend.

The internationalization of ethnobiological research and training has resulted in new directions in theory and application, enriching the field. In India, the tradition of conducting ethnobotanical inventories in various tribal areas has continued, but it is now supplemented by innovative approaches to studying the harvest of nontimber forest products in joint forest management schemes and practical strategies to create community biodiversity registers. Researchers in China have contributed studies on ecological succession in swidden fields, marketing of useful plants, and analysis of agroforestry practices. Equally impressive are developments in Mexico, where ethnobiologists have focused on the management of anthropogenic and natural ecosystems as well as the process of domestication of botanical resources.

III. THE NEW SYNTHESIS

The proliferation of labels and orientations for the study of local biological knowledge and practice is likely to continue. Consensus on precise definitions will be difficult to achieve because researchers are approaching the field from a variety of academic disciplines, bringing with them a wealth of new concepts and methods. Despite this dynamic development, there is some semblance of an agreement on a typology for the discipline.

Distinct approaches to gathering empirical data on the reciprocal interactions between people and biological organisms will continue to be referred to by terms such as economic botany, ethnobotany, ethnoecology, and ethnozoology. Ethnobiology is becoming the preferred term for an integrative discipline that draws on all these approaches to analyze traditional biological knowledge and practices throughout the world. It is unified by a central theory that local peoples' systematic knowledge and management of biological organisms and ecosystems can be classified as biological sciences, covalidated by qualitative and quantitative research methods. Perceived in this way, ethnobiology blends conventional studies carried out by economic botanists, ethnobotanists, ethnozoologists, and ethnoscientists that present a limited vision of local people's interaction with the natural environment. This provides an opportunity for reintegration of various disciplines of natural science, counteracting the impact of the reductionism of global scientific knowledge.

Ethnobiology seeks not only to integrate these various lines of scientific research but also to focus them on supporting community development and biodiversity conservation, which are clearly multidisciplinary endeavors. American ethnobiologist Darrell Posey argued for this style of "advocacy" or "applied ethnobiology," whose goal is to reform the economic, environmental, and social policies that are at the root of many problems which affect people in rural and urban settings. This approach has stimulated a new generation of researchers to blend scientific research with an awareness of political and ecological problems, including the loss of biotic and genetic resources, indigenous struggles for land and resource rights, and negative aspects of globalization. The ultimate goal is sustainable development, as defined by cultural, ecological, and economic parameters.

The yen for integration goes beyond creating an interdisciplinary, applied field of study. Although ethnobiologists have tended to concentrate on the empirical side of local biological knowledge, there is renewed interest in symbolic and other interpretive approaches that could give a broader view of how people perceive their natural surroundings. Many ethnobiologists advocate adopting a participatory approach through which the entire study, from research design to application of the results, is conceived as a collaborative effort between local people and researchers. These developments represent a significant achievement for ethnobiology, a field that has always sought to blend perspectives from many cultural traditions.

See Also the Following Articles

AGRICULTURE, TRADITIONAL • HISTORICAL AWARENESS OF BIODIVERSITY • HUNTER-GATHERER SOCIETIES, ECOLOGICAL IMPACT OF • INDIGENOUS PEOPLES, BIODIVERSITY AND • SOCIAL AND CULTURAL FACTORS

Bibliography

Alexiades, M. N. (1996). *Selected Guidelines for Ethnobotanical Research: A Field Manual.* New York Botanical Garden, New York.

Atran, S. (1990). *Cognitive Foundations of Natural History: Towards an Anthropology of Science.* Cambridge Univ. Press, Cambridge, UK.

Balée, W. (1994). *Footprints of the Forest. Ka'apor Ethnobotany—The Historical Ecology of Plant Utilization by an Amazonian People.* Columbia Univ. Press, New York.

Balick, M. J., and Cox, P. A. (1996). *Plants, People and Culture: The Science of Ethnobotany.* Scientific American, New York.

Berlin, B. (1992). *Ethnobiological Classification: Principles of Categorization of Plants and Animals in Traditional Societies.* Princeton Univ. Press, Princeton, NJ.

Clément, D. (1998). L'Ethnobiologie/ethnobiology. *Anthropologica* 40, 7–34.

Cotton, C. M. (1996). *Ethnobotany. Principles and Applications.* Wiley, London.

Diamond, J. (1998). *Guns, Germs and Steel: A Short History of Everybody for the Last 13,000 Years.* Vintage, London.

Ellen, R. F., Parkes, P. S. C., and Bicker, A. (Eds.) (2000). *Indigenous Environmental Knowledge and Its Transformations,* Studies in Environmental Anthropology. Harwood, Amsterdam.

Ford, R. I. (Ed.) (1978). The nature and status of ethnobotany, Anthropological Papers No. 67. Univ. of Michigan, Museum of Anthropology, Ann Arbor.

Johns, T. (1990). *With Bitter Herbs They Shall Eat It: Chemical Ecology and the Origins of Human Diet and Medicine.* Univ. of Arizona Press, Tucson.

Lévi-Strauss, C. (1966). *The Savage Mind.* Univ. of Chicago Press, Chicago.

Martin, G. (1995). *Ethnobotany.* Chapman & Hall, London.

Plotkin, M., and Famolare, L. (1992). *Sustainable Harvest and Marketing of Rain Forest Products.* Island Press, Washington, D.C.

Posey, D. (Ed.) (1999). *Cultural and Spiritual Values of Biodiversity.* Intermediate Technology, London.

Prance, G. T., Chadwick, D. J., and Marsh, J. (1994). *Ethnobotany and the Search for New Drugs.* Wiley, Chichester, UK.

Schultes, R. E., and Raffauf, R. F. (1990). *The Healing Forest: Medicinal and Toxic Plants of the Northwest Amazonia.* Dioscorides Press, Portland, OR.

Simpson, B. B., and Connor-Ogorzaly, M. (1995). *Economic Botany: Plants in Our World,* 2nd ed. McGraw-Hill, New York.

EUKARYOTES, ORIGIN OF

Dorion Sagan* and Lynn Margulis[†]
*Sciencewriters and [†] University of Massachusetts

I. Overview
II. Differences between Prokaryotes and Eukaryotes
III. Symbiosis as an Explanatory Mechanism
IV. Eukaryote Origins via Endosymbiosis
V. Serial Endosymbiosis Theory

EUKARYOTES, ORGANISMS WITH NUCLEATED CELLS (plants, animals, fungi, and protoctists), evolved symbiogenetically from the fusion of an archaebacterium and a eubacterium. This process resulted in the integration of the two prokaryotes' genomes to form chromosomes within a membrane-bounded nucleus.

GLOSSARY

endosymbiosis Condition of one organism living inside another; includes intracellular symbiosis (endocytobiosis) and extracellular symbiosis.
microtubule Hollow cellular structure that is 24–25 nm wide and made of the protein tubulin; it is the main component of centrioles, kinetosomes, the mitotic spindle, and the undulipodium or eukaryotic flagellum.
protoctists All eukaryotic organisms besides plants, animals, and fungi; includes the traditional groups protozoa, algae, and slime molds.
spirochetes Helically shaped gram-negative bacteria with flagella in the periplasmic space between the two cell membranes.
symbiogenesis Production of new organelles, cells, tissues, organs, or species by the symbiotic integration of two different organisms.
undulipodium Cilium or eukaryotic flagellum composed of a $[9(2) + 2]$ microtubular axoneme.

I. OVERVIEW

This article discusses the origin of eukaryotes, the only life-forms on Earth that are not bacteria. Eukaryotes, the one- or many-celled beings whose cells contain nuclei, can be classified unambiguously into one of four groups (or "kingdoms"): plants, animals, fungi, and protoctists. Bacteria (kingdom Monera), by contrast, are made of prokaryotic cells, whether or not their bodies are composed of one or many cells. Protoctists (kingdom Protoctista), which may be an unfamiliar term, include all those nucleated organisms that are not plants, animals, or fungi. Among the live present-day protoctists are free-living nucleated organisms such as single-celled amoebae and *Paramecium* and huge multicellular seaweeds such as the giant kelp. The smaller members of the kingdom, the protists, are made of cells that resemble the tissue cells typical of plants and animals. Protoctista includes composite organisms—we call them the "water neithers" because they are all aquatic but are neither plant nor animal (nor fungus). All green, brown, and red seaweeds are protoctists, but so are water molds, slime molds, slime nets, diatoms, ciliates, and many other even more obscure members of this kingdom. The aquatic photosynthesizers—seaweeds, diatoms, golden yellow algae, and the like—differ from the plants, members of the kingdom Plantae, in that, unlike plants, no protoctist develops from embryos.

There is evolutionary continuity between large familiar organisms made of nucleated (eukaryotic) cells and the nucleated free-living single-celled organisms: the protists (this informal term includes all single- and few-celled protoctists). For over 100 years, and even today, unfortunately, these microorganisms are known as "protozoans." Protozoa, a word derived from Greek roots meaning "first" and "animal," is misleading because "Protozoa" are in no way animals and indeed the early uni- and multi-cellular eukaryotes—the protists—evolved not only into animals but into modern protoctists, plants, and fungi as well.

The greatest morphological difference in all of modern life is that between prokaryotic organisms—bacteria—and eukaryotes—all the rest of life. From unicellular pond ciliates to blue whales and redwood trees, the eukaryotes show fundamental resemblances to each other. At a cell level, the difference between a nucleated amoeba cell and a bacterial cell is far greater than the difference between that of an amoeba, a human epidermal cell, and a sperm cell from the male sex organ of a ginkgo tree. The difference between prokaryotes and eukaryotes is profound: it is comparable to the gap between the Kitty Hawk flying machine of the Wright brothers and the Concorde jet liner. Unlike the evolution of humans from apelike predecessors, which is bridged by the fossil bones of ape–human intermediates (australopithecine and *Homo* species like *Australopithecus afarensis, Homo erectus, Homo ergastor*, and *H. sapiens neandertalensis*), the evolution of eukaryotes from prokaryotes appears disjunct.

It is almost miraculous that organisms composed of the larger, more complex cells appear in the fossil record. The earliest eukaryotes are fossils known as acritarchs, dated by radioactive igneous intrusives and stratigraphic correlation to be 1200 million years old. The only evidence for any life prior to the body and trace fossils of extinct, presumed eukaryotes is for bacteria—both as single microfossils and as microbial communities that left fossil remains. Bacterial life is ancient. Evidence for the earliest bacteria places them roughly contemporaneous with Earth's formation of a solid crust, almost four billion years ago. So the big question is: How, from bacteria or prokaryotes, did eukaryotes, nucleated cells, evolve? The answer seems to be: by

symbiosis, or the living together of different kinds of bacteria. The evolutionary process of symbiogenesis—the sequence of events that occur when symbiotic partners merge to form new cells, tissues, organs, or organisms—was crucial in bridging this evolutionary gap. The union of different types of bacteria to form amoebalike cells that became ancestral to the rest of the protoctists, fungi, plants, and animals is described here. This event spawned a revolution in biodiversity.

II. DIFFERENCES BETWEEN PROKARYOTES AND EUKARYOTES

One model of the origin of nucleated cells is depicted in Fig. 1. Three distinct symbiotic mergers (blue triangles) are hypothesized.

The first microbes to be discovered were protists, today called eukaryotic microorganisms. The pond water microbes studied by Dutch draper turned microscopist Antoni van Leeuwenhoek included *Mesodinium rubrum* and green algae. Most prokaryotic cells, evolutionary predecessors to protists, are up to a thousand times smaller than nucleated cells such as those of an amoeba or an animal tissue cell. The term eukaryote literally means "true kernel," from Greek roots; prokaryote comes from words meaning "before kernel." The kernel in question is the nucleus, which contains chromosomes of varying numbers. These numbers are not necessarily correlated to complexity. A dandelion, for example, has many more chromosomes than the forty-six found in all human cells except sperm and egg, which have twenty-three chromosomes each, and blood cells. Mammalian blood cells even lose their nuclei in the process of development and of course eventually die, being unable to reproduce.

Just as an individual understands the world on the basis of his or her own experience, so have biologists traditionally understood life on the basis of large, familiar organisms. Close inspection of the putative prokaryotic ancestors of all eukaryotes from single-celled yeasts to giant sequoias reveals significant and great differences. Prokaryotes lack true chromosomes. They never have membrane-bounded nuclei. The DNA of bacteria floats free in strands that have little or no protein and often the cells bear DNA in tiny rings called plasmids. Bacterial ("chromonemal") DNA, unlike that of eukaryotes, is not packaged into the winding protein-coated chromosome structures. In eukaryotic cells, DNA complexed with protein and capable of staining a deep red color lines up and separates in an intricate pattern (called mitosis) as the cell divides.

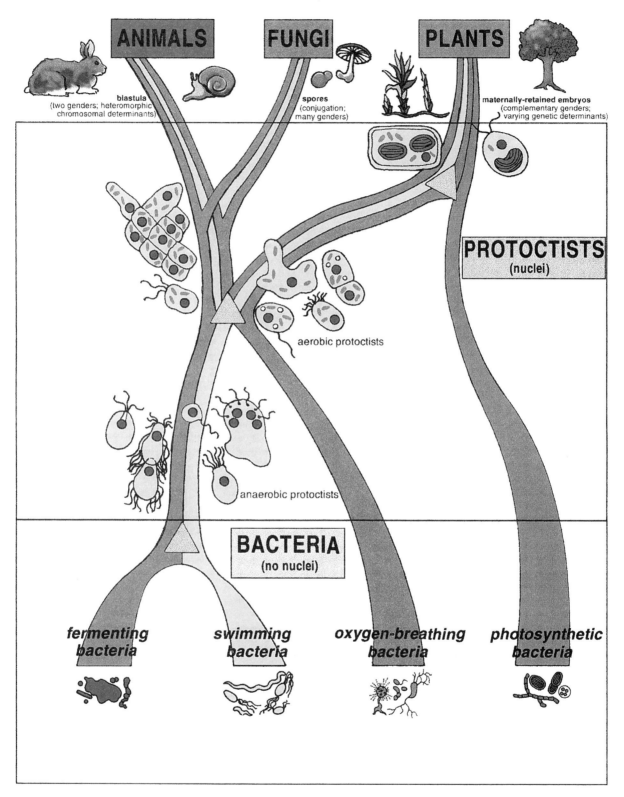

FIGURE 1   Origin of nucleated cells.

Although standard bacteriology texts refer to species of bacteria and assign them binomials (e.g., *Escherichia coli, Bacillus anthracis, Chromatium pfennigii*), the life-habits of bacteria defy any definition of species, especially the zoological. If a species is defined as a group of organisms whose "members can breed with each other," only one species of bacterium lives on Earth today! In eukaryotes, at least plants and animals, species are maintained by sexual reproduction. Sex cells reminiscent of ancestral protist eukaryotes recognize each other, come together, and fuse to form new cells now with twice the number of chromosomes as the sex cells. These new cells, after fertilization in plants and animals, grow by mitotic cell division (reproduction) to form an embryo. Embryos continue to grow and differentiate into adult plants or tissues and organs of adult animals. Since a bacterium can inject from a few to virtually all of its genes into any other entirely different bacterium, the standard species concept applied to them has no meaning. A bacterium after receipt of genes from another bacterium is a recombinant. This new and different bacterium can then reproduce in its recombined state indefinitely. Sexually-preserved species borders are an imposed fiction in the case of bacteria, because every minute recombination is rampant. Sorin Sonea and Maurice Panisset note that if bacteria had been discovered on Mars instead of on Earth, their astounding natural history would have led scientists to classify them as radically different from all other life-forms.

Other common features of eukaryotes are also lacking in bacteria. Eukaryotes harbor small membrane-bounded bodies called organelles inside their cells. A most important set of eukaryotic organelles, in addition to the ever-present nucleus, are the mitochondria. These are oxygen-using inclusions, respiring entities that provide the cell with energy. All algal and plant cells have another set of organelles that contain chlorophyll: plastids. These are called chloroplasts if they are green. Plastids are named rhodoplasts or phaeoplasts if found inside certain red or brown protoctists (e.g., in coralline red algae or brown kelp). Plastids contain many different pigments, but these always include chlorophyll *a*, one of the chemical requisites of oxygenic photosynthesis. Chloroplasts mediate the conversion of solar energy into intracellular fuel in algae and plants. They are, of course, absent in animals, whose metabolic fuel is derived from energetic compounds in the environment—food. Another general difference between prokaryotes and eukaryotes is that, with a couple of significant exceptions, eukaryotic cells are aerobes: they depend on oxygen for cell respiration.

The fossil record of oxidized minerals in Earth's crust clearly tells us that free oxygen gas was not present in significant quantities until well after life's origins. Many modern forms of bacteria—such as the methane-generating archaebacteria that coexist with cellulose-degrading bacteria in the special grass-digesting "stomachs" (the rumen) of bison, cows, and water buffaloes—are anaerobes. They are poisoned by oxygen. Today's eukaryotes somehow evolved from earlier life-forms that were intolerant of oxygen. Indeed, modern-day eukaryotes that lack mitochondria, such as pelomyxids, retortamonads, pyrsonymphids, trichomonads, calonymphids, and hypermastigotes, can be interpreted as "living fossils." These swimming protists are legacies of the missing links between the tiny gene-trading prokaryotes and the sexual species-forming eukaryotes.

III. SYMBIOSIS AS AN EXPLANATORY MECHANISM

Symbiosis was defined in 1873 by the German botanist Anton de Bary as the "living together of differently named organisms." To the biologist it refers specifically to organisms of different species that remain in more than casual contact. Bacteria of different kinds are certainly recognizable by their morphological and physiological traits rather than by their genetic and sexual behaviors. They need not be confused by inappropriate comparison to the familiar, sexually reproductive plants and animals who can easily be assigned to species proper. Biologists distinguish epibiosis, the symbiotic contact of skin or outer coating, from endosymbiosis, a kind of Jonah-in-the-whale association in which one organism enters, and sometimes fully merges with, the entered organism. Symbiogenesis is a kind of permanent mating. One might be tempted to accuse nature of bestiality were it not for the fact that the "unions" in question far transcend fleeting sexual "mistakes." The Russian naturalist Konstantin Merezhkovsky (1855–1921) was among those of the last century to suggest that symbiosis leads not only to fleeting exploitative (parasitism) or mutually food-providing (symbiotrophic) relationships, but also to the formation of entirely new organs and organisms. He coined the term "symbiogenesis" and described many of the most conspicuous examples.

The origin of eukaryotic cells, which represents such a "jump" in both the fossil record and the composition of extant life, is best explained by Merezhkovsky's symbiogenesis. Today, genetic, comparative morphological, and other lines of evidence suggest that this kind of "innovation by association" was the modus vivendi that transformed gene-trading bacteria into cells with nuclei, mitochondria, and, in the case of algae and plants, plastids. In life's long history, natural selection acting on symbiogenesis created new species, not only of eukaryotic cells but also of their descendants: sexually reproducing lineages of multicellular bodies. We think that only eukaryotic organisms with their multiple integrated genetic systems form true species.

But how can mere living together lead to the formation of new species? A variety of modern-day organismic associations help to explain this. Epibionts, such as the plovers that feed atop crocodiles by cleaning their teeth, are examples of symbiotic partners. They are compensated, if one considers their location and delicious potential edibility, for their services. Episymbiosis often evolves because organisms, which can be viewed as open thermodynamic systems, "leak"—that is, they incompletely make use of food and energy reserves. This leakage allows other organisms to move in. Human skin, even the surface of the human eye and its lashes, serves as a breeding ground for mites and microorganisms. Such normal human accoutrements as bacteria and protoctists lead to health problems if their populations grow too much. Yet their removal is also dangerous, in part because they keep each other in check. The fungus *Candida*, for example, which we associate with yeast infections, is normally kept in check by bacteria, which is why yogurt, which contains *Acidophilus*, is sometimes prescribed as a culinary prophylactic for the ailment.

Endosymbiosis seems disturbingly intrusive, but it leads to radical transformation. An example of endosymbiosis would be the human gut bacteria that metabolize vitamin B_{12}, a vitamin that humans, unaided by bacteria, are unable to synthesize. A striking example of symbiosis is the seaside worm, *Convoluta roscoffensis*, which dwells on the beaches of northern France, southern England, and southwest Spain. To the casual observer this organism is a seaweed: it looks like green scum. During rough seas or stormy weather—or even when a biologist stomps the surrounding ground—the "seaweed" uses its muscles to burrow into the sand. The "seaweed" is really composed of tiny worms whose mouths are closed. They live by direct internal absorption of nutrients provided by the gardens of algae within their cells. The endosymbiotic feeling, as often is the case, is mutual: the host organisms, the worms, receive free nutrients, whereas the resident, eaten but undigested, receive a protective enclosure. The living translucent "greenhouse" of a mobile being results: the green

animal eludes environmental danger at a moment's notice.

Endosymbiosis seems to be a species-creating phenomenon. The freshwater coelenterate *Hydra*, for example, is transparent, brown, or green, depending on whether its cells contain photosynthetic inclusions full of the green alga *Chlorella*. Symbiotic associations between other animal hosts abound. Green, food-producing algae or cyanobacteria live inside giant clams, snails, and even in the hair shafts of polar bears. These minute tubes provide a warm, snug locale for arctic photosynthesizers.

Photosynthetic alliances grow and evolve from relatively casual epibiosis to permanent endosymbiotic mergers. The photosynthesizers that dwell on sun-illumined surfaces of other living beings are ingested, but in their struggle to resist digestion they continue to metabolize. Their continued metabolism provides the would-be predator with additional food—the evolutionary equivalent of a free lunch.

A somewhat different mode for the establishment of endosymbiosis is that of infection. Microbes in the millions are normally and nonpathologically present on and in plant and animal (including human) bodies. Intimate associations among distinct organisms are not necessarily detrimental. But of course relationships degenerate. When one rapid grower enters and exploits another, it naturally produces wastes and may upset a delicate internal balance. Infection can evolve into tolerable truce and permanent productive association if the rapidly growing invader moderates its behavior and growth in response to the "objections" of the would-be invaded.

A striking example of endosymbiosis-by-infection was documented by Korean-American biologist Professor Kwang Jeon at the University of Tennessee, Knoxville. Jeon's collection of amoebae were tainted by a sample sent to him by colleagues that contained amoebae infected with bacteria. Most of the new amoebae died. All of the survivors of this natural selection scourge eventually transformed into what was de facto a new species of amoeba. The transformed amoebae retained an internal collection of relatively slow-growing endosymbiotic bacteria. Some 40,000 bacteria resided in each amoeba! Jeon showed that the new amoebae, now an amoeba–bacterial complex, depended on what had once been a fatal infection. To test this hypothesis he replaced the nuclei from bacteria-containing amoebae with "normal" nuclei from amoebae that were never infected by the newly acquired bacteria. He found that members of the "new" amoeba species invariably died unless they were "rescued" at about three days postoperative by a reimplantation of the formerly pathogenic bacteria. The nuclei, now accustomed to the presence of hordes of cytoplasmic bacteria, required their continued presence for existence. A chronic deleterious association had become a condition for life; the bacterial "infection" was now an intracellular requirement, and by definition the pathogen had become an organelle.

IV. EUKARYOTE ORIGINS VIA ENDOSYMBIOSIS

RNA sequence comparisons of molecules that contain some 1400 nucleotide base pairs sharply divide bacteria into two groups. Professor Carl Woese (at the University of Illinois, Urbana) christened these two groups of organisms—both prokaryotes—"archaebacteria" and "eubacteria." A comparison between Woese's phylogeny that ignores symbiosis and is based on molecular criteria and ours based on symbiogenesis is shown in Fig. 2. Both are of course highly simplified.

As their name suggests, the archaebacteria are putatively a more ancient lineage. Their metabolism appears to have remained relatively unchanged from the early days of an Earth that lacked atmospheric oxygen. The archaebacteria include three great subgroups: methane-producing, thermacidophilic, and salt-tolerant prokaryotes. The first two derive their energy in environments free of oxygen gas and produce wastes that do not include O_2. Some, such as *Thermoplasma* and *Sulfolobus*, inhabit boiling-hot water like the natural hot springs of Yellowstone National Park. The third group of archaebacteria, the halophiles, tolerate conditions of extreme salinity that are prohibitive to the growth not only of larger organisms but of other bacteria. Their genetic similarity, environmental tolerance, and ability to live in the absence of free oxygen gas point to an ancient history for the archaebacteria.

The eubacteria include all the rest, and the vast majority, of prokaryotes. Although extremely metabolically diverse, these bacteria include forms adept at oxygenic photosynthesis, ammonia oxidation, ammonification, manganese oxidation and reduction, iron oxidation and reduction, sulfur oxidation and reduction, fermentation, and nitrate respiration. The eubacteria possess RNA sequences, lipid composition, and other traits that distinguish them from the archaebacteria.

Intriguingly, however, archaebacterial RNAs show more kinship to the RNAs of eukaryotes than they do to those of other bacteria. This is only one of a slew of clues pointing to a shared ancestry between archaebacteria and modern eukaryotes. Compounds much like the histone proteins that form the packaging of chromosomes, for example, are found in *Thermoplasma*. Yet there are perhaps even more features tying us organisms made of eukaryotic cells to the eubacteria. Most striking are the relatively narrow metabolic repertoire of all familiar large organisms, the oxidative respiration of all plants and animals and fungi, and the photosynthesis of plants. The photosynthetic protoctists such as kelp and seaweed have exactly the same system of food and energy gathering as do the cyanobacteria. The highly circumscribed metabolism of all eukaryotes is well represented in the bacterial world, a realm that evolved so many other kinds of metabolism.

Plants and algae share oxygen-producing photosynthesis with cyanobacteria, and almost all eukaryotes (animals, plants, fungi, and protoctists) share their intracellular oxygen-using abilities with respiring bacteria. Many other traits, such as fermentation of organic compounds into butyric acid and the oxidation of hydrogen sulfide to sulfate, remain impossible for us self-designated "higher" organisms. (Perhaps "larger" is a more accurate term!)

The metabolic similarities of eukaryotes to a very few kinds of bacteria (prokaryotes), combined with the near absence of intermediate forms either living or in the fossil record, suggest an endosymbiotic, bacterial origin for eukaryotic cells. Other evidence is even more provocative. The oxygen-using organelles, the membrane-bounded mitochondria, that are found in nearly all eukaryotic organisms are strikingly similar to respiring bacteria in the details of their chemistry. They directly reproduce as do respiring bacteria. Their own DNA is separate from the DNA of the chromosomes in the nucleus of eukaryotic cells. These same traits apply to the plastids of algae and plants, bacteria-sized photosynthetic cell inclusions. They too have their own DNA and their own bacterial way of reproducing that is entirely different from the mitosis typical of eukaryotic cell division. The clincher, however, is that stretches of DNA in mitochondria and chloroplasts are statistically far more like stretches in the DNA of certain free-living forms of bacteria than they are like the DNA of the cell in which they reside. Indeed, no reasonable alternatives to the evolution of eukaryotes via bacterial endosymbioses of the forerunners to mitochondria and plastids exist, although the extent, the number of symbiotic events, and the nature of the details remain a matter of scientific investigation and debate.

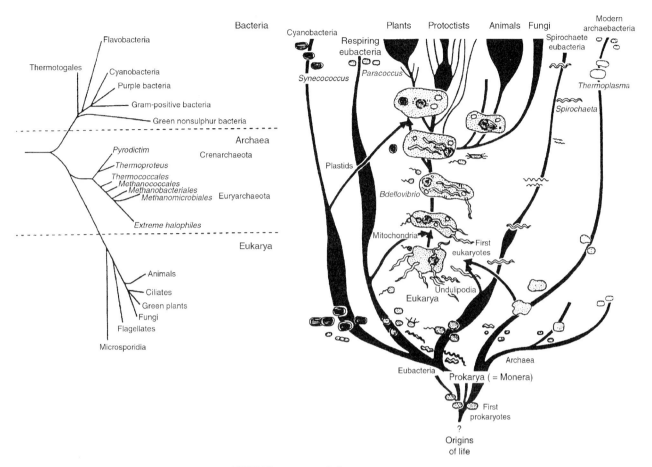

FIGURE 2   Woese's phylogeny versus ours.

V. SERIAL ENDOSYMBIOSIS THEORY

More than one explanation has been proffered to cover the evolutionary gap from prokaryotes to eukaryotes. Clearly, however, symbiosis—and not just a one-time symbiosis—appears to be involved. All organisms are open thermodynamic systems exchanging matter, energy, and information with their environment. The notion of a totally isolated, independent organism may work as a mathematical assumption or a Platonic idealization, but in the real world of responsive biological activity both genetic exchange and metabolic evolution present untold opportunities for organismic interaction. One organism sheltering another, predator–prey relations, and genetic exchanges lead to alliances of sexuality and symbiosis. Nonetheless, in obeisance to Darwin, traditional evolutionary theory permits only a series of minute and gradual changes that are touted to generate all propagating mutant descendants. Symbi-

ogenesis, the sudden appearance of new organisms by permanent association of different kinds of partners, is not usually what jumps to mind when one considers the evolution of new species. It should.

Serial endosymbiosis theory posits a succession of up to four evolutionary associations that we think led to modern-day cells. The eukaryotic cells of plants are the result of at least two and perhaps three separate evolutionary partnerships among at least three or perhaps four kinds of bacteria. One was the photosynthetic bacteria that conferred photosynthetic abilities upon the ancestors to plants—latter-day cyanobacteria that today exist as the plastid organelles of plant cells. The nucleated cells of animals are the result of at least one, perhaps two separate evolutionary partnerships among at least two or perhaps three kinds of bacteria.

To understand the evolutionary origins of nucleated cells, it helps to return to Earth in its geological infancy. In the Lower Archean eon, about four billion years ago, no protoctists, plants, animals, or fungi existed, only the metabolically diverse bacteria. We can provisionally trace the evolution of metabolic diversity by comparison with pathways in modern-day organisms. The first bacteria may have been fermenters because such organisms would need only complex organic compounds preexisting in the environment, rather than other organisms, on which to feed in order to grow. Among the earliest life-forms were photosynthetic cells, green and purple bacteria that much later yielded most of the real estate of the sunlit Earth's surface to algae and plants. Comparative analyses suggest that these ancient purple bacteria, like their cousins today, used hydrogen sulfide (H_2S) rather than water (H_2O) to make their organic (carbon–hydrogen) bodies. The hydrogen sulfide, spewed from volcanoes, would have been more plentiful on the early, more tectonically active Earth. But as Earth cooled and tectonic activity subsided, sulfide became less available than that ubiquitous solvent water. Mutations appear to have allowed some bacteria to alter the chemical reactions of their metabolism such that they could now, using the energy of sunlight, break the hydrogen–oxygen bonds of water to make their bodies. Life of course existed in water since its inception. But the use of water as a metabolic resource led necessarily to an entropic waste—oxygen. Powered by sunlight, early life discarded oxygen, which reacts strongly with the carbon–hydrogen compounds of living bodies, into the atmosphere. The rock record of rust, oxidized iron, and uranium oxides demonstrates that enormous quantities of oxygen did not begin to build up in Earth's atmosphere until about two billion years ago.

Life's difference from the universe around it, and its tendency to evolve toward greater levels of complexity, reflects its status as an open, entropy-generating system. The use of energy and the inevitable production of waste as a result lead to changes in life as it metabolically stays the same and reproduces. Reproduction of course is an extension of metabolism. But life also changes because of the toxic effects that its growth has on itself. The cyanobacteria that first mutated to use the hydrogen in water for their electron donor were also the first to be poisoned by oxygen toxicity. Among the prolific anaerobic organisms living at the surface were also rapidly swimming bacteria including spirochetes, many kinds of which dwell today in symbiotic alliance with other organisms in the hindguts (swollen intestines) of wood-eating termites.

Just as the Industrial Revolution changed human civilization in part due to new systems of railway, steamboat, and eventually airplane transportation, fast-moving bacteria are posited in Serial Endosymbiosis Theory (SET) to have conferred major changes upon early cells.

The most provisional part of the evolutionary story of the symbiotic eubacterial origin of eukaryotic cells is the part that explains the origin of intracellular movement. The association of fast-moving eubacteria, perhaps spirochetes (speedy corkscrew-shaped burrowing feeders upon and even permanently attached to other organisms), with larger archaebacteria is postulated to be the most far-reaching but integral part of the story.

Intracellular mobility—the "cytoskeletal system"—of all eukaryotic cells is remarkable for its extent, complexity, and ubiquity. Cells with nuclei not only undergo the elaborate alignment and separation of reproduced chromosomes during mitotic cell division, they also have characteristic motility structures such as the mitotic spindle with its microtubules. Microtubules, 24 nm in diameter and varying in length, are not found in prokaryotes. Among the many microtubule-based structures such as sperm tails, cilia, and moving spines are the mitotic spindles themselves. The mitotic spindle, crucial for eukaryotic cell reproduction, seems to appear out of nothing to become superbly organized and to separate chromosomes during cell division. Eukaryotic cells are famous for the incessant streaming of their cytoplasm. The intracellular activities are reminiscent of a crowded train station filmed in time-lapse photography. Many kinds of nucleated cells (of plant sperm, animal tissue, and myriad protoctists) bear special motility structures that display a ninefold symmetry in cross section. These, whether cilia, sperm tail, or sensory hair, are called "undulipodia." The whipping tail used in cell propulsion, sensory detection, or propulsion of eggs along the fallopian tubes of mammals are all examples of undulipodia. In cross section the shafts of these motile cell structures have a characteristic "[9(2)+2]" pattern. As seen with the electron microscope, nine pairs of tubes surround one central pair. This cross section structure, 250 nm in diameter, is found to be nearly universal in eukaryotic organisms, from the whipping tails of the sperm of ferns, mosses, cycads, and trees (yes, some trees do produce sperm) to the cilia that propel mastigote protists and ciliates through water. The structure is absent in all prokaryotes whether archaebacteria or eubacteria.

The existence of amitochondriate mastigotes—undulipodiated unicells that lack mitochondria and that are, in addition, poisoned by oxygen—suggests that the original symbiotic merger was between some anaerobic "host" bacterium and some kind of rapid swimmer bacterium. As has often been witnessed in modern-day symbiotic partnerships, the rapid swimmer would have been attracted to the food leaking out of the larger bacterium and the partnership would eventually have been selected in the evolution of permanent motile associations. Ciliates today often differ only in the placement of their undulipodia. Moreover, the microsurgical grafting of patterned patches of undulipodia from one ciliate to another leads to altered patterns in which the ciliate reproduces without benefit of a sexual partner. Although DNA is absent in the motility organelle of waving undulipodia itself, many other facts suggest a possible symbiotic origin. Grafted patterns of altered undulipodia "breed true." The sheer resemblance of whipping undulipodia—not easily differentiable in some organisms, such as *Mixotricha paradoxa*, from actual attached spirochetes—to fast-moving, free-living bacteria provides a clue.

Early evolution may have featured a great proliferation of mastigote (undulipodiated) cells resulting from a successful union of anaerobic archaebacteria and spirochetes or spirochete-like organisms in which the DNA of the motile form entered the newly evolved nucleus as it formed. The quick-swimming partnership complex would have conferred a greater ability to acquire food on the merger. The greater access to new food by the larger, now more quickly moving archaebacterial–eubacterial complex portended a great future. As often occurs in evolution, association led to new opportunity and power. Eventually, the greatest, most integrated partnerships formed unified genetic entities. These genetic entities went on to add new members in the evolution of new cells, protoctist lineages ancestral to plants, fungi, and animals.

Much more and better-quality evidence exists for later symbioses. Whether or not swimming eubacteria merged with archaebacteria to form mastigotes such as ciliates, these undulipodiated cells were very likely invaded or infected, some two billion years ago, by oxygen-using bacteria. These oxygen-using or respiring bacteria, similar to free-living forms today—some of which are not only oxygen-respiring but also invaders of other cells (e.g., *Bdellovibrio*)—proliferated in the aftermath of the evolution of oxygenic photosynthesis. The gaseous waste from using water as a source for hydrogen was the production, eventually at a planetary level, of the highly reactive oxygen gas (O_2). This gas naturally, and sometimes fatally, burns cell tissue. Many organisms were forced to retreat to the muds, out of harm's way. Some hung on until animals such as the ancestors to insects evolved. Hiding like soldiers in a tank or submarine, the oxygen-poisoned anaerobes today proliferate only in biological enclosures, such as sulfurous muds, the termite hindgut, the cow rumen, or other natural protective settings.

The respirers, however, turned the metabolic negative of volatile oxygen into the metabolic positive of a new, more efficient source of energy. Bustling about, funneling potentially dangerous redox reaction into their own growth, the ancestors to mitochondria kept the harmful gas away from others by using it themselves. This quality led to the greatest burst of biodiversity of all: the fabulous protoctist proliferation of respirers, both alone and in associations with other organisms. The symbiotically evolved amoebae of Jeon, mentioned earlier, were infected with respirers. Merged with amitochondriate, anaerobic mastigotes, the respiring forerunners to modern mitochondria powered the metabolism of the cells of all the major macroscopic lineages: protoctists, fungi, plants, and animals.

The final major symbiosis—for which, as in the case of mitochondria, there is good genetic evidence—occurred in the ancestors to algae and plants. Taxonomically, the major difference between algae and plants is that plants form from a fertilized nucleus that grows into an embryo that is retained in maternal tissue. By definition the embryo is a multicellular eukaryote that develops from a fertilization event, the fusion of two nuclei from complementarily gendered parents. Algal growth is more helter-skelter, amorphous, and never embryonic. The archaebacteria, perhaps with symbiotic spirochete offspring providing them varied means of mobility, merged with respiring bacteria. Today mitochondria, although they cannot be grown on their own, appear sometimes to revert back to their earlier, freely reproducing lifestyle; in cancer cells, for example, mitochondria often reproduce rampantly. The nucleus itself does not appear to be the direct result of cell symbioses. The nucleus may, however, represent the intracellular equivalent of a "government" that evolved to deal with the first merger of separate cell lineages shown as the lowest triangle in Fig. 1.

After the original archaebacterium–mitochondria module was established, the new nucleated cells had to feed. Many fed on photosynthetic bacteria. Bacteria do not have immune systems, and "eaten" food, like Jonah in the whale, continued to live and sometimes even reproduced. Although perhaps highly perilous, for starving mastigotes with symbiotically derived mitochondria it might also have been a godsend. With all the competition to find food, those who produced their own, or merged with photosynthetic organisms capable of production of their own food, would have been at a distinct advantage under the usual conditions of starvation. The algae and its derivative plant lineage are traceable to eukaryotes that symbiotically acquired bacterial photosynthesizers.

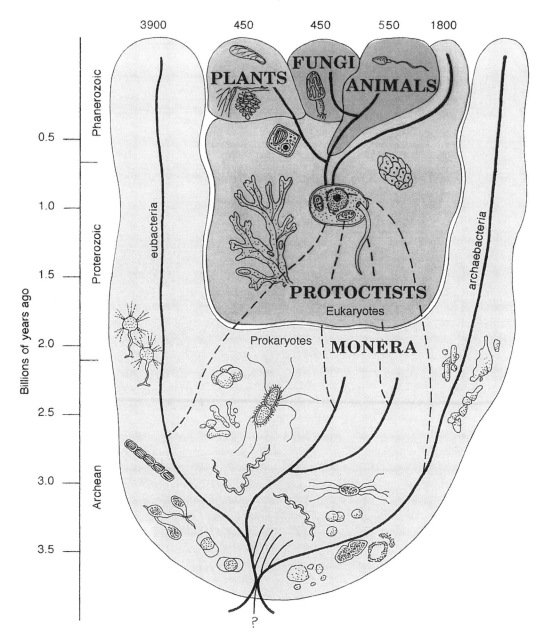

FIGURE 3 Classification based on symbiogenesis.

Today all familiar visible organisms are eukaryotes with nucleated, mitochondriate cells. Whether at the zoo, the forest, or the green prairie, we witness not isolated individuals but an evolutionary fact of life: hundreds of millions of years of symbiotic interliving. The simplest amoeba is no independent cell, rather it represents the outcome of an evolutionary flirtation with death and microbial diversity. A large animal, such as a hedgehog or woman, is a highly organized, differen-

tiated clone of amoeba-like protist cells. An overall classification of life based on this view of the symbiogenetic origin of eukaryotes is shown in Fig. 3. Biodiversity is obvious, widespread, and crucial to our global environment. The existing major phyla, classes, genera, and species resulted not only from mutation accumulation but also by symbiotic alliance and permanent merger. Microbial proliferation and diversification behind the production of new species and large evolutionary transitions in the past are still with us as the understory of life on Earth.

See Also the Following Articles

ARCHAEA, ORIGIN OF • BACTERIAL GENETICS • BIODIVERSITY, EVOLUTION AND • DIVERSITY, MOLECULAR LEVEL • MICROBIAL BIODIVERSITY • NUCLEIC ACID BIODIVERSITY • ORIGIN OF LIFE, THEORIES OF

Bibliography

Folch, R. (2000). *Biosfera*, 11 vols. Enciclopedia Catalana, Barcelona, Spain (*Encyclopedia of the Biosphere*, English language version *in preparation* with the Gale Group, Farmington Hills, Michigan).

Margulis, L. (1993). *Symbiosis in Cell Evolution*, 2nd ed. W. H. Freeman, New York.

Margulis, L., and Schwartz, K. V. (1998). *Five Kingdoms: An Illustrated Guide to the Phyla of Life on Earth*. W. H. Freeman, New York.

Woese, C. R., Kandler, O., and Wheelis, M. L. (1990). *Proc. Nat. Acad. Sci. USA* 87, 4576–4579.

EUROPE, ECOSYSTEMS OF

Ladislav Mucina
University of Stellenbosch

I. Introduction
II. Ecosystem Diversity: Concepts and Approaches
III. Biodiversity of Habitats and Vegetation
IV. Challenges for the Future

GLOSSARY

biome A large-scale complex of ecosystems sharing similar climate and vegetation structure.

ecological biodiversity Variety of biotic communities (plant, animal, and microbial communities) and their complexes (ecosystems, landscapes, and biomes).

formation Large-scale (subcontinental or continental) vegetation complex defined primarily on the basis of a combination of dominating life-forms (hence, vegetation structure).

habitat A place of dwelling of a biotic community (or a complex thereof) showing particular combination of ecological factors occupying certain area in a certain period of time.

orobiome Mountain range characterized by particular climatic pattern and characteristic sequence of vegetation zones.

syntaxon (plural syntaxa) A category of vegetation typology based on the floristic–sociological approach (known also as the Braun-Blanquet approach); the basic syntaxon rank is "association,"

which further groups into "alliances," which group into "orders," and orders group into "classes."

vegetation megazone Large-scale vegetation complex characteristic for a zonobiome.

vegetation survey Product of research activity aimed at the description and classification of vegetation cover on various levels of complexity in a certain geographic area using various field and data-evaluation methods.

zonobiome Broad ecological topographical unit characterized by a certain climatic pattern.

THE PROBLEM OF biodiversity of European ecosystems is approached from the point of view of biomes and zonobiomes, flora-based vegetation typology (phytosociological tradition to classification of vegetation), and habitat classifications. The concepts of biome (based on a combination of dominating life-form and climate) and zonobiomes (by H. Walter) are compared for Europe. The most effective classification of ecosystems in Europe is based on the floristic–sociological approach, which has deep roots in European tradition and is widely used throughout Europe. This approach has yielded a hierarchical system of vegetation units which allow insight into the diversity of vegetation types of this continent. Basic features of diversity of vegetation types (syntaxa) are briefly outlined, and the importance of vegetation classification for delimitation and calibration of habitat classification systems in Europe is discussed. The tasks for further research and socioeconomic implications of the diversity of ecosystems in Europe are outlined.

I. INTRODUCTION

Hardly any place in the world supports in a relatively small area such a variety of countries, peoples, cultures, histories, languages, political views, cuisine, and types of cheese and wine as does Europe. There might be more languages and people in India, higher species diversity of plants in the tropical rain forest of Columbia, or more diverse cuisine in China, but there is only one Europe.

Europe's nature is showing many faces, both pristine ones in the form of tracks of tundra and forests and created (or mis-created) ones—those carrying the signs of the hand of man. Europe is a patchwork of ecosystems connected by an intricate net of gradients of ecological factors ranging from those controlling continental patterns of vegetation megazones and biomes to local gradients creating small-scale mosaics of biotic communities.

This article discusses the variability of European ecosystems and a major framework of biodiversity patterns at various scales of complexity.

Europe is a continent, a large chunk of land; therefore, I discuss this subject from the viewpoint of diversity of large-scale ecological units such as vegetation megazones and biomes. The terrestrial habitats (and their complexes) will dominate this discourse and I emphasize the diversity of vegetation assemblages, which are traditionally considered the core of structure and functioning of all terrestrial ecosystems.

Europe is approximately 10 million km² and spans 35–81°N of latitude and 60°E–10°W of longitude. Although Europe is the second smallest continent (7% of the world's land surface), the basic classification of ecosystem diversity can be discussed only at a very large scale of biome. Not only the limited extent of this article but also the extraordinary wealth of scientific knowledge collected over centuries and the diversity of the subject impose constraints on the level of detail and accuracy of this discourse.

II. ECOSYSTEM DIVERSITY: CONCEPTS AND APPROACHES

Ecosystem, often defined as the union of biotic community and its environment, presents an obvious difficult scaling problem. Ecosystems as real entities occur in space and time. Depending on conditions, one can recognize natural or artificial borders allowing classification of ecosystems, an exercise meant to set a framework for simplifying the complexity of ecosystems and featuring them in a synoptic way.

A. Classification of Vegetation

Most European ecosystems are obviously terrestrial ecosystems accompanied by semiterrestrial and aquatic ones. Vegetation (defined as the unity of plants occupying certain area in a certain time; Palmer and White, 1994) is the major biotic element of terrestrial ecosystems. Hence, the problem of classifying terrestrial ecosystems is a problem of classification of vegetation.

The approaches to the classification of vegetation are manifold and largely depend on criteria (and their weighting), aims, and means (Mucina, 1997a). Several basic approaches to classification of vegetation have emerged during the past century (Whittaker, 1978). Among these, the floristic–sociologic approach (Westhoff and van der Maarel, 1978; Dierschke, 1994) became a standard communication tool among European (and other) vegetation scientists involved in or harvesting fruits of vegetation classification exercises.

1. Floristic–Sociologic Approach to Vegetation Classification

The basis of the floristic–sociological approach to classification of vegetation is the notion of the total floristic composition and the notion that some species indicate the environmental conditions better than others. These are called diagnostic species (Braun-Blanquet, 1964). Vegetation stands are selected by following various criteria, among which the so-called representativity of the studied vegetation type is the leading one. This approach is aimed at recognition of plant communities—plant assemblages recurrent in space and in time and showing distinct floristic composition which reflects certain combinations of current and past habitat conditions. The basic vegetation unit of this approach is the so-called association, a theoretical construction that is a result of the abstraction classification process involving many vegetation stands. A hierarchy of vegetation units based on groups of species having similarity between plant communities and their groups is also a vital part of the floristic–sociological approach. Floristically and ecologically similar associations are grouped into alliances, the alliances are grouped into orders, and the orders are grouped into classes. The vegetation classes can be less formally grouped into divisions. Particular associations, alliances, orders, classes, and divisions are termed syntaxa (Westhoff and van der Maarel, 1978) of various ranks and build the syntaxonomic hierarchy. An elaborate, highly formalized system of rules has been introduced to govern the formation of names of the syntaxa.

2. Dominance Approach to Classification of Vegetation

Another approach to classification of vegetation emphasizes the role of the dominant species or dominant growth forms, marking a departure from using plant taxonomy as the currency. Without discussing this approach in great detail (Whittaker, 1978), the notion of formation is the core tool of the dominance approach. The formation is the vegetation component (thus, in terrestrial ecosystems it is the leading element) of the concept of biome. Although the floristic–sociological approach is well suited for classification and description of vegetation on small geographic scales, formation and biome are, for the same purpose, well suited on large scales (subcontinental and continental).

3. Classification of Habitats

Despite much critical challenging of some subjective points regarding criteria for selection of stands and criteria of weighting of species in the classification process and sometimes awkward nomenclature, the floristic–sociological approach is aimed at enhancing effectiveness of communication when addressing units of vegetation cover. It is therefore not surprising that it was the terminology and concepts of this approach that were adopted as the basis of the majority of the units of the habitat systems developed for various purposes by the former European Communities (now European Union) authorities.

The classification of habitats passed several stages of development, spanning CORINE habitat classification (Commission of European Communities, 1991), Palearctic habitat classification (Devilliers and Devilliers-Terschuren, 1996), and EUNIS3 habitat classification (Davies and Moss, 1998). Obviously, the purpose of habitat classification is a practical one. It should serve (and serves) important purposes by delimitation and evaluation of land-use characteristics, projection and management of nature reserves of other areas of special nature conservancy interests (e.g., the Natura 2000 network of Special Areas of Conservation), etc. Not surprisingly, such a habitat classification scheme is a legal standard within the borders of the European Union.

III. BIODIVERSITY OF HABITATS AND VEGETATION

A. Sources

The current vegetation landscape of Europe is a result of manifold forces forming and reforming the habitat networks during the past few thousands of years; undoubtedly, some features of this habitat and vegetation patchwork are very old.

One can hardly avoid using oversimplification when talking about the sources of community (including vegetation and animal communities) and ecosystem diversity of present-day Europe. I distinguish four major prerequisites to the community and ecosystem diversity (Table I):

1. Taxon diversity: Plant and animal communities are composed of individuals belonging to several (often many) taxa. Speciation and associated processes of within-taxon diversification in addition to migrations of taxa—either ancient, natural ones or current ones largely driven by man's activities—are the major sources of taxon diversity.

2. Habitat diversity: Habitats are "homes" of biotic communities and the greater the variety of habitats, the greater the variety of biotic communities populating them. The character of habitats is determined primarily by ecological factors—their character and dynamics. Regardless of whether one considers spatial or temporal scales, one can recognize geological, geomorphologic,

TABLE I

Prerequisites and Sources of the Ecosystem and Community Diversity

Prerequisite	Sources
Taxon diversity	Speciation
	Migrations
Habitat diversity	Geological processes
	Geomorphological processes
	Soil-formation processes
	Hydrologic dynamics
	Climate change
Diversity of biotic interactions	Competition and related negative interactions
	Facilitation and related positive interactions
Diversity of human interference	Disturbance and removal of habitats
	Creation of new habitats

and soil-formation processes accompanied by hydrological dynamics and climate changes as the major natural sources of the habitat diversity.

3. Diversity of biotic interactions: The individuals representing different taxa interact within the biotic communities in many ways. They may compete for resources, facilitate each other's growth and reproduction by a plethora of positive interactions, and may behave indifferently. Undoubtedly, the biotic interactions form the face of each community and thus contribute to ecological diversity.

4. Diversity of human interference: Man has made himself the center of the universe and from this point of view has also become the dominant source of the disturbance (up to complete removal) but also creation of new habitats and new biotic communities through manipulating ecological factors and facilitation of species migrations.

B. Patterns

1. Biomes of Europe

Europe has six (of the world's nine) zonobiomes as defined by Heinrich Walter (Walter and Breckle, 1991, pp. 22–25). These include, from south to north, Mediterranean zonobiome (IV), warm-temperate zonobiome (V), nemoral zonobiome (VI), continental zonobiome (VII), boreal zonobiome (VIII), and polar zonobiome (IX). Furthermore, many zonoecotones mediate between the particular zonobiomes and these are designated as IV–V, V–IV, V–VI, VI–VII, VII–VI, VII(III), VIII–VI, and IX–VIII. The zonoecotone III–IV, mediating between subtropical zonobiome (not represented on the continent) and Mediterranean zonobiome, is found on the Canary Islands.

The Walter's biome scheme, largely based on a combination of climatic characters, is currently under revision in light of more modern definitions of the biome concept involving the combination of climate and functional types (including traditional life-forms) using modeling to predict vegetation patterns (Prentice et al., 1992; Haxeltine and Prentice, 1996) or to reconstruct the patterns of the past (Prentice et al., 1996).

Following the combined approach to define biomes using climate and vegetation structure, one can tentatively recognize 10 biomes in Europe and the Canary Islands (Table II).

Surprisingly little attention has been devoted to the definition of extant European biomes in comparison with other parts of the world. This can to a great extent be attributed to traditions in classification of ecosystems in Europe. The dominance-based approach to classification of vegetation leading directly to definition of biomes when exercised on a large scale is the predominant approach in parts of the world in which there is a lack of advanced knowledge on flora. Europe is undoubtedly a region in which the flora-based approach has a long and firmly rooted tradition.

2. Diversity of Vegetation Types

To list all syntaxa described in Europe is impossible here. Recent account of the high-rank syntaxa (Mucina, 1997b; Rodwell et al., 1998) revealed that the European vegetation can be classified into 75 classes (Table III

TABLE II

Preliminary Scheme of European Biomes with Corresponding Walter's Zonobiomes

Biome	Zonobiome and zonoecotone	Comment
Continental Semidesert Biome	VII(III)	Around Caspian Sea
Subtropical Semidesert Biome	III–IV	Canary Islands, southern Spain?
Mediterranean Biome	IV; IV–V	Entire Mediterranean basin
Temperate Grassland (Steppe) Biome	VII; VI–VII	Including Pannonian basin
Temperate Laurisilva Forest Biome	V; V–VI	Macaronesia, Bay of Biscaya?
Temperate Rain Forest Biome	V	Possibly on Azores
Temperate Deciduous Forest Biome	VI; VII–VI	Most of western and central Europe
Boreal Evergreen Forest Biome	VIII; VIII–VII; VIII–VI	Scandinavia, Russia
Tundra Biome	IX in part; IX–VIII	Scandinavia, northern Russia
Arctic Desert Biome	IX in part	Northern Europe, Arctic islands

TABLE III

Survey of European Phytosociological Classes and Their Correspondence with the Biome and Orobiome Classification Scheme[d]

Class	Or	Al	Characteristics	Biome/zonality
Aquatic vegetation				
Lemnetea	1	2	Communities of floating cormophytes (pleustophytes)	All but T Azonal
Charetea fragilis	3	5	Communities of submerged ramified macroalgae	All but T Azonal
Potametea	3	7	Communities of freshwater, rooted, submerged, and floating macrophytes	All Azonal
Ruppietea maritimae	1	2	Communities of rooted, submerged macrophytes of brackish waters	B, N, M, CSD Azonal
Zosteretea	3	3	Communities of rooted, submerged macrophytes of shallow seas	Shallow sea Azonal
Vegetation of freshwater marshes and fens				
Isoeto-Littorelletea	2	8	Dwarf amphibious vegetation of oligotrophic water bodies, and dystrophic lakes and bog hollows	T, B, N, M, OA Azonal
Isoeto-Nanojuncetea	2	10	Dwarf amphibious vegetation of banks and bottoms of mesotrophic and eutrophic temporary water bodies	T, B, N, M, OA Azonal
Montio-Cardaminetea	1	8	Moss- and herb-rich vegetation of water springs and edges of fast-running high-mountain rapids	T, B, N, M, OM Azonal
Phragmiti-Magnocaricetea	3	12	Vegetation of reeds and sedge-dominated fresh-water and brackish swamps	T, B, N, OA Azonal
Scheuchzerio-Caricetea	2	6	Vegetation of transitional mires, fens and bog hollows	T, B, N, M, OA, OM Azonal
Oxycocco-Sphagnetea	3	9	Vegetation of ombrogenic raised bogs and wooded boreal bogs	T, B, (N), OA Azonal
Coastal vegetation				
Cakiletea maritimae	3	7	Strandline dwelling, short-lived nitrophilous vegetation	B, N, M Azonal
Saginetea maritimae	2	5	Dwarf pioneer vegetating populating loamy and sandy soils in habitats under salt-spray influence	B, N, M Intrazonal?
Agropyretea pungentis	1	1	Coastal, salt-sprayed grasslands on soft rocks and related vegetation of inland saline heavy loamy soils	M Intrazonal
Crithmo-Staticetea	3	12	Chasmophytic coastal vegetation under salt-spray influence	B, N, M Azonal
Thero-Salicornietea	2	7	Vegetation of maritime salt marshes dominated by annual succulents	B, N, M Azonal
Spartinetea maritimae	1	1	Pioneer salt marsh grassland swards dominated by Spartina	B, N, M Azonal
Salicornietea fruticosae	2	11	Mediterranean and thermo-Atlantic perennial salt marsh scrub	M Intrazonal
Juncetea maritimi	3	13	Perennial maritime meadows and related herb-rich salt marshes	B, N, M Azonal
Honckenyo-Elymetea	2	4	Vegetation of embryonic shifting dunes of Atlantic, North Sea, and Baltic coasts	B, (N) Azonal
Ammophiletea	2	8	Vegetation of white and gray coastal sand dunes of Atlantic–Mediterranean distribution	(B), N, M Azonal
Chasmophytic vegetation				
Asplenietea trichomanis	18	73	Chasmophytic vegetation of rock faces, fissures, and ledges	B, N, M, OA, OM Azonal

continues

continued

Class	Or	A1	Characteristics	Biome/zonality
Adiantetea	1	2	Chasmophytic, fern- and moss-rich communities of water-rilled habitats of the Mediterranean	M Intrazonal
Thlaspietea rotundifolii	13	49	Vegetation of screes, gravel riverbanks, and related habitats	B, N, M, OA, OM Azonal
Arctic and alpine vegetation				
Loiseleurio-Vaccinietea	1	4	Dwarf shrub alpine and subalpine heathland of boreal and arctic Europe	T: Zonal OA: intrazonal
Salicetea herbaceae	1	9	Snow-bed communities on siliceous substrates	T, OA Intrazonal?
Juncetea trifidi	7	21	Siliceous alpine and boreal grasslands	T, OA Intrazonal?
Carici-Kobresietea	3	3	Circumpolar cold steppe and field vegetation and alpine wind-swept Elyna grasslands	T: Zonal OA: intrazonal
Elyno-Seslerietea	5	19	Alpine and subalpine calcareous grasslands	(B), OA Intrazonal?
Mulgedio-Aconitetea	3	15	Subactic–subalpine to alpine tall-grass and tall-herb grasslands and related krummholz	B, N, OA, OM Azonal?
Synanthropic vegetation				
Oryzetea sativae	1	1	Weed communities of rice fields	N, M Azonal
Bidentetea tripartiti	1	2	Annual ruderal communities of periodically flooded, nutrient-rich riparian and related synanthropic habitats	B, N, M Azonal
Polygono-Poetea annuae	1	4	Short-lived, trampled, ruderal plant communities	B, N, M, OA Azonal
Stellarietea mediae	6	49	Annual, herb-rich ruderal and agrestal communities	B, N, M, OA Azonal
Artemisietea vulgaris	6	18	Perennial thistle-rich (sub)xerophilous ruderal communities of temperate and Mediterranean regions	B, N, M (CSD) Azonal
Galio-Urticetea	5	18	Nitrophilous synanthropic herb-rich communities of woodland and riparian fringes	B, N, M Azonal
Epilobietea angustifolii	1	4	Tall-herb and tall-grass communities of woodland clearings and related shrubbery	B, N, (M) Azonal
Temperate heathlands and grasslands				
Calluno-Ulicetea	3	16	Temperate and boreal heathlands and related grasslands on nutrient-poor soils	N: Intrazonal R: zonal
Koelerio-Corynephoretea	6	17	Grasslands and related short-lived, herb-rich plant communities of sandy, nutrient-poor soils of temperate and boreal Europe	(B), N Intrazonal
Molinio-Arrhenatheretea	9	39	Nutrient-rich, mesic (pastures, hay meadows, and lawns) and wet grasslands	B, N, (M), S, OA Azonal
Trifolio-Geranietea	2	5	Herb-rich fringe communities of temperate woodlands	B, N, (M) Azonal
Festuco-Brometea	12	64	Steppes, rocky steppes, and sandy grasslands of the temperate and subboreal regions	S: Zonal N: intrazonal
Puccinellio-Salicornietea	8	37	Continental (inland) salt marshes, salt pans, and salt steppes	(N), S, (CSD) Intrazonal?
Temperate and boreal woodlands and scrub				
Rhamno-Prunetea	3	17	Shrub mantle communities in regions of temperate deciduous woods	B, N, (M) Azonal
Salicetea purpureae	1	7	Willow and poplar riparian woods and scrub of temperate Europe	B, N, M Azonal

continues

continued

Class	Or	Al	Characteristics	Biome/zonality
Alnetea glutinosae	3	5	Alder and willow carrs	B, N, (M) Azonal
Querco-Fagetea	5	22	Deciduous mesic and subxerophilous woods of temperate and subboreal Europe	(B), N, (M) Zonal
Erico-Pinetea	1	7	Relict pine woods of oro(sub)mediterranean and montane temperate distribution over limestone	(N), OA, (OM) Intrazonal
Junipero-Pinetea	2	7	Oromediterranean dry juniper–pine woods and inner-alpine steppic pine woods and related scrub	(OA), OM Intrazonal
Vaccinio-Piceetea	2	9	Boreal needle woods	B, (N) Zonal
Oromediterranean grassland and scrub				
Festucetea indigestae	1	3	Oro-Iberian siliceous Festuca indigesta grasslands	OM Zonal
Saginetea piliferae	1	2	Cyrno-Sardian oromediterranean siliceous grasslands and herblands	OM Zonal
Festuco-Ononidetea	2	9	Dry basiphilous pastures at high altitudes of the Submediterranean and western Mediterranean	OA, OM Zonal?
Carlino-Genistetea	1	3	Cyrno-Sardian oromediterranean calciphilous grasslands and phrygana	OM Zonal
Rumici-Astragaletea	1	1	Sicilian oromediterranean scrub and related grasslands	OM Zonal
Daphno-Festucetea	2	7	Greek and Aegean oromediterranean calciphilous grasslands and phrygana	OM Zonal
Mediterranean vegetation				
Thero-Brachypodietea	4	15	Mediterranean terrestrial plant communities dominated by annual low-grown herbs and grasses	M Intrazonal
Lygeo-Stipetea	3	9	Mediterranean pseudosteppes and related perennial grasslands	M Intrazonal
Cytisetea scopario-striati	1	7	Thermomediterranean broom heathlands (retamal)	M Intrazonal
Cisto-Levanduletea	2	5	Low-grown Mediterranean scrub (macchia, matorral, garrigue, tomillar, and phrygana) on siliceous and ultramafic substrata	M Zonal
Rosmarinetea officinalis	6	16	Low-grown calciphilous Mediterranean scrub (maquis, matorral, garrigue, tomillar, and phrygana)	M: Zonal OM: intrazonal
Nerio-Tamaricetea	3	8	Mediterranean riparian scrub	M Intrazonal
Quercetea ilicis	2	16	Mediterranean broad-leaved forests and maquis	M Intrazonal
Semidesert vegetation				
Pegano-Salsoletea	3	10	Thermomediterranean and Macaronesian halonitrophilous semidesert scrub	M Intrazonal
Artemisietea lerchianae	2	2	Aralo-Caspian subhalophilous semideserts	SD Zonal
Macaronesian vegetation				
Zygophyllo-Polycarpetea	1	2	Macaronesian. halophilous coastal dune scrub	SSD Intrazonal
Kleinio-Euphorbietea	1	1	Macaronesian succulent scrub on lava beds (tabaibal and cardonal)	SSD Zonal
Oleo-Rhamnetea	2	2	Macaronesian matorral and related scrub	M Zonal

continues

continued

Class	Or	Al	Characteristics	Biome/zonality
Pruno hixae-Lauretea	2	7	Macaronesian laurisilva and related scrub	L
				Zonal
Cytiso-Pinetea	1	1	Canarian pine woods and related scrub	M
				Intrazonal
Spartocytisetea	1	1	Canarian high-mountain volcanic semideserts	SSD
				Zonal
Aeonio-Greenovietea	1	2	Macaronesian chasmophytic vegetation of exposed volcanic rocks	M, L, SSD
				Azonal

[a] The definition of Europe includes Europe proper, Iceland, Madeira, Azores, and the Canary Islands. For a full nomenclature, major species groups, and literature sources, see Mucina (1997a). Or, number of phytosociological orders; Al, number of phytosociological alliances. Biomes and orobiomes: T, Tundra Biome; B, Boreal Evergreen Forest Biome; N, Temperate Deciduous Forest Biome; M, Mediterranean Biome; CSD, Continental Semidesert Biome; SSD, Subtropical Semidesert Biome; S, Temperate Grassland (Steppe) Biome; L, Temperate Laurisilva Forest Biome; R, Temperate Rain Forest Biome; OA, Alpine Orobiome; OM, Mediterranean Orobiome. The abbreviations in parentheses indicate "in part" statements.

provides the major features of these classes), 223 orders, and 844 alliances. The number of associations is unknown, although one estimate suggests there might be as many as 2500–3000. For further details on the diversity of syntaxa occurring in various European countries, refer to the many national vegetation surveys (and transnational vegetation accounts) listed in Table IV.

Several interesting biodiversity patterns can be observed within the groups of syntaxa, including classes, orders, and alliances (Table III):

1. The Mediterranean region (supporting the Mediterranean biome and Mediterranean orobiome) shows a high concentration of classes and subordinated syntaxa. This can be partly ascribed to high alpha, beta, and gamma diversities (Cowling *et al.*, 1996) and partly to the insular character of many types of Mediterranean habitats (and their complexes). Not only are numerous true marine islands very abundant within the Mediterranean but also the mountain summits form an archipelago of their own, demonstrating their own rates of evolution and combination of ecological and biogeographical factors leading to specific plant assemblages.

2. The high-latitude regions (arctic and boreal zones) show lower diversity of vegetation types partly as a result of low extant alpha diversity that diminished dramatically by the last ice age, partly because of the large-scale occurrence of uniform habitat complexes, and partly due to adversity of climate.

3. A group of azonal (and intrazonal) rather than broadly distributed zonal classes shows the highest diversity of orders and alliances (reflecting high diversity of ecological–biogeographic patterns). Among those with a particular position are Asplenietea trichomanis

and Thlaspietea rotundifolii (classes comprising vegetation of special habitats such as rock faces, fissures, and screes), on the one hand, and classes of Festuco-Brometea, Koelerio-Corynephoretea, and Puccinellio-Salicornietea (comprising both primary and secondary dry grasslands, the latter on saline soils) on the other hand. The former two classes have insular character of distribution pattern: The communities are small scale, occupying ecologically isolated habitats and housing most European endemic species. The group of dry grassland classes have relatively large-scale distribution and comprise Europe's most species-rich plant communities. Russian steppes, for instance, can support more than 60 species per square meter. Large-scale distribution and rich species pools are prerequisites to high beta diversities manifested in a high number of vegetation types.

4. The synanthropic (man-made and man-controlled) vegetation of Europe is also very diverse in terms of the number of syntaxa. The classes of Stellarietea media, Artemisietea, and Galio-Urticetea comprise 49, 18, and 18 alliances, respectively. The same pattern applies to the Molinio-Arrhenatheretea, comprising 39 alliances of plant communities of secondary mesic grasslands. These numbers are indicative of high diversity of disturbance factors related to human activities such as agriculture, silviculture, building activities, and unintentional transport of seeds and fruits over continental borders. The flora of Europe was enriched by many alien species adding to the diversity of vegetation types.

5. Mediterranean classes such as Thero-Brachypodietea, Rosmarinetea officinalis, Quercetea ilicis, and, to some extent, Cisto-Micrometea also show high diversity with regard to the number of alliances. Still

TABLE IV

Major Monographs and Summary Accounts of Ecosystem, Habitat, and Vegetation Diversity in Europe[a]

Source	Country	Type[b]	Comment
All-Europe surveys of ecosystems, vegetation, and habitats			
Goodall (1977–1999)	World	SM	Accounts on major ecosystems of the world (including Europe)
Braun-Blanquet(1933–1940)	Europe	SM	Unfinished series of syntaxonomic monographs
Tüxen(1975–1981)	Europe	SM	Unfinished series of syntaxonomic monographs
Dierßen (1996)	Northern Europe	M	Detailed account on vegetation of northern Europe
Ellenberg (1998)	Central Europe	M	Last edition of a famous vegetation monograph
Ernst (1976)	Europe	M	Syntaxonomic survey of the Violetea calaminariae
Hartmann and Jahn (1967)	Central Europe	M	Woodland vegetation
Hartmann (1974)	Central Europe	M	Woodland vegetation
Horvat *et al.* (1974)	Southeastern Europe	M	Detailed account of vegetation of Southeastern Europe and the Balkans
Mayer (1984)	Europe	M	Woodland vegetation
Ozenda (1988)	Europe	M	Vegetation monograph; translated into English
Polunin and Walters (1985)	Europe	M	Popular account of major biomes and ecosystems of Europe
von Hübschman (1986)	Central Europe	M	Survey of moss communities
Commission of European Communities (1991)	Europe	R	CORINE habitat classification system
Davies and Moss (1998)	Europe	R	EUNIS habitat classification
Devilliers and Devilliers-Terschuren (1996)	Europe	R	Palearctic habitat classification
Påhlsson (1994)	Northern Europe	R	Nordic habitat classification system
Rodwell *et al.* (1998)	Europe	R	Syntaxonomic calibration of the EUNIS units
Klika and Hadač (1944)	Central Europe	C	Annotated checklist of syntaxa of Central Europe
Mucina (1997)	Europe	C	Conspectus of vegetation classes
Lohmeyer *et al.* (1962)	Europe	L	List of high-ranked syntaxa of Europe
National surveys			
Moravec (1998)	Czech Republic	SM	First volume of Czech Vegetation Survey
Mucina *et al.* (1993)	Austria	SM	Completed series in three volumes
Rodwell (1991–1998)	United Kingdom	SM	Completed series in five volumes
Schaminée *et al.* (1995–1999)	Netherlands	SM	Completed series in five volumes
Valachovic (1996 et seq.)	Slovakia	SM	Two volumes of the Slovak Vegetation Survey
Various editors (1996 et seq.)	Germany	SM	Four volumes of the German Vegetation Survey
Coldea (1997)	Romania	M	First volume of national survey
Fremstad (1997)	Norway	M	Monographic survey of major vegetation types
Moravec (1998)	Czech Republic	M	First volume of national survey
Peinado Lorca and Rivas-Martínez (1987)	Spain	M	Account of vegetation of regions of Spain
Pott (1995)	Germany	M	Second edition
Rašomavičius (1997)	Lithuania	M	First volume of national survey (meadows)
Soó (1980)	Hungary	M	A volume devoted to vegetation from a six-volume series
Szafer (1959)	Poland	M	Monographic survey of vegetation in two volumes
Tansley (1949)	United Kingdom	M	Monographic survey of vegetation in two volumes
Westhoff and Den Held (1969)	Netherlands	M	Brief, systematic descriptive account of vegetation units
Apostolova and Slavova (1997)	Bulgaria	C	Account of vegetation units described between 1891 and 1995
Borhidi (1996)	Hungary	C	Account of Hungarian vegetation units
Braun-Blanquet *et al.* (1952)	France	C	Account of units of the Mediterranean France
Julve (1993)	France	C	Annotated checklist of high-rank vegetation units of France
Korotkov *et al.* (1991)	USSR	C	Annotated checklist of vegetation units of the former USSR
Matuszkiewicz (1981)	Poland	C	Key to identification of vegetation units
Moravec (1983)	Czech Republic	C	Annotated (conservation aspects) checklist of syntaxa
Solomakha (1996)	Ukraine	C	Annotated account of vegetation units
Lakušič *et al.* (1977)	Bosnia	L	List of syntaxa of Bosnia and Herzegowina
Rivas-Martínez *et al.* (1998)	Spain and Portugal	L	List of syntaxa for entire Spain and continental Portugal
Soó (1971)	Hungary	L	Checklist of vegetation units described from Hungary
Zupančič (1986)	Yugoslavia	L	List of syntaxa of the former Yugoslavia

[a] A bibliography of the featured surveys and checklists can be requested from the author.

[b] M, single monographic treatment; SM, series of monographic treatments; R, report serving a government body; C, conspectus (short account of vegetation units including limited descriptive treatment); L, list of vegetation units.

higher diversity becomes obvious on the level of associations reflecting the relict character and diversity of disturbance regimes showing a fine-grain pattern in the Mediterranean.

6. The lowest diversity of vegetation types is encountered in the classes typifying vegetation of the freshwater, marine, and coastal vegetation. Often, only a few orders or alliances are found within classes such as Lemnetea minors, Potametea, Ruppietea, Spartinetea, Hockenyo-Elymetea, and Crithmo-Staticetea. However, on the whole, the nonterrestrial complex of habitats is very diverse, which is reflected in the high number of phytosociological classes.

3. Diversity of Habitat Types

Although vegetation forms, at least in the terrestrial ecosystems, the major component of these ecosystems, the habitat typology cannot reflect and is not reflecting only vegetation typology. For instance, as found by Rodwell *et al.* (1998), 60% of 277 units of the EUNIS3 habitat classification were characterized in phytosociological (syntaxonomic) terms, whereas the other units are largely abiotic and carry geomorphologic and hydrological features or they are associated with fauna or nonvascular plants. Table V shows that this is the case especially in EUNIS habitat classes (groups of similar habitats) such a class A (marine habitats) and class J (constructed, industrial and other artificial habitats).

The judgment on diversity of habitats appears more difficult than that based on classification of vegetation. First, there are many incommensurable variables to be considered by definition of a habitat type. Second, the spatial and temporal scaling of the habitat types

TABLE V

Simplified System of EUNIS3 Habitat Units in Relation to Syntaxonomic Units[a]

EUNIS3 habitat class/subclass	Cl	Or	Al
Marine habitats			
Littoral rocks	—	—	—
Littoral sediments	2	2	2
Coastal salt marshes and saline habitats	8	10	17
Infralittoral (shallow subtidal or nontidal) rocks	—	—	—
Infralittoral (shallow subtidal or nontidal) sediments	3	5	5
Circalittoral (deep subtidal or nontidal) rocks	—	—	—
Circalittoral (deep subtidal or nontidal) sediments	—	—	—
Deep seabed (more than 200 m depth)	—	—	—
Pelagic water body	—	—	—
Anoxic marine habitats	—	—	—
Tidal or permanently flooded sea caves	—	—	—
Coastal habitats			
Coastal dune and sand habitats	22	25	33
Coastal shingle habitats	2	4	4
Rock cliffs, ledges, and shores, including the supralittoral	3	3	3
Inland surface water habitats			
Surface standing waters	8	12	16
Surface running waters	2	2	3
Littoral zone of inland surface water bodies	9	9	9
Mire, bog, and fen habitats			
Raised and blanket bogs	2	4	7
Valley bogs, poor fens, and transition mires	4	5	11
Aapa, palsa, and polygon mires	—	—	—
Base-rich fens	3	3	3
Sedge and reedbeds, normally without freestanding water	3	5	5
Inland saline and brackish marshes and reedbeds	2	2	4

continues

continued

EUNIS3 habitat class/subclass	Cl	Or	Al
Grassland habitats			
Dry grasslands	17	43	75
Mesic grasslands	5	9	21
Seasonally wet and wet grasslands	6	10	29
Alpine and subalpine grasslands, tall herbs, and ferns	9	18	35
Moss- and lichen-dominated habitats	3	3	3
Inland saline grasslands	3	7	7
Heatlands and scrub habitats			
Arctic and temperate scrub habitats	8	9	10
Maquis, matorral, and submediterranean deciduous thickets	11	15	34
Arctic, alpine, and subalpine dwarf shrub habitats	6	9	13
Temperate heathland	3	4	9
Garrigues	10	12	13
Spiny Mediterranean heath (phrygana and hedgehog heaths)	12	15	17
Themo-Atlantic xerophytic habitats	3	3	4
Hedgerows	—	—	—
Shrub plantations	—	—	—
Woodland and forest habitats and other wooded land			
Broad-leaved deciduous and mixed woodland	13	22	50
Broad-leaved evergreen woodland	7	8	16
Coniferous woodland	13	18	33
Lines of trees, sparsely wooded land, early stage woodland, and coppice	—	—	—
Inland sparsely vegetated and unvegetated habitats			
Terrestrial underground case, cave systems, and water bodies	—	—	—
Screes	1	9	35
Inland cliffs and exposed rock habitats	7	19	22
Frost or ice-dominated habitats	—	—	—
Inland sedimentary and organic habitats with very sparse or no vegetation	1	1	1
Volcanic features	3	3	3
Regularly or recently cultivated habitats and gardens			
Arable land and market gardens	2	2	2
Gardens	3	2	2
Constructed, industrial, and other artificial habitats			
Buildings of cities, towns, and villages	2	2	2
Low-density buildings	—	—	—
Extractive industrial sites	—	—	—
Transport networks	—	—	—
Highly artificial man-made waters and associated structures	—	—	—
Waste deposits	—	—	—

[a] From Davies and Moss (1997) and Rodwell *et al.* (1998). Cl, Or, and Al: number of phytosociological classes, orders, and alliances, respectively.

is very complex. This complexity is reflected in all classification systems of habitats know until today. In order to cope with the complexity, the habitat classification systems are hierarchical, but often on the same level of hierarchy, thus mixing small-scale habitats with habitat complexes recognizable on the landscape level. Obviously, the calibration of the habitat units and the classification system leaves much room for improvement (Rodwell *et al.*, 1998; Waterton *et al.*, 1998).

IV. CHALLENGES FOR THE FUTURE

There are many perspectives on ecological diversity. I present one that is not the favorite of granting agencies in Europe (and perhaps also elsewhere): the perspective of the diversity of ecosystem types themselves.

The legitimate status of the concept of ecological diversity as a virtual part of biodiversity has been recognized since "biodiversity," a brushed-up concept, started to sweep research agendas throughout the world: This happened largely due to the catalyzing effect of the U.S. Strategy Conference on Biological Diversity (Ghilarov, 1996) and later "beatification" of biodiversity by the Rio World Summit and resulting documents of the convention. Ecological diversity (including diversity of communities, ecosystems, landscapes, and biomes) was recognized by international and national bodies as a subject worth studying (and supporting) in the name of developing technological tools for its protection and sustainable use.

In Europe, as elsewhere, taxonomic diversity is the leading topic of biodiversity research and conservation. However, due to deeply rooted research traditions and the awareness of national governments, surveys on higher levels of biological complexity have been given much attention in the past: Almost every European nation has a national vegetation map, and many have land-use and landscape-unit maps. Modern national vegetation surveys have been recently completed or are under way in many crucial European countries [see Table IV for a survey; see Mucina *et al.* (1994) and Rodwell *et al.* (1995) for progress reports].

With Europe growing together as a union of nations sharing political agendas and an economic future, the large-scale tools for ecological diversity conservation and sustainable use are emerging. These include pan-European habitat classification scheme EUNIS3, a vegetation map of Europe (a long-term cooperative project headed by Dr. U. Bohn, Bonn, Germany), and the European Vegetation Survey (a Working Group of the International Association for Vegetation Science). The fall of the Iron Curtain in 1989 and 1990 posed a new challenge for research agendas in ecological diversity—unification of scientific standards and application tools along the West–East gradient within Europe. There is still much progress to be made.

Also, because we need to know the flora and fauna (as well as their intraspecific variability down to the genetic structures), we need to survey biotic communities, habitats, ecosystems, and biomes. This need is not only fed by scientific curiosity but also by practical, technological needs to ensure effective conservation and use of biological diversity as a whole.

Acknowledgments

I am indebted to Milan Chytrý, Hartmut Dierschke, Julian Dring, Sandro Pignatti, John Rodwell, and Joop Schaminée for long and fruitful discussions on the diversity of European vegetation—our common passion. Roberto Canullo carefully read the penultimate manuscript version.

See Also the Following Articles

AFRICA, ECOSYSTEMS OF • ASIA, ECOSYSTEMS OF • AUSTRALIA, ECOSYSTEMS OF • NORTH AMERICA, PATTERNS OF BIODIVERSITY IN • SOUTH AMERICA, ECOSYSTEMS OF

Bibliography

Braun-Blanquet, J. (1964). *Pflanzensoziologie. Grundzüge der Vegetationskunde*, 3rd ed. Springer, Wien.
Commission of European Communities (1991). *CORINE Biotopes. The Design, Compilation and Use of an Inventory of Sites of Major Importance for Nature Conservation in the European Communities.* Office for Official Publications of the European Communities, Luxembourg.
Cowling, R. M., Rundel, P. W., Lamont, B. B., Arroyo, M. K., and Arianoutsou, M. (1996). Plant diversity in Mediterranean-climate regions. *Trends Ecol. Evol.* **11**, 362.
Davies, C. E., and Moss, D. (1998). EUNIS habitat classification: Final report. European Topic Centre on Nature Conservation, 1997 Work Programme Task 7.5.1. Report to the European Environmental Agency, December 1997.
Devilliers, P., and Devilliers-Terschuren, J. (1996). A classification of palearctic habitats, Nature and Environment No. 78. Council of Europe, Strasbourg.
Ghilarov, A. (1996). What does "biodiversity" mean—Scientific problem or convenient myth? *Trends Ecol. Evol.* **11**, 304.
Haxeltine, A., and Prentice, I. C. (1996). BIOME3: An equilibrium biosphere model based on ecophysiological constraints, resource availability and competition among plant functional types. *Global Biogeochem. Cycles* **10**, 693.
Mucina, L. (1997a). Classification of vegetation: Past, present and future. *J. Veg. Sci.* **8**, 751.
Mucina, L. (1997b). Conspectus of classes of European vegetation. *Fol. Geobot. Phytotax.* **32**, 117.
Mucina, L., Rodwell, J. S., Schaminée, J. H. J., and Dierschke, H. (1994). European Vegetation Survey: Current state of some national programmes. *J. Veg. Sci.* **4**, 429.
Palmer, M. W., and White, P. S. (1994). On the existence of ecological communities. *J. Veg. Sci.* **5**, 279.
Prentice, I. C., Cramer, W., Harrison, S. P., Leemans, R., Monserud, R. A., and Solomon, A. M. (1992). A global biome model based on plant physiology and dominance, soil properties and climate. *J. Biogeogr.* **19**, 117.
Prentice, I. C., Guiot, J., Huntley, B., Jolly, D., and Chaddadi, R.

(1996). Reconstructing biomes from palaeoecological data: A general method and its application to European pollen data at 0 and 6 ka. *Climate Dynamics* **12,** 185.

Rodwell, J. S., Pignatti, S., Mucina, L., and Schaminée, J. H. J. (1995). European Vegetation Survey: Update on progress. *J. Veg. Sci.* **6,** 759.

Rodwell, J., Dring, J., Pignatti, S., Schaminée, J., and Mucina, L. (1998). *Scientific Background to the EUNIS Habitat Classification: Phytosociological Relationships of EUNIS Habitats.* Unit of Vegetation Science, Lancaster University, Lancaster & European Topic Centre on Nature Conservation, Paris.

Walter, H., and Breckle, S.-W. (1991). *Ökologie der Erde. Band 1: Ökologische Grundlagen in globaler Sicht,* 2nd ed. Fischer, Stuttgart.

Waterton, C., Grove-White, R., Rodwell, J. S., and Wynne, B. (1995). CORINE: Databases and nature conservation. Report to World Wide Fund for Nature UK, Lancaster.

Westhoff, V., and van der Maarel, E. (1978). The Braun–Blanquet approach. In *Classification of Plant Communities* (R. H. Whittaker, ed.), pp. 287–399. Junk, The Hague.

Whittaker, R. H. (ed.) (1978). *Classification of Plant Communities.* Junk, The Hague.

EUTROPHICATION AND OLIGOTROPHICATION

JoAnn M. Burkholder
North Carolina State University

I. Introduction
II. Changes in Flora and Fauna
III. Oligotrophication: Reversing the Impacts
of Eutrophication

GLOSSARY

algae Primitive plantlike organisms that photosynthesize. These organisms can be unicellular, filamentous, or colonial microscopic forms (microalgae), or they can be macroscopic (macroalgae), consisting of a primitive plant body (thallus) that lacks vascular tissue in most species.

anoxia Condition in which there is almost no dissolved oxygen in the water (<0.1 mg of dissolved oxygen [DO]/L).

bloom Proliferation of algae in river, lake, estuarine, or marine waters. Older literature referred to a bloom as 5000 or more algal cells per liter (L) (or 5 cells per ml), although this density generally is too low to discolor the water. Algal "blooms" range in cell density from eukaryote blooms (e.g., dinoflagellates at 10^3 to 10^4 cells/ml, to cyanobacteria at 10^8 to 10^9 cells/ml).

eutrophic Trophic status of an aquatic ecosystem that is characterized by relatively low phytoplankton species diversity but high phytoplankton production (biomass as mean chlorophyll *a* ca. 15–40 μg/L), with the phytoplankton often dominated by cyanobacteria in lakes, and by dinoflagellates or other flagellates in estuaries); high nutrient concentrations and loadings (for example, in lakes, mean inorganic N ca. 1900 μg/L, mean total P ca. 80 μg/L), high decomposition in the bottom water and surface sediments (with abundant organic materials available for this process); and bottom-water dissolved oxygen deficits, sometimes with occasional to frequent fish kills. Eutrophic lakes typically are shallow with well-developed littoral zones (area where light penetration is sufficient to support growth of rooted plants), sometimes extending across most of the bottom area.

harmful algae Algae that are undesirable to humans because (a) they become too abundant in response to nutrient overenrichment and then, at night, use most or all of the oxygen in the water for their respiration, so that fish and other organisms suffocate or become seriously physiologically stressed; (b) they become too abundant in response to nutrient enrichment, and overgrow beds of desirable rooted vegetation so that the beneficial plants cannot receive enough light to survive; (c) they cause or promote disease in other plants or animals; or (d) they produce toxins that hurt or kill finfish, shellfish, or other higher trophic levels including humans. "Harmful algae" include prokaryotic cyanobacteria or blue-green algae. More recently, the term has been used to include organisms that are not photosynthetic, primitive plantlike organisms—for example, certain nontoxic animal-like dinoflagellates, which cause fish disease (e.g., *Amyloodinium ocellatum*), and toxic animal-like dinoflagellates (e.g., the toxic *Pfiesteria* complex), which do not have their own chloroplasts for photosynthesis, but which resemble plantlike di-noflagellates in appearance and other general features.

hypoxia Condition in which the water has depressed levels of oxygen that are too low to sustain healthy fish populations (usually considered as ≥0.1 to <2 [sometimes <4 or <4.5] mg DO/L). Note that hypoxic levels of 3–4 mg DO/L can stress or kill sensitive egg and larval stages of some finfish and shellfish species, and that many motile fish actively avoid hypoxic areas.

mesotrophic Trophic status of an aquatic ecosystem that is characterized by moderate phytoplankton production and moderate nutrient concentrations and loadings.

mixotrophy Form of nutrition involving both autotrophic (photosynthetic) and heterotrophic carbon acquisition.

oligotrophic Trophic status of an aquatic ecosystem that is characterized by relatively high phytoplankton species diversity but low phytoplankton production (chlorophyll a < ca. 10 [g/L]); low nutrient concentrations and loadings (for example, mean inorganic N < ca. 700 μg/L, mean total P < 10 μg/L); low decomposition (with little organic material available to decompose); and plentiful oxygen throughout the water column.

salt wedge Water from the ocean, with higher salt content, that moves into an estuary along the bottom of the water column, beneath less dense fresh water that has moved into the same area from a river. The salt content makes the ocean water heavier than the fresh (riverine) water, so that under calm conditions, the estuarine water becomes density-stratified. This "salt wedge" of bottom water can become somewhat isolated from the overlying fresh or less brackish water. The longer the period in which the total water column is not mixed by winds or storms, the more distinct the two water layers or strata become. Salt wedges most often develop in warm seasons when plantlike phytoplankton production is high in the surface waters and respiration by heterotrophs (bacteria, fungi, animals) is also high, especially in the lower water column and sediments. Nutrients typically are higher within the salt wedge than in the overlying water, because of decomposition processes. At the same time, the bottom-water salt wedge can become hypoxic or anoxic, underlying waters can be saturated or supersaturated with oxygen from phytoplankton photosynthesis.

trophic status Ranking system for aquatic ecosystems, based on the amount of organic production and nutrient (N,P) levels. The major component that is usually considered in assigning trophic status is phytoplankton production, but this can be misleading. For example, some lakes are classified as oligotrophic because water-column nutrients and phytoplankton production are low, despite the fact that benthic plant production (e.g., of rooted angiosperms) is high.

EUTROPHICATION IS the natural aging process of aquatic ecosystems, formerly used mostly in reference to the natural aging of lakes wherein a large, deep, nutrient-poor lake eventually becomes more nutrient-rich, more productive with plant and animal life, and slowly fills in to become a pond, then a marsh. More recently, this term has been used in abbreviated reference to "cultural" or accelerated eutrophication of lakes, rivers, estuaries, and coastal marine environments, wherein the natural eutrophication process is accelerated (often by hundreds or thousands of years) by human activities that add nutrients to the aquatic system. Oligotrophication is a reversal of the eutrophication process, in which waters become less nutrient-enriched and/or support less plant and animal production. This process can occur naturally, for example, when the inflow to a moderately productive lake is severed so that the lake slowly becomes an ombrotrophic, nutrient-poor system or bog. Oligotrophication can also occur from human intervention, for example, following diversion of sewage from a small lake and dredging/removal of nutrient-rich bottom sediments.

I. INTRODUCTION

A. Overview across Aquatic Habitats

From freshwater lakes to estuaries and marine coastal waters, human-derived sources of nutrient pollution have rapidly changed water quality and aquatic community structure within the past 200 years of industrialization and rapid population growth. The predominant theme throughout most of the world is increased nutrient enrichment or cultural eutrophication, rather than nutrient decreases or oligotrophication; thus, eutrophication will be emphasized here although both phenomena will be addressed. Despite advances in treatment of human sewage in some countries during the late 20th century, despite bans on use of phosphorus in certain domestic or industrial practices, despite declines in agricultural fertilizers in some geographic regions, and despite modest improvements in environmental education in localized areas, the massive recent increase in global human population growth has increased nutrient loadings to aquatic ecosystems (Fig. 1). These increases have been greatest in estuarine and coastal marine areas where population growth has been highest, and where nearly two-thirds of the people of the world now reside. As a reflection of this trend, estuaries have been reported to receive more nutrient inputs per unit surface area than any other type of aquatic ecosystem.

Nutrients are essential for primary production by phytoplankton, benthic micro- and macroalgae, and aquatic angiosperms, which directly or indirectly support aquatic food webs. In freshwaters, phosphorus is the least abundant among the nutrients needed in large quantity (macronutrients) by biological organisms. Thus, it is the first element that becomes limiting to biological productivity in many freshwater systems. Phosphorus is also the nutrient that limits plant growth in many tropical coastal marine waters. In temperate and polar coastal marine environments, nitrogen is the most important nutrient that first limits primary production; and at the estuarine interface between marine and freshwater habitats, both N and P can "colimit" plant production, especially in late winter-spring seasons of high precipitation and accompanying high inorganic N inputs. Other nutrients, notably silica and iron, can also sometimes significantly influence the outcome of species dominance and the structure and abundance of phytoplankton communities under cultural eutrophication.

In moderation, nutrients promote beneficial increases in phytoplankton and benthic algal production and, in turn, higher production of zooplankton, macroinvertebrates, finfish, and shellfish that utilize the plant production directly or indirectly for food. But when added in excess, nutrient pollution can cause overgrowth of micro- and macroalgae, leading to oxygen depletion in bottom waters and sometimes throughout the water column. Although fish kills can be an obvious sign of acute impacts from cultural eutrophication, associated subtle, chronic impacts may be more damaging to aquatic communities over the long term. Sustained loss in biodiversity is foremost among these chronic impacts, resulting from underlying mechanisms that scientists are only beginning to understand.

Cultural eutrophication promotes major shifts in the structure of both plant and animal communities, generally affecting dominant components of every trophic level from microbial decomposers to macrofauna. There is clear, compelling evidence of altered aquatic community structure and significantly reduced biodiversity from cultural eutrophication in many freshwater, estuarine, and marine ecosystems. Surface waters across the earth are now sustaining such impacts; even the open oceans are no longer sufficiently isolated to avoid nutrient pollution from atmospheric deposition. Similar trends have been demonstrated from the microfossil records of lakes and of estuaries, showing a dramatic increase in nutrients and associated organic carbon deposits and a sharp, sustained decrease in the diversity of aquatic species.

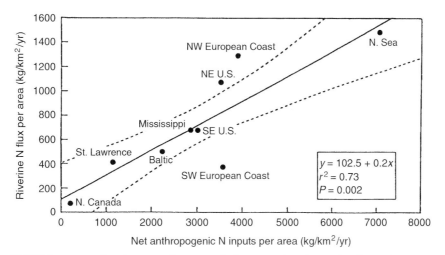

FIGURE 1 Export of total nitrogen from watersheds surrounding the North Atlantic Ocean, as a function of net anthropogenic inputs of nitrogen to their watersheds. Net anthropogenic inputs are defined as industrial N fertilizer + N fixation by legume crops + atmospheric inputs of oxidized N + net imports of N in food and feedstock. Reprinted from Vitousek et al. (1997), originally from Kluwer Academic Publishers, *Biogeochemistry* 35, 181–226, Howarth et al., Fig. 5a (1996) with kind permission from Kluwer Academic Publishers.

B. Factors Influencing Impacts from Nutrient Enrichment

Among the most important factors that influence the extent of adverse impacts from nutrient enrichment is water exchange rate, or the time required to completely replace the water in the system with new water. Waters that can "self-cleanse" or rapidly flush, such as fast-flowing rivers or run-of-river impoundments, are less sensitive to elevated nutrient loading than more contained or enclosed waters such as lakes and estuarine lagoons. Similarly, waters along open, wave-swept marine coasts are less sensitive to nutrient loading than quiet, sheltered upper embayments. The occurrence of other pollutants along with the nutrients—for example, suspended solids that can reduce light for growth of aquatic plants or toxic substances and harmful bacteria that can stress or kill some aquatic species—can exacerbate impacts from nutrient enrichment. The form of the nutrients can also be important. For example, swine wastes are much richer than untreated human sewage in organic nutrients (N, P, and C). Certain undesirable algae prefer organic nutrient forms over inorganic sources.

The timing and frequency of the source inputs are important, as well. During high-precipitation periods in winter, sewage is more effectively diluted and cold temperatures discourage the growth of many nuisance algae. Also, in some waterways, agriculture is the major source of nutrients annually, but during low-flow periods in warmer seasons, sewage can contribute half or more of the river volume and the loading of nutrients—at a time when the readily available nutrients stimulate noxious algal overgrowth of the system.

The initial aquatic community structure also influences the overall impacts of nutrient loads to aquatic ecosystems. Systems with low nutrient enrichment rely more on recycled nutrients than on introduction of new nutrient sources, and this may be a driving evolutionary force leading to greater specialization and diversity, as suggested by Howarth and others. Diverse communities, characteristic of oligotrophic or nutrient-depauperate waters, tend to be dominated by sensitive species that are limited to a narrow range of environmental conditions. Oligotrophic ecosystems are more sensitive (or less resistant), overall, to stress from nutrient enrichment. The most predictable outcomes are loss of sensitive species and increased abundance of generalist or opportunistic species that are more resistant to the undesirable water quality changes and other stresses effected by nutrient pollution. As an ecosystem becomes more eutrophic (that is, increasingly stressed by eutrophication), it becomes more resistant to further change.

Increasingly eutrophic lakes, lower rivers, and estuaries are commonly characterized by seasonal low oxygen stress. Hypoxia and anoxia develop when oxygen consumption exceeds supply, as a result of two major factors. The algae—often in densities of millions to billions of cells/ml of water—often are the first food web component to be stimulated by nutrient enrichment. They are net producers of oxygen through photosynthesis during the day, but at night they consume oxygen for respiration. Dense populations of algae in the upper and mid-depths of the water column can consume most of the available dissolved oxygen. Low-oxygen stress in the system also occurs because of increased decomposition. As plants and animals die and settle out over the growing season, their remains are decomposed by bacteria and fungi in oxygen-demanding processes that can rapidly deplete the oxygen from the lower water column, sometimes extending to mid-depths or shallower. Oxygen solubility decreases with increasing temperature. Thus, low-oxygen stress tends to be most pronounced in warmer seasons when algal biomass, decomposition rates, and respiratory rates (and oxygen requirements) of fish and other animals generally are high.

Low-oxygen conditions are exacerbated when bottom waters become somewhat isolated from the upper water column that receives oxygen (i) during the day from phytoplankton growth, and (ii) throughout the diel cycle from the overlying air as it diffuses into the water, and from phytoplankton growth (during the day). The isolating effect occurs when the water column becomes thermally (density) stratified, as in the bottom water layer (hypolimnion) of a stratified lake, or salinity (density) stratified, as in the lower water-column salt wedge in an estuary. As a result, sessile bottom-dwelling animals become physiologically stressed and may suffocate unless they can slow their metabolism until oxygen is replenished. Motile animals that rely on the bottom waters as critical nursery areas for their young, or as a refuge area to escape predation, can physically avoid the low-oxygen areas but may nonetheless suffer population declines because of critical habitat loss.

Organic matter in poorly treated sewage and animal waste contributes to low-oxygen stress in aquatic systems, as another source of material for decomposition or "biochemical oxygen demand." Chemical oxygen demand may further deplete oxygen, through oxidation of high levels of ammonia and other inorganic reduced compounds in the wastes. These oxygen-demanding processes often are relatively localized problems occurring, for example, near sewage outfalls or animal waste spills. In contrast, certain oxidized nutrients such as nitrate can be highly soluble in water and, therefore, can be transported considerable distances as shown, for example, by Mallin *et al.*'s work. This transport creates a time lag between the introduction of nutrients into one area and adverse impacts at some distance from the source. Oxygen consumption from decomposition of excess phytoplankton production thus can occur on a large scale that is difficult to track or to relate to one specific, original source.

II. CHANGES IN FLORA AND FAUNA

A. Microalgae

1. Species Shifts across Nutrient Gradients

Phytoplankton dominate the flora of oligotrophic systems (Fig. 2). They respond quickly to nutrient inputs because the tiny plantlike organisms are immersed in the enriched medium, in contact with it on all surfaces. Phytoplankton with optimal growth at elevated nutrient concentrations are especially stimulated or "selected for" by nutrient enrichment, and they eventually overgrow and replace species that grow best at lower nutrient levels. Along a nutrient enrichment gradient from oligotrophic to highly nutrient-enriched, the phytoplankton community structure gradually shifts from low abundance of many species and dominance by small flagellates and picoplankton (algae $\leq 2\ \mu$m in diameter), with energy flow channeled through a microbial loop of bacteria and small flagellates rather than directly up the food chain to herbivorous zooplankton, to high abundance of relatively few species consisting mainly of large cells or large colonies as well as seasonally abundant flagellates. As lakes, lower rivers, and estuaries become more eutrophic, diatom species with higher N and P optima predominate in colder periods, certain colonial green algae with high N optima are abundant in early summer, and dinoflagellates (Dinophyceae: *Gymnodinium* spp., *Peridinium* spp.) and filamentous and colonial cyanobacteria with high P optima dominate in late summer. If dissolved silica, needed by diatoms to make their cell walls, is limiting in colder seasons, flagellates (as examples, dinoflagellates, cryptomonads, euglenoids) become more abundant.

In shallow freshwater lakes, estuaries and lagoons where open-water habitat is limited, high phytoplankton densities occasionally occur in response to nutrient enrichment (for example, high flagellate densities that

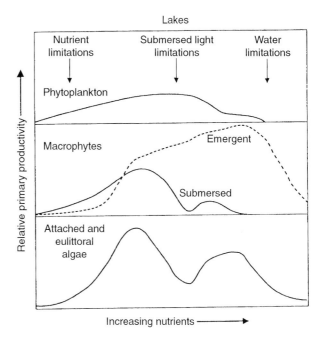

FIGURE 2 Generalized relationship of primary productivity of major plant groups in lakes under increasing nutrient enrichment. Nutrients are the primary resources limiting plant production under oligotrophic conditions, and phytoplankton (low in abundance or relative primary productivity, but high in species diversity) predominate over other plant groups. As eutrophication progresses with increasing nutrient enrichment, light becomes more important than nutrients as the primary resource limiting plant production. In mesotrophic conditions prior to the onset of light limitation, submersed macrophytes can dominate, but they eventually are eliminated because of low light availability and because emergent plants predominate. In late stages of the aging process, lakes become increasingly shallow and gradually function as wetlands. Plant production is then limited primarily by the availability of water. Reprinted from Wetzel (1983), with permission; originally from Wetzel (1979).

historically occurred in response to excrement from duck farms in western Long Island Sound). More commonly, benthic algae known as epiphytes (growing on the leaves of submersed aquatic vegetation or SAV), or floating "drift" macroalgae (in sheltered estuarine and marine coastal embayments), are rapidly stimulated by the nutrient increases and restrict light for underlying plants (Figs. 2 and 3). Increased nutrient enrichment generally promotes a decrease in the species diversity of benthic, photosynthetic algae, and a shift in dominance from diatoms to filamentous green algae or coccoid cyanobacteria. The combination of low light and rich organic substrates, characteristic of planktonic and (moreso) benthic habitats in such systems, also selects for mixotrophic microalgae, which become abundant and diverse under moderately eutrophic conditions.

Many regions of the world have landscapes domi-

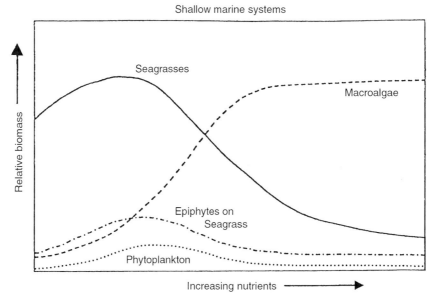

FIGURE 3 Generalized shift in biomass of major plant groups with increasing nutrient inputs to shallow marine ecosystems. Occasionally the phytoplankton dominate, but usually macroalgae dramatically increase, while submersed rooted plants decline through competition for light and/ or nitrogen. Reprinted with permission from Harlin (1993), © CRC Press, Boca Raton, FL.

nated by run-of-river impoundment or reservoirs, rather than natural lakes. Similar trends in impacts from nutrient loading on aquatic communities can occur over time in reservoirs as in lakes, with the exception of depauperate rooted macrophyte populations (discussed later) in reservoirs with variable depth imposed by controlled drawdown involved in potable water supply management or flood control. Impacts of cultural eutrophication in reservoirs can be mitigated by rapid water exchange (weeks to months, rather than years in most natural lakes). Moreover, the high turbidity from sediment loading/resuspension, characteristic of many reservoirs and estuaries as well as some natural lakes, often makes available light—rather than nutrients—the primary resource limiting the productivity of phytoplankton and other flora in the systems. Nutrients act as secondary factors controlling plant growth under such conditions. Such systems often can sustain higher nutrient (especially P) loading than clear, natural lakes while supporting less phytoplankton biomass because of their features of more rapid water exchange and light limitation from high suspended sediments.

"Bottom-up" control by nutrients interacts with grazing pressure from zooplankton and other fauna in "top-down" trophic-level effects. Such effects are well documented in freshwater lakes and streams, and can also be operative in estuaries and coastal marine waters.

Many field and laboratory studies have demonstrated that proliferation of phytoplankton and benthic microalgae under nutrient enrichment (for example, seagrass epiphytes as described by Neckles and colleagues) can be significantly reduced during periods when grazers are abundant. Thus, herbivores can sometimes alleviate eutrophication impacts by holding algal production in check. Decreased grazing pressure can allow higher algal biomass to develop in response to excess nutrients, but decreases in grazing can help promote algal blooms only where nutrient inputs are sufficiently high to support such blooms. Therefore, grazing pressure is regarded as a secondary factor controlling phytoplankton production under cultural eutrophication.

2. Long-Term Human Influence

Long-term data sets have provided two lines of compelling evidence in support of major impacts from cultural eutrophication on phytoplankton assemblage structure in aquatic ecosystems, with potentially serious ecological and economic ramifications. First, the geological record in sediment cores taken from many freshwater lakes and estuaries clearly shows that long-term major shifts have occurred under increased nutrient enrichment—progressing from a balance between species-rich planktonic (mostly centric) and benthic (mostly pennate) diatom assemblages, to dominance by planktonic diatoms with low species diversity (Fig. 4). Planktonic diatoms are selected for because they are the early recipients of water-column nutrient sources. Benthic diatoms encounter decreased light because of the overlying plankton growth, eventually leading to decreased production.

The second line of evidence concerns living phytoplankton assemblages with species shifts that have been related to changes in supply ratios of the two most limiting nutrients. Under nutrient-limiting conditions, species with similar optima in other environmental factors (for example, temperature or light) that maintain faster growth compete more successfully for the available nutrient resource (Figs. 5–7). Elegant work by Tilman and colleagues examined the response of freshwater diatom species to shifts in Si : P ratios. Rhee and colleagues extended these concepts to controlling influences of N : P ratios. Low molecular N : P ratios (7–15 in those studies; up to ca. 29 : 1 by weight in Smith's work) favored tested blue-green and diatom species, whereas higher N : P ratios favored green algae. Many green algae grow optimally at high N_i concentrations, whereas many blue-greens and dinoflagellates have a high P_i requirement. Diatoms, unlike the other algae, require major supplies of silica as well as N and P. When Si : N and Si : P ratios are high, the available Si favors growth of diatoms that can effectively compete for N and P resources. However, as Si : N and Si : P ratios decrease, silica becomes limiting for diatom growth, and more N and P remain available for growth of flagellates and other algae that do not require silica. These concepts were extended to natural lakes by Tilman and colleagues (for diatoms and other algae), Smith (for blue-greens), and others.

Silica is only slowly (years) made available for new diatom growth, through dissolution of dead diatoms and other natural silica sources. In contrast, N and P cycles are rapidly affected by anthropogenic inputs. As N and P enrichment increase, the Si : N and Si : P supply ratios are depressed. Eutrophication can effect a decrease in dissolved Si abundance by initially stimulating high growth of diatoms, to the point that they deplete the dissolved silica pool needed by developing diatom populations in subsequent seasons. Long-term data sets on estuarine, coastal, and freshwater phytoplankton communities indicate that shifts to dominance by flagellated algae or blue-greens—including some harmful bloom-forming species—have coincided with decreased abundance of diatoms and decreased Si : N and Si : P ratios. Such trends have been documented by Smayda and others in temperate and sub-Arctic waters such as the Great Lakes, the Gulf of Mexico near the

mouth of the Mississippi River, New England coastal waters, Chesapeake Bay estuarine waters, the Black Sea, and the coasts of northern Europe.

In oligotrophic tropical marine waters, mixotrophic dinoflagellates tend to predominate and symbiotic interactions are common in both planktonic and benthic communities. Little is known about impacts of eutrophication on phytoplankton community structure in such systems, but it is hypothesized that increased nutrient enrichment would shift the community structure to higher proportions of flagellated photosynthetic algae with less reliance on heterotrophy or symbiosis. Trace metals such as iron have also been shown to be limiting to phytoplankton growth in some estuarine and marine waters, where N and P are at levels that would otherwise be expected to support more algal production.

As eutrophication progresses, the previously described shift in temperate-zone phytoplankton community structure from certain diatoms to other diatom species, flagellates, and cyanobacteria causes subtle, but important, undesirable changes for the food web that can adversely affect secondary production. For example, as reviewed by Kilham and colleagues, some diatom species produce high quantities of certain lipids that are essential for zooplankton reproduction. Algal species that may replace these diatoms under increasingly eutrophic conditions do not produce these lipids, or produce much fewer of them. Analogous phenomena occur in fresh waters, estuaries, and coastal environments. For example, Starr and colleagues reported that spawning of green sea urchins and blue mussels apparently is triggered by a heat-stable metabolite that is released in high abundance by certain species of phytoplankton, especially certain diatoms such as *Skeletonema costatum* (Bacillariophyceae). This substance is not produced, or is produced in much lower quantities, by flagellated algae that replace these diatoms under cultural eutrophication.

3. Harmful Algal Blooms and Anthropogenic Nutrient Enrichment

Among the algal species favored by nutrient enrichment are noxious forms such as cyanobacteria that are toxic to zooplankton, fish and wildlife in fresh waters and certain estuaries (Baltic Sea and Australia), and certain dinoflagellates that are toxic to finfish and shellfish in estuarine and marine coastal areas worldwide (Table I). For example, the two known toxic *Pfiesteria* species have been most active in waters degraded by poorly treated sewage, swine effluent spills, and other excessive nutrient inputs, and these organisms have been experimentally stimulated by nutrient enrichment. However,

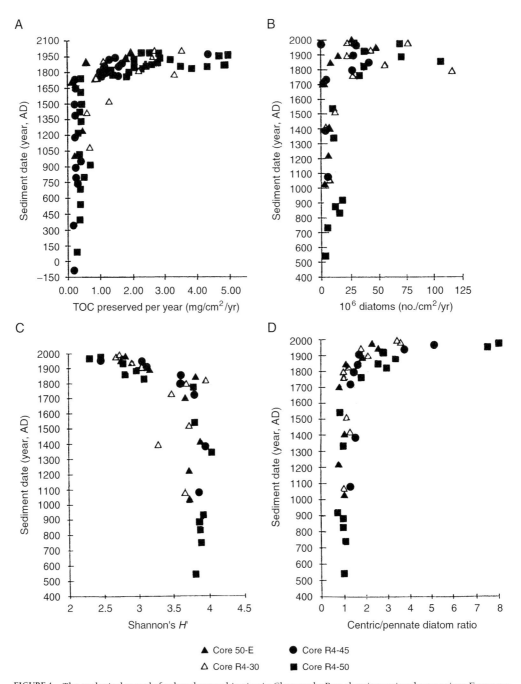

FIGURE 4   The geological record of cultural eutrophication in Chesapeake Bay, showing major changes since European settlement. The data from four sediment cores are graphed by the average date assigned to each sediment sample (depth layer) according to radiocarbon and pollen methods. (a) Total organic carbon (TOC, indicating total system productivity) preserved over time (historic record from A.D. −150 to A.D. 1990). Modified from Cooper and Brush (1991), with permission. (b) Diatom cell numbers per year (A.D. 400 to A.D. 1990). (c) Diatom community diversity calculated as Shannon's H (A.D. 400 to A.D. 1990). (d) Centric/pennate diatom ratios (A.D. 400 to A.D. 1990). TOC, diatom numbers, and the centric/pennate diatom ratios all showed a significant and abrupt increase following the time of European settlement in the 1700s. The total diatom community diversity, in contrast, significantly decreased post-1700s, relative to pre-1700s diversity. Graphs b–d were modified with permission from Cooper © (1995) American Association for the Advancement of Science.

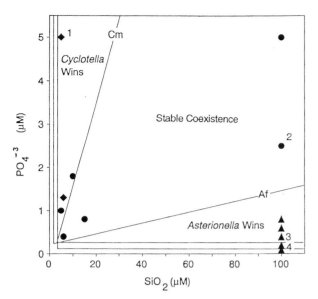

FIGURE 5 The outcome of competition between diatoms for limiting nutrients (silica or phosphate) across a gradient of Si : P ratio comparing two freshwater planktonic diatoms, *Cyclotella meneghiniana* and *Asterionella formosa*, under growth-limiting conditions (temperature, light, and other conditions held constant). These chemostat culture experiments indicate conditions in which *A. formosa* is dominant because it can outgrow *C. meneghiniana* at very low P (as micromolar concentrations or micromoles per liter of phosphate, μM PO_4^{-3}); conditions wherein the two species can coexist (circles) because one is limited by P, and the other is limited by Si (as silicon dioxide concentrations, μM SiO_2); and conditions wherein *C. meneghiniana* can dominate because it can outgrow *A. formosa* at lower Si. Reprinted from Tilman (1982), with permission.

other species of harmful algae are found in oligotrophic waters and do not appear to be favored by elevated nutrients. The planktonic Gulf Coast red tide dinoflagellate, *Gymnodinium breve*, and certain benthic toxic dinoflagellates that inhabit coral reefs (*Gambierdiscus toxicus, Prorocentrum lima*) are among various harmful species that have been reported to achieve optimal growth in low-nutrient conditions.

Despite their economic impacts worldwide, remarkably little is known about the nutritional ecology of most species of estuarine and marine toxic, parasitic, and other harmful algae. Overgeneralizations should be avoided that prematurely negate the potential for stimulation of various other harmful/toxic estuarine and marine algae by nutrient enrichment. As additional species are thoroughly evaluated, it will be important to consider that the concentration of a nutrient at any given point in time may not be correlated with its actual bioavailability, and that phytoplankton can grow for long periods on internally stored (luxury-consumed)

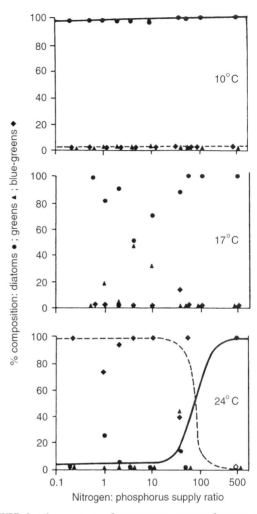

FIGURE 6 The outcome of competition among diatoms, green algae, and blue-green algae for limiting nutrients (nitrogen or phosphorus) across a gradient of N : P ratios under silica-replete (nonlimiting) conditions, showing additional influence of temperature. The comparison includes representative species of freshwater lake diatoms (circles), green algae (triangles), and blue-greens (cyanobacteria; diamonds) in chemostat culture under growth-limiting conditions for N and P (but not for Si). Note that at the coldest temperature indicative of conditions during spring and fall seasons in this north temperate lake, the cold-optimal diatoms outcompeted the greens and blue-greens across all N : P ratios. That is, temperature, rather than N or P, was the most important condition limiting the growth of greens and blue-greens relative to diatoms. The midrange temperature represented a transitional area where some growth could occur for species representing each group. The highest temperature favored warm-optimal blue-greens (that tend to have high P optima), especially at low N : P ratios when P was more abundant. Reprinted from Harper (1992), with permission; originally modified from Tilman *et al.* (1986).

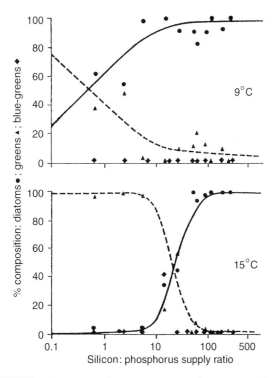

FIGURE 7 The outcome of competition among diatoms, green algae, and blue-green algae for limiting nutrients (phosphorus or silica) across a gradient of P : Si, showing the influence of temperature. The outcome of competition is indicated among freshwater lake diatoms (circles), green algae (triangles), and blue-greens (cyanobacteria; diamonds) in chemostat culture under growth-limiting conditions for Si (needed only by the cold-optimal diatoms) and P at two temperatures. At the lower temperature, cold-optimal diatoms dominated across all Si : P ratios. However, at the higher temperature they were able to dominate over green algae only at high Si : P ratios, when their Si requirements were alleviated by the relatively high Si. The blue-green species were able to outcompete the other groups only at one transitional Si : P ratio even at the higher temperature, which was still suboptimal for these organisms. Reprinted from Harper (1992), with permission; originally modified from Tilman *et al.* (1986).

nutrient pools. The luxury consumption and subsequent use of inorganic N, P, and C; the cell budgets for these nutrients; and the role of heterotrophy (including indirect stimulation by nutrient enrichment, through increased growth of microbial prey) in supplying carbon remain to be examined for most harmful estuarine and marine species.

B. Macroalgae

1. Freshwater Assemblages

The major flora of certain aquatic habitats are macroalgae, and they include some notorious aquatic "weeds"

that respond to cultural eutrophication. Temperate-zone freshwater, oligotrophic hard-water lakes may contain abundant benthic charophytes and, rarely, a few species of small brown macroalgae (Phaeophyceae). Oligotrophic soft-water lakes may have populations of small, benthic red macroalgae (Rhodophyceae), especially near spring-fed areas with bubbling carbon dioxide. As eutrophication progresses, populations of certain filamentous green algae and cyanobacteria that would otherwise be considered as microalgae become visually common, especially under high P enrichment. For example, *Cladophora glomerata* (Chlorophyceae) forms masses of long hairlike growth (dominant during the 1960s in the littoral zone of the west basin of Lake Erie); and cyanobacteria (or blue-green algae, Cyanophyceae; for example, *Anabaena, Aphanizomenon, Microcystis*) form tufts or amorphous masses of greenish, bluish-green, or reddish gray slime.

Macroalgae are the dominant autotrophs of midorder streams where hard substrata are available for colonization and light is moderate to high, especially after leaf fall in colder seasons. Stream-inhabiting macroalgae are often much more diverse and abundant than lake macroalgal floras. Nutrient-poor (often soft-water) stream segments are colonized by diverse, abundant cold-optimal chrysophytes (e.g., *Eunotia pectinalis* [Bacillariophyceae], *Tetrasporopsis* [Chrysophyceae]), green algae (*Spirogyra, Tetraspora, Oedogonium*), red algae (*Batrachospermum, Paralemanea* [Rhodophyceae]), or cyanobacterial mat formers (*Phormidium*). As eutrophication progresses, soft waters may develop larger populations of filamentous green algae such as certain species of *Oedogonium* and *Mougeotia* or dense growth of colonial *Hydrodictyon* and *Stigeoclonium* (also known to tolerate high metal concentrations that may be found in poorly treated sewage). Hard-water streams under high P enrichment may become choked with massive growth of *Cladophora* spp.

2. Estuarine and Marine Assemblages

Macroalgae also dominate the flora of many shallow estuaries, lagoons, and upper embayments, coral reefs, and rocky intertidal/subtidal habitats (Fig. 3). Nutrient enrichment leads to reduction in the diversity of macroalgae and associated fauna. Opportunistic species of green algae within the genera *Enteromorpha* (also found in saltwater lakes such as the Great Salt Lake of Utah), *Ulva* and *Cladophora*, and within the brown algal genus *Ectocarpus* have rapid growth rates and proliferate because they can more quickly take advantage of the ele-

TABLE I

Harmful Estuarine and Coastal Marine Microalgae That Have Been Linked to Anthropogenic Nutrient Enrichment

Harmful species	Link to cultural eutrophication
Chattonella antiqua	Bloomed under cumulative high loading of poorly treated sewage and other wastes, coinciding with human population growth (Japan; fish kills, toxic; Lam and Ho, 1989).
Chrysochromulina polylepis	Toxic outbreaks followed change in nutrient supply ratios from cumulative increased nutrient loading (Europe; fish kills, toxic; Kaas, *et al.*, 1991; Maestrini and Granéli, 1991).
Gymnodinium mikimotoi	Bloomed under cumulative high loading of poorly treated sewage and other wastes, coinciding with human population growth (Japan, as *G. nagasakiense*; fish kills, PSP; Lam and Ho, 1989).
Gonyaulax polygramma	Bloomed under cumulative high loading of poorly treated sewage and other wastes, coinciding with human population growth (Japan; fish kills from oxygen depletion; Lam and Ho, 1989).
Noctiluca scintillans	Bloomed under cumulative high loading of poorly treated sewage and other wastes, coinciding with human population growth (Japan; fish kills from oxygen depletion; Lam and Ho, 1989).
Nodularia spumigena	Blooms followed change in nutrient supply ratios from cumulative increased nutrient loading by sewage, agricultural wastes (Baltic Sea—Rinne *et al.*, 1981; estuary in Australia—Hillman *et al.*, 1990).
Pfiesteria piscicida, P. shumwayae sp. nov.	Most kills (with highest cell densities) have occurred in P- and N-enriched estuaries (e.g., near phosphate mining, sewage inputs, or animal waste spills); between kill events can prey upon flagellated algae that are stimulated by inorganic nutrients; bloomed 1 week after a major swine effluent lagoon rupture (with extremely high phosphorus and ammonium) into an estuary, in a location where high abundance of these dinoflagellates had not been documented for at least the previous 1.5 yr; highly correlated with phytoplankton biomass in other entrophic estuaries (mid-Atlantic and southeastern United States; fish kills, epizootics; Burkholder *et al.*, 1995a; Glasgow *et al.*, 1995; Burkholder and Glasgow, 1997).
Phaeocystis spp.	Bloomed following cumulative high loading of poorly treated sewage (Europe; fish—*Phaeocystis pouchetii* [Hariot] Lagerheim); blooms were correlated with altered N/P ratios from cumulative increased nutrient loading (*P. pouchetii*; Hallegraeff, 1993; Riegman *et al.*, 1993); bloomed 1 week after a major swine effluent lagoon rupture into a eutrophic estuary (*Phaeocystis globosa* Scherffel, along with *Pfiesteria* spp.; southeastern United States; Burkholder *et al.*, 1996).
Prorocentrum minimum	Bloomed under cumulative high loading of poorly treated sewage and other wastes, coinciding with human population growth (Japan; fish kills, toxic; Lam and Ho, 1989); blooms coincide with cumulative high loading of N from sewage, agricultural runoff, atmospheric loading, etc. (southeastern United States; Mallin, 1994).
Prymnesium parvum	Toxic outbreaks usually have occurred under eutrophic conditions (fish kills; Edvardsen and Paasche, 1997).
Pseudo-nitzschia multiseries, other Pseudo-nitzschia spp.	Have occurred with sewage and other wastes (Canada; ASP) (Smith *et al.*, 1990); consistent seasonal blooms in the Mississippi and Atchafalya River plume areas, associated with hypereutrophic conditions (Q. Dortch *et al.*, *unpublished data*) and in Prince Edward Island, Canada following anthropogenic nutrient loading and drought (Smith *et al.*, 1990).

Note that many of the known harmful estuarine and marine microalgae and heterotrophic or animal-like dinoflagellates also have been shown to be stimulated by N or P enrichment in culture, expected since they are photosynthetic. Also note that blooms of the toxic *Pseudo-nitzschia* complex have not been associated with cultural eutrophication in the northwestern United States. Reprinted from Burkholder (1998), with permission.

vated nutrient levels and shade out other species. Repeated, pulsed nutrient enrichment to temporarily elevated levels—characteristic of many eutrophic waters—is known to encourage these rapidly growing species, whereas more slowly growing perennial species tend to store large quantities of nutrients (for example, N). Macroalgal overgrowth in brackish/marine habitats commonly has been related to elevated inorganic N. Ammonium is more readily used (less energetically

costly) than nitrate, but either N_j form has significantly stimulated growth of opportunistic macroalgae in field research and mesocosm experiments. Phosphate can stimulate macroalgal growth when N is abundant.

For example, in Boston Harbor, Massachusetts, sea lettuce (*Ulva*) formed dense populations for many years near outfalls of poorly treated sewage. The massive seaweed growth reduced water flow, smothered shellfish, and affected shoreline areas with an intolerable stench of hydrogen sulfide from decomposition of senescent/dead material. In Peel/Harvey Estuary, Australia, dense mats of *Cladophora*, other green macroalgae, and cyanobacteria developed in response to P loading. In seagrass meadows worldwide, such overgrowth has been documented to reduce light and promote declines in various seagrass species. Along the intertidal area of the Baltic Sea, sewage inputs have been related to overgrowth of formerly dominant brown seaweeds (*Fucus* spp.) by opportunistic macroalgae.

In subtidal outfalls, nutrient enrichment from sewage and other wastes has stimulated overgrowth by red algae such as certain species of *Polysiphonia*. The exotic green macroalga *Caulerpa taxifolia* is a subtidal, subtropical "weed" that has invaded colder waters of the Mediterranean, and sewage appears to further stimulate its robust growth. This organism produces metabolites that discourage predation; sea urchins typically avoid consuming *C. taxifolia* and starve to death if it is the only available food. Subtidal forests of giant kelp (*Macrocystis pyrifera*) died back and failed to reproduce during the 1960s and 1970s near outfalls of sewage discharged from Los Angeles. The kelp bed loss apparently resulted from toxic substances in the poorly treated sewage.

Sensitive oligotrophic coastal marine coral reefs have been overgrown and smothered by macroalgae after nutrient input from sewage. In Kaneohe Bay, Hawaii, for example, phytoplankton near sewage outfalls first removed the N_i and were transported to the central area of the coral reef where they decomposed and released the N_i for stimulation of the opportunistic green macroalga, *Dictyosphaeria cavernosa*. Inorganic nutrient concentrations are frequently below detection in natural coral reef waters. Lapointe's elegant work indicated that even minor sewage-related increases in N_i (≥ 1 μM) and P_i (0.1–0.2 μM soluble reactive phosphate) off the coast of Jamaica, together with decreased herbivory from natural and fishing-related disturbance, were sufficient to stimulate blooms of the green "eutrophic indicator" macroalga, *Chaetomorpha linum* and cyanobacteria that have overgrown coral reefs in that area.

Reduced light from such excessive nutrient-stimulated macroalgal (and sometimes phytoplankton) growth affects coral growth by decreasing the productivity of zooxanthellae, the symbiotic photosynthetic dinoflagellates inside the coral tissue that provide much of the corals' nutrition. Nutrient overenrichment also can shift species dominance within the coral community. As the corals are outcompeted for space, filter-feeding taxa such as sponges may be stimulated by increased phytoplankton food resources, and they can

replace corals as dominant species. Sedimentation of decomposing phytoplankton, and other disturbances such as destructive techniques for fish harvesting, can exacerbate the impacts of cultural eutrophication on coral growth and survival.

C. Aquatic Macrophytes

1. Freshwater Communities

Floating, submersed, and emergent aquatic vascular plants, commonly called macrophytes, respond to nutrient enrichment in different ways because the competitive forces that they encounter vary with plant growth habit. *Rooted plants*—including those with specialized floating leaves, submersed growth, or emergent growth—generally do not compete with phytoplankton and other algae for nutrients, because they obtain most nutrients (except carbon, taken from the water) from the nutrient-rich sediment, with leaf uptake of nutrients from the less enriched overlying water as a secondary nutrient source. In contrast, *floating plants* at the water surface, such as the Lemnaceae (duckweed family) and the exotic weed *Eichhornia crassipes* (water hyacinth), must compete with suspended algae for N and P, but rely upon the overlying air for their carbon. Floating submersed plants must compete with suspended algae for all major nutrients. They tend to have well-developed root systems to aid in nutrient acquisition, and in shallow waters with sparse open-water habitat, they can outcompete phytoplankton for light by forming dense surface populations that reduce or eliminate available light in the underlying water. However, submersed rooted aquatic vegetation (SAV) is usually eliminated in highly eutrophic systems, primarily through light reduction by phytoplankton and other algal overgrowth, as described by Phillips and colleagues, Wetzel, and others (Fig. 2).

Like stream-inhabiting macroalgae, rooted macrophytes in river systems reach maximal abundance and species diversity in midorder segments. But, in contrast to lake-inhabiting macroalgae, submersed freshwater macrophyte communities attain highest species richness and abundance in mesotrophic lakes, often with dominance by *Potamogeton* spp. Light plays a major role in progressive SAV decline as lakes become increasingly eutrophic. Nutrient enrichment can stimulate macrophyte growth through luxury uptake and tight recycling, sometimes without an increase in phytoplankton. Over time, however, epiphytic algae take advantage of the greatly increased surface area for colonization afforded by the macrophytes, and they can both severely shade the underlying plants and restrict their carbon acquisition.

Increased macrophyte detritus and epiphyte production stimulates growth of invertebrates and, in turn, can lead to increased abundance of cyprinids (for example, bluegill sunfish—*Lepomis macrochiris*) and other invertebrate-feeding fish. These fish cause increased turbidity, both through direct sediment disturbance during feeding and indirectly through reduction of zooplankton density, which relieves grazing pressure on the phytoplankton. Macrophytes generally die back in autumn and then sprout from seeds or perennating belowground structures (rhizomes, tubers, etc.) in the following spring. As the lake turbidity increases, light limitation prevents this spring regrowth. Under accelerated eutrophication, dramatic declines in macrophyte population typically occur over a relatively short period (several years). The loss of most submersed macrophytes is often a critical turning point in the eutrophication of a lake. Without the habitat that provided cover for fish and substrata for littoral-zone invertebrates, cyprinid fish increase their grazing pressure on zooplankton, and their removal of zooplankton effectively decreases grazing pressure on the phytoplankton. The increased anoxic/hypoxic conditions from yet-higher phytoplankton production cause further reductions in benthic invertebrate species and abundance, leading to more intense cyprinid grazing pressure on zooplankton. Thus, highly eutrophic lakes contain dense phytoplankton, food-limited (stunted) cyprinids, and little else.

Emergent macrophytes at the littoral fringe of lakes, rivers, estuaries, and marine coasts can maintain enhanced growth and biomass over a wide range of nutrient inputs. Under certain conditions with sustained high nutrient loading, however, emergent macrophytes can decline. The increased nutrients stimulate increased growth, greater stem density and, often, accelerated longitudinal growth, leading to higher intraspecific competition for light and a weakening of the stems. The elevated nutrients also promote higher algal growth and increased detritus production (and sediment anoxia), and higher growth of epiphytes on the submersed portion of the macrophyte stems. The increased weight makes the plants less well anchored, leading to dieback at the water's edge. Elevated N_i also promotes reduction in supporting vascular tissues, and loss of stem strength.

2. Estuarine and Marine Macrophyte Communities

Brackish waters are colonized by rooted, mostly freshwater species with moderate salt tolerance (as examples, certain *Potamogeton* spp., *Valisneria americana, Zanichellia* spp.). In such habitats, light is often the primary resource limiting growth. Species with broader salt tolerance such as *Ruppia maritima* also can be abundant. Only about 50 species of angiosperms, mostly close relatives of freshwater *Potamogeton* spp., have the salt tolerance needed to thrive in marine habitats. Nearly all of these "seagrasses" grow in muddy substrata of shallow coastal lagoons and quiet embayments. Estuarine and marine macrophytes tend to be highly sensitive to light reduction and, thus, susceptible to eutrophication-related turbidity from algal (phytoplankton, epiphyte, and macroalgae) overgrowth and sediment loading/resuspension. Such shading causes gradual dieback and loss of most SAV, promoting dramatic declines, in turn, in the diversity and abundance of many plant and animal species that depend on the habitat provided by these plants.

More sensitive SAV species are replaced by others that are tolerant of eutrophic conditions. For example, among subtropical seagrasses, field observations and limited experiments have indicated that turtle grass (*Thalassia testudinum*) is more sensitive to eutrophication than shoal grass (*Halodule wright-ii*) and manatee grass (*Syringodium filiforme*). Other regions may not have additional seagrass species available to replace more sensitive species and, even where such species are present, they typically offer less desirable fish nursery habitat than the former dominant. Thus, in seagrass meadows under nutrient ove-enrichment, more oligotrophic seagrass species are replaced by less sensitive species when available. As eutrophication progresses the seagrasses are eliminated, and rapidly growing macroalgae or phytoplankton become the dominant flora (Fig. 3).

Although light reduction is considered the major mechanism for seagrass decline under cultural eutrophication, excessive nutrients can act independently of light to promote seagrass loss (Fig. 8). The dominant north temperate species, eelgrass (*Zostera marina*), apparently lacks a physiological mechanism to inhibit nitrate uptake through its leaves, as indicated by the research of Pregnall, Burkholder, and colleagues. Most plants take up nitrate during the day with energy from photosynthesis. In contrast, *Z. marina* takes up water-column nitrate day or night if it becomes available, as shown by Touchette and colleagues (Fig. 9). This species probably evolved in N_i-poor coastal waters, and sustained nitrate uptake under temporary enrichment may have developed as a once highly advantageous competitive strategy. However, as coastal waters have become more eutrophic from sewage, septic effluent

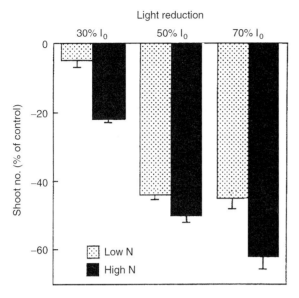

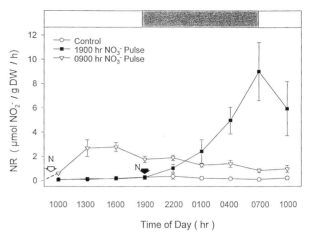

FIGURE 8 The effects of water-column nitrate enrichment and light reduction on shoot production of the seagrass, *Zostera marina*. From author's outdoor mesocosm experiments, indicated as the percent decrease from shoot production of control plants that did not receive water-column nitrate additions or light reduction (except that plants in controls and treatments all received an additional 30% light reduction for 3 hr at 0900, 1200, and 1500 hr on a 3-day rotation using neutral density screens to simulate conditions during high tide). Treatments were imposed for 10 weeks during the fall growing season for *Z. marina*. Controls were maintained at ambient natural light (except during simulated high tide) and nitrate (<30 μg NO_3^-N/L). Treatments included low N (at 50 μg NO_3^-N/L, added daily as a pulse of enrichment) and high N (at 100 μg NO_3^-N/L) at each of three imposed light levels as 30, 50, or 70% reduction of ambient surface light (I_o, accomplished using neutral density shades, with additional shading at simulated high tide as noted). *Z. marina* in all treatments with water-column nitrate enrichment declined in shoot production relative to shoot production of control plants, and the nitrate inhibition effect was exacerbated by light reduction (means $\pm$ 1 standard error; $P < 0.05$, n = 3). These effects were not caused by algal overgrowth, which was maintained at low levels in controls and all treatments throughout the experiment.

FIGURE 9 The response of the seagrass, *Zostera marina*, to pulsed water-column nitrate enrichment in light and dark periods. Plant nitrate uptake is indicated as leaf activity of the enzyme (nitrate reductase [NR], used to actively take up nitrate) of previously unenriched shoots. A spike of nitrate (110 μg NO_3^-N/L) was added in the morning (white arrow) or, to a subsample of plants from the same population, at night (black arrow). Nitrate reductase activity (plotted as micromoles of nitrite product produced per gram dry weight of plant leaf tissue per hour) indicated that *Z. marina* took up nitrate day or night, whenever a pulse was detected (means $\pm$ 1 standard error). In fact, maximal NR activity was significantly higher when nitrate was added during the dark period. Reprinted from Touchette *et al.* (2000), with permission.

leachate, and other anthropogenic sources, sustained uptake of water-column nitrate likely has become a disadvantage. Nitrate enrichment to the sediments, under control by an abundant microbial consortium, does not cause a similar effect and, instead, can be mildly stimulatory.

Nitrate uptake is a metabolically expensive process, requiring high cellular energy. Research by Touchette and colleagues indicated that sustained water-column nitrate uptake by *Z. marina* can promote severe internal carbon imbalances, apparently from the need to shunt C skeletons from photosynthesis for use in high amino acid synthesis to prevent internal accumulation of toxic

products such as ammonia. The physiological mechanism of an internal "carbon drain" from sustained nitrate uptake has been documented (for algae) by Turpin and colleagues. A common trait of *Z. marina* shoots under excessive water-column nitrate enrichment is structurally weakened growing regions, perhaps analogous to the above-mentioned loss of stem strength that has been reported in certain freshwater emergent macrophytes under nitrate enrichment. Excessive N_i enrichment has also promoted seagrass attack by pathogens (for example, the slime mold *Labrynthula zosteroides*), hypothesized to occur because N and C are internally shunted to amino acid production rather than to production of alkyloids and other antimicrobial compounds.

Another seagrass that has been examined for the nitrate inhibition phenomenon, *Halodule wrightii*, and certain macroalgae (for example, *Ulva lactuca*) have shown depressed growth in response to nitrate enrichment, although at much higher N levels (ca. 1.4 mg NO_3^-N/L, pulsed daily for 4 to 5 weeks) than for *Z. marina* (50–110 μg NO_3^-N/L, pulsed daily for 5 to 8 weeks). *Ruppia maritima* is stimulated by high water-column nitrate but inhibited by elevated N_i as ammonia;

and *Z. marina* has been experimentally inhibited by high ammonia levels as well. In mesocosm experiments, light reduction has been shown to exacerbate the inhibitory effects of water-column nitrate enrichment on shoot production in *Z. marina*. Warm temperatures also exacerbate water-column nitrate enrichment impacts on root growth of this seagrass, suggesting that warming trends in climate change may be expected to interact with eutrophication to adversely affect this beneficial habitat species.

D. Microfauna

1. Freshwater Communities

Whereas plants and mixotrophic algae respond directly to nutrient enrichment, animals are generally indirect recipients of eutrophication impacts. Nonetheless, all are significantly affected by nutrient control of the quality (including size, taste, and ease of filtering) and quantity of their algal, bacterial or detrital food. Freshwater zooplankton consist of three main groups: (i) protozoa (mainly ciliates and flagellates) and (ii) rotifers form the "microzooplankton" (maximum dimension ca. 45 μm to less than 70 μm); and (iii) crustaceans form the "macrozooplankton" (mainly cladocera, and calanoid and cyclopoid copepods, ca. 70 μm to less than 500 μm).

Protozoa diversity and abundance increase with the amount of available organic matter that becomes directly or indirectly available through eutrophication until the fauna are restricted by low-oxygen conditions. Oligotrophic protozoan plankton can be dominated in number by small-bodied ciliates, whereas large ciliates generally contribute most of the biomass (especially members of the Oligotrichida such as *Plagiopyla nasuta* and *Paramecium trichium*). Mesotrophic and eutrophic lakes are more commonly dominated in both number and biomass by small-bodied species (e.g., member of the Scuticociliatida). Protozoans are relatively sparse in oligotrophic lakes in comparison to the species-rich and abundant fauna that develops in moderately eutrophic waters. Aggregates of bacteria and detritus form in the water column as organic materials decompose, and these microhabitats can support protozoa that rival shallow benthic communities in diversity. Subtropical lakes show similar trends, but have been reported to support higher ciliate abundance and diversity than temperate lakes at a given trophic status.

Heterotrophic flagellates in the water column are bacterivores, whereas larger ciliates and heliozoa commonly consume mixed populations of heterotrophic and autotrophic flagellates and other algae. In low-oxygen waters, heliozoan amoebae and ciliates such as *Col-*

eps and *Euplotes* retain endosymbiont green algae (zoochlorellae) that photosynthesize and generate oxygen for their hosts. As nutrient enrichment increases, bottom waters and surface sediments become more organically enriched; but the physiological demands on protozoans and other microfauna such as certain rotifers are also greater. Survival generally depends on tolerance of low-oxygen and often co-occurring acidic conditions. In organically overloaded, hypoxic sediments, protozoa species diversity is very low relative to that in well-aerated sites. Bacteria and dissolved organic matter form the major food sources, and are consumed by species such as *Paramecium, Chilomonas,* and *Astasia.* Anoxic sediments become colonized by specialized protozoa such as pelobionts and diplomonads, and by certain ciliates and amoebae.

Rotifers include many algal herbivores as well as detritivores and a few carnivorous species. They generally attain highest species diversity and abundance, along with increasing small-celled phytoplankton, in moderately eutrophic lakes and lower rivers. These microzooplankton tend to be favored in abundance over macrozooplankton under increasing eutrophication. However, rotifer species diversity declines as eutrophication progresses to phytoplankton dominance by cyanobacteria, especially large mucilaginous colonial species that clog the filtering apparati of the animals. Rotifers are adversely affected as well by the toxins from certain cyanobacteria that are seasonally characteristic of eutrophic lakes.

Among macrozooplankton, cyclopoid copepods (raptorial feeders) generally feed most efficiently on larger "particles" (for example, algae with biovolume $\leq 1,000$ μm^3), whereas calanoid copepods consume particles ≤ 100 μm^3, and cladocerans eat small particles at ca. 10 μm^3. Changes in biodiversity under cultural eutrophication depend, to a large extent, on differences among these species in efficiencies of feeding at certain particle (algal or detrital) concentrations and size ranges, and on the responses of life history stages to food limitation. Nutrient enrichment alters food particle size and abundance which, in turn, leads to competitive species displacement. Larger-bodied individuals are often more efficient filter feeders than smaller fauna. However, larger species may not be competitively superior because their juveniles tend to be more vulnerable to food limitation than adults of smaller species. Smaller species withstand food depletion by reducing metabolism, growth, and egg production. Moreover, their specific food ingestion rate (milligrams food ingested per milligram of zooplankton biomass) is usually higher than that of larger zooplankton.

Calanoid copepods (for example, *Eudiaptomus*) often dominate the microfauna of oligotrophic and meso-trophic lakes. These organisms feed efficiently on larger algal cells in those systems, with higher ingestion efficiencies at low food density. As algal prey densities increase under eutrophication, larger cladoceran species become abundant under low to moderate predation by fish, whereas high predation tends to select for smaller cladocerans such as *Bosmina* or *Ceriodaphnia* over *Daphnia*. The larger cladoceran species generally have higher population growth rates than calanoid copepods. Their population cycles during warmer months have been shown to be mainly controlled by the relative proportions of edible and inedible algae, temperature, and predation. Like rotifers in the presence of abundant food, cladocerans can reproduce parthenogenically, thus allowing for rapid growth rates. They are more efficient filter feeders at moderate to high algal densities because they ingest more food for the same amount of energy expended, relative to ingestion under food-limited conditions.

Macrozooplankton biomass increases under eutrophication in both temperate and tropical lakes, with increasing dominance by small-bodied species. High-efficiency bacterial feeders are selected for as bacterial abundances increase under accelerated eutrophication. However, abundant co-occuring cyanobacteria with copious mucilage clog the filtering apparati of large-bodied microzooplankton. Moreover, some cyanobacteria are directly toxic to these fauna. In highly eutrophic lakes, the more selective feeding of calanoid copepods, and the seizing behavior of herbivorous cyclopoid copepods, may afford one or both groups competitive advantage. Smaller cladocerans, such as the littoral-zone chydorid *Chydorus,* may become abundant during dense cyanobacteria blooms because of their high efficiency in filtering the extremely small, solitary bacteria and blue-green prokaryote particles associated with the blooms while avoiding filter apparatus clogging by large, mucilaginous cyanobacteria colonies. Eutrophication can cause other impacts on zooplankton species, apart from changes in available food. For example, as bottom waters become increasingly hypoxic, survival is depressed for zooplankton eggs that sink to the bottom as part of the life cycle. Thus, nutrient enrichment can impair zooplankton recruitment as an indirect impact.

2. Estuarine and Marine Communities

Impacts of anthropogenic nutrient loading have, understandably, been more difficult to generalize in the complex flow/water exchange environments of estuaries and from large-scale marine environments with high physical and biological variability. Analyses by Michell and others have indicated that nutrients generally enhance phytoplankton biomass and carnivores depress herbivore biomass, but that the couplings between trophic levels (for example, phytoplankton and zooplankton) are weaker than in more "closed" systems (lakes and coastal lagoons).

Brackish and marine habitats differ from fresh waters in having fewer rotifers or cladocera, more extensive representation by protozoans (especially ciliates and foraminiferans), and often-abundant planktonic nauplii (young life history stages) of sessile adult fauna ranging from mollusks and malacostracan crustaceans to vertebrates. Holoplankton spend their entire lives in the water column. Among these are microzooplankton including larval forms of certain macrozooplankton, as well as tintinnid and nonloricate ciliated protozoans, heterotrophic flagellates, and amoebae. In benthic habitats, the "meiofauna" (similar in size to microzooplankton) include nematodes, harpacticoid copepods, many turbellarians, and several minor phyla with diverse feeding habits and lifestyles, mostly acting as consumers of other microbes as prey. Macrozooplankton include copepods, especially of the genus *Acartia*, other cyclopoid copepods and planktonic harpacticoids, noncopepod crustaceans (especially carideans and mysids), and chaetognaths (arrow worms). Meroplankton, which spend only part of their lives in the plankton as larval stages, may include immature forms of benthic invertebrates and tunicates; eggs, larvae, and juveniles of shrimp, crabs, and fish; and sexual stages of hydrozoan and scyphozoan cnidarians (jellyfishes).

The microzooplankton are important phytoplankton herbivores among estuarine and marine zooplankton. They are strongly influenced by changing phytoplankton food quality and abundance under eutrophication, with trends that are somewhat analogous to those described for freshwater microzooplankton. Zooplankton species diversity is highest in moderately nutrient-enriched waters, but significantly declines as nutrient enrichment becomes more excessive, with accompanying shifts to dominance of unpalatable algal species and pronounced bottom-water oxygen deficits. For example, in the Peel-Harvey Estuarine System of Australia, large populations of calanoid copepods *Sulcanus* and *Gladioferans* grazed winter diatom blooms but were rapidly eliminated when the noxious (toxic) filamentous cyanobacterium, *Nodularia spumigena*, became abundant. Predation of carnivorous macrozooplankton by planktivorous fish such as menhaden (e.g., young life history stages of Atlantic menhaden, *Brevoortia tyran-*

nus) can locally favor small-bodied zooplankton. Such fish become more abundant in localized areas such as small tributaries because they are known to track plankton blooms. Later in the growing season, these fish switch to phytoplankton prey and thus, directly compete with herbivorous zooplankton for food resources.

Under high nutrient enrichment, changes in food quality/availability also adversely affect larval stages of many species represented in the meroplankton. Tunicates can increase at the expense of planktonic crustacea, and gelatinous meroplankton (cnidarians, especially jellyfish) can increase at the expense of fish. Jellyfish are also favored under increasing eutrophication. They are predators on zooplankton and juvenile fish, and they are avoided by most pelagic fish that do not consume them. Low-oxygen conditions in bottom-water habitat decrease survival of zooplankton eggs and of eggs and larvae of many meroplankton species that settle to the bottom sediments as part of their life cycle.

E. Invertebrate Macrofauna

The freshwater zoobenthos, or bottom macrofauna (>0.5 mm, or 500 μm), consist mostly of insect larvae (with terrestrial adult stages), crustacea, worms, and mollusks. Eutrophication in the littoral zone of lakes and rivers, with accompanying organic pollution, causes similar impacts on the invertebrate macrofauna of these systems. In early stages of eutrophication, oligochaetes, chironomids, gastropods, and sphaerids increase, and mayfly nymphs such as *Hexagenia* decrease. As eutrophication continues, the major change that coincides with progressive hypoxia/anoxia in overlying waters and surface sediments is the decline and then disappearance of additional oxygen-sensitive species such as stonefly (*Diura* spp.) and certain other taxa of mayfly nymphs (e.g., *Baetis, Rhithrogena*) as well as certain caddisfly larvae (e.g., *Rhyacophila, Hydropsyche*) and bivalve mollusks.

Eutrophication increases organic matter and bacterial decay and depresses oxygen concentrations while increasing settlement of organic detritus to benthic organisms such as certain flatworms (e.g., *Polycelis*) that use the detritus for food. In the profundal sediments underlying deeper waters of lakes, the macroinvertebrate biomass increases but is comprised of low oxygen-tolerant species such as certain chironomid larvae and oligochaete worms (e.g., the oligocheate *Tubifex tubifex*, which, in one study, survived, grew, and reproduced under continuous anoxia for 10 months). The species diversity of tubificid oligochaetes decreases with advanced eutrophication and organic enrichment, and as-

sociated oxygen deficits. However, if oxygen is periodically available, the rich food supply in combination with the lack of more oxygen-sensitive competitors allows robust growth. Declines in chironomid communities occur most rapidly in the change between nutrient-poor to moderately nutrient-enriched (mesotrophic) waters; and numbers of oligochaetes (relative to chironomids) increase as organic enrichment increases. In the intermediate zone between the littoral and profundal, the increased supply of "fresh" littoral detritus is consumed by large-bodied detritivores such as the bivalve mollusk *Dreissena*.

Estuarine and marine coastal invertebrate macrofauna, including mollusks, polycheates, decapods, and other crustacea, and nemerteans, spend their adult lives buried beneath the sediment surface. They are highly diverse with food acquisition as filter feeders, nonselective deposit feeders, selective deposit feeders, or raptorial/other predators, making generalizations difficult. Increasing organic matter from/accompanying nutrient enrichment tends to cause similar impacts as in freshwater systems, namely, an increase in macroinvertebrate abundance under moderately eutrophic conditions, and a decline in species diversity and abundance as eutrophication and associated hypoxia/anoxia progress. Organisms that burrow into anaerobic sediments must be able to gain access to oxygen in the overlying water. Increasing eutrophication leads to elimination of burrowing organisms as the anaerobic zone moves closer to the sediment surface. In highly overenriched areas, only worms such as *Capitella* may survive. Oxygen does not need to be completely absent for damage to occur—hypoxic waters with 3.0 to 4.3 mg DO/L have been related to mortality of some benthic invertebrate species and to loss of habitat for shellfish species such as lobster that require higher oxygen availability.

Changes in food quality/quantity also can reduce the species diversity and abundance of estuarine and coastal marine invertebrate macrofauna under highly eutrophic conditions. For example, an extensive early study by Filice in San Francisco Bay demonstrated that in domestic sewage outfall areas, few species survived. Species diversity was higher but still depressed in surrounding areas that received dilute sewage. However, in those areas, some species (e.g., the clams *Gemma gemma, Mya arenaria,* and *Macoma inconspicua;* the polychaete worm *Polydora uncata,* and the barnacle *Balanus improvisus*) apparently took advantage of new energy and material resources, and became highly abundant relative to abundance in control areas without sewage influence. In the Great South Bay complex of

Long Island, New York, previously mentioned duck farms along the bay tributaries fertilized the water with nutrients and organic wastes, and stimulated dense blooms of the small algae, *Nannochloris* sp. and *Stichococcus* sp. (densities greater than 10^6 cells/ml). These algae were very different from the previous phytoplankton community in the area, which consisted of mixed species that are needed to support oysters. Following these changes, oyster populations significantly declined, apparently because they were unable to thrive on a diet consisting only of these small algae.

F. Vertebrate Macrofauna

Eutrophication initially reduces and then eliminates sensitive lake fishes (e.g., salmonids and coregonids), by eliminating oxygen-replete, colder bottom-water habitat and well-oxygenated spawning areas. Fish can avoid low-oxygen waters, but the cold-optimal species encounter warmer waters as they are forced to move from deep areas into the shallows. Most temperate-zone fishes breed in the littoral. Their eggs are more vulnerable to short-term (e.g., nightly) oxygen deficits and to the low-oxygen microenvironment of increased detritus. Thus, the critical habitat of these sensitive species is destroyed by nutrient overenrichment, and spawning and recruitment are depressed.

Analogous impacts occur in rivers and estuaries. For example, in parts of the Baltic Sea, cod eggs laid in well-oxygenated surface waters die when they sink to anoxic bottom waters. Oxygen levels in the bottom waters of the Baltic's deep basins are negatively correlated with juvenile codfish abundance. Hypoxia in estuaries has been linked to depressed survival of larval fish, mortality of certain benthic invertebrates used as fish prey, and loss of habitat for mobile species of finfish such as cod that require high oxygen availability. Hypoxia and anoxia represent a growing problem for many estuaries and coastal marine waters, such as the Chesapeake Bay, the Baltic Sea, the Black Sea, the Pamlico Estuary, Long Island Sound, the North Sea, and the Gulf of Mexico by the mouth of the Mississippi River.

Overall population sizes and biomass of fish usually increase with nutrient overenrichment, which is sometimes regarded as a beneficial effect in early to midstages of eutrophication. However, dominance shifts from species such as lake trout to cyprinids, bullheads (*Ictalurus* spp.), and other coarse fish that can tolerate low oxygen concentrations (Fig. 10). In lakes with well-developed littoral zones of rooted aquatic vegetation, high predation pressure from piscivorous fish such as largemouth

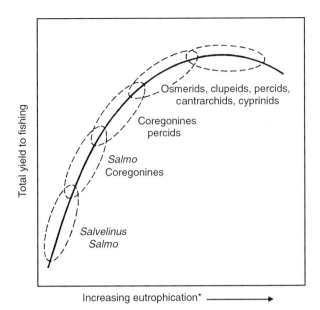

FIGURE 10 Trends in fish yields and taxonomic composition with increasing eutrophication, indicating total yield (as fish caught) of representative taxonomic groups from various North American lakes. Modified from Harper (1992), with permission; originally from Colby *et al.* (1972). Note that Harper used an *x*-axis labeled "morphoedaphic index," which is calculated considering both phosphorus content (indicator of eutrophication) and lake physical characteristics (for example, depth).

bass (*Micropterus salmoides*) or pike (e.g., *Esox lucius*) maintains low biomass of cyprinid fish and, thus, low predation pressure on large-bodied zooplankton. These large zooplankton, in turn, control phytoplankton biomass through strong grazing pressure, so that there is high visibility for visually feeding piscivorous fish and abundant light for macrophyte growth.

As eutrophication progresses, the littoral-zone macrophytes disappear and the carrying capacity of the lake for piscivorous fish is reduced. Thus, (zoo-) planktivorous fish are freed from high predation. Their larvae tend to selectively feed on the largest herbivorous zooplankton species (based on the size-efficiency hypothesis of Brooks and Dodson), causing an overall reduction in herbivore size. Smaller herbivores cannot exert enough grazing pressure to control blooms of cyanobacteria and other noxious algae, which proliferate in eutrophic lakes and lower rivers in the absence of the large-bodied zooplankton. Turbidity increases from higher cyprinid feeding activities and from loss of the littoral-zone "filtering" effect of the macrophytes, further impairing visual predation by remaining piscivorous fish. Photosynthetic activity from the high algal biomass can elevate the pH to levels that adversely affect gill function in sensitive fish species.

Similar impacts on fish populations from loss of desirable littoral zone species have been documented in estuarine and coastal waters. For example, along the Swedish coast of the Baltic Sea, excessive nutrient enrichment has been related to increases in nuisance cyanobacteria and other filamentous species, as mentioned, and a decrease in formerly dominant *Fucus vesiculosus* (Phaeophyceae), probably because of reduced light availability. This seaweed had provided critical habitat for herring spawns and the shift to dominance by filamentous macroalgae led to decreased egg hatch in the herring populations.

Although aquatic birds and other vertebrates can move among aquatic environments with relative independence, some general effects of eutrophication on these macrofauna also have been described. Increased macrophyte vegetation under moderate nutrient enrichment in mesotrophic systems has been linked to increased numbers of herbivorous waterfowl such as moorhen (*Gallinula chloropus*) and whooper swan (*Cignus cignus*). Some piscivorous birds such as grebes and herons have also increased. As eutrophication progresses, however, bird species may decline following undesirable changes in food and habitat. For example, pochards (*Aythya ferina*) decreased on Lake Constance as their main food supply, the macroalga *Chara*, declined under nutrient enrichment. General macrophyte habitat loss under nutrient overenrichment has also been linked to the gradual disappearance of many waterfowl such as black and mute swans (*Cygnus atratus, C. olor strepera*), Canada geese (*Branta canadensis*), coots (*Fulica atra*), teal (*Anas crecca*), and gadwells (*Anas strepera*). Sensitive life history stages of amphibians (for example, frog eggs) have been killed in increasing hypoxic events within littoral zones of eutrophic lakes.

III. OLIGOTROPHICATION— REVERSING THE IMPACTS OF EUTROPHICATION

It is possible to reverse nutrient loading impacts, at least to some extent, by reducing the nutrient inputs. This phenomenon can occur naturally when, for example, a major storm such as a hurricane causes high flooding or sedimentation that effectively re-routes a nutrient-rich tributary away from a receiving lake or estuary (such as an oxbow lake that is cut off from the main river); when natural flooding destroys a dam and eliminates a run-of-river impoundment; or when a raised bog system becomes isolated from nutrient-rich stream and groundwater sources. As nutrient loading declines, species diversity generally increases while the system production decreases. In ombrotrophic bog systems, there is also a shift over time to more acid-tolerant species, as the system's sole source of new water is low-pH precipitation.

Human influences are more pervasive than the natural forces that can contribute to oligotrophication. Cultural eutrophication is an acceleration of a long-term, natural process. Resilience, defined as the rate of recovery to the predisturbance state, depends on the initial status of the system in the natural eutrophication process and on the degree of nutrient loading sustained. Ecosystem resilience is generally considered to increase with increased nutrient loading rate and to decrease with increasing food chain length. Recent analysis by Cottingham and Carpenter suggested a modification of this relationship for pelagic food webs of north temperate lakes. In that analysis, planktivore-dominated (short-length) food webs were more resilient at baseline P loading rates of 0.1 to 1.0 μg/L/day. Piscivore-dominated (long-length) food webs were more resilient at high baseline P loading rates (2.0 μg/L/day), apparently because the additional nutrients were incorporated into the biota more rapidly.

Eutrophic systems, dominated by "generalist or opportunist" species that are insensitive to the adverse impacts from nutrient overenrichment, are considered to be relatively resistant to further change or stress as mentioned. Morever, the adverse impacts of cultural eutrophication in such systems can be partially reversed in a relatively short period of time (several years). Less nutrient enriched systems (for example, early phases of moderate nutrient enrichment or mesotrophy) have more specialized species, more opportunity for biotic adjustment of elemental cycles, and tighter coupling among element cycles. The diverse communities of such systems, comprised of more sensitive species, are relatively sensitive to further stress imposed by nutrient enrichment under accelerated eutrophication; and reversal or "recovery" to a "pristine" oligotrophic state is difficult to accomplish.

The degree of "restoration" success also depends on certain physical features of the system, especially mean depth and flushing rate. Aquatic ecosystems tend to retain most nutrients from loading events in the bottom sediments. In deep lakes, most of the nutrient inputs from the previous years are unavailable to phytoplankton of the next growing season because the lower water column remains permanently stratified and, thus, isolated from the upper water column where most of the viable phytoplankton occur. In shallow systems, wind and internal currents mix the water column and surficial sediments completely and frequently over an annual cycle, so that some portion of the nutrients from the relatively rich bottom waters and sediments repeatedly becomes available for phytoplankton growth. Reservoirs, rivers, well-flushed estuaries and wave-swept marine coastal waters are usually less sensitive to nutrient loading and more easily "reversed" in the eutrophication process than lakes or poorly flushed coastal lagoons with long water exchange times (on the order of months to years) because the latter systems cannot flush the nutrient-laden water through and "self-cleanse."

As other, more practical considerations, the extent to which the acceleration of the eutrophication process can be reversed would be expected to depend on the feature of the aquatic ecosystem that is targeted for improvement. Nutrient reductions would exert the most direct effects on plant communities such as freshwater/estuarine phytoplankton or estuarine/coastal macroalgae. Thus, a goal of reduced incidence of algal blooms may be more rapidly achieved than that of improved fish communities with growth that is indirectly rather than directly affected by the nutrient inputs. The economic feasibility of controls that can be exerted on anthropogenic nutrient sources is also important. Pragmatically, the highest prognosis for success would be expected for natural waters that are affected mostly by sewage and other point (pipe) source nutrient dischargers, because they are much easier to control than nonpoint (diffuse) sources.

One of the most famous early reports of successful reversal of cultural eutrophication involved removing sewage discharges from Lake Washington within metropolitan Seattle, Washington, in the United States. This large, deep lake (128 km², maximum depth 59 m, mean depth 18 m) historically had shown water quality degradation in response to sewage inputs. In 1922 a diversion was created to carry the raw sewage from 30 outfalls away from the lake (into nearby Puget Sound). Algal blooms and fish kills soon abated. However, in 1930 sewage effluents began to be discharged into the lake from treatment plants in outlying communities, and noxious cyanobacteria blooms and hypoxia again increased. About 76 million liters of sewage without inorganic nutrient removal were discharged daily into the lake by 1962. Nearly a decade of effort by a courageous

limnologist, Dr. T. Edmonson, led the city to support zero discharge of sewage into the lake by 1968. By 1970 the lake's phytoplankton growth had decreased to levels that had not been seen since the early 1950s, and the improvements in water quality and aesthetics were heralded by the citizenry as a "rapid and remarkable" recovery. This recovery was aided by the fact that the lake is relatively deep and that the major nutrient sources to be controlled were sewage pipes. It is unlikely that the degree of recovery reversed the water quality and aquatic communities to a circa 1930, more pristine status. However, the partial recovery (partial oligotrophication) represented a major improvement.

Other partial reversals of cultural eutrophication have been reported worldwide. In freshwaters they mostly have been achieved by targeting P reductions. In estuaries, both P and N have been reduced or "co-managed" for best results. Studies in north temperate estuaries by Fisher, Chesterikoff, and colleagues indicate that if only P, but not N, is removed from sewage inputs to upstream fresh waters, the P removal can actually exacerbate N-related eutrophication problems in downstream estuaries. The freshwater P reductions decrease riverine algal growth that, if present, would also have removed a substantial portion of the N_i in the sewage before it reached the estuary. Soluble nitrate, in particular, is transported downstream at higher concentrations than if freshwater algal blooms had been available to consume it, and the increased N_i is thus available to stimulate higher phytoplankton growth in the receiving estuary.

In large lakes, partial reversals can achieve highly desirable results. For example, the degradation of fish communities in Lake Erie reached its most extreme level in the 1960s from a combination of eutrophication, overexploitation of fishery resources, extensive habitat modification, and other pollution. Beginning in the 1970s, fishery management strategies and pollution abatement programs contributed to a dramatic reversal. Lake Erie walleye fisheries rebounded to world-class status, and point-source P loading significantly declined, especially after sewage treatment was improved at the major point source discharger to the west basin of the Lake (the Detroit metropolitan wastewater treatment plant) and after mandated use of detergents without phosphate. The P reductions effected a dramatic decrease in the abundance of nuisance phytoplankton species and of zooplankton biomass, as well as a decline in the abundance of pollution-tolerant oligochaetes and an overall shift in macroinvertebrates to more pollution-intolerant taxa. Similarly, in the Bay of Quinte on Lake Ontario, P loading reductions after 1977 led to a decline in the abundance and biomass of oligochaete worms, sphaeriid mollusks, isopod crustacea, and some chironomids. Dominance in both chironomid and oligochaete communities shifted to species less tolerant of eutrophic conditions.

Nutrient reductions have also led to success stories in estuaries and coastal lagoons. For example, industrial point source nutrient reductions to the Seto Inland Sea in Japan during the 1980s promoted a noticeable decline in the frequency and magnitude of toxic dinoflagellate blooms. Nutrient loadings to Cockburn Sound, Australia, from industrial point sources, and reduction in nutrient loadings to Chesapeake Bay from sewage point sources, decreased phytoplankton blooms and promoted an increase in seagrass meadow habitat. Sewage discharges to the shallow estuary, Mumford Cove, in Connecticut (United States) were rerouted to another waterway in the late 1980s, and within two years massive nuisance blooms of the macroalga, *Ulva lactuca*, were eliminated. Thus, partial "oligotrophication" from removal of sewage can have fairly rapid, positive results even in shallow systems where the sediments could provide some degree of nutrient replenishment to the overlying water.

From an overall ecosystem standpoint, the extent to which aquatic habitats can be "fully reversed" in the cultural eutrophication process depends on a more complex "endpoint" than variables such as reductions in nuisance algal growth, recovery of beneficial macrophyte beds, or increases in desirable fish species. The full range of chronic impacts from variables that frequently accompany nutrient enrichment are poorly understood. For example, scientists only recently determined that low-level water-column nitrate enrichment could inhibit the growth of certain sensitive seagrasses as a direct, subtle, and potentially serious physiological effect. Scientists only recently reported, as well, that nutrients can act to indirectly stimulate toxic, animal-like organisms such as *Pfiesteria*, mediated through stimulation of algal prey that are consumed by *Pfiesteria* species. The National Research Council has reported that there are more than 100 different enteric pathogens in sewage including viruses, bacteria, and parasites that cause waterborne diseases in humans. The fate of many of these organisms and their impacts on aquatic life have not been examined, but available evidence indicates that some of them can survive for months to years in the bottom sediments of lakes, rivers, and estuaries. Thus, many challenges of cultural eutrophication remain to be resolved before the full extent of its impacts on aquatic ecosystems, and the success of reversing those impacts, can be evaluated.

See Also the Following Articles

COASTAL BEACH ECOSYSTEMS • ESTUARINE ECOSYSTEMS • LAKE AND POND ECOSYSTEMS • PLANKTON, STATUS AND ROLE OF • RIVER ECOSYSTEMS • SEAGRASSES

Bibliography

Burkholder, J. M. (1998). Implications of harmful microalgae and heterotrophic dinoflagellates in management of sustainable marine fisheries. *Ecological Applications* 8(1) Suppl., S37–S62.

Burkholder, J. M., Mason, K. M., and Glasgow, H. B. Jr. (1992). Water-column nitrate enrichment promotes decline of eelgrass *Zostera marina*: Evidence from seasonal mesocosm experiments. *Marine Ecology Progress Series* 81, 163–178.

*Carpenter, S. R. (1988). Complex interactions in lake communities. Springer-Verlag, New York.

Chesterikoff, A., Garban, B., Billen, G., and Poulin, M. (1992). Inorganic nitrogen dynamics in the River Seine downstream from Paris (France). *Biogeochemistry* 17, 147–164.

Colby, P. J., Spangler, G. R., Hurley, D. A., and McCombie, A. M. (1972). Effects of eutrophication on salmonid communities in oligotrophic lakes. *Journal of the Fisheries Research Board of Canada* 29, 975–983.

*Cooper, S. R. (1995). Chesapeake Bay watershed historical land use: impact on water quality and diatom communities. *Ecological Applications* 5, 703–723.

Cooper, S. R., and Brush, G. S. (1991). Long-term history of Chesapeake Bay anoxia. *Science* 254, 992–996.

Cottingham, K. L., and Carpenter, S. R. (1994). Predictive indices of ecosystem resilience in models of north temperate lakes. *Ecology* 75, 2127–2138.

Edmondson, W. T. (1970). Phosphorus, nitrogen, and algae in Lake Washington after diversion of sewage. *Science* 169, 690–691.

Filice, F. P. (1959). The effect of wastes on the distribution of bottom invertebrates in the San Francisco Bay estuary. *Wasmann Journal of Biology* 17, 1–17.

Fisher, T. R., Peele, E. R., Ammerman, J. W., and Harding, L. W. Jr. (1992). Nutrient limitation of phytoplankton in Chesapeake Bay. *Marine Ecology Progress Series* 82, 51–63.

*Hallegraeff, G. M. (1993). A review of harmful algal blooms and their apparent global increase. *Phycologia* 32, 79–99.

*Harlin, M. M. (1993). Changes in major plant groups following nutrient enrichment. In *Eutrophic Shallow Estuaries and Lagoons* (A. J. McComb, Ed.), pp. 173–187. CRC Press, Boca Raton.

*Harper, D. (1992) *Eutrophication of Freshwaters – Principles, Problems and Restoration*. Chapman & Hall, New York.

*Hecky, P. E., and Kilham, P. (1988). Nutrient limitation of phytoplankton in freshwater and marine environments: A review of recent evidence on the effects of enrichment. *Limnology and Oceanography* 33, 796–822.

*Howarth, R. W. (1988). Nutrient limitation of net primary production in marine ecosystems. *Annual Review of Ecology and Systematics* 19, 89–110.

Kilham, S. S., Kreeger, D. A., Goulden, C. E., and Lynn, S. G. (1997). Effects of algal food quality on fecundity and population growth rates of *Daphnia*. *Freshwater Biology* 38, 639–647.

* Asterisk indicates comprehensive secondary references on aspects of this topic; other references were cited in the text, as well.

*Lapointe, B. E. (1997). Nutrient thresholds for bottom-up control of macroalgal blooms on coral reefs in Jamaica and southeast Florida. *Limnology & Oceanography* **42**, 1119–1131.

Mallin, M. A., Paerl, H. W., Rudek, J., and Bats, P. W. (1993). Regulation of estuarine primary production by watershed rainfall and river flow. *Marine Ecology Progress Series* **93**, 199–203.

Michell, F. (1999). Eutrophication, fisheries, and consumer-resource dynamics in marine pelagic ecosystems. *Science* **285**, 1396–1398.

National Research Council (1993). *Managing Wastewater in Coastal Urban Areas*. National Academy Press, Washington, DC.

Neckles, H. A., Wetzel, R. L., and Orth, R. J. (1993). Relative effects of nutrient enrichment and grazing on epiphyte-macrophyte (*Zostera marina* L.) dynamics. *Oecologia* **93**, 285–293.

Phillips, G. L., Eminson, D., and Moss, B. (1978). A mechanism to account for macrophyte decline in progressively eutrophicated freshwaters. *Aquatic Botany* **4**, 103–126.

*Rhee, G.-Y. (1982). Effects of environmental factors on phytoplankton growth. In *Advances in Microbial Ecology* (K. C. Marshall, Ed.), pp. 33–74. Plenum Press, New York.

Rhyther, J. H. (1989). Historical perspective of phytoplankton blooms on Long Island and the green tides of the 1950s. In *Novel Phytoplankton Blooms, Coastal and Estuarine Studies No. 35* (E. M. Cosper, V. M. Bricelj, and E. J. Carpenter, Eds.), pp. 375–383. Springer-Verlag, New York.

Schelske, C. L., Stoermer, E. F., Fahnenstiel, G. L., and Haibach, M. (1986). Phosphorus enrichment, silica utilization, and biogeochemical silica depletion in the Great Lakes. *Canadian Journal of Fisheries and Aquatic Sciences* **43**, 407–415.

*Schindler, D. W. (1987). Determining ecosystem responses to anthropogenic stress. *Canadian Journal of Fisheries & Aquatic Sciences* **44** (Suppl. 1), 6–25.

*Short, F. T., and Wyllie-Echeverria, S. (1996). Natural and human-induced disturbance of seagrasses. *Environmental Conservation* **23**, 17–27.

Smayda, T. J. (1989). Primary production and the global epidemic of phytoplankton blooms in the sea: a linkage? In *Novel Phytoplankton Blooms. Coastal and Estuarine Studies No. 35* (E. M. Cosper, V. M. Bricelj, and E. J. Carpenter, Eds.), pp. 449–484. Springer-Verlag, New York.

Smith, V. H. (1983). Low nitrogen to phosphorus ratios favor dominance by blue-green algae in lake phytoplankton. *Science* **221**, 669–671.

Starr, M., Himmelman, J. H., and Therriault, J.-C. (1990). Direct coupling of marine invertebrate spawning with phytoplankton blooms. *Science* **247**, 1701–1704.

Tilman, D. (1982). *Resource Competition and Community Structure*. Princeton University Press, Princeton.

Tilman, D., Kiesling, R., Sterner, R., Kilham, S. S., and Johnson, F. A. (1986). Green, blue-green, and diatom algae: taxonomic differences in competitive ability for phosphorus, silicon, and nitrogen. *Archives für Hydrobiologie* **106**, 473–485.

*Tilman, D., Kilham, S. S., and Kilham, P. (1982). Phytoplankton community ecology: The role of limiting nutrients. *Annual Review of Ecology and Systematics* **13**, 349–372.

Touchette, B. W. (2000). Review of nitrogen and phosphorus metabolism in seagrasses. *Journal of Experimental Marine Biology and Ecology* (in press).

Turpin, D. H. (1991). Effects of inorganic N availability on algal photosynthesis and carbon metabolism. *Journal of Psychology* **27**, 14–20.

Vitousek, P. M., Aber, J., Howarth, R. W., Likens, G. E., Matson, P. A., Schindler, D. W., Schlesinger, W. H., and Tilman, G. D. (1997). Human alteration of the global nitrogen cycle: Causes and consequences. *Ecological Applications* **7**, 737–750.

Wetzel, R. G. (1979). The role of the littoral zone and detritus in lake metabolism. *Archives für Hydrobiologie Beih. Ergebn. Limnologie* **13**, 145–161.

Wetzel, R. G. (1983). *Limnology*. Saunders College Publishing, Philadelphia.

EVOLUTION, THEORY OF

Catherine L. Craig
Harvard University, Tufts University

I. A Brief History of Evolutionary Thought
II. The Universal "Tree" of Life
III. Structure of Evolutionary Biology and How
     It Is Studied

GLOSSARY

evolution Descent, with modification from one (or at
most, a few) original ancestors
evolutionary theory A body of statements about the
general laws, principles, or causes of evolution
lateral transfer Genetic information passed between
organisms through means other than inter-breeding.
phylogenetic tree A hypothesis for describing the history and relationships among living species.

EVOLUTION, DESCENT WITH MODIFICATION, is a
scientific fact. Evolutionary theory is a coherent body
of interconnected statements, based on reasoning and
evidence, that describes the processes of branching of
lineages and changes within lineages (including extinction and gene transfer). Nothing in biology makes
sense unless it is studied in an evolutionary context
(Dobshansky, 1973), and all fields of biology provide
insight into the processes of evolution and mechanisms
of evolutionary change. Therefore, the goal of evolu-

tionary biology is to determine patterns of ancestor-
descendent relationships among organisms in time and
space, understand the processes that gave rise to them,
and discern how the specific attributes of organisms
originated and changed through time.

I. A BRIEF HISTORY OF EVOLUTIONARY THOUGHT

The development of evolutionary theory to date can be
characterized by four major periods of expansion that
reflect increasingly sophisticated means of documenting and quantifying natural variation at the molecular,
cellular, and organismal level. The most significant development in evolutionary theory occurred between
1859 and 1930, the Darwinian period. The most significant insight of this time was recognition of the common ancestry of organisms and the force of natural
selection. The next period of expansion in evolutionary
thought occurred between 1930 and 1960 when Darwinian theory was reconciled with modern genetics,
systematics, and paleontology. The current period of
expansion in evolutionary biology, 1960 to the present,
began when Lewontin and Hubby discovered the enormous amount of natural variation at the protein and
nucleotide level. Currently, the molecular revolution
has refocused evolutionary studies on decoding the genetic ciphers that explain evolution. The most fundamental concept in evolutionary biology, the concept

that all organisms can be divided into discrete species, is being challenged as are long-held notions about adaptation and the relationships among species. The current tree of life divides organisms into three kingdoms, the Eukarya, the Archaea, and the Bacteria.

This chapter is organized into two sections. The first section outlines the periods of expansion of evolutionary thought; the second section outlines the structure of evolutionary biology, explains how evolution is studied, and discusses recent empirical works that highlight new findings.

A. Darwinian Evolution: A Paradigm Shift in Evolutionary Thought

The word evolution is derived from the Latin word *evolutio,* or "unrolling." Prior to Darwin, classical thought argued that God created all species according a chain of being or Scala Naturae. The chain of being followed God's plan, a gradation from inanimate objects to animate forms of life, plants, invertebrates, and finally humans. Because the Scala Naturae was an expression of God's design, all objects were perfect, permanent, and unchanging: no new forms of life could have arisen or become extinct since the time of Creation. Therefore, the role of natural science was to catalog God's creations to make manifest His wisdom. Based on this view, Linnaeus established a universal framework for classifying similar species into similar genera that were thought to reflect God's design but that did not imply any genealogical connectedness. These ideas culminated in the most significant pre-Darwian theory of evolution, put forth by Jean Baptiste Lamarck (1744–1829). Lamarck proposed that organisms evolved through adaptation to the physical environment. He introduced the theory of "organic progression" in which species originated continually via independent events of spontaneous generation and subsequently evolved "up" the scale of nature (Fig. 1a). Lamarck envisioned two hierarchical chains of life, one for plants and one for animals, and a world where extinction was not possible. Lamarck viewed changes in organisms to result from the use and disuse of organs and that these acquired characteristics were then passed to descendants. Extinction was not possible and, therefore, species identified from fossils were still present, although modified, through adaptation.

Charles Darwin (1809–1882) founded the current field of evolution. His most important contribution to the theory of evolution was the concept that all species diverged from a common ancestor and that natural selection is the mechanism driving speciation. Prior to Darwin, natural scientists accepted the idea that species struggle against one another for existence. Darwin's experience as a naturalist and the writings of Malthus, "Essay of Population Growth" (1798), helped him see that individuals, not species, struggle for life. Malthus's essay argued that the rate of human population growth was greater than the rate of the increase of food. Therefore, if reproduction is untethered, famine would drive animal populations to extinction. The realization that individuals with superior characteristics would live to reproduce, while individuals of the same species with inferior characteristics would not, triggered Darwin's belief in evolution. The struggle for existence, that Darwin believed resulted in the differential reproduction of individuals, is defined as natural selection (Fig. 1b).

B. The Synthetic Theory of Evolution

By the 1870s most scientists accepted the theory of evolution by common descent but, for at least 60 years after the publication of *The Origin of Species,* there was no consensus that natural selection was the mechanism by which evolution was achieved. The next period of expansion in evolutionary thought occurred in the 1930s and 1940s, when contributions of geneticists, systematists, and paleontologists reconciled Darwinian theory with the facts of genetics (Mayr and Provine, 1980). A modern understanding of genetics, together with the mathematical theory of population genetics by R. A. Fisher (1890–1962), J. B. S. Haldane (1892–1964), and Sewall Wright (1889–1988), demonstrated that mutation and natural selection were causes of adaptive evolution. The synthesis of studies in taxonomy and genetic principles built by E. Mayr (b. 1904) and G. L. Stebbins (b. 1906) provided evidence that evolution occurred gradually. The integration of paleontology with population genetics by G. G. Simpson (1902–1984) demonstrated that the synthetic theory of evolution was fully consistent with the fossil record. Alternative theories of the mechanisms of evolutionary change that had been formulated since Darwin but prior to the evolutionary synthesis, such as Neolamarckism and the view that biological systems were creative and purposeful, were demonstrated to be inconsistent with the fossil record. Therefore, the major achievement of the evolutionary synthesis was to fully integrate genetics and Darwinian evolution and to argue that the major features of evolution could be accounted for by within species processes, such as mutation, recombination, and natural selection (Futuyma, 1998).

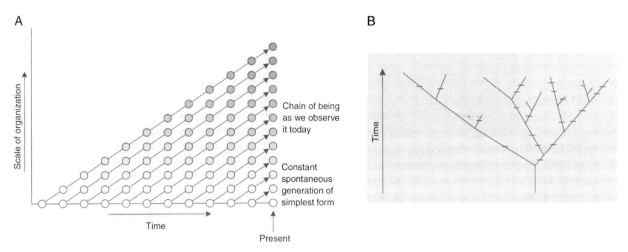

FIGURE 1 Proposed trees of ancestor–descendant relationships visualized by Lamark (A) and Darwin (B). (A) Jean Baptiste Lamark proposed that organisms originated spontaneously, continuously and evolved "up" a chain of natural being. (B) Darwin proposed that all species evolved from a common ancestor through the mechanism of natural selection (from Futuyma, 1998).

C. The Discovery of Protein and Nucleotide Diversity

Two schools of thought emerged during the synthetic period: the selectionist school from genetics and the mathematical theory of population genetics, and the adaptionist school from paleontology and systematics. The selectionist view is that there is no limit to an organism's variability and hence ability to evolve. Furthermore, all aspects and characteristics of organisms are variable and able to change rapidly. No qualitatively new phenomena results from increased organismal complexity and adaptations reflect the sum of infinitely variable genotypes. In contrast, the adaptationist school put forth the concept of "Bauplan," or the view that groups of organisms represent broad schemes of organization and that variations among groups reflect some degree of functional and developmental constraint. In its most extreme, the adaptationist school views organisms as relatively static and only able to change over long periods of time after populations of species have been isolated.

In the early 1960s, evolutionary population genetics became the central discipline of the study evolutionary processes. The most important contribution made during that period was the discovery of the vast amount of unexplained nucleotide and protein diversity and the subsequent explosion of mathematical theory to explain the role of mutation, drift and natural selection in the evolution of populations. The resulting, deepened understanding of the molecular basis of life and genetics revealed those questions that needed to be answered to reconcile the theory of evolutionary population genetics with concepts of speciation that are based on shared patterns of organismal organization and design.

D. The Molecular Revolution and Current Evolutionary Thought

The foundation of evolution, descent with modification from a common ancestor, as stated by Darwin, remains unchanged. The molecular revolution, however, continues to transform our perspective on evolutionary thought by providing molecular tools to dissect the pathways by which characters and organisms are modified and molecular tools to probe the mechanisms by which molecules, genes, and genomic architecture interact. The unfolding complexity of these genetic pathways continues to reveal an increasing number of insights that are both exciting as well as disturbing to some. One of these is the deep genetic similarity between what are otherwise very different organisms, for example, houseflies and humans. We now know that a substantial part of genetic and biological innovation is the result of the simple mechanism of gene duplication in which genes are subsequently coopted for new functions. Furthermore, the chimeric aspect of genomes shows that organisms from the archaeal lineages (the methanogens or methane-producing bacteria and bacteria that live in extreme environments), bacterial lineages (all remaining single cell organisms), and eukaryote lineages (all multicellular organisms) all contain genes from multiple sources that are likely to have been obtained by the process of lateral transfer, not interbreeding (vertical transfer) or even *de novo* (Fig. 2). In fact, the lateral transfer of genetic material is so extensive that some investigators no longer consider the Archeae and Eubacteria discrete or closed systems of organization but recognize them as fluid and reticulated biological systems (Doolittle, 1999).

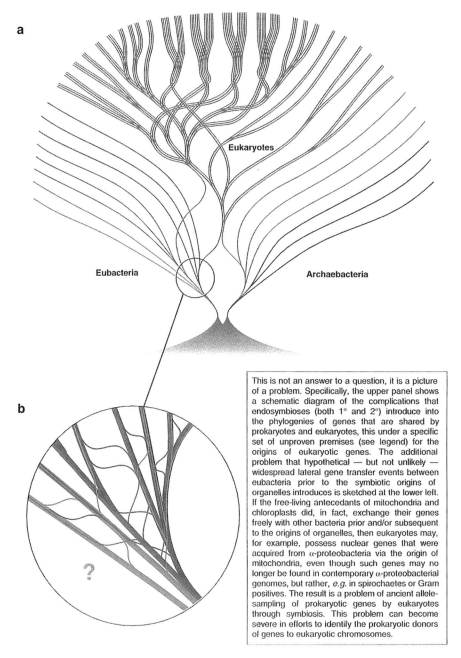

This is not an answer to a question, it is a picture of a problem. Specifically, the upper panel shows a schematic diagram of the complications that endosymbioses (both 1° and 2°) introduce into the phylogenies of genes that are shared by prokaryotes and eukaryotes, this under a specific set of unproven premises (see legend) for the origins of eukaryotic genes. The additional problem that hypothetical — but not unlikely — widespread lateral gene transfer events between eubacteria prior to the symbiotic origins of organelles introduces is sketched at the lower left. If the free-living antecedants of mitochondria and chloroplasts did, in fact, exchange their genes freely with other bacteria prior and/or subsequent to the origins of organelles, then eukaryotes may, for example, possess nuclear genes that were acquired from α-proteobacteria via the origin of mitochondria, even though such genes may no longer be found in contemporary α-proteobacterial genomes, but rather, *e.g.* in spirochaetes or Gram positives. The result is a problem of ancient allele-sampling of prokaryotic genes by eukaryotes through symbiosis. This problem can become severe in efforts to identify the prokaryotic donors of genes to eukaryotic chromosomes.

FIGURE 2 (a) The three kingdoms of living organisms. The origin of the Eukaryotes, all multicellular organisms, is thought to lie somewhere between the Archaebacteria (methane-producing bacteria and bacteria living in extreme environments) and the Eubacteria (all remaining single-celled organisms). Lateral transfer (b) of genetic information among all three lineages may be extensive and its implications a matter of active debate (from BioEssays 21.2, W. Martin, © 1999. Reprinted by permission of Wiley-Liss, Inc., a subsidiary of John Wiley & Sons, Inc.).

One effect of this more pluralistic view of organisms and the complexity of genetic systems revealed through the molecular revolution has been to refocus evolutionary thinking from the inheritance of fixed characters to the inheritance of systems of organization and control. Evidence for this perspective comes from integrative studies of the fossil record and molecular developmental biology. For example, current evidence from paleontology shows that by the end of the Ediacaran, 520 Ma, all of the major clades of animals, or "stem" groups of animals, had diverged, and since that time, only one of group of organisms, the Vendobionta (preserved as impressions, casts and molds; the status of this group is controversial), has gone extinct (Fig. 3). Divergence of the Bilateria (organisms with bilateral symmetry, i.e., the Chordata and Echinodermata; the Mollusca, Annelida, Platyhelminthes, and Brachiopoda; the Arthropoda, Nematoda, and Priapula) from the sponges, cnidarians, and ctenophores resulted in all of the animal taxa recognized today. Molecular evidence shows that early-diverging Bilateria taxa display both ancestral and derived characters that include similar gene content, development, and morphology (Knoll

and Carroll, 1999). All bilaterians are characterized by a modern body plan (rostral-caudal body axis) and an extensive cluster of homeobox genes (*Hox* genes encode a sequence of 60 amino acids that bind to DNA and control its expression) that control interacting networks of developmental regulators and key structural genes (Knoll and Carroll, 1999). The different degrees and types of organization that characterize the modern descendants of these groups are the result of a mosaic of genetic and morphological phenotypes whose components evolve at different rates.

Understanding the developmental basis of organismal variation provides insight into the potential for evolutionary change, as well as its pattern. For example, the two major groups of bilatarians—the Protosomes (illustrated primarily in Drosophila melanogaster) and the Deuterostomes (illustrated primarily by vertebrates)—share a common genetic regulatory repertoire. This means that the common ancestor of the arthropods and chordates would have had all the genes that they share, as well as all of the morphological characters that the shared genes regulate: photoreception organs, appendages, a heart, and the propensity for body seg-

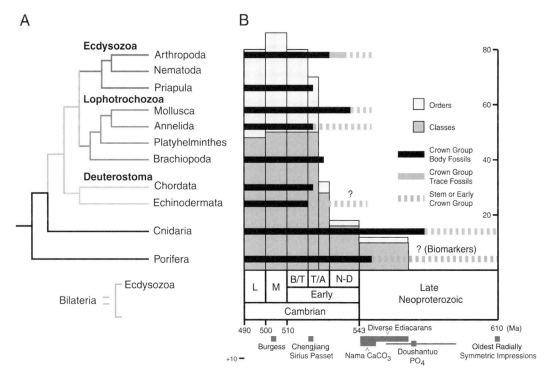

FIGURE 3 Animal diversity across the Peterozoic–Cambrian transition. By 520 Ma, all major clades of animals had diverged. All bilaterians are characterized by a rostral–caudal body plan and extensive cluster of homeobox genes. The descendents of these groups are mosaics of genetic and morphological phenotypes whose components evolve at different rates (Reprinted with permission from Knoll and Carroll, Early Animal Evolution © 1999, American Association for the Advancement of Science).

mentation (Knoll and Carroll, 1999). Through the replication and multiplication of regulatory networks, such as the *Hox* gene complex, taxa diverged as the complexity and the hierarchy of gene interactions increased. All of the proteins that cause the differentiation of body organization, however, did not and hence were present over 530 million years ago (Marin and Baker, 1998). Therefore, periods of great diversification in evolution, such as the Cambrian explosion, record the differentiation of crown (recently evolved) groups within clades that diverged as a result of an earlier radiation of genetic regulatory systems among stem (ancestral) taxa (Knoll and Carroll, 1999).

II. THE UNIVERSAL "TREE" OF LIFE

The current phylogenetic tree, based on small subunit RNA as well as whole genome-based phylogenetic analysis, distributes species among three groups of primary relatedness or domains: the Bacteria, the Eukarya, and the Archaea. The root of the Eukarya is between the Bacteria and Archaea (Fitz-Gibbon and House, 1999). Exploring the relationships among genes, regulatory processes, and the species that are contained in them has led to the concept of a phylogeny (the evolutionary history of an organism) as a diffuse cloud of gene histories that may make it difficult to organize organisms into Linnaean systems. Therefore, one current perspective of phylogenies or the histories of lineages are as models of the change in the interbreeding probabilities between organisms and models of genetic potentialities of organisms through time (Maddison, 1997).

III. STRUCTURE OF EVOLUTIONARY BIOLOGY AND HOW IT IS STUDIED

The are two central questions in evolutionary biology: The first question asks, what is the history of life, as defined by the history of the species, their origins, and extinction, and what are the steps through organismal features evolved? The second question asks, what are the causes of the history life? While the major principles of evolutionary biology and major causes have been established, new questions are continually posed and many long-standing questions remain to be fully answered.

The processes of evolution are both random and determinate. Studies in nine subdisciplines, as well as syntheses among these, contribute to understanding evolution and its mechanisms: evolutionary ecology,

behavioral evolution, evolutionary paleontology, evolutionary developmental biology, evolutionary physiology and morphology, evolutionary systematic biology, evolutionary genetics, molecular evolution, and human evolution. Research in all of these fields has accelerated in the past 20 years as the focus of evolutionary studies continues to shift.

A. Evolutionary Ecology

The most important questions in evolutionary ecology address the role the environment plays in speciation and extinction. During the synthetic period of evolution, the dominant view of speciation was that geographic barriers would develop between two populations prohibiting gene flow between them. Even though the populations might live in identical environments, they would diverge gradually as a result of the accumulation of random mutations. In contrast, however, some researchers have proposed that the barriers spawning species can also be ecological. In this case, selection, as opposed to isolation, is the factor that drives speciation. This implies that speciation is the product of an event, not a process. Both modes have received support in recent studies.

Peterson *et al.* defined and compared the ecological niches of birds (21 pairs), mammals (11 pairs), and butterflies (5 pairs) in an area of active speciation and population differentiation, the Isthmus of Tehuantepec. They tested the degree to which the ecological characteristics of one taxon were able to predict the geographic distribution of its putative sister taxon and vice versa. They then computed a genetic algorithm to produce a set of decision rules in ecological space (a model of the fundamental niche) that were projected onto maps to predict potential geographic distributions of the taxa. Across the 37 species pairs they found that a taxon on one side of the Isthmus could more accurately predict the distribution of its sister taxon on the opposite side of the Isthmus of Tehuantepec than the distribution of any of the other species. They concluded that because the species on each side of the Isthmus had been isolated for several million years, the sister species that evolved on each side of the barrier were the result of random genetic events but not natural selection (Peterson, Soberon, and Sanchez-Cordero, 1999).

An alternative view of the speciation process is that ecological variation, such as climate variation or differences in food resources can act as barriers, and can select for adaptation. This may result in individuals that are morphologically different and are unwilling to mate with each other even though they have never been physically separated. The most famous example of adaptive radiation and how ecological niche can drive the evolution within a population is seen in studies of Darwin's finches. While journeying on the *Beagle*, Darwin collected 14 different species of finches from the Galapagos Islands. The different species of finches showed a degree of variation in beak size, beak shape, and body size that usually characterizes differences among families of birds. Nevertheless, the entire radiation of the finches is believed to have occurred in less than 3 million years. Using microsatellites (multilocus genetic markers with high mutation rates), Petren, Grant, and Grant (1999) found that 13 of the species of Darwin's finch that they studied were monophyletic (derived from the same common ancestor). Their many differences in phenotypic traits, including beak size and shape, body size and plumage, are derived from an ancestor with a relatively long pointed beak that is associated with an insectivorous diet. From this ancestor, ground finches evolved blunted beaks that are efficient for crushing seeds while tree finches evolved beaks that allowed greater biting strength at the tip. Furthermore, once a novel beak evolved, body size and beak size changed rapidly and allometrically (Petren *et al.*, 1999). They concluded that divergence among the finches reflects strong selection for ecological type.

B. Behavioral Evolution

Behavior is studied at two levels: (a) the neural, hormonal, and developmental mechanisms that underlie the adaptive differences and the historical pathways leading to a current behavior and (b) the selective processes shaping behavioral evolution. When put in an evolutionary context, the importance of a behavior is measured through its effects on animal reproductive success. One particularly clear example is when the status or dominance rank of an animal that lives in a group or community correlates with the animal's reproductive success. For example, a recent analysis of 30 years of data gathered on the chimpanzee community at the Gombe Stream Research Center, Tanzania, has shown that female chimpanzees, who are solitary foragers, do not display an obvious linear hierarchy. Female status, however, affects her reproductive performance in several ways: offspring of high-ranking females survive longer, the age at which offspring reach sexual maturity is sooner when mothers are high-ranking, and average female life span is longer among high-ranking females (Fig. 4). Furthermore, a female's dominance correlates with age until she reaches 21 where the rank she has achieved strongly predicts her rank for the next 10 years. Understanding the effects of this kind

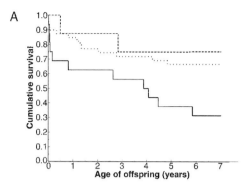

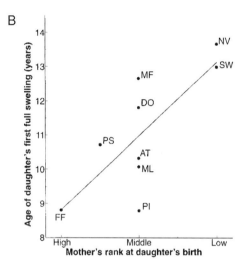

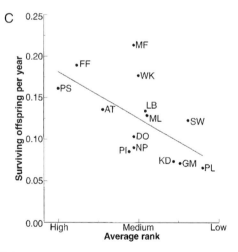

FIGURE 4 (A) Kaplan-Meier cumulative survival of offspring of females of high (-----), medium (·····), and low (—) rank. Shows that infants born to high-ranking females survive longer than infants born to low-ranking females. (B) Female offspring of high-ranking females mature sooner than the offspring of low-ranking females. (C) High-ranking females survive longer than low-ranking females (reprinted with permission from Pusey *et al.*, The influence of dominance rank on the reproductive success of female chimpanzees © 1997, American Association for the Advancement of Science).

of reproductive skew is important to the genetic diversity of species. When populations become small and isolated, genetic diversity of the population is likely to be reduced by the successful reproduction of a few dominant individuals (Pusey, Williams, and Goodall, 1997).

C. Evolutionary Paleontology

Paleontology addresses broadscale evolutionary patterns by tracing origins and fates of lineages and major groups, changes in characteristics and relationships of evolving lineages, and temporal variations in species diversity through the fossil record. Paleontologists, however, can only assess these changes through shifts in animal size. Hence paleontology cannot address the diversity or effects of the evolution of organismal physiology, only its pattern.

Research in evolutionary paleontology has become increasingly interdisciplinary and as a result four key, interrelated research questions have emerged: (a) What are the rules that govern biodiversity dynamics and do they apply at all temporal and spatial scales? (b) Why are major evolutionary innovations unevenly distributed in space and time? (c) How does the biosphere respond to environmental perturbations at global and region scales? (d) How have biological systems influenced the physical and chemical nature of the earth's surface, and vice versa (Jablonski, 1999)? For example, mass extinctions are not only important for the species that they eliminate, but for the evolutionary diversification and ecological restructuring that occur in their aftermath (Jablonski, 1998). Five major mass extinctions in the Metozoan fossil record were followed by rapid recovery of global taxonomic diversity and the radiation of new or previously minor groups. In a recent study, Jablonski examined the end-Cretaceous biotic recovery patterns among marine mollusks in four geographic provinces: the U.S. Gulf Coast, northern Europe, Northern Africa, and Pakistan-northern India. Although extinction intensities and selective effects were similar in the molluskan faunas of all four areas, the regions differed in the dynamics of diversifying clades and in the proportion of local taxa and invaders in the post-extinction biotas (Jablonski, 1998). In particular, three of the four regions lacked a rapid expansion and decline of "bloom taxa," as well as proportionally fewer invaders early in the recovery phase. Applying this result—contrary to the existing paradigm where global compendia of taxa are amassed to provide broad taxonomic analyses—Jablonski showed that distinct biogeographic regions are semiautonomous in their re-

sponse to environmental crises. While some investigators have proposed that the differential success of individual clades was due to intrinsic properties of the clade, Jablonski's studies show that biogeographic, environmental, and paleoecologic context are critical to any understanding how diversity changes.

D. Evolutionary Developmental Biology

Developmental, evolutionary biology reveals how changes in genetic information at the DNA level or genotype are translated into changes in organismal morphology or phenotype. For example, the molecular phylogenies of 18S ribosomal sequences show that the Bilateria should be reorganized into three different clades: the Deuterostoma (chordates and echinoderms), the Lophotrochozoa (molluscs, annelids, platyhelminths, and barchiopods), and the Ecdysozoa (chelicerates, crustacea, myriapods, insects, and onychophora). The result of this reordering, however, is that early evolving groups in derived clades must have displayed both ancestral and derived characters in combination (Knoll and Carroll, 1999).

Developmental biology shows that the apparent discrepancies between molecular and morphological data are resolved when the mechanisms that regulate gene expression are understood. For example, a marked trend in arthropod evolution has been an increase in body segment and appendage diversity. *Hox* genes play three major roles in the evolution of arthropods: (a) the entire arthropod clade contains the same set of *Hox* genes that are responsible for the recent diversity in Cambrian animals, (b) the increase in segment diversity is correlated with changes in the relative domains of *Hox gene*, and (c) changes in the morphology of homologous appendages are correlated with changes in the array of genes that are regulated by the same *Hox* gene (Knoll and Carroll, 1999). Therefore, most arthropod body plans are the result of very similar complexes of *Hox* genes. In addition, the correlation between diversification of *Hox* gene expression and patterns of evolution suggest that the diversification of bilaterian body plans is primarily due to the evolution of developmental regulatory systems, not changes in the genes themselves (Knoll and Carroll, 1999).

E. Evolutionary Physiology and Morphology

Evolutionary physiology and morphology determine how biochemical, physiological, and anatomical aspects of organisms affect their ability to adapt to new environments and ways of life. Contemporary studies focus on asking how form and function relate to each other during evolution, why some species are tolerant of broad physical ranges and others are not, and if there is more than one way for an animal to adapt to its environment. One of the most interesting discoveries of this field is that even when the fossils of ancestors are similar to their descendants, the descendent's biochemistries may be diverse and able to adapt to almost any physiological need.

For example, the macroevolutionary importance of plant chemistry on herbivore host shifts is critical to understanding the evolution of insect-plant interactions (Becerra, 1997). Members of the New World, monophyletic genus of the beetle, *Blepharida* (Chrysomelidaea: Alticinae), feed mainly on *Bursera* (Burseraceae). While plants in the *Bursera* genus produce an array of terpenes that are toxic or repellent to most herbivores, they only decrease *Blepharida* survival and growth rate. To determine the importance of plant chemistry to the phylogenetic diversification of the beetles and their feeding behaviors, Becerra (1997) constructed a dendogram of *Bursear* species based on their chemical similarity. She found that most clades of *Bursera* include plants that are in different chemical groups suggesting that their chemical similarity is partially independent of

plant phylogeny. She also found that *Blepharida* rarely shifted between chemically dissimilar plants and, in fact, subclades of *Blepharida* appear to have colonized species of plants that belonged to only one chemical group even if they shifted between hosts belonging to different subclades several times (Fig. 5). Therefore, comparison of plant phylogeny and plant chemical variation indicate a greater influence of host plant chemistry than host plant phylogeny in the evolution of the Blepharida and Bursera interaction.

F. Evolutionary Systematic Biology

Evolutionary biology is based on the principle that all organisms share a common history. Evolutionary systematics attempts to organize history by cataloging species, determining their genealogical or phylogenetic relationships, and classifying them into inclusive or hierarchical groups. Systematic studies prior to 1980 were largely based on morphological comparisons among taxa. Recent research, however, has reanalyzed and broadened past work using molecular tools. DNA analyses provide independent estimates of the age of phylogenetic groups while extending and revising our understanding of organismal relationships.

For example, corals are Cnidarians, the sister taxa of the Bilatarians, and evolved at least 600 million years

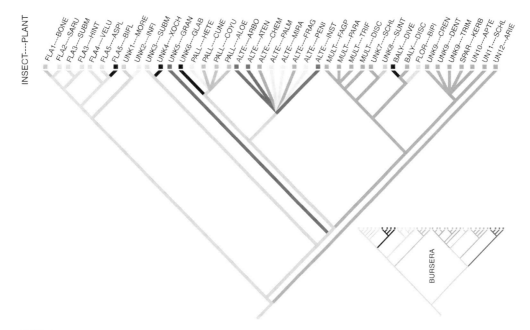

FIGURE 5 Comparison of *Blepharida* and *Bursera* phylogenies. The eight major clades of *Bursera* are traced on a phylogeny of *Blepharida*. These data show that plant chemistry has had a greater influence on the *Blepharida* and *Bursera* interactions than has host plant phylogeny (reprinted with permission from Becerra, Insects on Plants © 1997 American Association for the Advancement of Science).

ago. The sclaractinian corals (reef-building corals) have a continuous fossil record from the mid-Triassic (about 160 million years ago) and support some the world's most diverse marine communities. Despite this detail, their skeletal variability and morphology have made it difficult to understand the relationships among coral families and suborders (Romano and Palumbi, 1996). Molecular phylogenetic analysis of mitochondrial 16S ribosomal RNA, however, has helped to clarify coral evolution and revealed that sclaractinians originated 80 millions years earlier than they appear in the fossil record. The mitochondrial data show a deep split in the coral lineages that predates skeleton formation and thus is invisible in the fossil record. Using a combined molecular and traditional analyses, Roman and Palumbi suggest that the sclaractinian skeleton evolved repeatedly and converged during evolution. This is important because if the selective events that led to the multiple convergences in the sclaractinians can be identified, we will be able to understand the selective events that led to the complex, marine ecosystems that the corals support (Romano and Palumbi, 1996).

G. Human Evolution

Evolutionary biology is studied from two different viewpoints: a conceptual view, such as theoretical population genetics, or an organismal view, such as *Homo*. Anthropologists and biologists studying human evolution use the same principles, concepts, and tools as other biologists. Some investigators focus on the genetic processes that affect contemporary human populations, such as human genetics and medical genetics. Other investigators draw information from evolutionary systematics, paleontology, genetics, ecology, and animal behavior. One example of an organismal study that taps information from many fields of anthropological research explores the hypothesis that cooking has played a critical role in the evolution of human social systems. Using evidence from the fossil record, Wrangham *et al.* (1999) propose that cooking first appeared among hominoids 1.9 million years ago. They propose that the observed increase in female body mass was the effect of increased food energy that is released when foods are cooked, and that smaller tooth size reflected the reduced digestive effort that results when foods are partially broken down before eating. Furthermore, foods that were cooked and made more digestible resulted in increased nutrient accessibility. Food that was gathered, accumulated and stored until cooked became a valuable and defensible resource that changed the economics hominoid foraging. Females fitness, limited

by access to energetic resources, could be enhanced if the females formed bonds with food-guards or males willing to defend them. Females could have competed for the best food-guards through extended sexual attractiveness. Wrangham *et al.* (1999) suggest that cooking was responsible for the evolution of human social systems where pair-bonds are embedded within multimale and multifemale communities and are supported by strong, mutual, and frequently competing sexual conflicts (Wrangham, Jones, Laden, Pilbeam, and Conklin-Brittain, 1999).

H. Evolutionary Genetics

Evolutionary genetics, which includes population genetics, uses both molecular and classical genetic methods to understand the origin and effects of mutation and recombination. It employs empirical studies and theoretical methods to uncover the roles of genetic drift, gene flow, and natural selection to predict and interpret evolutionary change. Perhaps the most important finding in this field is the large amount of genetic variation in protein phenotypes that seem to have no immediate selective value. Furthermore, recombinant DNA technology has revealed that there are large amounts of previously undetected polymorphism at the nucleotide level that do not affect changes in amino acid sequence. Constancy in protein sequence in the face of nucleotide polymorphism shows that selection is acting directly at the DNA level preventing the evolution of deleterious proteins (Kreitman, 1983).

I. Molecular Evolution

Springing directly from the molecular revolution, molecular evolution focuses on causes of evolutionary change at the level of the gene, protein, and genome. It takes the position that evolution occurs on diverse scales of time. For example, antibodies recognize and distinguish among specific molecular patterns of antigens. Because antigens are structurally diverse, the repertoire of antibodies must be large enough to protect organisms from a wide range of pathogens and toxic agents. This diversity is achieved through three, germline, gene segments, which combine to generate antibody types. These are further altered through somatic mutation (not inherited) that increases the antigen's affinity and specificity as the immune response proceeds (French, Laskov, and Scharff, 1989; Tonegawa, 1983).

Wedemayer *et al.* (1997) compared the structure of the germline Fab fragments, and its complex with hapten, to the corresponding crystal structure of the affinity matured antibody 48G7. Combinatorial association between CDR1 and CDR2 regions in the primary antibody repertoire allows antibody combining sites to be more diverse than the flanking, germline-encoded sites. However, rather than a few large changes occurring at the active site, antibody maturation depends on small additive changes, many of which result from somatic mutations, to reconfigure the active site. By mapping structural changes due to nine amino acids, Wedemayer *et al.* (1997) showed that somatic mutations resulted in 30,000 times higher affinity for hapten. None of these, however, directly contacted the hapten; only two were within 5.5 A of the hapten binding site and all others were at least 10 A away. Instead, the mutations resulted in reorganized hydrogen bonding networks on the molecule's surface, shifting the backbone conformation of the antibody protein.

See Also the Following Articles

BIODIVERSITY, EVOLUTION AND • DARWIN, CHARLES • GENES, DESCRIPTION OF • PHYLOGENY

Bibliography

Becerra, J. X. (1997). Insects on plants: macroevolutionary chemical trends in host use. *Science* 276, 253–256.

Dobshansky, T. (1973). Nothing in biology makes sense except in the light of evolution. *Am. Biol. Teach.* 35, 125–129.

Doolilttle, F. (1999). Phylogenetic classification and the Universal Tree. *Science* 284, 2124–2128.

Fitz-Gibbon, S. T., and House, C. H. (1999). Whole genome-based phylogenetic analysis of free-living microorganisms. *Nucleic Acids Research* 27, 4218–4222.

French, D. L., Laskov, R., and Scharff, M. D. (1989). The role of somatic hypermutation in the generation of antibody diversity. *Science* 244, 1152–1157.

Futuyma, D. J. (1998). *Evolutionary Biology.* Sinauer, Sunderland.

Jablonski, D. (1998). Geographic variation in the molluscan recovery from the end-Cretaceous extinction. *Science* 279, 1327–1330.

Jablonski, D. (1999). The future of the fossil record. *Science* 284, 2114–2116.

Knoll, A. H., and Carroll, S. B. (1999). Early animal evolution: Emerging views from comparative biology and geology. *Science* 284, 2129–2137.

Kreitman, M. (1983). Nucleotide polymorphism at the alcohol dehydrogenase locus of Drosophila melanogaster. *Nature* 304, 414–417.

Maddison, W. P. (1997). Gene trees in species trees. *Systematic Biology* 46, 523–536.

Marin, I., and Baker, B. S. (1998). The evolutionary dynamics of sex determination. *Science* 281, 990–994.

Martin, W. (1999). Mosaic bacterial chromosomes: a challenge en route to a tree of genomes. *Bioessays* 21, 99–104.

Mayr, E., and Provine, W. (Eds.) (1980). The evolutionary synthesis: Perspectives on the unification of biology. Harvard University Press, Cambridge.

Peterson, A. T., Soberon, J., and Sanchez-Cordero, V. (1999). Conservatism of ecological niches in evolutionary time. *Science* 285, 1265–1267.

Petren, K., Grant, B. R., and Grant, P. R. (1999). A phylogeny of Darwin's finches based on microsatellite DNA length variation. *Proceedings of the Royal Society of London, B, February*, 321–329.

Pusey, A., Williams, J., and Goodall, J. (1997). The influence of dominance rank on the reproductive success of female chimpanzees. *Science* 277, 774–775.

Romano, S., and Palumbi, S. (1996). Evolution of scleractinian corals inferred from molecular systematics. *Science* 271, 640–642.

Tonegawa, S. (1983). Somatic generation of antibody diversity. *Nature* 302, 575–581.

Wedemayer, G. J., Patten, P. A., Wang, L. H., Schultz, P. G., and Stevens, R. C. (1997). Structural insights into the evolution of an antibody combining site. *Science* 276, 1665–1669.

Wrangham, R. W., Jones, J. H., Laden, G., Pilbeam, D., and Conklin-Brittain, N. (1999). The raw and the stolen: Cooking and the ecology of human origins. *Current Anthropology* 40, 567–594.

EX SITU, IN SITU CONSERVATION

Nigel Maxted
University of Birmingham

I. Introduction
II. Why Does Biodiversity Need Conservation?
III. What Threatens Biodiversity?
IV. What Is Biodiversity Conservation?
V. Methods of Conservation
VI. Sustainable and Integrated Biodiversity Conservation

GLOSSARY

complementary conservation Application of a range of conservation techniques (including *ex situ* and *in situ*) to conserve the target taxon, one technique acting as a backup to another. The degree of emphasis placed on each technique depends on the conservation aims, the type of species being conserved, the resources available, and whether the species has utilization potential.

conservation Maintenance of the diversity of living organisms, their habitats, and the interrelationships among organisms and their environment.

ecogeography Analysis of a species' ecological, geographical, and taxonomic characteristics to assist in the formulation of collection and conservation priorities.

effective population size (N_e) Number of conserved individuals that would undergo the same amount of random genetic drift as the actual population.

ex situ **conservation** Conservation of the components of biological diversity outside their natural habitats.

gene pool The total genetic diversity found within an individual species or species group.

genetic erosion Loss of genetic diversity from a species, often caused by anthropogenic factors.

in situ **conservation** Conservation of ecosystems and natural habitats and the maintenance and recovery of viable populations of species in their natural surroundings and, in the case of domesticates or cultivated species, in the surroundings where they have developed their distinctive properties.

keystone species Usually the dominant species within a habitat that tend to define it physiognomically and ecologically, for example, by determining nutrient and water cycling.

target taxon Species or species group that the conservation action is focused upon.

EX SITU AND IN SITU CONSERVATION approaches focus on the protection of species outside their natural habitats or in their natural surroundings, respectively. This article outlines the basic concepts of biodiversity conservation by discussing why conservation is needed, the major threats to animal and plant species, and how we can set about ensuring that the diversity of living organisms is maintained at the ecosystem, species, and genetic levels. The importance to humankind of the

Convention on Biological Diversity is also considered. A simple model for biodiversity conservation is proposed and discussed. This involves the selection of target taxa for conservation, gene pool concepts, ecogeographic surveys, preliminary survey missions, clarification of conservation objectives and field exploration, the two basic conservation strategies (_ex situ_ and _in situ_) and the range of conservation techniques, and ways that conservation is often linked to some form of utilization. Finally, the critical need of adopting a sustainable and integrated approach to conservation is emphasized, so that we can endeavor to pass to future generations the levels of biodiversity that we inherited from our forebears.

I. INTRODUCTION

The conservation of biodiversity is of critical importance now, because that very diversity is under threat of extinction and erosion, but also because it can be of direct and indirect benefit to humankind. Biodiversity benefits humans through the exploitation of animals and plants in agriculture and horticulture, the development of medicinal drugs, and the pivotal roles played by species in the functioning of all natural ecosystems. Biodiversity is also valuable for ethical, aesthetic, and recreational reasons. The fundamental importance of these issues to humankind was paramount at the United Nations Conference on the Environment and Development (UNCED) held in Rio de Janeiro, Brazil, in 1992. The Convention on Biological Diversity (CBD), which has subsequently been ratified by 175 countries (as of January, 1999), has as its objectives

> the conservation of biological diversity, the sustainable use of its components and the fair and equitable sharing of the benefits arising out of the utilisation of genetic resources. (Article 1: Objectives; Convention on Biological Diversity, 1992)

II. WHY DOES BIODIVERSITY NEED CONSERVATION?

Estimates of the total number of species in the world vary, and it is difficult to estimate rates of species extinction, but the consensus view summarized by Lugo

(1988) was that 15–20% of all species could become extinct by the turn of the century. It is even more difficult, if not impossible, to estimate precise levels of genetic erosion, that is, the loss of genetic diversity from extant species. However, the loss of genetic diversity must always be faster than the loss of species because there will be some genetic erosion from the species that remain extant. Loss of any genetic diversity means that the affected plants or animals will not be able to adapt to changing conditions quite so readily; for example, potato cultivars with a narrow genetic base were unable to withstand the infection of late potato blight (_Phytophthora infestans_) in the late 1840s in Ireland, the crop was devastated, and millions were forced to emigrate or starve. Although genetic erosion cannot be quantified accurately, it seems likely that virtually all species are currently suffering loss of genetic variation to varying degrees. Therefore, using Lugo's figures as a starting point, it may be estimated that 25–35% of plant and animal genetic diversity could possibly be lost over the 12 year period leading up to the year 2000 (Maxted _et al.,_ 1997a).

Of course loss of biodiversity also occurs at other levels than species, for habitats and ecological communities can also be degraded or destroyed. For example, virtually all of the natural grasslands in the United States have been lost since 1942 (Spellerberg, 1996), over 90% of natural wetlands in New Zealand have been lost since European settlement, and Fernside (1990) estimated that world forest loss was proceeding at 20,000 km^2 per year. In a similar study, FAO and UNEP (1991) found that annual rates of forest loss had increased from 113,000 km^2 to 169,000 km^2 per year between 1980 and 1990 in 76 countries.

It is important to realize, however, that both species extinction and genetic erosion can be natural events, just as species and genetic evolution are natural; nature is and it seems has always been dynamic. Yet the current levels of species extinction and genetic erosion are dramatically higher than the so-called background levels that existed hundreds, thousands, and millions of years ago. Humankind now has the capacity to drastically alter the world environment in ways that were not previously possible, and these anthropogenic changes are undoubtedly increasing the rates of species and genetic extinction.

III. WHAT THREATENS BIODIVERSITY?

The threat to biodiversity as a result of anthropogenic changes is not universal for all species. Some species are in greater danger of genetic erosion (or even of complete extinction) than others. These dangers must be evaluated carefully, so that those exposed to the highest risk can be given higher priority for conservation. However, it must also be borne in mind that levels of threat often change rapidly and unexpectedly. Thus, an area may suddenly come under the threat of industrial development, road-building, or logging. Diamond (1989), Gomez-Campo *et al.* (1992), WCMC (1992), and WRI *et al.* (1992) discuss in detail the kind of events that may lead to genetic erosion. These may be broadly grouped under the general headings of

- destruction, degradation, and fragmentation of natural habitats;
- overexploitation and incidental take;
- introduction of exotic species that compete with, prey on, or hybridize with native species;
- human socioeconomic change and upheaval;
- changes in agricultural practices and land use; and
- calamities, both natural and man-made.

It is valuable to establish a system by which the relative threat of genetic erosion can be assessed objectively. Guarino (1995) proposed a model for estimating the threat of genetic erosion that a taxon (wild or culti-

vated) faces in a particular region. The model may be used without having to visit the region involved, providing some background data are available on the taxon and the area. The model is based on scoring a number of parameters, such as the abundance of the taxon, the level of agricultural development, and the proximity and intensity of various types of human activity to the populations being studied. The higher the score, the greater the risk of genetic erosion and therefore the higher priority for *in situ* or *ex situ* conservation.

The International Union for Conservation of Nature and Natural Resources (IUCN, now known as the World Conservation Union) has developed a system of categories of conservation status, which is based on detailed knowledge of the population dynamics and genetics of the species concerned, the so-called IUCN Red Data List Categories (IUCN, 1994) (Fig. 1). Table I shows how some of these categories are determined. Using these categories, for example, a "critical" species would be assigned higher conservation priority than a "vulnerable" species. However, IUCN has tended to focus their attention almost exclusively on species extinction rather than genetic erosion within individual gene pools, and the latter may be of equal importance in terms of loss of biodiversity.

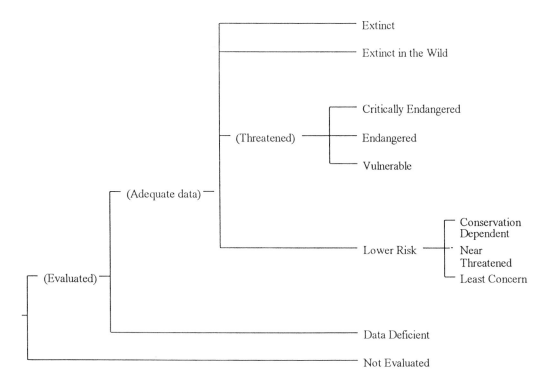

FIGURE 1 Structure of the IUCN categories of threat. (From IUCN, 1994.)

TABLE I

Threatened Category Thresholds[a]

Criteria	Main numerical thresholds		
	Critical	Endangered	Vulnerable
A. Rapid decline	>80% over 10 year or 3 generations	>50% over 10 years or 3 generations	>50% over 20 years or 5 generations
B. Small range (fragmented, declining, or fluctuating)	Extent of occurrence <100 km² or area of occupancy <10 km²	Extent of occurrence <5000 km² or area of occupancy <500 km²	Extent of occurrence <20,000 km² or area of occupancy <2000 km²
C. Small population (declining)	<250 mature individuals	<2500 mature individuals	<10,000 mature individuals
D1. Very small population	<50 mature individuals	<250 mature individuals	<1000 mature individuals
D2. Very small range	—	—	<100 km² or <5 locations
E. Unfavorable population viability analysis	Probability of extinction >50% within 5 years.	Probability of extinction >20% within 20 years	Probability of extinction >10% within 100 years

[a] From Department of the Environment (1996).

IV. WHAT IS IODIVERSITY CONSERVATION?

Spellerberg and Hardes (1992) stated that biodiversity conservation aims to "maintain the diversity of living organisms, their habitats and the interrelationships between organisms and their environment." But how is this to be achieved?

Conservationists must clearly define and understand the processes involved, and then attempt to develop practical techniques to achieve this objective. When undertaking a particular conservation exercise, a conservationist must use his or her knowledge of genetics, ecology, geography, taxonomy, and many other disciplines to understand and manage the biodiversity that is being conserved.

It is important to stress that conservation is not just about individual plant and animal species, but includes all aspects of biodiversity from ecosystems (a community of organisms and its abiotic environment), through communities (a collection of species found in a common environment or habitat), species, and populations, to genetic diversity within species. In recent years there has been a differentiation between conservation at the ecosystem and at the genetic levels, and these may be referred to as ecological and genetic conservation, respectively.

Ecological conservation focuses on the conservation of whole communities; although the survival of individuals and the extinction of particular species are a major concern, both are viewed in the larger context of overall community health. This form of whole-community conservation is exemplified by the 'Man and the Biosphere' program (UNESCO, 1996), which established a network of biosphere reserves representing distinct biomes and ecosystems throughout the world. The clear emphasis was on conservation of ecosystems: this program believed that individual species should be conserved as a component of ecosystems rather than on a species-by-species basis.

Genetic conservation focuses more explicitly on particular taxa and attempts to conserve the full range of genetic (allelic) variation within those taxa. The aim of this form of conservation is often utilitarian, for genetic diversity conservation is often linked intimately to human utilization. However, Maxted *et al.* (1997b) stressed the following points: first, species usefulness does not have to be defined in the strictest utilitarian sense, because plants and animals considered of aesthetic value are equally worthy of receiving human value as those of immediate use to plant or animal breeders; and second, in many cases individual species cannot be conserved without conserving the communities in which they naturally occur. So the distinction between the two basic forms of conservation is in practice blurred and may be viewed as artificial and of limited semantic importance, because the conservation of ecosystems and species are intimately linked. Just as it is difficult to focus conservation effort on the generality of the entire ecosystem, in practice conservationists need to focus on something more tangible that can be monitored and managed, even if only as an exemplar for the ecosystem as a whole.

So to undertake effective conservation, species interactions must be understood as far as possible. Even if the conservation target is a population of a single species, no populations survive in isolation. They exist within a community or ecosystem and interact with other species and the abiotic environment. Obvious examples of interactions include pollinators, seed dispersers, microbial symbionts, herbivores (whether natural or introduced by humans), predators, and pathogens. Thus, even when applying genetic _in situ_ conservation, the maintenance of genetic diversity will have to be considered within the context of whole-ecosystem conservation.

The so-called keystone species are important in this context, for these species contribute significantly to the structure of a community or its processes; they are the dominant species. The removal of a keystone species renders other members of a community vulnerable to extinction. Tropical trees that produce a rich food resource in the form of fruit or seeds, for example, can be considered keystone species, as they provide a vital food source for a diverse array of mammals and birds. Generally keystone species play an important part in interactions between different trophic levels, whether they are predators, herbivores, mutualists such as pollinators, or decomposers. So when considering the conservation of any particular species within an ecosystem, one must identify the inherent interactions and ensure their maintenance if the conservation project is to be successful and sustainable.

V. METHODS OF CONSERVATION

There is a need to develop appropriate methodologies for biodiversity conservation, particularly in the tropics. The tropical regions of the world have the highest levels of biodiversity, but their fauna and flora are the least well known and are most under threat. Also, tropical nations have few conservationists and often they are insufficiently trained; furthermore, the resources available for conservation activities are relatively limited. To address these issues, the CBD asks nations to:

> Promote and encourage research which contributes to the conservation and sustainable use of biological diversity, particularly in developing countries. (Article 12; CBD, 1992)

> Improve capability for determining scientific research priorities . . . via transparent, "user friendly" risk evaluation methodologies. (Chapter 35 of Agenda 21; CBD, 1992)

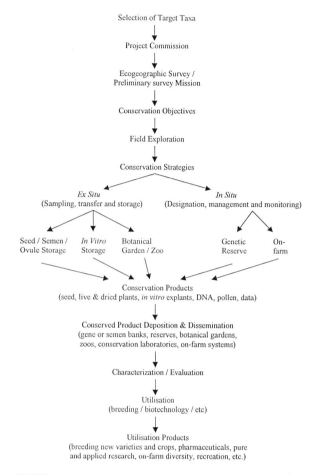

FIGURE 2 Model of biodiversity conservation. (Adapted from Maxted _et al.,_ 1997b.)

Maxted _et al._ (1997b) responded to this requirement—to clarify and enhance the methodologies and research programs that currently enable scientists to classify, conserve, manage, and utilize biodiversity—by proposing a model for plant and animal genetic diversity conservation (Fig. 2). The raw materials of genetic conservation are genes within gene pools, which represent the total diversity of genetic material of the particular taxon being conserved. The product of the gene pool (seeds, ovules, etc.) is either preserved or utilized as genetic diversity. The processes that link the raw matter and the utilized gene pool represent genetic conservation.

A. Selection of Target Taxa

Conservation activities will always be limited by the financial, temporal, and technical resources available. Conservation of ecosystems or species has a cost and the effort expended is directly related to how much society values that species and is therefore willing to pay. It is impossible to actively conserve or monitor all species, so it is important to make the most efficient and effective selection of species on which to focus conservation efforts. This choice should be objective and based on logical, scientific, and economic principles related to the perceived value of the species. Maxted *et al.* (1997a) discussed the sort of factors that provide a species with "value": current conservation status, potential economic use, threat of genetic erosion, genetic distinction, ecogeographic distinction, national or conservation agency priorities, biological importance, cultural importance, relative cost of conservation, conservation sustainability, and ethical and aesthetic considerations.

Rarely will one of the these factors on its own lead to a taxon being given conservation priority. More commonly, all or a range of these factors will be assessed for a particular taxon and then it will be given a certain level of national, regional, or world conservation priority. If the overall score passes a threshold level or is higher than those of competing taxa, the taxon will be conserved; it will then be either collected and conserved *ex situ,* an appropriate reserve will be established, or on-farm conservation will be proposed. Having listed the factors that affect the selection of target taxa in terms of value to society, those related to potential economic use will commonly be given higher comparative value, especially in economically poorer economies where income generation is of the highest priority. This anthropocentric and utilitarian view in the selection of conservation priorities may offend some conservationists, but when financial resources for conservation are, and are likely to remain, limited, and when men, women, and children are still suffering from malnutrition in many parts of the world, there appears to be no practical or ethical alternative to giving those species of most direct use to people the highest value and thus the highest conservation priority.

B. Project Commission

In practice, once taxa are selected for conservation, the actual conservation activities are necessarily preceded by some form of commission statement. This is likely to establish the objectives of the conservation, specify the target taxa and target areas, state how the material is to be utilized and where the conserved material is to be safely duplicated, and perhaps indicate which conservation techniques are to be employed. A clear,

concise commission statement will help to focus subsequent conservation activities.

Who writes the commission? The commission statement may be written by those contracting the conservation or those who actually undertake the conservation work. The commission may vary in taxonomic and geographic coverage, from a systematic collection program for a single species throughout its geographic range to a range of target taxa from a restricted location, for example, onion (*Allium*) species of Central Asia, large cat species worldwide, or chickpeas (*Cicer*) from the Western Tien Shen. In each case, however, a particular group of taxa from a defined geographical area must be considered to be insufficiently conserved (either *in situ* or *ex situ*), of sufficient actual or potential use, and/or endangered to warrant active conservation.

C. Ecogeographic Survey and Preliminary Survey Mission

Once the target taxon or group of taxa have been selected and delimited, the conservationist begins to amass and synthesize fundamental biological data to help formulate an appropriate conservation strategy. The synthesis and analysis of these data enables the conservationist to make vital decisions concerning, for example, which taxa to be included in the target group, where to find these taxa, which combination of *ex situ* and *in situ* conservation to use, what sampling strategy to adopt, where to store the germplasm and site the reserve or what captive breeding program would be most successful. If the basic biological data for a particular species, for example, the close lentil relative *Lens orientalis,* indicate that the species has been previously found on stony slopes at the edge of the Fergana valley in Uzbekistan, then further material of this species is likely to be currently found under similar constraints and is less likely to be found in different habitat types or in far distant regions.

The process of collating and analyzing geographical, ecological, and taxonomic data for use in designing conservation strategies is referred to as ecogeography and was defined by Maxted *et al.* (1995) as

> an ecological, geographical and taxonomic information gathering and synthesis process for a particular taxon. The results are predictive and can be used to assist in the formulation of collection and conservation priorities.

Ecogeographic studies involve the use of large and

complex data sets obtained from the literature and from the compilation of passport data associated with herbarium specimens and germplasm accessions. These data are synthesized to produce three basic products: the database, which contains the raw data for each taxon; the conspectus, which summarizes the data for each taxon; and the report, which discusses the contents of the database and conspectus, as well as proposing appropriate conservation strategies. Ecogeographic techniques enhance the efficiency of crop relative and wild species conservation because they enable the conservationist to identify clearly the geographical regions and ecological niches that the taxon inhabits, and so not only identify areas with high numbers of target taxa, but also areas that contain high taxonomic or genotypic diversity of taxa, uniqueness of habitat, economic or breeding importance, and so on.

If the available ecogeographic data for the target taxon are limited, the conservationist will not have sufficient background biological knowledge to formulate an effective conservation strategy. In this case it would be necessary to undertake an initial survey mission to gather the novel ecogeographic data required on which to base the actual strategy. The survey mission may be in the form of "coarse grid sampling," which involves traveling throughout a likely target region and sampling sites at relatively wide intervals over the whole region. The precise size of the interval between sites depends on the level of environmental diversity across the region, but it may involve sampling every 1–50 km. The population samples and data collected during this mission can then be used to formulate further conservation priorities and to develop an appropriate strategy, thus providing the same result as the ecogeographic survey for groups that are better biologically understood.

D. Conservation Objectives

The products of the ecogeographic survey or survey mission provide a basis for the conservationist to formulate future conservation priorities and strategies for the target taxon. Within the target area, zones of particular interest may be identified, for example, areas with high concentrations of diverse taxa, low or very high rainfall, or high frequency of saline soils or extremes of altitude or exposure. In general, areas with very distinctive characteristics are likely to contain plants with distinct genes or genotypes. If a taxon is found throughout a particular region, then the researcher can use the ecogeographic data to positively select a series of diverse habitats to designate as reserves. If a taxon has been found at one

location but not at another with similar ecogeographic conditions, then the ecogeographer may suggest that these similar locations should be searched. Within the target taxon, specific variants may be identified that warrant conservation priority, for example, species that have previously unrecognized utilization potential, populations that are particularly in danger of genetic erosion, or those that had not previously been conserved.

The conservationist must set out a clear, concise statement of the proposed conservation strategy for the target taxon and, if appropriate, prioritize actions. These may have been established in the project commission, but if not the conservationist should undertake the task. This should answer questions such as: Which populations require conservation? Can local farmers play a part in conservation activities? Do population levels require close monitoring? Should a national or international collecting team be directed to collect the priority target taxa? What conservation strategy or strategies are appropriate? What combination of conservation techniques is appropriate or is a more detailed study required before any of these questions can be answered?

E. Field Exploration

Once the conservation objectives have been clarified, whichever conservation strategy is to be applied, the ecogeographic information is used to locate and identify the general locality of the animal or plant populations that are to be conserved. The ecogeographic data will rarely be sufficiently comprehensive to precisely locate actual populations. Therefore, the preparatory element of conservation activities will be followed by field exploration, during which actual populations are located. Ideally, populations of the target taxon that contain the maximum amount of genetic diversity in the minimum number of populations will be identified, but how is this goal to be achieved? Commonly there will be too much diversity in both crops and wild species to conserve all their alleles, even if these were known then or at some future time. Thus the conservationist must attempt to conserve the range of diversity that best reflects the total genetic diversity of the species. How many animals or plants must be sampled, which specimens and what pattern of sampling is appropriate? To answer these specific questions the conservationist should know the amount of genetic variation within and between populations, local population structure, the breeding system, taxonomy and ecogeographic requirements of the target taxon, and many other biological details. Some of this information will be supplied following the ecogeographic survey, but some will remain unavailable. Therefore, the practice of field exploration will be modified depending on the biological information on the target taxon and target area that is available.

For a botanical project, the field botanist should select populations if they are found on the periphery of the target taxon's distribution or those that contain morphological or ecological variants. Atypical populations or those growing under atypical conditions may possess genes or alleles that are unknown or extremely rare in the target taxon's center of diversity, and this material possibly contains genetic variation that is of special use to breeders (e.g., disease or pest resistance or adaptation to soil or climate that is unknown in the crop itself).

F. Conservation Strategies

There are two basic conservation strategies, each composed of various techniques, that the conservationist can adopt to conserve genetic diversity once it has been located. The two strategies are *ex situ* and *in situ* conservation. Article 2 of the Convention on Biological Diversity (CBD, 1992) provides the following definition of these categories:

> *Ex situ* conservation means the conservation of components of biological diversity outside their natural habitats.

> *In situ* conservation means the conservation of ecosystems and natural habitats and the maintenance and recovery of viable populations of species in their natural surroundings and, in the case of domesticates or cultivated species, in the surroundings where they have developed their distinctive properties.

There is an obvious fundamental difference between these two strategies: *ex situ* conservation involves the sampling, transfer, and storage of target taxa from the target area, whereas *in situ* conservation involves the designation, management, and monitoring of target taxa where they are encountered. Because of this fundamental difference, there is little overlap between the two strategies. The two basic conservation strategies may be further subdivided into the following specific techniques, which are discussed in Sections V,G and V,H:

Ex Situ
    Seed/embryo storage
    In vitro storage

    Semen/ovule/pollen/DNA storage
    Field gene bank/livestock parks
    Botanic/zoological garden
In Situ
    Genetic reserve
    On-farm

G. *Ex Situ* Techniques

In *ex situ* conservation, genetic variation is maintained away from its original location and samples of a species, subspecies, or variety are taken and conserved either as living collections of plants or animals in field gene banks, botanic or zoological gardens, and arboreta, or as samples of seed, semen, ovules, tubers, tissue explants, pollen, or DNA maintained under special artificial conditions.

1. Seed/Embryo Storage Conservation

Ex situ seed/embryo collection and storage is the most convenient and widely used method of genetic conservation (Fig. 3). Seeds and embryos are the natural dispersal, storage, or generative organs for the majority of species. This technique involves collecting samples from individuals or populations and then transferring them to a gene bank for storage, usually at sub-zero temperatures. The procedure used for the bulk of orthodox-seeded plant species is to dry the seeds or embryos to a suitable moisture content (5–6%) before freezing at −20°C, but this method is only suitable for species that can be dried and stored at low temperature without losing viability. The advantages of this technique are that it is efficient and reproducible, and feasible for short-, medium-, and long-term secure storage. However, the disadvantages are that there are problems in storing recalcitrant-seeded plant species. The latter species cannot be dried and frozen in the way used for orthodox seeds, because they rarely produce seed or are normally clonally propagated.

FIGURE 3   Collecting seed from Tajikistan for *ex situ* conservation.

2. Botanical/Zoological Garden Conservation

Historically, botanical or zoological gardens were often associated with physic or medicinal gardens or displays of single specimens of zoological curiosities, and as such they did not attempt to reflect the genetic diversity of the species. These gardens now hold living collections of species that were collected in a particular location and moved to the garden to be conserved. The advantage of this method is that gardens do not have the same constraints as many other conservation agencies; they have the freedom to focus on wild species that may otherwise not be given sufficient priority for conservation. Yet there are two disadvantages to this technique. The first is that the number of species that can be genetically conserved in a botanical or zoological garden will always be limited because of the available space. The majority of these gardens are located in urban areas in temperate countries, and at their present sites most expansion would be prohibitively expensive. The majority of botanical and animal diversity is located in tropical climates, yet because most botanical and zoological gardens are in temperate countries, the collections must be kept in expensive greenhouses or other facilities, which also limits the space available. The second disadvantage is related to the first, namely, very few individuals of each species can be held, and this severely restricts the range of genetic diversity found in the wild that is protected. However, if the target species is very near extinction and only one or two specimens remain extant, this objection of course does not hold.

3. *In Vitro* Conservation

In vitro conservation involves the maintenance of explants in a sterile, pathogen-free environment, and it is widely used for vegetatively propagated and recalcitrant-seeded species. This method offers an alternative to field gene banks. It involves the establishment of tissue cultures of accessions on nutrient agar and their storage under controlled conditions of either slow or suspended growth. The main advantage is that it offers a solution to the long-term conservation problems of recalcitrant, sterile, or clonally propagated species. The main disadvantages are the risk of somaclonal variation, the need to develop individual maintenance protocols for the majority of species, and the relatively high-level technology and high cost required. The best answer for cheap, long-term *in vitro* conservation in the future may be cryopreservation (Hoyt, 1988), that is, the storage of frozen tissue cultures at very low temperatures, for example, in liquid nitrogen at −196°C. If this technique can be perfected to reduce the damage caused by freezing and thawing, it may be possible to preserve materials indefinitely.

4. Field Gene Bank/Livestock Park Conservation

The conservation of germplasm in field gene banks or livestock parks involves the collecting of plant or animal specimens from one location and the transfer and conservation at a second site. It has traditionally been the method for recalcitrant- (whose seeds cannot be dried and frozen without loss of viability) or sterile-seeded plant species or for those species for which it is preferable to store clonal material. Field gene banks are commonly used for species such as cocoa, rubber, coconut, mango, coffee, banana, cassava, sweet potato, and yam. Livestock parks or rare breed centers, as distinct from zoos, emphasize captive breeding programs and therefore genetic conservation. The advantages of field gene banks and livestock parks are that the species are easily accessible for utilization and evaluation can be undertaken while the material is being conserved. The disadvantages are that the material is restricted in terms of genetic diversity, is susceptible to pests, disease, and vandalism, and may require large areas of land.

5. Pollen/Semen/Ovule/DNA Conservation

The storage of pollen grains is possible under appropriate conditions that allow their subsequent use for crossing with living plant material. It may also be possible in the future to regenerate haploid plants routinely from pollen cultures, but no generalized protocols have been developed yet. The development of artificial insemination techniques in recent years has made semen and ovule storage routine, especially for domesticated animals. The storage of DNA under prescribed conditions can easily and inexpensively be achieved given the appropriate level of technology, but the regeneration of entire organisms from DNA cannot be envisaged at present, although single or small numbers of genes could subsequently be utilized. The advantage of pollen storage is that it is low cost and simple, but the disadvantage is that only paternal material would be conserved, and with DNA storage there are problems with subsequent gene isolation, cloning, and transfer.

H. *In Situ* Techniques

In situ techniques involve the maintenance of genetic variation at the location where it is encountered, either in the wild or in traditional farming systems. The majority of existing nature reserves and natural parks were established to conserve animals or to protect aesthetically beautiful landscapes, but even today few have plant conservation as their primary goal, let alone the genetic conservation of species (Hoyt, 1988).

1. Genetic Reserve Conservation

Conservation of wild species in a genetic reserve involves the location, designation, management, and monitoring of genetic diversity in a particular natural location (Fig. 4). This technique is the most appropriate for the bulk of wild species, because it can, when the management regime is minimal, be relatively inexpensive. Whether dealing with plants or animals, the objective is to contain the minimum number of individuals that can maintain genetic diversity within the species. If too few individuals are protected, genetic diversity will decline over time, and if too many are protected, resources may be wasted in managing the large population. To guide such efforts, conservationists will need to estimate the effective population size (N_e), that is, the number of conserved individuals that would undergo the same amount of random genetic drift as the actual population. Genetic reserves are appropriate for animals as well as for orthodox and non-orthodox seeded plant species, because numerous taxa can be protected in a single reserve that allows the continued evolution of species. However, the disadvantages are that the conserved material is not immediately available for human exploitation and, if the management regime is minimal, little characterization or evaluation data may be available. In the latter case, the reserve manager may even be unaware of the complete specific composition of the reserve that he or she is managing.

FIGURE 4 Surveying plant populations for an *in situ* reserve in Turkey.

2. On-Farm Conservation

Farmer-based conservation involves the maintenance of traditional crop or animal breeds or cultivation systems by farmers within traditional agricultural systems. On traditional farms, what are generally known as "land-races" of plants are sown and harvested, and each season the farmers keep a proportion of harvested seed for re-sowing. Traditional breeds of domestic animal are maintained by inter-breeding within and between local village stocks. Thus the land-race or breed is highly adapted to the local environment and is likely to contain locally adapted alleles that may prove useful for specific breeding programs. Home garden plant conservation is a closely related variant of on-farm conservation of land-races but on a smaller scale. It involves the cultivation of more species-diverse material in home, kitchen, backyard, or door-yard gardens. These home gardens focus on medicinal, flavoring, and vegetable species (e.g., tomatoes, peppers, digitalis, mint, thyme, parsley). The overall advantage of the on-farm technique is that it ensures the maintenance of highly adapted land-races and breeds and those wild species that traditional agriculture often depends on. However, these land-races or traditional breeds may yield less than their modern counterparts, and so traditional farmers may require some subsidy and possibly monitoring to ensure continued farming. It should be noted that contemporary economic forces tend to act against the continued farming of ancient land-races and breeds, which are currently suffering rapid genetic erosion; many face imminent extinction. A back-up system of *ex situ* conservation is therefore essential, as discussed in Section VI.

I. Community-Based Conservation

When applying all of the conservation techniques discussed here, professional conservationists have often failed to appreciate the role that local communities have successfully played in conserving animal and plant diversity within their local environment. It is now generally accepted that the present-day wealth of domesticated and non-domesticated biodiversity would not exist were it not for the conscious effort of local communities over millennia to conserve biodiversity in all its forms. For example, indigenous farmers in the Andes maintain a gene pool of over 3000 varieties of potatoes representing eight cultivated species, and in Papua New Guinea approximately 5000 varieties of sweet potato are cultivated, with a single farmer growing up to 20 varieties in one garden (McNeely *et al.*, 1995). For wild species, Prance *et al.* (1995) showed that four groups of Amazonian Indians use up to 79% of the tree species in their home ranges, and Milliken *et al.* (1992) in a similar study found that 81% of tree and vine species were utilized; this number rises to 86% when other categories of plants are added from literature sources.

So local communities have had and continue to have an essential role in biodiversity conservation. They not only continue to conserve by using traditional practices for their own future direct and indirect benefit, but also increasingly work in collaboration with professional conservationists to conserve broad-based biodiversity for the benefit of their host countries and humankind as a whole. Specifically, collaboration involving conservationists and local communities increases the overall efficiency of "professional" conservation, because local communities have a broader local knowledge base concerning the animal and plant species found in their area. Local communities are therefore able to assist in the development of a more practical, focused, and hopefully efficient approach to locally targeted conservation. The employment of a collaborative approach also empowers local people and engenders increased pride in native biodiversity and its conservation. In this way, rather than deferring responsibility to outside science-based experts, they can retain environmental responsibility and take greater pride in maintaining their own environment.

J. Conservation Products

The products of conservation activities are primarily conserved germplasm (seed, embryos, semen, and ovules), live plants and animals, dried plants, cultures, and conservation data. *Ex situ* conserved orthodox seed or animal semen and ovules are commonly held in gene or semen banks at sub-zero temperatures and, for seed, low moisture content to prolong their life. Live plants or animals are conserved in genetic reserves, field gene banks, botanical or zoological gardens, or parks and research laboratories. Germplasm that is stored in a suspended form, such as tissue, pollen, or DNA, is kept as cultures in specialist laboratory facilities. Dried voucher plant specimens are held in herbaria and linked to specific samples of germplasm, and are as much as possible made representative of the conserved populations. Conserved material is ideally associated with a range of passport data, which detail the taxonomic, geographical, and ecological provenance of the material. All passport data should be entered into a database and made available for the management of the material, the formulation of future conservation priorities and strategies, and any exploitation. The various conservation products, where they are stored, and where they should be duplicated are presented in Table II.

K. Conserved Product Dissemination

The conservation products are either maintained in their original environment or deposited in a range of *ex situ* storage facilities. Whether the germplasm, voucher specimens, or passport data are conserved *in situ* or *ex situ*, to ensure its safety it should ideally be duplicated in more than one location. The distribution of duplicate

TABLE II

Conservation Products and Their Storage and Duplication Sites

Conservation product	Storage site	Duplication site
Plant germplasm (seed, vegetative organs, etc.)	Gene bank	National, regional, and international gene banks, duplication with other conservation techniques
Animal germplasm (semen, ovules, eggs, embryos, etc.)	Gene bank	National, regional, and international gene banks, duplication with other conservation techniques
Live plants	Field gene bank, botanical garden, genetic reserve, on-farm	Duplication with other conservation techniques, e.g., gene bank storage of seed
Live animals	Zoological garden, genetic reserve	Duplication with other conservation techniques, e.g., storage of germplasm
Dried plants or preserved animals	Herbarium or museum	National, regional, and international herbaria or museums
Explants or plantlets	Tissue culture	Duplication with other conservation techniques, e.g., gene bank storage of seed
DNA and pollen	Various cultures	Duplication with other conservation techniques, e.g., gene bank storage of seed
Conservation data	Conservation database	Duplication with other national, regional, and international conservation agencies

sets of material avoids accidental loss of the material due to fire, economic or political difficulties, warfare, or other unforeseen circumstances. Duplication of the data is relatively easy from the conservation database, and copies should be held by the commissioning agency, relevant host country institutes, and other interested parties.

L. Biodiversity Utilization

As discussed earlier, there should be an intimate linkage between conservation and utilization. The products of conservation, whether they be "living" or "suspended," should be made available for use by humankind. Conservation can be seen as the safe-keeping of preserved material, so that the material is available at a future date. In certain cases the material can be used directly, say in the selection of forage accessions or local domesticated animal breeds, where little breeding is undertaken. The conserved material may also be used in reintroduction programs where the traditional breed or land-race has been lost locally owing to civil unrest or the application of perverse government incentives that encourage the alteration of traditional practices.

More commonly, the first stage of utilization will involve the recording of genetically controlled characteristics (characterization) and the plant material may be grown out under diverse environmental conditions to evaluate and screen for drought or salt tolerance, or be deliberately infected with diseases or pests to screen for particular biotic resistance (evaluation). The biotechnologist will be screening for single genes, which when located may be transferred into a host organism to generate more rapid growth, for example. The biochemist (bioprospector) will be screening for particular chemical products that may be of use to the pharmaceutical industry. The products of utilization are therefore numerous, including new varieties, new crops, improved breeds, and pharmaceuticals as well as more nebulous but equally valuable products such as a beautiful or more diverse environment for human recreational activities.

VI. SUSTAINABLE AND INTEGRATED BIODIVERSITY CONSERVATION

Having discussed in detail what constitutes biodiversity and how that biodiversity can be conserved, the point should be made that any biodiversity conservation program should be sustainable and integrated. Each conservation technique has its advantages and disadvantages. The two strategies of *ex situ* and *in situ* conservation should not be seen as alternatives or in opposition to one another, but rather as being complementary, as stated in Article 9 of the Convention on Biological Diversity (CBD, 1992). One conservation strategy or technique can act as a backup to another, the degree of emphasis placed on each depending on the conservation aims, the type of species being conserved, the resources available, and whether the species has utilization potential. The efficacy of adopting an integrated approach to conservation, or as some have called it a "holistic" approach (Withers, 1993), is now well established as the only sustainable option. Therefore, when formulating an overall conservation strategy for a species, conservationists should think in terms of applying a combination of the different techniques available, including both *in situ* as well as *ex situ* techniques, where the different methodologies complement each other. It may be helpful to think of the various techniques as pieces in a jig-saw puzzle that will complete the overall conservation strategy and thus ensure the maintenance of plant or animal genetic diversity.

The adoption of an integrated approach requires the conservationist to consider the characteristics and needs of the particular gene pool being conserved, and then to assess which of the strategies or combination of techniques offers the most appropriate option to maintain genetic diversity within that taxon. To formulate the conservation strategy, the conservationist may also need to address not only biological questions but also the practical and political ones: What are the species' storage characteristics? What do we know about its breeding success in captivity? Do we want to store the germplasm over the short, medium, or long term? How important is the species? Where is the species located and how accessible is it/does it need to be? Are there legal issues relating to access? How good is the infrastructure of the established reserves? What backup is necessary and/or desirable? How does the species conservation strategy fit within the local community development program? Given answers to these questions, the appropriate combination of techniques to conserve the gene pool can be applied in a pragmatic and balanced manner. The integration of conservation and community development is an important point to stress. We cannot expect local communities to altruistically forgo development for the benefit of a more abstract greater good, and so the practical application of the conservation strategy may need to be a compromise between scientific protocols and meeting the needs and desires of local people. Therefore the actual combination of techniques will be formulated afresh for each species or group of species, demonstrating the flexibility of the integrated approach.

Sustainability in the sense of continuance is a fundamental concept for conservation. Whether seed, semen, or embryos are collected for *ex situ* conservation in a gene bank, animals are incorporated into a captive breeding program, or a habitat is designated as a reserve, each option has a financial cost and it would be a waste of limited conservation funds not to ensure that the conservation project is sustainable at least in the medium term. If the species or genetic material is lost from a reserve, the resources expended on establishing the reserve would have been wasted and the cost of rehabilitating populations using materials stored *ex situ* would have to be considered. The latter option is commonly expensive and may require extensive research to ensure that the reintroduced animals or plants do not likewise go extinct. Unfortunately, many conservation projects are funded on a short-term basis, so it is essential that an effective project exit strategy is developed so the conservation program itself is sustainable.

Not only is it necessary to integrate the different conservation strategies and techniques, and to involve the local community in a sustainable conservation project, but it is also important to integrate the different potential agencies involved. This is particularly true for large *in situ* reserve projects where the project may naturally span national borders and professional disciplines. In these cases, the project team must ensure that the local, provincial, national, regional, and international conservation agencies, as well as professionals from the different disciplines involved such as environmentalists, foresters, agriculturalists, and politicians, work together to promote the success of the conservation project.

See Also the Following Articles

ENDANGERED ECOSYSTEMS • ENVIRONMENTAL ETHICS • GENE BANKS • GENETIC DIVERSITY • KEYSTONE SPECIES • SPECIES INTERACTIONS

Bibliography

Convention on Biological Diversity. (1992). *Convention on Biological Diversity: Text and Annexes*, pp. 1–34. Secretariat of the Convention on Biological Diversity, Montreal.
Department of the Environment. (1996). *Towards a Methodology for Costing Biodiversity Targets in the UK*. Department of the Environment, London.
Diamond, J. (1989). Overview of recent extinctions. In *Conservation for the Twenty-first Century* (D. Western and M. Pearl, eds.), pp. 37–41. Oxford University Press, Oxford, United Kingdom.
FAO and UNEP. (1991). *Tropical Forest Resources Assessment Project*. FAO, Rome.
Fernside, P. M. (1990). The rate and extent of deforestation in Brazilian Amazonia. *Environ. Conservation* **17**, 213–226.
Gomez-Campo,C., *et al.* (1992). *Libro Rojo de Especies Vegetales Amenazadas de Espana Peninsular e Islas Balneares*. Ministerio de Agricultura y Alimentacion, Madrid.
Guarino, L. (1995). Assessing the threat of genetic erosion. In *Collecting Plant Genetic Diversity: Technical Guidelines* (L. Guarino, V. Ramanatha Rao, and R. Reid, eds.), pp. 67–74. CAB International, Wallingford, United Kingdom.
Hoyt, E. (1988). *Conserving the Wild Relatives of Crops*. IBPGR/IUCN/WWF, Rome.
IUCN. (1994). *IUCN Red List Categories*. IUCN Species Survival Commission, Gland, Switzerland.
Lugo, A. E. (1988). Estimating reductions in the diversity of tropical forest species. In *Biodiversity* (E. O. Wilson, ed.), pp. 58–70. National Academy Press, Washington, D.C.
Maxted, N., van Slageren, M. W., and Rihan, J. (1995). Ecogeographic surveys. In *Collecting Plant Genetic Diversity: Technical Guidelines* (L. Guarino, V. Ramanatha Rao, and R. Reid, eds.), pp. 255–286. CAB International, Wallingford, United Kingdom.
Maxted, N., Hawkes, J. G., Guarino, L., and Sawkins, M. (1997a). The selection of taxa for plant genetic conservation. *Genetic Resources and Crop Evolution* **44**, 337–348.
Maxted, N., Ford-Lloyd, B. V., and Hawkes, J. G. (1997b). *Plant Genetic Conservation: The In Situ Approach*. Chapman & Hall, London.
McNeely, J. A., Gadgil, M., Leveque, C., Padoch, C., and Redford, K. (1995). Human influences on biodiversity. In *Global Biodiversity Assessment* (V. H. Heywood, ed.), pp. 711–821. Cambridge University Press, Cambridge, United Kingdom.
Milliken, W., Miller, R. P., Pollard, S. R., and Wandelli, E. V. (1992). *Ethnobotany of the Waimiri Atroari Indians of Brazil*. Royal Botanic Gardens, Kew, United Kingdom.
Prance, G. T., Balée, W., Boom, B. M., and Carneiro, R. L. (1995). Quantitative ethnobotany and the case for conservation in Amazonia. In *Ethnobotany: Evolution of a Discipline* (R. Evans-Schultes and S. von Reis, eds.), pp. 157–174. Dioscorides Press, Portland, Oregon.
Spellerberg, I. F. (1996). *Conservation Biology*. Longman Group Ltd., Harlow, United Kingdom.
Spellerberg, I. F., and Hardes, S. R. (1992). *Biological Conservation*. Cambridge University Press, Cambridge, United Kingdom.
UNESCO. (1996). *Biosphere Reserves: The Seville Strategy and the Statutory Framework of the World Network*. UNESCO, Paris.
World Conservation Monitoring Centre (WCMC). (1992). *Global Diversity: Status of the Earth's Living Resources*. Chapman & Hall, London.
Withers, L. A. (1993). Conservation methodologies with particular reference to *in vitro* conservation. In *Proceedings of the Asian Sweet Potato Germplasm Network Meeting, Guangzhou, China*, pp. 102–109. CIP, Manila.
WRI, IUCN, and UNEP. (1992). *Global Biodiversity Strategy*. World Resources Institute, Washington, D.C.

EXTINCTION, CAUSES OF

Richard B. Primack
Boston University

I. An Expanding Human Population
II. Habitat Destruction
III. Habitat Fragmentation
IV. Habitat Degradation and Pollution
V. Global Climate Change
VI. Overexploitation
VII. Exotic Species
VIII. Disease
IX. Multiple Factors

GLOSSARY

air pollution Lowering of air quality due to release of toxic materials by factories, automobiles, fires, and other human activities.

disease Infections by parasitic organisms that can cause weakness, decreased reproduction, and death.

exotic species Species that occurs outside of its natural range owing directly or indirectly to human activity.

global climate change Current and predicted changes in global temperature, rainfall, and other aspects of weather due to increased human production of carbon dioxide and other greenhouse gases.

habitat fragmentation Process by which a continuous area of habitat is divided into two or more fragments by roads, farms, fences, logging, and other human activities.

overexploitation Harvesting of a natural resource, such as fish or timber, at a rate more rapidly than it can be naturally replenished.

water pollution Lowering of water quality due to input of sewage, pesticides, agricultural run-off, and industrial wastes that can result in harm to aquatic plants and animals.

IF SPECIES AND NATURAL COMMUNITIES ARE ADAPTED TO LOCAL ENVIRONMENTAL CONDITIONS, why should they be faced with extinction? Shouldn't species and communities be able to persist in the same places that they have for thousands of years? Why are species going extinct now? The answers to these questions have become clear in recent decades: massive disturbances caused by people have altered, degraded, and destroyed the natural landscape on a vast scale, driving species and even communities to the point of extinction. Current rates of human-induced species extinctions are around 1000 times greater than past natural rates of extinction. The balance between natural rates of speciation and extinction have been drastically upset by human activities. The process of evolution will eventually create new species, but it will take thousands, if not millions, of years for these species to develop. And numerous unique species, such as pandas, elephants, and cheetahs, will be gone forever.

I. AN EXPANDING HUMAN POPULATION

The major threats to biological diversity that result from human activity are habitat destruction, habitat fragmentation, habitat degradation (including air and water pollution), the overexploitation of species for human use, the introduction of exotic species, and the increased spread of disease (Table I). Most threatened species face at least two or more of these threats, speeding their way toward extinction and hindering efforts to protect them. Typically, these threats develop so rapidly and on such a large scale that species are not able to adapt genetically to the changes or to disperse to a more hospitable location. These threats will continue to increase in the coming decades as the human population increases, as development and overexploitation continue, as the remaining natural habitats disappear, and as the global climate continues to change.

These threats to biological diversity are all caused by an ever-increasing use of the world's natural resources by an expanding human population. In Europe, only 15% of the land area remains unmodified by human activities, and the amount of many specific habitat types remaining is below 10%. The greatest destruction of biological communities has occurred during the last

150 years, during which the human population grew from 1 billion in 1850, to 2 billion in 1930, to 6 billion in 1999. World population will reach an estimated 10 billion by the year 2050 (Fig. 1). Human numbers have increased because birth rates have remained high while mortality rates have declined, particularly during the last century, as a result of both modern medical discoveries (specifically the control of disease) and the presence of more reliable food supplies. Population growth has slowed in the industrialized countries of the world, but it is still high in many areas of tropical Africa, Latin America, and Asia, where the greatest biological diversity is also found.

People use natural resources, such as fuelwood, wild meat, and wild plants, and convert vast amounts of natural habitat for agricultural and residential purposes. Because some degree of resource use is inevitable, population growth is partially responsible for the loss of biological diversity. All else being equal, more people equals less biodiversity. Some scientists have argued strongly that controlling the size of the human population is the key to protecting biological diversity. However, population growth is not the only cause of species extinction and habitat destruction: overconsumption of resources is also responsible. The rise of industrial capitalism and materialistic modern societies has greatly accelerated demands for natural resources, par-

TABLE I

Factors Responsible for Some Extinctions and Threatened Extinctions[a]

Group	Percentage due to each cause[b]					
	Habitat loss	Overexploitation[c]	Species introduction	Predators	Other	Unknown
Extinctions						
Mammals	19	23	20	1	1	36
Birds	20	11	22	0	2	37
Reptiles	5	32	42	0	0	21
Fishes	35	4	30	0	4	48
Threatened Extinctions[d]						
Mammals	68	54	6	8	12	—
Birds	58	30	28	2	1	—
Reptiles	53	63	17	3	6	—
Amphibians	77	29	14	—	3	—
Fishes	78	12	28	—	2	—

[a] From Reid and Miller (1989), based on data from various sources.

[b] These values represent the percentage of species that are influenced by the given factor. Some species may be influenced by more than one factor, thus the rows may exceed 100%.

[c] Overexploitation includes commercial, sport, and subsistence hunting, as well as live animal capture for any purpose.

[d] Threatened species and subspecies include those given in the IUCN categories critically endangered, endangered, and vulnerable.

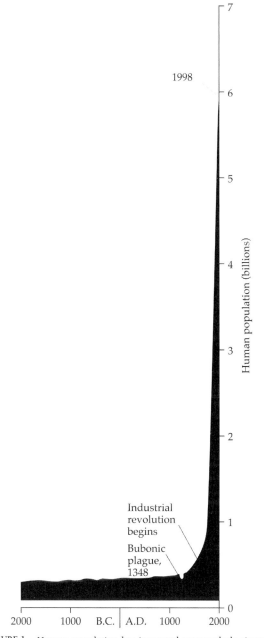

FIGURE 1 Human population has increased spectacularly since the seventeenth century. At current growth rates, the population will double in less than 40 years.

A. Unequal Use of Natural Resources

In many countries there is extreme inequality in the distribution of wealth, with a small percentage of the population controlling, owning, and consuming much of the wealth and natural resources such as good farmland, livestock, and timber resources. As a result, poor, rural people with no land or resources of their own are forced by necessity to destroy biological communities and hunt endangered species to extinction. Large landowners and business interests that force local farmers off their land is a common pattern in many countries of the developing world, a pattern often backed up by the government, the police, and the army. Political instability, lawlessness, and war also displace farmers into remote, undeveloped areas where they feel safer; most practice shifting cultivation, a form of agriculture involving cutting down forest, burning the plant material, and planting crops in the nutrient-rich ash, because it is the simplest way to make a living when they may be obliged to move again within a short time. The landless farmers and their families often exploit natural resources in their surroundings just to stay alive; often these resources are components of species-rich biological communities.

The responsibility for the destruction of biological diversity in species-rich tropical areas also lies in the unequal use of natural resources worldwide (Fig. 2). People in industrialized countries (and the wealthy minority in the developing countries) consume a disproportionate share of the world's energy, minerals, wood products, and food. Each year the average U.S. citizen uses 43 times more petroleum products, 34 times more aluminum, and 386 times more paper products than the average citizen of India. Wealthy countries and individuals leave a widespread "environmental footprint" in which their excessive patterns of consumption affect a wide area of the world that must supply their needs. This excessive consumption of resources is not sustainable in the long run. If this pattern is adopted by the expanding middle class in the developing world, it will cause massive environmental disruption.

B. Large-Scale Development Projects

In many cases, the factors causing habitat destruction, particularly in the developing world, are the large industrial and commercial activities associated with a global economy—such as mining, cattle ranching, commercial fishing, forestry, plantation agriculture, manufacturing, and dam construction—and initiated with the goal of making a profit. Many of these projects are funded by

ticularly in the developed countries. Inefficient and wasteful use and overconsumption of natural resources are major causes of the decline in biological diversity. For example, if less paper products were used and more paper was recycled, then there would be less need to cut down forests to manufacture new paper.

FIGURE 2 Citizens of the wealthy, developed countries of the world often criticize the poorer, developing nations for a lack of sound environmental policies but seem unwilling to acknowledge that their own excessive consumption of resources is a major part of the problem. (Cartoon by Scott Willis, *San Jose Mercury News.*)

national governments and international development banks and are touted as sources of jobs, commodities, and tax revenues. Others are initiated and funded by large multinational corporations. However, this exploitation of natural resources often is neither efficient nor cost-effective because the emphasis in these industries is on short-term gain, often at the expense of the long-term sustainability of the natural resources, and generally with little regard for the local people who depend on the resources.

II. HABITAT DESTRUCTION

Increasing human populations and their activities use even greater proportions of the world's terrestrial and marine environments and associated natural resources, resulting in the inevitable destruction of species, genetic variation, habitats, and ecosystem processes.

A. Habitat Loss

Habitat loss is the primary threat to the majority of vertebrate species currently facing extinction, a generalization that is certain to be true for threatened invertebrates, plants, and fungi as well. In many countries of the world, particularly on islands and in locations where human population density is high, most of the original habitat has been destroyed. More than 50% of the wildlife habitat has been destroyed in 49 of 61 Old World tropical countries. In tropical Asia, fully 65% of the primary forest habitat has been lost, with particularly high rates of destruction reported for Bangladesh (96%), Sri Lanka (86%), Vietnam (76%), and India (78%). Sub-Saharan Africa has similarly lost a total of about 65% of its forests, with losses being most severe in Rwanda (80%), Gambia (89%), and Ghana (82%). Two biologically rich nations, Zimbabwe and the Democratic Republic of Congo (formerly Zaire), are relatively better off, still having about half of their forests, although it is too soon to say how the recent civil war in the latter country has harmed its wildlife population. Present rates of deforestation vary considerably among countries, with particularly high annual rates of 1.5–2% for tropical countries such as Vietnam, Paraguay, Mexico, Cote d'Ivoire, and Costa Rica. In the Mediterranean region, which has been densely populated by people for thousands of years, only 10% of the original forest cover remains.

For many important wildlife species, the majority of habitat in their original range has been destroyed, and very little of the remaining habitat is protected. For certain Asian primates, such as the Javan gibbon, more than 95% of the original habitat has been destroyed, and some of these species are protected on less than 2% of their original range. The orangutan, a great ape that lives in Sumatra and Borneo, has lost 63% of its range and is protected in only 2% of its range. Such losses of habitat inevitably lead to the loss of species.

B. Rain Forest Loss

The destruction of tropical rain forests has come to be synonymous with the loss of species. Tropical rain forests occupy 7% of the Earth's land surface, but they are estimated to contain over 50% of its species. Therefore, the loss of these forests will result in the extinction of vast numbers of species. These evergreen to partly evergreen forests occur in frost-free areas below about 1800 m in altitude and have at least 100 mm (4 inches) of rain per month in most years. These forests are characterized by a great richness of species and a complexity of species interaction and specialization unparalleled in any other community. The original extent of tropical rain forests and related moist forests has been estimated at 16 million km^2, based on current patterns of rainfall and temperature. Less than half of this expanse was still present as of 1990, with the rate of destruction accelerating in the last decade. On a global scale, most rain forest destruction results from small-scale cultivation of crops and collection of firewood by poor farmers, most of whom have moved to forest areas to practice shifting cultivation out of desperation and poverty.

Other major causes include commercial logging in clear-cutting and selective logging operations, clearing for cattle ranches, clearing for cash-crop plantations (oil palm, cocoa, rubber, etc.) plus road building, and mining. At the current rate of destruction (around 140,000 km^2 per year), there will be no large blocks of tropical forest left after the year 2040, except in the relatively small national parks and protected areas, and a few remote areas of the Brazilian Amazon, central Africa, and the islands of Borneo and Papua New Guinea (Fig. 3). The situation is actually more grim than these projections indicate because the world's population is still increasing, and poverty is on the rise in many developing tropical countries, putting ever greater demands on the dwindling supply of rain forest.

C. Other Threatened Habitats

The plight of the tropical rain forests is perhaps the most widely publicized case of habitat destruction, but other habitats are also in grave danger.

1. Tropical Dry Forests

Tropical deciduous forests contain a large number of species, which in some places rival the diversity of the tropical rain forest. The land occupied by tropical deciduous forests is more suitable for agriculture and cattle ranching than the land occupied by tropical rain forests. Moderate seasonal rainfall, in the range of 250 to 2000 mm per year, allows mineral nutrients to be retained in the soil where they can be taken up by plants. Consequently, human population density is five times greater in dry forest areas of Central America than in adjacent rain forests. Today, the Pacific Coast of Central America has less than 2% of its original extent of deciduous dry forest remaining, and many species of this community have been eliminated or are threatened with extinction.

2. Grasslands

Temperate grasslands are another habitat type that has been almost completely destroyed by human activity with a consequent loss of species. It is relatively easy to convert large areas of grassland to farmland and cattle ranches. Illinois and Indiana, for example, originally contained 15 million ha (37 million acres) of tall-grass prairie, but now only 1400 ha (3500 acres) of this habitat—one ten-thousandth of the original area—remain undisturbed; the rest has been converted to farmland. This remaining area of prairie is fragmented and widely scattered across the landscape. Widespread efforts are being made to restore prairies in many areas of the world. Though such efforts are to be encouraged, it will not be possible to bring back the species that have already been lost.

3. Wetlands and Aquatic Habitats

Wetlands are critical habitats for fish, aquatic invertebrates, aquatic plants, and birds. They are also a resource for flood control, drinking water, and power production. Many aquatic species in lakes and streams have limited distributions that make them especially vulnerable to extinction. Wetlands are often filled in or drained for development, or they are altered by channelization of watercourses and dams. Many wetlands have been degraded by chemical pollution and siltation. When this happens, the aquatic species are fated with extinction. All of these factors are currently affecting the Florida Everglades, one of the premiere wildlife refuges in the United States, which is now on the verge of ecological collapse. During the last 200 years, over half of the wetlands in the United States have been destroyed, resulting in 40–50% of the freshwater snail species in the southeastern United

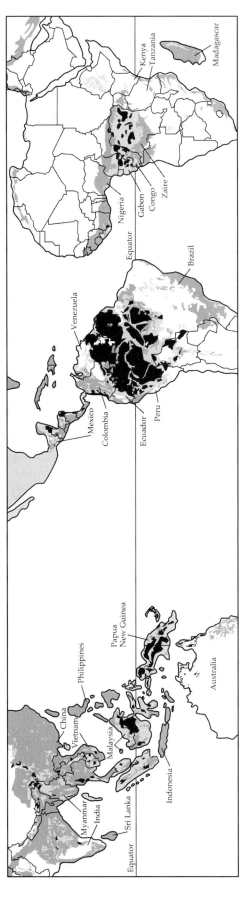

FIGURE 3 Tropical rain forests are found predominantly in wet, equatorial regions of the Americas, Africa, and Asia. Eight thousand years ago, tropical forests covered the entire shaded area, but human activities have resulted in the loss of a great deal of forest cover, shown here in the darkest shade. In the lighter-shaded area, forests remain, but they are secondary forests that have grown back following cutting; plantation forest such as rubber and teak; or forests degraded by logging and fuelwood collection. Only in the regions shown in black are there still blocks of intact natural tropical forest large enough to support all of their resident biodiversity. (After Bryant *et al.*, 1997.)

States becoming either extinct or endangered. More than 97% of the vernal pools in California's San Diego County have been destroyed; these unusual wetlands fill up with water in the winter and dry out in the summer, and support a unique endemic biota. When the habitat is damaged, many sensitive aquatic species are not able to survive.

4. Mangroves

Mangrove forests are among the most important wetland communities in tropical areas. Composed of species that are among the few woody plants able to tolerate salt water, mangrove forests occupy coastal areas with saline or brackish water, typically where there are muddy bottoms. Such habitats are similar to those occupied by salt marshes in the temperate zone. Mangroves are extremely important breeding grounds and feeding areas for shrimp and fish. They also play an important role in reducing storm damage. Despite their great economic value, mangroves are often harvested for timber and charcoal production and cleared for coastal development. In recent years, mangroves have been increasingly cleared for rice cultivation and commercial shrimp hatcheries, particularly in Southeast Asia, where as much as 15% of the mangrove area has been removed for aquaculture. The loss of mangroves is extensive in some parts of South and Southeast Asia; the percentage of mangroves lost is particularly high for India (85%), Thailand (87%), Pakistan (78%), and Bangladesh (73%).

5. Coral Reefs

Tropical coral reefs contain an estimated one-third of the ocean's fish species in only 0.2% of its surface area. Already 10% of all coral reefs have been destroyed, and as many as 30% more could be destroyed in the next few decades. The most severe destruction is taking place in the Philippines, where a staggering 90% of the reefs are dead or dying. The main culprits are pollution, which either kills the coral directly or allows excessive growth of algae; sedimentation following the removal of forests; overharvesting of fish, clams, and other animals; and, finally, fishermen blasting with dynamite and releasing cyanide and other poisons to collect the few remaining living creatures.

Extensive loss of coral reefs is expected within the next 40 years in tropical East Asia, around Madagascar and East Africa, and throughout the Caribbean (Fig. 4). In the Caribbean, a combination of overfishing, hurricane damage, water pollution, and disease is responsible for a dramatic decline of a large proportion of the coral reefs and their replacement by fleshy macroalgae. Elkhorn and staghorn corals, which were formerly common and gave structure to the community, have already become rare in many locations.

D. Desertification

Many biological communities in seasonally dry climates are degraded into man-made deserts by human activities, a process known as desertification. These commu-

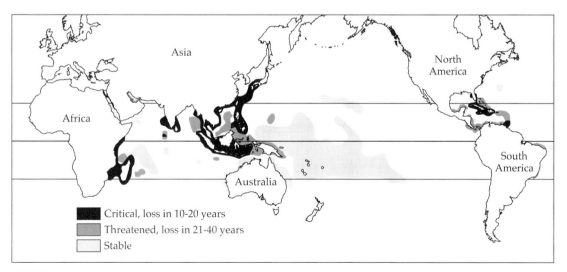

FIGURE 4   Extensive areas of coral will be damaged or destroyed by human activity over the next 40 years unless conservation measures can be implemented. (After Wilkinson from Weber, P. D. 1993. Reviving coral reefs. In L. R. Brown (ed), State of the World 1993. Norton, New York.)

nities include tropical grasslands, scrub, and deciduous forest, as well as temperate shrublands, such as those found in the Mediterranean region, southwestern Australia, South Africa, Chile, and southern California. While these areas initially may support agriculture, repeated cultivation, especially during dry and windy years, often leads to soil erosion and loss of water-holding capacity in the soil. Land may also be chronically overgrazed by domestic livestock, such as cattle, sheep, and goats, and woody plants may be cut down for fuel. The result is a progressive and largely irreversible degradation of the biological community and the loss of soil cover. Ultimately, the region takes on the appearance of a desert, and the original species of the area are lost.

III. HABITAT FRAGMENTATION

In addition to outright destruction, habitats that formerly occupied wide unbroken areas are now often divided into pieces by roads, fields, farms, houses, industries, fences, powerlines, and a broad range of other human activities. Habitat fragmentation is the process whereby a large, continuous area of habitat is both reduced in area and divided into two or more fragments. When habitat is destroyed, a patchwork of habitat fragments may be left behind. These fragments are often isolated from one another by a highly modified or degraded landscape (Fig. 5). Fragmentation almost always occurs during a severe reduction in habitat area, but it can also occur when area is reduced to only a minor degree if the original habitat is divided by roads, railroads, canals, power-lines, fences, oil pipelines, fire lanes, or other barriers to the free movement of species. The island model of biogeography is applicable to this situation: the fragments may be considered habitat islands in an inhospitable human-dominated sea. Habitat fragments differ from the original habitat in two important ways: (1) fragments have a greater amount of edge for the area of habitat and (2) the center of each habitat fragment is closer to an edge.

A. Barriers to Dispersal

Fragmentation may limit a species' potential for dispersal and colonization. Many bird, mammal, and insect species of the forest interior will not cross even very short distances of open area. In many species this is due to the high risk of predation in edge and open habitats. Habitat fragmentation creates barriers to the normal dispersal and colonization processes. In an undisturbed environment, seeds, spores, and animals move passively and actively across the landscape. When they arrive in a suitable but unoccupied area, new populations begin to develop at that site. Over time, populations of a species may build up and go extinct on a local scale as the species disperses from one suitable site to another and the biological community undergoes succession. Habitat fragmentation limits the ability of species to disperse to new habitats, and consequently the species may gradually die out.

Habitat fragmentation also reduces the foraging ability of individual animals. Many animal species, either as individuals or social groups, need to move freely across the landscape to feed on widely scattered resources. However, fences and other barriers may prevent the natural migration of animals, such as wildebeest or bison, forcing them to overgraze an unsuitable habitat, eventually leading to starvation of the animals and degradation of the habitat. Barriers to dispersal can restrict the ability of widely scattered species to find mates, leading to a loss of reproductive potential for many animal species. Plants also may have reduced seed production if butterflies and bees are less able to migrate among habitat fragments to pollinate flowers.

B. Edge Effects

Habitat fragmentation often changes the microenvironment at the fragment edge, resulting in increased light levels, higher daytime temperatures, higher wind speeds, and lower humidity. Each of these edge effects can have a significant impact on the vitality and composition of the species in the fragment. Species sensitive to humidity such as amphibians, many insects, and herbaceous plants will be eliminated from the forest fragments. Also, increased wind, lower humidity, and higher daytime temperatures make fires more likely in forest fragments. Fires may spread into habitat fragments from nearby agricultural fields that are being burned regularly, as in sugarcane harvesting, or from the irregular activities of farmers practicing shifting cultivation. In the process, many species will be eliminated.

C. Interspecific Interactions

Habitat fragmentation also increases the vulnerability of the fragment to invasion by exotic and native pest species. Omnivorous native animals, such as raccoons, skunks, and blue jays, and introduced animals, such as

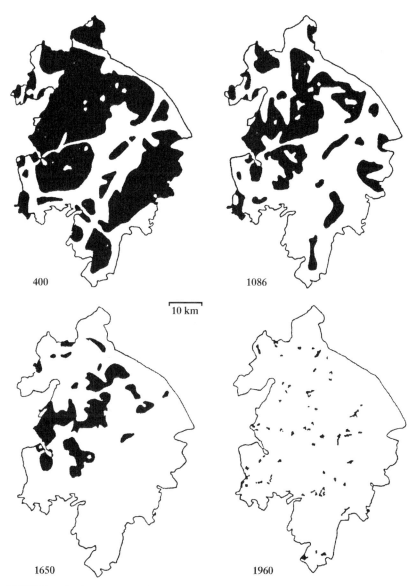

FIGURE 5   The forested areas of Warwickshire, England (shown in black), were fragmented and reduced in area over the centuries from A.D. 400 to 1960.

rats, may increase in population size along forest edges, where they can eat foods found in both undisturbed and disturbed habitats. These aggressive feeders may seek out the nests of interior forest birds, often preventing successful reproduction for many bird species hundreds of meters from the nearest forest edge. Weedy plant species and exotic herbivores can eliminate native plant species along the edges of fragments, and disease can similarly spread into the interior of habitat fragments.

IV. HABITAT DEGRADATION AND POLLUTION

Even when a habitat is unaffected by overt destruction or fragmentation, the communities and species in that habitat can be profoundly affected by human activities. Biological communities can be damaged and species locally extirpated or driven to extinction by external factors that do not change the structure of dominant plants in the community, so that the damage is not immediately apparent. For example, in temperate deciduous forests, physical degradation of a habitat might be caused by frequent, uncontrolled ground fires; these fires might not kill the mature trees, but the rich perennial wildflower community and insect fauna on the forest floor would gradually become impoverished. Keeping too many cattle in grassland communities gradually changes the biological community, often eliminating many native species and favoring exotic species that can tolerate grazing. Frequent boating and diving among coral reefs degrade the community, as fragile species are crushed by divers' flippers, boat hulls, and anchors. The most subtle form of environmental degradation is pollution, commonly caused by pesticides, sewage, fertilizer run-off from agricultural fields, industrial chemicals and wastes, emissions from factories and automobiles, and sediment deposits from eroded hillsides. The general effects of pollution on water quality, air quality, and even the global climate are cause for great concern, not only because of the threats to biological diversity, but also for their effects on human health.

A. Pesticides

The dangers of pesticides were brought to the world's attention in 1962 by Rachel Carson's influential book *Silent Spring*. Carson described a process known as biomagnification through which DDT (dichlorodiphenyltrichloroethane) and other organochlorine pesticides become concentrated as they ascend the food chain. These pesticides, at the time widely used on crop plants to kill insects and sprayed on water bodies to kill mosquito larvae, were harming wildlife populations, especially birds that ate large amounts of insects, fish, or other animals exposed to DDT and its by-products. Birds with high levels of concentrated pesticides in their tissues, particularly raptors such as hawks and eagles, became weak and tended to lay eggs with abnormally thin shells that cracked during incubation. As a result of failure to raise young and the outright death of many adults, populations of these birds showed dramatic declines throughout the world.

Recognition of this situation in the 1970s led many industrialized countries to ban the use of DDT and other stable pesticides. The ban eventually allowed the partial recovery of many bird populations, most notably peregrine falcons, ospreys, and bald eagles. Nevertheless, the continuing use of these classes of chemicals in other countries is still cause for concern, not only for endangered animal species, but for the potential long-term effects on people, particularly the workers who handle these chemicals in the field and the consumers of agricultural products treated with these chemicals.

B. Water Pollution

Water pollution has serious consequences for human populations: it destroys important food sources and contaminates drinking water with chemicals that can cause immediate and long-term harm to human health. In the broader picture, water pollution often severely damages aquatic communities. Rivers, lakes, and oceans are used as open sewers for industrial wastes and residential sewage. Pesticides, herbicides, oil products, heavy metals (such as mercury, lead, and zinc), detergents, and industrial wastes directly kill organisms living in aquatic environments. Even if the organisms are not killed outright, these chemicals can make the aquatic environment so inhospitable that species can no longer thrive. In contrast to a dump in the terrestrial environment, which has primarily local effects, toxic wastes in aquatic environments diffuse over a wide area. Toxic chemicals, even at very low levels, can be concentrated to lethal levels by aquatic organisms. Many aquatic environments are naturally low in essential minerals, such as nitrates and phosphates, and aquatic species have adapted to the natural absence of minerals by developing the ability to process large volumes of water and to concentrate these minerals. When these species process polluted water, they concentrate toxic chemicals along with the essential minerals, which eventually poison the plant or animal. Species that feed on these aquatic species then ingest these concentrations of toxic chemicals.

Essential minerals that are beneficial to plant and animal life can become harmful pollutants at high levels. Human sewage, agricultural fertilizers, detergents, and industrial processes often release large amounts of nitrates and phosphates into aquatic systems, initiating the process of cultural eutrophication. Although small amounts of these nutrients can stimulate plant and animal growth, high concentrations often result in thick "blooms" of algae at the surface on ponds, lakes, and coastal areas. These algal blooms may be so dense that they outcompete other plankton species and shade out bottom-dwelling plant species. As the algal mat becomes thicker, its lower layers sink to the bottom and die. The bacteria and fungi that decompose the dying algae grow in response to this added sustenance and consequently absorb all of the oxygen in the water. Without oxygen, much of the remaining animal life dies off, sometimes visibly in the form of masses of dead fish floating on the water's surface. The result is a greatly impoverished and simplified community consisting of only those species tolerant of polluted water and low oxygen levels. The spreading "dead zone" where the Mississippi River enters the Gulf of Mexico is an example of the dire consequences of water pollution.

C. Air Pollution

In the past, people assumed that the atmosphere was so vast that materials released into the air would be widely dispersed and their effects would be minimal. But today several types of air pollution are so widespread that they damage whole ecosystems.

1. Acid Rain

Acid rain is created when nitrates and sulfates released into the air by the burning of fossil fuels combine with atmospheric water to form acids that fall as rain. Acid rain lowers the pH of soil moisture and water bodies such as ponds and lakes. Increased acidity alone damages many plant and animal species: acid rain has been blamed for the death of large numbers of trees in Europe and North America. As the acidity of water bodies increases, many fish either fail to spawn or die outright (Fig. 6). Both increased acidity and water pollution are two likely factors behind the dramatic decline in amphibian populations throughout the world.

2. Ozone and Smog

Automobiles, power plants, and other industrial activities release hydrocarbons and nitrogen oxides as waste products. In the presence of sunlight, these chemicals react with the atmosphere to produce ozone and other secondary chemicals, collectively called photochemical smog. Although ozone in the upper atmosphere is important in filtering out ultraviolet radiation, high concentrations of ozone at ground level damage plant tissues and make them brittle, harming biological communities and reducing agricultural productivity. Ozone and smog are detrimental to both people and animals when inhaled, so controlling air pollution benefits both people and biological diversity.

3. Effects on Lichens

Even when communities are not destroyed by air pollution, species composition may be altered as more susceptible species are eliminated. Lichens, symbiotic organisms composed of fungi and algae that can survive in some of the harshest natural environments, are particularly susceptible to air pollution. Because each lichen species has distinct levels of tolerance to air pollution, the composition of the lichen community can be used as a biological indicator of the level of air pollution.

4. Ozone Depletion and Ultraviolet Radiation

As a result of human use of chlorofluorocarbons (CFCs) and other ozone-depleting chemicals, the atmospheric ozone layer has been significantly reduced. Ozone plays an important role in filtering out harmful ultraviolet radiation in sunlight. With less atmospheric ozone, more solar ultraviolet radiation reaches the Earth's surface. In humans, exposure to this UV radiation increases the risk of skin cancer. This UV radiation will possibly have a significant, negative impact on animals and plants exposed to direct sunlight, for example, amphibian eggs at the water surface.

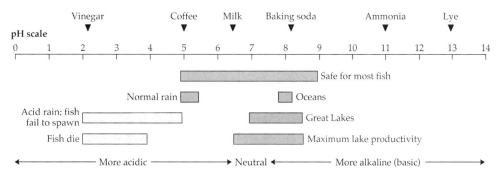

FIGURE 6   The pH scale, indicating ranges at which acidity becomes lethal to fish. Studies indicate that fish, amphibians, and invertebrates are indeed disappearing from heavily acidified lakes. (After Cox, G. W. 1993. *Conservation Ecology.* W. C. Brown, Dubuque, IA, based on data from the U.S. Fish and Wildlife Service.)

V. GLOBAL CLIMATE CHANGE

Scientists are now intensively studying atmospheric carbon dioxide, methane, and other "greenhouse" gases that are transparent to light but that absorb heat. During the past 100 years, global levels of carbon dioxide (CO_2), methane, and other trace gases have been steadily increasing, primarily as a result of burning coal, oil, and natural gas. Clearing forests to create farmland and burning firewood for heating and cooking also contribute to rising concentrations of CO_2. Carbon dioxide concentration in the atmosphere has increased from 290 parts per million (ppm) to 350 ppm over the last 100 years, and it is projected to double somewhere in the latter half of the twenty-first century. Increased concentration of these gases in the atmosphere will almost certainly cause a rise in global temperatures.

A. Prediction of Climate Change

Many scientists believe that these increased levels of greenhouse gases have affected the world's climate already, and that these effects will increase in the future. The best evidence seems to suggest that world climate has warmed by between 0.3° and 0.6° Celsius (°C) over the last 100 years. Predicting future weather patterns is extremely complex and difficult, even with all of the available weather data, simulation models, and super-

computers. However, the consensus among leading meteorologists is that the world climate will increase in temperature by an additional 1° to 3.5°C over the next century as a result of increased levels of carbon dioxide and other gases (Fig. 7). The increase could be even greater if carbon dioxide levels rise faster than predicted; it could be slightly less if all countries agreed to reduce their emissions of greenhouse gases. The increase in temperature will be greatest at high latitudes and over large continents. Many scientists also predict an increase in extreme weather events, such as hurricanes, flooding, and regional drought, associated with this warming.

B. Extinctions and Climate Change

In the northern temperature zone, species will need to migrate 200–300 km northward over the next century to remain in an area with the same climate conditions. It seems likely that many species will be unable to disperse rapidly enough to track the changing climate. Habitat fragmentation caused by human activities may further slow or prevent many species from migrating to new sites where suitable habitat exists. Many species of limited distribution and/or poor dispersal ability will undoubtedly go extinct, with widely distributed, easily dispersed species being favored in the new communities. Endemic mammals that are restricted to isolated

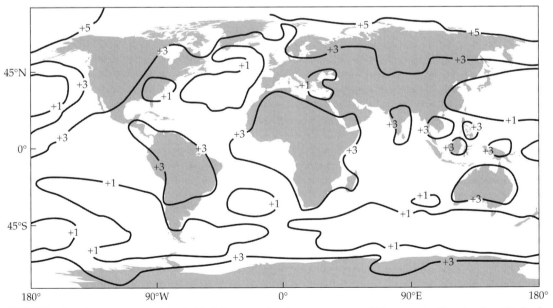

FIGURE 7   Complex computer models of global climate predict that temperatures will increase significantly when CO_2 levels double, which is projected to occur in the middle- to late twenty-first century. Predicted temperature increases, shown in °C, are greatest over continents and at high latitudes (i.e., closer to the poles). (After Intergovernmental Panel on Global Climate Change (IPPC). 1996)

mountain peaks or fish species found in a single lake are examples of species that will not easily be able to cross inhospitable terrain to reach a new, suitable location. The best hope for many species will be to migrate higher on mountain slopes or to disperse along valleys, rivers, and coastlines that are aligned north to south.

Concerns about global climate change, as important as they are, should not, however, divert our attention away from the massive habitat destruction that is the principal current cause of species extinction. The preservation of intact communities and the restoration of degraded communities are the most important and immediate priorities for conservation.

VI. OVEREXPLOITATION

A. Exploitation in Traditional Societies

People have always hunted and harvested the food and other resources they need in order to survive. As long as human populations were small and the methods of collection simple, people could sustainably harvest and hunt the plants and animals in their environment. In traditional societies, restrictions were often imposed to prevent overexploitation of natural resources. For example, the rights to specific harvesting territories were rigidly controlled; hunting in certain areas was banned; there were often prohibitions against taking females, juveniles, and undersized individuals; certain seasons of the year and times of the day were closed for harvesting; and certain efficient methods of harvesting were not allowed. These restrictions, which allowed traditional societies to exploit communal resources on a long-term, sustainable basis, are almost identical to the rigid fishing restrictions imposed on and proposed for many fisheries in industrialized nations. Among the most highly developed restrictions were those of the traditional or artisan societies of Micronesia and Polynesia. However, there are also numerous cases of large bird and mammal species being hunted to extinction by traditional people using simple methods of hunting.

B. Exploitation in Modern Societies

As human populations have increased, their use of the environment has escalated and their methods of harvesting have become dramatically more efficient, leading to an almost complete depletion of large animals from many biological communities, leaving strangely "empty" habitats. Guns are now used instead of blow-

pipes, spears, or arrows for hunting in the tropical rain forests and savannas. Powerful motorized fishing boats and enormous "factory ships" catch fish from the world's oceans. Small-scale local fishermen now have outboard motors on their canoes and boats, allowing them to harvest wider areas more rapidly.

In much of the world today resources are exploited opportunistically. If a market exists for a product, local people will search their environment to find and sell it. Whether people are poor and hungry or rich and greedy, they will use whatever methods are available to secure that product. Sometimes traditional groups will sell the rights to a resource, such as a forest or mining area, for cash to buy desired goods. In rural areas, the traditional controls that regulate the extraction of natural products have generally weakened. Where there has been substantial human migration, civil unrest, or war, controls may no longer exist. As a result, species are exploited to the point of extinction.

1. Trade in Wildlife

The legal and illegal trade in wildlife is responsible for the decline of many species. Worldwide trade in wildlife is valued at over $10 billion per year, not including timber and edible fish. One of the most pervasive examples of this is the international trade in furs, in which hunted species, such as the chinchilla (*Chinchilla* spp.), vicuña (*Vicugna vicugna*), giant otter (*Pteronura brasiliensis*), and numerous cat species, have been reduced to low numbers. Overharvesting of butterflies by insect collectors, of orchids, cacti, and other plants by horticulturists, of marine mollusks by shell collectors, and of tropical fish for aquarium hobbyists are further examples of whole biological communities being targeted to supply an enormous international demand (Table II). It has been estimated that 500 to 600 million tropical fish are sold worldwide for the aquarium market, and many times that number are killed during collection and shipping.

Besides a surprisingly large legal trade, billions of dollars are involved in the illegal trade of wildlife. A black market links poor local people, smugglers, corrupt customs officials, rogue dealers, and wealthy buyers who don't question the sources that they buy from. This trade has many of the same characteristics, the same practices, and sometimes the same criminal players as the illegal trade in drugs and weapons. Confronting these illegal activities has become a job for international law enforcement agencies. Clearly, people involved in the illegal trade of wildlife do not worry about species going extinct, unless it affects their profits.

TABLE II

Major Targeted Groups in the Worldwide Trade in Wildlife[a]

Group	Number traded each year[b]	Comments
Primates	25,000–30,000	Mostly used for biomedical research; also for pets, zoos, circuses, and private collections.
Birds	2–5 million	Zoos and pets. Mostly perching birds, but also legal and illegal trade in parrots.
Reptiles	2–3 million	Zoos and pets. Also 10–15 million raw skins. Reptile extracts used in some 50 million manufactured products (mainly from the wild but increasingly from farms).
Ornamental fish	500–600 million	Most saltwater tropical fish come from the wild and can be caught using illegal methods that damage other wildlife and the surrounding coral reef.
Reef corals	1000–2000 tons	Reefs are being destructively mined to provide aquarium decor and coral jewelry.
Orchids	9–10 million	Approximately 10% of the international trade comes from the wild, sometimes deliberately mislabeled to avoid regulations.
Cacti	7–8 million	Approximately 15% of the traded cacti come from the wild, with smuggling a major problem.

[a] Data from Fitzgerald (1989) and Hemley 1994.

[b] Numbers refer to the number of individuals unless otherwise specified.

2. Overfishing

In the North Atlantic, one species after another has been overfished to the point of diminishing return. The Atlantic bluefin tuna, for example, has experienced a 90% population decline over the past ten years. Similar grim scenarios can be recounted for other large fish prized for their flesh and for sport, such as the swordfish (*Xiphias gladius*). One of the most dramatic cases of overexploitation in recent years involves sharks. Over the last ten years, the fishing industry has been exploiting shark fisheries of the North Atlantic at a rate approximately 60% higher than the sharks can sustain in order to supply a growing international market. The boom in demand for shark meat and shark fins comes at a time when populations of many commercial fish species are severely depleted, so shark fishing has become a lucrative alternative for fishermen. But most shark species have a relatively slow reproductive cycle. As a result, the populations of many sharks are declining dramatically, and some species may soon go extinct.

Another striking example is the enormous increase in demand for seahorses (*Hippocampus* sp.) in China, which is tied to the nation's economic development. The Chinese use dried seahorses in their traditional medicine because it resembles a dragon and is believed to have a variety of healing powers. Around 20 tons of seahorses are consumed in China per year—roughly 6 million animals. Seahorse populations throughout the world are being decimated to supply this ever-increasing demand.

VII. EXOTIC SPECIES

Humans have radically altered patterns of species distribution by deliberately or accidentally transporting species throughout the world. The extent of this modern movement of human-transported species is unprecedented on a geological scale and has been described by Elton (1958) as "one of the great historical convulsions of the world's flora and fauna." Many areas of the world are strongly affected by exotic species. The United States currently has over 70 species of exotic fish, 80 species of exotic mollusks, 2000 species of exotic plants, and 2000 species of exotic insects. Exotic perennials completely dominate many North American wetlands: purple loosestrife (*Lythrum salicaria*) from Europe dominates marshes in eastern North America, while Japanese honeysuckle (*Lonicera japonica*) forms dense tangles in bottomlands of the southeastern United States.

The effects of these exotic insects on the native insect fauna can be devastating. At some localities in the southern United States, the diversity of insect species has declined by 40% following the invasion of exotic fire ants. The fire ants either directly attack and consume other insect species or outcompete them. Many bird species have shown dramatic declines once fire ants enter their habitat, again because of direct attack as well as loss of insect prey.

The great majority of exotic species do not become established or dominant because the new environment is not suitable to their needs. However, a certain percentage of species do establish themselves in their new homes, and many of these become abundant at the expense of native species. Exotic species are often able to thrive because their populations are not held in check by any of the local parasites or predators. These exotic species may displace native species through competition for limited resources, they may kill and eat native species to the point of extinction, or they may alter the habitat so that many natives are no longer able to persist. Exotic plant species frequently displace native species because they are better suited to the new conditions created by people, such as increased fire and introduced grazing animals.

Exotic species are considered to be the most serious threat facing the biota of the United States national park system. While the effects of habitat degradation, fragmentation, and pollution can potentially be corrected and reversed in a matter of years or decades as long as the original species are present, well-established exotic species may be impossible to remove from communities. They may have built up such large numbers and become so widely dispersed and so thoroughly integrated in the community that eliminating them may be extraordinarily difficult and expensive.

A. Exotic Species on Islands

Island species are particularly vulnerable to exotic species, because they have evolved in the absence of mainland herbivores and predators. The introduction of one exotic species to an island may cause the local extinction

FIGURE 8 The brown tree snake (*Boiga irregularis*) has been introduced onto many Pacific islands, where it devastates populations of endemic birds. This adult snake has just swallowed a bird. (Photograph by Julie Savidge.)

of numerous native species. Two examples illustrate the effects of introduced species on the biota of islands.

On Santa Catalina Island off the coast of California, 48 native plant species have been eliminated, primarily due to grazing by introduced goats, pigs, and deer. One-third of the plant species currently found on the island are exotics. Removal of goats from part of the island has led to the reappearance of several native plant species.

Birds of the Pacific islands are especially vulnerable to the effects of exotic species. For example, the brown tree snake (*Boiga irregularis*; Fig. 8) has been introduced onto a number of Pacific islands where it is devastating endemic bird populations. The snake eats eggs, nestlings, and adult birds; on Guam alone, the brown tree snake has driven ten endemic bird species to the point of extinction. Recent visitors have remarked on the absence of birdsong: "between the silence and the cobwebs, the rain forests of Guam have taken on the aura of a tomb" (Jaffe, 1994).

B. Exotic Species in Aquatic Habitats

Exotic species can have severe effects on vulnerable freshwater communities, in particular lakes and isolated stream systems. There has been a long history of introducing exotic commercial and sport fish species into lakes. Many introductions have been accidental. Often these exotic fish are larger and more aggressive than the native fish fauna, and they may eventually drive the local fish to extinction. Aquatic plants, invertebrates, and disease organisms can also become aggressive exotics outside their normal range.

Two examples illustrate the effects of exotic species on aquatic species. The freshwater fish fauna of the island nation of Madagascar has extremely high levels of endemism, with 14 of its 23 genera found nowhere else. Recent surveys of freshwater habitats were able to locate only 5 of the known native freshwater fish species of the island. Introduced fish dominate all of the freshwater habitats, and have completely replaced native species on the central plateau.

One of the most alarming recent invasions in North America was the arrival in 1988 of the Eurasian zebra mussel (*Dreissena polymorpha*) in the Great Lakes. Within two years zebra mussels had reached densities of 700,000 individuals per square meter in parts of Lake Erie, choking out native mussel species in the process. Zebra mussels have been found in the Detroit, Cumberland, and Tennessee Rivers; as it spreads south, this exotic species is causing enormous economic damage to fisheries, dams, power plants, water treatment facilities, and boats, as well as devastating the aquatic communities it encounters.

VIII. DISEASE

Disease caused by internal parasites is a natural control mechanism that reduces populations when they reach high densities. However, levels of disease can often increase in populations as a result of human activity. When animals are confined to habitat fragments at abnormally high densities, disease may spread more easily among individuals. Also, animals under stress or weakened by living in a degraded or polluted environment may be more susceptible to disease. Furthermore, as areas are fragmented by human activities, disease can spread more easily from domestic animals into wild populations. At Tanzania's Serengeti National Park, at least 25% of the lions have recently been killed by canine distemper, a viral disease apparently contracted from one or more of the 30,000 domestic dogs living near the park. For endangered species, such outbreaks can do phenomenal harm: the last population of black-footed ferrets known to occur in the wild was destroyed by canine distemper virus.

Diseases transported by people to new parts of the world can decimate species: North American chestnut trees (*Castanea dentata*), once common throughout the eastern United States, have been virtually obliterated by an ascomycete fungus carried on Chinese chestnut trees imported to New York City. Introduced fungal diseases are also eliminating elm trees (*Ulmus americana*) and flowering dogwoods (*Cornus florida*) from these forests (Fig. 9). Introduced diseases have particularly powerful adverse effects on endemic island species. Many endemic Hawaiian birds have been decimated and even driven to extinction by introduced avian malaria protozoans spreading from introduced bird species by introduced mosquitoes.

IX. MULTIPLE FACTORS

A combination of factors acting simultaneously or sequentially can overwhelm a species, as illustrated by the case of the large freshwater mussel *Margaritifera auricularia*. This species was formerly known from Western Europe to Morocco, but now it occurs in only one river and its adjoining canals in Catalonia, Spain. Its attractive shell and pearls have been used as ornaments by humans as far back as the Neolithic Age. The main reason for its decline, overcollecting, originally led to its disappearance from rivers in Central Europe in the fifteenth and sixteenth centuries, while pollution, destruction of freshwater habitats, and overcollecting continued to reduce its range in recent times. The mus-

FIGURE 9 Populations of flowering dogwood (*Cornus florida*) are declining in eastern North American forests because of anthracnose disease caused by the introduced fungus *Discula destructiva*. (Photograph by Jonathan P. Evans.)

sel is also affected by the loss of other species, since its larval stage needs to attach to certain species of fish to complete its life cycle. Unless strict conservation measures are implemented to prevent overcollecting, control water quality, maintain fish stocks, and protect the habitat, this culturally important species will soon be extinct. Such comprehensive conservation strategies are often needed to deal with the multiple threats to species.

Threats to biological diversity come from a number of different directions, but their underlying cause is the same: the magnitude of destructive human activity. It is often easy to blame a group of poor, rural people or a certain industry for the destruction of biological diversity, but the real challenge is to understand the local, national, and international linkages that promote the destruction and to find viable alternatives. These alternatives must include stabilizing the size of the human population, finding a livelihood for rural people that does not damage the environment, providing incentives and penalties that will convince industries to value the environment, and restricting trade in products that are obtained by damaging the environment. Yet an equally important part of the solution is to increase the willingness of wealthy and middle-class people in both developed and less-developed countries to reduce their consumption of the world's resources and to pay fair prices for products that are produced in a sustainable, nondestructive manner.

See Also the Following Articles

DEFORESTATION AND LAND CLEARING • DESERTIFICATION • EXTINCTION, RATES OF • EXTINCTIONS, MODERN INSTANCES OF • INTRODUCED SPECIES, EFFECT AND DISTRIBUTION • MASS EXTINCTIONS • PESTICIDES, USE AND EFFECTS OF • POLLUTION, OVERVIEW

Bibliography

Birkeland, C. (ed.). (1997). *The Life and Death of Coral Reefs.* Chapman and Hall, New York.

Bryant, D., Nelson, D., and Tangley, L. (1997). *The Last Frontier Forests: Ecosystems and Economies on the Edge.* World Resources Institute, Washington, D.C.

Drake, J. A., and Mooney, H. A. (eds.). (1989). *Biological Invasions: A Global Perspective.* John Wiley & Sons, Chichester, United Kingdom.

Elton, C. S. (1958). *The Ecology of Invasions.* John Wiley & Sons, New York.

Fitzgerald, S. (1989). *International Wildlife Trade: Whose Business Is It?* World Wildlife Fund, Washington, D.C.

Hemley, G. (ed.). (1994). *International Wildlife Trade: A CITES Sourcebook.* Island Press, Washington, D.C.

Intergovernmental Panel on Climate Change (IPCC). (1996). *Climate Change 1995: The Science of Climate Change.* Cambridge University Press, New York.

Jaffe, M. (1994). *And No Birds Sing.* Simon and Schuster, New York.

Laurance, W. F., and Bierregaard, R. O., Jr. (eds.). (1997). *Tropical Forest Remnants: Ecology, Management and Conservation of Fragmental Communities.* University of Chicago Press, Chicago.

Matthiessen, P. (1959). *Wildlife in America.* Viking Press, New York.

Meyer, W. B., and Turner, B. L. (eds.). (1994). *Changes in Land Use and Land Cover: A Global Perspective.* Cambridge University Press, New York.

Primack, R. (1998). *Essentials of Conservation Biology,* 2nd ed. Sinauer Associates, Sunderland, Massachusetts.

Reid, W. V., and Miller, K. R. (1989). *Keeping Options Alive: The Scientific Basis for Conserving Biodiversity.* World Resources Institute, Washington, D.C.

Simberloff, D., Schmitz, D. C., and Brown, T. C. (eds.). (1997). *Strangers in Paradise: Impact and Management of Nonindigenous Species in Florida.* Island Press, Washington, D.C.

Whitmore, T. C., and Sayer, J. A. (1992). *Tropical Deforestation and Species Extinction.* Chapman and Hall, London.

EXTINCTION, RATES OF

Jeffrey S. Levinton
State University of New York

GLOSSARY

background extinction A distinctly lower rate of extinction, more typical of most of the fossil record.
extinction rate The number or proportion of taxa becoming extinct per unit time or after an important geological temporal boundary.
mass extinction An extinction occurring over a short period of time that is of large magnitude, wide biogeographic impact, and involves the extinction of many taxonomically and ecologically distant groups.

MANY SPECIES ARE THREATENED by impending extinction and attempts have been made to assess population declines and to enact policies of recognizing endangerment by means of simple rules of thumb, such as the International Conservation Union's rule of three successive years of 80% decline. It is difficult to develop a measure of extinction rates of entire floras or faunas, if only because we usually have scant knowledge of the species pool before the impact. This is particularly a problem in species-rich tropical habitats, where cryptic species abound yet have not been identified completely. Surveys of especially rich faunas in tropical wet forests

are only being organized now, and the decline of the areal extent of these forests in recent years makes it nearly impossible to measure extinction rates, except by means of indirect estimates of species—area relationships. The same applies to species-rich marine communities such as coral reefs.

I. MEASURING EXTINCTION IN THE FOSSIL RECORD

A. Why the Fossil Record?

Extinction is very much the domain of the paleontologist. We believe that we are now possibly living through a mass extinction caused by human disturbance of high diversity tropical habitats. But we know very little about the extinction of species, except by fairly obvious mechanisms such as hunting. Can knowing that the dodo or the passenger pigeon was hunted to extinction help us very much with understanding climatically induced changes in key structural groups such as forest trees and reef corals, and their dependent species? Could such spotty knowledge be used to extrapolate to the broad sweep of geological time? Paleontological data has the advantage of large banks of "before and after" data on biodiversity. Its weakness, however, is in associating extinctions with unique causes, as we shall see.

Invasions have caused extensive extinctions on oceanic islands, particularly when alien predators overwhelmed small populations of endemic species in a matter of decades. Extinction on larger time scales, even over hundreds of years, is much more difficult to track. Unfortunately, the time scale for larger-scale changes over 100,000 years or more is probably unapproachable by the neontologist, who can only observe "normal" extinction, and we don't have much understanding of what occurs normally. It may be that fine-scale studies of the fossil record may eventually give us more insight into species-level extinction than neontological studies ever will. After all, the durations of animal species' life spans range from the order of 10^5 (land vertebrates) to 10^6 to 10^7 years (marine species). Even with millions of living species we are not likely to be able to document many cases of typical extinction of living animal species. The fossil record is probably our only hope of a model for study of extinction rates, especially on the scale of ocean basins and continents.

While its coverage of the total potential living biota is poor, the fossil record affords us a more complete glimpse of extinction rates of a number of readily fossilizable groups, both marine and terrestrial. We have a reasonably complete database that has stabilized over the years and one can readily trace extinctions across geological time horizons. The fossil record moreover gives us a deeper insight into what extinction really means. After all, we would like to produce a prospectus of the biological future of living communities following an extinction. Does the loss of a species have a disproportional importance, resulting in the extinction of many associated species? Following an extinction event, is there enough redundancy for the surviving species to evolve a new diverse fauna? Does the extinction of certain species cause the snowballing of a larger extinction event? With some judicious reasoning we can infer the answers to some of these questions with the use of the fossil record.

B. Measures and Types of Extinction Rates

Many extinctions in the fossil record appear to be precipitous and occur over short time periods of hundreds of thousands to a few million years. Impacts of extraterrestrial objects may have caused changes in a year or less. Such lengths of time are short when you consider the length of the record of the Phanerozoic era (545 million years).

It is possible to quantify the extent of the extinction with the following data:

1. The total pool of taxa before the extinction
2. The number of taxa that became extinct;

3. The time span over which the extinction occurs

We could calculate the number of taxa that become extinct or the percent of the former pool of taxa that became extinct. Time could be measured in years, but often we only have segments of relative geological time units such as geological stages.[1] In many parts of the fossil record, the absolute time represented by a stage is not accurately known, and different geological stages are often of great difference in temporal extent.

Charles Lyell developed an ingenious technique to estimate extinction rate by charting the gradual diminution of living species as one went back in geological time. This type of analysis can give longevities and extinction rates. Such Lyellian curves demonstrate, for example, that the diminution of bivalve species on the Pacific coast of the United States is at a steady pace whereas a more precipitous extinction occurred in the Atlantic.

C. Problems in Measuring Extinction Rates

1. Taxon-Level Bias

We typically think of extinction rate as a measure of the loss of species. To create a database for paleontology, the species level is very difficult to trust with any degree of confidence; most paleontologists tend to trust the genus and higher taxonomic levels in identifications. In recent years more and more effort has been directed toward accounting for the ranges of all named species in the fossil record, but most analyses have been done at the family or genus level. The large-scale database we now employ owes its existence to the dedicated work of David Raup and especially the late Jack Sepkoski, who continuously sought to produce a more and more complete database of the geological ranges of all fossil groups. Initially, the compilation was at the level of taxonomic order but subsequent analyses have moved down the taxonomic hierarchy to the family and generic levels.

Can extinctions of higher-level taxa be used to estimate species-level extinctions? To estimate species richness using numbers of higher-level taxa (e.g., orders), we assume that taxonomic diversity at higher taxonomic levels is correlated with species richness, but

---

[1] A set of stages comprises a geological series, and a set of series comprises a geological period (e.g., Cambrian, Cretaceous).

they are not necessarily correlated in this manner. This can be seen clearly where changes in ratios of one taxonomic level to another occur over broad spans of time. If the ratios change then higher taxonomic units might be flawed estimators of changes in species diversity. For example, the ratio of taxonomic orders to families decreased significantly from the Paleozoic to the Mesozoic era.

Over short periods of time the number of taxa at a higher taxonomic level (e.g., level of family) might have a regular relationship with a lower taxonomic level, such as species. In order to estimate species-level extinction from family-level extinction, David M. Raup used a *rarefaction technique* based on the sampling curve that relates the number of species collected at random to the number of families recovered.

The rarefaction approach is the best we have so far. Nevertheless, we must be careful in applying it. The biggest problem is the potential change in the relationship over geological time. For example, the ratio of families to species decreases by a factor of two from the Mesozoic to the Cenozoic, and other cases are known of changing ratios of species to genera. Selective extinction can also bias our conclusions. For example, certain families may be much more prone to extinction, owing to their presence in a particularly vulnerable habitat (e.g., coral reefs during a cooling event). This might overestimate total extinction, if these are added to a larger species list. It is also difficult to get sufficient data to calculate good rarefaction curves for all but the most abundant fossil groups.

2. Biased Preservation and Convergence

Estimates of extinction rates may be biased by preservation and abundance at the time of extinction. Preservation of appropriate habitats during an extinction may be greatly reduced. Thus a species might have survived, but there are no opportunities to see it because its usual facies of occurrence has not been preserved.

A common change in probability of preservation takes place when a systematic change in rock preservation occurs, as in the reduction of deposition during a regression phase of the sea,[2] as at the end of the Permian and just before the end of the Cretaceous. Suppose the ranges of a group of species all ended at the very terminus of the Cretaceous. A gradual reduction of deposition would, by sampling error alone, give the impression of a gradual disappearance of the fossil species. Even if deposition does not decline, previously rarer

species would be difficult to sample for presence during a general decline in abundance during extinction, just because we would be unlikely to find them. These biases have come to be known as the "Signor-Lipps effect," or "backward smearing," because a sharp extinction might appear to be gradual from fossil sampling. Only abundant forms would be sufficiently "findable" that we could assess their total geological range with confidence, especially up to the time of their extinction.

3. Accurate Estimate of Fossil Ranges

To estimate extinction rates, one must have an accurate accounting of the geological ranges of species. Then at any time horizon one would be able to account for the number of taxa that disappear from below to above the horizon.

Preservation and rock distributions may strongly bias our perception of geological ranges of taxa. All geological ranges of fossil groups are incomplete, owing to lack of appropriate preserved habitats and poor preservation. As incompleteness (or gaps) between fossil occurrences in a vertical section increases, it stands to reason that the actual temporal range of a taxon is greater than the record would indicate. The number of gaps in preservation may also be combined with a fossil recovery potential curve, which might correct for a change in the probability of preservation during the history of the taxa in question. On the grand scale, the volume of rock correlates positively with the total number of fossil taxa recovered in both marine and terrestrial environments; this suggests that incompleteness of preservation of environments may give us a false impression of true diversity.

4. Extinction Must be Compared with Origination Rates

If extinction occurs over a very short period of time, one can count the number of species before and after an event and assume that the decline can be explained by extinction alone. But speciation may be occurring continuously, which means that a decline in species richness may just as easily stem from a drop in speciation rate as an increase in extinction. Alternatively, speciation might keep pace with extinction, resulting in no loss of biodiversity. When the speciation and extinction rates are equal, an equilibrium exists, meaning the number of taxa remains constant. At the end of the Devonian, for example, large speciation rates balanced high extinction rates. But speciation rates collapsed during the latest Frasnian (Upper Devonian), which precipitated a severe reduction of marine species diversity. A dramatic extinction of mollusk species oc-

---

[2] Regression refers to a lowering of sea level in a given area; transgression refers to a rise in sea level.

curred in the subtropics of the western north Atlantic at the end of the Pliocene (ca. 3 million years ago), but this loss was more than compensated by the origin of new species.

5. Pseudoextinction

In many cases, paleontologists have followed lineages through a geological column and have named successions of species, which are recognized by a variety of character transformations. Thus, even though a lineage may not become extinct, the morphological changes result in an arbitrary extinction or pseudoextinction. Pseudoextinctions are a significant fraction of the total disappearances of taxonomic names from one geological horizon to the next.

6. Other Biases

Paleontologists are accustomed to dealing with a wide range of problems in preservation. Occasionally, an exquisitely preserved fossil biota, such as the Middle Cambrian Burgess Shale, demonstrates that most of the remaining fossil record has not preserved a wide variety of soft-bodied species and even a number of skeletonized taxa. Such unevenness of preservation also works at smaller scales and therefore preservation strongly biases our estimates of diversity. Monographic studies of the fossil record are also uneven and descriptions of species are often strongly correlated with the intensity of study by specialists, either between fossil groups or between time horizons. Recent studies have attempted to correct for these problems by normalizing diversity estimates by the number of monographs produced for a given group at a given time.

II. MASS EXTINCTION

A. Definitions and Identification of Mass Extinction

Strong temporal changes in taxon turnover were quantified first by paleontologist Norman Newell, who found peaks of activity in the Ordovician, Carboniferous, and Jurassic. Declines in standing taxon richness were simultaneous and relatively rapid among distantly related taxa, although increases were not so obviously coordinated. Figure 1 shows five conspicuous and precipitous drops in diversity, the most dramatic occurring at the end of the Permian. David M. Raup and John J. Sepkoski, Jr., analyzed overall extinction rate of marine taxa at the family level and found that four events fell outside of a one-sided 99% confidence interval from

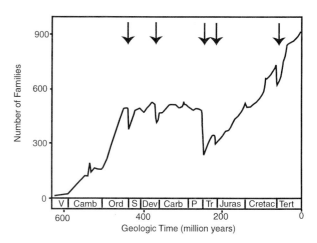

FIGURE 1 Change in numbers of marine and terrestrial taxonomic families throughout Phanerozoic time. Arrows point to times of mass extinction. From Sepkoski (1984).

the mean extinction rate trend: Ashgillean (Upper Ordovician), Frasnian (Late Devonian), Guadalupe-Dzhulfian (Late Permian), and Maastrichtian (Late Cretaceous). The Norian (Upper Triassic) fails this test but its widespread occurrence forces us to include it in the "big five" (Table I). A recent compilation of stratigraphic ranges of a wide variety of taxa produced by a wide variety of specialists yielded a similar overall pattern (Fig. 2). Inspection of extinction rates revealed peaks in the same times identified by Sepkoski, and, indeed, by paleontologists traditionally.

The big drops, *mass extinctions*, are to be distinguished from *background extinction*, which refers to the remainder and overwhelming majority of extinctions. While the big five are conspicuous, other mass extinctions have been recognized. About half of the marine genera disappeared in the Lower Cambrian and archaeocyathid reefs were decimated, perhaps owing to widespread marine anoxia. Another possible anoxic event caused a major extinction at the Cenomanian-Turonian boundary (Upper Cretaceous), although lowered productivity and global cooling may have contributed.

Statistical tests of mass extinction do not inspire confidence, because they combine many taxonomic groups of complicated taxonomic structure, reify them to independent data points, and usually analyze them using the assumptions of parametric statistics. Because the groups are enmeshed in a phylogenetic tree structure, it is not easy perform such analyses. Distributions of extinction events are usually skewed toward many events of extinction rates of a few percent. There may be as many as 12 mass extinctions.

TABLE I

Percent Extinction[a] at the Five Major Mass Extinctions in the Fossil Record[b]

Mass extinction	Families		Genera	
	Observed extinction	Calculated species-level extinction	Observed extinction	Calculated species-level extinction
End-Ordovician 439 Ma	26	84	60	85
Late Devonian 367 Ma	22	79	57	83
End-Permian 243 Ma	51	95	82	95
End-Triassic (Norian) 208 Ma	22	79	53	80
End-Cretaceous 65 Ma	16	70	47	76

[a] Extinctions/standing taxon richness $\times$ 100.
[b] After Jablonski, 1994.

It may seem inappropriate to fix on mass extinctions, which could be atypical end members, but if we cannot characterize these events, will we be able to explain the smaller extinctions that were far more common in the history of life? If mass extinctions are more or less larger-scale or even global versions of what might happen on a more local scale (extinctions stemming from, e.g., local tectonism, anoxia, regional sea level change), then maybe we can extrapolate what we learn about them to smaller scales, and vice versa. If we focused on times of heightened turnover in taxon richness that are confined to basins we might see mass extinctions writ small. There is some reason, however, to believe that the big five were distinctive and the effects of extinctions during these times transcended those of more mundane times.

If an extinction event is a statistical outlier, then how could we justify a separate category, requiring perhaps a set of extinction mechanisms that differ qualitatively from background extinction? This question opens up a can of worms. There has been a good deal of debate about what a statistical outlier really is and whether mass extinctions are really different from lesser periods of extinction. If we assembled extinction rates

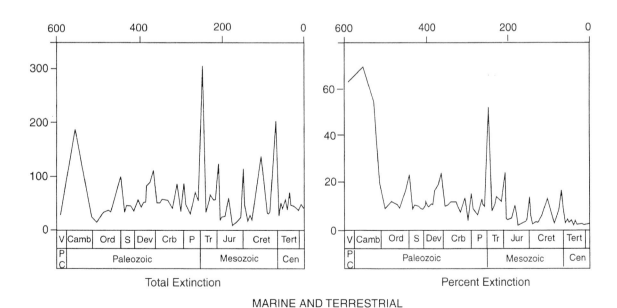

Total Extinction   Percent Extinction

MARINE AND TERRESTRIAL

FIGURE 2 A recent analysis of extinction rate of combined marine and fossil taxonomic families, based on recent compilations of a broad range of specialists. From Benton (1995).

into a frequency distribution curve we might argue that the mass extinctions sit squarely on the tails of some expected probability distribution, or a kill curve. The distribution of risk of marine genera consists of groups with mainly low risk, with some of much higher risk. With no mechanistic model in mind, it is not clear whether the somewhat bumpy distribution of extinction rates of marine genera is smooth or discontinuous, and we have no idea of the distribution for species. If we have to wait 100 million years for a mass extinction, are we waiting for an intense version of the same stuff or a truly distinctive event?

In the context of the time that they occur, mass extinctions are clear and major drops in taxon richness, distinct from extinction rates in the time periods before and after. The Permian appears as a sharp trough after an early Paleozoic time of expansion and then stabilization of numbers of taxa. The same can be said for the end-Cretaceous extinction. But numbers really tell only part of the story. A criterion based on a high extinction rate alone would stretch the confidence we have in our statistical assessments too far, and should only be a means of screening for candidates. To qualify, mass extinction events must have the following features:

1. The number of taxa becoming extinct is significantly greater than times of other extinctions.
2. The decline is concentrated in a small fraction of the Phanerozoic, for example, less than a geological series, or at most a few million years.
3. The extinction is broad-based taxonomically, affecting many distantly related taxa that have not arisen in the same time period as the decline.
4. The extinction affects many different biomes, perhaps not equally. For example, a mass extinction would not be confined to epibenthos on hard surfaces of a region, as opposed to coeval soft-bottom benthos living in the same area.
5. The extinction is geographically widespread, most likely global, in extent.
6. Mass extinction may be caused by mechanisms qualitatively different from background extinctions, but it may be caused merely by "much more of the same" (e.g., anoxia, climate change, unusually large extraterrestrial impacts).
7. Mass extinction affects taxa differently than during other extinctions; small-scale influences on the degree of extinction, such as geographic range are swamped during mass extinctions. David Jablonski found that during ordinary extinctions species with greater geographic ranges were less prone to become extinct.
8. The recovery period following the mass extinction marks the rise of either new taxonomic groups, the expansion of formerly rare groups, or complete reorganizations of ecosystem structure.

Table I demonstrates two important issues in quantifying the degree of extinction. First, the extinction event is characterized as a loss. But the nature of loss is unclear, as it may result from declining speciation rates, increasing extinction rates, or both. In the case of mass extinctions, speciation usually declines and extinction rate increases precipitously. Table I shows the startling results: About 95% of the marine (readily fossilizable) species became extinct at the end of the Permian, and the others of the "big five" took similarly big hits. The Permian was also bad for families (ca. 50% loss), but the others hovered around 20%. The lower loss of families in many extinctions suggests that there might be survivors of many families that could recover and proliferate following the extinction.

A compilation of major changes in both terrestrial and marine fossil groups, coordinated by Michael Benton, reveals some important features of the fossil record (Fig. 2). For one thing, extinction rate can be very high, with not much overall effect on total diversity, since originations may be high or even higher. Thus if one considers extinction rate separately, the Cambrian must be added to our roster of mass extinctions. In terms of percentage extinction, it looms over the rest of the fossil record, even the Permian. There is a possibility, however, that this extinction is more apparent than real. Some have argued that poor Upper Cambrian preservation biases our perception of Cambrian diversity. The analysis reveals a number of extinction peaks not easily seen in a plot of diversity alone. Most notable are strong extinctions in the Carboniferous, Jurassic, and mid-Cretaceous periods.

The quantitative aspects of mass extinction should not obscure some of the major qualitative effects, causing irreversible changes in the world's biota and therefore major reorganizations of the structure of life. At the end of the Permian, marine communities were reorganized completely owing to the end of dominance by brachiopods and the extinction of long-abundant forms such as trilobites. The end of the Mesozoic witnessed the demise of the long abundant carnivorous ammonites, leaving the modern world with a pitiful representation of the former glory of externally shelled cephalopods. Bivalves such as the inoceramids and rudists dominated Cretaceous shallow-water seas but they disappeared. Of course the dinosaurs also became extinct at this time, which presaged the evolution and diversification of the modern orders of mammals.

It is often difficult to estimate the time span over which extinction occurred, owing to strong uncertainty in the time span of geological stages, which is often the crucial level at which extinction is assessed. This is particularly a problem in the early and middle part of the Paleozoic, where time estimates of stage lengths are difficult to estimate with certainty. Within stages, rates are especially difficult because of uneven rates of sedimentation, which makes it invalid to assume a linear relationship between meters of geological section and time. For example, during a final flood stage of a transgression of the sea, sedimentation rates are often very low, and much time may be compressed in very little geological section.

The estimate of extinction is plagued by other factors as well. Worst of all, the incompleteness of the fossil record imposes biases that cloud an assessment of the tempo and degree of a mass extinction. During the Permian, an enormous regression of the sea resulted in the deposition of few marine deposits and correspondingly few marine fossils. As a result, a number of taxa appeared to become extinct, but they reappear in the Triassic, much as Lazarus was raised from the dead. *Lazarus taxa* may be explained by poor preservation in all facies, resulting in poor sampling of species, even those that really existed during the low point at the end of the extinction. Even if appropriate sedimentary rocks are widespread, reductions in population size may make any species more resistant to successful sampling. Alternatively, there may be localized havens in which such taxa may survive, but these refuge environments might escape preservation. In any case, extinction is overestimated. The problem is compounded by so-called *Elvis species,* which evolve after a mass extinction and converge by means of natural selection to resemble premass extinction morphotypes, much as Elvis impersonators now clutter the landscape, at least if you frequent Las Vegas. In Triassic reefs, sponges may be mistaken for Permian taxa, but they are unrelated.

Preservation and rock distributions may strongly bias our perception of geological ranges of taxa. A common change in probability of preservation occurs when a systematic change in rock preservation occurs, as in the reduction of deposition during a regression phase, as at the end of the Permian and just before the end of the Cretaceous. Suppose the ranges of a group of species all ended at the very terminus of the Cretaceous. A gradual reduction of deposition would, by sampling error alone, give the impression of a gradual disappearance of rarer fossil species. Thus a sudden ending of many taxa can be made to appear gradual merely by a gradual reduction of percent preservation.

Patterns of taxon survivorship of mollusks indicate that mass extinction may or may not be a qualitatively different phenomenon from background extinction. David Jablonski found that during normal periods in the Cretaceous period, extinctions of mollusks were correlated with planktotrophic larval development and geographic range, while clade survivorship was positively correlated with species richness. During the end-Cretaceous event, however, none of these held, and clade survival was correlated only with the geographic extent of the clade. After the event, the correlations found previously again obtained. In the end-Cretaceous extinction of planktonic diatoms, however, a different pattern emerged, as diatom species with benthic resting stages survived far better than those with no resting stages. Foraminifera species with specialized morphologies and larger size were eliminated and simpler morphologies were favored. Sea urchins suffered extensive extinctions across the K-T boundary, but bulk sediment processors and shallow water herbivorous species suffered more extinction than omnivores or selective deposit feeders, which suggests a relationship between high extinction and starvation. Here, properties that normally would be related to survival of individuals can be extrapolated to taxon survival. During the Permian mass extinction, gastropod success did not differ especially from periods of more subdued extinction, but groups with planktotrophic larvae and geographically restricted groups suffered more than average (Erwin, 1993). Thus, as extinction intensity increases, some qualitative changes may emerge for some taxa and biogeographic/dispersal properties, but not for all.

B. Causes of Mass Extinctions

Mass extinctions are associated in time with major environmental changes. The problem, of course, is, that other times of no mass extinction also mark times of environmental change, and it is fair to say that we could not easily predict all mass extinctions with nonfossil data alone. If environmental forcing, which transcends the abilities of species to survive or adapt, is a major cause of mass extinction, what are the factors? We can list them, but finding smoking guns is often another matter.

1. Impact or a series of impacts of extraterrestrially derived objects
2. Volcanism
3. Climate change

4. Lowering of sea level, which reduces available habitats for marine species
5. Anoxia, especially transgressive spread of deep-anoxic waters onto the continental shelves

These causes stem more from associations in time between inferred geological events and extinctions, and not from a solid model linking environmental change to extinction. The best example of the latter is the Permian mass extinction. Figure 3 illustrates a scenario of environmental change that may have triggered the extinction. The vast marine regression may have been the driving force behind a variety of environmental changes, including a rise in carbon dioxide, which led to increased temperature and oceanic anoxia. At the end of the Permian, sea level dropped, perhaps about 200 m, which was followed by a transgressive rise of sea level in the Lower Triassic of similar magnitude in just 2 my. Seasonality and reduction of habitat complexity during the regression may also have begotten environmental instability, beyond the adaptive ranges of a number of specialized groups. Volcanism may be a minor contribution to climate change at the end of the Permian, because calculations preclude much of a change in the large ^{13}C deviations at this time, owing to outgassing. Douglas Erwin likened this multicomponent explanation to *Murder on the Orient Express* by Agatha Christie, where twelve culprits are ultimately found to have conspired to murder the victim. Great for murder mysteries but maddening for science. Even this cast of characters ignores the hypothesis of global cooling triggered by glaciation, but this may be discounted as glacial evidence can be dated much before the extinction begins. Paleontologists Norman Newell and Anthony Hallam have implicated sea level change in a number of extinctions throughout the Mesozoic, but they too are often combined with other events, such as bolide impacts, anoxia, and temperature change.

C. The Pace of Mass Extinctions

The end of the Cretaceous is not the most dramatic mass extinction in the Phanerozoic (Figs. 1 and 2). At the time, however, both major terrestrial and marine elements were lost, the fauna was sufficiently modern to be understood ecologically, and some of our favorites,

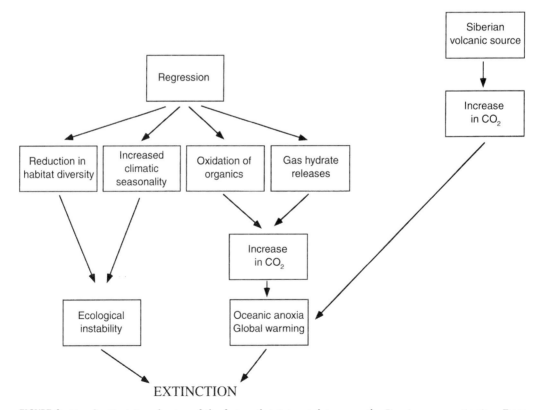

FIGURE 3 Douglas Erwin's reckoning of the factors that interacted to cause the Permian mass extinction. From Erwin (1993).

such as dinosaurs and ammonites, bit the dust. Luis and Walter Alvarez and colleagues set off a debate that has yet to flag by suggesting that a massive asteroid impact caused the extinctions by blanketing the earth with dust spread along ballistic trajectories outside the atmosphere. Such catastrophes had been suggested before by paleontologists, but here was the first tangible evidence.

The fact of an end-Cretaceous impact is supported by a worldwide anomaly of high concentrations of the element iridium in rocks just at the end-Cretaceous boundary (K-T boundary). Although there still is some controversy about this, extraordinarily high iridium concentrations indicate an extraterrestrial origin for some of the material in the rock. Shock structures on quartz crystals suggests that an enormous crater should be present on a continent. A possible piece of ejecta has been found in a core in the Pacific at the K-T boundary, which indicates that the bolide was likely a typical metal- and sulphide-rich carbonaceous chondrite rather than deriving from cometary materials.

The site of impact has probably been located in the megacrater at Chixulub in the Yucatan of Mexico. The crater harbors an armory of smoking guns, including shocked breccia clasts similar to shocked rock fragments found worldwide, tektite-like glasses, a pronounced iridium anomaly and a radiometrically estimated geological age of 65.2 Ma, which match ages of worldwide K-T boundary samples with tektites. The crater suggests a bolide of some 10 km in diameter. If the impact were at an angle, presumably more material would be spattered into the atmosphere, but it is clear from the worldwide iridium anomaly that winds could have spread the calamity throughout the earth.

We at present can only speculate the possible biological consequences. The dust cloud would exist for a time sufficient to severely disrupt climate by cutting off all light, and temperature might have been expected to drop precipitously. A stable oxygen isotope anomaly at the boundary gives evidence for a sudden temperature change. The impact should therefore have affected all organisms dependent on light and warm temperatures. Deep-water forms not so dependent on light or warm temperature, such as nuculid bivalves, would be expected to survive. Alternatively, a hot plume emanating from the impact site could have accelerated the production of nitric acid, causing a worldwide shower of acid rain that might have poisoned the upper ocean.

Can extraterrestrial impacts be used to explain other major extinctions? The results are mixed. Positive and negative evidence for an iridium anomaly has been found for the Frasnian-Famennian (Late Devonian)

mass extinction, but George McGhee and colleagues found that the extinction itself was spread over at least 7 my and climatic effects are evident. Solid evidence for impacts unfortunately postdate the Frasnian-Famennian event. There seems to be no iridium anomaly associated with the terminal Ordovician extinction. A similar iridium anomaly in sediments of 34 my of age occurs simultaneously with the disappearance of five dominant radiolarian species, and at the general time of a mammalian extinction, but the larger picture of biotic change across this boundary is gradual, with no suggestion of a catastrophe. We are therefore left with the end-Cretaceous extinction to consider.

The Alvarez theory has one strong and other weaker predictions. (a) Extinctions must follow or coincide with the impact. (b) One might also expect many groups to die off instantaneously, but a less catastrophic change in temperature and light might have a prolonged effect. (c) Finally, groups more prone to light stress or temperature increase would be more vulnerable (e.g., phytoplankton versus deep-water deposit-feeding benthos), because an impact might likely spew dust into the atmosphere, lowering world temperature. The response of the sensitive groups should be geologically instantaneous.

As in most other mass extinctions the end of the Cretaceous was preferential as to organisms affected. Groups associated strongly (Foraminifera, coccolithophorids) or weakly (ammonites) with the water column suffered the most strongly, while benthic forms (e.g., bivalve mollusks) generally suffered less. Members of food webs less dependent on plant material (marine deposit feeders, scavengers, stream inhabitants, and small insectivorous mammals) suffered less than strict herbivores. The relative success of sediment-feeding invertebrates relative to suspension feeders may be due more to their occurrence in deeper waters.

There is also an apparent thermal bias in extinction at the K-T boundary: S. M. Stanley found that mollusks and Foraminifera in the tropical Tethyan sea suffered large-scale extinction, and were replaced by higher-latitude contemporaries. The question of timing is more confusing. Coccolithophores and nonglobigerinoid Foraminifera disappeared so precipitously (and simultaneously with the iridium anomaly) that chalks give way to clastic sediments in a knife-edge contact in several sections. In the chalk of Denmark, the Maastrichtian fauna, dominated by brachiopods, disappears abruptly, with no prior warning in terms of reduced diversity or early extinction of specialized forms. The sediments above the chalk are clayey and indicative of anoxic conditions. They also have a spike of iridium.

Turbidity, loss of an appropriate sediment, and anoxia may all have contributed to the abrupt extinction. Radiolitid and hippuritid rudists bit the dust during a period of flourishing radiation. The current evidence suggests that Cretaceous vertebrates also bit the dust at the boundary, but preservation is probably too spotty to tell whether it is sudden in any respect.

Unfortunately, the larger story is not nearly so simple. Some fossil groups, including land plants, inoceramid bivalves, and ammonites, experienced major extinctions or reduced speciation rates several million years before the impact occurred. The rudistids, cone-shaped bivalves that often formed Cretaceous reefs died relatively suddenly, but well before the very end of the Cretaceous. Most embarrassing for the impact theory, the freshwater biotas seem to have emerged unscathed. One would have thought that organisms such as turtles and crocodiles would be most vulnerable to a major extraterrestrial impact.

The dinosaurs represent an interesting case. The dinosaur fauna of the late Maastrichtian included fewer than 20 species in 15 genera and 10 families, chiefly in the North American western interior. There is no good evidence, however, that the dinosaurs were declining steadily toward this low number in the last 9 million years of the Cretaceous. It may well be that, while the dinosaur fauna was but a remnant at the end of the Cretaceous, their demise was nevertheless caused by the impact.

The biogeography of extinction in the end-Cretaceous provides some insight. Maastrichtian planktonic foraminifera disappeared suddenly at the K-T boundary in middle and low latitudes. At high latitudes, however, a number of groups survive unscathed into the Danian, the beginning of the Paleocene epoch. The rudistid bivalves, associated largely with tropical and subtropical waters became extinct toward the end of the Cretaceous, but otherwise there are no differences with latitude in bivalve mollusk extinctions.

The sharpness of the boundary for any group is clouded by the imperfections of preservation. If the ranges of certain groups fail to extend all the way to the K-T boundary, then it is possible that perfect preservation would have given us a far different picture. Many western Tethyan ammonites appear to become extinct below the boundary, but Charles Marshall and Peter Ward demonstrated that the confidence limits of a number of lineages allow the possibilities that poor preservation is the reason why some fossil ranges fail to continue right to the K-T boundary. Of course, this does not necessarily prove that the species became extinct at this time, they could have become extinct before or after.

The analysis does exclude being sure that the geological range should be read literally. N. MacLeod performed a similar analysis with Upper Cretaceous foraminifera and found out the ranges are compatible with a sudden extinction. They also, however, are compatible with many other possible scenarios. Some groups penetrate the boundary and it appears that there is good evidence that the fossils were not reworked up into the Danian by erosion and bioturbation.[3]

In a mass extinction we would love to have a single cause to explain extinctions, but, like the Permian, there may be several interacting and a succession of climatic changes that caused a range of extinctions at the end of the Cretaceous.

1. In the late Maastrichtian, sea level decreased by 150 to 200 m, making a hypothesis of increased terrestrial seasonality compatible with the ultimate disappearance. Just below the K-T boundary sea level suddenly rises.
2. Temperature dropped.
3. The Deccan volcanics in India, enormous in scope, probably spewed a variety of substances that strongly affected the atmosphere and might produce effects resembling those of a bolide impact, but the activity probably predated the K-T boundary.

The evidence suggests that there may be more than one process at work in causing major extinctions. The evidence from land plants, dinosaurs, and many mollusks points to a change in conditions well before the Cretaceous-Tertiary boundary. Most notable are groups such as inoceramid bivalves and some of the ammonites that seem to be tracking an environmental deterioration before the termination of the Maastrichtian, or final Cretaceous stage. This would violate the fundamental prediction that extraterrestrial events should be followed by, not preceded by, extinctions. But the iridium layer, plus its associated faunal disappearances, cannot be reconciled with any hypothesis of gradual climatic deterioration. There clearly was a sudden extinction at the K-T boundary, but it was focused on lower latitude microplankton; we can only speculate about suddenness of extinction in other groups. We are left with a compound hypothesis, at least for the proximate cause of the totality of end-Cretaceous extinctions.

Complexity also characterizes the Permian extinction. It makes sense that sea-level drop was an important

---

[3] The Danian is the earliest part of the Paleocene epoch, which followed the K-T boundary.

factor, operating by the effect of reduced habitable area and reduced environmental heterogeneity, but we have to reconcile the precipitous change in sea-level coverage in the last stage of the Permian, with the pattern of extinction, which was initiated earlier than the sea-level drop. Also, the extinction seemed to concentrate on certain ecological groups, particularly tropical forms at the end, but it affected high-latitude groups earlier in the Permian. Most discouraging of all, what if the extraterrestrial influence occurred in a series of impacts, as has been argued to be possible in the Frasnian-Famennian (end-Devonian) extinction, rather than as one big bang? Without a series of signals (e.g., a series of definitive iridium spikes of extraterrestrial origin), such a hypothesis is very speculative.

In conclusion, there is credible evidence for the role of an extraterrestrial object in an extinction in the end-Cretaceous. While other extinction factors may have been at work in the Late Cretaceous, extinctions occurring before the time corresponding to an extraterrestrial iridium anomaly do not falsify the impact hypothesis. At any time in the record, *some* groups must be declining. Even if we accept the impact as a source of some extinction, the Cretaceous still appears to be a compound event, as environmental change and the pace of extinction both accelerate before the K-T boundary. Isn't it bad luck that a large asteroid happened to smack into the earth just as sea level was changing as much as it did in the whole of the Phanerozoic? Several other mass extinctions also seem to be complicated, and were probably associated with changes in climate and sea level that are probably interrelated. Anoxia may also be an important cause of mass extinctions, as witnessed by the extensive development of black shales in certain periods. What is lacking at present is a credible evaluation of the relative effects of these factors.

D. Biogeographic Aspects of Extinction

Some of the Phanerozoic changes in taxon richness can be related to a large degree to changes in the degree of *provinciality*. If we take the species area effect as given and constant, it is easy to calculate that, as one large province is divided in two, reductions of area are more than compensated by the increased total species richness as long as dispersal between areas is limited. Thus a temporal increase in provinciality results in an increase of total biodiversity. A reduction results in a decrease. Major worldwide deterioration of climate (e.g., worldwide cooling, increase of seasonality) might be an example of a time when provinciality might decrease.

The Silurian period, for example, was one of extreme cosmopolitanism, with one province of approximately 90 articulate brachiopod genera in the North Silurian Realm. In the Ludlow (Upper Silurian) two provinces can be delineated, with about 90 genera in each province. In the Devonian, Arthur Boucot found that the number of provinces increased to six; the total numbers of articulate brachiopod genera increased to about 350 on average. During the Frasnian (Middle Devonian), this provinciality decreased relatively suddenly, and generic richness returned to 93.

The onset of the Permian extinction was also marked by a decrease in numbers of provinces, and the Early Triassic marks a nadir of provinciality in the Phanerozoic. During the end of the Paleozoic, geographically restricted bivalve genera succumbed before more widespread genera, suggesting that the overall environmental change was filtering out those forms that define provinciality in the first place. Norman Newell argued that the extinction was related to the major fall in sea level. Shallow marine seas were reduced from a coverage of 40% of their possible extent in the Early Permian to less than 15% in the latest Permian and then expanded to 34% in the Early Triassic. James Valentine and Eldredge Moores speculated that reduced rates of sea-floor spreading may have been responsible for a lowering of ridge activity, depression of deep-sea bottoms, and the consequential large-scale marine regression. The significant reduction in area, coupled with continental assembly of Pangaea at the end of the Permian, may have increased extinction rates and would have homogenized the fauna due to the possible presence of more intershelf dispersal possibilities. In contrast, the Pleistocene reduction of area covered by the sea was far lower, and on the basis of area alone the modest marine extinctions are therefore not surprising from this point of view. Area reduction itself might not be a potent agent of extinction. Sea-level drops would hardly affect the shallow water habitat distribution of oceanic islands, where most modern families are widely distributed. Sea-level drop may just be a correlate of another change.

The changing spatial relationships generated by continental drift and sea level fluctuations must have had important influences on climate. James Valentine's theory of climate change generated by continental assembly and fragmentation attempted to relate climate and sea level to sea-floor spreading. Periods of continental assembly were envisioned as times when interior continental climates were severe, affecting the continental shelf faunas. In contrast, times of fragmentation were times when the continents' climate was more moderate due to ameliorating marine conditions; this permitted the buildup of shallow water diversity. While the post-Permian expansion may fit this pattern, evidence from the Paleozoic does not seem to show an increase in continental fragmentation during the early-mid Paleozoic. Indeed, the continents were maximally fragmented and arrayed along the equator during the Cambrian. Continental drift and arrangement nevertheless has had profound effects on climate and probably extinction. During the Ordovician and Silurian periods, Gondwana drifted southward from its Cambrian position at the equator and came to rest on the geographic south pole. This coincides with the Late Ordovician glacial tillites that have been found in North Africa, and a large reduction in the degree of marine provinciality relative to the early Ordovician. In the Cenozoic, the spatial arrangements of the continents about the Pacific and Atlantic Ocean made for a quite different climatic history. The North Atlantic was a more enclosed basin and was far more severely affected by the late Cenozoic polar cooling. The Pleistocene initiated severe enough climates to cause a major molluskan extinction in the southeastern United States Shelf, while Pacific American faunas showed no increased extinction.

The effects of increasing access between biogeographic realms can be illustrated by the large-scale interchange of mammals between North and South America after the Pliocene establishment of the Isthmus of Panama, following the disappearance of the Bolivar Trough marine barrier. Before the interchange there was long-term stability in numbers of mammalian families. As a probable result of North America's initial higher taxon richness, more taxa moved from north to south than in the reverse direction. In South America, where taxon richness now exceeded previous "steady-state" levels by more than 50%, there was about a 70% increase in extinction rates. Descendants of the North American invaders participated in an evolutionary radiation, resulting ultimately in an overall richness higher than previous levels. Mammalian diversity is now higher in South America, in contrast to the situation previous to the exchange. This suggests that area does have an effect on regulating diversity, but evolutionary changes can impose a significant overprint on diversity.

E. Periodicity in Extinction, or Just Ups and Downs?

Periodicity of extinction or climatic change predicted by astronomical or geophysical theories would be the most convincing way to establish a terrestrial or extraterrestrial cause of extinction. If extinctions are measur-

ably periodic, it may be that only one credible cyclic theory would fit the available pattern. The precedent for such an approach lies with the longstanding theories of the periodicity of Pleistocene glaciations. The Yugoslav astronomer Milankovitch theorized that Pleistocene glacial advances and retreats might be regulated by changes in high latitude insolation, caused by cyclic changes in the earth's orbital eccentricity, tilt, and time of perihelion. A power spectrum analysis of temporal changes of abundance of Pleistocene planktonic fossils in oceanic cores corresponded well to climate changes estimated by stable oxygen isotopes and to periodicity peaks predicted by the Milankovitch theory.

A number of studies in recent years have taken up this theme and related these cycles to sedimentary cycles, including some of the classic midcontinent alternations of carbonate and mudstone. Many of these cycles occurred during times when there was no significant amount of continental glaciation, and represent transgressive-regressive cycles.[4] For example, sedimentary cycles in the lacustrine Early Mesozoic supergroup correspond to periodicities of approximately 25,000, 44,000, 100,000, 130,000, and 400,000 years. These periodicities, in turn, correspond to those expected from celestial processes, such as the precession of the equinoxes, the obliquity cycle, and the eccentricity cycle. Cyclic processes such as the precession of the equinoxes may have driven continental heating cycles that rearranged wind and climate.

Milankovitch climatic rhythms also appear in mid-Cretaceous black shale sedimentary cycles. These cycles consist of alternations of carbonate and shale, with intervals of highly oxidized (red) and highly reduced (black) strata. They are particularly interesting, as they occur in marine sequences and must have reflected periods of ocean bottom anoxia, alternating with vigorous bottom mixing and high productivity in the water column. On an even smaller scale, El Niño-La Niña cycles and the North Atlantic Oscillation, whose forcing mechanisms of periodicity are not well understood, are known to cause cycles of benthic abundance in coastal communities and in small bays and fjords, such as the Swedish Gullmar Fjord.

The earth's history has been dominated by large-scale changes in climate, arrangement of continents, volcanism, and sea level. Alfred G. Fischer developed a theory connecting physical conditions with the overall pattern of Phanerozoic life. Global sea level was relatively high in both the mid-Paleozoic and Mesozoic.

_____

[4] Rises and falls of sea.

Periods of continental breakup, when dispersed and thinner continents resulted in smaller ocean basins, would be associated with higher sea levels. Periods of continental aggregation, when continental crust was bunched up due to collisions and ocean basins, were therefore more commodious, which resulted in lower stands of sea level. The temporal variation in granite emplacement matches the sea-level curve. This suggests a causal link between active continental fragmentation, volcanism, and sea level, an environmental condition of obvious importance to the world marine biota.

Fischer speculated on the presence of a causal connection between changes in terrestrial vulcanism and global climate through the greenhouse effect (Fig. 4). Increased volcanism may have liberated carbon dioxide into the atmosphere. As these periods were of higher sea-level stand, erosion would have been minimal, and loss of CO_2 in weathering would be suppressed. During times of low sea level, low volcanism would reduce the liberation of CO_2, and increased weathering would consume CO_2. Thus, the mid-Paleozoic amelioration

was associated with high CO_2, which, in turn, caused a greenhouse effect and an increase of surface temperature. The end of the Paleozoic witnessed the termination of such conditions, and an "icehouse effect" resulted in a deterioration of climate mainly at high latitudes. It is not clear whether these cited fluctuations are irregular temporal changes or regular oscillations.

In a time of extraordinary paleontological excitement, that, regrettably, has passed, David M. Raup and John J. Sepkoski reported a periodicity of about 26 my in the occurrence of extinction peaks of taxonomic families. Analyses of genera produced an even better periodic signal. To consider an extinction important, Raup and Sepkoski used a threshold level of 2%; minor variations on this criterion change the periodicity to an average time of as much as 30 million years between peaks, but the time between specific peaks varies substantially.

A number of celestial cycles have been suggested to explain the cyclicity. Of course, the most interesting ones are those that would cause rains of extraterrestrial objects on the earth or major changes in climate. Thus far, no theory works very well and it has also been suggested that a number of random models can explain the presence of cycles. This issue has not been settled yet, and if the cycles are real there is great hope that they can be related to an extraterrestrial source. Right now, the variation in extent of extinction and the average time between extinctions are not definitive enough to corroborate any models, which usually involve impacts of extraterrestrial objects such as asteroids or comets. An independent survey of extinctions supervised by Michael Benton failed to corroborate the presence of extinction periodicity.

III. BACKGROUND EXTINCTION AND TURNOVER

A. Normal Extinction?

The great spans of geological time between mass extinctions also witnessed significant appearances and extinctions, but at lower frequencies. The temporal pattern of such extinctions is not clear. Arthur Boucot suggested that periods of several million years are often dominated by a set of ecologically distinct species, whose coordinated extinction might be followed by the invasion or evolution of a new set of ecologically similar forms. In recent years, a number of studies have shown that turnover (including extinction) is common at this tem-

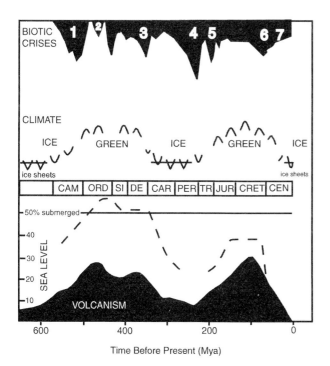

FIGURE 4 "Supercycles" of the Phanerozoic, postulated by Fischer. Sea level curve is superposed on a diagram of granite emplacement, times of glaciation, times of biotic crises (numbered) as determined by N. D. Newell, and a general estimate of climate, characterized by either icehouse (*I*) or greenhouse (*G*) conditions. Modified from Fischer (1984).

poral scale, but the exact pattern is variable. In some cases, periods of relatively low extinction are punctuated by high rates of a wide range of distantly related taxa, which are correlated with local environmental change such as basinal sea level change. Other studies however show rather high extinction and appearance rates with no punctuations in extinction. This is a field that needs to be explored with far more data collection and studies of environmental reconstruction before we can conclude anything.

At this smaller scale of extinction some of the same mechanisms as mass extinction may be in effect. Sea-level rise and fall, climatic change, and other factors can operate on a smaller scale to cause extinction. At this scale, however, biological factors may be of great importance. Biological factors in extinction might include the following:

1. Competitive displacement by an invader
2. Elimination by an overwhelming predator or herbivore
3. Spread of disease

Some of these factors may have been involved in a worldwide extinction of the so-called megafauna, a group of mammals and large flightless birds that disappeared at the end of the last glacial advance. The extinction involved large marsupial mammals and large flightless birds, the renowned Irish elk, large elephantine forms, saber-toothed marsupial, and placental cats, among others. It is possible that human hunting is the cause of these extinctions, but mobile human populations may also have brought novel diseases as they spread around the planet.

B. Declining Background Extinction

Sepkoski's extraction from the fossil database of statistical entities known as evolutionary faunas produced the fascinating result that the so-called evolutionary faunas (EF) are less and less prone to extinction as we approach the present. In the Ashgillean and Frasnian extinctions, for example, the more ancient Cambrian EF suffers more than the Paleozoic EF. In the Permian and Norian extinctions, the relatively older Paleozoic EF suffers more than the Modern EF. The successive evolutionary faunas also have progressively lower turnover (appearance plus extinction), which may make for increasing stability.

A class of distinctly lower family-level extinction rates decline during a long Paleozoic period of fairly constant taxon richness. In order to keep a steady state,

a decline in extinction must be matched by an overall decline of originations, which has also been found. Why should family-level extinction and origination rates decline over geological time? There is no simple solution, although it is tempting to believe that taxa over time have evolved more and more resistance to extinction, including reduced competition with other groups. Given the vagaries of extinction and the fact that extinction is usually an overwhelming process, driven by habitat loss, widespread marine anoxia, and other factors, this idea appears to be far fetched (Fig. 5).

There may an explanation to declining extinction that is a bit more mundane. Extinction may have declined as a result of the ratio of species numbers per families, which has been increasing steadily since the Mesozoic. If a family's representation in the world biota increases in numbers of species and its consequent ecological and geographic coverage, then the probability of family-level extinction may decline.

This explanation still does not provide a satisfactory answer to the decline in originations, which also decline over the same time period. Such a decline implies a long-term reduction in the production of novelties sufficient to define taxonomic families. In other words, the rate of origin of morphological diversity has decelerated over time. Two concomitant processes may have contributed to this decline in origins of basic morphological diversity. A general filling in of resource space may have made it difficult for wholly new forms to take root later and spread. Our world may very well be the tangled bank conceived by Darwin.

To summarize, extinction in the fossil record reveals the following main points:

1. Mass extinctions caused worldwide precipitous losses of species over a wide variety of taxonomic groups and habitats.

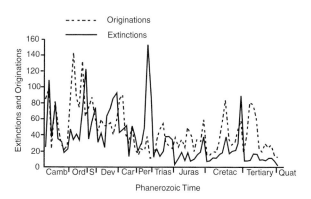

FIGURE 5 Total numbers of marine animal family originations and extinctions per geologic stage. From Hoffman and Ghiold (1985).

2. Environmental change in mass extinctions overwhelm the ability of a species to survive by profound changes in the environment over the whole species' range.

3. The major mass extinctions eliminated a very large majority of the species on the planet and caused major reorganizations of the world's biota.

4. In many cases, extinction appears to be selective with regard to ecological characters of the groups that survive, but in others one cannot identify any traits that make one group more resistant to extinction than others.

5. A distinctly lower level of extinction can be extracted from an analysis of the fossil data. This type of extinction might be a smaller version of mass extinctions, with relatively low rates of extinction punctuated by major regional environmental change driven by sea level changes. Lower-level extinctions rates appear to decline toward the present day. The explanation for this is unclear, but it may just be an artifact of the way taxonomic groups are classified into species. More recent taxonomic families have more species and more species-level extinctions would therefore be required to see the loss of a fossil family that occurred closer to the present day.

See Also the Following Articles

BIODIVERSITY, ORIGIN OF • EXTINCTION, CAUSES OF • EXTINCTIONS, MODERN EXAMPLES OF • FOSSIL RECORD • MASS EXTINCTIONS, CONCEPT OF • MASS EXTINCTIONS, NOTABLE EXAMPLES OF

Bibliography

Benton, M. J. (1995). Diversity and extinction in the history of life. *Science* 268, 252–258.

Erwin, D. H. (1993). *The Great Paleozoic Crisis*. Columbia University Press, New York.

Hallam, A., and Wignall, P. B. (1997). *Mass Extinctions and Their Aftermath*. Oxford University Press, Oxford.

Hoffman, A., and Ghiold, J. (1985). Randomness in the pattern of "mass extinctions," and "waves of origination." *Geological Magazine* 122, 1–4.

Jablonski, D. (1994). Extinctions in the fossil record. *Philosophical Transactions of the Royal Society, Series B* 344, pp. 11–17.

Levinton, J. S. (2000). *Genetics, Paleontology, and Macroevolution*, 2nd ed. Cambridge University Press, New York.

McGhee, G. R., Jr. (1996). *The Late Devonian Mass Extinction*. Columbia University Press. New York.

Raup, David M. (1986). *The Nemesis Affair*. W. W. Norton, New York.

Raup, David M. (1991). *Extinction: Bad Genes or Bad Luck?* W. W. Norton, New York.

Sepkoski, J. J., Jr. (1984). A kinetic model of Phanerozoic taxonomic diversity. III. Post-Paleozoic families and mass extinctions. *Paleobiology* 10, 246–267.

EXTINCTIONS, MODERN EXAMPLES OF

Gábor L. Lövei
Danish Institute of Agricultural Sciences

GLOSSARY

extinction Disappearance of the last living individual of a species. Extinction can be "local" if it concerns a definite population or location; we speak of "extinction in the wild" when the only individuals alive of a species are in captivity and of "global extinction" when no living individual remains of a species.

extinction cascade A chain of extinctions triggered by the extinction of a particular species on which many others depend. Species affected by other species are directly (parasites that live only on that species) or indirectly (predators that rely heavily on the species for food) linked with the extinct species through ecological links. Most species support other ones: a number of specialist herbivores can depend on a plant species for food or many parasites are host specific (can only parasitize one species). When these supporting species die out, the dependent spe-

cies also go extinct. This can trigger a chain of extinctions, termed an "extinction cascade." For example, when the passenger pigeon died, at least two feather lice parasites followed them into extinction. When more than one species dies out at the same time (this by definition must happen when the host of an obligate parasite dies out), the term "coextinction" is also used.

first-contact extinctions A wave of extinction of species native to a continent or island, following the first arrival of humans to that area.

living dead A term coined by the American tropical biologist Daniel Janzen, denoting the last living individuals of a species destined to extinction. By definition, extinction happens when the last individual of a species dies. In reality, however, extinction of a species can be certain even earlier. Most species need both male and female to reproduce. if there are no fertile individuals of one sex, the species is doomed even if several individuals are still alive. Similarly, below a certain population size, a species cannot form a self-sustaining population, and its numbers dwindle. The decline may take many years but its course cannot be easily altered.

metapopulation A series of populations belonging to the same species that are connected via regular migration to each other's habitat patches. An important recent realization is that most species exist as metapopulations and that this is probably the original, "natural" state of all species.

pseudo-extinction Extinction of local populations is sometimes erroneously termed as "pseudo-extinction." This is misleading because global extinction proceeds through the stepping stones of extinctions of local populations. There is no fundamental distinction between the extinction of a local population and the extinction of a species other than the species becomes extinct when the last local population dies out.

proximate cause(s) of extinction The actual immediate agent(s) that cause(s) a species to become extinct.

recolonization The reappearance of a species in an area where it has earlier been present, then went extinct.

species life span The time between the first record of a species in the fossil record to its disappearance. This time span is typically in the range of millions of years.

ultimate cause of extinction Being rare (few in numbers) and of limited distribution are precursors to extinction. The causes leading to rarity are the ultimate causes of extinction.

NO SPECIES LIVES FOREVER, AND EXTINCTION IS THE ULTIMATE FATE OF ALL LIVING SPECIES. The fossil record indicates that a recent extinction wave affecting terrestrial vertebrates was parallel with the arrival of modern humans to areas formerly uninhabited by them. These modern instances of extinction started at around 40,000 years ago. On continents, large mammals (especially those >50 kg body mass) were affected, while on islands, the impacts were mainly felt by birds. The causes of these extinctions are not well known but hunting, habitat alteration, and the introduction of nonnative species have caused extinctions. Our knowledge about extinctions is very incomplete, due to bias in research by taxonomy (vertebrate groups are better studied), geography (northern areas have received more attention), habitat (terrestrial habitats are better known than marine ones), biological reasons (certain groups do not fossilize), and methodological problems (methods of excavation and identification). Consequently, we can only crudely estimate the current rate of extinction. Even so it is evident that humans generated a new mass extinction, affecting all species in all habitats, and by the time it has run its course, it will potentially surpass the previous five mass extinction events in the history of earth. This article only deals with examples of extinction in the Quaternary period (from the final period of the last Ice Age, 10,000 years ago).

I. SPECIES LIFE SPANS

The life span of a species can vary widely but no species lives forever. Fossil records indicate that the average life span of an invertebrate species is about 11 million years, while mammal species live for about 1 million years (Table I). As a consequence, species existing today form only a small fraction of species that have ever lived. If we assume that the average life span of a species is 5 to 10 million years, and multicellular organisms have been on earth for a period of 600 million years, the plant and animal species currently living are not more than 1 to 2% of all those that have ever lived. For marine invertebrates, an estimated 95% of the species that had ever existed are today extinct. Extinction is thus the natural fate of all living species.

Extinction can and does happen at any time, and one can say that extinction is occurring continuously. Most of these extinctions are of local populations. For many species, a landscape contains several suitable habitat patches but not all patches are occupied at all times. Species constantly recolonize unoccupied patches and go extinct in others. The local populations in these patches form a kind of network called a "metapopulation." There is constant migration among the habitat patches, some of them (source patches) producing surplus individuals that colonize other patches; others are not so productive (sink patches). When a metapopulation cannot produce enough individuals to compensate for mortality, the species becomes regionally extinct.

TABLE I

Estimates of Species' Life Spans, from Origination to Extinction

Group	Estimated life span, million years
Dinoflagellates	13
All invertebrates	11
Cenozoic bivalves	10
Diatoms	8
Planktonic foraminifera	7
Echinoderms	6
Marine invertebrates	5–10
Marine animals	4–5
Cenozoic mammals	1–2
Mammals	1
All fossil groups	0.5–5

From May *et al.* (1995).

This results in a range contraction as regular migration between metapopulations does not occur.

Species may become globally rare and subsequently go extinct at any time. The "background rate of extinction" is estimated to be in the magnitude of 1 to 10 species/year through the geological periods. However, significant extinctions in earth's history occurred in clusters. During the last 500 million years, there were five such "mass extinctions," wiping out large proportions of the then-living species. The fossil record related to these has been intensively studied and hotly debated, but without producing an accepted interpretation about the causes of these mass extinctions. Extinctions affecting a more restricted group of species have also occurred on a smaller scale. Within those groups the extinctions were significant.

The most recent of such events commenced during the late Quaternary, about 100,000 years before present (yBP), and started to intensify about 40,000 yBP. Since then, on different continents and also on islands, at different times, several hundreds of land vertebrates, mostly large species (>50 kg body mass) have gone extinct. This extinction wave has not yet ended.

II. EXAMPLES, POSTGLACIAL

A. North American Extinctions

In late glacial North America (called the Wisconsin glaciation period, ending about 10,000 years ago), 71% of midlatitude mammal genera were lost, while in Alaska, the same loss was 56%. This is the opposite that would be expected from environmental conditions—we expect that if climate is the cause of these extinctions, more northerly species would be more severely affected. According to their trophic position, 71% of the herbivores, 67% of the bears, and 50% of the dogs and cats became extinct. Many of these have lived through cycles of glacial and interglacial periods, and extinction was not biased toward either older or newer genera. Environmental changes are therefore thought unlikely to have caused these extinctions. On the contrary, general conditions were at their worst during the period preceding the extinctions, 20,000 to 18,000 yBP. Conditions for large mammals have improved afterward, notably between 18,000 and 7000 yBP, when most extinctions occurred.

The postglacial extinctions are generally connected to the appearance of humans in the regions affected. The North American continent has suffered numerous avian extinctions during the end of the last glacial: 19 genera of birds became extinct during this period. In spite of taxonomic problems as well as scarce records (ten of these birds are only known from the single area of the Rancho La Brea tar pits in metropolitan Los Angeles in the United States), we can generalize that most of these extinct birds were large to very large by avian standards, and the loss of most or all of them can be attributed to ecological dependency on large mammals that also went extinct during the same time. The largest group of these extinct birds were raptors: condors, eagles, accipitrid vultures, and caracaras. Extant hunting birds, including eagles, feed on carrion as well as live prey that they themselves captured, so it is safe to assume the same way of life for these birds. The disappearance of large mammals must have resulted in a significant reduction of the available food base. As a consequence, many of them became extinct. A similar extinction cascade can be observed in the only remaining continent with diverse large ungulate fauna, Africa: where game becomes scarce or disappears, vultures and eagles also disappear. This points to the ecological plausibility of this hypothesis. Two other birds, *Panandris* and *Pyelorhamphus*, related to the North American icterids (Icteridae) of today, are thought to have been in a commensalist association with large herbivores—the "cowbirds" of the Pleistocene—and followed their hosts into extinctions. In Africa, there are several further groups of songbirds associated with large mammals, such as oxpeckers and drongos. It is likely that a variety of commensalist relationships also existed in the New World, and these must have been lost with the disappearance of most large mammals in North America.

B. Australian Extinctions

Australia, until the end of the last glacial, had a fauna of monotremes and marsupials that was as diverse as the placental faunas of other continents. In contrast to those, however, the Australian fauna was rich at the species if not the genus level, and it was not subjected to any significant intercontinental faunal exchange. This lead to a homogeneous fauna that seems to have been unable to withstand ecological stress. During the late Pleistocene, many species went extinct. This loss was comparable, in numbers of species, to extinctions on other continents. For example, while there existed only 15 genera versus the 32 in North America, the number of extinct species is about 60 in Australia and 51 in North America. All 19 species of marsupials heavier than 100 kg, and 22 of the 38 species that are 10 to 100 kg have become extinct. Three reptiles and the ostrich-sized *Genyornis newtoni* have met the same fate. A few of the extinct animals are depicted on Figure 1. The largest reptile was the varanid lizard *Megalania prisca*, which, at 7 m long, was probably a top predator. Among the extinct monotremes were large echnidas, such as *Zaglossus hacketti*, which was 1 m in length and 0.5 m in height. This seems to have been a proportionately large version of the small living echnida. Among the marsupials, there were large carnivores: a large morph of the tiger cat (present on one island until European contact), or the leopard-sized *Thylacoleo carnifex*, a marsupial lion named "giant killer possum." One species of the koalas, *Phascolarctos sirtoni*, which was about 30% larger than the living koala, also survived until the very late Pleistocene. The living koala is the only survivor of a diverse family that had its peak in the Tertiary.

The large and varied superfamily of Diprodontoidea

has totally disappeared during the late Pleistocene, completing a longer sequence of decline. By the last glacial, only two families were represented. Some species of the Palorchestidae were beasts that resembled a giant kangaroo but had a tapir-like trunk and huge, curved claws. The cowlike *Zygomaturus trilobus* had a 2 m long body, a huge, broad head, and a narrow, upturned snout. Judging from the frequent fossil remains, it was widely distributed in coastal and mountain Australia. Another browser was the large, slow *Diprotodon optatum* that had a feeding apparatus suggesting that it was browsing tough, succulents and shrubs. The kangaroos (family Macropodidae) today are the largest group of marsupials still living in Australia, although their diversity, too, was seriously reduced by the beginning of the Holocene. From the *Macropus* genus itself, at least eight species died out. Some of these were small, like today's wallabies, but *Macropus titan* and *M. ferragus* were real

FIGURE 1   A bestiary of most extinct late Pleistocene Australian vertebrate species. The silhouettes are drawn to scale, with the scale indicated by the human silhouette. The species, from left to right, are: Row 1: *Palorchestes azeal, Zygomaturus trilobus, Diprotodon optatum, D. minor, Euowenia grata.* Row 2: *Thylacoleo carnifex, Ramsayia curvirostris, Phascolonus gigas, Phascolomys major, P. medius, Vombatus hacketti, Phascolarctos stirtoni, Propleopus oscillans.* Row 3: *Proctoptodon goliah, P. rapha, P. pusio, Sthenturus maddocki, S. brownei, S. occidentalis, S. orientalis.* Row 4: *S. gilli, S. atlas, S. tindalei, S. pales, S. oreas, S. andersoni, Troposodon minor, Wallabaia indra.* Row 5: *Protemnodon roechus, P. anak, P. brehus, Macropus ferragus, M. birdselli, M. siva, M. titan.* Row 6: *M. rama, M. thor, M. piltonensis, M. gouldi, M. stirtoni, Sarcophilus naliarius, Zaglossus hacketti, Z. ramsayi.* Row 7: *Progura naracoortensis, P. gallinacea, Genyornis newtoni, Megalania prisca, Wonambi naracoortensis.* Reproduced with permission from a paper by P. Murray in *Quaternary Extinctions.* © University of Arizona Press, Tucson.

giants, the latter reaching a height of more than 2.5 m and a body mass of 250 kg. If these species have a living relative, these are about 25% or more smaller.

Although Australia has been inhabited by humans since at least 50,000 yBP, the extinctions were generally not believed to be linked to their presence. Recent evidence based on more exact radiocarbon dating of bird and egg remains casts doubt on this belief, and human predation seems the probable cause also in Australian extinctions.

C. African Extinctions

When examples of modern extinctions are discussed, Africa is often ignored. Some have called Africa "the living Pleistocene" because this is the only continent where a diverse and abundant fauna remained that bears any resemblance to the Ice Age. However, Africa has had its postglacial extinctions and as this is the continent with the longest period of human occupation, the analysis of these is potentially very important.

The most impressive examples of such extinctions come from northern Africa. This region, after the dry, hyperarid period between 40,000 and 12,000 yBP, experienced a moist period during which the fauna included, among other species, the African elephant, white rhinoceros, a zebra, warthog, giraffe, blue wildebeest, hartebeest, eland, roan antelope, and a species of reedbuck. Between 5000 and 4000 yBP, this moist period changed again, and this fauna disappeared from most of the Sahara but survived in the Maghreb area of North Africa. Some species were certainly lost, though: the Atlantic gazelle (*Gazella atlantica*), Thomas' camel (*Camelus thomasi*), the giant North African deer (*Megalocerus algericus*), and the long-horned North African buffalo (*Pelorovis antiquus*).

A similar cycle can be observed in southern Africa, although well dated records are missing from most of this region. In the Cape zone, however, the total disappearance of the long-horned buffalo, giant hartebeest (*Megalotragus priscus*), the giant Cape horse (*Equus capensis*), and the southern springbok (*Antidorcas australis*) happened around 12,000 to 9500 yBP. Their extinction in southern Africa was, at least partially, related to climate-driven environmental change.

During this period, however, there was a dramatic change in artifacts throughout the continent, indicating a very significant shift in human cultures, and with this, probably of hunting techniques and efficiency. Analyses of archaeological sites support the hypothesis of increased hunting proficiency: more remains of individuals in their prime age were found as well as rela-

tively more bones of "dangerous game." In North Africa, the appearance of domesticated animals may also have contributed to the decline.

The pace and extent of late Pleistocene and Holocene extinctions in Africa parallel that of Eurasia, where there was no sudden and massive extinction wave such as in North America. The only significant difference between these continents and the Americas is the length of human occupation. After Africa, Eurasia has the longest period of human presence, about 700,000 y. North America did not have a previous history of human habitation, and the pattern of extinctions is entirely different. This continent was swept by waves of extinction during the late Pleistocene/early Holocene, coinciding with the migration through the Bering Strait and then southward of anatomically modern humans. During this period, a very large proportion of the extant fauna disappeared in what is geologically and evolutionarily very short period of time.

III. EXAMPLES, FIRST-CONTACT EXTINCTIONS

On small islands all over the world, many species, especially birds, went extinct during the past 10,000 years (note that islands have an impoverished mammal fauna to start with, due to dispersal problems). The other common feature of these extinctions was that there was no taxonomic replacement of lost species. These extinctions are so tightly correlated with the arrival of humans that they were termed "first contact extinctions" (FCEs). In the Americas, FCEs occurred between 12,000 and 10,500 yBP, on the West Indies between 7000 and 5500 yBP, and on Madagascar between 2000 and 500 yBP (MacPhee and Marx, 1997).

These FCEs can take as little as 1 year on small islands and up to 1500 years on large islands and continents. On the Commander Islands, east of the Kamchatka Peninsula in the northern Pacific Ocean, humans arrived in 1741. Steller's sea cow (*Hydrodamalis gigas*) was extinct by 1768. On the Mascarenes Islands, humans arrived A.D. 1600, and the major extinctions terminated around 1900. In New Zealand, the first human colonists arrived at around A.D. 1000, and the first major episode of extinctions terminated by A.D. 1500. In the Mediterranean, humans colonized all the major islands between 10,000 and 4,000 yBP, and this also coincides with the extinctions of several endemic species, such as pigmy elephants, rhinoceroses, and hippopotamuses.

A. Madagascar

The extinction of large mammals and birds on the island of Madagascar during the Holocene was of similar magnitude than the earlier, Quaternary extinctions in North America, Australia, and New Zealand. Today's Madagascar fauna is a pale shadow of a once-diverse assemblage of spectacular species, a "magnificent bestiary" (R. E. Dewar). Humans colonized Madagascar only in historical times. The earliest dated archaeological site is from about 500 A.D. The extinctions started to happen not much later. It is generally agreed that these extinctions were caused by human activities, opinion only differs in what type of activity this was.

Seven of the 17 primate genera have disappeared completely, and two more lost its largest species. The extinct lemuroids were all large, and probably diurnal. The largest of these, *Megadalapis edwardsi,* had males with a body mass between 50 and 100 kg. Members of the smallest extinct genus (*Mesopropithecus*) was about as large as the largest living species, the indri (*Indri indri*). Several of these species had ways of locomotion that are unknown among today's living primates: walking on the ground on four legs (*Hadropithecus*), arm-swinging (*Paleopropithecus*), and vertical climbing similar to that of the koala bear (*Megadalapis*).

The other group that was severely affected is the large, flightless birds, ratites, commonly known as elephant birds. They are classified into two genera and 6 to 12 species. The largest of them (*Aepyornis maximus*) had a height of nearly 3 m and resembled a massive ostrich. The smallest, *Mullerornis betsilei,* was about half this size. These species are thought to have been terrestrial grazers—this ecological group is otherwise only represented by the pigmy hippo (*Hippopotamus lemerlei*). This species, together with a large viverrid (*Cryptoprocta spelea*) and an endemic aadrvark (*Plesiorycteropus madagascariensis*), is also extinct. C. *spelea* was the largest known carnivore in Madagascar and resembled a short-legged puma so much that earlier it was classified into the cat family (Felidae). The only reptiles that went extinct were giant land tortoises. The two species had carapace lengths of 80 cm and 120 cm, respectively, and were important consumers of ground vegetation.

B. The Pacific Islands

Humans have gradually colonized the world, and have relatively recently arrived to several oceanic islands. The colonization history of the Pacific Ocean islands is relatively well studied. The human expansion across the Pacific, starting from Southeast Asia, was accompanied by a wave of bird extinctions on all the islands that humans reached. The time span of this varied due to facts such as distance from neighboring islands, area, and terrain, but the scale of the human-driven extinctions is huge. On all the Hawaiian Islands, the number of endemic species known from fossil records only exceeds the number of living endemic species (i.e., more than half the endemic species were lost after human arrival). Every island in Oceania had, on average, an estimated 10 species lost. The total number of islands is about 800, so the loss of species or populations total 8000. Rails have especially suffered. All Oceanic islands studied so far have had one to four endemic species of rails, and thus an estimated total of about 2000 species, equaling 25% of the global species richness, was lost during human colonization of the Pacific. Most of these species were flightless forest dwellers.

C. New Zealand

The New Zealand archipelago lies in the South Pacific Ocean, between the latitudes of 30° and 47°. It is composed of two large and about 300 smaller islands. This land was originally part of the ancient supercontinent of Gondwanaland, which also included Antarctica and the other southern hemisphere continents. New Zealand separated from Australia about 75 million yBP. To this day, its flora and fauna contain elements from this common landmass before it broke into separate continents. Because of its relatively large size and isolation, evolution took on a prosperous and original course, resulting in a high degree of endemism. New Zealand's fauna was characterized by birds and a few reptiles. Before the arrival of humans, the only mammals were marine species and two small bats. There were no large predatory vertebrates, and in their absence, birds prospered. Many species arriving there with the power of flight had become unable to fly, but the best-known birds, the ratites, were originally flightless. The New Zealand ratites belong to two orders: the Apterygiformes (kiwis) and the Dinornithidiformes (moas). They have probably been separated from their relatives since the Cretaceous. The kiwis remained small, unobtrusive, and nocturnal and survived into the present.

Moas diversified into many species; current opinion accepts the existence of 12 moa species. All were ostrich-like, flightless, herbivorous birds with a considerable size range. The largest of them, *Dinornis giganteus,* was about 2 m tall and up to 250 kg in body mass, and the smallest, *Megalapteryx didinus,* was about the size of a large turkey, with an estimated live mass of 25 kg. In contrast to earlier opinion, it seems that most species were inhabiting forests, not grasslands. No species of the moas are left, and this is one of the best-known examples of large-scale, human-caused extinctions.

Polynesians have successfully colonized New Zealand about 1000 to 800 yBP and although we have no reliable record of moa densities before or after this period, nor do we know about moa evolutionary history in earlier ages, it is well documented that the cause of their demise was that the Maori have intensively hunted all species. Archaeological sites with large amount of moa bones are found all over New Zealand, some covering many hectares. The most detailed research on them was conducted on the eastern side of the South Island, and these convincingly demonstrate that man was a voracious hunter of moas: their nests were robbed, and their carcasses were probably utilized in a wasteful way. Dogs and rats introduced by man have probably also played a role in the extermination of moas, especially the smaller species. Moa hunting became intensive about one century after the arrival of Maori, coinciding with a rapid growth of the human population. Within a few centuries, hunting and forest burning accelerated the decline so that by about 400 yBP moas had become so scarce that they were no longer systematically hunted. Continued habitat destruction, sporadic hunting, and probably predation by feral dogs continued to destroy birds, and none were left by the time of European settlement.

The extinct bird species that have never been seen by Europeans include not only the moas but about 20 other bird species. These were often flightless (79% of all extinct species), ground nesting (89%), diurnal (96%), and larger than the closest surviving relative (71%). Fifteen of these were endemic to New Zealand, and five were very similar to living Australian relatives. No less than four of the fifteen were rails, thus echoing the extinction patterns of the Pacific islands (see earlier). Other birds lost include a flightless goose (*Cnemiornis calcitrans*), a giant rail (*Aptornis otidiformis*), a swan (*Cygnus sumnerensis*), and several flying birds. A coextinction with the moas was the extinction of the giant eagle, *Harpagornis moorei*, that was the largest known flying bird, probably preying on moa. After the extinction of its prey, or possibly even earlier, when the prey became rare, the predator disappeared.

IV. EXAMPLES, HISTORICAL EXTINCTIONS (1600–)

Since A.D.1500, during the "modern era," extinctions were closely correlated with the European expansion,

starting with the discovery of America in 1492. The time span of resulting extinctions differ by species and the area affected, but it gradually expended to include all areas and habitats of the earth.

A. New Zealand

During the European period of occupation in New Zealand (although "discovered" by the Dutch seafarer Abel Tasman in 1642, colonization of New Zealand did not start until about 1840), at least five further bird species have become extinct. There is no doubt that the environmental changes brought by Europeans in about 200 years exceeds those caused by the Polynesian occupants during the preceding centuries. This difference, however, is not due to intent but due to the difference in technology. The impact of the initial colonization in terms of extinctions is larger and more obvious because the Polynesians arrived to predator-free islands.

One of the recently exterminated species is the Stephen Island wren (*Xenicus lyallii*), the only known flightless songbird. Stephen Island is a small island in the Cook Straight, between the North and the South Islands of New Zealand. The first specimen of this bird was brought to the lightkeeper's house by his cat. Described as a new species to science, it was exterminated by the same cat within one year (1894). No person has ever seen a live specimen.

The catastrophic impact of predator invasion is exemplified by another New Zealand story, the rat invasion of Big South Cape Island. Big South Cape Island lies south of the South Island, and was known to harbour several endangered species when in 1964, ship rats (*Rattus rattus*) got on shore from a shipwreck. In two years' time, the rats reached very high densities, and four species of birds endemic to New Zealand, one native bat species (greater short-tailed bat, *Mysticina tuberculata robusta*), and numerous invertebrates became extinct. Other species were removed from the island, and thus, for example, the South Island saddleback (*Philesturnus carunculatus*), a thrush-sized bird, survived.

Many more species of birds, reptiles, amphibians, sea mammals, and invertebrates have also suffered a reduction of their former range. Typically, they became extinct on the main islands, surviving only on offshore ones, that were frequently but accidentally free of introduced mammals. For example, the tuatara, *Sphenodon punctatus* (with its sister species *S. guentheri*), the only living relative of the dinosaurs, has been found in early archaeological sites on the main islands. Today it only survives on a few offshore islands. It did not survive on islands where Polynesian rats (*Rattus exulans*) are present, but can be common on rat-free islands. Another example of on-islands-only species is the little spotted kiwi (*Apteryx oweni*), which had only one self-sustaining population on Kapiti Island near Wellington, and the recently discovered, undescribed tusked weta (a relative of grasshoppers).

B. Hawaiian Islands

The Hawaiian Islands are a group of volcanic islands, in distant isolation from any other land mass, in the middle of the Pacific Ocean. They were reached by Polynesian settlers at around 500 A.D. These islands have had a very diverse and unique fauna and flora, and as elsewhere, especially the vertebrates were seriously decimated. The best documented examples are again the birds. The extinct species include flightless geese, ibises, rails, a long-legged owl, a sea eagle, honeycreepers, and crows. Further, there is a group of species that have living populations on one island or another, but not on the one where they were found as subfossils.

The patterns of extinction are strikingly similar to New Zealand, except that there are no large numbers of songbirds reported from New Zealand. Man-induced changes in Hawaii may have been more extreme, or New Zealand originally did not have many songbirds.

The prehistorically extinct birds of the Hawaiian Islands include 1 species of petrel, at least 10, mostly flightless species of geese, 3 flightless ibises, 8 of rails, 3 of long-legged owls, 1 *Accipiter*, 2 large crows, 1 large meliphagid, and 15 species of Hawaiian honeycreepers, relatives of finches.

In the early 1980s, 82 endemic bird species were known from the Hawaiian Islands. Fifty-three of these became extinct prehistorically (before 1778 when Captain Cook discovered the islands). Area and elevation show significant positive correlation with the number of fossil and historically recorded bird species. On Molokai, the smallest of the 5 largest islands, with 676 km sq. of area and 1515 m a.s.l., there are 21 fossil and 9 historically known species. On Hawaii, the largest and highest (10646 km sq., 4206 m a.s.l.), 3 fossil and 23 historic species are known—although more fossil species are expected after more excavations are done.

Significant paleozoological findings are accumulating and it is difficult to draw a reliable and comprehensive picture about the original fauna of the Hawaiian Islands as well as a proper assessment of the extent and nature of extinctions. However, what we know now indicates that the effect of the human as exterminator, direct or indirect, of the fauna of this island archipelago is much more significant than earlier thought. Authorities claim that our previous knowledge of the prehuman fauna was so poor and extant species richness patterns are so pale remains of the original faunas that ecological and biogeographical studies using recently collected data are critically weakened.

C. Extinction Paradoxes

Interestingly, current extinction rates seem to be lowest in areas with a long history of human habitation. Plant extinction rates in areas with Mediterranean climate are low, ranging from 1% of all species in West Australia to 0.15% in the Mediterranean itself. Current threats to plants are one order magnitude larger: 10.2 to 15.2% of species are considered threatened. The suspected cause of the current low extinction rate is a "recording error": many of the extinctions occurring in the "prebotanical age." Indeed, the current extinction rates are lowest where agricultural cultivation has been the longest, 8000 to 6000 yBP. This is consistent with the view that most vulnerable species will have been lost by the time when botanical investigations started.

Similarly, the proportion of bird fauna extinct in the Pacific islands is inversely proportional to the length of human habitation of these islands: 80% of Hawaii's bird fauna is recently extinct or endangered against 10% on Vaunatu. Hawaii has been inhabited for about 1500 years and Vanuatu for 4000 years. Pimm and coworkers (in Lawton and May 1995) argue that the sensitive species have been eliminated by first colonists before record keeping began, and thus we have no direct evidence of first-contact extinctions in the Pacific.

V. EXAMPLES, EXTINCTION CASCADES

A species' "ecological environment" almost always includes other organisms that are essential for the species' survival. Species are connected through trophic links—they eat each other. Other vital ecological links include pollination, dispersal of seeds, and providing habitat. For example, bees, birds, and bats pollinate flowers, birds, and mammals disperse seeds, and trees provide nesting holes for birds. The extinction of a species can have reverberating consequences, affecting other species that are, directly (such as obligate parasites) or indirectly (such as shared predators), linked with the extinct species through such ecological links. Most species support other ones: a number of specialist herbivores can depend on a plant species for food, or many parasites are host specific (can only parasitize one species). When these supporting species die out, the dependent species also go extinct. This can trigger a chain of extinctions, termed an "extinction cascade."

With the death of the last passenger pigeon, a female named Martha in the zoo in Cincinnati, Ohio, in 1914, at least two species of feather lice, which were obligate parasites of this species, must also have perished, although there is no mention of this in any list of extinct species.

All moa, a group of 12 species of ratites of different size, went extinct not long after Polynesians settled in New Zealand. The largest known raptor, the giant eagle (*Harpagornis moorei*) also followed them into extinction. As there are large middens with thousands of moa bones at several sites in New Zealand but these do not contain bones of the eagle, it is thought that the eagle was not a victim of persecution or hunting, but became extinct after its food base, the formerly very common moa disappeared.

Likewise, several bird species that went extinct in North America at the end of the last Ice Age are suspected to have died out in an extinction cascade (see earlier).

VI. PROBLEMS IN OUR UNDERSTANDING OF EXTINCTIONS

We are aware that our knowledge of the actual extent of even recent extinctions is very fragmented and in-complete. There is, in other words, a huge difference between documented and real extinctions. This is due to a series of reasons. Some of these can be overcome with the development of science, but several of them results from the organisms' biology.

A. How Many Species Are There?

We do not know, even to an order of magnitude, how many species we share the earth with. Estimates of global species richness range from 3 to 80 million species. This has an obvious consequence for the estimation of extinction rates: 1000 species is a different relative share of a global total of 3 versus 80 million species.

B. Record Keeping Is Insufficient/Inadequate

The documentation of extinction is also uneven, both geographically (mostly from islands and northern temperate region) and taxonomically (higher organisms better reported). Since 1600, only 485 animal and 584 plant species are listed as extinct (Table II). We strongly suspect that even among vertebrate groups, documented extinctions are serious underestimates. For example, on the Solomon islands, where 164 bird species have been recorded, 12 have not been seen this century,

TABLE II

Species in Major Taxa That Have Become Extinct Since 1600 or Are Threatened with Extinction

	Total number of species				
	Extinct since 1600	Listed as threatened	Described (thousands)	% Extinct	% Threatened
Animals					
Molluscs	191	354	100	0.2	0.4
Crustaceans	4	126	40	0.01	0.3
Insects	61	873	1,000	0.006	0.09
Vertebrates	229	2,212	47	0.5	5
Fishes	29	452	24	0.1	2
Amphibians	2	59	3	0.1	2
Reptiles	23	167	6	0.4	3
Birds	116	1 029	9.5	1	11
Mammals	59	505	4.5	1	11
Total	485	3,565	1,400	0.04	0.3
Plants					
Gymnosperms	2	242	0.8	0.3	30
Dicotyledons	120	17,474	190	0.06	9
Monocotyleadons	462	4,421	52	0.9	9
Palms	4	925	2.8	0.1	33
Total	584	22,137	240	0.2	9

From May *et al.* (1995). Reproduced with permission.

but only 1 is listed as extinct. In Malaysia, a 4-year search for 266 freshwater fish species reported in the last century found only 122, yet few are recorded as extinct.

The current method of documenting extinct species is not entirely biologically valid. A species becomes officially extinct with the death of the last living individual. A species may be destined to extinction long before this happens. If mortality surpasses reproductive success, a species may get onto an "extinction trajectory"—numbers will continuously decrease without reversal, but it will take many years or decades until all individuals perish. Likewise, if there are no reproductively successful pairs remaining, the species has no hope of surviving, even though not all individuals are dead yet. These species are termed the "living dead."

Sometimes species can go through a "genetic bottleneck" when populations become so small that genetic variability practically disappears. Two such cases are known: the cheetah (*Acinonyx jubatus*), a fast-running predator in the cat family, must have gone through such a population crisis some thousand years ago. Today all living cheetahs are genetically virtually identical. The black robin (*Petroica traversi*), a small, endemic songbird on the Chatham Islands of New Zealand, had only one fertile female in the 1980s. In one of the success stories of today's conservation, this species was brought back from the brink of extinction—but genetic variability of the species is much reduced. Without human intervention, this species would have become extinct.

We also have to consider that so many species are recorded from only one location (for example, up to 40% of beetle species described in the Natural History Museum collection, London, United Kingdom) that it is difficult to assess anything but local extinctions.

C. Uneven Recording Effort

Documented insect extinctions are 100 times less than among vertebrate fauna. The difference is even larger if we consider that the number of insects is certainly much larger than we know today. There are 100 times more vertebrate than invertebrate taxonomists, and they are also 10 times more than the number of taxonomists of flowering plants. This uneven attention by humankind to different groups is also evident from the rate of describing new species. This is only 0.03 to 0.05% new species/year for birds. In tropical areas, 1 of every 100 plant specimens is new for science; for insects, fungi, and marine macrofauna, this can reach 20 to 80%.

D. Taxonomic and Habitat Bias

Our knowledge is also very biased by habitats and taxonomic relationships. We know much more about forests than seas, and while most of the mammal or bird species of the world are known, this cannot be said of other important and species-rich groups of organisms like fungi, nematodes, or arthropods.

The great taxonomic bias in our records is well exemplified in the 61 extinct insect species: 33 of these are butterflies and moths. These groups do not constitute more than half of all insect species (they are more likely to be about 25% maximum); their prevalence merely reflects that they are much better studied than other insect groups. It is perhaps real that 51 of these are island species but not that 42 of them are from Hawaii. Similarly, of the 10 continental extinct species, 9 are from North America. This indicates the distribution of researchers, not the real distribution of threatened or extinct species.

E. Geographical Bias

Our knowledge is particularly scant in areas of the earth with the highest biological diversity, the tropics, and thus any changes are much better documented in the northern temperate regions where only a minority of the global biodiversity can be found.

The geographical bias and variability reported in the literature include patterns that are real, while others imaginary. Sixty-one percent of all recorded animal extinctions are from islands—this is probably a real pattern. The numerical preponderance of the Pacific Islands is due to both their large numbers and recent human colonization. However, in the pattern that two-thirds of recent animal extinctions are from North America and the Caribbean, 20% from Australia is certainly an artifact. Similarly, all 45 plant extinctions in Africa are from the Cape flora, and two-thirds of continental plant extinctions are from North America and Australia. The rarity of such records from South America, Asia, and Africa is surely an artifact.

F. Methodological Obstacles

Some of these artifacts are historical and irreparable. Further inaccuracies result from the fact that much of extinction information is gathered by paleozoology. We have never witnessed these extinctions, and only remains of these extinct organisms are found. There are special difficulties in studying and interpreting fossil or subfossil material. Just to mention one, the screens used for sieving soil when excavating animal bones have been, until recently, too coarse to retain bones of small bird species. As a natural consequence, our knowledge of the true extent of bird extinction is grossly biased by this because there are many more small than large species of birds (just as in other groups of organisms).

G. Inherent, Biological Problems

Most of our fossils are from marine organisms, because they often have calcareous body parts that fossilize well. Fossilization on dry land is different: some groups and some climatic regions (insects in tropical climates) are simply not amenable to fossilization.

Birds are better known than other organisms because their skeletons fossilize better and their taxonomy, generally, is better known. In contrast, the original vegetation of the Hawaiian Island lowlands is a matter of conjecture as they were largely destroyed before botanists arrived to collect there. Entomologists can only speculate what the effect of this deforestation could have been for the arthropods as very few insects are preserved under Pacific island conditions.

VII. THEORETICAL ASPECTS OF EXTINCTION

The first step in the extinction process for a species is to become rare. It is conceptually useful to distinguish between *ultimate* causes of extinction (what causes species to be rare and thus vulnerable to extinction in the first case) and *proximate* causes of extinction (what is the actual cause of extinction). These latter generally include demographic and environmental stochasticity (random, large fluctuations in density and environmental conditions), genetic deterioration, and social dysfunction, although their respective importance is not well understood. Ultimate causes include hunting, habitat destruction, invasion by introduced species, and pollution.

Two *general tendencies* are relevant for the study of extinctions:

1. Species that are widespread tend to be abundant as well, but the causes of this positive correlation are not well understood. This also means that species most at risk from extinction (those that are sensitive to proximate causes) have small geographic ranges, because they will also be locally rare. This double jeopardy may be serious when populations and ranges are artificially reduced by ultimate causes of extinction.

2. The distribution of tropical species is generally more restricted than that of temperate species. Smaller ranges have been documented for tropical than temperate-region trees, mollusks, crustacea (crabs and relatives), fish, amphibians, reptiles, and mammals. A related trend is that average population densities of individual species increase from the equator toward the poles (proven for invertebrates, mammals, and birds). This fits with the first trend listed and is also consistent with a decline in range sizes toward the tropics.

As a consequence, disproportionately more tropical than temperate species are threatened with extinction. Of the 1029 threatened bird species, 442 live in tropical forests, more than twice the number of species living in wetlands, the next most threatened habitat category.

Most (direct as well as circumstantial) evidence indicates that most of the recent extinctions were caused by humans. Climate change has been invoked in some cases but the evidence for this is not strong. The actual form of human impact can be overhunting, habitat destruction, or introduction (voluntary or accidental) of nonnative species, mostly of predators (cats, dogs, rats) or browsing herbivores (pigs, goats, sheep). It was also suggested that humans have spread an extremely virulent pathogen, causing a "hyperdisease." In mollusks, birds, and mammals that went extinct since 1600 and have a known cause, 23% was due to hunting, 36% to habitat destruction, and 39% due to the introduction of exotic organisms. Once again, our knowledge is rather sketchy: in mammals that became extinct since 1600, only 30% have an established proximate cause of extinction.

A. Introductions as a Threat to Species

Introduced species have often been implicated in the extinction of native species. Many introduced species, however, have had no detectable effect on species in their new environments. However, the massive spread of organisms by humans to other areas of the globe may increase local diversity, but will result in large losses in global biodiversity. In order to understand the danger that pan-mixing of the earth's fauna and flora signify, let us consider a thought experiment in island biogeography. Species richness of an island is largely determined by its area: the larger the area, the more species the island contains. The same applies for continents. For examples, mammal species richness is related to the size of the individual continents. The resulting correlation allows to extrapolate the global species richness. A supercontinent, with an area equal to the total dry land on earth would support about 2000 mammal species. Currently, there are about 4200 mammal species. Therefore geographical isolation allowed evolution to generate nearly twice the biodiversity that could otherwise, on the basis of habitat area alone, be expected. As today human-assisted invasion is becoming a more and more prevalent biogeographic phenomenon, it is inevitable that more extinctions are predicted, with possibly catastrophic consequences for biodiversity.

Often there are more than one cause of extinctions. For example, the kokako (*Callaeas cinerea*), an endemic wattlebird in New Zealand, became extinct in most of its former distribution range (and is on the brink of global extinction) due to a combination of factors. These include the contraction and fragmentation of its original forest habitat plus the effects of introduced predators, mainly the European stoat (*Mustela erminea*) and the Australian brushtail possum (*Trichosurus vulpecula*).

B. Insights from Population Dynamics

A further difficulty to understand extinction is that the actual process of extinction is also very imperfectly documented. Only a few documented examples exist that link population decline to changes in species distributions. The stepping stones of global extinction are local extinctions, so it is logical to assume that as a species becomes more restricted and rare, its distribution range will become fragmented and gradually smaller. The European fir tree (*Abies* sp.) decline was indeed accompanied by population range fragmentation. The skipper butterfly (*Hesperia comma*) in Britain has crashed in this century. This process left scattered and highly fragmented populations by the 1950s. The same happened with many bird species in New Zealand (kokako, kaka *Nestor meridionalis*, a large parrot or the weka *Gallirallus australis,* a flightless rail).

The sensitivity of fragmented populations is underlined by a trend seen in the success of reintroduction attempts. Seventy-six percent of 133 documented reintroductions into former "core" areas, while only 48% of 54 translocations to periphery or beyond succeeded. However, not all species shows a similar range dynamics in the process of becoming rare. The Kirtland's warbler (*Dendroica kirtlandii*), a small insectivorous bird living in North American forests, withdrew into the historical center of its range during a recent 60% population collapse.

C. Special Traits Related to Density

While population densities typically fluctuate widely, some species are naturally rare. The study of rarity holds promise to understand processes related to extinction, although only vague clues are available today. It would be important to know, for example, if naturally and anthropogenically rare species are equally sensitive to proximate causes of extinction.

In plants, locally rare and geographically restricted species have lower levels of self-incompatibility and poorer dispersal abilities. Rare plants are overrepresented in certain families (Scrophulariaceae, Lamiaceae) and underrepresented in others (Rosaceae), at least in North America. This may indicate that there are some biological traits and adaptations that are shared by rare species.

Populations of large-bodied species fluctuate less than smaller-bodied taxa (although the measurement of population variability is not as easy as the concept suggests), yet body size is not a useful predictor of risk to extinction. In birds, body size was not a useful predictor of rates of population increase or decrease in a global sample of threatened species from 12 families at various trophic levels.

One important but counterintuitive fact is that trophic position has no consistent effect on extinction. It is difficult to detect a consistent tendency for more frequent extinction of species at higher trophic levels, fossil or extant. This is complicated by the difficulty in separating body size and trophic position (species at higher trophic levels are mostly large). Top predators, in other words, are not more prone to extinction than consumers at other levels.

It seems that large-bodied species are vulnerable to ultimate causes of extinction (hunting, habitat destruction) but less so to proximate causes (their populations fluctuate less).

D. Time Factor

The preceding important determinants are thought to vary in ecological time, 10 to 1000 years. However, as no species lives forever, there may be processes that are operating in evolutionary time. If range and abundance are also species-specific characteristics, some species will be more extinction prone (i.e., naturally rare and restricted in distribution) than others.

Among songbirds on West Indian islands, older taxa occur on fewer islands, have more restricted habitat distributions and have reduced population densities. However, body size : abundance plots within tribes of birds have more positive relationships than expected in taxonomically ancient tribes; large-bodied species in old tribes are more common than small-bodied ones. This could be the product of differential extinctions of large-bodied rare species over time.

Bivalves and gastropods in Cretaceous fossils achieved characteristic range sizes early in their history, and this changed little thereafter. In this group, locally endemic species have much higher chances of extinction than more cosmopolitan genera.

VIII. THE PRESENT: A FULL-FLEDGED MASS EXTINCTION

There are very few documented cases of extinction of lower organisms. This is an inevitable consequence of our ignorance of the degree of biodiversity of those groups. Given this, it is not surprising that the current rate of extinction can only be roughly guessed. It is extremely probable that the rate of recent extinctions is several magnitudes higher than the "background extinction rates" and probably surpasses any similar mass extinction events in the earth's history.

Given the above deficiencies, we can only estimate current rates of extinction. Among the comparatively well-studied birds and mammals, the documented extinctions this century numbers about 100 species. There are a total of 14,000 species of these classes, so the documented extinction rate this century is about 1%. This translates to an expected average life span of a bird or mammal species of around 10,000 years. This is 100 to 1000 times shorter than the average life span calculated from the fossil record.

Three different methods for predicting impending extinction rates suggest future life spans of birds and mammals of 200 to 400 years if current trends continue. These impending extinction rates are at least 10,000 times higher than background rates in the fossil record.

All evidence suggests that a sixth mass extinction event in the history of earth is underway. While the total effect cannot yet be guessed, we know that the sixth mass extinction will be unique in the earth's history. It will be the first resulting not from environmental changes but the extraordinary population growth and the activities of one species. Our species now uses an estimated 25 to 50% of terrestrial net primary productivity. This is without precedent, and will make the coming extinction qualitatively different from all previous mass extinctions. We know enough to realize the gravity of the problem. We need a more elaborate understanding of the phenomenon, how it affects different groups and geographic locations, as our conservation actions will become more and more critical for the future of life on earth.

See Also the Following Articles

EXTINCTION, CAUSES OF • EXTINCTION, RATES OF • LATENT EXTINCTIONS ("LIVING DEAD") • MAMMALS, LATE QUATERNARY, EXTINCTIONS OF • MARINE MAMMALS, EXTINCTIONS OF • MASS EXTINCTIONS, CONCEPT OF

Bibliography

Goodman, S. M., and Paterson, B. D. (Eds.) (1997). *Natural Change and Human Impact in Madagascar.* Smithsonian Institute Press, Washington, D.C.
Groombridge, B. (Ed.) (1992). *Global Biodiversity: Status of the Earth's Living Resources.* Chapman & Hall, London.
Groombridge, B. (Ed.) (1993, 1994). *IUCN Red List of Threatened Species.* IUCN, Gland, Switzerland.
Jablonski, D. 1994. Extinctions in the fossil record. *Philosophical Transactions of the Royal Society of London,* Series B344, 63–76.
Lawton, J. H., and May, R. M. (Eds.) (1995). *Extinction Rates.* Oxford University Press, Oxford, UK.
MacPhee, R. D. E., and Marx, P. A. (1997). The 40,000-year plague. Humans, hyperdisease, and first-contact extinctions. In *Natural Change and Human Impact in Madagascar* (S. M. Goodman and B. D. Patterson, Eds.), pp. 169–217. Smithsonian Institution Press, Washington, DC.
Martin, P. S. and Klein, R. G. (Eds.). (1984). *Quaternary Extinctions. A Prehistoric Revolution.* University of Arizona Press, Tucson, Arizona.
Martin, R. D. (1993). Primate origins: Plugging the gaps. *Nature* 363, 223–234.
May, R. M., Lawton, J. H., and Stork, N. E. (1995). Assessing extinction rates. In *Extinction Rates* (J. H. Lawton and R. M. May, Eds.), pp. 1–24. Oxford University Press, Oxford, UK.
Raup, D. M. (1978). Cohort analysis of generic survivorship. *Paleobiology* 4, 1–15.

FIRES, ECOLOGICAL EFFECTS OF

William Bond
University of Cape Town

GLOSSARY

fire regime The type of fire, mean and variance in fire frequency, intensity, severity, season, and areal extent of a burn in an ecosystem.
general circulation model (GCM) Computer models developed to simulate global climate and widely used for global climate change predictions.
serotiny Seeds stored on the plant with dispersal triggered by fire.

FIRE IS BOTH A NATURAL and anthropogenic disturbance influencing the distribution, structure and functioning of terrestrial ecosystems around the world. Many plants, and some animals, depend on fire for their continued existence. Others, such as rainforest species, are extremely intolerant of burning and need protection from fire. The properties of fire are changing as natural landscapes become fragmented and human influence becomes all pervasive. A sound understanding of its effects is an essential prerequisite for effectively managing this widespread ecological process.

I. FIRE IN EARTH HISTORY

Fire is an enormously influential disturbance over very large areas of land. Vegetation burns because the earth's atmosphere contains sufficient oxygen (>13%) to support combustion. Should oxygen levels rise above 30%, fires would be so frequent that dense forest vegetation, even in persistently wet climates, would be incinerated. There is an almost continuous record of fossil charcoal during the past 350 million years indicating that the atmosphere supported combustion for most of terrestrial plant evolution. Oxygen levels reached maxima in the Upper Carboniferous, 300 million years ago, when abundant fossil charcoal indicates frequent fires. At this and other times, frequent fires may have played a significant part in the ecology of palaeo-ecosystems.

Contemporary ecosystems contain a mix of ancient and modern plants with varied response to burning. Ancient fire-dependent elements include conifers such as the giant sequoias of California, whose association with fire has roots in the days of the dinosaurs. Many of the least fire-tolerant elements occur in angiosperm-dominated temperate and tropical forests of early Tertiary origin. Grasslands are the most flammable vegetation that has existed on the planet. Tropical (C4) grasses occupy one-fifth of the land surface but first appeared only 7 million years ago. Charcoal from marine sediments increased dramatically during the past 10 million years as fire-promoting ecosystems, especially grasslands and savannas, began to spread.

The current extent of fire-prone ecosystems has been drastically altered by humans. Hominids have used fire for perhaps as long as 1 million years. Regular use of fire, indicated by the frequency of domestic hearths in archeological sites, only became common in the past 100,000 years associated with the evolution of modern humans. Fire was widely used to promote the growth of edible plants in hunter–gatherer communities and to attract animals to hunting grounds. Fire was (and is) a tool used by farmers to clear new lands and to prepare sites for swidden-type farming. Increased intensity of human use of fire by crop farmers has led to the expansion of fire-prone systems in many parts of the world during the past few thousand years.

II. WORLD BIOMES AND FIRE INCIDENCE AND INFLUENCE

Fires are rare only at the extremes of the climatic continuum. Neither the most humid tropical and temperate forests nor the driest deserts experienced fire as a major factor. However, between these two extremes, fire has influenced the extent and composition of a great diversity of ecosystems, including boreal forests, dry conifer forests, many grasslands (especially those dominated by tall grasses), temperate woodlands, tropical savannas, Mediterranean-type shrublands, heathlands, and eucalypt woodlands. Forests with mast-flowering bamboo understories are also prone to burning after the bamboos flower and die, creating massive fuel loads. Humans have changed landscape patterns of burning. Even humid tropical forests are beginning to burn as a result of logging. All these biomes experience fires of widely differing frequency and severity which help shape ecosystem structure and function.

Given the wide geographic extent of ecosystems that burn, fire influences the distribution and abundance of many species. Some ecosystems are dominated by species that depend on fire to complete their life cycles. Others are dominated by species that tolerate burning but have no direct dependence on fire. Ecosystems that seldom or never burn, except when disturbed by human activity, contain mixtures of species that fortuitously tolerate burning and species extremely intolerant of fire. The impact of burning on biodiversity varies greatly among these different types of ecosystems and species-response patterns.

III. SPECIES RESPONSE TO BURNING

A. Plants

Fire has influenced some types of vegetation for so long that it has not only affected distribution of species but also led to the evolution of fire-dependent life histories. Burning triggers different stages in plant life cycles, including flowering, seed dispersal, and seed germination in fire-dependent plants. Perennial grasses and herbs, including orchids, lilies, and other bulb plants, flower prolifically after they have been burnt, often as a facultative response to higher light, water, and nutrient availability. The African fire lilies, *Cyrtanthus* spp., flower only after fire. In these and other species, flowering is stimulated by constituents of smoke such as ethylene (e.g., *Xanthorrhea*, the Australian "grass-trees"). Burning stimulates seed release from species with serotinous cone-like structures which store seeds on the plant for years between fires. Serotiny is common in conifers of the boreal forests and Mediterranean-climate regions and also among diverse groups of flowering plants in Australia and South Africa. Some species are polymorphic for the trait, with serotinous forms increasing in populations that regularly experience large severe burns. Burning also stimulates seed germination in plants that store dormant seeds in the soil. Heat-stimulated seed germination is common in many legumes and other groups with hard seed coats (e.g., members of Rhamnaceae). Thick seed coats prevent imbibition of water until cracked by the heat of a fire. Smoke-stimulated seed germination has been reported in fire-prone shrublands of South Africa, Australia, and California and also in some grassland species. Nitrogen dioxide, released in large quantities in smoke, cues seed germination in *Emmenanthe pendulifera*, a chaparral annual. Species with fire-stimulated flowering, seed release, or seed germination depend on fires for populations to expand.

Vegetative features of plants affect tolerance to burning. Thick bark and the ability to resprout allow many plants to survive burning. Neither feature is necessarily a fire adaptation. Sprouting is a common fire survival mechanism, either from the root stock or from branches above the ground. Some species posses large swollen burls or lignotubers which are thought to act as bud banks or storage reserves. Paradoxically, many woody plants in fire-dependent vegetation cannot resprout and are killed by fire. These nonsprouting plants often have higher seed production and higher seedling growth than related sprouting species. In some lineages, sprouting is the ancestral feature and loss of sprouting is viewed as an adaptive response to fire. Non-sprouting shrubs are particularly common in chaparral and similar shrublands and require fire to release seeds from serotinous cones or to stimulate germination. Among trees in fire-prone forests, many conifers do not sprout and a few eucalypts are also killed by fire. Nonsprouting species are particularly prone to local extinction if recruitment fails after burning.

Grasses are among the most fire resistant of all plant growth forms. The buds of new shoots are insulated by either layers of leaf sheaths or the soil where species have underground rhizomes. Grasslands recover from burning more rapidly than woody plants and can carry very frequent fires (1–3 years) in productive sites. Although many grasslands burn readily, few species have an obligate dependence on burning. Fire-stimulated flowering is rare but has been reported in many mostly temperate tussock grass species, including species of *Chionochloa* in New Zealand. Several very widespread warm-climate (C4) grasses (e.g., *Themeda triandra* and *Andropogon gerardi*) decline rapidly in the absence of burning. These species become locally extinct if fires are suppressed for more than a decade since litter accumulates, shades, and kills the grass plants.

Some woody plants have developed a remarkable ability to tolerate very frequent grassland fires. These create a particularly hostile environment for juvenile stages. Tree seedlings survive by rapidly acquiring the ability to sprout, and juveniles slowly develop food reserves in swollen roots and eventually produce bolting stems that place foliage above flame height. This peculiar life history occurs in several pine species (e.g., *Pinus palustris*) growing in grassy habitats and also in many savanna trees. A tropical Asian species, *Pinus merkusii*, produces persistent sprouting juveniles in frequently burnt populations but nonsprouting forms in infrequently burnt island populations. Juvenile stages of savanna trees can tolerate repeated burning, suffering repeated killing of the stem parts, for decades before they die or escape the flame zone to become adults.

The occurrence of species with flammable morphologies in fire-prone environments has led to the suggestion that flammability has evolved to promote burning. In most cases, features that promote flammability have probably evolved for alternative functions. However, some woody plants accumulate highly flammable fuel by retaining dead branches and require fire for recruitment. In western North America, serotinous pine species retain dead branches and recruit seedlings only after crown fires have destroyed the population. These species appear to both promote the spread of crown fires and benefit from fire-stimulated reproduction. This strategy contrasts strongly with self-pruning, thick-barked pines (e.g., *P. ponderosa*) which tolerate frequent surface fires but are restricted to productive sites which support rapid juvenile growth.

B. Animals

The direct effects of fire on wildlife are often surprisingly small. Agile animals flee to unburnt refugia, often moving across the fireline to places of safety. Soil is a very effective insulator so that many animals survive in crevices and cracks or in burrows in the soil. Mortalities of large mobile vertebrates, including humans, do occur but only in the most severe fires. Reptiles and slow-moving invertebrates can suffer higher mortalities and their carcasses provide a food source to scavenging birds and other creatures in the first few days after a burn. The threatened bald ibis of South Africa makes extensive use of recently burnt grasslands, as does the endangered whooping crane in its Texan winter feeding grounds.

The indirect effects of burning are generally far more important, especially changes in habitat attributes and the successional mosaic initiated by forest fires. A large crown fire in a forest causes drastic structural change and local extinction of all faunal elements that depend on unburnt forest habitat. Post-burn stages are colonized by a new suite of species. Different successional stages support different suites of animals. Even frequently burnt grasslands, such as those of the South African highveld, have distinct bird assemblages which turn over with successive years of regrowth after burning.

The pattern of fires across a landscape imposes a mosaic of patches of different successional ages. The size and configuration of patches influence rates of local extinction and patch re-colonization (meta-population structure). For example, nectar-feeding birds in shrublands of Australia and South Africa lose their food source, shrubby members of the Protea family, after a burn and have to seek unburnt stands for food. The landscape configuration of old stands with flowering proteas, and young stands with immature shrubs, produces a highly mobile bird assemblage. Changes in the spatial pattern of fires may change extinction risks in different faunal elements. Some species of Australian honeyeaters are threatened with extinction because changes in the fire regime no longer produce the right mix of mature and immature populations of nectar plants. The decline of many bird species in Australian savannas has been attributed to the development of homogeneous landscapes through the systematic burning of very large tracts of land. Smaller, patchier burns prevailed under aboriginal burning practices. The extinction of many Australian small to medium-sized (50 g–5 kg) mammals since European settlement (23 species) has been attributed to similar changes in the fire regime, but predation by feral animals is a major factor.

Recent fires in the lowland rain forests of Sumatra, caused by human impacts on forest structure, have had devastating effects on the forest fauna. Primary forest specialists, including squirrels, hornbills, and other fruit-eating and frugivorous birds, and some primate species disappear altogether from burnt and adjacent forests. The increasing risk of fire in humid tropical forests poses serious threats to survival of the forest fauna in addition to those caused by direct forest clearing.

IV. DETERMINANTS OF FIRE

The incidence of natural fires depends on much the same combination of ingredients as a campfire: dry fuel, flammable material (finely branched plant material that dries quickly), continuity of fuels, and a source of ignition. Suitable weather conditions and winds to carry the fire and pre-heat flammable material are also key factors. The moisture content of plant matter is critical. Dead matter has the lowest moisture content, whereas live leaves will burn more easily if their moisture content is low. The shape, size, and arrangement of plant parts influence moisture content and flammability. Plants with narrow leaves or thin branches dry rapidly and burn readily. Ecosystems that accumulate slow-decomposing litter are highly flammable. Leaves with large amounts of oils, fats, waxes, and terpenes also burn readily. Volatile substances enhance burning because they are released from leaves, burn fiercely, and thus dry and heat adjacent material.

Because fire depends very much on moisture content of vegetation, climate and weather conditions exert marked influences on the timing of burns. The length of a warm, dry period needed for ignition depends greatly on vegetation properties. A few days of hot, dry weather are sufficient to dry tall grasslands enough to sustain a fire, whereas months of extreme hot and dry conditions are needed for fire to burn pristine humid tropical forests. Therefore, large fires in woody ecosystems are generally associated with rare drought events, such as those produced by El Niño conditions. In contrast, large fires in arid grassy ecosystems are limited by the availability of fuel and are more common after high rainfall years. Fires generate their own heat that, under hot, dry, and windy conditions, creates a positive feedback, increasing the area burnt and the severity of the fire by drying the vegetation before the fire front.

A source of ignition is needed to start fires. Today, most fires are ignited by humans, except in sparsely populated regions. Lightning fires are still common in many landscapes. Lightning often accounts for large numbers of fires but small areas burnt because many are dowsed by subsequent rain or controlled by fire crews attuned to the risk of thunderstorm fires. Rockfalls also ignite fires, accounting for the spate of burns often triggered by earthquakes. Volcanic activity is of local significance, especially on islands in which lightning is unusual.

The contingent requirements of ignition, suitable weather conditions, and contiguous flammable vegetation make for a high degree of uncertainty in predicting landscape fire patterns.

A. The Fire Regime

The ecological effects of fire depend on the fire regime rather than the occurrence of a single fire. The fire regime is produced by the combined effects of climate, fuel properties of vegetation, and ignition frequency. It is characterized by the type of fire, mean and variance in fire frequency, intensity, severity, season, and areal extent of a burn. Types of fire include ground fires that burn the organic layers of the soil, surface fires that burn just above the ground, and crown fires that burn in the canopies of trees. Ground fires occur only in peaty soils in which they can be extremely damaging, destroying roots and completely altering soil properties. Crown fires predominate in short vegetation, including grasslands, shrublands, and low conifer forests in boreal and Mediterranean climate regions. Surface fires are common in most woodlands and forests. When crown fires do occur in forests, they produce massive stand-replacing fires. Fire exclusion in national parks, such as the giant sequoia forests of California, has led to an increase in young trees which now act as bridging fuels turning surface to crown fires with extremely damaging consequences.

Fire frequency is estimated from maps of fires, records of fire scars, or patterns of charcoal deposition in sediments. The mean and variance of fire return interval are important descriptors of historical disturbance patterns. Changes in fire frequency are an important cause of changes in ecosystem structure and function. Fire intensity is measured as energy released per meter of fire front. It is widely used in fire fighting operations. Fire severity is a distinct but related concept defined as the ecological impact of a single fire and usually estimated from the amount of plant biomass consumed. A fast-moving fire that consumes little biomass and a slow-moving fire that consumes more can have the same fire-line intensity but different severity. Fire severity is highly variable, literally depending on the weather during a burn, wind conditions, and the pre-burn condition of the vegetation. Fire season is largely dictated by the moisture content of flammable biomass. Where the vegetation dries out quickly, fires can burn in almost any season. Seasonal timing of burns can cause significant changes in species composition and ecosystem structure. Continuity of flammable vegetation, especially at the landscape scale, strongly influences the spread of fires. Habitat fragmentation can lead to a reduction in fire frequency of isolated fire-prone ecosystems or an increase in fire-excluding forests surrounded by flammable vegetation. Land abandonment in some countries has led to successional changes producing large, contiguous, highly flammable vegetation. In Spain, reduction of pastoral activities has led to the conversion of grasslands to highly flammable shrublands. This process has contributed to an increase in the area burnt annually from a few thousand hectares in the 1960s to hundreds of thousands of hectares in the 1990s. The increase in fires in southern European countries during this period has not been matched in southern Mediterranean countries in which rural depopulation is not as advanced.

V. ECOLOGICAL EFFECTS OF FIRE

A. Ecosystem Structure

The consequences of increasing fire frequency and severity for ecosystem structure are

to reduce vegetation height (tall forests to shorter ones and woodlands to shrublands);
to replace woody vegetation by grasslands;
to promote flammable species or communities (low litter decomposition rates, more xeromorphic leaves, and finer twigs/branches); and
to reduce biomass.

There are no reliable global estimates of the percentage reduction of plant biomass attributable to burning. Both tropical and temperate landscapes contain mixtures of fire-prone grassland or shrubland communities and closed forests which tend to exclude fire. Successional replacement of the fire-prone communities by fire-excluding forest elements frequently occurs when fires are suppressed. These changes can be very rapid. In southern Africa, forests have replaced both grasslands and savanna woodlands after 10–30 years of fire suppression. However, counterexamples exist in which fire suppression does not lead to changes in savanna/ forest boundaries. Stable boundaries often coincide with different soil types, with forests occurring on the better drained soils.

Changes from fire-excluding to fire-promoting ecosystem structures may also be rapid. In the Brazilian Amazon, fires in closed-canopy forests spread as a "thin, slowly creeping ribbon of flames a few tens of centimeters in height." Despite the low severity of an initial fire, structural changes causing an opening of the canopy dries out the understory and contributes to an increase in flammable understory biomass, increasing the risk of a second fire. Weedy vines and grasses quickly colonize twice-burned forests, further adding to the flammable biomass. Positive feedback of this kind is estimated to reduce a forest to vegetation resembling recently abandoned farmland in 20–30 years.

B. Ecosystem Function

The immediate effect of fire is gaseous loss of carbon and nitrogen from material that burns. More biomass burns in more severe fires and the nutrient losses are greater. Strong winds accompanying fire lead to losses of phosphorus and cations blown away in ash. Cation nutrients in ash tend to be mobile and in a plant-available form. Their presence leads to increases in soil pH—large increases in acid forest soils and smaller increases in neutral or alkaline soils in grasslands or savannas. Increased solar radiation, decreased evaporation, and higher pH lead to increased microbial activity, increased rates of mineralization, and increased availability of nutrients after a burn. After a chaparral burn, for example, nitrate increased more than 20-fold relative to unburnt controls. Short-term increases in nutrient availability can be offset by long-term decreases where fire frequencies are high and inputs to the system between fires are not high enough to replace losses. Severe fires can lead to nitrogen shortages. Many ecosystems have nitrogen-fixing organisms as major components of postburn vegetation which replace nitrogen losses in a few years.

Fire can lead to changes in ecosystem processes at landscape scales. The reduction in biomass caused by burning and changes in soil properties lead to temporary hydrological changes in patterns of stream flow. Severe fires can lead to changes in soil erosion. The Yellowstone fire of 1988 led to significant increases in sediment loads and altered the geomorphology of river systems. Debris spread over a distance of 12 km in one valley bottom.

C. Species and Populations

At the local scale, and within fire-prone communities, species respond to differences in fire frequency, season, and intensity. Variation in the fire interval is an important determinant of population status. In shrublands, the effect of fire on population growth depends on key demographic attributes of the species. Population size of non-sprouting species fluctuates more than that of sprouting species, and local extinction is not uncommon after a single fire. Species that are slow to mature are particularly vulnerable where populations are burnt before they have first flowered and set seed. Populations are also negatively affected where intervals between fires exceed the life span of a species or its seedbank. Many grasses are also sensitive to variation in fire frequency. In Australia, South Africa, and Southeast Asia, *Themeda triandra* dominates many frequently burnt grasslands. Populations may decline to local extinction if fires are excluded for longer than a decade. Manipulation of the fire interval is a key tool for influencing biodiversity of vegetation stands. Information on the reproductive status of plants, especially the size of the viable seedbank, at different postburn stages has been extensively used to help determine optimum fire frequencies to maintain particular species in fire-prone woody ecosystems.

Recovery of plant populations also depends on the unique combination of circumstances on the day a fire burns. These "event-dependent" effects can be as important as fire frequency in influencing biodiversity in some ecosystems. Fire season has a marked influence on recruitment of serotinous members of the Proteaceae in Western Australia and South African fynbos. Spring burns can reduce protea populations to less than one-tenth of their preburn density, whereas autumn burns can result in a 10-fold or greater increase in plant densities. Fire season also influences recovery of sprouting plants where the size of root reserves varies seasonally, affecting the vigor of resprouting. Manipulation of fire season is sometimes the only effective tool for managing densities of sprouting shrubs. Stem density of clonal species of hazel (*Corylus* spp.) in the understory of temperate forests increased 5-fold with four successive spring burns but was halved by successive summer burns. In Zambian woodlands, annual burns in the early dry season caused a 10-fold increase in tree seedlings and halved the adult mortality rate over a decade. Grassland composition can also be very sensitive to fire season. In a long-term burning experiment in the Kansas prairies, late-spring burns caused a halving of *Andropogon scoparius* biomass relative to fires burnt a few weeks

earlier. The effects of fire season on species and ecosystem recovery are poorly known in many ecosystems. They may be unimportant where the fire season is short because of climate constraints.

By definition, severe fires cause the most extreme biomass losses in ecosystems. In eucalypt and conifer forests, intense crown fires kill all aboveground plant growth. Where trees are incapable of sprouting, these fires cause complete replacement of canopy trees. Sprouting plants, especially shallow-rooted species, can be killed by intense burns. Fire intensity is an important factor in savanna ecology. Where grass growth is sufficient to carry fires at frequent intervals, burning kills the aboveground parts of trees. The amount of dieback depends on the intensity of the burn. In mesic savannas, fires are so frequent and so intense that juvenile trees may spend decades trapped in the grass layer. The frequency and intensity of fire are important determinants of tree biomass (and habitat structure) in these ecosystems.

The effects of fire severity on recruitment from seeds vary among species. Legumes and other plants with hard, dormant seeds in fire-prone shrublands do not germinate unless a burn heats the soil sufficiently. For example, the Australian shrub, *Acacia suaveolens*, will not germinate unless soils are heated to more than 50°C. Variation in fire intensity can directly affect species composition in fire-prone shrublands, incinerating some seeds and stimulating germination in others. Key species, such as legumes, may fail to germinate after low-intensity fires that are applied for safety reasons.

The relative sensitivity of plants to fire season and fire severity varies among species. This makes general prediction of population trends under different fire cycles difficult without species-specific information. In Mediterranean shrublands, species with diverse responses to frequency, season, and intensity of burn occur in the same community suggesting a history of fires that vary in these factors. It is a considerable conservation challenge to incorporate variability into fire regimes to maintain the full diversity of species.

VI. INTERACTIONS BETWEEN FIRE AND OTHER ECOLOGICAL PROCESSES

A. Fire and Herbivory

Fire interacts with, and is influenced by, other agents of disturbance that can also influence ecosystem structure on a large scale. Herbivores influence the distribution and biomass of plant parts and therefore the attributes of fire regimes. Heavily grazed savanna grasslands do not burn. Persistent heavy grazing often leads to an increase in tree densities because of the reduction in fire frequency. In Africa, elephants open up woodlands, enhancing grass growth which promotes more frequent severe fires. The combination of elephants and grass fires can cause a marked reduction in tree densities. In miombo woodlands of Zimbabwe, changes in woodland structure under the combined influence of elephants and fire markedly reduced bird diversity and led to local extinction of four endemic woodland bird species.

Insect herbivory also influences fire regimes, especially in northern ecosystems. In Balsam fir (*Abies balsamea*) and red spruce (*Picea rubens*) forests of eastern North America, fires are rare. Large-scale tree mortality is caused by spruce budworm outbreaks which inhibit the spread of fire by changing vegetation structure and fuel properties. In general, however, plants that provide fuel for burning make poor food for herbivores and vice versa. Fires burn readily where decomposition is slow, causing dead matter to accumulate. Slow decomposition is associated with high C:N ratios, high fiber contents, and high leaf-specific weight, all of which inhibit herbivore food intake. Thus, the most fire-prone vegetation tends to be least edible and vice versa.

B. Fire and Landscape Fragmentation

The pattern of fire in a landscape is sensitive to the contiguity of flammable vegetation. Landscape fragmentation can have major impacts on the fire regime, in turn affecting survival of species within fragments. Species losses from North American prairie fragments during a 50-year period were greater among plants of early postburn environments. Similar patterns of local extinction occur in fynbos fragments surrounded by nonflammable forests in South Africa. The causes of species loss in both prairies and fynbos is attributed to infrequent burning due to isolation of the fragments. Small forest patches in extensive areas of fire-prone shrublands or grasslands are also more likely to suffer local extinction of species intolerant of burning. In the tropical forests of Guyana, fire-tolerant forest tree species with thick, fissured bark and small seeds are unusually common near savanna boundaries and human settlements.

C. Fire and Invasives

The interplay between fire and invasive species can cause alarming ecosystem transformation. Direct effects of invasive plants can be minor relative to the indirect effects on fuel properties and fire regimes. Grass invasion into woody ecosystems has particularly damaging consequences. In Hawaii, invasion of tall non-native grasses has transformed the unique forests to grasslands entirely alien to the island ecosystem by fuelling frequent fires. In South America, invasion of logged-over tropical forests by fire-promoting grasses can cause elimination of the forest ecosystem and its rapid replacement by the alien grass. In southwestern Australia, species-rich heathlands are fire dependent but they have also been invaded by non-native grasses which burn so frequently that the heathlands are transformed into a species-poor savanna with scattered relictual trees. The reverse pattern, that of invasion of fire-prone grasslands by plants that do not burn easily, may also be a problem. For example, *Lantana camara* is invading fire-dependent grasslands in South Africa but burns much less readily than the native vegetation.

VII. MANAGING FIRES

Because wildfires are such a widespread feature of world vegetation, managing fires for particular objectives, including conservation of biodiversity, is a major concern. Attitudes and actions to wildfire are never neutral. Fires are actively suppressed in some ecosystems but purposely ignited in others. Wildfires commonly arouse public and media concern. The spread of the urban interface as more people flock to cities has increased threats of wildfire to people and properties. Smoke management has become a health and safety issue. Enormous effort and expense are often expended on fire management. In the former USSR, fires in protected areas were suppressed regardless of the ignition source using hundreds of planes and more than 8000 air-borne fire-fighters. Appropriate fire management policies continue to be a central management headache in protected areas as different as Yellowstone National Park in a coniferous ecosystem and Kruger National Park in African savannas. The expense of maintaining fire fighting teams consumes a significant fraction of conservation agency resources. Litigation, when fires move beyond the borders of protected areas, can also strain conservation budgets. Arson burns are not uncommon, sometimes as a protest against state authority. Fire management consumes a great deal of time and is a major expense in fire-prone ecosystems.

A. Fire Management Policies

There is no consensus on how to manage fires in protected areas or outside them. For the first half of the twentieth century, complete fire suppression was a common policy, and it still is in many parks. Suppression policies have slowly changed, partly because of their cost, partly because they do not work, and partly because of changes in ecological thinking. Fire suppression leads to the build-up of dead biomass in fire-prone ecosystems which produce very severe fires when they do burn. Suppression policies were also relaxed following the discovery of fire-dependent features of plants. In the Cape fynbos, the beautiful marsh rose (*Orothamnus zeyheri*) declined to a handful of plants before managers realized that the species had an obligate dependence on fire to stimulate germination of its seeds. There has also been increasing recognition of disturbance as a natural process in ecosystems. Outside of rain forests, complete fire suppression is rarely the aim of fire management. However, decades of suppression have led to pronounced changes in ecosystem structure that now pose major challenges regarding how to re-introduce fires without causing more problems. Many conifer forests were maintained as open parklands by frequent surface fires. Fire suppression has allowed numerous young trees to establish, creating "bridging fuels" and a switch from surface to crown fires which destroy the whole forest structure. In some savannas, fire suppression led to invasion of grasslands by forest species in a process that is very difficult to reverse except by costly manual clearing of trees to restore large mammal habitat.

Prescription burns are fires intentionally lit for management purposes. Safety for the fire crew, and for property, is always an important consideration. Therefore, prescribed burns often cause significant changes to the fire regime, especially to season and intensity but also to fire frequency. Changes to the fire regime can lead to significant population declines in sensitive species such as legumes, which depend on intense heat for seed germination.

Prescription burning calls for clear management objectives. Because fire is so influential a force on community structure and composition, decisions have to be made regarding the desirable objective and what kind of burning pattern should be used to achieve it. In the South African savanna parks, fire policy changed from complete fire exclusion to prescribed burning at fixed intervals (promoting large mammals) to fire regimes that "maximize biodiversity." The intention of the latter is to create and sustain a landscape mosaic of different successional ages or habitat types that can maintain viable populations of most species. What to do in practice is more difficult.

Another commonly considered policy is to re-create

"natural" burning regimes. This policy permits fires that have been ignited by lightning, rockfalls, or other non-human agents but suppresses fires of human origin. It is difficult to apply in practice because vegetation has been fragmented by roads, buildings, croplands, and the like so that fires can no longer spread unhindered across the landscape. Re-creating natural fire regimes has also been criticized for not including aboriginal influences on fire regimes which may have significantly influenced landscape patterns over millennia. A variation on natural burning policies is therefore to re-create aboriginal burning practices to try to reproduce pre-agricultural landscapes. Implementation of these policies is constrained by lack of biological knowledge, techniques for fire management, and safety-related considerations.

In practice, fire management increasingly involves a mix of fire suppression, prescription burns, and controlled natural fires depending on safety concerns and conservation objectives in different parts of the landscape.

VIII. FIRE AND GLOBAL CHANGE

A. Effects of Global Change on Fire Ecology

The indirect effects of climate change on disturbances such as fire and herbivory, and thereby on ecosystems, is likely to be as great or greater than direct influences of climate change in many fire-prone ecosystems. The pace of vegetation change in the western Canadian Arctic, an area subject to particularly large temperature change, appears to depend on fires. The treeline is moving northward with each successive fire because trees are limited more by suitable seed beds for regeneration than by climate constraints on growth. Ground fires that remove peat layers provide the necessary seed bed to cause trees to advance.

It is difficult to predict global change impacts on disturbance regimes. The multi-causal origins of fire create considerable uncertainty. Even in large tracts of undisturbed land such as the boreal forests, it is still difficult to balance the effects of changes in vegetation type, ignition (more convectional storms), length of the fire season, and fire weather to predict how fire regimes may change. Models that predict climate change impacts on fire regime combine physical models for fire spread, GCM predictions of future climate parameters, and models for future plant growth. Predictions suggest a higher incidence of fire, for example, in chaparral, with larger and more intense burns because of higher temperatures and higher fuel accumulation rates. However, instantaneous fire weather effects and extreme events drive contemporary fire patterns and are still poorly predicted by GCMs. In boreal forests, despite nearly a century of temperature increase, fires have declined in areas remote from human intervention. Boreal forests are predicted to burn more frequently and more intensely under global warming and elevated CO_2. The difference between predicted and observed patterns is partly due to an increase in precipitation associated with increasing temperatures.

B. Fire as a Source of Greenhouse Gases

Vegetation fires may contribute significantly to global climate change. Annual gross amounts of carbon released into the atmosphere from global savanna and forest fires (1990s) are estimated to be 1.7–4.1 pg compared with release of carbon from fossil fuel burning of 5 or 6 pg per year. If biomes remain stable in their distribution, most of the carbon released from burning will be taken up again in new plant growth. However, there is increasing concern that increasing fires will release stored carbon more permanently into the atmosphere because of more permanent ecological changes. Increased fires, and a change from surface to ground fires, in boreal forests could release 46–54 pg carbon into the atmosphere according to some predictions. The increasing frequency of fire in humid tropical forests is also of concern. Tropical forests are estimated to store one-fifth of the world's carbon. If burning of tropical forests continues, very large areas could be converted to flammable secondary scrub or grassland releasing this carbon into the atmosphere.

Fire is also important as a source of aerosols. Aerosols decrease regional and global irradiation through backscattering of incoming solar radiation. It has been estimated that the net effect of aerosols of pyrogenic origin is to reduce global irradiation by 2 W m^{-2}, causing a net global cooling effect of approximately 2°C. The possible magnitude of these effects has led to intensive efforts to quantify the frequency of vegetation fires using dedicated satellites and measurements of gaseous and aerosol emissions from fires. Public concern regarding atmospheric impacts of burning is leading to public pressure regarding the use of fire for conservation purposes. This may have positive effects if it leads to adoption of logging practices that reduce fire hazard in humid forests. Public pressure to suppress fires could have negative impacts on fire-prone ecosystem and their fire-dependent species.

See Also the Following Articles

CARBON CYCLE • CLIMATE CHANGE AND ECOLOGY, SYNERGISM OF • DISTURBANCE, MECHANISMS OF • FARMLAND AND RANGE ECOLOGY • GREENHOUSE EFFECT

Bibliography

Agee, J. K. (1993). *Fire Ecology of Pacific Northwest Forests.* Island Press, Washington, DC.

Bond, W. J., and van Wilgen, B. W. (1996). *Fire and Plants,* Population and Community Biology Series, Vol. 14. Chapman & Hall, London.

Clark, J. S., Cachier, H., Goldammer, J. G., and Stocks, B. J. (Eds.) 1997). *Sediment Records of Biomass Burning and Global Change.* Springer-Verlag, Berlin.

Crutzen, P. J., and Goldammer, J. G. (Eds.) (1993). *Fire in the Environment: The Ecological, Atmospheric, and Climatic Importance of Vegetation Fires.* Wiley, Chichester, UK.

Gill, A. M., Moore, P. H. R., and Martin, W. K. (1994). *Bibliography of Fire Ecology in Australia (Including Fire Science and Fire Management),* 4th Ed. NSW National Parks and Wildlife Service, Hurstville, Australia.

Goldammer, J. G. (Ed.) (1990). *Fire in the Tropical Biota. Ecosystem Processes and Global Challenges,* Ecological Studies Vol. 84. Springer-Verlag, Berlin.

Johnson, E. A. (1992). *Fire and Vegetation Dynamics: Studies from the North American Boreal Forest.* Cambridge Univ. Press, Cambridge, UK.

Keeley, J. E. (1995). *Bibliography on Fire Ecology and General Biology of Mediterranean-Type Ecosystems. Vol. I. California. Vol. II. Australia, Chile, the Mediterranean Basin and South Africa.* International Association of Wildland Fires, Fairfield, WA. (Includes electronic format)

Moreno, J. M., and Oechel, W. C. (1994). *The Role of Fire in Mediterranean-Type Ecosystems,* Ecological Studies Vol. 107. Springer-Verlag, Berlin.

Pyne, S. J. (1997). *World Fire: The Culture of Fire on Earth..* University of Washington Press, Seattle, WA.

Uhl, C., Hammond, D. S., ter Steeger, H., Schulze, M. D., Lefebvre, P., Schlesinger, P., Kinnaird, M. F., and O'Brien, T. G. (1998). Issues in international conservation: Perspectives on wildfire in the humid tropics. *Conserv. Biol.* **12,** 942–956.

Whelan, R. J. (1995). *The Ecology of Fire.* Cambridge Univ. Press, Cambridge, UK.

World Wide Web site: *http://www.csu.edu.au/firenet/.*

FISH, BIODIVERSITY OF

Gene S. Helfman
University of Georgia

I. Taxonomic Diversity
II. Geographic Diversity

GLOSSARY

adipose fin A small, fleshy fin without supporting spines or rays, set far back on the dorsal surface of many catfishes, characins, salmons, and other groups.

ancestral The taxon from which descendant species are derived, often synonymized as primitive or generalized. Ancestral traits or conditions are those which appear in an ancestor.

depauperate Of low diversity, lacking in species; opposite of speciose.

derived Later-appearing taxa within a lineage, often synonymized as advanced or specialized. Derived traits are those which appear in a descendant species and are changed from the ancestral condition.

endemic Restricted or native to a geographically defined area.

extant Living; opposite of extinct.

extirpated Locally extinct.

Gondwana A large supercontinent in what is now the Southern Hemisphere that separated during the Mesozoic, forming the modern continents of Australia, Antarctica, South America, and Africa.

species flock A large group of closely related, endemic species in a small area; all are descended from a single ancestor.

speciose Of high diversity, having many species in a group or area.

taxon (plural = taxa) A group of evolutionarily related species.

A FISH IS AN AQUATIC VERTEBRATE that (usually) has scales, fins, and gills (Box 1 and Fig. 1). Fishes constitute more than 25,000 of the known 48,000 species of living vertebrates and are divided taxonomically into three major groupings: jawless fishes (agnathans), cartilaginous fishes (chondrichthyans), and bony fishes (osteichthyans). Fishes are found almost everywhere on Earth where water of reasonable integrity exists. Fish habitats include the deep sea to depths of 8000 m, high mountain streams and lakes to 5000 m altitude, and just about every aquatic habitat in between. Marine fishes make up 58% of all species, freshwater species make up 41%, and 1% of fishes move regularly between the ocean and fresh water. Tropical areas have the highest diversity. Fish biodiversity is threatened by a wide range of human activities, but habitat modification, overharvest, and introduced species are particularly injurious.

I. TAXONOMIC DIVERSITY

A. Overview

Approximately 25,000 known fish species inhabit the earth's oceans, estuaries, rivers, lakes, and streams. This

What Is a Fish and What Are Fishes?

Generally defined, a fish is a cold-blooded, aquatic-living chordate with fin-like appendages, a body covered with scales, and that breathes using gills (Fig. 1). Exceptions to and variations in all traits are common. Some sharks, tunas, and billfishes are warm-blooded. Fins can be unsegmented and spiny or segmented and soft rayed (and soft rays can be hardened). Tail types include (i) the heterocercal tail of sharks and sturgeons, in which the notochord or vertebral axis extends considerably into the upper tail lobe; (ii) the abbreviate heterocercal tail of gars and bowfin, in which the vertebral axis extends only slightly into the upper lobe; (iii) the leptocercal or diphycercal tail of lungfishes, the coelacanth, and rattails, in which vertebrae or tail rays extend through the middle of the tail, forming a pointed tail; and (iv) the homocercal tail of most advanced bony fishes, in which the vertebral column ends at the tail base (the urostyle) and fin rays form a symmetrical, two-lobed tail. Scales also vary in terms of the number of layers of bony material that constitute them and the extent of spiny projections that cover their surface; more primitive fishes generally have heavier scales, and more advanced fishes have lighter scales, often with more projections. Scale types include the placoid scales of sharks, the ganoid scales of gars and bichirs (and sturgeons), the cycloid scales of lower teleosts, and the ctenoid scales of higher teleosts.

To ichthyologists, "fish" refers to one or more individuals of a single species, whereas "fishes" refers to more than one species, regardless of how many individuals are involved. Hence, this article is about fishes.

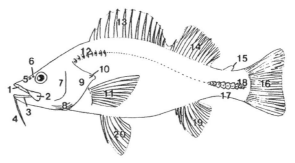

FIGURE 1 Anatomical features of a hypothetical fish, including some common features that are measured during fish identification. 1, premaxilla; 2, maxilla; 3, dentary; 4, barbel; 5, snout; 6, nostril; 7, preopercle; 8, branchiostegals; 9, opercle; 10, opercular spine; 11, pectoral fin; 12, lateral line; 13, first part of (or spinous) dorsal fin; 14, second part of (or soft) dorsal fin; 15, adipose fin; 16, caudal fin; 17, caudal peduncle; 18, lateral scutes; 19, anal fin; 20, pelvic or ventral fin (reproduced with permission from Greenfield and Thomerson, 1997).

incredible diversity exceeds that of all other vertebrate groups combined. It forms a wealth of biological wonder for ichthyologists (fish scientists), but such large numbers can be overwhelming to someone unfamiliar with the many taxonomic groups and their names. However, the different taxonomic groups are logically arranged according to well-studied evolutionary relationships, with those more closely related to groups that evolved earlier in geologic time (so-called ancestral or primitive taxa) placed earlier in lists and those groups

that evolved relatively recently (derived or advanced taxa) placed later in lists.

The most primitive of the living fishes are the jawless fishes. These arguably include 22 species of marine lancelets, and less arguably about 85 species of marine hagfishes and freshwater or anadromous (migratory between freshwater and marine) lampreys. Cartilaginous sharks, skates, rays, and chimaeras include approximately 10 orders, 43 families, and 850 species of almost entirely marine, relatively large-bodied predators. Skates and rays are more diverse than sharks, constituting about 55% of all cartilaginous fishes. Chimaeras are cartilaginous fishes distantly related to sharks that consist of 3 families and 31 species.

Bony fishes make up the vast majority of living fish species, exceeding all other groups in species, habitat, reproductive, and feeding diversity. Bony fishes vary in length from the 8-mm pygmy gobies to the 12-m long oarfish, 900-kg marlin, and 1000-kg-plus ocean sunfish. There are perhaps 45 orders and 435 families among the approximately 24,000 species of bony fishes, ranging from the relatively primitive lungfishes, coelacanth, sturgeons, bichirs, gars, and bowfin to the more advanced teleostean (higher bony fish) groups that include bonytongues, eels and tarpons, herring-like fishes and the so-called true teleostean groups of minnows, salmons, various deep-sea taxa, and cods. The spiny-rayed teleosts are the most evolutionarily advanced of the fishes and include mullets, silversides, scorpion fishes, perch-like fishes, tunas, flatfishes, and triggerfishes.

Phylum Chordata
Subphylum Cephalochordata; order Amphioxiformes (lancelets, two families, 22 species, marine, tropical and temperate)
Subphylum Vertebrata
Superclass Agnatha
Class Myxini; order Myxiniformes (living hagfishes; one family, 43 species, temperate marine)
Class Cephalaspidomorphi; order Petromyzontiformes (living lampreys; one family, 41 species, temperate fresh water and anadromous)

B. Cephalochordates

Lancelets may not be fishes because they lack scales, fins, and gills. However, their evolutionary and anatomical affinities are similar to what most workers believe characterized the ancestors of fishes, and lancelets are studied primarily by ichthyologists. Lancelets possess a dorsal nerve tube, in common with all vertebrates. They also have a notochord, which is a cartilaginous rod that runs the length of the body and is shared with all embryonic vertebrates as well as with the adults of many primitive living fishes. Lancelets are small (to 3 cm), slender organisms that as adults occupy sandy, usually shallow bottoms in all major tropical and temperate oceans. A commercial fishery for lancelets involving bottom dredging exists in southern China. The fishery is in apparent decline.

C. Agnathans

The first fishes lacked jaws. Modern jawless fishes—hagfishes and lampreys—look approximately similar, with slippery, eel-like bodies and jawless heads. The fossil record, however, indicates that they have been separated evolutionarily for hundreds of millions of years. Hence, most similarities are due to convergent or parallel evolution.

1. Hagfishes

Hagfishes, known also as slime eels or slime hags, produce copious mucus from many pairs of slime glands. A disturbed 2-ft-long hagfish can fill a 5-gallon bucket with slime. However, a hagfish covered in its own slime will suffocate. To rid itself of slime, a hagfish ties a knot in its tail and passes the knot forward until the slime is pushed off. Hagfishes are nocturnal predators on small invertebrates but are better known for their scavenging behavior, which involves burrowing into a dead or dying fish and consuming the prey from the inside.

Hagfishes occur almost worldwide in temperate and cold-temperate oceans, usually in water deeper than 30 m. Hagfishes can reach high densities, upwards of 0.5/m^2, on soft-bottom marine areas, which is the most abundant habitat type in the world. Hence, hagfishes could be ecologically important as predators and scavengers. They are also common in the diets of seals and sea lions. Hagfishes are commercially important because their hides are the popular "eelskin" of wallets, purses, and briefcases. Overfishing has depleted hagfish stocks in Korea and Japan, and new fisheries are being exploited, and probably overexploited, in the eastern Pacific and western Atlantic. This unfortunate chain of events has characterized marine fisheries worldwide. Drastic reductions in hagfish populations brought on by overfishing could potentially disrupt a widespread ecosystem; not enough is known about hagfish ecology to predict these impacts.

2. Lampreys

Lampreys also have a notochord rather than vertebrae. Lampreys go through a long-lived larval phase; the free-living, blind, toothless larva lives in silty stream-beds in which it filters microscopic organisms from the water for up to 7 years before transforming into an adult. In brook lampreys, larvae transform into non-feeding adults, live for 6 months, spawn, and die. Other species transform into feeding adults and live for 1–3 years as parasites on other fishes. They rasp holes in their host's skin and live off its body fluids. Accidental introduction of the parasitic marine lamprey into the Great Lakes of North America has contributed to the decline of lake trout, whitefishes, and blue pike.

Lampreys are cool-water species (30° north and south latitude or higher). Most lampreys live in fresh water, but some parasitic species are anadromous. Brook lampreys typically live in headwater streams, an ecosystem type frequently disrupted by human activities. Hence, several U.S. brook lampreys are imperiled. North America's smallest lamprey, the Miller Lake lamprey, was poisoned into extinction because it parasitized introduced trout in its only habitat—Miller Lake, Oregon (Fig. 2).

D. Cartilaginous Fishes

Elasmobranchs (sharks, skates, and rays) include about 800 species that live in all the world's oceans; a few live in fresh water. Elasmobranchs are characterized by a cartilaginous skeleton hardened by calcium deposits and usually five (sometimes six or seven) gill slits. They lack lungs or gas bladders but instead have large, oil-

FIGURE 2   The extinct Miller's Lake lamprey (reproduced with permission from Miller *et al.*, 1989).

filled livers which may aid in buoyancy. Their teeth and pedestal-like placoid scales develop from the same embryonic structures. Teeth are continually lost and replaced, and a shark may produce as many as 30,000 teeth during its lifetime. All elasmobranchs have internal fertilization, and many bear live young that are nourished by the mother via a complex umbilicus and placental structure analogous to that found in mammals. Slow growth, late maturation, and low reproductive output make many elasmobranchs exceptionally vulnerable to human exploitation.

Class Chondrichthyes: cartilaginous fishes
Subclass Elasmobranchii: shark-like fishes
Superorder Euselachii: modern sharks and rays
Order Heterodontiformes: 1 family, 8 species, bullhead and horn sharks
Order Orectolobiformes: 7 families, 31 species, including wobbegongs, nurse, and whale sharks
Order Carcharhiniformes: 8 families, 210 species, including catsharks, requiem, and hammerhead sharks
Order Lamniformes: 7 families, 16 species, including sand tiger, megamouth, thresher, basking, and mackerel sharks
Order Hexanchiformes: 2 families, 5 species, frill and cow sharks
Order Squaliformes: 4 families, 74 species, including sleeper and dogfish sharks
Order Squatiniformes: 1 family, 12 species, angel sharks
Order Pristiophoriformes: 1 family, 5 species, sawsharks
Order Rajiformes: 13 families, 456 species, including sawfishes, electric rays, guitarfishes, skates, stingrays, eagle rays, and manta rays

1. Sharks

Approximately 350 shark-like species are alive today (Fig. 3A). Sharks are generally large (>1 m), predatory fishes. The diverse requiem or ground sharks (carcharhiniforms) include the tiger, gray reef, bull, blue, lemon, and hammerhead sharks. Lamniform mackerel sharks are primarily offshore, pelagic inhabitants. The squaliform dogfishes, the second largest shark order, are most successful and abundant in the North Atlantic, North Pacific, and deep-sea regions.

Sharks range in size from the 15-g, 16- to 20-cm dwarf dogshark to the 12,000+ -kg, 12+ -m long whale shark, the largest fish in the world. White sharks as large as 6 m and 3324 kg are known; larger individuals are suspected. Other large sharks include basking sharks (9.8 m), great hammerheads (5.5 m), Greenland sharks (6.4 m), tiger sharks (5.9 m), and megamouth (5.4 m).

Sharks inhabit all oceans except the Antarctic. Depth records for sharks are held by the Portuguese shark at 3690 m and an unidentified dogfish at 4050 m. A few carcharhinid sharks enter fresh water; bull sharks have been captured 4200 km up the Amazon River and 1200 km up the Mississippi River. Large pelagic sharks may cross entire ocean basins; blue sharks have been tracked across the North Atlantic Ocean and back, a distance of 16,000 km.

Most sharks are predatory on large prey, but three of the largest sharks—the basking, megamouth, and whale sharks—feed on zooplankton. A small, 40-cm-long, midwater species, the cookie-cutter shark, is an ectoparasite on the sides of tunas, dolphins, whales, an occasional megamouth shark, and even rubber sonar domes of nuclear submarines. Some sharks use structures other than jaw teeth to capture prey. Thresher sharks use the long upper lobe of their tails to stun schooling prey. Sawsharks (and rajiform sawfishes) have elongate, blade-like snouts studded with lateral teeth which they slash laterally to disable prey. Hammerheads use their broadened hammer-shaped head to pin stingrays against the bottom before biting chunks out of the rays' wings.

Sharks are sensitive to chemicals, able to detect 1 part fish extract per 10 billion parts seawater. Sharks have good vision, although they tend to be slightly myopic (farsighted). Sharks are also highly sensitive to sounds, including infrasonic sound below 10 Hz, and can localize the direction from which underwater

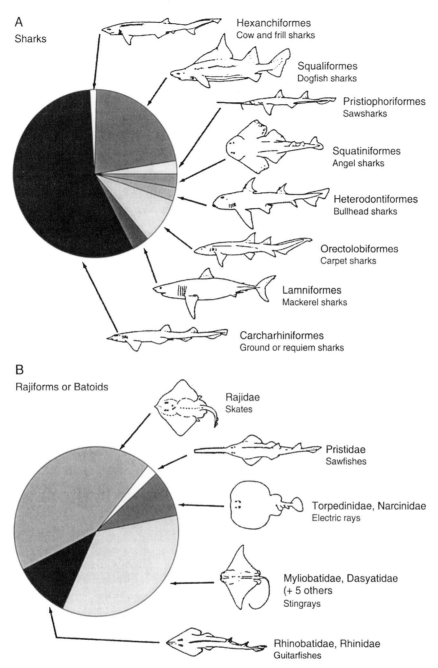

FIGURE 3   Taxonomic distribution and representative orders of (A) sharks and (B) skates and rays (modified with permission from Compagno, 1990).

sounds originate. Sharks can also locate prey by detecting the weak electric fields prey emit. This electrosensitivity may also allow them to navigate using the earth's geomagnetic fields. Many sharks have relatively large brains, with brain to body weight ratios comparable to those of some birds and mammals.

Sharks grow slowly and live long: spiny dogfish live 70–100 years and lemon sharks 50–60 years. Sharks also reproduce slowly. Lemon sharks may not mature until they are 24 years old, and spiny dogfish may not mature until they are 35 years old. Sharks produce relatively few, large young with a long gestation period. Clutch size varies from 1 or 2 live young (sand tigers, threshers, and makos) to 300 in the whale shark. Gestation periods average 9–12 months but may be 2 years in spiny dogfish and 3.5 years in basking sharks. Bullhead and nurse sharks lay eggs, but most sharks bear live young. The requiem sharks have the most advanced developmental pattern, in which an umbilical cord connects mother and embryo, transporting nutrients and oxygen to the embryo and carrying metabolic wastes to the mother.

Because many sharks mature late, reproduce at long intervals, and have low reproductive output, shark populations are easily and frequently overfished. Thresher sharks, school sharks, spiny dogfish, porbeagles, basking sharks, bull sharks, and soupfin sharks are all examples of shark stocks that have been overexploited. White sharks are protected in Australia and South Africa; white sharks and whale sharks are listed as endangered by the International Union for the Conservation of Nature.

2. Skates and Rays

The rajiform skates and rays are 450 species of mostly benthic (bottom-living), mostly marine forms (Fig. 3B). In skates and rays, the pectoral fins are fused to the sides of the head and the five gill slits are under the head. Skates are most diverse in deep water and at high latitudes, whereas stingrays are most diverse in tropical, inshore waters. Some rajiforms live much or all of their lives in fresh water. Largetooth sawfish frequently swim up rivers in Central and South America. Two stingray families contain entirely freshwater species—the river stingrays of South America and several species in the large stingray family Dasyatidae. The latter inhabit African, Southeast Asian, and New Guinea rivers.

Skates and rays feed mostly on benthic invertebrates, except for the huge (up to 6-m wide) manta rays, which capture small crustaceans and fishes in the water column. Torpedo rays stun prey with powerful electrical discharges (50 V and 50 A = 1 kW output). Rajiforms reproduce by either laying eggs (skates) or bearing live young (rays). Embryonic skates develop inside the "mermaid purse" egg cases for as much as 15 weeks.

Skates in some locales are actually increasing in number because of overexploitation of competing bony fishes, such as cod in the North Atlantic. However, the giant barndoor skate of the northwest Atlantic and its relative, the common skate of the northeast Atlantic, are caught incidental to bottom trawling for bony fishes; they have been seriously depleted and may face extinction. Largetooth sawfish in Nicaraguan lakes have been drastically overfished.

3. Chimaeras

Class Chondrichthyes
Subclass Holocephali
Order Chimaeriformes: 3 families, 31 species, chimaeras

Chimaeras, also known as rat- or rabbitfishes, share a cartilaginous skeleton and other features with elasmobranchs. They differ by having (i) the upper jaw permanently attached to the braincase, (ii) continually growing tooth plates in the jaws instead of replaceable teeth, (iii) a single gill flap instead of five or more gill slits, and (iv) no scales. Chimaeras swim by flapping their pectoral fins and by undulating their bodies. All chimaeras are egg-layers, the egg being protected by a horny shell. Adult chimaeras range in size from 60 to 200 cm. Chimaeras are cool-water, marine fishes that live in shallow to moderate depths between 80 and 2600 m, where they usually swim just above the bottom. Chimaeras eat predominantly hard-bodied benthic invertebrates, which they crush with their tooth plates. Surprisingly little is known about their general biology and natural history.

E. Bony Fishes

Modern bony fishes, often referred to as Osteichthyes (literally "bony fishes"), consist of seven major taxonomic groups. The first six, primitive taxa belong to groups that were much more diverse during Paleozoic and Mesozoic eras. Several of these primitive but modern fishes are classified as "bony" even though they have cartilaginous skeletons. Their skeletal condition is actually an advanced, specialized trait; their immediate ancestors were bony.

Grade Teleostomi (more commonly Osteichthyes)
Class Sarcopterygii
Sub- (or infra-) class Dipnoi
Order Ceratodontiformes: 1 family and 1 species, the Australian lungfish
Order Lepidosireniformes: 2 families, 5 species, South American and African lungfishes
Subclass Coelacanthimorpha
Order Coelacanthiformes: 1 family and 2 species, the coelacanths
Class Actinopterygii
Subclass Chondrostei
Order Acipenseriformes: 2 families, 26 species, sturgeons and paddlefishes

Order (or subclass) Polypteriformes: 1 family, 11 species, bichirs and reedfish
Subclass Neopterygii
Order Lepisosteiformes: 1 family, 7 species, gars
Order Amiiformes: 1 family and species, the bowfin

1. Lungfishes

Lungfishes today are represented by three families and six species, all located on the former Gondwanan continents of Australia, South America, and Africa. Lungfishes lack jaw teeth but have unusual toothplates on the mouth roof and floor. The Australian species, *Neoceratodus forsteri*, is limited to four river systems of northeastern Australia (Fig. 4A). It is large (1 m), with large scales, flipper-like fins, a broad tail, and a single lung. It crushes benthic crustaceans, mollusks, and small fishes with its tooth plates. It can, but does not have to, breathe atmospheric oxygen. Its young lack external gills. *Neoceratodus* populations have declined dramatically and the fish is protected; species recovery efforts include transplantation into several Queensland reservoirs and rivers.

The one South American and four African lungfishes have eel-like bodies; slender, almost-filamentous paired fins; lack scales; have paired lungs; have larvae with external gills; and must breathe air to survive. The four African species occur across central and south Africa, often in swampy areas that frequently experience drought. When a swamp dries up, African lungfishes dig a burrow and can wait 4 years for rains to return. The South American species occurs in swampy regions of the Amazon and Parana river basins. Comparatively little is known about its biology.

2. The Coelacanths

Coelacanths were thought to have gone extinct at the end of the Cretaceous, 65 million years ago, until a live one was trawled up in 1938 off South Africa (Fig. 4B). Today, a small, endangered population of 200–600 coelacanths lives at 100- to 500-m depths off two small volcanic islands in the Comore Archipelago, between Madagascar and Mozambique. In 1998, another species was discovered at similar depths in northern Indonesia. The living coelacanths have fleshy pectoral, pelvic, anal, and second dorsal fins (= the lobed fins that define the class Sarcopterygii); a symmetrical, three-lobed tail with a central extension; hollow neural spines (hence "coelacanth" or "hollow spines"); a unique unconstricted notochord; a joint in the dorsal braincase that aids jaw opening; relatively large, thick, bony scales; and live young.

Coelacanths are large (to 180 cm, 95 kg), old (probably 40–50 years), and produce relatively few, live young (5–26 young per clutch). The gestation period is about 13 months. Replacement rate in the population is therefore slow. Coelacanths are captured primarily as bycatch in the hook-and-line fishery for oilfish, and it is unlikely they can sustain even the current by-catch rate of 5–10 animals per year. Counts from small submarines indicate the Comoran species is declining. The Comoran government has outlawed its capture, and trade in coelacanths is outlawed by the Convention on the International Trade in Endangered Species (CITES).

3. Sturgeons and Paddlefishes

The most primitive actinopterygian (ray-finned) fishes are the chondrostean sturgeons and paddlefishes. All 24 species of sturgeons live in the Northern Hemisphere. All spawn in fresh water, although some species move seasonally between marine and fresh water. North American freshwater species include the lake sturgeon and three river sturgeons. Anadromous species include the Atlantic and white sturgeons, the latter being an occupant of west coast bays and rivers. White sturgeons attain the largest size of any North American freshwater fish (3.8 m, 630 kg). The world's largest freshwater fish is the beluga sturgeon of eastern Europe and Asia, *Huso huso*, at 8.6 m and 1300 kg.

Sturgeons have four barbels ahead of a ventrally located mouth, five rows of large bony shields on an otherwise scaleless body, and a heterocercal tail (Fig. 4C). They are exceptionally long-lived (118 years for beluga and 70–80 years for white sturgeon), mature slowly (as late as 30 years old), and spawn infrequently (every 3–5 years). They migrate up rivers to spawn in clean sand and gravel areas; hence, dam building and siltation of rivers both impede their reproduction. A spawning female can be worth thousands of dollars for her caviar alone, and many sturgeon stocks have been reduced 99% from historical levels. The shortnose sturgeon of Atlantic coastal rivers is listed as endangered, and lake sturgeon have been extirpated from a large part of their native range. Three species endemic to the Aral Sea may be extinct due to extensive drying of that once huge body of water, and several species in the former Soviet Union are fished mercilessly.

Two species of paddlefishes occur in large rivers of North America and China. Paddlefishes also have a heterocercal tail, unconstricted notochord, largely cartilaginous skeleton, and scaleless body. The most distinctive feature of the paddlefishes is their paddle or spoonbill, which is flat and rounded in the North American paddlefish and elongate and almost spear-like in the

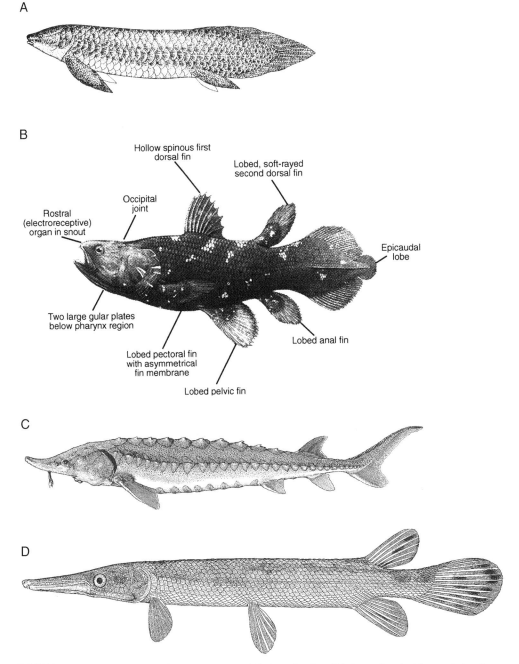

FIGURE 4 Living representatives of primitive groups. (A) Australian lungfish (reproduced with permission from Norman, 1905); (B) the living coelacanth, showing anatomic features (reproduced with permission from Musick *et al.* 1991); (C) Atlantic sturgeon (reproduced with permission from Vladykov and Greeley, 1963); and (D) alligator gar (reproduced with permission from Suttkus, 1963).

Chinese paddlefish. Paddlefishes are water column swimmers that feed on zooplankton and fishes. The paddle of the North American species may help direct food into its mouth, but abundant electroreceptor cells on the paddle suggest an additional, unknown function. North American paddlefish may live 30 years and grow to be 2.2 m in length and 83 kg in weight. Late maturation (at 10 years of age), infrequent spawning (every

2–5 years), lack of clean gravel spawning habitat, and overfishing have all contributed to population reductions and range contraction. The exceedingly rare and poorly studied Chinese paddlefish is larger, reaching more than 3 m in length. This paddlefish may be the most endangered fish in China because of overfishing, habitat destruction, pollution, and dam construction that blocks spawning migrations in the Yangtze River basin where it lives.

4. Bichirs

The polypteriform bichirs and ropefish of west and central tropical Africa are difficult to assign to any particular taxonomic group. Their gilled larvae, lobe-like fins, thick ganoid scales, and modified heterocercal tail suggest affinities with several living groups, particularly the chondrostean sturgeons and paddlefishes. However, their unique dorsal, caudal, and paired fins and unusual chromosomes place them apart from all extant groups. Taxonomists usually place them either in the Chondrostei or in their own subclass, the Brachiopterygii.

Ten species of polypteriforms are called bichirs (*Polypterus*) and an 11th, elongate species is the reedfish or ropefish, *Erpetoichthyes*. Bichirs reach 120 cm in length and ropefish 90 cm. All are predatory and inhabit shallow, swampy regions, in which they breathe atmospheric oxygen with their paired lungs. Bichirs, also called "flagfins," have unique dorsal and pectoral fins. Each dorsal finlet consists of a vertical spine with attached horizontal rays, looking like a flagpole with streaming banners. In more usual ray-finned fishes, dorsal fin rays emerge vertically from the body of the fish. The pectoral fin has an internal arrangement that involves a wishbone-shaped, flattened plate, again unlike that in any other fish. Bichirs are not particularly well studied; no evidence exists to suggest they are imperiled.

5. Gars

The lepisosteiform gars are seven species of elongate, predaceous fishes that occur in eastern North America and Central America (Fig. 4D). They typically inhabit backwater areas of lakes and rivers, such as oxbows and bayous, and breathe atmospheric oxygen using a highly vascularized gas bladder. Gars have bony skeletons, but their vertebral centra are unique, being convex anteriorly and concave posteriorly. In most fishes, the vertebrae are concave on both surfaces. Gars have an abbreviate heterocercal tail and hinged, diamond-shaped, interlocking ganoid (heavy and bony) scales. Gars are the only freshwater fish in North America with poisonous eggs. Alligator gars can be 3 m long and weigh 140 kg. In recent years, alligator gars have come under intense commercial fishing, and concern for their well-being is increasing.

6. The Bowfin

The bowfin, *Amia calva*, is the only living member of its genus, family, and order. The bowfin has the abbreviate heterocercal tail and spiral valve intestine of the gars but also has teleost-like biconcave vertebrae as well as cycloid scales, a relatively light scale type also possessed by many teleosts. The bowfin's head is exceptionally bony and the throat is covered by a distinctive large bone, the gular plate. Bowfin swim slowly forwards or backwards by passing undulations back and forth along their long dorsal fin. Bowfin occur throughout much of the eastern North America in backwater, often swampy areas; they also have a highly vascularized gas bladder which functions as a lung. They are relatively large and robust (to 1 m and 9 kg) and predatory on anything that moves. Bowfin males guard the young vigorously until they are relatively large (10 cm).

7. Teleosts

The division Teleostei ("perfect bone") contains most living fishes. Teleosts are not only taxonomically diverse but also ecologically diverse, occupying every aquatic habitat type and niche imaginable. The 23,600 living teleostean species are placed in 4064 genera, 426 families, and 38 orders. This incredible diversity is generally organized into four taxonomic subdivisions that reflect patterns of evolution that date back to the Mesozoic. These four main subdivisions are the osteoglossomorphs (bony tongues), elopomorphs (eels and tarpon), clupeomorphs (herrings), and the euteleosts, with the latter group containing the vast majority of modern bony fishes.

a. Class Actinopterygii

Subclass Neopterygii
Division Teleostei
Subdivision Osteoglossomorpha
Order Osteoglossiformes: 6 families, 217 species, including bonytongues, mooneyes, African knifefishes, and elephant fishes

Osteoglossiforms derive their name "bonytongue" from the teeth on their tongue that forms part of their bite. These freshwater fishes occur on all major continents except Europe. The arapaima of South America is one of the world's largest freshwater fishes, reaching a length of 2.5 m. The African mormyrid elephant fishes produce and detect weak electric fields, have large cerebellums, and have a brain size : body weight ratio comparable to that of humans (Fig. 5A). Two species, the mooneye and goldeye, occur in major river systems of northern North America.

b. Subdivision Elopomorpha

Order Elopiformes: 2 families, 8 species, ladyfishes and tarpons
Order Albuliformes: 3 families, 29 species, including bonefishes and spiny eels
Order Anguilliformes: 15 families, 738 species, including freshwater, moray, cutthroat, and conger eels
Order Saccopharyngiformes: 26 species, including swallower and gulper eels

Elopomorphs all have ribbon-shaped "leptocephalus" larvae. The Atlantic tarpon is a highly prized gamefish that reaches a length of 2.5 m and a mass of 150 kg (Fig. 5B). Albuliform bonefishes are also popular gamefishes that occupy sandy flats in shallow tropical waters. The 15 families of anguilliform "true" eels are distinguished from the approximately 45 other families of "eel-like" fishes that have independently evolved an elongate body. Anguillid eels are catadromous, spawning at sea but spending most of their lives in fresh water. Muraenid moray eels and their relatives are marine, tropical and warm temperate, predatory species. Synaphobranchid cutthroat eels include an endoparasitic species, the snubnose parasitic eel, which has been found in the heart of a mako shark. The saccopharyngiform deep-sea gulper and swallower eels have giant mouths but lack many head bones, scales, and fins found in most other fishes.

c. Subdivision Clupeomorpha

Order Clupeiformes: 5 families, 357 species, including anchovies and herrings

Clupeomorphs are small, schooling, silvery, pelagic marine and occasionally freshwater feeders on zooplankton and phytoplankton. Herrings, round herrings, shads, alewives, sprats, sardines, pilchards, and menhadens are extremely important commercial species. Anchovies range in size from a 2-cm Brazilian species to a piscivorous, riverine, 37-cm New Guinea anchovy. The largest clupeids are the Indo-Pacific chirocentrid wolf herrings, which reach a length of 1 m and have

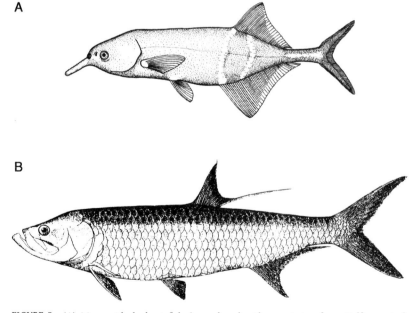

FIGURE 5 (A) Mormyrid elephant fish (reproduced with permission from Helfman *et al.*, 1997); (B) Atlantic tarpon (reproduced with permission from Hildebrand, 1963).

fang-like jaw teeth. Anadromous shads, alewives, and herrings occasionally establish landlocked populations in rivers, lakes, and reservoirs.

d. Subdivision Euteleostei

Approximately 95% of the teleosts are placed in the Euteleostei or "true" teleosts. This subdivision contains 391 families, 3795 genera, and 22,262 species—all in nine superorders.

i. Superorder Ostariophysi

Order Gonorhynchiformes: 4 families, 35 species, including milkfish
Order Cypriniformes: 5 families, 2660 species, including minnows, barbs, algae eaters, suckers, and loaches
Order Characiformes: 10 families, 1340 species, including freshwater hatchet fishes, tetras, and characins
Order Siluriformes: 34 families, 2405 species, including North American freshwater, airbreathing, electric, sea, upside-down, parasitic, callichthyid armored, and suckermouth armored catfishes
Order Gymnotiformes: 6 families, 62 species, including glass, ghost, and naked-back knifefishes and electric eel

Rivers, lakes, and streams worldwide are dominated numerically and ecologically by members of the superorder Ostariophysi. Ostariophysans include milkfish, minnows, carps, barbs, suckers, loaches, piranhas, tetras, catfishes, and electric eels. Two distinctive traits unite this otherwise disparate assemblage: (i) the Weberian apparatus, which is a series of modified anterior vertebrae that link the gas bladder to the inner ear and aid in hearing, and (ii) production and reaction to chemical alarm substances that are released when a fish is injured and lead to a stereotyped escape response in school members.

The gonorhynchiform milkfish, *Chanos chanos*, is an important food fish in the Indo-Pacific region and is often cultured in brackish fishponds. The Cypriniformes make up the largest order in the superorder. The Cyprinidae is the largest family of freshwater fishes and contains 2000 species of minnows, shiners, carps, barbs, barbels, gudgeons, chubs, dace, squawfishes, tench, rudd, bitterlings, bream, southeast Asian "sharks" (redtail black shark and bala shark), goldfish, koi (domesticated common carp), danios, and rasboras. Cyprinids are most diverse in Southeast Asia, followed by Africa, North America (270 species), and Europe, but are missing from South America and Australia. The

world's smallest freshwater fish is a Burmese cyprinid, *Danionella translucida,* which matures at 10 mm. The largest minnow in North America is the endangered, piscivorous Colorado squawfish, *Ptychocheilus lucius.*

Gyrinocheilid algae eaters scrape algae off rocks in swift, flowing waters. The catostomid suckers (i.e., buffaloes, quillback, carpsuckers, blue sucker, redhorses, jumprocks, and the extinct harelip sucker) include 70 species of North American fishes, with 1 species in eastern China. Loaches (Cobitidae) are 110 species of predominantly Eurasian stream fishes, including popular aquarium fishes such as the kuhli, clown, and skunk loaches, the weatherfishes, and the golden dojo. Weatherfishes (*Misgurnus*) become restless when barometric pressure decreases preceding a storm.

The characiforms are a speciose group of primarily tropical ostariophysans characterized (usually) by a rayless adipose fin and mouths armed with replacement dentition (e.g., piranhas). Body size ranges from very small (13 mm) tetras to large (1.5-m-long) tiger-fishes. Numerous aquarium fishes are included (headstanders, freshwater hatchet fishes, blind characins, pencil fishes, tetras, and silver dollars), as well as important food fishes (*Prochilodus, Colossoma,* and *Brycon*). Most characins (ca. 1150 species) are South American, about 200 are African, a few live in Central America, and 1 species, the Mexican tetra *Astyanax mexicanus,* extends naturally into southwestern Texas.

Catfishes (Siluriformes) are surprisingly diverse, with 34 families and more than 2400 species (Fig. 6A). Catfishes usually have barbels ("whiskers") and sometimes toxic spiny fins, and they are almost entirely freshwater, nocturnal, and benthic. Catfishes are most diverse in South America (e.g., loricariid suckermouth armored catfishes, 550 species; pimelodid long-whiskered catfishes, 300 species). Large species include the European wels (5 m, 330 kg), the Asian Mekong catfish (3 m, 300 kg), and a 3-m-long whiskered pimelodid of South America. The largest catfishes in North America are the flathead and blue catfishes at about 1.5 m and 50–68 kg. Some small catfishes are notable, such as the parasitic catfishes (Trichomycteridae) of South America, which normally parasitize the gills of fishes but are known to swim up the urethra of bathers and lodge there, necessitating surgical removal.

The gymnotiform South American knifefishes are unusual ostariophysans that produce and receive weak electric impulses. They have elongate, compressed bodies; an extremely long anal fin; and electrogenic tissue usually derived from modified muscle cells. Their electrical output is constant at high frequencies, whereas the osteoglossomorph mormyrids produce a pulsed,

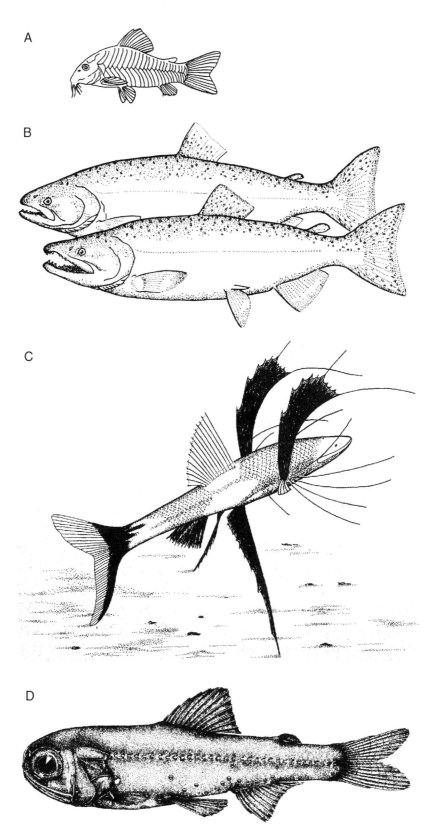

FIGURE 6   (A) Callichthyid armored catfish (reproduced with permission from Burgess, 1989);
(B) chinook salmon (female above and male below) (reproduced with permission from Scott
and Crossman, 1973); (C) tripodfish (reproduced with permission from Helfman *et al.*, 1997);
(D) lantern fish (reproduced with permission from Nafpaktitis *et al.*, 1977).

low-frequency output. Both groups detect objects that disrupt their electric fields. The 2-m-long electric eel (*Electrophorus*) produces a weak field for electrolocation and strong pulses for stunning prey or deterring predators.

ii. *Superorder Protacanthopterygii*

Order Esociformes: 2 families, 10 species, pikes and mudminnows
Order Osmeriformes: 13 families, 236 species, including barreleyes, smelts, salamander fish, and galaxiids
Order Salmoniformes: 1 family, 66 species, whitefishes, graylings, chars, trouts, and salmons

Protacanthopterygians are a mixed agglomeration of marine, freshwater, and diadromous fishes. Esociform pikes, pickerels, and mudminnows are Northern Hemisphere predators; the northern pike has the largest geographical distribution of any Northern Hemisphere fish, occurring across the northern portions of North America, Europe, and Asia. Mudminnows can survive winters in high-latitude lakes by breathing from air bubbles trapped under the ice. Osmeriforms are small, silvery, elongate, water-column dwelling fishes. Osmerids include commercially important species such as capelins, eulachons, Asian ayu, and smelts. The order also includes the Southern Hemisphere lepidogalaxiid salamander fish and the galaxiids. Salamander fish inhabit seasonal ponds of southwestern Australia, burying in drying mud and reemerging with the next rains. Galaxiids have suffered numerous extirpations and extinctions as a result of the stocking of nonnative trouts.

Salmoniforms are important commercially, ecologically, and aesthetically. Whitefishes and ciscoes are relatively large-scaled, zooplanktivorous salmonids of high-latitude North American and Eurasian lakes. Several North American species have been decimated due to introduced predators, competitors, and parasitic lampreys. Graylings are riverine fishes with a flowing dorsal fin. At least one species, the Michigan grayling, is extinct. The subfamily Salmoninae contains seven Eurasian and North American genera. The Siberian taimen, *Hucho taimen,* is the world's largest salmonid at 2 m and 70 kg. North American Salmoninae include the chars (lake, brook, and bull trout, Arctic char, and dolly varden). Arctic char live farther north than any other freshwater fish. The remaining salmonines are the Atlantic basin salmon and trout (e.g., Atlantic salmon and European brown trout), and the 11 species of Pacific basin trouts and salmons in the genus *Oncorhynchus,* 2 of which are endemic to Japan (Fig. 6B). Pacific trouts

and salmons include golden, cutthroat, and gila trouts and the spectacularly anadromous coho, chinook, chum, pink, and sockeye salmons, some of which undergo oceanic migrations of thousands of kilometers before returning to their birth river to spawn and die. The actual number of genetically distinct races of Pacific salmons is unknown because many stocks are reproductively isolated in small river systems. Evidence suggests that as many as 1000 stocks exist, 106 of which have gone extinct and an additional 314 of which are imperiled.

iii. *Superorder Stenopterygii*

Order Stomiiformes: 4 families, 321 species, including bristlemouths, marine hatchet fishes, and barbeled dragonfishes
Order Ateleopodiformes: 1 family, 12 species, jellynose fishes

Stenopterygians are deep-sea fishes, often with long teeth and large mouths. Gonostomatid bristlemouths may be the most abundant and widely distributed vertebrates on Earth. Idiacanthine black dragonfishes have a larva with eyes at the ends of elongate stalks.

iv. *Superorder Cyclosquamata*

Order Aulopiformes: 13 families, 219 species, including telescope fishes, tripod fishes, lizard fishes, and lancet fishes

Cyclosquamates are also deep-sea forms, including the bizarre giganturid telescope fishes with large tubular eyes, a huge mouth, flexible teeth, and an expandable stomach. Deep-sea tripod fishes have long pectoral, pelvic, and caudal rays that they use for resting on soft sediments of the deep ocean floor (Fig. 6C). Shallow representatives are the synodontid lizard fishes, which are common benthic predators on coral reefs worldwide. Alepisaurid lancet fishes are large (to 2 m) mesopelagic predators with a sail-like dorsal fin of unknown function.

v. *Superorder Scopelomorpha*

Order Myctophiformes: 2 families, 240 species, including lantern fishes

Scopelomorphs include the abundant, commercially important lantern fishes, which are identified based on species-specific photophore (light organ) patterns. Lantern fishes occur at middle depths from the Arctic to the Antarctic. They are important in the diets of many fishes as well as of marine mammals (Fig. 6D).

vi. Superorder Lampridiomorpha

Order Lampridiformes: 7 families, 19 species, including opahs, tube-eye, ribbonfishes, and oarfishes

Lampridiforms are generally open-water, oceanic fishes. Opahs are relatively large (1.8 m, 70 kg), oval-shaped, pelagic predators on squids and other fishes. The 30-cm-long tube-eye (*Stylephorus*) can increase the volume of its mouth 40-fold during feeding—a record among vertebrates. The elongate oarfish, *Regalecus*, may attain 12-m length and is the longest living teleost. It has a bluish-silvery body, scarlet head crest, and deep red fins. It is thought to be responsible for many "sea serpent" sightings.

vii. Superorder Polymixiomorpha

Order Polymixiiformes: 1 family, 5 species, beardfishes

viii. Superorder Paracanthopterygii

Order Percopsiformes: 3 families, 9 species, trout-perches, pirate perch, and cavefishes
Order Ophidiiformes: 5 families, 355 species, including pearlfishes, cusk-eels, and viviparous brotulas
Order Gadiformes: 12 families, 482 species, including rattails, hakes, and cods
Order Batrachoidiformes: 1 family, 69 species, toad-fishes
Order Lophiiformes: 16 families, 297 species, including goosefishes, frogfishes, handfishes, batfishes, and deep-sea anglerfishes

Paracanthopterygians are primarily benthic, marine, nocturnally active fishes; many live in the deep sea or in caves. Percopsiforms are small (<20 cm), freshwater fishes, most of which live in eastern North America. The anus of the swamp-dwelling aphrododerid pirate perch is located in the throat region of adults for functionally mysterious reasons. Amblyopsid cavefishes are often blind and scaleless forms highly adapted for cave life.

Ophidiiforms often live in holes or even inside other animals. Carapid pearlfishes live inside the body cavities of starfishes, sea cucumbers, clams, and sea squirts; some feed on the internal organs of their hosts. Ophidiid

and bythitid cusk eels and brotulas include blind cave species in freshwater systems of Caribbean and Galapagos Islands as well as coral reef species that hide deep within crevices. The neobythitine cusk eel, *Abyssobrotula galatheae*, holds the depth record for a fish at 8370 m in the Puerto Rico Trench.

The gadiforms include the cods, haddocks, hakes, pollocks, and whitings, which are some of the world's most important commercial fishes. True cods (Gadinae) have three dorsal fins and two anal fins. Many species have chin barbels. The burbot, *Lota lota*, of high-latitude, Northern Hemisphere lakes is the only freshwater species in the group. The commercially important Atlantic cod is the largest species (1.8 m, 90 kg), but fish more than 10 kg are rare due to drastic overfishing (Fig. 7A). One of the world's largest food fisheries is for North Pacific walleye pollock.

Batrachoidiforms include the midshipmen, which have hundreds of photophores, an unusual trait for a shallow dweller. Many batrachoids produce sounds by vibrating their gas bladders. Venomous toadfishes have dorsal and opercular spines which can inject a powerful toxin. Three South American toadfishes are restricted to fresh water.

Lophiiforms are a diverse and often bizarre-looking order of marine fishes that include benthic, shallow-water forms as well as highly modified, open-water, deep-sea forms. Many use a modified first dorsal spine as a lure for catching smaller fish. The meter-long western North Atlantic goosefish, *Lophius americanus*, has a huge mouth with long, recurved teeth which it uses to catch fishes and even diving seabirds. Antennariid frogfishes also rest on the bottom or walk over it with their pectoral and pelvic fins (Fig. 7B). The ogcocephalid batfishes walk on their pectorals, but they can also swim via jet propulsion by shooting water out their round, backward-facing opercular openings. The ceratioid anglerfishes include 11 families of strange-appearing bathypelagic predators, many of which have very small males that fuse to and become parasitic on the larger females. The endemic Australian handfishes include a Tasmanian species, the spotted handfish, that was once common but is now critically endangered due possibly to egg predation by an introduced starfish.

e. Superorder Acanthopterygii

Most bony fishes belong to a single superorder, the Acanthopterygii, which contains about 13,500 species in 251 families. Two small and one large taxonomic groupings, called series, are recognized, with the vast majority in the third series, the Percomorpha.

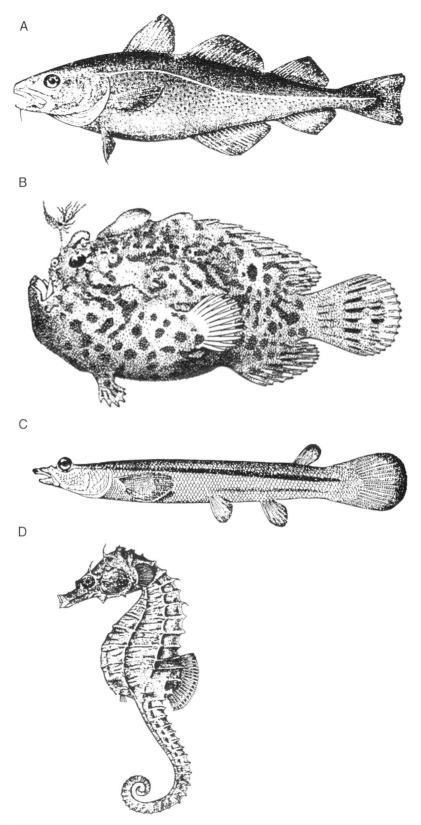

FIGURE 7 (A) Atlantic cod (reproduced with permission from Norman, 1905); (B) bloody frogfish (reproduced with permission from Heller and Snodgrass, 1903); (C) atherinomorph four-eyed fish (reproduced with permission from Norman, 1905); (D) lined seahorse (reproduced with permission from Norman, 1905).

i. Series Mugilomorpha

Order Mugiliformes: 1 family, 66 species, mullets

The mullets are a family of nearshore, marine and freshwater fishes of considerable economic importance. Many mullets feed on organic silt and minute plants, an unusual food type among fishes.

ii. Series Atherinomorpha

Order Atheriniformes: 8 families, 285 species, including rainbow fishes and silversides

Order Beloniformes: 5 families, 191 species, including needlefishes, flying fishes, and halfbeaks

Order Cyprinodontiformes: 8 families, 807 species, including topminnows, killifishes, livebearers, and pupfishes

Atherinomorphs are shallow-water, marine or freshwater fishes that live near the surface. Many atherinomorphs bear live young. Atheriniforms include the melanotaeniid rainbow fishes of Australia and New Guinea, in which males have brighter colors and longer fins than females, traits that make them popular aquarium species. Silversides are widespread, freshwater and marine schooling fishes and include the grunions of southern and Baja California, which ride waves up beaches on dark nights to spawn in wet sand biweekly during the summer. Beloniforms are predominantly silvery, marine fishes active at and sometimes above the surface of the water. The lower lobe of the tail in flying fishes is relatively long and is used to scull rapidly during takeoff. Many cyprinodontiforms, although basically freshwater fishes, tolerate considerable salinity and hence occur in streams on isolated oceanic islands (Fig. 7C). The rivulines of South America and Africa live only 1 year, laying eggs which survive in the dried bottoms of pools and which hatch with the next season's rains. *Rivulus marmoratus* of south Florida and the West Indies is the only truly hermaphroditic fish, fertilizing its own eggs. The livebearers include the mollies, platys, guppies, and swordtails of the aquarium trade. Some species are all female. Many cyprinodontid pupfishes are tolerant of extreme water conditions and consequently can live in saltmarsh and desert conditions. However, they cannot tolerate total desiccation, which has endangered many desert species that have to compete with humans for water. The Devil's Hole pupfish has the smallest known range of any fish species—one shallow shelf in a single spring in Death Valley, California. Other pupfish relatives inhabit Lake Titicaca, which at an elevation of 4570 m in the Andes Mountains is the highest natural lake with fishes.

iii. Series Percomorpha

Order Stephanoberyciformes: 9 families, 86 species, including whale fishes

Order Beryciformes: 7 families, 123 species, including flashlight fishes, roughies, and squirrelfishes

Order Zeiformes: 6 families, 39 species, including dories and boarfishes

Order Gasterosteiformes: 11 families, 257 species, including sticklebacks, pipefishes, seahorses, trumpet fishes, and shrimp fishes

Order Synbranchiformes: 3 families, 87 species, including swamp and tiretrack eels

Order Scorpaeniformes; 25 families, 1271 species, including scorpion fishes, rockfishes, sea robins, sablefishes, greenlings, sculpins, Baikal oilfishes, and lumpfishes

Order Perciformes (9300 species)

Suborder Percoidea: 71 families, 2860 species, including snooks, temperate basses, sea basses, centrarchid sunfishes, black basses, darters, perches, cardinal fishes, bluefishes, remoras, dolphin fishes, jacks, pompanos, snappers, grunts, croakers, drums, goatfishes, archerfishes, butterfly fishes, and angelfishes

Suborder Elassomatoidei: 1 family, 6 species, pygmy sunfishes

Suborder Labroidei : 6 families, 2200 species, including cichlids, surfperches, damselfishes, wrasses, and parrotfishes

Suborder Zoarcoidei: 9 families, 318 species, including eel pouts, gunnels, and wolffishes

Suborder Notothenioidei: 5 families, 122 species, including icefishes and Antarctic dragonfishes

Suborder Trachinoidei: 13 families, 212 species, including sand lances, weeverfishes, and stargazers

Suborder Blennioidei: 6 families, 732 species, including clinids and blennies

Suborder Icosteoidei: 1 family and species, ragfish

Suborder Gobiesocoidei: 1 family, 120 species, clingfishes

Suborder Callionymoidei: 2 families, 137 species, dragonets

Suborder Gobioidei: 8 families, 2120 species, including sleepers and gobies

Suborder Kurtoidei: 1 family, 2 species, nurseryfishes

Suborder Acanthuroidei: 6 families, 125 species, including spadefishes, scats, rabbitfishes, moorish idol, and surgeonfishes

Suborder Scombrolabracoidei: 1 family and species, the scombrolabracid

Suborder Scombroidei: 6 families, 136 species, including barracudas, mackerels, tunas, swordfish, and billfishes

Suborder Stromateoidei: 6 families, 65 species, including driftfishes and butterfishes

Suborder Anabantoidei: 5 families, 81 species, gouramis

Suborder Channoidei: 1 family, 21 species, snakeheads

Order Pleuronectiformes: 11 families, 570 species, including flounders and soles

Order Tetraodontiformes: 9 families, 339 species, including triggerfishes, boxfishes, trunkfishes, cowfishes, puffers, porcupine fishes, burrfishes, and ocean sunfishes

The percomorphs constitute by far the largest taxonomic group of fishes, far too many to deal with in any detail. What follows is a very brief overview of some of the more interesting orders, suborders, and families.

Beryciforms are shallow- to moderate-depth, often red, almost always nocturnal fishes, including the reef-dwelling squirrelfishes. Also included is the commercially important orange roughy, *Hoplostethus atlanticus,* of high-latitude, southern ocean regions. Orange roughies are being overexploited because they are slow growing and long-lived, taking more than 20 years to mature and reaching ages more than 100 years. Zeiforms include commercial species such as the European John Dory, *Zeus faber.*

Gasterosteiforms are small marine, estuarine, and freshwater fishes with dermal armor plating. Sticklebacks are well-studied fishes that frequently form distinct, isolated populations characterized by unusual spines, plates, and behavior. The suborder Syngnathoidei includes the bizarre pegasid seamoths and syngnathid pipefishes, sea dragons, and seahorses. Syngnathid pipefishes and seahorses are the only vertebrates in which the female "impregnates" the male by laying eggs in his brood pouch, which he then fertilizes and raises until hatching. Seahorses are heavily overfished for medicinal uses and the aquarium trade (Fig. 7D).

Synbranchiforms are primarily freshwater, eel-like, often air-breathing fishes. Swamp eels have recently been introduced into the southeastern United States and are of major concern as potential invading predators in river systems.

Scorpaeniforms are predominantly marine fishes, with spiny heads and sometimes venomous fin spines (i.e., stonefishes, scorpion fishes, and lionfishes). The sebastine rockfishes are a diverse, commercially impor-

tant, long-lived, and overfished group of the temperate North Pacific. Hexagrammid greenlings are littoral zone and kelp-associated fishes endemic to the North Pacific, including the highly edible lingcod, *Ophiodon elongatus.* The suborder Cottoidei includes many freshwater species, including the cottid sculpins of North American streams and the highly divergent (pelagic and livebearing) comephorid oilfishes of Lake Baikal in northern Asia. The cabezon of the Pacific coast of North America is unusual in having toxic eggs, whereas the lumpfish of the North Atlantic produces valuable caviar and has consequently been overfished.

The Perciformes are the largest order of percomorphs, with 148 families and 9300 species, including most marine and freshwater fishes of littoral (inshore and shallow-water) zones. Perciforms reach their greatest diversity on coral reefs, but they are also highly diverse in rivers, streams, and lakes. Coral reef perciforms include six of the eight largest fish families (gobies, wrasses, sea basses, blennies, damselfishes, and cardinal fishes). Two other large families, cichlids and croakers, are characteristic of tropical lakes and nearshore temperate marine habitats, respectively.

The largest perciform suborder is the Percoidei. Many percoids are bass-like fishes. Centropomids are large predatory fishes and include the snooks of tropical America, the barramundi of Australia, and the Nile perch of Africa (Fig. 8A). Nile perch are an introduced predator in Lake Victoria, in which they are thought to have extinguished perhaps hundreds of endemic cichlids. Moronid temperate basses include the striped and white bass of North America. The sea bass family Serranidae is one of the largest families (450 species) that range in size from small (<5 cm) anthiines and hamlets to giant groupers and jewfish (3 m long, 400 kg). Sea basses also include commercially important hinds, coneys, gag, and scamp (Fig. 8B). Many serranids are hermaphroditic, usually starting as female and then later becoming male, although some hamlets are both male and female simultaneously.

The Centrarchidae contains the sunfishes, crappies, rockbasses, and black basses of North American fresh waters (Fig. 8C). Largemouth bass have been introduced extensively as a sport fish. At least 162 species make up the family Percidae, 150 of which occur in North America. Percids include the yellow and Eurasian perch, walleye and sauger (= pikeperches), and about 150 species of small, stream-dwelling, spectacularly colored, and often imperiled darters.

Apogonid cardinal fishes are diverse (320 species), small (<10 cm), nocturnal coral reef fishes. Cardinal fishes mouthbrood their eggs, an unusual trait among

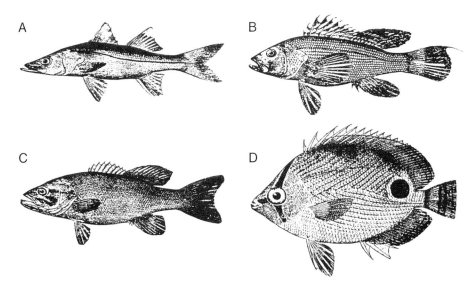

FIGURE 8 Representative percoids: (A) a centropomid snook; (B) a serranid black sea bass; (C) a centrarchid smallmouth black bass; (D) a chaetodontid four-eye butterfly fish (A, C, and D, reproduced with permission from Norman, 1905); B, reproduced with permission from Bigelow and Schroeder, 1953).

reef species. The voracious, schooling pomatomid blue-fish, *Pomatomus saltatrix,* occurs in most temperate and semitropical oceans except the eastern Pacific. In echeneid remoras or sharksuckers, the modified first dorsal fin forms a suction disk that is used to cling to various hosts. The coryphaenid dolphin fishes or mahi mahis are two species of open-water, pelagic predators that often associate with floating structure. They are also one of the few marine pelagic fishes that is successfully aquacultured. Carangid scads, jacks, pompanos, and amberjacks are a large family (140 species) of tropical nearshore predators.

Lutjanid snappers include 125 species of generally carnivorous marine fishes. Snappers are usually near-bottom dwellers (i.e., gray snapper, red snapper, and mangrove snapper), but some are water-column zooplanktivores (vermilion snapper).

Haemulidae grunts are moderate-sized coral reef fishes and are most diverse in the New World tropics. They are seen most often in their daytime resting schools around coral heads; at night they disperse to feed in surrounding reef and grass areas. Sciaenid croakers (270 species) have chin barbels and a muscularized gas bladder used for sound production. Sciaenids are a widespread family that is particularly diverse in the southeastern United States, but representatives occur widely in tropical marine and freshwater habitats. Species include red drum (spot tail bass), black drum, croakers, weakfish, sea trouts, kingfishes, white sea

bass, corbinas, and the endangered Mexican totoaba, one of the few unquestionably imperiled marine fishes. The range of the freshwater drum, *Aplodinotus grunniens,* includes much of eastern North America into Central America. Another bottom-oriented family is the tropical reef-dwelling mullid goatfishes, which have movable, muscularized chin barbels. Toxotid archerfishes are Indo-Pacific, brackish-water fishes well-known for their ability to shoot droplets of water that knock insects out of overhanging vegetation.

Two closely related, colorful reef families are the chaetodontid butterfly fishes (114 species) and the pomacanthid angelfishes (74 species). Both families are most diverse in the Indo-Pacific region. Butterfly fishes feed on coral polyps, small invertebrates, tube worms, or zooplankton, whereas angelfishes eat attached invertebrates such as sponges, tunicates, and anthozoans (Fig. 8D).

Elassomatoid pygmy sunfishes are a suborder of miniature (20- to 45-mm), colorful freshwater swamp dwellers of the southeastern United States.

Several very speciose families belong to the suborder Labroidei. Tropical marine families include the pomacentrid damselfishes (315 species), which are small, colorful, usually herbivorous, territorial reef dwellers. Some are zooplanktivores (*Chromis* and anemonefishes). Several occupy temperate regions (e.g., the garibaldi of California). The largest labroid family is the mostly tropical, reef-dwelling wrasses, Labridae

(500 species). Wrasses range in size from the 5-cm-long cleaner wrasses, slippery dicks, and blueheads to the 2.3-m-long Maori wrasse of the Indo-Pacific, which is hunted unmercifully for the live-fish restaurant trade. Cool temperate species include the California sheephead (Fig. 9A) and senorita, western Atlantic tautog and cunner, and the eastern Atlantic cuckoo wrasse. Many wrasses change sex from female to a more colorful male. The scarid parrotfishes (83 species) are almost exclusively coral reef dwellers, best known for their fused parrot-like teeth that are used for biting off algal and coral pieces, which are then crushed in the massive pharyngeal (throat) jaws. The embiotocid surfperches are 24 species of deep-bodied, temperate (mostly eastern Pacific) fishes associated with kelp beds and rocky reefs. They are live-bearers, feeding on zooplankton or small invertebrates. The largest labroid family is the freshwater cichlids, with more than 1300 species. Cichlids are chiefly tropical, South American and African fishes, with a few species that occur further north (the Rio Grande cichlid is found in south Texas). Central and South American species include freshwater angelfishes, discus, oscars, convict cichlids, and peacock bass (Fig. 9B). Most cichlids occur in Africa, where they are particularly speciose in the African Great Lakes and are threatened by introduced predators such as Nile perch. African tilapias and other cichlids have been deliberately or accidentally introduced into Florida, California, and Hawaii as a by-product of aquaculture activities.

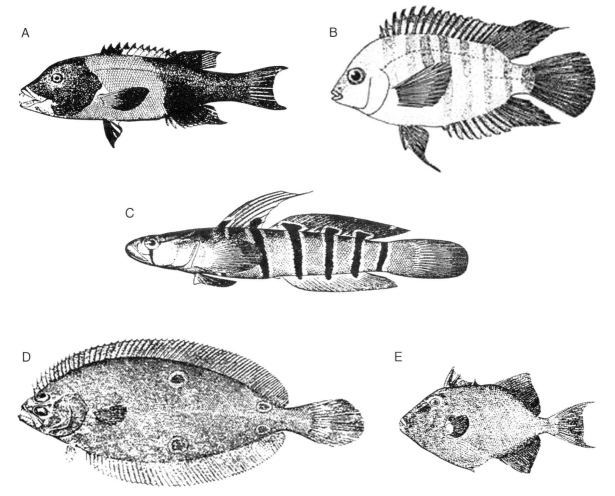

FIGURE 9   (A) California sheephead (reproduced with permission from Norman, 1905); (B) blue-eye cichlid (reproduced with permission from Greenfield and Thomerson, 1997); (C) Japanese goby (reproduced with permission from Norman, 1905); (D) four-spot flounder (reproduced with permission from Bigelow and Schroeder, 1953); (E) gray triggerfish (reproduced with permission from Bigelow and Schroeder, 1953).

Fishes of the suborder Zoarcoidei are all eel-like, bottom-living, marine, cool- to cold-water species. They range in size from the small, intertidal pricklebacks and gunnels to the live-bearing eel pouts, some of which live 3000 m below the surface. The large (to 2.5 m long) anarhicadid wolffishes and wolf eels of shallow North Pacific and Atlantic waters are anatomically and ecologically similar to moray eels.

Icefishes (suborder Notothenioidei) are mostly Antarctic, mostly benthic fishes that live under the ice and have antifreeze compounds in their blood. The crocodile icefishes lack red blood cells, hemoglobin, and myoglobin, and hence have colorless blood and flesh.

Trachinoids are marine, generally benthic fishes with a tendency to bury themselves in sand. Ammodytid sand lances are small, elongate, and abundant zooplankton feeders that spend their nights buried. Trachinid weeverfishes occur in the eastern Atlantic and Mediterranean and have venomous opercular and dorsal spines. Uranoscopid stargazers emit strong pulses of electricity (up to 50 V) from highly modified eye muscles.

Blennioids are small marine fishes that usually associate with structure. Chaenopsid pikeblennies and tubeblennies often live inside corals and worm tubes. Combtooth blennies are diverse (345 species) small fishes in tropical and subtropical waters; they scrape algae with their comb-like teeth.

The suborder Icosteoidei consists of the peculiar North Pacific ragfish, *Icosteus aenigmaticus*, which has a largely uncalcified, cartilaginous skeleton and is a preferred food of sperm whales. The small, marine and freshwater gobiesocoid clingfishes are shallow-water and even amphibious fishes often found in high-energy wave zones. Their pelvic fins are modified into a sucking disc.

Gobioids are usually small, benthic, often abundant fishes. The eleotrid sleepers are small to medium (to 60 cm) estuarine and stream fishes in tropical and subtropical areas, often on islands. The gobiid gobies are incredibly diverse (>1875 species). Many gobies have a suction disk formed by fused pelvic fins (Fig. 9C). The family includes the amphibious mudskippers. Gobies range in size from tiny pygmy gobies (i.e., 8- to 10-mm adults) to a comparative giant, the western Atlantic violet goby, which is a purplish eel-like fish 50 cm long. Round gobies have recently been introduced into North American Great Lakes from southern Europe and are spreading rapidly; other introduced goby species are now extremely common in San Francisco Bay.

The suborder Acanthuroidei includes the ephippid spadefishes of Atlantic reefs and beaches as well as the Indo-Pacific scatophagid scats, which get their name from feeding on the feces of other animals. Rabbitfishes are Indo-Pacific reef, grassbed, and estuarine herbivores that are convergent in many ways with some butterfly fishes, a description that also applies to the moorish idol (Zanclidae). The 72 species of usually herbivorous acanthurid surgeonfishes, unicornfishes, and tangs have a knifeblade on the caudal peduncle.

The suborder Scombroidei contains some of the largest and most spectacular marine fishes. Twenty species of barracudas inhabit tropical and subtropical oceans almost worldwide. The gempylid snake mackerels (23 species) are pelagic and deep-water predators, including the cosmopolitan oilfish, *Ruvettus pretiosus*, a large (1.8 m, 45 kg) predator of moderate depths. An active fishery for oilfish in the Comoro Islands captures endangered coelacanths as by-catch. The scombrid mackerels and tunas are quintessential open-sea predators, with streamlined bodies and a physiology geared to a high-speed lifestyle. They range from relatively small, 50-cm mackerels to giant bluefin tuna (4 m, 500 kg). Most are schooling fishes of tremendous commercial importance. The temperate and warm-temperate xiphiid swordfish and the more tropical istiophorid sailfishes, spearfishes, and marlins have an elongate upper jaw bone that forms the bill. It is used as a spear, a cutlass, or a billy. Swordfish grow to 530 kg, whereas blue and black marlin grow to 900 kg. Swordfish have been heavily overfished, particularly in the Atlantic.

Labyrinth fishes (Suborder Anabantoidei) have an auxiliary breathing structure in the gill chamber for aerial respiration. Anabantid climbing gouramis are African and Asian freshwater fishes that can move across wet ground and reportedly up wet tree trunks. The kissing gourami is the sole member of the family Helostomatidae. The belontiid gouramis, fighting fishes (bettas), and paradise fishes have elongate pelvic fin rays that serve as feelers. Bettas (Siamese fighting fish) have been bred to battle like fighting cocks, placing them among the few fishes that have been cultured for purposes other than food, appearance, or research.

Pleuronectiform flatfishes are distinctive, compressed, benthic fishes that have both eyes on the same side of the head (Fig. 9D). Many flatfishes are important commercially (e.g., dab, flounders, halibuts, plaice, sole, tonguefishes, turbots, and whiffs). Paralichthyids include the summer flounder and California halibut, the latter reaching 1.5 m and 30 kg. The pleuronectid right-eye flounders include the Atlantic and Pacific halibuts. Pacific halibuts may live 40 years and attain lengths of 3 m and masses of 200 kg. The fishery for Pacific halibut in the North Pacific is a well-regulated, sustainable enterprise.

The most advanced bony fishes are in the order Tetraodontiformes, an almost entirely marine order of medium-sized fishes with thick, leathery skin and with scales often modified into spines or bony plates. In balistoid triggerfishes and filefishes, the long, rigid first dorsal spine is locked erect by an interaction with the shorter, second spine (Fig. 9E). Ostraciid boxfishes are encased in a triangular or rectangular bony box, with just the fins and caudal peduncle emerging. Puffers and ocean sunfishes lack true teeth. Instead, the jaw bone has a cutting edge that looks like separated teeth or is fused into a parrot-like beak. Diodontid porcupine fishes inflate their body by filling the stomach with water, a process that also helps erect and interlock their body spines. Tetraodontid puffers concentrate a powerful and potentially fatal toxin, tetraodotoxin, in their viscera, which adds to the allure of eating puffers in licensed *Fugu* restaurants in Japan. The ocean sunfish, *Mola mola,* is one of the world's heaviest fishes at 1000–2,000 kg, producing as many as 300 million eggs. All this biomass is supported on a diet of jellyfishes.

II. GEOGRAPHIC DIVERSITY

A. Overview

Fishes occur just about everywhere water occurs as long as water is in its liquid state, is available through most of the year, and remains below 40°C. A major zoogeographic distinction can be made between marine and freshwater fishes, with substantial overlap occurring where intermediate salinities occur. Many fishes are restricted to pure fresh water (little or no salinity), many are restricted to normal oceanic salinity (about 35 parts per 1000 salt in water), some occur in both habitats at different times of their lives or of the year, and some occur and are even restricted to areas of intermediate salinity, such as estuaries.

In terms of numbers, about 58% of all fishes are marine and 41% live in fresh water, with the remaining 1% moving regularly between the two salinity designations (Fig. 10 and Box 2). Among the 10,250 freshwater species, 80% are primary or obligatory freshwater fishes and are intolerant of even moderate salinities. The remaining 20% can tolerate some salinity and hence inhabit upper estuarine areas or can cross through nearshore ocean regions to move from one river basin to another. Among the 14,500 marine fishes, the vast majority (69%) live in shallow, warm areas such as coral reefs. The remaining marine species are divided fairly evenly among shallow, cold, deep, open-ocean and deep-bottom areas (about 10% each). About 2% of marine fishes live in near-surface, open-sea (pelagic) habitats. The approximately 160 diadromous species that live in different salinity regions at different times of their lives are divided among three groups. Anadromous fishes (54%) live most of their lives in the ocean but then migrate to fresh water to spawn; this group includes lampreys, sturgeons, herrings, and salmons. Catadro-

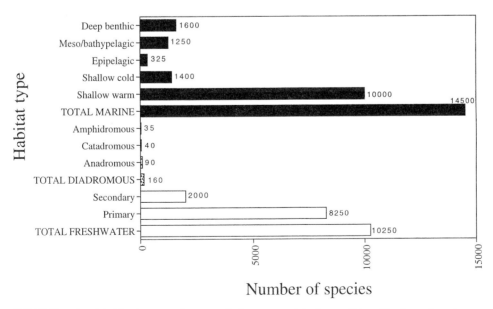

FIGURE 10 The global biodiversity of marine, diadromous, and freshwater fishes. Numbers of species are given for each habitat type and its subdivisions.

Why Are There so Many Freshwater Fishes?

The high global diversity of freshwater fishes (Fig. 10) is at first surprising. Fresh waters comprise only about 0.009% of the earth's water, which means that almost half of all fish species live in less than 1% of the world's water. This 7500-fold discrepancy in biodiversity per unit volume is probably best explained by the relative productivity and isolation of freshwater bodies. Most freshwater habitats are relatively shallow and receive ample sunlight and nutrients running off from adjacent land. Hence, freshwater habitats are relatively productive and capable of sustaining abundant life. Most of the ocean, in contrast, is deep, dark, and nutrient poor. Given that 81% of marine diversity occurs in shallow regions, a relationship between water depth and diversity in fresh waters is not surprising. In addition to the influence of available food is the comparative isolation of most freshwater habitats. Lakes are often created and affected by climatic and geologic forces (e.g., drought, floods, landslides, earthquakes, and uplifts) that separate them from other systems. Every lake can be relatively isolated from other lakes, which means that genetically distinct populations can evolve into new species and that little genetic mixing occurs between lakes. Small streams are separated from each other by larger rivers, which are barriers to the movement of small fishes, and large rivers are separated by oceans. As a result, freshwater habitats are perfect for the speciation process. Oceans, in contrast, are largely continuous habitats that are connected by currents, and ocean fishes typically produce larvae that float for several weeks or months on these currents. Hence, genetic exchange is common and opportunities for speciation are not as great. Oceanic basins have relatively distinct faunas, but connectedness within basins discourages the kind of genetic isolation needed for speciation of the sort seen in lakes and rivers.

mous fishes (25%) spend most of their lives in fresh water and migrate to the sea to spawn; included here are river eels, mullets, and temperate basses. Amphidromous fishes (21%) move between fresh and salt water, but migration to the spawning habitat occurs long before the fishes actually spawn; examples include gobies, sleepers, and galaxiids.

B. Freshwater Diversity

The world's freshwater habitats occur in six major zoogeographic regions or realms that correspond approximately to continental distributions, with important exceptions (Fig. 11). Each region has a fairly distinct fish fauna (again with some exceptions and shared elements).

1. The Nearctic Region

The Nearctic region consists of subtropical, tropical, temperate, and arctic North America. The region stretches from the Mexican Plateau to northern Canada and Alaska. The Nearctic contains 14 families of primary freshwater fishes, with about 950 species. The most diverse families are minnows, suckers, North American catfishes, perches (and darters), and sunfishes. Other important families include the lampreys, gars, salmons (many of which are anadromous), and whitefishes; sculpins, which are freshwater species in a primarily marine family (= "marine derivatives"); pickerels and mudminnows; killifishes; and livebearers. The Nearctic is further subdivided into three subregions: the Arctic-Atlantic (with six provinces), the Pacific (with seven provinces), and the Mexican Transition subdivision. Eleven major river systems drain the region; major lakes are abundant, the largest being the five Laurentian Great Lakes (Ontario, Erie, Huron, Michigan, and Superior).

2. The Neotropical Region

The Neotropical region contains South America and Middle America. It is the most speciose region of the world in terms of freshwater fishes, with 32 families and more than 2500 species. Particularly diverse groups include the colorful characiforms (1200 species of tetras, piranhas, characins, and freshwater hatchet fishes), 13 families and 1300 species of catfishes, 6 families and 62 species of gymnotiform South American electric knifefishes, and 150 species of cichlids. Several secondary freshwater and marine derivative groups are included: freshwater stingrays, herrings, silversides, needlefishes, killifishes, and croakers. Many species remain to be discovered and described, particularly in South America. The Neotropical region has been further divided into eight subdivisions with fairly distinctive faunas. Eight major river systems drain the region; major lakes include Lake Titicaca, the world's highest fish-containing lake.

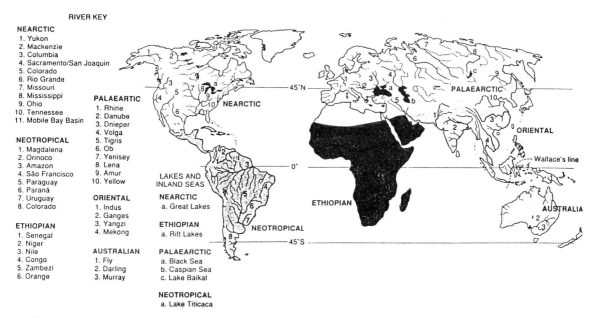

RIVER KEY

NEARCTIC
1. Yukon
2. Mackenzie
3. Columbia
4. Sacramento/San Joaquin
5. Colorado
6. Rio Grande
7. Missouri
8. Mississippi
9. Ohio
10. Tennessee
11. Mobile Bay Basin

PALAEARTIC
1. Rhine
2. Danube
3. Dnieper
4. Volga
5. Tigris
6. Ob
7. Yenisey
8. Lena
9. Amur
10. Yellow

NEOTROPICAL
1. Magdalena
2. Orinoco
3. Amazon
4. São Francisco
5. Paraguay
6. Paraná
7. Uruguay
8. Colorado

ORIENTAL
1. Indus
2. Ganges
3. Yangzi
4. Mekong

ETHIOPIAN
1. Senegal
2. Niger
3. Nile
4. Congo
5. Zambezi
6. Orange

AUSTRALIAN
1. Fly
2. Darling
3. Murray

LAKES AND INLAND SEAS

NEARCTIC
a. Great Lakes

ETHIOPIAN
a. Rift Lakes

PALAEARCTIC
a. Black Sea
b. Caspian Sea
c. Lake Baikal

NEOTROPICAL
a. Lake Titicaca

FIGURE 11   Major freshwater realms. Numbers indicate large river drainages, and letters show locations of larger lakes and inland seas. Neotropical, Ethiopian, Oriental, and Australian realms are shaded (reproduced with permission from Bond, 1996).

3. The Palearctic Region

The Palearctic region encompasses Eurasia, including Europe, northern Africa, and Asia north of the Oriental region. Twenty-seven families and about 550 species of temperate freshwater fishes occur in the region, dominated by minnows and loaches but also perches, pickerels, sturgeons, salmons, sculpins (including the Lake Baikal endemics), and 10 species of catfishes in four families. Diversity is greater in the southeastern than other parts of the region and also increases to the south, as is also the case in the Nearctic region. The Palearctic and Nearctic regions share numerous families and genera (sturgeons, paddlefishes, minnows, smelts, salmons, pikes, mudminnows, and perches) but only a few species (i.e., northern pike, longnose sucker, burbot, threespine stickleback, and fourhorn sculpin). The region is sometimes subdivided into six subregions based on faunal groupings. Ten major river systems drain the region; major lakes include the Black and Caspian Seas and Lake Baikal, the world's oldest and deepest lake.

4. The African or Ethiopian Region

The African or Ethiopian region is second to the Neotropics in freshwater fish diversity, with 47 families and more than 2000 species of primary and secondary freshwater fishes. The African region includes all the African continent south of the Sahara Desert, plus the large island of Madagascar with its endemic fauna. Half of the fishes are in the superorder Ostariophysi, including 300 minnows, 190 characiforms, and 360 catfishes in six families. Other diverse groups include killifishes and topminnows, elephant fishes and other osteoglossiforms, and cichlids. As many as 1000 cichlid fishes may occur in the three African Great Lakes of Lake Victoria, Lake Tanganyika, and Lake Malawi, with more cichlids in smaller surrounding lakes and rivers. Four lungfishes and all 11 polypteriform bichirs occur in Africa. Ten to 12 zoogeographic provinces are recognized, with six major river drainages, and numerous lakes including the African Great Lakes.

5. The Oriental Region

The Oriental region includes eastern Iran, India and Sri Lanka, China south of the Yangtze River, Southeast Asia, and the large island regions of Taiwan, the Philippines, and the East Indies/Indo-Malayan Archipelago. The Oriental region contains 43 families of primary and secondary freshwater fishes. Most diverse are the minnows, loaches, and 12 families of catfishes; clariid walking catfishes and bagrid catfishes are particularly diverse. Other important groups include algae eaters, river loaches, snakeheads, spiny eels, labyrinth fishes and gouramis, a few cichlids, and archerfishes. The Oriental region shares many families with the Palearctic to the north and the Ethiopian to the west but few with the Australian region to the southeast. The Oriental is often subdivided into two major subregions: Peninsular India with more than 700 species and Southeast Asia with more than 1000 species. Each subregion has two major river drainage systems; large lakes are uncommon. Southeast Asia is sometimes divided further into five zoogeographic regions.

6. The Australian Region

The Australian region (New Guinea, Australia, New Zealand, and Oceania) is relatively depauperate in true freshwater fishes, and in fact all but three of the freshwater fishes in the region are members of families obviously derived from marine groups. The northwestern border of the region, and the practical limit of primary freshwater fishes, is dramatically delineated by an ocean boundary that lies southeast of Java, Borneo, and Sulawesi and is known as Wallace's or Weber's Line. Nineteen families and about 210 species occur primarily in fresh water in the region, but only the Australian lungfish and 2 species of bonytongue saratogas are true freshwater fishes (another 33 families and 150 species of marine fishes frequently enter fresh waters in Australia). Other important families, many with species endemic to specific regions, include lampreys, river eels, herrings, two families of catfishes, southern smelts and graylings, salamander fish, galaxiids, silversides, rainbow fishes, barramundi, grunters, glassfishes, temperate perches or basses, sleepers and gobies, and torrentfishes (in New Zealand). One major river system occurs on New Guinea (the Fly) and two on Australia (Darling and Murray); permanent, large lakes are rare.

C. Marine Diversity

Delimiting zoogeographic regions in the world's oceans is complicated by depth, currents, and geographic locales; different faunal breaks occur depending on nearshore, pelagic, or deep-sea environments. The greatest fish diversity and the greatest geographic differentiation occur in nearshore, continental shelf (to about 100-m depth) regions. These regions are separated by continents, by large expanses of open ocean, and by currents that differ in temperature from that of the region in question. Temperature zones divide the seas into tropical, temperate, boreal, and polar regions (Fig. 12). In addition, different faunal groupings apply to pelagic fishes and to fishes of the deep sea.

1. The Indo-West Pacific Region

The Indo-West Pacific region includes shallow tropical seas that extend from South Africa and the Red Sea eastwards through the Indo-Malayan area and Australia to Hawaii and Easter Island; it also includes Micronesia, Melanesia, and Polynesia. The Indo-West Pacific is by far the most species-rich marine area, containing 3000 to 4000 tropical fish species, and is similarly diverse in sea snakes and many invertebrate taxa such as reef-building and soft corals, mollusks, tube worms, and echinoderms (Box 3). It is considered the center of evolution for many of the common coral reef fish families that occur in other tropical regions. Only a few families are endemic to the Indo-West Pacific (e.g., sillaginid whitings and rabbitfishes). Common families include moray eels, squirrelfishes, sea basses, grunts,

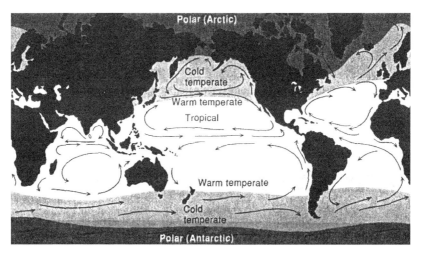

FIGURE 12   Major oceanic temperature regions and surface current patterns (reproduced with permission from Bond, 1996).

Box 3

The Deep Reef

Coral reef fishes—those associated with tropical, shallow waters in which coral reefs grow—are the most diverse group of fishes on the planet. Fish diversity is generally correlated with coral species diversity and coral coverage; 2000 fish species, and many other taxa, occur on the Great Barrier Reef of Australia, on which 500 species of coral also occur. However, most estimates of diversity of coral reef fishes are based on species found only in shallow water less than 50 m deep, where reef-building corals are abundant and where ichthyologists equipped with scuba gear can collect. An underappreciated and barely explored component of the diversity of reef fishes involves species that live in the deeper (50–150 m) "twilight zone" portion of the reef face, which has only recently been made accessible by the use of specialized, mixed-gas rebreathing equipment. The fish fauna of this region consists of species that seldom occur above 75 m. Many new species and even a few new genera have been discovered recently from the deep reef environment in families that are common in shallower water (e.g., wrasses, sea basses, damselfishes, angelfishes, and gobies). Most of the recent collecting using advanced diving techniques has occurred in deep reef areas of only Indo-Polynesia (e.g., Rarotonga, Palau, and New Guinea), but nearly 100 new species have already been found (Fig. 13). If we can extrapolate from the few deep reef areas that have been surveyed, our estimates of the number of fishes inhabiting "coral reefs" will have to be increased considerably. Also, fishes are just part of this barely explored, intermediate depth zone. How much of marine biodiversity are we missing?

FIGURE 13 The "Dr. Seuss fish"—*Belonoperca pylei*, a 6-cm long, brilliantly colored new species of sea bass from the deep reef zone of Rarotonga. The head and back are bright yellow, the body is pink, and the spots are orange (photo by Richard Pyle).

2. The Eastern Pacific Region

The Eastern Pacific region, with approximately 800 fish species, runs from southern Baja California to Ecuador, its northern and southern limits defined by the cold-water California and Peru currents. Despite its location in the Pacific Ocean, the Eastern Pacific is faunistically more similar to the tropical Atlantic, containing many species that are almost indistinguishable from Atlantic forms. The two oceans mixed before the Panamanian Isthmus formed and the two areas still share 12 species, despite 3 million years of physical separation. Most families are less diverse here than in the Western Atlantic, with the exceptions of sea catfishes, croakers, and herrings. Dactyloscopid sand stargazers occur here and in the Atlantic, but not in the Indo-West Pacific. Sixty-two Indo-West Pacific species have managed to cross the Eastern Pacific Barrier. Three provinces—Mexican, Panamanian, and Galapagos—are recognized.

3. The Western Atlantic Region

The Western Atlantic region is the second most diverse oceanic area, containing 1200 fish species. It includes Bermuda (which, although at 32°N, sits in the tropical Gulf Stream), southern Florida, the Bahamas Bank, the Caribbean Sea, and tropical and temperate portions of South America. Most of the families that occur in the Indo-West Pacific also occur in the Western Atlantic; a few families are more diverse here, such as grunts and toadfishes. Strong currents of warm water separate the Western Atlantic fauna from colder waters along much of its boundaries. It is subdivided into Caribbean, Brazilian, and West Indian provinces.

4. The Eastern Atlantic Region

The Eastern Atlantic region is a relatively small region along the west coast of Africa from Senegal to Angola and extending out to oceanic islands such as Ascension

snappers, cardinal fishes, butterfly fishes, angelfishes, damselfishes, wrasses, parrotfishes, surgeonfishes, gobies, triggerfishes, and boxfishes. Barriers to movement of Indo-West Pacific fishes are cool waters to the north and south and a vast expanse of open, deep water to the east known as the Eastern Pacific Barrier. Eight different provinces have been recognized in the Indo-West Pacific, separating it into a huge Indo-Polynesian region and seven smaller areas.

and St. Helena. Tropical marine fishes here are limited by cool-water currents impinging from both the north and the south as well as by substantial freshwater run-off and sediments from several major west African rivers, all factors which discourage coral reef growth. The region contains "only" 500 shore fishes; most coral reef families occur but are represented by only a few species. Porgies are particularly diverse. No subdivisions are recognized, with the possible exception of the warm-temperate Mediterranean. The Mediterranean Sea contains 540 species, with many species in the same families as those in the Eastern Atlantic. The Mediterranean has the dubious distinction of being the most heavily invaded tropical marine area in the world, with at least 52 alien fish species having moved in from the Red Sea (an Indo-West Pacific subregion) via the human-made Suez Canal.

5. The Arctic Region

The Arctic region encompasses high-latitude (above 60°N) waters of both the Pacific and Atlantic Oceans. It reaches from Nunivak Island, Alaska, northward and across the polar region to Newfoundland and Norway in the northern Atlantic. Of the two polar areas, the Arctic is more diverse. Successful groups include skates, herrings, salmons, smelts, cods, eelpouts, greenlings, sculpins, poachers, snailfishes, pricklebacks, wolffishes, gunnels, and right-eye flounders. Diversity within many of these groups is greater in the Pacific than in the Atlantic portions of the region. A total of 415 species occur here. Distribution of many of these families appears to be limited by temperature, with warmer waters and currents to the south determining species' boundaries.

6. The Antarctic Region

The Antarctic region (above 60°S) has its own distinctive fauna that is restricted to Antarctic waters and the surrounding Southern Ocean, including the cold waters of Australia, New Zealand, and nearby oceanic islands. Forty-nine families and 274 species occur here, 13 families and 174 species of which are identified as Antarctic continent species. A particularly successful group is the notothenioid icefishes and relatives, which account for 55% of antarctic species. Families include bovichthyids, cod and crocodile icefishes, plunderfishes, and dragonfishes. Non-notothenioids include skates, snailfishes, eelpouts, lantern fishes, eel cods, deep-sea cods, and southern flounders.

7. Temperate Regions

To the north of the Indo-West Pacific lie cooler temperate waters with their own characteristic fish faunas.

This area can be divided into four fairly distinct regions according to location and temperature: Japanese warm temperate and Californian warm temperate regions and Eastern and Western boreal regions. The warm temperate areas (from about Hong Kong to Tokyo in the west and from lower Baja California to central California in the east) contain a fauna that fluctuates seasonally, as tropical species move north in the summer and boreal species move south in the winter. Notable families to the west include lizard-fishes, flying fishes, mullets, jacks, sea basses, and croakers and to the east include endemic silversides, sea basses, croakers, damselfish, wrasses, and flatfishes. The more northerly boreal regions (approximately north from central California in the east and Korea in the west) contain similar families but different species. Important families include migratory salmonids, sculpins, rockfishes, snailfishes, greenlings, gunnels, pricklebacks, and right-eyed flounders. In the southern Pacific, cold water currents create at least three temperate faunal regions, with centers of distribution around Australia, New Zealand, and South America.

In the Atlantic, three northern temperate areas occur: the western and eastern Atlantic boreal regions and the Atlantic warm temperate or Carolinian region. The boreal regions (Newfoundland to Cape Hatteras in the west and British Isles to northern Europe and Scandinavia in the east) share a fauna of salmonids, cods, sticklebacks, poachers, sculpins, wolf-fishes, and right-eyed flounders, with occasional strays from more southerly waters during warm months. The Carolinian extends from Cape Hatteras south to Florida and also to the Gulf of Mexico, with southern Florida housing tropical species. Common groups are clupeids, sea robins, pipefishes, silversides, needlefishes, killifishes, croakers, left-eyed flounders, and puffers. Temperate faunas also occur in the southern Atlantic, but their areas and diversity are less than those of the southern temperate faunas of the Pacific Ocean. Two recognized regions are the eastern South American and southern African warm temperate regions. The former region's fauna includes sea catfishes, croakers, herrings, gobies, scorpion fishes, and sea basses; the latter area has many colder water members of Indo-West Pacific families.

8. Pelagic Regions

The 350 species of pelagic fishes occur in ocean surface waters to a depth of 200 m. This habitat type can be divided into 10 different regions based on faunal differences, with more joining of Southern Hemisphere areas because of the relative lack of large land masses. These regions are Arctic and Antarctic polar, North Pacific cold temperate, North Pacific warm temperate, tropical

Indo-Pacific, North Atlantic boreal, North Atlantic warm temperate, Atlantic tropical, southern warm temperate, and southern cold temperate. Many of the world's most important fisheries species occur in pelagic regions, including numerous sharks, sardines and herrings, salmons, codfishes, pollocks, hakes, haddocks, sauries, mackerels, and tunas. In addition, about 100 species of mostly pelagic fishes have a worldwide distribution. This group includes pelagic sharks (white, whale, tiger, and perhaps megamouth), swordfish, and ocean sunfish.

9. The Deep Sea

Waters deeper than 200 m are as much a zoogeographic as a habitat entity. The deep sea is generally divided into three major regions based on depth: the open-water mesopelagic (200–1000 m) and bathypelagic (1000–4000 m) regions and the bottom-associated benthal (200–1000 m) region. Benthal fishes are further divided into benthopelagic fishes that hover just above the bottom and benthic fishes that rest in contact with the bottom. Deep-sea fishes are most common in these regions between 40°N and 40°S latitude, approximately between San Francisco and Melbourne in the Pacific and between New York City and the Cape of Good Hope in the Atlantic. Abyssal and hadal (trench) regions deeper than 4000 m are relatively depauperate.

Each region has a characteristic and relatively diverse bony fish fauna consisting of fishes from many different taxonomic groups. The mesopelagic region worldwide contains about 750 species in seven different superorders and nine orders. Despite their lack of relatedness, deep-sea fishes share many anatomical and physiological features, suggesting independent, convergent evolution of adaptations to deep-water existence. Mesopelagic fishes typically have photophores (light organs) on their silvery bodies; have relatively large, often-tubular eyes; undergo daily migrations to surface waters to feed at night; and have large mouths and long teeth. Common names of mesopelagic fishes reflect these traits: barreleyes, bristlemouths, dragonfishes, sabertooth fishes, lantern fishes, tube-eyes, and swallowers.

The bathypelagic region is the largest habitat space on the earth, accounting for 88% of oceanic volume. The five superorders, nine orders, and 200 species in the cold, dark bathypelagic region share some traits with mesopelagic fishes but possess them in the extreme. Photophores are concentrated on lures used to attract prey; eyes are often small; mouths are extremely large and teeth very long; stomachs are expandable; bodies are black; and body musculature, bones, and scales are greatly reduced. These traits reflect greater habitat space and increasingly rare feeding opportuni-

ties with increasing depth, which select for an increasing need to conserve energy and to be able to take advantage of feeding opportunities. Again, bathypelagic fishes have names indicative of their adaptations: sawtooth eels, gulper eels, swallower eels, dragonfishes, anglerfishes, seadevils, and fangtooths.

Benthal fishes include about 1000 bony fish species in four superorders and nine orders, plus chimaeras and squaloid sharks. Different families inhabit bottom compared to open-water regions. Benthal fishes include greeneyes, tripod fishes, hakes, grenadiers, cusk-eels, batfishes, snailfishes, and eelpouts. Although diversity decreases below 1000 m, grenadiers and rattails live between 1000 and 4000 m, tripod fish have been found to 6000 m and snailfishes to 7000 m, and some cusk-eels have been found as deep as 8000 m.

Although some differences in species composition occur in different ocean basins or in association with different water masses, deep-sea species are relatively cosmopolitan, occurring in several different oceans. One trend is for fishes to occur deeper at lower latitudes, such that species that are bathypelagic near the equator may be mesopelagic at middle latitudes and even epipelagic at the poles.

Websites

www.fishbase.org (a remarkable site with information on fishes and fisheries worldwide)

www.fisheries.org/ (website of the American Fisheries Society)

www.fishlinkcentral.com (a clearinghouse for the aquarium trade)

www.fws.gov/r9endspp/endspp.html (U.S. Fish and Wildlife Service, Endangered Species homepage)

www.utexas.edu/depts/asih/ (website of the American Society of Ichthyologists and Herpetologists)

www.wcmc.org.uk (World Conservation Monitoring Centre; lists information on imperiled species worldwide)

See Also the Following Articles

FISH CONSERVATION • FISH STOCKS • LAKE AND POND ECOSYSTEMS • OCEAN ECOSYSTEMS • PELAGIC ECOSYSTEMS

Bibliography

Berra, T. M. (1981). *An Atlas of Distribution of the Freshwater Fish Families of the World.* Univ. of Nebraska Press, Lincoln.
Bigelow and Schroeder (1953). Fishes of the Gulf of Maine. *Fish Bull. Fish Wildlife Serv.* 53 (*Fish Bull.* 74).

Bond, C. E. (1996). *Biology of Fishes,* 2nd ed. Saunders, Fort Worth, TX.

Bone, Q., Marshall, N. B., and Blaxter, J. H. S. (1995). *Biology of Fishes,* 2nd ed. Blackie, London.

Briggs, J. C. (1995). *Global Biogeography.* Elsevier, Amsterdam.

Burgess (1989). *An Atlas of Freshwater and Marine Catfishes.* TFH Publishers, Neptune City, New Jersey.

Compagno, L. J. V. (1984). FAO species catalogue. Vol. 4: Sharks of the world. An annotated and illustrated catalogue of shark species known to date. FAO Fisheries Synopsis No. 125. Food and Agriculture Organization, Rome.

Compagno, L. J. V. (1990). *Environ. Biol. Fish* 28, 33–75.

Eschmeyer, W. N. (Ed.) (1998). *Catalog of Fishes.* California Academy of Sciences, San Francisco.

Greenfield, and Thomerson (1997). *Fishes of the Continental Waters of Belize.* Univ Press of Florida, Gainesville, Florida.

Helfman, G. S., Collette, B. B., and Facey, D. E. (1997). *The Diversity of Fishes.* Blackwell Science, Malden, MA.

Heller and Snodgrass (1903). Papers from the Hopkins Stanford Galapagos Expedition, 1898–1899. XV. New fishes. *Proc. Wash. Acad. Sci.* 5, 189–229.

Hildebrand (1963). Family Elopidae. *In:* Fishes of the Western North Atlantic, Part 7, Salmon, trouts, and others. *Mem. Sears Foundation Mar. Res.* 1, 111–145.

Miller *et al.* (1989). Extinctions of North American Fishes during the past century. *Fisheries* 14(6), 22–38.

Moyle, P. B., and Cech, J. J., Jr. (1996). *Fishes: An Introduction to Ichthyology,* 3rd ed. Prentice-Hall, Upper Saddle River, NJ.

Musick *et al.* (1991). The biology of *Latimeria chalumnae* and evolution of coelacanths. *Environ. Biol. Fish* 32, 1–435.

Nafpaktitis *et al.* (1977). Family Myctophidae. *In:* Fishes of the Western North Atlantic. *Mem. Sears Foundation Mar. Res.* 1(7), 13–265.

Nelson, J. S. (1994). *Fishes of the World,* 3rd ed. Wiley, New York.

Norman, J. R. (1905). A history of fishes.

Paxton, J. R., and Eschmeyer, W. N. (Ed.) (1998). *Encyclopedia of Fishes,* 2nd ed. Academic Press, San Diego.

Scott and Crossman (1973). Freshwater fishes of Canada. *Fish Res. Board Can. Bull.* 184, 1–920.

Suttkus (1963). Order Lepisostei. *In:* Fishes of the Western North Atlantic, Part 3, Soft-rayed bony fishes. *Mem. Sears Foundation Mar. Res.* 1, 61–88.

Vladykov, and Greeley (1963). Order Acipenseroidei. *In:* Fishes of the Western North Atlantic, Part 3, Soft-rayed bony fishes. *Mem. Sears Foundation Mar. Res.* 1(3), 24–60.

FISH CONSERVATION

Carl Safina
National Audubon Society

I. Introduction
II. The World Ocean Fishing Situation
III. Other Factors Affecting Fish Conservation
IV. Marine Fish Conservation in a Biodiversity Context
V. The Course Ahead

GLOSSARY

aquaculture Commercial farming of aquatic organisms, including seaweeds, raising captive-bred fish, and raising wild-born fish in captivity. Mariculture refers specifically to marine (saltwater) aquaculture.

ballast Water taken in by ships to balance them after they have unloaded their cargo.

bycatch Any living thing caught unintentionally in fishing gear; sometimes called bykill because so many such creatures are discarded dead. About one-fourth of the total world catch is bykill.

depletion In fisheries, reduction to population levels low enough to reduce or threaten future productivity.

ecological integrity The naturally evolved numerical and functional relationships among species and their environment in a given area.

fishery A collective effort to gather, collect, or catch wild aquatic wildlife or plants for recreational or commercial purposes. Fisheries extract large num-

bers of wild fish, sea urchins, corals, seaweeds, shrimp, snails, clams, scallops, squids, turtles, whales, and other creatures.

harvest To gather a crop. "Harvest" is an appropriate word for farming operations, including fish farming, but not for catching or collecting wild animals or plants. This term is widely misused in industry public relations to make the extraction of wild fish, natural stands of trees, and other wild organisms seem like agriculture, though nothing is planted or nurtured and these things are merely taken for profit. For wildlife, including wild fish, appropriate words include, among others: catch, fish for, take, extract, land, gather, and collect.

high seas Parts of the ocean outside national boundaries, usually beyond 200 miles of any nation's coast.

keystone species Species whose removal causes a chain or "cascade" of ecological effects among other species.

marine reserves Designated areas where no fishing, mining, or other consumptive use is allowed, usually for purposes of replenishing nearby fishing grounds or maintaining normal evolution, growth, and fecundity.

overfishing Extracting marine organisms faster than they can reproduce.

pollution Introduction of substances in quantities that are threatening to living resources, biological processes, and human health and activities.

OVERFISHING AND HABITAT DEGRADATION have driven many fish populations to historic lows. Poor fisheries management, increasing human population pressures, and habitat deterioration from several factors have caused this situation. But many fish are resilient and can recover within a decade or two if given viable habitat and a respite from overfishing.

I. INTRODUCTION

An ocean of water covers more than 70% of Earth, and travelers from another world might more logically assume that this planet would be named Ocean. Ninety-seven percent of Earth's water is in the ocean (2% is locked up as ice, and 1% is in surface freshwater or groundwater). Moreover, 99% of the living space on the planet in which life can exist—the "biosphere"—is in the sea. The basis of most life in the sea are the single-celled plants, or "phytoplankton," that create food from sunlight (through photosynthesis) and drift in the upper 1% of the ocean's volume. Plants attached to the bottom can only live in shallow coastal areas where they can get enough sunlight. Most life in the other 99% of the sea relies on food coming from that thin upper layer. Of 33 living animal phyla (the category that reflects the different basic body plans of living things), 32 are found in the ocean, 15 exclusively so. Only one is exclusively found on land (the Onycophora, or velvet worms). This article focuses on the world's ocean fishes. (For a synopsis of biodiversity issues facing freshwater fish, see Box 1.)

II. THE WORLD OCEAN FISHING SITUATION

Many factors affect life in the sea, but fishing has caused the largest changes and is the major current agent of ecological disturbance in the ocean. The United Nations' Food and Agriculture Organization has stated that modern fishing "is globally nonsustainable."

In the twentieth century, annual landings of wild ocean fish increased 25-fold, from 3 million metric tons to a plateau of over 80 million metric tons (Fig. 1). In the 1950s and 1960s, fishing technologies exploded as fishing fleets adapted war technologies such as radar, sonar, and LORAN to peaceful efforts of food gathering. Radar allowed boats to keep fishing in total fog, and sonar could find schools of fish deep beneath the sea's surface. LORAN turned the trackless sea into a grid and allowed boats to find and return to precise spots

---

Box 1

Synopsis of Freshwater Biodiversity Issues

Worldwide, lakes and rivers contain at least 8400 fish species, roughly 40% of known fish species on Earth, and almost 20% of all vertebrates (freshwaters support almost one-quarter of the planet's known biodiversity, in only 0.01% of the planet's water; however, the oceans hold vastly larger populations).

Lakes are isolated habitats, leading to a high rate of evolution of species of fish and other animals. For instance, three-quarters of the 2000 plant and animal species in Russia's Lake Baikal are found nowhere else. The lakes of Africa's Rift Valley have produced explosive speciation—99% of the 500 cichlid fish in Lake Malawi, for instance, live nowhere else. And Lake Tanganyika, the least-species-rich lake in the Rift Valley lake chain, has 25% more species of freshwater fishes than all of Europe.

Freshwaters are being degraded and species eliminated at a rate probably comparable to that occurring in tropical rain forests. Habitat loss and introduced species are the two greatest problems for freshwater biodiversity. Introduced predatory fish have already wiped out nearly 70% of Lake Victoria's cichlid fishes, and threaten to soon reduce its unique fish biota by 90%. In the United States half the rivers and streams are significantly polluted, and 98% of U.S. rivers (outside Alaska) are blocked by dams. Consequently, 20% of fish species and more than half the mussels in U.S. freshwaters are endangered or have become extinct, in contrast with only 7% of the mammals and birds in the United States.

Of the world's estimated 9000 freshwater fish species, about 1800 (20%) are in serious decline or have gone extinct. Ecologist Norman Myers has called the freshwater fish situation "the greatest extinction spasm of vertebrates in recent times."

---

where fish gather. Satellite-generated maps faxed directly to boats in midocean now track movements of water temperature fronts, showing where to find the fish.

Overfishing, first recognized in the early 1900s, is now prevalent in most major fishing areas. Not all of the catch is used for food. About a third of the world's catch becomes fertilizer, animal feeds, and industrial

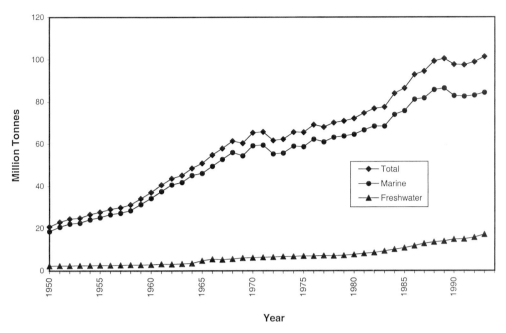

FIGURE 1   World fish landings (includes aquaculture).

products, and about a quarter of the catch is unwanted and shoveled overboard.

The annual rate of increase of world fish landings is now approaching zero. But many fisheries reached the limit decades earlier than the peak in global landings. Indeed, most regions in the Atlantic, Mediterranean, Pacific, and even Antarctic have declining catches. In some regions where catches peaked as long ago as the early 1970s, catches have declined by half. Only in the Indian Ocean has the total catch been barely increasing, but there is little room for more growth there. Declines in some individual fisheries have been catastrophic. Newfoundland cod, which supported one of the world's largest fisheries for over 400 years, crashed 99% between the early 1980s and early 1990s. Many other fish populations have declined by 80% or more. Some major fishing grounds are now closed in hopes that they will recover from exhaustion. Most important marine resources are considered fully exploited or overfished, and no major untapped fishery resources remain in the world. Humanity's former faith in the sea's inexhaustibility was wrong.

A. Social and Economic Concerns

Worldwide, about 200 million people depend on fishing for their livelihoods. Marine fishing supplies almost 20% of all animal protein consumed, and in Asia more than one billion people rely on fish as their main source of animal protein.

In the last few years, many thousands of fishers around the world have experienced severe decline or loss of their source of income. Depletion caused primarily by industrial overfishing and exacerbated by coastal habitat degradation threatens tens of millions of jobs, as well as major food sources. As preferred species are fished out, and less-preferred fish are targeted, prices rise for these less-preferred fish. Increases in price eventually removes the species from the tables of the poor. Distant markets and increasingly globalized trade networks allow and encourage regional overfishing or irresponsible aquaculture that exhausts local resources, then move on to tap new source areas. This results in serial depletion and habitat degradation along faraway, out-of-sight-out-of-mind shores. For example: bluefin tuna are depleted in the Atlantic and off Australia and New Zealand because of demand in Tokyo, where they become the world's most expensive sushi; appetites in the Northern Hemisphere support destructive shrimp farms in the tropics; and shark populations are declining in many places around the world because of demand for their fins, primarily in China.

B. Management Problems

In most of the world, intense political lobbying by the fishing industry usually causes fishery managers to disregard scientific assessments and scientific advice to bring catches within sustainable limits.

While no group of fishes is exempt, Atlantic bottomfishes such as cod, haddock, pollock, flounders, and Atlantic halibut have been among the worst managed. Various species of groupers and wild salmon are depleted throughout most of their ranges. And the world's giant fishes—the tunas, billfishes (like swordfish and marlins), and sharks—among the most magnificent of the ocean's wildlife, are also among the most mistreated. In much of the world, management in these fisheries is lacking or ineffective. Sharks, subject to intense overfishing in many regions, are easily depleted because they are generally slow growing with low reproductive rates. Fishing annually kills roughly 50 million sharks worldwide, mostly for just their fins, which are prized as a thickener in "shark fin soup." Only 3 of 26 major shark-fishing countries have management and research programs. Overfishing has driven virtually all large fishes in the Atlantic to all-time lows. The Atlantic tuna commission's mismanagement has resulted in systematic overfishing of swordfish, white marlin, blue marlin, bigeye tuna, and bluefin tuna; those populations have declined 60–90%.

In many other cases, poor monitoring or outright cheating enlarge the gap between fishery science and fishing activities. In one 7-year period, foreign fleets fishing the Grand Banks removed *sixteen times* the quotas that had been set for cod, flounders, and redfish by the Northwest Atlantic Fisheries Organization. A seized Spanish boat kept two log books—one showing actual catches and one for reporting to authorities. Illegal fishing is rampant in the Southern Ocean (around Antarctica), particularly for toothfish, which are marketed as Chilean sea bass (this fishery kills large numbers of seabirds as well.)

No country can be called generally successful in fisheries management. Most are hardly trying. International cooperation on migratory populations and fishing outside national waters—on the "high seas"—is particularly difficult, and in some cases a country unhappy with the restrictions of an agreement simply ignored them outright or through lack of enforcement.

C. Fish Fights: International Fishing Conflicts

Fish conflicts, many of them violent, have erupted around the world. Norway, Iceland, the United States, Canada, Indonesia, Taiwan, Nicaragua, Russia, China, the Philippines, Japan, France, Britain, Spain, and many others have been involved in international fishing disputes. In disputed waters off the Philippines, Chinese vessels were seized and scores of crew members were jailed. In the Galápagos, troops were deployed after fishers seized a biological station that was trying to stop rampant illegal fishing inside a marine reserve. In France, fishers protesting fish imports rioted, causing $4.5 million in damage. NATO allies have pointed guns at each other in disputes over cod, and a Russian Navy boat fatally shot Chinese poachers. More trouble is likely as hungry boats scour the oceans for dwindling resources.

D. Fish as Commodities/Fish as Wildlife

Marine creatures are the only wild animals still hunted commercially on a large scale. And we usually treat fish as mere commodities, forgetting that fish are wildlife. For instance, we speak of a population of fish as "stock," as in "the New England groundfish stock." We would not speak of the "Serengeti mammal stock" or the neighborhood "woodpecker stock." By preventing us from seeing fish as wildlife in natural ecological communities, the language used helps facilitate their overexploitation.

Probably the most abused word throughout fisheries is "harvest." "Harvest," applied to wild populations, is an industry public-relations word—the intent is to make activities that take wild living things from natural communities seem like farming. Until recently, humanity has distinguished between hunting and gathering wild things that we do not grow and agriculture, where we raise and harvest crops that we do grow. Nowadays, many people try to blur this distinction by using "harvest" to describe hunting and gathering operations as a way to put a better face on taking large quantities of individuals from wild populations, usually for profit. But harvest is an agricultural word—it means to reap a plant crop that has been sown and nurtured. We usually don't even say we've "harvested" agricultural animals like cows, pigs, and chickens when we slaughter them for meat, even though we raise them for those purposes. So why should we say that we "harvest" wild animals like cod or sharks? Industry people even speak of "harvesting" whales to avoid using the obviously correct word, "hunting," for the pursuing and killing of wild mammals, in this case the largest animals that have ever lived. When a wild fish or whale or a naturally growing tree is simply taken from a wild community, "harvest" is the wrong word. Appropriate words include: catch, fish for, take, extract, land, gather, collect (e.g., oysters), boat (i.e., they boated 2000 salmon), cut, or hunt.

E. Intensity and Limits of Fisheries

Humans exert tremendous pressure on the seas, using about a third of all "primary production"—the food energy made by plants from sunlight—in coastal areas. The coasts, where nutrients continually enter the marine environment primarily from the land, are the most productive parts of the oceans by far. The coastal areas, representing only 10% of the oceans' total area, produce 90% of the world catch.

Many industrialized fishing ships deploy large-scale fishing gear such as long-lines that are 50 *miles* long with thousands of baited hooks, drift nets 40 miles long (now subject to a United Nations ban), and bag-shaped trawl nets sometimes large enough to engulf twelve Boeing 747 jetliners.

Between 1970 and 1990, the world's industrialized fishing fleet increased at twice the rate of the catch, doubling in number and tonnage. The fleets of the world now wield twice the fishing power needed to catch what the oceans can produce. If the fleets had not grown at all since1970, they would be able to catch the same number of fish. Fishing has either reached or exceeded natural limits virtually everywhere in the ocean.

F. Kinds of Overfishing

Overfishing is the main reason fish populations have declined. Overfishing means catching marine creatures faster than they can reproduce. "Fisheries" extract large numbers of wild fish, shrimp, snails, clams, scallops, squids, sea urchins, corals, seaweeds, turtles, whales, and other kinds of creatures. Different kinds of overfishing have been identified (see Box 2), and various factors can influence a species' vulnerability to overfishing (see Box 3).

Overfishing causes: (1) massive depletion of many species; (2) loss of spawners, thus fewer young produced and increased risk of reproductive failure in times of poor environmental conditions (e.g., unusual ocean temperatures); (3) declines in average sizes of fish and other marine creatures; (4) loss of genetic diversity; (5) genetic change toward less desirable characteristics like smaller size potential; (6) disruption of natural communities; and (7) disruption of human communities. Simultaneous overfishing of many species leads to functional loss of species or species groups. "Keystone" species are those whose removal causes a chain or "cascade" of ecological effects. For example, removal of algae-eating fishes can cause coral reefs to be killed by algae overgrowth.

Until recently, one of the main assumptions in fish-

Box 2

Different Kinds of Overfishing Affect Many Species

"Growth overfishing" is when too many fish are caught while very small. In species like groupers that change sex with age, a critical shortage of one sex can result.

"Recruitment overfishing" is reproductive failure due to depletion of breeders.

"Demographic overfishing" turns a population with many age groups into a population with only one or two significant age classes doing most of the breeding, making the population vulnerable to years when natural fluctuations cause poor survival of young.

"Genetic overfishing" is when intense fishing changes a population's gene pool.

"Serial overfishing" is the depletion of one targeted species after another.

"Ecosystem overfishing" causes great changes in species composition and functional loss of key species, and can result in long-term community changes.

Overfishing related directly to the increasing human population is sometimes called "Malthusian overfishing."

---

ery management was that the number of young produced is generally not related to the number of spawners, and many practicing managers who were taught this in school still believe it. The reasoning was that environmental factors cause a different percentage of young to survive from year to year regardless of the number of eggs laid. But because any surviving fraction (say, 1%) of a large number of eggs is a higher number of survivors than the same percentage of a small number of eggs, the assumption that numbers of young are not related to numbers of parents is illogical. In 1996, Ransom Myers and Nicholas Barrowman proved this assumption wrong. They studied nearly 400 data sets from different species and asked the following questions: (1) Are the largest groups of young fish entering a fishery produced by the highest populations of spawners? (2) Are the smallest groups of young fish entering a fishery produced by the lowest populations of spawners? (3) Are above-average groups of young fish entering a fishery produced by above-average populations of spawners? Their findings?: "The answer to all three questions is almost always 'yes.'"

Consideration of These Factors Can Help Us Assess the Vulnerability of a Species to Overfishing

1. Inherent vulnerability: Does the species have low growth rate, low fecundity, high catchability (especially while immature), vulnerable behavior (e.g., spawns in groups), vulnerability to environmental changes, increasing vulnerability due to changes in fishing technology, or a poorly understood life cycle?

2. Human-related environmental risk: Does the species suffer destruction of key habitats, widespread effects of pollution, or conflicts with introduced species?

3. Is the population large or small, lightly or heavily exploited?

4. Is there management? Does it involve long-term conservation and sustainability as a goal? Does it benefit from independent and objective scientific advice? Does it have effective mechanisms for monitoring and enforcement and data collection?

Box 4
Dolphin-Safe's Downside

After several Pacific dolphin populations fell 50 to 80% due to drowning in tuna nets, the United States adopted a "dolphin-safe policy" in 1990, which barred the import of tuna caught by setting nets around dolphins (yellowfin tuna and dolphin swim together in the eastern tropical Pacific). Consequently, dolphin kills dropped 99%, but other bycatch increased. The average bycatch from 1000 net sets around dolphin herds includes 500 dolphins, 52 billfish, 10 sea turtles, and no sharks. In contrast, average bycatch from 1000 sets around floating objects (where tuna also gather), the preferred fishing method in the wake of the "dolphin-safe" regulations, includes only 2 dolphin, but also 654 billfish, 102 sea turtles, and 13,958 sharks. In addition, large numbers of the tuna caught under floating objects are juveniles.

When fishing is reduced, many depleted populations can recover. When fishing ceased in the North Sea during the world wars, fish populations increased significantly. On the East Coast of North America, a spectacular recovery of striped bass followed strict limits on fishing. Subsequent limits on fishing for several other species also resulted in increases in those populations. Protecting fish until they reach average spawning age is one of the simplest, fastest, most effective ways to allow recovery.

G. Bycatch

Every fishery catches unintended, unwanted creatures, called bycatch or bykill. Bycatch includes nontarget and juvenile fishes, seabirds, marine mammals, and any other creature that the fishers are not trying to catch. Bykill currently threatens several species of dolphins and albatrosses with extinction.

By weight, about 20% of the animals caught worldwide are discarded. By number, a much higher percentage are thrown away, because usually it is smaller fish in the catch that are discarded. In many fisheries, bycatch exceeds target catch. Shrimp trawling incurs more bykill than any other fishery. Discarded bykill outnumbers shrimp catch by anywhere from 2-to-1 up to 8-to-1 or even more. Shrimp trawls are the largest human source of mortality to adult sea turtles. The Gulf of Mexico shrimp fishery annually kills and discards 12 *million* juvenile snappers and 2800 metric tonnes of mostly juvenile sharks. But waste is not the only issue. Using everything caught would not fix the biological effects of killing millions of young fishes, as well as breeding-age seabirds, turtles, dolphins, and other mammals.

Lost fishing gear also catches, kills, and wastes sea life (this is called "ghostfishing"). Gill nets are frequently lost. For instance, the drift gill net (drift net) fishery of the North Pacific, which set approximately 30,000 *miles* of netting per night before the United Nations banned it, lost roughly 6000 miles of netting annually. A Norwegian study concluded that lost gill nets continued killing fish for at least 7 years; a New England study found them still catching fish 3 years after they were lost. Lost fishing gear frequently tangles seabirds and marine mammals even in areas as seemingly "remote" as the Antarctic. (See Boxes 4 and 5.)

H. Subsidies and Economics

Because there are too many boats, fisheries are said to be "overcapitalized," and are generally unprofitable. Worldwide, to catch $70 billion dollars worth of fish, fisheries incur total costs of $124 billion per year. Government subsidies, variously estimated at $22 to $54

Albatrocities

Bykill from commercial fishing operations is the most serious threat facing albatross populations. Each year over one hundred *million* hooks are set on longlines in the Southern Ocean. Each longline is up to 80 *miles* long and can bear 3000 baited hooks. Albatrosses often congregate in the same areas where boats fish. As the line is being played out from the moving vessel, albatrosses sometimes pick up the bait just before the weight of the sinking line pulls them underwater. Work by New Zealand ornithologist Dr. Sandy Bartle and colleagues shows population declines of 50 to 80% in some albatross colonies in the last several decades.

million, plug those huge deficits. Such massive subsidies arise from governments' efforts to preserve employment. Subsidies include fuel tax exemptions, price controls, low-interest loans, free access, and outright grants for gear or infrastructure.

Subsidies are responsible for building more fishing power than the resources can support, which increases political and social pressure for continued high catches. But high subsidies can only lead to greater and greater economic distress and further depletion. The ecologist Norman Myers concludes, "Subsidies are far and away the principal cause of overfishing."

Fishery depletion currently costs the U.S. economy $8 billion and 300,000 jobs each year. Widespread depletion has driven seafood prices higher and faster than those for other meats. Whereas export prices for pork in 1991 were 55% above the 1975 price, and for beef 75% above, those for marine fish were 335% higher. Increased demand for fish will drive prices even higher.

Many studies conclude that what are needed are reductions in fishing power—fewer and smaller-scale boats, each employing more people per unit of horsepower—if we are to begin to rebuild the fish populations.

III. OTHER FACTORS AFFECTING FISH CONSERVATION

As discussed so far, fishing is currently the major agent of change in the oceans, but a fuller compilation of the major threats to marine biodiversity includes:

- fishing operations, which cause depletion, catch large numbers of nontarget species, degrade seafloor and coral reef habitats, and cause major changes in living communities throughout the world's oceans;
- distant markets, which exert tremendous pull on resources from around the world, yet are relatively buffered from the economic, social, and ecological effects of local depletion because they can switch to new sources;
- pollution from those chemicals that cause toxic effects and hormonal disruption;
- pollution from excessive nutrients, which increase the frequency and severity of harmful algal blooms such as red tides, brown tides, and the single-celled fish-killing organism called Pfiesteria, and cause oxygen depletion in the waters;
- physical habitat alterations, including development, wetland alteration, and on-land deforestation, which sends clouds of sediments into rivers and coral reefs;
- introductions of species to areas where they are not native, causing competition, predation, and spread of disease;
- aquaculture or "fish farming," which usually entails extensive habitat damage, pollution, and introductions of alien species and diseases;
- debris, which tangles, traps, or is ingested by large numbers of marine fishes, mammals, birds, and other creatures;
- dams in rivers, which prevent fishes such as certain salmon, sturgeons, herring, and others from reaching spawning areas, or kill their young on their way to the sea;
- atmospheric warming, which stresses corals, lowers overall ocean productivity, melts polar ice, threatens to raise sea level and inundate coastal wetlands and low islands, and may change ocean current patterns, resulting in changes in abundance and distributions of marine wildlife; and
- ozone thinning, which lowers ocean productivity by killing significant amounts of plankton.

We should bear in mind that gravity takes things from land to sea. Almost everything that enters a storm drain goes directly into the ocean, from used oil to litter. The less chemicals people use on their lawns and farms, the less we will swim in and eat in our seafood. More oil enters the ocean from urban runoff, including automobile leaks, than from large tanker spills. About three-quarters of ocean pollution comes from the land (Table I).

TABLE I

TABLE I

Sources of Marine Pollution

Offshore oil production	1%	
Maritime transportation (shipping)	12%	
Dumping	10%	
Subtotal of ocean-based sources		23%
Runoff and land-based discharges	44%	
Atmospheric deposition into oceans	33%	
Subtotal of land-based sources		77%
Total		100%

A. Human Population Growth

By 2050, 7–10 billion people will be joining you for dinner, and the seas will feel their impact directly. Each year the world adds 100 million new human appetites. To keep pace for just the next decade, we will have to find 20 million more tons of seafood. A third of the world catch is now used for animal feed and fertilizer, but even if it were simply eaten by people, it would maintain present consumption levels per person for only 20 years. The widespread recoveries of wild fish in the face of increasing demand seem unlikely, and improved conservation would not be able to keep pace with population growth anyway. We have hit the limit of the oceans' natural capacity to feed us.

Coastal habitat disappears in proportion to population growth, and growth in coastal areas is four times the U.S. national average. Sixty-five percent of major cities are located along the coasts, and half to three-quarters of the human race will soon live within 50 miles of the sea. Diminished habitat means diminished productivity.

B. Habitat Issues

Human alteration of aquatic habitats is a major cause of fishes' decline—*the* major reason in freshwater, and the second-largest factor (after fishing) in estuarine and continental shelf habitats. Marine habitat is physically altered by development, aquaculture, mining, and fishing activities. Development often causes: extensive filling and diking of critical wetland breeding and nursery areas; dredging to deepen shallow estuaries for shipping; shoreline stabilization; and dams that alter normal patterns of sedimentation, erosion, and water flow. Dams also block migrations to and from spawning areas, and divert to agriculture, industries, and homes the freshwater flows needed by wetland and estuarine creatures. Aquaculture often destroys local wetlands and

mangrove areas. Mining removes coral and minerals, often adding toxic chemicals and sediments that destroy seafloor creatures. Agriculture and deforestation put large amounts of sediments into coastal waters, smothering spawning beds, seagrass flats, and coral reefs. Trawl nets and dredges dragged for fish, clams, and other species extensively damage seafloor structures and living communities. To catch fish from coral reefs, where nets are impractical, fishers often use explosives or cyanide that fragment or poison corals.

Coastal habitat losses entail major costs. In the Chesapeake estuary since 1970, hickory shad has declined 96%, American shad 66%, alewife and blueback herring 92%, and oysters 96% as a result of habitat alteration. Destruction of half the world's mangroves has eliminated an estimated 5 million tons of annual catch, which is about 5% of total world landings. In the Pacific Northwest of the United States and Canada, intensive deforestation and water diversion destroyed thousands of river miles of salmon spawning and nursery habitat, eliminating hundreds of salmon runs and thousands of related jobs.

Coral reefs support Earth's most diverse fish communities. Coral mining (for construction), fishing with explosives and poisons, destruction from boat anchors and ships running aground, and silt from farms and clear-cut logging cause major damage to coral-dependent communities. Sediment flowing off deforested land is a major threat to coral reefs, killing corals by clogging them, blocking sunlight, and preventing larval settlement.

C. Introduced Species

Aquaculture, the pet trade, sport fishing, and shipping intentionally and accidentally introduce species to new parts of the world. When ships empty their cargo, they often rebalance by taking on water. In the water are larvae, eggs, living cells, and small animals. When the ships arrive and discharge this "ballast" water, these hitchhiking species—it is estimated that more than 3000 species are in motion in ships on any given day—may be introduced to a part of the world where they have never existed. Often they arrive without the predators, competitors, or pathogens that limit them at home. Sometimes, organisms are introduced to places where the native species cannot escape or compete with them.

Invasive species greatly alter ecological communities. In San Francisco Bay, a tiny Chinese clam displaced 95% of the bottom community, and now consumes so much plankton that native plankton-eating and, in turn, predatory fish may be affected. In eastern Europe's Black

Sea, introduction of an American jellyfish reduced plankton biomass by 90% in less than a decade, causing small fishes like anchovies to crash. Reef fishes intentionally introduced into Hawaiian waters for sport fishing have affected local fish communities. The list of examples is long, and future invasions are inevitable. In one study, 5 of 80 ships entering Australia carried toxic phytoplankton, which can produce red tides and cause shellfish to become poisonous.

Even transporting salmon relatively short distances from one river system to the next for aquaculture or into hatcheries, as often happens, can move diseases into new areas, or expose fish to habitats where they are doomed by diseases for which they have no immunity.

D. Aquaculture

Many are counting on aquaculture—farming aquatic organisms—to fill the gap between nature's bounty and the hunger of people. But it is by no means clear that aquaculture actually produces more food.

Aquaculture is exploding. For example, shrimp farming increased sevenfold from the mid-1980s to the mid-1990s, with global value now exceeding $6 billion annually. Aquaculture now produces a fifth of humanity's fish. In freshwater, aquaculture now exceeds wild catch. In the oceans it produces about half the salmon and shrimp in commerce.

Aquaculture will do no more to save wild fish than poultry farms do to save wild birds. Most saltwater fish farming causes environmental degradation, displaces local people reliant on wild fish and their natural habitats but provides few jobs, does not produce long-term income growth in impoverished communities, provides food mainly for rich northern countries, and is driven by short-term economics. Aquacultural products are often expensive, and commonly exported to rich countries rather than used locally to ease the hunger of poor people.

Crowded fish farming operations pollute by releasing pesticides, antibiotics, and oxygen-robbing feed and feces into semienclosed waterways. Shrimp farming has a record of boom and bust caused by pollution and disease outbreaks. In Asia, shrimp ponds rarely remain viable for more than 5 to 10 years before being abandoned because of contamination. Thousands of Atlantic salmon have escaped in the Pacific Northwest. The state of Washington listed the Atlantic salmon as a "pollutant" because of its potential to cause problems for the several native Pacific species of salmon. Atlantic salmon have been discovered breeding in the wild in British Columbia. Even in the Atlantic, escaped Atlantic

salmon interbreeding with wild salmon cause genetic degradation, because wild salmon are genetically adapted to specific characteristics in the rivers where they spawn. Moreover, ponds and other open systems attract wild fish-eating birds such as cormorants and herons, often bringing them in range of farmers' guns.

Most importantly, most aquaculture consumes vast quantities of wild fish as feed—and is thus a net *loss* of food from the oceans. Most believe that aquaculture contributes to the supply of human food, and this is true for the few herbivorous species. But fish are not cabbages; they don't grow on sunlight. Most fish and shrimp are carnivorous, and feeding them uses two to four times as much wild fish as the farms produce. Farming such species actually contributes to the ocean's depletion.

Aquaculture does not take pressure off the seas. Wild larvae of species that do not breed in captivity are sometimes intensively caught for pen-rearing. Increased shrimp and salmon farming has not lessened fishing intensity even for those species, and has brought trouble to wild shrimp and salmon through diseases and habitat destruction.

Aquaculture can be done responsibly but as usually done now it destroys naturally productive habitats that support ocean fisheries and biodiverse wild communities, and introduces alien species, parasites, and diseases that threaten local biodiversity. Responsible aquaculture enterprises should not be sited where natural habitat is affected, its wastewater should be treated, and it should focus on herbivorous species in closed systems to prevent escape.

E. Pollutants

Pollution is the introduction of substances in quantities that are threatening to living resources and human health and activities. Chemical pollution can cause spectacular mass mortalities or cause subtle changes in population composition, impaired sexual development and reproductive success, impaired growth rates, deteriorated seafood quality, tumors and other diseases, and outbreaks of harmful algal blooms such as red tides, *Pfiesteria*, or normally innocuous algae that overgrow corals or deplete waters of oxygen. Chemical pollutants tend to concentrate in surface waters, where larvae and eggs also concentrate. Chemicals enter the seas from sewage, industrial outfall, agricultural runoff, ocean dumping, aquaculture, accidental spills, and from the air (acid rain has significantly damaged freshwater fishes in North America and Europe). Major chemical pollutants include insecticides and herbicides, detergents, PCBs, elements such as chlorine and heavy metals, petroleum products (nearly half the gasoline/oil mixture run through 2-cycle outboard engines exits unburned), mining wastes, fuel ash, radioactive materials, and excessive nutrients from sewage, farm animals, and fertilizers.

When algal blooms caused by excessive nutrients use up those nutrients and die, their subsequent decomposition can rob the water of oxygen, suffocating marine creatures. The surprising fact that fertilizer or other nutrients in excessive amounts can kill aquatic organisms is sometimes called "the paradox of enrichment."

About three-quarters of marine pollution comes from the land, creating an asymmetry whereby the people causing the pollution do not feel its effects directly or immediately, and the people feeling it (e.g., fishers) cannot directly affect its origin. That is one reason why laws are necessary. In the United States clean water legislation has helped keep some fisheries economically viable by controlling pollutants.

F. Global Atmospheric Change and Fish Conservation

Major atmospheric changes, including ozone thinning and global warming, have significant implications for marine life. Earth's climate has changed dramatically as ice ages came and went, but these changes took thousands of years, allowing life-forms long periods in which to adapt. Human-caused climate change is happening much faster, and some habitats and species will probably not change or move fast enough to survive.

The burning of fossil fuels has increased atmospheric levels of carbon dioxide, methane, and other heat-trapping gases, and most scientists agree that this is intensifying the "greenhouse effect," the warming of Earth's atmosphere. Oceans slow the buildup of greenhouse gases by absorbing about a third of the carbon produced by burning, but substantial climate change is likely over the next few decades. The effects of such change are not straightforward, but planetary warming is melting polar ice caps, causing population declines in ice-dependent krill (the shrimplike key prey species for Antarctic fishes, marine mammals, penguins, and other seabirds), and raising sea levels throughout the world. This rise is expected to cause some flooding and loss to coastal marshes, mangrove areas, low-lying islands, and their critical associated fish-nursery habitats, as well as to some cities.

Warming is likely to alter ocean currents. Such changes may intensify the recurrent El Niño phenome-

non, which changes sea temperatures and exerts a major influence on the survival and distribution of oceanic populations. In parts of the California Current, planktonic animals declined 70% over the last 40 years, a decline many scientists think may be linked to global warming. Other possible changes include alteration of currents such as the Gulf Stream and Kuroshio Current, and weakening of major upwellings off South America and Africa. Such changes would affect the production, distribution, and survival of fishes and other creatures in the world's oceans, and alter the fish availability among countries.

Certain human-generated compounds such as chlorofluorocarbons rise into the stratosphere and destroy the ozone that shields Earth from the sun's ultraviolet radiation (UV). The biologically damaging UV is called UV-B, and it penetrates many meters below the sea surface. It has increased under the Antarctic "ozone hole" and elsewhere. Also, recent volcanic eruptions (such as that of Mt. Pinatubo in the Philippines) have reduced total atmospheric ozone by as much as 10%, leading to increased UV-B of up to 20% in latitudes as far from the poles as Florida and the Bahamas. This has caused significant damage to plankton (small drifting animals and plants, including fish eggs and fish larvae), corals, and bottom-living organisms.

IV. MARINE FISH CONSERVATION IN A BIODIVERSITY CONTEXT

Biodiversity is the diversity of living things, and can be thought of as occurring on three levels: genetic diversity, which is the genetic variability among members of the same species; species diversity, the variety of species found in a community or ecosystem; and community diversity, the varieties of biological communities on Earth. We will use these categories to examine the fish conservation issues introduced above.

A. Marine Fish Biodiversity

The greater the diversity of habitats, and the more isolated habitats are from each other, the greater the diversity of species adapted to those habitats. (Though freshwater covers only 1% of Earth and ocean fish populations are vastly larger, 40% of the world's fish species live in freshwaters because they are so isolated from each other.) Because most ocean habitat diversity is coastal, so is most ocean fish diversity—there may be 10 times as many coastal fish species (roughly 13,000) as truly open-ocean fishes (about 1200 species).

B. Genetic Diversity within Species

Conservation of biodiversity is virtually never an official management goal in fisheries. Far from it; if we wanted to exterminate the fish, we could hardly have gone about it differently than we have gone about fishing.

Fishing can induce genetic changes through unintentional selective breeding because of intensive, nonrandom fishing mortality. In intentional artificial selection (as for farm animals) we choose and breed individuals with traits that we desire. But in fishing, we remove from the breeding population those individuals with the traits we like.

Fishing activities can cause evolution. Evolution is a change in the relative frequency of genes. It need *not* result in "progress" or new species; it merely shapes species for better survival under prevailing, often changing, conditions (if change is extensive, new species can result). Fishing is so intense in so many places that many fishes have a higher chance of dying by being caught by people than by natural causes (for instance, in some populations 80–90% of adult cod or salmon are killed by fishing; in such situations the probability of survival becomes vanishingly small).

Fishing can alter inherited characteristics of a population in two ways: (1) by applying a selective pressure that kills individuals based on certain traits, such as size or age; and (2) by applying a random pressure so intense that the population is depleted low enough to lose genes from the pool. Note that fishing *can* do these things, and it does both of them sometimes, but not all fishing results in genetic changes; whether it does is usually determined by fishing intensity.

Genes related to size, growth rate, and age of sexual maturity are most likely to be affected by fishing. "Growth overfishing" prevents fish from reaching large size, thus incidentally selecting *for* fish genetically preprogrammed to mature and breed younger and at smaller size than normal. Fish that reproduce pass their genes on. The genes of fish that get killed before breeding will begin dying out. Fish genetically programmed to breed at slightly younger age than average are also likely to have reduced growth potential. Thus, intense fishing that allows few individuals to live long will allow early-maturing fish to make a disproportionately large contribution to the next generation, and allow normally maturing fish to make a disproportionately small contribution. In a study modeling the contribution of three age groupings of cod to overall reproduction, the late-maturing subgroup made virtually no contribution to reproduction after only 7 years of intense fishing. Such a scenario indicates the likelihood of artificially driven evolution in wild populations. So fishing can unintentionally breed a population of younger-reproducing, shorter-lived, smaller fish.

Probably most intensive fisheries create conditions necessary for genetic change, but genetic changes can be difficult to confirm. The reduced average sizes and first-breeding ages caused by killing most fish before they live long do not necessarily reflect accompanying genetic change. Whether genetic or not, examples abound of population-level effects of overfishing. In a rockfish (*Sebastes alutus*), lightly exploited populations had a modal age of 30 years, and 73% of the fish were over age 20; heavily exploited populations had a modal age of 12 years, with only 7% of the fish over age 20.

Average sizes of many U.S. Southeast species declined about 75% between the early 1970s and the mid-1980s: red porgies went from 2.6 pounds to 1.3 pounds, red snapper from 18.0 to 4.4 pounds, snowy grouper from 17.6 to 4.4 pounds, speckled hind from 19.1 to 6.6 pounds, scamp from 10.1 to 3.3 pounds, and gag grouper from 18.0 to 4.4 pounds. For species such as groupers, which change sex in older age, intense removal of larger individuals can dramatically affect sex ratios and mate supply, greatly lowering the population's reproduction. Jim Bohnsack of the National Marine Fisheries Service says that if almost all the big ones are taken and the little ones left, "The result is a race of miniatures."

Overfishing reduced the California sardine population's age structure from five spawning age classes to only two, and two consecutive years of poor oceanic conditions led to spawning failure and collapse. However, even when vast populations of naturally small, short-lived fishes such as sardines, anchovies, and herrings collapsed to one-thousandth of their former numbers, high enough numbers have remained to maintain biodiversity. Yet in some animal populations, older or larger individuals are more likely to carry more "heterozygosity"—differing forms of particular genes. Larger individuals often spend more time on breeding areas than younger ones, so fisheries targeting spawning groups, such as those for groupers, orange roughy, and many others, can subject the most genetically diverse individuals to the most intense mortality. This kind of genetic diversity loss is believed by some to be a common consequence of heavy exploitation, even without reducing populations to near-extinction levels.

Theory suggests that genetic change from fishing is likely to be common, and several studies have found convincing evidence. Perhaps the species best-studied before and after commercial exploitation is a long-lived fish called the orange roughy. Off New Zealand, only

6 years of heavy commercial exploitation reduced populations 70% and significantly reduced genetic diversity within those populations, probably because older individuals in this very long-lived species were more heterozygous.

In one North Sea cod population, intensive fishing reduced the chances of a young cod surviving to breeding age by 95%. Over several decades, average age of sexual maturation declined from about 10 to 7–8 years, apparently through genetic change. Some populations of flounder and haddock also show convincing evidence of genetic change. In a population of Atlantic salmon, average age at maturity had dropped after two decades of heavy fishing. In a study of chinook salmon, average spawning age fell by 2 years. Other studied salmon showed other changes.

In sum, some studies found no apparent changes attributable to fishing, some showed changes that were inconclusively genetic, and other studies showed genetic changes that were driven by intense fishing.

C. Species Diversity

Human activities significantly change species composition and abundance, and predator–prey and competitive relationships. Naturally evolved numerical and functional relationships among species are sometimes referred to as the community's "ecological integrity."

One indication that many regions' ecological integrity is greatly diminished is the changing world catch. Since1950, fisheries, forced to work lower on the food chain as they deplete large fish, have shown a gradual transition from catching mostly large, long-lived, fish-eating bottomfish such as cod toward catching small, plankton-eating, open-ocean fishes like herring, as well as short-lived, low-on-the-food-chain invertebrates like squid. Daniel Pauly, the eminent ecologist who helped discover this trend, remarked, "If things go unchecked, we might end up with a marine junkyard dominated by plankton."

Various marine communities have been changed by overfishing. A common pattern—particularly acute in coral reef systems but also seen in kelp communities and elsewhere—is selective removal first of the largest species, then of large individuals of smaller species. People removed most large groupers and sea basses from many coral reefs and kelp forests, and manatees, dugongs, and sea turtles from seagrass communities, before scientists ever studied these habitats. Though no one really understands how those communities are evolved to function, they were certainly very different.

Since Columbus landed in the Bahamas, for instance, Caribbean sea turtle populations have declined 99%.

Depletion of fish populations changes patterns of abundance, distribution, and competition among plants, urchins, corals, sponges, tunicates, and other creatures. Alarming declines in North Pacific seabirds, Steller's sea lions, and other wildlife may be caused by heavy fishing. When fishing depleted Barents Sea herring and capelin, cod failed for lack of food. Squid sometimes *increase* following the collapse of their overfished predators. Off New England and maritime Canada, overfishing shifted communities dominated by cod, haddock, and flounder to domination by spiny dogfish (a small shark) and skates (a type of ray).

But not even skates are safe. Fishing has driven the commonest skate in the northeast Atlantic, *Raja batis*, to local extinction in the Irish Sea. The barndoor skate was one of the Northwest Atlantic's largest and most numerous skates in the1950s, appearing in 10% of research samples. Once estimated to number more than half a million off Newfoundland alone, not a single barndoor has been caught in that area since the 1970s, and if trends continue this skate could become the first well-documented extinction in a fully marine fish.

1. Vulnerability of Marine Fishes to Extinction

Few fully marine creatures, and perhaps no fully marine fish species, have gone extinct because of human activities. Many marine fishes have large distributions and a greater chance for replacement by neighboring populations if they are locally wiped out.

But the important thing is to prevent, not document, endangerment and extinctions. The time to save a species is when it is still common. Waiting until biodiversity declines, until populations falter, and until more species names get added to various endangered lists will only ensure three things: (1) more species will get into serious trouble, (2) the value to humanity from extremely important natural resources such as fishes will be seriously diminished, creating more of the economic, social, and nutritional problems already suffered by people in too many communities throughout the world, and (3) fixing the problems will be much harder, more expensive, and prone to failure.

Numerous fish have already been listed on the U.S. Endangered Species List, on the U.S. Overfished Species List, and under the world Convention on International Trade in Endangered Species (CITES) and the World Conservation Union's "Red List." The Red List contains over 100 marine fish that have suffered major and rapid population declines or significant local extinctions (extirpations).

Most vulnerable are the anadromous species—those that breed in freshwater but mature in salt water—such as salmon and sturgeons, because they suffer overfishing plus the effects of habitat disruption in rivers that are heavily dammed, subject to intensive agricultural water withdrawals, or subject to massive siltation from destructive logging practices. Hundreds of North American salmon runs are already extinct, and the trends for sturgeons and totoaba (a large fish of the croaker family that is dependent on Colorado River flow into the Gulf of California) suggest that biological extinction may become a real possibility.

Groupers are also vulnerable; they are generally slow-growing fishes whose spawning groups are easily targeted by fishers. Long-lived, slowly reproducing species such as large sharks and sawfishes are particularly vulnerable, and many such species are now depleted and threatened. Several large sharks, including the great white, are now totally protected in some regions, an acknowledgment of their threatened status.

No doubt endangerment to fishes is underappreciated simply because their populations are not well studied. Most of the 100-plus species listed on the World Conservation Union's "Red List" got there simply because the London Zoological Society and World Wide Fund for Nature convened a workshop to compare known information about fish population changes against the Union's listing criteria. Similarly, the barndoor skate became endangered without anyone realizing it, simply because no one had analyzed decades of already-existing data.

Not just fish are affected. Wild abalone populations off California have been greatly reduced by fishing, and the white abalone is almost certain to go extinct in the wild because probably too few individuals are left to breed effectively. Who would have thought that fishing would be intense enough to drive a snail to extinction?

2. Kinds of Extinction

Overfishing and habitat degradation can lead to a four-step process of extinction, in which people suffer the major effects long before the last animals vanish. From social, economic, and genetic standpoints, extinction is a process rather than an event; fishing economies usually go extinct and gene pools lose diversity long before the last fish dies.

The first stage of the extinction process is depletion, wherein the fish population is reduced below its most productive level; fish are smaller and fewer than they could be, and less likely to spawn a strong year class of young fish. Consequently, the population's ability to support fishing is reduced, and profit margins decline. Other members of the ecosystem may experience food shortages or an unnatural relaxation of predation pressure or competition.

The next stage is ecological extinction, where the animal's population drops so low that the species no longer effectively functions as prey, predator, or competitor in the community. If it is an important "keystone species" affecting many others, entire marine communities may undergo profound shifts in numerical and functional relationships, and species less valued by people may increase. Animals that are ecologically extinct—essentially removed—from the southern California kelp community, for example, include sea otters, black sea bass, white sea bass, large groupers, all the large kelp bass, sheephead, rays, flatfish, rock fish, lobsters, abalones, sea urchins, sea cucumbers, and others.

Another stage, commercial extinction, occurs when the species becomes too rare to be fished profitably. During this phase, fishery managers sometimes close large areas to fishing, as off New England and Newfoundland, or in the salmon fisheries off the Pacific U.S. coast.

If commercial and ecological extinction are not reversed, total extinction may become a possibility. But by the time total extinction becomes an issue, all the other practical effects of the creature's disappearance from commerce and from the ecosystem have already been suffered by people. One cannot ignore the community effects of overfishing on the top predators—humans. Through dislocation, malnutrition in remote local villages, job loss, and loss of social identity, humans suffer the major effects of overfishing long before the animals themselves completely vanish.

To date, total extinctions are rare in the oceans, but this may not always remain so. We may unfortunately see more fish on endangered species lists in the future. If we do, it may not be too late to save them. But it will likely be too late to save the fishing jobs and coastal communities that once depended on them. A recovery vision should always include regaining naturally evolved numerical and functional relationships among species.

D. Ecosystem Diversity and Fishes

Though the ocean may seem like one big bathtub, to the creatures in it and the people making a living from it, the sea is a mosaic of habitat types. Water circulation patterns can create discrete habitats because temperature alone can be a boundary for sea life. Currents also cause drifting seaweeds to accumulate in places like the Sargasso "Sea," an area in the middle of the Atlantic characterized by vast mats of sargassum weed used as habitat by many creatures. The sediments of the seafloor vary in texture, from boulders to fine silt, all creating different habitats. The area between high and low tides—the intertidal zone—likewise contains particular habitats for specially adapted creatures that live on rocks, in sand, or in tide pools. Along coasts, shallow "estuaries," where freshwater from rivers mixes with salty seawater, have their own suite of habitats: seagrass flats, salt marshes, bays between outer beaches and mainlands, and tidal inlets. Tropical and subtropical shores often support mangrove forests growing in salt or brackish water. Near warm shorelines, coral reefs support the most spectacular diversity of fishes and other creatures in the sea.

The distribution pattern of these major habitat types is what we speak of with the term "ecosystem diversity." Just as fishery management has generally ignored genetics, it has also generally ignored ecosystem considerations. Fishing activities, aquaculture, coastal development (for instance, damming rivers and filling wetlands), and atmospheric warming have resulted in changes in the distribution and functioning of habitats upon which marine communities rely, that is, changes in ecosystem diversity.

Aquaculture has significantly altered coastal and estuarine habitats in many parts of the world. The conversion of mangrove habitats into fish and shrimp farms over extensive areas is a form of ecosystem depletion. Coastal development has also destroyed habitat in ecologically significant areas. For instance, about half of the wetlands in the United States have been lost to practices like filling, which has led to a significant shift in the distribution of aquatic habitats. Atmospheric warming is both melting Antarctic ice and stressing coral reefs. This is causing change in these habitats on a global scale.

Coral reefs and continental shelves are two major kinds of habitats experiencing profound changes due to human activities. Small increases in temperatures in tropical seas have been implicated in coral "bleaching," whereby corals lose the helpful algal cells living inside them, turn white, and may die. Widespread death of corals, of course, has potentially severe implications for coral-dependent fish communities.

Coral reefs are the oldest and largest structures built by living things. When scientists first surveyed them in the1950s and 1960s, most coral reefs were poised on the edge of profound change in the composition and abundance of species and the functioning of coral reef communities. For example, Caribbean overfishing had already depleted herbivorous fishes to very low levels when algae-eating sea urchins crashed from a disease. Without herbivores, algae overgrew corals, killing them over large areas, and altering the community. Conversely, heavy fishing off East Africa depleted fish that prey on coral-eating sea urchins. With their predators largely fished-out, the urchin populations exploded to densities that were 100 times normal, and proceeded to inflict significant damage to reefs. The depletion of predator fishes has also been implicated in destructive outbreaks of the coral-eating sea star known as the crown-of-thorns. Such ecological changes affect large communities of organisms.

Fishing often kills corals directly when people use explosives to kill fish, or pound the coral with rocks on lines to scare fish toward nets, or use cyanide to stun fishes. All of these methods cause extensive destruction of coral reefs, through either breakage or, in the case of cyanide, death to corals by poisoning.

Coral reef systems are perhaps more sensitive to environmental change than most other marine ecosystems. For various reasons, species of algae may overtake a damaged reef prevent corals from recovering or recolonizing. Many coral reefs are now degraded by human activities, their fish communities modified, and the living corals replaced by algae growing on dead coral heads. Some areas of the Caribbean lost 50–90% of their live corals in the decade from the late 1980s to the late 1990s, because of overfishing, diseases, algal overgrowth, sewage overfertilization, and smothering silt running off deforested land areas. Some scientists believe these changes may be very long-lasting.

Aside from coral reef areas, continental shelves nearly everywhere are affected by bottom trawling—the towing of large bag-shaped nets or shellfish dredges over the seafloor. Trawling is the fishing method that takes half of the world catch. The vast majority of the world's seabed is encrusted or honeycombed by living things, and trawling causes enough incidental damage of nontarget invertebrates, plants, shellfishes, and fishes to cause major community changes over large areas. Trawling crushes, kills, exposes, and removes these living sources of nourishment and hiding places, making life dangerous for young fishes and almost certainly lowering the habitat's capacity to produce abundant fish populations. In one study of a scallop fishery, only 12% of the scallops in an area were actually caught, but many of those not caught were crushed by scallop dredges.

Until just a few years ago, trawlers were unable to work on high-relief, bouldery, or rubble-strewn bottom habitats without risking hanging up and losing their nets and gear. But nowadays, new kinds of trawls make every kind of seabed—whether silt, sand, clay, gravel, cobble, boulder, rock reef, worm reef, mussel bed, seagrass flat, sponge bottom, or coral reef—vulnerable to trawling. Furthermore, trawls hit many areas repeatedly each year. The seafloor structures that juvenile fishes rely on are often easily destroyed by bottom trawls, which kill or disperse the organisms that create them. Our focus on extracting edible fishes at all costs has effectively blindered fishery managers to the essential food and shelter that these fish themselves require.

Trawling is not uniformly bad for all species or all bottom habitats, and in fact some species do better in trawled than in undisturbed habitats (just as a few species do better in logging clear-cut areas). But most species are not helped by trawling, and some communities are seriously damaged. Recovery of certain bottom communities could require up to a century even if trawling was stopped today. Writing in the journal *Nature*, Daniel Pauly and Villy Christensen observed that, "Contrary to some terrestrial systems such as rainforests, of which large undisturbed tracts still exist,... the overwhelming bulk of the world's trawlable shelves are impacted by fishing, leaving few sanctuaries where biomasses and biodiversity remain high."

V. THE COURSE AHEAD

A. Marine Protected Areas

In times past, fisheries benefited from de facto refuges: undiscovered locations or places too deep or remote to probe. But now, because of new technologies, when fish run they can't hide. Facing this reality, some scientists have proposed that simply closing some areas to fishing and resource extraction, to allow them to function naturally, while leaving adjacent areas open, is the best way to manage fishing. These closed areas have been suggested especially where precise estimates on population sizes and sustainable catch levels are lacking and where gathering lots of data or intensive management and monitoring are impractical.

Such "marine reserves" allow fish to go through their normal patterns of growth, abundance, reproduction, genetic diversity, and community structure. Marine reserves can (1) protect critical spawning adults, (2) maintain natural size and age structure, (3) maintain genetic diversity, (4) prevent "serial overfishing," where one species and then the next and the next are depleted, (5) maintain natural communities, while allowing fish-

ing to continue nearby, and (6) provide scientific areas for the study of more naturally functioning systems. Few such reserves exist; indeed, fishing is even allowed in all U.S. National Marine "Sanctuaries." Marine reserves have been established in several parts of the world, and they can increase fish yields in adjacent areas as populations recover, fish reach their normally evolved maximum size, and their offspring begin wandering outside the reserve. In the Philippines, catches adjacent to one reserve tripled within two years of the reserve's designation. In a small Caribbean reserve, overall biomass of commercially important species increased 60% and snappers increased 220% in two years; but groupers did not increase, probably because severe depletion eliminated any source of larvae.

Eggs and larvae of fishes and other marine creatures may drift many miles from their source, so reserves must either be self-sustaining sources of fish or areas where incoming juveniles can grow and reproduce. Ideally, reserve networks should be designed to maintain genetic and community diversity over large regional areas. Large adults are disproportionately fecund—a single female red snapper 24 inches (61 cm) long produces as many eggs (about 9 million) as *200* 16-inch (41 cm) females. One study estimated that if 20% of the red snapper's Gulf of Mexico habitat was protected, total egg production would be 1200% greater than under current fishing pressure. Total fish catch would increase even though a fifth of the area was off-limits to fishing.

For more migratory fish, seasonal and area closures of areas such as spawning and nursery grounds could be helpful. For example, longlines catch mostly immature swordfish in certain areas. Those areas should be closed to long-lining during times where breeders or juvenile fish concentrate. Reserves are generally opposed by fishers at first, but in New Zealand they worked so well that 10 years after their establishment nearly 80% of fishers wanted *more* reserves.

B. Other Solutions, and Reasons for Optimism

The current situation is poor, but *not* bleak. There are reasons for optimism. For one, we know enough about many human-induced problems, especially overfishing and habitat alteration, to fix them. Clearly one of the most important things that could be done for overfishing and bycatch is to remove the artificial tax breaks, supports, and all the other subsidies that are propping up fisheries incapable of existing off the resources. And many fishes (though not sharks) have high reproductive

Smuggling Whales

Though whales are mammals, not fishes, whaling is sometimes considered a "fishery" and is similar to fishing in many aspects.

"Are the whale products available today exclusively from species hunted or traded in accordance with international treaties? A recent spot check of Japanese retail markets shows that they are not and that the existence of legal whaling serves as a cover for the sale of illegal whale products." That was the question and the answer reported by two marine biologists in the journal *Science* in 1994.

Japan legally kills 300 minke whales in Antarctic waters each year, ostensibly for "scientific" purposes. But the "samples" go to food markets, and the markets serve as continued incentive to hunt whales illegally. Iceland recently quit the International Whaling Commission to evade the ban on whaling, and Norway simply resumed whaling for minkes in nearby waters.

Sometimes the meat labeled "minke whale" is really from endangered species and from parts of the world where the great whales are supposed to be protected after being hunted to near extinction. While only minke whale meat can be legally sold in Japan, DNA tests of fourteen samples purchased in commercial markets in 1993 indicated that "minke" meat included minke, humpback, and finback whales from the North Pacific, North Atlantic, Australian region, and Antarctic, as well as dolphins. "We were surprised by the large number of species and the fact that they came from all over the world," wrote Steve Palumbi, now at Harvard University.

Smuggling whales is not difficult. Recently 260 tons of whale meat labeled as Norwegian shrimp was seized in Russia en route to Japan. This is why the arguments about sustainable hunting of minke whales are flawed and sometimes fraudulent, and why any whaling may prevent recovery of endangered whales in the vast areas of the world's oceans where they are still vanishingly rare.

The Atlantic gray whale was apparently hunted to extinction, but whales stand a good chance of recovery when effectively protected. This has been proven by increases of several great whale species in the Northwest Atlantic, humpbacks that breed in Hawaii, blue whales off California, and the recent full recovery of the Pacific gray whale—one of the great successes of the U.S. Endangered Species Act.

potential and may be able to repopulate relatively quickly if fishing and habitat degradation are controlled. Where this has been done, protected populations have often shown the ability to increase rapidly.

Some areas should be closed to trawling even if left open to stationary fishing gear such as traps or hook-and-line, which do not destroy habitat. New Zealand and Australia have closed areas to bottom trawling, as have some U.S. states. Trawls should be allowed on shallow sandy bottoms that are relatively resistant to disturbance, and barred from harder, higher-relief, and deeper bottoms where their damage is much more serious. Government and industry should create incentives for developing fishing gear that does not degrade the very habitat upon which the fishing communities depend. Intelligently designed financial incentives for encouraging new problem-solving technology could tap fishers' inherent inventiveness in constructive ways.

Reducing the problems of species introductions will require methods for safely sterilizing ship ballast water, and controlling aquacultural movements of species and their hitchhikers and pathogens.

Public interest and commitment regarding the oceans are much higher in just the last few years than ever before, and conservationists *are* discovering that fish are wildlife. Consumers hold tremendous power to direct change, but they have hardly been informed about how they can "vote with their wallets" in the marketplace to favor sustainable fishing. Initiatives like the Audubon Seafood Lover's Almanac and the Marine Stewardship Council (the latter is a global voluntary incentive program wherein participating seafood producers receive an on-pack logo saying this product is sustainably produced) are beginning to help empower consumers. Much more can be done in this area.

Many discussions of fisheries highlight the need for more information in order to manage wisely. But one does not need to fully understand ecosystems to limit obviously detrimental human behavior. Most of what limits our ability to manage is political resistance and lack of enforcement—better knowledge will not necessarily lead to better management (see Box 6). Reducing the undue influence of the irresponsible segments of the fishing industry in management arenas, and their incessant political pressure to allow higher and unsustainable catches, is a necessary first step.

At present, in almost all parts of the marine realm, exploitation reigns and conservation begs indulgence. The burden of proof must shift, so that conservation and restoration are the rule and exploitation becomes an effectively regulated privilege. Otherwise, the kingdom of the sea will sink deeper into poverty, creating more and more problems and hunger along the world's increasingly crowded coasts. Of course, human over-population and overconsumption drive or worsen each of the problems discussed here.

Anyone interested in conservation can join organizations working in this area, write letters to agency and government officials, speak at meetings and hearings, and write opinion pieces and letters to editors of various publications. Making your voice heard is the most important thing.

The solutions to overfishing and marine habitat destruction depend on political will, but politics reflects public opinion, and that means that politics can be changed. Already, because of changing political "sea-scapes," several formerly unlikely long shots have been achieved—including the global ban on drift netting, the strengthening of U.S. fisheries law toward ending overfishing and hastening the rebuilding of depleted populations, several new international fishing treaties and agreements, and increasing acceptance of the "precautionary principle" that says we should act conservatively, consider future needs, and so avoid irreversible change. Only by increasing our political involvement and effectiveness will we be able to improve the ecological health and protect the biodiversity of the world's oceans. Conservation itself is political activity in the service of living nature.

See Also the Following Articles

AQUACULTURE • COASTAL BEACH ECOSYSTEMS • ESTUARINE ECOSYSTEMS • FISH, BIODIVERSITY OF • FISH STOCKS • GREENHOUSE EFFECT • KEYSTONE SPECIES • MARINE ECOSYSTEMS, HUMAN IMPACT ON • REEF ECOSYSTEMS • SALMON

Bibliography

Bohnsack, J. (1996).Marine reserves, zoning, and the future of fishery management. *Fisheries* 21, 14–16.
Carlton, J. T. (1996). Marine bioinvasions: The alteration of marine ecosystems by nonindigenous species. *Oceanography* 9, 36–43.
Casey, J. M., and R. A. Myers. (1998). Near extinction of a large, widely distributed fish. *Science* 281, 690–692.
Jackson, J. (1997). Reefs since Columbus. *Coral Reefs* 16, S23–S32.
Myers, N. (1997). *Perverse Subsidies; Their Nature, Scale and Impacts.* The MacArthur Foundation, Chicago.
Myers, R. A., and N. J. Barrowman. (1996). Is fish recruitment related to spawner abundance? *Fishery Bull.* 94, 707–724.
Naylor, R., *et al.* (1998). Nature's subsidies to shrimp and salmon farming. *Science* 282, 883–884.
Norse, E. A. (1993). *Global Marine Biological Diversity.* Island Press, Washington, D.C.
Pauly, D., and V. Christensen. (1995). Primary production required to sustain global fisheries. *Nature* 374, 255–257.
Pauly, D., V. Christensen,, J. Dalsgaard,, R. Froese,, and F. Torres, Jr. (1998). Fishing down marine food webs. *Science* 279, 860–863.
Roberts, C. M. (1995). Effects of fishing on the ecosystem structure of coral reefs. *Conservation Biol.* 9, 988–995.
Safina, C. (1993). Bluefin tuna in the West Atlantic: Negligent management, and the making of an endangered species. *Conservation Biol.* 7, 229–234.
Safina, C. (1994). Where have all the fishes gone? *Issues Sci. Technol.* 10 (Spring), 37–43.
Safina, C. (1995). The world's imperiled fish. *Sci. Amer.*, November, pp. 46–53.
Safina, C. (1998). *Song for the Blue Ocean.* Henry Holt, New York.
Smith, R. C., *et al.* (1992). Ozone depletion: Ultraviolet radiation and phytoplankton biology in Antarctic waters. *Science* 255, 952–959.
Sutherland, W. J. (1990). Evolution and fisheries. *Nature* 344, 814–815.
Watling, L., and E. Norse, *et al.* (1998). Special Section: Effects of mobile fishing gear on marine benthos. *Conservation Biol.* 12, 1178–1240.

FISH STOCKS

Daniel Pauly* and Rainer Froese[†]
*Fisheries Centre, University of British Columbia, and [†]International Center for Living Aquatic Resources Management, Manila

I. Major Adaptations of Fishes
II. Respiratory Constraints to Growth and Related Processes
III. Distribution of Exploited Fish Stocks
IV. Ecosystem Impacts of Fisheries
V. Managing Fish Biodiversity Information
VI. Preserving Fish Biodiversity

GLOSSARY

biomass Collective weight or mass of all the members of a given population or stock at a given time, or, on the average, over a certain time period.

bioquads Occurrence record of organisms, serving as key units for biodiversity research and consisting of four elements (species names, location, time, and source).

catches The fish (or other aquatic organisms) of a given stock killed during a certain period by the operation of fishing gear(s). This definition implies that fish not landed, that is, discarded at sea, or killed by lost gear ("ghost fishing"), should be counted as part of the catch of a fishery.

ecosystem Area where a set of species interact in characteristic fashion, and generate among them biomass flows that are stronger than those linking that area to adjacent ones.

recruitment Entry of juvenile fish into the (adult) stock. Recruitment is distinguished from reproduction, because the eggs and larvae that result from

fish spawning usually suffer tremendous and largely unpredictable mortalities, thus uncoupling spawning from recruitment.

trophic level A number indicating the position of a species within an ecosystem though the number of steps linking it to the plants. By definition, plants are TL = 1, herbivores are TL = 2, and so on. Note that trophic levels do not need to be whole numbers; intermediate values occur among omnivorous consumers.

FISH STOCKS ARE POPULATIONS OF "FISH," THAT IS, VERTEBRATES WITH GILLS AND FINS, SUBJECTED TO EXPLOITATION BY HUMANS. Populations are components of species, inhabiting part of their overall range, and usually having little genetic exchange with adjacent populations. The major adaptations determining the spatial distribution of fish stock biomass pertain to the anatomy, reproductive biology, and respiratory physiology of the species to which the stocks belong. Also, fishing has become increasingly important to the biodiversity of fish, either through its direct impacts (changes of stock size and age structure, and overall biomass reductions, down to extirpation of populations), or by modifying the ecosystems in which they are embedded. Research devoted to monitoring the biodiversity of fish (or other organisms) must be able to handle large amounts of suitably formatted distributional information, here defined as consisting of "bioquads." Management regimes aiming at preserving fish biodiversity will have to include much stricter regulation of fishing and the establishment of no-take areas.

I. MAJOR ADAPTATIONS OF FISHES

A. Anatomy and Physiology

With about 25,000 recognized species in over 500 families, fish are the most diverse vertebrate group. However, their watery habitat, while failing to protect them from modern fishing gear, makes it difficult to fully appreciate this diversity, and the extent to which it is now threatened. It is even more difficult for us, as air breathers, to perceive the constraints under which fish, as water breathers, were forced to evolve.

Water is an extremely dense medium, 775 times heavier and 55 times more viscous than air. Also, water contains 30 times less oxygen than air, and this oxygen diffuses 300,000 times more slowly than in air. These physical constraints, which shaped all early life-forms, including the jawless predecessors of the fish, the agnathans, are best visualized by describing the major evolutionary trends leading from agnathans to modern fish (Fig. 1A).

The first of these trends was the evolution of jaws from the first upper and lower gill arches of agnathans. This built on the intimate connection, in the most primitive vertebrates, between the feeding apparatus (i.e.,

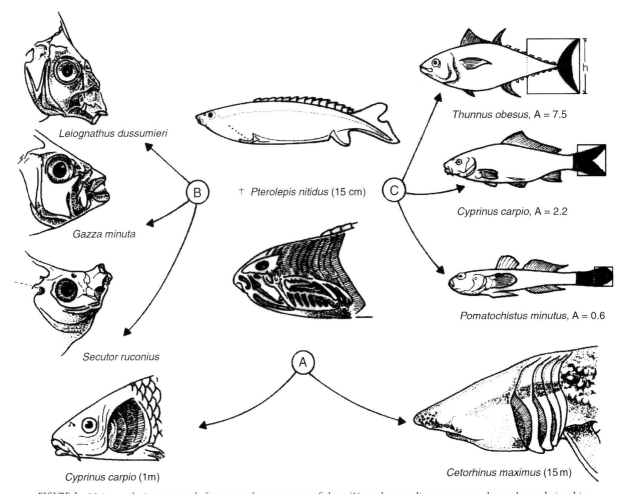

FIGURE 1 Major evolutionary trends from agnathans to extant fishes. (Note that no direct ancestor–descendant relationships are implied among the groups depicted.) (A) Trends toward larger gills; (B) trends toward efficient jaws; (C) trends toward effective paired and unpaired fins. [Note the aspect ratio of the caudal fin, defined by $A = h^2/s$, where h is the height and s the surface (in black) of the caudal fin, and of which high values define fast, large-gilled continuous swimmers, and conversely for low values.]

the mouth) and the respiratory apparatus (i.e., the gills adjacent to slits on both sides of the anterior part of the alimentary canal). Water-breathing invertebrates lack this close connection between feeding and breathing, one reason why even the largest among them (giant squids) cannot reach the mass of the largest fish (20 metric tons, for the whale shark *Rhincodon typus*).

The reorganization of the head of early fish allowed larger gills to evolve, which allowed the higher metabolic rates required for swimming in open waters. This transition was assisted by the gradual loss of the heavy armor protecting the slow, bottom-slurping agnathans. The fine "teeth" covering the bodies of sharks are vestiges of this armor.

Fast swimming in open water required better fins, both for propulsion and for steering. Propulsion is provided in most fish by oscillations of a caudal fin whose aspect ratio (Fig. 1C) gradually increased toward tunas and other derived, fast-swimming groups with very large gills. Steering, on the other hand, is provided by dorsal, pectoral, and anal fins. These fins are stiffened for precise action by hard, bony rays in the most derived fish, the teleosts, whose evolutionary success was further enhanced by a complexly built protrusile mouth that enables capture of a wide range of food items (Fig. 1B).

Subtle anatomical changes in fish can thus create more niches for increasing the numbers of specialists, which then occupy increasing numbers of closely packed niches. Ecosystems in which these changes have run for long periods, undisturbed by physical changes, therefore contain very large numbers of fish species. Their numbers are even larger in areas such as the Great Lakes of Africa and the tropical Indo-Pacific, where changes of water levels have repeatedly isolated basins and subpopulations, thereby accelerating species differentiation (Fig. 2).

B. Reproduction and Recruitment

Though many ancient fishes such as sharks and rays or the coelacanth *Latimeria chalumnae* practice internal fertilization and produce few large eggs or live offspring, most recently evolved fishes produce numerous small eggs that are fertilized externally and develop as part of the plankton, without parental care. The larvae that emerge from those eggs, after less than one day in warm tropical waters and up to two weeks (and more for larger eggs) in cold temperate waters, are usually elongated, as befit small, finless zooplankton feeders.

The average zooplankton concentrations that these

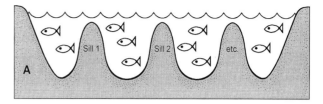
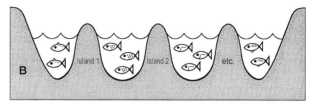
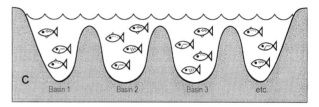

FIGURE 2   Schematic representation of how changes in water level can multiply, by creating isolated subpopulations, the number of species occurring in a given area. Such a mechanism, driven by repeated climatic changes, is thought to explain the large number of fish species in Southeast Asian marine waters and in the Great Lakes of Africa.

larvae encounter, even during spawning seasons attuned with zooplankton production cycles, are usually far too low to allow survival of fish larvae, and the overwhelming majority of such larvae perish. Those that tend to survive usually happened to have hatched within plankton-rich water layers. These layers are usually a few centimeters thick and last for only a few days of calm, between wind-driven or other mixing events, such as storms or upwelling pulses, that enrich surface waters with nutrients from deeper waters. This implies that large biomasses of fish can build up only when and where the local oceanographic conditions take the form of "triads" defined by (1) nutrient enrichment, such as generated by wind-driven mixing, (2) high plankton concentration, such as generated by various mechanisms including fronts, and (3) retention of larvae, required to prevent these weak swimmers from drifting away from suitable habitat. In pelagic fishes that build high biomass, for example, the anchovies and sardines in coastal upwelling systems off northwestern and southwestern Africa, Peru, and California, these triads occur only when the coastal winds range from 4 to 6 m per second. Weaker winds do not generate enough enrichment, and stronger winds disperse the larvae offshore.

Fish have developed several strategies to deal with the uncertain recruitment that results from the triad requirements. One is being small, short-lived, and capable of quickly building up large biomass under favorable environmental conditions. The other is being large, long-lived, and capable of weathering long series of recruitment failures through repeated spawning by old, large, and highly fecund adults. An example of the former strategy is provided by the Peruvian anchovy *Engraulis ringens*, whereas the northern cod, *Gadus morhua*, provides an example of the latter. Yet another strategy is to reduce the dependence on environmental conditions by various forms of parental care, such as nesting and guarding (e.g., in catfishes, family Clariidae), mouth-brooding (e.g., in cardinal fishes, family Apogonidae), and live-bearing (e.g., in ocean perches, genus *Sebastes*).

Another important feature of fish stocks is that, contrary to earlier assumptions of homogeneity, most appear to consist of well-differentiated individuals, each aiming to reproduce at the very place where it was hatched. Or, put differently: most migratory fish tend to "home." This behavior, well documented only in Pacific and Atlantic salmon (*Oncorhynchus* and *Salmo*, respectively), implies that individual fish, when reproducing, do not seek "optimal" sites, but rather spawn as close as possible to the site at which they hatched, and to which they are imprinted. This tendency to either stay in or return to a certain area makes it difficult for fish stocks to rebuild once they have been decimated by local overfishing or pollution.

II. RESPIRATORY CONSTRAINTS TO GROWTH AND RELATED PROCESSES

A. Basic Geometrical Constraints

Fish growth, as in other animals, requires both food and oxygen, the latter being required to synthesize the substance (adenosine triphosphate or ATP) that serves as fuel to all organisms. For oxygen to be metabolically available, it must be inside the fish body, that is, it must have passed though its gills. Thus, since oxygen cannot be stored inside the fish body (contrary to food, which can be stored as gut contents and as fat), the metabolic and growth rate of fish are largely proportional to the surface area of their gills. So fish that quickly reach large sizes have gills with large surface areas (as in tunas), and conversely in slow-growing fishes (like groupers). Moreover, gill area per unit of body mass declines with size, because the two-dimensional gill area cannot keep up with the three-dimensional increase of body mass. Hence larger fish dispose of relatively less oxygen to supply their metabolism, the reason why they ultimately stop growing. Also, environmental factors that tend to increase metabolic rate—especially elevated temperatures, but also including other form of stress—have the effect of reducing the maximum size that the fish of a given population can reach (Figs. 3A and 3B). This is why tropical fish tend to be smaller than their respective cold-water relatives. A similar mechanism explains the nearly constant relationship in fish between size at first maturity and maximum size (Figs. 3C and 3D).

B. Adaptation to Respiratory Constraints

Fish have evolved various strategies and tactics to overcome respiratory constraints. One strategy, illustrated in Fig. 1B, is to evolve large gills, a route taken by numerous open-water ("pelagic") species, culminating in tunas (Fig. 4).

Another strategy is the evolution of life cycles in which the juveniles migrate to deeper, cooler waters as they grow and then, upon maturing, produce eggs that quickly float up to the warmer surface layers, out of reach of the often cannibalistic adults. Such typical cycles are completed by an onshore drift of the larvae to coastal areas, and productive shallow nurseries for the early, voracious juveniles, which again migrate into deeper waters as they grow.

A tactic to accommodate metabolic stress, which is particularly useful in areas with strong seasonal temperature oscillations, is for the feeding adults to store fat during the warmer part of the season (late summer to early fall). Fat requires far less oxygen for maintenance than protein of muscle and other tissues. As temperature declines, the accumulated fat is converted into other tissues, notably gonads, whose contents are shed in spring, thus reducing body mass when temperatures again start to increase. These cycles, which use fat as protection against respiratory stress, are the reason why temperate fish tend to contain more muscle and visceral fat than tropical species, where temperatures, although high, do not fluctuate much in the course of a year.

Another tactic that delays respiratory stress is associated with ontogenetic shifts in diet composition. Here, the young fish feed on a diffuse, small prey (e.g., invertebrate zooplankton), while the adults, via their sheer size, can capture energy-rich prey such as other fish, which are acquired at lesser cost by the predator.

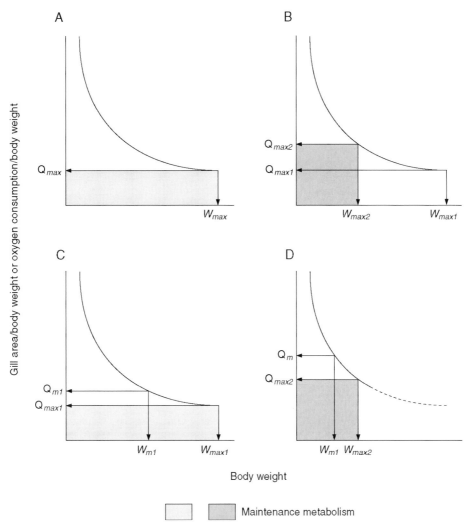

FIGURE 3 Schematic representation of the relationships linking, in fish, respiratory area (and hence metabolic rate), maximum body size, and size at first maturity. (A) As body size increases, gill area per body weight decreases, down to a level when it suffices only for maintenance metabolism. This defines the maximum size that can be reached. (B) Any environmental factor increasing oxygen demand for maintenance (such as elevated temperature) reduces the maximum size that fish can reach. (C) The relative metabolic rate at first maturity (Q_m) is necessarily higher than that associated with maximum size (Q_{max}). (D) An evolved, near constancy of the ratio Q_m/Q_{max} (about 1.4 from guppy to tuna) ensures that fish destined to remain small (as in case B) also spawn at smaller sizes.

C. Relationships between Growth and Mortality

Whichever strategy and tactic fish use to grow, more time will be needed in large species than in small fish for the size at first reproduction to be reached. Large sizes thus imply, other things being equal, more time during which the growing fish may become the prey of some predator. Hence the evolution of large fish was coupled with a reduction of their relative vulnerability to various predators, mainly by their ability to grow quickly through "small-size" stages in which mortality is highest. Fish capable of reaching large size and that have a high longevity also have low rates of natural mortality (Fig. 5). Hence fishing tends to have a stronger impact on species with low natural mortality, such as sharks or rockfishes. Because these are often the top predators, their reduction tends to disrupt the food webs in which they are embedded.

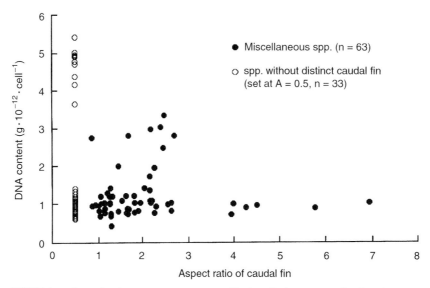

FIGURE 4 Relationship between DNA contents of body cells (a measure of cell size) versus caudal fin aspect in fish. Note triangular patterns, indicating that active fish with high aspect ratios are limited to small cells (which are metabolically more active than large cells), whereas more sluggish fish may have either small or large cells. Based on records in *FishBase 98*.

III. DISTRIBUTION OF EXPLOITED FISH STOCKS

A. Overall Distribution Ranges

Although mostly confined to water, fish occur in a wider range of habitats than any other vertebrate or invertebrate group. Thus, fish range from the upper reaches of streams in high mountain ranges (e.g., many river loaches, Balitoridae) to the mouths of temperate and tropical rivers (e.g., many gray mullets, Mugilidae). In the marine realm, fish range from the intertidal to the ocean's abyss, both as predators in their desert-like expanses (e.g., skipjack tuna, *Katsuwonus pelamis*) or as components of the rich, newly discovered deep-sea vent ecosystems (e.g., some live-bearing brotulas, Bythitidae). Environmental adaptations include the ability to deal with an enormous range of pressures (from

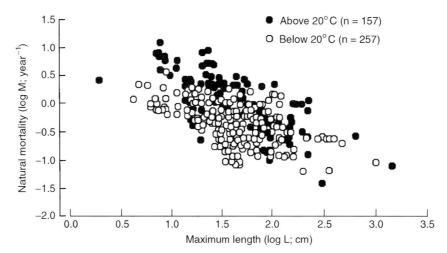

FIGURE 5 Relationships between maximum length, temperature, and rate of natural mortality (*M*) in fish, based on records in *FishBase 98*.

about one to hundreds of atmospheres), temperatures (from −1.8°C in polar waters to about 40°C in hot springs, tolerated by some tilapias), and salinities (from close to distilled water preferred by the discus fish, *Symphysodon discus*, of Amazonia to about 10%, e.g., in West African hypersaline coastal lagoons inhabited by the blackchin tilapia, *Sarotherodon melanotheron*), to list only three environmental factors. No single fish species or family, however, spans more than small fractions of these ranges. Rather, these various adaptations are exhibited by a bewildering variety of forms, ranging from minute gobies that are fully grown at close to 1 cm (e.g., *Mystichthys luzonensis*) to the 15 m reached by whale sharks (*Rhincodon typus*). These two species, incidentally, are exploited for food in the Philippines. The former, despite its turnover rate, is in danger of extinction in the small lake where it is endemic because of overfishing and pollution. The latter will be extirpated if the new directed, export-oriented fishery for this slow-growing fish continues.

B. Adaptations to Open-Ocean Habitats

Fish have different strategies to deal with the low production of the oceans. Tuna have adopted a high-energy strategy, wherein their tightly packed schools quickly move from one food patch to the other, essentially hopping from one "oasis" to the next and minimizing the time spent in the intervening desert-like expanses. Others, notably the lantern fishes (Myctophidae), occur in scattered populations that, at dawn, migrate from 1000 m down to the surface waters, and back again at dusk. These different strategies imply very different biomasses: tens of millions of metric tons for the major tuna species (prior to their recent depletion by various longline, purse seine, and other fisheries) against an estimated global biomass of one billion metric tons for the lantern fish and associated communities. The latter number is often viewed as a promising figure, from which various estimates of potential yields have been derived. Most of these estimates, however, do not consider the extremely dilute nature of this biomass (usually less than 1 g per metric ton of water).

C. Shelf Communities

1. Definition of Neritic Stocks

Most fish stocks are neritic, that is, occur above the continental shelves, the productive areas of shallow waters (down to 200 m) around the continents, from which about 90% of the world marine fisheries catches

are extracted. Shelves may have rocky or soft (sandy or muddy) substrates, and usually support two weakly connected fish communities, one species-rich and consisting of bottom or "demersal" fishes, the other consisting of fewer species of open-water or "pelagic" fishes. The fish of demersal communities are those exhibiting the specialized fins and mouths mentioned earlier, enabling utilization of distinctive food sources, particularly on reefs in both temperate and tropical regions.

On coral reefs, this fine partitioning of resources culminates in hundreds of fish species sharing a single reef, with dozens of specialists for each of its food resource types, from the filamentous algae consumed, for example, by damselfishes (Pomacentridae), the encrusting algae consumed by parrot fishes (Scaridae), the coral themselves, consumed by butterfly fishes (Chaetodontidae), to the small invertebrates consumed by, for example, wrasses (Labridae). A vast array of predators such as groupers (Serranidae) and sharks (Carcharhinidae) regulate the number of these smaller fishes. Hard-bottom shelves and, in tropical areas, the coral reefs that occur down to 30 m are also exploited wherever they occur. The fishing gear used over hard bottoms are mainly traps and handlines (the latter both sport and commercial), which are rather selective gears that would have relatively minor impacts were it not for their excessive numbers.

2. Demersal Fish Stocks

The demersal fish living in, on, or just above shelf soft bottoms consist of specialized flatfishes and rays and numerous generalized teleosts feeding on bottom invertebrates (the zoobenthos) and smaller fishes. The complex communities thus formed can reach very high biomass, at shallow depth in the tropics (20–50 m) and deeper in colder waters. In the warm waters of the tropics, bacteria induce a quick remineralization of the dead organic matter (detritus) falling out of the lighted part of the water column. This allows very little detritus to become available for consumption by the zoobenthos. In cold water, on the other hand, the short but intensive burst of algal production occurring in the spring is consumed only partly by the zooplankton of the upper water layers. Most of the remainder is consumed as detritus after falling down to the sea bottom as "marine snow." Thus, cold-water soft-bottom communities can occur in very deep waters, down to the shelf slopes (200–300 m) and well beyond. Indeed, the latest trend in fisheries "development" is the exploitation of deep-sea stocks of cod-like fish (order Gadiformes), orange roughy (*Hoplostethus atlanticus*), and other fish, down to depths of 1000 m or more, through ventures that even in principle could never be managed so as to achieve sustainability.

Wherever they occur, soft-bottom shelves are nowadays invariably subjected to bottom trawling, a very unselective fishing method that is environmentally damaging. This involves dragging a heavy, chain-studded net over the sea bottom and "catching," that is, removing all that it encounters. Not surprisingly, this procedure has often been compared to harvesting crops with a bulldozer. Trawler catches thus consist of targeted species (usually shrimps in the tropics and subtropics) plus a vast number of nontarget species, often the juveniles of demersals with large adult sizes, and literally parts of the habitat of bottom-fishes, notably sessile invertebrates and chunks of reefs lifted from the sea bottom. Nontarget species and debris are then discarded, and it is therefore trawlers that contribute most to the global discarding problem. Presently, about 30 million metric tons of various fish species are discarded; this is a very high discard rate when compared to the 90 million metric tons that appear in global landing statistics.

The contribution of trawlers to habitat destruction, including conversion of richly structured bottom habitats into featureless expanses of mud, is well recognized, and can only be compared in terms of scale with global deforestation and the ensuing trend toward desertification. Only recently has the impact on biodiversity of this mode of fishing begun to be evaluated in systematic fashion. The information so far available indicates high impacts and a tendency for small generalized fish and invertebrates to replace larger specialized fish, a trend that amplifies the food web effects to be described later.

3. Pelagic Fish Stocks

The pelagic communities over most shelf areas previously consisted of both major and minor stocks and stocklets of herrings, sardines (Clupeidae), anchovies (Engraulidae), and their relatives, and of their predators, notably mackerels and tunas (Scombridae) and various jacks (Carangidae). In many parts of the world, pelagic fisheries have eliminated the minor stocks and stocklets, and now depend wholly on annual recruitment to the remaining major stocks. The overfishing of old, highly fecund adults in these remaining stocks explains much of their volatility. Indeed, the present emphasis of much fisheries research on "variability" is thus devoted largely to a secondary phenomenon created by the fishery itself. It is true, however, that pelagic stocks, feeding lower in the food web, often closely track environmental changes, such as the decline of the Peruvian anchovy *Engraulis ringens* during El Niño

events, and their subsequent rebuilding, mainly from recruits produced off northern Chile.

Pelagic fish tend to form tightly structured, dense schools, which protects them from predators and facilitates detection and herding of scattered food patches. The fisheries rely on this behavior when deploying purse seines, which can surround and catch such schools in one go, often with associated predators such as dolphins. Large pelagics such as billfish (Xiphiidae and Istiophoridae) are caught by arrays of longlines, set by the thousands along shelf edges, which also capture, besides the target species, large amounts of by-catch (notably sharks). These sharks were previously left on the spot, but are now finned before the carcasses are discarded. Longlines are indeed as unselective as the now banned giant driftnets that, in the 1980s, erected "walls of death" that were hundreds of kilometers long across the migratory routes of fish in the North Pacific and the Atlantic.

4. Overall Status of Neritic Stocks

When combined, the demersal and pelagic fisheries of shelves and adjacent waters represent major threats to fish biodiversity. Particularly endangered are groupers and other slow-growing bottomfish, and pelagics such as bluefin tuna and various species of sharks and billfish.

Besides the fisheries, one factor contributing to this endangerment is the traditional separation of research devoted to fisheries management ("stock assessments") from that devoted to conservation and to ecosystem research. Both lines of research are separated institutionally, in terms of their methods and publication outlets, and in terms of what they perceive as their mandates. Overcoming this separation is crucial if fish biodiversity is to be maintained in the face of the onslaught by fisheries. Key needs are the development of tools and concepts for integrating information on fish biodiversity and ecosystem function with the knowledge gained through a century of applied, single-species fisheries research. Before considering these, however, evidence for fisheries impacts on ecosystems will be presented.

IV. ECOSYSTEM IMPACTS OF FISHERIES

A. Historical Trends

The earliest fishing gear so far identified by archeologists are bone harpoons that were recovered, along with other evidence of systematic fishing, from a site 90,000 years old, in the present-day Democratic Republic of Congo (formerly Zaire). Tellingly, the main species that was targeted appears to have been a now extinct, very large freshwater catfish.

This pattern of fisheries exterminating the stocks upon which they originally relied, then moving on to other species, is now understood to be common. This contradicts earlier perceptions of the ocean's quasi-inexhaustible resources, as expressed among others by such Victorian grandees as the geologist Charles Lyell and the zoologist Thomas Huxley. They were misled by the then prevailing abundance of various stocks of coastal fish (notably herring, *Clupea harengus*), and by what may be called "Lamarck's Fallacy": the notion that "animals living in the waters, especially in seawater ... are protected from the destruction of their species by Man. Their multiplication is so rapid and their means of evading pursuit or traps are so great that there is no likelihood of his being able to destroy the entire species in any of these animals."

The industrialization of the fisheries, first in Northern Europe and then in North America at the end of the nineteenth century, quickly showed these predictions to be wrong. Most coastal stocklets of herring and other small pelagics were extirpated, and faded even from memory, therein soon followed, after the introduction of bottom trawling, by coastal stocks of demersal fishes.

The practical response to this was the introduction of bigger boats with bigger engines, fishing farther offshore. Another response was the creation of research bodies (such as the International Council for the Exploration of the Sea, founded in 1902) to assess the reason why the resources were declining. Also, several countries (notably Norway and the United States) initiated costly programs wherein juvenile cod and other fish were raised in hatcheries and then thrown into the sea, in the vain hope that they would replenish the stocks rather than be eaten by happy predators (which they were).

B. Emergence of the Sustainability Concept

The First World War put an end to the stocking programs. It also established that a strong reduction of fishing effort, as caused by the drafting of fishers and vessels into the war effort, and the spiking of major fishing grounds by underwater mines (thus creating the first marine protected areas), would lead to a recovery of depleted fish stocks. Yet the Second World War, and another demonstration of stocks rebuilding themselves when subjected to less fishing, was required for the

notion of sustainable fishing to establish itself. This notion implies that some appropriate level of fishing effort (number of vessels or gear, mesh size) exists such that catches (or "yield") can be maintained at high levels—hence the concept of "maximum sustainable yield" or MSY. This led to the emergence of "fish population dynamics" and "stock assessments," wherein mathematical models of single-species fish stocks and of their response to targeted fishing became the mainstay of fisheries research. R. J. Beverton, S. J. Holt, and J. A. Gulland in England, W. E. Ricker in Canada, and W. E. Schaefer in the United States proposed most of these still-used models during an extremely creative period lasting from the early 1950s to the mid-1970s.

Yet in spite of these advances, the fisheries never became sustainable. One obvious reason was that, given a resource to which access was essentially open, the fisheries never could limit their collective effort at the level supposed to generate MSY. Rather, effort levels increased well beyond that, permitting some fleet owners to increase their stakes even as the aggregate "rent" from the fisheries declined. Recent trends toward subsidization of offshore and distant water fleets, driven by international competition, have aggravated these economic issues, enabling commercial profits to be gained even from strongly overexploited stocks. These developments are so widespread that they have rendered obvious the impacts which fisheries have on ecosystems.

C. Fishing Down Marine Food Webs

The ecosystem impacts of fisheries are due mainly to the fact that the targeted fish function as part of food webs, both as consumers and as prey. Within food webs, the fish of different species occupy distinct trophic levels (TL), each defining a step away from plants, which have a definitional TL of 1. Thus, fish feeding on planktonic algae have TL = 2, fish feeding on herbivorous zooplankton have TL = 3, and so on. It is important here to recognize that most fish tend to have intermediate TL values (2.7, 3.5, 4.1, etc.), reflecting the catholic nature of their diet.

Fisheries, by removing biomass from one of several fish stocks, necessarily modify food webs, thus forcing predators of the targeted species to shift toward available alternative prey, if any. Such adjustments were previously not distinguishable from natural fluctuations. They have gradually become highly visible, however, because they change the mean trophic level of the landings extracted from different stocks. Moreover, the changes induced by fishing are not of a random nature, with decreases in one area matched by increases in another. Rather, they are directed, with a clear downward trend (Fig. 6A), due to the link between growth and natural mortality mentioned in Section II. Thus, in large fish, even a low level of fishing mortality generated by a well-managed fishery will quickly exceed the low level of total mortality (i.e., natural + fishing mortality) that can be accommodated by the stock. By-catch species are even more endangered because the fishing will not stop as their numbers dwindle until they are eradicated, as has happened with rays in the Irish Sea. The trend of mean trophic level resulting from this (see Fig. 6A), reflecting a phenomenon now known as "fishing down marine food webs," provides a clear indication that, globally, fisheries generate levels of effort well past those required for sustainability, however defined. Indeed, other indices can be used to indicate that global changes have occurred in the composition of global fisheries landings, and in the structure of the ecosystems from which these landings are extracted (Fig. 6B).

Fisheries-induced modification of the structure of marine and freshwater ecosystems has strong indirect impacts on fish biodiversity, in addition to the direct impacts of reducing the biomass of the target and associated stocks by a factor of 10 or more, as is usually the case. Incorporating these indirect effects in fisheries stock assessments has proven to be difficult so far. This

is true for objective reasons (ecosystems are complex, and their behavior under exploitation, due to the large number of stocks to be considered, is difficult to simulate) and for subjective reasons (notably a perceived lack of suitable field data on these many stocks).

The recent development of robust ecosystem simulation tools should allow the first of these issues to be addressed. Overcoming the second not only involves pointing out the existence of suitable data, often lost in the "gray literature," but in making such data available in suitable format to all who are aware of the need for a transition from single-species to ecosystem-based fisheries assessments. This brings us to the issues related to the standardization, dissemination, and uses of biodiversity information.

V. MANAGING FISH BIODIVERSITY INFORMATION

A. Biodiversity as a Conceptual Challenge

There is a widespread perception that the main obstacle to the conservation of fish stocks and of fish biodiversity is "lack of data," a notion strengthened by public statements of biologists worried about the lack of funding for relevant research. However, simple lack of data cannot be the problem, not after the 250 years since Lin-

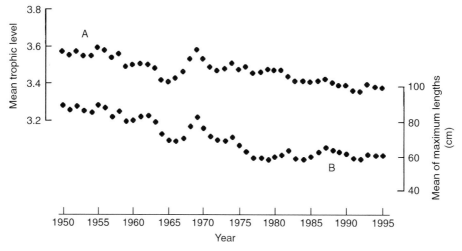

FIGURE 6 Trend, for the Northeast Atlantic from 1950 to 1996, of two indices of sustained fishing, based on landings originally compiled by the Food and Agriculture Organization of the United Nations, and other data in *FishBase 98*. (A) Trend in the mean trophic level of fisheries landings. (B) Trend in the mean maximum size of fish species in the landings. Note parallel declines, indicative of structural changes in the ecosystem from which the landings are extracted. Similar trends occur throughout the marine and freshwater fisheries of the world.

naeus created the taxonomic standards required for biodiversity research, 100 years of applied fisheries research, and at least 50 years of advances in ecosystem research. Rather, the problem here is the fragmentation of the database collected so far. Indeed, many studies conducted in recent years on the status of various stocks fail to consider previous knowledge on their relative abundance and distribution, and thus contribute to shifting baselines, wherein only the most recent and usually low estimates are used as reference for conservation or rebuilding efforts.

One reason for this reluctance of biologists to consolidate existing data into comprehensive, global databases may be due in part to the perception that biological data are too difficult to standardize, or are useless once standardized. Addressing these issues will be a key task of biodiversity research, and we now present a few ideas related to this.

There is consensus that the objects of biodiversity research are genes, populations, species, and ecosystems. However, there is little consensus as to what distinguishes biodiversity from the existing disciplines of fisheries biology, ecology, biogeography, population genetics, or taxonomy. As a result, the array of data being claimed to be essential for biodiversity studies reads like a composite list of the data traditionally used in the older disciplines, with few attempts at integration or prioritization. Such integration and prioritization are possible, however, by giving emphasis, in biodiversity studies, to data that are: (1) relevant to current research issues (e.g., richness, rarity, distinctiveness, representativeness, threat, function, and utility of species); (2) part of the data traditionally collected in taxonomy, biogeography, population genetics, and ecology; (3) widely available, in sufficient quantity; (4) pertinent to past, present, and most likely future trends; (5) easy to collect; (6) easy to standardize; (7) easy to verify; and (8) suggestive of new lines of research.

B. Bioquads as Key Biodiversity Data Sets

A minimum core of biodiversity information that fulfills these eight criteria is provided by "bioquads" (from "quads," short for quadriads), consisting of: (a) the scientific name of a taxon, usually a biological species or other evolutionarily significant unit; (b) the locality where a specimen of this taxon has been encountered; (c) the date (time) of the encounter; and (d) the authority or source reporting (a)–(c).

Of the research items mentioned under criterion (1),

richness (number of species encountered) is derived directly from the bioquads from a given area. Distinctiveness (how much the species encountered differ from each other) is derived from the classification of these species into higher taxa such as families, orders, and classes. Representativeness (how closely an area represents a predefined ecosystem type) is derived by comparing observed species composition with the typical composition of the ecosystem type under study. The utility of species to humans can be derived from published, or local knowledge, or from catches in the case of fish. Status of threat can be derived from trends in the distribution area defined by bioquads. Rarity can be estimated from the number of bioquads available for a species in a given area, standardized by sampling effort.

Taxonomists have made a conscious effort to systematically compile data of this sort in specimen collections, and to publish them in original species descriptions and revisions. As a result, bioquad-type data are readily available in enormous numbers (about 10 million for fish alone) in museum collections, survey reports, historical photos and films, and other forms (criterion 3). While museum collections go back over 200 years, some literature contains verifiable records that date back to antiquity (criterion 4). Also, archeological data reach back to the dawn of modern humanity (see the earlier record pertaining to giant catfish).

Numerous scientific surveys and projects also continuously collect contemporary bioquads. Other sources are the commercial fisheries and the many laypersons whose hobby is to observe and sometimes to collect fish and other wildlife. These activities are most likely to continue in the foreseeable future (criterion 5). An increasing number of the preceding data sources are available in computer-readable form (criteria 3, 5, and 6).

Efforts do exist to standardize the elements of the bioquad (criterion 6). For example, the Species 2000 Initiative has embarked on the task of providing a standard reference list of the valid names of the known 1.75 million species sharing Earth with humans (see the website www.sp2000.org). Geographical coordinates and the international date and time format are obvious standards for items (2) and (3), although there remains a need for a global gazetteer to deal efficiently with localities reported without coordinates, and there is a need for standards to deal with date and time ranges. On the other hand, standards exist for sources such as printed publications, databases, photos, films, and personal communications. Many of these were considered when developing FishBase, a computerized database on the biology, ecology, and uses of fish containing a vast number of bioquads (see the following).

The necessary verification (criterion 7) of millions of data points can only be done automatically. Basically, a computer can verify a scientific name against a standard list, compare the indicated locality and date against the established range of a species, and judge the reliability of a source, for example, by the number of outliers it has reported previously. Procedures will have to be established, however, on how to deal with the different types of outliers, some of which may represent valid new information.

An important consideration is how fast a research agenda based on bioquads will be exhausted (criterion 8). Important here is the ability of well-structured relational databases to interlink independently developed data sets. Thus, the scientific name links to all available information on a species, including taxonomy, systematics, genetics, biology, ecology, and human uses. The locality connects to all available information on surrounding environments, including province, country, continent, habitat, ecosystem, and tectonic plate. The combination of species, locality, and date points to a population or stock. Date and time in connection with the locality can be used to infer a wide range of environmental conditions, from local temperatures to current fisheries legislation. The source relates to the human dimension, such as persons and institutions working on certain species groups or in a certain area, representing the scientific interface between humans and the other species (Fig. 7).

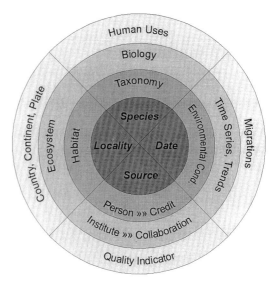

FIGURE 7 Interrelationships of the elements of biodiversity, articulated through the four elements of bioquads (species, location, times, and source).

bioquads as defined previously. Important here is that a new original of this graph is generated on the fly, from currently available bioquads, every time the relevant routine of FishBase is evoked, and that each of its "dots" can be clicked to verify the four elements of the underlying bioquad.

C. Databases as Tools for Management of Biodiversity Information

Two major initiatives presently exist to assemble and make widely available, for research on fish biodiversity, the information presently held by various institutions (notably museums). One is NEODAT, which makes accessible on the Internet about 400,000 bioquad records pertaining to freshwater fish of the Neotropics (NEODAT; www.fowler.acnatsci.org). The other is FishBase, an ongoing international collaborative project dedicated to assembling the estimated 10 million existing fish bioquads and to combining them with other, standardized biological information on fish. The intention here is to provide a global relational database, addressing head-on the data fragmentation issue mentioned earlier (see www.fishbase.org).

Figure 8 shows the geographic distribution of Nile tilapia, *Oreochromis niloticus*, through dots representing

VI. PRESERVING FISH BIODIVERSITY

A. Traditional Approaches to Stock Management

None of the foregoing considerations will help, however, if fisheries are allowed to continue undermining their resource base, which they will if fisheries management continues to rely on the panoply of approaches so far deployed. These traditional approaches include, among other things: (1) mesh size restriction; (2) restriction on the amount and/or species of fish that may be legally landed; (3) effort limitation, for example, through caps on the vessel tonnage that may deployed; and (4) seasonal closures.

Besides being extremely hard to enforce, these approaches—which are invariably conceived in the context of single-species assessments—fail to address the ecosystem effects mentioned earlier. Thus, mesh sizes

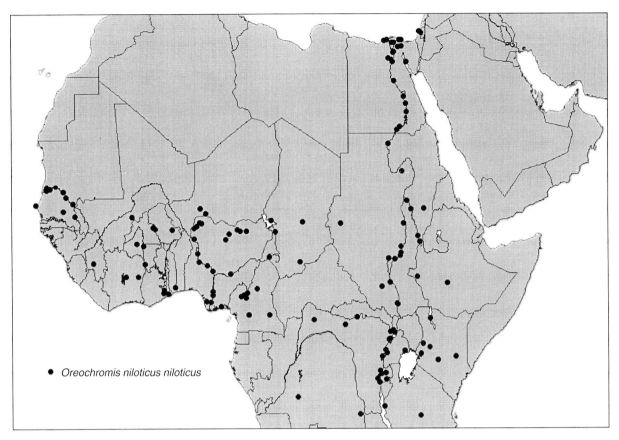

FIGURE 8 Distribution of Nile tilapia (*Oreochromis niloticus*) based on 425 bioquads contributed by the Musée Royal de l'Afrique Centrale, Tervuren, Belgium, and other sources. In the computerized version of this graph, each dot can be "clicked" to reveal the four elements of the underlying bioquad, thus allowing identification of outliers, temporal trends, etc.

above a certain limit, meant to protect the young of a given species, do not prevent associated species form being caught. Indeed, when combined with restrictions on total allowable catch (TAC), and on the landing of bycatch (as is often the case), mesh size restrictions become the very reason for discarding both the young of targeted species and the nontarget species. Limits on nominal fishing effort are subverted by technological developments, such as improved gears and navigation instruments (e.g., GPS), which increase the catching power of fishing vessels. Thus, government-run vessel retirement schemes often end up subsidizing the modernization of fishing fleets. Finally, seasonal closure of various areas usually has negligible ecological impacts, because the fishing effort expended during the open season is sufficient for the sea bottom to be scraped up numerous times by trawlers, and for the stocks of long-lived fishes to be severely impacted.

B. Marine Protected Areas

There is an emerging consensus among fisheries scientists and conservationists that the only fisheries management tool that will allow the recovery of damaged stock and ecosystems is the establishment of Marine Protected Areas (MPAs), including permanent No-Take zones as their core. Such core zones are easy to enforce—at least relative to the task of enforcing mesh sizes or TACs. Also, technology-driven increases of fishing effort can be ignored, and there is assurance that the long-lived organisms of seafloors and their associated fish communities can gradually return to a semblance of their original configurations. However, much research will have to be devoted to identifying the optimal size and location of MPAs, particularly for migratory stocks.

Still, traditional fisheries management, aimed at limiting effective fishing effort, will have to continue around MPAs, lest they become marine larders or fish-attracting rather than fish-producing zones from which resources are drained by fisheries operating at their very periphery.

Finally, the social context of fisheries will have to change: fisheries do not harvest crops they have sown. Rather, they exploit the natural productivity of wildlife; thus there are inherent limits to global fish catches, and future fisheries will not meet the demand of an ever-increasing human population. Indeed, the massive ecosystem changes already described indicate that these limits have been reached in most parts of the world, and that sustainable fisheries must be embedded in some form of ecosystem management.

See Also the Following Articles

ADAPTATION • FISH, BIODIVERSITY OF • FISH CONSERVATION • MARINE ECOSYSTEMS

Bibliography

Froese, R., and D. Pauly (eds.). (1998). *FishBase 98: Concepts, Design and Data Sources*. ICLARM, Manila. [Distributed with two CD-ROMs; also see the website www.fishbase.org.]

Hawksworth, D. L., P. M. Kirk, and S. D. Clarke (eds.). (1997). *Biodiversity Information: Needs and Options*. CAB International, Wallingford, United Kingdom.

Miller, R. I. (ed.). (1994). *Mapping the Diversity of Nature*. Chapman & Hall, London.

Mooney, P. (ed.). (1998). Ecosystem management for sustainable fisheries. *Ecol. Appl.*, Supplement to 8(1).

National Research Council. (1999). *Sustaining Marine Fisheries*. National Academy Press, Washington, D.C.

Nelson, J. (1994). *Fishes of the World*, 3rd ed. John Wiley & Sons, New York.

Pauly, D. (1994). *On the Sex of Fishes and the Gender of Scientists: Essays in Fisheries Science*. Chapman & Hall, London.

Paxton, J. R., and W. N. Eschmeyer (eds.). (1998). *Encyclopedia of Fishes*. Academic Press, San Diego.

Reaka-Kudla, M. L., D. E. Wilson, and E. O. Wilson (eds.). *Biodiversity II. Understanding and Protecting our Biological Resources*. Joseph Henry Press, Washington, D.C.

FLIES, GNATS, AND MOSQUITOES

Brian V. Brown
Natural History Museum of Los Angeles County

I. Introduction
II. Major Subdivisions
III. Life History Diversity
IV. Conservation Biology of Diptera

GLOSSARY

Diptera Group of insects to which the flies, including gnats and mosquitoes, belong.
larva (pl. larvae) Immature stage of flies, often called maggots.

THE DIPTERA, or true flies, are a well-established, monophyletic group of insects with more than 124,000 extant species. They are found worldwide and interact with the environment at almost all trophic levels as scavengers, filter-feeders, herbivores, predators, parasitoids, and parasites.

I. INTRODUCTION

The insect Order Diptera, with more than 124,000 currently described, extant species, ranks as one of the worlds largest groups of organisms. Along with the other insect megadiversity groups—Coleoptera (beetles; 350,000 species), Lepidoptera (butterflies and moths; 120,000 species), and Hymenoptera (ants, bees, wasps, and sawflies; 130,000 species)—they form the largest aggregation of species on the planet. Each of these megadiversity groups has more species assigned to it than to any other group of organisms except plants (300,000 species). Like other insects, most species of Diptera are still undescribed, and the actual number could range as high as 1 million or more species.

Diptera are found on every continent, including Antarctica. The relative percentage of Diptera within the insect fauna rises with latitude (and elevation) as other, less cold-adapted taxa are lost; for instance, over one-half of all insect species recorded from the Canadian high arctic are Diptera. The extant species of all biogeographical regions have been cataloged (Table I), as have the fossil Diptera. Some catalogs are badly outdated and the number of species described for each region is often more indicative of the amount of taxonomic activity directed at a given fauna than the true diversity. For instance, the Palearctic Region is the best studied and has more described species than the relatively poorly known, but presumably more diverse, Neotropical Region. Therefore, conclusions about dipteran biogeography based on these numbers will be highly inaccurate.

Diptera are a well-established monophyletic group, with the most obvious defining character being the reduction of the hind wings to small, club-shaped organs called halteres. The insect orders considered to be most closely related to Diptera are Siphonaptera (fleas) and Mecoptera (scorpionflies), although recent molecular

Region	Number of species
Nearctic	25,000
Neotropical	20,000
Palearctic	29,000
Afrotropical	16,318
Australasian	15,764
Oriental	15,964
Antarctica	60

Compiled from the various catalogs.

research indicates that the highly aberrant Strepsiptera (twisted-wing parasites) may be their closest relatives.

Fossil Diptera are common in the amber faunas of the world, which however extend back in time only to the early Cretaceous. The oldest dipteran fossil is believed to be from at least the mid-Triassic.

Like the other megadiversity groups, the Diptera are holometabolous insects, with a separate egg, larva, pupa, and adult stage. Generally, most feeding takes place in the larval stage, whereas the adult is usually specialized for reproduction and dispersal. Some adult Diptera are voracious feeders, however, requiring substantial supplemental feeding to mature their eggs or power their flight.

Flies are common in natural, disturbed, and urban habitats. Larvae are found on land and in freshwater; there are relatively few marine or brackish water species. Some species are synanthropic and have been transported around the world with human activities. Diptera are the most important vectors of human and animal disease, and a few plant-feeding species have become agricultural pests. The life history of most species of flies, however, is unknown. The group is so large, and there are so many undescribed species, that the science of dipterology is still in its relative infancy.

II. MAJOR SUBDIVISIONS

Table II presents a general list of the families of Diptera and some of the higher taxa that contain them. This list is based on only one of several possible classifications, however, and cannot be considered the final word in dipteran groupings. Common names, where well estab-

lished, are given in Table II. Many families lack common names, including the second largest family in the order (Tachinidae). That a group with more described species than the mammals lacks a common name is a good indication of the lack of general appreciation for the importance and ubiquity of the Diptera. The number of described species in each family is given, but in many instances these numbers are badly out of date and should be considered a bare minimum. Certainly, in most families of Diptera there are a large number of undescribed species awaiting the attention of specialists. It is estimated that we have described only 10% or fewer of the species in many families.

The Diptera are traditionally organized into three suborders: the Nematocera, the orthorrhaphous Brachycera, and the cyclorrhaphous Brachycera. Of these, it is likely that only the Cyclorrhapha is a monophyletic, or natural, group. Although a consensus on the higher relationships within the Diptera is not yet available, it is generally agreed that the "Nematocera" is a paraphyletic assemblage of relatively primitive dipteran families and that some subgroup of this assemblage is more closely related to the Brachycera than to other nematocerans. Similarly, within the Brachycera, a group widely considered to be monophyletic, the orthorrhaphous families are now believed to be a paraphyletic assemblage relative to the Cyclorrhapha. Finally, within the Cyclorrhapha is a group of flies traditionally called the Aschiza (here called the lower Cyclorrhapha), which is probably paraphyletic with respect to the monophyletic Schizophora. The details of the phylogeny of the Diptera has been reviewed extensively by Yeates and Wiegmann and their discussion of various groups is highly recommended for understanding the current status of dipteran phylogeny.

Within the Diptera, higher taxa show repeated patterns of relatively primitive, nondiverse grades of organization with relatively highly derived, speciose sister taxa. Thus the lower Diptera are collectively much less diverse than the Brachycera, with the notable exception of the incredibly large family Tipulidae. Within the Brachycera, the Muscomorpha is by far the largest infraorder and the Schizophora has four times more species than the lower Cyclorrhapha. The numbers are less disparate in the Schizophora, with about 26,000 acalypterates and 19,000 calypterates, but the monophyly of the acalypterates is highly contentious, making such a comparison questionable.

The distribution of species among the families of Diptera is extremely divergent, with the largest number found in the Tipulidae and the Tachinidae. Together,

TABLE II
Simplified List of Major Subgroups and Families of Diptera

	Common name	Number of species
Lower Diptera (=Nematocerous groups)		
Infraorder Ptychopteromorpha		
Ptychopteridae	Phantom crane files	61
Tanyderidae		42
Infraorder Culicomorpha		
Superfamily Culicoidea		
Culicidae	Mosquitoes	3,000
Dixidae	Meniscus midges	175
Corethrellidae		61
Chaoboridae	Phantom midges	50
Superfamily Chironomoidea		
Ceratopogonidae	Biting midges	5,360
Chironomidae	Midges	5,000
Simuliidae	Black files	1,475
Thaumaleidae		121
Infraorder Blephariceromorpha		
Blephariceridae	Net-winged midges	300
Deuterophlebiidae	Mountain midges	14
Nymphomyiidae		7
Infraorder Bibionomorpha		
Axymyiidae		5
Bibionidae	March flies	675
Cecidomyiidae	Gall midges	4,600
Mycetophilidae	Fungus gnats	3,000
Pachyneuridae		4
Sciaridae	Dark-winged fungus gnats	1,000
Infraorder Tipulomorpha		
Tipulidae	Crane flies	14,000
Trichoceridae	Winter crane flies	110
Infraorder Psychodomorpha		
Anisopodidae	Wood gnats	100
Perissomatidae		5
Psychodidae	Sand flies, moth flies	2700
Scatopsidae		255
Synneuridae		4
Brachycera (an unranked taxon)		
Infraorder Xylophagomorpha		
Xylophagidae		111
Infraorder Tabanomorpha		
Athericidae		90
Pelecorhynchidae		49
Rhagionidae	Snipe flies	520
Tabanidae	Horseflies, deer flies	3,000
Vermileonidae	Ant lions	31

continues

continued

	Common name	Number of species
Infraorder Stratiomyomorpha		
Pantophthalmidae		20
Stratiomyidae	Soldier flies	2,500
Xylomyidae		107
Infraorder Muscomorpha		
Superfamily Nemestrinoidea		
Acroceridae	Small-headed flies	500
Nemestrinidae	Tangle-veined flies	250
Superfamily Asiloidea		
Apioceridae	Flower-loving flies	145
Asilidae	Robber flies	5,600
Bombyliidae	Bee flies	4,800
Mydidae	Mydas flies	353
Scenopinidae	Window flies	330
Therevidae	Stiletto flies	850
Superfamily Empidoidea		
Dolichopodide	Long-legged flies	5,100
Empididae	Dance flies	3,500
Cyclorrhapha (an unranked taxon of Muscomorpha)		
"Lower Cyclorrhapha" (=Aschiza of previous authors)		
Ironomyiidae		1
Lonchopteridae	Spear-winged flies	35
Opetiidae		3
Phoridae	Humpbacked flies, scuttle flies	3,200
Pipunculidae	Big-headed flies	600
Platypezidae	Flat-footed flies	215
Sciadoceridae		2
Syrphidae	Hover flies, flower flies	5,800
Schizophora		
"Acalypterates" (possibly not monophyletic)		
Superfamily Nerioidea		
Cypselosomatidae		30
Micropezidae	Stilt-legged flies	500
Neriidae	Cactus flies	110
Superfamily Diopsoidea		
Diopsidae	Stalk-eyed flies	153
Megamerinidae		13
Nothybidae		8
Psilidae	Rust flies	200
Somatiidae		7
Strongylophthalmyiidae		27
Syringogastridae		8
Tanypezidae		18
Superfamily Conopoidea		
Conopidae	Thick-headed flies	800
Superfamily Tephritoidea		
Lonchaeidae		700
Otitidae		800

continues

continued

	Common name	Number of species
Pallopteridae		52
Piophilidae	Skipper flies	71
Platystomatidae		1,000
Pyrgotidae		200
Richardiidae		170
Tachiniscidae		3
Tephritidae	Fruit flies	4,000
Superfamily Lauxanioidea		
Celyphidae	Beetle flies	90
Chamaemyiidae	Aphid flies, silver flies	250
Eurychoromyiidae		1
Lauxaniidae		1,550
Superfamily Sciomyzoidea		
Coelopidae	Seaweed flies	30
Dryomyzidae		30
Helosciomyzidae		23
Sciomyzidae	Marsh flies	515
Ropalomeridae		30
Sepsidae	Black scavenger flies	240
Superfamily Opomyzoidea		
Suprafamily Clusioinea		
Acartophthalmidae		3
Clusiidae		217
Suprafamily Agromyzoinea		
Agromyzidae	Leafminer flies	3,500
Fergusoninidae		25
Odiniidae		50
Suprafamily Opomyzoinea		
Anthomyzidae		49
Marginidae		2
Opomyzidae		40
Suprafamily Asteioinea		
Asteiidae		100
Aulacigastridae		25
Neurochaetidae	Upside-down flies	5
Periscelididae		19
Teratomyzidae		5
Xenasteiidae		8
Superfamily Carnoidea		
Australimyzidae		5
Braulidae	Bee lice	7
Canacidae	Beach flies	113
Carnidae		41
Chloropidae	Frit flies	2,000
Cryptochaetidae		25
Milichiidae		190
Tethinidae		126

continues

continued

	Common name	Number of species
Superfamily Sphaeroceroidea		
Chyromyidae		40
Heleomyzidae (s.l.)		520
Mormotomyiidae		1
Sphaeroceridae	Lesser dung flies	2,500
Superfamily Ephydroidea		
Camillidae		11
Curtonotidae		36
Diastatidae		22
Drosophilidae	Vinegar flies, pomace flies	2,900
Ephydridae	Shore flies	1,300
Calyptratae		
Superfamily Hippoboscoidea		
Glossinidae	Tsetse flies	22
Hippoboscidae	Louse flies	200
Nycteribiidae	Bat flies	260
Streblidae	Bat flies	220
Superfamily Muscoidea		
Anthomyiidae		1,100
Fanniidae		265
Muscidae		3,880
Scathophagidae		250
Superfamily Oestroidea		
Calliphoridae	Blow flies	1,000
Mystacinobiidae		1
Oestridae	Bot flies, warble flies	42
Rhinophoridae		100
Sarcophagidae	Flesh flies	2,500
Tachinidae		9,200
TOTAL		124,390

Modified from Yeates and Wiegmann (1999) and McAlpine (1989). Most important taxa are in bold. Approximate number of described species from various sources.

these families account for almost 20% of the species in the order.

The Tipulidae, or crane flies, are elongate, long-legged, somewhat fragile flies that are found nearly everywhere on earth. The adults are sometimes found at lights but are most often seen resting on vegetation. The larvae have a variety of habitats, including terrestrial—in soil, mosses, and decaying wood—freshwater, and intertidal. They are scavengers, herbivores, or predators.

In contrast to the eclectic habits of the Tipulidae, larvae of Tachinidae have a single way of life: parasitism. All known species are internal parasitoids of other ar-

thropods, mostly other insects and, within the insects, mostly larval Lepidoptera. Eggs are either laid on or in the host or are broadcast in suitable areas. Larvae hatch from broadcast eggs and wait in ambush for hosts. Some tachinids produce microtype eggs that are designed to be ingested by the hosts; these eggs hatch inside the host and penetrate the gut wall to enter the body cavity. Adult tachinids are stout, bristly, housefly-like flies that are seen frequently on vegetation or flowers.

There are another 21 families of Diptera that can be considered large, possessing about 2000 or more described species. Together, these 23 largest families (including tipulids and tachinids) comprise about

100,000 species, or approximately 80% of the Order Diptera. In order of decreasing number of species, these other 21 large families are as follows:

• Syrphidae: hover flies, flower flies (5800 spp.). Adults of this family often are brightly colored mimics of Hymenoptera (bees and wasps). The larvae are saprophagous, predatory, or herbivorous. Some saprophagous species are called rat-tailed maggots because of their elongate posterior breathing tubes. Many of the predatory species live exposed on plants, feeding on aphids.

• Asilidae: robber flies (5600 spp.). The adults are voracious predators that usually attack their prey while in flight, stabbing them with their heavily sclerotized, swordlike proboscis and injecting digestive fluids. The larvae are also predatory.

• Ceratopogonidae: biting midges (5360 spp.). The tiny adult females of this family require blood meals to mature their eggs. Some species bite vertebrates, other feed on large insects, and still others are predatory on small insects. The larvae are predacious, living in damp terrestrial or freshwater habitats.

• Dolichopodidae: long-legged flies (5100 spp.). These flies are often metallic green in color and are found commonly on undergrowth or on tree trunks. The larvae and adults are predatory.

• Chironomidae: midges (5000 spp.). These flies are among the most abundant benthic invertebrates in freshwater environments; some are also terrestrial or intertidal. The adult males often form enormous mating swarms.

• Bombyliidae (including Mythicomyiidae): bee flies (4800 spp.). Species of this family are most diverse in dry areas, including deserts, and are nearly absent from tropical rain forests. The larvae are parasitoids of the immature stages of various other insects or predatory on grasshopper egg pods.

• Cecidomyiidae: gall midges (4600 spp.). Cecidomyiids are generally considered one of the largest, yet most poorly known groups of Diptera. Larvae often form galls on various plants, but there are non-gallforming species and many mycophagous forms as well.

• Tephritidae: fruit flies (4000 spp.). Larval tephritids are phytophagous, attacking a wide variety of plants and sometimes forming galls. The adults often have color patterns on their wings.

• Muscidae: houseflies and relatives (3880 spp.). A few extremely well-known muscid species are synanthropic, especially *Musca domestica* and *Stomoxys calcitrans*. Larvae of most species live in decaying organic material, where they are either saprophagous or

predatory. Some species are saprophagous in early larval instars, becoming carnivorous later.

• Agromyzidae: leaf-miner flies (3500 spp.). Along with some other insects, larvae of many species of agromyzids feed within plant leaves, excavating the distinctive, light-colored tunnels called mines. The adults are small, usually dark-colored flies; some species are marked with yellow.

• Empididae: dance flies (3500 spp.). Adults of this family are found mostly in damp terrestrial habitats, often near water. Common and diverse in temperate regions, they are less prevalent in tropical lowland forests. Larvae and adults of both sexes are predatory; larvae of a few species are parasitoids of caddisfly larvae.

• Phoridae: humpbacked flies, scuttle flies (3200 spp.). Phorids are common and diverse nearly everywhere except Antarctica. Larvae can be predators, parasitoids, true parasites, herbivores, or scavengers. Many species are associated with social insects.

• Tabanidae: horseflies, deer flies (3000 spp.). Adult female tabanids are well-known blood feeders, although some species feed only on nectar. The larvae are predatory, usually found near water.

• Culicidae: mosquitoes (3000 spp.). Adult females bite vertebrates to obtain blood meals, and they often transmit diseases. Larvae are aquatic.

• Mycetophilidae: fungus gnats (3000 spp.). Adults are found mostly in humid, forested areas. The larvae feed mainly on fungi, although some species spin webs to capture insect prey and a few species are parasitoids of flatworms.

• Drosophilidae: vinegar flies, pomace flies (2900 spp.). This family is best known for *Drosophila melanogaster*, the ubiquitous model organism for genetic research. Although adults are commonly found around overripe fruit or on mushrooms, the larvae of this family have a variety of lifestyles, from saprophagy to parasitism and predation.

• Psychodidae: moth flies (2700 spp.). The scavenging larvae of these flies are usually found in moist conditions, in soil, rotting wood, or other decaying vegetation. Adults of one subfamily are blood feeders and transmit the disease leishmaniasis to humans.

• Sarcophagidae: flesh flies (2500 spp.). Sarcophagid larvae are saprophagous, parasitoids, predators or commensals in the nests of solitary Hymenoptera. The adults of some species are associated with filth.

• Sphaeroceridae: lesser dung flies (2500 spp.). Larvae and adults of this family are commonly found on dung, carrion, and decaying organic material.

• Stratiomyidae: soldier flies (2500 spp.). The adults of this family are often brightly colored and conspicuous on flowers. The larvae are often found near water, in decaying organic material, or under bark. They are saprophagous, herbivorous, or predatory.

• Chloropidae: frit flies (2000 spp.). Larvae of chloropids have nearly every conceivable way of life, from scavenging to predation, parasitism and herbivory, although there are many more plant feeders in this family than the similarly diverse Phoridae.

A. Nematocerous Families

The nematocerous families are the relatively primitive members of the Diptera, characterized by long, unconsolidated antennae consisting of many segments. There are 26 families in this group and approximately 40,000 species included. Many are associated with aquatic habitats and some, such as mosquitoes, black flies, and biting midges, are voracious blood feeders.

B. Brachycera

Most Diptera belong to the Brachycera, a group characterized by the reduction or fusion of antennal segments to eight or fewer and by modifications to the larval head and mouthparts. With about 80,000 described species, this group contains many of the best known flies, such as houseflies and fruit flies.

The lower Brachycera includes several lineages constituting approximately 30,000 species. The larvae of most species are predatory, although there are a few parasitoid groups as well. The most familiar are the large families Asilidae, Bombyliidae, Tabanidae, Dolichopodidae, and Empididae.

The Cyclorrhapha, with about 50,000 species, includes a few primitive lineages and the Schizophora. The primitive groups are relatively small, with the exceptions of the large families Phoridae and Syrphidae. Within the Schizophora, there are a plethora of smaller acalypterate families that are rare, but many, such as the Tephritidae, Agromyzidae, Drosophilidae, Sphaeroceridae, and Ephydridae, are abundant and commonly encountered. The Calyptratae includes the familiar houseflies, flesh flies, and blow flies as well as the speciose parasitoids of the Tachinidae. Also included are a number of mammal parasites, including some truly bizarre, spider-like bat parasites in the Nycteribiidae and Streblidae.

III. LIFE HISTORY DIVERSITY

A. Larvae

The larvae of flies are abundant and widespread in most terrestrial and aquatic habitats. They are often encountered in soil (including sand), dead wood, dung, carrion, decaying vegetation, and among the refuse of social insect colonies. Relatively few are found living exposed on vegetation, including some herbivorous tipulids and many aphidophagous syrphids. Aquatic forms are found on the bottom and in the water column of lakes, streams, and ponds. Larval Diptera are predators, scavengers, herbivores, parasitoids, or even true parasites.

1. Scavengers

This is the lifestyle most commonly associated with Diptera larvae, especially those of the Brachycera. More than half of the 128 families recognized herein have larvae that feed on decaying organic material or organic detritus. Most receive their nutrition from bacteria and other microorganisms of decay, not from the main substance on which they are found. They concentrate these organisms and other suspended particles with a sievelike pharyngeal filter.

Among the most obvious terrestrial scavengers are the larvae of blow flies (Calliphoridae) that are found on newly dead animal carcasses. Early stages of decay are characterized by large numbers of calliphorid and muscid larvae, followed later by drosophilids, fanniids, phorids, piophilids, sepsids, sphaerocerids, and others as decay proceeds. The fauna of buried carrion is different, with the calliphorids largely excluded. Instead, the muscid genus *Muscina*, various phorids, and sphaerocerids predominate. The fauna of carrion immersed in water has also been studied and found to differ from that on dry land. There has been considerable study of the succession of scavenging Diptera larvae on dead animals, leading to the potential for their use in forensic entomology.

Decaying vegetable matter is also rich in scavenging Diptera, especially muscids, sphaerocerids, sciarids, and others. Some muscids are obligate thermophiles, requiring the heat generated by the decay of large piles of compost. Decaying seashore vegetation supports coelopids, sphaerocerids, and anthomyiids. Rotting fruit is the food of drosophilids, stratiomyids, and some phorids.

In aquatic environments, many larval Diptera feed on small organic particles in the water or on the substrate. Larvae of Culicidae and Simuliidae filter particles from the water with their brushlike labral fans.

2. Herbivores

Diptera that feed on living plants (including algae and fungi) are found in 37 of the 128 families recognized herein. Some, such as Agromyzidae, Anthomyiidae, Bibionidae, Cecidomyiidae, Chloropidae, Phoridae, Psilidae, Sciaridae, Tephritidae, and Tipulidae, include species considered to be pests to human agriculture. Fly larvae attack all parts of plants, including fruits, flowers, stems, leaves (as leaf-miners), and roots.

3. Predators

Predators are organisms that kill more than a single host organism for their feeding. There are 35 families with this way of life, including nearly all of the non-cyclorrhaphan Brachycera. Most have extremely active larvae that attack other invertebrates as their major food source. The larvae of tabanids have been known to kill frogs, an interesting reversal of the usual chain of events. Many predatory dipteran larvae are beneficial to humans in controlling insects considered to be pests. Examples include syrphids attacking aphids and predatory muscid larvae that kill larvae of other muscids, such as houseflies.

4. Parasitoids

Parasitoids develop on and kill a single host. Twenty-two families of Diptera have this way of life, which was reviewed recently by Feener and Brown. All species of the second largest family of Diptera, the Tachinidae, are parasitoids. All dipteran parasitoids attack other invertebrates, usually other insects, but unusual hosts include terrestrial flatworms, mollusks, earthworms, millipedes, spiders, and scorpions.

5. Parasites

Parasites feed on a single host, but do not normally kill it. True parasites include the bot flies, whose larvae live under the skin or in the nasal cavities of various mammals, including humans. Some bot fly larvae live in the stomachs and alimentary tracts of horses, elephants, zebras, and rhinos. Other parasites of vertebrates are calliphorids, chloropids, piophilids, and muscids. True parasites of invertebrates are less well known but occur in a few families, such as Phoridae.

6. No Free-Living Larvae

The larvae of the four families of the Hippoboscoidea—Glossinidae, Hippoboscidae, Nycteribiidae and Streblidae—are retained in the female abdomen and nourished by secretions of the accessory glands. They are deposited by females as fully mature, third-instar larvae, which quickly pupariate.

B. Adults

Most adult Diptera receive the majority of their nutrition as larvae and do not feed extensively. Many need carbohydrates to power their flight, however, and feed on the nectar in flowers, or on honeydew, the sweet secretions of Homoptera. Some flies require nitrogen for nourishing their eggs and are thus avid flower visitors in search of pollen. Flower foraging for nectar and pollen makes Diptera adults important pollinators of plants.

At least some parasitoids feed on the hemolymph of their hosts after oviposition, and many dipteran species feed on dead insects, carrion, dung, or rotting vegetation. A few adult Diptera, including Deuterophebiidae and some Oestridae, have vestigial mouthparts and do not feed.

Adults of a number of families have species that are well-known blood feeders, especially most or all species of Ceratopogonidae, Culicidae, Glossinidae, Hippoboscidae, Nycteribiidae, Simuliidae, Streblidae, and Tabanidae. Other families with fewer blood-feeding species are Muscidae, Rhagionidae, and Athericidae.

Some families of Diptera have species that are predatory as adults. This lifestyle is especially well developed in the Asilidae, Dolichopodidae, and Empididae but also occurs in some other families, such as Muscidae and Phoridae.

C. Special Associations

1. Aquatic Diptera

The larvae of many nematocerous families, as well as some Brachycera, are found in freshwater habitats. Among the most consistently aquatic forms are those families in the Culicomorpha and Blephariceromorpha, including such well-known families as Culicidae (mosquitoes), Simuliidae (black flies), and Chironomidae (midges). Chironomidae in particular can be exceedingly abundant, and various species assemblages are often used for assessment of water quality. The larvae of Simuliidae and Blephariceromorpha are found almost exclusively in clean, running water, with the Deuterophlebiidae and Blephariceridae being especially adapted to fast-flowing streams. Larval Culicidae are often found in ephemeral ponds that appear after snowmelt in temperate regions and after heavy rains elsewhere.

2. Phytotelmata

Phytotelmata are structures of plants that allow accumulations of water. They occur in various parts of plants, including leaves, leaf axils (especially of bromeliads and bananas), stems (especially of bamboos), fruits, and specialized structures (such as pitchers in pitcher plants). Among the approximately 20 families of Diptera that utilize phytotelmata, the Culicidae are the most prominent, being the most regular and most numerous larvae present. The other most frequently encountered families are Chironomidae, Ceratopogonidae, and Psychodidae, but further research has found that other families, such as Phoridae and Tipulidae, might also be common. Our knowledge of the fauna of phytotelmata is still fragmentary and much further work needs to be directed at these habitats.

The insect trapping structures of pitcher plants (Nepenthaceae and Sarraceniaceae) frequently harbor dipteran larvae. For instance, a distinctive fauna of about 80 species of aquatic Diptera is specialized to live in the carnivorous pitcher plants of the genus *Nepenthes* in Southeast Asia. The species found in this habitat are mostly from aquatic groups, such as mosquitoes, but some are from families more usually found in terrestrial habitats—Calliphoridae, Chloropidae, and Phoridae, for example. Nearly all feed on the trapped, drowned insects, or filter microorganisms associated with the decay of such insects from the water.

3. Social Insects

Social insects are the ants, bees, wasps, and termites that have organized societies. They have one or a few females responsible for all the egg laying, while other members of the colony (usually sterile females) gather food and do other tasks. Diptera are associated with social insects as scavengers, predators, parasitoids, and parasites. In the following treatment, only those species intimately associated with social insects are discussed, whereas opportunistic, generalized predators, such as robber flies (Asilidae), will be ignored.

Most flies known to be associated with social insects are brachycerans. Among the few nematocerous records of such associations is the astonishing observation that some Southeast Asian mosquito adults (genus *Malaya*) feed on honeydew carried in the mouthparts of ant workers.

Among the Brachycera associated with social insects, the most diverse are the Phoridae. These small flies are commonly found in association with ants, especially army ants (both New World and Old World), termites (especially in Africa and Southeast Asia), and, to a lesser extent, social bees and wasps. Many are parasitoids, laying their eggs inside the bodies of their adult hosts, usually ants, termites, and stingless bees; some are also known to parasitize the immature stages of ants. Many are scavengers, living in the refuse piles of the large colonies of army ants and leaf-cutter ants, where they are joined by scavenging larvae of other families. Some phorid larvae are predatory on ant brood, as are the larvae of some syrphids. The females of phorid species living in social insect nests are often remarkably modi-

fied, with reduced wings, eyes, and body sclerotization. One remarkable genus from Southeast Asia has adult females that mimic the larvae of their army ant hosts. Other females are heavily armored and have a rounded, teardrop shaped body form. This limuoid body form allows the females to escape damage when accosted by aggressive host ants, bees, or termites.

Other dipteran larvae found in social insect nests belong to the families Braulidae, Calliphoridae, Fanniidae, Sarcophagidae, Sphaeroceridae, and Syrphidae. Most are scavengers, although some are predatory.

Army ant raids are spectacular tropical phenomena that provide a number of opportunities for dipteran associates. Tachinids and conopids hover or perch near the raid front, darting down to parasitize the crickets, cockroaches, and other insects flushed by the foraging ants. Closer to the ground, parasitic phorids dart at their respective hosts: either the army ants themselves or other ants that are victims of the ant raids. Sarcophagids swarm on the leaves of nearby undergrowth, feeding on the droppings of birds attracted to the insects flushed out by the ant raids. Often, the raid front is best identified by listening for the loudest buzzing from the activity of hundreds of flies.

4. Kleptoparasites

Many Diptera exploit other insects or invertebrates that sequester or reserve food for long periods of time. By waiting before feeding, or by feeding slowly, these hosts provide a window of opportunity for larvae or adults of flies. The relationship between the host and kleptoparasite is often developed to such an extent that the flies live permanently associated with their food provider. Phoresy (transportation of the kleptoparasite by the host) often occurs in these associations.

A commonly observed example of kleptoparasitism and phoresy in Diptera is the association of large spiders and flies. Spider webs provide a continuous supply of food, and often a source of stored insect carcasses for later feeding. This warehouse is exploited by a number of fly families, including Ceratopogonidae, Cecidomyiidae, Milichiidae, Chloropidae, Lonchaeidae, Phoridae, and Empididae. In many instances, the flies perch on the bodies of the spider, waiting for their next meal to arrive.

Another, much more restricted association occurs between dung beetles of the family Scarabaeidae and flies of the family Sphaeroceridae. The flies ride on the bodies of the dung beetles, waiting until the beetles find and bury feces for their own larvae to feed upon. The flies briefly hop off the beetles, lay their own eggs on the dung, and then rejoin their food-providing host.

Kleptoparasites also occur in the nests of social Hymenoptera, where fly larvae are fed by deceived workers or attack the provisions left for a developing bee, wasp, or ant larva. A full review of kleptoparasitism is given by Sivinski *et al.*

5. Swarming

Swarms of adult Diptera are a common sight in most environments. Usually these are aggregations of males, allowing females to easily find a mate. Often, swarms are seen in the canopy of forests, under overhanging branches, in sunlight "pools" in forests, near fire towers emerging above tree level, or at the summit of tall hills or mountains ("hill-topping").

Most nematocerous families engage in swarming, as do about 15 families of Brachycera. Flies that engage in this behavior often have associated structural modifications, including a well-developed anal lobe of the wing and enlarged compound eyes. The development of the anal lobe probably allows for better maneuvering and hovering within the swarm, whereas the increased size of the eye allows male flies to more precisely place themselves, and assess their place, within the swarm. As females choose dominant males within the swarm, based on their relative position, assessment of position is critical for male mating success.

Species of dance flies (subfamily Empidinae) have elaborated on this basic pattern. Males of many species catch prey (usually smaller flies and other soft-bodied insects) and carry them in the swarm. They offer the prey to the nonhunting females as "nuptial gifts" to be fed upon during mating. A few species have reversed the trend of male-dominated swarms: instead, females form the aggregation, which prey-bearing males visit to selectively mate with the most desirable individuals (such female-dominated swarms are also known for some Phoridae). In some dance flies, the prey items themselves have changed, with males presenting already fed-upon prey, inanimate objects such as plant seeds, or even inedible bodily secretions. In these species, the giving of nuptial gifts apparently has become ritualized, losing all of its functional basis.

IV. CONSERVATION BIOLOGY OF DIPTERA

Because of the perception of almost all Diptera as disease-carrying filth flies, there is little public sympathy or interest in their conservation. Also, most Diptera are extremely poorly known, and the study of many families is still contingent on the recognition and description of the many undescribed species.

An exception to these statements is the Syrphidae, a group that is popular among amateur collectors, especially in Europe. There, Red Lists of endangered species exist for many regions, including Britain and parts of the mainland. When species that are at the edge of their distribution are eliminated from consideration, it appears that most endangered syrphids are either saproxylic or associated with wetlands.

Saproxylic insects of all orders are considered to be among the most endangered insects in Europe. These are species that are obligately associated with rotting wood, a habitat that is largely absent from young, even-aged tree plantations or from older forests that are sanitized or managed by removal of dead trees and wood. Most of the saproxylic habitats are afforded by old-growth forests with adequate numbers of injured, bleeding, moribund, or dead trees available. Different species of insects are associated with different types of decay, including whether or not the trees are still standing. Other variables include whether the trees are large or small, whether they are exposed to light and low humidity or are shaded throughout the day, and how long the process of decay has proceeded. Rot holes (some of which contain water), sap runs, dead branches on otherwise healthy trees, and loosened bark are all microhabitats that have specialized insect (including Diptera) faunas. The saproxylic faunas of other regions have not been assessed, but as deforestation proceeds throughout the world, Diptera associated with this habitat will likely be threatened.

Wetlands have been greatly reduced in many areas of the world, in many instances for the express purpose of eliminating biting fly vectors of disease. The loss of such habitats, however, also eliminates populations of other aquatic Diptera that have important ecological roles in the environment.

Another group of Diptera whose conservation needs are relatively well known is the endangered fauna of Hawaii. There are several species of *Drosophila* that have been proposed for listing as endangered species.

Because of their poor public image, Diptera are usually not considered flagship species for conservation projects. In the southwestern United States, however, an endangered species of mydid fly has been used to spearhead efforts to halt the final destruction of an endangered habitat in the Los Angeles area. The Delhi Sands Giant Flower-Loving Fly (*Rhaphiomidas terminatus abdominalis*) is the largest and most identifiable of a number of threatened taxa that live in this habitat, which is reduced to just a few acres in extent. The other subspecies of this taxon, *R. t. terminatus*, previously went extinct when its habitat in the coastal dunes near the Los Angeles International Airport was almost completely destroyed by urbanization. Other species of *Rhaphiomidas* are also considered endangered, as are other Diptera living in small, isolated, sandy habitats.

In general, the conservation status of Diptera on a worldwide scale is unknown. Undoubtedly many species are lost to deforestation, but only those in a few well-known groups (like syrphids) or in habitats that are of interest to some people (wetlands, sand dunes) have been studied. The situation is unlikely to change until we know much more about the systematics of flies.

See Also the Following Articles

BEETLES • BUTTERFLIES • HYMENOPTERA • INSECTS, OVERVIEW • MOTHS

Bibliography

Cumming, J. M. (1994). Sexual selection and the evolution of dance fly mating systems (Diptera: Empididae: Empidinae). *Canadian Entomologist* **126**, 907–920.

Evenhuis, N. L. (1994). *Catalogue of the Fossil Flies*. Backhuys, Leiden.

Feener, D. H. Jr., and Brown, B. V. (1997). Diptera as parasitoids. *Annual Review of Entomology* **42**, 73–97.

Ferrar, P. (1987). *A guide to the breeding habits and immature stages of Diptera Cyclorrhapha*. Entomonograph 8, parts 1 & 2. E. J. Brill/Scandinavian Science Press. Leiden, Copenhagen.

McAlpine, J. F. (ed.) (1989). *Manual of Nearctic Diptera, Volume 3*. Agriculture Canada Monograph No. 32.

McAlpine, J. F., and Munroe, D. D. (1968). Swarming of lonchaeid flies and other insects, with descriptions of four new species of Lonchaeidae (Diptera). *Canadian Enomologist* **100**, 1154–1178.

Sivinski, J., Marshall, S., and Petersson, E. (1999). Kleptoparasitism and phoresy in the Diptera. *Florida Entomologist* **82**, 179–197.

Smith, K. G. V. (1989). An introduction to the immature stages of British flies: Diptera larvae, with notes on eggs, puparia and pupae. *Handbooks for the Identification of British Insects* **10** (14).

Yeates, D. K., and Wiegmann, B. M. (1999). Congruence and controversy: Toward a higher-level phylogeny of Diptera. *Annual Review of Entomology* **44**, 397–428.

ISBN 0-12-226867-9

90038

9 780122 268670